## ■ TABLE 1.7
Approximate Physical Properties of Some Common Gases at Standard Atmospheric Pressure (BG Units)

| Gas | Temperature (°F) | Density, $\rho$ (slugs/ft³) | Specific Weight, $\gamma$ (lb/ft³) | Dynamic Viscosity, $\mu$ (lb·s/ft²) | Kinematic Viscosity, $\nu$ (ft²/s) | Gas Constant,[a] $R$ (ft·lb/slug·°R) | Specific Heat Ratio,[b] $k$ |
|---|---|---|---|---|---|---|---|
| Air (standard) | 59 | 2.38 E − 3 | 7.65 E − 2 | 3.74 E − 7 | 1.57 E − 4 | 1.716 E + 3 | 1.40 |
| Carbon dioxide | 68 | 3.55 E − 3 | 1.14 E − 1 | 3.07 E − 7 | 8.65 E − 5 | 1.130 E + 3 | 1.30 |
| Helium | 68 | 3.23 E − 4 | 1.04 E − 2 | 4.09 E − 7 | 1.27 E − 3 | 1.242 E + 4 | 1.66 |
| Hydrogen | 68 | 1.63 E − 4 | 5.25 E − 3 | 1.85 E − 7 | 1.13 E − 3 | 2.466 E + 4 | 1.41 |
| Methane (natural gas) | 68 | 1.29 E − 3 | 4.15 E − 2 | 2.29 E − 7 | 1.78 E − 4 | 3.099 E + 3 | 1.31 |
| Nitrogen | 68 | 2.26 E − 3 | 7.28 E − 2 | 3.68 E − 7 | 1.63 E − 4 | 1.775 E + 3 | 1.40 |
| Oxygen | 68 | 2.58 E − 3 | 8.31 E − 2 | 4.25 E − 7 | 1.65 E − 4 | 1.554 E + 3 | 1.40 |

[a]Values of the gas constant are independent of temperature.
[b]Values of the specific heat ratio depend only slightly on temperature.

## ■ TABLE 1.8
Approximate Physical Properties of Some Common Gases at Standard Atmospheric Pressure (SI Units)

| Gas | Temperature (°C) | Density, $\rho$ (kg/m³) | Specific Weight, $\gamma$ (N/m³) | Dynamic Viscosity, $\mu$ (N·s/m²) | Kinematic Viscosity, $\nu$ (m²/s) | Gas Constant,[a] $R$ (J/kg·K) | Specific Heat Ratio,[b] $k$ |
|---|---|---|---|---|---|---|---|
| Air (standard) | 15 | 1.23 E + 0 | 1.20 E + 1 | 1.79 E − 5 | 1.46 E − 5 | 2.869 E + 2 | 1.40 |
| Carbon dioxide | 20 | 1.83 E + 0 | 1.80 E + 1 | 1.47 E − 5 | 8.03 E − 6 | 1.889 E + 2 | 1.30 |
| Helium | 20 | 1.66 E − 1 | 1.63 E + 0 | 1.94 E − 5 | 1.15 E − 4 | 2.077 E + 3 | 1.66 |
| Hydrogen | 20 | 8.38 E − 2 | 8.22 E − 1 | 8.84 E − 6 | 1.05 E − 4 | 4.124 E + 3 | 1.41 |
| Methane (natural gas) | 20 | 6.67 E − 1 | 6.54 E + 0 | 1.10 E − 5 | 1.65 E − 5 | 5.183 E + 2 | 1.31 |
| Nitrogen | 20 | 1.16 E + 0 | 1.14 E + 1 | 1.76 E − 5 | 1.52 E − 5 | 2.968 E + 2 | 1.40 |
| Oxygen | 20 | 1.33 E + 0 | 1.30 E + 1 | 2.04 E − 5 | 1.53 E − 5 | 2.598 E + 2 | 1.40 |

[a]Values of the gas constant are independent of temperature.
[b]Values of the specific heat ratio depend only slightly on temperature.

# 7th edition

# Fundamentals of
# Fluid
# Mechanics

## Bruce R. Munson
Department of Aerospace Engineering
Iowa State University
Ames, Iowa

## Theodore H. Okiishi
Department of Mechanical Engineering
Iowa State University
Ames, Iowa

## Wade W. Huebsch
Department of Mechanical and Aerospace Engineering
West Virginia University
Morgantown, West Virginia

## Alric P. Rothmayer
Department of Aerospace Engineering
Iowa State University
Ames, Iowa

**WILEY**

John Wiley & Sons, Inc.

Executive Publisher:  Don Fowley
Senior Editor and Product Designer:  Jennifer Welter
Content Manager:  Kevin Holm
Senior Content Editor:  Wendy Ashenberg
Creative Director:  Harry Nolan
Senior Designer:  Madelyn Lesure
Executive Marketing Manager:  Christopher Ruel
Editorial Operations Manager:  Melissa Edwards
Photo Researcher:  Sheena Goldstein
Assistant Editor:  Samantha Mandel
Senior Production Editor:  John Curley
Media Specialist:  Lisa Sabatini
Production Management Services:  Ingrao Associates/Suzanne Ingrao
Cover Design:  Madelyn Lesure
Cover Photo:  Graham Jeffery/Sensitive Light

This book was set in 10/12 Times Roman by Aptara®, Inc., and printed and bound by R.R. Donnelley/Jefferson City. The cover was printed by R.R. Donnelley/Jefferson City.

This book is printed on acid free paper. ∞

Founded in 1807, John Wiley & Sons, Inc. has been a valued source of knowledge and understanding for more than 200 years, helping people around the world meet their needs and fulfill their aspirations. Our company is built on a foundation of principles that include responsibility to the communities we serve and where we live and work. In 2008, we launched a Corporate Citizenship Initiative, a global effort to address the environmental, social, economic, and ethical challenges we face in our business. Among the issues we are addressing are carbon impact, paper specifications and procurement, ethical conduct within our business and among our vendors, and community and charitable support. For more information, please visit our website: www.wiley.com/go/citizenship.

*Library of Congress Cataloging-in-Publication Data*

Munson, Bruce Roy, 1940-

  Fundamentals of fluid mechanics / Bruce R. Munson, Theodore H. Okiishi, Wade W. Huebsch, Alric P. Rothmayer—7th edition.

     pages cm

  Includes indexes.

  ISBN 978-1-118-11613-5

1. Fluid mechanics—Textbooks.  I. Okiishi, T. H. (Theodore Hisao), 1939- II. Huebsch, Wade W. III. Rothmayer, Alric P., 1959- IV. Title.

  TA357.M86 2013

  532–dc23

                                                                        2012011618

ISBN   978-1-118-11613-5 (Main Book)
ISBN   978-1-118-39971-2 (Binder-Ready Version)

Printed in the United States of America

10  9  8  7  6  5  4  3  2

# About the Authors

**Bruce R. Munson**, Professor Emeritus of Engineering Mechanics at Iowa State University, received his B.S. and M.S. degrees from Purdue University and his Ph.D. degree from the Aerospace Engineering and Mechanics Department of the University of Minnesota in 1970.

Prior to joining the Iowa State University faculty in 1974, Dr. Munson was on the mechanical engineering faculty of Duke University from 1970 to 1974. From 1964 to 1966, he worked as an engineer in the jet engine fuel control department of Bendix Aerospace Corporation, South Bend, Indiana.

Dr. Munson's main professional activity has been in the area of fluid mechanics education and research. He has been responsible for the development of many fluid mechanics courses for studies in civil engineering, mechanical engineering, engineering science, and agricultural engineering and is the recipient of an Iowa State University Superior Engineering Teacher Award and the Iowa State University Alumni Association Faculty Citation.

He has authored and coauthored many theoretical and experimental technical papers on hydrodynamic stability, low Reynolds number flow, secondary flow, and the applications of viscous incompressible flow. He is a member of The American Society of Mechanical Engineers.

**Ted H. Okiishi**, Professor Emeritus of Mechanical Engineering at Iowa State University, joined the faculty there in 1967 after receiving his undergraduate and graduate degrees from that institution.

From 1965 to 1967, Dr. Okiishi served as a U.S. Army officer with duty assignments at the National Aeronautics and Space Administration Lewis Research Center, Cleveland, Ohio, where he participated in rocket nozzle heat transfer research, and at the Combined Intelligence Center, Saigon, Republic of South Vietnam, where he studied seasonal river flooding problems.

Professor Okiishi and his students have been active in research on turbomachinery fluid dynamics. Some of these projects have involved significant collaboration with government and industrial laboratory researchers, with two of their papers winning the ASME Melville Medal (in 1989 and 1998).

Dr. Okiishi has received several awards for teaching. He has developed undergraduate and graduate courses in classical fluid dynamics as well as the fluid dynamics of turbomachines.

He is a licensed professional engineer. His professional society activities include having been a vice president of The American Society of Mechanical Engineers (ASME) and of the American Society for Engineering Education. He is a Life Fellow of The American Society of Mechanical Engineers and past editor of its *Journal of Turbomachinery*. He was recently honored with the ASME R. Tom Sawyer Award.

**Wade W. Huebsch**, Associate Professor in the Department of Mechanical and Aerospace Engineering at West Virginia University, received his B.S. degree in aerospace engineering from San Jose State University where he played college baseball. He received his M.S. degree in mechanical engineering and his Ph.D. in aerospace engineering from Iowa State University in 2000.

Dr. Huebsch specializes in computational fluid dynamics research and has authored multiple journal articles in the areas of aircraft icing, roughness-induced flow phenomena, and boundary layer flow control. He has taught both undergraduate and graduate courses in fluid mechanics and has developed a new undergraduate course in computational fluid dynamics. He has received multiple teaching awards such as Outstanding Teacher and Teacher of the Year from the College of Engineering and Mineral Resources at WVU as well as the Ralph R.

Teetor Educational Award from SAE. He was also named as the Young Researcher of the Year from WVU. He is a member of the American Institute of Aeronautics and Astronautics, the Sigma Xi research society, the Society of Automotive Engineers, and the American Society of Engineering Education.

**Alric P. Rothmayer**, Professor of Aerospace Engineering at Iowa State University, received his undergraduate and graduate degrees from the Aerospace Engineering Department at the University of Cincinnati, during which time he also worked at NASA Langley Research Center and was a visiting graduate research student at the Imperial College of Science and Technology in London. He joined the faculty at Iowa State University (ISU) in 1985 after a research fellowship sponsored by the Office of Naval Research at University College in London.

Dr. Rothmayer has taught a wide variety of undergraduate fluid mechanics and propulsion courses for over 25 years, ranging from classical low and high speed flows to propulsion cycle analysis.

Dr. Rothmayer was awarded an ISU Engineering Student Council Leadership Award, an ISU Foundation Award for Early Achievement in Research, an ISU Young Engineering Faculty Research Award, and a National Science Foundation Presidential Young Investigator Award. He is an Associate Fellow of the American Institute of Aeronautics and Astronautics (AIAA), and was chair of the 3rd AIAA Theoretical Fluid Mechanics Conference.

Dr. Rothmayer specializes in the integration of Computational Fluid Dynamics with asymptotic methods and low order modeling for viscous flows. His research has been applied to diverse areas ranging from internal flows through compliant tubes to flow control and aircraft icing. In 2001, Dr. Rothmayer won a NASA Turning Goals into Reality (TGIR) Award as a member of the Aircraft Icing Project Team, and also won a NASA Group Achievement Award in 2009 as a member of the LEWICE Ice Accretion Software Development Team. He was also a member of the SAE AC-9C Aircraft Icing Technology Subcommittee of the Aircraft Environmental Systems Committee of SAE and the Fluid Dynamics Technical Committee of AIAA.

# Preface

This book is intended for junior and senior engineering students who are interested in learning some fundamental aspects of fluid mechanics. We developed this text to be used as a first course. The principles considered are classical and have been well-established for many years. However, fluid mechanics education has improved with experience in the classroom, and we have brought to bear in this book our own ideas about the teaching of this interesting and important subject. This seventh edition has been prepared after several years of experience by the authors using the previous editions for introductory courses in fluid mechanics. On the basis of this experience, along with suggestions from reviewers, colleagues, and students, we have made a number of changes in this edition. The changes (listed below, and indicated by the word *New* in descriptions in this preface) are made to clarify, update, and expand certain ideas and concepts.

## New to This Edition

In addition to the continual effort of updating the scope of the material presented and improving the presentation of all of the material, the following items are new to this edition. With the widespread use of new technologies involving the web, DVDs, digital cameras and the like, there is an increasing use and appreciation of the variety of visual tools available for learning. As in recent editions, this fact has been addressed in the new edition by continuing to include additional new illustrations, graphs, photographs, and videos.

**Illustrations:** *New* illustrations and graphs have been added to this edition, as well as updates to past ones. The book now contains nearly 1600 illustrations. These illustrations range from simple ones that help illustrate a basic concept or equation to more complex ones that illustrate practical applications of fluid mechanics in our everyday lives.

**Photographs:** This edition has also added *new* photographs throughout the book to enhance the text. The total number of photographs now exceeds 300. Some photos involve situations that are so common to us that we probably never stop to realize how fluids are involved in them. Others involve new and novel situations that are still baffling to us. The photos are also used to help the reader better understand the basic concepts and examples discussed. Combining the illustrations, graphs and photographs, the book has approximately 1900 visual aids.

**Videos:** The video library has been enhanced by the addition of 19 *new* video segments directly related to the text material, as well as multiple updates to previous videos (i.e. same topic with an updated video clip). In addition to being strategically located at the appropriate places within the text, they are all listed, each with an appropriate thumbnail photo, in the video index. They illustrate many of the interesting and practical applications of real-world fluid phenomena. There are now 175 videos.

**Examples:** The book contains 5 *new* example problems that involve various fluid flow fundamentals. Some of these examples also incorporate *new* PtD (Prevention through Design) discussion material. The PtD project, under the direction of the National Institute for Occupational Safety and Health, involves, in part, the use of textbooks to encourage the proper design and use of workday equipment and material so as to reduce accidents and injuries in the workplace.

**Problems and Problem Types:** Approximately 30% *new* homework problems have been added for this edition, with a total number of 1484 problems in the text (additional problems in *WileyPLUS*). Also, *new* multiple-choice concept questions (developed by Jay Martin and John Mitchell of the University of Wisconsin-Madison) have been added at the beginning of each Problems section. These questions test the students' knowledge of basic chapter concepts. This edition has also significantly improved the homework problem integration with the

*WileyPLUS* course management system. *New* icons have been introduced in the Problems section to help instructors and students identify which problems are available to be assigned within *WileyPLUS* for automatic grading, and which problems have tutorial help available.

**Author:** A *new* co-author was brought on board for this edition. We are happy to welcome Dr. Alric P. Rothmayer.

### Within *WileyPLUS*:

*New* **What an Engineer Sees** animations demonstrate an engineer's perspective of everyday objects, and relates the transfer of theory to real life through the solution of a problem involving that everyday object.

*New* **Office-Hours Videos** demonstrate the solution of selected problems, focusing specifically on those areas in which students typically run into difficulty, with video and voiceover. Over 700 **homework problems from the text** that can be assigned for automatic feedback and grading (34 new for the 7th edition). Including 65 **GO (Guided Online) Tutorial** problems (26 new for this edition).

## Key Features

### Illustrations, Photographs, and Videos

(© Photograph courtesy of Pend Oreille Public Utility District.)

Fluid mechanics has always been a "visual" subject—much can be learned by viewing various aspects of fluid flow. In this new edition we have made several changes to reflect the fact that with new advances in technology, this visual component is becoming easier to incorporate into the learning environment, for both access and delivery, and is an important component to the learning of fluid mechanics. Thus, *new* photographs and illustrations have been added to the book. Some of these are within the text material; some are used to enhance the example problems; and some are included as margin figures of the type shown in the left margin to more clearly illustrate various points discussed in the text. In addition, *new* video segments have been added, bringing the total number of video segments to 175. These video segments illustrate many interesting and practical applications of real-world fluid phenomena. Each video segment is identified at the appropriate location in the text material by a video icon and thumbnail photograph of the type shown in the left margin. The full video library is shown in the video index at the back of the book. Each video segment has a separate associated text description of what is shown in the video. There are many homework problems that are directly related to the topics in the videos.

### Examples

*V1.9 Floating razor blade*

One of our aims is to represent fluid mechanics as it really is—an exciting and useful discipline. To this end, we include analyses of numerous everyday examples of fluid-flow phenomena to which students and faculty can easily relate. In the seventh edition there are 5 new examples and a total of 164 examples that provide detailed solutions to a variety of problems. Some of the new examples incorporate Prevention through Design (PtD) material. Many of the examples illustrate what happens if one or more of the parameters is changed. This gives the user a better feel for some of the basic principles involved. In addition, many of the examples contain *new* photographs of the actual device or item involved in the example. Also, all of the examples are outlined and carried out with the problem solving methodology of "Given, Find, Solution, and Comment" as discussed on page 5 in the "Note to User" before Example 1.1.

### Fluids in the News

The set of approximately 60 short "Fluids in the News" stories reflect some of the latest important, and novel, ways that fluid mechanics affects our lives. Many of these problems have homework problems associated with them.

## Homework Problems

A set of more than 1480 homework problems (approximately 30% *new* to this edition) stresses the practical application of principles. The problems are grouped and identified according to topic. An effort has been made to include several *new*, easier problems at the start of each group. The following types of problems are included:

1) new conceptual multiple-choice problems
2) "standard" problems,
3) computer problems,
4) discussion problems,
5) supply-your-own-data problems,
6) review problems with solutions,
7) problems based on the "Fluids in the News" topics,
8) problems based on the fluid videos,

9) Excel-based lab problems,
10) "Lifelong learning" problems,
11) problems that require the user to obtain a photograph/image of a given flow situation and write a brief paragraph to describe it,
12) simple CFD problems to be solved using ANSYS Academic CFD Software,
13) Fundamental of Engineering (FE) exam questions available on book website.

*Lab Problems*—There are 30 extended, laboratory-type problems that involve actual experimental data for simple experiments of the type that are often found in the laboratory portion of many introductory fluid mechanics courses. The data for these problems are provided in Excel format.

*Lifelong Learning Problems*—Each chapter has lifelong learning problems that involve obtaining additional information about various new state-of-the-art fluid mechanics topics and writing a brief report about this material.

*Review Problems*—There is a set of 186 review problems covering most of the main topics in the book. Complete, detailed solutions to these problems can be found in the *Student Solutions Manual and Study Guide for Fundamentals of Fluid Mechanics*, by Munson et al. (© 2013 John Wiley and Sons, Inc.).

## Well-Paced Concept and Problem-Solving Development

Since this is an introductory text, we have designed the presentation of material to allow for the gradual development of student confidence in fluid problem solving. Each important concept or notion is considered in terms of simple and easy-to-understand circumstances before more complicated features are introduced. Many pages contain a brief summary (a highlight) sentence that serves to prepare or remind the reader about an important concept discussed on that page.

Several brief components have been added to each chapter to help the user obtain the "big picture" idea of what key knowledge is to be gained from the chapter. A brief Learning Objectives section is provided at the beginning of each chapter. It is helpful to read through this list prior to reading the chapter to gain a preview of the main concepts presented. Upon completion of the chapter, it is beneficial to look back at the original learning objectives to ensure that a satisfactory level of understanding has been acquired for each item. Additional reinforcement of these learning objectives is provided in the form of a Chapter Summary and Study Guide at the end of each chapter. In this section a brief summary of the key concepts and principles introduced in the chapter is included along with a listing of important terms with which the student should be familiar. These terms are highlighted in the text. A list of the main equations in the chapter is included in the chapter summary.

## System of Units

Two systems of units continue to be used throughout most of the text: the International System of Units (newtons, kilograms, meters, and seconds) and the British Gravitational System (pounds, slugs, feet, and seconds). About one-half of the examples and homework problems are in each set of units. The English Engineering System (pounds, pounds mass, feet, and seconds) is used in the discussion of compressible flow in Chapter 11. This usage is standard practice for the topic.

## Topical Organization

In the first four chapters the student is made aware of some fundamental aspects of fluid motion, including important fluid properties, regimes of flow, pressure variations in fluids at rest and in motion, fluid kinematics, and methods of flow description and analysis. The Bernoulli equation is introduced in Chapter 3 to draw attention, early on, to some of the interesting effects of fluid motion on the distribution of pressure in a flow field. We believe that this timely consideration of elementary fluid dynamics increases student enthusiasm for the more complicated material that follows. In Chapter 4 we convey the essential elements of kinematics, including Eulerian and Lagrangian mathematical descriptions of flow phenomena, and indicate the vital relationship between the two views. For teachers who wish to consider kinematics in detail before the material on elementary fluid dynamics, Chapters 3 and 4 can be interchanged without loss of continuity.

Chapters 5, 6, and 7 expand on the basic analysis methods generally used to solve or to begin solving fluid mechanics problems. Emphasis is placed on understanding how flow phenomena are described mathematically and on when and how to use infinitesimal and finite control volumes. The effects of fluid friction on pressure and velocity distributions are also considered in some detail. A formal course in thermodynamics is not required to understand the various portions of the text that consider some elementary aspects of the thermodynamics of fluid flow. Chapter 7 features the advantages of using dimensional analysis and similitude for organizing test data and for planning experiments and the basic techniques involved.

Owing to the growing importance of computational fluid dynamics (CFD) in engineering design and analysis, material on this subject is included in Appendix A. This material may be omitted without any loss of continuity to the rest of the text. This introductory CFD overview includes examples and problems of various interesting flow situations that are to be solved using ANSYS Academic CFD software.

Chapters 8 through 12 offer students opportunities for the further application of the principles learned early in the text. Also, where appropriate, additional important notions such as boundary layers, transition from laminar to turbulent flow, turbulence modeling, and flow separation are introduced. Practical concerns such as pipe flow, open-channel flow, flow measurement, drag and lift, the effects of compressibility, and the fluid mechanics fundamentals associated with turbomachines are included.

Students who study this text and who solve a representative set of the exercises provided should acquire a useful knowledge of the fundamentals of fluid mechanics. Faculty who use this text are provided with numerous topics to select from in order to meet the objectives of their own courses. More material is included than can be reasonably covered in one term. All are reminded of the fine collection of supplementary material. We have cited throughout the text various articles and books that are available for enrichment.

# Student and Instructor Resources

*Student Solutions Manual and Study Guide,* by Munson et al. (© 2013 John Wiley and Sons, Inc.)—This short paperback book is available as a supplement for the text. It provides detailed solutions to the Review Problems and a concise overview of the essential points of most of the main sections of the text, along with appropriate equations, illustrations, and worked examples. This supplement is available through *WileyPLUS*, your local bookstore, or you may purchase it on the Wiley website at www.wiley.com/college/munson.

**Student Companion Site**—The student section of the book website at www.wiley.com/college/munson contains the assets listed below. Access is free-of-charge.

Video Library                                  Comprehensive Table of Conversion Factors
Review Problems with Answers                    CFD Driven Cavity Example
Lab Problems

**Instructor Companion Site**—The instructor section of the book website at www.wiley.com/college/munson contains the assets in the Student Companion Site, as well as the following, which are available only to professors who adopt this book for classroom use:

- Instructor Solutions Manual, containing complete, detailed solutions to all of the problems in the text.
- Figures from the text, appropriate for use in lecture slides.

These instructor materials are password-protected. Visit the Instructor Companion Site to register for a password.

*WileyPLUS*. *WileyPLUS* combines the complete, dynamic online text with all of the teaching and learning resources you need, in one easy-to-use system. This edition offers a much tighter integration between the book and *WileyPLUS*. The instructor assigns *WileyPLUS*, but students decide how to buy it: they can buy the new, printed text packaged with a *WileyPLUS* registration code at no additional cost or choose digital delivery of *WileyPLUS*, use the online text and integrated read, study, and practice tools, and save off the cost of the new book.

*WileyPLUS* offers today's engineering students the interactive and visual learning materials they need to help them grasp difficult concepts—and apply what they've learned to solve problems in a dynamic environment. A robust variety of examples and exercises enable students to work problems, see their results, and obtain instant feedback including hints and reading references linked directly to the online text.

Contact your local Wiley representative, or visit www.wileyplus.com for more information about using *WileyPLUS* in your course.

## Acknowledgments

We wish to express our gratitude to the many persons who provided suggestions for this and previous editions through reviews and surveys. In addition, we wish to express our appreciation to the many persons who supplied photographs and videos used throughout the text. Finally, we thank our families for their continued encouragement during the writing of this seventh edition.

Working with students over the years has taught us much about fluid mechanics education. We have tried in earnest to draw from this experience for the benefit of users of this book. Obviously we are still learning, and we welcome any suggestions and comments from you.

BRUCE R. MUNSON
THEODORE H. OKIISHI
WADE W. HUEBSCH
ALRIC P. ROTHMAYER

# Featured in this Book

# 5 Finite Control Volume Analysis

CHAPTER OPENING PHOTO: Wind turbine farms (this is the Middelgrunden Offshore Wind Farm in Denmark) are becoming more common. Finite control volume analysis can be used to estimate the amount of energy transferred between the moving air and each turbine rotor. (Photograph courtesy of Siemens Wind Power.)

## Learning Objectives

After completing this chapter, you should be able to:

- select an appropriate finite control volume to solve a fluid mechanics problem.
- apply conservation of mass and energy and Newton's second law of motion to the contents of a finite control volume to get important answers.
- know how velocity changes and energy transfers in fluid flows are related to forces and torques.
- understand why designing for minimum loss of energy in fluid flows is so important.

***Learning Objectives***
at the beginning of each chapter focus students' attention as they read the chapter.

***Photographs, Illustrations, and Videos***
(more than 2000 in the 7th edition) help readers connect theory to the physical world.

***Summary Sentences***
emphasize important points for the reader.

***Fluids in the News***
relate brief stories involving current, sometimes novel, applications of fluids phenomena. Many are associated with homework problems.

V9.3 Human aerodynamic wake

*Most familiar flows involve large Reynolds numbers.*

**Figure 9.6** Character of the steady, viscous flow past a circular cylinder: (*a*) low Reynolds number flow, (*b*) moderate Reynolds number flow, (*c*) large Reynolds number flow.

field. Since the shear stress (i.e., viscous effect) is the product of the fluid viscosity and the velocity gradient, it follows that viscous effects are confined to the boundary layer and wake regions.

The characteristics described in Figs. 9.5 and 9.6 for flow past a flat plate and a circular cylinder are typical of flows past streamlined and blunt bodies, respectively. The nature of the flow depends strongly on the Reynolds number. (See Ref. 31 for many examples illustrating this behavior.) Most familiar flows are similar to the large Reynolds number flows depicted in Figs. 9.5c and 9.6c, rather than the low Reynolds number flow situations. (See the photograph at the beginning of Chapters 7 and 11.) In the remainder of this chapter we will investigate more thoroughly these ideas and determine how to calculate the forces on immersed bodies.

## EXAMPLE 9.2    Characteristics of Flow Past Objects

**GIVEN** It is desired to experimentally determine the various characteristics of flow past a car as shown in Fig E9.2. The following tests could be carried out: (a) $U = 20$ mm/s flow of glycerin past a scale model that is 34-mm tall, 100-mm long, and 40-mm wide, (b) $U = 20$ mm/s air flow past the same scale model, or (c) $U = 25$ m/s airflow past the actual car, which is 1.7-m tall, 5-m long, and 2-m wide.

**FIND** Would the flow characteristics for these three situations be similar? Explain.

**Figure E9.2**
© Stanford University, with permission.

## EXAMPLE 2.2 Incompressible and Isothermal Pressure-Depth Variations

**GIVEN** In 2010, the world's tallest building, the Burj Khalifa skyscraper, was completed and opened in the United Arab Emirates. The final height of the building, which had remained a secret until completion, is 2717 ft (828 m).

**FIND** (a) Estimate the ratio of the pressure at the 2717-ft top of the building to the pressure at its base, assuming the air to be at a common temperature of 59 °F. (b) Compare the pressure calculated in part (a) with that obtained by assuming the air to be incompressible with $\gamma = 0.0765$ lb/ft$^3$ at 14.7 psi (abs) (values for air at standard sea level conditions).

### SOLUTION

(a) For the assumed isothermal conditions, and treating air as a compressible fluid, Eq. 2.10 can be applied to yield

$$\frac{p_2}{p_1} = \exp\left[ -\frac{g(z_2 - z_1)}{RT_0} \right]$$

$$= \exp\left\{ -\frac{(32.2 \text{ ft/s}^2)(2717 \text{ ft})}{(1716 \text{ ft} \cdot \text{lb/slug} \cdot {}^\circ\text{R})[(59 + 460){}^\circ\text{R}]} \right\}$$

$$= 0.906 \qquad \text{(Ans)}$$

(b) If the air is treated as an incompressible fluid we can apply Eq. 2.5. In this case

$$p_2 = p_1 - \gamma(z_2 - z_1)$$

or

$$\frac{p_2}{p_1} = 1 - \frac{\gamma(z_2 - z_1)}{p_1}$$

$$= 1 - \frac{(0.0765 \text{ lb/ft}^3)(2717 \text{ ft})}{(14.7 \text{ lb/in.}^2)(144 \text{ in.}^2/\text{ft}^2)} = 0.902 \qquad \text{(Ans)}$$

**COMMENTS** Note that there is little difference between the two results. Since the pressure difference between the bottom and top of the building is small, it follows that the variation in fluid density is small and, therefore, the compressible

■ **Figure E2.2** (Figure courtesy of Emaar Properties, Dubai, UAE.)

fluid and incompressible fluid analyses yield essentially the same result.

We see that for both calculations the pressure decreases by approximately 10% as we go from ground level to the top of this tallest building. It does not require a very large pressure difference to support a 2717-ft-tall column of fluid as light as air. This result supports the earlier statement that the changes in pressures in air and other gases due to elevation changes are very small, even for distances of hundreds of feet. [...]ences between the top and bottom of a [...] gas, or in a gas storage tank, are negligib[...] volved are very small.

---

## Example Problems

provide detailed solutions for interesting, real-world situations, with a consistent methodology and comments.

---

## 2.13 Chapter Summary and Study Guide

In this chapter the pressure variation in a fluid at rest is considered, along with some important consequences of this type of pressure variation. It is shown that for incompressible fluids at rest the pressure varies linearly with depth. This type of variation is commonly referred to as hydrostatic pressure distribution. For compressible fluids at rest the pressure distribution will not generally be hydrostatic, but Eq. 2.4 remains valid and can be used to determine the pressure distribution if additional information about the variation of the specific weight is specified. The distinction between absolute and gage pressure is discussed along with a consideration of barometers for the measurement of atmospheric pressure.

Pressure-measuring devices called manometers, which utilize static liquid columns, are analyzed in detail. A brief discussion of mechanical and electronic pressure gages is also included. Equations for determining the magnitude and location of the resultant fluid force acting on a plane surface in contact with a static fluid are developed. A general approach for determining the magnitude and location of the resultant fluid force acting on a curved surface in contact with a static fluid is described. For submerged or floating bodies the concept of the buoyant force and the use of Archimedes' principle are reviewed.

The following checklist provides a study guide for this chapter. When your study of the entire chapter and end-of-chapter exercises has been completed, you should be able to

*Pascal's law*
*surface force*
*body force*
*incompressible fluid*
*hydrostatic pressure*
*  distribution*
*pressure head*
*compressible fluid*
*U.S. standard*
*  atmosphere*
*absolute pressure*
*gage pressure*
*vacuum pressure*
*barometer*
*manometer*
*Bourdon pressure*
*  gage*
*center of pressure*
*buoyant force*
*Archimedes' principle*
*center of buoyancy*

■ write out meanings of the terms listed here in the margin and understand each of the related concepts. These terms are particularly important and are set in *italic, bold, and color* type in the text.
■ calculate the [...]
■ calculate the [...] if the variatio[...]
■ use the conce[...] ments using [...]
■ determine the [...] plane surface.

| | | |
|---|---|---|
| Pressure gradient in a stationary fluid | $\dfrac{dp}{dz} = -\gamma$ | (2.4) |
| Pressure variation in a stationary incompressible fluid | $p_1 = \gamma h + p_2$ | (2.7) |
| Hydrostatic force on a plane surface | $F_R = \gamma h_c A$ | (2.18) |
| Location of hydrostatic force on a plane surface | $y_R = \dfrac{I_{xc}}{y_c A} + y_c$ | (2.19) |
| | $x_R = \dfrac{I_{xyc}}{y_c A} + x_c$ | (2.20) |
| Buoyant force | $F_B = \gamma \mathcal{V}$ | (2.22) |
| Pressure gradient in rigid-body motion | $-\dfrac{\partial p}{\partial x} = \rho a_x, \quad -\dfrac{\partial p}{\partial y} = \rho a_y, \quad -\dfrac{\partial p}{\partial z} = \gamma + \rho a_z$ | (2.24) |
| Pressure gradient in rigid-body rotation | $\dfrac{\partial p}{\partial r} = \rho r \omega^2, \quad \dfrac{\partial p}{\partial \theta} = 0, \quad \dfrac{\partial p}{\partial z} = -\gamma$ | (2.30) |

---

## Chapter Summary and Study Guide

at the end of each chapter helps students focus their study and summarizes key equations.

---

### Review Problems

Go to Appendix G (*WileyPLUS* or the book's web site, www.wiley.com/college/munson) for a set of review problems with answers. Detailed solutions can be found in the *Student Solution Manual and Study Guide for Fundamentals of Fluid Mechanics*, by Munson et al. © 2013 John Wiley and Sons, Inc.).

### Conceptual Questions

**1.1C** The correct statement for the definition of density is

a) Density is the mass per unit volume.
b) Density is the volume per unit mass.
c) Density is the weight per unit volume.
d) Density is the weight divided by gravity.
e) Density is the mass divided by the weight.

**1.2C** Given the following equation where *p* is pressure in lb/ft$^2$, $\gamma$ is the specific weight in lb/ft$^3$, *V* is the magnitude of velocity in

c) Useful only for very low density gases.
d) Indicates that two solids in contact will not slip if the joining force is large.

**1.4C** In fluids, the shearing strain rate $\dfrac{du}{dy}$ for a Newtonian fluid has dimensions of:

a) L/T$^2$.          c) L$^2$/T.
b) 1/T.              d) L$^2$/T$^2$.

**1.5C** The laminar velocity profile for a Newtonian fluid is shown

---

## Review Problems

covering the main topics in the book give readers another opportunity to practice. Complete, detailed solutions to these problems are in *WileyPLUS* and the *Student Solutions Manual and Study Guide for Fundamentals of Fluid Mechanics*, by Munson et. al. (© John Wiley & Sons, Inc.).

---

## Conceptual Questions

help evaluate student conceptual understanding (more available in *WileyPLUS*).

| | |
|---|---|
| (WILEY) | Problem available in *WileyPLUS* at instructor's discretion. |
| (WILEY GO) | Tutoring problem available in *WileyPLUS* at instructor's discretion. |
| ▶ | Problem is related to a chapter video available in *WileyPLUS*. |
| ✳ | Problem to be solved with aid of programmable calculator or computer. |
| † | Open-ended problem that requires critical thinking. These problems require various assumptions to provide the necessary input data. There are not unique answers to these problems. |

## WileyPLUS Icons

in the problems section identify problems that are available to instructors to assign in *WileyPLUS* for automatic grading.

## Homework Problems

at the end of each chapter (over 1400 for the 7th edition) stress the practical applications of fluid mechanics principles.

## Lab Problems

in *WileyPLUS* and on the book website provide actual data in Excel format for experiments of the type found in many introductory fluid mechanics labs.

## Lifelong Learning Problems

require readers to obtain information from other sources about new state-of-the-art topics, and to write a brief report.

## FE Exam Problems

give students practice with problems similar to those on the FE Exam.

## CFD Problems

in select chapters can be solved using ANSYS Academic CFD software.

■ Figure P5.121

**5.122** ▶ (WILEY) Water is pumped from a tank, point ( of a water plant aerator, point (2), as shown in **Video V** P5.122 at a rate of 3.0 ft³/s. **(a)** Determine the power t adds to the water if the head loss from (1) to (2) where **(b)** Determine the head loss from (2) to the bottom column, point (3), if the average velocity at (3) is $V_3$ =

■ Figure P5.122

**5.123** (WILEY) Water is to be moved from one large reserv at a higher elevation as indicated in Fig. P5.123. The los energy associated with 2.5 ft³/s being pumped from s (2) is loss = $61\bar{V}^2/2$ ft²/s², where $\bar{V}$ is the average velo in the 8-in. inside-diameter piping involved. Determin of shaft power required.

■ Figure P5.123

**5.124** A $\frac{3}{4}$-hp motor is required by a 24-in.-diameter stream of air hav Determine the aerodynamic efficien

### ■ Lab Problems

**7.1LP** This problem involves the time that it takes water to drain from two geometrically similar tanks. To proceed with this problem, go to *WileyPLUS* or the book's web site, www.wiley.com/college/munson.

**7.2LP** This problem involves determining the frequency of vortex shedding from a circular cylinder as water flows past it. To proceed with this problem, go to *WileyPLUS* or the book's web site, www.wiley.com/college/munson.

**7.3LP** This problem involves the determination of the head loss for flow through a valve. To proceed with this problem, go to *WileyPLUS* or the book's web site, www.wiley.com/college/munson.

**7.4LP** This problem involves the calibration of a rotameter. To proceed with this problem, go to *WileyPLUS* or the book's web site, www.wiley.com/college/munson.

### ■ Lifelong Learning Problems

**7.1LL** Microfluidics is the study of fluid flow in fabricated devices at the micro scale. Advances in microfluidics have enhanced the ability of scientists and engineers to perform laboratory experiments using miniaturized devices known as a "lab-on-a-chip." Obtain information about a lab-on-a-chip device that is available commercially and investigate its capabilities. Summarize your findings in a brief report.

**7.2LL** For some types of aerodynamic wind tunnel testing, it is difficult to simultaneously match both the Reynolds number and Mach number between model and prototype. Engineers have developed several potential solutions to the problem including pressurized wind tunnels and lowering the temperature of the flow. Obtain information about cryogenic wind tunnels and explain the advantages and disadvantages. Summarize your findings in a brief report.

### ■ FE Exam Problems

Sample FE (Fundamental of Engineering) exam questions for fluid mechanics are provided in *WileyPLUS* or on the book's web site, www.wiley.com/college/munson.

### ■ Computational Fluid Dynamics (CFD)

The CFD problems associated with this chapter have been developed for use with the ANSYS Academic CFD software package that is associated with this text. See *WileyPLUS* or the book's web site (www.wiley.com/college/munson) for additional details.

**7.1CFD** This CFD problem involves investigation of the Reynolds number significance in fluid dynamics through the simulation of flow past a cylinder. To proceed with this problem, go to *WileyPLUS* or the book's web site, www.wiley.com/college/munson.

**There are additional CFD problems located in *WileyPLUS*.**

## Student Solutions Manual and Study Guide

a brief paperback book by Munson et al. (©2013 John Wiley & Sons, Inc.), contains detailed solutions to the Review Problems, and a study guide with brief summary and sample problems with solutions for major sections of the book.

# Contents

## 1 Introduction  1

*Learning Objectives*  1
1.1 Some Characteristics of Fluids  3
1.2 Dimensions, Dimensional
    Homogeneity, and Units  4
    1.2.1 Systems of Units  7
1.3 Analysis of Fluid Behavior  11
1.4 Measures of Fluid Mass and Weight  11
    1.4.1 Density  11
    1.4.2 Specific Weight  12
    1.4.3 Specific Gravity  12
1.5 Ideal Gas Law  12
1.6 Viscosity  14
1.7 Compressibility of Fluids  20
    1.7.1 Bulk Modulus  20
    1.7.2 Compression and Expansion
          of Gases  21
    1.7.3 Speed of Sound  22
1.8 Vapor Pressure  23
1.9 Surface Tension  24
1.10 A Brief Look Back in History  27
1.11 Chapter Summary and Study Guide  29
References  30
Review Problems  31
Conceptual Questions  31
Problems  31

## 2 Fluid Statics  40

*Learning Objectives*  40
2.1 Pressure at a Point  40
2.2 Basic Equation for Pressure Field  42
2.3 Pressure Variation in a Fluid at Rest  43
    2.3.1 Incompressible Fluid  44
    2.3.2 Compressible Fluid  47
2.4 Standard Atmosphere  49
2.5 Measurement of Pressure  50
2.6 Manometry  52
    2.6.1 Piezometer Tube  52
    2.6.2 U-Tube Manometer  53
    2.6.3 Inclined-Tube Manometer  56
2.7 Mechanical and Electronic Pressure-
    Measuring Devices  57
2.8 Hydrostatic Force on a Plane Surface  59

2.9 Pressure Prism  65
2.10 Hydrostatic Force on a Curved Surface  68
2.11 Buoyancy, Flotation, and Stability  70
    2.11.1 Archimedes' Principle  70
    2.11.2 Stability  73
2.12 Pressure Variation in a Fluid with
    Rigid-Body Motion  74
    2.12.1 Linear Motion  75
    2.12.2 Rigid-Body Rotation  77
2.13 Chapter Summary and Study Guide  79
References  80
Review Problems  80
Conceptual Questions  81
Problems  81

## 3 Elementary Fluid Dynamics—The Bernoulli Equation  101

*Learning Objectives*  101
3.1 Newton's Second Law  101
3.2 $\mathbf{F} = m\mathbf{a}$ along a Streamline  104
3.3 $\mathbf{F} = m\mathbf{a}$ Normal to a Streamline  108
3.4 Physical Interpretation  110
3.5 Static, Stagnation, Dynamic,
    and Total Pressure  113
3.6 Examples of Use of the Bernoulli
    Equation  117
    3.6.1 Free Jets  118
    3.6.2 Confined Flows  120
    3.6.3 Flowrate Measurement  126
3.7 The Energy Line and the Hydraulic
    Grade Line  131
3.8 Restrictions on Use of the
    Bernoulli Equation  134
    3.8.1 Compressibility Effects  134
    3.8.2 Unsteady Effects  136
    3.8.3 Rotational Effects  138
    3.8.4 Other Restrictions  139
3.9 Chapter Summary and Study Guide  139
References  141
Review Problems  141
Conceptual Questions  141
Problems  141

# 4 *Fluid Kinematics* 157

*Learning Objectives* 157

**4.1** The Velocity Field 157
    4.1.1 Eulerian and Lagrangian Flow Descriptions 160
    4.1.2 One-, Two-, and Three-Dimensional Flows 161
    4.1.3 Steady and Unsteady Flows 162
    4.1.4 Streamlines, Streaklines, and Pathlines 162

**4.2** The Acceleration Field 166
    4.2.1 The Material Derivative 166
    4.2.2 Unsteady Effects 169
    4.2.3 Convective Effects 169
    4.2.4 Streamline Coordinates 173

**4.3** Control Volume and System Representations 175

**4.4** The Reynolds Transport Theorem 176
    4.4.1 Derivation of the Reynolds Transport Theorem 178
    4.4.2 Physical Interpretation 183
    4.4.3 Relationship to Material Derivative 183
    4.4.4 Steady Effects 184
    4.4.5 Unsteady Effects 184
    4.4.6 Moving Control Volumes 186
    4.4.7 Selection of a Control Volume 187

**4.5** Chapter Summary and Study Guide 188
References 189
Review Problems 189
Conceptual Questions 189
Problems 190

# 5 *Finite Control Volume Analysis* 199

*Learning Objectives* 199

**5.1** Conservation of Mass—The Continuity Equation 200
    5.1.1 Derivation of the Continuity Equation 200
    5.1.2 Fixed, Nondeforming Control Volume 202
    5.1.3 Moving, Nondeforming Control Volume 208
    5.1.4 Deforming Control Volume 210

**5.2** Newton's Second Law—The Linear Momentum and Moment-of-Momentum Equations 213
    5.2.1 Derivation of the Linear Momentum Equation 213
    5.2.2 Application of the Linear Momentum Equation 214
    5.2.3 Derivation of the Moment-of-Momentum Equation 228
    5.2.4 Application of the Moment-of-Momentum Equation 229

**5.3** First Law of Thermodynamics—The Energy Equation 236
    5.3.1 Derivation of the Energy Equation 236
    5.3.2 Application of the Energy Equation 239
    5.3.3 Comparison of the Energy Equation with the Bernoulli Equation 243
    5.3.4 Application of the Energy Equation to Nonuniform Flows 249
    5.3.5 Combination of the Energy Equation and the Moment-of-Momentum Equation 252

**5.4** Second Law of Thermodynamics—Irreversible Flow 253

**5.5** Chapter Summary and Study Guide 253
References 254
Review Problems 255
Conceptual Questions 255
Problems 255

# 6 *Differential Analysis of Fluid Flow* 276

*Learning Objectives* 276

**6.1** Fluid Element Kinematics 277
    6.1.1 Velocity and Acceleration Fields Revisited 278
    6.1.2 Linear Motion and Deformation 278
    6.1.3 Angular Motion and Deformation 279

**6.2** Conservation of Mass 282
    6.2.1 Differential Form of Continuity Equation 282
    6.2.2 Cylindrical Polar Coordinates 285
    6.2.3 The Stream Function 285

**6.3** Conservation of Linear Momentum 288
    6.3.1 Description of Forces Acting on the Differential Element 289
    6.3.2 Equations of Motion 291

**6.4** Inviscid Flow   292
  6.4.1 Euler's Equations of Motion   292
  6.4.2 The Bernoulli Equation   292
  6.4.3 Irrotational Flow   294
  6.4.4 The Bernoulli Equation for
       Irrotational Flow   296
  6.4.5 The Velocity Potential   296
**6.5** Some Basic, Plane Potential Flows   286
  6.5.1 Uniform Flow   300
  6.5.2 Source and Sink   301
  6.5.3 Vortex   303
  6.5.4 Doublet   306
**6.6** Superposition of Basic, Plane Potential
    Flows   308
  6.6.1 Source in a Uniform
       Stream—Half-Body   308
  6.6.2 Rankine Ovals   311
  6.6.3 Flow around a Circular Cylinder   313
**6.7** Other Aspects of Potential Flow
    Analysis   318
**6.8** Viscous Flow   319
  6.8.1 Stress-Deformation Relationships   319
  6.8.2 The Navier–Stokes Equations   320
**6.9** Some Simple Solutions for Viscous,
    Incompressible Fluids   321
  6.9.1 Steady, Laminar Flow between
       Fixed Parallel Plates   322
  6.9.2 Couette Flow   324
  6.9.3 Steady, Laminar Flow in
       Circular Tubes   326
  6.9.4 Steady, Axial, Laminar Flow
       in an Annulus   329
**6.10** Other Aspects of Differential Analysis   331
  6.10.1 Numerical Methods   331
**6.11** Chapter Summary and Study Guide   332
References   333
Review Problems   334
Conceptual Questions   334
Problems   334

**7 Dimensional Analysis,
Similitude, and Modeling**   346

*Learning Objectives*   346
**7.1** Dimensional Analysis   347
**7.2** Buckingham Pi Theorem   349
**7.3** Determination of Pi Terms   350
**7.4** Some Additional Comments about
    Dimensional Analysis   355
  7.4.1 Selection of Variables   355

  7.4.2 Determination of Reference
       Dimensions   356
  7.4.3 Uniqueness of Pi Terms   358
**7.5** Determination of Pi Terms by
    Inspection   359
**7.6** Common Dimensionless Groups
    in Fluid Mechanics   360
**7.7** Correlation of Experimental Data   364
  7.7.1 Problems with One Pi Term   365
  7.7.2 Problems with Two or More
       Pi Terms   366
**7.8** Modeling and Similitude   368
  7.8.1 Theory of Models   368
  7.8.2 Model Scales   372
  7.8.3 Practical Aspects of Using
       Models   372
**7.9** Some Typical Model Studies   374
  7.9.1 Flow through Closed Conduits   374
  7.9.2 Flow around Immersed Bodies   377
  7.9.3 Flow with a Free Surface   381
**7.10** Similitude Based on Governing
    Differential Equations   384
**7.11** Chapter Summary and Study Guide   387
References   388
Review Problems   388
Conceptual Questions   389
Problems   389

**8 Viscous Flow in Pipes**   400

*Learning Objectives*   400
**8.1** General Characteristics of Pipe Flow   401
  8.1.1 Laminar or Turbulent Flow   402
  8.1.2 Entrance Region and Fully
       Developed Flow   405
  8.1.3 Pressure and Shear Stress   406
**8.2** Fully Developed Laminar Flow   407
  8.2.1 From $\mathbf{F} = m\mathbf{a}$ Applied Directly
       to a Fluid Element   407
  8.2.2 From the Navier–Stokes
       Equations   411
  8.2.3 From Dimensional Analysis   413
  8.2.4 Energy Considerations   414
**8.3** Fully Developed Turbulent Flow   416
  8.3.1 Transition from Laminar to
       Turbulent Flow   416
  8.3.2 Turbulent Shear Stress   418
  8.3.3 Turbulent Velocity Profile   422
  8.3.4 Turbulence Modeling   426
  8.3.5 Chaos and Turbulence   426

**8.4**    Dimensional Analysis of Pipe Flow    426
    8.4.1    Major Losses    427
    8.4.2    Minor Losses    432
    8.4.3    Noncircular Conduits    442
**8.5**    Pipe Flow Examples    445
    8.5.1    Single Pipes    445
    8.5.2    Multiple Pipe Systems    455
**8.6**    Pipe Flowrate Measurement    459
    8.6.1    Pipe Flowrate Meters    459
    8.6.2    Volume Flowmeters    464
**8.7**    Chapter Summary and Study Guide    465
References    467
Review Problems    468
Conceptual Questions    468
Problems    468

**9 Flow Over Immersed Bodies**    480

*Learning Objectives*    480
**9.1**    General External Flow Characteristics    481
    9.1.1    Lift and Drag Concepts    482
    9.1.2    Characteristics of Flow Past an Object    485
**9.2**    Boundary Layer Characteristics    489
    9.2.1    Boundary Layer Structure and Thickness on a Flat Plate    489
    9.2.2    Prandtl/Blasius Boundary Layer Solution    493
    9.2.3    Momentum Integral Boundary Layer Equation for a Flat Plate    497
    9.2.4    Transition from Laminar to Turbulent Flow    502
    9.2.5    Turbulent Boundary Layer Flow    504
    9.2.6    Effects of Pressure Gradient    507
    9.2.7    Momentum Integral Boundary Layer Equation with Nonzero Pressure Gradient    511
**9.3**    Drag    512
    9.3.1    Friction Drag    513
    9.3.2    Pressure Drag    514
    9.3.3    Drag Coefficient Data and Examples    516
**9.4**    Lift    528
    9.4.1    Surface Pressure Distribution    528
    9.4.2    Circulation    537
**9.5**    Chapter Summary and Study Guide    541
References    542
Review Problems    543
Conceptual Questions    543
Problems    544

**10 Open-Channel Flow**    554

*Learning Objectives*    554
**10.1**    General Characteristics of Open-Channel Flow    555
**10.2**    Surface Waves    556
    10.2.1    Wave Speed    556
    10.2.2    Froude Number Effects    559
**10.3**    Energy Considerations    561
    10.3.1    Specific Energy    562
    10.3.2    Channel Depth Variations    565
**10.4**    Uniform Depth Channel Flow    566
    10.4.1    Uniform Flow Approximations    566
    10.4.2    The Chezy and Manning Equations    567
    10.4.3    Uniform Depth Examples    570
**10.5**    Gradually Varied Flow    575
**10.6**    Rapidly Varied Flow    576
    10.6.1    The Hydraulic Jump    577
    10.6.2    Sharp-Crested Weirs    582
    10.6.3    Broad-Crested Weirs    585
    10.6.4    Underflow Gates    587
**10.7**    Chapter Summary and Study Guide    589
References    590
Review Problems    591
Conceptual Questions    591
Problems    591

**11 Compressible Flow**    601

*Learning Objectives*    601
**11.1**    Ideal Gas Relationships    602
**11.2**    Mach Number and Speed of Sound    607
**11.3**    Categories of Compressible Flow    610
**11.4**    Isentropic Flow of an Ideal Gas    614
    11.4.1    Effect of Variations in Flow Cross-Sectional Area    615
    11.4.2    Converging–Diverging Duct Flow    617
    11.4.3    Constant Area Duct Flow    631
**11.5**    Nonisentropic Flow of an Ideal Gas    631
    11.5.1    Adiabatic Constant Area Duct Flow with Friction (Fanno Flow)    631
    11.5.2    Frictionless Constant Area Duct Flow with Heat Transfer (Rayleigh Flow)    642
    11.5.3    Normal Shock Waves    648
**11.6**    Analogy between Compressible and Open-Channel Flows    655
**11.7**    Two-Dimensional Compressible Flow    657

**11.8** Chapter Summary and Study Guide   658
References   661
Review Problems   662
Conceptual Questions   662
Problems   662

## 12 *Turbomachines*   667

*Learning Objectives*   667
**12.1** Introduction   668
**12.2** Basic Energy Considerations   669
**12.3** Basic Angular Momentum
     Considerations   673
**12.4** The Centrifugal Pump   675
    12.4.1   Theoretical Considerations   676
    12.4.2   Pump Performance
          Characteristics   680
    12.4.3   Net Positive Suction Head
          (NPSH)   682
    12.4.4   System Characteristics and
          Pump Selection   684
**12.5** Dimensionless Parameters and
     Similarity Laws   688
    12.5.1   Special Pump Scaling Laws   690
    12.5.2   Specific Speed   691
    12.5.3   Suction Specific Speed   692
**12.6** Axial-Flow and Mixed-Flow Pumps   693
**12.7** Fans   695
**12.8** Turbines   695
    12.8.1   Impulse Turbines   696
    12.8.2   Reaction Turbines   704
**12.9** Compressible Flow Turbomachines   707
    12.9.1   Compressors   708
    12.9.2   Compressible Flow Turbines   711
**12.10** Chapter Summary and Study Guide   713
References   715
Review Problems   715
Conceptual Questions   715
Problems   716

*A Computational Fluid
Dynamics*   725

*B Physical Properties of
Fluids*   737

*C Properties of the U.S.
Standard Atmosphere*   742

*D Compressible Flow Graphs
for an Ideal Gas (k = 1.4)*   744

*E Comprehensive Table of
Conversion Factors*
*See www.wiley.com/college/munson or
WileyPLUS for this material.*

*F CFD Problems and Tutorials*
*See www.wiley.com/college/munson or
WileyPLUS for this material.*

*G Review Problems*
*See www.wiley.com/college/munson or
WileyPLUS for this material.*

*H Lab Problems*
*See www.wiley.com/college/munson or
WileyPLUS for this material.*

*I CFD Driven Cavity Example*
*See www.wiley.com/college/munson or
WileyPLUS for this material.*

*Answers*   ANS-1

*Index*   I-1

*Video Index*   VI-1

# 1 Introduction

CHAPTER OPENING PHOTO: The nature of air bubbles rising in a liquid is a function of fluid properties such as density, viscosity, and surface tension. (Left: air in oil; right: air in soap.) (Photographs copyright 2007 by Andrew Davidhazy, Rochester Institute of Technology.)

## Learning Objectives

After completing this chapter, you should be able to:

- determine the dimensions and units of physical quantities.
- identify the key fluid properties used in the analysis of fluid behavior.
- calculate common fluid properties given appropriate information.
- explain effects of fluid compressibility.
- use the concepts of viscosity, vapor pressure, and surface tension.

Fluid mechanics is the discipline within the broad field of applied mechanics that is concerned with the behavior of liquids and gases at rest or in motion. It covers a vast array of phenomena that occur in nature (with or without human intervention), in biology, and in numerous engineered, invented, or manufactured situations. There are few aspects of our lives that do not involve fluids, either directly or indirectly.

The immense range of different flow conditions is mind-boggling and strongly dependent on the value of the numerous parameters that describe fluid flow. Among the long list of parameters involved are (1) the physical size of the flow, $\ell$; (2) the speed of the flow, $V$; and (3) the pressure, $p$, as indicated in the figure in the margin for a light aircraft parachute recovery system. These are just three of the important parameters that, along with many others, are discussed in detail in various sections of this book. To get an inkling of the range of some of the parameter values involved and the flow situations generated, consider the following.

(Photograph courtesy of CIRRUS Design Corporation.)

- Size, $\ell$

Every flow has a characteristic (or typical) length associated with it. For example, for flow of fluid within pipes, the pipe diameter is a characteristic length. Pipe flows include

1

the flow of water in the pipes in our homes, the blood flow in our arteries and veins, and the airflow in our bronchial tree. They also involve pipe sizes that are not within our everyday experiences. Such examples include the flow of oil across Alaska through a 4-foot-diameter, 799-mile-long pipe and, at the other end of the size scale, the new area of interest involving flow in nano scale pipes whose diameters are on the order of $10^{-8}$ m. Each of these pipe flows has important characteristics that are not found in the others.

Characteristic lengths of some other flows are shown in Fig. 1.1$a$.

■ Speed, $V$

As we note from The Weather Channel, on a given day the wind speed may cover what we think of as a wide range, from a gentle 5-mph breeze to a 100-mph hurricane or a 250-mph tornado. However, this speed range is small compared to that of the almost imperceptible flow of the fluid-like magma below the Earth's surface that drives the continental drift motion of the tectonic plates at a speed of about $2 \times 10^{-8}$ m/s or the hypersonic airflow past a meteor as it streaks through the atmosphere at $3 \times 10^4$ m/s.

Characteristic speeds of some other flows are shown in Fig. 1.1$b$.

■ Pressure, $p$

The pressure within fluids covers an extremely wide range of values. We are accustomed to the 35 psi (lb/in.$^2$) pressure within our car's tires, the "120 over 70" typical blood pressure reading, or the standard 14.7 psi atmospheric pressure. However, the large 10,000 psi pressure in the hydraulic ram of an earth mover or the tiny $2 \times 10^{-6}$ psi pressure of a sound wave generated at ordinary talking levels are not easy to comprehend.

Characteristic pressures of some other flows are shown in Fig. 1.1$c$.

*V1.1 Mt. St. Helens eruption*

*V1.2 E. coli swimming*

■ **Figure 1.1**   Characteristic values of some fluid flow parameters for a variety of flows: ($a$) object size, ($b$) fluid speed, ($c$) fluid pressure.

The list of fluid mechanics applications goes on and on. But you get the point. Fluid mechanics is a very important, practical subject that encompasses a wide variety of situations. It is very likely that during your career as an engineer you will be involved in the analysis and design of systems that require a good understanding of fluid mechanics. Although it is not possible to adequately cover all of the important areas of fluid mechanics within one book, it is hoped that this introductory text will provide a sound foundation of the fundamental aspects of fluid mechanics.

## 1.1 Some Characteristics of Fluids

One of the first questions we need to explore is—what is a fluid? Or we might ask—what is the difference between a solid and a fluid? We have a general, vague idea of the difference. A solid is "hard" and not easily deformed, whereas a fluid is "soft" and is easily deformed (we can readily move through air). Although quite descriptive, these casual observations of the differences between solids and fluids are not very satisfactory from a scientific or engineering point of view. A closer look at the molecular structure of materials reveals that matter that we commonly think of as a solid (steel, concrete, etc.) has densely spaced molecules with large intermolecular cohesive forces that allow the solid to maintain its shape, and to not be easily deformed. However, for matter that we normally think of as a liquid (water, oil, etc.), the molecules are spaced farther apart, the intermolecular forces are smaller than for solids, and the molecules have more freedom of movement. Thus, liquids can be easily deformed (but not easily compressed) and can be poured into containers or forced through a tube. Gases (air, oxygen, etc.) have even greater molecular spacing and freedom of motion with negligible cohesive intermolecular forces, and as a consequence are easily deformed (and compressed) and will completely fill the volume of any container in which they are placed. Both liquids and gases are fluids.

*Both liquids and gases are fluids.*

---

**F l u i d s   i n   t h e   N e w s**

**Will what works in air work in water?** For the past few years a San Francisco company has been working on small, maneuverable submarines designed to travel through water using wings, controls, and thrusters that are similar to those on jet airplanes. After all, water (for submarines) and air (for airplanes) are both fluids, so it is expected that many of the principles governing the flight of airplanes should carry over to the "flight" of winged submarines. Of course, there are differences. For example, the submarine must be designed to withstand external pressures of nearly 700 pounds per square inch greater than that inside the vehicle. On the other hand, at high altitude where commercial jets fly, the exterior pressure is 3.5 psi rather than standard sea-level pressure of 14.7 psi, so the vehicle must be pressurized internally for passenger comfort. In both cases, however, the design of the craft for minimal drag, maximum lift, and efficient thrust is governed by the same fluid dynamic concepts.

---

Although the differences between solids and fluids can be explained qualitatively on the basis of molecular structure, a more specific distinction is based on how they deform under the action of an external load. Specifically, *a fluid is defined as a substance that deforms continuously when acted on by a shearing stress of any magnitude.* A shearing stress (force per unit area) is created whenever a tangential force acts on a surface as shown by the figure in the margin. When common solids such as steel or other metals are acted on by a shearing stress, they will initially deform (usually a very small deformation), but they will not continuously deform (flow). However, common fluids such as water, oil, and air satisfy the definition of a fluid—that is, they will flow when acted on by a shearing stress. Some materials, such as slurries, tar, putty, toothpaste, and so on, are not easily classified since they will behave as a solid if the applied shearing stress is small, but if the stress exceeds some critical value, the substance will flow. The study of such materials is called *rheology* and does not fall within the province of classical fluid mechanics. Thus, all the fluids we will be concerned with in this text will conform to the definition of a fluid given previously.

Although the molecular structure of fluids is important in distinguishing one fluid from another, it is not yet practical to study the behavior of individual molecules when trying to describe the behavior of fluids at rest or in motion. Rather, we characterize the behavior by considering the average, or macroscopic, value of the quantity of interest, where the average is evaluated over a small volume containing a large number of molecules. Thus, when we say that the velocity at a certain point in a fluid is so much, we are really indicating the average velocity of the molecules in a small volume surrounding the point. The volume is small compared with the physical dimensions of the system of interest, but large compared with the average distance between molecules. Is this a reasonable way to describe the behavior of a fluid? The answer is generally yes, since the spacing between molecules is typically very small. For gases at normal pressures and temperatures, the spacing is on the order of $10^{-6}$ mm, and for liquids it is on the order of $10^{-7}$ mm. The number of molecules per cubic millimeter is on the order of $10^{18}$ for gases and $10^{21}$ for liquids. It is thus clear that the number of molecules in a very tiny volume is huge and the idea of using average values taken over this volume is certainly reasonable. We thus assume that all the fluid characteristics we are interested in (pressure, velocity, etc.) vary continuously throughout the fluid—that is, we treat the fluid as a *continuum*. This concept will certainly be valid for all the circumstances considered in this text. One area of fluid mechanics for which the continuum concept breaks down is in the study of rarefied gases such as would be encountered at very high altitudes. In this case the spacing between air molecules can become large and the continuum concept is no longer acceptable.

## 1.2 Dimensions, Dimensional Homogeneity, and Units

Since in our study of fluid mechanics we will be dealing with a variety of fluid characteristics, it is necessary to develop a system for describing these characteristics both *qualitatively* and *quantitatively*. The qualitative aspect serves to identify the nature, or type, of the characteristics (such as length, time, stress, and velocity), whereas the quantitative aspect provides a numerical measure of the characteristics. The quantitative description requires both a number and a standard by which various quantities can be compared. A standard for length might be a meter or foot, for time an hour or second, and for mass a slug or kilogram. Such standards are called **units,** and several systems of units are in common use as described in the following section. The qualitative description is conveniently given in terms of certain *primary quantities,* such as length, $L$, time, $T$, mass, $M$, and temperature, $\Theta$. These primary quantities can then be used to provide a qualitative description of any other *secondary quantity*: for example, area $\doteq L^2$, velocity $\doteq LT^{-1}$, density $\doteq ML^{-3}$, and so on, where the symbol $\doteq$ is used to indicate the *dimensions* of the secondary quantity in terms of the primary quantities. Thus, to describe qualitatively a velocity, $V$, we would write

*Fluid characteristics can be described qualitatively in terms of certain basic quantities such as length, time, and mass.*

$$V \doteq LT^{-1}$$

and say that "the dimensions of a velocity equal length divided by time." The primary quantities are also referred to as **basic dimensions.**

For a wide variety of problems involving fluid mechanics, only the three basic dimensions, $L$, $T$, and $M$ are required. Alternatively, $L$, $T$, and $F$ could be used, where $F$ is the basic dimensions of force. Since Newton's law states that force is equal to mass times acceleration, it follows that $F \doteq MLT^{-2}$ or $M \doteq FL^{-1}T^2$. Thus, secondary quantities expressed in terms of $M$ can be expressed in terms of $F$ through the relationship above. For example, stress, $\sigma$, is a force per unit area, so that $\sigma \doteq FL^{-2}$, but an equivalent dimensional equation is $\sigma \doteq ML^{-1}T^{-2}$. Table 1.1 provides a list of dimensions for a number of common physical quantities.

All theoretically derived equations are **dimensionally homogeneous**—that is, the dimensions of the left side of the equation must be the same as those on the right side, and all additive separate terms must have the same dimensions. We accept as a fundamental premise that all equations describing physical phenomena must be dimensionally homogeneous. If this were not true, we would be attempting to equate or add unlike physical quantities, which would not make sense. For example, the equation for the velocity, $V$, of a uniformly accelerated body is

$$V = V_0 + at \tag{1.1}$$

**■ Table 1.1**
**Dimensions Associated with Common Physical Quantities**

| | FLT System | MLT System | | FLT System | MLT System |
|---|---|---|---|---|---|
| Acceleration | $LT^{-2}$ | $LT^{-2}$ | Power | $FLT^{-1}$ | $ML^2T^{-3}$ |
| Angle | $F^0L^0T^0$ | $M^0L^0T^0$ | Pressure | $FL^{-2}$ | $ML^{-1}T^{-2}$ |
| Angular acceleration | $T^{-2}$ | $T^{-2}$ | Specific heat | $L^2T^{-2}\Theta^{-1}$ | $L^2T^{-2}\Theta^{-1}$ |
| Angular velocity | $T^{-1}$ | $T^{-1}$ | Specific weight | $FL^{-3}$ | $ML^{-2}T^{-2}$ |
| Area | $L^2$ | $L^2$ | Strain | $F^0L^0T^0$ | $M^0L^0T^0$ |
| Density | $FL^{-4}T^2$ | $ML^{-3}$ | Stress | $FL^{-2}$ | $ML^{-1}T^{-2}$ |
| Energy | $FL$ | $ML^2T^{-2}$ | Surface tension | $FL^{-1}$ | $MT^{-2}$ |
| Force | $F$ | $MLT^{-2}$ | Temperature | $\Theta$ | $\Theta$ |
| Frequency | $T^{-1}$ | $T^{-1}$ | Time | $T$ | $T$ |
| Heat | $FL$ | $ML^2T^{-2}$ | Torque | $FL$ | $ML^2T^{-2}$ |
| Length | $L$ | $L$ | Velocity | $LT^{-1}$ | $LT^{-1}$ |
| Mass | $FL^{-1}T^2$ | $M$ | Viscosity (dynamic) | $FL^{-2}T$ | $ML^{-1}T^{-1}$ |
| Modulus of elasticity | $FL^{-2}$ | $ML^{-1}T^{-2}$ | Viscosity (kinematic) | $L^2T^{-1}$ | $L^2T^{-1}$ |
| Moment of a force | $FL$ | $ML^2T^{-2}$ | Volume | $L^3$ | $L^3$ |
| Moment of inertia (area) | $L^4$ | $L^4$ | Work | $FL$ | $ML^2T^{-2}$ |
| Moment of inertia (mass) | $FLT^2$ | $ML^2$ | | | |
| Momentum | $FT$ | $MLT^{-1}$ | | | |

where $V_0$ is the initial velocity, $a$ the acceleration, and $t$ the time interval. In terms of dimensions the equation is

$$LT^{-1} \doteq LT^{-1} + LT^{-2}T$$

and thus Eq. 1.1 is dimensionally homogeneous.

Some equations that are known to be valid contain constants having dimensions. The equation for the distance, $d$, traveled by a freely falling body can be written as

$$d = 16.1t^2 \tag{1.2}$$

and a check of the dimensions reveals that the constant must have the dimensions of $LT^{-2}$ if the equation is to be dimensionally homogeneous. Actually, Eq. 1.2 is a special form of the well-known equation from physics for freely falling bodies,

$$d = \frac{gt^2}{2} \tag{1.3}$$

in which $g$ is the acceleration of gravity. Equation 1.3 is dimensionally homogeneous and valid in any system of units. For $g = 32.2$ ft/s$^2$ the equation reduces to Eq. 1.2 and thus Eq. 1.2 is valid only for the system of units using feet and seconds. Equations that are restricted to a particular system of units can be denoted as *restricted homogeneous equations,* as opposed to equations valid in any system of units, which are *general homogeneous equations.* The preceding discussion indi-

*General homogeneous equations are valid in any system of units.*

cates one rather elementary, but important, use of the concept of dimensions: the determination of one aspect of the generality of a given equation simply based on a consideration of the dimensions of the various terms in the equation. The concept of dimensions also forms the basis for the powerful tool of *dimensional analysis,* which is considered in detail in Chapter 7.

*Note to the users of this text.*   All of the examples in the text use a consistent problem-solving methodology, which is similar to that in other engineering courses such as statics. Each example highlights the key elements of analysis: *Given, Find, Solution,* and *Comment.*

The *Given* and *Find* are steps that ensure the user understands what is being asked in the problem and explicitly list the items provided to help solve the problem.

The *Solution* step is where the equations needed to solve the problem are formulated and the problem is actually solved. In this step, there are typically several other tasks that help to set up the solution and are required to solve the problem. The first is a drawing of the problem; where appropriate, it is always helpful to draw a sketch of the problem. Here the relevant geometry and coordinate system to be used as well as features such as control volumes, forces and pressures, velocities, and mass flow rates are included. This helps in gaining a visual understanding of the problem. Making appropriate assumptions to solve the problem is the second task. In a realistic engineering problem-solving environment, the necessary assumptions are developed as an integral part of the solution process. Assumptions can provide appropriate simplifications or offer useful constraints, both of which can help in solving the problem. Throughout the examples in this text, the necessary assumptions are embedded within the **Solution** step, as they are in solving a real-world problem. This provides a realistic problem-solving experience.

The final element in the methodology is the **Comment.** For the examples in the text, this section is used to provide further insight into the problem or the solution. It can also be a point in the analysis at which certain questions are posed. For example: Is the answer reasonable, and does it make physical sense? Are the final units correct? If a certain parameter were changed, how would the answer change? Adopting this type of methodology will aid in the development of problem-solving skills for fluid mechanics, as well as other engineering disciplines.

---

## EXAMPLE 1.1   Restricted and General Homogeneous Equations

**GIVEN**   A liquid flows through an orifice located in the side of a tank as shown in Fig. E1.1. A commonly used equation for determining the volume rate of flow, $Q$, through the orifice is

$$Q = 0.61\, A\sqrt{2gh}$$

where $A$ is the area of the orifice, $g$ is the acceleration of gravity, and $h$ is the height of the liquid above the orifice.

**FIND**   Investigate the dimensional homogeneity of this formula.

*(a)*                          *(b)*

■ **Figure E1.1**

### SOLUTION

The dimensions of the various terms in the equation are $Q =$ volume/time $\doteq L^3T^{-1}$, $A =$ area $\doteq L^2$, $g =$ acceleration of gravity $\doteq LT^{-2}$, and $h =$ height $\doteq L$.

These terms, when substituted into the equation, yield the dimensional form:

$$(L^3T^{-1}) \doteq (0.61)(L^2)(\sqrt{2})(LT^{-2})^{1/2}(L)^{1/2}$$

or

$$(L^3T^{-1}) \doteq [0.61\sqrt{2}](L^3T^{-1})$$

It is clear from this result that the equation is dimensionally homogeneous (both sides of the formula have the same dimensions of $L^3T^{-1}$), and the number $0.61\sqrt{2}$ is dimensionless.

If we were going to use this relationship repeatedly, we might be tempted to simplify it by replacing $g$ with its standard value of 32.2 ft/s$^2$ and rewriting the formula as

$$Q = 4.90\, A\sqrt{h} \qquad (1)$$

A quick check of the dimensions reveals that

$$L^3T^{-1} \doteq (4.90)(L^{5/2})$$

and, therefore, the equation expressed as Eq. 1 can only be dimensionally correct if the number 4.90 has the dimensions of $L^{1/2}T^{-1}$. Whenever a number appearing in an equation or formula has dimensions, it means that the specific value of the number will depend on the system of units used. Thus, for the case being considered with feet and seconds used as units, the number 4.90 has units of ft$^{1/2}$/s. Equation 1 will only give the correct value for $Q$ (in ft$^3$/s) when $A$ is expressed in square feet and $h$ in feet. Thus, Eq. 1 is a *restricted* homogeneous equation, whereas the original equation is a *general* homogeneous equation that would be valid for any consistent system of units.

**COMMENT**   A quick check of the dimensions of the various terms in an equation is a useful practice and will often be helpful in eliminating errors—that is, as noted previously, all physically meaningful equations must be dimensionally homogeneous. We have briefly alluded to units in this example, and this important topic will be considered in more detail in the next section.

## 1.2.1 Systems of Units

In addition to the qualitative description of the various quantities of interest, it is generally necessary to have a quantitative measure of any given quantity. For example, if we measure the width of this page in the book and say that it is 10 units wide, the statement has no meaning until the unit of length is defined. If we indicate that the unit of length is a meter, and define the meter as some standard length, a unit system for length has been established (and a numerical value can be given to the page width). In addition to length, a unit must be established for each of the remaining basic quantities (force, mass, time, and temperature). There are several systems of units in use, and we shall consider three systems that are commonly used in engineering.

***International System (SI).*** In 1960 the Eleventh General Conference on Weights and Measures, the international organization responsible for maintaining precise uniform standards of measurements, formally adopted the *International System of Units* as the international standard. This system, commonly termed SI, has been widely adopted worldwide and is widely used (although certainly not exclusively) in the United States. It is expected that the long-term trend will be for all countries to accept SI as the accepted standard and it is imperative that engineering students become familiar with this system. In SI the unit of length is the meter (m), the time unit is the second (s), the mass unit is the kilogram (kg), and the temperature unit is the kelvin (K). Note that there is no degree symbol used when expressing a temperature in kelvin units. The kelvin temperature scale is an absolute scale and is related to the Celsius (centigrade) scale (°C) through the relationship

$$K = °C + 273.15$$

Although the Celsius scale is not in itself part of SI, it is common practice to specify temperatures in degrees Celsius when using SI units.

The force unit, called the newton (N), is defined from Newton's second law as

$$1 \text{ N} = (1 \text{ kg})(1 \text{ m/s}^2)$$

*In mechanics it is very important to distinguish between weight and mass.*

Thus, a 1-N force acting on a 1-kg mass will give the mass an acceleration of $1 \text{ m/s}^2$. Standard gravity in SI is $9.807 \text{ m/s}^2$ (commonly approximated as $9.81 \text{ m/s}^2$) so that a 1-kg mass weighs 9.81 N under standard gravity. Note that weight and mass are different, both qualitatively and quantitatively! The unit of *work* in SI is the joule (J), which is the work done when the point of application of a 1-N force is displaced through a 1-m distance in the direction of a force. Thus,

$$1 \text{ J} = 1 \text{ N} \cdot \text{m}$$

The unit of *power* is the watt (W) defined as a joule per second. Thus,

$$1 \text{ W} = 1 \text{ J/s} = 1 \text{ N} \cdot \text{m/s}$$

Prefixes for forming multiples and fractions of SI units are given in Table 1.2. For example, the notation kN would be read as "kilonewtons" and stands for $10^3$ N. Similarly, mm would be read as "millimeters" and stands for $10^{-3}$ m. The centimeter is not an accepted unit of length in

■ Table 1.2
**Prefixes for SI Units**

| Factor by Which Unit Is Multiplied | Prefix | Symbol | Factor by Which Unit Is Multiplied | Prefix | Symbol |
|---|---|---|---|---|---|
| $10^{15}$ | peta | P | $10^{-2}$ | centi | c |
| $10^{12}$ | tera | T | $10^{-3}$ | milli | m |
| $10^{9}$ | giga | G | $10^{-6}$ | micro | $\mu$ |
| $10^{6}$ | mega | M | $10^{-9}$ | nano | n |
| $10^{3}$ | kilo | k | $10^{-12}$ | pico | p |
| $10^{2}$ | hecto | h | $10^{-15}$ | femto | f |
| $10$ | deka | da | $10^{-18}$ | atto | a |
| $10^{-1}$ | deci | d | | | |

the SI system, so for most problems in fluid mechanics in which SI units are used, lengths will be expressed in millimeters or meters.

**British Gravitational (BG) System.** In the BG system the unit of length is the foot (ft), the time unit is the second (s), the force unit is the pound (lb), and the temperature unit is the degree Fahrenheit (°F) or the absolute temperature unit is the degree Rankine (°R), where

$$°R = °F + 459.67$$

The mass unit, called the *slug,* is defined from Newton's second law (force = mass × acceleration) as

$$1 \text{ lb} = (1 \text{ slug})(1 \text{ ft }/\text{s}^2)$$

This relationship indicates that a 1-lb force acting on a mass of 1 slug will give the mass an acceleration of 1 ft/s$^2$.

The weight, $\mathcal{W}$ (which is the force due to gravity, $g$), of a mass, $m$, is given by the equation

$$\mathcal{W} = mg$$

and in BG units

$$\mathcal{W}(\text{lb}) = m(\text{slugs}) \, g(\text{ft}/\text{s}^2)$$

Since Earth's standard gravity is taken as $g = 32.174$ ft/s$^2$ (commonly approximated as 32.2 ft/s$^2$), it follows that a mass of 1 slug weighs 32.2 lb under standard gravity.

*Two systems of units that are widely used in engineering are the British Gravitational (BG) System and the International System (SI).*

---

## F l u i d s   i n   t h e   N e w s

**How long is a foot?** Today, in the United States, the common length *unit* is the *foot,* but throughout antiquity the unit used to measure length has quite a history. The first length units were based on the lengths of various body parts. One of the earliest units was the Egyptian cubit, first used around 3000 B.C. and defined as the length of the arm from elbow to extended fingertips. Other measures followed, with the foot simply taken as the length of a man's foot. Since this length obviously varies from person to person it was often "standardized" by using the length of the current reigning royalty's foot. In 1791 a special French commission proposed that a new universal length unit called a meter (metre) be defined as the distance of one-quarter of the Earth's meridian (north pole to the equator) divided by 10 million. Although controversial, the meter was accepted in 1799 as the standard. With the development of advanced technology, the length of a meter was redefined in 1983 as the distance traveled by light in a vacuum during the time interval of 1/299,792,458 s. The foot is now defined as 0.3048 meter. Our simple rulers and yardsticks indeed have an intriguing history.

---

**English Engineering (EE) System.** In the EE system, units for force *and* mass are defined independently; thus special care must be exercised when using this system in conjunction with Newton's second law. The basic unit of mass is the pound mass (lbm), and the unit of force is the pound (lb).[1] The unit of length is the foot (ft), the unit of time is the second (s), and the absolute temperature scale is the degree Rankine (°R). To make the equation expressing Newton's second law dimensionally homogeneous we write it as

$$\mathbf{F} = \frac{m\mathbf{a}}{g_c} \tag{1.4}$$

where $g_c$ is a constant of proportionality, which allows us to define units for both force and mass. For the BG system, only the force unit was prescribed and the mass unit defined in a consistent manner such that $g_c = 1$. Similarly, for SI the mass unit was prescribed and the force unit defined in a consistent manner such that $g_c = 1$. For the EE system, a 1-lb force is defined as that force which gives a 1 lbm a standard acceleration of gravity, which is taken as 32.174 ft/s$^2$. Thus, for Eq. 1.4 to be both numerically and dimensionally correct

$$1 \text{ lb} = \frac{(1 \text{ lbm})(32.174 \text{ ft/s}^2)}{g_c}$$

---

[1]It is also common practice to use the notation, lbf, to indicate pound force.

**Figure 1.2** Comparison of SI, BG, and EE units.

so that

$$g_c = \frac{(1 \text{ lbm})(32.174 \text{ ft/s}^2)}{(1 \text{ lb})}$$

With the EE system, weight and mass are related through the equation

$$\mathcal{W} = \frac{mg}{g_c}$$

where $g$ is the local acceleration of gravity. Under conditions of standard gravity ($g = g_c$) the weight in pounds and the mass in pound mass are numerically equal. Also, since a 1-lb force gives a mass of 1 lbm an acceleration of 32.174 ft /s$^2$ and a mass of 1 slug an acceleration of 1 ft /s$^2$, it follows that

$$1 \text{ slug} = 32.174 \text{ lbm}$$

*In this text we will primarily use the BG system and SI for units.* The EE system is used very sparingly, and only in those instances where convention dictates its use, such as for the compressible flow material in Chapter 11. Approximately one-half the problems and examples are given in BG units and one-half in SI units. We cannot overemphasize the importance of paying close attention to units when solving problems. It is very easy to introduce huge errors into problem solutions through the use of incorrect units. Get in the habit of using a *consistent* system of units throughout a given solution. It really makes no difference which system you use as long as you are consistent; for example, don't mix slugs and newtons. If problem data are specified in SI units, then use SI units throughout the solution. If the data are specified in BG units, then use BG units throughout the solution. The relative sizes of the SI, BG, and EE units of length, mass, and force are shown in Fig. 1.2.

Tables 1.3 and 1.4 provide conversion factors for some quantities that are commonly encountered in fluid mechanics. For convenient reference these tables are reproduced on the inside of the back cover. Note that in these tables (and others) the numbers are expressed by using computer exponential notation. For example, the number 5.154 E + 2 is equivalent to $5.154 \times 10^2$ in scientific notation, and the number 2.832 E − 2 is equivalent to $2.832 \times 10^{-2}$. More extensive tables of conversion factors for a large variety of unit systems can be found in Appendix E.

*When solving problems it is important to use a consistent system of units, e.g., don't mix BG and SI units.*

**■ Table 1.3**

**Conversion Factors from BG and EE Units to SI Units**

(See inside of back cover.)

**■ Table 1.4**

**Conversion Factors from SI Units to BG and EE Units**

(See inside of back cover.)

## EXAMPLE 1.2  BG and SI Units

**GIVEN**  A tank of liquid having a total mass of 36 kg rests on a support in the equipment bay of the Space Shuttle.

**FIND**  Determine the force (in newtons) that the tank exerts on the support shortly after lift off when the shuttle is accelerating upward as shown in Fig. E1.2a at 15 ft /s².

### SOLUTION

A free-body diagram of the tank is shown in Fig. E1.2b, where $\mathcal{W}$ is the weight of the tank and liquid, and $F_f$ is the reaction of the floor on the tank. Application of Newton's second law of motion to this body gives

$$\sum \mathbf{F} = m\,\mathbf{a}$$

or

$$F_f - \mathcal{W} = ma \tag{1}$$

where we have taken upward as the positive direction. Since $\mathcal{W} = mg$, Eq. 1 can be written as

$$F_f = m\,(g + a) \tag{2}$$

Before substituting any number into Eq. 2, we must decide on a system of units, and then be sure all of the data are expressed in these units. Since we want $F_f$ in newtons, we will use SI units so that

$$F_f = 36 \text{ kg} \left[ 9.81 \text{ m /s}^2 + (15 \text{ ft /s}^2)(0.3048 \text{ m /ft}) \right]$$
$$= 518 \text{ kg} \cdot \text{m /s}^2$$

Since 1 N = 1 kg · m /s², it follows that

$$F_f = 518 \text{ N} \qquad \text{(downward on floor)} \qquad \textbf{(Ans)}$$

■ **Figure E1.2a**  (Photograph courtesy of NASA.)

The direction is downward since the force shown on the free-body diagram is the force of the support *on the tank* so that the force the tank exerts *on the support* is equal in magnitude but opposite in direction.

**COMMENT**  As you work through a large variety of problems in this text, you will find that units play an essential role in arriving at a numerical answer. Be careful! It is easy to mix units and cause large errors. If in the above example the acceleration had been left as 15 ft/s² with $m$ and $g$ expressed in SI units, we would have calculated the force as 893 N and the answer would have been 72% too large!

■ **Figure E1.2b**

### F l u i d s   i n   t h e   N e w s

**Units and space travel**  A NASA spacecraft, the Mars Climate Orbiter, was launched in December 1998 to study the Martian geography and weather patterns. The spacecraft was slated to begin orbiting Mars on September 23, 1999. However, NASA officials lost communication with the spacecraft early that day and it is believed that the spacecraft broke apart or overheated because it came too close to the surface of Mars. Errors in the maneuvering commands sent from earth caused the Orbiter to sweep within 37 miles of the surface rather than the intended 93 miles. The subsequent investigation revealed that the errors were due to a simple mix-up in *units*. One team controlling the Orbiter used SI units, whereas another team used BG units. This costly experience illustrates the importance of using a consistent system of units.

## 1.3    Analysis of Fluid Behavior

The study of fluid mechanics involves the same fundamental laws you have encountered in physics and other mechanics courses. These laws include Newton's laws of motion, conservation of mass, and the first and second laws of thermodynamics. Thus, there are strong similarities between the general approach to fluid mechanics and to rigid-body and deformable-body solid mechanics. This is indeed helpful since many of the concepts and techniques of analysis used in fluid mechanics will be ones you have encountered before in other courses.

The broad subject of fluid mechanics can be generally subdivided into *fluid statics,* in which the fluid is at rest, and *fluid dynamics,* in which the fluid is moving. In the following chapters we will consider both of these areas in detail. Before we can proceed, however, it will be necessary to define and discuss certain fluid *properties* that are intimately related to fluid behavior. It is obvious that different fluids can have grossly different characteristics. For example, gases are light and compressible, whereas liquids are heavy (by comparison) and relatively incompressible. A syrup flows slowly from a container, but water flows rapidly when poured from the same container. To quantify these differences, certain fluid properties are used. In the following several sections, the properties that play an important role in the analysis of fluid behavior are considered.

## 1.4    Measures of Fluid Mass and Weight

### 1.4.1  Density

*The density of a fluid is defined as its mass per unit volume.*

The ***density*** of a fluid, designated by the Greek symbol $\rho$ (rho), is defined as its mass per unit volume. Density is typically used to characterize the mass of a fluid system. In the BG system, $\rho$ has units of slugs/ft$^3$ and in SI the units are kg/m$^3$.

The value of density can vary widely between different fluids, but for liquids, variations in pressure and temperature generally have only a small effect on the value of $\rho$. The small change in the density of water with large variations in temperature is illustrated in Fig. 1.3. Tables 1.5 and 1.6 list values of density for several common liquids. The density of water at 60 °F is 1.94 slugs/ft$^3$ or 999 kg/m$^3$. The large difference between those two values illustrates the importance of paying attention to units! Unlike liquids, the density of a gas is strongly influenced by both pressure and temperature, and this difference will be discussed in the next section.

The *specific volume, v,* is the *volume* per unit mass and is therefore the reciprocal of the density—that is,

$$v = \frac{1}{\rho}$$

(1.5)

This property is not commonly used in fluid mechanics but is used in thermodynamics.

■ **Figure 1.3**   **Density of water as a function of temperature.**

■ **Table 1.5**
**Approximate Physical Properties of Some Common Liquids (BG Units)**

| (See inside of front cover.) |
| --- |

■ **Table 1.6**
**Approximate Physical Properties of Some Common Liquids (SI Units)**

| (See inside of front cover.) |
| --- |

### 1.4.2 Specific Weight

*Specific weight is weight per unit volume; specific gravity is the ratio of fluid density to the density of water at a certain temperature.*

The ***specific weight*** of a fluid, designated by the Greek symbol $\gamma$ (gamma), is defined as its *weight* per unit volume. Thus, specific weight is related to density through the equation

$$\gamma = \rho g \tag{1.6}$$

where $g$ is the local acceleration of gravity. Just as density is used to characterize the mass of a fluid system, the specific weight is used to characterize the weight of the system. In the BG system, $\gamma$ has units of lb/ft$^3$ and in SI the units are N/m$^3$. Under conditions of standard gravity ($g = 32.174$ ft/s$^2$ = 9.807 m/s$^2$), water at 60 °F has a specific weight of 62.4 lb/ft$^3$ and 9.80 kN/m$^3$. Tables 1.5 and 1.6 list values of specific weight for several common liquids (based on standard gravity). More complete tables for water can be found in Appendix B (Tables B.1 and B.2).

### 1.4.3 Specific Gravity

13.55

Water

Mercury

1

The ***specific gravity*** of a fluid, designated as *SG*, is defined as the ratio of the density of the fluid to the density of water at some specified temperature. Usually the specified temperature is taken as 4 °C (39.2 °F), and at this temperature the density of water is 1.94 slugs/ft$^3$ or 1000 kg/m$^3$. In equation form, specific gravity is expressed as

$$SG = \frac{\rho}{\rho_{\text{H}_2\text{O@4 °C}}} \tag{1.7}$$

and since it is the *ratio* of densities, the value of *SG* does not depend on the system of units used. For example, the specific gravity of mercury at 20 °C is 13.55. This is illustrated by the figure in the margin. Thus, the density of mercury can be readily calculated in either BG or SI units through the use of Eq. 1.7 as

$$\rho_{\text{Hg}} = (13.55)(1.94 \text{ slugs/ft}^3) = 26.3 \text{ slugs/ft}^3$$

or

$$\rho_{\text{Hg}} = (13.55)(1000 \text{ kg/m}^3) = 13.6 \times 10^3 \text{ kg/m}^3$$

It is clear that density, specific weight, and specific gravity are all interrelated, and from a knowledge of any one of the three the others can be calculated.

## 1.5   Ideal Gas Law

Gases are highly compressible in comparison to liquids, with changes in gas density directly related to changes in pressure and temperature through the equation

$$\rho = \frac{p}{RT} \tag{1.8}$$

where $p$ is the absolute pressure, $\rho$ the density, $T$ the absolute temperature,[2] and $R$ is a gas constant. Equation 1.8 is commonly termed the ***ideal*** or *perfect **gas law,*** or the *equation of state* for

---

[2]We will use $T$ to represent temperature in thermodynamic relationships although $T$ is also used to denote the basic dimension of time.

an ideal gas. It is known to closely approximate the behavior of real gases under normal conditions when the gases are not approaching liquefaction.

Pressure in a fluid at rest is defined as the normal force per unit area exerted on a plane surface (real or imaginary) immersed in a fluid and is created by the bombardment of the surface with the fluid molecules. From the definition, pressure has the dimension of $FL^{-2}$ and in BG units is expressed as lb/ft² (psf) or lb/in.² (psi) and in SI units as N/m². In SI, 1 N/m² defined as a *pascal,* abbreviated as Pa, and pressures are commonly specified in pascals. The pressure in the ideal gas law must be expressed as an **absolute pressure,** denoted (abs), which means that it is measured relative to absolute zero pressure (a pressure that would only occur in a perfect vacuum). Standard sea-level atmospheric pressure (by international agreement) is 14.696 psi (abs) or 101.33 kPa (abs). For most calculations these pressures can be rounded to 14.7 psi and 101 kPa, respectively. In engineering it is common practice to measure pressure relative to the local atmospheric pressure, and when measured in this fashion it is called **gage pressure.** Thus, the absolute pressure can be obtained from the gage pressure by adding the value of the atmospheric pressure. For example, as shown by the figure in the margin on the next page, a pressure of 30 psi (gage) in a tire is equal to 44.7 psi (abs) at standard atmospheric pressure. Pressure is a particularly important fluid characteristic and it will be discussed more fully in the next chapter.

*In the ideal gas law, absolute pressures and temperatures must be used.*

# $E$XAMPLE 1.3   Ideal Gas Law

**GIVEN**   The compressed air tank shown in Fig. E1.3a has a volume of 0.84 ft³. The temperature is 70 °F and the atmospheric pressure is 14.7 psi (abs).

**FIND**   When the tank is filled with air at a gage pressure of 50 psi, determine the density of the air and the weight of air in the tank.

## $S$OLUTION

The air density can be obtained from the ideal gas law (Eq. 1.8)

$$\rho = \frac{p}{RT}$$

so that

$$\rho = \frac{(50 \text{ lb/in.}^2 + 14.7 \text{ lb/in.}^2)(144 \text{ in.}^2/\text{ft}^2)}{(1716 \text{ ft} \cdot \text{lb/slug} \cdot \text{°R})[(70 + 460)\text{°R}]}$$

$$= 0.0102 \text{ slugs/ft}^3 \qquad \textbf{(Ans)}$$

Note that both the pressure and temperature were changed to absolute values.

■ **Figure E1.3a**   **(Photograph courtesy of Jenny Products, Inc.)**

The weight, $\mathcal{W}$, of the air is equal to

$$\mathcal{W} = \rho g \times (\text{volume})$$
$$= (0.0102 \text{ slug/ft}^3)(32.2 \text{ ft/s}^2)(0.84 \text{ ft}^3)$$
$$= 0.276 \text{ slug} \cdot \text{ft/s}^2$$

so that since 1 lb = 1 slug · ft /s²

$$\mathcal{W} = 0.276 \text{ lb} \qquad \textbf{(Ans)}$$

**COMMENT**   By repeating the calculations for various values of the pressure, $p$, the results shown in Fig. E1.3b are obtained. Note that doubling the gage pressure does not double the amount of air in the tank, but doubling the absolute pressure does. Thus, a scuba diving tank at a gage pressure of 100 psi does not contain twice the amount of air as when the tank reads 50 psi.

(50 psi, 0.276 lb)

$\mathcal{W}$, lb

$p$, psi

■ **Figure E1.3b**

■ **Table 1.7**

**Approximate Physical Properties of Some Common Gases at Standard Atmospheric Pressure (BG Units)**

(See inside of front cover.)

■ **Table 1.8**

**Approximate Physical Properties of Some Common Gases at Standard Atmospheric Pressure (SI Units)**

(See inside of front cover.)

The gas constant, $R$, which appears in Eq. 1.8, depends on the particular gas and is related to the molecular weight of the gas. Values of the gas constant for several common gases are listed in Tables 1.7 and 1.8. Also in these tables the gas density and specific weight are given for standard atmospheric pressure and gravity and for the temperature listed. More complete tables for air at standard atmospheric pressure can be found in Appendix B (Tables B.3 and B.4).

## 1.6    Viscosity

*V1.3 Viscous fluids*

*V1.4 No-slip condition*

*Real fluids, even though they may be moving, always "stick" to the solid boundaries that contain them.*

The properties of density and specific weight are measures of the "heaviness" of a fluid. It is clear, however, that these properties are not sufficient to uniquely characterize how fluids behave since two fluids (such as water and oil) can have approximately the same value of density but behave quite differently when flowing. Apparently, some additional property is needed to describe the "fluidity" of the fluid.

To determine this additional property, consider a hypothetical experiment in which a material is placed between two very wide parallel plates as shown in Fig. 1.4a. The bottom plate is rigidly fixed, but the upper plate is free to move. If a solid, such as steel, were placed between the two plates and loaded with the force $P$ as shown, the top plate would be displaced through some small distance, $\delta a$ (assuming the solid was mechanically attached to the plates). The vertical line $AB$ would be rotated through the small angle, $\delta\beta$, to the new position $AB'$. We note that to resist the applied force, $P$, a shearing stress, $\tau$, would be developed at the plate–material interface, and for equilibrium to occur, $P = \tau A$ where $A$ is the effective upper plate area (Fig. 1.4b). It is well known that for elastic solids, such as steel, the small angular displacement, $\delta\beta$ (called the shearing strain), is proportional to the shearing stress, $\tau$, that is developed in the material.

What happens if the solid is replaced with a fluid such as water? We would immediately notice a major difference. When the force $P$ is applied to the upper plate, it will move continuously with a velocity, $U$ (after the initial transient motion has died out) as illustrated in Fig. 1.5. This behavior is consistent with the definition of a fluid—that is, if a shearing stress is applied to a fluid it will deform continuously. A closer inspection of the fluid motion between the two plates would reveal that the fluid in contact with the upper plate moves with the plate velocity, $U$, and the fluid in contact with the bottom fixed plate has a zero velocity. The fluid between the two plates moves with velocity $u = u(y)$ that would be found to vary linearly, $u = Uy/b$, as illustrated in Fig. 1.5. Thus, a *velocity gradient, $du/dy$*, is developed in the fluid between the plates. In this particular case the velocity gradient is a constant since $du/dy = U/b$, but in more complex flow situations, such

■ **Figure 1.4** (*a*) Deformation of material placed between two parallel plates. (*b*) Forces acting on upper plate.

(a)      (b)

44.7 — 30

14.7 — 0

0 — –14.7
(abs)    (gage)
   $p$, psi

■ **Figure 1.5** **Behavior of a fluid placed between two parallel plates.**

as that shown by the photograph in the margin, this is not true. The experimental observation that the fluid "sticks" to the solid boundaries is a very important one in fluid mechanics and is usually referred to as the ***no-slip condition.*** All fluids, both liquids and gases, satisfy this condition.

In a small time increment, $\delta t$, an imaginary vertical line $AB$ in the fluid would rotate through an angle, $\delta\beta$, so that

$$\tan\delta\beta \approx \delta\beta = \frac{\delta a}{b}$$

Since $\delta a = U\,\delta t$, it follows that

$$\delta\beta = \frac{U\,\delta t}{b}$$

We note that in this case, $\delta\beta$ is a function not only of the force $P$ (which governs $U$) but also of time. Thus, it is not reasonable to attempt to relate the shearing stress, $\tau$, to $\delta\beta$ as is done for solids. Rather, we consider the *rate* at which $\delta\beta$ is changing and define the ***rate of shearing strain, $\dot{\gamma}$,*** as

$$\dot{\gamma} = \lim_{\delta t\to 0} \frac{\delta\beta}{\delta t}$$

which in this instance is equal to

$$\dot{\gamma} = \frac{U}{b} = \frac{du}{dy}$$

A continuation of this experiment would reveal that as the shearing stress, $\tau$, is increased by increasing $P$ (recall that $\tau = P/A$), the rate of shearing strain is increased in direct proportion—that is,

$$\tau \propto \dot{\gamma}$$

or

$$\tau \propto \frac{du}{dy}$$

This result indicates that for common fluids such as water, oil, gasoline, and air the shearing stress and rate of shearing strain (velocity gradient) can be related with a relationship of the form

$$\tau = \mu\,\frac{du}{dy} \tag{1.9}$$

*V1.5 Capillary tube viscometer*

*Dynamic viscosity is the fluid property that relates shearing stress and fluid motion.*

where the constant of proportionality is designated by the Greek symbol $\mu$ (mu) and is called the ***absolute viscosity, dynamic viscosity,*** or simply the *viscosity* of the fluid. In accordance with Eq. 1.9, plots of $\tau$ versus $du/dy$ should be linear with the slope equal to the viscosity as illustrated in Fig. 1.6. The actual value of the viscosity depends on the particular fluid, and for a particular fluid the viscosity is also highly dependent on temperature as illustrated in Fig. 1.6 with the two curves for water. Fluids for which the shearing stress is *linearly* related to the rate of shearing strain (also referred to as rate of angular deformation) are designated as ***Newtonian fluids*** after I. Newton (1642–1727). Fortunately, most common fluids, both liquids and gases, are Newtonian. A more general formulation of Eq. 1.9 which applies to more complex flows of Newtonian fluids is given in Section 6.8.1.

■ **Figure 1.6** Linear variation of shearing stress with rate of shearing strain for common fluids.

---

**F l u i d s   i n   t h e   N e w s**

**An extremely viscous fluid** Pitch is a derivative of tar once used for waterproofing boats. At elevated temperatures it flows quite readily. At room temperature it feels like a solid—it can even be shattered with a blow from a hammer. However, it is a liquid. In 1927 Professor Parnell heated some pitch and poured it into a funnel. Since that time it has been allowed to flow freely (or rather, drip slowly) from the funnel. The flowrate is quite small. In fact, to date only seven drops have fallen from the end of the funnel, although the eighth drop is poised ready to fall "soon." While nobody has actually seen a drop fall from the end of the funnel, a beaker below the funnel holds the previous drops that fell over the years. It is estimated that the pitch is about 100 billion times more viscous than water.

---

*For non-Newtonian fluids, the apparent viscosity is a function of the shear rate.*

Fluids for which the shearing stress is not linearly related to the rate of shearing strain are designated as **non-Newtonian fluids.** Although there is a variety of types of non-Newtonian fluids, the simplest and most common are shown in Fig. 1.7. The slope of the shearing stress versus rate of shearing strain graph is denoted as the *apparent viscosity, $\mu_{ap}$.* For Newtonian fluids the apparent viscosity is the same as the viscosity and is independent of shear rate.

For *shear thinning fluids* the apparent viscosity decreases with increasing shear rate—the harder the fluid is sheared, the less viscous it becomes. Many colloidal suspensions and polymer solutions are shear thinning. For example, latex paint does not drip from the brush because the shear rate is small and the apparent viscosity is large. However, it flows smoothly onto the wall because the thin layer of paint between the wall and the brush causes a large shear rate and a small apparent viscosity.

■ **Figure 1.7** Variation of shearing stress with rate of shearing strain for several types of fluids, including common non-Newtonian fluids.

$\tau < \tau_{yield}$

$\tau > \tau_{yield}$

*V1.6 Non-Newtonian behavior*

For *shear thickening fluids* the apparent viscosity increases with increasing shear rate—the harder the fluid is sheared, the more viscous it becomes. Common examples of this type of fluid include water–corn starch mixture and water–sand mixture ("quicksand"). Thus, the difficulty in removing an object from quicksand increases dramatically as the speed of removal increases.

The other type of behavior indicated in Fig. 1.7 is that of a *Bingham plastic,* which is neither a fluid nor a solid. Such material can withstand a finite, nonzero shear stress, $\tau_{yield}$, the yield stress, without motion (therefore, it is not a fluid), but once the yield stress is exceeded it flows like a fluid (hence, it is not a solid). Toothpaste and mayonnaise are common examples of Bingham plastic materials. As indicated in the figure in the margin, mayonnaise can sit in a pile on a slice of bread (the shear stress less than the yield stress), but it flows smoothly into a thin layer when the knife increases the stress above the yield stress.

From Eq. 1.9 it can be readily deduced that the dimensions of viscosity are $FTL^{-2}$. Thus, in BG units viscosity is given as lb · s/ft$^2$ and in SI units as N · s/m$^2$. Values of viscosity for several common liquids and gases are listed in Tables 1.5 through 1.8. A quick glance at these tables reveals the wide variation in viscosity among fluids. Viscosity is only mildly dependent on pressure and the effect of pressure is usually neglected. However, as previously mentioned, and as illustrated in Fig. 1.8, viscosity is very sensitive to temperature. For example, as the temperature of water changes from 60 to 100 °F the density decreases by less than 1%, but the viscosity decreases by about 40%. It is thus clear that particular attention must be given to temperature when determining viscosity.

Figure 1.8 shows in more detail how the viscosity varies from fluid to fluid and how for a given fluid it varies with temperature. It is to be noted from this figure that the viscosity of liquids decreases with an increase in temperature, whereas for gases an increase in temperature causes an increase in viscosity. This difference in the effect of temperature on the viscosity of liquids and gases can again be traced back to the difference in molecular structure. The liquid molecules are closely spaced, with strong cohesive forces between molecules, and the resistance to relative motion between adjacent layers of fluid is related to these intermolecular forces. As

■ **Figure 1.8** Dynamic (absolute) viscosity of some common fluids as a function of temperature.

the temperature increases, these cohesive forces are reduced with a corresponding reduction in resistance to motion. Since viscosity is an index of this resistance, it follows that the viscosity is reduced by an increase in temperature. In gases, however, the molecules are widely spaced and intermolecular forces negligible. In this case, resistance to relative motion arises due to the exchange of momentum of gas molecules between adjacent layers. As molecules are transported by random motion from a region of low bulk velocity to mix with molecules in a region of higher bulk velocity (and vice versa), there is an effective momentum exchange that resists the relative motion between the layers. As the temperature of the gas increases, the random molecular activity increases with a corresponding increase in viscosity.

The effect of temperature on viscosity can be closely approximated using two empirical formulas. For gases the *Sutherland equation* can be expressed as

*Viscosity is very sensitive to temperature.*

$$\mu = \frac{CT^{3/2}}{T + S} \tag{1.10}$$

where $C$ and $S$ are empirical constants, and $T$ is absolute temperature. Thus, if the viscosity is known at two temperatures, $C$ and $S$ can be determined. Or, if more than two viscosities are known, the data can be correlated with Eq. 1.10 by using some type of curve-fitting scheme.

For liquids an empirical equation that has been used is

$$\mu = De^{B/T} \tag{1.11}$$

where $D$ and $B$ are constants and $T$ is absolute temperature. This equation is often referred to as *Andrade's equation*. As was the case for gases, the viscosity must be known at least for two temperatures so the two constants can be determined. A more detailed discussion of the effect of temperature on fluids can be found in Ref. 1.

---

# $E$XAMPLE 1.4    Viscosity and Dimensionless Quantities

**GIVEN**  A dimensionless combination of variables that is important in the study of viscous flow through pipes is called the *Reynolds number*, Re, defined as $\rho VD/\mu$ where, as indicated in Fig. E1.4, $\rho$ is the fluid density, $V$ the mean fluid velocity, $D$ the pipe diameter, and $\mu$ the fluid viscosity. A Newtonian fluid having a viscosity of $0.38 \ \text{N} \cdot \text{s/m}^2$ and a specific gravity of 0.91 flows through a 25-mm-diameter pipe with a velocity of 2.6 m/s.

**FIND**  Determine the value of the Reynolds number using (a) SI units and (b) BG units.

## SOLUTION

**(a)**  The fluid density is calculated from the specific gravity as

$$\rho = SG \ \rho_{\text{H}_2\text{O@4 °C}} = 0.91 \ (1000 \ \text{kg/m}^3) = 910 \ \text{kg/m}^3$$

■ **Figure E1.4**

and from the definition of the Reynolds number

$$\text{Re} = \frac{\rho VD}{\mu} = \frac{(910 \ \text{kg/m}^3)(2.6 \ \text{m/s})(25 \ \text{mm})(10^{-3} \ \text{m/mm})}{0.38 \ \text{N} \cdot \text{s/m}^2}$$

$$= 156 \ (\text{kg} \cdot \text{m/s}^2)/\text{N}$$

However, since $1 \ \text{N} = 1 \ \text{kg} \cdot \text{m/s}^2$ it follows that the Reynolds number is unitless—that is,

$$\text{Re} = 156 \tag{Ans}$$

The value of any dimensionless quantity does not depend on the system of units used if all variables that make up the quantity are expressed in a consistent set of units. To check this, we will calculate the Reynolds number using BG units.

**(b)**  We first convert all the SI values of the variables appearing in the Reynolds number to BG values by using the conversion factors from Table 1.4. Thus,

$$\rho = (910 \ \text{kg/m}^3)(1.940 \times 10^{-3}) = 1.77 \ \text{slugs/ft}^3$$

$$V = (2.6 \ \text{m/s})(3.281) = 8.53 \ \text{ft/s}$$

$$D = (0.025 \ \text{m})(3.281) = 8.20 \times 10^{-2} \ \text{ft}$$

$$\mu = (0.38 \ \text{N} \cdot \text{s/m}^2)(2.089 \times 10^{-2}) = 7.94 \times 10^{-3} \ \text{lb} \cdot \text{s/ft}^2$$

and the value of the Reynolds number is

$$Re = \frac{(1.77 \text{ slugs/ft}^3)(8.53 \text{ ft/s})(8.20 \times 10^{-2} \text{ ft})}{7.94 \times 10^{-3} \text{ lb} \cdot \text{s/ft}^2}$$

$$= 156 \, (\text{slug} \cdot \text{ft/s}^2)/\text{lb} = 156 \qquad \textbf{(Ans)}$$

since 1 lb = 1 slug $\cdot$ ft/s$^2$.

**COMMENTS** The values from part (a) and part (b) are the same, as expected. Dimensionless quantities play an important role in fluid mechanics, and the significance of the Reynolds number as well as other important dimensionless combinations will be discussed in detail in Chapter 7. It should be noted that in the Reynolds number it is actually the ratio $\mu/\rho$ that is important, and this is the property that is defined as the kinematic viscosity.

---

## EXAMPLE 1.5 Newtonian Fluid Shear Stress

**GIVEN** The velocity distribution for the flow of a Newtonian fluid between two fixed wide, parallel plates (see Fig. E1.5a) is given by the equation

$$u = \frac{3V}{2}\left[1 - \left(\frac{y}{h}\right)^2\right]$$

where $V$ is the mean velocity. The fluid has a viscosity of 0.04 lb $\cdot$ s/ft$^2$. Also, $V = 2$ ft/s and $h = 0.2$ in.

**FIND** Determine: (a) the shearing stress acting on the bottom wall, and (b) the shearing stress acting on a plane parallel to the walls and passing through the centerline (midplane).

### SOLUTION

For this type of parallel flow the shearing stress is obtained from Eq. 1.9,

$$\tau = \mu \frac{du}{dy} \qquad \textbf{(1)}$$

Thus, if the velocity distribution $u = u(y)$ is known, the shearing stress can be determined at all points by evaluating the velocity gradient, $du/dy$. For the distribution given

$$\frac{du}{dy} = -\frac{3Vy}{h^2} \qquad \textbf{(2)}$$

**(a)** Along the bottom wall $y = -h$ so that (from Eq. 2)

$$\frac{du}{dy} = \frac{3V}{h}$$

and therefore the shearing stress is

$$\tau_{\substack{\text{bottom} \\ \text{wall}}} = \mu\left(\frac{3V}{h}\right) = \frac{(0.04 \text{ lb} \cdot \text{s/ft}^2)(3)(2 \text{ ft/s})}{(0.2 \text{ in.})(1 \text{ ft/12 in.})}$$

$$\doteq 14.4 \text{ lb/ft}^2 \text{ (in direction of flow)} \qquad \textbf{(Ans)}$$

This stress creates a drag on the wall. Since the velocity distribution is symmetrical, the shearing stress along the upper wall would have the same magnitude and direction.

**(b)** Along the midplane where $y = 0$ it follows from Eq. 2 that

$$\frac{du}{dy} = 0$$

and thus the shearing stress is

$$\tau_{\text{midplane}} = 0 \qquad \textbf{(Ans)}$$

■ **Figure E1.5a**

**COMMENT** From Eq. 2 we see that the velocity gradient (and therefore the shearing stress) varies linearly with $y$ and in this particular example varies from 0 at the center of the channel to 14.4 lb/ft$^2$ at the walls. This is shown in Fig. E1.5b. For the more general case the actual variation will, of course, depend on the nature of the velocity distribution.

■ **Figure E1.5b**

---

Quite often viscosity appears in fluid flow problems combined with the density in the form

$$\nu = \frac{\mu}{\rho}$$

*Kinematic viscosity is defined as the ratio of the absolute viscosity to the fluid density.*

This ratio is called the **kinematic viscosity** and is denoted with the Greek symbol $\nu$ (nu). The dimensions of kinematic viscosity are $L^2/T$, and the BG units are ft$^2$/s and SI units are m$^2$/s. Values of kinematic viscosity for some common liquids and gases are given in Tables 1.5 through 1.8. More extensive tables giving both the dynamic and kinematic viscosities for water and air can be found in Appendix B (Tables B.1 through B.4), and graphs showing the variation in both dynamic and kinematic viscosity with temperature for a variety of fluids are also provided in Appendix B (Figs. B.1 and B.2).

Although in this text we are primarily using BG and SI units, dynamic viscosity is often expressed in the metric CGS (centimeter-gram-second) system with units of dyne · s/cm$^2$. This combination is called a *poise,* abbreviated P. In the CGS system, kinematic viscosity has units of cm$^2$/s, and this combination is called a *stoke,* abbreviated St.

## 1.7    Compressibility of Fluids

### 1.7.1 Bulk Modulus

An important question to answer when considering the behavior of a particular fluid is how easily can the volume (and thus the density) of a given mass of the fluid be changed when there is a change in pressure? That is, how compressible is the fluid? A property that is commonly used to characterize compressibility is the **bulk modulus,** $E_v$, defined as

$$E_v = -\frac{dp}{d\forall/\forall} \tag{1.12}$$

where $dp$ is the differential change in pressure needed to create a differential change in volume, $d\forall$, of a volume $\forall$. This is illustrated by the figure in the margin. The negative sign is included since an increase in pressure will cause a decrease in volume. Since a decrease in volume of a given mass, $m = \rho\forall$, will result in an increase in density, Eq. 1.12 can also be expressed as

$$E_v = \frac{dp}{d\rho/\rho} \tag{1.13}$$

The bulk modulus (also referred to as the *bulk modulus of elasticity*) has dimensions of pressure, $FL^{-2}$. In BG units, values for $E_v$ are usually given as lb/in.$^2$ (psi) and in SI units as N/m$^2$ (Pa). Large values for the bulk modulus indicate that the fluid is relatively incompressible—that is, it takes a large pressure change to create a small change in volume. As expected, values of $E_v$ for common liquids are large (see Tables 1.5 and 1.6). For example, at atmospheric pressure and a temperature of 60 °F it would require a pressure of 3120 psi to compress a unit volume of water 1%. This result is representative of the compressibility of liquids. Since such large pressures are required to effect a change in volume, we conclude that liquids can be considered as *incompressible* for most practical engineering applications. As liquids are compressed the bulk modulus increases, but the bulk modulus near atmospheric pressure is usually the one of interest. The use of bulk modulus as a property describing compressibility is most prevalent when dealing with liquids, although the bulk modulus can also be determined for gases.

*V1.7 Water balloon*

## F l u i d s   i n   t h e   N e w s

**This water jet is a blast** Usually liquids can be treated as incompressible fluids. However, in some applications the *compressibility* of a liquid can play a key role in the operation of a device. For example, a water pulse generator using compressed water has been developed for use in mining operations. It can fracture rock by producing an effect comparable to a conventional explosive such as gunpowder. The device uses the energy stored in a water-filled accumulator to generate an ultrahigh-pressure water pulse ejected through a 10- to 25-mm-diameter discharge valve. At the ultrahigh pressures used (300 to 400 MPa, or 3000 to 4000 atmos-

pheres), the water is compressed (i.e., the volume reduced) by about 10 to 15%. When a fast-opening valve within the pressure vessel is opened, the water expands and produces a jet of water that upon impact with the target material produces an effect similar to the explosive force from conventional explosives. Mining with the water jet can eliminate various hazards that arise with the use of conventional chemical explosives, such as those associated with the storage and use of explosives and the generation of toxic gas by-products that require extensive ventilation. (See Problem 1.110.)

### 1.7.2 Compression and Expansion of Gases

When gases are compressed (or expanded), the relationship between pressure and density depends on the nature of the process. If the compression or expansion takes place under constant temperature conditions (*isothermal process*), then from Eq. 1.8

$$\frac{p}{\rho} = \text{constant} \tag{1.14}$$

If the compression or expansion is frictionless and no heat is exchanged with the surroundings (*isentropic process*), then

$$\frac{p}{\rho^k} = \text{constant} \tag{1.15}$$

where $k$ is the ratio of the specific heat at constant pressure, $c_p$, to the specific heat at constant volume, $c_v$ (i.e., $k = c_p/c_v$). The two specific heats are related to the gas constant, $R$, through the equation $R = c_p - c_v$. As was the case for the ideal gas law, the pressure in both Eqs. 1.14 and 1.15 must be expressed as an absolute pressure. Values of $k$ for some common gases are given in Tables 1.7 and 1.8 and for air over a range of temperatures, in Appendix B (Tables B.3 and B.4). The pressure–density variations for isothermal and isentropic conditions are illustrated in the margin figure.

*The value of the bulk modulus depends on the type of process involved.*

With explicit equations relating pressure and density, the bulk modulus for gases can be determined by obtaining the derivative $dp/d\rho$ from Eq. 1.14 or 1.15 and substituting the results into Eq. 1.13. It follows that for an isothermal process

$$E_v = p \tag{1.16}$$

and for an isentropic process,

$$E_v = kp \tag{1.17}$$

Note that in both cases the bulk modulus varies directly with pressure. For air under standard atmospheric conditions with $p = 14.7$ psi (abs) and $k = 1.40$, the isentropic bulk modulus is 20.6 psi. A comparison of this figure with that for water under the same conditions ($E_v = 312,000$ psi) shows that air is approximately 15,000 times as compressible as water. It is thus clear that in dealing with gases, greater attention will need to be given to the effect of compressibility on fluid behavior. However, as will be discussed further in later sections, gases can often be treated as incompressible fluids if the changes in pressure are small.

---

## EXAMPLE 1.6  Isentropic Compression of a Gas

**GIVEN** A cubic foot of air at an absolute pressure of 14.7 psi is compressed isentropically to $\frac{1}{2}$ ft³ by the tire pump shown in Fig. E1.6*a*.

**FIND** What is the final pressure?

### SOLUTION

For an isentropic compression

$$\frac{p_i}{\rho_i^k} = \frac{p_f}{\rho_f^k}$$

where the subscripts $i$ and $f$ refer to initial and final states, respectively. Since we are interested in the final pressure, $p_f$, it follows that

$$p_f = \left(\frac{\rho_f}{\rho_i}\right)^k p_i$$

■ **Figure E1.6*a***

As the volume, $V$. is reduced by one-half, the density must double, since the mass, $m = \rho V$. of the gas remains constant. Thus, with $k = 1.40$ for air

$$p_f = (2)^{1.40}(14.7 \text{ psi}) = 38.8 \text{ psi (abs)} \qquad \textbf{(Ans)}$$

**COMMENT** By repeating the calculations for various values of the ratio of the final volume to the initial volume, $V_f/V_i$, the results shown in Fig. E1.6b are obtained. Note that even though air is often considered to be easily compressed (at least compared to liquids), it takes considerable pressure to significantly reduce a given volume of air as is done in an automobile engine where the compression ratio is on the order of $V_f/V_i = 1/8 = 0.125$.

■ **Figure E1.6b**

### 1.7.3 Speed of Sound

*The velocity at which small disturbances propagate in a fluid is called the speed of sound.*

Another important consequence of the compressibility of fluids is that disturbances introduced at some point in the fluid propagate at a finite velocity. For example, if a fluid is flowing in a pipe and a valve at the outlet is suddenly closed (thereby creating a localized disturbance), the effect of the valve closure is not felt instantaneously upstream. It takes a finite time for the increased pressure created by the valve closure to propagate to an upstream location. Similarly, a loudspeaker diaphragm causes a localized disturbance as it vibrates, and the small change in pressure created by the motion of the diaphragm is propagated through the air with a finite velocity. The velocity at which these small disturbances propagate is called the *acoustic velocity* or the ***speed of sound,*** *c*. It will be shown in Chapter 11 that the speed of sound is related to changes in pressure and density of the fluid medium through the equation

$$c = \sqrt{\frac{dp}{d\rho}} \qquad \textbf{(1.18)}$$

or in terms of the bulk modulus defined by Eq. 1.13

$$c = \sqrt{\frac{E_v}{\rho}} \qquad \textbf{(1.19)}$$

Since the disturbance is small, there is negligible heat transfer and the process is assumed to be isentropic. Thus, the pressure–density relationship used in Eq. 1.18 is that for an isentropic process.

For gases undergoing an isentropic process, $E_v = kp$ (Eq. 1.17) so that

$$c = \sqrt{\frac{kp}{\rho}}$$

and making use of the ideal gas law, it follows that

$$c = \sqrt{kRT} \qquad \textbf{(1.20)}$$

*V1.8 As fast as a speeding bullet*

Thus, for ideal gases the speed of sound is proportional to the square root of the absolute temperature. For example, for air at 60 °F with $k = 1.40$ and $R = 1716 \text{ ft} \cdot \text{lb/slug} \cdot °R$, it follows that $c = 1117 \text{ ft/s}$. The speed of sound in air at various temperatures can be found in Appendix B (Tables B.3 and B.4). Equation 1.19 is also valid for liquids, and values of $E_v$ can be used to determine the speed of sound in liquids. For water at 20 °C, $E_v = 2.19 \text{ GN/m}^2$ and $\rho = 998.2 \text{ kg/m}^3$ so that $c = 1481 \text{ m/s}$ or 4860 ft/s. As shown by the figure in the margin, the speed of sound is much higher in water than in air. If a fluid were truly incompressible ($E_v = \infty$)

the speed of sound would be infinite. The speed of sound in water for various temperatures can be found in Appendix B (Tables B.1 and B.2).

---

## $E$XAMPLE 1.7    Speed of Sound and Mach Number

**GIVEN**   A jet aircraft flies at a speed of 550 mph at an altitude of 35,000 ft, where the temperature is $-66\ ^\circ$F and the specific heat ratio is $k = 1.4$.

**FIND**   Determine the ratio of the speed of the aircraft, $V$, to that of the speed of sound, $c$, at the specified altitude.

### $S$OLUTION

From Eq. 1.20 the speed of sound can be calculated as

$$c = \sqrt{kRT}$$
$$= \sqrt{(1.40)(1716\ \text{ft lb/slug }^\circ\text{R})(-66 + 460)\ ^\circ\text{R}}$$
$$= 973\ \text{ft/s}$$

Since the air speed is

$$V = \frac{(550\ \text{mi/hr})(5280\ \text{ft/mi})}{(3600\ \text{s/hr})} = 807\ \text{ft/s}$$

the ratio is

$$\frac{V}{c} = \frac{807\ \text{ft/s}}{973\ \text{ft/s}} = 0.829 \qquad \textbf{(Ans)}$$

■ **Figure E1.7**

**COMMENT**   This ratio is called the *Mach number,* Ma. If Ma $< 1.0$ the aircraft is flying at *subsonic* speeds, whereas for Ma $> 1.0$ it is flying at *supersonic* speeds. The Mach number is an important dimensionless parameter used in the study of the flow of gases at high speeds and will be further discussed in Chapters 7 and 11.

By repeating the calculations for different temperatures, the results shown in Fig. E1.7 are obtained. Because the speed of sound increases with increasing temperature, for a constant airplane speed, the Mach number decreases as the temperature increases.

---

## 1.8    Vapor Pressure

*A liquid boils when the pressure is reduced to the vapor pressure.*

It is a common observation that liquids such as water and gasoline will evaporate if they are simply placed in a container open to the atmosphere. Evaporation takes place because some liquid molecules at the surface have sufficient momentum to overcome the intermolecular cohesive forces and escape into the atmosphere. If the container is closed with a small air space left above the surface, and this space evacuated to form a vacuum, a pressure will develop in the space as a result of the vapor that is formed by the escaping molecules. When an equilibrium condition is reached so that the number of molecules leaving the surface is equal to the number entering, the vapor is said to be saturated and the pressure that the vapor exerts on the liquid surface is termed the ***vapor pressure,*** $p_v$. Similarly, if the end of a completely liquid-filled container is moved as shown in the figure in the margin without letting any air into the container, the space between the liquid and the end becomes filled with vapor at a pressure equal to the vapor pressure.

Since the development of a vapor pressure is closely associated with molecular activity, the value of vapor pressure for a particular liquid depends on temperature. Values of vapor pressure for water at various temperatures can be found in Appendix B (Tables B.1 and B.2), and the values of vapor pressure for several common liquids at room temperatures are given in Tables 1.5 and 1.6.

*Boiling,* which is the formation of vapor bubbles within a fluid mass, is initiated when the absolute pressure in the fluid reaches the vapor pressure. As commonly observed in the kitchen, water at standard atmospheric pressure will boil when the temperature reaches 212 °F (100 °C)—that is, the vapor pressure of water at 212 °F is 14.7 psi (abs). However, if we attempt to boil water at a higher elevation, say 30,000 ft above sea level (the approximate elevation of Mt. Everest), where the atmospheric pressure is 4.37 psi (abs), we find that boiling will start when the temperature is about 157 °F. At this temperature the vapor pressure of water is 4.37 psi (abs). For the U.S. Standard Atmosphere (see Section 2.4), the boiling temperature is a function of altitude as shown in the figure in the margin. Thus, boiling can be induced at a given pressure acting on the fluid by raising the temperature, or at a given fluid temperature by lowering the pressure.

An important reason for our interest in vapor pressure and boiling lies in the common observation that in flowing fluids it is possible to develop very low pressure due to the fluid motion, and if the pressure is lowered to the vapor pressure, boiling will occur. For example, this phenomenon may occur in flow through the irregular, narrowed passages of a valve or pump. When vapor bubbles are formed in a flowing fluid, they are swept along into regions of higher pressure where they suddenly collapse with sufficient intensity to actually cause structural damage. The formation and subsequent collapse of vapor bubbles in a flowing fluid, called *cavitation*, is an important fluid flow phenomenon to be given further attention in Chapters 3 and 7.

*In flowing liquids it is possible for the pressure in localized regions to reach vapor pressure, thereby causing cavitation.*

## 1.9 Surface Tension

*V1.9 Floating razor blade*

At the interface between a liquid and a gas, or between two immiscible liquids, forces develop in the liquid surface that cause the surface to behave as if it were a "skin" or "membrane" stretched over the fluid mass. Although such a skin is not actually present, this conceptual analogy allows us to explain several commonly observed phenomena. For example, a steel needle or a razor blade will float on water if placed gently on the surface because the tension developed in the hypothetical skin supports it. Small droplets of mercury will form into spheres when placed on a smooth surface because the cohesive forces in the surface tend to hold all the molecules together in a compact shape. Similarly, discrete bubbles will form in a liquid. (See the photograph at the beginning of Chapter 1.)

These various types of surface phenomena are due to the unbalanced cohesive forces acting on the liquid molecules at the fluid surface. Molecules in the interior of the fluid mass are surrounded by molecules that are attracted to each other equally. However, molecules along the surface are subjected to a net force toward the interior. The apparent physical consequence of this unbalanced force along the surface is to create the hypothetical skin or membrane. A tensile force may be considered to be acting in the plane of the surface along any line in the surface. The intensity of the molecular attraction per unit length along any line in the surface is called the **surface tension** and is designated by the Greek symbol $\sigma$ (sigma). For a given liquid the surface tension depends on temperature as well as the other fluid it is in contact with at the interface. The dimensions of surface tension are $FL^{-1}$ with BG units of lb/ft and SI units of N/m. Values of surface tension for some common liquids (in contact with air) are given in Tables 1.5 and 1.6 and in Appendix B (Tables B.1 and B.2) for water at various temperatures. As indicated by the figure in the margin, the value of the surface tension decreases as the temperature increases.

## F l u i d s   i n   t h e   N e w s

**Walking on water** Water striders are insects commonly found on ponds, rivers, and lakes that appear to "walk" on water. A typical length of a water strider is about 0.4 in., and they can cover 100 body lengths in one second. It has long been recognized that it is *surface tension* that keeps the water strider from sinking below the surface. What has been puzzling is how they propel themselves at such a high speed. They can't pierce the water surface or they would sink. A team of mathematicians and engineers from the Massachusetts Institute of Technology (MIT) applied conventional flow visualization techniques and high-speed video to examine in detail the movement of the water striders. They found that each stroke of the insect's legs creates dimples on the surface with underwater swirling vortices sufficient to propel it forward. It is the rearward motion of the vortices that propels the water strider forward. To further substantiate their explanation, the MIT team built a working model of a water strider, called Robostrider, which creates surface ripples and underwater vortices as it moves across a water surface. Waterborne creatures, such as the water strider, provide an interesting world dominated by surface tension. (See Problem 1.131.)

■ **Figure 1.9** Forces acting on one-half of a liquid drop.

The pressure inside a drop of fluid can be calculated using the free-body diagram in Fig. 1.9. If the spherical drop is cut in half (as shown), the force developed around the edge due to surface tension is $2\pi R\sigma$. This force must be balanced by the pressure difference, $\Delta p$, between the internal pressure, $p_i$, and the external pressure, $p_e$, acting over the circular area, $\pi R^2$. Thus,

$$2\pi R\sigma = \Delta p\,\pi R^2$$

or

$$\Delta p = p_i - p_e = \frac{2\sigma}{R} \tag{1.21}$$

It is apparent from this result that the pressure inside the drop is greater than the pressure surrounding the drop. (Would the pressure on the inside of a bubble of water be the same as that on the inside of a drop of water of the same diameter and at the same temperature?)

Among common phenomena associated with surface tension is the rise (or fall) of a liquid in a capillary tube. If a small open tube is inserted into water, the water level in the tube will rise above the water level outside the tube, as is illustrated in Fig. 1.10a. In this situation we have a liquid–gas–solid interface. For the case illustrated there is an attraction (adhesion) between the wall of the tube and liquid molecules which is strong enough to overcome the mutual attraction (cohesion) of the molecules and pull them up the wall. Hence, the liquid is said to *wet* the solid surface.

The height, $h$, is governed by the value of the surface tension, $\sigma$, the tube radius, $R$, the specific weight of the liquid, $\gamma$, and the *angle of contact, $\theta$*, between the fluid and tube. From the free-body diagram of Fig. 1.10b we see that the vertical force due to the surface tension is equal to $2\pi R\sigma \cos\theta$ and the weight is $\gamma\pi R^2 h$, and these two forces must balance for equilibrium. Thus,

$$\gamma\pi R^2 h = 2\pi R\sigma \cos\theta$$

so that the height is given by the relationship

$$h = \frac{2\sigma \cos\theta}{\gamma R} \tag{1.22}$$

The angle of contact is a function of both the liquid and the surface. For water in contact with clean glass $\theta \approx 0°$. It is clear from Eq. 1.22 that the height is inversely proportional to the tube radius, and therefore, as indicated by the figure in the margin, the rise of a liquid in a tube as a result of capillary action becomes increasingly pronounced as the tube radius is decreased.

If adhesion of molecules to the solid surface is weak compared to the cohesion between molecules, the liquid will not wet the surface and the level in a tube placed in a nonwetting liquid will actually be depressed, as shown in Fig. 1.10c. Mercury is a good example of a nonwetting liquid when it is in contact with a glass tube. For nonwetting liquids the angle of contact is greater than 90°, and for mercury in contact with clean glass $\theta \approx 130°$.

■ **Figure 1.10** Effect of capillary action in small tubes. (*a*) Rise of column for a liquid that wets the tube. (*b*) Free-body diagram for calculating column height. (*c*) Depression of column for a nonwetting liquid.

### EXAMPLE 1.8 Capillary Rise in a Tube

**GIVEN** Pressures are sometimes determined by measuring the height of a column of liquid in a vertical tube.

**FIND** What diameter of clean glass tubing is required so that the rise of water at 20 °C in a tube due to capillary action (as opposed to pressure in the tube) is less than $h = 1.0$ mm?

### SOLUTION

From Eq. 1.22

$$h = \frac{2\sigma \cos \theta}{\gamma R}$$

so that

$$R = \frac{2\sigma \cos \theta}{\gamma h}$$

For water at 20 °C (from Table B.2), $\sigma = 0.0728$ N/m and $\gamma = 9.789$ kN/m$^3$. Since $\theta \approx 0°$ it follows that for $h = 1.0$ mm,

$$R = \frac{2(0.0728 \text{ N/m})(1)}{(9.789 \times 10^3 \text{ N/m}^3)(1.0 \text{ mm})(10^{-3} \text{ m/mm})}$$

$$= 0.0149 \text{ m}$$

and the minimum required tube diameter, $D$, is

$$D = 2R = 0.0298 \text{ m} = 29.8 \text{ mm} \qquad \textbf{(Ans)}$$

**COMMENT** By repeating the calculations for various values of the capillary rise, $h$, the results shown in Fig. E1.8 are obtained.

Note that as the allowable capillary rise is decreased, the diameter of the tube must be significantly increased. There is always some capillarity effect, but it can be minimized by using a large enough diameter tube.

■ **Figure E1.8**

(Photograph copyright 2007 by Andrew Davidhazy, Rochester Institute of Technology.)

Surface tension effects play a role in many fluid mechanics problems, including the movement of liquids through soil and other porous media, flow of thin films, formation of drops and bubbles, and the breakup of liquid jets. For example, surface tension is a main factor in the formation of drops from a leaking faucet, as shown in the photograph in the margin. Surface phenomena associated with liquid–gas, liquid–liquid, and liquid–gas–solid interfaces are exceedingly complex, and a more detailed and rigorous discussion of them is beyond the scope of this text. Fortunately, in many fluid mechanics problems, surface phenomena, as characterized by surface tension, are not important, since inertial, gravitational, and viscous forces are much more dominant.

### F l u i d s   i n   t h e   N e w s

**Spreading of oil spills** With the large traffic in oil tankers there is great interest in the prevention of and response to oil spills. As evidenced by the famous *Exxon Valdez* oil spill in Prince William Sound in 1989, oil spills can create disastrous environmental problems. A more recent example of this type of catastrophe is the oil spill that occurred in the Gulf of Mexico in 2010. It is not surprising that much attention is given to the rate at which an oil spill spreads. When spilled, most oils tend to spread horizontally into a smooth and slippery surface, called a slick. There are many factors that influence the ability of an oil slick to spread, including the size of the spill, wind speed and direction, and the physical properties of the oil. These properties include *surface tension, specific gravity,* and *viscosity.* The higher the surface tension the more likely a spill will remain in place. Since the specific gravity of oil is less than one, it floats on top of the water, but the specific gravity of an oil can increase if the lighter substances within the oil evaporate. The higher the viscosity of the oil, the greater the tendency to stay in one place.

## 1.10    A Brief Look Back in History

Before proceeding with our study of fluid mechanics, we should pause for a moment to consider the history of this important engineering science. As is true of all basic scientific and engineering disciplines, their actual beginnings are only faintly visible through the haze of early antiquity. But we know that interest in fluid behavior dates back to the ancient civilizations. Through necessity there was a practical concern about the manner in which spears and arrows could be propelled through the air, in the development of water supply and irrigation systems, and in the design of boats and ships. These developments were, of course, based on trial-and-error procedures without any knowledge of mathematics or mechanics. However, it was the accumulation of such empirical knowledge that formed the basis for further development during the emergence of the ancient Greek civilization and the subsequent rise of the Roman Empire. Some of the earliest writings that pertain to modern fluid mechanics are those of Archimedes (287–212 B.C.), a Greek mathematician and inventor who first expressed the principles of hydrostatics and flotation. Elaborate water supply systems were built by the Romans during the period from the fourth century B.C. through the early Christian period, and Sextus Julius Frontinus (A.D. 40–103), a Roman engineer, described these systems in detail. However, for the next 1000 years during the Middle Ages (also referred to as the Dark Ages), there appears to have been little added to further understanding of fluid behavior.

As shown in Fig. 1.11, beginning with the Renaissance period (about the fifteenth century) a rather continuous series of contributions began that forms the basis of what we consider to be the science of fluid mechanics. Leonardo da Vinci (1452–1519) described through sketches and writings many different types of flow phenomena. The work of Galileo Galilei (1564–1642) marked the beginning of experimental mechanics. Following the early Renaissance period and during the seventeenth and eighteenth centuries, numerous significant contributions were made. These include theoretical and mathematical advances associated with the famous names of Newton, Bernoulli, Euler, and d'Alembert. Experimental aspects of fluid mechanics were also advanced during this period, but unfortunately the two different approaches, theoretical and experimental, developed along separate paths. *Hydrodynamics* was the term associated with the theoretical or mathematical study of idealized, frictionless fluid behavior, with the term *hydraulics* being used to describe the applied or experimental aspects of real fluid behavior, particularly the behavior of water. Further contributions and refinements were made to both theoretical hydrodynamics and experimental hydraulics during the nineteenth century, with the general differential equations describing fluid motions that are used in modern fluid mechanics being developed in this period. Experimental hydraulics became more of a science, and many of the results of experiments performed during the nineteenth century are still used today.

At the beginning of the twentieth century, both the fields of theoretical hydrodynamics and experimental hydraulics were highly developed, and attempts were being made to unify the two. In 1904 a classic paper was presented by a German professor, Ludwig Prandtl (1875–1953), who introduced the concept of a "fluid boundary layer," which laid the foundation for the unification of the theoretical and experimental aspects of fluid mechanics. Prandtl's idea was that for flow next to

■ **Figure 1.11**   **Time line of some contributors to the science of fluid mechanics.**

a solid boundary a thin fluid layer (boundary layer) develops in which friction is very important, but outside this layer the fluid behaves very much like a frictionless fluid. This relatively simple concept provided the necessary impetus for the resolution of the conflict between the hydrodynamicists and the hydraulicists. Prandtl is generally accepted as the founder of modern fluid mechanics.

Also, during the first decade of the twentieth century, powered flight was first successfully demonstrated with the subsequent vastly increased interest in *aerodynamics*. Because the design of aircraft required a degree of understanding of fluid flow and an ability to make accurate predictions of the effect of airflow on bodies, the field of aerodynamics provided a great stimulus for the many rapid developments in fluid mechanics that took place during the twentieth century.

As we proceed with our study of the fundamentals of fluid mechanics, we will continue to note the contributions of many of the pioneers in the field. Table 1.9 provides a chronological

*The rich history of fluid mechanics is fascinating, and many of the contributions of the pioneers in the field are noted in the succeeding chapters.*

Leonardo da Vinci

Isaac Newton

Daniel Bernoulli

Ernst Mach

■ **Table 1.9**

**Chronological Listing of Some Contributors to the Science of Fluid Mechanics Noted in the Text**[a]

**ARCHIMEDES** (287– 212 B.C.)
Established elementary principles of buoyancy and flotation.

**SEXTUS JULIUS FRONTINUS** (A.D. 40–103)
Wrote treatise on Roman methods of water distribution.

**LEONARDO da VINCI** (1452–1519)
Expressed elementary principle of continuity; observed and sketched many basic flow phenomena; suggested designs for hydraulic machinery.

**GALILEO GALILEI** (1564–1642)
Indirectly stimulated experimental hydraulics; revised Aristotelian concept of vacuum.

**EVANGELISTA TORRICELLI** (1608–1647)
Related barometric height to weight of atmosphere, and form of liquid jet to trajectory of free fall.

**BLAISE PASCAL** (1623–1662)
Finally clarified principles of barometer, hydraulic press, and pressure transmissibility.

**ISAAC NEWTON** (1642–1727)
Explored various aspects of fluid resistance— inertial, viscous, and wave; discovered jet contraction.

**HENRI de PITOT** (1695–1771)
Constructed double-tube device to indicate water velocity through differential head.

**DANIEL BERNOULLI** (1700–1782)
Experimented and wrote on many phases of fluid motion, coining name "hydrodynamics"; devised manometry technique and adapted primitive energy principle to explain velocity-head indication; proposed jet propulsion.

**LEONHARD EULER** (1707–1783)
First explained role of pressure in fluid flow; formulated basic equations of motion and so-called Bernoulli theorem; introduced concept of cavitation and principle of centrifugal machinery.

**JEAN le ROND d'ALEMBERT** (1717–1783)
Originated notion of velocity and acceleration components, differential expression of continuity, and paradox of zero resistance to steady nonuniform motion.

**ANTOINE CHEZY** (1718–1798)
Formulated similarity parameter for predicting flow characteristics of one channel from measurements on another.

**GIOVANNI BATTISTA VENTURI** (1746–1822)
Performed tests on various forms of mouthpieces— in particular, conical contractions and expansions.

**LOUIS MARIE HENRI NAVIER** (1785–1836)
Extended equations of motion to include "molecular" forces.

**AUGUSTIN LOUIS de CAUCHY** (1789–1857)
Contributed to the general field of theoretical hydrodynamics and to the study of wave motion.

**GOTTHILF HEINRICH LUDWIG HAGEN** (1797–1884)
Conducted original studies of resistance in and transition between laminar and turbulent flow.

**JEAN LOUIS POISEUILLE** (1799–1869)
Performed meticulous tests on resistance of flow through capillary tubes.

**HENRI PHILIBERT GASPARD DARCY** (1803–1858)
Performed extensive tests on filtration and pipe resistance; initiated open-channel studies carried out by Bazin.

**JULIUS WEISBACH** (1806–1871)
Incorporated hydraulics in treatise on engineering mechanics, based on original experiments; noteworthy for flow patterns, nondimensional coefficients, weir, and resistance equations.

**WILLIAM FROUDE** (1810–1879)
Developed many towing-tank techniques, in particular the conversion of wave and boundary layer resistance from model to prototype scale.

**ROBERT MANNING** (1816–1897)
Proposed several formulas for open-channel resistance.

**GEORGE GABRIEL STOKES** (1819–1903)
Derived analytically various flow relationships ranging from wave mechanics to viscous resistance— particularly that for the settling of spheres.

**ERNST MACH** (1838–1916)
One of the pioneers in the field of supersonic aerodynamics.

**■ Table 1.9 (continued)**

Osborne Reynolds

Ludwig Prandtl

**OSBORNE REYNOLDS** (1842–1912)
Described original experiments in many fields—cavitation, river model similarity, pipe resistance—and devised two parameters for viscous flow; adapted equations of motion of a viscous fluid to mean conditions of turbulent flow.

**JOHN WILLIAM STRUTT, LORD RAYLEIGH** (1842–1919)
Investigated hydrodynamics of bubble collapse, wave motion, jet instability, laminar flow analogies, and dynamic similarity.

**VINCENZ STROUHAL** (1850–1922)
Investigated the phenomenon of "singing wires."

**EDGAR BUCKINGHAM** (1867–1940)
Stimulated interest in the United States in the use of dimensional analysis.

**MORITZ WEBER** (1871–1951)
Emphasized the use of the principles of similitude in fluid flow studies and formulated a capillarity similarity parameter.

**LUDWIG PRANDTL** (1875–1953)
Introduced concept of the boundary layer and is generally considered to be the father of present-day fluid mechanics.

**LEWIS FERRY MOODY** (1880–1953)
Provided many innovations in the field of hydraulic machinery. Proposed a method of correlating pipe resistance data that is widely used.

**THEODOR VON KÁRMÁN** (1881–1963)
One of the recognized leaders of twentieth century fluid mechanics. Provided major contributions to our understanding of surface resistance, turbulence, and wake phenomena.

**PAUL RICHARD HEINRICH BLASIUS** (1883–1970)
One of Prandtl's students who provided an analytical solution to the boundary layer equations. Also demonstrated that pipe resistance was related to the Reynolds number.

*a*Used by permission of IIHR—Hydroscience & Engineering, The University of Iowa.

listing of some of these contributors and reveals the long journey that makes up the history of fluid mechanics. This list is certainly not comprehensive with regard to all past contributors but includes those who are mentioned in this text. As mention is made in succeeding chapters of the various individuals listed in Table 1.9, a quick glance at this table will reveal where they fit into the historical chain.

It is, of course, impossible to summarize the rich history of fluid mechanics in a few paragraphs. Only a brief glimpse is provided, and we hope it will stir your interest. References 2 to 5 are good starting points for further study, and in particular Ref. 2 provides an excellent, broad, easily read history. Try it—you might even enjoy it!

## 1.11    Chapter Summary and Study Guide

This introductory chapter discussed several fundamental aspects of fluid mechanics. Methods for describing fluid characteristics both quantitatively and qualitatively are considered. For a quantitative description, units are required, and in this text, two systems of units are used: the British Gravitational (BG) system (pounds, slugs, feet, and seconds) and the International (SI) System (newtons, kilograms, meters, and seconds). For the qualitative description the concept of dimensions is introduced in which basic dimensions such as length, $L$, time, $T$, and mass, $M$, are used to provide a description of various quantities of interest. The use of dimensions is helpful in checking the generality of equations, as well as serving as the basis for the powerful tool of dimensional analysis discussed in detail in Chapter 7.

Various important fluid properties are defined, including fluid density, specific weight, specific gravity, viscosity, bulk modulus, speed of sound, vapor pressure, and surface tension. The ideal gas law is introduced to relate pressure, temperature, and density in common gases, along with a brief discussion of the compression and expansion of gases. The distinction between absolute and gage pressure is introduced and this important idea is explored more fully in Chapter 2.

*fluid*
*units*
*basic dimensions*
*dimensionally*
  *homogeneous*
*density*
*specific weight*
*specific gravity*
*ideal gas law*
*absolute pressure*
*gage pressure*
*no-slip condition*
*rate of shearing strain*
*absolute viscosity*
*Newtonian fluid*
*non-Newtonian fluid*
*kinematic viscosity*
*bulk modulus*
*speed of sound*
*vapor pressure*
*surface tension*

The following checklist provides a study guide for this chapter. When your study of the entire chapter and end-of-chapter exercises has been completed you should be able to

■ write out meanings of the terms listed here in the margin and understand each of the related concepts. These terms are particularly important and are set in *italic, **bold,** and color* type in the text.

■ determine the dimensions of common physical quantities.

■ determine whether an equation is a general or restricted homogeneous equation.

■ use both BG and SI systems of units.

■ calculate the density, specific weight, or specific gravity of a fluid from a knowledge of any two of the three.

■ calculate the density, pressure, or temperature of an ideal gas (with a given gas constant) from a knowledge of any two of the three.

■ relate the pressure and density of a gas as it is compressed or expanded using Eqs. 1.14 and 1.15.

■ use the concept of viscosity to calculate the shearing stress in simple fluid flows.

■ calculate the speed of sound in fluids using Eq. 1.19 for liquids and Eq. 1.20 for gases.

■ determine whether boiling or cavitation will occur in a liquid using the concept of vapor pressure.

■ use the concept of surface tension to solve simple problems involving liquid–gas or liquid–solid–gas interfaces.

Some of the important equations in this chapter are:

| | | |
|---|---|---|
| Specific weight | $\gamma = \rho g$ | **(1.6)** |
| Specific gravity | $SG = \dfrac{\rho}{\rho_{H_2O@4\,°C}}$ | **(1.7)** |
| Ideal gas law | $\rho = \dfrac{p}{RT}$ | **(1.8)** |
| Newtonian fluid shear stress | $\tau = \mu \dfrac{du}{dy}$ | **(1.9)** |
| Bulk modulus | $E_v = -\dfrac{dp}{d\forall/\forall}$ | **(1.12)** |
| Speed of sound in an ideal gas | $c = \sqrt{kRT}$ | **(1.20)** |
| Capillary rise in a tube | $h = \dfrac{2\sigma \cos \theta}{\gamma R}$ | **(1.22)** |

# References

1. Reid, R. C., Prausnitz, J. M., and Sherwood, T. K., *The Properties of Gases and Liquids,* 3rd Ed., McGraw-Hill, New York, 1977.

2. Rouse, H. and Ince, S., *History of Hydraulics,* Iowa Institute of Hydraulic Research, Iowa City, 1957, Dover, New York, 1963.

3. Tokaty, G. A., *A History and Philosophy of Fluid Mechanics,* G. T. Foulis and Co., Ltd., Oxfordshire, Great Britain, 1971.

4. Rouse, H., *Hydraulics in the United States 1776–1976,* Iowa Institute of Hydraulic Research, Iowa City, Iowa, 1976.

5. Garbrecht, G., ed., *Hydraulics and Hydraulic Research—A Historical Review,* A. A. Balkema, Rotterdam, Netherlands, 1987.

6. Brenner, M. P., Shi, X. D., Eggens, J., and Nagel, S. R., *Physics of Fluids,* Vol. 7, No. 9, 1995.

7. Shi, X. D., Brenner, M. P., and Nagel, S. R., *Science,* Vol. 265, 1994.

| | |
|---|---|
| (WILEY) | Problem available in *WileyPLUS* at instructor's discretion. |
| (WILEY GO) | Tutoring problem available in *WileyPLUS* at instructor's discretion. |
| ▶ | Problem is related to a chapter video available in *WileyPLUS*. |
| ✳ | Problem to be solved with aid of programmable calculator or computer. |
| † | Open-ended problem that requires critical thinking. These problems require various assumptions to provide the necessary input data. There are not unique answers to these problems. |

## Review Problems

Go to Appendix G (*WileyPLUS* or the book's web site, www.wiley.com/college/munson) for a set of review problems with answers. Detailed solutions can be found in the *Student Solution* *Manual and Study Guide for Fundamentals of Fluid Mechanics,* by Munson et al. © 2013 John Wiley and Sons, Inc.).

## Conceptual Questions

**1.1C** The correct statement for the definition of density is

a) Density is the mass per unit volume.

b) Density is the volume per unit mass.

c) Density is the weight per unit volume.

d) Density is the weight divided by gravity.

e) Density is the mass divided by the weight.

**1.2C** Given the following equation where $p$ is pressure in lb/ft$^2$, $\gamma$ is the specific weight in lb/ft$^3$, $V$ is the magnitude of velocity in ft/s, $g$ is in ft/s$^2$, and $z$ is height in feet. If values are substituted into the equation, will the correct value of C be determined?

$$\frac{p}{\gamma} + \frac{V^2}{2g} + z = C$$

a) Yes if the constant C has units of ft.

b) Yes if the constant C is dimensionless.

c) No, the equation cannot produce the correct value of C.

d) Yes if the constant C has units of ft and the specific weight is multiplied by the conversion factor from lbm to lbf.

**1.3C** The no-slip condition is:

a) An experimental observation that the velocity of a fluid in contact with a solid surface is equal to the velocity of the surface.

b) Valid only for liquids.

c) Useful only for very low density gases.

d) Indicates that two solids in contact will not slip if the joining force is large.

**1.4C** In fluids, the shearing strain rate $\dfrac{du}{dy}$ for a Newtonian fluid has dimensions of:

a) L/T$^2$.       c) L$^2$/T.

b) 1/T.       d) L$^2$/T$^2$.

**1.5C** The laminar velocity profile for a Newtonian fluid is shown below.

Which figure best describes the variation of shear stress with distance from the plate?

**Additional conceptual questions are available in *WileyPLUS* at the instructor's discretion.**

## Problems

**Note: Unless specific values of required fluid properties are given in the problem statement, use the values found in the tables on the inside of the front cover. Answers to the even-numbered problems are listed at the end of the book. The Lab Problems as well as the videos that accompany problems can be accessed in WileyPLUS or the book's web site, www.wiley.com/college/munson.**

### Section 1.2 Dimensions, Dimensional Homogeneity, and Units

**1.1** The force, $F$, of the wind blowing against a building is given by $F = C_D \rho V^2 A/2$, where $V$ is the wind speed, $\rho$ the density of the air, $A$ the cross-sectional area of the building, and $C_D$ is a constant termed the drag coefficient. Determine the dimensions of the drag coefficient.

**1.2** Determine the dimensions, in both the *FLT* system and the *MLT* system, for **(a)** the product of mass times velocity, **(b)** the product of force times volume, and **(c)** kinetic energy divided by area.

**1.3** Verify the dimensions, in both the *FLT* and *MLT* systems, of the following quantities which appear in Table 1.1: **(a)** volume, **(b)** acceleration, **(c)** mass, **(d)** moment of inertia (area), and **(e)** work.

**1.4** Determine the dimensions, in both the *FLT* system and the *MLT* system, for **(a)** the product of force times acceleration, **(b)** the product of force times velocity divided by area, and **(c)** momentum divided by volume.

**1.5** Verify the dimensions, in both the *FLT* and *MLT* systems, of the following quantities which appear in Table 1.1: **(a)** angular velocity, **(b)** energy, **(c)** moment of inertia (area), **(d)** power, and **(e)** pressure.

**1.6** Verify the dimensions, in both the *FLT* system and the *MLT* system, of the following quantities which appear in Table 1.1: **(a)** frequency, **(b)** stress, **(c)** strain, **(d)** torque, and **(e)** work.

**1.7** If $u$ is a velocity, $x$ a length, and $t$ a time, what are the dimensions (in the *MLT* system) of **(a)** $\partial u/\partial t$, **(b)** $\partial^2 u/\partial x \partial t$, and **(c)** $\int (\partial u/\partial t)\, dx$?

**1.8** Verify the dimensions, in both the *FLT* system and the *MLT* system, of the following quantities which appear in Table 1.1: **(a)** acceleration, **(b)** stress, **(c)** moment of a force, **(d)** volume, and **(e)** work.

**1.9** If $p$ is a pressure, $V$ a velocity, and $\rho$ a fluid density, what are the dimensions (in the *MLT* system) of **(a)** $p/\rho$, **(b)** $pV\rho$, and **(c)** $p/\rho V^2$?

**1.10** If $P$ is a force and $x$ a length, what are the dimensions (in the *FLT* system) of **(a)** $dP/dx$, **(b)** $d^3P/dx^3$, and **(c)** $\int P\, dx$?

**1.11** ⟨WILEY⟩ If $V$ is a velocity, $\ell$ a length, and $\nu$ a fluid property (the kinematic viscosity) having dimensions of $L^2T^{-1}$, which of the following combinations are dimensionless: **(a)** $V\ell\nu$, **(b)** $V\ell/\nu$, **(c)** $V^2\nu$, **(d)** $V/\ell\nu$?

**1.12** If $V$ is a velocity, determine the dimensions of $Z$, $\alpha$, and $G$, which appear in the dimensionally homogeneous equation

$$V = Z(\alpha - 1) + G$$

**1.13** The volume rate of flow, $Q$, through a pipe containing a slowly moving liquid is given by the equation

$$Q = \frac{\pi R^4 \Delta p}{8\mu\ell}$$

where $R$ is the pipe radius, $\Delta p$ the pressure drop along the pipe, $\mu$ a fluid property called viscosity $(FL^{-2}T)$, and $\ell$ the length of pipe. What are the dimensions of the constant $\pi/8$? Would you classify this equation as a general homogeneous equation? Explain.

**1.14** According to information found in an old hydraulics book, the energy loss per unit weight of fluid flowing through a nozzle connected to a hose can be estimated by the formula

$$h = (0.04 \text{ to } 0.09)(D/d)^4 V^2/2g$$

where $h$ is the energy loss per unit weight, $D$ the hose diameter, $d$ the nozzle tip diameter, $V$ the fluid velocity in the hose, and $g$ the acceleration of gravity. Do you think this equation is valid in any system of units? Explain.

**1.15** ⟨WILEY⟩ The pressure difference, $\Delta p$, across a partial blockage in an artery (called a *stenosis*) is approximated by the equation

$$\Delta p = K_v \frac{\mu V}{D} + K_u \left(\frac{A_0}{A_1} - 1\right)^2 \rho V^2$$

where $V$ is the blood velocity, $\mu$ the blood viscosity $(FL^{-2}T)$, $\rho$ the blood density $(ML^{-3})$, $D$ the artery diameter, $A_0$ the area of the unobstructed artery, and $A_1$ the area of the stenosis. Determine the dimensions of the constants $K_v$ and $K_u$. Would this equation be valid in any system of units?

**1.16** Assume that the speed of sound, $c$, in a fluid depends on an elastic modulus, $E_v$, with dimensions $FL^{-2}$, and the fluid density, $\rho$, in the form $c = (E_v)^a(\rho)^b$. If this is to be a dimensionally homogeneous equation, what are the values for $a$ and $b$? Is your result consistent with the standard formula for the speed of sound? (See Eq. 1.19.)

**1.17** ⟨WILEY⟩ A formula to estimate the volume rate of flow, $Q$, flowing over a dam of length, $B$, is given by the equation

$$Q = 3.09\, BH^{3/2}$$

where $H$ is the depth of the water above the top of the dam (called the head). This formula gives $Q$ in ft³/s when $B$ and $H$ are in feet. Is the constant, 3.09, dimensionless? Would this equation be valid if units other than feet and seconds were used?

**1.18** The force, $P$, that is exerted on a spherical particle moving slowly through a liquid is given by the equation

$$P = 3\pi\mu DV$$

where $\mu$ is a fluid property (viscosity) having dimensions of $FL^{-2}T$, $D$ is the particle diameter, and $V$ is the particle velocity. What are the dimensions of the constant, $3\pi$? Would you classify this equation as a general homogeneous equation?

**†1.19** Cite an example of a restricted homogeneous equation contained in a technical article found in an engineering journal in your field of interest. Define all terms in the equation, explain why it is a restricted equation, and provide a complete journal citation (title, date, etc.).

**1.20** ⟨WILEY⟩ Make use of Table 1.3 to express the following quantities in SI units: **(a)** 10.2 in./min, **(b)** 4.81 slugs, **(c)** 3.02 lb, **(d)** 73.1 ft/s², **(e)** 0.0234 lb · s/ft².

**1.21** ⟨WILEY⟩ Make use of Table 1.4 to express the following quantities in BG units: **(a)** 14.2 km, **(b)** 8.14 N/m³, **(c)** 1.61 kg/m³, **(d)** 0.0320 N · m/s, **(e)** 5.67 mm/hr.

**1.22** ⟨WILEY⟩ Express the following quantities in SI units: **(a)** 160 acres, **(b)** 15 gallons (U.S.), **(c)** 240 miles, **(d)** 79.1 hp, **(e)** 60.3 °F.

**1.23** For Table 1.3 verify the conversion relationships for: **(a)** area, **(b)** density, **(c)** velocity, and **(d)** specific weight. Use the basic conversion relationships: 1 ft = 0.3048 m; 1 lb = 4.4482 N; and 1 slug = 14.594 kg.

**1.24** For Table 1.4 verify the conversion relationships for: **(a)** acceleration, **(b)** density, **(c)** pressure, and **(d)** volume flowrate. Use the basic conversion relationships: 1 m = 3.2808 ft; 1N = 0.22481 lb; and 1 kg = 0.068521 slug.

**1.25** ⟨WILEY⟩ Water flows from a large drainage pipe at a rate of 1200 gal/min. What is this volume rate of flow in **(a)** m³/s, **(b)** liters/min, and **(c)** ft³/s?

**1.26** Dimensionless combinations of quantities (commonly called dimensionless parameters) play an important role in fluid mechanics. Make up five possible dimensionless parameters by using combinations of some of the quantities listed in Table 1.1.

**1.27** ⟨WILEY GO⟩ An important dimensionless parameter in certain types of fluid flow problems is the *Froude number* defined as $V/\sqrt{g\ell}$, where $V$ is a velocity, $g$ the acceleration of gravity, and $\ell$ a length. Determine the value of the Froude number for $V = 10$ ft/s, $g = 32.2$ ft/s², and $\ell = 2$ ft. Recalculate the Froude number using SI units for $V$, $g$, and $\ell$. Explain the significance of the results of these calculations.

### Section 1.4 Measures of Fluid Mass and Weight

**1.28** Obtain a photograph/image of a situation in which the density or specific weight of a fluid is important. Print this photo and write a brief paragraph that describes the situation involved.

**1.29** A tank contains 500 kg of a liquid whose specific gravity is 2. Determine the volume of the liquid in the tank.

**1.30** ⟨WILEY⟩ Clouds can weigh thousands of pounds due to their liquid water content. Often this content is measured in grams per cubic meter (g/m³). Assume that a cumulus cloud occupies a volume of one cubic kilometer, and its liquid water content is 0.2 g/m³. **(a)** What is the volume of this cloud in cubic miles? **(b)** How much does the water in the cloud weigh in pounds?

**1.31** (WILEY) A tank of oil has a mass of 25 slugs. **(a)** Determine its weight in pounds and in newtons at the Earth's surface. **(b)** What would be its mass (in slugs) and its weight (in pounds) if located on the moon's surface where the gravitational attraction is approximately one-sixth that at the Earth's surface?

**1.32** (WILEY) A certain object weighs 300 N at the Earth's surface. Determine the mass of the object (in kilograms) and its weight (in newtons) when located on a planet with an acceleration of gravity equal to 4.0 ft/s$^2$.

**1.33** (WILEY) The density of a certain type of jet fuel is 775 kg/m$^3$. Determine its specific gravity and specific weight.

**1.34** ▶ (WILEY) A *hydrometer* is used to measure the specific gravity of liquids. (See **Video V2.8**.) For a certain liquid, a hydrometer reading indicates a specific gravity of 1.15. What is the liquid's density and specific weight? Express your answer in SI units.

**1.35** The specific weight of a certain liquid is 85.3 lb/ft$^3$. Determine its density and specific gravity.

**1.36** (WILEY) An open, rigid-walled, cylindrical tank contains 4 ft$^3$ of water at 40 °F. Over a 24-hour period of time the water temperature varies from 40 to 90 °F. Make use of the data in Appendix B to determine how much the volume of water will change. For a tank diameter of 2 ft, would the corresponding change in water depth be very noticeable? Explain.

**†1.37** Estimate the number of pounds of mercury it would take to fill your bathtub. List all assumptions and show all calculations.

**1.38** A mountain climber's oxygen tank contains 1 lb of oxygen when he begins his trip at sea level where the acceleration of gravity is 32.174 ft/s$^2$. What is the weight of the oxygen in the tank when he reaches the top of Mt. Everest where the acceleration of gravity is 32.082 ft/s$^2$? Assume that no oxygen has been removed from the tank; it will be used on the descent portion of the climb.

**1.39** (WILEY.GO) The information on a can of pop indicates that the can contains 355 mL. The mass of a full can of pop is 0.369 kg, while an empty can weighs 0.153 N. Determine the specific weight, density, and specific gravity of the pop and compare your results with the corresponding values for water at 20 °C. Express your results in SI units.

**\*1.40** (WILEY) The variation in the density of water, $\rho$, with temperature, $T$, in the range 20 °C $\leq T \leq$ 50 °C, is given in the following table.

| Density (kg/m$^3$) | 998.2 | 997.1 | 995.7 | 994.1 | 992.2 | 990.2 | 988.1 |
|---|---|---|---|---|---|---|---|
| Temperature (°C) | 20 | 25 | 30 | 35 | 40 | 45 | 50 |

Use these data to determine an empirical equation of the form $\rho = c_1 + c_2 T + c_3 T^2$ which can be used to predict the density over the range indicated. Compare the predicted values with the data given. What is the density of water at 42.1 °C?

**1.41** (WILEY) If 1 cup of cream having a density of 1005 kg/m$^3$ is turned into 3 cups of whipped cream, determine the specific gravity and specific weight of the whipped cream.

**1.42** A liquid when poured into a graduated cylinder is found to weigh 8 N when occupying a volume of 500 ml (milliliters). Determine its specific weight, density, and specific gravity.

**†1.43** The presence of raindrops in the air during a heavy rainstorm increases the average density of the air–water mixture. Estimate by what percent the average air–water density is greater than that of just still air. State all assumptions and show calculations.

## Section 1.5 Ideal Gas Law

**1.44** Determine the mass of air in a 2 m$^3$ tank if the air is at room temperature, 20 °C, and the absolute pressure within the tank is 200 kPa (abs).

**1.45** (WILEY) Nitrogen is compressed to a density of 4 kg/m$^3$ under an absolute pressure of 400 kPa. Determine the temperature in degrees Celsius.

**1.46** (WILEY) The temperature and pressure at the surface of Mars during a Martian spring day were determined to be −50 °C and 900 Pa, respectively. **(a)** Determine the density of the Martian atmosphere for these conditions if the gas constant for the Martian atmosphere is assumed to be equivalent to that of carbon dioxide. **(b)** Compare the answer from part (a) with the density of the Earth's atmosphere during a spring day when the temperature is 18 °C and the pressure 101.6 kPa (abs).

**1.47** (WILEY) A closed tank having a volume of 2 ft$^3$ is filled with 0.30 lb of a gas. A pressure gage attached to the tank reads 12 psi when the gas temperature is 80 °F. There is some question as to whether the gas in the tank is oxygen or helium. Which do you think it is? Explain how you arrived at your answer.

**1.48** A tire having a volume of 3 ft$^3$ contains air at a gage pressure of 26 psi and a temperature of 70 °F. Determine the density of the air and the weight of the air contained in the tire.

**1.49** (WILEY) A compressed air tank contains 5 kg of air at a temperature of 80 °C. A gage on the tank reads 300 kPa. Determine the volume of the tank.

**1.50** (WILEY) A rigid tank contains air at a pressure of 90 psia and a temperature of 60 °F. By how much will the pressure increase as the temperature is increased to 110 °F?

**1.51** The density of oxygen contained in a tank is 2.0 kg/m$^3$ when the temperature is 25 °C. Determine the gage pressure of the gas if the atmospheric pressure is 97 kPa.

**1.52** The helium-filled blimp shown in Fig. P1.52 is used at various athletic events. Determine the number of pounds of helium within it if its volume is 68,000 ft$^3$ and the temperature and pressure are 80 °F and 14.2 psia, respectively.

■ **Figure P1.52**

**\*1.53** Develop a computer program for calculating the density of an ideal gas when the gas pressure in pascals (abs), the temperature in degrees Celsius, and the gas constant in J/kg · K are specified. Plot the density of helium as a function of temperature from 0 °C to 200 °C and pressures of 50, 100, 150, and 200 kPa (abs).

## Section 1.6 Viscosity (also see Lab Problems 1.1LP and 1.2LP)

**1.54** Obtain a photograph/image of a situation in which the viscosity of a fluid is important. Print this photo and write a brief paragraph that describes the situation involved.

**1.55** For flowing water, what is the magnitude of the velocity gradient needed to produce a shear stress of 1.0 N/m²?

**1.56** Make use of the data in Appendix B to determine the dynamic viscosity of glycerin at 85 °F. Express your answer in both SI and BG units.

**1.57** Make use of the data in Appendix B to determine the dynamic viscosity of mercury at 75 °F. Express your answer in BG units.

**1.58** ▶ WILEY○ One type of *capillary-tube viscometer* is shown in **Video V1.5** and in Fig. P1.58. For this device the liquid to be tested is drawn into the tube to a level above the top etched line. The time is then obtained for the liquid to drain to the bottom etched line. The kinematic viscosity, $\nu$, in m²/s is then obtained from the equation $\nu = KR^4t$ where $K$ is a constant, $R$ is the radius of the capillary tube in mm, and $t$ is the drain time in seconds. When glycerin at 20 °C is used as a calibration fluid in a particular viscometer, the drain time is 1430 s. When a liquid having a density of 970 kg/m³ is tested in the same viscometer the drain time is 900 s. What is the dynamic viscosity of this liquid?

Glass strengthening bridge

Etched lines

Capillary tube

■ **Figure P1.58**

**1.59** ▶ WILEY○ The viscosity of a soft drink was determined by using a capillary tube viscometer similar to that shown in Fig. P1.58 and **Video V1.5**. For this device the kinematic viscosity, $\nu$, is directly proportional to the time, $t$, that it takes for a given amount of liquid to flow through a small capillary tube. That is, $\nu = Kt$. The following data were obtained from regular pop and diet pop. The corresponding measured specific gravities are also given. Based on these data, by what percent is the absolute viscosity, $\mu$, of regular pop greater than that of diet pop?

|        | Regular pop | Diet pop |
|--------|-------------|----------|
| $t$(s) | 377.8       | 300.3    |
| SG     | 1.044       | 1.003    |

**1.60** Determine the ratio of the dynamic viscosity of water to air at a temperature of 60 °C. Compare this value with the corresponding ratio of kinematic viscosities. Assume the air is at standard atmospheric pressure.

**1.61** WILEY○ The viscosity of a certain fluid is $5 \times 10^{-4}$ poise. Determine its viscosity in both SI and BG units.

**1.62** The kinematic viscosity and specific gravity of a liquid are $3.5 \times 10^{-4}$ m²/s and 0.79, respectively. What is the dynamic viscosity of the liquid in SI units?

**1.63** A liquid has a specific weight of 59 lb/ft³ and a dynamic viscosity of 2.75 lb·s/ft². Determine its kinematic viscosity.

**1.64** The kinematic viscosity of oxygen at 20 °C and a pressure of 150 kPa (abs) is 0.104 stokes. Determine the dynamic viscosity of oxygen at this temperature and pressure.

**\*1.65** ▶ Fluids for which the shearing stress, $\tau$, is not linearly related to the rate of shearing strain, $\dot{\gamma}$, are designated as non-Newtonian fluids. Such fluids are commonplace and can exhibit unusual behavior, as shown in **Video V1.6**. Some experimental data obtained for a particular non-Newtonian fluid at 80 °F are shown below.

| $\tau$(lb/ft²)       | 0 | 2.11 | 7.82 | 18.5 | 31.7 |
|----------------------|---|------|------|------|------|
| $\dot{\gamma}$ (s⁻¹) | 0 | 50   | 100  | 150  | 200  |

Plot these data and fit a second-order polynomial to the data using a suitable graphing program. What is the apparent viscosity of this fluid when the rate of shearing strain is 70 s⁻¹? Is this apparent viscosity larger or smaller than that for water at the same temperature?

**1.66** ▶ WILEY○ Water flows near a flat surface and some measurements of the water velocity, $u$, parallel to the surface, at different heights, $y$, above the surface are obtained. At the surface $y = 0$. After an analysis of the data, the lab technician reports that the velocity distribution in the range $0 < y < 0.1$ ft is given by the equation

$$u = 0.81 + 9.2y + 4.1 \times 10^3 y^3$$

with $u$ in ft/s when $y$ is in ft. **(a)** Do you think that this equation would be valid in any system of units? Explain. **(b)** Do you think this equation is correct? Explain. You may want to look at **Video 1.4** to help you arrive at your answer.

**1.67** WILEY○ Calculate the Reynolds numbers for the flow of water and for air through a 4-mm-diameter tube, if the mean velocity is 3 m/s and the temperature is 30 °C in both cases (see Example 1.4). Assume the air is at standard atmospheric pressure.

**1.68** SAE 30 oil at 60 °F flows through a 2-in.-diameter pipe with a mean velocity of 5 ft/s. Determine the value of the Reynolds number (see Example 1.4).

**1.69** For air at standard atmospheric pressure the values of the constants that appear in the Sutherland equation (Eq. 1.10) are $C = 1.458 \times 10^{-6}$ kg/(m·s·K$^{1/2}$) and $S = 110.4$ K. Use these values to predict the viscosity of air at 10 °C and 90 °C and compare with values given in Table B.4 in Appendix B.

**\*1.70** Use the values of viscosity of air given in Table B.4 at temperatures of 0, 20, 40, 60, 80, and 100 °C to determine the constants $C$ and $S$ which appear in the Sutherland equation (Eq. 1.10). Compare your results with the values given in Problem 1.69. (*Hint:* Rewrite the equation in the form

$$\frac{T^{3/2}}{\mu} = \left(\frac{1}{C}\right)T + \frac{S}{C}$$

and plot $T^{3/2}/\mu$ versus $T$. From the slope and intercept of this curve, $C$ and $S$ can be obtained.)

**1.71** ▶ The viscosity of a fluid plays a very important role in determining how a fluid flows. (See **Video V1.3**.) The value of the viscosity depends not only on the specific fluid but also on the fluid temperature. Some experiments show that when a liquid, under the action of a constant driving pressure, is forced with a low velocity, $V$, through a small horizontal tube, the velocity is given by the equation $V = K/\mu$. In this equation $K$ is a constant for a given tube and pressure, and $\mu$ is the dynamic viscosity. For a particular

liquid of interest, the viscosity is given by Andrade's equation (Eq. 1.11) with $D = 5 \times 10^{-7}$ lb·s/ft$^2$ and $B = 4000$ °R. By what percentage will the velocity increase as the liquid temperature is increased from 40 °F to 100 °F? Assume all other factors remain constant.

**\*1.72** Use the value of the viscosity of water given in Table B.2 at temperatures of 0, 20, 40, 60, 80, and 100 °C to determine the constants $D$ and $B$ which appear in Andrade's equation (Eq. 1.11). Calculate the value of the viscosity at 50 °C and compare with the value given in Table B.2. (*Hint:* Rewrite the equation in the form

$$\ln \mu = (B)\frac{1}{T} + \ln D$$

and plot $\ln \mu$ versus $1/T$. From the slope and intercept of this curve, $B$ and $D$ can be obtained. If a nonlinear curve-fitting program is available, the constants can be obtained directly from Eq. 1.11 without rewriting the equation.)

**1.73** For a certain liquid $\mu = 7.1 \times 10^{-5}$ lb·s/ft$^2$ at 40 °F and $\mu = 1.9 \times 10^{-5}$ lb·s/ft$^2$ at 150 °F. Make use of these data to determine the constants $D$ and $B$ which appear in Andrade's equation (Eq. 1.11). What would be the viscosity at 80 °F?

**1.74** WILEY GO For a parallel plate arrangement of the type shown in Fig. 1.5 it is found that when the distance between plates is 2 mm, a shearing stress of 150 Pa develops at the upper plate when it is pulled at a velocity of 1 m/s. Determine the viscosity of the fluid between the plates. Express your answer in SI units.

**1.75** Two flat plates are oriented parallel above a fixed lower plate as shown in Fig. P1.75. The top plate, located a distance $b$ above the fixed plate, is pulled along with speed $V$. The other thin plate is located a distance $cb$, where $0 < c < 1$, above the fixed plate. This plate moves with speed $V_1$, which is determined by the viscous shear forces imposed on it by the fluids on its top and bottom. The fluid on the top is twice as viscous as that on the bottom. Plot the ratio $V_1/V$ as a function of $c$ for $0 < c < 1$.

■ **Figure P1.75**

**1.76** ▶ There are many fluids that exhibit non-Newtonian behavior (see, for example, **Video V1.6**). For a given fluid the distinction between Newtonian and non-Newtonian behavior is usually based on measurements of shear stress and rate of shearing strain. Assume that the viscosity of blood is to be determined by measurements of shear stress, $\tau$, and rate of shearing strain, $du/dy$, obtained from a small blood sample tested in a suitable viscometer. Based on the data given below, determine if the blood is a Newtonian or non-Newtonian fluid. Explain how you arrived at your answer.

| $\tau$ (N/m$^2$) | 0.04 | 0.06 | 0.12 | 0.18 | 0.30 | 0.52 | 1.12 | 2.10 |
|---|---|---|---|---|---|---|---|---|
| $du/dy$ (s$^{-1}$) | 2.25 | 4.50 | 11.25 | 22.5 | 45.0 | 90.0 | 225 | 450 |

**1.77** WILEY The sled shown in Fig. P1.77 slides along on a thin horizontal layer of water between the ice and the runners. The horizontal force that the water puts on the runners is equal to 1.2 lb when the sled's speed is 50 ft/s. The total area of both runners in contact with the water is 0.08 ft$^2$, and the viscosity of the water is

$3.5 \times 10^{-5}$ lb·s/ft$^2$. Determine the thickness of the water layer under the runners. Assume a linear velocity distribution in the water layer.

■ **Figure P1.77**

**1.78** WILEY A 25-mm-diameter shaft is pulled through a cylindrical bearing as shown in Fig. P1.78. The lubricant that fills the 0.3-mm gap between the shaft and bearing is an oil having a kinematic viscosity of $8.0 \times 10^{-4}$ m$^2$/s and a specific gravity of 0.91. Determine the force $P$ required to pull the shaft at a velocity of 3 m/s. Assume the velocity distribution in the gap is linear.

■ **Figure P1.78**

**1.79** A piston having a diameter of 5.48 in. and a length of 9.50 in. slides downward with a velocity $V$ through a vertical pipe. The downward motion is resisted by an oil film between the piston and the pipe wall. The film thickness is 0.002 in., and the cylinder weighs 0.5 lb. Estimate $V$ if the oil viscosity is 0.016 lb·s/ft$^2$. Assume the velocity distribution in the gap is linear.

**1.80** WILEY A 10-kg block slides down a smooth inclined surface as shown in Fig. P1.80. Determine the terminal velocity of the block if the 0.1-mm gap between the block and the surface contains SAE 30 oil at 60 °F. Assume the velocity distribution in the gap is linear, and the area of the block in contact with the oil is 0.1 m$^2$.

■ **Figure P1.80**

**1.81** WILEY A layer of water flows down an inclined fixed surface with the velocity profile shown in Fig. P1.81. Determine the magnitude and direction of the shearing stress that the water exerts on the fixed surface for $U = 2$ m/s and $h = 0.1$ m.

■ **Figure P1.81**

**1.82** A thin layer of glycerin flows down an inclined, wide plate with the velocity distribution shown in Fig. P1.82. For $h = 0.3$ in. and $\alpha = 20°$, determine the surface velocity, $U$. Note that for equilibrium, the component of weight acting parallel to the plate surface must be balanced by the shearing force developed along the plate surface. In your analysis assume a unit plate width.

■ **Figure P1.82**

*****1.83** (WILEY) Standard air flows past a flat surface, and velocity measurements near the surface indicate the following distribution:

| $y$ (ft) | 0.005 | 0.01 | 0.02 | 0.04 | 0.06 | 0.08 |
|---|---|---|---|---|---|---|
| $u$ (ft/s) | 0.74 | 1.51 | 3.03 | 6.37 | 10.21 | 14.43 |

The coordinate $y$ is measured normal to the surface and $u$ is the velocity parallel to the surface. **(a)** Assume the velocity distribution is of the form

$$u = C_1 y + C_2 y^3$$

and use a standard curve-fitting technique to determine the constants $C_1$ and $C_2$. **(b)** Make use of the results of part (a) to determine the magnitude of the shearing stress at the wall ($y = 0$) and at $y = 0.05$ ft.

**1.84** (WILEY) A new computer drive is proposed to have a disc, as shown in Fig. P1.84. The disc is to rotate at 10,000 rpm, and the reader head is to be positioned 0.0005 in. above the surface of the disc. Estimate the shearing force on the reader head as a result of the air between the disc and the head.

■ **Figure P1.84**

**1.85** (WILEY) The space between two 6-in.-long concentric cylinders is filled with glycerin (viscosity = $8.5 \times 10^{-3}$ lb · s/ft²). The inner cylinder has a radius of 3 in. and the gap width between cylinders is 0.1 in. Determine the torque and the power required to rotate the inner cylinder at 180 rev/min. The outer cylinder is fixed. Assume the velocity distribution in the gap to be linear.

**1.86** (WILEY) A pivot bearing used on the shaft of an electrical instrument is shown in Fig. P1.86. An oil with a viscosity of $\mu = 0.010$ lb · s/ft² fills the 0.001-in. gap between the rotating shaft and the stationary base. Determine the frictional torque on the shaft when it rotates at 5000 rpm.

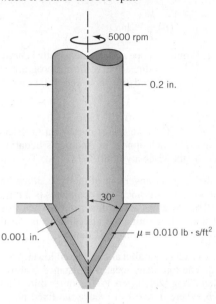

■ **Figure P1.86**

**1.87** The viscosity of liquids can be measured through the use of a *rotating cylinder viscometer* of the type illustrated in Fig. P1.87. In this device the outer cylinder is fixed and the inner cylinder is rotated with an angular velocity, $\omega$. The torque $\mathcal{T}$ required to develop $\omega$ is measured and the viscosity is calculated from these two measurements. **(a)** Develop an equation relating $\mu$, $\omega$, $\mathcal{T}$, $\ell$, $R_o$, and $R_i$. Neglect end effects and assume the velocity distribution in the gap is linear. **(b)** The following torque-angular velocity data were obtained with a rotating cylinder viscometer of the type discussed in part (a).

| Torque (ft · lb) | 13.1 | 26.0 | 39.5 | 52.7 | 64.9 | 78.6 |
|---|---|---|---|---|---|---|
| Angular velocity (rad/s) | 1.0 | 2.0 | 3.0 | 4.0 | 5.0 | 6.0 |

For this viscometer $R_o = 2.50$ in., $R_i = 2.45$ in., and $\ell = 5.00$ in. Make use of these data and a standard curve-fitting program to determine the viscosity of the liquid contained in the viscometer.

■ **Figure P1.87**

**1.88** One type of rotating cylinder viscometer, called a *Stormer* viscometer, uses a falling weight, $\mathcal{W}$, to cause the cylinder to rotate with an angular velocity, $\omega$, as illustrated in Fig. P1.88. For this device the viscosity, $\mu$, of the liquid is related to $\mathcal{W}$ and $\omega$ through the equation $\mathcal{W} = K\mu\omega$, where $K$ is a constant that depends only on the geometry (including the liquid depth) of the viscometer. The value of $K$ is usually determined by using a calibration liquid (a liquid of known viscosity).

**(a)** Some data for a particular Stormer viscometer, obtained using glycerin at 20 °C as a calibration liquid, are given below. Plot values of the weight as ordinates and values of the angular velocity as abscissae. Draw the best curve through the plotted points and determine $K$ for the viscometer.

| $\mathcal{W}$ (lb) | 0.22 | 0.66 | 1.10 | 1.54 | 2.20 |
|---|---|---|---|---|---|
| $\omega$ (rev/s) | 0.53 | 1.59 | 2.79 | 3.83 | 5.49 |

**(b)** A liquid of unknown viscosity is placed in the same viscometer used in part (a), and the data given below are obtained. Determine the viscosity of this liquid.

| $\mathcal{W}$ (lb) | 0.04 | 0.11 | 0.22 | 0.33 | 0.44 |
|---|---|---|---|---|---|
| $\omega$ (rev/s) | 0.72 | 1.89 | 3.73 | 5.44 | 7.42 |

■ **Figure P1.88**

**1.89** (WILEY) A 12-in.-diameter circular plate is placed over a fixed bottom plate with a 0.1-in. gap between the two plates filled with glycerin as shown in Fig. P1.89. Determine the torque required to rotate the circular plate slowly at 2 rpm. Assume that the velocity distribution in the gap is linear and that the shear stress on the edge of the rotating plate is negligible.

■ **Figure P1.89**

†**1.90** Vehicle shock absorbers damp out oscillations caused by road roughness. Describe how a temperature change may affect the operation of a shock absorber.

**1.91** (WILEY) Some measurements on a blood sample at 37 °C (98.6 °F) indicate a shearing stress of 0.52 N/m² for a corresponding rate of shearing strain of 200 s⁻¹. Determine the apparent viscosity of the blood and compare it with the viscosity of water at the same temperature.

## Section 1.7 Compressibility of Fluids

**1.92** Obtain a photograph/image of a situation in which the compressibility of a fluid is important. Print this photo and write a brief paragraph that describes the situation involved.

**1.93** (WILEY) A sound wave is observed to travel through a liquid with a speed of 1500 m/s. The specific gravity of the liquid is 1.5. Determine the bulk modulus for this fluid.

**1.94** A rigid-walled cubical container is completely filled with water at 40 °F and sealed. The water is then heated to 100 °F. Determine the pressure that develops in the container when the water reaches this higher temperature. Assume that the volume of the container remains constant and the value of the bulk modulus of the water remains constant and equal to 300,000 psi.

**1.95** In a test to determine the bulk modulus of a liquid it was found that as the absolute pressure was changed from 15 to 3000 psi the volume decreased from 10.240 to 10.138 in.³ Determine the bulk modulus for this liquid.

**1.96** (WILEY) Estimate the increase in pressure (in psi) required to decrease a unit volume of mercury by 0.1%.

**1.97** (WILEY) A 1-m³ volume of water is contained in a rigid container. Estimate the change in the volume of the water when a piston applies a pressure of 35 MPa.

**1.98** Determine the speed of sound at 20 °C in **(a)** air, **(b)** helium, and **(c)** natural gas (methane). Express your answer in m/s.

**1.99** Calculate the speed of sound in m/s for **(a)** gasoline, **(b)** mercury, and **(c)** seawater.

**1.100** (WILEY) Air is enclosed by a rigid cylinder containing a piston. A pressure gage attached to the cylinder indicates an initial reading of 25 psi. Determine the reading on the gage when the piston has compressed the air to one-third its original volume. Assume the compression process to be isothermal and the local atmospheric pressure to be 14.7 psi.

**1.101** Repeat Problem 1.100 if the compression process takes place without friction and without heat transfer (isentropic process).

**1.102** (WILEY) Carbon dioxide at 30 °C and 300 kPa absolute pressure expands isothermally to an absolute pressure of 165 kPa. Determine the final density of the gas.

**1.103** Oxygen at 30 °C and 300 kPa absolute pressure expands isothermally to an absolute pressure of 120 kPa. Determine the final density of the gas.

**1.104** (WILEY) Natural gas at 70 °F and standard atmospheric pressure of 14.7 psi (abs) is compressed isentropically to a new absolute pressure of 70 psi. Determine the final density and temperature of the gas.

**1.105** Compare the isentropic bulk modulus of air at 101 kPa (abs) with that of water at the same pressure.

*****1.106** Develop a computer program for calculating the final gage pressure of gas when the initial gage pressure, initial and final volumes, atmospheric pressure, and the type of process (isothermal or isentropic) are specified. Use BG units. Check your program against the results obtained for Problem 1.100.

**1.107** Often the assumption is made that the flow of a certain fluid can be considered as incompressible flow if the density of the fluid changes by less than 2%. If air is flowing through a tube such that the air pressure at one section is 9.0 psi and at a downstream section it is 8.6 psi at the same temperature, do you think that this flow could be considered an incompressible flow? Support your answer with the necessary calculations. Assume standard atmospheric pressure.

**1.108** (WILEY○) An important dimensionless parameter concerned with very high-speed flow is the *Mach number,* defined as $V/c$, where $V$ is the speed of the object such as an airplane or projectile, and $c$ is the speed of sound in the fluid surrounding the object. For a projectile traveling at 800 mph through air at 50 °F and standard atmospheric pressure, what is the value of the Mach number?

**1.109** Jet airliners typically fly at altitudes between approximately 0 to 40,000 ft. Make use of the data in Appendix C to show on a graph how the speed of sound varies over this range.

**1.110** (WILEY○) (See Fluids in the News article titled **"This water jet is a blast,"** Section 1.7.1.) By what percent is the volume of water decreased if its pressure is increased to an equivalent to 3000 atmospheres (44,100 psi)?

## Section 1.8 Vapor Pressure

**1.111** (WILEY○) During a mountain climbing trip it is observed that the water used to cook a meal boils at 90 °C rather than the standard 100 °C at sea level. At what altitude are the climbers preparing their meal? (See Tables B.2 and C.2 for data needed to solve this problem.)

**1.112** (WILEY○) When a fluid flows through a sharp bend, low pressures may develop in localized regions of the bend. Estimate the minimum absolute pressure (in psi) that can develop without causing cavitation if the fluid is water at 160 °F.

**1.113** A partially filled closed tank contains ethyl alcohol at 68 °F. If the air above the alcohol is evacuated, what is the minimum absolute pressure that develops in the evacuated space?

**1.114** (WILEY○) Estimate the minimum absolute pressure (in pascals) that can be developed at the inlet of a pump to avoid cavitation if the fluid is carbon tetrachloride at 20 °C.

**1.115** (WILEY○) When water at 70 °C flows through a converging section of pipe, the pressure decreases in the direction of flow. Estimate the minimum absolute pressure that can develop without causing cavitation. Express your answer in both BG and SI units.

**1.116** (WILEY○) At what atmospheric pressure will water boil at 35 °C? Express your answer in both SI and BG units.

## Section 1.9 Surface Tension

**1.117** Obtain a photograph/image of a situation in which the surface tension of a fluid is important. Print this photo and write a brief paragraph that describes the situation involved.

**1.118** (▶) When a 2-mm-diameter tube is inserted into a liquid in an open tank, the liquid is observed to rise 10 mm above the free surface of the liquid (see **Video V1.10**). The contact angle between the liquid and the tube is zero, and the specific weight of the liquid is $1.2 \times 10^4$ N/m³. Determine the value of the surface tension for this liquid.

**1.119** An open 2-mm-diameter tube is inserted into a pan of ethyl alcohol, and a similar 4-mm-diameter tube is inserted into a pan of water. In which tube will the height of the rise of the fluid column due to capillary action be the greatest? Assume the angle of contact is the same for both tubes.

**1.120** (WILEY○) Small droplets of carbon tetrachloride at 68 °F are formed with a spray nozzle. If the average diameter of the droplets is 200 $\mu$m, what is the difference in pressure between the inside and outside of the droplets?

**1.121** A 12-mm-diameter jet of water discharges vertically into the atmosphere. Due to surface tension the pressure inside the jet will be slightly higher than the surrounding atmospheric pressure. Determine this difference in pressure.

**1.122** Estimate the excess pressure inside a raindrop having a diameter of 3 mm.

**1.123** What is the difference between the pressure inside a soap bubble and atmospheric pressure for a 3-in.-diameter bubble? Assume the surface tension of the soap film to be 70% of that of water at 70 °F.

**1.124** (▶) (WILEY○) As shown in **Video V1.9**, surface tension forces can be strong enough to allow a double-edge steel razor blade to "float" on water, but a single-edge blade will sink. Assume that the surface tension forces act at an angle $\theta$ relative to the water surface as shown in Fig. P1.124. **(a)** The mass of the double-edge blade is $0.64 \times 10^{-3}$ kg, and the total length of its sides is 206 mm. Determine the value of $\theta$ required to maintain equilibrium between the blade weight and the resultant surface tension force. **(b)** The mass of the single-edge blade is $2.61 \times 10^{-3}$ kg, and the total length of its sides is 154 mm. Explain why this blade sinks. Support your answer with the necessary calculations.

**■ Figure P1.24**

**1.125** To measure the water depth in a large open tank with opaque walls, an open vertical glass tube is attached to the side of the tank. The height of the water column in the tube is then used as a measure of the depth of water in the tank. **(a)** For a true water depth in the tank of 3 ft, make use of Eq. 1.22 (with $\theta \simeq 0°$) to determine the percent error due to capillarity as the diameter of the glass tube is changed. Assume a water temperature of 80 °F. Show your results on a graph of percent error versus tube diameter, $D$, in the range 0.1 in. $< D <$ 1.0 in. **(b)** If you want the error to be less than 1%, what is the smallest tube diameter allowed?

**1.126** (▶) (WILEY○ GO) Under the right conditions, it is possible, due to surface tension, to have metal objects float on water. (See **Video V1.9**.) Consider placing a short length of a small diameter steel ($\gamma = 490$ lb/ft³) rod on a surface of water. What is the maximum diameter that the rod can have before it will sink? Assume that the surface tension forces act vertically upward. *Note:* A standard paper clip has a diameter of 0.036 in. Partially unfold a paper clip and see if you can get it to float on water. Do the results of this experiment support your analysis?

**1.127** (▶) (WILEY○) An open, clean glass tube, having a diameter of 3 mm, is inserted vertically into a dish of mercury at 20 °C (see **Video V1.10**). How far will the column of mercury in the tube be depressed?

**1.128** (▶) (WILEY○) An open, clean glass tube ($\theta = 0°$) is inserted vertically into a pan of water (see **Video V1.10**). What tube diameter is needed if the water level in the tube is to rise one tube diameter (due to surface tension)?

**1.129** (▶) (WILEY○) Determine the height that water at 60 °F will rise due to capillary action in a clean, $\frac{1}{4}$-in.-diameter tube (see **Video V1.10**). What will be the height if the diameter is reduced to 0.01 in.?

**1.130** Two vertical, parallel, clean glass plates are spaced a distance of 2 mm apart. If the plates are placed in water, how high will the water rise between the plates due to capillary action?

**1.131** (See Fluids in the News article titled "**Walking on water,**" Section 1.9.) **(a)** The water strider bug shown in Fig. P1.131 is supported on the surface of a pond by surface tension acting along the interface between the water and the bug's legs. Determine the minimum length of this interface needed to support the bug. Assume the bug weighs $10^{-4}$ N and the surface tension force acts vertically upwards. **(b)** Repeat part (a) if surface tension were to support a person weighing 750 N.

■ **Figure P1.131**

■ **Lab Problems**

**1.1LP** This problem involves the use of a Stormer viscometer to determine whether a fluid is a Newtonian or a non-Newtonian fluid. To proceed with this problem, go to Appendix H, which is located in *WileyPLUS* or on the book's web site, www.wiley.com/college/munson.

**1.2LP** This problem involves the use of a capillary tube viscometer to determine the kinematic viscosity of water as a function of temperature. To proceed with this problem, go to Appendix H, which is located in *WileyPLUS* or on the book's web site, www.wiley.com/college/munson.

■ **Lifelong Learning Problems**

**1.1 LL** Although there are numerous non-Newtonian fluids that occur naturally (quicksand and blood among them), with the advent of modern chemistry and chemical processing, many new manufactured non-Newtonian fluids are now available for a variety of novel applications. Obtain information about the discovery and use of newly developed non-Newtonian fluids. Summarize your findings in a brief report.

**1.2 LL** For years, lubricating oils and greases obtained by refining crude oil have been used to lubricate moving parts in a wide variety of machines, motors, and engines. With the increasing cost of crude oil and the potential for the reduced availability of it, the need for non-petroleum-based lubricants has increased considerably. Obtain information about non-petroleum-based lubricants. Summarize your findings in a brief report.

**1.3 LL** It is predicted that nano technology and the use of nano sized objects will allow many processes, procedures, and products that, as of now, are difficult for us to comprehend. Among new nano technology areas is that of nano scale fluid mechanics. Fluid behavior at the nano scale can be entirely different than that for the usual everyday flows with which we are familiar. Obtain information about various aspects of nano fluid mechanics. Summarize your findings in a brief report.

■ **FE Exam Problems**

Sample FE (Fundamentals of Engineering) exam question for fluid mechanics are provided in *WileyPLUS* or on the book's web site, www.wiley.com/college/munson.

# 2 Fluid Statics

CHAPTER OPENING PHOTO: *Floating iceberg:* An iceberg is a large piece of fresh water ice that originated as snow in a glacier or ice shelf and then broke off to float in the ocean. Although the fresh water ice is lighter than the salt water in the ocean, the difference in densities is relatively small. Hence, only about one ninth of the volume of an iceberg protrudes above the ocean's surface, so that what we see floating is literally "just the tip of the iceberg." (© oversnap/iStockphoto.)

## Learning Objectives

After completing this chapter, you should be able to:

- determine the pressure at various locations in a fluid at rest.
- explain the concept of manometers and apply appropriate equations to determine pressures.
- calculate the hydrostatic pressure force on a plane or curved submerged surface.
- calculate the buoyant force and discuss the stability of floating or submerged objects.

In this chapter we will consider an important class of problems in which the fluid is either at rest or moving in such a manner that there is no relative motion between adjacent particles. In both instances there will be no shearing stresses in the fluid, and the only forces that develop on the surfaces of the particles will be due to the pressure. Thus, our principal concern is to investigate pressure and its variation throughout a fluid and the effect of pressure on submerged surfaces. The absence of shearing stresses greatly simplifies the analysis and, as we will see, allows us to obtain relatively simple solutions to many important practical problems.

## 2.1 Pressure at a Point

As we briefly discussed in Chapter 1, the term *pressure* is used to indicate the normal force per unit area at a given point acting on a given plane within the fluid mass of interest. A question that immediately arises is how the pressure at a point varies with the orientation of the plane passing through the point. To answer this question, consider the free-body diagram, illustrated in Fig. 2.1,

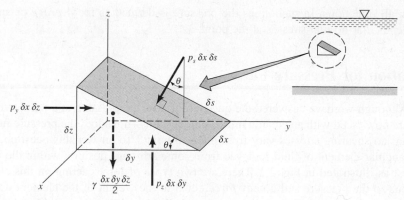

■ **Figure 2.1**    **Forces on an arbitrary wedge-shaped element of fluid.**

that was obtained by removing a small triangular wedge of fluid from some arbitrary location within a fluid mass. Since we are considering the situation in which there are no shearing stresses, the only external forces acting on the wedge are due to the pressure and the weight. For simplicity the forces in the $x$ direction are not shown, and the $z$ axis is taken as the vertical axis so the weight acts in the negative $z$ direction. Although we are primarily interested in fluids at rest, to make the analysis as general as possible, we will allow the fluid element to have accelerated motion. The assumption of zero shearing stresses will still be valid as long as the fluid element moves as a rigid body; that is, there is no relative motion between adjacent elements.

The equations of motion (Newton's second law, $\mathbf{F} = m\mathbf{a}$) in the $y$ and $z$ directions are, respectively,

$$\sum F_y = p_y\,\delta x\,\delta z - p_s\,\delta x\,\delta s \sin\theta = \rho\,\frac{\delta x\,\delta y\,\delta z}{2}\,a_y$$

$$\sum F_z = p_z\,\delta x\,\delta y - p_s\,\delta x\,\delta s \cos\theta - \gamma\,\frac{\delta x\,\delta y\,\delta z}{2} = \rho\,\frac{\delta x\,\delta y\,\delta z}{2}\,a_z$$

where $p_s$, $p_y$, and $p_z$ are the average pressures on the faces, $\gamma$ and $\rho$ are the fluid specific weight and density, respectively, and $a_y$, $a_z$ the accelerations. Note that a pressure must be multiplied by an appropriate area to obtain the force generated by the pressure. It follows from the geometry that

$$\delta y = \delta s \cos\theta \qquad \delta z = \delta s \sin\theta$$

so that the equations of motion can be rewritten as

$$p_y - p_s = \rho a_y \frac{\delta y}{2}$$

$$p_z - p_s = (\rho a_z + \gamma)\frac{\delta z}{2}$$

*The pressure at a point in a fluid at rest is independent of direction.*

Since we are really interested in what is happening at a point, we take the limit as $\delta x$, $\delta y$, and $\delta z$ approach zero (while maintaining the angle $\theta$), and it follows that

$$p_y = p_s \qquad p_z = p_s$$

or $p_s = p_y = p_z$. The angle $\theta$ was arbitrarily chosen so we can conclude that *the pressure at a point in a fluid at rest, or in motion, is independent of direction as long as there are no shearing stresses present*. This important result is known as **Pascal's law,** named in honor of Blaise Pascal (1623–1662), a French mathematician who made important contributions in the field of hydrostatics. Thus, as shown by the photograph in the margin, at the junction of the side and bottom of the beaker, the pressure is the same on the side as it is on the bottom. In Chapter 6 it will be shown that for moving fluids in which there is relative motion between particles (so that shearing stresses develop), the normal stress at a point, which corresponds to pressure in fluids at rest, is not necessarily the same in all directions. In such cases the pressure is defined as the *average* of any three mutually perpendicular normal stresses at the point.

## 2.2    Basic Equation for Pressure Field

Although we have answered the question of how the pressure at a point varies with direction, we are now faced with an equally important question—how does the pressure in a fluid in which there are no shearing stresses vary from point to point? To answer this question, consider a small rectangular element of fluid removed from some arbitrary position within the mass of fluid of interest as illustrated in Fig. 2.2. There are two types of forces acting on this element: *surface forces* due to the pressure and a *body force* equal to the weight of the element. Other possible types of body forces, such as those due to magnetic fields, will not be considered in this text.

If we let the pressure at the center of the element be designated as $p$, then the average pressure on the various faces can be expressed in terms of $p$ and its derivatives, as shown in Fig. 2.2. We are actually using a Taylor series expansion of the pressure at the element center to approximate the pressures a short distance away and neglecting higher order terms that will vanish as we let $\delta x$, $\delta y$, and $\delta z$ approach zero. This is illustrated by the figure in the margin. For simplicity the surface forces in the $x$ direction are not shown. The resultant surface force in the $y$ direction is

*The pressure may vary across a fluid particle.*

$$\delta F_y = \left( p - \frac{\partial p}{\partial y}\frac{\delta y}{2} \right) \delta x\, \delta z - \left( p + \frac{\partial p}{\partial y}\frac{\delta y}{2} \right) \delta x\, \delta z$$

or

$$\delta F_y = -\frac{\partial p}{\partial y}\, \delta x\, \delta y\, \delta z$$

Similarly, for the $x$ and $z$ directions the resultant surface forces are

$$\delta F_x = -\frac{\partial p}{\partial x}\, \delta x\, \delta y\, \delta z \qquad \delta F_z = -\frac{\partial p}{\partial z}\, \delta x\, \delta y\, \delta z$$

The resultant surface force acting on the element can be expressed in vector form as

$$\delta \mathbf{F}_s = \delta F_x \hat{\mathbf{i}} + \delta F_y \hat{\mathbf{j}} + \delta F_z \hat{\mathbf{k}}$$

■ **Figure 2.2**   **Surface and body forces acting on small fluid element.**

*The resultant surface force acting on a small fluid element depends only on the pressure gradient if there are no shearing stresses present.*

or

$$\delta \mathbf{F}_s = -\left( \frac{\partial p}{\partial x} \hat{\mathbf{i}} + \frac{\partial p}{\partial y} \hat{\mathbf{j}} + \frac{\partial p}{\partial z} \hat{\mathbf{k}} \right) \delta x \, \delta y \, \delta z \qquad (2.1)$$

where $\hat{\mathbf{i}}$, $\hat{\mathbf{j}}$, and $\hat{\mathbf{k}}$ are the unit vectors along the coordinate axes shown in Fig. 2.2. The group of terms in parentheses in Eq. 2.1 represents in vector form the *pressure gradient* and can be written as

$$\frac{\partial p}{\partial x} \hat{\mathbf{i}} + \frac{\partial p}{\partial y} \hat{\mathbf{j}} + \frac{\partial p}{\partial z} \hat{\mathbf{k}} = \nabla p$$

where

$$\nabla( \; ) = \frac{\partial( \; )}{\partial x} \hat{\mathbf{i}} + \frac{\partial( \; )}{\partial y} \hat{\mathbf{j}} + \frac{\partial( \; )}{\partial z} \hat{\mathbf{k}}$$

and the symbol $\nabla$ is the *gradient* or "del" vector operator. Thus, the resultant surface force per unit volume can be expressed as

$$\frac{\delta \mathbf{F}_s}{\delta x \, \delta y \, \delta z} = -\nabla p$$

Since the $z$ axis is vertical, the weight of the element is

$$-\delta \mathcal{W} \hat{\mathbf{k}} = -\gamma \, \delta x \, \delta y \, \delta z \, \hat{\mathbf{k}}$$

where the negative sign indicates that the force due to the weight is downward (in the negative $z$ direction). Newton's second law, applied to the fluid element, can be expressed as

$$\sum \delta \mathbf{F} = \delta m \, \mathbf{a}$$

where $\sum \delta \mathbf{F}$ represents the resultant force acting on the element, $\mathbf{a}$ is the acceleration of the element, and $\delta m$ is the element mass, which can be written as $\rho \, \delta x \, \delta y \, \delta z$. It follows that

$$\sum \delta \mathbf{F} = \delta \mathbf{F}_s - \delta \mathcal{W} \hat{\mathbf{k}} = \delta m \, \mathbf{a}$$

or

$$-\nabla p \, \delta x \, \delta y \, \delta z - \gamma \, \delta x \, \delta y \, \delta z \, \hat{\mathbf{k}} = \rho \, \delta x \, \delta y \, \delta z \, \mathbf{a}$$

and, therefore,

$$-\nabla p - \gamma \hat{\mathbf{k}} = \rho \mathbf{a} \qquad (2.2)$$

Equation 2.2 is the general equation of motion for a fluid in which there are no shearing stresses. We will use this equation in Section 2.12 when we consider the pressure distribution in a moving fluid. For the present, however, we will restrict our attention to the special case of a fluid at rest.

## 2.3 Pressure Variation in a Fluid at Rest

For a fluid at rest $\mathbf{a} = 0$ and Eq. 2.2 reduces to

$$\nabla p + \gamma \hat{\mathbf{k}} = 0$$

or in component form

$$\frac{\partial p}{\partial x} = 0 \qquad \frac{\partial p}{\partial y} = 0 \qquad \frac{\partial p}{\partial z} = -\gamma \qquad (2.3)$$

These equations show that the pressure does not depend on $x$ or $y$. Thus, as we move from point to point in a horizontal plane (any plane parallel to the $x$–$y$ plane), the pressure does not change. Since $p$ depends only on $z$, the last of Eqs. 2.3 can be written as the ordinary differential equation

$$\boxed{\dfrac{dp}{dz} = -\gamma} \tag{2.4}$$

Equation 2.4 is the fundamental equation for fluids at rest and can be used to determine how pressure changes with elevation. This equation and the figure in the margin indicate that the pressure gradient in the vertical direction is negative; that is, the pressure decreases as we move upward in a fluid at rest. There is no requirement that $\gamma$ be a constant. Thus, it is valid for fluids with constant specific weight, such as liquids, as well as fluids whose specific weight may vary with elevation, such as air or other gases. However, to proceed with the integration of Eq. 2.4 it is necessary to stipulate how the specific weight varies with $z$.

If the fluid is flowing (i.e., not at rest with $\mathbf{a} = 0$), then the pressure variation is usually much more complex than that given by Eq. 2.4. For example, the pressure distribution on your car as it is driven along the road varies in a complex manner with $x$, $y$, and $z$. This idea is covered in detail in Chapters 3, 6, and 9.

### 2.3.1 Incompressible Fluid

Since the specific weight is equal to the product of fluid density and acceleration of gravity ($\gamma = \rho g$), changes in $\gamma$ are caused by a change in either $\rho$ or $g$. For most engineering applications the variation in $g$ is negligible, so our main concern is with the possible variation in the fluid density. In general, a fluid with constant density is called an ***incompressible fluid.*** For liquids the variation in density is usually negligible, even over large vertical distances, so that the assumption of constant specific weight when dealing with liquids is a good one. For this instance, Eq. 2.4 can be directly integrated

$$\int_{p_1}^{p_2} dp = -\gamma \int_{z_1}^{z_2} dz$$

to yield

$$p_2 - p_1 = -\gamma(z_2 - z_1)$$

or

$$p_1 - p_2 = \gamma(z_2 - z_1) \tag{2.5}$$

*V2.1 Pressure on a car*

where $p_1$ and $p_2$ are pressures at the vertical elevations $z_1$ and $z_2$, as is illustrated in Fig. 2.3.

Equation 2.5 can be written in the compact form

$$p_1 - p_2 = \gamma h \tag{2.6}$$

or

$$p_1 = \gamma h + p_2 \tag{2.7}$$

where $h$ is the distance, $z_2 - z_1$, which is the depth of fluid measured downward from the location of $p_2$. This type of pressure distribution is commonly called a ***hydrostatic distribution,*** and Eq. 2.7

*V2.2 Demonstration of atmospheric pressure.*

■ **Figure 2.3** Notation for pressure variation in a fluid at rest with a free surface.

pA = 0

A = 1 in.²

23.1 ft

𝒲 = 10 lb

Water

pA = 10 lb

shows that in an incompressible fluid at rest the pressure varies linearly with depth. The pressure must increase with depth to "hold up" the fluid above it.

It can also be observed from Eq. 2.6 that the pressure difference between two points can be specified by the distance $h$ since

$$h = \frac{p_1 - p_2}{\gamma}$$

In this case $h$ is called the **pressure head** and is interpreted as the height of a column of fluid of specific weight $\gamma$ required to give a pressure difference $p_1 - p_2$. For example, a pressure difference of 10 psi can be specified in terms of pressure head as 23.1 ft of water ($\gamma = 62.4$ lb/ft³), or 518 mm of Hg ($\gamma = 133$ kN/m³). As illustrated by the figure in the margin, a 23.1-ft-tall column of water with a cross-sectional area of 1 in.² weighs 10 lb.

## F l u i d s   i n   t h e   N e w s

**Giraffe's blood pressure** A giraffe's long neck allows it to graze up to 6 m above the ground. It can also lower its head to drink at ground level. Thus, in the circulatory system there is a significant *hydrostatic pressure* effect due to this elevation change. To maintain blood to its head throughout this change in elevation, the giraffe must maintain a relatively high blood pressure at heart level—approximately two and a half times that in humans. To prevent rupture of blood vessels in the high-pressure lower leg

regions, giraffes have a tight sheath of thick skin over their lower limbs that acts like an elastic bandage in exactly the same way as do the g-suits of fighter pilots. In addition, valves in the upper neck prevent backflow into the head when the giraffe lowers its head to ground level. It is also thought that blood vessels in the giraffe's kidney have a special mechanism to prevent large changes in filtration rate when blood pressure increases or decreases with its head movement. (See Problem 2.17.)

When one works with liquids there is often a free surface, as is illustrated in Fig. 2.3, and it is convenient to use this surface as a reference plane. The reference pressure $p_0$ would correspond to the pressure acting on the free surface (which would frequently be atmospheric pressure), and thus if we let $p_2 = p_0$ in Eq. 2.7 it follows that the pressure $p$ at any depth $h$ below the free surface is given by the equation:

$$p = \gamma h + p_0 \tag{2.8}$$

As is demonstrated by Eq. 2.7 or 2.8, the pressure in a homogeneous, incompressible fluid at rest depends on the depth of the fluid relative to some reference plane, and it is *not* influenced by the *size* or *shape* of the tank or container in which the fluid is held. Thus, in Fig. 2.4

Liquid surface
($p = p_0$)

h

A                                                                B

Specific weight γ

■ **Figure 2.4**  Fluid pressure in containers of arbitrary shape.

the pressure is the same at all points along the line *AB*, even though the containers have very irregular shapes. The actual value of the pressure along *AB* depends only on the depth, *h*, the surface pressure, $p_0$, and the specific weight, $\gamma$, of the liquid in the container.

## EXAMPLE 2.1 Pressure–Depth Relationship

**GIVEN** Because of a leak in a buried gasoline storage tank, water has seeped in to the depth shown in Fig. E2.1. The specific gravity of the gasoline is $SG = 0.68$.

**FIND** Determine the pressure at the gasoline–water interface and at the bottom of the tank. Express the pressure in units of lb/ft², lb/in.², and as a pressure head in feet of water.

■ **Figure E2.1**

## SOLUTION

Since we are dealing with liquids at rest, the pressure distribution will be hydrostatic, and therefore the pressure variation can be found from the equation:

$$p = \gamma h + p_0$$

With $p_0$ corresponding to the pressure at the free surface of the gasoline, then the pressure at the interface is

$$
\begin{aligned}
p_1 &= SG\gamma_{H_2O}\, h + p_0 \\
&= (0.68)(62.4\ \text{lb/ft}^3)(17\ \text{ft}) + p_0 \\
&= 721 + p_0\ (\text{lb/ft}^2)
\end{aligned}
$$

If we measure the pressure relative to atmospheric pressure (gage pressure), it follows that $p_0 = 0$, and therefore

$$p_1 = 721\ \text{lb/ft}^2 \qquad \textbf{(Ans)}$$

$$\frac{p_1}{} = \frac{721\ \text{lb/ft}^2}{144\ \text{in.}^2/\text{ft}^2} = 5.01\ \text{lb/in.}^2 \qquad \textbf{(Ans)}$$

$$\frac{p_1}{\gamma_{H_2O}} = \frac{721\ \text{lb/ft}^2}{62.4\ \text{lb/ft}^3} = 11.6\ \text{ft} \qquad \textbf{(Ans)}$$

It is noted that a rectangular column of water 11.6 ft tall and 1 ft² in cross section weighs 721 lb. A similar column with a 1-in.² cross section weighs 5.01 lb.

We can now apply the same relationship to determine the pressure at the tank bottom; that is,

$$
\begin{aligned}
p_2 &= \gamma_{H_2O}\, h_{H_2O} + p_1 \\
&= (62.4\ \text{lb/ft}^3)(3\ \text{ft}) + 721\ \text{lb/ft}^2 \qquad \textbf{(Ans)} \\
&= 908\ \text{lb/ft}^2
\end{aligned}
$$

$$p_2 = \frac{908\ \text{lb/ft}^2}{144\ \text{in.}^2/\text{ft}^2} = 6.31\ \text{lb/in.}^2 \qquad \textbf{(Ans)}$$

$$\frac{p_2}{\gamma_{H_2O}} = \frac{908\ \text{lb/ft}^2}{62.4\ \text{lb/ft}^3} = 14.6\ \text{ft} \qquad \textbf{(Ans)}$$

**COMMENT** Observe that if we wish to express these pressures in terms of *absolute* pressure, we would have to add the local atmospheric pressure (in appropriate units) to the previous results. A further discussion of gage and absolute pressure is given in Section 2.5.

---

*The transmission of pressure throughout a stationary fluid is the principle upon which many hydraulic devices are based.*

The required equality of pressures at equal elevations throughout a system is important for the operation of hydraulic jacks (see Fig. 2.5a), lifts, and presses, as well as hydraulic controls on aircraft and other types of heavy machinery. The fundamental idea behind such devices and systems is demonstrated in Fig. 2.5b. A piston located at one end of a closed system filled with a liquid, such as oil, can be used to change the pressure throughout the system, and thus transmit an applied force $F_1$ to a second piston where the resulting force is $F_2$. Since the pressure $p$ acting on the faces of both pistons is the same (the effect of elevation changes is usually negligible for this type of hydraulic device), it follows that $F_2 = (A_2/A_1)F_1$. The piston area $A_2$ can be made much larger than $A_1$ and therefore a large mechanical advantage can be developed; that is, a small force applied at the smaller piston can be used to develop a large force at the larger piston. The applied force could be created manually through some type of mechanical device, such as a hydraulic jack, or through compressed air acting directly on the surface of the liquid, as is done in hydraulic lifts commonly found in service stations.

$F_2 = pA_2$    $F_1 = pA_1$

$A_2$    $A_1$

(a)    (b)

■ **Figure 2.5**  (a) Hydraulic jack, (b) Transmission of fluid pressure.

### 2.3.2  Compressible Fluid

We normally think of gases such as air, oxygen, and nitrogen as being **compressible fluids** because the density of the gas can change significantly with changes in pressure and temperature. Thus, although Eq. 2.4 applies at a point in a gas, it is necessary to consider the possible variation in $\gamma$ before the equation can be integrated. However, as was discussed in Chapter 1, the specific weights of common gases are small when compared with those of liquids. For example, the specific weight of air at sea level and 60 °F is 0.0763 lb/ft³, whereas the specific weight of water under the same conditions is 62.4 lb/ft³. Since the specific weights of gases are comparatively small, it follows from Eq. 2.4 that the pressure gradient in the vertical direction is correspondingly small, and even over distances of several hundred feet the pressure will remain essentially constant for a gas. This means we can neglect the effect of elevation changes on the pressure for stationary gases in tanks, pipes, and so forth in which the distances involved are small.

For those situations in which the variations in heights are large, on the order of thousands of feet, attention must be given to the variation in the specific weight. As is described in Chapter 1, the equation of state for an ideal (or perfect) gas is

$$\rho = \frac{p}{RT}$$

*If the specific weight of a fluid varies significantly as we move from point to point, the pressure will no longer vary linearly with depth.*

where $p$ is the absolute pressure, $R$ is the gas constant, and $T$ is the absolute temperature. This relationship can be combined with Eq. 2.4 to give

$$\frac{dp}{dz} = -\frac{gp}{RT}$$

and by separating variables

$$\int_{p_1}^{p_2} \frac{dp}{p} = \ln\frac{p_2}{p_1} = -\frac{g}{R}\int_{z_1}^{z_2}\frac{dz}{T} \tag{2.9}$$

where $g$ and $R$ are assumed to be constant over the elevation change from $z_1$ to $z_2$. Although the acceleration of gravity, $g$, does vary with elevation, the variation is very small (see Tables C.1 and C.2 in Appendix C), and $g$ is usually assumed constant at some average value for the range of elevation involved.

Before completing the integration, one must specify the nature of the variation of temperature with elevation. For example, if we assume that the temperature has a constant value $T_0$ over the range $z_1$ to $z_2$ (*isothermal* conditions), it then follows from Eq. 2.9 that

$$p_2 = p_1 \exp\left[ -\frac{g(z_2 - z_1)}{RT_0} \right] \qquad (2.10)$$

This equation provides the desired pressure–elevation relationship for an isothermal layer. As shown in the margin figure, even for a 10,000-ft altitude change, the difference between the constant temperature (isothermal) and the constant density (incompressible) results is relatively minor. For nonisothermal conditions a similar procedure can be followed if the temperature–elevation relationship is known, as is discussed in the following section.

---

## EXAMPLE 2.2   Incompressible and Isothermal Pressure–Depth Variations

**GIVEN** In 2010, the world's tallest building, the Burj Khalifa skyscraper, was completed and opened in the United Arab Emirates. The final height of the building, which had remained a secret until completion, is 2717 ft (828 m).

**FIND** (a) Estimate the ratio of the pressure at the 2717-ft top of the building to the pressure at its base, assuming the air to be at a common temperature of 59 °F. (b) Compare the pressure calculated in part (a) with that obtained by assuming the air to be incompressible with $\gamma = 0.0765$ lb/ft$^3$ at 14.7 psi (abs) (values for air at standard sea level conditions).

### SOLUTION

(a) For the assumed isothermal conditions, and treating air as a compressible fluid, Eq. 2.10 can be applied to yield

$$\frac{p_2}{p_1} = \exp\left[ -\frac{g(z_2 - z_1)}{RT_0} \right]$$

$$= \exp\left\{ -\frac{(32.2 \text{ ft/s}^2)(2717 \text{ ft})}{(1716 \text{ ft} \cdot \text{lb/slug} \cdot {}^\circ\text{R})[(59 + 460){}^\circ\text{R}]} \right\}$$

$$= 0.906 \qquad \textbf{(Ans)}$$

(b) If the air is treated as an incompressible fluid we can apply Eq. 2.5. In this case

$$p_2 = p_1 - \gamma(z_2 - z_1)$$

or

$$\frac{p_2}{p_1} = 1 - \frac{\gamma(z_2 - z_1)}{p_1}$$

$$= 1 - \frac{(0.0765 \text{ lb/ft}^3)(2717 \text{ ft})}{(14.7 \text{ lb/in.}^2)(144 \text{ in.}^2/\text{ft}^2)} = 0.902 \qquad \textbf{(Ans)}$$

**COMMENTS** Note that there is little difference between the two results. Since the pressure difference between the bottom and top of the building is small, it follows that the variation in fluid density is small and, therefore, the compressible

■ **Figure E2.2** (Figure courtesy of Emaar Properties, Dubai, UAE.)

fluid and incompressible fluid analyses yield essentially the same result.

We see that for both calculations the pressure decreases by approximately 10% as we go from ground level to the top of this tallest building. It does not require a very large pressure difference to support a 2717-ft-tall column of fluid as light as air. This result supports the earlier statement that the changes in pressures in air and other gases due to elevation changes are very small, even for distances of hundreds of feet. Thus, the pressure differences between the top and bottom of a horizontal pipe carrying a gas, or in a gas storage tank, are negligible since the distances involved are very small.

## 2.4 Standard Atmosphere

An important application of Eq. 2.9 relates to the variation in pressure in the earth's atmosphere. Ideally, we would like to have measurements of pressure versus altitude over the specific range for the specific conditions (temperature, reference pressure) for which the pressure is to be determined. However, this type of information is usually not available. Thus, a "standard atmosphere" has been determined that can be used in the design of aircraft, missiles, and spacecraft and in comparing their performance under standard conditions. The concept of a standard atmosphere was first developed in the 1920s, and since that time many national and international committees and organizations have pursued the development of such a standard. The currently accepted standard atmosphere is based on a report published in 1962 and updated in 1976 (see Refs. 1 and 2), defining the so-called **U.S. standard atmosphere,** which is an idealized representation of middle-latitude, year-round mean conditions of the earth's atmosphere. Several important properties for standard atmospheric conditions at *sea level* are listed in Table 2.1, and Fig. 2.6 shows the temperature profile for the U.S. standard atmosphere. As is shown in this figure, the temperature decreases with altitude in the region nearest the earth's surface (*troposphere*), then becomes essentially constant in the next layer (*stratosphere*), and subsequently starts to increase in the next layer. Typical events that occur in the atmosphere are shown in the figure in the margin.

Since the temperature variation is represented by a series of linear segments, it is possible to integrate Eq. 2.9 to obtain the corresponding pressure variation. For example, in the troposphere, which extends to an altitude of about 11 km (~36,000 ft), the temperature variation is of the form

$$T = T_a - \beta z \tag{2.11}$$

*The standard atmosphere is an idealized representation of mean conditions in the earth's atmosphere.*

### ■ Table 2.1
**Properties of U.S. Standard Atmosphere at Sea Level***

| Property | SI Units | BG Units |
|---|---|---|
| Temperature, $T$ | 288.15 K (15 °C) | 518.67 °R (59.00 °F) |
| Pressure, $p$ | 101.33 kPa (abs) | 2116.2 lb/ft² (abs) [14.696 lb/in.² (abs)] |
| Density, $\rho$ | 1.225 kg/m³ | 0.002377 slugs/ft³ |
| Specific weight, $\gamma$ | 12.014 N/m³ | 0.07647 lb/ft³ |
| Viscosity, $\mu$ | $1.789 \times 10^{-5}$ N · s/m² | $3.737 \times 10^{-7}$ lb · s/ft² |

*Acceleration of gravity at sea level = 9.807 m/s² = 32.174 ft/s².

■ **Figure 2.6** **Variation of temperature with altitude in the U.S. standard atmosphere.**

where $T_a$ is the temperature at sea level ($z = 0$) and $\beta$ is the *lapse rate* (the rate of change of temperature with elevation). For the standard atmosphere in the troposphere, $\beta = 0.00650$ K/m or 0.00357 °R/ft.

Equation 2.11 used with Eq. 2.9 yields

$$p = p_a\left(1 - \frac{\beta z}{T_a}\right)^{g/R\beta} \tag{2.12}$$

where $p_a$ is the absolute pressure at $z = 0$. With $p_a$, $T_a$, and $g$ obtained from Table 2.1, and with the gas constant $R = 286.9$ J/kg · K or 1716 ft · lb/slug · °R, the pressure variation throughout the troposphere can be determined from Eq. 2.12. This calculation shows that at the outer edge of the troposphere, where the temperature is −56.5 °C (−69.7 °F), the absolute pressure is about 23 kPa (3.3 psia). It is to be noted that modern jetliners cruise at approximately this altitude. Pressures at other altitudes are shown in Fig. 2.6, and tabulated values for temperature, acceleration of gravity, pressure, density, and viscosity for the U.S. standard atmosphere are given in Tables C.1 and C.2 in Appendix C.

## 2.5 Measurement of Pressure

*Pressure is designated as either absolute pressure or gage pressure.*

Since pressure is a very important characteristic of a fluid field, it is not surprising that numerous devices and techniques are used in its measurement. As is noted briefly in Chapter 1, the pressure at a point within a fluid mass will be designated as either an **absolute** pressure or a **gage** pressure. Absolute pressure is measured relative to a perfect vacuum (absolute zero pressure), whereas gage pressure is measured relative to the local atmospheric pressure. Thus, a gage pressure of zero corresponds to a pressure that is equal to the local atmospheric pressure. Absolute pressures are always positive, but gage pressures can be either positive or negative depending on whether the pressure is above atmospheric pressure (a positive value) or below atmospheric pressure (a negative value). A negative gage pressure is also referred to as a *suction* or *vacuum* **pressure**. For example, 10 psi (abs) could be expressed as −4.7 psi (gage), if the local atmospheric pressure is 14.7 psi, or alternatively 4.7 psi suction or 4.7 psi vacuum. The concept of gage and absolute pressure is illustrated graphically in Fig. 2.7 for two typical pressures located at points 1 and 2.

In addition to the reference used for the pressure measurement, the *units* used to express the value are obviously of importance. As is described in Section 1.5, pressure is a force per unit area, and the units in the BG system are lb/ft$^2$ or lb/in.$^2$, commonly abbreviated psf or psi, respectively. In the SI system the units are N/m$^2$; this combination is called the pascal and written as Pa (1 N/m$^2$ = 1 Pa). As noted earlier, pressure can also be expressed as the height of a column of liquid. Then the units will refer to the height of the column (in., ft, mm, m, etc.), and in addition, the liquid in the column must be specified ($H_2O$, Hg, etc.). For example, standard atmospheric pressure can be expressed as 760 mm Hg (abs). *In this text, pressures will be assumed to be gage pressures unless specifically designated absolute.* For example, 10 psi or 100 kPa would be gage pressures, whereas 10 psia or 100 kPa (abs) would refer to absolute pressures. It is to be

■ **Figure 2.7** **Graphical representation of gage and absolute pressure.**

■ **Figure 2.8** Mercury barometer.

noted that *pressure differences* are independent of the reference, so that no special notation is required in this case.

The measurement of atmospheric pressure is usually accomplished with a mercury **barometer,** which in its simplest form consists of a glass tube closed at one end with the open end immersed in a container of mercury as shown in Fig. 2.8. The tube is initially filled with mercury (inverted with its open end up) and then turned upside down (open end down), with the open end in the container of mercury. The column of mercury will come to an equilibrium position where its weight plus the force due to the vapor pressure (which develops in the space above the column) balances the force due to the atmospheric pressure. Thus,

$$p_{atm} = \gamma h + p_{vapor} \tag{2.13}$$

where $\gamma$ is the specific weight of mercury. For most practical purposes the contribution of the vapor pressure can be neglected since it is very small [for mercury, the fluid most commonly used in barometers, $p_{vapor} = 0.000023$ lb/in.$^2$ (abs) at a temperature of 68 °F], so that $p_{atm} \approx \gamma h$. It is conventional to specify atmospheric pressure in terms of the height, $h$, in millimeters or inches of mercury. Note that if water were used instead of mercury, the height of the column would have to be approximately 34 ft rather than 29.9 in. of mercury for an atmospheric pressure of 14.7 psia! This is shown to scale in the figure in the margin. The concept of the mercury barometer is an old one, with the invention of this device attributed to Evangelista Torricelli in about 1644.

---

## EXAMPLE 2.3 | Barometric Pressure

**GIVEN** A mountain lake has an average temperature of 10 °C and a maximum depth of 40 m. The barometric pressure is 598 mm Hg.

**FIND** Determine the absolute pressure (in pascals) at the deepest part of the lake.

### SOLUTION

The pressure in the lake at any depth, $h$, is given by the equation

$$p = \gamma h + p_0$$

where $p_0$ is the pressure at the surface. Since we want the absolute pressure, $p_0$ will be the local barometric pressure expressed in a consistent system of units; that is

$$\frac{p_{barometric}}{\gamma_{Hg}} = 598 \text{ mm} = 0.598 \text{ m}$$

and for $\gamma_{Hg} = 133$ kN/m$^3$

$$p_0 = (0.598 \text{ m})(133 \text{ kN/m}^3) = 79.5 \text{ kN/m}^2$$

From Table B.2, $\gamma_{H_2O} = 9.804$ kN/m$^3$ at 10 °C and therefore

$$p = (9.804 \text{ kN/m}^3)(40 \text{ m}) + 79.5 \text{ kN/m}^2$$
$$= 392 \text{ kN/m}^2 + 79.5 \text{ kN/m}^2$$
$$= 472 \text{ kPa (abs)} \qquad \textbf{(Ans)}$$

**COMMENT** This simple example illustrates the need for close attention to the units used in the calculation of pressure; that is, be sure to use a *consistent* unit system, and be careful not to add a pressure head (m) to a pressure (Pa).

**Weather, barometers, and bars** One of the most important indicators of weather conditions is *atmospheric pressure*. In general, a falling or low pressure indicates bad weather; rising or high pressure, good weather. During the evening TV weather report in the United States, atmospheric pressure is given as so many inches (commonly around 30 in.). This value is actually the height of the mercury column in a mercury *barometer* adjusted to sea level. To determine the true atmospheric pressure at a particular location, the elevation relative to sea level must be known. Another unit used by meteorologists to indicate atmospheric pressure is the *bar*, first used in weather reporting in 1914 and defined as $10^5$ N/m². The definition of a bar is probably related to the fact that standard sea-level pressure is $1.0133 \times 10^5$ N/m², that is, only slightly larger than one bar. For typical weather patterns, "sea-level equivalent" atmospheric pressure remains close to one bar. However, for extreme weather conditions associated with tornadoes, hurricanes, or typhoons, dramatic changes can occur. The lowest atmospheric sea-level pressure ever recorded was associated with a typhoon, Typhoon Tip, in the Pacific Ocean on October 12, 1979. The value was 0.870 bar (25.8 in. Hg). (See Problem 2.24.)

## 2.6    Manometry

*Manometers use vertical or inclined liquid columns to measure pressure.*

A standard technique for measuring pressure involves the use of liquid columns in vertical or inclined tubes. Pressure-measuring devices based on this technique are called **manometers.** The mercury barometer is an example of one type of manometer, but there are many other configurations possible depending on the particular application. Three common types of manometers include the piezometer tube, the U-tube manometer, and the inclined-tube manometer.

### 2.6.1 Piezometer Tube

The simplest type of manometer consists of a vertical tube, open at the top, and attached to the container in which the pressure is desired, as illustrated in Fig. 2.9. The figure in the margin shows an important device whose operation is based on this principle. It is a sphygmomanometer, the traditional instrument used to measure blood pressure.

Since manometers involve columns of fluids at rest, the fundamental equation describing their use is Eq. 2.8

$$p = \gamma h + p_0$$

which gives the pressure at any elevation within a homogeneous fluid in terms of a reference pressure $p_0$ and the vertical distance $h$ between $p$ and $p_0$. Remember that in a fluid at rest pressure will *increase* as we move *downward* and will decrease as we move *upward*. Application of this equation to the piezometer tube of Fig. 2.9 indicates that the pressure $p_A$ can be determined by a measurement of $h_1$ through the relationship

$$p_A = \gamma_1 h_1$$

where $\gamma_1$ is the specific weight of the liquid in the container. Note that since the tube is open at the top, the pressure $p_0$ can be set equal to zero (we are now using gage pressure), with the height

■ **Figure 2.9**   **Piezometer tube.**

■ **Figure 2.10** Simple U-tube manometer.

$h_1$ measured from the meniscus at the upper surface to point (1). Since point (1) and point $A$ within the container are at the same elevation, $p_A = p_1$.

Although the piezometer tube is a very simple and accurate pressure-measuring device, it has several disadvantages. It is suitable only if the pressure in the container is greater than atmospheric pressure (otherwise air would be sucked into the system), and the pressure to be measured must be relatively small so the required height of the column is reasonable. Also the fluid in the container in which the pressure is to be measured must be a liquid rather than a gas.

### 2.6.2 U-Tube Manometer

To overcome the difficulties noted previously, another type of manometer which is widely used consists of a tube formed into the shape of a U, as is shown in Fig. 2.10. The fluid in the manometer is called the *gage fluid*. To find the pressure $p_A$ in terms of the various column heights, we start at one end of the system and work our way around to the other end, simply utilizing Eq. 2.8. Thus, for the U-tube manometer shown in Fig. 2.10, we will start at point $A$ and work around to the open end. The pressure at points $A$ and (1) are the same, and as we move from point (1) to (2) the pressure will increase by $\gamma_1 h_1$. The pressure at point (2) is equal to the pressure at point (3), since the pressures at equal elevations in a continuous mass of fluid at rest must be the same. Note that we could not simply "jump across" from point (1) to a point at the same elevation in the right-hand tube since these would not be points within the same continuous mass of fluid. With the pressure at point (3) specified, we now move to the open end where the pressure is zero. As we move vertically upward the pressure decreases by an amount $\gamma_2 h_2$. In equation form these various steps can be expressed as

$$p_A + \gamma_1 h_1 - \gamma_2 h_2 = 0$$

*The contribution of gas columns in manometers is usually negligible since the weight of the gas is so small.*

and, therefore, the pressure $p_A$ can be written in terms of the column heights as

$$p_A = \gamma_2 h_2 - \gamma_1 h_1 \tag{2.14}$$

A major advantage of the U-tube manometer lies in the fact that the gage fluid can be different from the fluid in the container in which the pressure is to be determined. For example, the fluid in $A$ in Fig. 2.10 can be either a liquid or a gas. If $A$ does contain a gas, the contribution of the gas column, $\gamma_1 h_1$, is almost always negligible so that $p_A \approx p_2$, and in this instance Eq. 2.14 becomes

*V2.3 Blood pressure measurement*

$$p_A = \gamma_2 h_2$$

Thus, for a given pressure the height, $h_2$, is governed by the specific weight, $\gamma_2$, of the gage fluid used in the manometer. If the pressure $p_A$ is large, then a heavy gage fluid, such as mercury, can be used and a reasonable column height (not too long) can still be maintained. Alternatively, if the pressure $p_A$ is small, a lighter gage fluid, such as water, can be used so that a relatively large column height (which is easily read) can be achieved.

## EXAMPLE 2.4    Simple U-Tube Manometer

**GIVEN** A closed tank contains compressed air and oil ($SG_{oil} = 0.90$) as is shown in Fig. E2.4. A U-tube manometer using mercury ($SG_{Hg} = 13.6$) is connected to the tank as shown. The column heights are $h_1 = 36$ in., $h_2 = 6$ in., and $h_3 = 9$ in.

**FIND** Determine the pressure reading (in psi) of the gage.

■ **Figure E2.4**

## SOLUTION

Following the general procedure of starting at one end of the manometer system and working around to the other, we will start at the air–oil interface in the tank and proceed to the open end where the pressure is zero. The pressure at level (1) is

$$p_1 = p_{air} + \gamma_{oil}(h_1 + h_2)$$

This pressure is equal to the pressure at level (2), since these two points are at the same elevation in a homogeneous fluid at rest. As we move from level (2) to the open end, the pressure must decrease by $\gamma_{Hg}h_3$, and at the open end the pressure is zero. Thus, the manometer equation can be expressed as

$$p_{air} + \gamma_{oil}(h_1 + h_2) - \gamma_{Hg}h_3 = 0$$

or

$$p_{air} + (SG_{oil})(\gamma_{H_2O})(h_1 + h_2) - (SG_{Hg})(\gamma_{H_2O})h_3 = 0$$

For the values given

$$p_{air} = -(0.9)(62.4 \text{ lb/ft}^3)\left(\frac{36 + 6}{12}\text{ ft}\right)$$
$$+ (13.6)(62.4 \text{ lb/ft}^3)\left(\frac{9}{12}\text{ ft}\right)$$

so that

$$p_{air} = 440 \text{ lb/ft}^2$$

Since the specific weight of the air above the oil is much smaller than the specific weight of the oil, the gage should read the pressure we have calculated; that is,

$$p_{gage} = \frac{440 \text{ lb/ft}^2}{144 \text{ in.}^2/\text{ft}^2} = 3.06 \text{ psi} \qquad \textbf{(Ans)}$$

**COMMENTS** Note that the air pressure is a function of the height of the mercury in the manometer and the depth of the oil (both in the tank and in the tube). It is not just the mercury in the manometer that is important.

Assume that the gage pressure remains at 3.06 psi, but the manometer is altered so that it contains only oil. That is, the mercury is replaced by oil. A simple calculation shows that in this case the vertical oil-filled tube would need to be $h_3 = 11.3$ ft tall, rather than the original $h_3 = 9$ in. There is an obvious advantage of using a heavy fluid such as mercury in manometers.

---

*Manometers are often used to measure the difference in pressure between two points.*

The U-tube manometer is also widely used to measure the *difference* in pressure between two containers or two points in a given system. Consider a manometer connected between containers *A* and *B* as is shown in Fig. 2.11. The difference in pressure between *A* and *B* can be found

■ **Figure 2.11** Differential U-tube manometer.

by again starting at one end of the system and working around to the other end. For example, at A the pressure is $p_A$, which is equal to $p_1$, and as we move to point (2) the pressure increases by $\gamma_1 h_1$. The pressure at $p_2$ is equal to $p_3$, and as we move upward to point (4) the pressure decreases by $\gamma_2 h_2$. Similarly, as we continue to move upward from point (4) to (5) the pressure decreases by $\gamma_3 h_3$. Finally, $p_5 = p_B$, since they are at equal elevations. Thus,

$$p_A + \gamma_1 h_1 - \gamma_2 h_2 - \gamma_3 h_3 = p_B$$

Or, as indicated in the figure in the margin, we could start at B and work our way around to A to obtain the same result. In either case, the pressure difference is

$$p_A - p_B = \gamma_2 h_2 + \gamma_3 h_3 - \gamma_1 h_1$$

When the time comes to substitute in numbers, be sure to use a consistent system of units!

Capillarity due to surface tension at the various fluid interfaces in the manometer is usually not considered, since for a simple U-tube with a meniscus in each leg, the capillary effects cancel (assuming the surface tensions and tube diameters are the same at each meniscus), or we can make the capillary rise negligible by using relatively large bore tubes (with diameters of about 0.5 in. or larger; see Section 1.9). Two common gage fluids are water and mercury. Both give a well-defined meniscus (a very important characteristic for a gage fluid) and have well-known properties. Of course, the gage fluid must be immiscible with respect to the other fluids in contact with it. For highly accurate measurements, special attention should be given to temperature since the various specific weights of the fluids in the manometer will vary with temperature.

---

# EXAMPLE 2.5    U-Tube Manometer

**GIVEN**    As will be discussed in Chapter 3, the volume rate of flow, $Q$, through a pipe can be determined by means of a flow nozzle located in the pipe as illustrated in Fig. E2.5a. The nozzle creates a pressure drop, $p_A - p_B$, along the pipe that is related to the flow through the equation $Q = K\sqrt{p_A - p_B}$, where $K$ is a constant depending on the pipe and nozzle size. The pressure drop is frequently measured with a differential U-tube manometer of the type illustrated.

**FIND**    (a) Determine an equation for $p_A - p_B$ in terms of the specific weight of the flowing fluid, $\gamma_1$, the specific weight of the gage fluid, $\gamma_2$, and the various heights indicated. (b) For $\gamma_1 = 9.80$ kN/m³, $\gamma_2 = 15.6$ kN/m³, $h_1 = 1.0$ m, and $h_2 = 0.5$ m, what is the value of the pressure drop, $p_A - p_B$?

## SOLUTION

(a)    Although the fluid in the pipe is moving, the fluids in the columns of the manometer are at rest so that the pressure variation in the manometer tubes is hydrostatic. If we start at point A and move vertically upward to level (1), the pressure will decrease by $\gamma_1 h_1$ and will be equal to the pressure at (2) and at (3). We can now move from (3) to (4) where the pressure has been further reduced by $\gamma_2 h_2$. The pressures at levels (4) and (5) are equal, and as we move from (5) to B the pressure will increase by $\gamma_1(h_1 + h_2)$. Thus, in equation form

$$p_A - \gamma_1 h_1 - \gamma_2 h_2 + \gamma_1(h_1 + h_2) = p_B$$

or

$$p_A - p_B = h_2(\gamma_2 - \gamma_1)    \quad \textbf{(Ans)}$$

**COMMENT**    It is to be noted that the only column height of importance is the differential reading, $h_2$. The differential

■ **Figure E2.5a**

manometer could be placed 0.5 or 5.0 m above the pipe ($h_1 = 0.5$ m or $h_1 = 5.0$ m), and the value of $h_2$ would remain the same.

(b)    The specific value of the pressure drop for the data given is

$$p_A - p_B = (0.5 \text{ m})(15.6 \text{ kN/m}^3 - 9.80 \text{ kN/m}^3)$$
$$= 2.90 \text{ kPa}    \quad \textbf{(Ans)}$$

**COMMENT**    By repeating the calculations for manometer fluids with different specific weights, $\gamma_2$, the results shown in Fig. E2.5b are obtained. Note that relatively small pressure differences can be measured if the manometer fluid has nearly the same specific weight as the flowing fluid. It is the difference in the specific weights, $\gamma_2 - \gamma_1$, that is important.

Hence, by rewriting the answer as $h_2 = (p_A - p_B)/(\gamma_2 - \gamma_1)$ it is seen that even if the value of $p_A - p_B$ is small, the value of $h_2$ can be large enough to provide an accurate reading provided the value of $\gamma_2 - \gamma_1$ is also small.

■ **Figure E2.5b**

### 2.6.3 Inclined-Tube Manometer

To measure small pressure changes, a manometer of the type shown in Fig. 2.12 is frequently used. One leg of the manometer is inclined at an angle $\theta$, and the differential reading $\ell_2$ is measured along the inclined tube. The difference in pressure $p_A - p_B$ can be expressed as

$$p_A + \gamma_1 h_1 - \gamma_2 \ell_2 \sin\theta - \gamma_3 h_3 = p_B$$

or

$$p_A - p_B = \gamma_2 \ell_2 \sin\theta + \gamma_3 h_3 - \gamma_1 h_1 \qquad (2.15)$$

*Inclined-tube manometers can be used to measure small pressure differences accurately.*

where it is to be noted the pressure difference between points (1) and (2) is due to the *vertical* distance between the points, which can be expressed as $\ell_2 \sin\theta$. Thus, for relatively small angles the differential reading along the inclined tube can be made large even for small pressure differences. The inclined-tube manometer is often used to measure small differences in gas pressures so that if pipes $A$ and $B$ contain a gas, then

$$p_A - p_B = \gamma_2 \ell_2 \sin\theta$$

or

$$\ell_2 = \frac{p_A - p_B}{\gamma_2 \sin\theta} \qquad (2.16)$$

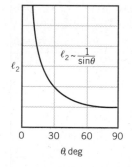

where the contributions of the gas columns $h_1$ and $h_3$ have been neglected. Equation 2.16 and the figure in the margin show that the differential reading $\ell_2$ (for a given pressure difference) of the inclined-tube manometer can be increased over that obtained with a conventional U-tube manometer by the factor $1/\sin\theta$. Recall that $\sin\theta \to 0$ as $\theta \to 0$.

■ **Figure 2.12** Inclined-tube manometer.

## 2.7    Mechanical and Electronic Pressure-Measuring Devices

*A Bourdon tube pressure gage uses a hollow, elastic, and curved tube to measure pressure.*

V2.4 Bourdon gage

Although manometers are widely used, they are not well suited for measuring very high pressures or pressures that are changing rapidly with time. In addition, they require the measurement of one or more column heights, which, although not particularly difficult, can be time consuming. To overcome some of these problems numerous other types of pressure-measuring instruments have been developed. Most of these make use of the idea that when a pressure acts on an elastic structure the structure will deform, and this deformation can be related to the magnitude of the pressure. Probably the most familiar device of this kind is the **Bourdon pressure gage,** which is shown in Fig. 2.13*a*. The essential mechanical element in this gage is the hollow, elastic curved tube (Bourdon tube) which is connected to the pressure source as shown in Fig. 2.13*b*. As the pressure within the tube increases the tube tends to straighten, and although the deformation is small, it can be translated into the motion of a pointer on a dial as illustrated. Since it is the difference in pressure between the outside of the tube (atmospheric pressure) and the inside of the tube that causes the movement of the tube, the indicated pressure is gage pressure. The Bourdon gage must be calibrated so that the dial reading can directly indicate the pressure in suitable units such as psi, psf, or pascals. A zero reading on the gage indicates that the measured pressure is equal to the local atmospheric pressure. This type of gage can be used to measure a negative gage pressure (vacuum) as well as positive pressures.

The *aneroid* barometer is another type of mechanical gage that is used for measuring atmospheric pressure. Since atmospheric pressure is specified as an absolute pressure, the conventional Bourdon gage is not suitable for this measurement. The common aneroid barometer contains a hollow, closed, elastic element that is evacuated so that the pressure inside the element is near absolute zero. As the external atmospheric pressure changes, the element deflects, and this motion can be translated into the movement of an attached dial. As with the Bourdon gage, the dial can be calibrated to give atmospheric pressure directly, with the usual units being millimeters or inches of mercury.

For many applications in which pressure measurements are required, the pressure must be measured with a device that converts the pressure into an electrical output. For example, it may be desirable to continuously monitor a pressure that is changing with time. This type of pressure-measuring device is called a *pressure transducer,* and many different designs are used. One possible type of transducer is one in which a Bourdon tube is connected to a linear variable differential transformer (LVDT), as is illustrated in Fig. 2.14. The core of the LVDT is connected to the free end of the Bourdon tube so that as a pressure is applied the resulting motion of the end of the tube moves the core through the coil and an output voltage develops. This voltage is a linear function of the pressure and could be recorded on an oscillograph or digitized for storage or processing on a computer.

(*a*)                                         (*b*)

■ **Figure 2.13**    (*a*) Liquid-filled Bourdon pressure gages for various pressure ranges. (*b*) Internal elements of Bourdon gages. The "C-shaped" Bourdon tube is shown on the left, and the "coiled spring" Bourdon tube for high pressures of 1000 psi and above is shown on the right. (Photographs courtesy of Weiss Instruments, Inc.)

■ **Figure 2.14** **Pressure transducer that combines a linear variable differential transformer (LVDT) with a Bourdon gage. (From Ref. 4, used by permission.)**

---

### F l u i d s   i n   t h e   N e w s

**Tire pressure warning** Proper tire inflation on vehicles is important for more than ensuring long tread life. It is critical in preventing accidents such as rollover accidents caused by underinflation of tires. The National Highway Traffic Safety Administration is developing a regulation regarding four-tire tire-pressure monitoring systems that can warn a driver when a tire is more than 25% underinflated. Some of these devices are currently in operation on select vehicles; it is expected that they will soon be required on all vehicles. A typical tire-pressure monitoring system fits within the tire and contains a *pressure transducer* (usually either a piezo-resistive or a capacitive-type transducer) and a transmitter that sends the information to an electronic control unit within the vehicle. Information about tire pressure and a warning when the tire is underinflated is displayed on the instrument panel. The environment (hot, cold, vibration) in which these devices must operate, their small size, and required low cost provide challenging constraints for the design engineer.

---

*It is relatively complicated to make accurate pressure transducers for the measurement of pressures that vary rapidly with time.*

One disadvantage of a pressure transducer using a Bourdon tube as the elastic sensing element is that it is limited to the measurement of pressures that are static or only changing slowly (quasistatic). Because of the relatively large mass of the Bourdon tube, it cannot respond to rapid changes in pressure. To overcome this difficulty, a different type of transducer is used in which the sensing element is a thin, elastic diaphragm that is in contact with the fluid. As the pressure changes, the diaphragm deflects, and this deflection can be sensed and converted into an electrical voltage. One way to accomplish this is to locate strain gages either on the surface of the diaphragm not in contact with the fluid, or on an element attached to the diaphragm. These gages can accurately sense the small strains induced in the diaphragm and provide an output voltage proportional to pressure. This type of transducer is capable of measuring accurately both small and large pressures, as well as both static and dynamic pressures. For example, strain-gage pressure transducers of the type shown in Fig. 2.15 are used to measure arterial blood pressure, which is a relatively small pressure that varies periodically with a fundamental frequency of about 1 Hz. The transducer is usually connected to the blood vessel by means of a liquid-filled, small-diameter tube called a pressure catheter. Although the strain-gage type of transducer can be designed to have very good frequency response (up to approximately 10 kHz), they become less sensitive at the higher frequencies since the diaphragm must be made stiffer to achieve the higher frequency response. As an alternative, the diaphragm can be constructed of a piezoelectric crystal to be used as both the elastic element and the sensor. When a pressure is applied to the crystal, a voltage develops because of the deformation of the crystal. This voltage is directly related to the applied pressure. Depending on the design, this type of transducer can be used to measure both very low and high pressures (up to approximately 100,000 psi) at high frequencies. Additional information on pressure transducers can be found in Refs. 3, 4, and 5.

*(a)*

*(b)*

■ **Figure 2.15** (*a*) Photograph of a typical pressure transducer with a male thread fitting in front of the diaphragm for system connection and an electrical connector in the rear of the device. (*b*) Schematic diagram of a typical pressure transducer device (male thread connector not shown). Deflection of the diaphragm due to pressure is measured with a silicon beam on which strain gages and an associated bridge circuit have been deposited.

## 2.8    Hydrostatic Force on a Plane Surface

*V2.5 Hoover dam*

When a surface is submerged in a fluid, forces develop on the surface due to the fluid. The determination of these forces is important in the design of storage tanks, ships, dams, and other hydraulic structures. For fluids at rest we know that the force must be *perpendicular* to the surface since there are no shearing stresses present. We also know that the pressure will vary linearly with depth as shown in Fig. 2.16 if the fluid is incompressible. For a horizontal surface, such as the bottom of a liquid-filled tank (Fig. 2.16*a*), the magnitude of the resultant force is simply $F_R = pA$, where $p$ is the uniform pressure on the bottom and $A$ is the area of the bottom. For the open tank shown, $p = \gamma h$. Note that if atmospheric pressure acts on both sides of the bottom, as is illustrated, the *resultant* force on the bottom is simply due to the liquid in the tank. Since the pressure is constant and uniformly distributed over the bottom, the resultant force acts through the centroid of the area as shown in Fig. 2.16*a*. As shown in Fig. 2.16*b*, the pressure on the ends of the tank is not uniformly distributed. Determination of the resultant force for situations such as this is presented as follows.

(a) Pressure on tank bottom                  (b) Pressure on tank ends

■ **Figure 2.16** (*a*) **Pressure distribution and resultant hydrostatic force on the bottom of an open tank.** (*b*) **Pressure distribution on the ends of an open tank.**

*The resultant force of a static fluid on a plane surface is due to the hydrostatic pressure distribution on the surface.*

For the more general case in which a submerged plane surface is inclined, as is illustrated in Fig. 2.17, the determination of the resultant force acting on the surface is more involved. For the present we will assume that the fluid surface is open to the atmosphere. Let the plane in which the surface lies intersect the free surface at 0 and make an angle $\theta$ with this surface as in Fig. 2.17. The $x$–$y$ coordinate system is defined so that 0 is the origin and $y = 0$ (i.e., the $x$ axis) is directed along the surface as shown. The area can have an arbitrary shape as shown. We wish to determine the direction, location, and magnitude of the resultant force acting on one side of this area due to the liquid in contact with the area. At any given depth, $h$, the force acting on $dA$ (the differential area of Fig. 2.17) is $dF = \gamma h \, dA$ and is perpendicular to the surface. Thus, the magnitude of the resultant force can be found by summing these differential forces over the entire surface. In equation form

$$F_R = \int_A \gamma h \, dA = \int_A \gamma y \sin \theta \, dA$$

■ **Figure 2.17** **Notation for hydrostatic force on an inclined plane surface of arbitrary shape.**

where $h = y \sin \theta$. For constant $\gamma$ and $\theta$

$$F_R = \gamma \sin \theta \int_A y \, dA \tag{2.17}$$

The integral appearing in Eq. 2.17 is the *first moment of the area* with respect to the $x$ axis, so we can write

$$\int_A y \, dA = y_c A$$

where $y_c$ is the $y$ coordinate of the centroid of area $A$ measured from the $x$ axis which passes through 0. Equation 2.17 can thus be written as

$$F_R = \gamma A y_c \sin \theta$$

*The magnitude of the resultant fluid force is equal to the pressure acting at the centroid of the area multiplied by the total area.*

or more simply as

$$\boxed{F_R = \gamma h_c A} \tag{2.18}$$

where, as shown by the figure in the margin, $h_c$ is the vertical distance from the fluid surface to the centroid of the area. Note that the magnitude of the force is independent of the angle $\theta$. As indicated by the figure in the margin, it depends only on the specific weight of the fluid, the total area, and the depth of the centroid of the area below the surface. In effect, Eq. 2.18 indicates that the magnitude of the resultant force is equal to the pressure at the centroid of the area multiplied by the total area. Since all the differential forces that were summed to obtain $F_R$ are perpendicular to the surface, the resultant $F_R$ must also be perpendicular to the surface.

Although our intuition might suggest that the resultant force should pass through the centroid of the area, this is not actually the case. The $y$ coordinate, $y_R$, of the resultant force can be determined by summation of moments around the $x$ axis. That is, the moment of the resultant force must equal the moment of the distributed pressure force, or

$$F_R y_R = \int_A y \, dF = \int_A \gamma \sin \theta \, y^2 \, dA$$

and, therefore, since $F_R = \gamma A y_c \sin \theta$

$$y_R = \frac{\int_A y^2 \, dA}{y_c A}$$

The integral in the numerator is the *second moment of the area (moment of inertia)*, $I_x$, with respect to an axis formed by the intersection of the plane containing the surface and the free surface ($x$ axis). Thus, we can write

$$y_R = \frac{I_x}{y_c A}$$

Use can now be made of the parallel axis theorem to express $I_x$ as

$$I_x = I_{xc} + A y_c^2$$

where $I_{xc}$ is the second moment of the area with respect to an axis passing through its *centroid* and parallel to the $x$ axis. Thus,

$$\boxed{y_R = \frac{I_{xc}}{y_c A} + y_c} \tag{2.19}$$

As shown by Eq. 2.19 and the figure in the margin, the resultant force does not pass through the centroid but for nonhorizontal surfaces is always *below* it, since $I_{xc}/y_c A > 0$.

The $x$ coordinate, $x_R$, for the resultant force can be determined in a similar manner by summing moments about the $y$ axis. Thus,

$$F_R x_R = \int_A \gamma \sin \theta \, xy \, dA$$

The resultant fluid force does not pass through the centroid of the area.

and, therefore,

$$x_R = \frac{\int_A xy \, dA}{y_c A} = \frac{I_{xy}}{y_c A}$$

where $I_{xy}$ is the product of inertia with respect to the $x$ and $y$ axes. Again, using the parallel axis theorem,[1] we can write

$$\boxed{x_R = \frac{I_{xyc}}{y_c A} + x_c}$$   (2.20)

where $I_{xyc}$ is the product of inertia with respect to an orthogonal coordinate system passing through the *centroid* of the area and formed by a translation of the $x$–$y$ coordinate system. If the submerged area is symmetrical with respect to an axis passing through the centroid and parallel to either the $x$ or $y$ axis, the resultant force must lie along the line $x = x_c$, since $I_{xyc}$ is identically zero in this case. The point through which the resultant force acts is called the **center of pressure.** It is to be noted from Eqs. 2.19 and 2.20 that as $y_c$ increases the center of pressure moves closer to the centroid of the area. Since $y_c = h_c/\sin \theta$, the distance $y_c$ will increase if the depth of submergence, $h_c$, increases, or, for a given depth, the area is rotated so that the angle, $\theta$, decreases. Thus, the hydrostatic force on the right-hand side of the gate shown in the margin figure acts closer to the centroid of the gate than the force on the left-hand side. Centroidal coordinates and moments of inertia for some common areas are given in Fig. 2.18.

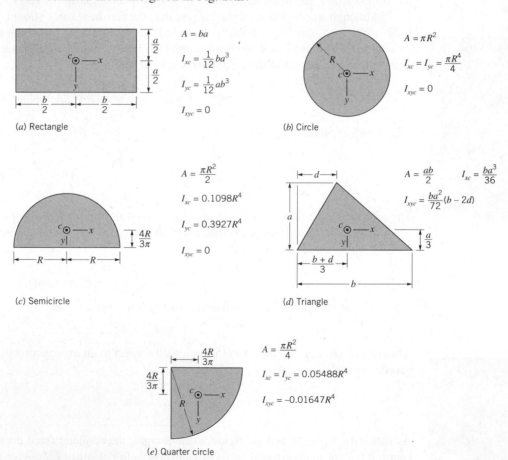

(a) Rectangle

$A = ba$

$I_{xc} = \frac{1}{12}ba^3$

$I_{yc} = \frac{1}{12}ab^3$

$I_{xyc} = 0$

(b) Circle

$A = \pi R^2$

$I_{xc} = I_{yc} = \frac{\pi R^4}{4}$

$I_{xyc} = 0$

(c) Semicircle

$A = \frac{\pi R^2}{2}$

$I_{xc} = 0.1098R^4$

$I_{yc} = 0.3927R^4$

$I_{xyc} = 0$

(d) Triangle

$A = \frac{ab}{2}$   $I_{xc} = \frac{ba^3}{36}$

$I_{xyc} = \frac{ba^2}{72}(b - 2d)$

(e) Quarter circle

$A = \frac{\pi R^2}{4}$

$I_{xc} = I_{yc} = 0.05488R^4$

$I_{xyc} = -0.01647R^4$

■ **Figure 2.18**   **Geometric properties of some common shapes.**

[1]Recall that the parallel axis theorem for the product of inertia of an area states that the product of inertia with respect to an orthogonal set of axes ($x$–$y$ coordinate system) is equal to the product of inertia with respect to an orthogonal set of axes parallel to the original set and passing through the centroid of the area, plus the product of the area and the $x$ and $y$ coordinates of the centroid of the area. Thus, $I_{xy} = I_{xyc} + Ax_cy_c$.

**The Three Gorges Dam** The Three Gorges Dam being constructed on China's Yangtze River will contain the world's largest hydroelectric power plant when in full operation. The dam is of the concrete gravity type, having a length of 2309 m with a height of 185 m. The main elements of the project include the dam, two power plants, and navigation facilities consisting of a ship lock and lift. The power plants will contain 26 Francis-type turbines, each with a capacity of 700 megawatts. The spillway section, which is the center section of the dam, is 483 m long with 23 bottom outlets and 22 surface sluice gates.

The maximum discharge capacity is 102,500 cu m per second. After more than 10 years of construction, the dam gates were finally closed, and on June 10, 2003, the reservoir had been filled to its interim level of 135 m. Due to the large depth of water at the dam and the huge extent of the storage pool, *hydrostatic pressure forces* have been a major factor considered by engineers. When filled to its normal pool level of 175 m, the total reservoir storage capacity is 39.3 billion cu m. All of the originally planned components of the project (except for the ship lift) were completed in 2008. (See Problem 2.111.)

---

## EXAMPLE 2.6   Hydrostatic Force on a Plane Circular Surface

**GIVEN**  The 4-m-diameter circular gate of Fig. E2.6a is located in the inclined wall of a large reservoir containing water ($\gamma = 9.80$ kN/m³). The gate is mounted on a shaft along its horizontal diameter, and the water depth is 10 m above the shaft.

**FIND**  Determine

**(a)**  the magnitude and location of the resultant force exerted on the gate by the water and

**(b)**  the moment that would have to be applied to the shaft to open the gate.

## SOLUTION

**(a)**  To find the magnitude of the force of the water we can apply Eq. 2.18,

$$F_R = \gamma h_c A$$

and since the vertical distance from the fluid surface to the centroid of the area is 10 m, it follows that

$$F_R = (9.80 \times 10^3 \text{ N/m}^3)(10 \text{ m})(4\pi \text{ m}^2)$$
$$= 1230 \times 10^3 \text{ N} = 1.23 \text{ MN} \qquad \textbf{(Ans)}$$

To locate the point (center of pressure) through which $F_R$ acts, we use Eqs. 2.19 and 2.20,

$$x_R = \frac{I_{xyc}}{y_c A} + x_c \qquad y_R = \frac{I_{xc}}{y_c A} + y_c$$

For the coordinate system shown, $x_R = 0$ since the area is symmetrical, and the center of pressure must lie along the diameter $A$-$A$. To obtain $y_R$, we have from Fig. 2.18

$$I_{xc} = \frac{\pi R^4}{4}$$

and $y_c$ is shown in Fig. E2.6b. Thus,

$$y_R = \frac{(\pi/4)(2 \text{ m})^4}{(10 \text{ m}/\sin 60°)(4\pi \text{ m}^2)} + \frac{10 \text{ m}}{\sin 60°}$$
$$= 0.0866 \text{ m} + 11.55 \text{ m} = 11.6 \text{ m}$$

■ **Figure E2.6a–c**

and the distance (along the gate) below the shaft to the center of pressure is

$$y_R - y_c = 0.0866 \text{ m} \qquad \textbf{(Ans)}$$

We can conclude from this analysis that the force on the gate due to the water has a magnitude of 1.23 MN and acts through a point along its diameter *A-A* at a distance of 0.0866 m (along the gate) below the shaft. The force is perpendicular to the gate surface as shown in Fig. E2.6b.

**COMMENT**  By repeating the calculations for various values of the depth to the centroid, $h_c$, the results shown in Fig. E2.6d are obtained. Note that as the depth increases, the distance between the center of pressure and the centroid decreases.

**(b)**  The moment required to open the gate can be obtained with the aid of the free-body diagram of Fig. E2.6c. In this diagram $\mathcal{W}$

is the weight of the gate and $O_x$ and $O_y$ are the horizontal and vertical reactions of the shaft on the gate. We can now sum moments about the shaft

$$\sum M_c = 0$$

and, therefore,

$$
\begin{aligned}
M &= F_R(y_R - y_c) \\
&= (1230 \times 10^3 \, \text{N})(0.0866 \, \text{m}) \\
&= 1.07 \times 10^5 \, \text{N} \cdot \text{m} \qquad \textbf{(Ans)}
\end{aligned}
$$

(10m, 0.0886 m)

■ **Figure E2.6d**

---

## EXAMPLE 2.7    Hydrostatic Pressure Force on a Plane Triangular Surface

**GIVEN** An aquarium contains seawater ($\gamma = 64.0 \, \text{lb/ft}^3$) to a depth of 1 ft as shown in Fig. E2.7a. To repair some damage to one corner of the tank, a triangular section is replaced with a new section as illustrated in Fig. E2.7b.

**FIND** Determine

**(a)** the magnitude of the force of the seawater on this triangular area, and

**(b)** the location of this force.

### SOLUTION

**(a)** The various distances needed to solve this problem are shown in Fig. E2.7c. Since the surface of interest lies in a vertical plane, $y_c = h_c = 0.9 \, \text{ft}$, and from Eq. 2.18 the magnitude of the force is

$$
\begin{aligned}
F_R &= \gamma h_c A \\
&= (64.0 \, \text{lb/ft}^3)(0.9 \, \text{ft})[(0.3 \, \text{ft})^2/2] = 2.59 \, \text{lb} \qquad \textbf{(Ans)}
\end{aligned}
$$

**COMMENT** Note that this force is independent of the tank length. The result is the same if the tank is 0.25 ft, 25 ft, or 25 miles long.

**(b)** The y coordinate of the center of pressure (CP) is found from Eq. 2.19,

$$y_R = \frac{I_{xc}}{y_c A} + y_c$$

and from Fig. 2.18

(b)

(c)

(d)

■ **Figure E2.7b–d**

■ **Figure E2.7a** (Photograph courtesy of Tenecor Tanks, Inc.)

$$I_{xc} = \frac{(0.3 \text{ ft})(0.3 \text{ ft})^3}{36} = \frac{0.0081}{36} \text{ ft}^4$$

so that

$$y_R = \frac{0.0081/36 \text{ ft}^4}{(0.9 \text{ ft})(0.09/2 \text{ ft}^2)} + 0.9 \text{ ft}$$

$$= 0.00556 \text{ ft} + 0.9 \text{ ft} = 0.906 \text{ ft} \qquad \textbf{(Ans)}$$

Similarly, from Eq. 2.20

$$x_R = \frac{I_{xyc}}{y_c A} + x_c$$

and from Fig. 2.18

$$I_{xyc} = \frac{(0.3 \text{ ft})(0.3 \text{ ft})^2}{72}(0.3 \text{ ft}) = \frac{0.0081}{72} \text{ ft}^4$$

so that

$$x_R = \frac{0.0081/72 \text{ ft}^4}{(0.9 \text{ ft})(0.09/2 \text{ ft}^2)} + 0 = 0.00278 \text{ ft} \qquad \textbf{(Ans)}$$

**COMMENT**   Thus, we conclude that the center of pressure is 0.00278 ft to the right of and 0.00556 ft below the centroid of the area. If this point is plotted, we find that it lies on the median line for the area as illustrated in Fig. E2.7d. Since we can think of the total area as consisting of a number of small rectangular strips of area $\delta A$ (and the fluid force on each of these small areas acts through its center), it follows that the resultant of all these parallel forces must lie along the median.

## 2.9     Pressure Prism

An informative and useful graphical interpretation can be made for the force developed by a fluid acting on a plane rectangular area. Consider the pressure distribution along a vertical wall of a tank of constant width $b$, which contains a liquid having a specific weight $\gamma$. Since the pressure must vary linearly with depth, we can represent the variation as is shown in Fig. 2.19a, where the pressure is equal to zero at the upper surface and equal to $\gamma h$ at the bottom. It is apparent from this diagram that the average pressure occurs at the depth $h/2$ and, therefore, the resultant force acting on the rectangular area $A = bh$ is

$$F_R = p_{av} A = \gamma \left(\frac{h}{2}\right) A$$

which is the same result as obtained from Eq. 2.18. The pressure distribution shown in Fig. 2.19a applies across the vertical surface, so we can draw the three-dimensional representation of the pressure distribution as shown in Fig. 2.19b. The base of this "volume" in pressure-area space is the plane surface of interest, and its altitude at each point is the pressure. This volume is called the *pressure prism,* and it is clear that the magnitude of the resultant force acting on the rectangular surface is equal to the volume of the pressure prism. Thus, for the prism of Fig. 2.19b the fluid force is

*The magnitude of the resultant fluid force is equal to the volume of the pressure prism and passes through its centroid.*

$$F_R = \text{volume} = \frac{1}{2}(\gamma h)(bh) = \gamma \left(\frac{h}{2}\right) A$$

where $bh$ is the area of the rectangular surface, $A$.

The resultant force must pass through the *centroid* of the pressure prism. For the volume under consideration the centroid is located along the vertical axis of symmetry of the surface and at a distance of $h/3$ above the base (since the centroid of a triangle is located at $h/3$ above its base). This result can readily be shown to be consistent with that obtained from Eqs. 2.19 and 2.20.

(a)

(b)

■ **Figure 2.19**   **Pressure prism for vertical rectangular area.**

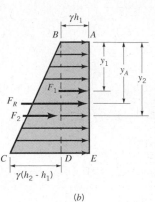

(a)

(b)

■ **Figure 2.20** Graphical representation of hydrostatic forces on a vertical rectangular surface.

This same graphical approach can be used for plane rectangular surfaces that do not extend up to the fluid surface, as illustrated in Fig. 2.20a. In this instance, the cross section of the pressure prism is trapezoidal. However, the resultant force is still equal in magnitude to the volume of the pressure prism, and it passes through the centroid of the volume. Specific values can be obtained by decomposing the pressure prism into two parts, *ABDE* and *BCD*, as shown in Fig. 2.20b. Thus,

$$F_R = F_1 + F_2$$

where the components can readily be determined by inspection for rectangular surfaces. The location of $F_R$ can be determined by summing moments about some convenient axis, such as one passing through *A*. In this instance

$$F_R y_A = F_1 y_1 + F_2 y_2$$

and $y_1$ and $y_2$ can be determined by inspection.

For inclined plane rectangular surfaces the pressure prism can still be developed, and the cross section of the prism will generally be trapezoidal, as is shown in Fig. 2.21. Although it is usually convenient to measure distances along the inclined surface, the pressures developed depend on the vertical distances as illustrated.

*The use of the pressure prism concept to determine the force on a submerged area is best suited for plane rectangular surfaces.*

The use of pressure prisms for determining the force on submerged plane areas is convenient if the area is rectangular so the volume and centroid can be easily determined. However, for other nonrectangular shapes, integration would generally be needed to determine the volume and centroid. In these circumstances it is more convenient to use the equations developed in the previous section, in which the necessary integrations have been made and the results presented in a convenient and compact form that is applicable to submerged plane areas of any shape.

The effect of atmospheric pressure on a submerged area has not yet been considered, and we may ask how this pressure will influence the resultant force. If we again consider the pressure distribution on a plane vertical wall, as is shown in Fig. 2.22a, the pressure varies from zero at the surface to $\gamma h$ at the bottom. Since we are setting the surface pressure equal to zero, we are using

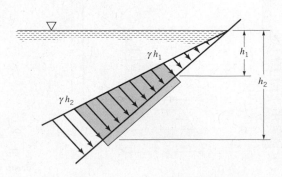

■ **Figure 2.21** Pressure variation along an inclined plane area.

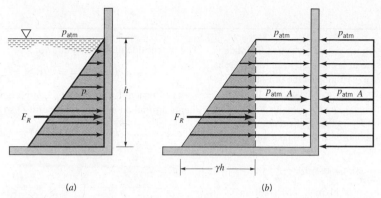

■ **Figure 2.22**   Effect of atmospheric pressure on the resultant force acting on a plane vertical wall.

*The resultant fluid force acting on a submerged area is affected by the pressure at the free surface.*

atmospheric pressure as our datum, and thus the pressure used in the determination of the fluid force is gage pressure. If we wish to include atmospheric pressure, the pressure distribution will be as is shown in Fig. 2.22b. We note that in this case the force on one side of the wall now consists of $F_R$ as a result of the hydrostatic pressure distribution, plus the contribution of the atmospheric pressure, $p_{atm}A$, where $A$ is the area of the surface. However, if we are going to include the effect of atmospheric pressure on one side of the wall, we must realize that this same pressure acts on the outside surface (assuming it is exposed to the atmosphere), so that an equal and opposite force will be developed as illustrated in the figure. Thus, we conclude that the *resultant* fluid force on the surface is that due only to the gage pressure contribution of the liquid in contact with the surface—the atmospheric pressure does not contribute to this resultant. Of course, if the surface pressure of the liquid is different from atmospheric pressure (such as might occur in a closed tank), the resultant force acting on a submerged area, $A$, will be changed in magnitude from that caused simply by hydrostatic pressure by an amount $p_s A$, where $p_s$ is the gage pressure at the liquid surface (the outside surface is assumed to be exposed to atmospheric pressure).

# EXAMPLE 2.8     Use of the Pressure Prism Concept

**GIVEN**   A pressurized tank contains oil ($SG = 0.90$) and has a square, 0.6-m by 0.6-m plate bolted to its side, as is illustrated in Fig. E2.8a. The pressure gage on the top of the tank reads 50 kPa, and the outside of the tank is at atmospheric pressure.

**FIND**   What is the magnitude and location of the resultant force on the attached plate?

■ **Figure E2.8**

## SOLUTION

The pressure distribution acting on the inside surface of the plate is shown in Fig. E2.8*b*. The pressure at a given point on the plate is due to the air pressure, $p_s$, at the oil surface and the pressure due to the oil, which varies linearly with depth as is shown in the figure. The resultant force on the plate (having an area $A$) is due to the components, $F_1$ and $F_2$, where $F_1$ and $F_2$ are due to the rectangular and triangular portions of the pressure distribution, respectively. Thus,

$$
\begin{aligned}
F_1 &= (p_s + \gamma h_1)\, A \\
&= [50 \times 10^3\ \text{N/m}^2 \\
&\quad + (0.90)(9.81 \times 10^3\ \text{N/m}^3)(2\ \text{m})](0.36\ \text{m}^2) \\
&= 24.4 \times 10^3\ \text{N}
\end{aligned}
$$

and

$$
\begin{aligned}
F_2 &= \gamma\left(\frac{h_2 - h_1}{2}\right) A \\
&= (0.90)(9.81 \times 10^3\ \text{N/m}^3)\left(\frac{0.6\ \text{m}}{2}\right)(0.36\ \text{m}^2) \\
&= 0.954 \times 10^3\ \text{N}
\end{aligned}
$$

The magnitude of the resultant force, $F_R$, is therefore

$$
F_R = F_1 + F_2 = 25.4 \times 10^3\ \text{N} = 25.4\ \text{kN} \qquad \textbf{(Ans)}
$$

The vertical location of $F_R$ can be obtained by summing moments around an axis through point $O$ so that

$$
F_R y_O = F_1(0.3\ \text{m}) + F_2(0.2\ \text{m})
$$

or

$$
\begin{aligned}
y_O &= \frac{(24.4 \times 10^3\ \text{N})(0.3\ \text{m}) + (0.954 \times 10^3\ \text{N})(0.2\ \text{m})}{25.4 \times 10^3\ \text{N}} \\
&= 0.296\ \text{m} \qquad \textbf{(Ans)}
\end{aligned}
$$

Thus, the force acts at a distance of 0.296 m above the bottom of the plate along the vertical axis of symmetry.

**COMMENT**   Note that the air pressure used in the calculation of the force was gage pressure. Atmospheric pressure does not affect the resultant force (magnitude or location), since it acts on both sides of the plate, thereby canceling its effect.

## 2.10   Hydrostatic Force on a Curved Surface

*V2.6 Pop bottle*

The equations developed in Section 2.8 for the magnitude and location of the resultant force acting on a submerged surface only apply to plane surfaces. However, many surfaces of interest (such as those associated with dams, pipes, and tanks) are nonplanar. The domed bottom of the beverage bottle shown in the figure in the margin shows a typical curved surface example. Although the resultant fluid force can be determined by integration, as was done for the plane surfaces, this is generally a rather tedious process and no simple, general formulas can be developed. As an alternative approach, we will consider the equilibrium of the fluid volume enclosed by the curved surface of interest and the horizontal and vertical projections of this surface.

For example, consider a curved portion of the swimming pool shown in Fig. 2.23*a*. We wish to find the resultant fluid force acting on section *BC* (which has a unit length perpendicular to the plane of the paper) shown in Fig. 2.23*b*. We first isolate a volume of fluid that is bounded by the surface of interest, in this instance section *BC*, the horizontal plane surface *AB*, and the vertical plane surface *AC*. The free-body diagram for this volume is shown in Fig. 2.23*c*. The magnitude and location of forces $F_1$ and $F_2$ can be determined from the relationships for planar surfaces. The weight, $\mathcal{W}$, is simply the specific weight of the fluid times the enclosed volume and acts through the center of gravity (CG) of the mass of fluid contained within the volume. The forces $F_H$ and $F_V$ represent the components of the force that the tank *exerts on the fluid.*

In order for this force system to be in equilibrium, the horizontal component $F_H$ must be equal in magnitude and collinear with $F_2$, and the vertical component $F_V$ equal in magnitude and collinear with the resultant of the vertical forces $F_1$ and $\mathcal{W}$. This follows since the three forces acting on the fluid mass ($F_2$, the resultant of $F_1$ and $\mathcal{W}$, and the resultant force that the tank exerts on the mass) must form a *concurrent* force system. That is, from the principles of statics, it is known that when a body is held in equilibrium by three nonparallel forces, they must be concurrent (their lines of action intersect at a common point) and coplanar. Thus,

$$
\begin{aligned}
F_H &= F_2 \\
F_V &= F_1 + \mathcal{W}
\end{aligned}
$$

and the magnitude of the resultant is obtained from the equation

$$
F_R = \sqrt{(F_H)^2 + (F_V)^2}
$$

■ **Figure 2.23**  Hydrostatic force on a curved surface. (Photograph courtesy of Intex Marketing, Ltd.)

The resultant $F_R$ passes through the point $O$, which can be located by summing moments about an appropriate axis. The resultant force of the fluid acting *on the curved surface BC* is equal and opposite in direction to that obtained from the free-body diagram of Fig. 2.23c. The desired fluid force is shown in Fig. 2.23d.

---

## *E*XAMPLE 2.9    Hydrostatic Pressure Force on a Curved Surface

**GIVEN**  A 6-ft-diameter drainage conduit of the type shown in Fig. E2.9a is half full of water at rest, as shown in Fig. E2.9b.

**FIND**  Determine the magnitude and line of action of the resultant force that the water exerts on a 1-ft length of the curved section *BC* of the conduit wall.

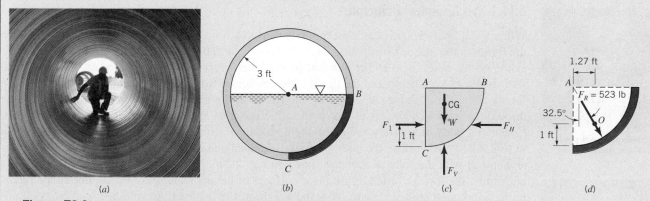

■ **Figure E2.9**  (Photograph courtesy of CONTECH Construction Products, Inc.)

## *S*OLUTION

We first isolate a volume of fluid bounded by the curved section *BC*, the horizontal surface *AB*, and the vertical surface *AC*, as shown in Fig. E2.9c. The volume has a length of 1 ft. The forces acting on the volume are the horizontal force, $F_1$, which acts on the vertical surface *AC*, the weight, $\mathcal{W}$, of the fluid contained within the volume, and the horizontal and vertical components of the force of the conduit wall on the fluid, $F_H$ and $F_V$, respectively.

The magnitude of $F_1$ is found from the equation

$$F_1 = \gamma h_c A = (62.4 \text{ lb/ft}^3)(\tfrac{3}{2} \text{ ft})(3 \text{ ft}^2) = 281 \text{ lb}$$

and this force acts 1 ft above $C$ as shown. The weight $\mathcal{W} = \gamma \mathcal{V}$, where $\mathcal{V}$ is the fluid volume, is

$$\mathcal{W} = \gamma \mathcal{V} = (62.4 \text{ lb/ft}^3)(9\pi/4 \text{ ft}^2)(1 \text{ ft}) = 441 \text{ lb}$$

and acts through the center of gravity of the mass of fluid, which according to Fig. 2.18 is located 1.27 ft to the right of *AC* as shown. Therefore, to satisfy equilibrium

$$F_H = F_1 = 281 \text{ lb} \qquad F_V = \mathcal{W} = 441 \text{ lb}$$

and the magnitude of the resultant force is

$$F_R = \sqrt{(F_H)^2 + (F_V)^2}$$
$$= \sqrt{(281 \text{ lb})^2 + (441 \text{ lb})^2} = 523 \text{ lb} \qquad \textbf{(Ans)}$$

The force the water exerts *on the conduit wall* is equal, but *opposite in direction,* to the forces $F_H$ and $F_V$ shown in Fig. E2.9c. Thus, the resultant force *on the conduit wall* is shown in Fig. E2.9d. This force acts through the point $O$ at the angle shown.

**COMMENT** An inspection of this result will show that the line of action of the resultant force passes through the center of the conduit. In retrospect, this is not a surprising result since at each point on the curved surface of the conduit the elemental force due to the pressure is normal to the surface, and each line of action must pass through the center of the conduit. It therefore follows that the resultant of this concurrent force system must also pass through the center of concurrence of the elemental forces that make up the system.

This same general approach can also be used for determining the force on curved surfaces of pressurized, closed tanks. If these tanks contain a gas, the weight of the gas is usually negligible in comparison with the forces developed by the pressure. Thus, the forces (such as $F_1$ and $F_2$ in Fig. 2.23$c$) on horizontal and vertical projections of the curved surface of interest can simply be expressed as the internal pressure times the appropriate projected area.

## F l u i d s   i n   t h e   N e w s

**Miniature, exploding pressure vessels** Our daily lives are safer because of the effort put forth by engineers to design safe, lightweight pressure vessels such as boilers, propane tanks, and pop bottles. Without proper design, the large *hydrostatic pressure forces on the curved surfaces* of such containers could cause the vessel to explode with disastrous consequences. On the other hand, the world is a friendlier place because of miniature pressure vessels that are designed to explode under the proper conditions—popcorn kernels. Each grain of popcorn contains a small amount of water within the special, impervious hull (pressure vessel) which, when heated to a proper temperature, turns to steam, causing the kernel to explode and turn itself inside out. Not all popcorn kernels have the proper properties to make them pop well. First, the kernel must be quite close to 13.5% water. With too little moisture, not enough steam will build up to pop the kernel; too much moisture causes the kernel to pop into a dense sphere rather than the light fluffy delicacy expected. Second, to allow the pressure to build up, the kernels must not be cracked or damaged.

## 2.11 Buoyancy, Flotation, and Stability

### 2.11.1 Archimedes' Principle

(Photograph courtesy of Cameron Balloons.)

When a stationary body is completely submerged in a fluid (such as the hot air balloon shown in the figure in the margin), or floating so that it is only partially submerged, the resultant fluid force acting on the body is called the ***buoyant force.*** A net upward vertical force results because pressure increases with depth and the pressure forces acting from below are larger than the pressure forces acting from above, as shown by the figure in the margin. This force can be determined through an approach similar to that used in the previous section for forces on curved surfaces. Consider a body of arbitrary shape, having a volume $\mathcal{V}$, that is immersed in a fluid as illustrated in Fig. 2.24$a$. We enclose the body in a parallelepiped and draw a free-body diagram of the parallelepiped with the body removed as shown in Fig. 2.24$b$. Note that the forces $F_1$, $F_2$, $F_3$, and $F_4$ are simply the forces exerted on the plane surfaces of the parallelepiped (for simplicity the forces in the $x$ direction are not shown), $\mathcal{W}$ is the weight of the shaded fluid volume (parallelepiped minus body), and $F_B$ is the force the body is exerting *on the fluid.* The forces on the vertical surfaces, such as $F_3$ and $F_4$, are all equal and cancel, so the equilibrium equation of interest is in the $z$ direction and can be expressed as

$$F_B = F_2 - F_1 - \mathcal{W} \tag{2.21}$$

If the specific weight of the fluid is constant, then

$$F_2 - F_1 = \gamma(h_2 - h_1)A$$

where $A$ is the horizontal area of the upper (or lower) surface of the parallelepiped, and Eq. 2.21 can be written as

$$F_B = \gamma(h_2 - h_1)A - \gamma[(h_2 - h_1)A - \mathcal{V}]$$

Simplifying, we arrive at the desired expression for the buoyant force

$$\boxed{F_B = \gamma \mathcal{V}} \tag{2.22}$$

V2.7 Cartesian Diver

■ **Figure 2.24**    Buoyant force on submerged and floating bodies.

where $\gamma$ is the specific weight of the fluid and $V$ is the volume of the body. The effects of the specific weight (or density) of the body as compared to that of the surrounding fluid are illustrated by the figure in the margin. The direction of the buoyant force, which is the force of the fluid *on the body,* is opposite to that shown on the free-body diagram. Therefore, the buoyant force has a magnitude equal to the weight of the fluid displaced by the body and is directed vertically upward. This result is commonly referred to as ***Archimedes' principle*** in honor of Archimedes (287–212 B.C.), a Greek mechanician and mathematician who first enunciated the basic ideas associated with hydrostatics.

    The location of the line of action of the buoyant force can be determined by summing moments of the forces shown on the free-body diagram in Fig. 2.24*b* with respect to some convenient axis. For example, summing moments about an axis perpendicular to the paper through point *D* we have

$$F_B y_c = F_2 y_1 - F_1 y_1 - W y_2$$

and on substitution for the various forces

$$V y_c = V_T y_1 - (V_T - V) y_2 \tag{2.23}$$

*Archimedes' principle states that the buoyant force has a magnitude equal to the weight of the fluid displaced by the body and is directed vertically upward.*

where $V_T$ is the total volume $(h_2 - h_1)A$. The right-hand side of Eq. 2.23 is the first moment of the displaced volume $V$ with respect to the *x–z* plane so that $y_c$ is equal to the *y* coordinate of the centroid of the volume $V$. In a similar fashion, it can be shown that the *x* coordinate of the buoyant force coincides with the *x* coordinate of the centroid. Thus, we conclude that the *buoyant force passes through the centroid of the displaced volume* as shown in Fig. 2.24*c*. The point through which the buoyant force acts is called the ***center of buoyancy.***

## F l u i d s   i n   t h e   N e w s

**Concrete canoes** A solid block of concrete thrown into a pond or lake will obviously sink. But if the concrete is formed into the shape of a canoe it can be made to float. Of, course the reason the canoe floats is the development of the *buoyant force* due to the displaced volume of water. With the proper design, this vertical force can be made to balance the weight of the canoe plus passengers—the canoe floats. Each year since 1988 a National Concrete Canoe Competition for university teams is jointly spon-

sored by the American Society of Civil Engineers and Master Builders Inc. The canoes must be 90% concrete and are typically designed with the aid of a computer by civil engineering students. Final scoring depends on four components: a design report, an oral presentation, the final product, and racing. In 2011 California Polytechnic State University, San Luis Obispo, won the national championship with its 208-pound, 20-foot-long canoe. (See Problem 2.147.)

*V2.8 Hydrometer*

These same results apply to floating bodies that are only partially submerged, as illustrated in Fig. 2.24*d*, if the specific weight of the fluid above the liquid surface is very small compared with the liquid in which the body floats. Since the fluid above the surface is usually air, for practical purposes this condition is satisfied.

In the derivations presented above, the fluid is assumed to have a constant specific weight, $\gamma$. If a body is immersed in a fluid in which $\gamma$ varies with depth, such as in a layered fluid, the magnitude of the buoyant force remains equal to the weight of the displaced fluid. However, the buoyant force does not pass through the centroid of the displaced volume, but rather, it passes through the center of gravity of the displaced volume.

---

## EXAMPLE 2.10 ■ Buoyant Force on a Submerged Object

**GIVEN** A Type I offshore life jacket (personal flotation device) of the type worn by commercial fishermen is shown in Fig. E2.10*a*. It is designed for extended survival in rough, open water. According to U.S. Coast Guard regulations, the life jacket must provide a minimum 22-lb net upward force on the user. Consider such a life jacket that uses a foam material with a specific weight of 2.0 lb/ft³ for the main flotation material. The remaining material (cloth, straps, fasteners, etc.) weighs 1.3 lb and is of negligible volume.

**FIND** Determine the minimum volume of foam needed for this life jacket.

### SOLUTION

A free-body diagram of the life jacket is shown in Fig. E2.10*b*, where $F_B$ is the buoyant force acting on the life jacket, $\mathcal{W}_F$ is the weight of the foam, $\mathcal{W}_S = 1.3$ lb is the weight of the remaining material, and $F_U = 22$ lb is the required force on the user. For equilibrium it follows that

$$F_B = \mathcal{W}_F + \mathcal{W}_S + F_U \qquad (1)$$

where from Eq. 2.22

$$F_B = \gamma_{\text{water}}\, V$$

Here $\gamma_{\text{water}} = 64.0$ lb/ft³ is the specific weight of seawater and $V$ is the volume of the foam. Also $\mathcal{W}_{\text{foam}} = \gamma_{\text{foam}}\, V$, where $\gamma_{\text{foam}} = 2.0$ lb/ft³ is the specific weight of the foam. Thus, from Eq. 1

$$\gamma_{\text{water}}\, V = \gamma_{\text{foam}}\, V + \mathcal{W}_S + F_U$$

■ **Figure E2.10*a***

or

$$\begin{aligned} V &= (\mathcal{W}_S + F_U)/(\gamma_{\text{water}} - \gamma_{\text{foam}}) \\ &= (1.3 \text{ lb} + 22 \text{ lb})/(64.0 \text{ lb/ft}^3 - 2.0 \text{ lb/ft}^3) \\ &= 0.376 \text{ ft}^3 \end{aligned} \qquad \textbf{(Ans)}$$

**COMMENTS** In this example, rather than using difficult-to-calculate hydrostatic pressure force on the irregularly shaped life jacket, we have used the buoyant force. The net effect of the pressure forces on the surface of the life jacket is equal to the upward buoyant force. Do not include both the buoyant force and the hydrostatic pressure effects in your calculations—use one or the other.

There is more to the proper design of a life jacket than just the volume needed for the required buoyancy. According to regulations, a Type I life jacket must also be designed so that it provides proper protection to the user by turning an unconscious person in the water to a face-up position as shown in Fig. E2.10*a*. This involves the concept of the stability of a floating object (see Section 2.11.2). The life jacket should also provide minimum interference under ordinary working conditions so as to encourage its use by commercial fishermen.

■ **Figure E2.10*b***

*V2.9 Atmospheric buoyancy*

The effects of buoyancy are not limited to the interaction between a solid body and a fluid. Buoyancy effects can also be seen within fluids alone, as long as a density difference exists. Consider the shaded portion of Fig. 2.24c to be a volume of fluid instead of a solid. This volume of fluid is submerged in the surrounding fluid and therefore has a buoyant force due to the fluid it displaces (like the solid). If this volume contains fluid with a density of $\rho_1$, then the downward force due to weight is $\mathcal{W} = \rho_1 g V$. In addition, if the surrounding fluid has the same density, then the buoyant force on the volume due to the displaced fluid will be $F_B = \gamma V = \rho_1 g \mathcal{V}$. As expected, in this case the weight of the volume is exactly balanced by the buoyant force acting on the volume so there is no net force. However, if the density of fluid in the volume is $\rho_2$, then $\mathcal{W}$ and $F_B$ will not balance and there will be a net force in the upward or downward direction depending on whether the density in the volume ($\rho_2$) is less than or greater than, respectively, the density of the surrounding fluid. Note that this difference can develop from two different fluids with different densities or from temperature differences within the same fluid causing density variations in space. For example, smoke from a fire rises because it is lighter (due to its higher temperature) than the surrounding air.

## F l u i d s   i n   t h e   N e w s

**Explosive lake** In 1986 a tremendous explosion of carbon dioxide ($CO_2$) from Lake Nyos, west of Cameroon, killed more than 1700 people and livestock. The explosion resulted from a buildup of $CO_2$ that seeped into the high-pressure water at the bottom of the lake from warm springs of $CO_2$-bearing water. The $CO_2$-rich water is heavier than pure water and can hold a volume of $CO_2$ more than five times the water volume. As long as the gas remains dissolved in the water, the stratified lake (i.e., pure water on top, $CO_2$ water on the bottom) is stable. But if some mechanism causes the gas

bubbles to nucleate, they rise, grow, and cause other bubbles to form, feeding a chain reaction. A related phenomenon often occurs when a pop bottle is shaken and then opened. The pop shoots from the container rather violently. When this set of events occurred in Lake Nyos, the entire lake overturned through a column of rising and expanding *buoyant* bubbles. The heavier-than-air $CO_2$ then flowed through the long, deep valleys surrounding the lake and asphyxiated human and animal life caught in the gas cloud. One victim was 27 km downstream from the lake.

*V2.10 Density differences in fluids*

### 2.11.2 Stability

Another interesting and important problem associated with submerged or floating bodies is concerned with the stability of the bodies. As illustrated by the figure in the margin, a body is said to be in a *stable equilibrium* position if, when displaced, it returns to its equilibrium position. Conversely, it is in an *unstable equilibrium* position if, when displaced (even slightly), it moves to a new equilibrium position. Stability considerations are particularly important for submerged or floating bodies since the centers of buoyancy and gravity do not necessarily coincide. A small rotation can result in either a restoring or overturning couple. For example, for the *completely* submerged body shown in Fig. 2.25, which has a center of gravity below the center of buoyancy, a rotation from its equilibrium position will create a restoring couple formed by the weight, $\mathcal{W}$, and the buoyant force, $F_B$, which causes the body to rotate back to its original position. Thus, for this configuration the body is stable. It is to be noted that as long as the center of gravity falls *below* the center of buoyancy, this will always be true; that is, the body is in a *stable equilibrium* position with respect to small rotations. However, as is illustrated in Fig. 2.26, if the center of gravity of the

Stable

Unstable

*The stability of a body can be determined by considering what happens when it is displaced from its equilibrium position.*

■ **Figure 2.25** Stability of a completely immersed body—center of gravity below centroid.

■ **Figure 2.26** Stability of a completely immersed body—center of gravity above centroid.

V2.11 Stability of a floating cube

$c$ = centroid of original displaced volume

$c'$ = centroid of new displaced volume

Restoring couple

Stable

■ **Figure 2.27**  **Stability of a floating body—stable configuration.**

Marginally stable

$c$ = centroid of original displaced volume

$c'$ = centroid of new displaced volume

Overturning couple

Unstable

■ **Figure 2.28**  **Stability of a floating body—unstable configuration.**

Very stable

© **RiverNorthPhotography/ iStockphoto**

V2.12 Stability of a model barge

completely submerged body is above the center of buoyancy, the resulting couple formed by the weight and the buoyant force will cause the body to overturn and move to a new equilibrium position. Thus, a completely submerged body with its center of gravity *above* its center of buoyancy is in an *unstable equilibrium* position.

For *floating* bodies the stability problem is more complicated, since as the body rotates the location of the center of buoyancy (which passes through the centroid of the displaced volume) may change. As is shown in Fig. 2.27, a floating body such as a barge that rides low in the water can be stable even though the center of gravity lies above the center of buoyancy. This is true since as the body rotates the buoyant force, $F_B$, shifts to pass through the centroid of the newly formed displaced volume and, as illustrated, combines with the weight, $\mathcal{W}$, to form a couple that will cause the body to return to its original equilibrium position. However, for the relatively tall, slender body shown in Fig. 2.28, a small rotational displacement can cause the buoyant force and the weight to form an overturning couple as illustrated.

It is clear from these simple examples that determining the stability of submerged or floating bodies can be difficult since the analysis depends in a complicated fashion on the particular geometry and weight distribution of the body. Thus, although both the relatively narrow kayak and the wide houseboat shown in the figures in the margin are stable, the kayak will overturn much more easily than the houseboat. The problem can be further complicated by the necessary inclusion of other types of external forces such as those induced by wind gusts or currents. Stability considerations are obviously of great importance in the design of ships, submarines, bathyscaphes, and so forth; such considerations play a significant role in the work of naval architects (see, for example, Ref. 6).

## 2.12    Pressure Variation in a Fluid with Rigid-Body Motion

Although in this chapter we have been primarily concerned with fluids at rest, the general equation of motion (Eq. 2.2)

$$-\nabla p - \gamma \hat{\mathbf{k}} = \rho \mathbf{a}$$

*Even though a fluid may be in motion, if it moves as a rigid body there will be no shearing stresses present.*

was developed for both fluids at rest and fluids in motion, with the only stipulation being that there were no shearing stresses present. Equation 2.2 in component form, based on rectangular coordinates with the positive $z$ axis being vertically upward, can be expressed as

$$-\frac{\partial p}{\partial x} = \rho a_x \qquad -\frac{\partial p}{\partial y} = \rho a_y \qquad -\frac{\partial p}{\partial z} = \gamma + \rho a_z \qquad (2.24)$$

A general class of problems involving fluid motion in which there are no shearing stresses occurs when a mass of fluid undergoes rigid-body motion. For example, if a container of fluid accelerates along a straight path, the fluid will move as a rigid mass (after the initial sloshing motion has died out) with each particle having the same acceleration. Since there is no deformation, there will be no shearing stresses and, therefore, Eq. 2.2 applies. Similarly, if a fluid is contained in a tank that rotates about a fixed axis, the fluid will simply rotate with the tank as a rigid body, and again Eq. 2.2 can be applied to obtain the pressure distribution throughout the moving fluid. Specific results for these two cases (rigid-body uniform motion and rigid-body rotation) are developed in the following two sections. Although problems relating to fluids having rigid-body motion are not, strictly speaking, "fluid statics" problems, they are included in this chapter because, as we will see, the analysis and resulting pressure relationships are similar to those for fluids at rest.

### 2.12.1  Linear Motion

We first consider an open container of a liquid that is translating along a straight path with a constant acceleration **a** as illustrated in Fig. 2.29. Since $a_x = 0$, it follows from the first of Eqs. 2.24 that the pressure gradient in the x direction is zero ($\partial p / \partial x = 0$). In the y and z directions

$$\frac{\partial p}{\partial y} = -\rho a_y \tag{2.25}$$

$$\frac{\partial p}{\partial z} = -\rho(g + a_z) \tag{2.26}$$

The change in pressure between two closely spaced points located at $y$, $z$, and $y + dy$, $z + dz$ can be expressed as

$$dp = \frac{\partial p}{\partial y} dy + \frac{\partial p}{\partial z} dz$$

or in terms of the results from Eqs. 2.25 and 2.26

$$dp = -\rho a_y \, dy - \rho(g + a_z) \, dz \tag{2.27}$$

Along a line of *constant* pressure, $dp = 0$, and therefore from Eq. 2.27 it follows that the slope of this line is given by the relationship

$$\frac{dz}{dy} = -\frac{a_y}{g + a_z} \tag{2.28}$$

This relationship is illustrated by the figure in the margin. Along a free surface the pressure is constant, so that for the accelerating mass shown in Fig. 2.29 the free surface will be inclined if $a_y \neq 0$. In addition, all lines of constant pressure will be parallel to the free surface as illustrated.

■ **Figure 2.29**  Linear acceleration of a liquid with a free surface.

*The pressure distribution in a fluid mass that is accelerating along a straight path is not hydrostatic.*

For the special circumstance in which $a_y = 0$, $a_z \neq 0$, which corresponds to the mass of fluid accelerating in the vertical direction, Eq. 2.28 indicates that the fluid surface will be horizontal. However, from Eq. 2.26 we see that the pressure distribution is not hydrostatic, but is given by the equation

$$\frac{dp}{dz} = -\rho(g + a_z)$$

For fluids of constant density this equation shows that the pressure will vary linearly with depth, but the variation is due to the combined effects of gravity and the externally induced acceleration, $\rho(g + a_z)$, rather than simply the specific weight $\rho g$. Thus, for example, the pressure along the bottom of a liquid-filled tank which is resting on the floor of an elevator that is accelerating upward will be increased over that which exists when the tank is at rest (or moving with a constant velocity). It is to be noted that for a *freely falling* fluid mass ($a_z = -g$), the pressure gradients in all three coordinate directions are zero, which means that if the pressure surrounding the mass is zero, the pressure throughout will be zero. The pressure throughout a "blob" of orange juice floating in an orbiting space shuttle (a form of free fall) is zero. The only force holding the liquid together is surface tension (see Section 1.9).

# EXAMPLE 2.11    Pressure Variation in an Accelerating Tank

**GIVEN**  The cross section for the fuel tank of an experimental vehicle is shown in Fig. E2.11. The rectangular tank is vented to the atmosphere, and the specific gravity of the fuel is $SG = 0.65$. A pressure transducer is located in its side as illustrated. During testing of the vehicle, the tank is subjected to a constant linear acceleration, $a_y$.

**FIND**  (a) Determine an expression that relates $a_y$ and the pressure (in lb/ft$^2$) at the transducer. (b) What is the maximum acceleration that can occur before the fuel level drops below the transducer?

■ **Figure E2.11**

## SOLUTION

(a)  For a constant horizontal acceleration the fuel will move as a rigid body, and from Eq. 2.28 the slope of the fuel surface can be expressed as

$$\frac{dz}{dy} = -\frac{a_y}{g}$$

since $a_z = 0$. Thus, for some arbitrary $a_y$, the change in depth, $z_1$, of liquid on the right side of the tank can be found from the equation

$$-\frac{z_1}{0.75 \text{ ft}} = -\frac{a_y}{g}$$

or

$$z_1 = (0.75 \text{ ft})\left(\frac{a_y}{g}\right)$$

Since there is no acceleration in the vertical, $z$, direction, the pressure along the wall varies hydrostatically as shown by Eq. 2.26. Thus, the pressure at the transducer is given by the relationship

$$p = \gamma h$$

where $h$ is the depth of fuel above the transducer, and therefore

$$p = (0.65)(62.4 \text{ lb/ft}^3)[0.5 \text{ ft} - (0.75 \text{ ft})(a_y/g)]$$

$$= 20.3 - 30.4 \frac{a_y}{g} \qquad \textbf{(Ans)}$$

for $z_1 \leq 0.5$ ft. As written, $p$ would be given in lb/ft$^2$.

(b)  The limiting value for $(a_y)_{max}$ (when the fuel level reaches the transducer) can be found from the equation

$$0.5 \text{ ft} = (0.75 \text{ ft})\left[\frac{(a_y)_{max}}{g}\right]$$

or

$$(a_y)_{max} = \frac{2g}{3}$$

and for standard acceleration of gravity

$$(a_y)_{max} = \tfrac{2}{3}(32.2 \text{ ft/s}^2) = 21.5 \text{ ft/s}^2 \qquad \textbf{(Ans)}$$

**COMMENT**  Note that the pressure in horizontal layers is not constant in this example since $\partial p / \partial y = -\rho a_y \neq 0$. Thus, for example, $p_1 \neq p_2$.

### 2.12.2  Rigid-Body Rotation

After an initial "start-up" transient, a fluid contained in a tank that rotates with a constant angular velocity $\omega$ about an axis as is shown in Fig. 2.30 will rotate with the tank as a rigid body. It is known from elementary particle dynamics that the acceleration of a fluid particle located at a distance $r$ from the axis of rotation is equal in magnitude to $r\omega^2$, and the direction of the acceleration is toward the axis of rotation, as is illustrated in the figure. Since the paths of the fluid particles are circular, it is convenient to use cylindrical polar coordinates $r$, $\theta$, and $z$, defined in the insert in Fig. 2.30. It will be shown in Chapter 6 that in terms of cylindrical coordinates the pressure gradient $\nabla p$ can be expressed as

$$\nabla p = \frac{\partial p}{\partial r}\,\hat{\mathbf{e}}_r + \frac{1}{r}\frac{\partial p}{\partial \theta}\,\hat{\mathbf{e}}_\theta + \frac{\partial p}{\partial z}\,\hat{\mathbf{e}}_z \tag{2.29}$$

Thus, in terms of this coordinate system

$$\mathbf{a}_r = -r\omega^2\,\hat{\mathbf{e}}_r \qquad \mathbf{a}_\theta = 0 \qquad \mathbf{a}_z = 0$$

and from Eq. 2.2

$$\frac{\partial p}{\partial r} = \rho r \omega^2 \qquad \frac{\partial p}{\partial \theta} = 0 \qquad \frac{\partial p}{\partial z} = -\gamma \tag{2.30}$$

These results show that for this type of rigid-body rotation, the pressure is a function of two variables $r$ and $z$, and therefore the differential pressure is

$$dp = \frac{\partial p}{\partial r}\,dr + \frac{\partial p}{\partial z}\,dz$$

or

$$dp = \rho r \omega^2\,dr - \gamma\,dz \tag{2.31}$$

On a horizontal plane ($dz = 0$), it follows from Eq. 2.31 that $dp/dr = \rho\omega^2 r$, which is greater than zero. Hence, as illustrated in the figure in the margin, because of centrifugal acceleration, the pressure increases in the radial direction.

Along a surface of constant pressure, such as the free surface, $dp = 0$, so that from Eq. 2.31 (using $\gamma = \rho g$)

$$\frac{dz}{dr} = \frac{r\omega^2}{g}$$

Integration of this result gives the equation for surfaces of constant pressure as

$$z = \frac{\omega^2 r^2}{2g} + \text{constant} \tag{2.32}$$

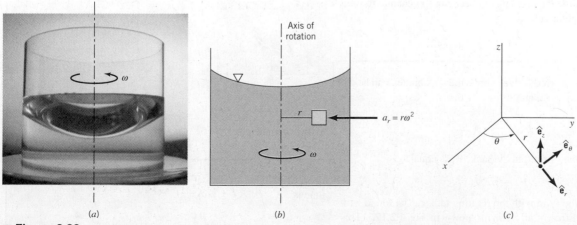

(a)      (b)      (c)

■ **Figure 2.30   Rigid-body rotation of a liquid in a tank. (Photograph courtesy of Geno Pawlak.)**

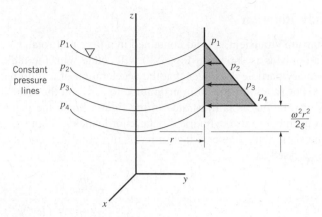

■ **Figure 2.31** Pressure distribution in a rotating liquid.

This equation reveals that these surfaces of constant pressure are parabolic, as illustrated in Fig. 2.31. Integration of Eq. 2.31 yields

*The free surface in a rotating liquid is curved rather than flat.*

$$\int dp = \rho\omega^2 \int r \, dr - \gamma \int dz$$

or

$$p = \frac{\rho\omega^2 r^2}{2} - \gamma z + \text{constant} \tag{2.33}$$

where the constant of integration can be expressed in terms of a specified pressure at some arbitrary point $r_0, z_0$. This result shows that the pressure varies with the distance from the axis of rotation, but at a fixed radius, the pressure varies hydrostatically in the vertical direction as shown in Fig. 2.31.

---

## EXAMPLE 2.12 Free Surface Shape of Liquid in a Rotating Tank

**GIVEN** It has been suggested that the angular velocity, $\omega$, of a rotating body or shaft can be measured by attaching an open cylinder of liquid, as shown in Fig. E2.12a, and measuring with some type of depth gage the change in the fluid level, $H - h_0$, caused by the rotation of the fluid.

**FIND** Determine the relationship between this change in fluid level and the angular velocity.

■ **Figure E2.12**

### SOLUTION

The height, $h$, of the free surface above the tank bottom can be determined from Eq. 2.32, and it follows that

$$h = \frac{\omega^2 r^2}{2g} + h_0$$

The initial volume of fluid in the tank, $\Psi_i$, is equal to

$$\Psi_i = \pi R^2 H$$

The volume of the fluid with the rotating tank can be found with the aid of the differential element shown in Fig. E2.12b. This

cylindrical shell is taken at some arbitrary radius, $r$, and its volume is

$$d\Psi = 2\pi rh \, dr$$

The total volume is, therefore,

$$\mathcal{V} = 2\pi \int_0^R r\left(\frac{\omega^2 r^2}{2g} + h_0\right) dr = \frac{\pi\omega^2 R^4}{4g} + \pi R^2 h_0$$

Since the volume of the fluid in the tank must remain constant (assuming that none spills over the top), it follows that

$$\pi R^2 H = \frac{\pi\omega^2 R^4}{4g} + \pi R^2 h_0$$

or

$$H - h_0 = \frac{\omega^2 R^2}{4g} \tag{Ans}$$

**COMMENT**    This is the relationship we were looking for. It shows that the change in depth could indeed be used to determine the rotational speed, although the relationship between the change in depth and speed is not a linear one.

---

### F l u i d s   i n   t h e   N e w s

**Rotating mercury mirror telescope**    A telescope mirror has the same shape as the parabolic free surface of a *liquid in a rotating tank*. The liquid mirror telescope (LMT) consists of a pan of liquid (normally mercury because of its excellent reflectivity) rotating to produce the required parabolic shape of the free surface mirror. With recent technological advances, it is possible to obtain the vibration-free rotation and the constant angular velocity necessary to produce a liquid mirror surface precise enough for astronomical use. Construction of the largest LMT, located at the University of British Columbia, has recently been completed. With a diameter of 6 ft and a rotation rate of 7 rpm, this mirror uses 30 liters of mercury for its 1-mm thick, parabolic-shaped mirror. One of the major benefits of a LMT (compared to a normal glass mirror telescope) is its low cost. Perhaps the main disadvantage is that a LMT can look only straight up, although there are many galaxies, supernova explosions, and pieces of space junk to view in any part of the sky. The next-generation LMTs may have movable secondary mirrors to allow a larger portion of the sky to be viewed. (See Problem 2.163.)

---

## 2.13  Chapter Summary and Study Guide

In this chapter the pressure variation in a fluid at rest is considered, along with some important consequences of this type of pressure variation. It is shown that for incompressible fluids at rest the pressure varies linearly with depth. This type of variation is commonly referred to as hydrostatic pressure distribution. For compressible fluids at rest the pressure distribution will not generally be hydrostatic, but Eq. 2.4 remains valid and can be used to determine the pressure distribution if additional information about the variation of the specific weight is specified. The distinction between absolute and gage pressure is discussed along with a consideration of barometers for the measurement of atmospheric pressure.

Pressure-measuring devices called manometers, which utilize static liquid columns, are analyzed in detail. A brief discussion of mechanical and electronic pressure gages is also included. Equations for determining the magnitude and location of the resultant fluid force acting on a plane surface in contact with a static fluid are developed. A general approach for determining the magnitude and location of the resultant fluid force acting on a curved surface in contact with a static fluid is described. For submerged or floating bodies the concept of the buoyant force and the use of Archimedes' principle are reviewed.

The following checklist provides a study guide for this chapter. When your study of the entire chapter and end-of-chapter exercises has been completed, you should be able to

*Pascal's law*
*surface force*
*body force*
*incompressible fluid*
*hydrostatic pressure*
  *distribution*
*pressure head*
*compressible fluid*
*U.S. standard*
  *atmosphere*
*absolute pressure*
*gage pressure*
*vacuum pressure*
*barometer*
*manometer*
*Bourdon pressure*
  *gage*
*center of pressure*
*buoyant force*
*Archimedes' principle*
*center of buoyancy*

- write out meanings of the terms listed here in the margin and understand each of the related concepts. These terms are particularly important and are set in *italic, bold, and color* type in the text.
- calculate the pressure at various locations within an incompressible fluid at rest.
- calculate the pressure at various locations within a compressible fluid at rest using Eq. 2.4 if the variation in the specific weight is specified.
- use the concept of a hydrostatic pressure distribution to determine pressures from measurements using various types of manometers.
- determine the magnitude, direction, and location of the resultant hydrostatic force acting on a plane surface.

- determine the magnitude, direction, and location of the resultant hydrostatic force acting on a curved surface.
- use Archimedes' principle to calculate the resultant hydrostatic force acting on floating or submerged bodies.
- analyze, based on Eq. 2.2, the motion of fluids moving with simple rigid-body linear motion or simple rigid-body rotation.

Some of the important equations in this chapter are:

| | | |
|---|---|---|
| Pressure gradient in a stationary fluid | $\dfrac{dp}{dz} = -\gamma$ | **(2.4)** |
| Pressure variation in a stationary incompressible fluid | $p_1 = \gamma h + p_2$ | **(2.7)** |
| Hydrostatic force on a plane surface | $F_R = \gamma h_c A$ | **(2.18)** |
| Location of hydrostatic force on a plane surface | $y_R = \dfrac{I_{xc}}{y_c A} + y_c$ | **(2.19)** |
| | $x_R = \dfrac{I_{xyc}}{y_c A} + x_c$ | **(2.20)** |
| Buoyant force | $F_B = \gamma \forall$ | **(2.22)** |
| Pressure gradient in rigid-body motion | $-\dfrac{\partial p}{\partial x} = \rho a_x, \quad -\dfrac{\partial p}{\partial y} = \rho a_y, \quad -\dfrac{\partial p}{\partial z} = \gamma + \rho a_z$ | **(2.24)** |
| Pressure gradient in rigid-body rotation | $\dfrac{\partial p}{\partial r} = \rho r \omega^2, \quad \dfrac{\partial p}{\partial \theta} = 0, \quad \dfrac{\partial p}{\partial z} = -\gamma$ | **(2.30)** |

# References

1. *The U.S. Standard Atmosphere, 1962*, U.S. Government Printing Office, Washington, DC, 1962.
2. *The U.S. Standard Atmosphere, 1976*, U.S. Government Printing Office, Washington, DC, 1976.
3. Benedict, R. P., *Fundamentals of Temperature, Pressure, and Flow Measurements*, 3rd Ed., Wiley, New York, 1984.
4. Dally, J. W., Riley, W. F., and McConnell, K. G., *Instrumentation for Engineering Measurements*, 2nd Ed., Wiley, New York, 1993.
5. Holman, J. P., *Experimental Methods for Engineers*, 4th Ed., McGraw-Hill, New York, 1983.
6. Comstock, J. P., ed., *Principles of Naval Architecture*, Society of Naval Architects and Marine Engineers, New York, 1967.
7. Hasler, A. F., Pierce, H., Morris, K. R., and Dodge, J., "Meteorological Data Fields 'In Perspective'," *Bulletin of the American Meteorological Society*, Vol. 66, No. 7, July 1985.

---

| | |
|---|---|
| (WILEY) | Problem available in *WileyPLUS* at instructor's discretion. |
| (WILEY GO) | Tutoring problem available in *WileyPLUS* at instructor's discretion. |
| ▶ | Problem is related to a chapter video available in *WileyPLUS*. |
| * | Problem to be solved with aid of programmable calculator or computer. |
| † | Open-ended problem that requires critical thinking. These problems require various assumptions to provide the necessary input data. There are not unique answers to these problems. |

# Review Problems

Go to Appendix G (*WileyPLUS* or the book's web site, www.wiley.com/college/munson) for a set of review problems with answers. Detailed solutions can be found in the *Student Solution Manual and Study Guide for Fundamentals of Fluid Mechanics*, by Munson et al. (© 2013 John Wiley and Sons, Inc.).

# Conceptual Questions

**2.1C** Two tubes connected to two water reservoirs are as shown below. The tube on the left is straight, and that on the right is a cone with the top area four times that of the base. The bottom area of the cone equals that of the straight tube. The height of the water is the same for both cases.

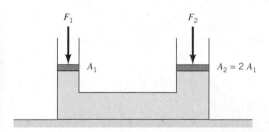

The relation between the pressures $p_1$ and $p_2$ at the base of the tubes is:

a) $p_2 = 4\,p_1$.    b) $p_2 = 2\,p_1$.    c) $p_2 = \dfrac{1}{2}\,p_1$.

d) $p_2 = p_1$.    e) $p_2 = \dfrac{1}{3}\,p_1$.

**2.2C** A system filled with a liquid is shown below. On the left there is a piston in a tube of cross-sectional area $A_1$ with a force $F_1$ applied to it, and on the right there is a piston in a tube of cross-sectional area $A_2$ that is twice that of $A_1$ and a force $F_2$. The pistons are weightless. The two liquid levels are the same.

The relation between the force $F_2$ and $F_1$ is

a) $F_2 = F_1$    b) $F_2 = 4\,F_1$    c) $F_2 = 2\,F_1$

d) $F_2 = F_1/2$    e) $F_2 = F_1/4$

**2.3C** For a fluid element at rest, the forces acting on the element are:

a) gravity, shear, and normal forces.

b) gravity and normal forces.

c) gravity and shear forces.

d) normal and shear forces.

**2.4C** A tank is filled with a liquid, and the surface is exposed to the atmosphere. Which of the following accurately represents the absolute pressure distribution on the right-hand side of the tank?

**2.5C** A wire is attached to a block of metal that is submerged in a tank of water as shown below. The graph that most correctly describes the relation between the force in the wire and time as the block is pulled slowly out of the water is

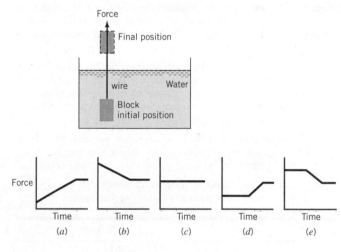

**Additional conceptual questions are available in *WileyPLUS* at the instructor's discretion.**

# Problems

**Note: Unless specific values of required fluid properties are given in the problem statement, use the values found in the tables on the inside of the front cover. Answers to the even-numbered problems are listed at the end of the book. The Lab Problems as well as the videos that accompany problems can be accessed in *WileyPLUS* or the book's web site, www.wiley.com/college/munson.**

## Section 2.3 Pressure Variation in a Fluid at Rest

**2.1** Obtain a photograph/image of a situation in which the fact that in a static fluid the pressure increases with depth is important. Print this photo and write a brief paragraph that describes the situation involved.

**2.2** A closed, 5-m-tall tank is filled with water to a depth of 4 m. The top portion of the tank is filled with air which, as indicated by a pressure gage at the top of the tank, is at a pressure of 20 kPa. Determine the pressure that the water exerts on the bottom of the tank.

**2.3** A closed tank is partially filled with glycerin. If the air pressure in the tank is 6 lb/in.² and the depth of glycerin is 10 ft, what is the pressure in lb/ft² at the bottom of the tank?

**2.4** Blood pressure is usually given as a ratio of the maximum pressure (systolic pressure) to the minimum pressure (diastolic pressure). As shown in **Video V2.3**, such pressures are commonly measured with a mercury manometer. A typical value for this ratio for a human would be 120/70, where the pressures are in mm Hg. **(a)** What would these pressures be in pascals? **(b)** If your car tire was inflated to 120 mm Hg, would it be sufficient for normal driving?

**2.5** (WILEY) An unknown immiscible liquid seeps into the bottom of an open oil tank. Some measurements indicate that the depth of the unknown liquid is 1.5 m and the depth of the oil (specific weight = 8.5 kN/m$^3$) floating on top is 5.0 m. A pressure gage connected to the bottom of the tank reads 65 kPa. What is the specific gravity of the unknown liquid?

**2.6** The water level in an open standpipe is 80 ft above the ground. What is the static pressure at a fire hydrant that is connected to the standpipe and located at ground level? Express your answer in psi.

**2.7** (WILEY) How high a column of SAE 30 oil would be required to give the same pressure as 700 mm Hg?

**2.8** What pressure, expressed in pascals, will a skin diver be subjected to at a depth of 40 m in seawater?

**2.9** (WILEY) Bathyscaphes are capable of submerging to great depths in the ocean. What is the pressure at a depth of 5 km, assuming that seawater has a constant specific weight of 10.1 kN/m$^3$? Express your answer in pascals and psi.

**2.10** For the great depths that may be encountered in the ocean the compressibility of seawater may become an important consideration. **(a)** Assume that the bulk modulus for seawater is constant and derive a relationship between pressure and depth which takes into account the change in fluid density with depth. **(b)** Make use of part (a) to determine the pressure at a depth of 6 km assuming seawater has a bulk modulus of $2.3 \times 10^9$ Pa and a density of 1030 kg/m$^3$ at the surface. Compare this result with that obtained by assuming a constant density of 1030 kg/m$^3$.

**2.11** Sometimes when riding an elevator or driving up or down a hilly road a person's ears "pop" as the pressure difference between the inside and outside of the ear is equalized. Determine the pressure difference (in psi) associated with this phenomenon if it occurs during a 150-ft elevation change.

**2.12** Develop an expression for the pressure variation in a liquid in which the specific weight increases with depth, $h$, as $\gamma = Kh + \gamma_0$, where $K$ is a constant and $\gamma_0$ is the specific weight at the free surface.

**\*2.13** In a certain liquid at rest, measurements of the specific weight at various depths show the following variation:

| $h$ (ft) | $\gamma$ (lb/ft$^3$) |
|:---:|:---:|
| 0 | 70 |
| 10 | 76 |
| 20 | 84 |
| 30 | 91 |
| 40 | 97 |
| 50 | 102 |
| 60 | 107 |
| 70 | 110 |
| 80 | 112 |
| 90 | 114 |
| 100 | 115 |

The depth $h = 0$ corresponds to a free surface at atmospheric pressure. Determine, through numerical integration of Eq. 2.4, the corresponding variation in pressure and show the results on a plot of pressure (in psf) versus depth (in feet).

**†2.14** Because of elevation differences, the water pressure in the second floor of your house is lower than it is in the first floor. For tall buildings this pressure difference can become unacceptable. Discuss possible ways to design the water distribution system in very tall buildings so that the hydrostatic pressure difference is within acceptable limits.

**\*2.15** Under normal conditions the temperature of the atmosphere decreases with increasing elevation. In some situations, however, a temperature inversion may exist so that the air temperature increases with elevation. A series of temperature probes on a mountain give the elevation–temperature data shown in the table below. If the barometric pressure at the base of the mountain is 12.1 psia, determine by means of numerical integration the pressure at the top of the mountain.

| Elevation (ft) | Temperature (°F) |
|:---:|:---:|
| 5000 | 50.1 (base) |
| 5500 | 55.2 |
| 6000 | 60.3 |
| 6400 | 62.6 |
| 7100 | 67.0 |
| 7400 | 68.4 |
| 8200 | 70.0 |
| 8600 | 69.5 |
| 9200 | 68.0 |
| 9900 | 67.1 (top) |

**†2.16** Although it is difficult to compress water, the density of water at the bottom of the ocean is greater than that at the surface because of the higher pressure at depth. Estimate how much higher the ocean's surface would be if the density of seawater were instantly changed to a uniform density equal to that at the surface.

**2.17** (See Fluids in the News article titled **"Giraffe's blood pressure,"** Section 2.3.1.) **(a)** Determine the change in hydrostatic pressure in a giraffe's head as it lowers its head from eating leaves 6 m above the ground to getting a drink of water at ground level as shown in Fig. P2.17. Assume the specific gravity of blood is $SG = 1$. **(b)** Compare the pressure change calculated in part **(a)** to the normal 120 mm of mercury pressure in a human's heart.

6 m

Water

■ **Figure P2.17**

### Section 2.4 Standard Atmosphere

**2.18** What would be the barometric pressure reading, in mm Hg, at an elevation of 4 km in the U.S. standard atmosphere? (Refer to Table C.2 in Appendix C.)

**2.19** An absolute pressure of 7 psia corresponds to what gage pressure for standard atmospheric pressure of 14.7 psia?

**2.20** (WILEY) Assume that a person skiing high in the mountains at an altitude of 15,000 ft takes in the same volume of air with each breath as she does while walking at sea level. Determine the ratio of the mass of oxygen inhaled for each breath at this high altitude compared to that at sea level.

**2.21** (WILEY) Pikes Peak near Denver, Colorado, has an elevation of 14,110 ft. (**a**) Determine the pressure at this elevation, based on Eq. 2.12. (**b**) If the air is assumed to have a constant specific weight of 0.07647 lb/ft³, what would the pressure be at this altitude? (**c**) If the air is assumed to have a constant temperature of 59 °F, what would the pressure be at this elevation? For all three cases assume standard atmospheric conditions at sea level (see Table 2.1).

**2.22** Equation 2.12 provides the relationship between pressure and elevation in the atmosphere for those regions in which the temperature varies linearly with elevation. Derive this equation and verify the value of the pressure given in Table C.2 in Appendix C for an elevation of 5 km.

**2.23** As shown in Fig. 2.6 for the U.S. standard atmosphere, the troposphere extends to an altitude of 11 km where the pressure is 22.6 kPa (abs). In the next layer, called the stratosphere, the temperature remains constant at −56.5 °C. Determine the pressure and density in this layer at an altitude of 15 km. Assume $g = 9.77$ m/s² in your calculations. Compare your results with those given in Table C.2 in Appendix C.

**2.24** (See Fluids in the News article titled **"Weather, barometers, and bars,"** Section 2.5.) The record low sea-level barometric pressure ever recorded is 25.8 in. of mercury. At what altitude in the standard atmosphere is the pressure equal to this value?

## Section 2.5 Measurement of Pressure

**2.25** On a given day, a barometer at the base of the Washington Monument reads 29.97 in. of mercury. What would the barometer reading be when you carry it up to the observation deck 500 ft above the base of the monument?

**2.26** Aneroid barometers can be used to measure changes in altitude. If a barometer reads 30.1 in. Hg at one elevation, what has been the change in altitude in meters when the barometer reading is 28.3 in. Hg? Assume a standard atmosphere and that Eq. 2.12 is applicable over the range of altitudes of interest.

**2.27** ▶ (WILEY) Bourdon gages (see **Video V2.4** and Fig. 2.13) are commonly used to measure pressure. When such a gage is attached to the closed water tank of Fig. P2.27 the gage reads 5 psi. What is the absolute air pressure in the tank? Assume standard atmospheric pressure of 14.7 psi.

**2.28** (WILEY) On the suction side of a pump, a Bourdon pressure gage reads 40 kPa vacuum. What is the corresponding absolute pressure if the local atmospheric pressure is 100 kPa (abs)?

**2.29** A Bourdon pressure gage attached to the outside of a tank containing air reads 77.0 psi when the local atmospheric pressure is 760 mm Hg. What will be the gage reading if the atmospheric pressure increases to 773 mm Hg?

## Section 2.6 Manometry

**2.30** Obtain a photograph/image of a situation in which the use of a manometer is important. Print this photo and write a brief paragraph that describes the situation involved.

**2.31** A water-filled U-tube manometer is used to measure the pressure inside a tank that contains air. The water level in the U-tube on the side that connects to the tank is 5 ft above the base of the tank. The water level in the other side of the U-tube (which is open to the atmosphere) is 2 ft above the base. Determine the pressure within the tank.

**2.32** (WILEY) A barometric pressure of 29.4 in. Hg corresponds to what value of atmospheric pressure in psia, and in pascals?

**2.33** (WILEY) For an atmospheric pressure of 101 kPa (abs) determine the heights of the fluid columns in barometers containing one of the following liquids: (**a**) mercury, (**b**) water, and (**c**) ethyl alcohol. Calculate the heights including the effect of vapor pressure and compare the results with those obtained neglecting vapor pressure. Do these results support the widespread use of mercury for barometers? Why?

**2.34** The closed tank of Fig. P.2.34 is filled with water and is 5 ft long. The pressure gage on the tank reads 7 psi. Determine: (**a**) the height, $h$, in the open water column, (**b**) the gage pressure acting on the bottom tank surface $AB$, and (**c**) the absolute pressure of the air in the top of the tank if the local atmospheric pressure is 14.7 psia.

■ **Figure P2.34**

**2.35** A mercury manometer is connected to a large reservoir of water as shown in Fig. P2.35. Determine the ratio, $h_w/h_m$, of the distances $h_w$ and $h_m$ indicated in the figure.

■ **Figure P2.27**

■ **Figure P2.35**

**2.36** A U-tube mercury manometer is connected to a closed pressurized tank as illustrated in Fig. P2.36. If the air pressure is 2 psi, determine the differential reading, $h$. The specific weight of the air is negligible.

■ **Figure P2.36**

**2.37** (WILEY) A U-tube manometer is connected to a closed tank containing air and water as shown in Fig. P2.37. At the closed end of the manometer the air pressure is 16 psia. Determine the reading on the pressure gage for a differential reading of 4 ft on the manometer. Express your answer in psi (gage). Assume standard atmospheric pressure and neglect the weight of the air columns in the manometer.

■ **Figure P2.37**

**2.38** Compartments A and B of the tank shown in Fig. P2.38 are closed and filled with air and a liquid with a specific gravity equal

■ **Figure P2.38**

to 0.6. Determine the manometer reading, $h$, if the barometric pressure is 14.7 psia and the pressure gage reads 0.5 psi. The effect of the weight of the air is negligible.

**2.39** (WILEY GO) A closed cylindrical tank filled with water has a hemispherical dome and is connected to an inverted piping system as shown in Fig. P2.39. The liquid in the top part of the piping system has a specific gravity of 0.8, and the remaining parts of the system are filled with water. If the pressure gage reading at $A$ is 60 kPa, determine (**a**) the pressure in pipe $B$, and (**b**) the pressure head, in millimeters of mercury, at the top of the dome (point $C$).

■ **Figure P2.39**

**2.40** Two pipes are connected by a manometer as shown in Fig. P2.40. Determine the pressure difference, $p_A - p_B$, between the pipes.

■ **Figure P2.40**

**2.41** An inverted open tank is held in place by a force $R$ as shown in Fig. P2.41. If the specific gravity of the manometer fluid is 2.5, determine the value of $h$.

■ **Figure P2.41**

**2.42** (WILEY) A U-tube manometer is connected to a closed tank as shown in Fig. P2.42. The air pressure in the tank is 0.50 psi and the liquid in the tank is oil ($\gamma = 54.0$ lb/ft$^3$). The pressure at point $A$ is 2.00 psi. Determine: **(a)** the depth of oil, $z$, and **(b)** the differential reading, $h$, on the manometer.

■ **Figure P2.42**

**2.43** For the inclined-tube manometer of Fig. P2.43, the pressure in pipe $A$ is 0.6 psi. The fluid in both pipes $A$ and $B$ is water, and the gage fluid in the manometer has a specific gravity of 2.6. What is the pressure in pipe $B$ corresponding to the differential reading shown?

■ **Figure P2.43**

**2.44** (WILEY) A flowrate measuring device is installed in a horizontal pipe through which water is flowing. A U-tube manometer is connected to the pipe through pressure taps located 3 in. on either side of the device. The gage fluid in the manometer has a specific weight of 112 lb/ft$^3$. Determine the differential reading of the manometer corresponding to a pressure drop between the taps of 0.5 lb/in.$^2$

**2.45** Small differences in gas pressures are commonly measured with a *micromanometer* of the type illustrated in Fig. P2.45. This device consists of two large reservoirs each having a cross-sectional area $A_r$ which are filled with a liquid having a specific weight $\gamma_1$ and connected by a U-tube of cross-sectional area $A_t$ containing a liquid of specific weight $\gamma_2$. When a differential gas

■ **Figure P2.45**

pressure $p_1 - p_2$, is applied, a differential reading, $h$, develops. It is desired to have this reading sufficiently large (so that it can be easily read) for small pressure differentials. Determine the relationship between $h$ and $p_1 - p_2$ when the area ratio $A_t/A_r$ is small, and show that the differential reading, $h$, can be magnified by making the difference in specific weights, $\gamma_2 - \gamma_1$, small. Assume that initially (with $p_1 = p_2$) the fluid levels in the two reservoirs are equal.

**2.46** (WILEY) The cylindrical tank with hemispherical ends shown in Fig. P2.46 contains a volatile liquid and its vapor. The liquid density is 800 kg/m$^3$, and its vapor density is negligible. The pressure in the vapor is 120 kPa (abs) and the atmospheric pressure is 101 kPa (abs). Determine: **(a)** the gage pressure reading on the pressure gage, and **(b)** the height, $h$, of the mercury, manometer.

■ **Figure P2.46**

**2.47** (WILEY) Determine the elevation difference, $\Delta h$, between the water levels in the two open tanks shown in Fig. P2.47.

■ **Figure P2.47**

**2.48** Water, oil, and an unknown fluid are contained in the vertical tubes shown in Fig. P2.48. Determine the density of the unknown fluid.

■ **Figure P2.48**

**2.49**  WILEY  For the configuration shown in Fig. P2.49 what must be the value of the specific weight of the unknown fluid? Express your answer in lb/ft³.

■ **Figure P2.49**

**2.50**  WILEY  An air-filled, hemispherical shell is attached to the ocean floor at a depth of 10 in. as shown in Fig. P2.50. A mercury barometer located inside the shell reads 765 mm Hg, and a mercury U-tube manometer designed to give the outside water pressure indicates differential reading of 735 mm Hg as illustrated. Based on these data, what is the atmospheric pressure at the ocean surface?

■ **Figure P2.50**

**2.51**  Water, oil, and salt water fill a tube as shown in Fig. P2.51. Determine the pressure at point 1 (inside the closed tube).

■ **Figure P2.51**

**\*2.52**  Both ends of the U-tube mercury manometer of Fig. P2.52 are initially open to the atmosphere and under standard atmospheric pressure. When the valve at the top of the right leg is open, the level of mercury below the valve is $h_1$. After the valve is closed, air pressure is applied to the left leg. Determine the relationship between the differential reading on the manometer and the applied gage pressure, $p_g$. Show on a plot how the differential reading varies with $p_g$ for $h_1 = 25, 50, 75,$ and 100 mm over the range $0 \leq p_g \leq 300$ kPa. Assume that the temperature of the trapped air remains constant.

■ **Figure P2.52**

**2.53**  The inverted U-tube manometer of Fig. P2.53 contains oil ($SG = 0.9$) and water as shown. The pressure differential between pipes $A$ and $B$, $p_A - p_B$, is $-5$ kPa. Determine the differential reading $h$.

■ **Figure P2.53**

**2.54**  An inverted U-tube manometer containing oil ($SG = 0.8$) is located between two reservoirs as shown in Fig. P2.54. The reservoir on the left, which contains carbon tetrachloride, is closed and pressurized to 8 psi. The reservoir on the right contains water and is open to the atmosphere. With the given data, determine the depth of water, $h$, in the right reservoir.

■ **Figure P2.54**

**2.55** Three different liquids with properties as indicated fill the tank and manometer tubes as shown in Fig. P2.55. Determine the specific gravity of Fluid 3.

Fluid 1: specific weight = 62.4 lb/ft³
Fluid 2: density = 1.60 slugs/ft³
Fluid 3: specific gravity = ?

■ **Figure P2.55**

**2.56** Determine the pressure of the water in pipe *A* shown in Fig. P2.56 if the gage pressure of the air in the tank is 2 psi.

■ **Figure P2.56**

**2.57** In Fig. P2.57 pipe *A* contains carbon tetrachloride ($SG = 1.60$) and the closed storage tank *B* contains a salt brine ($SG = 1.15$). Determine the air pressure in tank *B* if the pressure in pipe *A* is 25 psi.

■ **Figure P2.57**

**2.58** ⟨WILEY⟩ In Fig. P2.58 pipe *A* contains gasoline ($SG = 0.7$), pipe *B* contains oil ($SG = 0.9$), and the manometer fluid is mercury. Determine the new differential reading if the pressure in pipe *A* is decreased 25 kPa, and the pressure in pipe *B* remains constant. The initial differential reading is 0.30 m as shown.

■ **Figure P2.58**

**2.59** The mercury manometer of Fig. P2.59 indicates a differential reading of 0.30 m when the pressure in pipe *A* is 30-mm Hg vacuum. Determine the pressure in pipe *B*.

■ **Figure P2.59**

**2.60** ⟨WILEY⟩ The inclined differential manometer of Fig. P2.60 contains carbon tetrachloride. Initially the pressure differential between pipes *A* and *B*, which contain a brine ($SG = 1.1$), is zero as illustrated in the figure. It is desired that the manometer give a differential reading of 12 in. (measured along the inclined tube) for a pressure differential of 0.1 psi. Determine the required angle of inclination, $\theta$.

■ **Figure P2.60**

**2.61** The manometer fluid in the manometer of Fig. P2.61 has a specific gravity of 3.46. Pipes *A* and *B* both contain water. If the pressure in pipe *A* is decreased by 1.3 psi and the pressure in pipe *B* increases by 0.9 psi, determine the new differential reading of the manometer.

Water

2 ft

Water

1 ft

1 ft

Gage fluid
(SG = 3.46)

■ **Figure P2.61**

**2.62** (WILEY●) Determine the new differential reading along the inclined leg of the mercury manometer of Fig. P2.2, if the pressure in pipe A is decreased 10 kPa and the pressure in pipe B remains unchanged. The fluid in A has a specific gravity of 0.9 and the fluid in B is water.

A
SG = 0.9

Water

30°

100 mm

B

80 mm

50 mm

Mercury

■ **Figure P2.62**

**2.63** Determine the ratio of areas, $A_1/A_2$, of the two manometer legs of Fig. P2.63 if a change in pressure in pipe B of 0.5 psi gives a corresponding change of 1 in. in the level of the mercury in the right leg. The pressure in pipe A does not change.

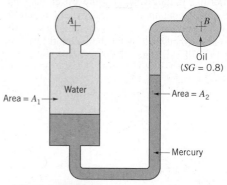

A

B

Oil
(SG = 0.8)

Water

Area = $A_1$

Area = $A_2$

Mercury

■ **Figure P2.63**

**2.64** (WILEY●) Determine the change in the elevation of the mercury in the left leg of the manometer of Fig. P2.64 as a result of an in-

Water

Oil (SG = 0.9)

18 in.

6 in.

12 in.

$\frac{1}{4}$-in.-diameter

30°

$\frac{1}{2}$-in.-diameter

Mercury

■ **Figure P2.64**

crease in pressure of 5 psi in pipe A while the pressure in pipe B remains constant.

**2.65** The U-shaped tube shown in Fig. 2.65 initially contains water only. A second liquid with specific weight, $\gamma$, less than water is placed on top of the water with no mixing occurring. Can the height, $h$, of the second liquid be adjusted so that the left and right levels are at the same height? Provide proof of your answer.

$\gamma$

$h$

$D_1 = 1.5\ D_2$

$D_2$

Water

■ **Figure P2.65**

**\*2.66** An inverted hollow cylinder is pushed into the water as is shown in Fig. P2.66. Determine the distance, $\ell$, that the water rises in the cylinder as a function of the depth, $d$, of the lower edge of the cylinder. Plot the results for $0 \le d \le H$, when $H$ is equal to 1 m. Assume the temperature of the air within the cylinder remains constant.

H

Water

d

D

$\ell$

Open end

■ **Figure P2.66**

## Section 2.8 Hydrostatic Force on a Plane Surface (also see Lab Problems 2.1LP, 2.2LP, 2.3LP, and 2.4LP)

**2.67** Obtain a photograph/image of a situation in which the hydrostatic force on a plane surface is important. Print this photo and write a brief paragraph that describes the situation involved.

**2.68** (WILEY●) The basic elements of a hydraulic press are shown in Fig. P2.68. The plunger has an area of 1 in.$^2$, and a force, $F_1$, can be applied to the plunger through a lever mechanism having a mechanical advantage of 8 to 1. If the large piston has an area of 150 in.$^2$, what load, $F_2$, can be raised by a force of 30 lb applied to the lever? Neglect the hydrostatic pressure variation.

$F_2$

Plunger

$F_1$

Hydraulic fluid

■ **Figure P2.68**

**2.69** A 0.3-m-diameter pipe is connected to a 0.02-m-diameter pipe, and both are rigidly held in place. Both pipes are horizontal with pistons at each end. If the space between the pistons is filled with water, what force will have to be applied to the larger piston to balance a force of 90 N applied to the smaller piston? Neglect friction.

**\*2.70** ▶ A Bourdon gage (see Fig. 2.13 and **Video V2.4**) is often used to measure pressure. One way to calibrate this type of gage is to use the arrangement shown in Fig. P2.70a. The container is filled with a liquid and a weight, $W$, placed on one side with the gage on the other side. The weight acting on the liquid through a 0.4-in.-diameter opening creates a pressure that is transmitted to the gage. This arrangement, with a series of weights, can be used to determine what a change in the dial movement, $\theta$, in Fig. P2.70b, corresponds to in terms of a change in pressure. For a particular gage, some data are given below. Based on a plot of these data, determine the relationship between $\theta$ and the pressure, $p$, where $p$ is measured in psi.

| $W$ (lb) | 0 | 1.04 | 2.00 | 3.23 | 4.05 | 5.24 | 6.31 |
|---|---|---|---|---|---|---|---|
| $\theta$ (deg.) | 0 | 20 | 40 | 60 | 80 | 100 | 120 |

(b)          (a)

■ **Figure P2.70**

**2.71** An inverted 0.1-m-diameter circular cylinder is partially filled with water and held in place as shown in Fig. P2.71. A force of 20 N is needed to pull the flat plate from the cylinder. Determine the air pressure within the cylinder. The plate is not fastened to the cylinder and has negligible mass.

■ **Figure P2.71**

**2.72** You partially fill a glass with water, place an index card on top of the glass, and then turn the glass upside down while holding the card in place. You can then remove your hand from the card and the card remains in place, holding the water in the glass. Explain how this works.

**2.73** ⓦ A piston having a cross-sectional area of 0.07 m² is located in a cylinder containing water as shown in Fig. P2.73. An open U-tube manometer is connected to the cylinder as shown. For $h_1 = 60$ mm and $h = 100$ mm, what is the value of the applied force, $P$, acting on the piston? The weight of the piston is negligible.

■ **Figure P2.73**

**2.74** ⓦ A 6-in.-diameter piston is located within a cylinder that is connected to a $\frac{1}{2}$-in.-diameter inclined-tube manometer as shown in Fig. P2.74. The fluid in the cylinder and the manometer is oil (specific weight = 59 lb/ft³). When a weight, $W$, is placed on the top of the cylinder, the fluid level in the manometer tube rises from point (1) to (2). How heavy is the weight? Assume that the change in position of the piston is negligible.

■ **Figure P2.74**

**2.75** A square gate (4 m by 4 m) is located on the 45° face of a dam. The top edge of the gate lies 8 m below the water surface. Determine the force of the water on the gate and the point through which it acts.

**2.76** A large, open tank contains water and is connected to a 6-ft-diameter conduit as shown in Fig. P2.76. A circular plug is used to seal the conduit. Determine the magnitude, direction, and location of the force of the water on the plug.

■ **Figure P2.76**

**2.77** A circular 2-m-diameter gate is located on the sloping side of a swimming pool. The side of the pool is oriented 60° relative to the horizontal bottom, and the center of the gate is located 3 m below the water surface. Determine the magnitude of the water force acting on the gate and the point through which it acts.

**2.78** A vertical rectangular gate is 8 ft wide and 10 ft long and weighs 6000 lb. The gate slides in vertical slots in the side of a reservoir containing water. The coefficient of friction between the slots and the gate is 0.03. Determine the minimum vertical force required to lift the gate when the water level is 4 ft above the top edge of the gate.

**2.79** A horizontal 2-m-diameter conduit is half filled with a liquid ($SG = 1.6$) and is capped at both ends with plane vertical surfaces. The air pressure in the conduit above the liquid surface is 200 kPa. Determine the resultant force of the fluid acting on one of the end caps, and locate this force relative to the bottom of the conduit.

**2.80** Concrete is poured into the forms as shown in Fig. P2.80 to produce a set of steps. Determine the weight of the sandbag needed to keep the bottomless forms from lifting off the ground. The weight of the forms is 85 lb, and the specific weight of the concrete is 150 lb/ft³.

■ **Figure P2.80**

**2.81** Forms used to make a concrete basement wall are shown in Fig. P2.81. Each 4-ft-long form is held together by four ties—two at the top and two at the bottom as indicated. Determine the tension in the upper and lower ties. Assume concrete acts as a fluid with a weight of 150 lb/ft³.

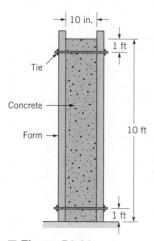

■ **Figure P2.81**

**2.82** ⓦILEY⊙ A structure is attached to the ocean floor as shown in Fig. P2.82. A 2-m-diameter hatch is located in an inclined wall and

■ **Figure P2.82**

hinged on one edge. Determine the minimum air pressure, $p_1$, within the container that will open the hatch. Neglect the weight of the hatch and friction in the hinge.

**2.83** An open rectangular container contains a liquid that has a specific weight that varies according to the equation $\gamma = c_1 + c_2 h$, where $c_1$ and $c_2$ are constants and $h$ is a vertical coordinate measured downward from the free surface. Derive an equation for the magnitude of the liquid force exerted on one wall of the container having a width $b$ and height $H$ and an equation that gives the vertical coordinate of this force.

**2.84** ⓦILEY⊙ A long, vertical wall separates seawater from fresh water. If the seawater stands at a depth of 7 m, what depth of freshwater is required to give a zero resultant force on the wall? When the resultant force is zero, will the moment due to the fluid forces be zero? Explain.

**2.85** A gate having the shape shown in Fig. P2.85 is located in the vertical side of an open tank containing water. The gate is mounted on a horizontal shaft. (**a**) When the water level is at the top of the gate, determine the magnitude of the fluid force on the rectangular portion of the gate above the shaft and the magnitude of the fluid force on the semicircular portion of the gate below the shaft. (**b**) For this same fluid depth determine the moment of the force acting on the semicircular portion of the gate with respect to an axis that coincides with the shaft.

■ **Figure P2.85**

**2.86** A pump supplies water under pressure to a large tank as shown in Fig. P2.86. The circular-plate valve fitted in the short discharge pipe on the tank pivots about its diameter $A$–$A$ and is held shut against the water pressure by a latch at $B$. Show that the force on the latch is independent of the supply pressure, $p$, and the height of the tank, $h$.

■ **Figure P2.86**

**2.87** (WILEY GO) A homogeneous, 4-ft-wide, 8-ft-long rectangular gate weighing 800 lb is held in place by a horizontal flexible cable as shown in Fig. P2.87. Water acts against the gate, which is hinged at point A. Friction in the hinge is negligible. Determine the tension in the cable.

■ **Figure P2.87**

**2.88** A rectangular gate 6 ft tall and 5 ft wide in the side of an open tank is held in place by the force $F$ as indicated in Fig. P.2.88. The weight of the gate is negligible, and the hinge at O is frictionless. (**a**) Determine the water depth, $h$, if the resultant hydrostatic force of the water acts 2.5 ft above the bottom of the gate, i.e., it is collinear with the applied force $F$. (**b**) For the depth of part (a), determine the magnitude of the resultant hydrostatic force. (**c**) Determine the force that the hinge puts on the gate under the above conditions.

■ **Figure P2.88**

†**2.89** Sometimes it is difficult to open an exterior door of a building because the air distribution system maintains a pressure difference between the inside and outside of the building. Estimate how big this pressure difference can be if it is "not too difficult" for an average person to open the door.

**2.90** An area in the form of an isosceles triangle with a base width of 6 ft and an attitude of 8 ft lies in the plane forming one wall of a tank that contains a liquid having a specific weight of 79.8 lb/ft³. The side slopes upward, making an angle of 60° with the horizontal. The base of the triangle is horizontal, and the vertex is above the base. Determine the resultant force the fluid exerts on the area when the fluid depth is 20 ft above the base of the triangular area. Show, with the aid of a sketch, where the center of pressure is located.

**2.91** Solve Problem 2.90 if the isosceles triangle is replaced with a right triangle having the same base width and altitude as the isosceles triangle.

**2.92** (WILEY) A vertical plane area having the shape shown in Fig. P2.92 is immersed in an oil bath (specific weight = 8.75 kN/m³). Determine the magnitude of the resultant force acting on one side of the area as a result of the oil.

■ **Figure P2.92**

**2.93** (WILEY GO) A 3-m-wide, 8-m-high rectangular gate is located at the end of a rectangular passage that is connected to a large open tank filled with water as shown in Fig. P2.93. The gate is hinged at its bottom and held closed by a horizontal force, $F_H$, located at the center of the gate. The maximum value for $F_H$ is 3500 kN. (**a**) Determine the maximum water depth, $h$, above the center of the gate that can exist without the gate opening. (**b**) Is the answer the same if the gate is hinged at the top? Explain your answer.

■ **Figure P2.93**

**2.94** Two square gates close two openings in a conduit connected to an open tank of water as shown in Fig. P2.94. When the water depth, $h$, reaches 5 m it is desired that both gates open at the same time. Determine the weight of the homogeneous horizontal gate and the horizontal force, $R$, acting on the vertical gate that is required to keep the gates closed until this depth is reached. The weight of the vertical gate is negligible, and both gates are hinged at one end as shown. Friction in the hinges is negligible.

■ **Figure P2.94**

**2.95** (WILEY) A gate having the cross section shown in Fig. 2.95 closes an opening 5 ft wide and 4 ft high in a water reservoir. The gate weighs 500 lb, and its center of gravity is 1 ft to the left of AC and 2 ft above BC. Determine the horizontal reaction that is developed on the gate at C.

■ **Figure P2.95**

**2.96** (WILEY) A gate having the cross section shown in Fig. P2.96 is 4 ft wide and is hinged at $C$. The gate weighs 18,000 lb, and its mass center is 1.67 ft to the right of the plane $BC$. Determine the vertical reaction at $A$ on the gate when the water level is 3 ft above the base. All contact surfaces are smooth.

■ **Figure P2.96**

**2.97** (WILEY) The massless, 4-ft-wide gate shown in Fig. P2.97 pivots about the frictionless hinge O. It is held in place by the 2000 lb counterweight, $\mathcal{W}$. Determine the water depth, $h$.

■ **Figure P2.97**

**\*2.98** A 200-lb homogeneous gate 10 ft wide and 5 ft long is hinged at point $A$ and held in place by a 12-ft-long brace as shown in Fig. P2.98. As the bottom of the brace is moved to the right, the water level remains at the top of the gate. The line of action of the

■ **Figure P2.98**

force that the brace exerts on the gate is along the brace. (**a**) Plot the magnitude of the force exerted on the gate by the brace as a function of the angle of the gate, $\theta$, for $0 \le \theta \le 90°$. (**b**) Repeat the calculations for the case in which the weight of the gate is negligible. Common on the result as $\theta \to 0$.

**2.99** (WILEY) An open tank has a vertical partition and on one side contains gasoline with a density $\rho = 700$ kg/m³ at a depth of 4 m, as shown in Fig. P2.99. A rectangular gate that is 4 m high and 2 m wide and hinged at one end is located in the partition. Water is slowly added to the empty side of the tank. At what depth, $h$, will the gate start to open?

■ **Figure P2.99**

**2.100** (WILEY) A 4-ft by 3-ft massless rectangular gate is used to close the end of the water tank shown in Fig. P2.100. A 200-lb weight attached to the arm of the gate at a distance $\ell$ from the frictionless hinge is just sufficient to keep the gate closed when the water depth is 2 ft, that is, when the water fills the semicircular lower portion of the tank. If the water were deeper, the gate would open. Determine the distance $\ell$.

■ **Figure P2.100**

**2.101** The rigid gate, $OAB$, of Fig. P2.101 is hinged at $O$ and rests against a rigid support at $B$. What minimum horizontal force, $P$, is required to hold the gate closed if its width is 3 m? Neglect the weight of the gate and friction in the hinge. The back of the gate is exposed to the atmosphere.

**2.102** (WILEY) A rectangular gate that is 2 m wide is located in the vertical wall of a tank containing water as shown in Fig. P2.102. It is desired to have the gate open automatically when the depth of water above the top of the gate reaches 10 m. (**a**) At what distance, $d$, should the frictionless horizontal shaft be located? (**b**) What is the magnitude of the force on the gate when it opens?

■ **Figure P2.101**

■ **Figure P2.102**

**2.103** (WILEY GO) A thin 4-ft-wide, right-angle gate with negligible mass is free to pivot about a frictionless hinge at point $O$, as shown in Fig. P2.103. The horizontal portion of the gate covers a 1-ft-diameter drain pipe that contains air at atmospheric pressure. Determine the minimum water depth, $h$, at which the gate will pivot to allow water to flow into the pipe.

■ **Figure P2.103**

**2.104** The inclined face $AD$ of the tank of Fig. P2.104 is a plane surface containing a gate $ABC$, which is hinged along line $BC$. The shape of the gate is shown in the plan view. If the tank contains water, determine the magnitude of the force that the water exerts on the gate.

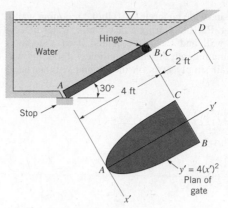

■ **Figure P2.104**

**2.105** (WILEY) An open rectangular tank is 2 m wide and 4 m long. The tank contains water to a depth of 2 m and oil ($SG = 0.8$) on top of the water to a depth of 1 m. Determine the magnitude and location of the resultant fluid force acting on one end of the tank.

**\*2.106** An open rectangular settling tank contains a liquid suspension that at a given time has a specific weight that varies approximately with depth according to the following data:

| $h$ (m) | $\gamma$ (N/m³) |
|---------|----------------|
| 0       | 10.0           |
| 0.4     | 10.1           |
| 0.8     | 10.2           |
| 1.2     | 10.6           |
| 1.6     | 11.3           |
| 2.0     | 12.3           |
| 2.4     | 12.7           |
| 2.8     | 12.9           |
| 3.2     | 13.0           |
| 3.6     | 13.1           |

The depth $h = 0$ corresponds to the free surface. Determine, by means of numerical integration, the magnitude and location of the resultant force that the liquid suspension exerts on a vertical wall of the tank that is 6 m wide. The depth of fluid in the tank is 3.6 m.

**2.107** (WILEY) The closed vessel of Fig. P2.107 contains water with an air pressure of 10 psi at the water surface. One side of the vessel contains a spout that is closed by a 6-in.-diameter circular gate that is hinged along one side as illustrated. The horizontal axis of the hinge is located 10 ft below the water surface. Determine the minimum torque that must be applied at the hinge to hold the gate shut. Neglect the weight of the gate and friction at the hinge.

■ **Figure P2.107**

**2.108** A 4-ft-tall, 8-in.-wide concrete (150 lb/ft³) retaining wall is built as shown in Fig. P2.108. During a heavy rain, water fills the space between the wall and the earth behind it to a depth $h$. Determine the maximum depth of water possible without the wall tipping over. The wall simply rests on the ground without being anchored to it.

■ **Figure P2.108**

**2.109** Water backs up behind a concrete dam as shown in Fig. 2.109. Leakage under the foundation gives a pressure distribution under the dam as indicated. If the water depth, $h$, is too great, the dam will topple over about its toe (point $A$). For the dimensions given, determine the maximum water depth for the following widths of the dam, $\ell = 20$, 30, 40, 50, and 60 ft. Base your analysis on a unit length of the dam. The specific weight of the concrete is 150 lb/ft³.

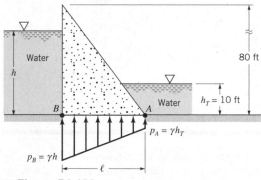

■ **Figure P2.109**

**2.110** ⟨WILEY⟩ The concrete dam of Fig. P2.110 weighs 23.6 kN/m³ and rests on a solid foundation. Determine the minimum coefficient of friction between the dam and the foundation required to keep the dam from sliding at the water depth shown. Assume no fluid uplift pressure along the base. Base your analysis on a unit length of the dam.

■ **Figure P2.110**

**2.111** (See Fluids in the News article titled **"The Three Gorges Dam,"** Section 2.8.) (**a**) Determine the horizontal hydrostatic force on the 2309-m-long Three Gorges Dam when the average depth of the water against it is 175 m. (**b**) If all of the 6.4 billion people on Earth were to push horizontally against the Three Gorges Dam, could they generate enough force to hold it in place? Support your answer with appropriate calculations.

### Section 2.10 Hydrostatic Force on a Curved Surface

**2.112** Obtain a photograph/image of a situation in which the hydrostatic force on a curved surface is important. Print this photo and write a brief paragraph that describes the situation involved.

**2.113** ⟨WILEY⟩ A 2-ft-diameter hemispherical plexiglass "bubble" is to be used as a special window on the side of an above-ground swimming pool. The window is to be bolted onto the vertical wall of the pool and faces outward, covering a 2-ft-diameter opening in the wall. The center of the opening is 4 ft below the surface. Determine the horizontal and vertical components of the force of the water on the hemisphere.

**2.114** Two round, open tanks containing the same type of fluid rest on a table top as shown in Fig. P2.114. They have the same bottom area, $A$, but different shapes. When the depth, $h$, of the liquid in the two tanks is the same, the pressure force of the liquids on the bottom of the two tanks is the same. However, the force that the table exerts on the two tanks is different because the weight in each of the tanks is different. How do you account for this apparent paradox?

■ **Figure P2.114**

**2.115** Two hemispherical shells are bolted together as shown in Fig. P2.115. The resulting spherical container, which weighs 300 lb, is filled with mercury and supported by a cable as shown. The container is vented at the top. If eight bolts are symmetrically located around the circumference, what is the vertical force that each bolt must carry?

■ **Figure P2.115**

**2.116** A 4-m-long curved gate is located in the side of a reservoir containing water as shown in Fig. P2.116. Determine the magnitude of the horizontal and vertical components of the force of the water on the gate. Will this force pass through point *A*? Explain.

■ **Figure P2.116**

**2.117** (WILEY) The 18-ft-long lightweight gate of Fig. P2.117 is a quarter circle and is hinged at *H*. Determine the horizontal force, *P*, required to hold the gate in place. Neglect friction at the hinge and the weight of the gate.

■ **Figure P2.117**

**2.118** ▶ The air pressure in the top of the 2-liter pop bottle shown in **Video V2.6** and Fig. P2.118 is 40 psi, and the pop depth is 10 in. The bottom of the bottle has an irregular shape with a diameter of 4.3 in. (**a**) If the bottle cap has a diameter of 1 in. what is the magnitude of the axial force required to hold the cap in place? (**b**) Determine the force needed to secure the bottom 2 in. of the bottle to its cylindrical sides. For this calculation assume the effect of the weight of the pop is negligible. (**c**) By how much does the weight of the pop increase the pressure 2 in. above the bottom? Assume the pop has the same specific weight as that of water.

■ **Figure P2.118**

**2.119** A tank wall has the shape shown in Fig. P2.119. Determine the horizontal and vertical components of the force of the water on a 4-ft length of the curved section *AB*.

■ **Figure P2.119**

**2.120** ▶ Hoover Dam (see **Video 2.5**) is the highest arch-gravity type of dam in the United States. A cross section of the dam is shown in Fig. P2.120(*a*). The walls of the canyon in which the dam is located are sloped, and just upstream of the dam the vertical plane shown in Figure P2.120(*b*) approximately represents the cross section of the water acting on the dam. Use this vertical cross section to estimate the resultant horizontal force of the water on the dam, and show where this force acts.

■ **Figure P2.120**

**2.121** (WILEY) A plug in the bottom of a pressurized tank is conical in shape, as shown in Fig. P2.121. The air pressure is 40 kPa, and the liquid in the tank has a specific weight of 27 kN/m³. Determine the magnitude, direction, and line of action of the force exerted on the curved surface of the cone within the tank due to the 40-kPa pressure and the liquid.

■ **Figure P2.121**

**2.122** (WILEY) The homogeneous gate shown in Fig. P2.122 consists of one quarter of a circular cylinder and is used to maintain a water depth of 4 m. That is, when the water depth exceeds 4 m, the gate

■ **Figure P2.122**

■ **Figure P2.128**

opens slightly and lets the water flow under it. Determine the weight of the gate per meter of length.

**2.123** (WILEY) The concrete (specific weight = 150 lb/ft³) seawall of Fig. P2.123 has a curved surface and restrains seawater at a depth of 24 ft. The trace of the surface is a parabola as illustrated. Determine the moment of the fluid force (per unit length) with respect to an axis through the toe (point A).

■ **Figure P2.123**

**2.124** A 12-in.-diameter pipe contains a gas under a pressure of 140 psi. If the pipe wall thickness is $\frac{1}{4}$ in., what is the average circumferential stress developed in the pipe wall?

**2.125** A cylindrical tank with its axis horizontal has a diameter of 2.0 m and a length of 4.0 m. The ends of the tank are vertical planes. A vertical 0.1-m-diameter pipe is connected to the top of the tank. The tank and the pipe are filled with ethyl alcohol to a level of 1.5 m above the top of the tank. Determine the resultant force of the alcohol on one end of the tank and show where it acts.

**2.126** Imagine the tank of Problem 2.125 split by a horizontal plane. Determine the magnitude of the resultant force of the alcohol on the bottom half of the tank.

**2.127** If the tank ends in Problem 2.125 are hemispherical, what is the magnitude of the resultant horizontal force of the alcohol on one of the curved ends?

**2.128** (WILEY) An open tank containing water has a bulge in its vertical side that is semicircular in shape as shown in Fig. P2.128. Determine the horizontal and vertical components of the force that the water exerts on the bulge. Base your analysis on a 1-ft length of the bulge.

**2.129** (WILEY) A closed tank is filled with water and has a 4-ft-diameter hemispherical dome as shown in Fig. 2.129. A U-tube manometer is connected to the tank. Determine the vertical force of the water on the dome if the differential manometer reading is 7 ft and the air pressure at the upper end of the manometer is 12.6 psi.

■ **Figure P2.129**

**2.130** (WILEY) A 3-m-diameter open cylindrical tank contains water and has a hemispherical bottom as shown in Fig. P2.130. Determine the magnitude, line of action, and direction of the force of the water on the curved bottom.

■ **Figure P2.130**

**2.131** Three gates of negligible weight are used to hold back water in a channel of width b as shown in Fig. P2.131. The force of the gate against the block for gate (b) is R. Determine (in terms of R) the force against the blocks for the other two gates.

### Section 2.11 Buoyancy, Flotation, and Stability

**2.132** Obtain a photograph/image of a situation in which Archimede's principle is important. Print this photo and write a brief paragraph that describes the situation involved.

**2.133** A freshly cut log floats with one fourth of its volume protruding above the water surface. Determine the specific weight of the log.

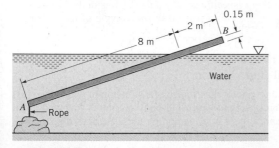

**■ Figure P2.131**

**2.134** A 3 ft × 3 ft × 3 ft wooden cube (specific weight = 37 lb/ft³) floats in a tank of water. How much of the cube extends above the water surface? If the tank were pressurized so that the air pressure at the water surface was increased by 1.0 psi (i.e., 1 psig), how much of the cube would extend above the water surface? Explain how you arrived at your answer.

**2.135** The homogeneous timber *AB* of Fig. P2.135 is 0.15 m by 0.35 m in cross section. Determine the specific weight of the timber and the tension in the rope.

**■ Figure P2.135**

**2.136** ⟨WILEY⟩ A river barge, whose cross section is approximately rectangular, carries a load of grain. The barge is 28 ft wide and 90 ft long. When unloaded, its draft (depth of submergence) is 5 ft. and with the load of grain the draft is 7 ft. Determine: (**a**) the unloaded weight of the barge, and (**b**) the weight of the grain.

**2.137** A tank of cross-sectional area *A* is filled with a liquid of specific weight $\gamma_1$ as shown in Fig. P2.137*a*. Show that when a cylinder of specific weight $\gamma_2$ and volume $V$ is floated in the liquid (see Fig. P2.137*b*), the liquid level rises by an amount $\Delta h = (\gamma_2/\gamma_1) \, V/A$.

**2.138** ⟨WILEY GO⟩ When the Tucurui Dam was constructed in northern Brazil, the lake that was created covered a large forest of valuable hardwood trees. It was found that even after 15 years underwater the trees were perfectly preserved and underwater logging was started. During the logging process a tree is selected, trimmed, and anchored with ropes to prevent it from shooting to the surface like a missile when cut. Assume that a typical large tree can be approximated as a truncated cone with a base diameter of 8 ft, a top diameter of 2 ft, and a height of 100 ft. Determine the resultant vertical force that the ropes must resist when the completely submerged tree is cut. The specific gravity of the wood is approximately 0.6.

**†2.139** Estimate the minimum water depth needed to float a canoe carrying two people and their camping gear. List all assumptions and show all calculations.

**2.140** ▶ An inverted test tube partially filled with air floats in a plastic water-filled soft drink bottle as shown in **Video V2.7** and Fig. P2.140. The amount of air in the tube has been adjusted so that it just floats. The bottle cap is securely fastened. A slight squeezing of the plastic bottle will cause the test tube to sink to the bottom of the bottle. Explain this phenomenon.

**■ Figure P2.140**

**2.141** An irregularly shaped piece of a solid material weighs 8.05 lb in air and 5.26 lb when completely submerged in water. Determine the density of the material.

**2.142** A 1-ft-diameter, 2-ft-long cylinder floats in an open tank containing a liquid having a specific weight $\gamma$. A U-tube manometer is connected to the tank as shown in Fig. P2.142. When the

**■ Figure P2.137**

**■ Figure P2.142**

pressure in pipe *A* is 0.1 psi below atmospheric pressure, the various fluid levels are as shown. Determine the weight of the cylinder. Note that the top of the cylinder is flush with the fluid surface.

**2.143** A 1-m-diameter cylindrical mass, *M*, is connected to a 2-m-wide rectangular gate as shown in Fig. P2.143. The gate is to open when the water level, *h*, drops below 2.5 m. Determine the required value for *M*. Neglect friction at the gate hinge and the pulley.

■ **Figure P2.143**

**2.144** The thin-walled, 1-m-diameter tank of Fig. P2.144 is closed at one end and has a mass of 90 kg. The open end of the tank is lowered into the water and held in the position shown by a steel block having a density of 7840 kg/m³. Assume that the air that is trapped in the tank is compressed at a constant temperature. Determine: (**a**) the reading on the pressure gage at the top of the tank, and (**b**) the volume of the steel block.

■ **Figure P2.144**

**2.145** ▶ ⓦ When a hydrometer (see Fig. P2.145 and **Video V2.8**) having a stem diameter of 0.30 in. is placed in water, the stem protrudes 3.15 in. above the water surface. If the water is

■ **Figure P2.145**

replaced with a liquid having a specific gravity of 1.10, how much of the stem would protrude above the liquid surface? The hydrometer weighs 0.042 lb.

**2.146** ⓦ A 2-ft-thick block constructed of wood (*SG* = 0.6) is submerged in oil (*SG* = 0.8) and has a 2-ft-thick aluminum (specific weight = 168 lb/ft³) plate attached to the bottom as indicated in Fig. P2.146. Determine completely the force required to hold the block in the position shown. Locate the force with respect to point *A*.

■ **Figure P2.146**

**2.147** ⓦ (See Fluids in the News article titled "**Concrete canoes,**" Section 2.11.1.) How much extra water does a 147-lb concrete canoe displace compared to an ultralightweight 38-lb Kevlar canoe of the same size carrying the same load?

**2.148** An iceberg (specific gravity 0.917) floats in the ocean (specific gravity 1.025). What percent of the volume of the iceberg is under water?

### Section 2.12 Pressure Variation in a Fluid with Rigid-Body Motion

**2.149** Obtain a photograph/image of a situation in which the pressure variation in a fluid with rigid-body motion is involved. Print this photo and write a brief paragraph that describes the situation involved.

**2.150** It is noted that while stopping, the water surface in a glass of water sitting in the cup holder of a car is slanted at an angle of 15° relative to the horizontal street. Determine the rate at which the car is decelerating.

**2.151** ⓦ An open container of oil rests on the flatbed of a truck that is traveling along a horizontal road at 55 mi/hr. As the truck slows uniformly to a complete stop in 5 s, what will be the slope of the oil surface during the period of constant deceleration?

**2.152** ⓦ A 5-gal, cylindrical open container with a bottom area of 120 in.² is filled with glycerin and rests on the floor of an elevator. (**a**) Determine the fluid pressure at the bottom of the container when the elevator has an upward acceleration of 3 ft/s². (**b**) What resultant force does the container exert on the floor of the elevator during this acceleration? The weight of the container is negligible. (Note: 1 gal = 231 in.³)

**2.153** An open rectangular tank 1 m wide and 2 m long contains gasoline to a depth of 1 m. If the height of the tank sides is 1.5 m, what is the maximum horizontal acceleration (along the long axis of the tank) that can develop before the gasoline would begin to spill?

**2.154** If the tank of Problem 2.153 slides down a frictionless plane that is inclined at 30° with the horizontal, determine the angle the free surface makes with the horizontal.

**2.155** (WILEY) A closed cylindrical tank that is 8 ft in diameter and 24 ft long is completely filled with gasoline. The tank, with its long axis horizontal, is pulled by a truck along a horizontal surface. Determine the pressure difference between the ends (along the long axis of the tank) when the truck undergoes an acceleration of 5 ft/s$^2$.

**2.156** The open U-tube of Fig. P2.156 is partially filled with a liquid. When this device is accelerated with a horizontal acceleration $a$, a differential reading $h$ develops between the manometer legs which are spaced a distance $\ell$ apart. Determine the relationship between $a$, $\ell$, and $h$.

■ **Figure P2.156**

**2.157** The U-tube of Fig. P2.157 is partially filled with water and rotates around the axis $a$–$a$. Determine the angular velocity that will cause the water to start to vaporize at the bottom of the tube (point $A$).

■ **Figure P2.157**

**2.158** The U-tube of Fig. P2.158 contains mercury and rotates about the off-center axis $a$–$a$. At rest, the depth of mercury in each leg is 150 mm as illustrated. Determine the angular velocity for which the difference in heights between the two legs is 75 mm.

■ **Figure P2.158**

**2.159** (WILEY) An open 1-m-diameter tank contains water at a depth of 0.7 m when at rest. As the tank is rotated about its vertical axis the center of the fluid surface is depressed. At what angular velocity will the bottom of the tank first be exposed? No water is spilled from the tank.

**2.160** An open, 2-ft-diameter tank contains water to a depth of 3 ft when at rest. If the tank is rotated about its vertical axis with an angular velocity of 180 rev/min, what is the minimum height of the tank walls to prevent water from spilling over the sides?

**2.161** A child riding in a car holds a string attached to a floating, helium-filled balloon. As the car decelerates to a stop, the balloon tilts backwards. As the car makes a right-hand turn, the balloon tilts to the right. On the other hand, the child tends to be forced forward as the car decelerates and to the left as the car makes a right-hand turn. Explain these observed effects on the balloon and child.

**2.162** (WILEY) A closed, 0.4-m-diameter cylindrical tank is completely filled with oil ($SG = 0.9$) and rotates about its vertical longitudinal axis with an angular velocity of 40 rad/s. Determine the difference in pressure just under the vessel cover between a point on the circumference and a point on the axis.

**2.163** (See Fluids in the News article titled **"Rotating mercury mirror telescope,"** Section 2.12.2.) The largest liquid mirror telescope uses a 6-ft-diameter tank of mercury rotating at 7 rpm to produce its parabolic-shaped mirror as shown in Fig. P2.163. Determine the difference in elevation of the mercury, $\Delta h$, between the edge and the center of the mirror.

■ **Figure P2.163**

■ **Lab Problems**

**2.1 LP** This problem involves the force needed to open a gate that covers an opening in the side of a water-filled tank. To proceed with this problem, go to Appendix H which is located in WileyPLUS or on the book's web site, www.wiley.com/college/munson.

**2.2 LP** This problem involves the use of a cleverly designed apparatus to investigate the hydrostatic, pressure force on a submerged rectangle. To proceed with this problem, go to Appendix H which is located in *WileyPLUS* or on the book's web site, www.wiley.com/college/munson.

**2.3 LP** This problem involves determining the weight needed to hold down an open-bottom box that has slanted sides when the box is filled with water. To proceed with this problem, go to Appendix H which is located in *WileyPLUS* or on the book's web site, www.wiley.com/college/munson.

**2.4 LP** This problem involves the use of a pressurized air pad to provide the vertical force to support a given load. To proceed with this problem, go to Appendix H which is located in *WileyPLUS* or on the book's web site, www.wiley.com/college/munson.

■ **Lifelong Learning Problems**

**2.1 LL** Although it is relatively easy to calculate the net hydrostatic pressure force on a dam, it is not necessarily easy to design and construct an appropriate, long-lasting, inexpensive dam. In fact, inspection of older dams has revealed that many of them are in peril of collapse unless corrective action is soon taken. Obtain information about the severity of the poor conditions of older dams throughout the country. Summarize your findings in a brief report.

**2.2 LL** Over the years the demand for high-quality, first-growth timber has increased dramatically. Unfortunately, most of the trees that supply such lumber have already been harvested. Recently, however, several companies have started to reclaim the numerous high-quality logs that sank in lakes and oceans during the logging boom times many years ago. Many of these logs are still in excellent condition. Obtain information about the use of fluid mechanics concepts in harvesting sunken logs. Summarize your findings in a brief report.

**2.3 LL** Liquid-filled manometers and Bourdon tube pressure gages have been the mainstay for measuring pressure for many, many years. However, for many modern applications these tried-and-true devices are not sufficient. For example, various new uses need small, accurate, inexpensive pressure transducers with digital outputs. Obtain information about some of the new concepts used for pressure measurement. Summarize your findings in a brief report.

■ **FE Exam Problems**

Sample FE (Fundamentals of Engineering) exam questions for fluid mechanics are provided in *WileyPLUS* or on the book's web site, www.wiley.com/college/munson.

# 3 Elementary Fluid Dynamics— The Bernoulli Equation

CHAPTER OPENING PHOTO: Flow past a blunt body: On any object placed in a moving fluid there is a stagnation point on the front of the object where the velocity is zero. This location has a relatively large pressure and divides the flow field into two portions—one flowing to the left, and one flowing to the right of the body. (Dye in water.)

## Learning Objectives

After completing this chapter, you should be able to:

- discuss the application of Newton's second law to fluid flows.
- explain the development, uses, and limitations of the Bernoulli equation.
- use the Bernoulli equation (stand-alone or in combination with the continuity equation) to solve simple flow problems.
- apply the concepts of static, stagnation, dynamic, and total pressures.
- calculate various flow properties using the energy and hydraulic grade lines.

*The Bernoulli equation may be the most used and abused equation in fluid mechanics.*

In this chapter we investigate some typical fluid motions (fluid dynamics) in an elementary way. We will discuss in some detail the use of Newton's second law ($\mathbf{F} = m\mathbf{a}$) as it is applied to fluid particle motion that is "ideal" in some sense. We will obtain the celebrated Bernoulli equation and apply it to various flows. Although this equation is one of the oldest in fluid mechanics and the assumptions involved in its derivation are numerous, it can be used effectively to predict and analyze a variety of flow situations. However, if the equation is applied without proper respect for its restrictions, serious errors can arise. Indeed, the Bernoulli equation is appropriately called the most used and the most abused equation in fluid mechanics.

A thorough understanding of the elementary approach to fluid dynamics involved in this chapter will be useful on its own. It also provides a good foundation for the material in the following chapters where some of the present restrictions are removed and "more nearly exact" results are presented.

## 3.1 Newton's Second Law

As a fluid particle moves from one location to another, it usually experiences an acceleration or deceleration. According to Newton's second law of motion, the net force acting on the fluid particle under consideration must equal its mass times its acceleration,

$$\mathbf{F} = m\mathbf{a}$$

In this chapter we consider the motion of inviscid fluids. That is, the fluid is assumed to have zero viscosity. If the viscosity is zero, then the thermal conductivity of the fluid is also zero and there can be no heat transfer (except by radiation).

In practice there are no inviscid fluids, since every fluid supports shear stresses when it is subjected to a rate of strain displacement. For many flow situations the viscous effects are relatively small compared with other effects. As a first approximation for such cases it is often possible to ignore viscous effects. For example, often the viscous forces developed in flowing water may be several orders of magnitude smaller than forces due to other influences, such as gravity or pressure differences. For other water flow situations, however, the viscous effects may be the dominant ones. Similarly, the viscous effects associated with the flow of a gas are often negligible, although in some circumstances they are very important.

We assume that the fluid motion is governed by pressure and gravity forces only and examine Newton's second law as it applies to a fluid particle in the form:

(Net pressure force on particle) + (net gravity force on particle) =
(particle mass) × (particle acceleration)

The results of the interaction between the pressure, gravity, and acceleration provide numerous useful applications in fluid mechanics.

*Inviscid fluid flow is governed by pressure and gravity forces.*

To apply Newton's second law to a fluid (or any other object), we must define an appropriate coordinate system in which to describe the motion. In general the motion will be three-dimensional and unsteady so that three space coordinates and time are needed to describe it. There are numerous coordinate systems available, including the most often used rectangular $(x, y, z)$ and cylindrical $(r, \theta, z)$ systems shown by the figures in the margin. Usually the specific flow geometry dictates which system would be most appropriate.

In this chapter we will be concerned with two-dimensional motion like that confined to the $x$–$z$ plane as is shown in Fig. 3.1$a$. Clearly we could choose to describe the flow in terms of the components of acceleration and forces in the $x$ and $z$ coordinate directions. The resulting equations are frequently referred to as a two-dimensional form of the *Euler equations* of motion in rectangular Cartesian coordinates. This approach will be discussed in Chapter 6.

As is done in the study of dynamics (Ref. 1), the motion of each fluid particle is described in terms of its velocity vector, **V**, which is defined as the time rate of change of the position of the particle. The particle's velocity is a vector quantity with a magnitude (the speed, $V = |\mathbf{V}|$) and direction. As the particle moves about, it follows a particular path, the shape of which is governed by the velocity of the particle. The location of the particle along the path is a function of where the particle started at the initial time and its velocity along the path. If it is **steady flow** (i.e., nothing changes with time at a given location in the flow field), each successive particle that passes through a given point [such as point (1) in Fig. 3.1$a$] will follow the same path. For such cases the path is a fixed line in the $x$–$z$ plane. Neighboring particles that pass on either side of point (1) follow their own paths, which may be of a different shape than the one passing through (1). The entire $x$–$z$ plane is filled with such paths.

For steady flows each particle slides along its path, and its velocity vector is everywhere tangent to the path. The lines that are tangent to the velocity vectors throughout the flow field are called **streamlines.** For many situations it is easiest to describe the flow in terms of the

Rectangular

Cylindrical

■ **Figure 3.1**  (*a*) Flow in the *x*–*z* plane. (*b*) Flow in terms of streamline and normal coordinates.

"streamline" coordinates based on the streamlines as are illustrated in Fig. 3.1b. The particle motion is described in terms of its distance, $s = s(t)$, along the streamline from some convenient origin and the local radius of curvature of the streamline, $\mathcal{R} = \mathcal{R}(s)$. The distance along the streamline is related to the particle's speed by $V = ds/dt$, and the radius of curvature is related to the shape of the streamline. In addition to the coordinate along the streamline, $s$, the coordinate normal to the streamline, $n$, as is shown in Fig. 3.1b, will be of use.

To apply Newton's second law to a particle flowing along its streamline, we must write the particle acceleration in terms of the streamline coordinates. By definition, the acceleration is the time rate of change of the velocity of the particle, $\mathbf{a} = d\mathbf{V}/dt$. For two-dimensional flow in the $x$–$z$ plane, the acceleration has two components—one along the streamline, $a_s$, the streamwise acceleration, and one normal to the streamline, $a_n$, the normal acceleration.

The streamwise acceleration results from the fact that the speed of the particle generally varies along the streamline, $V = V(s)$. For example, in Fig. 3.1a the speed may be 50 ft/s at point (1) and 100 ft/s at point (2). Thus, by use of the chain rule of differentiation, the $s$ component of the acceleration is given by $a_s = dV/dt = (\partial V/\partial s)(ds/dt) = (\partial V/\partial s)V$. We have used the fact that speed is the time rate of change of distance, $V = ds/dt$. Note that the streamwise acceleration is the product of the rate of change of speed with distance along the streamline, $\partial V/\partial s$, and the speed, $V$. Since $\partial V/\partial s$ can be positive, negative, or zero, the streamwise acceleration can, therefore, be positive (acceleration), negative (deceleration), or zero (constant speed).

The normal component of acceleration, the centrifugal acceleration, is given in terms of the particle speed and the radius of curvature of its path. Thus, $a_n = V^2/\mathcal{R}$, where both $V$ and $\mathcal{R}$ may vary along the streamline. These equations for the acceleration should be familiar from the study of particle motion in physics (Ref. 2) or dynamics (Ref. 1). A more complete derivation and discussion of these topics can be found in Chapter 4.

Thus, the components of acceleration in the $s$ and $n$ directions, $a_s$ and $a_n$, are given by

$$a_s = V \frac{\partial V}{\partial s}, \qquad a_n = \frac{V^2}{\mathcal{R}} \tag{3.1}$$

where $\mathcal{R}$ is the local radius of curvature of the streamline, and $s$ is the distance measured along the streamline from some arbitrary initial point. In general there is acceleration along the streamline (because the particle speed changes along its path, $\partial V/\partial s \neq 0$) and acceleration normal to the streamline (because the particle does not flow in a straight line, $\mathcal{R} \neq \infty$). Various flows and the accelerations associated with them are shown in the figure in the margin. As discussed in Section 3.6.2, for incompressible flow the velocity is inversely proportional to the streamline spacing. Hence, converging streamlines produce positive streamwise acceleration. To produce this acceleration there must be a net, nonzero force on the fluid particle.

To determine the forces necessary to produce a given flow (or conversely, what flow results from a given set of forces), we consider the free-body diagram of a small fluid particle as is shown in Fig. 3.2. The particle of interest is removed from its surroundings, and the reactions of the surroundings on the particle are indicated by the appropriate forces present, $\mathbf{F}_1$, $\mathbf{F}_2$, and so forth. For the present case, the important forces are assumed to be gravity and pressure. Other forces,

$a_s = a_n = 0$

$a_s > 0$

$a_s < 0$

$a_n > 0$

$a_s > 0, a_n > 0$

■ **Figure 3.2** **Isolation of a small fluid particle in a flow field. (Photo courtesy of Diana Sailplanes.)**

such as viscous forces and surface tension effects, are assumed negligible. The acceleration of gravity, $g$, is assumed to be constant and acts vertically, in the negative $z$ direction, at an angle $\theta$ relative to the normal to the streamline.

## 3.2    F = *m*a along a Streamline

Consider the small fluid particle of size $\delta s$ by $\delta n$ in the plane of the figure and $\delta y$ normal to the figure as shown in the free-body diagram of Fig. 3.3. Unit vectors along and normal to the streamline are denoted by $\hat{s}$ and $\hat{n}$, respectively. For steady flow, the component of Newton's second law along the streamline direction, $s$, can be written as

$$\sum \delta F_s = \delta m\, a_s = \delta m\, V \frac{\partial V}{\partial s} = \rho\, \delta\forall\, V \frac{\partial V}{\partial s} \tag{3.2}$$

where $\sum \delta F_s$ represents the sum of the $s$ components of all the forces acting on the particle, which has mass $\delta m = \rho\, \delta\forall$, and $V\, \partial V/\partial s$ is the acceleration in the $s$ direction. Here, $\delta\forall = \delta s\, \delta n\, \delta y$ is the particle volume. Equation 3.2 is valid for both compressible and incompressible fluids. That is, the density need not be constant throughout the flow field.

The gravity force (weight) on the particle can be written as $\delta W = \gamma\, \delta\forall$, where $\gamma = \rho g$ is the specific weight of the fluid (lb/ft³ or N/m³). Hence, the component of the weight force in the direction of the streamline is

$$\delta W_s = -\delta W \sin \theta = -\gamma\, \delta\forall \sin \theta$$

If the streamline is horizontal at the point of interest, then $\theta = 0$, and there is no component of particle weight along the streamline to contribute to its acceleration in that direction.

*In a flowing fluid the pressure varies from one location to another.*

As is indicated in Chapter 2, the pressure is not constant throughout a stationary fluid $(\nabla p \neq 0)$ because of the fluid weight. Likewise, in a flowing fluid the pressure is usually not constant. In general, for steady flow, $p = p(s, n)$. If the pressure at the center of the particle shown in Fig. 3.3 is denoted as $p$, then its average value on the two end faces that are perpendicular to the streamline are $p + \delta p_s$ and $p - \delta p_s$. Since the particle is "small," we can use a one-term Taylor series expansion for the pressure field (as was done in Chapter 2 for the pressure forces in static fluids) to obtain

$$\delta p_s \approx \frac{\partial p}{\partial s} \frac{\delta s}{2}$$

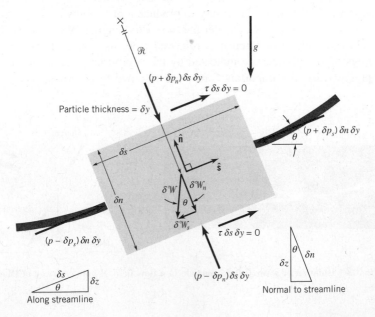

■ **Figure 3.3** Free-body diagram of a fluid particle for which the important forces are those due to pressure and gravity.

Thus, if $\delta F_{ps}$ is the net pressure force on the particle in the streamline direction, it follows that

$$\delta F_{ps} = (p - \delta p_s)\,\delta n\,\delta y - (p + \delta p_s)\,\delta n\,\delta y = -2\,\delta p_s\,\delta n\,\delta y$$

$$= -\frac{\partial p}{\partial s}\,\delta s\,\delta n\,\delta y = -\frac{\partial p}{\partial s}\,\delta \mathcal{V}$$

Note that the actual level of the pressure, $p$, is not important. What produces a net pressure force is the fact that the pressure is not constant throughout the fluid. The nonzero pressure gradient, $\nabla p = \partial p/\partial s\,\hat{\mathbf{s}} + \partial p/\partial n\,\hat{\mathbf{n}}$, is what provides a net pressure force on the particle. Viscous forces, represented by $\tau\,\delta s\,\delta y$, are zero, since the fluid is inviscid.

Thus, the net force acting in the streamline direction on the particle shown in Fig. 3.3 is given by

$$\sum \delta F_s = \delta \mathcal{W}_s + \delta F_{ps} = \left(-\gamma \sin\theta - \frac{\partial p}{\partial s}\right)\delta \mathcal{V} \tag{3.3}$$

By combining Eqs. 3.2 and 3.3, we obtain the following equation of motion along the streamline direction:

$$-\gamma \sin\theta - \frac{\partial p}{\partial s} = \rho V\frac{\partial V}{\partial s} = \rho a_s \tag{3.4}$$

We have divided out the common particle volume factor, $\delta \mathcal{V}$, that appears in both the force and the acceleration portions of the equation. This is a representation of the fact that it is the fluid density (mass per unit volume), not the mass, per se, of the fluid particle that is important.

The physical interpretation of Eq. 3.4 is that a change in fluid particle speed is accomplished by the appropriate combination of pressure gradient and particle weight along the streamline. For fluid static situations this balance between pressure and gravity forces is such that no change in particle speed is produced—the right-hand side of Eq. 3.4 is zero, and the particle remains stationary. In a flowing fluid the pressure and weight forces do not necessarily balance—the force unbalance provides the appropriate acceleration and, hence, particle motion.

# EXAMPLE 3.1    Pressure Variation along a Streamline

**GIVEN**   Consider the inviscid, incompressible, steady flow along the horizontal streamline $A-B$ in front of the sphere of radius $a$, as shown in Fig. E3.1a. From a more advanced theory of flow past a sphere, the fluid velocity along this streamline is

$$V = V_0\left(1 + \frac{a^3}{x^3}\right)$$

as shown in Fig. E3.1b.

**FIND**   Determine the pressure variation along the streamline from point $A$ far in front of the sphere ($x_A = -\infty$ and $V_A = V_0$) to point $B$ on the sphere ($x_B = -a$ and $V_B = 0$).

(a)

(b)

(c)

(d)

■ **Figure E3.1**

## Solution

Since the flow is steady and inviscid, Eq. 3.4 is valid. In addition, since the streamline is horizontal, $\sin\theta = \sin 0° = 0$ and the equation of motion along the streamline reduces to

$$\frac{\partial p}{\partial s} = -\rho V \frac{\partial V}{\partial s} \tag{1}$$

With the given velocity variation along the streamline, the acceleration term is

$$V\frac{\partial V}{\partial s} = V\frac{\partial V}{\partial x} = V_0\left(1 + \frac{a^3}{x^3}\right)\left(-\frac{3V_0 a^3}{x^4}\right)$$

$$= -3V_0^2\left(1 + \frac{a^3}{x^3}\right)\frac{a^3}{x^4}$$

where we have replaced $s$ by $x$ since the two coordinates are identical (within an additive constant) along streamline $A$–$B$. It follows that $V\,\partial V/\partial s < 0$ along the streamline. The fluid slows down from $V_0$ far ahead of the sphere to zero velocity on the "nose" of the sphere ($x = -a$).

Thus, according to Eq. 1, to produce the given motion the pressure gradient along the streamline is

$$\frac{\partial p}{\partial x} = \frac{3\rho a^3 V_0^2 (1 + a^3/x^3)}{x^4} \tag{2}$$

This variation is indicated in Fig. E3.1c. It is seen that the pressure increases in the direction of flow ($\partial p/\partial x > 0$) from point $A$ to point $B$. The maximum pressure gradient ($0.610\,\rho V_0^2/a$) occurs just slightly ahead of the sphere ($x = -1.205a$). It is the pressure gradient that slows the fluid down from $V_A = V_0$ to $V_B = 0$ as shown in Fig. E3.1b.

The pressure distribution along the streamline can be obtained by integrating Eq. 2 from $p = 0$ (gage) at $x = -\infty$ to pressure $p$ at location $x$. The result, plotted in Fig. E3.1d, is

$$p = -\rho V_0^2\left[\left(\frac{a}{x}\right)^3 + \frac{(a/x)^6}{2}\right] \tag{Ans}$$

**COMMENT** The pressure at $B$, a stagnation point since $V_B = 0$, is the highest pressure along the streamline ($p_B = \rho V_0^2/2$). As shown in Chapter 9, this excess pressure on the front of the sphere (i.e., $p_B > 0$) contributes to the net drag force on the sphere. Note that the pressure gradient and pressure are directly proportional to the density of the fluid, a representation of the fact that the fluid inertia is proportional to its mass.

## F l u i d s   i n   t h e   N e w s

**Incorrect raindrop shape** The incorrect representation that raindrops are teardrop shaped is found nearly everywhere—from children's books to weather maps on the Weather Channel. About the only time raindrops possess the typical teardrop shape is when they run down a windowpane. The actual shape of a falling raindrop is a function of the size of the drop and results from a balance between surface tension forces and the air pressure exerted on the falling drop. Small drops with a radius less than about 0.5 mm have a spherical shape because the surface tension effect (which is inversely proportional to drop size) wins over the increased pressure, $\rho V_0^2/2$, caused by the motion of the drop and exerted on its bottom. With increasing size, the drops fall faster and the increased pressure causes the drops to flatten. A 2-mm drop, for example, is flattened into a hamburger bun shape. Slightly larger drops are actually concave on the bottom. When the radius is greater than about 4 mm, the depression of the bottom increases and the drop takes on the form of an inverted bag with an annular ring of water around its base. This ring finally breaks up into smaller drops.

Streamline
$n = $ constant
$n$   $s$   $p = p(s)$

Equation 3.4 can be rearranged and integrated as follows. First, we note from Fig. 3.3 that along the streamline $\sin\theta = dz/ds$. Also we can write $V\,dV/ds = \frac{1}{2}d(V^2)/ds$. Finally, along the streamline the value of $n$ is constant ($dn = 0$) so that $dp = (\partial p/\partial s)\,ds + (\partial p/\partial n)\,dn = (\partial p/\partial s)\,ds$. Hence, as indicated by the figure in the margin, along a given streamline $p(s, n) = p(s)$ and $\partial p/\partial s = dp/ds$. These ideas combined with Eq. 3.4 give the following result valid along a streamline

$$-\gamma\frac{dz}{ds} - \frac{dp}{ds} = \frac{1}{2}\rho\frac{d(V^2)}{ds}$$

This simplifies to

$$dp + \frac{1}{2}\rho d(V^2) + \gamma\,dz = 0 \qquad \text{(along a streamline)} \tag{3.5}$$

*For steady, inviscid flow the sum of certain pressure, velocity, and elevation effects is constant along a streamline.*

which, for constant acceleration of gravity, can be integrated to give

$$\int\frac{dp}{\rho} + \frac{1}{2}V^2 + gz = C \qquad \text{(along a streamline)} \tag{3.6}$$

where $C$ is a constant of integration to be determined by the conditions at some point on the streamline.

*V3.2 Balancing ball*

In general it is not possible to integrate the pressure term because the density may not be constant and, therefore, cannot be removed from under the integral sign. To carry out this integration we must know specifically how the density varies with pressure. This is not always easily determined. For example, for a perfect gas the density, pressure, and temperature are related according to $\rho = p/RT$, where $R$ is the gas constant. To know how the density varies with pressure, we must also know the temperature variation. For now we will assume that the density and specific weight are constant (incompressible flow). The justification for this assumption and the consequences of compressibility will be considered further in Section 3.8.1 and more fully in Chapter 11.

With the additional assumption that the density remains constant (a very good assumption for liquids and also for gases if the speed is "not too high"), Eq. 3.6 assumes the following simple representation for steady, inviscid, incompressible flow.

$$p + \tfrac{1}{2}\rho V^2 + \gamma z = \text{constant along streamline} \tag{3.7}$$

*V3.3 Flow past a biker*

This is the celebrated **Bernoulli equation**—a very powerful tool in fluid mechanics. In 1738 Daniel Bernoulli (1700–1782) published his *Hydrodynamics* in which an equivalent of this famous equation first appeared. To use it correctly we must constantly remember the basic assumptions used in its derivation: (1) viscous effects are assumed negligible, (2) the flow is assumed to be steady, (3) the flow is assumed to be incompressible, and (4) the equation is applicable along a streamline. In the derivation of Eq. 3.7, we assume that the flow takes place in a plane (the $x$–$z$ plane). In general, this equation is valid for both planar and nonplanar (three-dimensional) flows, provided it is applied along the streamline.

We will provide many examples to illustrate the correct use of the Bernoulli equation and will show how a violation of the basic assumptions used in the derivation of this equation can lead to erroneous conclusions. The constant of integration in the Bernoulli equation can be evaluated if sufficient information about the flow is known at one location along the streamline.

---

# EXAMPLE 3.2   The Bernoulli Equation

**GIVEN**   Consider the flow of air around a bicyclist moving through still air with velocity $V_0$, as is shown in Fig. E3.2.

**FIND**   Determine the difference in the pressure between points (1) and (2).

## SOLUTION

In a coordinate fixed to the ground, the flow is unsteady as the bicyclist rides by. However, in a coordinate system fixed to the bike, it appears as though the air is flowing steadily toward the bicyclist with speed $V_0$. Since use of the Bernoulli equation is restricted to steady flows, we select the coordinate system fixed to the bike. If the assumptions of Bernoulli's equation are valid (steady, incompressible, inviscid flow), Eq. 3.7 can be applied as follows along the streamline that passes through (1) and (2)

$$p_1 + \tfrac{1}{2}\rho V_1^2 + \gamma z_1 = p_2 + \tfrac{1}{2}\rho V_2^2 + \gamma z_2$$

We consider (1) to be in the free stream so that $V_1 = V_0$ and (2) to be at the tip of the bicyclist's nose and assume that $z_1 = z_2$ and $V_2 = 0$ (both of which, as is discussed in Section 3.4, are reasonable assumptions). It follows that the pressure at (2) is greater than that at (1) by an amount

$$p_2 - p_1 = \tfrac{1}{2}\rho V_1^2 = \tfrac{1}{2}\rho V_0^2 \qquad \textbf{(Ans)}$$

**COMMENTS**   A similar result was obtained in Example 3.1 by integrating the pressure gradient, which was known because

**■ Figure E3.2**

the velocity distribution along the streamline, $V(s)$, was known. The Bernoulli equation is a general integration of **F** = *m***a**. To determine $p_2 - p_1$, knowledge of the detailed velocity distribution is not needed—only the "boundary conditions" at (1) and (2) are required. Of course, knowledge of the value of $V$ along the streamline is needed to determine the pressure at points between (1) and (2). Note that if we measure $p_2 - p_1$ we can determine the speed, $V_0$. As discussed in Section 3.5, this is the principle on which many velocity-measuring devices are based.

If the bicyclist were accelerating or decelerating, the flow would be unsteady (i.e., $V_0 \neq$ constant) and the above analysis would be incorrect since Eq. 3.7 is restricted to steady flow.

The difference in fluid velocity between two points in a flow field, $V_1$ and $V_2$, can often be controlled by appropriate geometric constraints of the fluid. For example, a garden hose nozzle is designed to give a much higher velocity at the exit of the nozzle than at its entrance where it is attached to the hose. As is shown by the Bernoulli equation, the pressure within the hose must be larger than that at the exit (for constant elevation, an increase in velocity requires a decrease in pressure if Eq. 3.7 is valid). It is this pressure drop that accelerates the water through the nozzle. Similarly, an airfoil is designed so that the fluid velocity over its upper surface is greater (on the average) than that along its lower surface. From the Bernoulli equation, therefore, the average pressure on the lower surface is greater than that on the upper surface. A net upward force, the lift, results.

## 3.3  F = ma Normal to a Streamline

*V3.4 Hydrocyclone separator*

In this section we will consider application of Newton's second law in a direction normal to the streamline. In many flows the streamlines are relatively straight, the flow is essentially one-dimensional, and variations in parameters across streamlines (in the normal direction) can often be neglected when compared to the variations along the streamline. However, in numerous other situations valuable information can be obtained from considering **F** = *m***a** normal to the streamlines. For example, the devastating low-pressure region at the center of a tornado can be explained by applying Newton's second law across the nearly circular streamlines of the tornado.

We again consider the force balance on the fluid particle shown in Fig. 3.3 and the figure in the margin. This time, however, we consider components in the normal direction, $\hat{\mathbf{n}}$, and write Newton's second law in this direction as

$$\sum \delta F_n = \frac{\delta m\, V^2}{\mathcal{R}} = \frac{\rho\, \delta \forall\, V^2}{\mathcal{R}} \tag{3.8}$$

where $\sum \delta F_n$ represents the sum of $n$ components of all the forces acting on the particle and $\delta m$ is particle mass. We assume the flow is steady with a normal acceleration $a_n = V^2/\mathcal{R}$, where $\mathcal{R}$ is the local radius of curvature of the streamlines. This acceleration is produced by the change in direction of the particle's velocity as it moves along a curved path.

We again assume that the only forces of importance are pressure and gravity. The component of the weight (gravity force) in the normal direction is

$$\delta \mathcal{W}_n = -\delta \mathcal{W} \cos \theta = -\gamma\, \delta \forall \cos \theta$$

*To apply* **F** = *m***a** *normal to streamlines, the normal components of force are needed.*

If the streamline is vertical at the point of interest, $\theta = 90°$, and there is no component of the particle weight normal to the direction of flow to contribute to its acceleration in that direction.

If the pressure at the center of the particle is $p$, then its values on the top and bottom of the particle are $p + \delta p_n$ and $p - \delta p_n$, where $\delta p_n = (\partial p/\partial n)(\delta n/2)$. Thus, if $\delta F_{pn}$ is the net pressure force on the particle in the normal direction, it follows that

$$\delta F_{pn} = (p - \delta p_n)\, \delta s\, \delta y - (p + \delta p_n)\, \delta s\, \delta y = -2\, \delta p_n\, \delta s\, \delta y$$

$$= -\frac{\partial p}{\partial n}\, \delta s\, \delta n\, \delta y = -\frac{\partial p}{\partial n}\, \delta \forall$$

*V3.5 Aircraft wing tip vortex*

Hence, the net force acting in the normal direction on the particle shown in Fig 3.3 is given by

$$\sum \delta F_n = \delta \mathcal{W}_n + \delta F_{pn} = \left( -\gamma \cos \theta - \frac{\partial p}{\partial n} \right) \delta \forall \tag{3.9}$$

By combining Eqs. 3.8 and 3.9 and using the fact that along a line normal to the streamline $\cos \theta = dz/dn$ (see Fig. 3.3), we obtain the following equation of motion along the normal direction:

$$-\gamma \frac{dz}{dn} - \frac{\partial p}{\partial n} = \frac{\rho V^2}{\mathcal{R}} \tag{3.10a}$$

*Weight and/or pressure can produce curved streamlines.*

The physical interpretation of Eq. 3.10 is that a change in the direction of flow of a fluid particle (i.e., a curved path, $\mathcal{R} < \infty$) is accomplished by the appropriate combination of pressure gradient and particle weight normal to the streamline. A larger speed or density or a smaller radius of curvature of the motion requires a larger force unbalance to produce the motion. For example, if gravity is neglected (as is commonly done for gas flows) or if the flow is in a horizontal ($dz/dn = 0$) plane, Eq. 3.10 becomes

$$\frac{\partial p}{\partial n} = -\frac{\rho V^2}{\mathcal{R}} \tag{3.10b}$$

*V3.6 Free vortex*

This indicates that the pressure increases with distance away from the center of curvature ($\partial p/\partial n$ is negative since $\rho V^2/\mathcal{R}$ is positive—the positive $n$ direction points toward the "inside" of the curved streamline). Thus, the pressure outside a tornado (typical atmospheric pressure) is larger than it is near the center of the tornado (where an often dangerously low partial vacuum may occur). This pressure difference is needed to balance the centrifugal acceleration associated with the curved streamlines of the fluid motion. (See Fig. E6.6a in Section 6.5.3.)

---

# EXAMPLE 3.3    Pressure Variation Normal to a Streamline

**GIVEN** Shown in Figs. E3.3a,b are two flow fields with circular streamlines. The velocity distributions are

$$V(r) = (V_0/r_0)r \quad \text{for case } (a)$$

and

$$V(r) = \frac{(V_0 r_0)}{r} \quad \text{for case } (b)$$

where $V_0$ is the velocity at $r = r_0$.

**FIND** Determine the pressure distributions, $p = p(r)$, for each, given that $p = p_0$ at $r = r_0$.

## SOLUTION

We assume the flows are steady, inviscid, and incompressible with streamlines in the horizontal plane ($dz/dn = 0$). Because the streamlines are circles, the coordinate $n$ points in a direction opposite that of the radial coordinate, $\partial/\partial n = -\partial/\partial r$, and the radius of curvature is given by $\mathcal{R} = r$. Hence, Eq. 3.10b becomes

$$\frac{\partial p}{\partial r} = \frac{\rho V^2}{r}$$

For case (a) this gives

$$\frac{\partial p}{\partial r} = \rho(V_0/r_0)^2 r$$

whereas for case (b) it gives

$$\frac{\partial p}{\partial r} = \frac{\rho(V_0 r_0)^2}{r^3}$$

For either case the pressure increases as $r$ increases since $\partial p/\partial r > 0$. Integration of these equations with respect to $r$, starting with a known pressure $p = p_0$ at $r = r_0$, gives

$$p - p_0 = (\rho V_0^2/2)[(r/r_0)^2 - 1] \tag{Ans}$$

**Figure E3.3**

for case (a) and

$$p - p_0 = (\rho V_0^2/2)[1 - (r_0/r)^2] \tag{Ans}$$

for case (b). These pressure distributions are shown in Fig. E3.3c.

**COMMENT** The pressure distributions needed to balance the centrifugal accelerations in cases (a) and (b) are not the same because the velocity distributions are different. In fact, for case (a) the pressure increases without bound as $r \to \infty$, whereas for case (b) the pressure approaches a finite value as $r \to \infty$. The streamline patterns are the same for each case, however.

Physically, case (a) represents rigid-body rotation (as obtained in a can of water on a turntable after it has been "spun up") and case (b) represents a free vortex (an approximation to a tornado, a hurricane, or the swirl of water in a drain, the "bathtub vortex"). See Fig. E6.6 for an approximation of this type of flow.

---

*The sum of pressure, elevation, and velocity effects is constant across streamlines.*

If we multiply Eq. 3.10 by $dn$, use the fact that $\partial p/\partial n = dp/dn$ if $s$ is constant, and integrate across the streamline (in the $n$ direction) we obtain

$$\int \frac{dp}{\rho} + \int \frac{V^2}{\mathcal{R}} \, dn + gz = \text{constant across the streamline} \tag{3.11}$$

To complete the indicated integrations, we must know how the density varies with pressure and how the fluid speed and radius of curvature vary with $n$. For incompressible flow the density is constant and the integration involving the pressure term gives simply $p/\rho$. We are still left, however, with the integration of the second term in Eq. 3.11. Without knowing the $n$ dependence in $V = V(s, n)$ and $\mathcal{R} = \mathcal{R}(s, n)$ this integration cannot be completed.

Thus, the final form of Newton's second law applied across the streamlines for steady, inviscid, incompressible flow is

$$p + \rho \int \frac{V^2}{\mathcal{R}} \, dn + \gamma z = \text{constant across the streamline} \tag{3.12}$$

As with the Bernoulli equation, we must be careful that the assumptions involved in the derivation of this equation are not violated when it is used.

## 3.4 Physical Interpretation

In the previous two sections, we developed the basic equations governing fluid motion under a fairly stringent set of restrictions. In spite of the numerous assumptions imposed on these flows, a variety of flows can be readily analyzed with them. A physical interpretation of the equations will be of help in understanding the processes involved. To this end, we rewrite Eqs. 3.7 and 3.12 here and interpret them physically. Application of $\mathbf{F} = m\mathbf{a}$ along and normal to the streamline results in

$$p + \tfrac{1}{2}\rho V^2 + \gamma z = \text{constant along the streamline} \tag{3.13}$$

and

$$p + \rho \int \frac{V^2}{\mathcal{R}} \, dn + \gamma z = \text{constant across the streamline} \tag{3.14}$$

$p + \frac{1}{2} \rho V^2 + \gamma z$ = constant

$p + \rho \int \frac{V^2}{\mathcal{R}} dn + \gamma z$ = constant

as indicated by the figure in the margin.

The following basic assumptions were made to obtain these equations: The flow is steady, and the fluid is inviscid and incompressible. In practice none of these assumptions is exactly true.

A violation of one or more of the above assumptions is a common cause for obtaining an incorrect match between the "real world" and solutions obtained by use of the Bernoulli equation. Fortunately, many "real-world" situations are adequately modeled by the use of Eqs. 3.13 and 3.14 because the flow is nearly steady and incompressible and the fluid behaves as if it were nearly inviscid.

The Bernoulli equation was obtained by integration of the equation of motion along the "natural" coordinate direction of the streamline. To produce an acceleration, there must be an unbalance of the resultant forces, of which only pressure and gravity were considered to be important. Thus, there are three processes involved in the flow—mass times acceleration (the $\rho V^2/2$ term), pressure (the $p$ term), and weight (the $\gamma z$ term).

Integration of the equation of motion to give Eq. 3.13 actually corresponds to the work–energy principle often used in the study of dynamics [see any standard dynamics text (Ref. 1)]. This principle results from a general integration of the equations of motion for an object in a way very similar to that done for the fluid particle in Section 3.2. With certain assumptions, a statement of the work–energy principle may be written as follows:

> The work done on a particle by all forces acting on the particle is equal to the change of the kinetic energy of the particle.

The Bernoulli equation is a mathematical statement of this principle.

As the fluid particle moves, both gravity and pressure forces do work on the particle. Recall that the work done by a force is equal to the product of the distance the particle travels times the component of force in the direction of travel (i.e., work $= \mathbf{F} \cdot \mathbf{d}$). The terms $\gamma z$ and $p$ in Eq. 3.13 are related to the work done by the weight and pressure forces, respectively. The remaining term, $\rho V^2/2$, is obviously related to the kinetic energy of the particle. In fact, an alternate method of deriving the Bernoulli equation is to use the first and second laws of thermodynamics (the energy and entropy equations), rather than Newton's second law. With the appropriate restrictions, the general energy equation reduces to the Bernoulli equation. This approach is discussed in Section 5.4.

An alternate but equivalent form of the Bernoulli equation is obtained by dividing each term of Eq. 3.7 by the specific weight, $\gamma$, to obtain

*The Bernoulli equation can be written in terms of heights called heads.*

$$\frac{p}{\gamma} + \frac{V^2}{2g} + z = \text{constant on a streamline}$$

Each of the terms in this equation has the units of energy per weight ($LF/F = L$) or length (feet, meters) and represents a certain type of head.

The elevation term, $z$, is related to the potential energy of the particle and is called the ***elevation head***. The pressure term, $p/\gamma$, is called the ***pressure head*** and represents the height of a column of the fluid that is needed to produce the pressure $p$. The velocity term, $V^2/2g$, is the ***velocity head*** and represents the vertical distance needed for the fluid to fall freely (neglecting friction) if it is to reach velocity $V$ from rest. The Bernoulli equation states that the sum of the pressure head, the velocity head, and the elevation head is constant along a streamline.

---

## $E$XAMPLE 3.4    Kinetic, Potential, and Pressure Energy

**GIVEN**   Consider the flow of water from the syringe shown in Fig. E3.4a. As indicated in Fig. E3.4b, a force, $F$, applied to the plunger will produce a pressure greater than atmospheric at point (1) within the syringe. The water flows from the needle, point (2), with relatively high velocity and coasts up to point (3) at the top of its trajectory.

**FIND**   Discuss the energy of the fluid at points (1), (2), and (3) by using the Bernoulli equation.

|  | **Energy Type** | | |
| Point | Kinetic $\rho V^2/2$ | Potential $\gamma z$ | Pressure $p$ |
| --- | --- | --- | --- |
| 1 | Small | Zero | Large |
| 2 | Large | Small | Zero |
| 3 | Zero | Large | Zero |

■ **Figure E3.4**

## SOLUTION

If the assumptions (steady, inviscid, incompressible flow) of the Bernoulli equation are approximately valid, it then follows that the flow can be explained in terms of the partition of the total energy of the water. According to Eq. 3.13, the sum of the three types of energy (kinetic, potential, and pressure) or heads (velocity, elevation, and pressure) must remain constant. The table above indicates the relative magnitude of each of these energies at the three points shown in the figure.

The motion results in (or is due to) a change in the magnitude of each type of energy as the fluid flows from one location to another. An alternate way to consider this flow is as follows. The pressure gradient between (1) and (2) produces an acceleration to eject the water from the needle. Gravity acting on the particle between (2) and (3) produces a deceleration to cause the water to come to a momentary stop at the top of its flight.

**COMMENT**    If friction (viscous) effects were important, there would be an energy loss between (1) and (3) and for the given $p_1$ the water would not be able to reach the height indicated in the figure. Such friction may arise in the needle (see Chapter 8 on pipe flow) or between the water stream and the surrounding air (see Chapter 9 on external flow).

---

## F l u i d s   i n   t h e   N e w s

**Armed with a water jet for hunting** Archerfish, known for their ability to shoot down insects resting on foliage, are like submarine water pistols. With their snout sticking out of the water, they eject a high-speed water jet at their prey, knocking it onto the water surface where they snare it for their meal. The barrel of their water pistol is formed by placing their tongue against a groove in the roof of their mouth to form a tube. By snapping shut their gills, water is forced through the tube and directed with the tip of their tongue. The archerfish can produce a *pressure head* within their gills large enough so that the jet can reach 2 to 3 m. However, it is accurate to only about 1 m. Recent research has shown that archerfish are very adept at calculating where their prey will fall. Within 100 milliseconds (a reaction time twice as fast as a human's), the fish has extracted all the information needed to predict the point where the prey will hit the water. Without further visual cues it charges directly to that point.

---

A net force is required to accelerate any mass. For steady flow the acceleration can be interpreted as arising from two distinct occurrences—a change in speed along the streamline and a change in direction if the streamline is not straight. Integration of the equation of motion along the streamline accounts for the change in speed (kinetic energy change) and results in the Bernoulli equation. Integration of the equation of motion normal to the streamline accounts for the centrifugal acceleration ($V^2/\mathcal{R}$) and results in Eq. 3.14.

*The pressure variation across straight streamlines is hydrostatic.*

When a fluid particle travels along a curved path, a net force directed toward the center of curvature is required. Under the assumptions valid for Eq. 3.14, this force may be either gravity or pressure, or a combination of both. In many instances the streamlines are nearly straight ($\mathcal{R} = \infty$) so that centrifugal effects are negligible and the pressure variation across the streamlines is merely hydrostatic (because of gravity alone), even though the fluid is in motion.

---

## EXAMPLE 3.5    Pressure Variation in a Flowing Stream

**GIVEN**    Water flows in a curved, undulating waterslide as shown in Fig. E3.5a. As an approximation to this flow, consider

■ **Figure E3.5b**

the inviscid, incompressible, steady flow shown in Fig. E3.5b. From section A to B the streamlines are straight, while from C to D they follow circular paths.

**FIND**    Describe the pressure variation between points (1) and (2) and points (3) and (4).

■ **Figure E3.5a**    **(Photo courtesy of Schlitterbahn® Waterparks.)**

## SOLUTION

With the above assumptions and the fact that $\mathcal{R} = \infty$ for the portion from $A$ to $B$, Eq. 3.14 becomes

$$p + \gamma z = \text{constant}$$

The constant can be determined by evaluating the known variables at the two locations using $p_2 = 0$ (gage), $z_1 = 0$, and $z_2 = h_{2-1}$ to give

$$p_1 = p_2 + \gamma(z_2 - z_1) = p_2 + \gamma h_{2-1} \qquad \textbf{(Ans)}$$

Note that since the radius of curvature of the streamline is infinite, the pressure variation in the vertical direction is the same as if the fluid were stationary.

However, if we apply Eq. 3.14, between points (3) and (4), we obtain (using $dn = -dz$)

$$p_4 + \rho \int_{z_3}^{z_4} \frac{V^2}{\mathcal{R}}(-dz) + \gamma z_4 = p_3 + \gamma z_3$$

With $p_4 = 0$ and $z_4 - z_3 = h_{4-3}$ this becomes

$$p_3 = \gamma h_{4-3} - \rho \int_{z_3}^{z_4} \frac{V^2}{\mathcal{R}} dz \qquad \textbf{(Ans)}$$

To evaluate the integral, we must know the variation of $V$ and $\mathcal{R}$ with $z$. Even without this detailed information we note that the integral has a positive value. Thus, the pressure at (3) is less than the hydrostatic value, $\gamma h_{4-3}$, by an amount equal to $\rho \int_{z_3}^{z_4} (V^2/\mathcal{R}) \, dz$. This lower pressure, caused by the curved streamline, is necessary to accelerate the fluid around the curved path.

**COMMENT**    Note that we did not apply the Bernoulli equation (Eq. 3.13) across the streamlines from (1) to (2) or (3) to (4). Rather we used Eq. 3.14. As is discussed in Section 3.8, application of the Bernoulli equation across streamlines (rather than along them) may lead to serious errors.

## 3.5    Static, Stagnation, Dynamic, and Total Pressure

*Each term in the Bernoulli equation can be interpreted as a form of pressure.*

A useful concept associated with the Bernoulli equation deals with the stagnation and dynamic pressures. These pressures arise from the conversion of kinetic energy in a flowing fluid into a "pressure rise" as the fluid is brought to rest (as in Example 3.2). In this section we explore various results of this process. Each term of the Bernoulli equation, Eq. 3.13, has the dimensions of force per unit area—psi, lb/ft$^2$, N/m$^2$. The first term, $p$, is the actual thermodynamic pressure of the fluid as it flows. To measure its value, one could move along with the fluid, thus being "static" relative to the moving fluid. Hence, it is normally termed the ***static pressure.*** Another way to measure the static pressure would be to drill a hole in a flat surface and fasten a piezometer tube as indicated by the location of point (3) in Fig. 3.4. As we saw in Example 3.5, the pressure in the flowing fluid at (1) is $p_1 = \gamma h_{3-1} + p_3$, the same as if the fluid were static. From the manometer considerations of Chapter 2, we know that $p_3 = \gamma h_{4-3}$. Thus, since $h_{3-1} + h_{4-3} = h$ it follows that $p_1 = \gamma h$.

The third term in Eq. 3.13, $\gamma z$, is termed the *hydrostatic pressure,* in obvious regard to the hydrostatic pressure variation discussed in Chapter 2. It is not actually a pressure but does represent the change in pressure possible due to potential energy variations of the fluid as a result of elevation changes.

The second term in the Bernoulli equation, $\rho V^2/2$, is termed the ***dynamic pressure.*** Its interpretation can be seen in Fig. 3.4 by considering the pressure at the end of a small tube inserted into the flow and pointing upstream. After the initial transient motion has died out, the liquid will fill the tube to a height of $H$ as shown. The fluid in the tube, including that at its tip, (2), will be stationary. That is, $V_2 = 0$, or point (2) is a ***stagnation point.***

If we apply the Bernoulli equation between points (1) and (2), using $V_2 = 0$ and assuming that $z_1 = z_2$, we find that

$$p_2 = p_1 + \tfrac{1}{2}\rho V_1^2$$

■ **Figure 3.4**  Measurement of static and stagnation pressures.

Stagnation point

(a)

(b)

■ **Figure 3.5** Stagnation points.

*V3.7 Stagnation point flow*

Hence, the pressure at the stagnation point is greater than the static pressure, $p_1$, by an amount $\rho V_1^2/2$, the dynamic pressure.

It can be shown that there is a stagnation point on any stationary body that is placed into a flowing fluid. Some of the fluid flows "over" and some "under" the object. The dividing line (or surface for three-dimensional flows) is termed the *stagnation streamline* and terminates at the stagnation point on the body. (See the photograph at the beginning of the chapter.) For symmetrical objects (such as a baseball) the stagnation point is clearly at the tip or front of the object as shown in Fig. 3.5a. For other flows such as a water jet against a car as shown in Fig. 3.5b, there is also a stagnation point on the car.

If elevation effects are neglected, the *stagnation pressure,* $p + \rho V^2/2$, is the largest pressure obtainable along a given streamline. It represents the conversion of all of the kinetic energy into a pressure rise. The sum of the static pressure, hydrostatic pressure, and dynamic pressure is termed the *total pressure,* $p_T$. The Bernoulli equation is a statement that the total pressure remains constant along a streamline. That is,

$$p + \tfrac{1}{2}\rho V^2 + \gamma z = p_T = \text{constant along a streamline} \tag{3.15}$$

Again, we must be careful that the assumptions used in the derivation of this equation are appropriate for the flow being considered.

---

### F l u i d s   i n   t h e   N e w s

**Pressurized eyes** Our eyes need a certain amount of internal pressure in order to work properly, with the normal range being between 10 and 20 mm of mercury. The pressure is determined by a balance between the fluid entering and leaving the eye. If the pressure is above the normal level, damage may occur to the optic nerve where it leaves the eye, leading to a loss of the visual field termed glaucoma. Measurement of the pressure within the eye can be done by several different noninvasive types of instruments, all of which measure the slight deformation of the eyeball when a force is put on it. Some methods use a physical probe that makes contact with the front of the eye, applies a known force, and measures the deformation. One noncontact method uses a calibrated "puff" of air that is blown against the eye. The *stagnation pressure* resulting from the air blowing against the eyeball causes a slight deformation, the magnitude of which is correlated with the pressure within the eyeball. (See Problem 3.28.)

---

Knowledge of the values of the static and stagnation pressures in a fluid implies that the fluid speed can be calculated. This is the principle on which the *Pitot-static tube* is based [H. de Pitot (1695–1771)]. As shown in Fig. 3.6, two concentric tubes are attached to two pressure gages (or a differential gage) so that the values of $p_3$ and $p_4$ (or the difference $p_3 - p_4$) can be determined. The center tube measures the stagnation pressure at its open tip. If elevation changes are negligible,

$$p_3 = p + \tfrac{1}{2}\rho V^2$$

**■ Figure 3.6**   The Pitot-static tube.

where $p$ and $V$ are the pressure and velocity of the fluid upstream of point (2). The outer tube is made with several small holes at an appropriate distance from the tip so that they measure the static pressure. If the effect of the elevation difference between (1) and (4) is negligible, then

$$p_4 = p_1 = p$$

By combining these two equations we see that

$$p_3 - p_4 = \tfrac{1}{2}\rho V^2$$

which can be rearranged to give

$$V = \sqrt{2(p_3 - p_4)/\rho} \tag{3.16}$$

*Pitot-static tubes measure fluid velocity by converting velocity into pressure.*

The actual shape and size of Pitot-static tubes vary considerably. A typical Pitot-static probe used to determine aircraft airspeed is shown in Fig. 3.7. (See Fig. E3.6a also.)

# F l u i d s   i n   t h e   N e w s

**Bugged and plugged Pitot tubes** Although a *Pitot tube* is a simple device for measuring aircraft airspeed, many airplane accidents have been caused by inaccurate Pitot tube readings. Most of these accidents are the result of having one or more of the holes blocked and, therefore, not indicating the correct pressure (speed). Usually this is discovered during takeoff when time to resolve the issue is short. The two most common causes for such a blockage are either that the pilot (or ground crew) has forgotten to remove the protective Pitot tube cover or that insects have built their nest within the tube where the standard visual check cannot detect it. The most serious accident (in terms of number of fatalities) caused by a blocked Pitot tube involved a Boeing 757 and occurred shortly after takeoff from Puerto Plata in the Dominican Republic. Incorrect airspeed data were automatically fed to the computer, causing the autopilot to change the angle of attack and the engine power. The flight crew became confused by the false indications; the aircraft stalled and then plunged into the Caribbean Sea killing all aboard.

*V3.8 Airspeed indicator*

**■ Figure 3.7**   Airplane Pitot-static probe. (*a*) Schematic, (*b*) Photograph. (Photograph courtesy of Aero-Instruments Co., LLC.)

## EXAMPLE 3.6 Pitot-Static Tube

**GIVEN** An airplane flies 200 mph at an elevation of 10,000 ft in a standard atmosphere as shown in Fig. E3.6a.

**FIND** Determine the pressure at point (1) far ahead of the airplane, the pressure at the stagnation point on the nose of the airplane, point (2), and the pressure difference indicated by a Pitot-static probe attached to the fuselage.

(2)

(1)
$V_1$ = 200 mph

Pitot-static tube

■ **Figure E3.6a** (Photo courtesy of Hawker Beechcraft Corporation.)

## SOLUTION

From Table C.1 we find that the static pressure at the altitude given is

$$p_1 = 1456 \text{ lb/ft}^2 \text{ (abs)} = 10.11 \text{ psia} \qquad \textbf{(Ans)}$$

Also the density is $\rho = 0.001756$ slug/ft$^3$.

If the flow is steady, inviscid, and incompressible and elevation changes are neglected, Eq. 3.13 becomes

$$p_2 = p_1 + \frac{\rho V_1^2}{2}$$

With $V_1 = 200$ mph = 293 ft/s and $V_2 = 0$ (since the coordinate system is fixed to the airplane), we obtain

$$p_2 = 1456 \text{ lb/ft}^2 + (0.001756 \text{ slugs/ft}^3)(293^2 \text{ ft}^2/\text{s}^2)/2$$

$$= (1456 + 75.4) \text{ lb/ft}^2 \text{ (abs)}$$

Hence, in terms of gage pressure

$$p_2 = 75.4 \text{ lb/ft}^2 = 0.524 \text{ psi} \qquad \textbf{(Ans)}$$

Thus, the pressure difference indicated by the Pitot-static tube is

$$p_2 - p_1 = \frac{\rho V_1^2}{2} = 0.524 \text{ psi} \qquad \textbf{(Ans)}$$

**COMMENTS** Note that it is very easy to obtain incorrect results by using improper units. Do not add lb/in.$^2$ and lb/ft$^2$. Recall that (slug/ft$^3$)(ft$^2$/s$^2$) = (slug · ft/s$^2$)/(ft$^2$) = lb/ft$^2$.

(200 mph, 0.951)

$p_1/p_2$

$V_1$, mph

■ **Figure E3.6b**

It was assumed that the flow is incompressible—the density remains constant from (1) to (2). However, since $\rho = p/RT$, a change in pressure (or temperature) will cause a change in density. For this relatively low speed, the ratio of the absolute pressures is nearly unity [i.e., $p_1/p_2 = (10.11 \text{ psia})/(10.11 + 0.524 \text{ psia}) = 0.951$], so that the density change is negligible. However, by repeating the calculations for various values of the speed, $V_1$, the results shown in Fig. E3.6b are obtained. Clearly at the 500- to 600-mph speeds normally flown by commercial airliners, the pressure ratio is such that density changes are important. In such situations it is necessary to use compressible flow concepts to obtain accurate results. (See Section 3.8.1 and Chapter 11.)

The Pitot-static tube provides a simple, relatively inexpensive way to measure fluid speed. Its use depends on the ability to measure the static and stagnation pressures. Care is needed to obtain these values accurately. For example, an accurate measurement of static pressure requires that none of the fluid's kinetic energy be converted into a pressure rise at the point of measurement. This requires a smooth hole with no burrs or imperfections. As indicated in Fig. 3.8, such imperfections can cause the measured pressure to be greater or less than the actual static pressure.

Also, the pressure along the surface of an object varies from the stagnation pressure at its stagnation point to values that may be less than the free stream static pressure. A typical pressure variation for a Pitot-static tube is indicated in Fig. 3.9. Clearly it is important that

*Accurate measurement of static pressure requires great care.*

$\frac{V}{p}$     $\frac{V}{p}$     $\frac{V}{p}$

(1)     (1)     (1)
$p_1 > p$     $p_1 < p$     $p_1 = p$

■ **Figure 3.8** Incorrect and correct design of static pressure taps.

■ **Figure 3.9**  Typical pressure distribution along a Pitot-static tube.

■ **Figure 3.10**  Cross section of a directional-finding Pitot-static tube.

the pressure taps be properly located to ensure that the pressure measured is actually the static pressure.

In practice it is often difficult to align the Pitot-static tube directly into the flow direction. Any misalignment will produce a nonsymmetrical flow field that may introduce errors. Typically, yaw angles up to 12 to 20° (depending on the particular probe design) give results that are less than 1% in error from the perfectly aligned results. Generally it is more difficult to measure static pressure than stagnation pressure.

One method of determining the flow direction and its speed (thus the velocity) is to use a directional-finding Pitot tube as is illustrated in Fig. 3.10. Three pressure taps are drilled into a small circular cylinder, fitted with small tubes, and connected to three pressure transducers. The cylinder is rotated until the pressures in the two side holes are equal, thus indicating that the center hole points directly upstream. The center tap then measures the stagnation pressure. The two side holes are located at a specific angle ($\beta = 29.5°$) so that they measure the static pressure. The speed is then obtained from $V = [2(p_2 - p_1)/\rho]^{1/2}$.

The above discussion is valid for incompressible flows. At high speeds, compressibility becomes important (the density is not constant) and other phenomena occur. Some of these ideas are discussed in Section 3.8, while others (such as shockwaves for supersonic Pitot-tube applications) are discussed in Chapter 11.

The concepts of static, dynamic, stagnation, and total pressure are useful in a variety of flow problems. These ideas are used more fully in the remainder of the book.

## 3.6    Examples of Use of the Bernoulli Equation

In this section we illustrate various additional applications of the Bernoulli equation. Between any two points, (1) and (2), on a streamline in steady, inviscid, incompressible flow the Bernoulli equation can be applied in the form

$$p_1 + \tfrac{1}{2}\rho V_1^2 + \gamma z_1 = p_2 + \tfrac{1}{2}\rho V_2^2 + \gamma z_2 \qquad \text{(3.17)}$$

Obviously, if five of the six variables are known, the remaining one can be determined. In many instances it is necessary to introduce other equations, such as the conservation of mass. Such considerations will be discussed briefly in this section and in more detail in Chapter 5.

■ **Figure 3.11** Vertical flow from a tank.

### 3.6.1 Free Jets

One of the oldest equations in fluid mechanics deals with the flow of a liquid from a large reservoir. A modern version of this type of flow involves the flow of coffee from a coffee urn as indicated by the figure in the margin. The basic principles of this type of flow are shown in Fig. 3.11 where a jet of liquid of diameter $d$ flows from the nozzle with velocity $V$. (A nozzle is a device shaped to accelerate a fluid.) Application of Eq. 3.17 between points (1) and (2) on the streamline shown gives

$$\gamma h = \tfrac{1}{2}\rho V^2$$

We have used the facts that $z_1 = h$, $z_2 = 0$, the reservoir is large ($V_1 \cong 0$) and open to the atmosphere ($p_1 = 0$ gage), and the fluid leaves as a ***"free jet"*** ($p_2 = 0$). Thus, we obtain

$$V = \sqrt{2\,\frac{\gamma h}{\rho}} = \sqrt{2gh} \tag{3.18}$$

which is the modern version of a result obtained in 1643 by Torricelli (1608–1647), an Italian physicist.

*The exit pressure for an incompressible fluid jet is equal to the surrounding pressure.*

The fact that the exit pressure equals the surrounding pressure ($p_2 = 0$) can be seen by applying $\mathbf{F} = m\mathbf{a}$, as given by Eq. 3.14, across the streamlines between (2) and (4). If the streamlines at the tip of the nozzle are straight ($\mathcal{R} = \infty$), it follows that $p_2 = p_4$. Since (4) is on the surface of the jet, in contact with the atmosphere, we have $p_4 = 0$. Thus, $p_2 = 0$ also. Since (2) is an arbitrary point in the exit plane of the nozzle, it follows that the pressure is atmospheric across this plane. Physically, since there is no component of the weight force or acceleration in the normal (horizontal) direction, the pressure is constant in that direction.

Once outside the nozzle, the stream continues to fall as a free jet with zero pressure throughout ($p_5 = 0$) and as seen by applying Eq. 3.17 between points (1) and (5), the speed increases according to

$$V = \sqrt{2g\,(h + H)}$$

where, as shown in Fig. 3.11, $H$ is the distance the fluid has fallen outside the nozzle.

Equation 3.18 could also be obtained by writing the Bernoulli equation between points (3) and (4) using the fact that $z_4 = 0$, $z_3 = \ell$. Also, $V_3 = 0$ since it is far from the nozzle, and from hydrostatics, $p_3 = \gamma(h - \ell)$.

As learned in physics or dynamics and illustrated in the figure in the margin, any object dropped from rest that falls through a distance $h$ in a vacuum will obtain the speed $V = \sqrt{2gh}$, the same as the water leaving the spout of the watering can shown in the figure in the margin on the next page. This is consistent with the fact that all of the particle's potential energy is converted to kinetic energy, provided viscous (friction) effects are negligible. In terms of heads, the elevation head at point (1) is converted into the velocity head at point (2). Recall that for the case shown in Fig. 3.11 the pressure is the same (atmospheric) at points (1) and (2).

For the horizontal nozzle of Fig. 3.12a, the velocity of the fluid at the centerline, $V_2$, will be slightly greater than that at the top, $V_1$, and slightly less than that at the bottom, $V_3$, due to the differences in elevation. In general, $d \ll h$ as shown in Fig. 3.12b and we can safely use the centerline velocity as a reasonable "average velocity."

■ **Figure 3.12**   Horizontal flow from a tank.

■ **Figure 3.13**   Vena contracta effect for a sharp-edged orifice.

*The diameter of a fluid jet is often smaller than that of the hole from which it flows.*

If the exit is not a smooth, well-contoured nozzle, but rather a flat plate as shown in Fig. 3.13, the diameter of the jet, $d_j$, will be less than the diameter of the hole, $d_h$. This phenomenon, called a *vena contracta* effect, is a result of the inability of the fluid to turn the sharp 90° corner indicated by the dotted lines in the figure.

Since the streamlines in the exit plane are curved ($\mathcal{R} < \infty$), the pressure across them is not constant. It would take an infinite pressure gradient across the streamlines to cause the fluid to turn a "sharp" corner ($\mathcal{R} = 0$). The highest pressure occurs along the centerline at (2) and the lowest pressure, $p_1 = p_3 = 0$, is at the edge of the jet. Thus, the assumption of uniform velocity with straight streamlines and constant pressure is not valid at the exit plane. It is valid, however, in the plane of the vena contracta, section $a-a$. The uniform velocity assumption is valid at this section provided $d_j \ll h$, as is discussed for the flow from the nozzle shown in Fig. 3.12.

The vena contracta effect is a function of the geometry of the outlet. Some typical configurations are shown in Fig. 3.14 along with typical values of the experimentally obtained *contraction coefficient*, $C_c = A_j / A_h$, where $A_j$ and $A_h$ are the areas of the jet at the vena contracta and the area of the hole, respectively.

# F l u i d s   i n   t h e   N e w s

**Cotton candy, glass wool, and steel wool** Although cotton candy and glass wool insulation are made of entirely different materials and have entirely different uses, they are made by similar processes. Cotton candy, invented in 1897, consists of sugar fibers. Glass wool, invented in 1938, consists of glass fibers. In a cotton candy machine, sugar is melted and then forced by centrifugal action to flow through numerous tiny *orifices* in a spinning "bowl." Upon emerging, the thin streams of liquid sugar cool very quickly and become solid threads that are collected on a stick or cone. Making glass wool in-

sulation is somewhat more complex, but the basic process is similar. Liquid glass is forced through tiny orifices and emerges as very fine glass streams that quickly solidify. The resulting intertwined flexible fibers, glass wool, form an effective insulation material because the tiny air "cavities" between the fibers inhibit air motion. Although steel wool looks similar to cotton candy or glass wool, it is made by an entirely different process. Solid steel wires are drawn over special cutting blades that have grooves cut into them so that long, thin threads of steel are peeled off to form the matted steel wool.

$$C_C = A_j/A_h = (d_j/d_h)^2$$

(a) Knife edge      (b) Well rounded

(c) Sharp edge      (d) Re-entrant

■ **Figure 3.14** **Typical flow patterns and contraction coefficients for various round exit configurations. (a) Knife edge, (b) Well rounded, (c) Sharp edge, (d) Re-entrant.**

### 3.6.2 Confined Flows

*The continuity equation states that mass cannot be created or destroyed.*

In many cases the fluid is physically constrained within a device so that its pressure cannot be pre-scribed a priori as was done for the free jet examples above. Such cases include nozzles and pipes of variable diameter for which the fluid velocity changes because the flow area is different from one section to another. For these situations it is necessary to use the concept of conservation of mass (the continuity equation) along with the Bernoulli equation. The derivation and use of this equation are discussed in detail in Chapters 4 and 5. For the needs of this chapter we can use a simplified form of the continuity equation obtained from the following intuitive arguments. Con-sider a fluid flowing through a fixed volume (such as a syringe) that has one inlet and one outlet as shown in Fig. 3.15a. If the flow is steady so that there is no additional accumulation of fluid within the volume, the rate at which the fluid flows into the volume must equal the rate at which it flows out of the volume (otherwise, mass would not be conserved).

The *mass flowrate* from an outlet, $\dot{m}$ (slugs/s or kg/s), is given by $\dot{m} = \rho Q$, where $Q$ (ft$^3$/s or m$^3$/s) is the **volume flowrate**. If the outlet area is $A$ and the fluid flows across this area (normal to the area) with an average velocity $V$, then the volume of the fluid crossing this area in a time interval $\delta t$ is $VA \, \delta t$, equal to that in a volume of length $V \, \delta t$ and cross-sectional area $A$ (see Fig. 3.15b). Hence, the vol-ume flowrate (volume per unit time) is $Q = VA$. Thus, $\dot{m} = \rho VA$. To conserve mass, the inflow rate must equal the outflow rate. If the inlet is designated as (1) and the outlet as (2), it follows that $\dot{m}_1 = \dot{m}_2$. Thus, conservation of mass requires

$$\rho_1 A_1 V_1 = \rho_2 A_2 V_2$$

If the density remains constant, then $\rho_1 = \rho_2$, and the above becomes the **continuity equation** for incompressible flow

$$\boxed{A_1 V_1 = A_2 V_2, \text{ or } Q_1 = Q_2} \tag{3.19}$$

For example, if as shown by the figure in the margin the outlet flow area is one-half the size of the inlet flow area, it follows that the outlet velocity is twice that of the inlet velocity, since

■ **Figure 3.15**   (*a*) Flow through a syringe. (*b*) Steady flow into and out of a volume.

$V_2 = A_1 V_1/A_2 = 2V_1$. Use of the Bernoulli equation and the flowrate equation (continuity equation) is demonstrated by Example 3.7.

---

# EXAMPLE 3.7   Flow from a Tank—Gravity Driven

**GIVEN**   A stream of refreshing beverage of diameter $d = 0.01$ m flows steadily from the cooler of diameter $D = 0.20$ m as shown in Figs. E3.7a and b.

**FIND**   Determine the flowrate, $Q$, from the bottle into the cooler if the depth of beverage in the cooler is to remain constant at $h = 0.20$ m.

■ **Figure E3.7**

## SOLUTION

For steady, inviscid, incompressible flow, the Bernoulli equation applied between points (1) and (2) is

$$p_1 + \tfrac{1}{2}\rho V_1^2 + \gamma z_1 = p_2 + \tfrac{1}{2}\rho V_2^2 + \gamma z_2 \qquad (1)$$

With the assumptions that $p_1 = p_2 = 0$, $z_1 = h$, and $z_2 = 0$, Eq. 1 becomes

$$\tfrac{1}{2}V_1^2 + gh = \tfrac{1}{2}V_2^2 \qquad (2)$$

Although the liquid level remains constant ($h = $ constant), there is an average velocity, $V_1$, across section (1) because of the flow from the tank. From Eq. 3.19 for steady incompressible flow, conservation of mass requires $Q_1 = Q_2$, where $Q = AV$. Thus, $A_1V_1 = A_2V_2$, or

$$\frac{\pi}{4}D^2 V_1 = \frac{\pi}{4}d^2 V_2$$

Hence,

$$V_1 = \left(\frac{d}{D}\right)^2 V_2 \qquad (3)$$

Equations 1 and 3 can be combined to give

$$V_2 = \sqrt{\frac{2gh}{1 - (d/D)^4}} = \sqrt{\frac{2(9.81 \text{ m/s}^2)(0.20 \text{ m})}{1 - (0.01 \text{ m}/0.20 \text{ m})^4}} = 1.98 \text{ m/s}$$

Thus,

$$Q = A_1V_1 = A_2V_2 = \frac{\pi}{4}(0.01 \text{ m})^2(1.98 \text{ m/s})$$

$$= 1.56 \times 10^{-4} \text{ m}^3\text{/s} \qquad \textbf{(Ans)}$$

**COMMENTS**   In this example we have not neglected the kinetic energy of the water in the tank ($V_1 \neq 0$). If the tank diameter is large compared to the jet diameter ($D \gg d$), Eq. 3 indicates that $V_1 \ll V_2$ and the assumption that $V_1 \approx 0$ would be reasonable. The error associated with this assumption can be seen by calculating the ratio of the flowrate assuming $V_1 \neq 0$, denoted $Q$, to that assuming $V_1 = 0$, denoted $Q_0$. This ratio, written as

$$\frac{Q}{Q_0} = \frac{V_2}{V_2|_{D=\infty}} = \frac{\sqrt{2gh/[1 - (d/D)^4]}}{\sqrt{2gh}} = \frac{1}{\sqrt{1 - (d/D)^4}}$$

is plotted in Fig. E3.7c. With $0 < d/D < 0.4$ it follows that $1 < Q/Q_0 \lesssim 1.01$, and the error in assuming $V_1 = 0$ is less than 1%. For this example with $d/D = 0.01 \text{ m}/0.20 \text{ m} = 0.05$, it follows that $Q/Q_0 = 1.000003$. Thus, it is often reasonable to assume $V_1 = 0$.

Note that this problem was solved using points (1) and (2) located at the free surface and the exit of the pipe, respectively. Although this was convenient (because most of the variables are known at those points), other points could be selected and the same result would be obtained. For example, consider points (1) and (3) as indicated in Fig. E3.7b. At (3), located sufficiently far from the tank exit, $V_3 = 0$ and $z_3 = z_2 = 0$. Also, $p_3 = \gamma h$ since the pressure is hydrostatic sufficiently far from the exit. Use of this information in the Bernoulli equation applied between (3) and (2) gives the exact same result as obtained using it between (1) and (2). The only difference is that the elevation head, $z_1 = h$, has been interchanged with the pressure head at (3), $p_3/\gamma = h$.

The fact that a kinetic energy change is often accompanied by a change in pressure is shown by Example 3.8.

## EXAMPLE 3.8   Flow from a Tank—Pressure Driven

**GIVEN**   Air flows steadily from a tank, through a hose of diameter $D = 0.03$ m, and exits to the atmosphere from a nozzle of diameter $d = 0.01$ m as shown in Fig. E3.8. The pressure in the tank remains constant at 3.0 kPa (gage) and the atmospheric conditions are standard temperature and pressure.

**FIND**   Determine
(a)   the flowrate and
(b)   the pressure in the hose.

$p_1 = 3.0$ kPa   $D = 0.03$ m   $d = 0.01$ m
(1)   Air   (2)   (3)   $Q$

■ **Figure E3.8a**

## SOLUTION

(a)   If the flow is assumed steady, inviscid, and incompressible, we can apply the Bernoulli equation along the streamline from (1) to (2) to (3) as

$$p_1 + \tfrac{1}{2}\rho V_1^2 + \gamma z_1 = p_2 + \tfrac{1}{2}\rho V_2^2 + \gamma z_2$$
$$= p_3 + \tfrac{1}{2}\rho V_3^2 + \gamma z_3$$

With the assumption that $z_1 = z_2 = z_3$ (horizontal hose), $V_1 = 0$ (large tank), and $p_3 = 0$ (free jet), this becomes

$$V_3 = \sqrt{\frac{2p_1}{\rho}}$$

and

$$p_2 = p_1 - \tfrac{1}{2}\rho V_2^2 \qquad (1)$$

The density of the air in the tank is obtained from the perfect gas law, using standard absolute pressure and temperature, as

$$\rho = \frac{p_1}{RT_1}$$

$$= \frac{(3.0 + 101) \text{ kN/m}^2 \times 10^3 \text{ N/kN}}{(286.9 \text{ N} \cdot \text{m/kg} \cdot \text{K})(15 + 273)\text{K}}$$

$$= 1.26 \text{ kg/m}^3$$

Thus, we find that

$$V_3 = \sqrt{\frac{2(3.0 \times 10^3 \text{ N/m}^2)}{1.26 \text{ kg/m}^3}} = 69.0 \text{ m/s}$$

or

$$Q = A_3 V_3 = \frac{\pi}{4} d^2 V_3 = \frac{\pi}{4} (0.01 \text{ m})^2 (69.0 \text{ m/s})$$

$$= 0.00542 \text{ m}^3/\text{s} \qquad \text{(Ans)}$$

**COMMENT**    Note that the value of $V_3$ is determined strictly by the value of $p_1$ (and the assumptions involved in the Bernoulli equation), independent of the "shape" of the nozzle. The pressure head within the tank, $p_1/\gamma = (3.0 \text{ kPa})/(9.81 \text{ m/s}^2)(1.26 \text{ kg/m}^3) = 243$ m, is converted to the velocity head at the exit, $V_2^2/2g = (69.0 \text{ m/s})^2/(2 \times 9.81 \text{ m/s}^2) = 243$ m. Although we used gage pressure in the Bernoulli equation ($p_3 = 0$), we had to use absolute pressure in the perfect gas law when calculating the density.

**(b)**    The pressure within the hose can be obtained from Eq. 1 and the continuity equation (Eq. 3.19)

$$A_2 V_2 = A_3 V_3$$

Hence,

$$V_2 = A_3 V_3 / A_2 = \left(\frac{d}{D}\right)^2 V_3$$

$$= \left(\frac{0.01 \text{ m}}{0.03 \text{ m}}\right)^2 (69.0 \text{ m/s}) = 7.67 \text{ m/s}$$

and from Eq. 1

$$p_2 = 3.0 \times 10^3 \text{ N/m}^2 - \tfrac{1}{2}(1.26 \text{ kg/m}^3)(7.67 \text{ m/s})^2$$

$$= (3000 - 37.1)\text{N/m}^2 = 2963 \text{ N/m}^2 \qquad \text{(Ans)}$$

**COMMENTS**    In the absence of viscous effects, the pressure throughout the hose is constant and equal to $p_2$. Physically, the decreases in pressure from $p_1$ to $p_2$ to $p_3$ accelerate the air and increase its kinetic energy from zero in the tank to an intermediate value in the hose and finally to its maximum value at the nozzle exit. Since the air velocity in the nozzle exit is nine

■ **Figure E3.8b**

■ **Figure E3.8c**

times that in the hose, most of the pressure drop occurs across the nozzle ($p_1 = 3000 \text{ N/m}^2$, $p_2 = 2963 \text{ N/m}^2$, and $p_3 = 0$).

Since the pressure change from (1) to (3) is not too great [i.e., in terms of absolute pressure $(p_1 - p_3)/p_1 = 3.0/101 = 0.03$], it follows from the perfect gas law that the density change is also not significant. Hence, the incompressibility assumption is reasonable for this problem. If the tank pressure were considerably larger or if viscous effects were important, application of the Bernoulli equation to this situation would be incorrect.

By repeating the calculations for various nozzle diameters, $d$, the results shown in Figs. E3.8b,c are obtained. The flowrate increases as the nozzle is opened (i.e., larger $d$). Note that if the nozzle diameter is the same as that of the hose ($d = 0.03$ m), the pressure throughout the hose is atmospheric (zero gage).

---

**F l u i d s   i n   t h e   N e w s**

**Hi-tech inhaler**    The term *inhaler* often brings to mind a treatment for asthma or bronchitis. Work is underway to develop a family of inhalation devices that can do more than treat respiratory ailments. They will be able to deliver medication for diabetes and other conditions by spraying it to reach the bloodstream through the lungs. The concept is to make the spray droplets fine enough to penetrate to the lungs' tiny sacs, the alveoli, where exchanges between blood and the outside world take place. This is accomplished by use of a laser-machined *nozzle* containing an array of very fine holes that cause the liquid to divide into a mist of micron-scale droplets. The device fits the hand and accepts a disposable strip that contains the medicine solution sealed inside a blister of laminated plastic and the nozzle. An electrically actuated piston drives the liquid from its reservoir through the nozzle array and into the respiratory system. To take the medicine, the patient breathes through the device and a differential pressure transducer in the inhaler senses when the patient's breathing has reached the best condition for receiving the medication. At that point, the piston is automatically triggered.

In many situations the combined effects of kinetic energy, pressure, and gravity are important. Example 3.9 illustrates this.

---

# EXAMPLE 3.9  Flow in a Variable Area Pipe

**GIVEN**  Water flows through a pipe reducer as is shown in Fig. E3.9. The static pressures at (1) and (2) are measured by the inverted U-tube manometer containing oil of specific gravity, $SG$, less than one.

**FIND**  Determine the manometer reading, $h$.

## SOLUTION

With the assumptions of steady, inviscid, incompressible flow, the Bernoulli equation can be written as

$$p_1 + \tfrac{1}{2}\rho V_1^2 + \gamma z_1 = p_2 + \tfrac{1}{2}\rho V_2^2 + \gamma z_2$$

The continuity equation (Eq. 3.19) provides a second relationship between $V_1$ and $V_2$ if we assume the velocity profiles are uniform at those two locations and the fluid incompressible:

$$Q = A_1 V_1 = A_2 V_2$$

By combining these two equations we obtain

$$p_1 - p_2 = \gamma(z_2 - z_1) + \tfrac{1}{2}\rho V_2^2\left[1 - (A_2/A_1)^2\right] \tag{1}$$

This pressure difference is measured by the manometer and can be determined by using the pressure–depth ideas developed in Chapter 2. Thus,

$$p_1 - \gamma(z_2 - z_1) - \gamma\ell - \gamma h + SG\,\gamma h + \gamma\ell = p_2$$

or

$$p_1 - p_2 = \gamma(z_2 - z_1) + (1 - SG)\gamma h \tag{2}$$

As discussed in Chapter 2, this pressure difference is neither merely $\gamma h$ nor $\gamma(h + z_1 - z_2)$.

Equations 1 and 2 can be combined to give the desired result as follows:

$$(1 - SG)\gamma h = \frac{1}{2}\rho V_2^2\left[1 - \left(\frac{A_2}{A_1}\right)^2\right]$$

**■ Figure E3.9**

or since $V_2 = Q/A_2$

$$h = (Q/A_2)^2\,\frac{1 - (A_2/A_1)^2}{2g(1 - SG)} \tag{Ans}$$

**COMMENT**  The difference in elevation, $z_1 - z_2$, was not needed because the change in elevation term in the Bernoulli equation exactly cancels the elevation term in the manometer equation. However, the pressure difference, $p_1 - p_2$, depends on the angle $\theta$, because of the elevation, $z_1 - z_2$, in Eq. 1. Thus, for a given flowrate, the pressure difference, $p_1 - p_2$, as measured by a pressure gage would vary with $\theta$, but the manometer reading, $h$, would be independent of $\theta$.

---

*V3.10 Venturi channel*

In general, an increase in velocity is accompanied by a decrease in pressure. For example, the velocity of the air flowing over the top surface of an airplane wing is, on the average, faster than that flowing under the bottom surface. Thus, the net pressure force is greater on the bottom than on the top—the wing generates a lift.

If the differences in velocity are considerable, the differences in pressure can also be considerable. For flows of gases, this may introduce compressibility effects as discussed in Section 3.8 and Chapter 11. For flows of liquids, this may result in **cavitation,** a potentially dangerous situation that results when the liquid pressure is reduced to the vapor pressure and the liquid "boils."

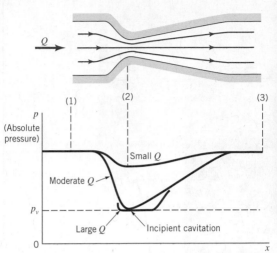

■ **Figure 3.16**  Pressure variation and cavitation in a variable area pipe.

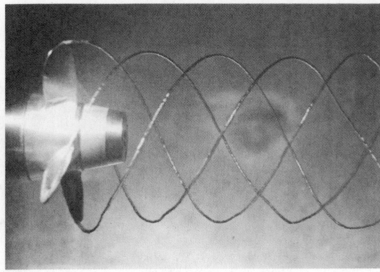

■ **Figure 3.17**  Tip cavitation from a propeller. (Photograph courtesy of The Pennsylvania State University, Applied Research Laboratory, Garfield Thomas Water Tunnel.)

*Cavitation occurs when the pressure is reduced to the vapor pressure.*

As discussed in Chapter 1, the vapor pressure, $p_v$, is the pressure at which vapor bubbles form in a liquid. It is the pressure at which the liquid starts to boil. Obviously this pressure depends on the type of liquid and its temperature. For example, water, which boils at 212 °F at standard atmospheric pressure, 14.7 psia, boils at 80 °F if the pressure is 0.507 psia. That is, $p_v = 0.507$ psia at 80 °F and $p_v = 14.7$ psia at 212 °F. (See Tables B.1 and B.2.)

One way to produce cavitation in a flowing liquid is noted from the Bernoulli equation. If the fluid velocity is increased (for example, by a reduction in flow area as shown in Fig. 3.16), the pressure will decrease. This pressure decrease (needed to accelerate the fluid through the constriction) can be large enough so that the pressure in the liquid is reduced to its vapor pressure. A simple example of cavitation can be demonstrated with an ordinary garden hose. If the hose is "kinked," a restriction in the flow area in some ways analogous to that shown in Fig. 3.16 will result. The water velocity through this restriction will be relatively large. With a sufficient amount of restriction the sound of the flowing water will change—a definite "hissing" sound is produced. This sound is a result of cavitation.

*Cavitation can cause damage to equipment.*

In such situations boiling occurs (though the temperature need not be high), vapor bubbles form, and then they collapse as the fluid moves into a region of higher pressure (lower velocity). This process can produce dynamic effects (imploding) that cause very large pressure transients in the vicinity of the bubbles. Pressures as large as 100,000 psi (690 MPa) are believed to occur. If the bubbles collapse close to a physical boundary they can, over a period of time, cause damage to the surface in the cavitation area. Tip cavitation from a propeller is shown in Fig. 3.17. In this case the high-speed rotation of the propeller produced a corresponding low pressure on the propeller. Obviously, proper design and use of equipment are needed to eliminate cavitation damage.

---

# *E*XAMPLE 3.10    Siphon and Cavitation

**GIVEN**  A liquid can be siphoned from a container as shown in Fig. E3.10*a*, provided the end of the tube, point (3), is below the free surface in the container, point (1), and the maximum elevation of the tube, point (2), is "not too great." Consider water at 60 °F being siphoned from a large tank through a constant-diameter hose as shown in Fig. E3.10*b*. The end of the siphon is 5 ft below the bottom of the tank, and the atmospheric pressure is 14.7 psia.

**FIND**  Determine the maximum height of the hill, $H$, over which the water can be siphoned without cavitation occurring.

## SOLUTION

If the flow is steady, inviscid, and incompressible we can apply the Bernoulli equation along the streamline from (1) to (2) to (3) as follows:

$$p_1 + \tfrac{1}{2}\rho V_1^2 + \gamma z_1 = p_2 + \tfrac{1}{2}\rho V_2^2 + \gamma z_2$$
$$= p_3 + \tfrac{1}{2}\rho V_3^2 + \gamma z_3 \qquad (1)$$

With the tank bottom as the datum, we have $z_1 = 15$ ft, $z_2 = H$, and $z_3 = -5$ ft. Also, $V_1 = 0$ (large tank), $p_1 = 0$ (open tank), $p_3 = 0$ (free jet), and from the continuity equation $A_2 V_2 = A_3 V_3$, or because the hose is constant diameter, $V_2 = V_3$. Thus, the speed of the fluid in the hose is determined from Eq. 1 to be

$$V_3 = \sqrt{2g(z_1 - z_3)} = \sqrt{2(32.2\ \text{ft/s}^2)[15 - (-5)]\ \text{ft}}$$
$$= 35.9\ \text{ft/s} = V_2$$

Use of Eq. 1 between points (1) and (2) then gives the pressure $p_2$ at the top of the hill as

$$p_2 = p_1 + \tfrac{1}{2}\rho V_1^2 + \gamma z_1 - \tfrac{1}{2}\rho V_2^2 - \gamma z_2$$
$$= \gamma(z_1 - z_2) - \tfrac{1}{2}\rho V_2^2 \qquad (2)$$

From Table B.1, the vapor pressure of water at 60 °F is 0.256 psia. Hence, for incipient cavitation the lowest pressure in the system will be $p = 0.256$ psia. Careful consideration of Eq. 2 and Fig. E3.10b will show that this lowest pressure will occur at the top of the hill. Since we have used gage pressure at point (1) ($p_1 = 0$), we must use gage pressure at point (2) also. Thus, $p_2 = 0.256 - 14.7 = -14.4$ psi and Eq. 2 gives

$$(-14.4\ \text{lb/in.}^2)(144\ \text{in.}^2/\text{ft}^2)$$
$$= (62.4\ \text{lb/ft}^3)(15 - H)\text{ft} - \tfrac{1}{2}(1.94\ \text{slugs/ft}^3)(35.9\ \text{ft/s})^2$$

or

$$H = 28.2\ \text{ft} \qquad \textbf{(Ans)}$$

For larger values of $H$, vapor bubbles will form at point (2) and the siphon action may stop.

**COMMENTS** Note that we could have used absolute pressure throughout ($p_2 = 0.256$ psia and $p_1 = 14.7$ psia) and obtained the same result. The lower the elevation of point (3), the larger the flowrate and, therefore, the smaller the value of $H$ allowed.

We could also have used the Bernoulli equation between (2) and (3), with $V_2 = V_3$, to obtain the same value of $H$. In this case it would not have been necessary to determine $V_2$ by use of the Bernoulli equation between (1) and (3).

The above results are independent of the diameter and length of the hose (provided viscous effects are not important). Proper design of the hose (or pipe) is needed to ensure that it will not collapse due to the large pressure difference (vacuum) between the inside and outside of the hose.

■ **Figure E3.10a**

■ **Figure E3.10b**

By using the fluid properties listed in Table 1.5 and repeating the calculations for various fluids, the results shown in Fig. E3.10c are obtained. The value of $H$ is a function of both the specific weight of the fluid, $\gamma$, and its vapor pressure, $p_v$.

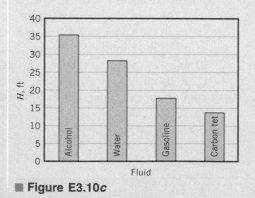

■ **Figure E3.10c**

### 3.6.3 Flowrate Measurement

Many types of devices using principles involved in the Bernoulli equation have been developed to measure fluid velocities and flowrates. The Pitot-static tube discussed in Section 3.5 is an example. Other examples discussed below include devices to measure flowrates in pipes and

**Figure 3.18**  Typical devices for measuring flowrate in pipes.

conduits and devices to measure flowrates in open channels. In this chapter we will consider "ideal" *flowmeters*—those devoid of viscous, compressibility, and other "real-world" effects. Corrections for these effects are discussed in Chapters 8 and 10. Our goal here is to understand the basic operating principles of these simple flowmeters.

An effective way to measure the flowrate through a pipe is to place some type of restriction within the pipe as shown in Fig. 3.18 and to measure the pressure difference between the low-velocity, high-pressure upstream section (1) and the high-velocity, low-pressure downstream section (2). Three commonly used types of flowmeters are illustrated: the *orifice meter*, the *nozzle meter*, and the *Venturi meter*. The operation of each is based on the same physical principles—an increase in velocity causes a decrease in pressure. The difference between them is a matter of cost, accuracy, and how closely their actual operation obeys the idealized flow assumptions.

We assume the flow is horizontal ($z_1 = z_2$), steady, inviscid, and incompressible between points (1) and (2). The Bernoulli equation becomes

$$p_1 + \tfrac{1}{2}\rho V_1^2 = p_2 + \tfrac{1}{2}\rho V_2^2$$

(The effect of nonhorizontal flow can be incorporated easily by including the change in elevation, $z_1 - z_2$, in the Bernoulli equation.)

If we assume the velocity profiles are uniform at sections (1) and (2), the continuity equation (Eq. 3.19) can be written as

$$Q = A_1 V_1 = A_2 V_2$$

*The flowrate varies as the square root of the pressure difference across the flowmeter.*

where $A_2$ is the small ($A_2 < A_1$) flow area at section (2). Combination of these two equations results in the following theoretical flowrate

$$Q = A_2 \sqrt{\frac{2(p_1 - p_2)}{\rho[1 - (A_2/A_1)^2]}} \tag{3.20}$$

Thus, as shown by the figure in the margin, for a given flow geometry ($A_1$ and $A_2$) the flowrate can be determined if the pressure difference, $p_1 - p_2$, is measured. The actual measured flowrate, $Q_{\text{actual}}$, will be smaller than this theoretical result because of various differences between the "real world" and the assumptions used in the derivation of Eq. 3.20. These differences (which are quite consistent and may be as small as 1 to 2% or as large as 40%, depending on the geometry used) can be accounted for by using an empirically obtained discharge coefficient as discussed in Section 8.6.1.

## EXAMPLE 3.11 Venturi Meter

**GIVEN** Kerosene ($SG = 0.85$) flows through the Venturi meter shown in Fig. E3.11a with flowrates between 0.005 and 0.050 m³/s.

**FIND** Determine the range in pressure difference, $p_1 - p_2$, needed to measure these flowrates.

■ **Figure E3.11a**

*(figure labels: Kerosene, SG = 0.85; $D_1 = 0.1$ m; (2) • $D_2 = 0.06$ m; (1); Q; 0.005 m³/s ≤ Q ≤ 0.050 m³/s)*

## SOLUTION

If the flow is assumed to be steady, inviscid, and incompressible, the relationship between flowrate and pressure is given by Eq. 3.20. This can be rearranged to give

$$p_1 - p_2 = \frac{Q^2 \rho [1 - (A_2/A_1)^2]}{2 A_2^2}$$

With the density of the flowing fluid

$$\rho = SG\, \rho_{H_2O} = 0.85(1000 \text{ kg/m}^3) = 850 \text{ kg/m}^3$$

and the area ratio

$$A_2/A_1 = (D_2/D_1)^2 = (0.06 \text{ m}/0.10 \text{ m})^2 = 0.36$$

the pressure difference for the smallest flowrate is

$$p_1 - p_2 = (0.005 \text{ m}^3\text{/s})^2 (850 \text{ kg/m}^3) \frac{(1 - 0.36^2)}{2\,[(\pi/4)(0.06 \text{ m})^2]^2}$$

$$= 1160 \text{ N/m}^2 = 1.16 \text{ kPa}$$

Likewise, the pressure difference for the largest flowrate is

$$p_1 - p_2 = (0.05)^2(850) \frac{(1 - 0.36^2)}{2[(\pi/4)(0.06)^2]^2}$$

$$= 1.16 \times 10^5 \text{ N/m}^2 = 116 \text{ kPa}$$

Thus,

$$1.16 \text{ kPa} \leq p_1 - p_2 \leq 116 \text{ kPa} \qquad \textbf{(Ans)}$$

**COMMENTS** These values represent the pressure differences for inviscid, steady, incompressible conditions. The ideal results presented here are independent of the particular flowmeter geometry—an orifice, nozzle, or Venturi meter (see Fig. 3.18).

It is seen from Eq. 3.20 that the flowrate varies as the square root of the pressure difference. Hence, as indicated by the numerical results and shown in Fig. E3.11b, a 10-fold increase in flowrate requires a 100-fold increase in pressure difference. This nonlinear relationship can cause difficulties when measuring flowrates over a wide range of values. Such measurements would require pressure transducers with a wide range of operation. An alternative is to use two flowmeters in parallel—one for the larger and one for the smaller flowrate ranges.

*(graph: vertical axis $p_1 - p_2$, kPa from 0 to 120; horizontal axis Q, m³/s from 0 to 0.05; labels (0.05 m³/s, 116 kPa) and (0.005 m³/s, 1.16 kPa))*

■ **Figure E3.11b**

---

Other flowmeters based on the Bernoulli equation are used to measure flowrates in open channels such as flumes and irrigation ditches. Two of these devices, the *sluice gate* and the *sharp-crested weir*, are discussed below under the assumption of steady, inviscid, incompressible flow. These and other open-channel flow devices are discussed in more detail in Chapter 10.

Sluice gates like those shown in Fig. 3.19a are often used to regulate and measure the flowrate in open channels. As indicated in Fig. 3.19b, the flowrate, $Q$, is a function of the water depth upstream, $z_1$, the width of the gate, $b$, and the gate opening, $a$. Application of the Bernoulli equation and continuity equation between points (1) and (2) can provide a good approximation to the actual flowrate obtained. We assume the velocity profiles are uniform sufficiently far upstream and downstream of the gate.

■ **Figure 3.19**  Sluice gate geometry. (Photograph courtesy of Plasti-Fab, Inc.)

Thus, we apply the Bernoulli equation between points on the free surfaces at (1) and (2) to give

$$p_1 + \tfrac{1}{2}\rho V_1^2 + \gamma z_1 = p_2 + \tfrac{1}{2}\rho V_2^2 + \gamma z_2$$

Also, if the gate is the same width as the channel so that $A_1 = bz_1$ and $A_2 = bz_2$, the continuity equation gives

$$Q = A_1 V_1 = bV_1 z_1 = A_2 V_2 = bV_2 z_2$$

With the fact that $p_1 = p_2 = 0$, these equations can be combined and rearranged to give the flowrate as

> *The flowrate under a sluice gate depends on the water depths on either side of the gate.*

$$Q = z_2 b \sqrt{\frac{2g(z_1 - z_2)}{1 - (z_2/z_1)^2}} \tag{3.21}$$

In the limit of $z_1 \gg z_2$ this result simply becomes

$$Q = z_2 b \sqrt{2gz_1}$$

This limiting result represents the fact that if the depth ratio, $z_1/z_2$, is large, the kinetic energy of the fluid upstream of the gate is negligible and the fluid velocity after it has fallen a distance $(z_1 - z_2) \approx z_1$ is approximately $V_2 = \sqrt{2gz_1}$.

The results of Eq. 3.21 could also be obtained by using the Bernoulli equation between points (3) and (4) and the fact that $p_3 = \gamma z_1$ and $p_4 = \gamma z_2$ since the streamlines at these sections are straight. In this formulation, rather than the potential energies at (1) and (2), we have the pressure contributions at (3) and (4).

The downstream depth, $z_2$, not the gate opening, $a$, was used to obtain the result of Eq. 3.21. As was discussed relative to flow from an orifice (Fig. 3.14), the fluid cannot turn a sharp 90° corner. A vena contracta results with a contraction coefficient, $C_c = z_2/a$, less than 1. Typically $C_c$ is approximately 0.61 over the depth ratio range of $0 < a/z_1 < 0.2$. For larger values of $a/z_1$ the value of $C_c$ increases rapidly.

---

## ***E*XAMPLE 3.12**  Sluice Gate

**GIVEN**  Water flows under the sluice gate shown in Fig. E3.12a.    **FIND**  Determine the approximate flowrate per unit width of the channel.

## SOLUTION

Under the assumptions of steady, inviscid, incompressible flow, we can apply Eq. 3.21 to obtain $Q/b$, the flowrate per unit width, as

$$\frac{Q}{b} = z_2 \sqrt{\frac{2g(z_1 - z_2)}{1 - (z_2/z_1)^2}}$$

In this instance $z_1 = 5.0$ m and $a = 0.80$ m so the ratio $a/z_1 = 0.16 < 0.20$, and we can assume that the contraction coefficient is approximately $C_c = 0.61$. Thus, $z_2 = C_c a = 0.61$ $(0.80$ m$) = 0.488$ m and we obtain the flowrate

$$\frac{Q}{b} = (0.488 \text{ m}) \sqrt{\frac{2(9.81 \text{ m/s}^2)(5.0 \text{ m} - 0.488 \text{ m})}{1 - (0.488 \text{ m}/5.0 \text{ m})^2}}$$

$$= 4.61 \text{ m}^2/\text{s} \qquad \textbf{(Ans)}$$

**COMMENT** If we consider $z_1 \gg z_2$ and neglect the kinetic energy of the upstream fluid, we would have

$$\frac{Q}{b} = z_2 \sqrt{2gz_1} = 0.488 \text{ m} \sqrt{2(9.81 \text{ m/s}^2)(5.0 \text{ m})}$$

$$= 4.83 \text{ m}^2/\text{s}$$

In this case the difference in $Q$ with or without including $V_1$ is not too significant because the depth ratio is fairly large $(z_1/z_2 = 5.0/0.488 = 10.2)$. Thus, it is often reasonable to neglect the kinetic energy upstream from the gate compared to that downstream of it.

By repeating the calculations for various flow depths, $z_1$, the results shown in Fig. E3.12b are obtained. Note that the

■ **Figure E3.12a**

■ **Figure E3.12b**

flowrate is not directly proportional to the flow depth. Thus, for example, if during flood conditions the upstream depth doubled from $z_1 = 5$ m to $z_1 = 10$ m, the flowrate per unit width of the channel would not double but would increase only from 4.61 m²/s to 6.67 m²/s.

Another device used to measure flow in an open channel is a *weir*. A typical rectangular, sharp-crested weir is shown in Fig. 3.20. For such devices the flowrate of liquid over the top of the weir plate is dependent on the weir height, $P_w$, the width of the channel, $b$, and the head, $H$, of the water above the top of the weir. Application of the Bernoulli equation can provide a simple approximation of the flowrate expected for these situations, even though the actual flow is quite complex.

Between points (1) and (2) the pressure and gravitational fields cause the fluid to accelerate from velocity $V_1$ to velocity $V_2$. At (1) the pressure is $p_1 = \gamma h$, while at (2) the pressure is essentially atmospheric, $p_2 = 0$. Across the curved streamlines directly above the top of the weir plate (section $a$–$a$), the pressure changes from atmospheric on the top surface to some maximum value within the fluid stream and then to atmospheric again at the bottom surface. This distribution is indicated in Fig. 3.20. Such a pressure distribution, combined with the streamline curvature and gravity, produces a rather nonuniform velocity profile across this section. This velocity distribution can be obtained from experiments or a more advanced theory.

■ **Figure 3.20** Rectangular, sharp-crested weir geometry.

For now, we will take a very simple approach and assume that the weir flow is similar in many respects to an orifice-type flow with a free streamline. In this instance we would expect the average velocity across the top of the weir to be proportional to $\sqrt{2gH}$ and the flow area for this rectangular weir to be proportional to $Hb$. Hence, it follows that

$$Q = C_1 Hb \sqrt{2gH} = C_1 b \sqrt{2g} \, H^{3/2}$$

where $C_1$ is a constant to be determined.

Simple use of the Bernoulli equation has provided a method to analyze the relatively complex flow over a weir. The correct functional dependence of $Q$ on $H$ has been obtained ($Q \sim H^{3/2}$, as indicated by the figure in the margin), but the value of the coefficient $C_1$ is unknown. Even a more advanced analysis cannot predict its value accurately. As is discussed in Chapter 10, experiments are used to determine the value of $C_1$.

---

# EXAMPLE 3.13    Weir

**GIVEN**    Water flows over a triangular weir, as is shown in Fig. E3.13.

**FIND**    Based on a simple analysis using the Bernoulli equation, determine the dependence of the flowrate on the depth $H$. If the flowrate is $Q_0$ when $H = H_0$, estimate the flowrate when the depth is increased to $H = 3H_0$.

■ **Figure E3.13**

## SOLUTION

With the assumption that the flow is steady, inviscid, and incompressible, it is reasonable to assume from Eq. 3.18 that the average speed of the fluid over the triangular notch in the weir plate is proportional to $\sqrt{2gH}$. Also, the flow area for a depth of $H$ is $H[H \tan (\theta/2)]$. The combination of these two ideas gives

$$Q = AV = H^2 \tan \frac{\theta}{2} (C_2 \sqrt{2gH}) = C_2 \tan \frac{\theta}{2} \sqrt{2g} \, H^{5/2} \quad \textbf{(Ans)}$$

where $C_2$ is an unknown constant to be determined experimentally.

Thus, an increase in the depth by a factor of three (from $H_0$ to $3H_0$) results in an increase of the flowrate by a factor of

$$\frac{Q_{3H_0}}{Q_{H_0}} = \frac{C_2 \tan(\theta/2) \sqrt{2g} \, (3H_0)^{5/2}}{C_2 \tan(\theta/2) \sqrt{2g} \, (H_0)^{5/2}}$$

$$= 15.6 \quad \textbf{(Ans)}$$

**COMMENT**    Note that for a triangular weir the flowrate is proportional to $H^{5/2}$, whereas for the rectangular weir discussed above, it is proportional to $H^{3/2}$. The triangular weir can be accurately used over a wide range of flowrates.

---

# 3.7    The Energy Line and the Hydraulic Grade Line

As was discussed in Section 3.4, the Bernoulli equation is actually an energy equation representing the partitioning of energy for an inviscid, incompressible, steady flow. The sum of the various energies of the fluid remains constant as the fluid flows from one section to another. A useful interpretation of the Bernoulli equation can be obtained through use of the concepts of the **hydraulic grade line** (HGL) and the **energy line** (EL). These ideas represent a geometrical interpretation of a flow and can often be effectively used to better grasp the fundamental processes involved.

*The hydraulic grade line and energy line are graphical forms of the Bernoulli equation.*

For steady, inviscid, incompressible flow the total energy remains constant along a streamline. The concept of "head" was introduced by dividing each term in Eq. 3.7 by the specific weight, $\gamma = \rho g$, to give the Bernoulli equation in the following form

$$\frac{p}{\gamma} + \frac{V^2}{2g} + z = \text{constant on a streamline} = H \quad \textbf{(3.22)}$$

■ **Figure 3.21**  Representation of the energy line and the hydraulic grade line.

Each of the terms in this equation has the units of length (feet or meters) and represents a certain type of head. The Bernoulli equation states that the sum of the pressure head, the velocity head, and the elevation head is constant along a streamline. This constant is called the *total head*, $H$, and is shown in the figure in the margin.

The energy line is a line that represents the total head available to the fluid. As shown in Fig. 3.21, the elevation of the energy line can be obtained by measuring the stagnation pressure with a Pitot tube. (A Pitot tube is the portion of a Pitot-static tube that measures the stagnation pressure. See Section 3.5.) The stagnation point at the end of the Pitot tube provides a measurement of the total head (or energy) of the flow. The static pressure tap connected to the piezometer tube shown, on the other hand, measures the sum of the pressure head and the elevation head, $p/\gamma + z$. This sum is often called the *piezometric head*. The static pressure tap does not measure the velocity head.

According to Eq. 3.22, the total head remains constant along the streamline (provided the assumptions of the Bernoulli equation are valid). Thus, a Pitot tube at any other location in the flow will measure the same total head, as is shown in the figure. The elevation head, velocity head, and pressure head may vary along the streamline, however.

*Under the assumptions of the Bernoulli equation, the energy line is horizontal.*

The locus of elevations provided by a series of Pitot tubes is termed the energy line, EL. The locus provided by a series of piezometer taps is termed the hydraulic grade line, HGL. Under the assumptions of the Bernoulli equation, the energy line is horizontal. If the fluid velocity changes along the streamline, the hydraulic grade line will not be horizontal. If viscous effects are important (as they often are in pipe flows), the total head does not remain constant due to a loss in energy as the fluid flows along its streamline. This means that the energy line is no longer horizontal. Such viscous effects are discussed in Chapters 5 and 8.

The energy line and hydraulic grade line for flow from a large tank are shown in Fig. 3.22. If the flow is steady, incompressible, and inviscid, the energy line is horizontal and at the elevation of the liquid in the tank (since the fluid velocity in the tank and the pressure on the surface

■ **Figure 3.22**  The energy line and hydraulic grade line for flow from a tank.

■ **Figure 3.23**  Use of the energy line and the hydraulic grade line.

are zero). The hydraulic grade line lies a distance of one velocity head, $V^2/2g$, below the energy line. Thus, a change in fluid velocity due to a change in the pipe diameter results in a change in the elevation of the hydraulic grade line. At the pipe outlet the pressure head is zero (gage), so the pipe elevation and the hydraulic grade line coincide.

*For flow below (above) the hydraulic grade line, the pressure is positive (negative).*

The distance from the pipe to the hydraulic grade line indicates the pressure within the pipe, as is shown in Fig. 3.23. If the pipe lies below the hydraulic grade line, the pressure within the pipe is positive (above atmospheric). If the pipe lies above the hydraulic grade line, the pressure is negative (below atmospheric). Thus, a scale drawing of a pipeline and the hydraulic grade line can be used to readily indicate regions of positive or negative pressure within a pipe.

---

**E**XAMPLE 3.14    **Energy Line and Hydraulic Grade Line**

**GIVEN**  Water is siphoned from the tank shown in Fig. E3.14 through a hose of constant diameter. A small hole is found in the hose at location (1) as indicated.

**FIND**  When the siphon is used, will water leak out of the hose, or will air leak into the hose, thereby possibly causing the siphon to malfunction?

■ **Figure E3.14**

**S**OLUTION

Whether air will leak into or water will leak out of the hose depends on whether the pressure within the hose at (1) is less than or greater than atmospheric. Which happens can be easily determined by using the energy line and hydraulic grade line concepts. With the assumption of steady, incompressible, inviscid flow it follows that the total head is constant—thus, the energy line is horizontal.

Since the hose diameter is constant, it follows from the continuity equation ($AV = $ constant) that the water velocity in the hose is constant throughout. Thus, the hydraulic grade line is a constant distance, $V^2/2g$, below the energy line as shown in Fig. E3.14. Since the pressure at the end of the hose is atmospheric, it follows that the hydraulic grade line is at the same elevation as the end of the hose outlet. The fluid within the hose at any point above the hydraulic grade line will be at less than atmospheric pressure.

Thus, air will leak into the hose through the hole at point (1).    **(Ans)**

**COMMENT**  In practice, viscous effects may be quite important, making this simple analysis (horizontal energy line) incorrect. However, if the hose is "not too small diameter," "not too long," the fluid "not too viscous," and the flowrate "not too large," the above result may be very accurate. If any of these assumptions are relaxed, a more detailed analysis is required (see Chapter 8). If the end of the hose were closed so that the flowrate were zero, the hydraulic grade line would coincide with the energy line ($V^2/2g = 0$ throughout), the pressure at (1) would be greater than atmospheric, and water would leak through the hole at (1).

---

The above discussion of the hydraulic grade line and the energy line is restricted to ideal situations involving inviscid, incompressible flows. Another restriction is that there are no "sources" or "sinks" of energy within the flow field. That is, there are no pumps or turbines involved. Alterations in the energy line and hydraulic grade line concepts due to these devices are discussed in Chapters 5 and 8.

## 3.8 Restrictions on Use of the Bernoulli Equation

Proper use of the Bernoulli equation requires close attention to the assumptions used in its derivation. In this section we review some of these assumptions and consider the consequences of incorrect use of the equation.

### 3.8.1 Compressibility Effects

One of the main assumptions is that the fluid is incompressible. Although this is reasonable for most liquid flows, it can, in certain instances, introduce considerable errors for gases.

In the previous section, we saw that the stagnation pressure, $p_{stag}$, is greater than the static pressure, $p_{static}$, by an amount $\Delta p = p_{stag} - p_{static} = \rho V^2/2$, provided that the density remains constant. If this dynamic pressure is not too large compared with the static pressure, the density change between two points is not very large and the flow can be considered incompressible. However, since the dynamic pressure varies as $V^2$, the error associated with the assumption that a fluid is incompressible increases with the square of the velocity of the fluid, as indicated by the figure in the margin. To account for compressibility effects, we must return to Eq. 3.6 and properly integrate the term $\int dp/\rho$ when $\rho$ is not constant.

A simple, although specialized, case of compressible flow occurs when the temperature of a perfect gas remains constant along the streamline—isothermal flow. Thus, we consider $p = \rho RT$, where $T$ is constant. (In general, $p$, $\rho$, and $T$ will vary.) For steady, inviscid, isothermal flow, Eq. 3.6 becomes

$$RT \int \frac{dp}{p} + \frac{1}{2} V^2 + gz = \text{constant}$$

where we have used $\rho = p/RT$. The pressure term is easily integrated and the constant of integration evaluated if $z_1$, $p_1$, and $V_1$ are known at some location on the streamline. The result is

$$\frac{V_1^2}{2g} + z_1 + \frac{RT}{g} \ln\left(\frac{p_1}{p_2}\right) = \frac{V_2^2}{2g} + z_2 \tag{3.23}$$

*The Bernoulli equation can be modified for compressible flows.*

Equation 3.23 is the inviscid, isothermal analog of the incompressible Bernoulli equation. In the limit of small pressure difference, $p_1/p_2 = 1 + (p_1 - p_2)/p_2 = 1 + \varepsilon$, with $\varepsilon \ll 1$ and Eq. 3.23 reduces to the standard incompressible Bernoulli equation. This can be shown by use of the approximation $\ln(1 + \varepsilon) \approx \varepsilon$ for small $\varepsilon$. The use of Eq. 3.23 in practical applications is restricted by the inviscid flow assumption, since (as is discussed in Section 11.5) most isothermal flows are accompanied by viscous effects.

A much more common compressible flow condition is that of isentropic (constant entropy) flow of a perfect gas. Such flows are reversible adiabatic processes—"no friction or heat transfer"—and are closely approximated in many physical situations. As discussed fully in Chapter 11, for isentropic flow of a perfect gas the density and pressure are related by $p/\rho^k = C$, where $k$ is the specific heat ratio and $C$ is a constant. Hence, the $\int dp/\rho$ integral of Eq. 3.6 can be evaluated as follows. The density can be written in terms of the pressure as $\rho = p^{1/k} C^{-1/k}$ so that Eq. 3.6 becomes

$$C^{1/k} \int p^{-1/k}\, dp + \frac{1}{2} V^2 + gz = \text{constant}$$

The pressure term can be integrated between points (1) and (2) on the streamline and the constant $C$ evaluated at either point ($C^{1/k} = p_1^{1/k}/\rho_1$ or $C^{1/k} = p_2^{1/k}/\rho_2$) to give the following:

$$C^{1/k} \int_{p1}^{p2} p^{-1/k}\, dp = C^{1/k}\left(\frac{k}{k-1}\right)\left[p_2^{(k-1)/k} - p_1^{(k-1)/k}\right]$$

$$= \left(\frac{k}{k-1}\right)\left(\frac{p_2}{\rho_2} - \frac{p_1}{\rho_1}\right)$$

Thus, the final form of Eq. 3.6 for compressible, isentropic, steady flow of a perfect gas is

$$\left(\frac{k}{k-1}\right)\frac{p_1}{\rho_1} + \frac{V_1^2}{2} + gz_1 = \left(\frac{k}{k-1}\right)\frac{p_2}{\rho_2} + \frac{V_2^2}{2} + gz_2 \tag{3.24}$$

The similarities between the results for compressible isentropic flow (Eq. 3.24) and incompressible isentropic flow (the Bernoulli equation, Eq. 3.7) are apparent. The only differences are the factors of $[k/(k-1)]$ that multiply the pressure terms and the fact that the densities are different ($\rho_1 \neq \rho_2$). In the limit of "low-speed flow" the two results are exactly the same, as is seen by the following.

We consider the stagnation point flow of Section 3.5 to illustrate the difference between the incompressible and compressible results. As is shown in Chapter 11, Eq. 3.24 can be written in dimensionless form as

$$\frac{p_2 - p_1}{p_1} = \left[\left(1 + \frac{k-1}{2}\mathrm{Ma}_1^2\right)^{k/k-1} - 1\right] \quad \text{(compressible)} \tag{3.25}$$

where (1) denotes the upstream conditions and (2) the stagnation conditions. We have assumed $z_1 = z_2$, $V_2 = 0$, and have denoted $\mathrm{Ma}_1 = V_1/c_1$ as the upstream *Mach number*—the ratio of the fluid velocity to the speed of sound, $c_1 = \sqrt{kRT_1}$.

A comparison between this compressible result and the incompressible result is perhaps most easily seen if we write the incompressible flow result in terms of the pressure ratio and the Mach number. Thus, we divide each term in the Bernoulli equation, $\rho V_1^2/2 + p_1 = p_2$, by $p_1$ and use the perfect gas law, $p_1 = \rho RT_1$, to obtain

$$\frac{p_2 - p_1}{p_1} = \frac{V_1^2}{2RT_1}$$

Since $\mathrm{Ma}_1 = V_1/\sqrt{kRT_1}$ this can be written as

$$\frac{p_2 - p_1}{p_1} = \frac{k\mathrm{Ma}_1^2}{2} \quad \text{(incompressible)} \tag{3.26}$$

Equations 3.25 and 3.26 are plotted in Fig. 3.24. In the low-speed limit of $\mathrm{Ma}_1 \to 0$, both of the results are the same. This can be seen by denoting $(k-1)\mathrm{Ma}_1^2/2 = \tilde{\varepsilon}$ and using the binomial expansion, $(1 + \tilde{\varepsilon})^n = 1 + n\tilde{\varepsilon} + n(n-1)\tilde{\varepsilon}^2/2 + \cdots$, where $n = k/(k-1)$, to write Eq. 3.25 as

*For small Mach numbers the compressible and incompressible results are nearly the same.*

$$\frac{p_2 - p_1}{p_1} = \frac{k\mathrm{Ma}_1^2}{2}\left(1 + \frac{1}{4}\mathrm{Ma}_1^2 + \frac{2-k}{24}\mathrm{Ma}_1^4 + \cdots\right) \quad \text{(compressible)}$$

For $\mathrm{Ma}_1 \ll 1$ this compressible flow result agrees with Eq. 3.26. The incompressible and compressible equations agree to within about 2% up to a Mach number of approximately $\mathrm{Ma}_1 = 0.3$. For larger Mach numbers the disagreement between the two results increases.

■ **Figure 3.24** Pressure ratio as a function of Mach number for incompressible and compressible (isentropic) flow.

Thus, a rule of thumb is that the flow of a perfect gas may be considered as incompressible provided the Mach number is less than about 0.3. In standard air ($T_1 = 59$ °F, $c_1 = \sqrt{kRT_1} = 1117$ ft/s) this corresponds to a speed of $V_1 = \text{Ma}_1 c_1 = 0.3(1117 \text{ ft/s}) = 335$ ft/s $= 228$ mi/hr. At higher speeds, compressibility may become important.

---

# EXAMPLE 3.15  Compressible Flow—Mach Number

**GIVEN**  The jet shown in Fig. E3.15 flies at Mach 0.82 at an altitude of 10 km in a standard atmosphere.

**FIND**  Determine the stagnation pressure on the leading edge of its wing if the flow is incompressible; and if the flow is compressible isentropic.

## SOLUTION

From Tables 1.8 and C.2 we find that $p_1 = 26.5$ kPa (abs), $T_1 = -49.9$ °C, $\rho = 0.414$ kg/m$^3$, and $k = 1.4$. Thus, if we assume incompressible flow, Eq. 3.26 gives

$$\frac{p_2 - p_1}{p_1} = \frac{k\text{Ma}_1^2}{2} = 1.4 \frac{(0.82)^2}{2} = 0.471$$

or

$$p_2 - p_1 = 0.471(26.5 \text{ kPa}) = 12.5 \text{ kPa} \qquad \textbf{(Ans)}$$

On the other hand, if we assume isentropic flow, Eq. 3.25 gives

$$\frac{p_2 - p_1}{p_1} = \left\{ \left[ 1 + \frac{(1.4 - 1)}{2} (0.82)^2 \right]^{1.4/(1.4-1)} - 1 \right\}$$
$$= 0.555$$

or

$$p_2 - p_1 = 0.555 (26.5 \text{ kPa}) = 14.7 \text{ kPa} \qquad \textbf{(Ans)}$$

**COMMENT**  We see that at Mach 0.82 compressibility effects are of importance. The pressure (and, to a first approximation, the

■ **Figure E3.15**  (© RobHowarth/iStockphoto)

lift and drag on the airplane; see Chapter 9) is approximately $14.7/12.5 = 1.18$ times greater according to the compressible flow calculations. This may be very significant. As discussed in Chapter 11, for Mach numbers greater than 1 (supersonic flow) the differences between incompressible and compressible results are often not only quantitative but also qualitative.

Note that if the airplane were flying at Mach 0.30 (rather than 0.82) the corresponding values would be $p_2 - p_1 = 1.670$ kPa for incompressible flow and $p_2 - p_1 = 1.707$ kPa for compressible flow. The difference between these two results is about 2%.

---

### 3.8.2  Unsteady Effects

Another restriction of the Bernoulli equation (Eq. 3.7) is the assumption that the flow is steady. For such flows, on a given streamline the velocity is a function of only $s$, the location along the streamline. That is, along a streamline $V = V(s)$. For unsteady flows the velocity is also a function of time, so that along a streamline $V = V(s, t)$. Thus when taking the time derivative of the velocity to obtain the streamwise acceleration, we obtain $a_s = \partial V/\partial t + V \partial V/\partial s$ rather than just $a_s = V \partial V/\partial s$ as is true for steady flow. For steady flows the acceleration is due to the change in velocity resulting from a change in position of the particle (the $V \partial V/\partial s$ term), whereas for unsteady flow there is an additional contribution to the acceleration resulting from a change in velocity with time at a fixed location (the $\partial V/\partial t$ term). These effects are discussed in detail in Chapter 4. The net effect is that the inclusion of the unsteady term, $\partial V/\partial t$, does not allow the equation of motion to be easily integrated (as was done to obtain the Bernoulli equation) unless additional assumptions are made.

*The Bernoulli equation can be modified for unsteady flows.*

The Bernoulli equation was obtained by integrating the component of Newton's second law (Eq. 3.5) along the streamline. When integrated, the acceleration contribution to this equation, the $\frac{1}{2}\rho d(V^2)$ term, gave rise to the kinetic energy term in the Bernoulli equation. If the steps leading to Eq. 3.5 are repeated with the inclusion of the unsteady effect $(\partial V/\partial t \neq 0)$, the following is obtained:

$$\rho \frac{\partial V}{\partial t} ds + dp + \frac{1}{2}\rho d(V^2) + \gamma \, dz = 0 \qquad \text{(along a streamline)}$$

For incompressible flow this can be easily integrated between points (1) and (2) to give

$$p_1 + \frac{1}{2}\rho V_1^2 + \gamma z_1 = \rho \int_{s_1}^{s_2} \frac{\partial V}{\partial t} ds + p_2 + \frac{1}{2}\rho V_2^2 + \gamma z_2 \qquad \text{(along a streamline)} \qquad \textbf{(3.27)}$$

*V3.11 Oscillations in a U-tube*

Equation 3.27 is an unsteady form of the Bernoulli equation valid for unsteady, incompressible, inviscid flow. Except for the integral involving the local acceleration, $\partial V/\partial t$, it is identical to the steady Bernoulli equation. In general, it is not easy to evaluate this integral because the variation of $\partial V/\partial t$ along the streamline is not known. In some situations the concepts of "irrotational flow" and the "velocity potential" can be used to simplify this integral. These topics are discussed in Chapter 6.

---

## EXAMPLE 3.16    Unsteady Flow—U-Tube

**GIVEN**   An incompressible, inviscid liquid is placed in a vertical, constant diameter U-tube as indicated in Fig. E3.16. When released from the nonequilibrium position shown, the liquid column will oscillate at a specific frequency.

**FIND**   Determine this frequency.

■ **Figure E3.16**

## SOLUTION

The frequency of oscillation can be calculated by use of Eq. 3.27 as follows. Let points (1) and (2) be at the air–water interfaces of the two columns of the tube and $z = 0$ correspond to the equilibrium position of these interfaces. Hence, $p_1 = p_2 = 0$ and if $z_2 = z$, then $z_1 = -z$. In general, $z$ is a function of time, $z = z(t)$. For a constant diameter tube, at any instant in time the fluid speed is constant throughout the tube, $V_1 = V_2 = V$, and the integral representing the unsteady effect in Eq. 3.27 can be written as

$$\int_{s_1}^{s_2} \frac{\partial V}{\partial t} ds = \frac{dV}{dt}\int_{s_1}^{s_2} ds = \ell \frac{dV}{dt}$$

where $\ell$ is the total length of the liquid column as shown in the figure. Thus, Eq. 3.27 can be written as

$$\gamma(-z) = \rho\ell \frac{dV}{dt} + \gamma z$$

Since $V = dz/dt$ and $\gamma = \rho g$, this can be written as the second-order differential equation describing simple harmonic motion

$$\frac{d^2z}{dt^2} + \frac{2g}{\ell}z = 0$$

which has the solution $z(t) = C_1 \sin(\sqrt{2g/\ell}\, t) + C_2 \cos (\sqrt{2g/\ell}\, t)$. The values of the constants $C_1$ and $C_2$ depend on the initial state (velocity and position) of the liquid at $t = 0$. Thus, the liquid oscillates in the tube with a frequency

$$\omega = \sqrt{2g/\ell} \qquad \textbf{(Ans)}$$

**COMMENT**   This frequency depends on the length of the column and the acceleration of gravity (in a manner very similar to the oscillation of a pendulum). The period of this oscillation (the time required to complete an oscillation) is $t_0 = 2\pi\sqrt{\ell/2g}$.

---

In a few unsteady flow cases, the flow can be made steady by an appropriate selection of the coordinate system. Example 3.17 illustrates this.

## $E$XAMPLE 3.17 Unsteady or Steady Flow

**GIVEN** A submarine moves through seawater ($SG = 1.03$) at a depth of 50 m with velocity $V_0 = 5.0$ m/s as shown in Fig. E3.17.

**FIND** Determine the pressure at the stagnation point (2).

## $S$OLUTION

In a coordinate system fixed to the ground, the flow is unsteady. For example, the water velocity at (1) is zero with the submarine in its initial position, but at the instant when the nose, (2), reaches point (1) the velocity there becomes $\mathbf{V}_1 = -V_0\hat{\mathbf{i}}$. Thus, $\partial\mathbf{V}_1/\partial t \neq 0$ and the flow is unsteady. Application of the steady Bernoulli equation between (1) and (2) would give the incorrect result that $p_1 = p_2 + \rho V_0^2/2$. According to this result, the static pressure is greater than the stagnation pressure—an incorrect use of the Bernoulli equation.

We can either use an unsteady analysis for the flow (which is outside the scope of this text) or redefine the coordinate system so that it is fixed on the submarine, giving steady flow with respect to this system. The correct method would be

$$p_2 = \frac{\rho V_1^2}{2} + \gamma h = [(1.03)(1000)\ \text{kg/m}^3]\ (5.0\ \text{m/s})^2/2$$
$$+ (9.80 \times 10^3\ \text{N/m}^3)(1.03)(50\ \text{m})$$

**Figure E3.17**

$$= (12,900 + 505,000)\ \text{N/m}^2$$
$$= 518\ \text{kPa} \qquad \textbf{(Ans)}$$

similar to that discussed in Example 3.2.

**COMMENT** If the submarine were accelerating, $\partial V_0/\partial t \neq 0$, the flow would be unsteady in either of the above coordinate systems and we would be forced to use an unsteady form of the Bernoulli equation.

---

Some unsteady flows may be treated as "quasisteady" and solved approximately by using the steady Bernoulli equation. In these cases the unsteadiness is "not too great" (in some sense), and the steady flow results can be applied at each instant in time as though the flow were steady. The slow draining of a tank filled with liquid provides an example of this type of flow.

### 3.8.3 Rotational Effects

*Care must be used in applying the Bernoulli equation across streamlines.*

Another of the restrictions of the Bernoulli equation is that it is applicable along the streamline. Application of the Bernoulli equation across streamlines (i.e., from a point on one streamline to a point on another streamline) can lead to considerable errors, depending on the particular flow conditions involved. In general, the Bernoulli constant varies from streamline to streamline. However, under certain restrictions this constant is the same throughout the entire flow field. Example 3.18 illustrates this fact.

## $E$XAMPLE 3.18 Use of Bernoulli Equation across Streamlines

**GIVEN** Consider the uniform flow in the channel shown in Fig. E3.18a. The liquid in the vertical piezometer tube is stationary.

**FIND** Discuss the use of the Bernoulli equation between points (1) and (2), points (3) and (4), and points (4) and (5).

## $S$OLUTION

If the flow is steady, inviscid, and incompressible, Eq. 3.7 written between points (1) and (2) gives

$$p_1 + \tfrac{1}{2}\rho V_1^2 + \gamma z_1 = p_2 + \tfrac{1}{2}\rho V_2^2 + \gamma z_2$$
$$= \text{constant} = C_{12}$$

**Figure E3.18**

Since $V_1 = V_2 = V_0$ and $z_1 = z_2 = 0$, it follows that $p_1 = p_2 = p_0$ and the Bernoulli constant for this streamline, $C_{12}$, is given by

$$C_{12} = \tfrac{1}{2}\rho V_0^2 + p_0$$

Along the streamline from (3) to (4) we note that $V_3 = V_4 = V_0$ and $z_3 = z_4 = h$. As was shown in Example 3.5, application of $\mathbf{F} = m\mathbf{a}$ across the streamline (Eq. 3.12) gives $p_3 = p_1 - \gamma h$ because the streamlines are straight and horizontal. The above facts combined with the Bernoulli equation applied between (3) and (4) show that $p_3 = p_4$ and that the Bernoulli constant along this streamline is the same as that along the streamline between (1) and (2). That is, $C_{34} = C_{12}$, or

$$p_3 + \tfrac{1}{2}\rho V_3^2 + \gamma z_3 = p_4 + \tfrac{1}{2}\rho V_4^2 + \gamma z_4 = C_{34} = C_{12}$$

Similar reasoning shows that the Bernoulli constant is the same for any streamline in Fig. E3.18. Hence,

$$p + \tfrac{1}{2}\rho V^2 + \gamma z = \text{constant throughout the flow}$$

Again from Example 3.5 we recall that

$$p_4 = p_5 + \gamma H = \gamma H$$

If we apply the Bernoulli equation across streamlines from (4) to (5), we obtain the incorrect result $H = p_4/\gamma + V_4^2/2g$. The correct result is $H = p_4/\gamma$.

From the above we see that we can apply the Bernoulli equation across streamlines (1)–(2) and (3)–(4) (i.e., $C_{12} = C_{34}$) but not across streamlines from (4) to (5). The reason for this is that while the flow in the channel is "irrotational," it is "rotational" between the flowing fluid in the channel and the stationary fluid in the piezometer tube. Because of the uniform velocity profile across the channel, it is seen that the fluid particles do not rotate or "spin" as they move. The flow is "irrotational." However, as seen in Fig. E3.18b, there is a very thin shear layer between (4) and (5) in which adjacent fluid particles interact and rotate or "spin." This produces a "rotational" flow. A more complete analysis would show that the Bernoulli equation cannot be applied across streamlines if the flow is "rotational" (see Chapter 6).

*V3.12 Flow over a cavity*

As is suggested by Example 3.18, if the flow is "irrotational" (i.e., the fluid particles do not "spin" as they move), it is appropriate to use the Bernoulli equation across streamlines. However, if the flow is "rotational" (fluid particles "spin"), use of the Bernoulli equation is restricted to flow along a streamline. The distinction between irrotational and rotational flow is often a very subtle and confusing one. These topics are discussed in more detail in Chapter 6. A thorough discussion can be found in more advanced texts (Ref. 3).

### 3.8.4  Other Restrictions

Another restriction on the Bernoulli equation is that the flow is inviscid. As is discussed in Section 3.4, the Bernoulli equation is actually a first integral of Newton's second law along a streamline. This general integration was possible because, in the absence of viscous effects, the fluid system considered was a conservative system. The total energy of the system remains constant. If viscous effects are important the system is nonconservative (dissipative) and energy losses occur. A more detailed analysis is needed for these cases. Such material is presented in Chapter 5.

*The Bernoulli equation is not valid for flows that involve pumps or turbines.*

The final basic restriction on use of the Bernoulli equation is that there are no mechanical devices (pumps or turbines) in the system between the two points along the streamline for which the equation is applied. These devices represent sources or sinks of energy. Since the Bernoulli equation is actually one form of the energy equation, it must be altered to include pumps or turbines, if these are present. The inclusion of pumps and turbines is covered in Chapters 5 and 12.

In this chapter we have spent considerable time investigating fluid dynamic situations governed by a relatively simple analysis for steady, inviscid, incompressible flows. Many flows can be adequately analyzed by use of these ideas. However, because of the rather severe restrictions imposed, many others cannot. An understanding of these basic ideas will provide a firm foundation for the remainder of the topics in this book.

## 3.9    Chapter Summary and Study Guide

In this chapter, several aspects of the steady flow of an inviscid, incompressible fluid are discussed. Newton's second law, $\mathbf{F} = m\mathbf{a}$, is applied to flows for which the only important forces are those due to pressure and gravity (weight)—viscous effects are assumed negligible. The result is the often-used Bernoulli equation, which provides a simple relationship among pressure, elevation, and velocity variations along a streamline. A similar but less often used equation is also obtained to describe the variations in these parameters normal to a streamline.

The concept of a stagnation point and the corresponding stagnation pressure is introduced, as are the concepts of static, dynamic, and total pressure and their related heads.

*steady flow*
*streamline*
*Bernoulli equation*
*elevation head*
*pressure head*
*velocity head*
*static pressure*
*dynamic pressure*
*stagnation point*
*stagnation pressure*
*total pressure*
*Pitot-static tube*
*free jet*
*volume flowrate*
*continuity equation*
*cavitation*
*flowmeter*
*hydraulic grade line*
*energy line*

Several applications of the Bernoulli equation are discussed. In some flow situations, such as the use of a Pitot-static tube to measure fluid velocity or the flow of a liquid as a free jet from a tank, a Bernoulli equation alone is sufficient for the analysis. In other instances, such as confined flows in tubes and flowmeters, it is necessary to use both the Bernoulli equation and the continuity equation, which is a statement of the fact that mass is conserved as fluid flows.

The following checklist provides a study guide for this chapter. When your study of the entire chapter and end-of-chapter exercises has been completed, you should be able to

■ write out meanings of the terms listed here in the margin and understand each of the related concepts. These terms are particularly important and are set in ***italic, bold, and color*** type in the text.

■ explain the origin of the pressure, elevation, and velocity terms in the Bernoulli equation and how they are related to Newton's second law of motion.

■ apply the Bernoulli equation to simple flow situations, including Pitot-static tubes, free jet flows, confined flows, and flowmeters.

■ use the concept of conservation of mass (the continuity equation) in conjunction with the Bernoulli equation to solve simple flow problems.

■ apply Newton's second law across streamlines for appropriate steady, inviscid, incompressible flows.

■ use the concepts of pressure, elevation, velocity, and total heads to solve various flow problems.

■ explain and use the concepts of static, stagnation, dynamic, and total pressures.

■ use the energy line and the hydraulic grade line concepts to solve various flow problems.

■ explain the various restrictions on use of the Bernoulli equation.

Some of the important equations in this chapter are:

| | | |
|---|---|---|
| Streamwise and normal acceleration | $a_s = V\dfrac{\partial V}{\partial s}, \qquad a_n = \dfrac{V^2}{\mathcal{R}}$ | **(3.1)** |
| Force balance along a streamline for steady inviscid flow | $\displaystyle\int \dfrac{dp}{\rho} + \dfrac{1}{2}V^2 + gz = C \qquad$ (along a streamline) | **(3.6)** |
| Bernoulli equation | $p + \tfrac{1}{2}\rho V^2 + \gamma z = $ constant along streamline | **(3.7)** |
| Pressure gradient normal to streamline for inviscid flow in absence of gravity | $\dfrac{\partial p}{\partial n} = -\dfrac{\rho V^2}{\mathcal{R}}$ | **(3.10b)** |
| Force balance normal to a streamline for steady, inviscid, incompressible flow | $p + \rho \displaystyle\int \dfrac{V^2}{\mathcal{R}}\, dn + \gamma z = $ constant across the streamline | **(3.12)** |
| Velocity measurement for a Pitot-static tube | $V = \sqrt{2\,(p_3 - p_4)/\rho}$ | **(3.16)** |
| Free jet | $V = \sqrt{2\,\dfrac{\gamma h}{\rho}} = \sqrt{2gh}$ | **(3.18)** |
| Continuity equation | $A_1 V_1 = A_2 V_2,$ or $Q_1 = Q_2$ | **(3.19)** |
| Flowmeter equation | $Q = A_2 \sqrt{\dfrac{2(p_1 - p_2)}{\rho[1 - (A_2/A_1)^2]}}$ | **(3.20)** |
| Sluice gate equation | $Q = z_2 b \sqrt{\dfrac{2g(z_1 - z_2)}{1 - (z_2/z_1)^2}}$ | **(3.21)** |
| Total head | $\dfrac{p}{\gamma} + \dfrac{V^2}{2g} + z = $ constant on a streamline $= H$ | **(3.22)** |

## References

1. Riley, W. F., and Sturges, L. D., *Engineering Mechanics: Dynamics,* 2nd Ed., Wiley, New York, 1996.
2. Tipler, P. A., *Physics,* Worth, New York, 1982.
3. Panton, R. L., *Incompressible Flow,* Wiley, New York, 1984.

| | |
|---|---|
|  | Problem available in *WileyPLUS* at instructor's discretion. |
|  | Tutoring problem available in *WileyPLUS* at instructor's discretion. |
|  | Problem is related to a chapter video available in *WileyPLUS*. |
| ✳ | Problem to be solved with aid of programmable calculator or computer. |
| † | Open-ended problem that requires critical thinking. These problems require various assumptions to provide the necessary input data. There are not unique answers to these problems. |

## Review Problems

Go to Appendix G (*WileyPLUS* or the book's web site, www. wiley.com/college/munson) for a set of review problems with answers. Detailed solutions can be found in the *Student Solution Manual and Study Guide for Fundamentals of Fluid Mechanics,* by Munson et al. (© 2013 John Wiley and Sons, Inc.).

## Conceptual Questions

**3.1C** In low-speed flow, the following device is called a diffuser:

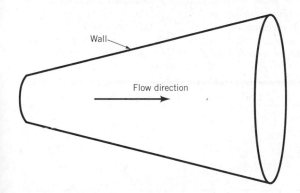

The fluid velocity:

a) increases in the direction of flow.

b) decreases in the direction of flow.

c) stays the same in the direction of flow.

d) None of the above.

**3.2C** Consider constant altitude, steady flow along a streamline with a flow that satisfies the assumptions necessary for Bernoulli's equation. Which of the following have a constant value along the streamline?

a) Internal energy    c) Stagnation or total pressure

b) Static or local pressure    d) Dynamic or velocity pressure

**3.3C** The stagnation pressure for a fluid is

a) the pressure due to the velocity of the fluid.

b) not defined unless the fluid is moving.

c) the total pressure when the fluid is brought to rest.

d) the sum of the static and dynamic pressures.

**3.4C** Pitot tubes are placed in two ducts in which air is flowing and the static pressures of the flows are equal. The stagnation pressure tap and the static pressure tap are connected to manometers as shown. The manometer reading for duct A is height h, and that for duct B is 2h. The relation between the velocity $V_A$ in duct A and the velocity $V_B$ in duct B is

a) $V_B$ equals $V_A/2$.

b) $V_B$ equals $\sqrt{2}V_A$.

c) $V_B$ equals $4V_A$.

d) $V_B$ equals $2V_A$.

e) $V_B$ equals $V_A$.

**Additional conceptual questions are available in *WileyPLUS* at the instructor's discretion.**

## Problems

**Note: Unless specific values of required fluid properties are given in the problem statement, use the values found in the tables on the inside of the front cover. Answers to the even-numbered problems are listed at the end of the book. The Lab Problems as well as the videos that accompany problems can be accessed in *WileyPLUS* or the book's web site, www.wiley.com/college/munson.**

### Section 3.2 F = *m*a along a Streamline

**3.1** Obtain a photograph/image of a situation that can be analyzed by use of the Bernoulli equation. Print this photo and write a brief paragraph that describes the situation involved.

**3.2** Air flows steadily along a streamline from point (1) to point (2) with negligible viscous effects. The following conditions are measured: At point (1) $z_1 = 2$ m and $p_1 = 0$ kPa; at point (2) $z_2 = 10$ m, $p_2 = 20$ N/m$^2$, and $V_2 = 0$. Determine the velocity at point (1).

**3.3** (WILEY) Water flows steadily through the variable area horizontal pipe shown in Fig. P3.3. The centerline velocity is given by $\mathbf{V} = 10(1 + x)\,\hat{\mathbf{i}}$ ft/s, where $x$ is in feet. Viscous effects are neglected. **(a)** Determine the pressure gradient, $\partial p/\partial x$, (as a function of $x$) needed to produce this flow. **(b)** If the pressure at section (1) is 50 psi, determine the pressure at (2) by (i) integration of the pressure gradient obtained in **(a)**, (ii) application of the Bernoulli equation.

■ **Figure P3.3**

**3.4** (WILEY) What pressure gradient along the streamline, $dp/ds$, is required to accelerate water in a horizontal pipe at a rate of 30 m/s$^2$?

**3.5** (WILEY GO) At a given location the airspeed is 20 m/s and the pressure gradient along the streamline is 100 N/m$^3$. Estimate the airspeed at a point 0.5 m farther along the streamline.

**3.6** (WILEY) What pressure gradient along the streamline, $dp/ds$, is required to accelerate water upward in a vertical pipe at a rate of 30 ft/s$^2$? What is the answer if the flow is downward?

**3.7** The Bernoulli equation is valid for steady, inviscid, incompressible flows with constant acceleration of gravity. Consider flow on a planet where the acceleration of gravity varies with height so that $g = g_0 - cz$, where $g_0$ and $c$ are constants. Integrate "$\mathbf{F} = m\mathbf{a}$" along a streamline to obtain the equivalent of the Bernoulli equation for this flow.

**3.8** An incompressible fluid flows steadily past a circular cylinder as shown in Fig. P3.8. The fluid velocity along the dividing streamline ($-\infty \le x \le -a$) is found to be $V = V_0(1 - a^2/x^2)$, where $a$ is the radius of the cylinder and $V_0$ is the upstream velocity. **(a)** Determine the pressure gradient along this streamline. **(b)** If the upstream pressure is $p_0$, integrate the pressure gradient to obtain the pressure $p(x)$ for $-\infty \le x \le -a$. **(c)** Show from the result of part **(b)** that the pressure at the stagnation point ($x = -a$) is $p_0 + \rho V_0^2/2$, as expected from the Bernoulli equation.

■ **Figure P3.8**

**3.9** Consider a compressible liquid that has a constant bulk modulus. Integrate "$\mathbf{F} = m\mathbf{a}$" along a streamline to obtain the equivalent of the Bernoulli equation for this flow. Assume steady, inviscid flow.

### Section 3.3 F = ma Normal to a Streamline

**3.10** Obtain a photograph/image of a situation in which Newton's second law applied across the streamlines (as given by Eq. 3.12)

is important. Print this photo and write a brief paragraph that describes the situation involved.

**3.11** (WILEY) Air flows along a horizontal, curved streamline with a 20 ft radius with a speed of 100 ft/s. Determine the pressure gradient normal to the streamline.

**3.12** (WILEY) Water flows around the vertical two-dimensional bend with circular streamlines and constant velocity as shown in Fig. P3.12. If the pressure is 40 kPa at point (1), determine the pressures at points (2) and (3). Assume that the velocity profile is uniform as indicated.

■ **Figure P3.12**

**3.13** (WILEY) Water flows around the vertical two-dimensional bend with circular streamlines as is shown in Fig. P3.13. The pressure at point (1) is measured to be $p_1 = 25$ psi and the velocity across section $a–a$ is as indicated in the table. Calculate and plot the pressure across section $a–a$ of the channel [$p = p(z)$ for $0 \le z \le 2$ ft].

| $z$ (ft) | $V$ (ft/s) |
|---|---|
| 0 | 0 |
| 0.2 | 8.0 |
| 0.4 | 14.3 |
| 0.6 | 20.0 |
| 0.8 | 19.5 |
| 1.0 | 15.6 |
| 1.2 | 8.3 |
| 1.4 | 6.2 |
| 1.6 | 3.7 |
| 1.8 | 2.0 |
| 2.0 | 0 |

■ **Figure P3.13**

**3.14**  Water in a container and air in a tornado flow in horizontal circular streamlines of radius $r$ and speed $V$ as shown in **Video V3.6** and Fig. P3.14. Determine the radial pressure gradient, $\partial p/\partial r$, needed for the following situations: **(a)** The fluid is water with $r = 3$ in. and $V = 0.8$ ft/s. **(b)** The fluid is air with $r = 300$ ft and $V = 200$ mph.

■ **Figure P3.14**

**3.15** Air flows smoothly over the hood of your car and up past the windshield. However, a bug in the air does not follow the same path; it becomes splattered against the windshield. Explain why this is so.

## Section 3.5 Static, Stagnation, Dynamic, and Total Pressure

**3.16** Obtain a photograph/image of a situation in which the concept of the stagnation pressure is important. Print this photo and write a brief paragraph that describes the situation involved.

**3.17** At a given point on a horizontal streamline in flowing air, the static pressure is $-2.0$ psi (i.e., a vacuum) and the velocity is 150 ft/s. Determine the pressure at a stagnation point on that streamline.

**†3.18** A drop of water in a zero-g environment (as in the International Space Station) will assume a spherical shape as shown in Fig. P3.18a. A raindrop in the cartoons is typically drawn as in Fig. P3.18b. The shape of an actual raindrop is more nearly like that shown in Fig. 3.18c. Discuss why these shapes are as indicated.

*(a)*          *(b)*          *(c)*

■ **Figure P3.18**

**3.19** When an airplane is flying 200 mph at 5000-ft altitude in a standard atmosphere, the air velocity at a certain point on the wing is 273 mph relative to the airplane. **(a)** What suction pressure is developed on the wing at that point? **(b)** What is the pressure at the leading edge (a stagnation point) of the wing?

**3.20** Some animals have learned to take advantage of the Bernoulli effect without having read a fluid mechanics book. For example, a typical prairie dog burrow contains two entrances—a flat front door and a mounded back door as shown in Fig. P3.20. When the wind blows with velocity $V_0$ across the front door, the average velocity across the back door is greater than $V_0$ because of the mound. Assume the air velocity across the back door is $1.07V_0$. For a wind velocity of 6 m/s, what pressure differences,

$p_1 - p_2$, are generated to provide a fresh airflow within the burrow?

■ **Figure P3.20**

**3.21** A loon is a diving bird equally at home "flying" in the air or water. What swimming velocity under water will produce a dynamic pressure equal to that when it flies in the air at 40 mph?

**†3.22** Estimate the pressure on your hand when you hold it in the stream of air coming from the air hose at a filling station. List all assumptions and show calculations. Warning: Do not try this experiment; it can be dangerous!

**3.23** A person holds her hand out of an open car window while the car drives through still air at 65 mph. Under standard atmospheric conditions, what is the maximum pressure on her hand? What would be the maximum pressure if the "car" were an Indy 500 racer traveling 220 mph?

**3.24** A Pitot-static tube is used to measure the velocity of helium in a pipe. The temperature and pressure are 40 °F and 25 psia. A water manometer connected to the Pitot-static tube indicates a reading of 2.3 in. Determine the helium velocity. Is it reasonable to consider the flow as incompressible? Explain.

**3.25** A Bourdon-type pressure gage is used to measure the pressure from a Pitot tube attached to the leading edge of an airplane wing. The gage is calibrated to read in miles per hour at standard sea level conditions (rather than psi). If the airspeed meter indicates 150 mph when flying at an altitude of 10,000 ft, what is the true airspeed?

**†3.26** Estimate the force of a hurricane strength wind against the side of your house. List any assumptions and show all calculations.

**3.27** A 40-mph wind blowing past your house speeds up as it flows up and over the roof. If elevation effects are negligible, determine **(a)** the pressure at the point on the roof where the speed is 60 mph if the pressure in the free stream blowing toward your house is 14.7 psia. Would this effect tend to push the roof down against the house, or would it tend to lift the roof? **(b)** Determine the pressure on a window facing the wind if the window is assumed to be a stagnation point.

**3.28** (See Fluids in the News article titled "**Pressurized eyes**," Section 3.5.) Determine the air velocity needed to produce a stagnation pressure equal to 10 mm of mercury.

## Section 3.6.1 Free Jets

**3.29** Water flows through a hole in the bottom of a large, open tank with a speed of 8 m/s. Determine the depth of water in the tank. Viscous effects are negligible.

**†3.30** Estimate the pressure needed at the pumper truck in order to shoot water from the street level onto a fire on the roof of a five-story building. List all assumptions and show all calculations.

**3.31** Water flows from the faucet on the first floor of the building shown in Fig. P3.31 with a maximum velocity of 20 ft/s. For steady inviscid flow, determine the maximum water velocity from the basement faucet and from the faucet on the second floor (assume each floor is 12 ft tall).

■ **Figure P3.31**

**3.32** Laboratories containing dangerous materials are often kept at a pressure slightly less than ambient pressure so that contaminants can be filtered through an exhaust system rather than leaked through cracks around doors, etc. If the pressure in such a room is 0.1 in. of water below that of the surrounding rooms, with what velocity will air enter the room through an opening? Assume viscous effects are negligible.

**†3.33** The "supersoaker" water gun shown in Fig. P3.33 can shoot more than 30 ft in the horizontal direction. Estimate the minimum pressure, $p_1$, needed in the chamber in order to accomplish this. List all assumptions and show all calculations.

■ **Figure P3.33**

**3.34** (WILEY) Streams of water from two tanks impinge upon each other as shown in Fig. P3.34. If viscous effects are negligible and point $A$ is a stagnation point, determine the height $h$.

■ **Figure P3.34**

**3.35** Several holes are punched into a tin can as shown in Fig. P3.35. Which of the figures represents the variation of the water velocity as it leaves the holes? Justify your choice.

■ **Figure P3.35**

**3.36** (WILEY) Water flows from a pressurized tank, through a 6-in.-diameter pipe, exits from a 2-in.-diameter nozzle, and rises 20 ft above the nozzle as shown in Fig. P3.36. Determine the pressure in the tank if the flow is steady, frictionless, and incompressible.

■ **Figure P3.36**

**3.37** (WILEY) An inviscid, incompressible liquid flows steadily from the large pressurized tank shown in Fig. P.3.37. The velocity at the exit is 40 ft/s. Determine the specific gravity of the liquid in the tank.

■ **Figure P3.37**

## Section 3.6.2  Confined Flows (also see Lab Problems 3.1LP and 3.3LP)

**3.38** Obtain a photograph/image of a situation that involves a confined flow for which the Bernoulli and continuity equations are important. Print this photo and write a brief paragraph that describes the situation involved.

**3.39** Air flows steadily through a horizontal 4-in.-diameter pipe and exits into the atmosphere through a 3-in.-diameter nozzle. The velocity at the nozzle exit is 150 ft/s. Determine the pressure in the pipe if viscous effects are negligible.

**3.40** For the pipe enlargement shown in Fig. P3.40, the pressures at sections (1) and (2) are 56.3 and 58.2 psi, respectively. Determine the weight flowrate (lb/s) of the gasoline in the pipe.

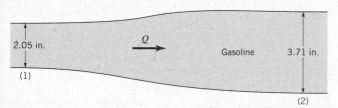

■ **Figure P3.40**

**3.41** WILEY GO A fire hose nozzle has a diameter of $1\frac{1}{8}$ in. According to some fire codes, the nozzle must be capable of delivering at least 250 gal/min. If the nozzle is attached to a 3-in.-diameter hose, what pressure must be maintained just upstream of the nozzle to deliver this flowrate?

**3.42** ▶ WILEY Water flowing from the 0.75-in.-diameter outlet shown in **Video V8.15** and Fig. P3.42 rises 2.8 in. above the outlet. Determine the flowrate.

■ **Figure P3.42**

**3.43** ▶ Pop (with the same properties as water) flows from a 4-in.-diameter pop container that contains three holes as shown in Fig. P3.43 (see **Video 3.9**). The diameter of each fluid stream is 0.15 in., and the distance between holes is 2 in. If viscous effects are negligible and quasi-steady conditions are assumed, determine the time at which the pop stops draining from the top hole. Assume the pop surface is 2 in. above the top hole when $t = 0$. Compare your results with the time you measure from the video.

■ **Figure P3.43**

**3.44** Water flows steadily through the large tanks shown in Fig. P3.44. Determine the water depth, $h_A$.

■ **Figure P3.44**

**3.45** Water (assumed inviscid and incompressible) flows steadily in the vertical variable-area pipe shown in Fig. P3.45. Determine the flowrate if the pressure in each of the gages reads 50 kPa.

■ **Figure P3.45**

**3.46** WILEY Air is drawn into a wind tunnel used for testing automobiles as shown in Fig. P3.46. **(a)** Determine the manometer reading, $h$, when the velocity in the test section is 60 mph. Note that there is a 1-in. column of oil on the water in the manometer. **(b)** Determine the difference between the stagnation pressure on the front of the automobile and the pressure in the test section.

■ **Figure P3.46**

**3.47** Natural gas (methane) flows from a 3-in.-diameter gas main, through a 1-in.-diameter pipe, and into the burner of a furnace at a rate of 100 ft³/hour. Determine the pressure in the gas main if the pressure in the 1-in. pipe is to be 6 in. of water greater than atmospheric pressure. Neglect viscous effects.

**3.48** WILEY Small-diameter, high-pressure liquid jets can be used to cut various materials as shown in Fig. P3.48. If viscous effects are negligible, estimate the pressure needed to produce a 0.10-mm-diameter water jet with a speed of 700 m/s. Determine the flowrate.

■ **Figure P3.48**

**3.49** Water (assumed frictionless and incompressible) flows steadily from a large tank and exits through a vertical, constant diameter pipe as shown in Fig. P3.49. The air in the tank is pressurized to 50 kN/m². Determine **(a)** the height $h$, to which the water rises, **(b)** the water velocity in the pipe, and **(c)** the pressure in the horizontal part of the pipe.

■ **Figure P3.49**

**3.50** (WILEY) Water (assumed inviscid and incompressible) flows steadily with a speed of 10 ft/s from the large tank shown in Fig. P3.50. Determine the depth, $H$, of the layer of light liquid (specific weight = 50 lb/ft³) that covers the water in the tank.

■ **Figure P3.50**

**3.51** Water flows through the pipe contraction shown in Fig. P3.51. For the given 0.2-m difference in manometer level, determine the flowrate as a function of the diameter of the small pipe, $D$.

■ **Figure P3.51**

**3.52** (WILEY) Water flows through the pipe contraction shown in Fig. P3.52. For the given 0.2-m difference in the manometer level, determine the flowrate as a function of the diameter of the small pipe, $D$.

■ **Figure P3.52**

**3.53** (WILEY) A 0.15-m-diameter pipe discharges into a 0.10-m-diameter pipe. Determine the velocity head in each pipe if they are carrying 0.12 m³/s of kerosene.

**3.54** (WILEY) Carbon tetrachloride flows in a pipe of variable diameter with negligible viscous effects. At point $A$ in the pipe the pressure and velocity are 20 psi and 30 ft/s, respectively. At location $B$ the pressure and velocity are 23 psi and 14 ft/s. Which point is at the higher elevation and by how much?

**\*3.55** Water flows from a 20-mm-diameter pipe with a flowrate $Q$ as shown in Fig. P3.55. Plot the diameter of the water stream, $d$, as a function of distance below the faucet, $h$, for values of $0 \leq h \leq 1$ m and $0 \leq Q \leq 0.004$ m³/s. Discuss the validity of the one-dimensional assumption used to calculate $d = d(h)$, noting, in particular, the conditions of small $h$ and small $Q$.

■ **Figure P3.55**

**3.56** Water flows upward through a variable area pipe with a constant flowrate, $Q$, as shown in Fig. P3.56. If viscous effects are negligible, determine the diameter, $D(z)$, in terms of $D_1$ if the pressure is to remain constant throughout the pipe. That is, $p(z) = p_1$.

■ **Figure P3.56**

**3.57** (WILEY) The circular stream of water from a faucet is observed to taper from a diameter of 20 mm to 10 mm in a distance of 50 cm. Determine the flowrate.

**3.58** (WILEY GO) Water is siphoned from the tank shown in Fig. P3.58. The water barometer indicates a reading of 30.2 ft. Determine the maximum value of $h$ allowed without cavitation occurring. Note that the pressure of the vapor in the closed end of the barometer equals the vapor pressure.

■ **Figure P3.58**

**3.59** Water is siphoned from a tank as shown in Fig. P3.59. Determine the flowrate and the pressure at point $A$, a stagnation point.

■ **Figure P3.59**

**3.60** A 50-mm-diameter plastic tube is used to siphon water from the large tank shown in Fig. P3.60. If the pressure on the outside of the tube is more than 30 kPa greater than the pressure within the tube, the tube

will collapse and siphon will stop. If viscous effects are negligible, determine the minimum value of $h$ allowed without the siphon stopping.

■ **Figure P3.60**

**3.61** Water is siphoned from the tank shown in Fig. P3.61. Determine the flowrate from the tank and the pressure at points (1), (2), and (3) if viscous effects are negligible.

■ **Figure P3.61**

**3.62** Redo Problem 3.61 if a 1-in-diameter nozzle is placed at the end of the tube.

**3.63** (WILEY) A smooth plastic, 10-m-long garden hose with an inside diameter of 20 mm is used to drain a wading pool as is shown in Fig. P3.63. If viscous effects are neglected, what is the flowrate from the pool?

■ **Figure P3.63**

**3.64** (WILEY) Water exits a pipe as a free jet and flows to a height $h$ above the exit plane as shown in Fig. P3.64. The flow is steady, incompressible, and frictionless. (a) Determine the height $h$. (b) Determine the velocity and pressure at section (1).

■ **Figure P3.64**

**3.65** (WILEY) Water flows steadily from a large, closed tank as shown in Fig. P3.65. The deflection in the mercury manometer is 1 in. and viscous effects are negligible. **(a)** Determine the volume flowrate. **(b)** Determine the air pressure in the space above the surface of the water in the tank.

■ **Figure P3.65**

**3.66** Carbon dioxide flows at a rate of 1.5 ft³/s from a 3-in. pipe in which the pressure and temperature are 20 psi (gage) and 120 °F into a 1.5-in. pipe. If viscous effects are neglected and incompressible conditions are assumed, determine the pressure in the smaller pipe.

**3.67** Oil of specific gravity 0.83 flows in the pipe shown in Fig. P3.67. If viscous effects are neglected, what is the flowrate?

■ **Figure P3.67**

**3.68** Water flows steadily through the variable area pipe shown in Fig. P3.68 with negligible viscous effects. Determine the manometer reading, $H$, if the flowrate is 0.5 m³/s and the density of the manometer fluid is 600 kg/m³.

■ **Figure P3.68**

**3.69** Determine the flowrate through the pipe in Fig. P3.69.

■ **Figure P3.69**

**3.70** The specific gravity of the manometer fluid shown in Fig. P3.70 is 1.07. Determine the volume flowrate, $Q$, if the flow is inviscid and incompressible and the flowing fluid is **(a)** water, **(b)** gasoline, or **(c)** air at standard conditions.

■ **Figure P3.70**

**3.71** (WILEY) Water flows steadily with negligible viscous effects through the pipe shown in Fig. P3.71. It is known that the 4-in.-diameter section of thin-walled tubing will collapse if the pressure within it becomes less than 10 psi below atmospheric pressure. Determine the maximum value that $h$ can have without causing collapse of the tubing.

■ **Figure P3.71**

**3.72** (WILEY) Helium flows through a 0.30-m-diameter horizontal pipe with a temperature of 20 °C and a pressure of 200 kPa (abs) at a rate of 0.30 kg/s. If the pipe reduces to 0.25-m-diameter, determine the pressure difference between these two sections. Assume incompressible, inviscid flow.

**3.73** (WILEY) Water is pumped from a lake through an 8-in. pipe at a rate of 10 ft³/s. If viscous effects are negligible, what is the pressure in the suction pipe (the pipe between the lake and the pump) at an elevation 6 ft above the lake?

**3.74** Air is drawn into a small open-circuit wing tunnel as shown in Fig. P3.74. Atmospheric pressure is 98.7 kPa (abs) and the temperature is 27 °C. If viscous effects are negligible, determine the pressure at the stagnation point on the nose of the airplane. Also determine the manometer reading, $h$, for the manometer attached to the static pressure tap within the test section of the wind tunnel if the air velocity within the test section is 50 m/s.

**Figure P3.74**

**Figure P3.78**

**3.75** Air flows through the device shown in Fig. P3.75. If the flowrate is large enough, the pressure within the constriction will be low enough to draw the water up into the tube. Determine the flowrate, $Q$, and the pressure needed at section (1) to draw the water into section (2). Neglect compressibility and viscous effects.

**Figure P3.75**

**3.76** ⟨WILEY⟩ Water flows steadily from the large open tank shown in Fig. 3.76. If viscous effects are negligible, determine **(a)** the flowrate, $Q$, and **(b)** the manometer reading, $h$.

**Figure P3.76**

**3.77** ⟨WILEY⟩ Water from a faucet fills a 16-oz glass (volume = 28.9 in.$^3$) in 20 s. If the diameter of the jet leaving the faucet is 0.60 in., what is the diameter of the jet when it strikes the water surface in the glass which is positioned 14 in. below the faucet?

**3.78** ▶ Air flows steadily through a converging–diverging rectangular channel of constant width as shown in Fig. 3.78 and **Video V3.10**. The height of the channel at the exit and the exit velocity are $H_0$ and $V_0$ respectively. The channel is to be shaped so that the distance, $d$, that water is drawn up into tubes attached to static pressure taps along the channel wall is linear with distance along the channel. That is $d = (d_{max}/L)\,x$, where $L$ is the channel length and $d_{max}$ is the maximum water depth (at the minimum channel height: $x = L$). Determine the height, $H(x)$, as a function of $x$ and the other important parameters.

***3.79** Water flows from a large tank and through a pipe of variable area as shown in Fig. 3.79. The area of the pipe is given by $A = A_0[1 - x(1 - x/\ell)/2\ell]$, where $A_0$ is the area at the beginning ($x = 0$) and end ($x = \ell$) of the pipe. Plot graphs of the pressure within the pipe as a function of distance along the pipe for water depths of $h = 1, 4, 10$, and 25 m.

**Figure P3.79**

**3.80** ⟨WILEY⟩ If viscous effects are neglected and the tank is large, determine the flowrate from the tank shown in Fig. P3.80.

**Figure P3.80**

**3.81** Water flows steadily downward in the pipe shown in Fig. P.3.81 with negligible losses. Determine the flowrate.

■ **Figure P3.81**

■ **Figure P3.84**

**3.82**  Water flows steadily from a large open tank and discharges into the atmosphere though a 3-in.-diameter pipe as shown in Fig. P3.82. Determine the diameter, $d$, in the narrowed section of the pipe at $A$ if the pressure gages at $A$ and $B$ indicate the same pressure.

■ **Figure P3.82**

**3.83**  (WILEY) Water flows from a large tank as shown in Fig. P3.83. Atmospheric pressure is 14.5 psia, and the vapor pressure is 1.60 psia. If viscous effects are neglected, at what height, $h$, will cavitation begin? To avoid cavitation, should the value of $D_1$ be increased or decreased? To avoid cavitation, should the value of $D_2$ be increased or decreased? Explain.

■ **Figure P3.83**

**3.84**  ▶ (WILEY) Water flows into the sink shown in Fig. P3.84 and **Video V5.1** at a rate of 2 gal/min. If the drain is closed, the water will eventually flow through the overflow drain holes rather than over the edge of the sink. How many 0.4-in.-diameter drain holes are needed to ensure that the water does not overflow the sink? Neglect viscous effects.

**3.85**  (WILEY) What pressure, $p_1$, is needed to produce a flowrate of 0.09 ft³/s from the tank shown in Fig. P3.85?

■ **Figure P3.85**

**3.86**  The vent on the tank shown in Fig. P3.86 is closed and the tank pressurized to increase the flowrate. What pressure, $p_1$, is needed to produce twice the flowrate of that when the vent is open?

■ **Figure P3.86**

**3.87**  (WILEY) Water is siphoned from the tank shown in Fig. P3.87. Determine the flowrate from the tank and the pressures at points (1), (2), and (3) if viscous effects are negligible.

■ **Figure P3.87**

■ **Figure P3.90**

**3.88** (WILEY) Water is siphoned from a large tank and discharges into the atmosphere through a 2-in.-diameter tube as shown in Fig. P3.88. The end of the tube is 3 ft below the tank bottom, and viscous effects are negligible. **(a)** Determine the volume flowrate from the tank. **(b)** Determine the maximum height, $H$, over which the water can be siphoned without cavitation occurring. Atmospheric pressure is 14.7 psia, and the water vapor pressure is 0.26 psia.

**3.91** (WILEY) JP-4 fuel ($SG = 0.77$) flows through the Venturi meter shown in Fig. P3.91 with a velocity of 15 ft/s in the 6-in. pipe. If viscous effects are negligible, determine the elevation, $h$, of the fuel in the open tube connected to the throat of the Venturi meter.

■ **Figure P3.88**

■ **Figure P3.91**

**3.89** (WILEY) Determine the manometer reading, $h$, for the flow shown in Fig. P3.89.

**3.92** (WILEY) Water, considered an inviscid, incompressible fluid, flows steadily as shown in Fig. P3.92. Determine $h$.

■ **Figure P3.89**

■ **Figure P3.92**

**3.90** Water flows steadily from the pipe shown in Fig. P3.90 with negligible viscous effects. Determine the maximum flowrate if the water is not to flow from the open vertical tube at $A$.

**3.93** (WILEY) Determine the flowrate through the submerged orifice shown in Fig. P3.93 if the contraction coefficient is $C_c = 0.63$.

■ **Figure P3.93**

**3.94** An ancient device for measuring time is shown in Fig. P3.94. The axisymmetric vessel is shaped so that the water level falls at a constant rate. Determine the shape of the vessel, $R = R(z)$, if the water level is to decrease at a rate of 0.10 m/hr and the drain hole is 5.0 mm in diameter. The device is to operate for 12 hr without needing refilling. Make a scale drawing of the shape of the vessel.

■ **Figure P3.94**

**3.95** A long water trough of triangular cross section is formed from two planks as is shown in Fig. P3.95. A gap of 0.1 in. remains at the junction of the two planks. If the water depth initially was 2 ft, how long a time does it take for the water depth to reduce to 1 ft?

■ **Figure P3.95**

**\*3.96** A spherical tank of diameter $D$ has a drain hole of diameter $d$ at its bottom. A vent at the top of the tank maintains atmospheric pressure at the liquid surface within the tank. The flow is quasi-steady and inviscid and the tank is full of water initially. Determine the water depth as a function of time, $h = h(t)$, and plot graphs of $h(t)$ for tank diameters of 1, 5, 10, and 20 ft if $d = 1$ in.

**†3.97** A small hole develops in the bottom of the stationary row-boat shown in Fig. P3.97. Estimate the amount of time it will take for the boat to sink. List all assumptions and show all calculations.

■ **Figure P3.97**

**3.98** When the drain plug is pulled, water flows from a hole in the bottom of a large, open cylindrical tank. Show that if viscous effects are negligible and if the flow is assumed to be quasisteady, then it takes 3.41 times longer to empty the entire tank than it does to empty the first half of the tank. Explain why this is so.

**\*3.99** The surface area, $A$, of the pond shown in Fig. P3.99 varies with the water depth, $h$, as shown in the table. At time $t = 0$ a valve is opened and the pond is allowed to drain through a pipe of diameter $D$. If viscous effects are negligible and quasisteady conditions are assumed, plot the water depth as a function of time from when the valve is opened ($t = 0$) until the pond is drained for pipe diameters of $D = 0.5, 1.0, 1.5, 2.0, 2.5$, and 3.0 ft. Assume $h = 18$ ft at $t = 0$.

■ **Figure P3.99**

| h (ft) | A [acres (1 acre = 43,560 ft²)] |
|--------|----------------------------------|
| 0 | 0 |
| 2 | 0.3 |
| 4 | 0.5 |
| 6 | 0.8 |
| 8 | 0.9 |
| 10 | 1.1 |
| 12 | 1.5 |
| 14 | 1.8 |
| 16 | 2.4 |
| 18 | 2.8 |

**3.100** (WILEY) Water flows through a horizontal branching pipe as shown in Fig. P3.100. Determine the pressure at section (3).

■ **Figure P3.100**

**3.101** Water flows through the horizontal Y-fitting shown in Fig. P3.101. If the flowrate and pressure in pipe (1) are $Q_1 = 2.3$ ft³/s and $p_1 = 50$ lb/ft³, determine the pressures $p_2$ and $p_3$, in pipes (2) and (3) under the assumption that the flowrate divides evenly between pipes (2) and (3).

■ **Figure P3.101**

**3.102** Water flows through the branching pipe shown in Fig. P3.102. If viscous effects are negligible, determine the pressure at section (2) and the pressure at section (3).

■ **Figure P3.102**

**3.103** ⟨WILEY⟩ Water flows through the horizontal branching pipe shown in Fig. P3.103 at a rate of 10 ft³/s. If viscous effects are negligible, determine the water speed at section (2), the pressure at section (3), and the flowrate at section (4).

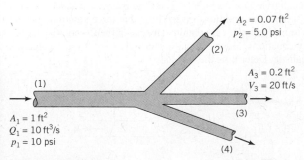

■ **Figure P3.103**

**3.104** ⟨WILEY⟩ Water flows from a large tank through a large pipe that splits into two smaller pipes as shown in Fig. P3.104. If viscous effects are negligible, determine the flowrate from the tank and the pressure at point (1).

■ **Figure P3.104**

**3.105** Air, assumed incompressible and inviscid, flows into the outdoor cooking grill through nine holes of 0.40-in. diameter as shown in Fig. P3.105. If a flowrate of 40 in.³/s into the grill is required to maintain the correct cooking conditions, determine the pressure within the grill near the holes.

■ **Figure P3.105**

**3.106** ⟨WILEY⟩ An air cushion vehicle is supported by forcing air into the chamber created by a skirt around the periphery of the vehicle as shown in Fig. P3.106. The air escapes through the 3-in. clearance between the lower end of the skirt and the ground (or water). Assume the vehicle weighs 10,000 lb and is essentially rectangular in shape, 30 by 65 ft. The volume of the chamber is large enough so that the kinetic energy of the air within the chamber is negligible. Determine the flowrate, $Q$, needed to support the vehicle. If the ground clearance were reduced to 2 in., what flowrate would be needed? If the vehicle weight were reduced to 5000 lb and the ground clearance maintained at 3 in., what flowrate would be needed?

■ **Figure P3.106**

**3.107** ⟨WILEY⟩ Water flows from the pipe shown in Fig. P3.107 as a free jet and strikes a circular flat plate. The flow geometry shown is axisymmetrical. Determine the flowrate and the manometer reading, $H$.

■ **Figure P3.107**

**3.108** (WILEY) A conical plug is used to regulate the airflow from the pipe shown in Fig. P3.108. The air leaves the edge of the cone with a uniform thickness of 0.02 m. If viscous effects are negligible and the flowrate is 0.50 m³/s, determine the pressure within the pipe.

■ **Figure P3.108**

**3.109** (WILEY) Water flows steadily from a nozzle into a large tank as shown in Fig. P3.109. The water then flows from the tank as a jet of diameter $d$. Determine the value of $d$ if the water level in the tank remains constant. Viscous effects are negligible.

■ **Figure P3.109**

**3.110** A small card is placed on top of a spool as shown in Fig. P3.110. It is not possible to blow the card off the spool by blowing air through the hole in the center of the spool. The harder one

blows, the harder the card "sticks" to the spool. In fact by blowing hard enough it is possible to keep the card against the spool with the spool turned upside down. (*Note*: It may be necessary to use a thumb tack to prevent the card from sliding from the spool.) Explain this phenomenon.

■ **Figure P3.110**

**3.111** Observations show that it is not possible to blow the table tennis ball from the funnel shown in Fig. P3.111a. In fact, the ball can be kept in an inverted funnel, Fig P3.111b, by blowing through it. The harder one blows through the funnel, the harder the ball is held within the funnel. Explain this phenomenon.

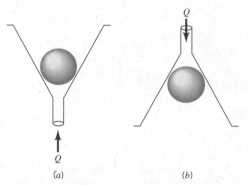

■ **Figure P3.111**

**3.112** Water flows down the sloping ramp shown in Fig. P3.112 with negligible viscous effects. The flow is uniform at sections (1) and (2). For the conditions given show that three solutions for the downstream depth, $h_2$, are obtained by use of the Bernoulli and continuity equations. However, show that only two of these solutions are realistic. Determine these values.

■ **Figure P3.112**

**3.113** (WILEY) Water flows in a rectangular channel that is 2.0 m wide as shown in Fig. P3.113. The upstream depth is 70 mm. The water surface rises 40 mm as it passes over a portion where the channel bottom rises 10 mm. If viscous effects are negligible, what is the flowrate?

**Figure P3.113**

### Section 3.6.3 Flowrate Measurement (also see Lab Problems 3.2LP and 3.4LP)

**3.114** Obtain a photograph/image of a situation that involves some type of flowmeter. Print this photo and write a brief paragraph that describes the situation involved.

**3.115** (WILEY) A Venturi meter with a minimum diameter of 3 in. is to be used to measure the flowrate of water through a 4-in.-diameter pipe. Determine the pressure difference indicated by the pressure gage attached to the flowmeter if the flowrate is 0.5 ft³/s and viscous effects are negligible.

**3.116** (WILEY) Determine the flowrate through the Venturi meter shown in Fig. P3.116 if ideal conditions exist.

**Figure P3.116**

**3.117** (WILEY GO) For what flowrate through the Venturi meter of Problem 3.116 will cavitation begin if $p_1 = 275$ kPa gage, atmospheric pressure is 101 kPa (abs), and the vapor pressure is 3.6 kPa (abs)?

**3.118** What diameter orifice hole, $d$, is needed if under ideal conditions the flowrate through the orifice meter of Fig. P3.118 is to be 30 gal/min of seawater with $p_1 - p_2 = 2.37$ lb/in.²? The contraction coefficient is assumed to be 0.63.

**Figure P3.118**

**3.119–** ▶ A weir (see **Video V10.13**) of trapezoidal cross section is used to measure the flowrate in a channel as shown in Fig. P3.119. If the flowrate is $Q_0$ when $H = \ell/2$, what flowrate is expected when $H = \ell$?

**Figure P3.119**

**3.120** The flowrate in a water channel is sometimes determined by use of a device called a Venturi flume. As shown in Fig. P3.120, this device consists simply of a bump on the bottom of the channel. If the water surface dips a distance of 0.07 m for the conditions shown, what is the flowrate per width of the channel? Assume the velocity is uniform and viscous effects are negligible.

**Figure P3.120**

**3.121** (WILEY) Water flows under the inclined sluice gate shown in Fig. P3.121. Determine the flowrate if the gate is 8 ft wide.

**Figure P3.121**

### Section 3.7 The Energy Line and the Hydraulic Grade Line

**3.122** (WILEY) Water flows in a vertical pipe of 0.15-m diameter at a rate of 0.2 m³/s and a pressure of 200 kPa at an elevation of 25 m. Determine the velocity head and pressure head at elevations of 20 and 55 m.

**3.123** (WILEY) Draw the energy line and the hydraulic grade line for the flow of Problem 3.83.

**3.124** Draw the energy line and hydraulic grade line for the flow shown in Problem 3.71.

### Section 3.8 Restrictions on Use of the Bernoulli Equation

**3.125** Obtain a photograph/image of a flow in which it would not be appropriate to use the Bernoulli equation. Print this photo and write a brief paragraph that describes the situation involved.

**3.126** (WILEY) Listed below are typical flight speeds for two aircraft. For which of these conditions would it be reasonable to use the incompressible Bernoulli equation to study the aerodynamics associated with their flight? Explain.

| Aircraft | Flight speed, km/hr | |
| --- | --- | --- |
| | Cruise | Landing approach |
| Boeing 787 | 913 | 214 |
| F-22 fighter | 1960 | 250 |

**3.127** A meteorologist uses a Pitot-static tube to measure the wind speed in a tornado. Based on the damage caused by the storm, the tornado is rated as EF5 on the Enhanced Fujita Scale. This means that the wind speed is estimated to be in the range of 261 to 318 mph. Is it reasonable to use the incompressible Pitot-tube equation (Eq. 3.16) to determine the actual wind speed, or must compressible effects to taken into account? Explain.

■ **Lab Problems**

**3.1 LP** This problem involves the pressure distribution between two parallel circular plates. To proceed with this problem, go to Appendix H which is located in *WileyPLUS* or on the book's web site, www.wiley.com/college/munson.

**3.2 LP** This problem involves the calibration of a nozzle-type flowmeter. To proceed with this problem, go to Appendix H which is located in *WileyPLUS* or on the book's web site, www.wiley.com/college/munson.

**3.3 LP** This problem involves the pressure distribution in a two-dimensional channel. To proceed with this problem, go to Appendix H which is located in *WileyPLUS* or on the book's web site, www.wiley.com/college/munson.

**3.4 LP** This problem involves the determination of the flowrate under a sluice gate as a function of the water depth. To proceed with this problem, go to Appendix H which is located in *WileyPLUS* or on the book's web site, www.wiley.com/college/munson.

■ **Lifelong Learning Problems**

**3.1 LL** The concept of the use of a Pitot-static tube to measure the airspeed of an airplane is rather straightforward. However, the design and manufacture of reliable, accurate, inexpensive Pitot-static tube airspeed indicators is not necessarily simple. Obtain information about the design and construction of modern Pitot-static tubes. Summarize your findings in a brief report.

**3.2 LL** In recent years damage due to hurricanes has been significant, particularly in the southeastern United States. The low barometric pressure, high winds, and high tides generated by hurricanes can combine to cause considerable damage. According to some experts, in the coming years hurricane frequency may increase because of global warming. Obtain information about the fluid mechanics of hurricanes. Summarize your findings in a brief report.

**3.3 LL** Orifice, nozzle, or Venturi flowmeters have been used for a long time to predict accurately the flowrate in pipes. However, recently there have been several new concepts suggested or used for such flowrate measurements. Obtain information about new methods to obtain pipe flowrate information. Summarize your findings in a brief report.

**3.4 LL** Ultra-high-pressure, thin jets of liquids can be used to cut various materials ranging from leather to steel and beyond. Obtain information about new methods and techniques proposed for liquid jet cutting and investigate how they may alter various manufacturing processes. Summarize your findings in a brief report.

■ **FE Exam Problems**

Sample FE (Fundamentals of Engineering) exam questions for fluid mechanics are provided in *WileyPLUS* or on the book's web site, www.wiley.com/college/munson.

# 4 Fluid Kinematics

CHAPTER OPENING PHOTO: A vortex ring: The complex, three-dimensional structure of a smoke ring is indicated in this cross-sectional view. (Smoke in air.) [Photograph courtesy of R. H. Magarvey and C. S. MacLatchy (Ref. 4).]

## Learning Objectives

After completing this chapter, you should be able to:

- discuss the differences between the Eulerian and Lagrangian descriptions of fluid motion.
- identify various flow characteristics based on the velocity field.
- determine the streamline pattern and acceleration field given a velocity field.
- discuss the differences between a system and control volume.
- apply the Reynolds transport theorem and the material derivative.

*V4.1 Streaklines*

In this chapter we will discuss various aspects of fluid motion without being concerned with the actual forces necessary to produce the motion. That is, we will consider the *kinematics* of the motion—the velocity and acceleration of the fluid, and the description and visualization of its motion. The analysis of the specific forces necessary to produce the motion (the *dynamics* of the motion) will be discussed in detail in the following chapters. A wide variety of useful information can be gained from a thorough understanding of fluid kinematics. Such an understanding of how to describe and observe fluid motion is an essential step to the complete understanding of fluid dynamics.

## 4.1 The Velocity Field

In general, fluids flow. That is, there is a net motion of molecules from one point in space to another point as a function of time. As is discussed in Chapter 1, a typical portion of fluid contains so many molecules that it becomes totally unrealistic (except in special cases) for us to attempt to account for the motion of individual molecules. Rather, we employ the continuum hypothesis and consider fluids to be made up of fluid particles that interact with each other and with their surroundings. Each particle contains numerous molecules. Thus, we can describe the flow of a fluid in terms of the motion of fluid particles rather than individual molecules. This motion can be described in terms of the velocity and acceleration of the fluid particles.

The infinitesimal particles of a fluid are tightly packed together (as is implied by the continuum assumption). Thus, at a given instant in time, a description of any fluid property (such as density,

**157**

*V4.2 Velocity field*

*V4.3 Cylinder-velocity vectors*

■ **Figure 4.1** **Particle location in terms of its position vector.**

pressure, velocity, and acceleration) may be given as a function of the fluid's location. This representation of fluid parameters as functions of the spatial coordinates is termed a ***field representation*** of the flow. Of course, the specific field representation may be different at different times, so that to describe a fluid flow we must determine the various parameters not only as a function of the spatial coordinates ($x$, $y$, $z$, for example) but also as a function of time, $t$. Thus, to completely specify the temperature, $T$, in a room we must specify the temperature field, $T = T(x, y, z, t)$, throughout the room (from floor to ceiling and wall to wall) at any time of the day or night.

Shown in the margin figure is one of the most important fluid variables, the ***velocity field,***

$$\mathbf{V} = u(x, y, z, t)\hat{\mathbf{i}} + v(x, y, z, t)\hat{\mathbf{j}} + w(x, y, z, t)\hat{\mathbf{k}}$$

where $u$, $v$, and $w$ are the $x$, $y$, and $z$ components of the velocity vector. By definition, the velocity of a particle is the time rate of change of the position vector for that particle. As is illustrated in Fig. 4.1, the position of particle $A$ relative to the coordinate system is given by its *position vector,* $\mathbf{r}_A$, which (if the particle is moving) is a function of time. The time derivative of this position gives the *velocity* of the particle, $d\mathbf{r}_A/dt = \mathbf{V}_A$. By writing the velocity for all of the particles, we can obtain the field description of the velocity vector $\mathbf{V} = \mathbf{V}(x, y, z, t)$.

Since the velocity is a vector, it has both a direction and a magnitude. The magnitude of $\mathbf{V}$, denoted $V = |\mathbf{V}| = (u^2 + v^2 + w^2)^{1/2}$, is the speed of the fluid. (It is very common in practical situations to call $V$ velocity rather than speed, i.e., "the velocity of the fluid is 12 m/s.") As is discussed in the next section, a change in velocity results in an acceleration. This acceleration may be due to a change in speed and/or direction.

| F l u i d s i n t h e N e w s |
|---|

**Follow those particles** Superimpose two photographs of a bouncing ball taken a short time apart and draw an arrow between the two images of the ball. This arrow represents an approximation of the velocity (displacement/time) of the ball. The particle image velocimeter (PIV) uses this technique to provide the instantaneous *velocity field* for a given cross section of a flow. The flow being studied is seeded with numerous micron-sized particles that are small enough to follow the flow yet big enough to reflect enough light to be captured by the camera. The flow is illuminated with a light sheet from a double-pulsed laser. A digital camera captures both light pulses on the same image frame, allowing the movement of the particles to be tracked. By using appropriate computer software to carry out a pixel-by-pixel interrogation of the double image, it is possible to track the motion of the particles and determine the two components of velocity in the given cross section of the flow. By using two cameras in a stereoscopic arrangement, it is possible to determine all three components of velocity. (See Problem 4.18.)

# $E$XAMPLE 4.1 | Velocity Field Representation

**GIVEN** A velocity field is given by $\mathbf{V} = (V_0/\ell)(-x\hat{\mathbf{i}} + y\hat{\mathbf{j}})$ where $V_0$ and $\ell$ are constants.

**FIND** At what location in the flow field is the speed equal to $V_0$? Make a sketch of the velocity field for $x \geq 0$ by drawing arrows representing the fluid velocity at representative locations.

## SOLUTION

The $x$, $y$, and $z$ components of the velocity are given by $u = -V_0 x/\ell$, $v = V_0 y/\ell$, and $w = 0$ so that the fluid speed, $V$, is

$$V = (u^2 + v^2 + w^2)^{1/2} = \frac{V_0}{\ell}(x^2 + y^2)^{1/2} \tag{1}$$

The speed is $V = V_0$ at any location on the circle of radius $\ell$ centered at the origin $[(x^2 + y^2)^{1/2} = \ell]$ as shown in Fig. E4.1a. **(Ans)**

The direction of the fluid velocity relative to the $x$ axis is given in terms of $\theta = \arctan(v/u)$ as shown in Fig. E4.1b. For this flow

$$\tan \theta = \frac{v}{u} = \frac{V_0 y/\ell}{-V_0 x/\ell} = \frac{y}{-x}$$

Thus, along the $x$ axis ($y = 0$) we see that $\tan \theta = 0$, so that $\theta = 0°$ or $\theta = 180°$. Similarly, along the $y$ axis ($x = 0$) we obtain $\tan \theta = \pm\infty$ so that $\theta = 90°$ or $\theta = 270°$. Also, for $y = 0$ we find $\mathbf{V} = (-V_0 x/\ell)\hat{\mathbf{i}}$, while for $x = 0$ we have $\mathbf{V} = (V_0 y/\ell)\hat{\mathbf{j}}$,

indicating (if $V_0 > 0$) that the flow is directed away from the origin along the $y$ axis and toward the origin along the $x$ axis as shown in Fig. E4.1a.

By determining $\mathbf{V}$ and $\theta$ for other locations in the $x$–$y$ plane, the velocity field can be sketched as shown in the figure. For example, on the line $y = x$ the velocity is at a 45° angle relative to the $x$ axis ($\tan \theta = v/u = -y/x = -1$). At the origin $x = y = 0$ so that $\mathbf{V} = 0$. This point is a stagnation point. The farther from the origin the fluid is, the faster it is flowing (as seen from Eq. 1). By careful consideration of the velocity field it is possible to determine considerable information about the flow.

**COMMENT** The velocity field given in this example approximates the flow in the vicinity of the center of the sign shown in Fig. E4.1c. When wind blows against the sign, some air flows over the sign, some under it, producing a stagnation point as indicated.

(b)

(c)

(a)

■ **Figure E4.1**

Flow

Flow

*Either Eulerian or Lagrangian methods can be used to describe flow fields.*

The figure in the margin shows the velocity field (i.e., velocity vectors) at a given instant for flow past two square bars. It is possible to obtain much qualitative and quantitative information for complex flows by using plots such as this.

### 4.1.1 Eulerian and Lagrangian Flow Descriptions

There are two general approaches in analyzing fluid mechanics problems (or problems in other branches of the physical sciences, for that matter). The first method, called the *Eulerian method,* uses the field concept introduced above. In this case, the fluid motion is given by completely prescribing the necessary properties (pressure, density, velocity, etc.) as functions of space and time. From this method we obtain information about the flow in terms of what happens at fixed points in space as the fluid flows through those points.

A typical Eulerian representation of the flow is shown by the figure in the margin which involves flow past an airfoil at angle of attack. The pressure field is indicated by using a contour plot showing lines of constant pressure, with gray shading indicating the intensity of the pressure.

The second method, called the *Lagrangian method,* involves following individual fluid particles as they move about and determining how the fluid properties associated with these particles change as a function of time. That is, the fluid particles are "tagged" or identified, and their properties determined as they move.

The difference between the two methods of analyzing fluid flow problems can be seen in the example of smoke discharging from a chimney, as is shown in Fig. 4.2. In the Eulerian method one may attach a temperature-measuring device to the top of the chimney (point 0) and record the temperature at that point as a function of time. At different times there are different fluid particles passing by the stationary device. Thus, one would obtain the temperature, $T$, for that location ($x = x_0, y = y_0,$ and $z = z_0$) as a function of time. That is, $T = T(x_0, y_0, z_0, t)$. The use of numerous temperature-measuring devices fixed at various locations would provide the temperature field, $T = T(x, y, z, t)$. The temperature of a particle as a function of time would not be known unless the location of the particle were known as a function of time.

In the Lagrangian method, one would attach the temperature-measuring device to a particular fluid particle (particle $A$) and record that particle's temperature as it moves about. Thus, one would obtain that particle's temperature as a function of time, $T_A = T_A(t)$. The use of many such measuring devices moving with various fluid particles would provide the temperature of these fluid particles as a function of time. The temperature would not be known as a function of position unless the location of each particle were known as a function of time. If enough information in Eulerian form is available, Lagrangian information can be derived from the Eulerian data—and vice versa.

Example 4.1 provides an Eulerian description of the flow. For a Lagrangian description we would need to determine the velocity as a function of time for each particle as it flows along from one point to another.

In fluid mechanics it is usually easier to use the Eulerian method to describe a flow—in either experimental or analytical investigations. There are, however, certain instances in which the Lagrangian method is more convenient. For example, some numerical fluid mechanics calculations are based on determining the motion of individual fluid particles (based on the appropriate interactions among the particles), thereby describing the motion in Lagrangian terms. Similarly, in

Location 0:
$T = T(x_0, y_0, t)$

Particle $A$:
$T_A = T_A(t)$

■ **Figure 4.2** Eulerian and Lagrangian descriptions of temperature of a flowing fluid.

*V4.4 Follow the particles (experiment)*

*V4.5 Follow the particles (computer)*

*Most flow fields are actually three-dimensional.*

*V4.6 Flow past a wing*

some experiments individual fluid particles are "tagged" and are followed throughout their motion, providing a Lagrangian description. Oceanographic measurements obtained from devices that flow with the ocean currents provide this information. Similarly, by using X-ray opaque dyes, it is possible to trace blood flow in arteries and to obtain a Lagrangian description of the fluid motion. A Lagrangian description may also be useful in describing fluid machinery (such as pumps and turbines) in which fluid particles gain or lose energy as they move along their flow paths.

Another illustration of the difference between the Eulerian and Lagrangian descriptions can be seen in the following biological example. Each year thousands of birds migrate between their summer and winter habitats. Ornithologists study these migrations to obtain various types of important information. One set of data obtained is the rate at which birds pass a certain location on their migration route (birds per hour). This corresponds to an Eulerian description—"flowrate" at a given location as a function of time. Individual birds need not be followed to obtain this information. Another type of information is obtained by "tagging" certain birds with radio transmitters and following their motion along the migration route. This corresponds to a Lagrangian description— "position" of a given particle as a function of time.

### 4.1.2 One-, Two-, and Three-Dimensional Flows

Generally, a fluid flow is a rather complex three-dimensional, time-dependent phenomenon— $\mathbf{V} = \mathbf{V}(x, y, z, t) = u\hat{\mathbf{i}} + v\hat{\mathbf{j}} + w\hat{\mathbf{k}}$. In many situations, however, it is possible to make simplifying assumptions that allow a much easier understanding of the problem without sacrificing needed accuracy. One of these simplifications involves approximating a real flow as a simpler one- or two-dimensional flow.

In almost any flow situation, the velocity field actually contains all three velocity components ($u$, $v$, and $w$, for example). In many situations the ***three-dimensional flow*** characteristics are important in terms of the physical effects they produce. (See the photograph at the beginning of Chapter 4.) For these situations it is necessary to analyze the flow in its complete three-dimensional character. Neglect of one or two of the velocity components in these cases would lead to considerable misrepresentation of the effects produced by the actual flow.

The flow of air past an airplane wing provides an example of a complex three-dimensional flow. A feel for the three-dimensional structure of such flows can be obtained by studying Fig. 4.3, which is a photograph of the flow past a model wing; the flow has been made visible by using a flow visualization technique.

In many situations one of the velocity components may be small (in some sense) relative to the two other components. In situations of this kind it may be reasonable to neglect the smaller component and assume ***two-dimensional flow.*** That is, $\mathbf{V} = u\hat{\mathbf{i}} + v\hat{\mathbf{j}}$, where $u$ and $v$ are functions of $x$ and $y$ (and possibly time, $t$).

It is sometimes possible to further simplify a flow analysis by assuming that two of the velocity components are negligible, leaving the velocity field to be approximated as a ***one-dimensional flow*** field. That is, $\mathbf{V} = u\hat{\mathbf{i}}$. As we will learn from examples throughout the remainder of the book, although there are very few, if any, flows that are truly one-dimensional, there are

■ **Figure 4.3**  Flow visualization of the complex three-dimensional flow past a delta wing. (Photograph courtesy of T. T. Lim and Y. D. Cui, National University of Singapore.)

Solenoid off, valve closed

Spring
Solenoid    Diaphragm

Outlet ←

Inlet

Solenoid on, valve open

*V4.7 Flow types*

many flow fields for which the one-dimensional flow assumption provides a reasonable approximation. There are also many flow situations for which use of a one-dimensional flow field assumption will give completely erroneous results.

### 4.1.3 Steady and Unsteady Flows

In the previous discussion we have assumed *steady flow*—the velocity at a given point in space does not vary with time, $\partial \mathbf{V}/\partial t = 0$. In reality, almost all flows are unsteady in some sense. That is, the velocity does vary with time. It is not difficult to believe that *unsteady flows* are usually more difficult to analyze (and to investigate experimentally) than are steady flows. Hence, considerable simplicity often results if one can make the assumption of steady flow without compromising the usefulness of the results. Among the various types of unsteady flows are nonperiodic flow, periodic flow, and truly random flow. Whether or not unsteadiness of one or more of these types must be included in an analysis is not always immediately obvious.

An example of a nonperiodic, unsteady flow is that produced by turning off a faucet to stop the flow of water. Usually this unsteady flow process is quite mundane, and the forces developed as a result of the unsteady effects need not be considered. However, if the water is turned off suddenly (as with the electrically operated valve in a dishwasher shown in the figure in the margin), the unsteady effects can become important [as in the "water hammer" effects made apparent by the loud banging of the pipes under such conditions (Ref. 1)].

In other flows the unsteady effects may be periodic, occurring time after time in basically the same manner. The periodic injection of the air–gasoline mixture into the cylinder of an automobile engine is such an example. The unsteady effects are quite regular and repeatable in a regular sequence. They are very important in the operation of the engine.

---

**F l u i d s   i n   t h e   N e w s**

**New pulsed liquid-jet scalpel** High-speed liquid-jet cutters are used for cutting a wide variety of materials such as leather goods, jigsaw puzzles, plastic, ceramic, and metal. Typically, compressed air is used to produce a continuous stream of water that is ejected from a tiny nozzle. As this stream impacts the material to be cut, a high pressure (the stagnation pressure) is produced on the surface of the material, thereby cutting the material. Such liquid-jet cutters work well in air but are difficult to control if the jet must pass through a liquid as often happens

in surgery. Researchers have developed a new pulsed jet cutting tool that may allow surgeons to perform microsurgery on tissues that are immersed in water. Rather than using a steady water jet, the system uses *unsteady flow*. A high-energy electrical discharge inside the nozzle momentarily raises the temperature of the microjet to approximately 10,000 °C. This creates a rapidly expanding vapor bubble in the nozzle and expels a tiny fluid jet from the nozzle. Each electrical discharge creates a single, brief jet, which makes a small cut in the material.

---

In many situations the unsteady character of a flow is quite random. That is, there is no repeatable sequence or regular variation to the unsteadiness. This behavior occurs in *turbulent flow* and is absent from *laminar flow*. The "smooth" flow of highly viscous syrup onto a pancake represents a "deterministic" laminar flow. It is quite different from the turbulent flow observed in the "irregular" splashing of water from a faucet onto the sink below it. The "irregular" gustiness of the wind represents another random turbulent flow. The differences between these types of flows are discussed in considerable detail in Chapters 8 and 9.

It must be understood that the definition of steady or unsteady flow pertains to the behavior of a fluid property as observed at a fixed point in space. For steady flow, the values of all fluid properties (velocity, temperature, density, etc.) at any fixed point are independent of time. However, the value of those properties for a given fluid particle may change with time as the particle flows along, even in steady flow. Thus, the temperature of the exhaust at the exit of a car's exhaust pipe may be constant for several hours, but the temperature of a fluid particle that left the exhaust pipe five minutes ago is lower now than it was when it left the pipe, even though the flow is steady.

*V4.8 Jupiter red spot*

### 4.1.4 Streamlines, Streaklines, and Pathlines

Although fluid motion can be quite complicated, there are various concepts that can be used to help in the visualization and analysis of flow fields. To this end we discuss the use of streamlines, streaklines, and pathlines in flow analysis. The streamline is often used in analytical work, while the streakline and pathline are often used in experimental work.

A *streamline* is a line that is everywhere tangent to the velocity field. If the flow is steady, nothing at a fixed point (including the velocity direction) changes with time, so the streamlines are fixed lines in space. (See the photograph at the beginning of Chapter 6.) For unsteady flows the streamlines may change shape with time. Streamlines are obtained analytically by integrating the equations defining lines tangent to the velocity field. As illustrated in the margin figure, for two-dimensional flows the slope of the streamline, $dy/dx$, must be equal to the tangent of the angle that the velocity vector makes with the $x$ axis or

*V4.9 Streamlines*

$$\frac{dy}{dx} = \frac{v}{u} \tag{4.1}$$

If the velocity field is known as a function of $x$ and $y$ (and $t$ if the flow is unsteady), this equation can be integrated to give the equation of the streamlines.

For unsteady flow there is no easy way to produce streamlines experimentally in the laboratory. As discussed below, the observation of dye, smoke, or some other tracer injected into a flow can provide useful information, but for unsteady flows it is not necessarily information about the streamlines.

---

## EXAMPLE 4.2    Streamlines for a Given Velocity Field

**GIVEN**  Consider the two-dimensional steady flow discussed in Example 4.1, $\mathbf{V} = (V_0/\ell)(-x\hat{\mathbf{i}} + y\hat{\mathbf{j}})$.

**FIND**  Determine the streamlines for this flow.

### SOLUTION

Since

$$u = (-V_0/\ell)x \text{ and } v = (V_0/\ell)y \tag{1}$$

it follows that streamlines are given by solution of the equation

$$\frac{dy}{dx} = \frac{v}{u} = \frac{(V_0/\ell)y}{-(V_0/\ell)x} = -\frac{y}{x}$$

in which variables can be separated and the equation integrated to give

$$\int \frac{dy}{y} = -\int \frac{dx}{x}$$

or

$$\ln y = -\ln x + \text{constant}$$

Thus, along the streamline

$$xy = C, \quad \text{where } C \text{ is a constant} \tag{Ans}$$

By using different values of the constant $C$, we can plot various lines in the $x$–$y$ plane—the streamlines. The streamlines for $x \geq 0$ are plotted in Fig. E4.2. A comparison of this figure with Fig. E4.1a illustrates the fact that streamlines are lines tangent to the velocity field.

**COMMENT**  Note that a flow is not completely specified by the shape of the streamlines alone. For example, the streamlines for the flow with $V_0/\ell = 10$ have the same shape as those for the flow with $V_0/\ell = -10$. However, the direction of the flow is opposite for these two cases. The arrows in Fig. E4.2 representing the flow direction are correct for $V_0/\ell = 10$ since, from Eq. 1, $u = -10x$ and $v = 10y$. That is, the flow is from right to left. For $V_0/\ell = -10$ the arrows are reversed. The flow is from left to right.

■ **Figure E4.2**

*V4.10 Streaklines*

A *streakline* consists of all particles in a flow that have previously passed through a common point. Streaklines are more of a laboratory tool than an analytical tool. They can be obtained by taking instantaneous photographs of marked particles that all passed through a given location in the flow field at some earlier time. Such a line can be produced by continuously injecting marked fluid (neutrally buoyant smoke in air or dye in water) at a given location (Ref. 2). (See Fig. 9.1.) If the flow is steady, each successively injected particle follows precisely behind the previous one, forming a steady streakline that is exactly the same as the streamline through the injection point.

For unsteady flows, particles injected at the same point at different times need not follow the same path. An instantaneous photograph of the marked fluid would show the streakline at that instant, but it would not necessarily coincide with the streamline through the point of injection at that particular time nor with the streamline through the same injection point at a different time (see Example 4.3).

The third method used for visualizing and describing flows involves the use of *pathlines*. A pathline is the line traced out by a given particle as it flows from one point to another. The pathline is a Lagrangian concept that can be produced in the laboratory by marking a fluid particle (dying a small fluid element) and taking a time exposure photograph of its motion. (See the photograph at the beginning of Chapter 7.)

---

## F l u i d s   i n   t h e   N e w s

**Air bridge spanning the oceans** It has long been known that large quantities of material are transported from one location to another by airborne dust particles. It is estimated that 2 billion metric tons of dust are lifted into the atmosphere each year. Most of these particles settle out fairly rapidly, but significant amounts travel large distances. Scientists are beginning to understand the full impact of this phenomenon—it is not only the tonnage transported, but the type of material transported that is significant. In addition to the mundane inert material we all term "dust," it is now known that a wide variety of hazardous materials and organisms are also carried along these literal *particle paths*. Satellite images reveal the amazing rate by which desert soils and other materials are transformed into airborne particles as a result of storms that produce strong winds. Once the tiny particles are aloft, they may travel thousands of miles, crossing the oceans and eventually being deposited on other continents. For the health and safety of all, it is important that we obtain a better understanding of the air bridges that span the oceans and also understand the ramification of such material transport.

---

*For steady flow, streamlines, streaklines, and pathlines are the same.*

If the flow is steady, the path taken by a marked particle (a pathline) will be the same as the line formed by all other particles that previously passed through the point of injection (a streakline). For such cases these lines are tangent to the velocity field. Hence, pathlines, streamlines, and streaklines are the same for steady flows. For unsteady flows none of these three types of lines need be the same (Ref. 3). Often one sees pictures of "streamlines" made visible by the injection of smoke or dye into a flow as is shown in Fig. 4.3. Actually, such pictures show streaklines rather than streamlines. However, for steady flows the two are identical; only the nomenclature is incorrectly used.

---

## $E$XAMPLE 4.3 | Comparison of Streamlines, Pathlines, and Streaklines

**GIVEN** Water flowing from the oscillating slit shown in Fig. E4.3a produces a velocity field given by $\mathbf{V} = u_0 \sin[\omega(t - y/v_0)]\hat{\mathbf{i}} + v_0\hat{\mathbf{j}}$, where $u_0$, $v_0$, and $\omega$ are constants. Thus, the $y$ component of velocity remains constant ($v = v_0$), and the $x$ component of velocity at $y = 0$ coincides with the velocity of the oscillating sprinkler head $[u = u_0 \sin(\omega t)$ at $y = 0]$.

**FIND** (a) Determine the streamline that passes through the origin at $t = 0$; at $t = \pi/2\omega$.
(b) Determine the pathline of the particle that was at the origin at $t = 0$; at $t = \pi/2$.
(c) Discuss the shape of the streakline that passes through the origin.

## SOLUTION

(a) Since $u = u_0 \sin[\omega(t - y/v_0)]$ and $v = v_0$, it follows from Eq. 4.1 that streamlines are given by the solution of

$$\frac{dy}{dx} = \frac{v}{u} = \frac{v_0}{u_0 \sin[\omega(t - y/v_0)]}$$

in which the variables can be separated and the equation integrated (for any given time $t$) to give

$$u_0 \int \sin\left[\omega\left(t - \frac{y}{v_0}\right)\right] dy = v_0 \int dx$$

or

$$u_0(v_0/\omega) \cos\left[\omega\left(t - \frac{y}{v_0}\right)\right] = v_0 x + C \quad \text{(1)}$$

where $C$ is a constant. For the streamline at $t = 0$ that passes through the origin ($x = y = 0$), the value of $C$ is obtained from Eq. 1 as $C = u_0 v_0/\omega$. Hence, the equation for this streamline is

$$x = \frac{u_0}{\omega}\left[\cos\left(\frac{\omega y}{v_0}\right) - 1\right] \quad \text{(2) (Ans)}$$

Similarly, for the streamline at $t = \pi/2\omega$ that passes through the origin, Eq. 1 gives $C = 0$. Thus, the equation for this streamline is

$$x = \frac{u_0}{\omega}\cos\left[\omega\left(\frac{\pi}{2\omega} - \frac{y}{v_0}\right)\right] = \frac{u_0}{\omega}\cos\left(\frac{\pi}{2} - \frac{\omega y}{v_0}\right)$$

or

$$x = \frac{u_0}{\omega}\sin\left(\frac{\omega y}{v_0}\right) \quad \text{(3) (Ans)}$$

**COMMENT**   These two streamlines, plotted in Fig. E4.3b, are not the same because the flow is unsteady. For example, at the origin ($x = y = 0$) the velocity is $\mathbf{V} = v_0\hat{\mathbf{j}}$ at $t = 0$ and $\mathbf{V} = u_0\hat{\mathbf{i}} + v_0\hat{\mathbf{j}}$ at $t = \pi/2\omega$. Thus, the angle of the streamline passing through the origin changes with time. Similarly, the shape of the entire streamline is a function of time.

**(b)**   The pathline of a particle (the location of the particle as a function of time) can be obtained from the velocity field and the definition of the velocity. Since $u = dx/dt$ and $v = dy/dt$ we obtain

$$\frac{dx}{dt} = u_0\sin\left[\omega\left(t - \frac{y}{v_0}\right)\right] \quad \text{and} \quad \frac{dy}{dt} = v_0$$

The $y$ equation can be integrated (since $v_0 = $ constant) to give the $y$ coordinate of the pathline as

$$y = v_0 t + C_1 \quad \text{(4)}$$

where $C_1$ is a constant. With this known $y = y(t)$ dependence, the $x$ equation for the pathline becomes

$$\frac{dx}{dt} = u_0\sin\left[\omega\left(t - \frac{v_0 t + C_1}{v_0}\right)\right] = -u_0\sin\left(\frac{C_1\omega}{v_0}\right)$$

This can be integrated to give the $x$ component of the pathline as

$$x = -\left[u_0\sin\left(\frac{C_1\omega}{v_0}\right)\right]t + C_2 \quad \text{(5)}$$

where $C_2$ is a constant. For the particle that was at the origin ($x = y = 0$) at time $t = 0$, Eqs. 4 and 5 give $C_1 = C_2 = 0$. Thus, the pathline is

$$x = 0 \quad \text{and} \quad y = v_0 t \quad \text{(6) (Ans)}$$

Similarly, for the particle that was at the origin at $t = \pi/2\omega$, Eqs. 4 and 5 give $C_1 = -\pi v_0/2\omega$ and $C_2 = -\pi u_0/2\omega$. Thus, the pathline for this particle is

$$x = u_0\left(t - \frac{\pi}{2\omega}\right) \quad \text{and} \quad y = v_0\left(t - \frac{\pi}{2\omega}\right) \quad \text{(7)}$$

The pathline can be drawn by plotting the locus of $x(t)$, $y(t)$ values for $t \geq 0$ or by eliminating the parameter $t$ from Eq. 7 to give

$$y = \frac{v_0}{u_0}x \quad \text{(8) (Ans)}$$

**COMMENT**   The pathlines given by Eqs. 6 and 8, shown in Fig. E4.3c, are straight lines from the origin (rays). The pathlines and streamlines do not coincide because the flow is unsteady.

**(c)**   The streakline through the origin at time $t = 0$ is the locus of particles at $t = 0$ that previously ($t < 0$) passed through the origin. The general shape of the streaklines can be seen as follows. Each particle that flows through the origin travels in a straight line (pathlines are rays from the origin), the slope of which lies between $\pm v_0/u_0$ as shown in Fig. E4.3d. Particles passing through the origin at different times are located on different rays from the origin and at different distances from the origin. The net result is that a stream of dye continually injected at the origin (a streakline) would have the shape shown in Fig. E4.3d. Because of the unsteadiness, the streakline will vary with time, although it will always have the oscillating, sinuous character shown.

**COMMENT**   Similar streaklines are given by the stream of water from a garden hose nozzle that oscillates back and forth in a direction normal to the axis of the nozzle.

In this example neither the streamlines, pathlines, nor streaklines coincide. If the flow were steady, all of these lines would be the same.

■ **Figure E4.3(a), (b)**

■ Figure E4.3*(c), (d)*

## 4.2 The Acceleration Field

*V4.11 Pathlines*

*Acceleration is the time rate of change of velocity for a given particle.*

As indicated in the previous section, we can describe fluid motion by either (1) following individual particles (Lagrangian description) or (2) remaining fixed in space and observing different particles as they pass by (Eulerian description). In either case, to apply Newton's second law ($\mathbf{F} = m\mathbf{a}$) we must be able to describe the particle acceleration in an appropriate fashion. For the infrequently used Lagrangian method, we describe the fluid acceleration just as is done in solid body dynamics— $\mathbf{a} = \mathbf{a}(t)$ for each particle. For the Eulerian description we describe the *acceleration field* as a function of position and time without actually following any particular particle. This is analogous to describing the flow in terms of the velocity field, $\mathbf{V} = \mathbf{V}(x, y, z, t)$, rather than the velocity for particular particles. In this section we will discuss how to obtain the acceleration field if the velocity field is known.

The acceleration of a particle is the time rate of change of its velocity. For unsteady flows the velocity at a given point in space (occupied by different particles) may vary with time, giving rise to a portion of the fluid acceleration. In addition, a fluid particle may experience an acceleration because its velocity changes as it flows from one point to another in space. For example, water flowing through a garden hose nozzle under steady conditions (constant number of gallons per minute from the hose) will experience an acceleration as it changes from its relatively low velocity in the hose to its relatively high velocity at the tip of the nozzle.

### 4.2.1 The Material Derivative

Consider a fluid particle moving along its pathline as is shown in Fig. 4.4. In general, the particle's velocity, denoted $\mathbf{V}_A$ for particle $A$, is a function of its location and the time. That is,

$$\mathbf{V}_A = \mathbf{V}_A(\mathbf{r}_A, t) = \mathbf{V}_A[x_A(t), y_A(t), z_A(t), t]$$

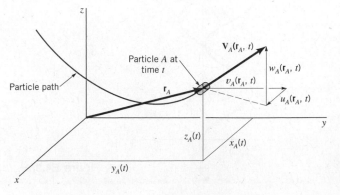

■ Figure 4.4
**Velocity and position of particle *A* at time *t*.**

where $x_A = x_A(t)$, $y_A = y_A(t)$, and $z_A = z_A(t)$ define the location of the moving particle. By definition, the acceleration of a particle is the time rate of change of its velocity. Since the velocity may be a function of both position and time, its value may change because of the change in time as well as a change in the particle's position. Thus, we use the chain rule of differentiation to obtain the acceleration of particle $A$, denoted $\mathbf{a}_A$, as

$$\mathbf{a}_A(t) = \frac{d\mathbf{V}_A}{dt} = \frac{\partial \mathbf{V}_A}{\partial t} + \frac{\partial \mathbf{V}_A}{\partial x}\frac{dx_A}{dt} + \frac{\partial \mathbf{V}_A}{\partial y}\frac{dy_A}{dt} + \frac{\partial \mathbf{V}_A}{\partial z}\frac{dz_A}{dt} \tag{4.2}$$

Using the fact that the particle velocity components are given by $u_A = dx_A/dt$, $v_A = dy_A/dt$, and $w_A = dz_A/dt$, Eq. 4.2 becomes

$$\mathbf{a}_A = \frac{\partial \mathbf{V}_A}{\partial t} + u_A\frac{\partial \mathbf{V}_A}{\partial x} + v_A\frac{\partial \mathbf{V}_A}{\partial y} + w_A\frac{\partial \mathbf{V}_A}{\partial z}$$

Since the above is valid for any particle, we can drop the reference to particle $A$ and obtain the acceleration field from the velocity field as

$$\mathbf{a} = \frac{\partial \mathbf{V}}{\partial t} + u\frac{\partial \mathbf{V}}{\partial x} + v\frac{\partial \mathbf{V}}{\partial y} + w\frac{\partial \mathbf{V}}{\partial z} \tag{4.3}$$

This is a vector result whose scalar components can be written as

$$a_x = \frac{\partial u}{\partial t} + u\frac{\partial u}{\partial x} + v\frac{\partial u}{\partial y} + w\frac{\partial u}{\partial z}$$

$$a_y = \frac{\partial v}{\partial t} + u\frac{\partial v}{\partial x} + v\frac{\partial v}{\partial y} + w\frac{\partial v}{\partial z} \tag{4.4}$$

and

$$a_z = \frac{\partial w}{\partial t} + u\frac{\partial w}{\partial x} + v\frac{\partial w}{\partial y} + w\frac{\partial w}{\partial z}$$

where $a_x$, $a_y$, and $a_z$ are the $x$, $y$, and $z$ components of the acceleration.

The above result is often written in shorthand notation as

*The material derivative is used to describe time rates of change for a given particle.*

$$\mathbf{a} = \frac{D\mathbf{V}}{Dt}$$

where the operator

$$\frac{D(\ )}{Dt} \equiv \frac{\partial(\ )}{\partial t} + u\frac{\partial(\ )}{\partial x} + v\frac{\partial(\ )}{\partial y} + w\frac{\partial(\ )}{\partial z} \tag{4.5}$$

is termed the **material derivative** or *substantial derivative*. An often-used shorthand notation for the material derivative operator is

$$\frac{D(\ )}{Dt} = \frac{\partial(\ )}{\partial t} + (\mathbf{V} \cdot \nabla)(\ ) \tag{4.6}$$

The dot product of the velocity vector, $\mathbf{V}$, and the gradient operator, $\nabla(\ ) = \partial(\ )/\partial x\,\hat{\mathbf{i}} + \partial(\ )/\partial y\,\hat{\mathbf{j}} + \partial(\ )/\partial z\,\hat{\mathbf{k}}$ (a vector operator) provides a convenient notation for the spatial derivative terms appearing in the Cartesian coordinate representation of the material derivative. Note that the notation $\mathbf{V} \cdot \nabla$ represents the operator $\mathbf{V} \cdot \nabla(\ ) = u\partial(\ )/\partial x + v\partial(\ )/\partial y + w\partial(\ )/\partial z$.

The material derivative concept is very useful in analysis involving various fluid parameters, not just the acceleration. The material derivative of any variable is the rate at which that variable changes with time for a given particle (as seen by one moving along with the fluid—the Lagrangian description). For example, consider a temperature field $T = T(x, y, z, t)$ associated with a given flow, like the flame shown in the figure in the margin. It may be of interest to determine the time rate of change of temperature of a fluid particle (particle $A$) as it moves through this temperature field. If the velocity, $\mathbf{V} = \mathbf{V}(x, y, z, t)$, is known, we can apply the chain rule to determine the rate of change of temperature as

$$\frac{dT_A}{dt} = \frac{\partial T_A}{\partial t} + \frac{\partial T_A}{\partial x}\frac{dx_A}{dt} + \frac{\partial T_A}{\partial y}\frac{dy_A}{dt} + \frac{\partial T_A}{\partial z}\frac{dz_A}{dt}$$

This can be written as

$$\frac{DT}{Dt} = \frac{\partial T}{\partial t} + u\frac{\partial T}{\partial x} + v\frac{\partial T}{\partial y} + w\frac{\partial T}{\partial z} = \frac{\partial T}{\partial t} + \mathbf{V} \cdot \nabla T$$

As in the determination of the acceleration, the material derivative operator, $D(\ )/Dt$, appears.

## EXAMPLE 4.4    Acceleration along a Streamline

**GIVEN**  An incompressible, inviscid fluid flows steadily past a tennis ball of radius $R$, as shown in Fig. E4.4a. According to a more advanced analysis of the flow, the fluid velocity along streamline $A$–$B$ is given by

$$\mathbf{V} = u(x)\hat{\mathbf{i}} = V_0\left(1 + \frac{R^3}{x^3}\right)\hat{\mathbf{i}}$$

where $V_0$ is the upstream velocity far ahead of the sphere.

**FIND**  Determine the acceleration experienced by fluid particles as they flow along this streamline.

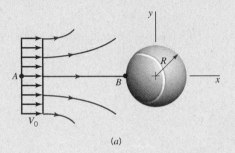

(a)

## SOLUTION

Along streamline $A$–$B$ there is only one component of velocity ($v = w = 0$) so that from Eq. 4.3

$$\mathbf{a} = \frac{\partial \mathbf{V}}{\partial t} + u\frac{\partial \mathbf{V}}{\partial x} = \left(\frac{\partial u}{\partial t} + u\frac{\partial u}{\partial x}\right)\hat{\mathbf{i}}$$

or

$$a_x = \frac{\partial u}{\partial t} + u\frac{\partial u}{\partial x}, \qquad a_y = 0, \qquad a_z = 0$$

Since the flow is steady, the velocity at a given point in space does not change with time. Thus, $\partial u/\partial t = 0$. With the given velocity distribution along the streamline, the acceleration becomes

$$a_x = u\frac{\partial u}{\partial x} = V_0\left(1 + \frac{R^3}{x^3}\right)V_0[R^3(-3x^{-4})]$$

or

$$a_x = -3(V_0^2/R)\frac{1 + (R/x)^3}{(x/R)^4} \qquad \textbf{(Ans)}$$

**COMMENTS**  Along streamline $A$–$B$ ($-\infty \le x \le -R$ and $y = 0$) the acceleration has only an $x$ component, and it is negative (a deceleration). Thus, the fluid slows down from its upstream

(b)

■ **Figure E4.4**

velocity of $\mathbf{V} = V_0\hat{\mathbf{i}}$ at $x = -\infty$ to its stagnation point velocity of $\mathbf{V} = 0$ at $x = -R$, the "nose" of the ball. The variation of $a_x$ along streamline $A$–$B$ is shown in Fig. E4.4b. It is the same result as is obtained in Example 3.1 by using the streamwise component of the acceleration, $a_x = V\,\partial V/\partial s$. The maximum deceleration occurs at $x = -1.205R$ and has a value of $a_{x,max} = -0.610\,V_0^2/R$. Note that this maximum deceleration increases with increasing velocity and decreasing size. As indicated in the following table, typical values of this deceleration can be quite large. For example, the $a_{x,max} = -4.08 \times 10^4$ ft/s$^2$ value for a pitched baseball is a deceleration approximately 1500 times that of gravity.

| Object | $V_0$ (ft/s) | $R$ (ft) | $a_{x,max}$ (ft/s²) |
|--------|--------------|----------|---------------------|
| Rising weather balloon | 1 | 4.0 | $-0.153$ |
| Soccer ball | 20 | 0.80 | $-305$ |
| Baseball | 90 | 0.121 | $-4.08 \times 10^4$ |
| Tennis ball | 100 | 0.104 | $-5.87 \times 10^4$ |
| Golf ball | 200 | 0.070 | $-3.49 \times 10^5$ |

In general, for fluid particles on streamlines other than $A$–$B$, all three components of the acceleration ($a_x$, $a_y$, and $a_z$) will be nonzero.

### 4.2.2  Unsteady Effects

*The local derivative is a result of the unsteadiness of the flow.*

*V4.12 Unsteady flow*

As is seen from Eq. 4.5, the material derivative formula contains two types of terms—those involving the time derivative $[\partial(\ )/\partial t]$ and those involving spatial derivatives $[\partial(\ )/\partial x, \partial(\ )/\partial y,$ and $\partial(\ )/\partial z]$. The time derivative portions are denoted as the *local derivative*. They represent effects of the unsteadiness of the flow. If the parameter involved is the acceleration, that portion given by $\partial \mathbf{V}/\partial t$ is termed the **local acceleration.** For steady flow the time derivative is zero throughout the flow field $[\partial(\ )/\partial t \equiv 0]$, and the local effect vanishes. Physically, there is no change in flow parameters at a fixed point in space if the flow is steady. There may be a change of those parameters for a fluid particle as it moves about, however.

If a flow is unsteady, its parameter values (velocity, temperature, density, etc.) at any location may change with time. For example, an unstirred ($\mathbf{V} = 0$) cup of coffee will cool down in time because of heat transfer to its surroundings. That is, $DT/Dt = \partial T/\partial t + \mathbf{V} \cdot \nabla T = \partial T/\partial t < 0$. Similarly, a fluid particle may have nonzero acceleration as a result of the unsteady effect of the flow. Consider flow in a constant diameter pipe as is shown in Fig. 4.5. The flow is assumed to be spatially uniform throughout the pipe. That is, $\mathbf{V} = V_0(t)\,\hat{\mathbf{i}}$ at all points in the pipe. The value of the acceleration depends on whether $V_0$ is being increased, $\partial V_0/\partial t > 0$, or decreased, $\partial V_0/\partial t < 0$. Unless $V_0$ is independent of time ($V_0 \equiv$ constant) there will be an acceleration, the local acceleration term. Thus, the acceleration field, $\mathbf{a} = \partial V_0/\partial t\,\hat{\mathbf{i}}$, is uniform throughout the entire flow, although it may vary with time ($\partial V_0/\partial t$ need not be constant). The acceleration due to the spatial variations of velocity ($u\,\partial u/\partial x, v\,\partial v/\partial y$, etc.) vanishes automatically for this flow, since $\partial u/\partial x = 0$ and $v = w = 0$. That is,

$$\mathbf{a} = \frac{\partial \mathbf{V}}{\partial t} + u\frac{\partial \mathbf{V}}{\partial x} + v\frac{\partial \mathbf{V}}{\partial y} + w\frac{\partial \mathbf{V}}{\partial z} = \frac{\partial \mathbf{V}}{\partial t} = \frac{\partial V_0}{\partial t}\,\hat{\mathbf{i}}$$

### 4.2.3  Convective Effects

$V_2 > V_1$

$V_1$

The portion of the material derivative (Eq. 4.5) represented by the spatial derivatives is termed the *convective derivative*. It represents the fact that a flow property associated with a fluid particle may vary because of the motion of the particle from one point in space where the parameter has one value to another point in space where its value is different. For example, the water velocity at the inlet of the garden hose nozzle shown in the figure in the margin is different (both in direction and speed) than it is at the exit. This contribution to the time rate of change of the parameter for the particle can occur whether the flow is steady or unsteady.

$V_0(t)$

$V_0(t)$

$x$

■ **Figure 4.5**    Uniform, unsteady flow in a constant diameter pipe.

Water heater

Hot
$T_{out} > T_{in}$

Pathline

$\frac{\partial T}{\partial t} = 0$

$\frac{DT}{Dt} \neq 0$

Cold
$T_{in}$

■ **Figure 4.6** Steady-state operation of a water heater. (Photograph courtesy of American Water Heater Company.)

*The convective derivative is a result of the spatial variation of the flow.*

It is due to the convection, or motion, of the particle through space in which there is a gradient $[\nabla(\ ) = \partial(\ )/\partial x\,\hat{\mathbf{i}} + \partial(\ )/\partial y\,\hat{\mathbf{j}} + \partial(\ )/\partial z\,\hat{\mathbf{k}}]$ in the parameter value. That portion of the acceleration given by the term $(\mathbf{V} \cdot \nabla)\mathbf{V}$ is termed the ***convective acceleration.***

As is illustrated in Fig. 4.6, the temperature of a water particle changes as it flows through a water heater. The water entering the heater is always the same cold temperature, and the water leaving the heater is always the same hot temperature. The flow is steady. However, the temperature, $T$, of each water particle increases as it passes through the heater—$T_{out} > T_{in}$. Thus, $DT/Dt \neq 0$ because of the convective term in the total derivative of the temperature. That is, $\partial T/\partial t = 0$, but $u\,\partial T/\partial x \neq 0$ (where $x$ is directed along the streamline), since there is a nonzero temperature gradient along the streamline. A fluid particle traveling along this nonconstant temperature path ($\partial T/\partial x \neq 0$) at a specified speed ($u$) will have its temperature change with time at a rate of $DT/Dt = u\,\partial T/\partial x$ even though the flow is steady ($\partial T/\partial t = 0$).

The same types of processes are involved with fluid accelerations. Consider flow in a variable area pipe as shown in Fig. 4.7. It is assumed that the flow is steady and one-dimensional with velocity that increases and decreases in the flow direction as indicated. As the fluid flows from section (1) to section (2), its velocity increases from $V_1$ to $V_2$. Thus, even though $\partial \mathbf{V}/\partial t = 0$ (steady flow), fluid particles experience an acceleration given by $a_x = u\,\partial u/\partial x$ (convective acceleration). For $x_1 < x < x_2$, it is seen that $\partial u/\partial x > 0$ so that $a_x > 0$—the fluid accelerates. For $x_2 < x < x_3$, it is seen that $\partial u/\partial x < 0$ so that $a_x < 0$—the fluid decelerates. This acceleration and deceleration are shown in the figure in the margin. If $V_1 = V_3$, the amount of acceleration precisely balances the amount of deceleration even though the distances between $x_2$ and $x_1$ and $x_3$ and $x_2$ are not the same.

The concept of the material derivative can be used to determine the time rate of change of any parameter associated with a particle as it moves about. Its use is not restricted to fluid mechanics alone. The basic ingredients needed to use the material derivative concept are the field description of the parameter, $P = P(x, y, z, t)$, and the rate at which the particle moves through that field, $\mathbf{V} = \mathbf{V}(x, y, z, t)$.

■ **Figure 4.7** Uniform, steady flow in a variable area pipe.

## EXAMPLE 4.5    Acceleration from a Given Velocity Field

**GIVEN**  Consider the steady, two-dimensional flow field discussed in Example 4.2.

**FIND**  Determine the acceleration field for this flow.

### SOLUTION

In general, the acceleration is given by

$$\mathbf{a} = \frac{D\mathbf{V}}{Dt} = \frac{\partial \mathbf{V}}{\partial t} + (\mathbf{V} \cdot \nabla)(\mathbf{V})$$

$$= \frac{\partial \mathbf{V}}{\partial t} + u\frac{\partial \mathbf{V}}{\partial x} + v\frac{\partial \mathbf{V}}{\partial y} + w\frac{\partial \mathbf{V}}{\partial z} \qquad (1)$$

where the velocity is given by $\mathbf{V} = (V_0/\ell)(-x\hat{\mathbf{i}} + y\hat{\mathbf{j}})$ so that $u = -(V_0/\ell)x$ and $v = (V_0/\ell)y$. For steady $[\partial(\ )/\partial t = 0]$, two-dimensional $[w = 0$ and $\partial(\ )/\partial z = 0]$ flow, Eq. 1 becomes

$$\mathbf{a} = u\frac{\partial \mathbf{V}}{\partial x} + v\frac{\partial \mathbf{V}}{\partial y}$$

$$= \left(u\frac{\partial u}{\partial x} + v\frac{\partial u}{\partial y}\right)\hat{\mathbf{i}} + \left(u\frac{\partial v}{\partial x} + v\frac{\partial v}{\partial y}\right)\hat{\mathbf{j}}$$

Hence, for this flow the acceleration is given by

$$\mathbf{a} = \left[\left(-\frac{V_0}{\ell}\right)(x)\left(-\frac{V_0}{\ell}\right) + \left(\frac{V_0}{\ell}\right)(y)(0)\right]\hat{\mathbf{i}}$$

$$+ \left[\left(-\frac{V_0}{\ell}\right)(x)(0) + \left(\frac{V_0}{\ell}\right)(y)\left(\frac{V_0}{\ell}\right)\right]\hat{\mathbf{j}}$$

or

$$a_x = \frac{V_0^2 x}{\ell^2}, \qquad a_y = \frac{V_0^2 y}{\ell^2} \qquad \textbf{(Ans)}$$

**COMMENTS**  The fluid experiences an acceleration in both the $x$ and $y$ directions. Since the flow is steady, there is no local acceleration—the fluid velocity at any given point is constant in time. However, there is a convective acceleration due to the change in velocity from one point on the particle's pathline to another. Recall that the velocity is a vector—it has both a magnitude and a direction. In this flow both the fluid speed (magnitude) and flow direction change with location (see Fig. E4.1a).

For this flow the magnitude of the acceleration is constant on circles centered at the origin, as is seen from the fact that

$$|\mathbf{a}| = (a_x^2 + a_y^2 + a_z^2)^{1/2} = \left(\frac{V_0}{\ell}\right)^2 (x^2 + y^2)^{1/2} \qquad (2)$$

■ **Figure E4.5**

Also, the acceleration vector is oriented at an angle $\theta$ from the $x$ axis, where

$$\tan\theta = \frac{a_y}{a_x} = \frac{y}{x}$$

This is the same angle as that formed by a ray from the origin to point $(x, y)$. Thus, the acceleration is directed along rays from the origin and has a magnitude proportional to the distance from the origin. Typical acceleration vectors (from Eq. 2) and velocity vectors (from Example 4.1) are shown in Fig. E4.5 for the flow in the first quadrant. Note that $\mathbf{a}$ and $\mathbf{V}$ are not parallel except along the $x$ and $y$ axes (a fact that is responsible for the curved pathlines of the flow) and that both the acceleration and velocity are zero at the origin $(x = y = 0)$. An infinitesimal fluid particle placed precisely at the origin will remain there, but its neighbors (no matter how close they are to the origin) will drift away.

## EXAMPLE 4.6    The Material Derivative

**GIVEN**  A fluid flows steadily through a two-dimensional nozzle of length $\ell$ as shown in Fig. E4.6a. The nozzle shape is given by

$$y/\ell = \pm 0.5/[1 + (x/\ell)]$$

If viscous and gravitational effects are negligible, the velocity field is approximately

$$u = V_0[1 + x/\ell], \quad v = -V_0 y/\ell \qquad (1)$$

and the pressure field is

$$p - p_0 = -(\rho V_0^2/2)[(x^2 + y^2)/\ell^2 + 2x/\ell]$$

where $V_0$ and $p_0$ are the velocity and pressure at the origin, $x = y = 0$. Note that the fluid speed increases as it flows through the nozzle. For example, along the centerline ($y = 0$), $V = V_0$ at $x = 0$ and $V = 2V_0$ at $x = \ell$.

**FIND** Determine, as a function of $x$ and $y$, the time rate of change of pressure felt by a fluid particle as it flows through the nozzle.

## SOLUTION

The time rate of change of pressure at any given, fixed point in this steady flow is zero. However, the time rate of change of pressure felt by a particle flowing through the nozzle is given by the material derivative of the pressure and is not zero. Thus,

$$\frac{Dp}{Dt} = \frac{\partial p}{\partial t} + u\frac{\partial p}{\partial x} + v\frac{\partial p}{\partial y} = u\frac{\partial p}{\partial x} + v\frac{\partial p}{\partial y} \quad (2)$$

where the $x$ and $y$ components of the pressure gradient can be written as

$$\frac{\partial p}{\partial x} = -\frac{\rho V_0^2}{\ell}\left(\frac{x}{\ell} + 1\right) \quad (3)$$

and

$$\frac{\partial p}{\partial y} = -\frac{\rho V_0^2}{\ell}\left(\frac{y}{\ell}\right) \quad (4)$$

Therefore, by combining Eqs. (1), (2), (3), and (4) we obtain

$$\frac{Dp}{Dt} = V_0\left(1 + \frac{x}{\ell}\right)\left(-\frac{\rho V_0^2}{\ell}\right)\left(\frac{x}{\ell} + 1\right) + \left(-V_0\frac{y}{\ell}\right)\left(-\frac{\rho V_0^2}{\ell}\right)\left(\frac{y}{\ell}\right)$$

or

$$\frac{Dp}{Dt} = -\frac{\rho V_0^3}{\ell}\left[\left(\frac{x}{\ell} + 1\right)^2 - \left(\frac{y}{\ell}\right)^2\right] \quad (5) \quad \textbf{(Ans)}$$

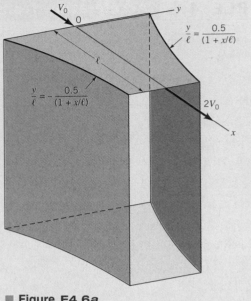

■ **Figure E4.6a**

**COMMENT** Lines of constant pressure within the nozzle are indicated in Fig. E4.6b, along with some representative streamlines of the flow. Note that as a fluid particle flows along its streamline, it moves into areas of lower and lower pressure. Hence, even though the flow is steady, the time rate of change of the pressure for any given particle is negative. This can be verified from Eq. (5) which, when plotted in Fig. E4.6c, shows that for any point within the nozzle $Dp/Dt < 0$.

■ **Figure E4.6b**

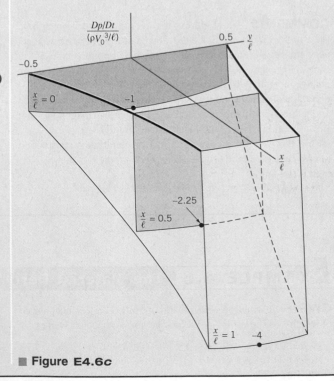

■ **Figure E4.6c**

### 4.2.4 Streamline Coordinates

*V4.13 Streamline coordinates*

In many flow situations, it is convenient to use a coordinate system defined in terms of the streamlines of the flow. An example for steady, two-dimensional flows is illustrated in Fig. 4.8. Such flows can be described either in terms of the usual $x$, $y$ Cartesian coordinate system (or some other system such as the $r$, $\theta$ polar coordinate system) or the streamline coordinate system. In the streamline coordinate system the flow is described in terms of one coordinate along the streamlines, denoted $s$, and the second coordinate normal to the streamlines, denoted $n$. Unit vectors in these two directions are denoted by $\hat{\mathbf{s}}$ and $\hat{\mathbf{n}}$, as shown in the figure. Care is needed not to confuse the coordinate distance $s$ (a scalar) with the unit vector along the streamline direction, $\hat{\mathbf{s}}$.

The flow plane is therefore covered by an orthogonal curved net of coordinate lines. At any point the $s$ and $n$ directions are perpendicular, but the lines of constant $s$ or constant $n$ are not necessarily straight. Without knowing the actual velocity field (hence, the streamlines) it is not possible to construct this flow net. In many situations, appropriate simplifying assumptions can be made so that this lack of information does not present an insurmountable difficulty. One of the major advantages of using the streamline coordinate system is that the velocity is always tangent to the $s$ direction. That is,

$$\mathbf{V} = V\,\hat{\mathbf{s}}$$

This allows simplifications in describing the fluid particle acceleration and in solving the equations governing the flow.

For steady, two-dimensional flow we can determine the acceleration as

$$\mathbf{a} = \frac{D\mathbf{V}}{Dt} = a_s\hat{\mathbf{s}} + a_n\hat{\mathbf{n}}$$

where $a_s$ and $a_n$ are the streamline and normal components of acceleration, respectively, as indicated by the figure in the margin. We use the material derivative because by definition the acceleration is the time rate of change of the velocity of a given particle as it moves about. If the streamlines

■ **Figure 4.8**
**Streamline coordinate system for two-dimensional flow.**

are curved, both the speed of the particle and its direction of flow may change from one point to another. In general, for steady flow both the speed and the flow direction are a function of location—$V = V(s, n)$ and $\hat{\mathbf{s}} = \hat{\mathbf{s}}(s, n)$. For a given particle, the value of $s$ changes with time, but the value of $n$ remains fixed because the particle flows along a streamline defined by $n = $ constant. (Recall that streamlines and pathlines coincide in steady flow.) Thus, application of the chain rule gives

$$\mathbf{a} = \frac{D(V\hat{\mathbf{s}})}{Dt} = \frac{DV}{Dt}\hat{\mathbf{s}} + V\frac{D\hat{\mathbf{s}}}{Dt}$$

or

$$\mathbf{a} = \left(\frac{\partial V}{\partial t} + \frac{\partial V}{\partial s}\frac{ds}{dt} + \frac{\partial V}{\partial n}\frac{dn}{dt}\right)\hat{\mathbf{s}} + V\left(\frac{\partial \hat{\mathbf{s}}}{\partial t} + \frac{\partial \hat{\mathbf{s}}}{\partial s}\frac{ds}{dt} + \frac{\partial \hat{\mathbf{s}}}{\partial n}\frac{dn}{dt}\right)$$

This can be simplified by using the fact that for steady flow nothing changes with time at a given point so that both $\partial V/\partial t$ and $\partial \hat{\mathbf{s}}/\partial t$ are zero. Also, the velocity along the streamline is $V = ds/dt$, and the particle remains on its streamline ($n = $ constant) so that $dn/dt = 0$. Hence,

$$\mathbf{a} = \left(V\frac{\partial V}{\partial s}\right)\hat{\mathbf{s}} + V\left(V\frac{\partial \hat{\mathbf{s}}}{\partial s}\right)$$

*The orientation of the unit vector along the streamline changes with distance along the streamline.*

The quantity $\partial \hat{\mathbf{s}}/\partial s$ represents the limit as $\delta s \to 0$ of the change in the unit vector along the streamline, $\delta\hat{\mathbf{s}}$, per change in distance along the streamline, $\delta s$. The magnitude of $\hat{\mathbf{s}}$ is constant ($|\hat{\mathbf{s}}| = 1$; it is a unit vector), but its direction is variable if the streamlines are curved. From Fig. 4.9 it is seen that the magnitude of $\partial\hat{\mathbf{s}}/\partial s$ is equal to the inverse of the radius of curvature of the streamline, $\mathcal{R}$, at the point in question. This follows because the two triangles shown ($AOB$ and $A'O'B'$) are similar triangles so that $\delta s/\mathcal{R} = |\delta\hat{\mathbf{s}}|/|\hat{\mathbf{s}}| = |\delta\hat{\mathbf{s}}|$, or $|\delta\hat{\mathbf{s}}/\delta s| = 1/\mathcal{R}$. Similarly, in the limit $\delta s \to 0$, the direction of $\delta\hat{\mathbf{s}}/\delta s$ is seen to be normal to the streamline. That is,

$$\frac{\partial\hat{\mathbf{s}}}{\partial s} = \lim_{\delta s \to 0}\frac{\delta\hat{\mathbf{s}}}{\delta s} = \frac{\hat{\mathbf{n}}}{\mathcal{R}}$$

Hence, the acceleration for steady, two-dimensional flow can be written in terms of its streamwise and normal components in the form

$$\mathbf{a} = V\frac{\partial V}{\partial s}\hat{\mathbf{s}} + \frac{V^2}{\mathcal{R}}\hat{\mathbf{n}} \quad \text{or} \quad a_s = V\frac{\partial V}{\partial s}, \qquad a_n = \frac{V^2}{\mathcal{R}} \tag{4.7}$$

The first term, $a_s = V\,\partial V/\partial s$, represents the convective acceleration along the streamline, and the second term, $a_n = V^2/\mathcal{R}$, represents centrifugal acceleration (one type of convective acceleration) normal to the fluid motion. These components can be noted in Fig. E4.5 by resolving the acceleration vector into its components along and normal to the velocity vector. Note that the unit vector $\hat{\mathbf{n}}$ is directed from the streamline toward the center of curvature. These forms of the acceleration were used in Chapter 3 and are probably familiar from previous dynamics or physics considerations.

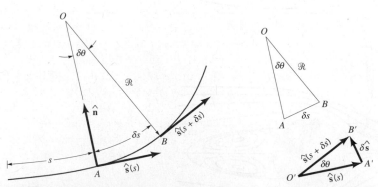

■ **Figure 4.9** Relationship between the unit vector along the streamline, $\hat{\mathbf{s}}$, and the radius of curvature of the streamline, $\mathcal{R}$.

## 4.3   Control Volume and System Representations

As is discussed in Chapter 1, a fluid is a type of matter that is relatively free to move and interact with its surroundings. As with any matter, a fluid's behavior is governed by fundamental physical laws that are approximated by an appropriate set of equations. The application of laws such as the conservation of mass, Newton's laws of motion, and the laws of thermodynamics forms the foundation of fluid mechanics analyses. There are various ways that these governing laws can be applied to a fluid, including the system approach and the control volume approach. By definition, a *system* is a collection of matter of fixed identity (always the same atoms or fluid particles), which may move, flow, and interact with its surroundings. A *control volume,* on the other hand, is a volume in space (a geometric entity, independent of mass) through which fluid may flow.

*Both control volume and system concepts can be used to describe fluid flow.*

A system is a specific, identifiable quantity of matter. It may consist of a relatively large amount of mass (such as all of the air in the Earth's atmosphere), or it may be an infinitesimal size (such as a single fluid particle). In any case, the molecules making up the system are "tagged" in some fashion (dyed red, either actually or only in your mind) so that they can be continually identified as they move about. The system may interact with its surroundings by various means (by the transfer of heat or the exertion of a pressure force, for example). It may continually change size and shape, but it always contains the same mass.

A mass of air drawn into an air compressor can be considered as a system. It changes shape and size (it is compressed), its temperature may change, and it is eventually expelled through the outlet of the compressor. The matter associated with the original air drawn into the compressor remains as a system, however. The behavior of this material could be investigated by applying the appropriate governing equations to this system.

One of the important concepts used in the study of statics and dynamics is that of the free-body diagram. That is, we identify an object, isolate it from its surroundings, replace its surroundings by the equivalent actions that they put on the object, and apply Newton's laws of motion. The body in such cases is our system—an identified portion of matter that we follow during its interactions with its surroundings. In fluid mechanics, it is often quite difficult to identify and keep track of a specific quantity of matter. A finite portion of a fluid contains an uncountable number of fluid particles that move about quite freely, unlike a solid that may deform but usually remains relatively easy to identify. For example, we cannot as easily follow a specific portion of water flowing in a river as we can follow a branch floating on its surface.

We may often be more interested in determining the forces put on a fan, airplane, or automobile by air flowing past the object than we are in the information obtained by following a given portion of the air (a system) as it flows along. Similarly, for the Space Shuttle launch vehicle shown in the figure in the margin, we may be more interested in determining the thrust produced than we are in the information obtained by following the highly complex, irregular path of the exhaust plume from the rocket engine nozzle. For these situations we often use the control volume approach. We identify a specific volume in space (a volume associated with the fan, airplane, or automobile, for example) and analyze the fluid flow within, through, or around that volume. In general, the control volume can be a moving volume, although for most situations considered in this book we will use only fixed, nondeformable control volumes. The matter within a control volume may change with time as the fluid flows through it. Similarly, the amount of mass within the volume may change with time. The control volume itself is a specific geometric entity, independent of the flowing fluid.

(Photograph courtesy of NASA.)

Examples of control volumes and *control surfaces* (the surface of the control volume) are shown in Fig. 4.10. For case (*a*), fluid flows through a pipe. The fixed control surface consists of the inside surface of the pipe, the outlet end at section (2), and a section across the pipe at (1). One portion of the control surface is a physical surface (the pipe), while the remainder is simply a surface in space (across the pipe). Fluid flows across part of the control surface but not across all of it.

Another control volume is the rectangular volume surrounding the jet engine shown in Fig. 4.10*b*. If the airplane to which the engine is attached is sitting still on the runway, air flows through this control volume because of the action of the engine within it. The air that was within the engine itself at time $t = t_1$ (a system) has passed through the engine and is outside of the control volume at a later time $t = t_2$ as indicated. At this later time other air (a different system) is within the engine. If the airplane is moving, the control volume is fixed relative to an observer on the airplane, but it

--- --- Control volume surface    System at time $t_1$    System at time $t_2 > t_1$

■ **Figure 4.10**  Typical control volumes: (*a*) fixed control volume, (*b*) fixed or moving control volume, (*c*) deforming control volume.

is a moving control volume relative to an observer on the ground. In either situation air flows through and around the engine as indicated.

The deflating balloon shown in Fig. 4.10*c* provides an example of a deforming control volume. As time increases, the control volume (whose surface is the inner surface of the balloon) decreases in size. If we do not hold onto the balloon, it becomes a moving, deforming control volume as it darts about the room. The majority of the problems we will analyze can be solved by using a fixed, nondeforming control volume. In some instances, however, it will be advantageous, in fact necessary, to use a moving, deforming control volume.

In many ways the relationship between a system and a control volume is similar to the relationship between the Lagrangian and Eulerian flow description introduced in Section 4.1.1. In the system or Lagrangian description, we follow the fluid and observe its behavior as it moves about. In the control volume or Eulerian description, we remain stationary and observe the fluid's behavior at a fixed location. (If a moving control volume is used, it virtually never moves with the system—the system flows through the control volume.) These ideas are discussed in more detail in the next section.

*The governing laws of fluid motion are stated in terms of fluid systems, not control volumes.*

All of the laws governing the motion of a fluid are stated in their basic form in terms of a system approach. For example, "the mass of a system remains constant," or "the time rate of change of momentum of a system is equal to the sum of all the forces acting on the system." Note the word system, not control volume, in these statements. To use the governing equations in a control volume approach to problem solving, we must rephrase the laws in an appropriate manner. To this end we introduce the Reynolds transport theorem in the following section.

## 4.4    The Reynolds Transport Theorem

We are sometimes interested in what happens to a particular part of the fluid as it moves about. Other times we may be interested in what effect the fluid has on a particular object or volume in space as fluid interacts with it. Thus, we need to describe the laws governing fluid motion using both system concepts (consider a given mass of the fluid) and control volume concepts (consider a given volume). To do this we need an analytical tool to shift from one representation to the other. The ***Reynolds transport theorem*** provides this tool.

All physical laws are stated in terms of various physical parameters. Velocity, acceleration, mass, temperature, and momentum are but a few of the more common parameters. Let $B$ represent any of these (or other) fluid parameters and $b$ represent the amount of that parameter per unit mass. That is,

$$B = mb$$

where $m$ is the mass of the portion of fluid of interest. For example, as shown by the figure in the margin, if $B = m$, the mass, it follows that $b = 1$. The mass per unit mass is unity. If $B = mV^2/2$, the kinetic energy of the mass, then $b = V^2/2$, the kinetic energy per unit mass. The parameters $B$ and $b$ may be scalars or vectors. Thus, if $\mathbf{B} = m\mathbf{V}$, the momentum of the mass, then $\mathbf{b} = \mathbf{V}$. (The momentum per unit mass is the velocity.)

The parameter $B$ is termed an *extensive property*, and the parameter $b$ is termed an *intensive property*. The value of $B$ is directly proportional to the amount of the mass being considered, whereas the value of $b$ is independent of the amount of mass. The amount of an extensive property that a system possesses at a given instant, $B_{sys}$, can be determined by adding up the amount associated with each fluid particle in the system. For infinitesimal fluid particles of size $\delta \forall$ and mass $\rho \, \delta \forall$,

| $B$ | $b = B/m$ |
| --- | --- |
| $m$ | $1$ |
| $m\mathbf{V}$ | $\mathbf{V}$ |
| $\frac{1}{2}mV^2$ | $\frac{1}{2}V^2$ |

this summation (in the limit of $\delta \forall \to 0$) takes the form of an integration over all the particles in the system and can be written as

$$B_{sys} = \lim_{\delta \forall \to 0} \sum_i b_i (\rho_i \, \delta \forall_i) = \int_{sys} \rho b \, d\forall$$

The limits of integration cover the entire system—a (usually) moving volume. We have used the fact that the amount of $B$ in a fluid particle of mass $\rho \, \delta \forall$ is given in terms of $b$ by $\delta B = b \rho \, \delta \forall$.

Most of the laws governing fluid motion involve the time rate of change of an extensive property of a fluid system—the rate at which the momentum of a system changes with time, the rate at which the mass of a system changes with time, and so on. Thus, we often encounter terms such as

$$\frac{dB_{sys}}{dt} = \frac{d\left( \int_{sys} \rho b \, d\forall \right)}{dt} \tag{4.8}$$

To formulate the laws into a control volume approach, we must obtain an expression for the time rate of change of an extensive property within a control volume, $B_{cv}$, not within a system. This can be written as

$$\frac{dB_{cv}}{dt} = \frac{d\left( \int_{cv} \rho b \, d\forall \right)}{dt} \tag{4.9}$$

*Differences between control volume and system concepts are subtle but very important.*

where the limits of integration, denoted by cv, cover the control volume of interest. Although Eqs. 4.8 and 4.9 may look very similar, the physical interpretation of each is quite different. Mathematically, the difference is represented by the difference in the limits of integration. Recall that the control volume is a volume in space (in most cases stationary, although if it moves it need not move with the system). On the other hand, the system is an identifiable collection of mass that moves with the fluid (indeed it is a specified portion of the fluid). We will learn that even for those instances when the control volume and the system momentarily occupy the same volume in space, the two quantities $dB_{sys}/dt$ and $dB_{cv}/dt$ need not be the same. The Reynolds transport theorem provides the relationship between the time rate of change of an extensive property for a system and that for a control volume—the relationship between Eqs. 4.8 and 4.9.

---

## $E$XAMPLE 4.7    Time Rate of Change for a System and a Control Volume

**GIVEN**    Fluid flows from the fire extinguisher tank shown in Fig. E4.7a.

**FIND**    Discuss the differences between $dB_{sys}/dt$ and $dB_{cv}/dt$ if $B$ represents mass.

## SOLUTION

With $B = m$, the system mass, it follows that $b = 1$ and Eqs. 4.8 and 4.9 can be written as

$$\frac{dB_{sys}}{dt} \equiv \frac{dm_{sys}}{dt} = \frac{d\left( \int_{sys} \rho \, d\forall \right)}{dt}$$

and

$$\frac{dB_{cv}}{dt} \equiv \frac{dm_{cv}}{dt} = \frac{d\left( \int_{cv} \rho \, d\forall \right)}{dt}$$

(a)

■ **Figure E4.7**
(© DNY59/iStockphoto)

Physically, these represent the time rate of change of mass within the system and the time rate of change of mass within the control volume, respectively. We choose our system to be the fluid within the tank at the time the valve was opened ($t = 0$) and the control volume to be the tank itself as shown in Fig. E4.7$b$. A short time after the valve is opened, part of the system has moved outside of the control volume as is shown in Fig. E4.7$c$. The control volume remains fixed. The limits of integration are fixed for the control volume; they are a function of time for the system.

Clearly, if mass is to be conserved (one of the basic laws governing fluid motion), the mass of the fluid in the system is constant, so that

$$\frac{d\left( \int_{\text{sys}} \rho \, d\Psi \right)}{dt} = 0$$

On the other hand, it is equally clear that some of the fluid has left the control volume through the nozzle on the tank. Hence, the amount of mass within the tank (the control volume) decreases with time, or

$$\frac{d\left( \int_{\text{cv}} \rho \, d\Psi \right)}{dt} < 0$$

The actual numerical value of the rate at which the mass in the control volume decreases will depend on the rate at which the fluid flows through the nozzle (i.e., the size of the nozzle and the speed and density of the fluid). Clearly the meanings of $dB_{\text{sys}}/dt$ and $dB_{\text{cv}}/dt$ are different. For this example, $dB_{\text{cv}}/dt < dB_{\text{sys}}/dt$. Other situations may have $dB_{\text{cv}}/dt \geq dB_{\text{sys}}/dt$.

■ **Figure E4.7**

### 4.4.1 Derivation of the Reynolds Transport Theorem

A simple version of the Reynolds transport theorem relating system concepts to control volume concepts can be obtained easily for the one-dimensional flow through a fixed control volume such as the variable area duct section shown in Fig. 4.11$a$. We consider the control volume to be that stationary volume within the duct between sections (1) and (2) as indicated in Fig. 4.11$b$. The system that we consider is that fluid occupying the control volume at some initial time $t$. A short time later, at time $t + \delta t$, the system has moved slightly to the right. The fluid particles that coincided with section (2) of the control surface at time $t$ have moved a distance $\delta \ell_2 = V_2 \, \delta t$ to the right, where $V_2$ is the velocity of the fluid as it passes section (2). Similarly, the fluid initially at section (1) has moved a distance $\delta \ell_1 = V_1 \, \delta t$, where $V_1$ is the fluid velocity at section (1). We assume the fluid flows across sections (1) and (2) in a direction normal to these surfaces and that $V_1$ and $V_2$ are constant across sections (1) and (2).

*The moving system flows through the fixed control volume.*

As is shown in Fig. 4.11$c$, the outflow from the control volume from time $t$ to $t + \delta t$ is denoted as volume II, the inflow as volume I, and the control volume itself as CV. Thus, the system at time $t$ consists of the fluid in section CV; that is, "SYS = CV" at time $t$. At time $t + \delta t$ the system consists of the same fluid that now occupies sections (CV − I) + II. That is, "SYS = CV − I + II" at time $t + \delta t$. The control volume remains as section CV for all time.

■ **Figure 4.11**   Control volume and system for flow through a variable area pipe.

If $B$ is an extensive parameter of the system, then the value of it for the system at time $t$ is

$$B_{sys}(t) = B_{cv}(t)$$

since the system and the fluid within the control volume coincide at this time. Its value at time $t + \delta t$ is

$$B_{sys}(t + \delta t) = B_{cv}(t + \delta t) - B_I(t + \delta t) + B_{II}(t + \delta t)$$

Thus, the change in the amount of $B$ in the system, $\delta B_{sys}$, in the time interval $\delta t$ divided by this time interval is given by

$$\frac{\delta B_{sys}}{\delta t} = \frac{B_{sys}(t + \delta t) - B_{sys}(t)}{\delta t} = \frac{B_{cv}(t + \delta t) - B_I(t + \delta t) + B_{II}(t + \delta t) - B_{sys}(t)}{\delta t}$$

By using the fact that at the initial time $t$ we have $B_{sys}(t) = B_{cv}(t)$, this ungainly expression may be rearranged as follows.

$$\frac{\delta B_{sys}}{\delta t} = \frac{B_{cv}(t + \delta t) - B_{cv}(t)}{\delta t} - \frac{B_I(t + \delta t)}{\delta t} + \frac{B_{II}(t + \delta t)}{\delta t} \tag{4.10}$$

*The time rate of change of a system property is a Lagrangian concept.*

In the limit $\delta t \to 0$, the left-hand side of Eq. 4.10 is equal to the time rate of change of $B$ for the system and is denoted as $DB_{sys}/Dt$. We use the material derivative notation, $D(\ )/Dt$, to denote this time rate of change to emphasize the Lagrangian character of this term. (Recall from Section 4.2.1 that the material derivative, $DP/Dt$, of any quantity $P$ represents the time rate of change of that quantity associated with a given fluid particle as it moves along.) Similarly, the quantity $DB_{sys}/Dt$ represents the time rate of change of property $B$ associated with a system (a given portion of fluid) as it moves along.

In the limit $\delta t \to 0$, the first term on the right-hand side of Eq. 4.10 is seen to be the time rate of change of the amount of $B$ within the control volume

$$\lim_{\delta t \to 0} \frac{B_{cv}(t + \delta t) - B_{cv}(t)}{\delta t} = \frac{\partial B_{cv}}{\partial t} = \frac{\partial\left(\int_{cv} \rho b\, d\Psi\right)}{\partial t} \tag{4.11}$$

$V_2$

The third term on the right-hand side of Eq. 4.10 represents the rate at which the extensive parameter $B$ flows from the control volume across the control surface. As indicated by the figure in the margin, during the time interval from $t = 0$ to $t = \delta t$ the volume of fluid that flows across section (2) is given by $\delta \Psi_{II} = A_2\, \delta \ell_2 = A_2 (V_2 \delta t)$. Thus, the amount of $B$ within region II, the outflow region, is its amount per unit volume, $\rho b$, times the volume

$$B_{II}(t + \delta t) = (\rho_2 b_2)(\delta \Psi_{II}) = \rho_2 b_2 A_2 V_2\, \delta t$$

where $b_2$ and $\rho_2$ are the constant values of $b$ and $\rho$ across section (2). Thus, the rate at which this property flows from the control volume, $\dot{B}_{out}$, is given by

$$\dot{B}_{out} = \lim_{\delta t \to 0} \frac{B_{II}(t + \delta t)}{\delta t} = \rho_2 A_2 V_2 b_2 \tag{4.12}$$

(2)
$t = 0$

$\delta \Psi_{II}$

$V_2 \delta t$

(2)
$t = \delta t$

Similarly, the inflow of $B$ into the control volume across section (1) during the time interval $\delta t$ corresponds to that in region I and is given by the amount per unit volume times the volume, $\delta \Psi_I = A_1\, \delta \ell_1 = A_1 (V_1\, \delta t)$. Hence,

$$B_I(t + \delta t) = (\rho_1 b_1)(\delta \Psi_I) = \rho_1 b_1 A_1 V_1\, \delta t$$

where $b_1$ and $\rho_1$ are the constant values of $b$ and $\rho$ across section (1). Thus, the rate of inflow of the property $B$ into the control volume, $\dot{B}_{in}$, is given by

$$\dot{B}_{in} = \lim_{\delta t \to 0} \frac{B_I(t + \delta t)}{\delta t} = \rho_1 A_1 V_1 b_1 \tag{4.13}$$

If we combine Eqs. 4.10, 4.11, 4.12, and 4.13, we see that the relationship between the time rate of change of $B$ for the system and that for the control volume is given by

$$\frac{DB_{sys}}{Dt} = \frac{\partial B_{cv}}{\delta t} + \dot{B}_{out} - \dot{B}_{in} \tag{4.14}$$

or

$$\frac{DB_{sys}}{Dt} = \frac{\partial B_{cv}}{\partial t} + \rho_2 A_2 V_2 b_2 - \rho_1 A_1 V_1 b_1 \tag{4.15}$$

This is a version of the Reynolds transport theorem valid under the restrictive assumptions associated with the flow shown in Fig. 4.11—fixed control volume with one inlet and one outlet having uniform properties (density, velocity, and the parameter $b$) across the inlet and outlet with the velocity normal to sections (1) and (2). Note that the time rate of change of $B$ for the system (the left-hand side of Eq. 4.15 or the quantity in Eq. 4.8) is not necessarily the same as the rate of change of $B$ within the control volume (the first term on the right-hand side of Eq. 4.15 or the quantity in Eq. 4.9). This is true because the inflow rate ($b_1 \rho_1 V_1 A_1$) and the outflow rate ($b_2 \rho_2 V_2 A_2$) of the property $B$ for the control volume need not be the same.

*The time derivative associated with a system may be different from that for a control volume.*

---

## EXAMPLE 4.8    Use of the Reynolds Transport Theorem

**GIVEN** Consider again the flow from the fire extinguisher shown in Fig. E4.7. Let the extensive property of interest be the system mass ($B = m$, the system mass, or $b = 1$).

**FIND** Write the appropriate form of the Reynolds transport theorem for this flow.

### SOLUTION

Again we take the control volume to be the fire extinguisher and the system to be the fluid within it at time $t = 0$. For this case there is no inlet, section (1), across which the fluid flows into the control volume ($A_1 = 0$). There is, however, an outlet, section (2). Thus, the Reynolds transport theorem, Eq. 4.15, along with Eq. 4.9 with $b = 1$ can be written as

$$\frac{Dm_{sys}}{Dt} = \frac{\partial \left( \int_{cv} \rho \, d\forall \right)}{\partial t} + \rho_2 A_2 V_2 \qquad (1) \quad \textbf{(Ans)}$$

**COMMENT** If we proceed one step further and use the basic law of conservation of mass, we may set the left-hand side of this equation equal to zero (the amount of mass in a system is constant) and rewrite Eq. 1 in the form

$$\frac{\partial \left( \int_{cv} \rho \, d\forall \right)}{\partial t} = -\rho_2 A_2 V_2 \qquad (2)$$

The physical interpretation of this result is that the rate at which the mass in the tank decreases in time is equal in magnitude but opposite to the rate of flow of mass from the exit, $\rho_2 A_2 V_2$. Note the units for the two terms of Eq. 2 (kg/s or slugs/s). If there were both an inlet and an outlet to the control volume shown in Fig. E4.7, Eq. 2 would become

$$\frac{\partial \left( \int_{cv} \rho \, d\forall \right)}{\partial t} = \rho_1 A_1 V_1 - \rho_2 A_2 V_2 \qquad (3)$$

In addition, if the flow were steady, the left-hand side of Eq. 3 would be zero (the amount of mass in the control would be constant in time) and Eq. 3 would become

$$\rho_1 A_1 V_1 = \rho_2 A_2 V_2$$

This is one form of the conservation of mass principle discussed in Section 3.6.2—the mass flowrates into and out of the control volume are equal. Other more general forms are discussed in Chapter 5.

---

Equation 4.15 is a simplified version of the Reynolds transport theorem. We will now derive it for much more general conditions. A general, fixed control volume with fluid flowing through it is shown in Fig. 4.12. The flow field may be quite simple (as in the above one-dimensional flow considerations), or it may involve a quite complex, unsteady, three-dimensional situation such as the flow through a human heart as illustrated by the figure in the margin. In any case we again consider the system to be the fluid within the control volume at the initial time $t$. A short time later a portion of the fluid (region II) has exited from the control volume, and additional fluid (region I, not part of the original system) has entered the control volume.

■ **Figure 4.12**  Control volume and system for flow through an arbitrary, fixed control volume.

■ **Figure 4.13**  Typical control volume with more than one inlet and outlet.

*The simplified Reynolds transport theorem can be easily generalized.*

We consider an extensive fluid property $B$ and seek to determine how the rate of change of $B$ associated with the system is related to the rate of change of $B$ within the control volume at any instant. By repeating the exact steps that we did for the simplified control volume shown in Fig. 4.11, we see that Eq. 4.14 is valid for the general case also, provided that we give the correct interpretation to the terms $\dot{B}_{\text{out}}$ and $\dot{B}_{\text{in}}$. In general, the control volume may contain more (or less) than one inlet and one outlet. A typical pipe system may contain several inlets and outlets as are shown in Fig. 4.13. In such instances we think of all inlets grouped together $(\text{I} = \text{I}_a + \text{I}_b + \text{I}_c + \cdots)$ and all outlets grouped together $(\text{II} = \text{II}_a + \text{II}_b + \text{II}_c + \cdots)$, at least conceptually.

The term $\dot{B}_{\text{out}}$ represents the net flowrate of the property $B$ from the control volume. Its value can be thought of as arising from the addition (integration) of the contributions through each infinitesimal area element of size $\delta A$ on the portion of the control surface dividing region II and the control volume. This surface is denoted $\text{CS}_{\text{out}}$. As is indicated in Fig. 4.14, in time $\delta t$ the volume of fluid that passes across each area element is given by $\delta\text{V} = \delta\ell_n\,\delta A$, where $\delta\ell_n = \delta\ell\cos\theta$ is the height (normal to the base, $\delta A$) of the small volume element, and $\theta$ is the angle between the velocity vector and the outward pointing normal to the surface, $\hat{\mathbf{n}}$. Thus, since $\delta\ell = V\,\delta t$, the amount of the property $B$ carried across the area element $\delta A$ in the time interval $\delta t$ is given by

$$\delta B = b\rho\,\delta\text{V} = b\rho(V\cos\theta\,\delta t)\,\delta A$$

The rate at which $B$ is carried out of the control volume across the small area element $\delta A$, denoted $\delta\dot{B}_{\text{out}}$, is

$$\delta\dot{B}_{\text{out}} = \lim_{\delta t\to 0}\frac{\rho b\,\delta\text{V}}{\delta t} = \lim_{\delta t\to 0}\frac{(\rho bV\cos\theta\,\delta t)\,\delta A}{\delta t} = \rho bV\cos\theta\,\delta A$$

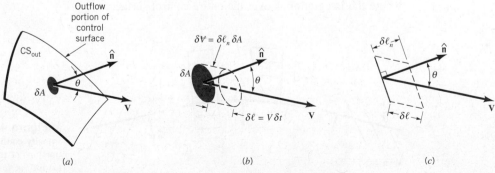

■ **Figure 4.14**  Outflow across a typical portion of the control surface.

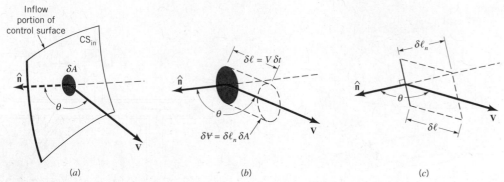

■ **Figure 4.15** Inflow across a typical portion of the control surface.

By integrating over the entire outflow portion of the control surface, $CS_{out}$, we obtain

$$\dot{B}_{out} = \int_{cs_{out}} d\dot{B}_{out} = \int_{cs_{out}} \rho b V \cos\theta \, dA$$

The quantity $V\cos\theta$ is the component of the velocity normal to the area element $\delta A$. From the definition of the dot product, this can be written as $V\cos\theta = \mathbf{V} \cdot \hat{\mathbf{n}}$. Hence, an alternate form of the outflow rate is

*The flowrate of a parameter across the control surface is written in terms of a surface integral.*

$$\dot{B}_{out} = \int_{cs_{out}} \rho b \mathbf{V} \cdot \hat{\mathbf{n}} \, dA \tag{4.16}$$

In a similar fashion, by considering the inflow portion of the control surface, $CS_{in}$, as shown in Fig. 4.15, we find that the inflow rate of $B$ into the control volume is

$$\dot{B}_{in} = -\int_{cs_{in}} \rho b V \cos\theta \, dA = -\int_{cs_{in}} \rho b \mathbf{V} \cdot \hat{\mathbf{n}} \, dA \tag{4.17}$$

We use the standard notation that the unit normal vector to the control surface, $\hat{\mathbf{n}}$, points out from the control volume. Thus, as is shown in Fig. 4.16, $-90° < \theta < 90°$ for outflow regions (the normal component of $\mathbf{V}$ is positive; $\mathbf{V} \cdot \hat{\mathbf{n}} > 0$). For inflow regions $90° < \theta < 270°$ (the normal component of $\mathbf{V}$ is negative; $\mathbf{V} \cdot \hat{\mathbf{n}} < 0$). The value of $\cos\theta$ is, therefore, positive on the $CV_{out}$ portions of the control surface and negative on the $CV_{in}$ portions. Over the remainder of the control surface, there is no inflow or outflow, leading to $\mathbf{V} \cdot \hat{\mathbf{n}} = V\cos\theta = 0$ on those portions. On such portions either $V = 0$ (the fluid "sticks" to the surface) or $\cos\theta = 0$ (the fluid "slides" along the surface without crossing it) (see Fig. 4.16). Therefore, the net flux (flowrate) of parameter $B$ across the entire control surface is

$$\dot{B}_{out} - \dot{B}_{in} = \int_{cs_{out}} \rho b \mathbf{V} \cdot \hat{\mathbf{n}} \, dA - \left(-\int_{cs_{in}} \rho b \mathbf{V} \cdot \hat{\mathbf{n}} \, dA\right)$$

$$= \int_{cs} \rho b \mathbf{V} \cdot \hat{\mathbf{n}} \, dA \tag{4.18}$$

where the integration is over the entire control surface.

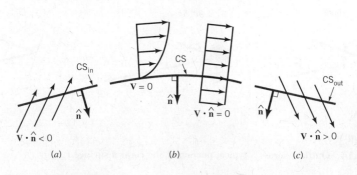

■ **Figure 4.16** Possible velocity configurations on portions of the control surface: (a) inflow, (b) no flow across the surface, (c) outflow.

By combining Eqs. 4.14 and 4.18 we obtain

$$\frac{DB_{sys}}{Dt} = \frac{\partial B_{cv}}{\partial t} + \int_{cs} \rho b \mathbf{V} \cdot \hat{\mathbf{n}}\, dA$$

This can be written in a slightly different form by using $B_{cv} = \int_{cv} \rho b\, d\mathcal{V}$ so that

$$\boxed{\frac{DB_{sys}}{Dt} = \frac{\partial}{\partial t} \int_{cv} \rho b\, d\mathcal{V} + \int_{cs} \rho b\, \mathbf{V} \cdot \hat{\mathbf{n}}\, dA} \qquad \textbf{(4.19)}$$

Equation 4.19 is the general form of the Reynolds transport theorem for a fixed, nondeforming control volume. Its interpretation and use are discussed in the following sections.

### 4.4.2  Physical Interpretation

The Reynolds transport theorem as given in Eq. 4.19 is widely used in fluid mechanics (and other areas as well). At first it appears to be a rather formidable mathematical expression—perhaps one to be steered clear of if possible. However, a physical understanding of the concepts involved will show that it is a rather straightforward, relatively easy-to-use tool. Its purpose is to provide a link between control volume ideas and system ideas.

The left side of Eq. 4.19 is the time rate of change of an arbitrary extensive parameter of a system. This may represent the rate of change of mass, momentum, energy, or angular momentum of the system, depending on the choice of the parameter $B$.

Because the system is moving and the control volume is stationary, the time rate of change of the amount of $B$ within the control volume is not necessarily equal to that of the system. The first term on the right side of Eq. 4.19 represents the rate of change of $B$ within the control volume as the fluid flows through it. Recall that $b$ is the amount of $B$ per unit mass, so that $\rho b\, d\mathcal{V}$ is the amount of $B$ in a small volume $d\mathcal{V}$. Thus, the time derivative of the integral of $\rho b$ throughout the control volume is the time rate of change of $B$ within the control volume at a given time.

Control surface

The last term in Eq. 4.19 (an integral over the control surface) represents the net flowrate of the parameter $B$ across the entire control surface. As illustrated by the figure in the margin, over a portion of the control surface this property is being carried out of the control volume ($\mathbf{V} \cdot \hat{\mathbf{n}} > 0$); over other portions it is being carried into the control volume ($\mathbf{V} \cdot \hat{\mathbf{n}} < 0$). Over the remainder of the control surface there is no transport of $B$ across the surface since $b\mathbf{V} \cdot \hat{\mathbf{n}} = 0$, because either $b = 0$, $\mathbf{V} = 0$, or $\mathbf{V}$ is parallel to the surface at those locations. The mass flowrate through area element $\delta A$, given by $\rho \mathbf{V} \cdot \hat{\mathbf{n}}\, \delta A$, is positive for outflow (efflux) and negative for inflow (influx). Each fluid particle or fluid mass carries a certain amount of $B$ with it, as given by the product of $B$ per unit mass, $b$, and the mass. The rate at which this $B$ is carried across the control surface is given by the area integral term of Eq. 4.19. This net rate across the entire control surface may be negative, zero, or positive depending on the particular situation involved.

### 4.4.3  Relationship to Material Derivative

In Section 4.2.1 we discussed the concept of the material derivative $D(\ )/Dt = \partial(\ )/\partial t + \mathbf{V} \cdot \nabla(\ ) = \partial(\ )/\partial t + u\, \partial(\ )/\partial x + v\, \partial(\ )/\partial y + w\, \partial(\ )/\partial z$. The physical interpretation of this derivative is that it provides the time rate of change of a fluid property (temperature, velocity, etc.) associated with a particular fluid particle as it flows. The value of that parameter for that particle may change because of unsteady effects [the $\partial(\ )/\partial t$ term] or because of effects associated with the particle's motion [the $\mathbf{V} \cdot \nabla(\ )$ term].

Careful consideration of Eq. 4.19 indicates the same type of physical interpretation for the Reynolds transport theorem. The term involving the time derivative of the control volume integral represents unsteady effects associated with the fact that values of the parameter within the control volume may change with time. For steady flow this effect vanishes—fluid flows through the control volume, but the amount of any property, $B$, within the control volume is constant in time. The term involving the control surface integral represents the convective effects associated with the flow of the system across the fixed control surface. The sum of these two terms gives the rate of change of the parameter $B$ for the system. This corresponds to the interpretation of the material derivative,

*The Reynolds transport theorem is the integral counterpart of the material derivative.*

$D(\ )/Dt = \partial(\ )/\partial t + \mathbf{V} \cdot \nabla(\ )$, in which the sum of the unsteady effect and the convective effect gives the rate of change of a parameter for a fluid particle. As is discussed in Section 4.2, the material derivative operator may be applied to scalars (such as temperature) or vectors (such as velocity). This is also true for the Reynolds transport theorem. The particular parameters of interest, $B$ and $b$, may be scalars or vectors.

Thus, both the material derivative and the Reynolds transport theorem equations represent ways to transfer from the Lagrangian viewpoint (follow a particle or follow a system) to the Eulerian viewpoint (observe the fluid at a given location in space or observe what happens in the fixed control volume). The material derivative (Eq. 4.5) is essentially the infinitesimal (or derivative) equivalent of the finite size (or integral) Reynolds transport theorem (Eq. 4.19).

### 4.4.4 Steady Effects

Consider a steady flow $[\partial(\ )/\partial t \equiv 0]$ so that Eq. 4.19 reduces to

$$\frac{DB_{\text{sys}}}{Dt} = \int_{\text{cs}} \rho b \mathbf{V} \cdot \hat{\mathbf{n}}\, dA \qquad (4.20)$$

In such cases if there is to be a change in the amount of $B$ associated with the system (nonzero left-hand side), there must be a net difference in the rate that $B$ flows into the control volume compared with the rate that it flows out of the control volume. That is, the integral of $\rho b \mathbf{V} \cdot \hat{\mathbf{n}}$ over the inflow portions of the control surface would not be equal and opposite to that over the outflow portions of the surface.

Control volume

Momentum out

(Photograph courtesy of NASA.)

Consider steady flow through the "black box" control volume that is shown in Fig. 4.17. If the parameter $B$ is the mass of the system, the left-hand side of Eq. 4.20 is zero (conservation of mass for the system as discussed in detail in Section 5.1). Hence, the flowrate of mass into the box must be the same as the flowrate of mass out of the box because the right-hand side of Eq. 4.20 represents the net flowrate through the control surface. On the other hand, assume the parameter $B$ is the momentum of the system. The momentum of the system need not be constant. In fact, according to Newton's second law the time rate of change of the system momentum equals the net force, $\mathbf{F}$, acting on the system. In general, the left-hand side of Eq. 4.20 will therefore be nonzero. Thus, the right-hand side, which then represents the net flux of momentum across the control surface, will be nonzero. The flowrate of momentum into the control volume need not be the same as the flux of momentum from the control volume. We will investigate these concepts much more fully in Chapter 5. They are the basic principles describing the operation of such devices as jet or rocket engines like the one shown in the figure in the margin.

For steady flows the amount of the property $B$ within the control volume does not change with time. The amount of the property associated with the system may or may not change with time, depending on the particular property considered and the flow situation involved. The difference between that associated with the control volume and that associated with the system is determined by the rate at which $B$ is carried across the control surface—the term $\int_{\text{cs}} \rho b \mathbf{V} \cdot \hat{\mathbf{n}}\, dA$.

*The Reynolds transport theorem involves both steady and unsteady effects.*

### 4.4.5 Unsteady Effects

Consider unsteady flow $[\partial(\ )/\partial t \neq 0]$ so that all terms in Eq. 4.19 must be retained. When they are viewed from a control volume standpoint, the amount of parameter $B$ within the system may change because the amount of $B$ within the fixed control volume may change with time

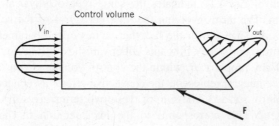

Control volume

$V_{\text{in}}$        $V_{\text{out}}$

$F$

■ **Figure 4.17** Steady flow through a control volume.

■ **Figure 4.18**    Unsteady flow through a constant diameter pipe.

*For some flow situations, certain portions of the Reynolds transport theorem are automatically zero.*

[the $\partial(\int_{cv} \rho b \, d\mathcal{V})/\partial t$ term] and because there may be a net nonzero flow of that parameter across the control surface (the $\int_{cs} \rho b \mathbf{V} \cdot \hat{\mathbf{n}} \, dA$ term).

For the special unsteady situations in which the rate of inflow of parameter $B$ is exactly balanced by its rate of outflow, it follows that $\int_{cs} \rho b \mathbf{V} \cdot \hat{\mathbf{n}} \, dA = 0$, and Eq. 4.19 reduces to

$$\frac{DB_{sys}}{Dt} = \frac{\partial}{\partial t} \int_{cv} \rho b \, d\mathcal{V} \qquad (4.21)$$

For such cases, any rate of change in the amount of $B$ associated with the system is equal to the rate of change of $B$ within the control volume. This can be illustrated by considering flow through a constant diameter pipe as is shown in Fig. 4.18. The control volume is as shown, and the system is the fluid within this volume at time $t_0$. We assume the flow is one-dimensional with $\mathbf{V} = V_0 \hat{\mathbf{i}}$, where $V_0(t)$ is a function of time, and that the density is constant. At any instant in time, all particles in the system have the same velocity. We let $\mathbf{B} =$ system momentum $= m\mathbf{V} = mV_0 \hat{\mathbf{i}}$, where $m$ is the system mass, so that $\mathbf{b} = \mathbf{B}/m = \mathbf{V} = V_0 \hat{\mathbf{i}}$, the fluid velocity. The magnitude of the momentum efflux across the outlet [section (2)] is the same as the magnitude of the momentum influx across the inlet [section (1)]. However, the sign of the efflux is opposite to that of the influx since $\mathbf{V} \cdot \hat{\mathbf{n}} > 0$ for the outflow and $\mathbf{V} \cdot \hat{\mathbf{n}} < 0$ for the inflow. Note that $\mathbf{V} \cdot \hat{\mathbf{n}} = 0$ along the sides of the control volume. Thus, with $\mathbf{V} \cdot \hat{\mathbf{n}} = -V_0$ on section (1), $\mathbf{V} \cdot \hat{\mathbf{n}} = V_0$ on section (2), and $A_1 = A_2$, we obtain

$$\int_{cs} \rho b \mathbf{V} \cdot \hat{\mathbf{n}} \, dA = \int_{cs} \rho(V_0 \hat{\mathbf{i}})(\mathbf{V} \cdot \hat{\mathbf{n}}) \, dA$$

$$= \int_{(1)} \rho(V_0 \hat{\mathbf{i}})(-V_0) \, dA + \int_{(2)} \rho(V_0 \hat{\mathbf{i}})(V_0) \, dA$$

$$= -\rho V_0^2 A_1 \hat{\mathbf{i}} + \rho V_0^2 A_2 \hat{\mathbf{i}} = 0$$

It is seen that for this special case Eq. 4.21 is valid. The rate at which the momentum of the system changes with time is the same as the rate of change of momentum within the control volume. If $V_0$ is constant in time, there is no rate of change of momentum of the system, and for this special case each of the terms in the Reynolds transport theorem is zero by itself.

Consider the flow through a variable area pipe shown in Fig. 4.19. In such cases the fluid velocity is not the same at section (1) as it is at (2). Hence, the efflux of momentum from the control volume is not equal to the influx of momentum, so that the convective term in Eq. 4.20 [the integral of $\rho \mathbf{V}(\mathbf{V} \cdot \hat{\mathbf{n}})$ over the control surface] is not zero. These topics will be discussed in considerably more detail in Chapter 5.

■ **Figure 4.19**    Flow through a variable area pipe.

■ **Figure 4.20** Example of a moving control volume.

### 4.4.6 Moving Control Volumes

For most problems in fluid mechanics, the control volume may be considered as a fixed volume through which the fluid flows. There are, however, situations for which the analysis is simplified if the control volume is allowed to move or deform. The most general situation would involve a control volume that moves, accelerates, and deforms. As one might expect, the use of these control volumes can become fairly complex.

A number of important problems can be most easily analyzed by using a nondeforming control volume that moves with a constant velocity. Such an example is shown in Fig. 4.20 in which a stream of water with velocity $\mathbf{V}_1$ strikes a vane that is moving with constant velocity $\mathbf{V}_0$. It may be of interest to determine the force, $\mathbf{F}$, that the water puts on the vane. Such problems frequently occur in turbines where a stream of fluid (water or steam, for example) strikes a series of blades that move past the nozzle. To analyze such problems it is advantageous to use a moving control volume. We will obtain the Reynolds transport theorem for such control volumes.

We consider a control volume that moves with a constant velocity as is shown in Fig. 4.21. The shape, size, and orientation of the control volume do not change with time. The control volume merely translates with a constant velocity, $\mathbf{V}_{cv}$, as shown. In general, the velocity of the control volume and that of the fluid are not the same, so that there is a flow of fluid through the moving control volume just as in the stationary control volume cases discussed in Section 4.4.2. The main difference between the fixed and the moving control volume cases is that it is the *relative velocity*, $\mathbf{W}$, that carries fluid across the moving control surface, whereas it is the *absolute velocity*, $\mathbf{V}$, that carries the fluid across the fixed control surface. The relative velocity is the fluid velocity relative to the moving control volume—the fluid velocity seen by an observer riding along on the control volume. The absolute velocity is the fluid velocity as seen by a stationary observer in a fixed coordinate system.

*The absolute and relative velocities differ by an amount equal to the control volume velocity.*

The difference between the absolute and relative velocities is the velocity of the control volume, $\mathbf{V}_{cv} = \mathbf{V} - \mathbf{W}$, or

$$\mathbf{V} = \mathbf{W} + \mathbf{V}_{cv} \tag{4.22}$$

Since the velocity is a vector, we must use vector addition as is shown in Fig. 4.22 to obtain the relative velocity if we know the absolute velocity and the velocity of the control volume. Thus, if the water leaves the nozzle in Fig. 4.20 with a velocity of $\mathbf{V}_1 = 100\hat{\mathbf{i}}$ ft/s and the vane has a velocity of $\mathbf{V}_0 = 20\hat{\mathbf{i}}$ ft/s (the same as the control volume), it appears to an observer riding on the vane that the water approaches the vane with a velocity of $\mathbf{W} = \mathbf{V} - \mathbf{V}_{cv} = 80\hat{\mathbf{i}}$ ft/s. In general, the absolute

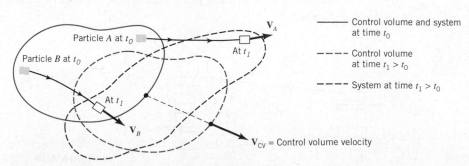

■ **Figure 4.21** Typical moving control volume and system.

$V_A$ = Absolute velocity of $A$

$V_B$

$V_{CV}$

$W_A$ = Velocity of $A$ relative to control volume

$W_B$ = Velocity of $B$ relative to control volume

$V_{CV}$

■ **Figure 4.22**   Relationship between absolute and relative velocities.

velocity, **V**, and the control volume velocity, $\mathbf{V}_{cv}$, will not be in the same direction so that the relative and absolute velocities will have different directions (see Fig. 4.22).

The Reynolds transport theorem for a moving, nondeforming control volume can be derived in the same manner that it was obtained for a fixed control volume. As is indicated in Fig. 4.23, the only difference that needs be considered is the fact that relative to the moving control volume the fluid velocity observed is the relative velocity, not the absolute velocity. An observer fixed to the moving control volume may or may not even know that he or she is moving relative to some fixed coordinate system. If we follow the derivation that led to Eq. 4.19 (the Reynolds transport theorem for a fixed control volume), we note that the corresponding result for a moving control volume can be obtained by simply replacing the absolute velocity, **V**, in that equation by the relative velocity, **W**. Thus, the Reynolds transport theorem for a control volume moving with constant velocity is given by

*The Reynolds transport theorem for a moving control volume involves the relative velocity.*

$$\frac{DB_{sys}}{Dt} = \frac{\partial}{\partial t} \int_{cv} \rho b \, d\mathbf{\Psi} + \int_{cs} \rho b \, \mathbf{W} \cdot \hat{\mathbf{n}} \, dA \qquad \textbf{(4.23)}$$

where the relative velocity is given by Eq. 4.22.

### 4.4.7 Selection of a Control Volume

Any volume in space can be considered as a control volume. It may be of finite size or it may be infinitesimal in size, depending on the type of analysis to be carried out. In most of our cases, the control volume will be a fixed, nondeforming volume. In some situations, we will consider control volumes that move with constant velocity. In either case, it is important that considerable thought go into the selection of the specific control volume to be used.

The selection of an appropriate control volume in fluid mechanics is very similar to the selection of an appropriate free-body diagram in dynamics or statics. In dynamics, we select the body in which we are interested, represent the object in a free-body diagram, and then apply the appropriate governing laws to that body. The ease of solving a given dynamics problem is often very dependent on the specific object that we select for use in our free-body diagram. Similarly, the ease of solving a given fluid mechanics problem is often very dependent on the choice of the control volume used. Only by practice can we develop skill at selecting the "best" control volume. None are "wrong," but some are "much better" than others.

Solution of a typical problem will involve determining parameters such as velocity, pressure, and force at some point in the flow field. It is usually best to ensure that this point is located on the control surface, not "buried" within the control volume. The unknown will then appear in the convective term (the surface integral) of the Reynolds transport theorem. If possible, the control

*V4.14 Control volume*

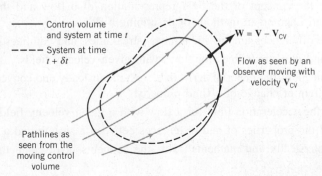

—— Control volume and system at time $t$

- - - - System at time $t + \delta t$

$\mathbf{W} = \mathbf{V} - \mathbf{V}_{CV}$

Flow as seen by an observer moving with velocity $\mathbf{V}_{CV}$

Pathlines as seen from the moving control volume

■ **Figure 4.23**   Control volume and system as seen by an observer moving with the control volume.

■ **Figure 4.24** Various control volumes for flow through a pipe.

surface should be normal to the fluid velocity so that the angle $\theta$ ($\mathbf{V} \cdot \hat{\mathbf{n}} = V \cos \theta$ as shown by the figure in the margin) in the flux terms of Eq. 4.19 will be 0 or 180°. This will usually simplify the solution process.

Figure 4.24 illustrates three possible control volumes associated with flow through a pipe. If the problem is to determine the pressure at point (1), the selection of the control volume (a) is better than that of (b) because point (1) lies on the control surface. Similarly, control volume (a) is better than (c) because the flow is normal to the inlet and exit portions of the control volume. None of these control volumes are wrong—(a) will be easier to use. Proper control volume selection will become much clearer in Chapter 5 where the Reynolds transport theorem is used to transform the governing equations from the system formulation into the control volume formulation, and numerous examples using control volume ideas are discussed.

# 4.5    Chapter Summary and Study Guide

*field representation*
*velocity field*
*Eulerian method*
*Lagrangian method*
*one-, two-, and three-*
    *dimensional flow*
*steady and unsteady*
    *flow*
*streamline*
*streakline*
*pathline*
*acceleration field*
*material derivative*
*local acceleration*
*convective acceleration*
*system*
*control volume*
*Reynolds transport*
    *theorem*

This chapter considered several fundamental concepts of fluid kinematics. That is, various aspects of fluid motion are discussed without regard to the forces needed to produce this motion. The concepts of a field representation of a flow and the Eulerian and Lagrangian approaches to describing a flow are introduced, as are the concepts of velocity and acceleration fields.

The properties of one-, two-, or three-dimensional flows and steady or unsteady flows are introduced along with the concepts of streamlines, streaklines, and pathlines. Streamlines, which are lines tangent to the velocity field, are identical to streaklines and pathlines if the flow is steady. For unsteady flows, they need not be identical.

As a fluid particle moves about, its properties (i.e., velocity, density, temperature) may change. The rate of change of these properties can be obtained by using the material derivative, which involves both unsteady effects (time rate of change at a fixed location) and convective effects (time rate of change due to the motion of the particle from one location to another).

The concepts of a control volume and a system are introduced, and the Reynolds transport theorem is developed. By using these ideas, the analysis of flows can be carried out using a control volume (a volume, usually fixed, through which the fluid flows), whereas the governing principles are stated in terms of a system (a flowing portion of fluid).

The following checklist provides a study guide for this chapter. When your study of the entire chapter and end-of-chapter exercises has been completed, you should be able to

- write out meanings of the terms listed here in the margin and understand each of the related concepts. These terms are particularly important and are set in ***italic, bold, and color*** type in the text.

- understand the concept of the field representation of a flow and the difference between Eulerian and Lagrangian methods of describing a flow.

- explain the differences among streamlines, streaklines, and pathlines.

- calculate and plot streamlines for flows with given velocity fields.

- use the concept of the material derivative, with its unsteady and convective effects, to determine time rate of change of a fluid property.

- determine the acceleration field for a flow with a given velocity field.

- understand the properties of and differences between a system and a control volume.

- interpret, physically and mathematically, the concepts involved in the Reynolds transport theorem.

Some of the important equations in this chapter are:

| | | |
|---|---|---|
| Equation for streamlines | $\dfrac{dy}{dx} = \dfrac{v}{u}$ | **(4.1)** |
| Acceleration | $\mathbf{a} = \dfrac{\partial \mathbf{V}}{\partial t} + u\dfrac{\partial \mathbf{V}}{\partial x} + v\dfrac{\partial \mathbf{V}}{\partial y} + w\dfrac{\partial \mathbf{V}}{\partial z}$ | **(4.3)** |
| Material derivative | $\dfrac{D(\ )}{Dt} = \dfrac{\partial(\ )}{\partial t} + (\mathbf{V} \cdot \nabla)(\ )$ | **(4.6)** |
| Streamwise and normal components of acceleration | $a_s = V\dfrac{\partial V}{\partial s}, \qquad a_n = \dfrac{V^2}{\mathscr{R}}$ | **(4.7)** |
| Reynolds transport theorem (restricted form) | $\dfrac{DB_{\text{sys}}}{Dt} = \dfrac{\partial B_{\text{cv}}}{\partial t} + \rho_2 A_2 V_2 b_2 - \rho_1 A_1 V_1 b_1$ | **(4.15)** |
| Reynolds transport theorem (general form) | $\dfrac{DB_{\text{sys}}}{Dt} = \dfrac{\partial}{\partial t}\displaystyle\int_{\text{cv}} \rho b\, d\Psi + \int_{\text{cs}} \rho b\, \mathbf{V} \cdot \hat{\mathbf{n}}\, dA$ | **(4.19)** |
| Relative and absolute velocities | $\mathbf{V} = \mathbf{W} + \mathbf{V}_{\text{cv}}$ | **(4.22)** |

# References

1. Streeter, V. L., and Wylie, E. B., *Fluid Mechanics*, 8th Ed., McGraw-Hill, New York, 1985.
2. Goldstein, R. J., *Fluid Mechanics Measurements*, Hemisphere, New York, 1983.
3. Homsy, G. M., et al., *Multimedia Fluid Mechanics* CD-ROM, 2nd Ed., Cambridge University Press, New York, 2007.
4. Magarvey, R. H., and MacLatchy, C. S., The Formation and Structure of Vortex Rings, *Canadian Journal of Physics*, Vol. 42, 1964.

| | |
|---|---|
| **WILEY** | Problem available in *WileyPLUS* at instructor's discretion. |
| **WILEY GO** | Tutoring problem available in *WileyPLUS* at instructor's discretion. |
| ▶ | Problem is related to a chapter video available in *WileyPLUS*. |
| ✳ | Problem to be solved with aid of programmable calculator or computer. |
| † | Open-ended problem that requires critical thinking. These problems require various assumptions to provide the necessary input data. There are not unique answers to these problems. |

# Review Problems

Go to Appendix G (*WileyPLUS* or the book's web site, www.wiley.com/college/munson) for a set of review problems with answers. Detailed solutions can be found in the *Student Solution Manual and Study Guide for Fundamentals of Fluid Mechanics,* by Munson et al. (© 2013 John Wiley and Sons, Inc.).

# Conceptual Questions

**4.1C** A velocity field is given by:

$$\mathbf{V} = 6x\hat{j} \text{ m/s}.$$

Velocity fields are plotted below, where the arrows indicate the magnitude and direction of the velocity vectors. The picture that best describes the field is:

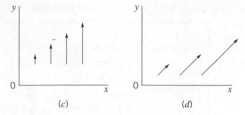

(a)          (b)          (c)          (d)

**4.2C** Consider the path of a football thrown by Brett Favre. Suppose that you are interested in determining how far the ball travels. Then you are interested in:

a) neither Lagrangian nor Eulerian views.

b) the Eulerian view of the field.

c) the Lagrangian or particle view of the football.

**4.3C** The following is true of the system (versus control volume) form of the basic equations (mass, momentum, angular momentum, and energy):

a) The equations apply to a fixed quantity of mass.

b) The equations are formulated in terms of extensive properties.

c) The equations apply to a fixed volume.

**Additional conceptual questions are available in *WileyPLUS* at the instructor's discretion.**

# Problems

**Note: Unless specific values of required fluid properties are given in the problem statement, use the values found in the tables on the inside of the front cover. Answers to the even-numbered problems are listed at the end of the book. The Lab Problems as well as the videos that accompany problems can be accessed in *WileyPLUS* or the book's web site, www.wiley. com/college/munson.**

### Section 4.1 The Velocity Field

**4.1** Obtain a photograph/image of a situation in which a fluid is flowing. Print this photo and draw in some lines to represent how you think some streamlines may look. Write a brief paragraph to describe the acceleration of a fluid particle as it flows along one of these streamlines.

**4.2** The velocity field of a flow is given by $\mathbf{V} = (3y + 2)\hat{\mathbf{i}} + (x - 8)\hat{\mathbf{j}} + 5z\hat{\mathbf{k}}$ ft/s , where $x$, $y$, and $z$ are in feet. Determine the fluid speed at the origin ($x = y = z = 0$) and on the $y$ axis ($x = z = 0$).

**4.3** The velocity field of a flow is given by $\mathbf{V} = 2x^2t\hat{\mathbf{i}} + [4y(t - 1) + 2x^2t]\hat{\mathbf{j}}$ m/s, where $x$ and $y$ are in meters and $t$ is in seconds. For fluid particles on the $x$ axis, determine the speed and direction of flow.

**4.4** A two-dimensional velocity field is given by $u = 1 + y$ and $v = 1$. Determine the equation of the streamline that passes through the origin. On a graph, plot this streamline.

**4.5** The velocity field of a flow is given by $\mathbf{V} = (5z - 3)\hat{\mathbf{i}} + (x + 4)\hat{\mathbf{j}} + 4y\hat{\mathbf{k}}$ ft/s, where $x$, $y$, and $z$ are in feet. Determine the fluid speed at the origin ($x = y = z = 0$) and on the $x$ axis ($y = z = 0$).

**4.6** A flow can be visualized by plotting the velocity field as velocity vectors at representative locations in the flow as shown in **Video V4.2** and Fig. E4.1. Consider the velocity field given in polar coordinates by $v_r = -10/r$, and $v_\theta = 10/r$. This flow approximates a fluid swirling into a sink as shown in Fig. P4.6. Plot the velocity field at locations given by $r = 1, 2,$ and 3 with $\theta = 0$, 30, 60, and 90°.

■ **Figure P4.6**

**4.7** The velocity field of a flow is given by $\mathbf{V} = 20y/(x^2 + y^2)^{1/2}\hat{\mathbf{i}} - 20x/(x^2 + y^2)^{1/2}\hat{\mathbf{j}}$ ft/s, where $x$ and $y$ are in feet. Determine the fluid speed at points along the $x$ axis; along the $y$ axis. What is the angle between the velocity vector and the $x$ axis at points $(x, y) = (5, 0), (5, 5),$ and $(0, 5)$?

**4.8** The components of a velocity field are given by $u = x + y$, $v = xy^3 + 16$, and $w = 0$. Determine the location of any stagnation points ($\mathbf{V} = 0$) in the flow field.

**4.9** The $x$ and $y$ components of velocity for a two-dimensional flow are $u = 6y$ ft/s and $v = 3$ ft/s, where $y$ is in feet. Determine the equation for the streamlines and sketch representative streamlines in the upper half plane.

**4.10** The velocity field of a flow is given by $u = -V_0y/(x^2 + y^2)^{1/2}$ and $v = V_0x/(x^2 + y^2)^{1/2}$, where $V_0$ is a constant. Where in the flow field is the speed equal to $V_0$? Determine the equation of the streamlines and discuss the various characteristics of this flow.

**4.11** A velocity field is given by $\mathbf{V} = x\hat{\mathbf{i}} + x(x - 1)(y + 1)\hat{\mathbf{j}}$, where $u$ and $v$ are in ft/s and $x$ and $y$ are in feet. Plot the streamline that passes through $x = 0$ and $y = 0$. Compare this streamline with the streakline through the origin.

**4.12** From time $t = 0$ to $t = 5$ hr radioactive steam is released from a nuclear power plant accident located at $x = -1$ mile and $y = 3$ miles. The following wind conditions are expected: $\mathbf{V} = 10\hat{\mathbf{i}} - 5\hat{\mathbf{j}}$ mph for $0 < t < 3$ hr, $\mathbf{V} = 15\hat{\mathbf{i}} + 8\hat{\mathbf{j}}$ mph for $3 < t < 10$ hr, and $\mathbf{V} = 5\hat{\mathbf{i}}$ mph for $t > 10$ hr. Draw to scale the expected streakline of the steam for $t = 3$, 10, and 15 hr.

**4.13** The $x$ and $y$ components of a velocity field are given by $u = x^2 y$ and $v = -xy^2$. Determine the equation for the streamlines of this flow and compare it with those in Example 4.2. Is the flow in this problem the same as that in Example 4.2? Explain.

**4.14** In addition to the customary horizontal velocity components of the air in the atmosphere (the "wind"), there often are vertical air currents (thermals) caused by buoyant effects due to uneven heating of the air as indicated in Fig. P4.14. Assume that the velocity field in a certain region is approximated by $u = u_0$, $v = v_0 (1 - y/h)$ for $0 < y < h$, and $u = u_0$, $v = 0$ for $y > h$. Plot the shape of the streamline that passes through the origin for values of $u_0/v_0 = 0.5$, 1, and 2.

■ **Figure P4.14**

**4.15** ▶ As shown in **Video V4.6** and Fig. P4.15, a flying airplane produces swirling flow near the end of its wings. In certain circumstances this flow can be approximated by the velocity field $u = -Ky/(x^2 + y^2)$ and $v = Kx/(x^2 + y^2)$, where $K$ is a constant depending on various parameters associated with the airplane (i.e., its weight, speed) and $x$ and $y$ are measured from the center of the swirl. **(a)** Show that for this flow the velocity is inversely proportional to the distance from the origin. That is, $V = K/(x^2 + y^2)^{1/2}$. **(b)** Show that the streamlines are circles.

■ **Figure P4.15**

**†4.16** For any steady flow the streamlines and streaklines are the same. For most unsteady flows this is not true. However, there are unsteady flows for which the streamlines and streaklines are the same. Describe a flow field for which this is true.

**4.17** A 10-ft-diameter dust devil that rotates one revolution per second travels across the Martian surface (in the $x$-direction) with a speed of 5 ft/s. Plot the pathline etched on the surface by a fluid particle 10 ft from the center of the dust devil for time $0 \le t \le 3$ s. The particle position is given by the sum of that for a stationary swirl $[x = 10 \cos(2\pi t), y = 10 \sin(2\pi t)]$ and that for a uniform velocity ($x = 5t, y = $ constant), where $x$ and $y$ are in feet and $t$ is in seconds.

**4.18** (See Fluids in the News article titled **"Follow those particles,"** Section 4.1.) Two photographs of four particles in a flow past a sphere are superposed as shown in Fig. P4.18. The time interval between the photos is $\Delta t = 0.002$ s. The locations of the particles, as determined from the photos, are shown in the table. **(a)** Determine the fluid velocity for these particles. **(b)** Plot a graph to compare the results of part **(a)** with the theoretical velocity which is given by $V = V_0(1 + a^3/x^3)$, where $a$ is the sphere radius and $V_0$ is the fluid speed far from the sphere.

| Particle | $x$ at $t = 0$ s (ft) | $x$ at $t = 0.002$ s (ft) |
|---|---|---|
| 1 | −0.500 | −0.480 |
| 2 | −0.250 | −0.232 |
| 3 | −0.140 | −0.128 |
| 4 | −0.120 | −0.112 |

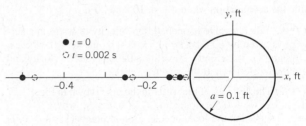

■ **Figure P4.18**

**†4.19** Pathlines and streaklines provide ways to visualize flows. Another technique would be to instantly inject a line of dye across streamlines and observe how this line moves as time increases. For example, consider the initially straight dye line injected in front of the circular cylinder shown in Fig. P4.19. Discuss how this dye line would appear at later times. How would you calculate the location of this line as a function of time?

■ **Figure P4.19**

### Section 4.2 The Acceleration Field

**4.20** (WILEY) A velocity field is given by $u = cx^2$ and $v = cy^2$, where $c$ is a constant. Determine the $x$ and $y$ components of the acceleration. At what point (points) in the flow field is the acceleration zero?

**4.21** (WILEY) Determine the acceleration field for a three-dimensional flow with velocity components $u = -x$, $v = 4x^2 y^2$, and $w = x - y$.

**4.22** A three-dimensional velocity field is given by $u = 2x$, $v = -y$, and $w = z$. Determine the acceleration vector.

**4.23** Water flows through a constant diameter pipe with a uniform velocity given by $\mathbf{V} = (8/t + 5)\hat{\mathbf{j}}$ m/s, where $t$ is in seconds. Determine the acceleration at time $t = 1$, 2, and 10 s.

**4.24** (WILEY) The velocity of air in the diverging pipe shown in Fig. P4.24 is given by $V_1 = 4t$ ft/s and $V_2 = 2t$ ft/s, where $t$ is in seconds. **(a)** Determine the local acceleration at points (1) and (2). **(b)** Is the average convective acceleration between these two points negative, zero, or positive? Explain.

$V_1 = 4t$ ft/s $\qquad$ $V_2 = 2t$ ft/s

(1)

(2)

■ **Figure P4.24**

**4.25** (WILEY) Water flows in a pipe so that its velocity triples every 20 s. At $t = 0$ it has $u = 5$ ft/s. That is, $\mathbf{V} = u(t)\hat{\mathbf{i}} = 5\,(3^{t/20})\hat{\mathbf{i}}$ ft/s. Determine the acceleration when $t = 0$, 10, and 20 s.

**4.26** (WILEY) When a valve is opened, the velocity of water in a certain pipe is given by $u = 10(1 - e^{-t})$, $v = 0$, and $w = 0$, where $u$ is in ft/s and $t$ is in seconds. Determine the maximum velocity and maximum acceleration of the water.

**4.27** (WILEY) The velocity of the water in the pipe shown in Fig. P4.27 is given by $V_1 = 0.50t$ m/s and $V_2 = 1.0t$ m/s, where $t$ is in seconds. Determine the local acceleration at points (1) and (2). Is the average convective acceleration between these two points negative, zero, or positive? Explain.

$V_1 = $
$0.50t$ m/s $\qquad$ $V_2 = $
$1.0t$ m/s

(2)

(1)

■ **Figure P4.27**

**4.28** (WILEY) A shock wave is a very thin layer (thickness $= \ell$) in a high-speed (supersonic) gas flow across which the flow properties (velocity, density, pressure, etc.) change from state (1) to state (2) as shown in Fig. P4.28. If $V_1 = 1800$ fps, $V_2 = 700$ fps, and $\ell = 10^{-4}$ in., estimate the average deceleration of the gas as it flows across the shock wave. How many $g$'s deceleration does this represent?

$V_1$ $\qquad$ $V_2$

$\ell$

Shock wave

■ **Figure P4.28**

†**4.29** Estimate the average acceleration of water as it travels through the nozzle on your garden hose. List all assumptions and show all calculations.

†**4.30** A stream of water from the faucet strikes the bottom of the sink. Estimate the maximum acceleration experienced by the water particles. List all assumptions and show calculations.

**4.31** (WILEY) As a valve is opened, water flows through the diffuser shown in Fig. P4.31 at an increasing flowrate so that the velocity along the centerline is given by $\mathbf{V} = u\hat{\mathbf{i}} = V_0(1 - e^{-ct})(1 - x/\ell)\,\hat{\mathbf{i}}$, where $u_0$, $c$, and $\ell$ are constants. Determine the acceleration as a function of $x$ and $t$. If $V_0 = 10$ ft/s and $\ell = 5$ ft, what value of $c$ (other than $c = 0$) is needed to make the acceleration zero for any $x$ at $t = 1$ s? Explain how the acceleration can be zero if the flowrate is increasing with time.

$y$

$\ell/2$

$u$ $\qquad$ $u = \frac{1}{2}V_0(1 - e^{-ct})$

$x$

$u = V_0(1 - e^{-ct})$

■ **Figure P4.31**

**4.32** The fluid velocity along the $x$ axis shown in Fig. P4.32 changes from 6 m/s at point $A$ to 18 m/s at point $B$. It is also known that the velocity is a linear function of distance along the streamline. Determine the acceleration at points $A$, $B$, and $C$. Assume steady flow.

$V_A = 6$ m/s $\qquad$ $V_B = 18$ m/s

$A$ $\qquad$ $C$ $\qquad$ $B$ $\quad$ $x$

0.05 m

0.1 m

■ **Figure P4.32**

**4.33** (WILEY) A fluid flows along the $x$ axis with a velocity given by $\mathbf{V} = (x/t)\,\hat{\mathbf{i}}$, where $x$ is in feet and $t$ in seconds. **(a)** Plot the speed for $0 \le x \le 10$ ft and $t = 3$ s. **(b)** Plot the speed for $x = 7$ ft and $2 \le t \le 4$ s. **(c)** Determine the local and convective acceleration. **(d)** Show that the acceleration of any fluid particle in the flow is zero. **(e)** Explain physically how the velocity of a particle in this unsteady flow remains constant throughout its motion.

**4.34** ▶ (WILEY) A hydraulic jump is a rather sudden change in depth of a liquid layer as it flows in an open channel as shown in Fig. P4.34 and **Video V10.12**. In a relatively short distance (thickness = $\ell$) the liquid depth changes from $z_1$ to $z_2$, with a corresponding change in velocity from $V_1$ to $V_2$. If $V_1$ = 1.20 ft/s, $V_2$ = 0.30 ft/s, and $\ell$ = 0.02 ft, estimate the average deceleration of the liquid as it flows across the hydraulic jump. How many $g$'s deceleration does this represent?

■ **Figure P4.34**

(a)

(b)

■ **Figure P4.38**

**4.35** ▶ (WILEY) A fluid particle flowing along a stagnation streamline, as shown in **Video V4.9** and Fig. P4.35, slows down as it approaches the stagnation point. Measurements of the dye flow in the video indicate that the location of a particle starting on the stagnation streamline a distance $s$ = 0.6 ft upstream of the stagnation point at $t$ = 0 is given approximately by $s = 0.6e^{-0.5t}$, where $t$ is in seconds and $s$ is in feet. (a) Determine the speed of a fluid particle as a function of time, $V_{particle}(t)$, as it flows along the streamline. (b) Determine the speed of the fluid as a function of position along the streamline, $V = V(s)$. (c) Determine the fluid acceleration along the streamline as a function of position, $a_s = a_s(s)$.

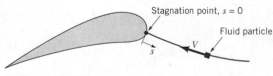

■ **Figure P4.35**

**4.36** (WILEY) A nozzle is designed to accelerate the fluid from $V_1$ to $V_2$ in a linear fashion. That is, $V = ax + b$, where $a$ and $b$ are constants. If the flow is constant with $V_1$ = 10 m/s at $x_1$ = 0 and $V_2$ = 25 m/s at $x_2$ = 1 m, determine the local acceleration, the convective acceleration, and the acceleration of the fluid at points (1) and (2).

**4.37** (WILEY) Repeat Problem 4.36 with the assumption that the flow is not steady, but at the time when $V_1$ = 10 m/s and $V_2$ = 25 m/s, it is known that $\partial V_1/\partial t$ = 20 m/s² and $\partial V_2/\partial t$ = 60 m/s².

**4.38** An incompressible fluid flows through the converging duct shown in Fig. P4.38a with velocity $V_0$ at the entrance. Measurements indicate that the actual velocity of the fluid near the wall of the duct along streamline A–F is as shown in Fig. P4.38b. Sketch the component of acceleration along this streamline, $a$, as a function of $s$. Discuss the important characteristics of your result.

**\*4.39** Air flows steadily through a variable area pipe with a velocity of $\mathbf{V} = u(x)\hat{\mathbf{i}}$ ft/s, where the approximate measured values of $u(x)$ are given in the table. Plot the acceleration as a function of $x$ for $0 \leq x \leq 12$ in. Plot the acceleration if the flowrate is increased by a factor of $N$ (i.e., the values of $u$ are increased by a factor of $N$) for $N$ = 2, 4, 10.

| $x$ (in.) | $u$ (ft/s) | $x$ (in.) | $u$ (ft/s) |
|---|---|---|---|
| 0 | 10.0 | 7 | 20.1 |
| 1 | 10.2 | 8 | 17.4 |
| 2 | 13.0 | 9 | 13.5 |
| 3 | 20.1 | 10 | 11.9 |
| 4 | 28.3 | 11 | 10.3 |
| 5 | 28.4 | 12 | 10.0 |
| 6 | 25.8 | 13 | 10.0 |

**\*4.40** (WILEY) As is indicated in Fig. P4.40, the speed of exhaust in a car's exhaust pipe varies in time and distance because of the periodic nature of the engine's operation and the damping effect with distance from the engine. Assume that the speed is given by $V = V_0[1 + ae^{-bx} \sin(\omega t)]$, where $V_0$ = 8 fps, $a$ = 0.05, $b$ = 0.2 ft$^{-1}$, and $\omega$ = 50 rad/s. Calculate and plot the fluid acceleration at $x$ = 0, 1, 2, 3, 4, and 5 ft for $0 \leq t \leq \pi/25$ s.

■ **Figure P4.40**

**4.41** Water flows down the face of the dam shown in Fig. P4.41. The face of the dam consists of two circular arcs with radii of 10 and 20 ft as shown. If the speed of the water along streamline A–B is approximately $V = (2gh)^{1/2}$, where the distance $h$ is as indicated, plot the normal acceleration as a function of distance along the streamline, $a_n = a_n(s)$.

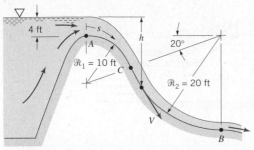

■ **Figure P4.41**

**4.42** WILEY Water flows over the crest of a dam with speed $V$ as shown in Fig. P4.42. Determine the speed if the magnitude of the normal acceleration at point (1) is to equal the acceleration of gravity, $g$.

■ **Figure P4.42**

**4.43** Water flows under the sluice gate shown in Fig. P4.43. If $V_1 = 3$ m/s, what is the normal acceleration at point (1)?

■ **Figure P4.43**

**4.44** For the flow given in Problem 4.41, plot the streamwise acceleration, $a_s$, as a function of distance, $s$, along the surface of the dam from $A$ to $C$.

**4.45** ▶ WILEY Assume that the streamlines for the wingtip vortices from an airplane (see Fig. P4.15 and **Video V4.6**) can be approximated by circles of radius $r$ and that the speed is $V = K/r$, where $K$ is a constant. Determine the streamline acceleration, $a_s$, and the normal acceleration, $a_n$, for this flow.

**4.46** WILEY A fluid flows past a sphere with an upstream velocity of $V_0 = 40$ m/s as shown in Fig. P4.46. From a more advanced theory it is found that the speed of the fluid along the front part of the sphere is $V = \frac{3}{2}V_0 \sin \theta$. Determine the streamwise and normal components of acceleration at point $A$ if the radius of the sphere is $a = 0.20$ m.

■ **Figure P4.46**

**4.47** WILEY The velocity components for steady flow through the nozzle shown in Fig. P4.47 are $u = -V_0 x/\ell$ and $v = V_0[1 + (y/\ell)]$, where $V_0$ and $\ell$ are constants. Determine the ratio of the magnitude of the acceleration at point (1) to that at point (2).

■ **Figure P4.47**

**4.48** Water flows through the curved hose shown in Fig. P4.48 with an increasing speed of $V = 10t$ ft/s, where $t$ is in seconds. For $t = 2$ s determine **(a)** the component of acceleration along the streamline, **(b)** the component of acceleration normal to the streamline, and **(c)** the net acceleration (magnitude and direction).

■ **Figure P4.48**

**\*4.49** WILEY A fluid flows past a circular cylinder of radius $a$ with an upstream speed of $V_0$ as shown in Fig. P4.49. A more advanced theory indicates that if viscous effects are negligible, the velocity of the fluid along the surface of the cylinder is given by $V = 2V_0 \sin \theta$. Determine the streamline and normal components of acceleration on the surface of the cylinder as a function of $V_0$, $a$, and $\theta$ and plot graphs of $a_s$ and $a_n$ for $0 \le \theta \le 90°$ with $V_0 = 10$ m/s and $a = 0.01, 0.10, 1.0,$ and $10.0$ m.

■ **Figure P4.49**

**4.50** Determine the $x$ and $y$ components of acceleration for the flow $u = c(x^2 - y^2)$ and $v = -2cxy$, where $c$ is a constant. If $c > 0$, is the particle at point $x = x_0 > 0$ and $y = 0$ accelerating or decelerating? Explain. Repeat if $x_0 < 0$.

**4.51** ⟨WILEY⟩ When the floodgates in a channel are opened, water flows along the channel downstream of the gates with an increasing speed given by $V = 4(1 + 0.1t)$ ft/s, for $0 \le t \le 20$ s, where $t$ is in seconds. For $t > 20$ s the speed is a constant $V = 12$ ft/s. Consider a location in the curved channel where the radius of curvature of the streamlines is 50 ft. For $t = 10$ s, determine (a) the component of acceleration along the streamline, (b) the component of acceleration normal to the streamline, and (c) the net acceleration (magnitude and direction). Repeat for $t = 30$ s.

**4.52** ⟨WILEY⟩ Water flows steadily through the funnel shown in Fig. P4.52. Throughout most of the funnel the flow is approximately radial (along rays from $O$) with a velocity of $V = c/r^2$, where $r$ is the radial coordinate and $c$ is a constant. If the velocity is 0.4 m/s when $r = 0.1$ m, determine the acceleration at points $A$ and $B$.

■ **Figure P4.52**

**4.53** ⟨WILEY⟩ Water flows though the slit at the bottom of a two-dimensional water trough as shown in Fig. P4.53. Throughout most of the trough the flow is approximately radial (along rays from $O$) with a velocity of $V = c/r$, where $r$ is the radial coordinate and $c$ is a constant. If the velocity is 0.04 m/s when $r = 0.1$ m, determine the acceleration at points $A$ and $B$.

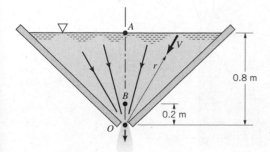

■ **Figure P4.53**

**4.54** ⟨WILEY⟩ Air flows from a pipe into the region between two parallel circular disks as shown in Fig. P4.54. The fluid velocity in the gap between the disks is closely approximated by $V = V_0R/r$, where $R$ is the radius of the disk, $r$ is the radial coordinate, and $V_0$ is the fluid velocity at the edge of the disk. Determine the acceleration for $r = 1$, 2, or 3 ft if $V_0 = 5$ ft/s and $R = 3$ ft.

■ **Figure P4.54**

**4.55** ⟨WILEY⟩ Air flows into a pipe from the region between a circular disk and a cone as shown in Fig. P4.55. The fluid velocity in the gap between the disk and the cone is closely approximated by $V = V_0R^2/r^2$, where $R$ is the radius of the disk, $r$ is the radial coordinate, and $V_0$ is the fluid velocity at the edge of the disk. Determine the acceleration for $r = 0.5$ and 2 ft if $V_0 = 5$ ft/s and $R = 2$ ft.

■ **Figure P4.55**

## Section 4.2.1 The Material Derivative

**4.56** ⟨WILEY⟩ Air flows steadily through a long pipe with a speed of $u = 50 + 0.5x$, where $x$ is the distance along the pipe in feet, and $u$ is in ft/s. Due to heat transfer into the pipe, the air temperature, $T$, within the pipe is $T = 300 + 10x$ °F. Determine the rate of change of the temperature of air particles as they flow past the section at $x = 5$ ft.

**4.57** A gas flows along the $x$ axis with a speed of $V = 5x$ m/s and a pressure of $p = 10x^2$ N/m², where $x$ is in meters. (a) Determine the time rate of change of pressure at the fixed location $x = 1$. (b) Determine the time rate of change of pressure for a fluid particle flowing past $x = 1$. (c) Explain without using any equations why the answers to parts (a) and (b) are different.

**4.58** ⟨WILEY⟩ Assume the temperature of the exhaust in an exhaust pipe can be approximated by $T = T_0(1 + ae^{-bx})[1 + c\cos(\omega t)]$, where $T_0 = 100$ °C, $a = 3$, $b = 0.03$ m$^{-1}$, $c = 0.05$, and $\omega = 100$ rad/s. If the exhaust speed is a constant 3 m/s, determine the time rate of change of temperature of the fluid particles at $x = 0$ and $x = 4$ m when $t = 0$.

**4.59** ⟨WILEY⟩ GO A bicyclist leaves from her home at 9 A.M. and rides to a beach 40 mi away. Because of a breeze off the ocean, the temperature at the beach remains 60 °F throughout the day. At the cyclist's home the temperature increases linearly with time, going from 60 °F at 9 A.M. to 80 °F by 1 P.M. The temperature is assumed to vary linearly as a function of position between the cyclist's home and the beach. Determine the rate of change of temperature observed by the cyclist for the following conditions: (a) as she pedals 10 mph through a town 10 mi from her home at 10 A.M.; (b) as she eats lunch at a rest stop 30 mi from her home at noon; (c) as she arrives enthusiastically at the beach at 1 P.M., pedaling 20 mph.

**4.60** ⟨WILEY⟩ The temperature distribution in a fluid is given by $T = 10x + 5y$, where $x$ and $y$ are the horizontal and vertical coordinates in meters and $T$ is in degrees centigrade. Determine the time rate of change of temperature of a fluid particle traveling (a) horizontally with $u = 20$ m/s, $v = 0$ or (b) vertically with $u = 0$, $v = 20$ m/s.

## Section 4.4 The Reynolds Transport Theorem

**4.61** Obtain a photograph/image of a situation in which a fluid is flowing. Print this photo and draw a control volume through which the fluid flows. Write a brief paragraph that describes how the fluid flows into and out of this control volume.

**4.62** (WILEY) Water flows through a duct of square cross section as shown in Fig. P4.62 with a constant, uniform velocity of $V = 20$ m/s. Consider fluid particles that lie along line $A$–$B$ at time $t = 0$. Determine the position of these particles, denoted by line $A'$–$B'$, when $t = 0.20$ s. Use the volume of fluid in the region between lines $A$–$B$ and $A'$–$B'$ to determine the flowrate in the duct. Repeat the problem for fluid particles originally along line $C$–$D$; along line $E$–$F$. Compare your three answers.

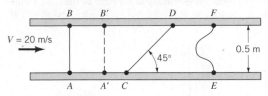

■ **Figure P4.62**

**4.63** (WILEY) Repeat Problem 4.62 if the velocity profile is linear from 0 to 20 m/s across the duct as shown in Fig. P4.63.

■ **Figure P4.63**

**4.64** ▶ (WILEY) In the region just downstream of a sluice gate, the water may develop a reverse flow region as is indicated in Fig. P4.64 and **Video V10.9**. The velocity profile is assumed to consist of two uniform regions, one with velocity $V_a = 10$ fps and the other with $V_b = 3$ fps. Determine the net flowrate of water across the portion of the control surface at section (2) if the channel is 20 ft wide.

■ **Figure P4.64**

**4.65** (WILEY) At time $t = 0$ the valve on an initially empty (perfect vacuum, $\rho = 0$) tank is opened and air rushes in. If the tank has a volume of $V_0$ and the density of air within the tank increases as $\rho = \rho_\infty(1 - e^{-bt})$, where $b$ is a constant, determine the time rate of change of mass within the tank.

**†4.66** From calculus, one obtains the following formula (Leibnitz rule) for the time derivative of an integral that contains time in both the integrand and the limits of the integration:

$$\frac{d}{dt}\int_{x_1(t)}^{x_2(t)} f(x, t)dx = \int_{x_1}^{x_2} \frac{\partial f}{\partial t} dx + f(x_2, t)\frac{dx_2}{dt} - f(x_1, t)\frac{dx_1}{dt}$$

Discuss how this formula is related to the time derivative of the total amount of a property in a system and to the Reynolds transport theorem.

**4.67** Air enters an elbow with a uniform speed of 10 m/s as shown in Fig. P4.67. At the exit of the elbow, the velocity profile is not uniform. In fact, there is a region of separation or reverse flow. The fixed control volume $ABCD$ coincides with the system at time $t = 0$. Make a sketch to indicate (a) the system at time $t = 0.01$ s and (b) the fluid that has entered and exited the control volume in that time period.

■ **Figure P4.67**

**4.68** ▶ (WILEY) A layer of oil flows down a vertical plate as shown in Fig. P4.68 with a velocity of $\mathbf{V} = (V_0/h^2)(2hx - x^2)\hat{\mathbf{j}}$ where $V_0$ and $h$ are constants. (a) Show that the fluid sticks to the plate and that the shear stress at the edge of the layer ($x = h$) is zero. (b) Determine the flowrate across surface $AB$. Assume the width of the plate is $b$. (Note: The velocity profile for laminar flow in a pipe has a similar shape. See **Video V6.13**.)

■ **Figure P4.68**

**4.69** (WILEY) Water flows in the branching pipe shown in Fig. P4.69 with uniform velocity at each inlet and outlet. The fixed control volume indicated coincides with the system at time $t = 20$ s. Make a sketch to indicate (a) the boundary of the system at time $t = 20.1$ s, (b) the fluid that left the control volume during that 0.1-s interval, and (c) the fluid that entered the control volume during that time interval.

■ **Figure P4.69**

■ **Figure P4.72**

**4.70** (WILEY) Two plates are pulled in opposite directions with speeds of 1.0 ft/s as shown in Fig. P4.70. The oil between the plates moves with a velocity given by $\mathbf{V} = 10\,y\hat{\mathbf{i}}$ ft/s, where $y$ is in feet. The fixed control volume $ABCD$ coincides with the system at time $t = 0$. Make a sketch to indicate (a) the system at time $t = 0.2$ s and (b) the fluid that has entered and exited the control volume in that time period.

**4.73** (WILEY) The wind blows across a field with an approximate velocity profile as shown in Fig. P4.73. Use Eq. 4.16 with the parameter $b$ equal to the velocity to determine the momentum flowrate across the vertical surface $A$–$B$, which is of unit depth into the paper.

■ **Figure P4.70**

■ **Figure P4.73**

**4.71** Water enters a 5-ft-wide, 1-ft-deep channel as shown in Fig. P4.71. Across the inlet the water velocity is 6 ft/s in the center portion of the channel and 1 ft/s in the remainder of it. Farther downstream the water flows at a uniform 2 ft/s velocity across the entire channel. The fixed control volume $ABCD$ coincides with the system at time $t = 0$. Make a sketch to indicate (a) the system at time $t = 0.5$ s and (b) the fluid that has entered and exited the control volume in that time period.

**4.74** Water flows from a nozzle with a speed of $V = 10$ m/s and is collected in a container that moves toward the nozzle with a speed of $V_{cv} = 2$ m/s as shown in Fig. P4.74. The moving control surface consists of the inner surface of the container. The system consists of the water in the container at time $t = 0$ and the water between the nozzle and the tank in the constant diameter stream at $t = 0$. At time $t = 0.1$ s what volume of the system remains outside of the control volume? How much water has entered the control volume during this time period? Repeat the problem for $t = 0.3$ s.

■ **Figure P4.71**

■ **Figure P4.74**

**4.72** (WILEY) Water flows through the 2-m-wide rectangular channel shown in Fig. P4.72 with a uniform velocity of 3 m/s. (a) Directly integrate Eq. 4.16 with $b = 1$ to determine the mass flowrate (kg/s) across section $CD$ of the control volume. (b) Repeat part (a) with $b = 1/\rho$, where $\rho$ is the density. Explain the physical interpretation of the answer to part (b).

■ **Lifelong Learning Problems**

**4.1LL** Even for the simplest flows it is often not easy to visually represent various flow field quantities such as velocity, pressure, or temperature. For more complex flows, such as those involving three-dimensional or unsteady effects, it is extremely difficult to "show the data." However, with the use of computers and appropriate software, novel methods are being devised to more effectively illustrate the structure of a given flow. Obtain information about methods used to present complex flow data. Summarize your findings in a brief report.

**4.2LL** For centuries people have obtained qualitative and quantitative information about various flow fields by observing the motion of objects or particles in a flow. For example, the speed of the current in a river can be approximated by timing how long it takes a stick to travel a certain distance. The swirling motion of a tornado can be observed by following debris moving within the tornado funnel. Recently, various high-tech methods using lasers and minute particles seeded within the flow have been developed to measure velocity fields. Such techniques include the laser doppler anemometer (LDA), the particle image velocimeter (PIV), and others. Obtain information about new laser-based techniques for measuring velocity fields. Summarize your findings in a brief report.

■ **FE Exam Problems**

Sample FE (Fundamentals of Engineering) exam questions for fluid mechanics are provided in *WileyPLUS* or on the book's web site, www.wiley.com/college/munson.

# 5 Finite Control Volume Analysis

## Learning Objectives

After completing this chapter, you should be able to:

■ select an appropriate finite control volume to solve a fluid mechanics problem.

■ apply conservation of mass and energy and Newton's second law of motion to the contents of a finite control volume to get important answers.

■ know how velocity changes and energy transfers in fluid flows are related to forces and torques.

■ understand why designing for minimum loss of energy in fluid flows is so important.

*Many fluid mechanics problems can be solved by using control volume analysis.*

To solve many practical problems in fluid mechanics, questions about the behavior of the contents of a finite region in space (a finite control volume) are answered. For example, we may be asked to estimate the maximum anchoring force required to hold a turbojet engine stationary during a test. Or we may be called on to design a propeller to move a boat both forward and backward. Or we may need to determine how much power it would take to move natural gas from one location to another many miles away.

The bases of finite control volume analysis are some fundamental laws of physics, namely, conservation of mass, Newton's second law of motion, and the first and second laws of thermodynamics. While some simplifying approximations are made for practicality, the engineering answers possible with the estimates of this powerful analysis method have proven valuable in numerous instances.

Conservation of mass is the key to tracking flowing fluid. How much enters and leaves a control volume can be ascertained.

Newton's second law of motion leads to the conclusion that forces can result from or cause changes in a flowing fluid's velocity magnitude and/or direction. Moment of force (torque) can result from or cause changes in a flowing fluid's moment of velocity. These forces and torques can be associated with work and power transfer.

The first law of thermodynamics is a statement of conservation of energy. The second law of thermodynamics identifies the loss of energy associated with every actual process. The mechanical energy equation based on these two laws can be used to analyze a large variety of steady, incompressible flows in terms of changes in pressure, elevation, speed, and of shaft work and loss.

Good judgment is required in defining the finite region in space, the control volume, used in solving a problem. What exactly to leave out of and what to leave in the control volume are important considerations. The formulas resulting from applying the fundamental laws to the contents of the control volume are easy to interpret physically and are not difficult to derive and use.

Because a finite region of space, a control volume, contains many fluid particles and even more molecules that make up each particle, the fluid properties and characteristics are often average values. Chapter 6 presents an analysis of fluid flow based on what is happening to the contents of an infinitesimally small region of space or control volume through which numerous molecules simultaneously flow (what we might call a point in space).

# 5.1 Conservation of Mass—The Continuity Equation

### 5.1.1 Derivation of the Continuity Equation

A system is defined as a collection of unchanging contents, so the ***conservation of mass*** principle for a system is simply stated as

Time rate of change of the system mass = 0

or

*The amount of mass in a system is constant.*

$$\frac{DM_{sys}}{Dt} = 0 \tag{5.1}$$

where the system mass, $M_{sys}$, is more generally expressed as

$$M_{sys} = \int_{sys} \rho \, d\mathcal{V} \tag{5.2}$$

and the integration is over the volume of the system. In words, Eq. 5.2 states that the system mass is equal to the sum of all the density-volume element products for the contents of the system.

For a system and a fixed, nondeforming control volume that are coincident at an instant of time, as illustrated in Fig. 5.1, the Reynolds transport theorem (Eq. 4.19) with $B = $ mass and $b = 1$ allows us to state that

$$\frac{D}{Dt} \int_{sys} \rho \, d\mathcal{V} = \frac{\partial}{\partial t} \int_{cv} \rho \, d\mathcal{V} + \int_{cs} \rho \mathbf{V} \cdot \hat{\mathbf{n}} \, dA \tag{5.3}$$

■ **Figure 5.1** System and control volume at three different instances of time. (*a*) System and control volume at time $t - \delta t$. (*b*) System and control volume at time $t$, coincident condition. (*c*) System and control volume at time $t + \delta t$.

or

$$
\begin{array}{ccc}
\text{Time rate of change} & \text{time rate of change} & \text{net rate of flow} \\
\text{of the mass of the} & = \text{of the mass of the} & + \text{of mass through} \\
\text{coincident system} & \text{contents of the coin-} & \text{the control} \\
 & \text{cident control volume} & \text{surface}
\end{array}
$$

In Eq. 5.3, we express the time rate of change of the system mass as the sum of two control volume quantities, the time rate of change of the mass of the contents of the control volume,

$$
\frac{\partial}{\partial t} \int_{cv} \rho \, d\Psi
$$

and the net rate of mass flow through the control surface,

$$
\int_{cs} \rho \mathbf{V} \cdot \hat{\mathbf{n}} \, dA
$$

Control surface

$V \cdot \hat{n} > 0$

$V \cdot \hat{n} < 0$

When a flow is steady, all field properties (i.e., properties at any specified point) including density remain constant with time and the time rate of change of the mass of the contents of the control volume is zero. That is,

$$
\frac{\partial}{\partial t} \int_{cv} \rho \, d\Psi = 0
$$

The integrand, $\mathbf{V} \cdot \hat{\mathbf{n}} \, dA$, in the mass flowrate integral represents the product of the component of velocity, $\mathbf{V}$, perpendicular to the small portion of control surface and the differential area, $dA$. Thus, $\mathbf{V} \cdot \hat{\mathbf{n}} \, dA$ is the volume flowrate through $dA$ and $\rho \mathbf{V} \cdot \hat{\mathbf{n}} \, dA$ is the mass flowrate through $dA$. Furthermore, as shown in the sketch in the margin, the sign of the dot product $\mathbf{V} \cdot \hat{\mathbf{n}}$ is "+" for flow *out* of the control volume and "−" for flow *into* the control volume since $\hat{\mathbf{n}}$ is considered positive when it points out of the control volume. When all of the differential quantities, $\rho \mathbf{V} \cdot \hat{\mathbf{n}} \, dA$, are summed over the entire control surface, as indicated by the integral

$$
\int_{cs} \rho \mathbf{V} \cdot \hat{\mathbf{n}} \, dA
$$

the result is the net mass flowrate through the control surface, or

$$
\int_{cs} \rho \mathbf{V} \cdot \hat{\mathbf{n}} \, dA = \sum \dot{m}_{out} - \sum \dot{m}_{in} \tag{5.4}
$$

where $\dot{m}$ is the mass flowrate (lbm/s, slug/s or kg/s). If the integral in Eq. 5.4 is positive, the net flow is out of the control volume; if the integral is negative, the net flow is into the control volume.

The control volume expression for conservation of mass, which is commonly called the ***continuity equation,*** for a fixed, nondeforming control volume is obtained by combining Eqs. 5.1, 5.2, and 5.3 to obtain

*The continuity equation is a statement that mass is conserved.*

$$
\boxed{\frac{\partial}{\partial t} \int_{cv} \rho \, d\Psi + \int_{cs} \rho \mathbf{V} \cdot \hat{\mathbf{n}} \, dA = 0} \tag{5.5}
$$

In words, Eq. 5.5 states that to conserve mass the time rate of change of the mass of the contents of the control volume plus the net rate of mass flow through the control surface must equal zero. Actually, the same result could have been obtained more directly by equating the rates of mass flow into and out of the control volume to the rates of accumulation and depletion of mass within the control volume (see Section 3.6.2). It is reassuring, however, to see that the Reynolds transport theorem works for this simple-to-understand case. This confidence will serve us well as we develop control volume expressions for other important principles.

An often-used expression for ***mass flowrate, $\dot{m}$,*** through a section of control surface having area $A$ is

$$
\boxed{\dot{m} = \rho Q = \rho A V} \tag{5.6}
$$

*Mass flowrate equals the product of density and volume flowrate.*

where $\rho$ is the fluid density, $Q$ is the volume flowrate (ft³/s or m³/s), and $V$ is the component of fluid velocity perpendicular to area $A$. Since

$$\dot{m} = \int_A \rho \mathbf{V} \cdot \hat{\mathbf{n}} \, dA$$

application of Eq. 5.6 involves the use of *representative* or average values of fluid density, $\rho$, and fluid velocity, $V$. For incompressible flows, $\rho$ is uniformly distributed over area $A$. For compressible flows, we will normally consider a uniformly distributed fluid density at each section of flow and allow density changes to occur only from section to section. The appropriate fluid velocity to use in Eq. 5.6 is the average value of the component of velocity normal to the section area involved. This average value, $\bar{V}$, defined as

$$\bar{V} = \frac{\displaystyle\int_A \rho \mathbf{V} \cdot \hat{\mathbf{n}} \, dA}{\rho \mathring{A}} \qquad (5.7)$$

is shown in the figure in the margin.

If the velocity, $\mathbf{V}$, is considered uniformly distributed (one-dimensional flow) over the section area, $A$, then

$$\bar{V} = \frac{\displaystyle\int_A \rho \mathbf{V} \cdot \hat{\mathbf{n}} \, dA}{\rho A} = V \qquad (5.8)$$

and the bar notation is not necessary (as in Example 5.1). When the flow is not uniformly distributed over the flow cross-sectional area, the bar notation reminds us that an average velocity is being used (as in Examples 5.2 and 5.4).

*V5.1 Sink flow*

### 5.1.2 Fixed, Nondeforming Control Volume

In many applications of fluid mechanics, an appropriate control volume to use is fixed and nondeforming. Several example problems that involve the continuity equation for fixed, nondeforming control volumes (Eq. 5.5) follow.

**E**XAMPLE 5.1 **Conservation of Mass—Steady, Incompressible Flow**

**GIVEN** A great danger to workers in confined spaces involves the consumption of breathable air (oxygen) and its replacement with other gases such as carbon dioxide ($CO_2$). To prevent this from happening, confined spaces need to be ventilated. Although there is no standard for air exchange rates, a complete change of the air every 3 minutes has been accepted by industry as providing effective ventilation.

A worker is performing maintenance in a small rectangular tank with a height of 10 ft and a square base 6 ft by 6 ft. Fresh air enters through an 8-in.-diameter hose and exits through a 4-in.-diameter port on the tank wall. The flow is assumed steady and incompressible.

**FIND** Determine

**(a)** the exchange rate needed (ft³/min) for this tank and

**(b)** the velocity of the air entering and exiting the tank at this exchange rate.

**S**OLUTION

**(a)** The necessary exchange rate, which will provide the flowrate entering and exiting the space, is based on the volume, $\mathcal{V}$, of the tank, where

$$\mathcal{V} = (10 \text{ ft})(6 \text{ ft})(6 \text{ ft}) = 360 \text{ ft}^3$$

Thus, 360 ft³ of air is needed to provide one complete air exchange. As described in the problem statement, to provide effective ventilation this must be done every 3 minutes.

Therefore, the required flowrate, $Q$, is

$$Q = \frac{360 \text{ ft}^3}{3 \text{ min}} = 120 \text{ cfm} \qquad \text{(Ans)}$$

**(b)**  The continuity equation, Eq. 5.5, can be used to calculate the velocities at the inlet and outlet. Thus,

$$\frac{\partial}{\partial t} \int_{cv} \rho \, dV + \sum \rho_{out} V_{out} A_{out} - \sum \rho_{in} V_{in} A_{in} = 0 \qquad \text{(1)}$$

We consider the volume within the tank to be the control volume, $A_{in}$ the cross-sectional area of the hose as it protrudes through the tank wall, and $A_{out}$ the area of the port in the tank wall. Since the flow is assumed steady and incompressible,

$$\frac{\partial}{\partial t} \int_{cv} \rho dV = 0$$

and

$$\rho_{out} = \rho_{in}$$

Thus, Eq. 1 reduces to

$$V_{out} A_{out} - V_{in} A_{in} = 0$$

or

$$V_{out} A_{out} = V_{in} A_{in} = Q \qquad \text{(2)}$$

which can be rearranged to solve for $V_{out}$ and $V_{in}$.

$$V_{out} = \frac{Q}{A_{out}} = \frac{120 \text{ ft}^2/\text{min}}{\left(\frac{\pi}{4}\right)\left(\frac{4}{12} \text{ ft}\right)^2} = 1380 \text{ ft/min} = 22.9 \text{ ft/s}$$

$$V_{in} = \frac{Q}{A_{in}} = \frac{120 \text{ ft}^2/\text{min}}{\left(\frac{\pi}{4}\right)\left(\frac{8}{12} \text{ ft}\right)^2} = 343 \text{ ft/min} = 5.73 \text{ ft/s} \qquad \text{(Ans)}$$

**COMMENTS**  In this example it is quite apparent how much the problem is simplified by assuming the flow is steady and incompressible. It is important for engineers to understand when such assumptions can be made. Also one can see how the velocity through an outlet/inlet is directly related to the flowrate and geometry of the outlet/inlet. For example, if the air velocity through the outlet causes too much dust to be stirred up, the velocity could be decreased by increasing the diameter of the outlet. The relationship between velocity and diameter is shown in Fig. E5.1. As expected, the velocity is inversely proportional to the square of the diameter.

As mentioned, there is no written standard for air exchange rates, but changing the air every 3 minutes has been accepted by industry. It is important for engineers to understand that individual industries and companies will have their own established safety precautions, depending on the nature of their work. As an engineer, one must heed these precautions. Safety is always a key component in the design process.

■ **Figure E5.1**

---

## EXAMPLE 5.2    Conservation of Mass—Steady, Compressible Flow

**GIVEN**  Air flows steadily between two sections in a long, straight portion of 4-in. inside diameter pipe as indicated in Fig. E5.2. The uniformly distributed temperature and pressure at each section are given. The average air velocity (nonuniform velocity distribution) at section (2) is 1000 ft/s.

**FIND**  Calculate the average air velocity at section (1).

■ **Figure E5.2**

### SOLUTION

The average fluid velocity at any section is that velocity which yields the section mass flowrate when multiplied by the section average fluid density and section area (Eq. 5.7). We relate the flows at sections (1) and (2) with the control volume designated with a dashed line in Fig. E5.2.

Equation 5.5 is applied to the contents of this control volume to obtain

$$\frac{\partial}{\partial t} \int_{cv} \rho \, dV + \int_{cs} \rho \mathbf{V} \cdot \hat{\mathbf{n}} \, dA = 0 \qquad \nearrow 0 \text{ (flow is steady)}$$

The time rate of change of the mass of the contents of this control volume is zero because the flow is steady. The control surface integral involves mass flowrates at sections (1) and (2) so that from Eq. 5.4 we get

$$\int_{cs} \rho \mathbf{V} \cdot \hat{\mathbf{n}} \, dA = \dot{m}_2 - \dot{m}_1 = 0$$

or

$$\dot{m}_1 = \dot{m}_2 \tag{1}$$

and from Eqs. 1, 5.6, and 5.7 we obtain

$$\rho_1 A_1 \overline{V}_1 = \rho_2 A_2 \overline{V}_2 \tag{2}$$

or since $A_1 = A_2$

$$\overline{V}_1 = \frac{\rho_2}{\rho_1} \overline{V}_2 \tag{3}$$

Air at the pressures and temperatures involved in this example problem behaves like an ideal gas. The ideal gas equation of state (Eq. 1.8) is

$$\rho = \frac{p}{RT} \tag{4}$$

Thus, combining Eqs. 3 and 4 we obtain

$$\overline{V}_1 = \frac{p_2 T_1 \overline{V}_2}{p_1 T_2}$$

$$= \frac{(18.4 \text{ psia})(540 \text{ °R})(1000 \text{ ft/s})}{(100 \text{ psia})(453 \text{ °R})} = 219 \text{ ft/s} \quad \textbf{(Ans)}$$

**COMMENT**  We learn from this example that the continuity equation (Eq. 5.5) is valid for compressible as well as incompressible flows. Also, nonuniform velocity distributions can be handled with the average velocity concept. Significant average velocity changes can occur in pipe flow if the fluid is compressible.

---

# EXAMPLE 5.3  Conservation of Mass—Two Fluids

**GIVEN**  The inner workings of a dehumidifier are shown in Fig. E5.3a. Moist air (a mixture of dry air and water vapor) enters the dehumidifier at the rate of 600 lbm/hr. Liquid water drains out of the dehumidifier at a rate of 3.0 lbm/hr. A simplified sketch of the process is provided in Fig. E5.3b.

**FIND**  Determine the mass flowrate of the dry air and the water vapor leaving the dehumidifier.

■ **Figure E5.3a**

■ **Figure E5.3b**

## SOLUTION

The unknown mass flowrate at section (2) is linked with the known flowrates at sections (1) and (3) with the control volume designated with a dashed line in Fig. E5.3b. The contents of the control volume are the air and water vapor mixture and the condensate (liquid water) in the dehumidifier at any instant.

Not included in the control volume are the fan and its motor, and the condenser coils and refrigerant. Even though the flow in the vicinity of the fan blade is unsteady, it is unsteady in a cyclical way. Thus, the flowrates at sections (1), (2), and (3) appear steady, and the time rate of change of the mass of the contents of the control volume may be considered equal to zero on a time-average basis. The application of Eqs. 5.4 and 5.5 to the control volume contents results in

$$\int_{cs} \rho \mathbf{V} \cdot \hat{\mathbf{n}} \, dA = -\dot{m}_1 + \dot{m}_2 + \dot{m}_3 = 0$$

or

$$\dot{m}_2 = \dot{m}_1 - \dot{m}_3 = 600 \text{ lbm/hr} - 3.0 \text{ lbm/hr}$$
$$= 597 \text{ lbm/hr} \qquad \textbf{(Ans)}$$

**COMMENT**    Note that the continuity equation (Eq. 5.5) can be used when there is more than one stream of fluid flowing through the control volume.

The answer is the same with a control volume that includes the cooling coils to be within the control volume. The continuity equation becomes

$$\dot{m}_2 = \dot{m}_1 - \dot{m}_3 + \dot{m}_4 - \dot{m}_5 \qquad \textbf{(1)}$$

where $\dot{m}_4$ is the mass flowrate of the cooling fluid flowing into the control volume, and $\dot{m}_5$ is the flowrate out of the control volume through the cooling coil. Since the flow through the coils is steady, it follows that $\dot{m}_4 = \dot{m}_5$. Hence, Eq. 1 gives the same answer as obtained with the original control volume.

---

# EXAMPLE 5.4    Conservation of Mass—Nonuniform Velocity Profile

**GIVEN**    Incompressible, laminar water flow develops in a straight pipe having radius $R$ as indicated in Fig. E5.4a. At section (1), the velocity profile is uniform; the velocity is equal to a constant value $U$ and is parallel to the pipe axis everywhere. At section (2), the velocity profile is axisymmetric and parabolic, with zero velocity at the pipe wall and a maximum value of $u_{max}$ at the centerline.

**FIND**

**(a)**    How are $U$ and $u_{max}$ related?

**(b)**    How are the average velocity at section (2), $\overline{V}_2$, and $u_{max}$ related?

■ **Figure E5.4a**

## SOLUTION

**(a)**    An appropriate control volume is sketched (dashed lines) in Fig. E5.4a. The application of Eq. 5.5 to the contents of this control volume yields

$$\frac{\partial}{\partial t} \int_{cv} \rho \, d\forall + \int_{cs} \rho \mathbf{V} \cdot \hat{\mathbf{n}} \, dA = 0 \qquad \overset{0 \text{ (flow is steady)}}{} \qquad \textbf{(1)}$$

At the inlet, section (1), the velocity is uniform with $V_1 = U$ so that

$$\int_{(1)} \rho \mathbf{V} \cdot \hat{\mathbf{n}} \, dA = -\rho_1 A_1 U \qquad \textbf{(2)}$$

At the outlet, section (2), the velocity is not uniform. However, the net flowrate through this section is the sum of flows through numerous small washer-shaped areas of size $dA_2 = 2\pi r \, dr$ as shown by the shaded area element in Fig. E5.4b. On each of

these infinitesimal areas the fluid velocity is denoted as $u_2$. Thus, in the limit of infinitesimal area elements, the summation is replaced by an integration, and the outflow through section (2) is given by

$$\int_{(2)} \rho \mathbf{V} \cdot \hat{\mathbf{n}} \, dA = \rho_2 \int_0^R u_2 2\pi r \, dr \qquad \textbf{(3)}$$

By combining Eqs. 1, 2, and 3, we get

$$\rho_2 \int_0^R u_2 2\pi r \, dr - \rho_1 A_1 U = 0 \qquad \textbf{(4)}$$

Since the flow is considered incompressible, $\rho_1 = \rho_2$. The parabolic velocity relationship for flow through section (2) is used in Eq. 4 to yield

$$2\pi u_{max} \int_0^R \left[ 1 - \left( \frac{r}{R} \right)^2 \right] r \, dr - A_1 U = 0 \qquad \textbf{(5)}$$

Integrating, we get from Eq. 5

$$2\pi u_{max} \left( \frac{r^2}{2} - \frac{r^4}{4R^2} \right)\Big|_0^R - \pi R^2 U = 0$$

■ **Figure E5.4b**

or

$$u_{max} = 2U \qquad \text{(Ans)}$$

**(b)** Since this flow is incompressible, we conclude from Eq. 5.7 that $U$ is the average velocity at all sections of the control volume. Thus, the average velocity at section (2), $\overline{V}_2$, is one-half the maximum velocity, $u_{max}$, there or

$$\overline{V}_2 = \frac{u_{max}}{2} \qquad \text{(Ans)}$$

**COMMENT** The relationship between the maximum velocity at section (2) and the average velocity is a function of the "shape" of the velocity profile. For the parabolic profile assumed in this example, the average velocity, $u_{max}/2$, is the actual "average" of the maximum velocity at section (2), $u_2 = u_{max}$, and the minimum velocity at that section, $u_2 = 0$. However, as shown in Fig. E5.4c, if the velocity profile is a different shape (nonparabolic), the average velocity is not necessarily one-half of the maximum velocity.

■ **Figure E5.4c**

---

# EXAMPLE 5.5 ■ Conservation of Mass—Unsteady Flow

**GIVEN** Construction workers in a trench of the type shown in Fig. E5.5a are installing a new waterline. The trench is 10 ft long, 5 ft wide, and 8 ft deep. As a result of being near an intersection, carbon dioxide from vehicle exhaust enters the trench at a rate of 10 ft³/min. Because carbon dioxide has a greater density than air, it will settle to the bottom of the trench and displace the air the workers need to breathe. Assume that there is negligible mixing between the air and carbon dioxide.

**FIND** **(a)** Estimate the time rate of change of the depth of carbon dioxide in the trench, $\partial h/\partial t$, in feet per minute at any instant.
**(b)** Calculate the time, $t_{h=6}$, it would take for the level of carbon dioxide to reach 6 ft, the approximate height to fully engulf the utility workers.

■ **Figure E5.5a**
**(Photo courtesy of Architect of the Capitol.)**

## SOLUTION

**(a)** We use the fixed, nondeforming control volume outlined with a dashed line in Fig. E5.5b. This control volume includes, at any instant, the carbon dioxide accumulated in the trench, some of the carbon dioxide flowing from the street into the trench, and some air. Application of Eqs. 5.5 and 5.6 to the contents of the control volume results in

$$\frac{\partial}{\partial t} \int_{\substack{air \\ volume}} \rho_{air} \, dV_{air} + \frac{\partial}{\partial t} \int_{\substack{co_2 \\ volume}} \rho_{co_2} \, dV_{co_2}$$
$$- \dot{m}_{co_2} + \dot{m}_{air} = 0 \quad (1)$$

where $\dot{m}_{co_2}$ and $\dot{m}_{air}$ are the mass flowrates of carbon dioxide into and air out of the control volume.

Recall that the mass, $dm$, of fluid contained in a small volume $dV$ is $dm = \rho \, dV$. Hence, the two integrals in Eq. 1 represent the total amount of air and carbon dioxide in the control volume, and

the sum of the first two terms is the time rate of change of mass within the control volume.

Note that the time rates of change of air mass and carbon dioxide mass are each not zero. Recognizing, however, that the air mass must be conserved, we know that the time rate of change of the mass of air in the control volume must be equal to the rate of air mass flow out of the control volume. Thus, applying Eqs. 5.5 and 5.6 to the air only and to the carbon dioxide only, we obtain

$$\frac{\partial}{\partial t} \int_{\substack{air \\ volume}} \rho_{air} \, dV_{air} + \dot{m}_{air} = 0$$

for air and

$$\frac{\partial}{\partial t} \int_{\substack{co_2 \\ volume}} \rho_{co_2} \, dV_{co_2} = \dot{m}_{co_2} \quad (2)$$

Control volume

$Q_{CO_2}$

8 ft

h

5 ft

10 ft

■ **Figure E5.5b**

for carbon dioxide. The volume of carbon dioxide in the control volume is given by

$$\int_{\substack{CO_2 \\ \text{volume}}} \rho_{CO_2} \, d\mathcal{V}_{CO_2} = \rho_{CO_2}[h(10 \text{ ft})(5 \text{ ft})] \quad (3)$$

Combining Eqs. 2 and 3, we obtain

$$\rho_{CO_2}(50 \text{ ft}^2)\frac{\partial h}{\partial t} = \dot{m}_{CO_2}$$

and, thus, since $\dot{m} = \rho Q$,

$$\frac{\partial h}{\partial t} = \frac{Q_{CO_2}}{50 \text{ ft}^2}$$

or

$$\frac{\partial h}{\partial t} = \frac{10 \text{ ft}^3/\text{min}}{50 \text{ ft}^2} = 0.20 \text{ ft/min} \quad \textbf{(Ans)}$$

**(b)** To find the time it will take to engulf the workers at a depth of $h = 6$ ft, we need only to divide the height the level needs to reach by the time rate of change of depth. That is,

$$t_{h=6} = \frac{6 \text{ ft}}{0.2 \text{ (ft/min)}} = 30 \text{ min} \quad \textbf{(Ans)}$$

**COMMENTS** As shown in this example, it would not take long for the air within the confined space of the trench to be displaced enough to become a danger to the workers. Furthermore, by the time workers feel the effects of an oxygen-deficient atmosphere, they may be unable to remove themselves from the dangerous space.

Note that the answer to part (b) can be easily obtained by realizing that the accumulated volume, $\mathcal{V}$, of a flow is the flowrate, $Q$, times the time it has been flowing, $t$. That is, $\mathcal{V} = Qt$. For this example, the volume of the carbon dioxide in the trench when it is 6 ft deep (neglecting the volume of the workers and equipment within the trench) is $\mathcal{V} = 5 \text{ ft} \times 10 \text{ ft} \times 6 \text{ ft} = 300 \text{ ft}^3$. Thus, with a flowrate of $Q = 10 \text{ ft}^3/\text{min}$, $t_{h=6} = 300 \text{ ft}^3/10 \text{ ft}^3/\text{min} = 30 \text{ min}$ in agreement with the foregoing answer.

The preceding example problems illustrate some important results of applying the conservation of mass principle to the contents of a fixed, nondeforming control volume. The dot product $\mathbf{V} \cdot \hat{\mathbf{n}}$ is "+" for flow out of the control volume and "−" for flow into the control volume. Thus, mass flowrate out of the control volume is "+" and mass flowrate in is "−." When the flow is steady, the time rate of change of the mass of the contents of the control volume

*The appropriate sign convention must be followed.*

*V5.2 Shop vac filter*

$$\frac{\partial}{\partial t} \int_{cv} \rho \, d\mathcal{V}$$

is zero and the net amount of mass flowrate, $\dot{m}$, through the control surface is, therefore, also zero

$$\sum \dot{m}_{out} - \sum \dot{m}_{in} = 0 \quad (5.9)$$

If the steady flow is also incompressible, the net amount of volume flowrate, $Q$, through the control surface is also zero:

$$\sum Q_{out} - \sum Q_{in} = 0 \quad (5.10)$$

An unsteady, but cyclical, flow can be considered steady on a time-average basis. When the flow is unsteady, the instantaneous time rate of change of the mass of the contents of the control volume is not necessarily zero and can be an important variable. When the value of

*V5.3 Flow through a contraction*

$$\frac{\partial}{\partial t} \int_{cv} \rho \, d\mathcal{V}$$

is "+," the mass of the contents of the control volume is increasing. When it is "−," the mass of the contents of the control volume is decreasing.

When the flow is uniformly distributed over the opening in the control surface (one-dimensional flow),

$$\dot{m} = \rho A V$$

where $V$ is the uniform value of the velocity component normal to the section area $A$. When the velocity is nonuniformly distributed over the opening in the control surface,

$$\dot{m} = \rho A \overline{V} \tag{5.11}$$

where $\overline{V}$ is the average value of the component of velocity normal to the section area $A$ as defined by Eq. 5.7.

For steady flow involving only one stream of a specific fluid flowing through the control volume at sections (1) and (2), as shown by the figure in the margin

$$\dot{m} = \rho_1 A_1 \overline{V}_1 = \rho_2 A_2 \overline{V}_2 \tag{5.12}$$

and for incompressible flow,

$$Q = A_1 \overline{V}_1 = A_2 \overline{V}_2 \tag{5.13}$$

For steady flow involving more than one stream of a specific fluid or more than one specific fluid flowing through the control volume,

$$\sum \dot{m}_{\text{in}} = \sum \dot{m}_{\text{out}}$$

The variety of example problems solved above should give the correct impression that the fixed, nondeforming control volume is versatile and useful.

---

**F l u i d s   i n   t h e   N e w s**

**New 1.6-gpf standards** Toilets account for approximately 40% of all indoor household water use. To conserve water, the new standard is 1.6 gallons of water per flush (gpf). Old toilets use up to 7 gpf; those manufactured after 1980 use 3.5 gpf. Neither is considered a low-flush toilet. A typical 3.2-person household in which each person flushes a 7-gpf toilet 4 times a day uses 32,700 gallons of water each year; with a 3.5-gpf toilet the amount is reduced to 16,400 gallons. Clearly the new 1.6-gpf toilets will save even more water. However, designing a toilet that flushes properly with such a small amount of water is not simple. Today there are two basic types involved: those that are gravity powered and those that are pressure powered. Gravity toilets (typical of most currently in use) have rather long cycle times. The water starts flowing under the action of gravity and the swirling vortex motion initiates the siphon action that builds to a point of discharge. In the newer pressure-assisted models, the *flowrate* is large, but the cycle time is short and the amount of water used is relatively small. (See Problem 5.33.)

---

### 5.1.3 Moving, Nondeforming Control Volume

*Some problems are most easily solved by using a moving control volume.*

It is sometimes necessary to use a nondeforming control volume attached to a moving reference frame. Examples include control volumes containing a gas turbine engine on an aircraft in flight, the exhaust stack of a ship at sea, and the gasoline tank of an automobile passing by.

As discussed in Section 4.4.6, when a moving control volume is used, the fluid velocity relative to the moving control volume (relative velocity) is an important flow field variable. The relative velocity, $\mathbf{W}$, is the fluid velocity seen by an observer moving with the control volume. The control volume velocity, $\mathbf{V}_{\text{cv}}$, is the velocity of the control volume as seen from a fixed coordinate system. The absolute velocity, $\mathbf{V}$, is the fluid velocity seen by a stationary observer in a fixed coordinate system. These velocities are related to each other by the vector equation

$$\mathbf{V} = \mathbf{W} + \mathbf{V}_{\text{cv}} \tag{5.14}$$

as illustrated by the figure in the margin. This is the same as Eq. 4.22, introduced earlier.

For a system and a moving, nondeforming control volume that are coincident at an instant of time, the Reynolds transport theorem (Eq. 4.23) for a moving control volume leads to

$$\frac{DM_{\text{sys}}}{Dt} = \frac{\partial}{\partial t} \int_{\text{cv}} \rho \, d\mathcal{V} + \int_{\text{cs}} \rho \mathbf{W} \cdot \hat{\mathbf{n}} \, dA \tag{5.15}$$

From Eqs. 5.1 and 5.15, we can get the control volume expression for conservation of mass (the continuity equation) for a moving, nondeforming control volume, namely,

$$\frac{\partial}{\partial t} \int_{\text{cv}} \rho \, d\mathcal{V} + \int_{\text{cs}} \rho \mathbf{W} \cdot \hat{\mathbf{n}} \, dA = 0 \tag{5.16}$$

Some examples of the application of Eq. 5.16 follow.

## EXAMPLE 5.6 Conservation of Mass—Compressible Flow with a Moving Control Volume

**GIVEN** An airplane moves forward at a speed of 971 km/hr as shown in Fig. E5.6a. The frontal intake area of the jet engine is 0.80 m$^2$ and the entering air density is 0.736 kg/m$^3$. A stationary observer determines that relative to the Earth, the jet engine exhaust gases move away from the engine with a speed of 1050 km/hr. The engine exhaust area is 0.558 m$^2$, and the exhaust gas density is 0.515 kg/m$^3$.

**FIND** Estimate the mass flowrate of fuel into the engine in kg/hr.

**■ Figure E5.6**

## SOLUTION

The control volume, which moves with the airplane (see Fig. E5.6b), surrounds the engine and its contents and includes all fluids involved at an instant. The application of Eq. 5.16 to these contents of the control volume yields

$$\frac{\partial}{\partial t}\int_{cv} \rho \, d\mathcal{V} + \int_{cs} \rho\mathbf{W}\cdot\hat{\mathbf{n}} \, dA = 0 \qquad (1)$$

0 (flow relative to moving control volume is considered steady on a time-average basis)

Assuming one-dimensional flow, we evaluate the surface integral in Eq. 1 and get

$$-\dot{m}_{\substack{\text{fuel}\\\text{in}}} - \rho_1 A_1 W_1 + \rho_2 A_2 W_2 = 0$$

or

$$\dot{m}_{\substack{\text{fuel}\\\text{in}}} = \rho_2 A_2 W_2 - \rho_1 A_1 W_1 \qquad (2)$$

We consider the intake velocity, $W_1$, relative to the moving control volume, as being equal in magnitude to the speed of the airplane, 971 km/hr. The exhaust velocity, $W_2$, also needs to be measured relative to the moving control volume. Since a fixed observer

noted that the exhaust gases were moving away from the engine at a speed of 1050 km/hr, the speed of the exhaust gases relative to the moving control volume, $W_2$, is determined as follows by using Eq. 5.14

$$V_2 = W_2 + V_{\text{plane}}$$

or

$$W_2 = V_2 - V_{\text{plane}} = 1050 \text{ km/hr} - (-971 \text{ km/hr})$$
$$= 2021 \text{ km/hr}$$

and is shown in Fig. E5.6b.

From Eq. 2,

$$\dot{m}_{\substack{\text{fuel}\\\text{in}}} = (0.515 \text{ kg/m}^3)(0.558 \text{ m}^2)(2021 \text{ km/hr})(1000 \text{ m/km})$$
$$- (0.736 \text{ kg/m}^3)(0.80 \text{ m}^2)(971 \text{ km/hr})(1000 \text{ m/km})$$
$$= (580{,}800 - 571{,}700) \text{ kg/hr}$$
$$\dot{m}_{\substack{\text{fuel}\\\text{in}}} = 9100 \text{ kg/hr} \qquad \textbf{(Ans)}$$

**COMMENT** Note that the fuel flowrate was obtained as the difference of two large, nearly equal numbers. Precise values of $W_2$ and $W_1$ are needed to obtain a modestly accurate value of $\dot{m}_{\text{fuel}}$.

# $E$XAMPLE 5.7 Conservation of Mass—Relative Velocity

**GIVEN** Water enters a rotating lawn sprinkler through its base at the steady rate of 1000 ml/s as sketched in Fig. E5.7. The exit area of each of the two nozzles is 30 mm².

**FIND** Determine the average speed of the water leaving the nozzle, relative to the nozzle, if

(a) the rotary sprinkler head is stationary,

(b) the sprinkler head rotates at 600 rpm, and

(c) the sprinkler head accelerates from 0 to 600 rpm.

■ **Figure E5.7**

## $S$OLUTION

(a) We specify a control volume that contains the water in the rotary sprinkler head at any instant. This control volume is nondeforming, but it moves (rotates) with the sprinkler head.

The application of Eq. 5.16 to the contents of this control volume for situation (a), (b), or (c) of the problem results in the same expression, namely,

$$\frac{\partial}{\partial t}\int_{cv}\rho\,d\Psi\;+\;\int_{cs}\rho\mathbf{W}\cdot\hat{\mathbf{n}}\,dA=0$$

0 flow is steady or the control volume is filled with an incompressible fluid

or

$$\sum\rho_{out}A_{out}W_{out}-\sum\rho_{in}A_{in}W_{in}=0 \qquad (1)$$

The time rate of change of the mass of water in the control volume is zero because the flow is steady and the control volume is filled with water.

Because there is only one inflow [at the base of the rotating arm, section (1)] and two outflows [the two nozzles at the tips of the arm, sections (2) and (3), each with the same area and fluid velocity], Eq. 1 becomes

$$\rho_2 A_2 W_2 + \rho_3 A_3 W_3 - \rho_1 A_1 W_1 = 0 \qquad (2)$$

Hence, for incompressible flow with $\rho_1 = \rho_2 = \rho_3$, Eq. 2 becomes

$$A_2 W_2 + A_3 W_3 - A_1 W_1 = 0$$

With $Q = A_1 W_1$, $A_2 = A_3$, and $W_2 = W_3$ it follows that

$$W_2 = \frac{Q}{2\,A_2}$$

or

$$W_2 = \frac{(1000\ \text{ml/s})(0.001\ \text{m}^3/\text{liter})(10^6\ \text{mm}^2/\text{m}^2)}{(1000\ \text{ml/liter})(2)(30\ \text{mm}^2)}$$

$$= 16.7\ \text{m/s} \qquad (\text{Ans})$$

(b), (c) The value of $W_2$ is independent of the speed of rotation of the sprinkler head and represents the average velocity of the water exiting from each nozzle with respect to the nozzle for cases (a), (b), and (c).

**COMMENT** The velocity of water discharging from each nozzle, when viewed from a stationary reference (i.e., $V_2$), will vary as the rotation speed of the sprinkler head varies since from Eq. 5.14,

$$V_2 = W_2 - U$$

where $U = \omega R$ is the speed of the nozzle and $\omega$ and $R$ are the angular velocity and radius of the sprinkler head, respectively.

---

When a moving, nondeforming control volume is used, the dot product sign convention used earlier for fixed, nondeforming control volume applications is still valid. Also, if the flow within the moving control volume is steady, or steady on a time-average basis, the time rate of change of the mass of the contents of the control volume is zero. Velocities seen from the control volume reference frame (relative velocities) must be used in the continuity equation. Relative and absolute velocities are related by a vector equation (Eq. 5.14), which also involves the control volume velocity.

*Care is needed to ensure that absolute and relative velocities are used correctly.*

### 5.1.4 Deforming Control Volume

Occasionally, a deforming control volume can simplify the solution of a problem. A deforming control volume involves changing volume size and control surface movement. Thus, the Reynolds transport theorem for a moving control volume can be used for this case, and Eqs. 4.23 and 5.1 lead to

$$\frac{DM_{sys}}{Dt} = \frac{\partial}{\partial t}\int_{cv} \rho \, d\mathcal{V} + \int_{cs} \rho \mathbf{W} \cdot \hat{\mathbf{n}} \, dA = 0 \qquad (5.17)$$

The time rate of change term in Eq. 5.17,

$$\frac{\partial}{\partial t}\int_{cv} \rho \, d\mathcal{V}$$

is usually nonzero and must be carefully evaluated because the extent of the control volume varies with time. The mass flowrate term in Eq. 5.17,

$$\int_{cs} \rho \mathbf{W} \cdot \hat{\mathbf{n}} \, dA$$

must be determined with the relative velocity, $\mathbf{W}$, the velocity referenced to the control surface. Since the control volume is deforming, the control surface velocity is not necessarily uniform and identical to the control volume velocity, $\mathbf{V}_{cv}$, as was true for moving, nondeforming control volumes.

*The velocity of the surface of a deforming control volume is not the same at all points on the surface.*

For the deforming control volume,

$$\mathbf{V} = \mathbf{W} + \mathbf{V}_{cs} \qquad (5.18)$$

where $\mathbf{V}_{cs}$ is the velocity of the control surface as seen by a fixed observer. The relative velocity, $\mathbf{W}$, must be ascertained with care wherever fluid crosses the control surface. Two example problems that illustrate the use of the continuity equation for a deforming control volume, Eq. 5.17, follow.

---

# EXAMPLE 5.8    Conservation of Mass—Deforming Control Volume

**GIVEN**   A syringe (Fig. E5.8a) is used to inoculate a cow. The plunger has a face area of 500 mm². The liquid in the syringe is to be injected steadily at a rate of 300 cm³/min. The leakage rate past the plunger is 0.10 times the volume flowrate out of the needle.

**FIND**   With what speed should the plunger be advanced?

## SOLUTION

The control volume selected for solving this problem is the deforming one illustrated in Fig. E5.8b. Section (1) of the control surface moves with the plunger. The surface area of section (1), $A_1$, is considered equal to the circular area of the face of the plunger, $A_p$, although this is not strictly true, since leakage occurs. The difference is small, however. Thus,

$$A_1 = A_p \qquad (1)$$

Liquid also leaves the needle through section (2), which involves fixed area $A_2$. The application of Eq. 5.17 to the contents of this control volume gives

$$\frac{\partial}{\partial t}\int_{cv} \rho \, d\mathcal{V} + \dot{m}_2 + \rho Q_{leak} = 0 \qquad (2)$$

Even though $Q_{leak}$ and the flow through section area $A_2$ are steady, the time rate of change of the mass of liquid in the shrinking control volume is not zero because the control volume

(a)

(© James McQuillan/iStockphoto.)

**Figure E5.8**

(b)

is getting smaller. To evaluate the first term of Eq. 2, we note that

$$\int_{cv} \rho \, d\mathcal{V} = \rho(\ell A_1 + \mathcal{V}_{needle}) \qquad (3)$$

where $\ell$ is the changing length of the control volume (see Fig. E5.8) and $\mathcal{V}_{needle}$ is the volume of the needle. From Eq. 3, we obtain

$$\frac{\partial}{\partial t}\int_{cv} \rho \, d\mathcal{V} = \rho A_1 \frac{\partial \ell}{\partial t} \qquad (4)$$

Note that

$$-\frac{\partial \ell}{\partial t} = V_p \tag{5}$$

where $V_p$ is the speed of the plunger sought in the problem statement. Combining Eqs. 2, 4, and 5 we obtain

$$-\rho A_1 V_p + \dot{m}_2 + \rho Q_{\text{leak}} = 0 \tag{6}$$

However, from Eq. 5.6, we see that

$$\dot{m}_2 = \rho Q_2 \tag{7}$$

and Eq. 6 becomes

$$-\rho A_1 V_p + \rho Q_2 + \rho Q_{\text{leak}} = 0 \tag{8}$$

Solving Eq. 8 for $V_p$ yields

$$V_p = \frac{Q_2 + Q_{\text{leak}}}{A_1} \tag{9}$$

Since $Q_{\text{leak}} = 0.1 Q_2$, Eq. 9 becomes

$$V_p = \frac{Q_2 + 0.1 Q_2}{A_1} = \frac{1.1 Q_2}{A_1}$$

and

$$V_p = \frac{(1.1)(300 \text{ cm}^3/\text{min})}{(500 \text{ mm}^2)} \left( \frac{1000 \text{ mm}^3}{\text{cm}^3} \right)$$

$$= 660 \text{ mm/min} \tag{Ans}$$

---

# EXAMPLE 5.9    Conservation of Mass—Deforming Control Volume

**GIVEN** Consider Example 5.5.

**FIND** Solve the problem of Example 5.5 using a deforming control volume that includes only the carbon dioxide accumulating in the trench.

## SOLUTION

For this deforming control volume, Eq. 5.17 leads to

$$\frac{\partial}{\partial t} \int_{\substack{\text{CO}_2 \\ \text{volume}}} \rho \, d\mathcal{V} + \int_{cs} \rho \mathbf{W} \cdot \hat{\mathbf{n}} \, dA = 0 \tag{1}$$

The first term of Eq. 1 can be evaluated as

$$\frac{\partial}{\partial t} \int_{\substack{\text{CO}_2 \\ \text{volume}}} \rho \, d\mathcal{V} = \frac{\partial}{\partial t} \left[ \rho h (10 \text{ ft})(5 \text{ ft}) \right]$$

$$= \rho (50 \text{ ft}^2) \frac{\partial h}{\partial t} \tag{2}$$

The second term of Eq. 1 can be evaluated as

$$\int_{cs} \rho \mathbf{W} \cdot \hat{\mathbf{n}} \, dA = -\rho \left( V_{\text{CO}_2} + \frac{\partial h}{\partial t} \right) A_{\text{CO}_2} \tag{3}$$

where $A_{\text{CO}_2}$ and $V_{\text{CO}_2}$ are the cross-sectional area and velocity of the carbon dioxide flowing from the street down into the trench. Note

that the velocity of the carbon dioxide entering the control surface and the velocity of the control surface are summed in Eq. 3 since they are in opposite directions and therefore, the relative velocity would be the two velocities combined (i.e. like the relative velocity of two cars passing each other going in opposite directions). The overall mass flow term is negative since the carbon dioxide is entering the control volume. Thus, from Eqs. 1, 2, and 3 we obtain

$$\frac{\partial h}{\partial t} = \frac{V_{\text{CO}_2} A_{\text{CO}_2}}{(50 \text{ ft}^2 - A_{\text{CO}_2})} = \frac{Q_{\text{CO}_2}}{(50 \text{ ft}^2 - A_{\text{CO}_2})}$$

or for $A_{\text{CO}_2} \ll 50 \text{ ft}^2$

$$\frac{\partial h}{\partial t} = \frac{10 \text{ ft}^3/\text{min}}{(50 \text{ ft}^2)} = 0.2 \text{ ft/min} \tag{Ans}$$

**COMMENT** Note that these results using a deforming control volume are the same as those obtained in Example 5.5 with a fixed control volume.

---

The conservation of mass principle is easily applied to the contents of a control volume. The appropriate selection of a specific kind of control volume (for example, fixed and nondeforming, moving and nondeforming, or deforming) can make the solution of a particular problem less complicated. In general, where fluid flows through the control surface, it is advisable to make the control surface perpendicular to the flow. In the sections ahead we learn that the conservation of mass principle is primarily used in combination with other important laws to solve problems.

## 5.2    Newton's Second Law—The Linear Momentum and Moment-of-Momentum Equations

*V5.4 Smokestack plume momentum*

### 5.2.1  Derivation of the Linear Momentum Equation

Newton's second law of motion for a system is

> Time rate of change of the     = sum of external forces
> linear momentum of the system    acting on the system

Since momentum is mass times velocity, the momentum of a small particle of mass $\rho\, d\forall$ is $\mathbf{V}\rho\, d\forall$. Thus, the momentum of the entire system is $\int_{sys}\mathbf{V}\rho\, d\forall$ and Newton's law becomes

$$\frac{D}{Dt}\int_{sys}\mathbf{V}\rho\, d\forall = \sum \mathbf{F}_{sys} \tag{5.19}$$

Any reference or coordinate system for which this statement is true is called *inertial*. A fixed coordinate system is inertial. A coordinate system that moves in a straight line with constant velocity and is thus without acceleration is also inertial. We proceed to develop the control volume formula for this important law. When a control volume is coincident with a system at an instant of time, the forces acting on the system and the forces acting on the contents of the coincident control volume (see Fig. 5.2) are instantaneously identical, that is,

*Forces acting on a flowing fluid can change its velocity magnitude and/or direction.*

$$\sum \mathbf{F}_{sys} = \sum \mathbf{F}_{\substack{\text{contents of the} \\ \text{coincident control volume}}} \tag{5.20}$$

Furthermore, for a system and the contents of a coincident control volume that is fixed and non-deforming, the Reynolds transport theorem [Eq. 4.19 with *b* set equal to the velocity (i.e., momentum per unit mass), and $B_{sys}$ being the system momentum] allows us to conclude that

*V5.5 Marine propulsion*

$$\frac{D}{Dt}\int_{sys}\mathbf{V}\rho\, d\forall = \frac{\partial}{\partial t}\int_{cv}\mathbf{V}\rho\, d\forall + \int_{cs}\mathbf{V}\rho\mathbf{V}\cdot\hat{\mathbf{n}}\, dA \tag{5.21}$$

or

> Time rate of change     time rate of change     net rate of flow
> of the linear          of the linear          of linear momentum
> momentum of the    =   momentum of the    +   through the
> system                 contents of the        control surface
>                        control volume

Equation 5.21 states that the time rate of change of system linear momentum is expressed as the sum of the two control volume quantities: the time rate of change of the *linear momentum of the contents of the control volume*, and the net rate of *linear momentum flow through the control surface*. As particles of mass move into or out of a control volume through the control surface, they carry linear momentum in or out. Thus, linear momentum flow should seem no more unusual than mass flow.

For a control volume that is fixed (and thus inertial) and nondeforming, Eqs. 5.19, 5.20, and 5.21 provide an appropriate mathematical statement of Newton's second law of motion as

$$\boxed{\frac{\partial}{\partial t}\int_{cv}\mathbf{V}\rho\, d\forall + \int_{cs}\mathbf{V}\rho\mathbf{V}\cdot\hat{\mathbf{n}}\, dA = \sum \mathbf{F}_{\substack{\text{contents of the} \\ \text{control volume}}}} \tag{5.22}$$

We call Eq. 5.22 the ***linear momentum equation***.

*V5.6 Force due to a water jet*

■ **Figure 5.2**    External forces acting on system and coincident control volume.

Flow out

Flow in

Control volume

$F_{\text{fluid out}}$

$\mathcal{W}$    $F_{\text{wall}}$

$F_{\text{fluid in}}$

In our application of the linear momentum equation, we initially confine ourselves to fixed, nondeforming control volumes for simplicity. Subsequently, we discuss the use of a moving but inertial, nondeforming control volume. We do not consider deforming control volumes and accelerating (noninertial) control volumes. If a control volume is noninertial, the acceleration components involved (for example, translation acceleration, Coriolis acceleration, and centrifugal acceleration) require consideration.

The forces involved in Eq. 5.22 are body and surface forces that act on what is contained in the control volume as shown in the sketch in the margin. The only body force we consider in this chapter is the one associated with the action of gravity. We experience this body force as weight, $\mathcal{W}$. The surface forces are basically exerted on the contents of the control volume by material just outside the control volume in contact with material just inside the control volume. For example, a wall in contact with fluid can exert a reaction surface force on the fluid it bounds. Similarly, fluid just outside the control volume can push on fluid just inside the control volume at a common interface, usually an opening in the control surface through which fluid flow occurs. An immersed object can resist fluid motion with surface forces.

The linear momentum terms in the momentum equation deserve careful explanation. We clarify their physical significance in the following sections.

### 5.2.2 Application of the Linear Momentum Equation

The linear momentum equation for an inertial control volume is a vector equation (Eq. 5.22). In engineering applications, components of this vector equation resolved along orthogonal coordinates, for example, $x$, $y$, and $z$ (rectangular coordinate system) or $r$, $\theta$, and $x$ (cylindrical coordinate system), will normally be used. A simple example involving steady, incompressible flow is considered first.

---

## $E$XAMPLE 5.10    Linear Momentum—Change in Flow Direction

**GIVEN** As shown in Fig. E5.10a, a horizontal jet of water exits a nozzle with a uniform speed of $V_1 = 10$ ft/s, strikes a vane, and is turned through an angle $\theta$. A similar situation is also shown in Video V5.6.

**FIND** Determine the anchoring force needed to hold the vane stationary if gravity and viscous effects are negligible.

### SOLUTION

We select a control volume that includes the vane and a portion of the water (see Figs. E5.10b, c) and apply the linear momentum equation to this fixed control volume. The only portions of the control surface across which fluid flows are section (1) (the entrance) and section (2) (the exit). Hence, the $x$ and $z$ components of Eq. 5.22 become

$$\frac{\partial}{\partial t} \int_{cv} u\,\rho\,d\mathcal{V}^{\,0 \text{ (flow is steady)}} + \int_{cs} u\,\rho\,\mathbf{V} \cdot \hat{\mathbf{n}}\,dA = \sum F_x$$

and

$$\frac{\partial}{\partial t} \int_{cv} w\,\rho\,d\mathcal{V}^{\,0 \text{ (flow is steady)}} + \int_{cs} w\,\rho\,\mathbf{V} \cdot \hat{\mathbf{n}}\,dA = \sum F_z$$

or

$$u_2\,\rho A_2 V_2 - u_1 \rho A_1 V_1 = \sum F_x \qquad (1)$$

and

$$w_2\,\rho A_2 V_2 - w_1 \rho A_1 V_1 = \sum F_z \qquad (2)$$

where $\mathbf{V} = u\hat{\mathbf{i}} + w\hat{\mathbf{k}}$, and $\sum F_x$ and $\sum F_z$ are the net $x$ and $z$ components of force acting on the contents of the control volume. Depending on the particular flow situation being considered and the coordinate system chosen, the $x$ and $z$ components of velocity, $u$ and $w$, can be positive, negative, or zero. In this example the flow is in the positive direction at both the inlet and the outlet.

The water enters and leaves the control volume as a free jet at atmospheric pressure. Hence, there is atmospheric pressure surrounding the entire control volume, and the net pressure force on the control volume surface is zero. If we neglect the weight of the water and vane, the only forces applied to the control volume contents are the horizontal and vertical components of the anchoring force, $F_{Ax}$ and $F_{Az}$, respectively.

With negligible gravity and viscous effects, and since $p_1 = p_2$, the speed of the fluid remains constant so that $V_1 = V_2 = 10$ ft/s (see the Bernoulli equation, Eq. 3.7). Hence, at section (1), $u_1 = V_1$, $w_1 = 0$, and at section (2), $u_2 = V_1 \cos\theta$, $w_2 = V_1 \sin\theta$.

By using this information, Eqs. 1 and 2 can be written as

$$V_1 \cos\theta\,\rho\,A_2 V_1 - V_1\,\rho\,A_1 V_1 = F_{Ax} \qquad (3)$$

and

$$V_1 \sin \theta \, \rho A_2 V_1 - 0 \, \rho A_1 V_1 = F_{Az} \quad \text{(4)}$$

Equations 3 and 4 can be simplified by using conservation of mass, which states that for this incompressible flow $A_1V_1 = A_2V_2$, or $A_1 = A_2$ since $V_1 = V_2$. Thus,

$$F_{Ax} = -\rho A_1 V_1^2 + \rho A_1 V_1^2 \cos \theta = -\rho A_1 V_1^2 (1 - \cos \theta) \quad \text{(5)}$$

and

$$F_{Az} = \rho A_1 V_1^2 \sin \theta \quad \text{(6)}$$

With the given data we obtain

$$F_{Ax} = -(1.94 \text{ slugs/ft}^3)(0.06 \text{ ft}^2)(10 \text{ ft/s})^2(1 - \cos \theta)$$
$$= -11.64(1 - \cos \theta) \text{ slugs} \cdot \text{ft/s}^2$$
$$= -11.64(1 - \cos \theta) \text{ lb} \quad \textbf{(Ans)}$$

and

$$F_{Az} = (1.94 \text{ slugs/ft}^3)(0.06 \text{ ft}^2)(10 \text{ ft/s})^2 \sin \theta$$
$$= 11.64 \sin \theta \text{ lb} \quad \textbf{(Ans)}$$

**COMMENTS**    The values of $F_{Ax}$ and $F_{Az}$ as a function of $\theta$ are shown in Fig. E5.10d. Note that if $\theta = 0$ (i.e., the vane does not turn the water), the anchoring force is zero. The inviscid fluid merely slides along the vane without putting any force on it. If $\theta = 90°$, then $F_{Ax} = -11.64$ lb and $F_{Az} = 11.64$ lb. It is necessary to push on the vane (and, hence, for the vane to push on the water) to the left ($F_{Ax}$ is negative) and up in order to change the direction of flow of the water from horizontal to vertical. This momentum change requires a force. If $\theta = 180°$, the water jet is turned back on itself. This requires no vertical force ($F_{Az} = 0$), but the horizontal force ($F_{Ax} = -23.3$ lb) is two times that required if $\theta = 90°$. This horizontal fluid momentum change requires a horizontal force only.

Note that the anchoring force (Eqs. 5, 6) can be written in terms of the mass flowrate, $\dot{m} = \rho A_1 V_1$, as

$$F_{Ax} = -\dot{m} V_1 (1 - \cos \theta)$$

and

$$F_{Az} = \dot{m} V_1 \sin \theta$$

In this example exerting a force on a fluid flow resulted in a change in its direction only (i.e., change in its linear momentum).

■ **Figure E5.10**

---

## F l u i d s   i n   t h e   N e w s

**Where the plume goes**  Commercial airliners have wheel brakes very similar to those on highway vehicles. In fact, antilock brakes now found on most new cars were first developed for use on airplanes. However, when landing, the major braking force comes from the engine rather than the wheel brakes. Upon touchdown, a piece of engine cowling translates aft and blocker doors drop down, directing the engine airflow into a honeycomb structure called a cascade. The cascade reverses the direction of the high-speed engine exhausts by nearly 180° so that it flows forward. As predicted by the *momentum equation*, the air passing through the engine produces a substantial braking force—the reverse thrust. Designers must know the flow pattern of the exhaust plumes to eliminate potential problems. For example, the plumes of hot exhaust must be kept away from parts of the aircraft where repeated heating and cooling could cause premature fatigue. Also the plumes must not reenter the engine inlet, blow debris from the runway in front of the engine, or envelop the vertical tail. (See Problem 5.78.)

# EXAMPLE 5.11 Linear Momentum—Weight, Pressure, and Change in Speed

**GIVEN** As shown in Fig. E5.11a, water flows through a nozzle attached to the end of a laboratory sink faucet with a flowrate of 0.6 liters/s. The nozzle inlet and exit diameters are 16 mm and 5 mm, respectively, and the nozzle axis is vertical. The mass of the nozzle is 0.1 kg. The pressure at section (1) is 464 kPa.

**FIND** Determine the anchoring force required to hold the nozzle in place.

## SOLUTION

The anchoring force sought is the reaction force between the faucet and nozzle threads. To evaluate this force, we select a control volume that includes the entire nozzle and the water contained in the nozzle at an instant, as is indicated in Figs. E5.11a and E5.11b. All of the vertical forces acting on the contents of this control volume are identified in Fig. E5.11b. The action of atmospheric pressure cancels out in every direction and is not shown. Gage pressure forces do not cancel out in the vertical direction and are shown. Application of the vertical or $z$ direction component of Eq. 5.22 to the contents of this control volume leads to

$$\overbrace{\frac{\partial}{\partial t} \int_{cv} w\rho \, d\!\!\!\forall}^{0 \text{ (flow is steady)}} + \int_{cs} w\rho \mathbf{V} \cdot \hat{\mathbf{n}} \, dA = F_A - \mathcal{W}_n - p_1 A_1$$
$$- \mathcal{W}_w + p_2 A_2 \qquad (1)$$

where $w$ is the $z$ direction component of fluid velocity, and the various parameters are identified in the figure.

Note that the positive direction is considered "up" for the forces. We will use this same sign convention for the fluid velocity, $w$, in Eq. 1. In Eq. 1, the dot product, $\mathbf{V} \cdot \hat{\mathbf{n}}$, is "+" for flow out of the control volume and "−" for flow into the control volume. For this particular example

$$\mathbf{V} \cdot \hat{\mathbf{n}} \, dA = \pm |w| \, dA \qquad (2)$$

with the "+" used for flow out of the control volume and "−" used for flow in. To evaluate the control surface integral in Eq. 1, we need to assume a distribution for fluid velocity, $w$, and fluid density, $\rho$. For simplicity, we assume that $w$ is uniformly distributed or constant, with magnitudes of $w_1$ and $w_2$ over cross-sectional areas $A_1$ and $A_2$. Also, this flow is incompressible so the fluid density, $\rho$, is constant throughout. Proceeding further we obtain for Eq. 1

$$(-\dot{m}_1)(-w_1) + \dot{m}_2(-w_2)$$
$$= F_A - \mathcal{W}_n - p_1 A_1 - \mathcal{W}_w + p_2 A_2 \qquad (3)$$

where $\dot{m} = \rho A V$ is the mass flowrate.

■ **Figure E5.11a**

$F_A$ = anchoring force that holds nozzle in place
$\mathcal{W}_n$ = weight of nozzle
$\mathcal{W}_w$ = weight of water contained in the nozzle
$p_1$ = gage pressure at section (1)
$A_1$ = cross section area at section (1)
$p_2$ = gage pressure at section (2)
$A_2$ = cross section area at section (2)
$w_1$ = z direction velocity at control volume entrance
$w_2$ = z direction velocity at control volume exit

■ **Figure E5.11b**

Note that $-w_1$ and $-w_2$ are used because both of these velocities are "down." Also, $-\dot{m}_1$ is used because it is associated with flow into the control volume. Similarly, $+\dot{m}_2$ is used because it is associated with flow out of the control volume. Solving Eq. 3 for the anchoring force, $F_A$, we obtain

$$F_A = \dot{m}_1 w_1 - \dot{m}_2 w_2 + \mathcal{W}_n + p_1 A_1 + \mathcal{W}_w - p_2 A_2 \quad (4)$$

From the conservation of mass equation, Eq. 5.12, we obtain

$$\dot{m}_1 = \dot{m}_2 = \dot{m} \quad (5)$$

which when combined with Eq. 4 gives

$$F_A = \dot{m}(w_1 - w_2) + \mathcal{W}_n + p_1 A_1 + \mathcal{W}_w - p_2 A_2 \quad (6)$$

It is instructive to note how the anchoring force is affected by the different actions involved. As expected, the nozzle weight, $\mathcal{W}_n$, the water weight, $\mathcal{W}_w$, and gage pressure force at section (1), $p_1 A_1$, all increase the anchoring force, while the gage pressure force at section (2), $p_2 A_2$, acts to decrease the anchoring force. The change in the vertical momentum flowrate, $\dot{m}(w_1 - w_2)$, will, in this instance, decrease the anchoring force because this change is negative ($w_2 > w_1$).

To complete this example, we use quantities given in the problem statement to quantify the terms on the right-hand side of Eq. 6.

From Eq. 5.6,

$$\dot{m} = \rho w_1 A_1 = \rho Q$$
$$= (999\ \text{kg/m}^3)(0.6\ \text{liter/s})(10^{-3}\ \text{m}^3/\text{liter})$$
$$= 0.599\ \text{kg/s} \quad (7)$$

and

$$w_1 = \frac{Q}{A_1} = \frac{Q}{\pi(D_1^2/4)}$$
$$= \frac{(0.6\ \text{liter/s})(10^{-3}\ \text{m}^3/\text{liter})}{\pi(16\ \text{mm})^2/4(1000^2\ \text{mm}^2/\text{m}^2)} = 2.98\ \text{m/s} \quad (8)$$

Also from Eq. 5.6,

$$w_2 = \frac{Q}{A_2} = \frac{Q}{\pi(D_2^2/4)}$$
$$= \frac{(0.6\ \text{liter/s})(10^{-3}\ \text{m}^3/\text{liter})}{\pi(5\ \text{mm})^2/4(1000^2\ \text{mm}^2/\text{m}^2)} = 30.6\ \text{m/s} \quad (9)$$

The weight of the nozzle, $\mathcal{W}_n$, can be obtained from the nozzle mass, $m_n$, with

$$\mathcal{W}_n = m_n g = (0.1\ \text{kg})(9.81\ \text{m/s}^2) = 0.981\ \text{N} \quad (10)$$

The weight of the water in the control volume, $\mathcal{W}_w$, can be obtained from the water density, $\rho$, and the volume of water, $V_w$, in the truncated cone of height $h$. That is,

$$\mathcal{W}_w = \rho V_w g$$

where

$$V_w = \tfrac{1}{12}\pi h(D_1^2 + D_2^2 + D_1 D_2)$$
$$= \frac{1}{12}\pi \frac{(30\ \text{mm})}{(1000\ \text{mm/m})}$$

$$\times \left[ \frac{(16\ \text{mm})^2 + (5\ \text{mm})^2 + (16\ \text{mm})(5\ \text{mm})}{(1000^2\ \text{mm}^2/\text{m}^2)} \right]$$
$$= 2.84 \times 10^{-6}\ \text{m}^3$$

Thus,

$$\mathcal{W}_w = (999\ \text{kg/m}^3)(2.84 \times 10^{-6}\ \text{m}^3)(9.81\ \text{m/s}^2)$$
$$= 0.0278\ \text{N} \quad (11)$$

The gage pressure at section (2), $p_2$, is zero since, as discussed in Section 3.6.1, when a subsonic flow discharges to the atmosphere as in the present situation, the discharge pressure is essentially atmospheric. The anchoring force, $F_A$, can now be determined from Eqs. 6 through 11 with

$$F_A = (0.599\ \text{kg/s})(2.98\ \text{m/s} - 30.6\ \text{m/s}) + 0.981\ \text{N}$$
$$+ (464\ \text{kPa})(1000\ \text{Pa/kPa})\frac{\pi(16\ \text{mm})^2}{4(1000^2\ \text{mm}^2/\text{m}^2)}$$
$$+ 0.0278\ \text{N} - 0$$

or

$$F_A = -16.5\ \text{N} + 0.981\ \text{N} + 93.3\ \text{N} + 0.0278\ \text{N}$$
$$= 77.8\ \text{N} \quad \textbf{(Ans)}$$

Since the anchoring force, $F_A$, is positive, it acts upward in the $z$ direction. The nozzle would be pushed off the pipe if it were not fastened securely.

**COMMENT**   The control volume selected above to solve problems such as these is not unique. The following is an alternate solution that involves two other control volumes—one containing only the nozzle and the other containing only the water in the nozzle. These control volumes are shown in Figs. E5.11c and E5.11d along with the vertical forces acting on the contents of each control volume. The new force involved, $R_z$, represents the interaction between the water and the conical inside surface of the nozzle. It includes the net pressure and viscous forces at this interface.

Application of Eq. 5.22 to the contents of the control volume of Fig. E5.11c leads to

$$F_A = \mathcal{W}_n + R_z - p_{\text{atm}}(A_1 - A_2) \quad (12)$$

■ **Figure E5.11c**

$(p_1 + p_{atm})A_1$

$w_1$

$\mathcal{W}_w$

$R_z$

(2)

$(p_2 + p_{atm})A_2$

$w_2$

■ **Figure E5.11d**

equal to the pressure times the projection of the surface area on a plane perpendicular to the axis of the nozzle. The projection of this area on a plane perpendicular to the $z$ direction is $A_1 - A_2$. The effect of the atmospheric pressure on the internal area (between the nozzle and the water) is already included in $R_z$ which represents the net force on this area.

Similarly, for the control volume of Fig. E5.11d we obtain

$$R_z = \dot{m}(w_1 - w_2) + \mathcal{W}_w + (p_1 + p_{atm})A_1 - (p_2 - p_{atm})A_2 \tag{13}$$

where $p_1$ and $p_2$ are gage pressures. From Eq. 13 it is clear that the value of $R_z$ depends on the value of the atmospheric pressure, $p_{atm}$, since $A_1 \neq A_2$. That is, we must use absolute pressure, not gage pressure, to obtain the correct value of $R_z$. From Eq. 13 we can easily identify which forces acting on the flowing fluid change its velocity magnitude and thus linear momentum.

By combining Eqs. 12 and 13 we obtain the same result for $F_A$ as before (Eq. 6):

$$F_A = \dot{m}(w_1 - w_2) + \mathcal{W}_n + p_1A_1 + W_w - p_2A_2$$

Note that although the force between the fluid and the nozzle wall, $R_z$, is a function of $p_{atm}$, the anchoring force, $F_A$, is not. That is, we were correct in using gage pressure when solving for $F_A$ by means of the original control volume shown in Fig. E5.11b.

The term $p_{atm}(A_1 - A_2)$ is the resultant force from the atmospheric pressure acting on the exterior surface of the nozzle (i.e., that portion of the surface of the nozzle that is not in contact with the water). Recall that the pressure force on a curved surface (such as the exterior surface of the nozzle) is

---

*V5.7 Air jet momentum flux*

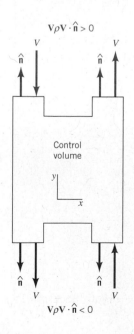

$\mathbf{V}\rho\mathbf{V}\cdot\hat{\mathbf{n}} > 0$

$V$    $V$

$\hat{\mathbf{n}}$    $\hat{\mathbf{n}}$

Control volume

$y$

$x$

$\hat{\mathbf{n}}$    $\hat{\mathbf{n}}$

$V$    $V$

$\mathbf{V}\rho\mathbf{V}\cdot\hat{\mathbf{n}} < 0$

Several important generalities about the application of the linear momentum equation (Eq. 5.22) are apparent in the example just considered.

1. When the flow is uniformly distributed over a section of the control surface where flow into or out of the control volume occurs, the integral operations are simplified. Thus, one-dimensional flows are easier to work with than flows involving nonuniform velocity distributions.

2. Linear momentum is directional; it can have components in as many as three orthogonal coordinate directions. Furthermore, along any one coordinate, the linear momentum of a fluid particle can be in the positive or negative direction and thus be considered as a positive or a negative quantity. In Example 5.11, only the linear momentum in the $z$ direction was considered (all of it was in the negative $z$ direction and was hence treated as being negative).

3. The flow of positive or negative linear momentum *into* a control volume involves a negative $\mathbf{V} \cdot \hat{\mathbf{n}}$ product. Momentum flow *out* of the control volume involves a positive $\mathbf{V} \cdot \hat{\mathbf{n}}$ product. The correct algebraic sign ($+$ or $-$) to assign to momentum flow ($\mathbf{V}\rho\mathbf{V} \cdot \hat{\mathbf{n}}\, dA$) will depend on the sense of the velocity ($+$ in positive coordinate direction, $-$ in negative coordinate direction) and the $\mathbf{V} \cdot \hat{\mathbf{n}}$ product ($+$ for flow out of the control volume, $-$ for flow into the control volume). This is shown in the figure in the margin. In Example 5.11, the momentum flow into the control volume past section (1) was a positive ($+$) quantity while the momentum flow out of the control volume at section (2) was a negative ($-$) quantity.

4. The time rate of change of the linear momentum of the contents of a nondeforming control volume (i.e., $\partial/\partial t \int_{cv} \mathbf{V}\rho\, d\mathcal{V}$) is zero for steady flow. The momentum problems considered in this text all involve steady flow.

5. If the control surface is selected so that it is perpendicular to the flow where fluid enters or leaves the control volume, the surface force exerted at these locations by fluid outside the control volume on fluid inside will be due to pressure, as shown by the figure in the margin on the next page. Furthermore, when subsonic flow exits from a control volume into the atmosphere, atmospheric pressure prevails at the exit cross section. In Example 5.11, the flow was subsonic and so we set the exit flow pressure at the atmospheric level. The continuity equation (Eq. 5.12) allowed us to evaluate the fluid flow velocities $w_1$ and $w_2$ at sections (1) and (2).

$p_{out}A_{out}$

$V_{out}$

(out)

Control volume

$V_{in}$ (in)

$p_{in}A_{in}$

*A control volume diagram is similar to a free-body diagram.*

6. The forces due to atmospheric pressure acting on the control surface may need consideration as indicated by Eq. 13 in Example 5.11 for the reaction force between the nozzle and the fluid. When calculating the anchoring force, $F_A$, the forces due to atmospheric pressure on the control surface cancel each other (for example, after combining Eqs. 12 and 13 the atmospheric pressure forces are no longer involved) and gage pressures may be used.

7. The external forces have an algebraic sign—positive if the force is in the assigned positive coordinate direction and negative otherwise.

8. Only external forces acting on the contents of the control volume are considered in the linear momentum equation (Eq. 5.22). If the fluid alone is included in a control volume, reaction forces between the fluid and the surface or surfaces in contact with the fluid [wetted surface(s)] will need to be in Eq. 5.22. If the fluid and the wetted surface or surfaces are within the control volume, the reaction forces between fluid and wetted surface(s) do not appear in the linear momentum equation (Eq. 5.22) because they are internal, not external forces. The anchoring force that holds the wetted surface(s) in place is an external force, however, and must therefore be in Eq. 5.22.

9. The force required to anchor an object will generally exist in response to surface pressure and/or shear forces acting on the control surface, to a change in linear momentum flow through the control volume containing the object, and to the weight of the object and the fluid contained in the control volume. In Example 5.11 the nozzle anchoring force was required mainly because of pressure forces and partly because of a change in linear momentum flow associated with accelerating the fluid in the nozzle. The weight of the water and the nozzle contained in the control volume influenced the size of the anchoring force only slightly.

## F l u i d s   i n   t h e   N e w s

**Motorized surfboard** When Bob Montgomery, a former professional surfer, started to design his motorized surfboard (called a jet board), he discovered that there were many engineering challenges to the design. The idea is to provide surfing to anyone, no matter where they live, near or far from the ocean. The rider stands on the device like a surfboard and steers it like a surfboard by shifting his or her body weight. A new, sleek, compact 45-hp engine and pump was designed to fit within the surfboard hull. Thrust is produced in response to the change in *linear momentum* of the water stream as it enters through the inlet passage and exits through an appropriately designed nozzle. Some of the fluid dynamic problems associated with designing the craft included one-way valves so that water does not get into the engine (at both the intake or exhaust ports), buoyancy, hydrodynamic lift, drag, thrust, and hull stability. (See Problem 5.79.)

To further demonstrate the use of the linear momentum equation (Eq. 5.22), we consider another one-dimensional flow example before moving on to other facets of this important equation.

## *E*XAMPLE 5.12    Linear Momentum—Pressure and Change in Flow Direction

**GIVEN** Water flows through a horizontal, 180° pipe bend as shown in Fig. E5.12a and illustrated in Fig. E5.12b. The flow cross-sectional area is constant at a value of 0.1 ft² through the bend. The magnitude of the flow velocity everywhere in the bend is axial and 50 ft/s. The absolute pressures at the entrance and exit of the bend are 30 psia and 24 psia, respectively.

**FIND** Calculate the horizontal (x and y) components of the anchoring force required to hold the bend in place.

(a)

■ **Figure E5.12**

# SOLUTION

Since we want to evaluate components of the anchoring force to hold the pipe bend in place, an appropriate control volume (see dashed line in Fig. E5.12b) contains the bend and the water in the bend at an instant. The horizontal forces acting on the contents of this control volume are identified in Fig. E5.12c. Note that the weight of the water is vertical (in the negative $z$ direction) and does not contribute to the $x$ and $y$ components of the anchoring force. All of the horizontal normal and tangential forces exerted on the fluid and the pipe bend are resolved and combined into the two resultant components, $F_{Ax}$ and $F_{Ay}$. These two forces act on the control volume contents, and thus for the $x$ direction, Eq. 5.22 leads to

$$\int_{cs} u\rho \mathbf{V} \cdot \hat{\mathbf{n}} \, dA = F_{Ax} \tag{1}$$

At sections (1) and (2), the flow is in the $y$ direction and therefore $u = 0$ at both cross sections. There is no $x$ direction momentum flow into or out of the control volume and we conclude from Eq. 1 that

$$F_{Ax} = 0 \tag{Ans}$$

For the $y$ direction, we get from Eq. 5.22

$$\int_{cs} v\rho \mathbf{V} \cdot \hat{\mathbf{n}} \, dA = F_{Ay} + p_1 A_1 + p_2 A_2 \tag{2}$$

For one-dimensional flow, the surface integral in Eq. 2 is easy to evaluate and Eq. 2 becomes

$$(+v_1)(-\dot{m}_1) + (-v_2)(+\dot{m}_2) = F_{Ay} + p_1 A_1 + p_2 A_2 \tag{3}$$

Note that the $y$ component of velocity is positive at section (1) but is negative at section (2). Also, the mass flowrate term is negative at section (1) (flow in) and is positive at section (2) (flow out). From the continuity equation (Eq. 5.12), we get

$$\dot{m} = \dot{m}_1 = \dot{m}_2 \tag{4}$$

and thus Eq. 3 can be written as

$$-\dot{m}(v_1 + v_2) = F_{Ay} + p_1 A_1 + p_2 A_2 \tag{5}$$

Solving Eq. 5 for $F_{Ay}$ we obtain

$$F_{Ay} = -\dot{m}(v_1 + v_2) - p_1 A_1 - p_2 A_2 \tag{6}$$

From the given data we can calculate the mass flowrate, $\dot{m}$, from Eq. 5.6 as

$$\dot{m} = \rho_1 A_1 v_1 = (1.94 \text{ slugs/ft}^3)(0.1 \text{ ft}^2)(50 \text{ ft/s})$$
$$= 9.70 \text{ slugs/s}$$

For determining the anchoring force, $F_{Ay}$, the effects of atmospheric pressure cancel, and thus gage pressures for $p_1$ and $p_2$ are appropriate. By substituting numerical values of variables into Eq. 6 and using the fact that 1 lb = 1 slug · ft/s², we get

$$F_{Ay} = -(9.70 \text{ slugs/s})(50 \text{ ft/s} + 50 \text{ ft/s})$$
$$- (30 \text{ psia} - 14.7 \text{ psia})(144 \text{ in.}^2/\text{ft}^2)(0.1 \text{ ft}^2)$$
$$- (24 \text{ psia} - 14.7 \text{ psia})(144 \text{ in.}^2/\text{ft}^2)(0.1 \text{ ft}^2)$$
$$F_{Ay} = -970 \text{ lb} - 220 \text{ lb} - 134 \text{ lb} = -1324 \text{ lb} \quad \text{(Ans)}$$

The negative sign for $F_{Ay}$ is interpreted as meaning that the $y$ component of the anchoring force is actually in the negative $y$ direction, not the positive $y$ direction as originally indicated in Fig. E5.12c.

**COMMENT** As with Example 5.11, the anchoring force for the pipe bend is independent of the atmospheric pressure. However, the force that the bend exerts on the fluid inside of it, $R_y$, depends on the atmospheric pressure. We can see this by using a control volume which surrounds only the fluid within the bend as shown in Fig. E5.12d. Application of the momentum equation to this situation gives

$$R_y = -\dot{m}(v_1 + v_2) - p_1 A_1 - p_2 A_2$$

where $p_1$ and $p_2$ must be in terms of absolute pressure because the force between the fluid and the pipe wall, $R_y$, is the complete pressure effect (i.e., absolute pressure). We see that forces exerted on the flowing fluid result in a change in its velocity direction (a change in linear momentum).

Thus, we obtain

$$R_y = -(9.70 \text{ slugs/s})(50 \text{ ft/s} + 50 \text{ ft/s})$$
$$- (30 \text{ psia})(144 \text{ in.}^2/\text{ft}^2)(0.1 \text{ ft}^2)$$
$$- (24 \text{ psia})(144 \text{ in.}^2/\text{ft}^2)(0.1 \text{ ft}^2) \tag{7}$$
$$= -1748 \text{ lb}$$

We can use the control volume that includes just the pipe bend (without the fluid inside it) as shown in Fig. E5.12e to

(b)

(c)

■ **Figure E5.12**   cont.

determine $F_{Ay}$, the anchoring force component in the $y$ direction necessary to hold the bend stationary. The $y$ component of the momentum equation applied to this control volume gives

$$F_{Ay} = R_y + p_{atm}(A_1 + A_2) \qquad (8)$$

where $R_y$ is given by Eq. 7. The $p_{atm}(A_1 + A_2)$ term represents the net pressure force on the outside portion of the control volume. Recall that the pressure force on the inside of the bend is ac-

counted for by $R_y$. By combining Eqs. 7 and 8 and using the fact that $p_{atm} = 14.7 \text{ lb/in.}^2 (144 \text{ in.}^2/\text{ft}^2) = 2117 \text{ lb/ft}^2$, we obtain

$$F_{Ay} = -1748 \text{ lb} + 2117 \text{ lb/ft}^2 (0.1 \text{ ft}^2 + 0.1 \text{ ft}^2)$$
$$= -1324 \text{ lb}$$

in agreement with the original answer obtained using the control volume of Fig. E5.12c.

(d)    (e)

■ **Figure E5.12**    cont.

*V5.8 Fire hose*

In Examples 5.10 and 5.12 the force exerted on a flowing fluid resulted in a change in flow direction only. This force was associated with constraining the flow, with a vane in Example 5.10, and with a pipe bend in Example 5.12. In Example 5.11 the force exerted on a flowing fluid resulted in a change in velocity magnitude only. This force was associated with a converging nozzle. Anchoring forces are required to hold a vane or conduit stationary. They are most easily estimated with a control volume that contains the vane or conduit and the flowing fluid involved. Alternately, two separate control volumes can be used, one containing the vane or conduit only and one containing the flowing fluid only.

---

**$E$XAMPLE 5.13**    **Linear Momentum—Pressure, Change in Speed, and Friction**

**GIVEN**    Air flows steadily between two cross sections in a long, straight portion of 4-in. inside diameter pipe as indicated in Fig. E5.13, where the uniformly distributed temperature and pressure at each cross section are given. If the average air velocity at section (2) is 1000 ft/s, we found in Example 5.2 that the average air velocity at section (1) must be 219 ft/s. Assume uniform velocity distributions at sections (1) and (2).

**FIND**    Determine the frictional force exerted by the pipe wall on the airflow between sections (1) and (2).

■ **Figure E5.13**

**SOLUTION**

The control volume of Example 5.2 is appropriate for this problem. The forces acting on the air between sections (1) and (2) are identified in Fig. E5.13. The weight of air is considered negligibly small. The reaction force between the wetted wall of the pipe and the flowing air, $R_x$, is the frictional force sought. Application of the axial component of Eq. 5.22 to this control volume yields

$$\int_{cs} u\rho \mathbf{V} \cdot \hat{\mathbf{n}} \, dA = -R_x + p_1 A_1 - p_2 A_2 \qquad (1)$$

The positive $x$ direction is set as being to the right. Furthermore, for uniform velocity distributions (one-dimensional flow), Eq. 1 becomes

$$(+u_1)(-\dot{m}_1) + (+u_2)(+\dot{m}_2) = -R_x + p_1 A_1 - p_2 A_2 \qquad (2)$$

From conservation of mass (Eq. 5.12) we get

$$\dot{m} = \dot{m}_1 = \dot{m}_2 \tag{3}$$

so that Eq. 2 becomes

$$\dot{m}(u_2 - u_1) = -R_x + A_2(p_1 - p_2) \tag{4}$$

Solving Eq. 4 for $R_x$, we get

$$R_x = A_2(p_1 - p_2) - \dot{m}(u_2 - u_1) \tag{5}$$

The equation of state gives

$$\rho_2 = \frac{p_2}{RT_2} \tag{6}$$

and the equation for area $A_2$ is

$$A_2 = \frac{\pi D_2^2}{4} \tag{7}$$

Thus, from Eqs. 3, 6, and 7

$$\dot{m} = \left(\frac{p_2}{RT_2}\right)\left(\frac{\pi D_2^2}{4}\right)u_2$$

English Engineering (EE) units are often used for this kind of flow. The gas constant, $R$, for air in EE units is

$$R = \frac{1716(\text{ft} \cdot \text{lb})/(\text{slug} \cdot {}^\circ\text{R})}{32.174(\text{lbm/slug})} = 53.3(\text{ft} \cdot \text{lb})/(\text{lbm} \cdot {}^\circ\text{R})$$

Hence, $\dot{m} = \dfrac{(18.4 \text{ psia})(144 \text{ in.}^2/\text{ft}^2)}{[53.3(\text{ft} \cdot \text{lb})/(\text{lbm} \cdot {}^\circ\text{R})](453 \,{}^\circ\text{R})}$

$$\times \frac{\pi(4 \text{ in.})^2}{4(144 \text{ in.}^2/\text{ft}^2)}(1000 \text{ ft/s}) = 9.57 \text{ lbm/s} \tag{8}$$

Thus, from Eqs. 5 and 8

$$R_x = \frac{\pi(4 \text{ in.})^2}{4}(100 \text{ psia} - 18.4 \text{ psia})$$
$$- (9.57 \text{ lbm})(1000 \text{ ft/s} - 219 \text{ ft/s})/$$
$$32.174(\text{lbm} \cdot \text{ft})/(\text{lb} \cdot \text{s}^2)$$
$$= 1025 \text{ lb} - 232 \text{ lb}$$

or

$$R_x = 793 \text{ lb} \qquad \textbf{(Ans)}$$

**COMMENT** For this compressible flow, the pressure difference drives the motion, which results in a frictional force, $R_x$, and an acceleration of the fluid (i.e., a velocity magnitude increase). For a similar incompressible pipe flow, a pressure difference results in fluid motion with a frictional force only (i.e., no change in velocity magnitude).

---

## EXAMPLE 5.14   Linear Momentum—Weight, Pressure, Friction, and Nonuniform Velocity Profile

**GIVEN** Consider the flow of Example 5.4 to be vertically upward as shown by Fig. E5.14.

**FIND** Develop an expression for the fluid pressure drop that occurs between sections (1) and (2).

### SOLUTION

A control volume (see dashed lines in Fig. E5.14) that includes only fluid from section (1) to section (2) is selected. The forces acting on the fluid in this control volume are identified in Fig. E5.14. The application of the axial component of Eq. 5.22 to the fluid in this control volume results in

$$\int_{cs} w\rho \mathbf{V} \cdot \hat{\mathbf{n}} \, dA = p_1 A_1 - R_z - \mathcal{W} - p_2 A_2 \tag{1}$$

where $R_z$ is the resultant force of the wetted pipe wall on the fluid. Further, for uniform flow at section (1), and because the flow at section (2) is out of the control volume, Eq. 1 becomes

$$(+w_1)(-\dot{m}_1) + \int_{A_2} (+w_2)\rho(+w_2 \, dA_2) = p_1 A_1 - R_z \\ - \mathcal{W} - p_2 A_2 \tag{2}$$

The positive direction is considered up. The surface integral over the cross-sectional area at section (2), $A_2$, is evaluated by using the parabolic velocity profile obtained in Example 5.4, $w_2 = 2w_1[1 - (r/R)^2]$, as

$$\int_{A_2} w_2 \rho w_2 \, dA_2 = \rho \int_0^R w_2^2 \, 2\pi r \, dr$$
$$= 2\pi\rho \int_0^R (2w_1)^2 \left[1 - \left(\frac{r}{R}\right)^2\right]^2 r \, dr$$

or

$$\int_{A_2} w_2 \rho w_2 \, dA_2 = 4\pi\rho w_1^2 \frac{R^2}{3} \tag{3}$$

Combining Eqs. 2 and 3 we obtain

$$-w_1^2 \rho\pi R^2 + \tfrac{4}{3}w_1^2 \rho\pi R^2 = p_1 A_1 - R_z - \mathcal{W} - p_2 A_2 \tag{4}$$

Solving Eq. 4 for the pressure drop from section (1) to section (2), $p_1 - p_2$, we obtain

$$p_1 - p_2 = \frac{\rho w_1^2}{3} + \frac{R_z}{A_1} + \frac{\mathcal{W}}{A_1} \qquad \textbf{(Ans)}$$

**COMMENT**    We see that the drop in pressure from section (1) to section (2) occurs because of the following:

1. The change in momentum flow between the two sections associated with going from a uniform velocity profile to a parabolic velocity profile, $\rho w_1^2/3$

2. Pipe wall friction, $R_z$

3. The weight of the water column, $\mathcal{W}$; a hydrostatic pressure effect

If the velocity profiles had been identically parabolic at sections (1) and (2), the momentum flowrate at each section would have been identical, a condition we call "fully developed" flow. Then the pressure drop, $p_1 - p_2$, would be due only to pipe wall friction and the weight of the water column. If in addition to being fully developed, the flow involved negligible weight effects (for example, horizontal flow of liquids or the flow of gases in any direction), the drop in pressure between any two sections, $p_1 - p_2$, would be a result of pipe wall friction only.

Note that although the average velocity is the same at section (1) as it is at section (2) $(\overline{V}_1 = \overline{V}_2 = w_1)$, the momentum flux across section (1) is not the same as it is across section (2). If it were, the left-hand side of Eq. (4) would be zero. For this nonuniform flow the momentum flux can be written in terms of the average velocity, $\overline{V}$, and the *momentum coefficient*, $\beta$, as

$$\beta = \frac{\int w \rho \mathbf{V} \cdot \hat{\mathbf{n}} \, dA}{\rho \overline{V}^2 A}$$

Hence the momentum flux can be written as

$$\int_{cs} w \rho \mathbf{V} \cdot \hat{\mathbf{n}} \, dA = -\beta_1 w_1^2 \rho \pi R^2 + \beta_2 w_1^2 \rho \pi R^2$$

where $\beta_1 = 1$ ($\beta = 1$ for uniform flow) and $\beta_2 = 4/3$ ($\beta > 1$ for any nonuniform flow).

■ **Figure E5.14**

---

## EXAMPLE 5.15    Linear Momentum—Thrust

**GIVEN**    A static thrust stand as sketched in Fig. E5.15 is to be designed for testing a jet engine. The following conditions are known for a typical test: Intake air velocity = 200 m/s; exhaust gas velocity = 500 m/s; intake cross-sectional area = 1 m²; intake static pressure = −22.5 kPa = 78.5 kPa (abs); intake static temperature = 268 K; exhaust static pressure = 0 kPa = 101 kPa (abs).

**FIND**    Estimate the nominal anchoring force for which to design.

## SOLUTION

The cylindrical control volume outlined with a dashed line in Fig. E5.15 is selected. The external forces acting in the axial direction are also shown. Application of the momentum equation (Eq. 5.22) to the contents of this control volume yields

$$\int_{cs} u \rho \mathbf{V} \cdot \hat{\mathbf{n}} \, dA = p_1 A_1 + F_{th} - p_2 A_2$$
$$- p_{atm}(A_1 - A_2) \qquad (1)$$

■ **Figure E5.15**

where the pressures are absolute. Thus, for one-dimensional flow, Eq. 1 becomes

$$(+u_1)(-\dot{m}_1) + (+u_2)(+\dot{m}_2) = (p_1 - p_{atm})A_1$$
$$- (p_2 - p_{atm})A_2 + F_{th} \qquad (2)$$

The positive direction is to the right. The conservation of mass equation (Eq. 5.12) leads to

$$\dot{m} = \dot{m}_1 = \rho_1 A_1 u_1 = \dot{m}_2 = \rho_2 A_2 u_2 \tag{3}$$

Combining Eqs. 2 and 3 and using gage pressure we obtain

$$\dot{m}(u_2 - u_1) = p_1 A_1 - p_2 A_2 + F_{th} \tag{4}$$

Solving Eq. 4 for the thrust force, $F_{th}$, we obtain

$$F_{th} = -p_1 A_1 + p_2 A_2 + \dot{m}(u_2 - u_1) \tag{5}$$

We need to determine the mass flowrate, $\dot{m}$, to calculate $F_{th}$, and to calculate $\dot{m} = \rho_1 A_1 u_1$, we need $\rho_1$. From the ideal gas equation of state

$$\rho_1 = \frac{p_1}{RT_1} = \frac{(78.5 \text{ kPa})(1000 \text{ Pa/kPa})[1(\text{N/m}^2)/\text{Pa}]}{(286.9 \text{ J/kg} \cdot \text{K})(268 \text{ K})(1 \text{ N} \cdot \text{m/J})}$$
$$= 1.02 \text{ kg/m}^3$$

Thus,

$$\dot{m} = \rho_1 A_1 u_1 = (1.02 \text{ kg/m}^3)(1 \text{ m}^2)(200 \text{ m/s})$$
$$= 204 \text{ kg/s} \tag{6}$$

Finally, combining Eqs. 5 and 6 and substituting given data with $p_2 = 0$, we obtain

$$F_{th} = -(1 \text{ m}^2)(-22.5 \text{ kPa})(1000 \text{ Pa/kPa})[1(\text{N/m}^2)/\text{Pa}]$$
$$+ (204 \text{ kg/s})(500 \text{ m/s} - 200 \text{ m/s})[1 \text{ N}/(\text{kg} \cdot \text{m/s}^2)]$$

or

$$F_{th} = 22{,}500 \text{ N} + 61{,}200 \text{ N} = 83{,}700 \text{ N} \tag{Ans}$$

**COMMENT** The force of the thrust stand on the engine is directed toward the right. Conversely, the engine pushes to the left on the thrust stand (or aircraft).

---

## F l u i d s   i n   t h e   N e w s

**Bow thrusters** In the past, large ships required the use of tugboats for precise maneuvering, especially when docking. Nowadays, most large ships (and many moderate to small ones as well) are equipped with bow thrusters to help steer in close quarters. The units consist of a mechanism (usually a ducted propeller mounted at right angles to the fore/aft axis of the ship) that takes water from one side of the bow and ejects it as a water jet on the other side. The *momentum flux* of this jet produces a starboard or port force on the ship for maneuvering. Sometimes a second unit is installed in the stern. Initially used in the bows of ferries, these versatile control devices have became popular in offshore oil servicing boats, fishing vessels, and larger oceangoing craft. They permit unassisted maneuvering alongside of oil rigs, vessels, loading platforms, fishing nets, and docks. They also provide precise control at slow speeds through locks, narrow channels, and bridges, where the rudder becomes very ineffective. (See Problem 5.80.)

---

## $E$XAMPLE 5.16   Linear Momentum—Nonuniform Pressure

**GIVEN** A sluice gate across a channel (Fig. E5.16a) of width $b$ is shown in the closed and open positions in Figs. E5.16b and E5.16c.

**FIND** Is the anchoring force required to hold the gate in place larger when the gate is closed or when it is open?

(a)

■ **Figure E5.16a**
(Photograph courtesy of Pend Oreille Public Utility District.)

### SOLUTION

We will answer this question by comparing expressions for the horizontal reaction force, $R_x$, between the gate and the water when the gate is closed and when the gate is open. The control volume used in each case is indicated with dashed lines in Figs. E5.16b and E5.16c.

When the gate is closed, the horizontal forces acting on the contents of the control volume are identified in Fig. E5.16d.

Application of Eq. 5.22 to the contents of this control volume yields

$$\int_{cs} u\rho \overset{0 \text{ (no flow)}}{\mathbf{V} \cdot \hat{\mathbf{n}}} \, dA = \tfrac{1}{2}\gamma H^2 b - R_x \tag{1}$$

Note that the hydrostatic pressure force, $\gamma H^2 b/2$, is used. From Eq. 1, the force exerted on the water by the gate (which is equal to the force necessary to hold the gate stationary) is

$$R_x = \tfrac{1}{2}\gamma H^2 b \qquad (2)$$

which is equal in magnitude to the hydrostatic force exerted on the gate by the water.

When the gate is open, the horizontal forces acting on the contents of the control volume are shown in Fig. E5.16e. Application of Eq. 5.22 to the contents of this control volume leads to

$$\int_{cs} u\rho \mathbf{V} \cdot \hat{\mathbf{n}}\, dA = \tfrac{1}{2}\gamma H^2 b - R_x - \tfrac{1}{2}\gamma h^2 b - F_f \qquad (3)$$

Note that because the water at sections (1) and (2) is flowing along straight, horizontal streamlines, the pressure distribution at those locations is hydrostatic, varying from zero at the free surface to $\gamma$ times the water depth at the bottom of the channel (see Chapter 3, Section 3.4). Thus, the pressure forces at sections (1) and (2) (given by the pressure at the centroid times the area) are $\gamma H^2 b/2$ and $\gamma h^2 b/2$, respectively. Also the frictional force between the channel bottom and the water is specified as $F_f$. The surface integral in Eq. 3 is nonzero only where there is flow across the control surface. With the assumption of uniform velocity distributions,

$$\int_{cs} u\rho \mathbf{V} \cdot \hat{\mathbf{n}}\, dA = (u_1)\rho(-u_1)Hb + (+u_2)\rho(+u_2)hb \qquad (4)$$

Thus, Eqs. 3 and 4 combine to form

$$-\rho u_1^2 Hb + \rho u_2^2 hb = \tfrac{1}{2}\gamma H^2 b - R_x - \tfrac{1}{2}\gamma h^2 b - F_f \qquad (5)$$

If $H \gg h$, the upstream velocity, $u_1$, is much less than $u_2$ so that the contribution of the incoming momentum flow to the control surface integral can be neglected and from Eq. 5 we obtain

$$R_x = \tfrac{1}{2}\gamma H^2 b - \tfrac{1}{2}\gamma h^2 b - F_f - \rho u_2^2 hb \qquad (6)$$

**■ Figure E5.16**

By using the continuity equation, $\dot{m} = \rho b H u_1 = \rho b h u_2$, Eq. 6 can be rewritten as

$$R_x = \tfrac{1}{2}\gamma H^2 b - \tfrac{1}{2}\gamma h^2 b - F_f - \dot{m}(u_2 - u_1) \qquad (7)$$

Hence, since $u_2 > u_1$, by comparing the expressions for $R_x$ (Eqs. 2 and 7), we conclude that the reaction force between the gate and the water (and therefore the anchoring force required to hold the gate in place) is smaller when the gate is open than when it is closed.    **(Ans)**

---

*The linear momentum equation can be written for a moving control volume.*

All of the linear momentum examples considered thus far have involved stationary and nondeforming control volumes which are thus inertial because there is no acceleration. A nondeforming control volume translating in a straight line at constant speed is also inertial because there is no acceleration. For a system and an inertial, moving, nondeforming control volume that are both coincident at an instant of time, the Reynolds transport theorem (Eq. 4.23) is

$$\frac{D}{Dt}\int_{sys} \mathbf{V}\rho\, d\forall = \frac{\partial}{\partial t}\int_{cv} \mathbf{V}\rho\, d\forall + \int_{cs} \mathbf{V}\rho \mathbf{W} \cdot \hat{\mathbf{n}}\, dA \qquad (5.23)$$

where $\mathbf{W}$ is the relative velocity. When we combine Eq. 5.23 with Eqs. 5.19 and 5.20, we get

$$\frac{\partial}{\partial t}\int_{cv} \mathbf{V}\rho\, d\forall + \int_{cs} \mathbf{V}\rho \mathbf{W} \cdot \hat{\mathbf{n}}\, dA = \sum \mathbf{F}_{\substack{\text{contents of the}\\ \text{control volume}}} \qquad (5.24)$$

When the equation relating absolute, relative, and control volume velocities (Eq. 5.14) is used with Eq. 5.24, the result is

$$\frac{\partial}{\partial t}\int_{cv} (\mathbf{W} + \mathbf{V}_{cv})\rho\, d\forall + \int_{cs} (\mathbf{W} + \mathbf{V}_{cv})\rho \mathbf{W} \cdot \hat{\mathbf{n}}\, dA = \sum \mathbf{F}_{\substack{\text{contents of the}\\ \text{control volume}}} \qquad (5.25)$$

*V5.9 Jellyfish*

*V5.10 Running on water*

For a constant control volume velocity, $\mathbf{V}_{cv}$, and steady flow in the control volume reference frame,

$$\frac{\partial}{\partial t} \int_{cv} (\mathbf{W} + \mathbf{V}_{cv}) \rho \, d\mathbf{V} = 0 \tag{5.26}$$

Also, for this inertial, nondeforming control volume

$$\int_{cs} (\mathbf{W} + \mathbf{V}_{cv}) \rho \mathbf{W} \cdot \hat{\mathbf{n}} \, dA = \int_{cs} \mathbf{W} \rho \mathbf{W} \cdot \hat{\mathbf{n}} \, dA + \mathbf{V}_{cv} \int_{cs} \rho \mathbf{W} \cdot \hat{\mathbf{n}} \, dA \tag{5.27}$$

For steady flow (on an instantaneous or time-average basis), Eq. 5.15 gives

$$\int_{cs} \rho \mathbf{W} \cdot \hat{\mathbf{n}} \, dA = 0 \tag{5.28}$$

Combining Eqs. 5.25, 5.26, 5.27, and 5.28, we conclude that the linear momentum equation for an inertial, moving, nondeforming control volume that involves steady (instantaneous or time-average) flow is

*The linear momentum equation for a moving control volume involves the relative velocity.*

$$\boxed{\int_{cs} \mathbf{W} \rho \mathbf{W} \cdot \hat{\mathbf{n}} \, dA = \sum \mathbf{F}_{\substack{\text{contents of the} \\ \text{control volume}}}} \tag{5.29}$$

Example 5.17 illustrates the use of Eq. 5.29.

---

# EXAMPLE 5.17   Linear Momentum—Moving Control Volume

**GIVEN**  A vane on wheels moves with constant velocity $\mathbf{V}_0$ when a stream of water having a nozzle exit velocity of $\mathbf{V}_1$ is turned 45° by the vane as indicated in Fig. E5.17a. Note that this is the same moving vane considered in Section 4.4.6 earlier. The speed of the water jet leaving the nozzle is 100 ft/s,

and the vane is moving to the right with a constant speed of 20 ft/s.

**FIND**  Determine the magnitude and direction of the force, **R**, exerted by the stream of water on the vane surface.

■ **Figure E5.17**

## SOLUTION

To determine the magnitude and direction of the force, **R**, exerted by the water on the vane, we apply Eq. 5.29 to the contents of the moving control volume shown in Fig. E5.17$b$. The forces acting on the contents of this control volume are indicated in Fig. E5.17$c$. Note that since the ambient pressure is atmospheric, all pressure forces cancel each other out. Equation 5.29 is applied to the contents of the moving control volume in component directions. For the $x$ direction (positive to the right), we get

$$\int_{cs} W_x \rho \, \mathbf{W} \cdot \hat{\mathbf{n}} \, dA = -R_x$$

or

$$(+W_1)(-\dot{m}_1) + (+W_2 \cos 45°)(+\dot{m}_2) = -R_x \quad (1)$$

where

$$\dot{m}_1 = \rho_1 W_1 A_1 \quad \text{and} \quad \dot{m}_2 = \rho_2 W_2 A_2.$$

For the vertical or $z$ direction (positive up) we get

$$\int_{cs} W_z \rho \mathbf{W} \cdot \hat{\mathbf{n}} \, dA = R_z - \mathcal{W}_w$$

or

$$(+W_2 \sin 45°)(+\dot{m}_2) = R_z - \mathcal{W}_w \quad (2)$$

We assume for simplicity that the water flow is frictionless and that the change in water elevation across the vane is negligible. Thus, from the Bernoulli equation (Eq. 3.7) we conclude that the speed of the water relative to the moving control volume, $W$, is constant or

$$W_1 = W_2$$

The relative speed of the stream of water entering the control volume, $W_1$, is

$$W_1 = V_1 - V_0 = 100 \text{ ft/s} - 20 \text{ ft/s} = 80 \text{ ft/s} = W_2$$

The water density is constant so that

$$\rho_1 = \rho_2 = 1.94 \text{ slugs/ft}^3$$

Application of the conservation of mass principle to the contents of the moving control volume (Eq. 5.16) leads to

$$\dot{m}_1 = \rho_1 W_1 A_1 = \rho_2 W_2 A_2 = \dot{m}_2$$

Combining results we get

$$R_x = \rho W_1^2 A_1 \, (1 - \cos 45°)$$

or

$$R_x = (1.94 \text{ slugs/ft}^3)(80 \text{ ft/s})^2(0.006 \text{ ft}^2)(1 - \cos 45°)$$
$$= 21.8 \text{ lb}$$

Also,

$$R_z = \rho W_1^2(\sin 45°)A_1 + \mathcal{W}_w$$

where

$$\mathcal{W}_w = \rho g A_1 \ell$$

Thus,

$$R_z = (1.94 \text{ slugs/ft}^3)(80 \text{ ft/s})^2(\sin 45°)(0.006 \text{ ft}^2)$$
$$+ (62.4 \text{ lb/ft}^3)(0.006 \text{ ft}^2)(1 \text{ ft})$$
$$= 52.6 \text{ lb} + 0.37 \text{ lb} = 53 \text{ lb}$$

Combining the components we get

$$R = \sqrt{R_x^2 + R_z^2} = [(21.8 \text{ lb})^2 + (53 \text{ lb})^2]^{1/2} = 57.3 \text{ lb}$$

The angle of **R** from the $x$ direction, $\alpha$, is

$$\alpha = \tan^{-1}\frac{R_z}{R_x} = \tan^{-1}(53 \text{ lb}/21.8 \text{ lb}) = 67.6°$$

The force of the water on the vane is equal in magnitude but opposite in direction from **R**; thus it points to the right and down at an angle of $67.6°$ from the $x$ direction and is equal in magnitude to 57.3 lb. **(Ans)**

**COMMENT** The force of the fluid on the vane in the $x$ direction, $R_x = 21.8$ lb, is associated with $x$-direction motion of the vane at a constant speed of 20 ft/s. Since the vane is not accelerating, this $x$-direction force is opposed mainly by a wheel friction force of the same magnitude. From basic physics we recall that the power this situation involves is the product of force and speed. Thus,

$$\mathcal{P} = R_x V_0$$
$$= \frac{(21.8 \text{ lb})(20 \text{ ft/s})}{550(\text{ft} \cdot \text{lb})/(\text{hp} \cdot \text{s})}$$
$$= 0.79 \text{ hp}$$

All of this power is consumed by friction.

---

*V5.11 Thrust from water nozzle*

It is clear from the preceding examples that a flowing fluid can be forced to

1. change direction
2. speed up or slow down
3. have a velocity profile change
4. do only some or all of the above
5. do none of the above

A net force on the fluid is required for achieving any or all of the first four above. The forces on a flowing fluid balance out with no net force for the fifth.

Typical forces considered in this book include

(a) pressure

(b) friction

(c) weight

and involve some type of constraint such as a vane, channel, or conduit to guide the flowing fluid. A flowing fluid can cause a vane, channel, or conduit to move. When this happens, power is produced.

The selection of a control volume is an important matter. For determining anchoring forces, consider including fluid and its constraint in the control volume. For determining force between a fluid and its constraint, consider including only the fluid in the control volume.

### 5.2.3 Derivation of the Moment-of-Momentum Equation[1]

In many engineering problems, the moment of a force with respect to an axis, namely, *torque*, is important. Newton's second law of motion has already led to a useful relationship between forces and linear momentum flow. The linear momentum equation can also be used to solve problems involving torques. However, by forming the moment of the linear momentum and the moment of the resultant force associated with each particle of fluid with respect to a point in an inertial coordinate system, we will develop a ***moment-of-momentum equation*** that relates *torques* and *angular momentum flow* for the contents of a control volume. When torques are important, the moment-of-momentum equation is often more convenient to use than the linear momentum equation.

*The angular momentum equation is derived from Newton's second law.*

Application of Newton's second law of motion to a particle of fluid yields

$$\frac{D}{Dt}(\mathbf{V}\rho\,\delta\!\!\!\forall) = \delta\mathbf{F}_{\text{particle}} \tag{5.30}$$

where $\mathbf{V}$ is the particle velocity measured in an inertial reference system, $\rho$ is the particle density, $\delta\!\!\!\forall$ is the infinitesimally small particle volume, and $\delta\mathbf{F}_{\text{particle}}$ is the resultant external force acting on the particle. If we form the moment of each side of Eq. 5.30 with respect to the origin of an inertial coordinate system, we obtain

$$\mathbf{r} \times \frac{D}{Dt}(\mathbf{V}\rho\,\delta\!\!\!\forall) = \mathbf{r} \times \delta\mathbf{F}_{\text{particle}} \tag{5.31}$$

where $\mathbf{r}$ is the position vector from the origin of the inertial coordinate system to the fluid particle (Fig. 5.3). We note that

$$\frac{D}{Dt}[(\mathbf{r} \times \mathbf{V})\rho\,\delta\!\!\!\forall] = \frac{D\mathbf{r}}{Dt} \times \mathbf{V}\rho\,\delta\!\!\!\forall + \mathbf{r} \times \frac{D(\mathbf{V}\rho\,\delta\!\!\!\forall)}{Dt} \tag{5.32}$$

and

$$\frac{D\mathbf{r}}{Dt} = \mathbf{V} \tag{5.33}$$

Thus, since

$$\mathbf{V} \times \mathbf{V} = 0 \tag{5.34}$$

by combining Eqs. 5.31, 5.32, 5.33, and 5.34, we obtain the expression

$$\frac{D}{Dt}[(\mathbf{r} \times \mathbf{V})\rho\,\delta\!\!\!\forall] = \mathbf{r} \times \delta\mathbf{F}_{\text{particle}} \tag{5.35}$$

■ **Figure 5.3** Inertial coordinate system.

---

[1]This section may be omitted, along with Sections 5.2.4 and 5.3.5, without loss of continuity in the text material. However, these sections are recommended for those interested in Chapter 12.

Equation 5.35 is valid for every particle of a system. For a system (collection of fluid particles), we need to use the sum of both sides of Eq. 5.35 to obtain

$$\int_{sys} \frac{D}{Dt}[(\mathbf{r} \times \mathbf{V})\rho \, d\Psi] = \sum (\mathbf{r} \times \mathbf{F})_{sys} \tag{5.36}$$

where

$$\sum \mathbf{r} \times \delta\mathbf{F}_{particle} = \sum (\mathbf{r} \times \mathbf{F})_{sys} \tag{5.37}$$

We note that

$$\frac{D}{Dt} \int_{sys} (\mathbf{r} \times \mathbf{V})\rho \, d\Psi = \int_{sys} \frac{D}{Dt}[(\mathbf{r} \times \mathbf{V})\rho \, d\Psi] \tag{5.38}$$

since the sequential order of differentiation and integration can be reversed without consequence. (Recall that the material derivative, $D(\;)/Dt$, denotes the time derivative following a given system; see Section 4.2.1.) Thus, from Eqs. 5.36 and 5.38 we get

$$\frac{D}{Dt} \int_{sys} (\mathbf{r} \times \mathbf{V})\rho \, d\Psi = \sum (\mathbf{r} \times \mathbf{F})_{sys} \tag{5.39}$$

or

Time rate of change of the $\quad$ = sum of external torques
moment-of-momentum of the system $\quad$ acting on the system

The sketch in the margin illustrates what torque, $\mathbf{T} = \mathbf{r} \times \mathbf{F}$, is. For a control volume that is instantaneously coincident with the system, the torques acting on the system and on the control volume contents will be identical:

$$\sum (\mathbf{r} \times \mathbf{F})_{sys} = \sum (\mathbf{r} \times \mathbf{F})_{cv} \tag{5.40}$$

Further, for the system and the contents of the coincident control volume that is fixed and nondeforming, the Reynolds transport theorem (Eq. 4.19) leads to

$$\frac{D}{Dt} \int_{sys} (\mathbf{r} \times \mathbf{V})\rho \, d\Psi = \frac{\partial}{\partial t} \int_{cv} (\mathbf{r} \times \mathbf{V})\rho \, d\Psi + \int_{cs} (\mathbf{r} \times \mathbf{V})\rho \mathbf{V} \cdot \hat{\mathbf{n}} \, dA \tag{5.41}$$

or

Time rate of change $\quad$ time rate of change $\quad$ net rate of flow
of the moment-of- $\quad$ of the moment-of- $\quad$ of the moment-of-
momentum of the $\quad$ = momentum of the $\quad$ + momentum through
system $\quad$ contents of the $\quad$ the control surface
$\quad$ control volume

For a control volume that is fixed (and therefore inertial) and nondeforming, we combine Eqs. 5.39, 5.40, and 5.41 to obtain the moment-of-momentum equation:

$$\boxed{\frac{\partial}{\partial t} \int_{cv} (\mathbf{r} \times \mathbf{V})\rho \, d\Psi + \int_{cs} (\mathbf{r} \times \mathbf{V})\rho \mathbf{V} \cdot \hat{\mathbf{n}} \, dA = \sum (\mathbf{r} \times \mathbf{F})_{\substack{\text{contents of the} \\ \text{control volume}}}} \tag{5.42}$$

*For a system, the rate of change of moment-of-momentum equals the net torque.*

An important category of fluid mechanical problems that is readily solved with the help of the moment-of-momentum equation (Eq. 5.42) involves machines that rotate or tend to rotate around a single axis. Examples of these machines include rotary lawn sprinklers, ceiling fans, lawn mower blades, wind turbines, turbochargers, and gas turbine engines. As a class, these devices are often called turbomachines.

### 5.2.4  Application of the Moment-of-Momentum Equation[2]

We simplify our use of Eq. 5.42 in several ways:

1. We assume that flows considered are one-dimensional (uniform distributions of average velocity at any section).

---

[2]This section may be omitted, along with Sections 5.2.3 and 5.3.5, without loss of continuity in the text material. However, these sections are recommended for those interested in Chapter 12.

*V5.12 Rotating lawn sprinkler*

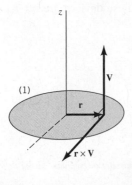

■ **Figure 5.4** (*a*) **Rotary water sprinkler.** (*b*) **Rotary water sprinkler, plane view.** (*c*) **Rotary water sprinkler, side view.**

2. We confine ourselves to steady or steady-in-the-mean cyclical flows. Thus,

$$\frac{\partial}{\partial t} \int_{cv} (\mathbf{r} \times \mathbf{V})\rho \, d\forall = 0$$

at any instant of time for steady flows or on a time-average basis for cyclical unsteady flows.

3. We work only with the component of Eq. 5.42 resolved along the axis of rotation.

*Change in moment of fluid velocity around an axis can result in torque and rotation around that same axis.*

Consider the rotating sprinkler sketched in Fig. 5.4. Because the direction and magnitude of the flow through the sprinkler from the inlet [section (1)] to the outlet [section (2)] of the arm changes, the water exerts a torque on the sprinkler head causing it to tend to rotate or to actually rotate in the direction shown, much like a turbine rotor. In applying the moment-of-momentum equation (Eq. 5.42) to this flow situation, we elect to use the fixed and nondeforming control volume shown in Fig. 5.4. This disk-shaped control volume contains within its boundaries the spinning or stationary sprinkler head and the portion of the water flowing through the sprinkler contained in the control volume at an instant. The control surface cuts through the sprinkler head's solid material so that the shaft torque that resists motion can be clearly identified. When the sprinkler is rotating, the flow field in the stationary control volume is cyclical and unsteady but steady in the mean. We proceed to use the axial component of the moment-of-momentum equation (Eq. 5.42) to analyze this flow.

The integrand of the moment-of-momentum flow term in Eq. 5.42,

$$\int_{cs} (\mathbf{r} \times \mathbf{V})\rho \mathbf{V} \cdot \hat{\mathbf{n}} \, dA$$

can be nonzero only where fluid is crossing the control surface. Everywhere else on the control surface this term will be zero because $\mathbf{V} \cdot \hat{\mathbf{n}} = 0$. Water enters the control volume axially through the hollow stem of the sprinkler at section (1). At this portion of the control surface, the component of $\mathbf{r} \times \mathbf{V}$ resolved along the axis of rotation is zero because as illustrated by the figure in the margin, $\mathbf{r} \times \mathbf{V}$ lies in the plane of section (1), perpendicular to the axis of rotation. Thus, there is no axial moment-of-momentum flow in at section (1). Water leaves the control volume through each of the two nozzle openings at section (2). For the exiting flow, the magnitude of the axial component of $\mathbf{r} \times \mathbf{V}$ is $r_2 V_{\theta 2}$, where $r_2$ is the radius from the axis of rotation to the nozzle centerline and $V_{\theta 2}$ is the value of the tangential component of the velocity of the flow exiting each nozzle as observed from a frame of reference attached to the fixed and nondeforming control volume. The fluid velocity measured relative to a fixed control surface is an absolute velocity, $\mathbf{V}$. The velocity of the nozzle exit flow as viewed from the nozzle is called the relative velocity, **W**. The absolute and relative velocities, **V** and **W**, are related by the vector relationship

$$\mathbf{V} = \mathbf{W} + \mathbf{U} \tag{5.43}$$

where **U** is the velocity of the moving nozzle as measured relative to the fixed control surface.

The cross product and the dot product involved in the moment-of-momentum flow term of Eq. 5.42,

$$\int_{cs} (\mathbf{r} \times \mathbf{V})\rho \mathbf{V} \cdot \hat{\mathbf{n}}\, dA$$

can each result in a positive or negative value. For flow into the control volume, $\mathbf{V} \cdot \hat{\mathbf{n}}$ is negative. For flow out, $\mathbf{V} \cdot \hat{\mathbf{n}}$ is positive. The relative directions of **r**, **V** and $\mathbf{r} \times \mathbf{V}$ as well as correct algebraic sign to assign the axis component of $\mathbf{r} \times \mathbf{V}$ can be ascertained by using the right-hand rule. The positive direction along the axis of rotation is the direction the thumb of the right-hand points when it is extended, and the remaining fingers are curled around the rotation axis in the positive direction of rotation as illustrated in Fig. 5.5. The direction of the axial component of $\mathbf{r} \times \mathbf{V}$ is similarly ascertained by noting the direction of the cross product of the radius from the axis of rotation, $r\hat{\mathbf{e}}_r$, and the tangential component of absolute velocity, $V_\theta\hat{\mathbf{e}}_\theta$. Thus, for the sprinkler of Fig. 5.4, we can state that

*The algebraic sign of $\mathbf{r} \times \mathbf{V}$ is obtained by the right-hand rule.*

$$\left[ \int_{cs} (\mathbf{r} \times \mathbf{V})\rho \mathbf{V} \cdot \hat{\mathbf{n}}\, dA \right]_{axial} = (-r_2 V_{\theta 2})(+\dot{m}) \tag{5.44}$$

where, because of mass conservation, $\dot{m}$ is the total mass flowrate through both nozzles. As was demonstrated in Example 5.7, the mass flowrate is the same whether the sprinkler rotates or not. The correct algebraic sign of the axial component of $\mathbf{r} \times \mathbf{V}$ can be easily remembered in the following way: if $\mathbf{V}_\theta$ and **U** are in the same direction, use $+$; if $\mathbf{V}_\theta$ and **U** are in opposite directions, use $-$.

The torque term $[\sum (\mathbf{r} \times \mathbf{F})_{\text{contents of the control volume}}]$ of the moment-of-momentum equation (Eq. 5.42) is analyzed next. Confining ourselves to torques acting with respect to the axis of rotation only, we conclude that the shaft torque is important. The net torque with respect to the axis of rotation associated with normal forces exerted on the contents of the control volume will be very small, if not zero. The net axial torque due to fluid tangential forces is also negligibly small for the control volume of Fig. 5.4. Thus, for the sprinkler of Fig. 5.4

$$\sum \left[ (\mathbf{r} \times \mathbf{F})_{\substack{\text{contents of the} \\ \text{control volume}}} \right]_{axial} = \mathbf{T}_{\text{shaft}} \tag{5.45}$$

Note that we have entered $T_{\text{shaft}}$ as a positive quantity in Eq. 5.45. This is equivalent to assuming that $T_{\text{shaft}}$ is in the same direction as rotation.

For the sprinkler of Fig. 5.4, the axial component of the moment-of-momentum equation (Eq. 5.42) is, from Eqs. 5.44 and 5.45

$$-r_2 V_{\theta 2}\dot{m} = T_{\text{shaft}} \tag{5.46}$$

We interpret $T_{\text{shaft}}$ being a negative quantity from Eq. 5.46 to mean that the shaft torque actually opposes the rotation of the sprinkler arms as shown in Fig. 5.4. The shaft torque, $T_{\text{shaft}}$, opposes rotation in all turbine devices.

(a)    (b)

■ **Figure 5.5    Right-hand rule convention.**

We could evaluate the **shaft power**, $\dot{W}_{shaft}$, associated with **shaft torque**, $T_{shaft}$, by forming the product of $T_{shaft}$ and the rotational speed of the shaft, $\omega$. [We use the notation that $W$ = work, $(\cdot) = d(\ )/dt$, and thus $\dot{W}$ = power.] Thus, from Eq. 5.46 we get

$$\dot{W}_{shaft} = T_{shaft}\,\omega = -r_2 V_{\theta 2}\dot{m}\,\omega \tag{5.47}$$

Since $r_2\omega$ is the speed of each sprinkler nozzle, $U$, we can also state Eq. 5.47 in the form

$$\dot{W}_{shaft} = -U_2 V_{\theta 2}\dot{m} \tag{5.48}$$

Shaft work per unit mass, $w_{shaft}$, is equal to $\dot{W}_{shaft}/\dot{m}$. Dividing Eq. 5.48 by the mass flowrate, $\dot{m}$, we obtain

$$w_{shaft} = -U_2 V_{\theta 2} \tag{5.49}$$

V5.13 Impulse-type lawn sprinkler

Negative shaft work as in Eqs. 5.47, 5.48, and 5.49 is work out of the control volume, that is, work done by the fluid on the rotor and thus its shaft.

The principles associated with this sprinkler example can be extended to handle most simplified turbomachine flows. The fundamental technique is not difficult. However, the geometry of some turbomachine flows is quite complicated.

Example 5.18 further illustrates how the axial component of the moment-of-momentum equation (Eq. 5.46) can be used.

---

# EXAMPLE 5.18 — Moment-of-Momentum—Torque

**GIVEN** Water enters a rotating lawn sprinkler through its base at the steady rate of 1000 ml/s as sketched in Fig. E5.18a. The exit area of each of the two nozzles is 30 mm$^2$ and the flow leaving each nozzle is in the tangential direction. The radius from the axis of rotation to the centerline of each nozzle is 200 mm.

**FIND** (a) Determine the resisting torque required to hold the sprinkler head stationary.

(b) Determine the resisting torque associated with the sprinkler rotating with a constant speed of 500 rev/min.

(c) Determine the speed of the sprinkler if no resisting torque is applied.

■ **Figure E5.18**

## SOLUTION

To solve parts **(a)**, **(b)**, and **(c)** of this example we can use the same fixed and nondeforming, disk-shaped control volume illustrated in Fig. 5.4. As indicated in Fig. E5.18a, the only axial torque considered is the one resisting motion, $T_{shaft}$.

**(a)** When the sprinkler head is held stationary as specified in part **(a)** of this example problem, the velocities of the fluid entering and leaving the control volume are shown in Fig. E5.18b. Equation 5.46 applies to the contents of this control volume. Thus,

$$T_{shaft} = -r_2 V_{\theta 2}\dot{m} \tag{1}$$

Since the control volume is fixed and nondeforming and the flow exiting from each nozzle is tangential,

$$V_{\theta 2} = V_2 \tag{2}$$

Equations 1 and 2 give

$$T_{shaft} = -r_2 V_2 \dot{m} \tag{3}$$

In Example 5.7, we ascertained that $V_2 = 16.7$ m/s. Thus, from Eq. 3 with

$$\dot{m} = Q\rho = \frac{(1000 \text{ ml/s})(10^{-3} \text{ m}^3/\text{liter})(999 \text{ kg/m}^3)}{(1000 \text{ ml/liter})}$$

$$= 0.999 \text{ kg/s}$$

we obtain

$$T_{\text{shaft}} = -\frac{(200 \text{ mm})(16.7 \text{ m/s})(0.999 \text{ kg/s})[1 \text{ (N/kg)}/(\text{m/s}^2)]}{(1000 \text{ mm/m})}$$

or

$$T_{\text{shaft}} = -3.34 \text{ N} \cdot \text{m} \qquad \textbf{(Ans)}$$

**(b)**   When the sprinkler is rotating at a constant speed of 500 rpm, the flow field in the control volume is unsteady but cyclical. Thus, the flow field is steady in the mean. The velocities of the flow entering and leaving the control volume are as indicated in Fig. E5.18c. The absolute velocity of the fluid leaving each nozzle, $V_2$, is from Eq. 5.43,

$$V_2 = W_2 - U_2 \qquad \textbf{(4)}$$

where

$$W_2 = 16.7 \text{ m/s}$$

as determined in Example 5.7. The speed of the nozzle, $U_2$, is obtained from

$$U_2 = r_2\omega \qquad \textbf{(5)}$$

Application of the axial component of the moment-of-momentum equation (Eq. 5.46) leads again to Eq. 3. From Eqs. 4 and 5,

$$V_2 = 16.7 \text{ m/s} - r_2\omega$$

$$= 16.7 \text{ m/s} - \frac{(200 \text{ mm})(500 \text{ rev/min})(2\pi \text{ rad/rev})}{(1000 \text{ mm/m})(60 \text{ s/min})}$$

or

$$V_2 = 16.7 \text{ m/s} - 10.5 \text{ m/s} = 6.2 \text{ m/s}$$

Thus, using Eq. 3, with $\dot{m} = 0.999$ kg/s (as calculated previously), we get

$$T_{\text{shaft}} = -\frac{(200 \text{ mm})(6.2 \text{ m/s}) \, 0.999 \text{ kg/s} \, [1 \text{ (N/kg)}/(\text{m/s}^2)]}{(1000 \text{ mm/m})}$$

or

$$T_{\text{shaft}} = -1.24 \text{ N} \cdot \text{m} \qquad \textbf{(Ans)}$$

**COMMENT**   Note that the resisting torque associated with sprinkler head rotation is much less than the resisting torque that is required to hold the sprinkler stationary.

**(c)**   When no resisting torque is applied to the rotating sprinkler head, a maximum constant speed of rotation will occur as demonstrated below. Application of Eqs. 3, 4, and 5 to the contents of the control volume results in

$$T_{\text{shaft}} = -r_2(W_2 - r_2\omega)\dot{m} \qquad \textbf{(6)}$$

For no resisting torque, Eq. 6 yields

$$0 = -r_2(W_2 - r_2\omega)\dot{m}$$

Thus,

$$\omega = \frac{W_2}{r_2} \qquad \textbf{(7)}$$

In Example 5.4, we learned that the relative velocity of the fluid leaving each nozzle, $W_2$, is the same regardless of the speed of rotation of the sprinkler head, $\omega$, as long as the mass flowrate of the fluid, $\dot{m}$, remains constant. Thus, by using Eq. 7 we obtain

$$\omega = \frac{W_2}{r_2} = \frac{(16.7 \text{ m/s})(1000 \text{ mm/m})}{(200 \text{ mm})} = 83.5 \text{ rad/s}$$

or

$$\omega = \frac{(83.5 \text{ rad/s})(60 \text{ s/min})}{2\pi \text{ rad/rev}} = 797 \text{ rpm} \qquad \textbf{(Ans)}$$

For this condition ($T_{\text{shaft}} = 0$), the water both enters and leaves the control volume with zero angular momentum.

**COMMENT**   Note that forcing a change in direction of a flowing fluid, in this case with a sprinkler, resulted in rotary motion and a useful "sprinkling" of water over an area.

By repeating the calculations for various values of the angular velocity, $\omega$, the results shown in Fig. E5.18d are obtained. It is seen that the magnitude of the resisting torque associated with rotation is less than the torque required to hold the rotor stationary. Even in the absence of a resisting torque, the rotor maximum speed is finite.

■ **Figure E5.18d**

When the moment-of-momentum equation (Eq. 5.42) is applied to a more general, one-dimensional flow through a rotating machine, we obtain

$$T_{\text{shaft}} = (-\dot{m}_{\text{in}})(\pm r_{\text{in}}V_{\theta\text{in}}) + \dot{m}_{\text{out}}(\pm r_{\text{out}}V_{\theta\text{out}}) \tag{5.50}$$

by applying the same kind of analysis used with the sprinkler of Fig. 5.4. The "−" is used with mass flowrate into the control volume, $\dot{m}_{\text{in}}$, and the "+" is used with mass flowrate out of the control volume, $\dot{m}_{\text{out}}$, to account for the sign of the dot product, $\mathbf{V} \cdot \hat{\mathbf{n}}$, involved. Whether "+" or "−" is used with the $rV_\theta$ product depends on the direction of $(\mathbf{r} \times \mathbf{V})_{\text{axial}}$. A simple way to determine the sign of the $rV_\theta$ product is to compare the direction of $V_\theta$ and the blade speed, $U$. As shown in the margin, if $V_\theta$ and $U$ are in the same direction, then the $rV_\theta$ product is positive. If $V_\theta$ and $U$ are in opposite directions, the $rV_\theta$ product is negative. The sign of the shaft torque is "+" if $T_{\text{shaft}}$ is in the same direction along the axis of rotation as $\omega$, and "−" otherwise.

The shaft power, $\dot{W}_{\text{shaft}}$, is related to shaft torque, $T_{\text{shaft}}$, by

$$\dot{W}_{\text{shaft}} = T_{\text{shaft}}\omega \tag{5.51}$$

Thus, using Eqs. 5.50 and 5.51 with a "+" sign for $T_{\text{shaft}}$ in Eq. 5.50, we obtain

$$\dot{W}_{\text{shaft}} = (-\dot{m}_{\text{in}})(\pm r_{\text{in}}\omega V_{\theta\text{in}}) + \dot{m}_{\text{out}}(\pm r_{\text{out}}\omega V_{\theta\text{out}}) \tag{5.52}$$

or since $r\omega = U$

$$\dot{W}_{\text{shaft}} = (-\dot{m}_{\text{in}})(\pm U_{\text{in}}V_{\theta\text{in}}) + \dot{m}_{\text{out}}(\pm U_{\text{out}}V_{\theta\text{out}}) \tag{5.53}$$

*When shaft torque and shaft rotation are in the same (opposite) direction, power is into (out of) the fluid.*

The "+" is used for the $UV_\theta$ product when $U$ and $V_\theta$ are in the same direction; the "−" is used when $U$ and $V_\theta$ are in opposite directions. Also, since $+T_{\text{shaft}}$ was used to obtain Eq. 5.53, when $\dot{W}_{\text{shaft}}$ is positive, power is into the fluid (for example, a pump), and when $\dot{W}_{\text{shaft}}$, is negative, power is out of the fluid (for example, a turbine).

The shaft work per unit mass, $w_{\text{shaft}}$, can be obtained from the shaft power, $\dot{W}_{\text{shaft}}$, by dividing Eq. 5.53 by the mass flowrate, $\dot{m}$. By conservation of mass,

$$\dot{m} = \dot{m}_{\text{in}} = \dot{m}_{\text{out}}$$

From Eq. 5.53, we obtain

$$w_{\text{shaft}} = -(\pm U_{\text{in}}V_{\theta\text{in}}) + (\pm U_{\text{out}}V_{\theta\text{out}}) \tag{5.54}$$

The application of Eqs. 5.50, 5.53, and 5.54 is demonstrated in Example 5.19. More examples of the application of Eqs. 5.50, 5.53, and 5.54 are included in Chapter 12.

# EXAMPLE 5.19    Moment-of-Momentum—Power

**GIVEN**  An air fan has a bladed rotor of 12-in. outside diameter and 10-in. inside diameter as illustrated in Fig. E5.19a. The height of each rotor blade is constant at 1 in. from blade inlet to outlet. The flowrate is steady, on a time-average basis, at 230 ft³/min and the absolute velocity of the air at blade inlet, $V_1$, is radial. The blade discharge angle is 30° from the tangential direction. The rotor rotates at a constant speed of 1725 rpm.

**FIND**  Estimate the power required to run the fan.

## SOLUTION

We select a fixed and nondeforming control volume that includes the rotating blades and the fluid within the blade row at an instant, as shown with a dashed line in Fig. E5.19a. The flow within this control volume is cyclical, but steady in the mean. The only torque we consider is the driving shaft torque, $T_{shaft}$. This torque is provided by a motor. We assume that the entering and leaving flows are each represented by uniformly distributed velocities and flow properties. Since shaft power is sought, Eq. 5.53 is appropriate. Application of Eq. 5.53 to the contents of the control volume in Fig. E5.19 gives

$$\dot{W}_{shaft} = -\dot{m}_1(\pm U_1 V_{\theta 1})^{\,0\,(\mathbf{V}_1 \text{ is radial})} + \dot{m}_2(\pm U_2 V_{\theta 2}) \quad (1)$$

From Eq. 1 we see that to calculate fan power, we need mass flowrate, $\dot{m}$, rotor exit blade velocity, $U_2$, and fluid tangential velocity at blade exit, $V_{\theta 2}$. The mass flowrate, $\dot{m}$, is easily obtained from Eq. 5.6 as

$$\dot{m} = \rho Q = \frac{(2.38 \times 10^{-3} \text{ slug/ft}^3)(230 \text{ ft}^3/\text{min})}{(60 \text{ s/min})}$$
$$= 0.00912 \text{ slug/s} \quad (2)$$

Often, problems involving fans are solved using English Engineering units. Since 1 slug = 32.174 lbm, we could have used as the density of air $\rho_{air} = (2.38 \times 10^{-3} \text{ slug/ft}^3)(32.174 \text{ lbm/slug}) = 0.0766 \text{ lbm/ft}^3$.

Then

$$\dot{m} = \frac{(0.0766 \text{ lbm/ft}^3)(230 \text{ ft}^3/\text{min})}{(60 \text{ s/min})} = 0.294 \text{ lbm/s}$$

The rotor exit blade speed, $U_2$, is

$$U_2 = r_2\omega = \frac{(6 \text{ in.})(1725 \text{ rpm})(2\pi \text{ rad/rev})}{(12 \text{ in./ft})(60 \text{ s/min})}$$
$$= 90.3 \text{ ft/s} \quad (3)$$

(b)

■ **Figure E5.19**

To determine the fluid tangential speed at the fan rotor exit, $V_{\theta 2}$, we use Eq. 5.43 to get

$$\mathbf{V}_2 = \mathbf{W}_2 + \mathbf{U}_2 \tag{4}$$

The vector addition of Eq. 4 is shown in the form of a "velocity triangle" in Fig. E5.19b. From Fig. E5.19b, we can see that

$$V_{\theta 2} = U_2 - W_2 \cos 30° \tag{5}$$

To solve Eq. 5 for $V_{\theta 2}$ we need a value of $W_2$, in addition to the value of $U_2$ already determined (Eq. 3). To get $W_2$, we recognize that

$$W_2 \sin 30° = V_{r2} \tag{6}$$

where $V_{r2}$ is the radial component of either $\mathbf{W}_2$ or $\mathbf{V}_2$. Also, using Eq. 5.6, we obtain

$$\dot{m} = \rho A_2 V_{r2} \tag{7}$$

or since

$$A_2 = 2\,\pi r_2 h \tag{8}$$

where $h$ is the blade height, Eqs. 7 and 8 combine to form

$$\dot{m} = \rho 2 \pi r_2 h V_{r2} \tag{9}$$

Taking Eqs. 6 and 9 together we get

$$W_2 = \frac{\dot{m}}{\rho 2 \pi r_2 h \sin 30°} = \frac{\rho Q}{\rho 2 \pi r_2 h \sin 30°} \tag{10}$$
$$= \frac{Q}{2 \pi r_2 h \sin 30°}$$

Substituting known values into Eq. 10, we obtain

$$W_2 = \frac{(230\ \text{ft}^3/\text{min})(12\ \text{in./ft})(12\ \text{in./ft})}{(60\ \text{s/min})2\pi(6\ \text{in.})(1\ \text{in.})\sin 30°}$$
$$= 29.3\ \text{ft/s}$$

By using this value of $W_2$ in Eq. 5 we get

$$V_{\theta 2} = U_2 - W_2 \cos 30°$$
$$= 90.3\ \text{ft/s} - (29.3\ \text{ft/s})(0.866) = 64.9\ \text{ft/s}$$

Equation 1 can now be used to obtain

$$\dot{W}_{\text{shaft}} = \dot{m}\,U_2 V_{\theta 2} = \frac{(0.00912\ \text{slug/s})(90.3\ \text{ft/s})(64.9\ \text{ft/s})}{[1\ (\text{slug} \cdot \text{ft/s}^2)/\text{lb}][550\ (\text{ft} \cdot \text{lb})/(\text{hp} \cdot \text{s})]}$$

with BG units.
    With EE units

$$\dot{W}_{\text{shaft}} = \frac{(0.294\ \text{lbm/s})(90.3\ \text{ft/s})(64.9\ \text{ft/s})}{[32.174\ (\text{lbm} \cdot \text{ft})/(\text{lb/s}^2)][550\ (\text{ft} \cdot \text{lb})/(\text{hp} \cdot \text{s})]}$$

In either case

$$\dot{W}_{\text{shaft}} = 0.097\ \text{hp} \tag{Ans}$$

**COMMENT** Note that the "+" was used with the $U_2 V_{\theta 2}$ product because $U_2$ and $V_{\theta 2}$ are in the same direction. This result, 0.097 hp, is the power that needs to be delivered through the fan shaft for the given conditions. Ideally, all of this power would go into the flowing air. However, because of fluid friction, only some of this power will produce useful effects (e.g., movement and pressure rise) on the air. How much useful effect depends on the efficiency of the energy transfer between the fan blades and the fluid.

# 5.3    First Law of Thermodynamics—The Energy Equation

### 5.3.1 Derivation of the Energy Equation

The *first law of thermodynamics* for a system is, in words,

| Time rate of increase of the total stored energy of the system | = | net time rate of energy addition by heat transfer into the system | + | net time rate of energy addition by work transfer into the system |

In symbolic form, this statement is

$$\frac{D}{Dt}\int_{\text{sys}} e\rho\,d\Psi = \left(\sum \dot{Q}_{\text{in}} - \sum \dot{Q}_{\text{out}}\right)_{\text{sys}} + \left(\sum \dot{W}_{\text{in}} - \sum \dot{W}_{\text{out}}\right)_{\text{sys}}$$

*The first law of thermodynamics is a statement of conservation of energy.*

or

$$\frac{D}{Dt}\int_{\text{sys}} e\rho\,d\Psi = (\dot{Q}_{\substack{\text{net} \\ \text{in}}} + \dot{W}_{\substack{\text{net} \\ \text{in}}})_{\text{sys}} \tag{5.55}$$

Some of these variables deserve a brief explanation before proceeding further. The total stored energy per unit mass for each particle in the system, $e$, is related to the internal energy per unit mass, $\breve{u}$, the kinetic energy per unit mass, $V^2/2$, and the potential energy per unit mass, $gz$, by the equation

$$e = \breve{u} + \frac{V^2}{2} + gz \tag{5.56}$$

The net *rate of heat transfer* into the system is denoted with $\dot{Q}_{\text{net in}}$, and the net rate of work transfer into the system is labeled $\dot{W}_{\text{net in}}$. Heat transfer and work transfer are considered "+" going into the system and "−" coming out.

Equation 5.55 is valid for inertial and noninertial reference systems. We proceed to develop the control volume statement of the first law of thermodynamics. For the control volume that is coincident with the system at an instant of time

$$(\dot{Q}_{\underset{\text{in}}{\text{net}}} + \dot{W}_{\underset{\text{in}}{\text{net}}})_{\text{sys}} = (\dot{Q}_{\underset{\text{in}}{\text{net}}} + \dot{W}_{\underset{\text{in}}{\text{net}}})_{\underset{\text{control volume}}{\text{coincident}}} \tag{5.57}$$

Furthermore, for the system and the contents of the coincident control volume that is fixed and nondeforming, the Reynolds transport theorem (Eq. 4.19 with the parameter $b$ set equal to $e$) allows us to conclude that

$$\frac{D}{Dt}\int_{\text{sys}} e\rho \, d\mathcal{V} = \frac{\partial}{\partial t}\int_{\text{cv}} e\rho \, d\mathcal{V} + \int_{\text{cs}} e\rho \mathbf{V} \cdot \hat{\mathbf{n}} \, dA \tag{5.58}$$

or in words,

| Time rate of increase of the total stored energy of the system | = | time rate of increase of the total stored energy of the contents of the control volume | + | net rate of flow of the total stored energy out of the control volume through the control surface |
|---|---|---|---|---|

Combining Eqs. 5.55, 5.57, and 5.58, we get the control volume formula for the first law of thermodynamics:

$$\boxed{\frac{\partial}{\partial t}\int_{\text{cv}} e\rho \, d\mathcal{V} + \int_{\text{cs}} e\rho \mathbf{V} \cdot \hat{\mathbf{n}} \, dA = (\dot{Q}_{\underset{\text{in}}{\text{net}}} + \dot{W}_{\underset{\text{in}}{\text{net}}})_{\text{cv}}} \tag{5.59}$$

*The energy equation involves stored energy and heat and work transfer.*

The total stored energy per unit mass, $e$, in Eq. 5.59 is for fluid particles entering, leaving, and within the control volume. Further explanation of the heat transfer and work transfer involved in this equation follows.

The heat transfer rate, $\dot{Q}$, represents all of the ways in which energy is exchanged between the control volume contents and surroundings because of a temperature difference. Thus, radiation, conduction, and/or convection are possible. As shown by the figure in the margin, heat transfer into the control volume is considered positive; heat transfer out is negative. In many engineering applications, the process is *adiabatic*; the heat transfer rate, $\dot{Q}$, is zero. The net heat transfer rate, $\dot{Q}_{\text{net in}}$, can also be zero when $\sum \dot{Q}_{\text{in}} - \sum \dot{Q}_{\text{out}} = 0$.

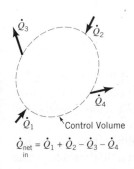

$$\dot{Q}_{\underset{\text{in}}{\text{net}}} = \dot{Q}_1 + \dot{Q}_2 - \dot{Q}_3 - \dot{Q}_4$$

The work transfer rate, $\dot{W}$, also called *power*, is positive when work is done on the contents of the control volume by the surroundings. Otherwise, it is considered negative. Work can be transferred across the control surface in several ways. In the following paragraphs, we consider some important forms of work transfer.

In many instances, work is transferred across the control surface by a moving shaft. In rotary devices such as turbines, fans, and propellers, a rotating shaft transfers work across that portion of the control surface that slices through the shaft. Even in reciprocating machines like positive displacement internal combustion engines and compressors that utilize piston-in-cylinder arrangements, a rotating crankshaft is used. Since work is the dot product of force and related displacement, rate of work (or power) is the dot product of force and related displacement per unit time. For a rotating shaft, the power transfer, $\dot{W}_{\text{shaft}}$, is related to the shaft torque that causes the rotation, $T_{\text{shaft}}$, and the angular velocity of the shaft, $\omega$, by the relationship

$$\dot{W}_{\text{shaft}} = T_{\text{shaft}}\omega$$

When the control surface cuts through the shaft material, the shaft torque is exerted by shaft material at the control surface. To allow for consideration of problems involving more than one shaft we use the notation

$$\dot{W}_{\underset{\text{net in}}{\text{shaft}}} = \sum \dot{W}_{\underset{\text{in}}{\text{shaft}}} - \sum \dot{W}_{\underset{\text{out}}{\text{shaft}}} \tag{5.60}$$

■ **Figure 5.6**  Simple, fully developed pipe flow.

$$u_1 = u_{max} \left[1 - \left(\frac{r}{R}\right)^2\right] \qquad u_2 = u_{max} \left[1 - \left(\frac{r}{R}\right)^2\right]$$

Work transfer can also occur at the control surface when a force associated with fluid normal stress acts over a distance. Consider the simple pipe flow illustrated in Fig. 5.6 and the control volume shown. For this situation, the fluid normal stress, $\sigma$, is simply equal to the negative of fluid pressure, $p$, in all directions; that is,

$$\sigma = -p \qquad (5.61)$$

This relationship can be used with varying amounts of approximation for many engineering problems (see Chapter 6).

$\dot{W} = \mathbf{F} \cdot \mathbf{V}$
$= FV \cos\theta$

The power transfer, $\dot{W}$, associated with a force $\mathbf{F}$ acting on an object moving with velocity $\mathbf{V}$ is given by the dot product $\mathbf{F} \cdot \mathbf{V}$. This is illustrated by the figure in the margin. Hence, the power transfer associated with normal stresses acting on a single fluid particle, $\delta\dot{W}_{normal\ stress}$, can be evaluated as the dot product of the normal stress force, $\delta\mathbf{F}_{normal\ stress}$, and the fluid particle velocity, $\mathbf{V}$, as

$$\delta\dot{W}_{normal\ stress} = \delta\mathbf{F}_{normal\ stress} \cdot \mathbf{V}$$

If the normal stress force is expressed as the product of local normal stress, $\sigma = -p$, and fluid particle surface area, $\hat{\mathbf{n}}\,\delta A$, the result is

$$\delta\dot{W}_{normal\ stress} = \sigma\hat{\mathbf{n}}\,\delta A \cdot \mathbf{V} = -p\hat{\mathbf{n}}\,\delta A \cdot \mathbf{V} = -p\mathbf{V} \cdot \hat{\mathbf{n}}\,\delta A$$

For all fluid particles on the control surface of Fig. 5.6 at the instant considered, power transfer due to fluid normal stress, $\dot{W}_{normal\ stress}$, is

$$\dot{W}_{normal\ stress} = \int_{cs} \sigma\mathbf{V} \cdot \hat{\mathbf{n}}\,dA = \int_{cs} -p\mathbf{V} \cdot \hat{\mathbf{n}}\,dA \qquad (5.62)$$

Note that the value of $\dot{W}_{normal\ stress}$ for particles on the wetted inside surface of the pipe is zero because $\mathbf{V} \cdot \hat{\mathbf{n}}$ is zero there. Thus, $\dot{W}_{normal\ stress}$ can be nonzero only where fluid enters and leaves the control volume. Although only a simple pipe flow was considered, Eq. 5.62 is quite general, and the control volume used in this example can serve as a general model for other cases.

*Work is transferred by rotating shafts, normal stresses, and tangential stresses.*

Work transfer can also occur at the control surface because of tangential stress forces. Rotating shaft work is transferred by tangential stresses in the shaft material. For a fluid particle, shear stress force power, $\delta\dot{W}_{tangential\ stress}$, can be evaluated as the dot product of tangential stress force, $\delta\mathbf{F}_{tangential\ stress}$, and the fluid particle velocity, $\mathbf{V}$. That is,

$$\delta\dot{W}_{tangential\ stress} = \delta\mathbf{F}_{tangential\ stress} \cdot \mathbf{V}$$

For the control volume of Fig. 5.6, the fluid particle velocity is zero everywhere on the wetted inside surface of the pipe. Thus, no tangential stress work is transferred across that portion of the control surface. Furthermore, if we select the control surface so that it is perpendicular to the fluid particle velocity, then the tangential stress force is also perpendicular to the velocity. Therefore, the tangential stress work transfer is zero on that part of the control surface. This is illustrated in the figure in the margin. Thus, in general, we select control volumes like the one of Fig. 5.6 and consider fluid tangential stress power transfer to be negligibly small.

Using the information we have developed about power, we can express the first law of thermodynamics for the contents of a control volume by combining Eqs. 5.59, 5.60, and 5.62 to obtain

$\delta\dot{W}_{tangential\ stress} = 0$

$$\frac{\partial}{\partial t} \int_{cv} e\rho\,d\mathcal{V} + \int_{cs} e\rho\mathbf{V} \cdot \hat{\mathbf{n}}\,dA = \dot{Q}_{net\atop in} + \dot{W}_{shaft\atop net\ in} - \int_{cs} p\mathbf{V} \cdot \hat{\mathbf{n}}\,dA \qquad (5.63)$$

When the equation for total stored energy (Eq. 5.56) is considered with Eq. 5.63, we obtain the *energy equation*:

$$\frac{\partial}{\partial t} \int_{cv} e\rho \, d\mathcal{V} + \int_{cs} \left( \check{u} + \frac{p}{\rho} + \frac{V^2}{2} + gz \right) \rho \mathbf{V} \cdot \hat{\mathbf{n}} \, dA = \dot{Q}_{\underset{in}{net}} + \dot{W}_{\underset{net\ in}{shaft}}$$   **(5.64)**

### 5.3.2 Application of the Energy Equation

In Eq. 5.64, the term $\partial/\partial t \int_{cv} e\rho \, d\mathcal{V}$ represents the time rate of change of the total stored energy, $e$, of the contents of the control volume. This term is zero when the flow is steady. This term is also zero in the mean when the flow is steady in the mean (cyclical).

In Eq. 5.64, the integrand of

$$\int_{cs} \left( \check{u} + \frac{p}{\rho} + \frac{V^2}{2} + gz \right) \rho \mathbf{V} \cdot \hat{\mathbf{n}} \, dA$$

can be nonzero only where fluid crosses the control surface ($\mathbf{V} \cdot \hat{\mathbf{n}} \neq 0$). Otherwise, $\mathbf{V} \cdot \hat{\mathbf{n}}$ is zero and the integrand is zero for that portion of the control surface. If the properties within parentheses, $\check{u}$, $p/\rho$, $V^2/2$, and $gz$, are all assumed to be uniformly distributed over the flow cross-sectional areas involved, the integration becomes simple and gives

$$\int_{cs} \left( \check{u} + \frac{p}{\rho} + \frac{V^2}{2} + gz \right) \rho \mathbf{V} \cdot \hat{\mathbf{n}} \, dA = \sum_{\underset{out}{flow}} \left( \check{u} + \frac{p}{\rho} + \frac{V^2}{2} + gz \right) \dot{m}$$

$$- \sum_{\underset{in}{flow}} \left( \check{u} + \frac{p}{\rho} + \frac{V^2}{2} + gz \right) \dot{m}$$   **(5.65)**

Furthermore, if there is only one stream entering and leaving the control volume, then

$$\int_{cs} \left( \check{u} + \frac{p}{\rho} + \frac{V^2}{2} + gz \right) \rho \mathbf{V} \cdot \hat{\mathbf{n}} \, dA =$$

$$\left( \check{u} + \frac{p}{\rho} + \frac{V^2}{2} + gz \right)_{out} \dot{m}_{out} - \left( \check{u} + \frac{p}{\rho} + \frac{V^2}{2} + gz \right)_{in} \dot{m}_{in}$$   **(5.66)**

Uniform flow as described above will occur in an infinitesimally small diameter streamtube as illustrated in Fig. 5.7. This kind of streamtube flow is representative of the steady flow of a particle of fluid along a pathline. We can also idealize actual conditions by disregarding nonuniformities in a finite cross section of flow. We call this one-dimensional flow, and although such uniform flow rarely occurs in reality, the simplicity achieved with the one-dimensional approximation often justifies its use. More details about the effects of nonuniform distributions of velocities and other fluid flow variables are considered in Section 5.3.4 and in Chapters 8, 9, and 10.

If shaft work is involved, the flow must be unsteady, at least locally (see Refs. 1 and 2). The flow in any fluid machine that involves shaft work is unsteady within that machine. For example, the velocity and pressure at a fixed location near the rotating blades of a fan are unsteady. However, upstream and downstream of the machine, the flow may be steady. Most often shaft work is associated with flow that is unsteady in a recurring or cyclical way. On a time-average basis for flow that is one-dimensional, cyclical, and involves only one stream of fluid entering and leaving

■ **Figure 5.7**   **Streamtube flow.**

$\dot{m}_{\text{out}}$

$\dot{m}_{\text{in}}$

$\dot{m}_{\text{in}} = \dot{m}_{\text{out}} = \dot{m}$

the control volume, such as the hair dryer shown in the figure in the margin, Eq. 5.64 can be simplified with the help of Eqs. 5.9 and 5.66 to form

$$\dot{m}\left[ \breve{u}_{\text{out}} - \breve{u}_{\text{in}} + \left(\frac{p}{\rho}\right)_{\text{out}} - \left(\frac{p}{\rho}\right)_{\text{in}} + \frac{V_{\text{out}}^2 - V_{\text{in}}^2}{2} + g(z_{\text{out}} - z_{\text{in}}) \right] = \dot{Q}_{\substack{\text{net} \\ \text{in}}} + \dot{W}_{\substack{\text{shaft} \\ \text{net in}}} \qquad (5.67)$$

We call Eq. 5.67 the *one-dimensional energy equation for steady-in-the-mean flow*. Note that Eq. 5.67 is valid for incompressible and compressible flows. Often, the fluid property called *enthalpy*, $\breve{h}$, where

$$\breve{h} = \breve{u} + \frac{p}{\rho} \qquad (5.68)$$

*The energy equation is sometimes written in terms of enthalpy.*

is used in Eq. 5.67. With enthalpy, the one-dimensional energy equation for steady-in-the-mean flow (Eq. 5.67) is

$$\dot{m}\left[ \breve{h}_{\text{out}} - \breve{h}_{\text{in}} + \frac{V_{\text{out}}^2 - V_{\text{in}}^2}{2} + g(z_{\text{out}} - z_{\text{in}}) \right] = \dot{Q}_{\substack{\text{net} \\ \text{in}}} + \dot{W}_{\substack{\text{shaft} \\ \text{net in}}} \qquad (5.69)$$

Equation 5.69 is often used for solving compressible flow problems. Examples 5.20 and 5.21 illustrate how Eqs. 5.67 and 5.69 can be used.

---

## $E$XAMPLE 5.20    Energy—Pump Power

**GIVEN** A pump delivers water at a steady rate of 300 gal/min as shown in Fig. E5.20. Just upstream of the pump [section (1)] where the pipe diameter is 3.5 in., the pressure is 18 psi. Just downstream of the pump [section (2)] where the pipe diameter is 1 in., the pressure is 60 psi. The change in water elevation across the pump is zero. The rise in internal energy of water, $\breve{u}_2 - \breve{u}_1$, associated with a temperature rise across the pump is 93 ft · lb/lbm. The pumping process is considered to be adiabatic.

**FIND** Determine the power (hp) required by the pump.

■ **Figure E5.20**

---

## $S$OLUTION

We include in our control volume the water contained in the pump between its entrance and exit sections. Application of Eq. 5.67 to the contents of this control volume on a time-average basis yields

0 (no elevation change)

$$\dot{m}\left[ \breve{u}_2 - \breve{u}_1 + \left(\frac{p}{\rho}\right)_2 - \left(\frac{p}{\rho}\right)_1 + \frac{V_2^2 - V_1^2}{2} + g(z_2 - z_1) \right]$$

0 (adiabatic flow)

$$= \dot{Q}_{\substack{\text{net} \\ \text{in}}} + \dot{W}_{\substack{\text{shaft} \\ \text{net in}}} \qquad (1)$$

We can solve directly for the power required by the pump, $\dot{W}_{\text{shaft net in}}$, from Eq. 1, after we first determine the mass flowrate, $\dot{m}$, the speed of flow into the pump, $V_1$, and the speed of the flow

out of the pump, $V_2$. All other quantities in Eq. 1 are given in the problem statement. From Eq. 5.6, we get

$$\dot{m} = \rho Q = \frac{(1.94 \text{ slugs/ft}^3)(300 \text{ gal/min})(32.174 \text{ lbm/slug})}{(7.48 \text{ gal/ft}^3)(60 \text{ s/min})}$$

$$= 41.8 \text{ lbm/s} \qquad (2)$$

Also from Eq. 5.6,

$$V = \frac{Q}{A} = \frac{Q}{\pi D^2/4}$$

so

$$V_1 = \frac{Q}{A_1} = \frac{(300 \text{ gal/min})4 (12 \text{ in./ft})^2}{(7.48 \text{ gal/ft}^3)(60 \text{ s/min})\pi (3.5 \text{ in.})^2}$$

$$= 10.0 \text{ ft/s} \qquad (3)$$

and

$$V_2 = \frac{Q}{A_2} = \frac{(300 \text{ gal/min})4\,(12 \text{ in./ft})^2}{(7.48 \text{ gal/ft}^3)(60 \text{ s/min})\pi\,(1 \text{ in.})^2}$$

$$= 123 \text{ ft/s} \qquad (4)$$

Substituting the values of Eqs. 2, 3, and 4 and values from the problem statement into Eq. 1 we obtain

$$\dot{W}_{\substack{\text{shaft} \\ \text{net in}}} = (41.8 \text{ lbm/s})\bigg[(93 \text{ ft·lb/lbm})$$

$$+ \frac{(60 \text{ psi})(144 \text{ in.}^2/\text{ft}^2)}{(1.94 \text{ slugs/ft}^3)(32.174 \text{ lbm/slug})}$$

$$-\frac{(18 \text{ psi})(144 \text{ in.}^2/\text{ft}^2)}{(1.94 \text{ slugs/ft}^3)(32.174 \text{ lbm/slug})}$$

$$+ \frac{(123 \text{ ft/s})^2 - (10.0 \text{ ft/s})^2}{2[32.174\,(\text{lbm·ft})/(\text{lb·s}^2)]}\bigg]$$

$$\times \frac{1}{[550(\text{ft·lb/s})/\text{hp}]} = 32.2 \text{ hp} \qquad \textbf{(Ans)}$$

**COMMENT**  Of the total 32.2 hp, internal energy change accounts for 7.09 hp, the pressure rise accounts for 7.37 hp, and the kinetic energy increase accounts for 17.8 hp.

---

# EXAMPLE 5.21    Energy—Turbine Power per Unit Mass of Flow

**GIVEN**  A steam turbine generator unit used to produce electricity is shown in Fig. E5.21a. Assume the steam enters a turbine with a velocity of 30 m/s and enthalpy, $\breve{h}_1$, of 3348 kJ/kg (see Fig. E5.21b). The steam leaves the turbine as a mixture of vapor and liquid having a velocity of 60 m/s and an enthalpy of 2550 kJ/kg. The flow through the turbine is adiabatic, and changes in elevation are negligible.

**FIND**  Determine the work output involved per unit mass of steam through-flow.

## SOLUTION

We use a control volume that includes the steam in the turbine from the entrance to the exit as shown in Fig. E5.21b. Applying Eq. 5.69 to the steam in this control volume we get

0 (elevation change is negligible)

0 (adiabatic flow)

$$\dot{m}\bigg[\breve{h}_2 - \breve{h}_1 + \frac{V_2^2 - V_1^2}{2} + g(z_2 - z_1)\bigg] = \dot{Q}_{\substack{\text{net} \\ \text{in}}} + \dot{W}_{\substack{\text{shaft} \\ \text{net in}}} \qquad (1)$$

The work output per unit mass of steam through-flow, $w_{\text{shaft net in}}$, can be obtained by dividing Eq. 1 by the mass flowrate, $\dot{m}$, to obtain

$$w_{\substack{\text{shaft} \\ \text{net in}}} = \frac{\dot{W}_{\substack{\text{shaft} \\ \text{net in}}}}{\dot{m}} = \breve{h}_2 - \breve{h}_1 + \frac{V_2^2 - V_1^2}{2} \qquad (2)$$

Since $w_{\text{shaft net out}} = -w_{\text{shaft net in}}$, we obtain

$$w_{\substack{\text{shaft} \\ \text{net out}}} = \breve{h}_1 - \breve{h}_2 + \frac{V_1^2 - V_2^2}{2}$$

or

$$w_{\substack{\text{shaft} \\ \text{net out}}} = 3348 \text{ kJ/kg} - 2550 \text{ kJ/kg}$$

$$+ \frac{[(30 \text{ m/s})^2 - (60 \text{ m/s})^2][1 \text{ J/(N·m)}]}{2[1\,(\text{kg·m})/(\text{N·s}^2)](1000 \text{ J/kJ})}$$

■ Figure E5.21a

■ Figure E5.21b

Thus,

$$w_{\substack{\text{shaft} \\ \text{net out}}} = 3348 \text{ kJ/kg} - 2550 \text{ kJ/kg} - 1.35 \text{ kJ/kg}$$

$$= 797 \text{ kJ/kg} \qquad \textbf{(Ans)}$$

**COMMENT**  Note that in this particular example, the change in kinetic energy is small in comparison to the difference in enthalpy involved. This is often true in applications involving steam turbines. To determine the power output, $\dot{W}_{\text{shaft}}$, we must know the mass flowrate, $\dot{m}$.

If the flow is steady throughout, one-dimensional, and only one fluid stream is involved, then the shaft work is zero and the energy equation is

$$\dot{m}\left[\breve{u}_{out} - \breve{u}_{in} + \left(\frac{p}{\rho}\right)_{out} - \left(\frac{p}{\rho}\right)_{in} + \frac{V_{out}^2 - V_{in}^2}{2} + g(z_{out} - z_{in})\right] = \dot{Q}_{\substack{net \\ in}} \tag{5.70}$$

*V5.14 Pelton wheel turbine*

We call Eq. 5.70 the *one-dimensional, steady-flow energy equation*. This equation is valid for incompressible and compressible flows. For compressible flows, enthalpy is most often used in the one-dimensional, steady-flow energy equation and, thus, we have

$$\dot{m}\left[\breve{h}_{out} - \breve{h}_{in} + \frac{V_{out}^2 - V_{in}^2}{2} + g(z_{out} - z_{in})\right] = \dot{Q}_{\substack{net \\ in}} \tag{5.71}$$

An example of the application of Eq. 5.70 follows.

## $E$XAMPLE 5.22    Energy—Temperature Change

**GIVEN** The 420-ft waterfall shown in Fig. E5.22*a* involves steady flow from one large body of water to another.

**FIND** Determine the temperature change associated with this flow.

### SOLUTION

To solve this problem, we consider a control volume consisting of a small cross-sectional streamtube from the nearly motionless surface of the upper body of water to the nearly motionless surface of the lower body of water as is sketched in Fig. E5.22*b*. We need to determine $T_2 - T_1$. This temperature change is related to the change of internal energy of the water, $\breve{u}_2 - \breve{u}_1$, by the relationship

$$T_2 - T_1 = \frac{\breve{u}_2 - \breve{u}_1}{\breve{c}} \tag{1}$$

■ **Figure E5.22*a***
[Photograph of Akaka Falls (Hawaii) courtesy of Scott and Margaret Jones.]

■ **Figure E5.22*b***

where $\breve{c} = 1$ Btu/(lbm · °R) is the specific heat of water. The application of Eq. 5.70 to the contents of this control volume leads to

$$\dot{m}\left[\breve{u}_2 - \breve{u}_1 + \left(\frac{p}{\rho}\right)_2 - \left(\frac{p}{\rho}\right)_1 + \frac{V_2^2 - V_1^2}{2} + g(z_2 - z_1)\right]$$
$$= \dot{Q}_{\substack{net \\ in}} \tag{2}$$

We assume that the flow is adiabatic. Thus $\dot{Q}_{net\,in} = 0$. Also,

$$\left(\frac{p}{\rho}\right)_1 = \left(\frac{p}{\rho}\right)_2 \tag{3}$$

because the flow is incompressible and atmospheric pressure prevails at sections (1) and (2). Furthermore,

$$V_1 = V_2 = 0 \qquad (4)$$

because the surface of each large body of water is considered motionless. Thus, Eqs. 1 through 4 combine to yield

$$T_2 - T_1 = \frac{g(z_1 - z_2)}{\check{c}}$$

so that with

$$\check{c} = [1 \text{ Btu/(lbm} \cdot {}^\circ\text{R})] (778 \text{ ft} \cdot \text{lb/Btu})$$

$$= [778 \text{ ft} \cdot \text{lb/(lbm} \cdot {}^\circ\text{R})]$$

$$T_2 - T_1 = \frac{(32.2 \text{ ft/s}^2)(420 \text{ ft})}{[778 \text{ ft} \cdot \text{lb/(lbm} \cdot {}^\circ\text{R})][32.2 \text{ (lbm} \cdot \text{ft)/(lb} \cdot \text{s}^2)]}$$

$$= 0.540 \,{}^\circ\text{R} \qquad \textbf{(Ans)}$$

**COMMENT**   Note that it takes a considerable change of potential energy to produce even a small increase in temperature.

A form of the energy equation that is most often used to solve incompressible flow problems is developed in the next section.

### 5.3.3 Comparison of the Energy Equation with the Bernoulli Equation

When the one-dimensional energy equation for steady-in-the-mean flow, Eq. 5.67, is applied to a flow that is steady, Eq. 5.67 becomes the one-dimensional, steady-flow energy equation, Eq. 5.70. The only difference between Eq. 5.67 and Eq. 5.70 is that shaft power, $\dot{W}_{\text{shaft net in}}$, is zero if the flow is steady throughout the control volume (fluid machines involve locally unsteady flow). If in addition to being steady, the flow is incompressible, we get from Eq. 5.70

$$\dot{m}\left[ \check{u}_{\text{out}} - \check{u}_{\text{in}} + \frac{p_{\text{out}}}{\rho} - \frac{p_{\text{in}}}{\rho} + \frac{V_{\text{out}}^2 - V_{\text{in}}^2}{2} + g(z_{\text{out}} - z_{\text{in}}) \right] = \dot{Q}_{\substack{\text{net} \\ \text{in}}} \qquad \textbf{(5.72)}$$

Dividing Eq. 5.72 by the mass flowrate, $\dot{m}$, and rearranging terms we obtain

$$\frac{p_{\text{out}}}{\rho} + \frac{V_{\text{out}}^2}{2} + gz_{\text{out}} = \frac{p_{\text{in}}}{\rho} + \frac{V_{\text{in}}^2}{2} + gz_{\text{in}} - \left( \check{u}_{\text{out}} - \check{u}_{\text{in}} - q_{\substack{\text{net} \\ \text{in}}} \right) \qquad \textbf{(5.73)}$$

where

$$q_{\substack{\text{net} \\ \text{in}}} = \frac{\dot{Q}_{\text{net in}}}{\dot{m}}$$

is the heat transfer rate per mass flowrate, or heat transfer per unit mass. Note that Eq. 5.73 involves energy per unit mass and is applicable to one-dimensional flow of a single stream of fluid between two sections or flow along a streamline between two sections.

If the steady, incompressible flow we are considering also involves negligible viscous effects (frictionless flow), then the Bernoulli equation, Eq. 3.7, can be used to describe what happens between two sections in the flow as

$$p_{\text{out}} + \frac{\rho V_{\text{out}}^2}{2} + \gamma z_{\text{out}} = p_{\text{in}} + \frac{\rho V_{\text{in}}^2}{2} + \gamma z_{\text{in}} \qquad \textbf{(5.74)}$$

where $\gamma = \rho g$ is the specific weight of the fluid. To get Eq. 5.74 in terms of energy per unit mass, so that it can be compared directly with Eq. 5.73, we divide Eq. 5.74 by density, $\rho$, and obtain

$$\frac{p_{\text{out}}}{\rho} + \frac{V_{\text{out}}^2}{2} + gz_{\text{out}} = \frac{p_{\text{in}}}{\rho} + \frac{V_{\text{in}}^2}{2} + gz_{\text{in}} \qquad \textbf{(5.75)}$$

A comparison of Eqs. 5.73 and 5.75 prompts us to conclude that

$$\check{u}_{\text{out}} - \check{u}_{\text{in}} - q_{\substack{\text{net} \\ \text{in}}} = 0 \qquad \textbf{(5.76)}$$

when the steady incompressible flow is frictionless. For steady incompressible flow with friction, we learn from experience (second law of thermodynamics) that

$$\check{u}_{\text{out}} - \check{u}_{\text{in}} - q_{\substack{\text{net} \\ \text{in}}} > 0 \qquad \textbf{(5.77)}$$

In Eqs. 5.73 and 5.75, we can consider the combination of variables

$$\frac{p}{\rho} + \frac{V^2}{2} + gz$$

as equal to *useful* or *available energy*. Thus, from inspection of Eqs. 5.73 and 5.75, we can conclude that $\breve{u}_{out} - \breve{u}_{in} - q_{net\ in}$ represents the **loss** of useful or available energy that occurs in an incompressible fluid flow because of friction. In equation form we have

$$\breve{u}_{out} - \breve{u}_{in} - q_{\substack{net \\ in}} = loss \tag{5.78}$$

For a frictionless flow, Eqs. 5.73 and 5.75 tell us that loss equals zero.

It is often convenient to express Eq. 5.73 in terms of loss as

$$\frac{p_{out}}{\rho} + \frac{V_{out}^2}{2} + gz_{out} = \frac{p_{in}}{\rho} + \frac{V_{in}^2}{2} + gz_{in} - loss \tag{5.79}$$

An example of the application of Eq. 5.79 follows.

---

# EXAMPLE 5.23    Energy—Effect of Loss of Available Energy

**GIVEN**   As shown in Fig. E5.23a, air flows from a room through two different vent configurations: a cylindrical hole in the wall having a diameter of 120 mm and the same diameter cylindrical hole in the wall but with a well-rounded entrance. The room pressure is held constant at 1.0 kPa above atmospheric pressure. Both vents exhaust into the atmosphere. As discussed in Section 8.4.2, the loss in available energy associated with flow through the cylindrical vent from the room to the vent exit is $0.5V_2^2/2$ where $V_2$ is the uniformly distributed exit velocity of air. The loss in available energy associated with flow through the rounded entrance vent from the room to the vent exit is $0.05V_2^2/2$, where $V_2$ is the uniformly distributed exit velocity of air.

**FIND**   Compare the volume flowrates associated with the two different vent configurations.

## SOLUTION

We use the control volume for each vent sketched in Fig. E5.23a. What is sought is the flowrate, $Q = A_2 V_2$, where $A_2$ is the vent exit cross-sectional area, and $V_2$ is the uniformly distributed exit velocity. For both vents, application of Eq. 5.79 leads to

$$\frac{p_2}{\rho} + \frac{V_2^2}{2} + \overset{\text{0 (no elevation change)}}{\cancel{gz_2}} = \frac{p_1}{\rho} + \underset{\text{0 }(V_1 \approx 0)}{\cancel{\frac{V_1^2}{2}}} + \cancel{gz_1} - {}_1loss_2 \tag{1}$$

where ${}_1loss_2$ is the loss between sections (1) and (2). Solving Eq. 1 for $V_2$ we get

$$V_2 = \sqrt{2\left[\left(\frac{p_1 - p_2}{\rho}\right) - {}_1loss_2\right]} \tag{2}$$

Since

$${}_1loss_2 = K_L \frac{V_2^2}{2} \tag{3}$$

where $K_L$ is the loss coefficient ($K_L = 0.5$ and $0.05$ for the two vent configurations involved), we can combine Eqs. 2 and 3 to get

$$V_2 = \sqrt{2\left[\left(\frac{p_1 - p_2}{\rho}\right) - K_L \frac{V_2^2}{2}\right]} \tag{4}$$

■ **Figure E5.23a**

Solving Eq. 4 for $V_2$ we obtain

$$V_2 = \sqrt{\frac{p_1 - p_2}{\rho[(1 + K_L)/2]}} \tag{5}$$

Therefore, for flowrate, $Q$, we obtain

$$Q = A_2 V_2 = \frac{\pi D_2^2}{4} \sqrt{\frac{p_1 - p_2}{\rho[(1 + K_L)/2]}} \qquad \text{(6)}$$

For the rounded entrance cylindrical vent, Eq. 6 gives

$$Q = \frac{\pi(120 \text{ mm})^2}{4(1000 \text{ mm/m})^2}$$

$$\times \sqrt{\frac{(1.0 \text{ kPa})(1000 \text{ Pa/kPa})[1(\text{N/m}^2)/(\text{Pa})]}{(1.23 \text{ kg/m}^3)[(1 + 0.05)/2][1(\text{N·s}^2)/(\text{kg·m})]}}$$

or

$$Q = 0.445 \text{ m}^3/\text{s} \qquad \textbf{(Ans)}$$

For the cylindrical vent, Eq. 6 gives us

$$Q = \frac{\pi(120 \text{ mm})^2}{4(1000 \text{ mm/m})^2}$$

$$\times \sqrt{\frac{(1.0 \text{ kPa})(1000 \text{ Pa/kPa})[1(\text{N/m}^2)/(\text{Pa})]}{(1.23 \text{ kg/m}^3)[(1 + 0.5)/2][1(\text{N·s}^2)/(\text{kg·m})]}}$$

or

$$Q = 0.372 \text{ m}^3/\text{s} \qquad \textbf{(Ans)}$$

**COMMENT**    By repeating the calculations for various values of the loss coefficient, $K_L$, the results shown in Fig. E5.23b are obtained. Note that the rounded entrance vent allows the passage of more air than does the cylindrical vent because the loss associated with the rounded entrance vent is less than that for the cylindrical one. For this flow the pressure drop, $p_1 - p_2$, has two purposes: (1) overcome the loss associated with the flow, and (2) produce the kinetic energy at the exit. Even if there were no loss (i.e., $K_L = 0$), a pressure drop would be needed to accelerate the fluid through the vent.

■ **Figure E5.23b**

An important group of fluid mechanics problems involves one-dimensional, incompressible, steady-in-the-mean flow with friction and shaft work. Included in this category are constant density flows through pumps, blowers, fans, and turbines. For this kind of flow, Eq. 5.67 becomes

$$\dot{m}\left[\check{u}_{\text{out}} - \check{u}_{\text{in}} + \frac{p_{\text{out}}}{\rho} - \frac{p_{\text{in}}}{\rho} + \frac{V_{\text{out}}^2 - V_{\text{in}}^2}{2} + g(z_{\text{out}} - z_{\text{in}})\right] = \dot{Q}_{\substack{\text{net} \\ \text{in}}} + \dot{W}_{\substack{\text{shaft} \\ \text{net in}}} \qquad \textbf{(5.80)}$$

Dividing Eq. 5.80 by mass flowrate and using the work per unit mass, $w_{\substack{\text{shaft} \\ \text{net in}}} = \dot{W}_{\substack{\text{shaft} \\ \text{net in}}}/\dot{m}$, we obtain

$$\frac{p_{\text{out}}}{\rho} + \frac{V_{\text{out}}^2}{2} + gz_{\text{out}} = \frac{p_{\text{in}}}{\rho} + \frac{V_{\text{in}}^2}{2} + gz_{\text{in}} + w_{\substack{\text{shaft} \\ \text{net in}}} - (\check{u}_{\text{out}} - \check{u}_{\text{in}} - q_{\substack{\text{net} \\ \text{in}}}) \qquad \textbf{(5.81)}$$

*The mechanical energy equation can be written in terms of energy per unit mass.*

If the flow is steady throughout, Eq. 5.81 becomes identical to Eq. 5.73, and the previous observation that $\check{u}_{\text{out}} - \check{u}_{\text{in}} - q_{\text{net in}}$ equals the loss of available energy is valid. Thus, we conclude that Eq. 5.81 can be expressed as

$$\boxed{\frac{p_{\text{out}}}{\rho} + \frac{V_{\text{out}}^2}{2} + gz_{\text{out}} = \frac{p_{\text{in}}}{\rho} + \frac{V_{\text{in}}^2}{2} + gz_{\text{in}} + w_{\substack{\text{shaft} \\ \text{net in}}} - \text{loss}} \qquad \textbf{(5.82)}$$

*V5.15 Energy transfer*

This is a form of the energy equation for steady-in-the-mean flow that is often used for incompressible flow problems. It is sometimes called the *mechanical energy* equation or the *extended Bernoulli* equation. Note that Eq. 5.82 involves energy per unit mass (ft · lb/slug $=$ ft$^2$/s$^2$ or N · m $=$ m$^2$/s$^2$).

According to Eq. 5.82, when the shaft work is into the control volume, as for example with a pump, a larger amount of loss will result in more shaft work being required for the same rise in available energy. Similarly, when the shaft work is out of the control volume (for example, a turbine), a larger loss will result in less shaft work out for the same drop in available energy. Designers spend a great deal of effort on minimizing losses in fluid flow components. The following examples demonstrate why losses should be kept as small as possible in fluid systems.

# EXAMPLE 5.24  Energy—Fan Work and Efficiency

**GIVEN** An axial-flow ventilating fan driven by a motor that delivers 0.4 kW of power to the fan blades produces a 0.6-m-diameter axial stream of air having a speed of 12 m/s. The flow upstream of the fan involves negligible speed.

**FIND** Determine how much of the work to the air actually produces useful effects, that is, fluid motion and a rise in available energy. Estimate the fluid mechanical efficiency of this fan.

## SOLUTION

We select a fixed and nondeforming control volume as is illustrated in Fig. E5.24. The application of Eq. 5.82 to the contents of this control volume leads to

$$\underset{\text{0 (atmospheric pressures cancel)}}{\qquad} \qquad \underset{0\,(V_1 \approx 0)}{\qquad}$$

$$w_{\substack{\text{shaft} \\ \text{net in}}} - \text{loss} = \left(\frac{p_2}{\rho} + \frac{V_2^2}{2} + g z_2\right) - \left(\frac{p_1}{\rho} + \frac{V_1^2}{2} + g z_1\right) \quad (1)$$

$$\underset{\text{0 (no elevation change)}}{\qquad}$$

where $w_{\text{shaft net in}} - \text{loss}$ is the amount of work added to the air that produces a useful effect. Equation 1 leads to

$$w_{\substack{\text{shaft} \\ \text{net in}}} - \text{loss} = \frac{V_2^2}{2} = \frac{(12 \text{ m/s})^2}{2[1(\text{kg·m})/(\text{N·s}^2)]}$$

$$= 72.0 \text{ N·m/kg} \qquad (2) \quad \textbf{(Ans)}$$

A reasonable estimate of *efficiency*, $\eta$, would be the ratio of amount of work that produces a useful effect, Eq. 2, to the amount of work delivered to the fan blades. That is,

$$\eta = \frac{w_{\substack{\text{shaft} \\ \text{net in}}} - \text{loss}}{w_{\substack{\text{shaft} \\ \text{net in}}}} \qquad (3)$$

To calculate the efficiency, we need a value of $w_{\text{shaft net in}}$, which is related to the power delivered to the blades, $\dot{W}_{\text{shaft net in}}$. We note that

$$w_{\substack{\text{shaft} \\ \text{net in}}} = \frac{\dot{W}_{\substack{\text{shaft} \\ \text{net in}}}}{\dot{m}} \qquad (4)$$

where the mass flowrate, $\dot{m}$, is (from Eq. 5.6)

$$\dot{m} = \rho A V = \rho \frac{\pi D_2^2}{4} V_2 \qquad (5)$$

For fluid density, $\rho$, we use 1.23 kg/m$^3$ (standard air) and, thus, from Eqs. 4 and 5 we obtain

■ **Figure E5.24**

$$w_{\substack{\text{shaft} \\ \text{net in}}} = \frac{\dot{W}_{\substack{\text{shaft} \\ \text{net in}}}}{(\rho \pi D_2^2/4)V_2}$$

$$= \frac{(0.4 \text{ kW})[1000 \text{ (Nm)}/(\text{skW})]}{(1.23 \text{ kg/m}^3)[(\pi)(0.6 \text{ m})^2/4](12 \text{ m/s})}$$

or

$$w_{\substack{\text{shaft} \\ \text{net in}}} = 95.8 \text{ N·m/kg} \qquad (6)$$

From Eqs. 2, 3, and 6 we obtain

$$\eta = \frac{72.0 \text{ N·m/kg}}{95.8 \text{ N·m/kg}} = 0.752 \qquad \textbf{(Ans)}$$

**COMMENT** Note that only 75% of the power that was delivered to the air resulted in useful effects and, thus, 25% of the shaft power is lost to air friction.

---

## F l u i d s   i n   t h e   N e w s

**Curtain of air** An air curtain is produced by blowing air through a long rectangular nozzle to produce a high-velocity sheet of air, or a "curtain of air." This air curtain is typically directed over a doorway or opening as a replacement for a conventional door. The air curtain can be used for such things as keeping warm air from infiltrating dedicated cold spaces, preventing dust and other contaminants from entering a clean environment, and even just keeping insects out of the workplace, still allowing people to enter or exit. A disadvantage over conventional doors is the added *power requirements* to operate the air curtain, although the advantages can outweigh the disadvantage for various industrial applications. New applications for current air curtain designs continue to be developed. For example, the use of air curtains as a means of road tunnel fire security is currently being investigated. In such an application, the air curtain would act to isolate a portion of the tunnel where fire has broken out and not allow smoke and fumes to infiltrate the entire tunnel system. (See Problem 5.127.)

If Eq. 5.82, which involves energy per unit mass, is multiplied by fluid density, $\rho$, we obtain

$$p_{\text{out}} + \frac{\rho V_{\text{out}}^2}{2} + \gamma z_{\text{out}} = p_{\text{in}} + \frac{\rho V_{\text{in}}^2}{2} + \gamma z_{\text{in}} + \rho w_{\substack{\text{shaft} \\ \text{net in}}} - \rho(\text{loss}) \qquad (5.83)$$

*V5.16 Water plant aerator*

where $\gamma = \rho g$ is the specific weight of the fluid. Equation 5.83 involves *energy per unit volume* and the units involved are identical with those used for pressure (ft · lb/ft$^3$ = lb/ft$^2$ or N · m/m$^3$ = N/m$^2$).

If Eq. 5.82 is divided by the acceleration of gravity, $g$, we get

$$\boxed{\frac{p_{\text{out}}}{\gamma} + \frac{V_{\text{out}}^2}{2g} + z_{\text{out}} = \frac{p_{\text{in}}}{\gamma} + \frac{V_{\text{in}}^2}{2g} + z_{\text{in}} + h_s - h_L} \qquad (5.84)$$

where

$$h_s = w_{\text{shaft net in}}/g = \frac{\dot{W}_{\substack{\text{shaft} \\ \text{net in}}}}{\dot{m}g} = \frac{\dot{W}_{\substack{\text{shaft} \\ \text{net in}}}}{\gamma Q} \qquad (5.85)$$

*The energy equation written in terms of energy per unit weight involves heads.*

is the **shaft work head** and $h_L = \text{loss}/g$ is the **head loss**. Equation 5.84 involves *energy per unit weight* (ft · lb/lb = ft or N · m/N = m). In Section 3.7, we introduced the notion of "head," which is energy per unit weight. Units of length (for example, ft, m) are used to quantify the amount of head involved. If a turbine is in the control volume, $h_s$ is negative because it is associated with shaft work out of the control volume. For a pump in the control volume, $h_s$ is positive because it is associated with shaft work into the control volume.

We can define a total head, $H$, as follows:

$$H = \frac{p}{\gamma} + \frac{V^2}{2g} + z$$

Then Eq. 5.84 can be expressed as

$$H_{\text{out}} = H_{\text{in}} + h_s - h_L$$

Some important possible values of $H_{\text{out}}$ in comparison to $H_{\text{in}}$ are shown in Fig. 5.8. Note that $h_L$ (head loss) always reduces the value of $H_{\text{out}}$, except in the ideal case when it is zero. Note also that $h_L$ lessens the effect of shaft work that can be extracted from a fluid. When $h_L = 0$ (ideal condition) the shaft work head, $h_s$, and the change in total head are the same. This head change is sometimes called ideal head change. The corresponding ideal shaft work head is the minimum required to achieve a desired effect. For work out, it is the maximum possible. Designers usually strive to minimize loss. In Chapter 12 we learn of one instance when minimum loss is sacrificed for survivability of fish coursing through a turbine rotor.

■ **Figure 5.8**    **Total-head change in fluid flows.**

# EXAMPLE 5.25   Energy—Head Loss and Power Loss

**GIVEN**  The pump shown in Fig. E5.25a adds 10 horsepower to the water as it pumps water from the lower lake to the upper lake. The elevation difference between the lake surfaces is 30 ft and the head loss is 15 ft.

**FIND**  Determine

**(a)**  the flowrate and

**(b)**  the power loss associated with this flow.

■ **Figure E5.25a**

## SOLUTION

**(a)**  The energy equation (Eq. 5.84) for this flow is

$$\frac{p_2}{\gamma} + \frac{V_2^2}{2g} + z_2 = \frac{p_1}{\gamma} + \frac{V_1^2}{2g} + z_1 + h_s - h_L \qquad (1)$$

where points 2 and 1 (corresponding to "out" and "in" in Eq. 5.84) are located on the lake surfaces. Thus, $p_2 = p_1 = 0$ and $V_2 = V_1 = 0$ so that Eq. 1 becomes

$$h_s = h_L + z_2 - z_1 \qquad (2)$$

where $z_2 = 30$ ft, $z_1 = 0$, and $h_L = 15$ ft. The pump head is obtained from Eq. 5.85 as

$$h_s = \dot{W}_{\text{shaft net in}}/\gamma\,Q$$
$$= (10\ \text{hp})(550\ \text{ft}\cdot\text{lb/s/hp})/(62.4\ \text{lb/ft}^3)\,Q$$
$$= 88.1/Q$$

where $h_s$ is in ft when $Q$ is in ft³/s.

Hence, from Eq. 2,

$$88.1/Q = 15\ \text{ft} + 30\ \text{ft}$$

or

$$Q = 1.96\ \text{ft}^3/\text{s} \qquad \text{(Ans)}$$

**COMMENT**  Note that in this example the purpose of the pump is to lift the water (a 30-ft head) and overcome the head loss

(a 15-ft head); it does not, overall, alter the water's pressure or velocity.

**(b)**  The power lost due to friction can be obtained from Eq. 5.85 as

$$\dot{W}_{\text{loss}} = \gamma\,Q h_L = (62.4\ \text{lb/ft}^3)(1.96\ \text{ft}^3/\text{s})(15\ \text{ft})$$
$$= 1830\ \text{ft}\cdot\text{lb/s}\ (1\ \text{hp}/550\ \text{ft}\cdot\text{lb/s})$$
$$= 3.33\ \text{hp} \qquad \text{(Ans)}$$

**COMMENTS**  The remaining 10 hp − 3.33 hp = 6.67 hp that the pump adds to the water is used to lift the water from the lower to the upper lake. This energy is not "lost," but it is stored as potential energy.

By repeating the calculations for various head losses, $h_L$, the results shown in Fig. E5.25b are obtained. Note that as the head loss increases, the flowrate decreases because an increasing portion of the 10 hp supplied by the pump is lost and, therefore, not available to lift the fluid to the higher elevation.

■ **Figure E5.25b**

A comparison of the energy equation and the Bernoulli equation has led to the concept of loss of available energy in incompressible fluid flows with friction. In Chapter 8, we discuss in detail some methods for estimating loss in incompressible flows with friction. In Section 5.4 and Chapter 11, we demonstrate that loss of available energy is also an important factor to consider in compressible flows with friction.

### 5.3.4  Application of the Energy Equation to Nonuniform Flows

The forms of the energy equation discussed in Sections 5.3.2 and 5.3.3 are applicable to one-dimensional flows, flows that are approximated with uniform velocity distributions where fluid crosses the control surface.

If the velocity profile at any section where flow crosses the control surface is not uniform, inspection of the energy equation for a control volume, Eq. 5.64, suggests that the integral

$$\int_{cs} \frac{V^2}{2} \rho \mathbf{V} \cdot \hat{\mathbf{n}} \, dA$$

*The kinetic energy coefficient is used to account for non-uniform flows.*

will require special attention. The other terms of Eq. 5.64 can be accounted for as already discussed in Sections 5.3.2 and 5.3.3.

For one stream of fluid entering and leaving the control volume, we can define the relationship

$$\int_{cs} \frac{V^2}{2} \rho \mathbf{V} \cdot \hat{\mathbf{n}} \, dA = \dot{m} \left( \frac{\alpha_{out} \overline{V}_{out}^2}{2} - \frac{\alpha_{in} \overline{V}_{in}^2}{2} \right)$$

where $\alpha$ is the **kinetic energy coefficient** and $\overline{V}$ is the average velocity defined earlier in Eq. 5.7. From the above we can conclude that

$$\frac{\dot{m} \alpha \overline{V}^2}{2} = \int_{A} \frac{V^2}{2} \rho \mathbf{V} \cdot \hat{\mathbf{n}} \, dA$$

for flow through surface area $A$ of the control surface. Thus,

$$\alpha = \frac{\int_{A} (V^2/2) \rho \mathbf{V} \cdot \hat{\mathbf{n}} \, dA}{\dot{m} \overline{V}^2/2} \tag{5.86}$$

$\alpha = 2$

Parabolic
(laminar)

$\alpha \approx 1.08$

Turbulent

$\alpha = 1$

Uniform

It can be shown that for any velocity profile, $\alpha \geq 1$, with $\alpha = 1$ only for uniform flow. Some typical velocity profile examples for flow in a conventional pipe are shown in the sketch in the margin. Therefore, for nonuniform velocity profiles, the energy equation on an energy per unit mass basis for the incompressible flow of one stream of fluid through a control volume that is steady in the mean is

$$\frac{p_{out}}{\rho} + \frac{\alpha_{out} \overline{V}_{out}^2}{2} + g z_{out} = \frac{p_{in}}{\rho} + \frac{\alpha_{in} \overline{V}_{in}^2}{2} + g z_{in} + w_{\substack{shaft \\ net\ in}} - loss \tag{5.87}$$

On an energy per unit volume basis we have

$$p_{out} + \frac{\rho \alpha_{out} \overline{V}_{out}^2}{2} + \gamma z_{out} = p_{in} + \frac{\rho \alpha_{in} \overline{V}_{in}^2}{2} + \gamma z_{in} + \rho w_{\substack{shaft \\ net\ in}} - \rho(loss) \qquad (5.88)$$

and on an energy per unit weight or head basis we have

$$\frac{p_{out}}{\gamma} + \frac{\alpha_{out} \overline{V}_{out}^2}{2g} + z_{out} = \frac{p_{in}}{\gamma} + \frac{\alpha_{in} \overline{V}_{in}^2}{2g} + z_{in} + \frac{w_{\substack{shaft \\ net\ in}}}{g} - h_L \qquad (5.89)$$

The following examples illustrate the use of the kinetic energy coefficient.

## EXAMPLE 5.26   Energy—Effect of Nonuniform Velocity Profile

**GIVEN** The small fan shown in Fig. E5.26 moves air at a mass flowrate of 0.1 kg/min. Upstream of the fan, the pipe diameter is 60 mm, the flow is laminar, the velocity distribution is parabolic, and the kinetic energy coefficient, $\alpha_1$, is equal to 2.0. Downstream of the fan, the pipe diameter is 30 mm, the flow is turbulent, the velocity profile is quite uniform, and the kinetic energy coefficient, $\alpha_2$, is equal to 1.08. The rise in static pressure across the fan is 0.1 kPa, and the fan motor draws 0.14 W.

**FIND** Compare the value of loss calculated: **(a)** assuming uniform velocity distributions, **(b)** considering actual velocity distributions.

■ **Figure E5.26**

## SOLUTION

Application of Eq. 5.87 to the contents of the control volume shown in Fig. E5.26 leads to

$$\frac{p_2}{\rho} + \frac{\alpha_2 \overline{V}_2^2}{2} + \overset{\overset{\text{0 (change in } gz \text{ is negligible)}}{\diagup}}{gz_2} = \frac{p_1}{\rho} + \frac{\alpha_1 \overline{V}_1^2}{2} + \overset{\diagup}{gz_1}$$
$$- loss + w_{\substack{shaft \\ net\ in}} \qquad (1)$$

or solving Eq. 1 for loss we get

$$loss = w_{\substack{shaft \\ net\ in}} - \left(\frac{p_2 - p_1}{\rho}\right) + \frac{\alpha_1 \overline{V}_1^2}{2} - \frac{\alpha_2 \overline{V}_2^2}{2} \qquad (2)$$

To proceed further, we need values of $w_{shaft\ net\ in}$, $\overline{V}_1$, and $\overline{V}_2$. These quantities can be obtained as follows. For shaft work

$$w_{\substack{shaft \\ net\ in}} = \frac{\text{power to fan motor}}{\dot{m}}$$

or

$$w_{\substack{shaft \\ net\ in}} = \frac{(0.14\ W)[(1\ N \cdot m/s)/W]}{0.1\ kg/min}(60\ s/min)$$
$$= 84.0\ N \cdot m/kg \qquad (3)$$

For the average velocity at section (1), $\overline{V}_1$, from Eq. 5.11 we obtain

$$\overline{V}_1 = \frac{\dot{m}}{\rho A_1}$$
$$= \frac{\dot{m}}{\rho(\pi D_1^2/4)} \qquad (4)$$

$$= \frac{(0.1\ kg/min)\ (1\ min/60\ s)\ (1000\ mm/m)^2}{(1.23\ kg/m^3)[\pi(60\ mm)^2/4]}$$
$$= 0.479\ m/s$$

For the average velocity at section (2), $\overline{V}_2$,

$$\overline{V}_2 = \frac{(0.1\ kg/min)\ (1\ min/60\ s)\ (1000\ mm/m)^2}{(1.23\ kg/m^3)[\pi(30\ mm)^2/4]}$$
$$= 1.92\ m/s \qquad (5)$$

**(a)** For the assumed uniform velocity profiles ($\alpha_1 = \alpha_2 = 1.0$), Eq. 2 yields

$$loss = w_{\substack{shaft \\ net\ in}} - \left(\frac{p_2 - p_1}{\rho}\right) + \frac{\overline{V}_1^2}{2} - \frac{\overline{V}_2^2}{2} \qquad (6)$$

Using Eqs. 3, 4, and 5 and the pressure rise given in the problem statement, Eq. 6 gives

$$loss = 84.0\ \frac{N \cdot m}{kg} - \frac{(0.1\ kPa)(1000\ Pa/kPa)(1\ N/m^2/Pa)}{1.23\ kg/m^3}$$

$$+ \frac{(0.479 \text{ m/s})^2}{2[1 \text{ (kg} \cdot \text{m)/(N} \cdot \text{s}^2)]} - \frac{(1.92 \text{ m/s})^2}{2[1 \text{ (kg} \cdot \text{m)/(N} \cdot \text{s}^2)]}$$

or

$$\text{loss} = 84.0 \text{ N} \cdot \text{m/kg} - 81.3 \text{ N} \cdot \text{m/kg}$$
$$+ 0.115 \text{ N} \cdot \text{m/kg} - 1.84 \text{ N} \cdot \text{m/kg}$$
$$= 0.975 \text{ N} \cdot \text{m/kg} \qquad \textbf{(Ans)}$$

**(b)**   For the actual velocity profiles ($\alpha_1 = 2$, $\alpha_2 = 1.08$), Eq. 1 gives

$$\text{loss} = w_{\substack{\text{shaft} \\ \text{net in}}} - \left(\frac{p_2 - p_1}{\rho}\right) + \alpha_1 \frac{\overline{V}_1^2}{2} - \alpha_2 \frac{\overline{V}_2^2}{2} \qquad \textbf{(7)}$$

If we use Eqs. 3, 4, and 5 and the given pressure rise, Eq. 7 yields

$$\text{loss} = 84 \text{ N} \cdot \text{m/kg} - \frac{(0.1 \text{ kPa})(1000 \text{ Pa/kPa})(1 \text{ N/m}^2/\text{Pa})}{1.23 \text{ kg/m}^3}$$

$$+ \frac{2(0.479 \text{ m/s})^2}{2[1 \text{ (kg} \cdot \text{m)/(N} \cdot \text{s}^2)]} - \frac{1.08(1.92 \text{ m/s})^2}{2[1 \text{ (kg} \cdot \text{m)/(N} \cdot \text{s}^2)]}$$

or

$$\text{loss} = 84.0 \text{ N} \cdot \text{m/kg} - 81.3 \text{ N} \cdot \text{m/kg}$$
$$+ 0.230 \text{ N} \cdot \text{m/kg} - 1.99 \text{ N} \cdot \text{m/kg}$$
$$= 0.940 \text{ N} \cdot \text{m/kg} \qquad \textbf{(Ans)}$$

**COMMENT**   The difference in loss calculated assuming uniform velocity profiles and actual velocity profiles is not large compared to $w_{\text{shaft net in}}$ for this fluid flow situation.

---

# $E$XAMPLE 5.27    Energy—Effect of Nonuniform Velocity Profile

**GIVEN**   Consider the flow situation of Example 5.14.

**FIND**   Apply Eq. 5.87 to develop an expression for the fluid pressure drop that occurs between sections (1) and (2). By compar-

ing the equation for pressure drop obtained presently with the result of Example 5.14, obtain an expression for loss between sections (1) and (2).

## SOLUTION

Application of Eq. 5.87 to the flow of Example 5.14 (see Fig. E5.14) leads to

$$0 \text{ (no shaft work)}$$

$$\frac{p_2}{\rho} + \frac{\alpha_2 \overline{w}_2^2}{2} + gz_2 = \frac{p_1}{\rho} + \frac{\alpha_1 \overline{w}_1^2}{2} + gz_1 - \text{loss} + w_{\substack{\text{shaft} \\ \text{net in}}} \qquad \textbf{(1)}$$

Solving Eq. 1 for the pressure drop, $p_1 - p_2$, we obtain

$$p_1 - p_2 = \rho \left[ \frac{\alpha_2 \overline{w}_2^2}{2} - \frac{\alpha_1 \overline{w}_1^2}{2} + g(z_2 - z_1) + \text{loss} \right] \qquad \textbf{(2)}$$

Since the fluid velocity at section (1), $w_1$, is uniformly distributed over cross-sectional area $A_1$, the corresponding kinetic energy coefficient, $\alpha_1$, is equal to 1.0. The kinetic energy coefficient at section (2), $\alpha_2$, needs to be determined from the velocity profile distribution given in Example 5.14. Using Eq. 5.86 we get

$$\alpha_2 = \frac{\int_{A_2} \rho w_2^3 \, dA_2}{\dot{m} \overline{w}_2^2} \qquad \textbf{(3)}$$

Substituting the parabolic velocity profile equation into Eq. 3 we obtain

$$\alpha_2 = \frac{\rho \int_0^R (2w_1)^3 [1 - (r/R)^2]^3 \, 2\pi r \, dr}{(\rho A_2 \overline{w}_2) \overline{w}_2^2}$$

From conservation of mass, since $A_1 = A_2$

$$w_1 = \overline{w}_2 \qquad \textbf{(4)}$$

Then, substituting Eq. 4 into Eq. 3, we obtain

$$\alpha_2 = \frac{\rho 8 \overline{w}_2^3 2\pi \int_0^R [1 - (r/R)^2]^3 r \, dr}{\rho \pi R^2 \overline{w}_2^3}$$

or

$$\alpha_2 = \frac{16}{R^2} \int_0^R [1 - 3(r/R)^2 + 3(r/R)^4 - (r/R)^6] r \, dr$$
$$= 2 \qquad \textbf{(5)}$$

Now we combine Eqs. 2 and 5 to get

$$p_1 - p_2 = \rho \left[ \frac{2.0 \overline{w}_2^2}{2} - \frac{1.0 \overline{w}_1^2}{2} + g(z_2 - z_1) + \text{loss} \right] \qquad \textbf{(6)}$$

However, from conservation of mass $\overline{w}_2 = \overline{w}_1 = \overline{w}$ so that Eq. 6 becomes

$$p_1 - p_2 = \frac{\rho \overline{w}^2}{2} + \rho g(z_2 - z_1) + \rho(\text{loss}) \qquad \textbf{(7)}$$

The term associated with change in elevation, $\rho g(z_2 - z_1)$, is equal to the weight per unit cross-sectional area, $\mathcal{W}/A$, of the water contained between sections (1) and (2) at any instant,

$$\rho g(z_2 - z_1) = \frac{\mathcal{W}}{A} \qquad \textbf{(8)}$$

Thus, combining Eqs. 7 and 8 we get

$$p_1 - p_2 = \frac{\rho \overline{w}^2}{2} + \frac{\mathcal{W}}{A} + \rho(\text{loss}) \qquad \textbf{(9)}$$

The pressure drop between sections (1) and (2) is due to:

1. The change in kinetic energy between sections (1) and (2) associated with going from a uniform velocity profile to a parabolic velocity profile.
2. The weight of the water column, that is, hydrostatic pressure effect.
3. Viscous loss.

Comparing Eq. 9 for pressure drop with the one obtained in Example 5.14 (i.e., the answer of Example 5.14), we obtain

$$\frac{\rho\bar{w}^2}{2} + \frac{\mathcal{W}}{A} + \rho(\text{loss}) = \frac{\rho\bar{w}^2}{3} + \frac{R_z}{A} + \frac{\mathcal{W}}{A} \quad (10)$$

or

$$\text{loss} = \frac{R_z}{\rho A} - \frac{\bar{w}^2}{6} \quad \textbf{(Ans)}$$

**COMMENT**   We conclude that while some of the pipe wall friction force, $R_z$, resulted in loss of available energy, a portion of this friction, $\rho A \bar{w}^2/6$, led to the velocity profile change.

### 5.3.5 Combination of the Energy Equation and the Moment-of-Momentum Equation[3]

If Eq. 5.82 is used for one-dimensional incompressible flow through a turbomachine, we can use Eq. 5.54, developed in Section 5.2.4 from the moment-of-momentum equation (Eq. 5.42), to evaluate shaft work. This application of both Eqs. 5.54 and 5.82 allows us to ascertain the amount of loss that occurs in incompressible turbomachine flows as is demonstrated in Example 5.28.

---

### $E$XAMPLE 5.28 Energy—Fan Performance

**GIVEN**   Consider the fan of Example 5.19.

**FIND**   Show that only some of the shaft power into the air is converted into useful effects. Develop a meaningful efficiency equation and a practical means for estimating lost shaft energy.

### SOLUTION

We use the same control volume used in Example 5.19. Application of Eq. 5.82 to the contents of this control volume yields

$$\frac{p_2}{\rho} + \frac{V_2^2}{2} + gz_2 = \frac{p_1}{\rho} + \frac{V_1^2}{2} + gz_1 + w_{\substack{\text{shaft} \\ \text{net in}}} - \text{loss} \quad (1)$$

As in Example 5.26, we can see with Eq. 1 that a "useful effect" in this fan can be defined as

$$\text{useful effect} = w_{\substack{\text{shaft} \\ \text{net in}}} - \text{loss}$$

$$= \left(\frac{p_2}{\rho} + \frac{V_2^2}{2} + gz_2\right) - \left(\frac{p_1}{\rho} + \frac{V_1^2}{2} + gz_1\right) \quad (2) \quad \textbf{(Ans)}$$

In other words, only a portion of the shaft work delivered to the air by the fan blades is used to increase the available energy of the air; the rest is lost because of fluid friction.

A meaningful efficiency equation involves the ratio of shaft work converted into a useful effect (Eq. 2) to shaft work into the air, $w_{\text{shaft net in}}$. Thus, we can express efficiency, $\eta$, as

$$\eta = \frac{w_{\substack{\text{shaft} \\ \text{net in}}} - \text{loss}}{w_{\substack{\text{shaft} \\ \text{net in}}}} \quad (3)$$

However, when Eq. 5.54, which was developed from the moment-of-momentum equation (Eq. 5.42), is applied to the contents of the control volume of Fig. E5.19, we obtain

$$w_{\substack{\text{shaft} \\ \text{net in}}} = +U_2 V_{\theta 2} \quad (4)$$

Combining Eqs. 2, 3, and 4, we obtain

$$\eta = \{[(p_2/\rho) + (V_2^2/2) + gz_2] - [(p_1/\rho) + (V_1^2/2) + gz_1]\}/U_2 V_{\theta 2} \quad (5) \quad \textbf{(Ans)}$$

Equation 5 provides us with a practical means to evaluate the efficiency of the fan of Example 5.19.

Combining Eqs. 2 and 4, we obtain

$$\text{loss} = U_2 V_{\theta 2} - \left[\left(\frac{p_2}{\rho} + \frac{V_2^2}{2} + gz_2\right) - \left(\frac{p_1}{\rho} + \frac{V_1^2}{2} + gz_1\right)\right] \quad (6) \quad \textbf{(Ans)}$$

**COMMENT**   Equation 6 provides us with a useful method of evaluating the loss due to fluid friction in the fan of Example 5.19 in terms of fluid mechanical variables that can be measured.

---

[3]This section may be omitted without loss of continuity in the text material. This section should not be considered without prior study of Sections 5.2.3 and 5.2.4. All of these sections are recommended for those interested in Chapter 12.

# 5.4    Second Law of Thermodynamics—Irreversible Flow

This section is available on *WileyPLUS*. Note that this entire section may be omitted without loss of continuity in the text material.

# 5.5    Chapter Summary and Study Guide

In this chapter the flow of a fluid is analyzed by using important principles including conservation of mass, Newton's second law of motion, and the first law of thermodynamics as applied to control volumes. The Reynolds transport theorem is used to convert basic system-oriented laws into corresponding control volume formulations.

The continuity equation, a statement of the fact that mass is conserved, is obtained in a form that can be applied to any flow—steady or unsteady, incompressible or compressible. Simplified forms of the continuity equation enable tracking of fluid everywhere in a control volume, where it enters, where it leaves, and within. Mass or volume flowrates of fluid entering or leaving a control volume and rate of accumulation or depletion of fluid within a control volume can be estimated.

The linear momentum equation, a form of Newton's second law of motion applicable to flow of fluid through a control volume, is obtained and used to solve flow problems. Net force results from or causes changes in linear momentum (velocity magnitude and/or direction) of fluid flowing through a control volume. Work and power associated with force can be involved.

The moment-of-momentum equation, which involves the relationship between torque and changes in angular momentum, is obtained and used to solve flow problems dealing with turbines (energy extracted from a fluid) and pumps (energy supplied to a fluid).

The steady-state energy equation, obtained from the first law of thermodynamics (conservation of energy), is written in several forms. The first (Eq. 5.69) involves power terms. The second form (Eq. 5.82 or 5.84) is termed the mechanical energy equation or the extended Bernoulli equation. It consists of the Bernoulli equation with extra terms that account for energy losses due to friction in the flow, as well as terms accounting for the work of pumps or turbines in the flow.

The following checklist provides a study guide for this chapter. When your study of the entire chapter and end-of-chapter exercises has been completed, you should be able to

*conservation of mass*
*continuity equation*
*mass flowrate*
*linear momentum equation*
*moment-of-momentum equation*
*shaft power*
*shaft torque*
*first law of thermodynamics*
*heat transfer rate*
*energy equation*
*loss*
*shaft work head*
*head loss*
*kinetic energy coefficient*

- ■ write out meanings of the terms listed here in the margin and understand each of the related concepts. These terms are particularly important and are set in *italic, **bold**, and color* type in the text.

- ■ select an appropriate control volume for a given problem and draw an accurately labeled control volume diagram.

- ■ use the continuity equation and a control volume to solve problems involving mass or volume flowrate.

- ■ use the linear momentum equation and a control volume, in conjunction with the continuity equation as necessary, to solve problems involving forces related to linear momentum change.

- ■ use the moment-of-momentum equation to solve problems involving torque and related work and power due to angular momentum change.

- ■ use the energy equation, in one of its appropriate forms, to solve problems involving losses due to friction (head loss) and energy input by pumps or extraction by turbines.

- ■ use the kinetic energy coefficient in the energy equation to account for nonuniform flows.

Some of the important equations in this chapter are:

| | | |
|---|---|---|
| Conservation of mass | $\dfrac{\partial}{\partial t} \displaystyle\int_{cv} \rho \, d\mathcal{V} + \int_{cs} \rho \mathbf{V} \cdot \hat{\mathbf{n}} \, dA = 0$ | (5.5) |
| Mass flowrate | $\dot{m} = \rho Q = \rho A V$ | (5.6) |

| | | |
|---|---|---|
| Average velocity | $\overline{V} = \dfrac{\displaystyle\int_A \rho \mathbf{V} \cdot \hat{\mathbf{n}}\, dA}{\rho A}$ | **(5.7)** |
| Steady-flow mass conservation | $\sum \dot{m}_{\text{out}} - \sum \dot{m}_{\text{in}} = 0$ | **(5.9)** |
| Moving control volume mass conservation | $\dfrac{\partial}{\partial t} \displaystyle\int_{\text{cv}} \rho\, d\forall + \int_{\text{cs}} \rho \mathbf{W} \cdot \hat{\mathbf{n}}\, dA = 0$ | **(5.16)** |
| Deforming control volume mass conservation | $\dfrac{DM_{\text{sys}}}{Dt} = \dfrac{\partial}{\partial t} \displaystyle\int_{\text{cv}} \rho\, d\forall + \int_{\text{cs}} \rho \mathbf{W} \cdot \hat{\mathbf{n}}\, dA = 0$ | **(5.17)** |
| Force related to change in linear momentum | $\dfrac{\partial}{\partial t} \displaystyle\int_{\text{cv}} \mathbf{V}\rho\, d\forall + \int_{\text{cs}} \mathbf{V}\rho \mathbf{V} \cdot \hat{\mathbf{n}}\, dA = \sum \mathbf{F}_{\substack{\text{contents of the} \\ \text{control volume}}}$ | **(5.22)** |
| Moving control volume force related to change in linear momentum | $\displaystyle\int_{\text{cs}} \mathbf{W}\rho \mathbf{W} \cdot \hat{\mathbf{n}}\, dA = \sum \mathbf{F}_{\substack{\text{contents of the} \\ \text{control volume}}}$ | **(5.29)** |
| Vector addition of absolute and relative velocities | $\mathbf{V} = \mathbf{W} + \mathbf{U}$ | **(5.43)** |
| Shaft torque from force | $\sum \left[ (\mathbf{r} \times \mathbf{F})_{\substack{\text{contents of the} \\ \text{control volume}}} \right]_{\text{axial}} = \mathbf{T}_{\text{shaft}}$ | **(5.45)** |
| Shaft torque related to change in moment-of-momentum (angular momentum) | $T_{\text{shaft}} = (-\dot{m}_{\text{in}})(\pm r_{\text{in}} V_{\theta\text{in}}) + \dot{m}_{\text{out}}(\pm r_{\text{out}} V_{\theta\text{out}})$ | **(5.50)** |
| Shaft power related to change in moment-of-momentum (angular momentum) | $\dot{W}_{\text{shaft}} = (-\dot{m}_{\text{in}})(\pm U_{\text{in}} V_{\theta\text{in}}) + \dot{m}_{\text{out}}(\pm U_{\text{out}} V_{\theta\text{out}})$ | **(5.53)** |
| First law of thermodynamics (Conservation of energy) | $\dfrac{\partial}{\partial t} \displaystyle\int_{\text{cv}} e\rho\, d\forall + \int_{\text{cs}} \left( \check{u} + \dfrac{p}{\rho} + \dfrac{V^2}{2} + gz \right) \rho \mathbf{V} \cdot \hat{\mathbf{n}}\, dA = \dot{Q}_{\substack{\text{net} \\ \text{in}}} + \dot{W}_{\substack{\text{shaft} \\ \text{net in}}}$ | **(5.64)** |
| Conservation of power | $\dot{m}\left[ \check{h}_{\text{out}} - \check{h}_{\text{in}} + \dfrac{V_{\text{out}}^2 - V_{\text{in}}^2}{2} + g(z_{\text{out}} - z_{\text{in}}) \right] = \dot{Q}_{\substack{\text{net} \\ \text{in}}} + \dot{W}_{\substack{\text{shaft} \\ \text{net in}}}$ | **(5.69)** |
| Conservation of mechanical energy | $\dfrac{p_{\text{out}}}{\rho} + \dfrac{V_{\text{out}}^2}{2} + gz_{\text{out}} = \dfrac{p_{\text{in}}}{\rho} + \dfrac{V_{\text{in}}^2}{2} + gz_{\text{in}} + w_{\substack{\text{shaft} \\ \text{net in}}} - \text{loss}$ | **(5.82)** |

# References

1. Eck, B., *Technische Stromungslehre*, Springer-Verlag, Berlin, Germany, 1957.
2. Dean, R. C., "On the Necessity of Unsteady Flow in Fluid Machines," *ASME Journal of Basic Engineering* 81D; 24–28, March 1959.
3. Moran, M. J., and Shapiro, H. N., *Fundamentals of Engineering Thermodynamics*, 6th Ed., Wiley, New York, 2008.

| | |
|---|---|
| WILEY○ | Problem available in *WileyPLUS* at instructor's discretion. |
| WILEY○ GO | Tutoring problem available in *WileyPLUS* at instructor's discretion. |
| ▶ | Problem is related to a chapter video available in *WileyPLUS*. |
| ∗ | Problem to be solved with aid of programmable calculator or computer. |
| † | Open-ended problem that requires critical thinking. These problems require various assumptions to provide the necessary input data. There are not unique answers to these problems. |

# Review Problems

Go to Appendix G (*WileyPLUS* or the book's web site, www. wiley.com/college/munson) for a set of review problems with answers. Detailed solutions can be found in the *Student Solution* *Manual and Study Guide for Fundamentals of Fluid Mechanics,* by Munson et al. (© 2013 John Wiley and Sons, Inc.).

# Conceptual Questions

**5.1C** A fluid flows steadily through a pipe with a uniform cross-sectional area. The density of the fluid decreases to half its initial value as it flows through the pipe. The correct statement about the average velocity $V$ is

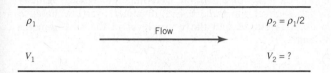

a) $V_2$ equals $2\,V_1$.    b) $V_2$ equals $V_1/2$.    c) $V_2$ equals $V_1$.
d) $V_2$ equals $V_1/4$.    e) $V_2$ equals $4\,V_1$.

**5.2C** Water flows steadily into and out of the system with four pipes shown below. The mass flowrate through three of the pipes in kg/s is indicated.

The mass flowrate $\dot{m}$ in the fourth pipe is
a) 13 kg/s.    b) 9 kg/s.    c) 6 kg/s.
d) 5 kg/s.    e) 4kg/s.

**5.3C** The nozzle on a fire hose is connected to the hose via a coupling. When the fire hose is in use with water flowing through it and the hose is stationary, the coupling is:

a) in equilibrium, so there is no force on the coupling.

b) in tension.

c) in compression.

**5.4C** Two fluid jets are pointed at surfaces as shown in the following figures. The fluids are incompressible, and the effects of gravity can be neglected. The mass flowrates and the velocities of the jets are identical. The cross-sectional areas of the jets do not change significantly as the fluid flows.

The correct statement regarding the horizontal forces is

a) $F_1$ equals $2\,F_2$.

b) $F_1$ is greater than 0 and $F_2$ equals 0.

c) $F_1$ equals $F_2/2$.

d) $F_1$ equals 0 and $F_2$ is greater than 0.

e) $F_1$ equals $F_2$.

**Additional conceptual questions are available in *WileyPLUS* at the instructor's discretion.**

# Problems

**Note: Unless specific values of required fluid properties are given in the problem statement, use the values found in the tables on the inside of the front cover. Answers to the even-numbered problems are listed at the end of the book. The Lab Problems as well as the videos that accompany problems can be accessed in *WileyPLUS* or the book's web site, www.wiley.com/college/munson.**

## Section 5.1.1 Derivation of the Continuity Equation

**5.1** Use the Reynolds transport theorem (Eq. 4.19) with $B$ = volume and, therefore, $b$ = volume/mass = 1/density to obtain the continuity equation for steady or unsteady incompressible flow through a fixed control volume: $\displaystyle\int_{cv} \mathbf{V} \cdot \hat{\mathbf{n}}\,dA = 0$.

**5.2** An incompressible fluid flows horizontally in the $x$–$y$ plane with a velocity given by

$$u = 30(y/h)^{1/2} \text{ m/s}, \text{ v} = 0$$

where $y$ and $h$ are in meters and $h$ is a constant. Determine the average velocity for the portion of the flow between $y = 0$ and $y = h$.

## Section 5.1.2 Fixed, Nondeforming Control Volume— Uniform Velocity Profile or Average Velocity

**5.3** Water flows steadily through the horizontal piping system shown in Fig. P5.3. The velocity is uniform at section (1), the mass flowrate is 10 slugs/s at section (2), and the velocity is nonuniform at section (3). **(a)** Determine the value of the quantity $\frac{D}{Dt}\int_{sys}\rho d\mathcal{V}$, where the system is the water contained in the pipe bounded by sections (1), (2), and (3). **(b)** Determine the mean velocity at section (2). **(c)** Determine, if possible, the value of the integral $\int_{(3)}\rho V \cdot \hat{n}\, dA$ over section (3). If it is not possible, explain what additional information is needed to do so.

(2)
10 slugs/s
Area = 0.3 ft²

15 ft/s

(1) Area = 0.7 ft²          (3) Area = 0.7 ft²

■ **Figure P5.3**

**5.4** (WILEY) Water flows out through a set of thin, closely spaced blades as shown in Fig. P5.4 with a speed of $V = 10$ ft/s around the entire circumference of the outlet. Determine the mass flowrate through the inlet pipe.

Inlet

0.08-ft diameter

0.1 ft

Blades

0.6 ft

60°

$V = 10$ ft/s

■ **Figure P5.4**

† **5.5** Estimate the rate (in gal/hr) that your car uses gasoline when it is being driven on an interstate highway. Determine how long it would take to empty a 12-oz soft-drink container at this flowrate. List all assumptions and show calculations.

**5.6** (WILEY) The pump shown in Fig. P5.6 produces a steady flow of 10 gal/s through the nozzle. Determine the nozzle exit diameter, $D_2$, if the exit velocity is to be $V_2 = 100$ ft/s.

Section (1)

Pump

Section (2)

$V_2$

$D_2$

■ **Figure P5.6**

**5.7** ▶ (WILEY) Water flows into a sink as shown in **Video V5.1** and Fig. P5.7 at a rate of 2 gal/min. Determine the average velocity through each of the three 0.4-in.-diameter overflow holes if the drain is closed and the water level in the sink remains constant.

Three 0.4-in.-diameter overflow holes

$Q = 2$ gal/min

Drain

■ **Figure P5.7**

**5.8** The wind blows through a 7 ft × 10 ft garage door opening with a speed of 5 ft/s as shown in Fig. P5.8. Determine the average speed, $V$, of the air through the two 3 ft × 4 ft openings in the windows.

$V$          $V$

3 ft          3 ft

16 ft          10 ft          5 ft/s

20°

22 ft

■ **Figure P5.8**

**5.9** The human circulatory system consists of a complex branching pipe network ranging in diameter from the aorta (largest) to the capillaries (smallest). The average radii and the number of these vessels are shown in the table. Does the average blood velocity increase, decrease, or remain constant as it travels from the aorta to the capillaries?

| Vessel | Average Radius, mm | Number |
|---|---|---|
| Aorta | 12.5 | 1 |
| Arteries | 2.0 | 159 |
| Arterioles | 0.03 | $1.4 \times 10^7$ |
| Capillaries | 0.006 | $3.9 \times 10^9$ |

**5.10** (WILEY) Air flows steadily between two cross sections in a long, straight section of 0.1-m-inside-diameter pipe. The static temperature and pressure at each section are indicated in Fig. P5.10. If the average air velocity at section (1) is 205 m/s, determine the average air velocity at section (2).

Section (1)    Section (2)

$p_1 = 77$ kPa (abs)    $p_2 = 45$ kPa (abs)
$T_1 = 268$ K    $T_2 = 240$ K
$V_1 = 205$ m/s

■ **Figure P5.10**

**5.11** ▶ (WILEY) A hydraulic jump (see **Video V10.11**) is in place downstream from a spillway as indicated in Fig. P5.11. Upstream of the jump, the depth of the stream is 0.6 ft and the average stream velocity is 18 ft/s. Just downstream of the jump, the average stream velocity is 3.4 ft/s. Calculate the depth of the stream, $h$, just downstream of the jump.

0.6 ft    $h$    3.4 ft/s
18 ft/s

■ **Figure P5.11**

**5.12** Water enters a rigid, sealed, cylindrical tank at a steady rate of 100 liters/hr and forces gasoline ($SG = 0.68$) out as is indicated in Fig. P5.12. What is the time rate of change of mass of gasoline contained in the tank?

Gasoline ($SG = 0.68$)

Water

■ **Figure P5.12**

**5.13** (WILEY) An evaporative cooling tower (see Fig. P5.13) is used to cool water from 110 to 80 °F. Water enters the tower at a rate of 250,000 lbm/hr. Dry air (no water vapor) flows into the tower at a rate of 151,000 lbm/hr. If the rate of wet airflow out of the tower is 156,900 lbm/hr, determine the rate of water evaporation in lbm/hr and the rate of cooled water flow in lbm/hr.

Wet air
$\dot{m} = 156,900$ lbm/hr

Warm water
$\dot{m} = 250,000$ lbm/hr

Dry air
$\dot{m} = 151,000$ lbm/hr

Cooled water

■ **Figure P5.13**

**5.14** (WILEY) At cruise conditions, air flows into a jet engine at a steady rate of 65 lbm/s. Fuel enters the engine at a steady rate of 0.60 lbm/s. The average velocity of the exhaust gases is 1500 ft/s relative to the engine. If the engine exhaust effective cross-sectional area is 3.5 ft$^2$, estimate the density of the exhaust gases in lbm/ft$^3$.

**5.15** (WILEY) Water at 0.1 m$^3$/s and alcohol ($SG = 0.8$) at 0.3 m$^3$/s are mixed in a $y$-duct as shown in Fig. 5.15. What is the average density of the mixture of alcohol and water?

Water and alcohol mix

Water
$Q = 0.1$ m$^3$/s

Alcohol ($SG = 0.8$)
$Q = 0.3$ m$^3$/s

■ **Figure P5.15**

**5.16** Oil having a specific gravity of 0.9 is pumped as illustrated in Fig. P5.16 with a water jet pump. The water volume flowrate is 1 m$^3$/s. The water and oil mixture has an average specific gravity of 0.95. Calculate the rate, in m$^3$/s, at which the pump moves oil.

Section (1)    Section (3)

Water
$Q_1 = 1$ m$^3$/s

Water and oil mix
($SG = 0.95$)

Section (2)

Oil ($SG = 0.9$)

■ **Figure P5.16**

**5.17** WILEY° Fresh water flows steadily into an open 55-gal drum initially filled with seawater. The fresh water mixes thoroughly with the seawater, and the mixture overflows out of the drum. If the fresh water flowrate is 10 gal/min, estimate the time in seconds required to decrease the difference between the density of the mixture and the density of fresh water by 50%.

## Section 5.1.2 Fixed, Nondeforming Control Volume— Nonuniform Velocity Profile

**5.18** WILEY° A water jet pump (see Fig. P5.18) involves a jet cross-sectional area of 0.01 m², and a jet velocity of 30 m/s. The jet is surrounded by entrained water. The total cross-sectional area associated with the jet and entrained streams is 0.075 m². These two fluid streams leave the pump thoroughly mixed with an average velocity of 6 m/s through a cross-sectional area of 0.075 m². Determine the pumping rate (i.e., the entrained fluid flowrate) involved in liters/s.

■ **Figure P5.18**

***5.19** To measure the mass flowrate of air through a 6-in.-inside-diameter pipe, local velocity data are collected at different radii from the pipe axis (see Table). Determine the mass flowrate corresponding to the data listed in the following table.

| r (in.) | Axial Velocity (ft/s) |
|---------|----------------------|
| 0 | 30 |
| 0.2 | 29.71 |
| 0.4 | 29.39 |
| 0.6 | 29.06 |
| 0.8 | 28.70 |
| 1.0 | 28.31 |
| 1.2 | 27.89 |
| 1.4 | 27.42 |
| 1.6 | 26.90 |
| 1.8 | 26.32 |
| 2.0 | 25.64 |
| 2.2 | 24.84 |
| 2.4 | 23.84 |
| 2.6 | 22.50 |
| 2.8 | 20.38 |
| 2.9 | 18.45 |
| 2.95 | 16.71 |
| 2.98 | 14.66 |
| 3.00 | 0 |

**5.20** WILEY° Two rivers merge to form a larger river as shown in Fig. P5.20. At a location downstream from the junction (before the two streams completely merge), the nonuniform velocity profile is as shown and the depth is 6 ft. Determine the value of V.

■ **Figure P5.20**

**5.21** ▶ WILEY° Various types of attachments can be used with the shop vac shown in **Video V5.2**. Two such attachments are shown in Fig. P5.21—a nozzle and a brush. The flowrate is 1 ft³/s. **(a)** Determine the average velocity through the nozzle entrance, $V_n$. **(b)** Assume the air enters the brush attachment in a radial direction all around the brush with a velocity profile that varies linearly from 0 to $V_b$ along the length of the bristles as shown in the figure. Determine the value of $V_b$.

■ **Figure P5.21**

**5.22** WILEY° An appropriate turbulent pipe flow velocity profile is

$$\mathbf{V} = u_c \left( \frac{R - r}{R} \right)^{1/n} \hat{\mathbf{i}}$$

where $u_c$ = centerline velocity, $r$ = local radius, $R$ = pipe radius, and $\hat{\mathbf{i}}$ = unit vector along pipe centerline. Determine the ratio of average velocity, $\bar{u}$, to centerline velocity, $u_c$, for **(a)** $n = 4$, **(b)** $n = 6$, **(c)** $n = 8$, **(d)** $n = 10$. Compare the different velocity profiles.

**5.23** WILEY°GO As shown in Fig. P5.23, at the entrance to a 3-ft-wide channel the velocity distribution is uniform with a velocity V. Further downstream the velocity profile is given by $u = 4y - 2y^2$, where $u$ is in ft/s and y is in ft. Determine the value of V.

■ **Figure P5.23**

**5.24** WILEY° An incompressible flow velocity field (water) is given as

$$\mathbf{V} = -\frac{1}{r}\hat{\mathbf{e}}_r + \frac{1}{r}\hat{\mathbf{e}}_\theta \text{ m/s}$$

where $r$ is in meters. **(a)** Calculate the mass flowrate through the cylindrical surface at $r = 1$ m from $z = 0$ to $z = 1$ m as shown in Fig. P5.24a. **(b)** Show that mass is conserved in the annular control volume from $r = 1$ m to $r = 2$ m and $z = 0$ to $z = 1$ m as shown in Fig. P5.24b.

(a)                            (b)

■ **Figure P5.24**

**5.25** Flow of a viscous fluid over a flat plate surface results in the development of a region of reduced velocity adjacent to the wetted surface as depicted in Fig. P5.25. This region of reduced flow is called a boundary layer. At the leading edge of the plate, the velocity profile may be considered uniformly distributed with a value $U$. All along the outer edge of the boundary layer, the fluid velocity component parallel to the plate surface is also $U$. If the $x$-direction velocity profile at section (2) is

$$\frac{u}{U} = \left(\frac{y}{\delta}\right)^{1/7}$$

develop an expression for the volume flowrate through the edge of the boundary layer from the leading edge to a location downstream at $x$ where the boundary layer thickness is $\delta$.

■ **Figure P5.25**

## Section 5.1.2 Fixed, Nondeforming Control Volume—Unsteady Flow

**5.26** ⓦⁱˡᵉʸ Air at standard conditions enters the compressor shown in Fig. P5.26 at a rate of 10 ft³/s. It leaves the tank through a 1.2-in.-diameter pipe with a density of 0.0035 slugs/ft³ and a uniform speed of 700 ft/s. **(a)** Determine the rate (slugs/s) at which the mass of air in the tank is increasing or decreasing. **(b)** Determine the average time rate of change of air density within the tank.

■ **Figure P5.26**

**5.27** ⓦⁱˡᵉʸ Estimate the time required to fill with water a cone-shaped container (see Fig. P5.27) 5 ft high and 5 ft across at the top if the filling rate is 20 gal/min.

■ **Figure P5.27**

**5.28** How long would it take to fill a cylindrical-shaped swimming pool having a diameter of 8 m to a depth of 1.5 m with water from a garden hose if the flowrate is 1.0 liter/s?

## Section 5.1.3 Moving, Nondeforming Control Volume

†**5.29** For an automobile moving along a highway, describe the control volume you would use to estimate the flowrate of air across the radiator. Explain how you would estimate the velocity of that air.

## Section 5.1.4 Deforming Control Volume

**5.30** ⓦⁱˡᵉʸ A hypodermic syringe (see Fig. P5.30) is used to apply a vaccine. If the plunger is moved forward at the steady rate of 20 mm/s and if vaccine leaks past the plunger at 0.1 of the volume flowrate out the needle opening, calculate the average velocity of the needle exit flow. The inside diameters of the syringe and the needle are 20 mm and 0.7 mm.

■ **Figure P5.30**

**5.31** ▶ ⓦⁱˡᵉʸ The Hoover Dam (see **Video V2.4**) backs up the Colorado River and creates Lake Mead, which is approximately 115 miles long and has a surface area of approximately 225 square miles. If during flood conditions the Colorado River flows into the lake at a rate of 45,000 cfs and the outflow from the dam is 8000 cfs, how many feet per 24-hour day will the lake level rise?

**5.32** ⓦⁱˡᵉʸ Storm sewer backup causes your basement to flood at the steady rate of 1 in. of depth per hour. The basement floor area is 1500 ft². What capacity (gal/min) pump would you rent to **(a)** keep the water accumulated in your basement at a constant level until the storm sewer is blocked off, and **(b)** reduce the water accumulation in your basement at a rate of 3 in./hr even while the backup problem exists?

**5.33** ⓦⁱˡᵉʸ (See Fluids in the News article "New 1.6-gpf standards," Section 5.1.2.) When a toilet is flushed, the water depth, $h$, in the tank as a function of time, $t$, is as given in the table. The size of the rectangular tank is 19 in. by 7.5 in. **(a)** Determine the volume of water used per flush, gpf. **(b)** Plot the flowrate for $0 \leq t \leq 6$ s.

| $t$ (s) | $h$ (in.) |
|---|---|
| 0 | 5.70 |
| 0.5 | 5.33 |
| 1.0 | 4.80 |
| 2.0 | 3.45 |
| 3.0 | 2.40 |
| 4.0 | 1.50 |
| 5.0 | 0.75 |
| 6.0 | 0 |

## Section 5.2.1 Derivation of the Linear Momentum Equation

**5.34** A fluid flows steadily in the $x$ direction through a control volume. Measurements indicate that to cause this flow the force acting on the contents of the control volume is 120 N in the negative $x$ direction. Determine the net rate of flow of linear momentum through the control surface.

**5.35** ⟨WILEY⟩ Consider the unsteady flow of a fluid in the $x$ direction through a control volume. The linear momentum of the fluid within the control volume is a function of time given by $200\, t\, \hat{\mathbf{i}}$ slug · ft/s, where $t$ is in seconds and $\hat{\mathbf{i}}$ is a unit vector in the $x$ direction. Measurements indicate that to cause this flow the force acting on the contents of the control volume is $40\, \hat{\mathbf{i}}$ lb. Determine the net rate of flow of linear momentum through the control surface.

## Section 5.2.2 Application of the Linear Momentum Equation (also see Lab Problems 5.1LP, 5.2LP, 5.3LP, and 5.4LP)

**5.36** ▶ ⟨WILEY⟩ A 10-mm-diameter jet of water is deflected by a homogeneous rectangular block (15 mm by 200 mm by 100 mm) that weighs 6 N as shown in **Video V5.6** and Fig. P5.36. Determine the minimum volume flowrate needed to tip the block.

■ **Figure P5.36**

†**5.37** When a baseball player catches a ball, the force of the ball on her glove is as shown as a function of time in Fig. P5.37. Describe how this situation is similar to the force generated by the deflection of a jet of water by a vane. *Note:* Consider many baseballs being caught in quick succession.

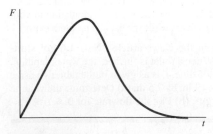

■ **Figure P5.37**

**5.38** ⟨WILEY⟩ Determine the anchoring force required to hold in place the conical nozzle attached to the end of the laboratory sink faucet shown in Fig. P5.38 when the water flowrate is 10 gal/min. The nozzle weight is 0.2 lb. The nozzle inlet and exit inside diameters are 0.6 and 0.2 in., respectively. The nozzle axis is vertical, and the axial distance between sections (1) and (2) is 1.2 in. The pressure at section (1) is 68 psi.

■ **Figure P5.38**

**5.39** ⟨WILEY⟩ Water flows through a horizontal, 180° pipe bend as is illustrated in Fig. P5.39. The flow cross-sectional area is constant at a value of 9000 mm². The flow velocity everywhere in the bend is 15 m/s. The pressures at the entrance and exit of the bend are 210 and 165 kPa, respectively. Calculate the horizontal ($x$ and $y$) components of the anchoring force needed to hold the bend in place.

■ **Figure P5.39**

**5.40** Water flows through a horizontal bend and discharges into the atmosphere as shown in Fig. P5.40. When the pressure gage reads 10 psi, the resultant $x$-direction anchoring force, $F_{Ax}$, in the horizontal plane required to hold the bend in place is shown on the figure. Determine the flowrate through the bend and the $y$-direction anchoring force, $F_{Ay}$, required to hold the bend in place. The flow is not frictionless.

■ **Figure P5.40**

**5.41** A free jet of fluid strikes a wedge as shown in Fig. P5.41. Of the total flow, a portion is deflected 30°; the remainder is not deflected. The horizontal and vertical components of force needed to hold the wedge stationary are $F_H$ and $F_V$, respectively. Gravity is negligible, and the fluid speed remains constant. Determine the force ratio, $F_H/F_V$.

**■ Figure P5.41**

**5.42** [WILEY] Water enters the horizontal, circular cross-sectional, sudden contraction nozzle sketched in Fig. P5.42 at section (1) with a uniformly distributed velocity of 25 ft/s and a pressure of 75 psi. The water exits from the nozzle into the atmosphere at section (2) where the uniformly distributed velocity is 100 ft/s. Determine the axial component of the anchoring force required to hold the contraction in place.

**■ Figure P5.42**

**†5.43** A truck carrying chickens is too heavy for a bridge that it needs to cross. The empty truck is within the weight limits; with the chickens it is overweight. It is suggested that if one could get the chickens to fly around the truck (i.e., by banging on the truck side) it would be safe to cross the bridge. Do you agree? Explain.

**5.44** ▶ [WILEY] Exhaust (assumed to have the properties of standard air) leaves the 4-ft-diameter chimney shown in **Video V5.4** and Fig. P5.44 with a speed of 6 ft/s. Because of the wind, after a few diameters downstream the exhaust flows in a horizontal direction with the speed of the wind, 15 ft/s. Determine the horizontal component of the force that the blowing wind exerts on the exhaust gases.

**■ Figure P5.44**

**5.45** Air flows steadily between two cross sections in a long, straight section on 12-in.-inside-diameter pipe. The static temperature and pressure at each section are indicated in Fig P5.45. If the average air velocity at section (2) is 320 m/s, determine the average air velocity at section (1). Determine the frictional force exerted by the pipe wall on the air flowing between sections (1) and (2). Assume uniform velocity distributions at each section.

Section (1)
$p_1 = 690$ kPa (abs)
$T_1 = 300$ K

Section (2)
$p_2 = 127$ kPa (abs)
$T_2 = 252$ K
$V_2 = 320$ m/s

**■ Figure P5.45**

**5.46** Water flows steadily from a tank mounted on a cart as shown in Fig. 5.46. After the water jet leaves the nozzle of the tank, it falls and strikes a vane attached to another cart. The cart's wheels are frictionless, and the fluid is inviscid. **(a)** Determine the speed of the water leaving the tank, $V_1$, and the water speed leaving the cart, $V_2$. **(b)** Determine the tension in rope A. **(c)** Determine the tension in rope B.

**■ Figure P5.46**

**5.47** [WILEY] Determine the magnitude and direction of the anchoring force needed to hold the horizontal elbow and nozzle combination shown in Fig. P5.47 in place. Atmospheric pressure is 100 kPa(abs). The gage pressure at section (1) is 100 kPa. At section (2), the water exits to the atmosphere.

$p_1 = 100$ kPa
$V_1 = 2$ m/s

**■ Figure P5.47**

**5.48** Water is added to the tank shown in Fig. P5.48 through a vertical pipe to maintain a constant (water) level. The tank is placed

**■ Figure P5.48**

on a horizontal plane which has a frictionless surface. Determine the horizontal force, $F$, required to hold the tank stationary. Neglect all losses.

**5.49** (WILEY GO) Water flows as two free jets from the tee attached to the pipe shown in Fig. P5.49. The exit speed is 15 m/s. If viscous effects and gravity are negligible, determine the $x$ and $y$ components of the force that the pipe exerts on the tee.

**■ Figure P5.49**

**5.50** A nozzle is attached to a vertical pipe and discharges water into the atmosphere as shown in Fig. P5.50. When the discharge is 0.1 m³/s, the gage pressure at the flange is 40 kPa. Determine the vertical component of the anchoring force required to hold the nozzle in place. The nozzle has a weight of 200 N, and the volume of water in the nozzle is 0.012 m³. Is the anchoring force directed upward or downward?

**■ Figure P5.50**

**5.51** (WILEY) The hydraulic dredge shown in Fig. P5.51 is used to dredge sand from a river bottom. Estimate the thrust needed from the propeller to hold the boat stationary. Assume the specific gravity of the sand/water mixture is $SG = 1.4$.

**■ Figure P5.51**

**5.52** (WILEY) A static thrust stand is to be designed for testing a specific jet engine, knowing the following conditions for a typical test.

$$\begin{aligned}
\text{intake air velocity} &= 700 \text{ ft/s}\\
\text{exhaust gas velocity} &= 1640 \text{ ft/s}\\
\text{intake cross section area} &= 10 \text{ ft}^2\\
\text{intake static pressure} &= 11.4 \text{ psia}\\
\text{intake static temperature} &= 480 \text{ °R}\\
\text{exhaust gas pressure} &= 0 \text{ psi}
\end{aligned}$$

estimate a nominal thrust to design for.

**5.53** A vertical jet of water leaves a nozzle at a speed of 10 m/s and a diameter of 20 mm. It suspends a plate having a mass of 1.5 kg as indicated in Fig. P5.53. What is the vertical distance $h$?

**■ Figure P5.53**

**5.54** (WILEY) A horizontal, circular cross-sectional jet of air having a diameter of 6 in. strikes a conical deflector as shown in Fig. P5.54. A horizontal anchoring force of 5 lb is required to hold the cone in place. Estimate the nozzle flowrate in ft³/s. The magnitude of the velocity of the air remains constant.

**■ Figure P5.54**

**5.55** (WILEY) A vertical, circular cross-sectional jet of air strikes a conical deflector as indicated in Fig. P5.55. A vertical anchoring force of 0.1 N is required to hold the deflector in place. Determine the mass (kg) of the deflector. The magnitude of velocity of the air remains constant.

**■ Figure P5.55**

**5.56** A vertical jet of water having a nozzle exit velocity of 15 ft/s with a diameter of 1 in. suspends a hollow hemisphere as indicated in Fig. P5.56. If the hemisphere is stationary at an elevation of 12 in., determine its weight.

12 in.

■ **Figure P5.56**

**5.57** Air flows into the atmosphere from a nozzle and strikes a vertical plate as shown in Fig. P5.57. A horizontal force of 12 N is required to hold the plate in place. Determine the reading on the pressure gage. Assume the flow to be incompressible and frictionless.

$p = ?$

12N

Area = 0.003 m$^2$

Area = 0.01 m$^2$

■ **Figure P5.57**

**5.58** WILEY• Water flows from a large tank into a dish as shown in Fig. P5.58. **(a)** If at the instant shown the tank and the water in it weigh $W_1$ lb, what is the tension, $T_1$, in the cable supporting the tank? **(b)** If at the instant shown the dish and the water in it weigh $W_2$ lb, what is the force, $F_2$, needed to support the dish?

$T_1$

Tank

10 ft

0.1-ft diameter

12 ft

2 ft

Dish

$F_2$

■ **Figure P5.58**

**5.59** WILEY• Two water jets of equal size and speed strike each other as shown in Fig. P5.59. Determine the speed, $V$, and direction, $\theta$, of the resulting combined jet. Gravity is negligible.

$V$

$\theta$

$V_2 = 10$ ft/s

0.1 ft

90°

0.1 ft

$V_1 = 10$ ft/s

■ **Figure P5.59**

**5.60** WILEY GO Assuming frictionless, incompressible, one-dimensional flow of water through the horizontal tee connection sketched in Fig. P5.60, estimate values of the $x$ and $y$ components of the force exerted by the tee on the water. Each pipe has an inside diameter of 1 m.

Section (3)

$z$

Section (2)

$Q_3 = 10$ m$^3$/s

$y$

$x$

Section (1)

$V_1 = 6$ m/s

$p_1 = 200$ kPa

■ **Figure P5.60**

**5.61** Water discharges into the atmosphere through the device shown in Fig. P5.61. Determine the $x$ component of force at the flange required to hold the device in place. Neglect the effect of gravity and friction.

$A = 0.4$ ft$^2$

30 ft/s

10 psi

Flange

20 ft/s

$A = 0.8$ ft$^2$

$A = 0.8$ ft$^2$

43°

$V$

■ **Figure P5.61**

**5.62** WILEY• Determine the magnitude of the horizontal component of the anchoring force required to hold in place the sluice gate shown in Fig. 5.62. Compare this result with the size of the horizontal component of the anchoring force required to hold in place the sluice gate when it is closed and the depth of water upstream is 10 ft.

■ **Figure P5.62**

**5.63** Water flows steadily into and out of a tank that sits on frictionless wheels as shown in Fig. P5.63. Determine the diameter $D$ so that the tank remains motionless if $F = 0$.

■ **Figure P5.63**

**5.64** WILEY◦ The rocket shown in Fig. P5.64, is held stationary by the horizontal force, $F_x$, and the vertical force, $F_z$. The velocity and pressure of the exhaust gas are 5000 ft/s and 20 psia at the nozzle exit, which has a cross section area of 60 in.$^2$. The exhaust mass flowrate is constant at 21 lbm/s. Determine the value of the restraining force $F_x$. Assume the exhaust flow is essentially horizontal.

■ **Figure P5.64**

**5.65** WILEY◦ GO A horizontal circular jet of air strikes a stationary flat plate as indicated in Fig. P5.65. The jet velocity is 40 m/s and the jet diameter is 30 mm. If the air velocity magnitude remains constant as the air flows over the plate surface in the directions shown, determine: **(a)** the magnitude of $F_A$, the anchoring force required to

■ **Figure P5.65**

hold the plate stationary; **(b)** the fraction of mass flow along the plate surface in each of the two directions shown; **(c)** the magnitude of $F_A$, the anchoring force required to allow the plate to move to the right at a constant speed of 10 m/s.

**5.66** Air discharges from a 2-in.-diameter nozzle and strikes a curved vane, which is in a vertical plane as shown in Fig. P5.66. A stagnation tube connected to a water U-tube manometer is located in the free air jet. Determine the horizontal component of the force that the air jet exerts on the vane. Neglect the weight of the air and all friction.

■ **Figure P5.66**

**5.67** WILEY◦ Water is sprayed radially outward over 180° as indicated in Fig. P5.67. The jet sheet is in the horizontal plane. If the jet velocity at the nozzle exit is 20 ft/s, determine the direction and magnitude of the resultant horizontal anchoring force required to hold the nozzle in place.

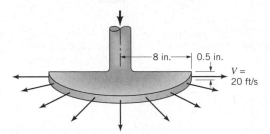

■ **Figure P5.67**

**5.68** WILEY◦ A sheet of water of uniform thickness ($h = 0.01$ m) flows from the device shown in Fig. P5.68. The water enters vertically through the inlet pipe and exits horizontally with a speed that varies linearly from 0 to 10 m/s along the 0.2-m length of the slit. Determine the $y$ component of anchoring force necessary to hold this device stationary.

■ **Figure P5.68**

**5.69** The results of a wind tunnel test to determine the drag on a body (see Fig. P5.69) are summarized below. The upstream [section (1)] velocity is uniform at 100 ft/s. The static pressures are given by $p_1 = p_2 = 14.7$ psia. The downstream velocity distribution, which is symmetrical about the centerline, is given by

$$u = 100 - 30\left(1 - \frac{|y|}{3}\right) \qquad |y| \le 3 \text{ ft}$$

$$u = 100 \qquad\qquad |y| > 3 \text{ ft}$$

where $u$ is the velocity in ft/s and $y$ is the distance on either side of the centerline in feet (see Fig. P5.69). Assume that the body shape does not change in the direction normal to the paper. Calculate the drag force (reaction force in $x$ direction) exerted on the air by the body per unit length normal to the plane of the sketch.

■ **Figure P5.69**

**5.70** WILEY A variable mesh screen produces a linear and axisymmetric velocity profile as indicated in Fig. P5.70 in the airflow through a 2-ft-diameter circular cross-sectional duct. The static pressures upstream and downstream of the screen are 0.2 and 0.15 psi and are uniformly distributed over the flow cross-sectional area. Neglecting the force exerted by the duct wall on the flowing air, calculate the screen drag force.

■ **Figure P5.70**

**5.71** Consider unsteady flow in the constant diameter, horizontal pipe shown in Fig. P5.71. The velocity is uniform throughout the entire pipe, but it is a function of time: $\mathbf{V} = u(t)\,\hat{\mathbf{i}}$. Use the $x$ component of the unsteady momentum equation to determine the pressure difference $p_1 - p_2$. Discuss how this result is related to $F_x = ma_x$.

■ **Figure P5.71**

**5.72** In a laminar pipe flow that is fully developed, the axial velocity profile is parabolic. That is,

$$u = u_c\left[1 - \left(\frac{r}{R}\right)^2\right]$$

as is illustrated in Fig. P5.72. Compare the axial direction momentum flowrate calculated with the average velocity, $\bar{u}$, with the axial direction momentum flowrate calculated with the nonuniform velocity distribution taken into account.

■ **Figure P5.72**

†**5.73** Water from a garden hose is sprayed against your car to rinse dirt from it. Estimate the force that the water exerts on the car. List all assumptions and show calculations.

**5.74** ▶ WILEY A Pelton wheel vane directs a horizontal, circular cross-sectional jet of water symmetrically as indicated in Fig. P5.74 and **Video V5.6**. The jet leaves the nozzle with a velocity of 100 ft/s. Determine the $x$-direction component of anchoring force required to **(a)** hold the vane stationary, **(b)** confine the speed of the vane to a value of 10 ft/s to the right. The fluid speed magnitude remains constant along the vane surface.

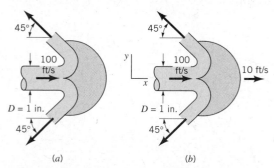

■ **Figure P5.74**

**5.75** ▶ WILEY The thrust developed to propel the jet ski shown in **Video V9.18** and Fig. P5.75 is a result of water pumped through the vehicle and exiting as a high-speed water jet. For the conditions shown in the figure, what flowrate is needed to produce a 300-lb thrust? Assume the inlet and outlet jets of water are free jets.

■ **Figure P5.75**

**5.76** Thrust vector control is a technique that can be used to greatly improve the maneuverability of military fighter aircraft. It consists of using a set of vanes in the exit of a jet engine to deflect the exhaust gases as shown in Fig. P5.76. **(a)** Determine the pitching moment (the moment tending to rotate the nose of the aircraft

$p_{out} = 0$
$V_{out} = 1500$ ft/s

20 ft

cg

Vane

$V_{in} = 300$ ft/s

$\theta = 8$ deg

$p_{in} = 0$

$\dot{m}_{in} = \dot{m}_{out} = 16$ slugs/s

■ **Figure P5.76**

up) about the aircraft's mass center (cg) for the conditions indicated in the figure. **(b)** By how much is the thrust (force along the centerline of the aircraft) reduced for the case indicated compared to normal flight when the exhaust is parallel to the centerline?

**5.77** The exhaust gas from the rocket shown in Fig. P5.77a leaves the nozzle with a uniform velocity parallel to the x axis. The gas is assumed to be discharged from the nozzle as a free jet. **(a)** Show that the thrust that is developed is equal to $\rho A V^2$, where $A = \pi D^2/4$. **(b)** The exhaust gas from the rocket nozzle shown in Fig. P5.77b is also uniform, but rather than being directed along the x axis, it is directed along rays from point 0 as indicated. Determine the thrust for this rocket.

(a)

(b)

■ **Figure P5.77**

**5.78** (WILEY⊙) (See Fluids in the News article titled **"Where the Plume goes,"** Section 5.2.2.) Air flows into the jet engine shown in Fig. P5.78 at a rate of 9 slugs/s and a speed of 300 ft/s. Upon landing, the engine exhaust exits through the reverse thrust mechanism with a speed of 900 ft/s in the direction indicated. Determine the reverse thrust applied by the engine to the airplane. Assume the inlet and exit pressures are atmospheric and that the mass flowrate of fuel is negligible compared to the air flowrate through the engine.

$V_3 = 900$ ft/s
(3)

4-ft diameter

(1)   30°

$V_1 = 300$ ft/s

(2)

$V_2 = 900$ ft/s

■ **Figure P5.78**

**5.79** (WILEY⊙) (See Fluids in the News article titled **"Motorized Surfboard,"** Section 5.2.2.) The thrust to propel the powered surfboard shown in Fig. P5.79 is a result of water pumped through the board that exits as a high-speed 2.75-in.-diameter jet. Determine

the flowrate and the velocity of the exiting jet if the thrust is to be 300 lb. Neglect the momentum of the water entering the pump.

■ **Figure P5.79**

**5.80** (WILEY⊙) (See Fluids in the News article titled **"Bow Thrusters,"** Section 5.2.2.) The bow thruster on the boat shown in Fig. P5.80 is used to turn the boat. The thruster produces a 1-m-diameter jet of water with a velocity of 10 m/s. Determine the force produced by the thruster. Assume that the inlet and outlet pressures are zero and that the momentum of the water entering the thruster is negligible.

$V = 10$ m/s

$D = 1$ m

■ **Figure P5.80**

**5.81** Water flows from a two-dimensional open channel and is diverted by an inclined plate as illustrated in Fig. P5.81. When the velocity at section (1) is 10 ft/s, what horizontal force (per unit width) is required to hold the plate in position? At section (1) the pressure distribution is hydrostatic, and the fluid acts as a free jet at section (2). Neglect friction.

Section (1)

20°

Plate

10 ft/s   4 ft

Section (2)

1.0 ft

■ **Figure P5.81**

**†5.82** If a valve in a pipe is suddenly closed, a large pressure surge may develop. For example, when the electrically operated shutoff valve in a dishwasher closes quickly, the pipes supplying the dishwasher may rattle or "bang" because of this large pressure pulse. Explain the physical mechanism for this "water hammer" phenomenon. How could this phenomenon be analyzed?

**5.83** (WILEY) A snowplow mounted on a truck clears a path 12 ft through heavy wet snow, as shown in Figure P5.83. The snow is 8 in. deep and its density is 10 lbm/ft$^3$. The truck travels at 30 mph. The snow is discharged from the plow at an angle of 45° from the direction of travel and 45° above the horizontal, as shown in Figure P5.83. Estimate the force required to push the plow.

■ **Figure P5.83**

## Section 5.2.3 Derivation of the Moment-of-Momentum Equation

**5.84** Describe a few examples (include photographs/images) of turbines where the force/torque of a flowing fluid leads to rotation of a shaft.

**5.85** Describe a few examples (include photographs/images) of pumps where a fluid is forced to move by "blades" mounted on a rotating shaft.

**5.86** An incompressible fluid flows outward through a blower as indicated in Fig. P5.86. The shaft torque involved, $T_{shaft}$, is estimated with the following relationship:

$$T_{shaft} = \dot{m} r_2 V_{\theta 2}$$

where $\dot{m}$ = mass flowrate through the blower, $r_2$ = outer radius of blower, and $V_{\theta 2}$ = tangential component of absolute fluid velocity leaving the blower. State the flow conditions that make this formula valid.

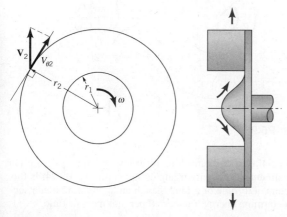

■ **Figure P5.86**

## Section 5.2.4 Application of the Moment-of-Momentum Equation

**5.87** (WILEY) Water enters a rotating lawn sprinkler through its base at the steady rate of 16 gal/min as shown in Fig. P5.87. The exit cross-sectional area of each of the two nozzles is 0.04 in.$^2$, and the flow leaving each nozzle is tangential. The radius from the axis of rotation to the centerline of each nozzle is 8 in. **(a)** Determine the resisting torque required to hold the sprinkler head stationary. **(b)** Determine the resisting torque associated with the sprinkler rotating with a constant speed of 500 rev/min. **(c)** Determine the angular velocity of the sprinkler if no resisting torque is applied.

■ **Figure P5.87**

**5.88** ▶ (WILEY GO) Five liters/s of water enter the rotor shown in **Video V5.10** and Fig. P5.88 along the axis of rotation. The cross-sectional area of each of the three nozzle exits normal to the relative velocity is 18 mm$^2$. How large is the resisting torque required to hold the rotor stationary? How fast will the rotor spin steadily if the resisting torque is reduced to zero and **(a)** $\theta = 0°$, **(b)** $\theta = 30°$, **(c)** $\theta = 60°$?

■ **Figure P5.88**

**5.89** (See Fluids in the News article titled **"Tailless Helicopters,"** Section 5.2.4.) Shown in Fig. P5.89 is a toy "helicopter" powered

■ **Figures P5.89**

by air escaping from a balloon. The air from the balloon flows radially through each of the three propeller blades and out through small nozzles at the tips of the blades. Explain physically how this flow can cause the rotation necessary to rotate the blades to produce the needed lifting force.

**5.90** A water turbine wheel rotates at the rate of 50 rpm in the direction shown in Fig. P5.90. The inner radius, $r_2$, of the blade row is 2 ft, and the outer radius, $r_1$, is 4 ft. The absolute velocity vector at the turbine rotor entrance makes an angle of 20° with the tangential direction. The inlet blade angle is 60° relative to the tangential direction. The blade outlet angle is 120°. The flowrate is 20 ft³/s. For the flow tangent to the rotor blade surface at inlet and outlet, determine an appropriate constant blade height, $b$, and the corresponding power available at the rotor shaft.

**■ Figure P5.90**

**5.91** (WILEY) A water turbine with radial flow has the dimensions shown in Fig. P5.91. The absolute entering velocity is 50 ft/s, and it makes an angle of 30° with the tangent to the rotor. The absolute exit velocity is directed radially inward. The angular speed of the rotor is 120 rpm. Find the power delivered to the shaft of the turbine.

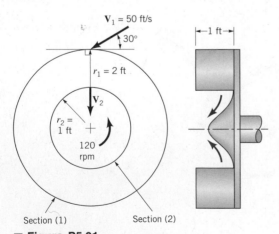

**■ Figure P5.91**

**5.92** A fan (see Fig. P5.92) has a bladed rotor of 12-in. outside diameter and 5-in. inside diameter and runs at 1725 rpm. The width of each rotor blade is 1 in. from blade inlet to outlet. The volume flowrate is steady at 230 ft³/min, and the absolute velocity of the air at blade inlet, $V_1$, is purely radial. The blade discharge angle is 30°

**■ Figure P5.92**

measured with respect to the tangential direction at the outside diameter of the rotor. **(a)** What would be a reasonable blade inlet angle (measured with respect to the tangential direction at the inside diameter of the rotor)? **(b)** Find the power required to run the fan.

**5.93** The radial component of velocity of water leaving the centrifugal pump sketched in Fig. P5.93 is 30 ft/s. The magnitude of the absolute velocity at the pump exit is 60 ft/s. The fluid enters the pump rotor radially. Calculate the shaft work required per unit mass flowing through the pump.

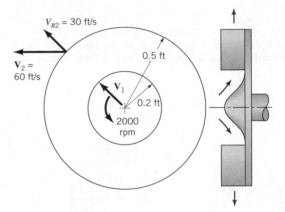

**■ Figure P5.93**

**5.94** An axial flow turbomachine rotor involves the upstream (1) and downstream (2) velocity triangles shown in Fig. P5.94. Is this turbomachine a turbine or a fan? Sketch an appropriate blade section and determine energy transferred per unit mass of fluid.

**■ Figure P5.94**

**5.95** An inward flow radial turbine (see Fig. P5.95) involves a nozzle angle, $\alpha_1$, of 60° and an inlet rotor tip speed, $U_1$, of 6 m/s. The ratio of rotor inlet to outlet diameters is 1.8. The absolute velocity leaving the rotor at section (2) is radial with a magnitude of

■ **Figure P5.95**

12 m/s. Determine the energy transfer per unit mass of fluid flowing through this turbine if the fluid is **(a)** air, **(b)** water.

5.96 A sketch of the arithmetic mean radius blade sections of an axial-flow water turbine stage is shown in Fig. P5.96. The rotor speed is 1000 rpm. **(a)** Sketch and label velocity triangles for the flow entering and leaving the rotor row. Use **V** for absolute velocity, **W** for relative velocity, and **U** for blade velocity. Assume flow enters and leaves each blade row at the blade angles shown. **(b)** Calculate the work per unit mass delivered at the shaft.

■ **Figure P5.96**

5.97 WILEY By using velocity triangles for flow upstream (1) and downstream (2) of a turbomachine rotor, prove that the shaft work in per unit mass flowing through the rotor is

$$w_{\text{shaft net in}} = \frac{V_2^2 - V_1^2 + U_2^2 - U_1^2 + W_1^2 - W_2^2}{2}$$

where V = absolute flow velocity magnitude, W = relative flow velocity magnitude, and U = blade speed.

Section 5.3.1 Derivation of the Energy Equation

5.98 Distinguish between shaft work and other kinds of work associated with a flowing fluid.

5.99 WILEY An incompressible fluid flows along a 0.20-m-diameter pipe with a uniform velocity of 3 m/s. If the pressure drop between the upstream and downstream sections of the pipe is 20 kPa, determine the power transferred to the fluid due to fluid normal stresses.

Section 5.3.2 Application of the Energy Equation—No Shaft Work and Section 5.3.3 Comparison of the Energy Equation with the Bernoulli Equation

5.100 A 100-ft-wide river with a flowrate of 2400 ft^3/s flows over a rock pile as shown in Fig. P5.100. Determine the direction of flow and the head loss associated with the flow across the rock pile.

■ **Figure P5.100**

5.101 WILEY A horizontal Venturi flow meter consists of a converging–diverging conduit as indicated in Fig. P5.101. The diameters of cross sections (1) and (2) are 6 and 4 in. The velocity and static pressure are uniformly distributed at cross sections (1) and (2). Determine the volume flowrate (ft^3/s) through the meter if $p_1 - p_2 =$ 3 psi, the flowing fluid is oil (ρ = 56 lbm/ft^3), and the loss per unit mass from (1) to (2) is negligibly small.

■ **Figure P5.101**

5.102 WILEY Oil (SG = 0.9) flows downward through a vertical pipe contraction as shown in Fig. P5.102. If the mercury manometer reading, h, is 100 mm, determine the volume flowrate for frictionless flow. Is the actual flowrate more or less than the frictionless value? Explain.

■ **Figure P5.102**

5.103 An incompressible liquid flows steadily along the pipe shown in Fig. P5.103. Determine the direction of flow and the head loss over the 6-m length of pipe.

■ **Figure P5.103**

**5.104** A siphon is used to draw water at 70 °F from a large container as indicated in Fig. P5.104. The inside diameter of the siphon line is 1 in. and the pipe centerline rises 3 ft above the essentially constant water level in the tank. Show that by varying the length of the siphon below the water level, $h$, the rate of flow through the siphon can be changed. Assuming frictionless flow, determine the maximum flowrate possible through the siphon. The limiting condition is the occurrence of cavitation in the siphon. Will the actual maximum flow be more or less than the frictionless value? Explain.

■ **Figure P5.104**

**5.105** (WILEY) A water siphon having a constant inside diameter of 3 in. is arranged as shown in Fig. P5.105. If the friction loss between A and B is $0.8V^2/2$, where $V$ is the velocity of flow in the siphon, determine the flowrate involved.

■ **Figure P5.105**

**5.106** Water flows through a valve (see Fig. P5.106) at the rate of 1000 lbm/s. The pressure just upstream of the valve is 90 psi and the pressure drop across the valve is 50 psi. The inside diameters of the valve inlet and exit pipes are 12 and 24 in. If the flow through the valve occurs in a horizontal plane, determine the loss in available energy across the valve.

■ **Figure P5.106**

**5.107** A gas expands through a nozzle from a pressure of 300 psia to a pressure of 5 psia. The enthalpy change involved, $\check{h}_1 - \check{h}_2$, is 150 Btu/lbm. If the expansion is adiabatic but with frictional effects and the inlet gas speed is negligibly small, determine the exit gas velocity.

**5.108** (WILEY) For the 180° elbow and nozzle flow shown in Fig. P5.108, determine the loss in available energy from section (1) to section (2). How much additional available energy is lost from section (2) to where the water comes to rest?

■ **Figure P5.108**

**5.109** An automobile engine will work best when the back pressure at the interface of the exhaust manifold and the engine block is minimized. Show how reduction of losses in the exhaust manifold, piping, and muffler will also reduce the back pressure. How could losses in the exhaust system be reduced? What primarily limits the minimization of exhaust system losses?

**5.110** (WILEY) (See Fluids in the News article titled **"Smart Shocks,"** Section 5.3.3.) A 200-lb force applied to the end of the piston of the shock absorber shown in Fig. P5.110 causes the two ends of the shock absorber to move toward each other with a speed of 5 ft/s. Determine the head loss associated with the flow of the oil through the channel. Neglect gravity and any friction force between the piston and cylinder walls.

■ **Figure P5.110**

## Section 5.3.2 Application of the Energy Equation— With Shaft Work

†**5.111** Based on flowrate and pressure rise information, estimate the power output of a human heart.

**5.112** (WILEY) What is the maximum possible power output of the hydroelectric turbine shown in Fig. P5.112?

50 m

1 m

6 m/s

Turbine

■ **Figure P5.112**

**5.113** (WILEY GO) Oil ($SG = 0.88$) flows in an inclined pipe at a rate of 5 ft³/s as shown in Fig. P5.113. If the differential reading in the mercury manometer is 3 ft, calculate the power that the pump supplies to the oil if head losses are negligible.

6 in.

$h$

$P$

Oil

12 in.

$H$

3 ft

■ **Figure P5.113**

**5.114** (WILEY) The pumper truck shown in Fig. P5.114 is to deliver 1.5 ft³/s to a maximum elevation of 60 ft above the hydrant. The

60 ft

10 psi
4-in.
diameter

Hydrant

■ **Figure P5.114**

pressure at the 4-in.-diameter outlet of the hydrant is 10 psi. If head losses are negligibly small, determine the power that the pump must add to the water.

**5.115** (WILEY) The hydroelectric turbine shown in Fig. P5.115 passes 8 million gal/min across a head of 600 ft. What is the maximum amount of power output possible? Why will the actual amount be less?

600 ft

Turbine

■ **Figure P5.115**

**5.116** (WILEY) A pump is to move water from a lake into a large, pressurized tank as shown in Fig. P5.116 at a rate of 1000 gal in 10 min or less. Will a pump that adds 3 hp to the water work for this purpose? Support your answer with appropriate calculations. Repeat the problem if the tank were pressurized to 3, rather than 2, atmospheres.

$p = 2$ atm

Air

20 ft

Pump

■ **Figure P5.116**

**5.117** (WILEY) Water is supplied at 150 ft³/s and 60 psi to a hydraulic turbine through a 3-ft inside-diameter inlet pipe as indicated in Fig. P5.117. The turbine discharge pipe has a 4-ft inside diameter. The static pressure at section (2), 10 ft below the turbine inlet, is 10-in. Hg vacuum. If the turbine develops 2500 hp, determine the power lost between sections (1) and (2).

Section (1)

$p_1 = 60$ psi
$Q = 150$ ft³/s
$D_1 = 3$ ft

Turbine

10 ft

$p_2 = 10$-in. Hg
vacuum
$D_2 = 4$ ft

Section (2)

■ **Figure P5.117**

**5.118** Water is pumped from the tank shown in Fig. P5.118a. The head loss is known to be 1.2 $V^2/2g$, where $V$ is the average velocity in the pipe. According to the pump manufacturer, the relationship between the pump head and the flowrate is as shown in Fig. P5.118b: $h_p = 20 - 2000\,Q^2$, where $h_p$ is in meters and $Q$ is in m³/s. Determine the flowrate, $Q$.

(a)                          (b)

■ **Figure P5.118**

**5.119** WILEY Water is pumped from the large tank shown in Fig. P5.119. The head loss is known to be equal to $4V^2/2g$ and the pump head is $h_p = 20 - 4Q^2$, where $h_p$ is in ft when $Q$ is in ft³/s. Determine the flowrate.

■ **Figure P5.119**

**5.120** WILEY Water flows by gravity from one lake to another as sketched in Fig. P5.120 at the steady rate of 80 gpm. What is the loss in available energy associated with this flow? If this same amount of loss is associated with pumping the fluid from the lower lake to the higher one at the same flowrate, estimate the amount of pumping power required.

■ **Figure P5.120**

**5.121** WILEY The turbine shown in Fig. P5.121 develops 100 hp when the flowrate of water is 20 ft³/s. If all losses are negligible, determine **(a)** the elevation $h$, **(b)** the pressure difference across the turbine, and **(c)** the flowrate expected if the turbine were removed.

■ **Figure P5.121**

**5.122** ▶ WILEY Water is pumped from a tank, point (1), to the top of a water plant aerator, point (2), as shown in **Video V5.16** and Fig. P5.122 at a rate of 3.0 ft³/s. **(a)** Determine the power that the pump adds to the water if the head loss from (1) to (2) where $V_2 = 0$ is 4 ft. **(b)** Determine the head loss from (2) to the bottom of the aerator column, point (3), if the average velocity at (3) is $V_3 = 2$ ft/s.

■ **Figure P5.122**

**5.123** WILEY Water is to be moved from one large reservoir to another at a higher elevation as indicated in Fig. P5.123. The loss of available energy associated with 2.5 ft³/s being pumped from sections (1) to (2) is loss = $61\bar{V}^2/2$ ft²/s², where $V$ is the average velocity of water in the 8-in. inside-diameter piping involved. Determine the amount of shaft power required.

■ **Figure P5.123**

**5.124** A $\frac{3}{4}$-hp motor is required by an air ventilating fan to produce a 24-in.-diameter stream of air having a uniform speed of 40 ft/s. Determine the aerodynamic efficiency of the fan.

**5.125** A pump moves water horizontally at a rate of 0.02 m³/s. Upstream of the pump where the pipe diameter is 90 mm, the pressure is 120 kPa. Downstream of the pump where the pipe diameter is 30 mm, the pressure is 400 kPa. If the loss in energy across the pump due to fluid friction effects is 170 N · m/kg, determine the hydraulic efficiency of the pump.

**5.126** (WILEY®) Water is to be pumped from the large tank shown in Fig. P5.126 with an exit velocity of 6 m/s. It was determined that the original pump (pump 1) that supplies 1 kW of power to the water did not produce the desired velocity. Hence, it is proposed that an additional pump (pump 2) be installed as indicated to increase the flowrate to the desired value. How much power must pump 2 add to the water? The head loss for this flow is $h_L = 250 \, Q^2$, where $h_L$ is in m when $Q$ is in m$^3$/s.

**■ Figure P5.126**

**5.127** (WILEY®) (See Fluids in the News article titled **"Curtain of Air,"** Section 5.3.3.) The fan shown in Fig. P5.127 produces an air curtain to separate a loading dock from a cold storage room. The air curtain is a jet of air 10 ft wide, 0.5 ft thick moving with speed $V = 30$ ft/s. The loss associated with this flow is loss = $K_L V^2/2$, where $K_L = 5$. How much power must the fan supply to the air to produce this flow?

**■ Figure P5.127**

## Section 5.3.3 Application of the Energy Equation— Combined with Linear Momentum

**5.128** If a $\frac{3}{4}$-hp motor is required by a ventilating fan to produce a 24-in. stream of air having a velocity of 40 ft/s as shown in Fig. P5.128, estimate **(a)** the efficiency of the fan and **(b)** the thrust of the supporting member on the conduit enclosing the fan.

**■ Figure P5.128**

**5.129** (WILEY®) Air flows past an object in a pipe of 2-m diameter and exits as a free jet as shown in Fig. P5.129. The velocity and pressure upstream are uniform at 10 m/s and 50 N/m$^2$, respectively. At the pipe exit the velocity is nonuniform as indicated. The shear stress along the pipe wall is negligible. **(a)** Determine the head loss associated with a particle as it flows from the uniform velocity upstream

of the object to a location in the wake at the exit plane of the pipe. **(b)** Determine the force that the air exerts on the object.

**■ Figure P5.129**

**5.130** ▶ Near the downstream end of a river spillway, a hydraulic jump often forms, as illustrated in Fig. P5.130 and **Video V10.11.** The velocity of the channel flow is reduced abruptly across the jump. Using the conservation of mass and linear momentum principles, derive the following expression for $h_2$,

$$h_2 = -\frac{h_1}{2} + \sqrt{\left(\frac{h_1}{2}\right)^2 + \frac{2V_1^2 h_1}{g}}$$

The loss of available energy across the jump can also be determined if energy conservation is considered. Derive the loss expression

$$\text{jump loss} = \frac{g(h_2 - h_1)^3}{4h_1 h_2}$$

**■ Figure P5.130**

**5.131** (WILEY®) Water flows steadily down the inclined pipe as indicated in Fig P5.131. Determine the following: **(a)** the difference in pressure $p_1 - p_2$, **(b)** the loss between sections (1) and (2), **(c)** the net axial force exerted by the pipe wall on the flowing water between sections (1) and (2).

**■ Figure P5.131**

**5.132** Water flows steadily in a pipe and exits as a free jet through an end cap that contains a filter as shown in Fig. P5.132. The flow is in a horizontal plane. The axial component, $R_y$, of the anchoring force needed to keep the end cap stationary is 60 lb. Determine the head loss for the flow through the end cap.

**■ Figure P5.132**

**5.133** (WILEY) When fluid flows through an abrupt expansion as indicated in Fig. P5.133, the loss in available energy across the expansion, $loss_{ex}$, is often expressed as

$$loss_{ex} = \left(1 - \frac{A_1}{A_2}\right)^2 \frac{V_1^2}{2}$$

where $A_1$ = cross-sectional area upstream of expansion, $A_2$ = cross-sectional area downstream of expansion, and $V_1$ = velocity of flow upstream of expansion. Derive this relationship.

**■ Figure P5.133**

**5.134** (WILEY) Two water jets collide and form one homogeneous jet as shown in Fig. P5.134. **(a)** Determine the speed, $V$, and direction, $\theta$, of the combined jet. **(b)** Determine the loss for a fluid particle flowing from (1) to (3), from (2) to (3). Gravity is negligible.

**■ Figure P5.134**

### Section 5.3.4  Application of the Energy Equation to Nonuniform Flows

**5.135** Water flows vertically upward in a circular cross-sectional pipe. At section (1), the velocity profile over the cross-sectional area is uniform. At section (2), the velocity profile is

$$\mathbf{V} = w_c \left(\frac{R - r}{R}\right)^{1/7} \hat{\mathbf{k}}$$

where $\mathbf{V}$ = local velocity vector, $w_c$ = centerline velocity in the axial direction, $R$ = pipe inside radius, and $r$ = radius from pipe axis. Develop an expression for the loss in available energy between sections (1) and (2).

**5.136** (WILEY) A small fan moves air at a mass flowrate of 0.004 lbm/s. Upstream of the fan, the pipe diameter is 2.5 in., the flow is laminar, the velocity distribution is parabolic, and the kinetic energy coefficient, $\alpha_1$, is equal to 2.0. Downstream of the fan, the pipe diameter is 1 in., the flow is turbulent, the velocity profile is quite flat, and the kinetic energy coefficient, $\alpha_2$, is equal to 1.08. If the rise in static pressure across the fan is 0.015 psi and the fan shaft draws 0.00024 hp, compare the value of loss calculated: **(a)** assuming uniform velocity distributions, **(b)** considering actual velocity distributions.

### Section 5.3.5  Combination of the Energy Equation and the Moment-of-Momentum Equation

**5.137** (WILEY) Air enters a radial blower with zero angular momentum. It leaves with an absolute tangential velocity, $V_\theta$, of 200 ft/s. The rotor blade speed at rotor exit is 170 ft/s. If the stagnation pressure rise across the rotor is 0.4 psi, calculate the loss of available energy across the rotor and the rotor efficiency.

**5.138** Water enters a pump impeller radially. It leaves the impeller with a tangential component of absolute velocity of 10 m/s. The impeller exit diameter is 60 mm, and the impeller speed is 1800 rpm. If the stagnation pressure rise across the impeller is 45 kPa, determine the loss of available energy across the impeller and the hydraulic efficiency of the pump.

**5.139** (WILEY) Water enters an axial-flow turbine rotor with an absolute velocity tangential component, $V_\theta$, of 15 ft/s. The corresponding blade velocity, $U$, is 50 ft/s. The water leaves the rotor blade row with no angular momentum. If the stagnation pressure drop across the turbine is 12 psi, determine the hydraulic efficiency of the turbine.

**5.140** An inward flow radial turbine (see Fig. P5.140) involves a nozzle angle, $\alpha_1$, of 60° and an inlet rotor tip speed, $U_1$, of 30 ft/s. The ratio of rotor inlet to outlet diameters is 2.0. The radial component of velocity remains constant at 20 ft/s through the rotor, and the flow leaving the rotor at section (2) is without angular momentum. If the flowing fluid is water and the stagnation pressure drop across the rotor is 16 psi, determine the loss of available energy across the rotor and the hydraulic efficiency involved.

**■ Figure P5.140**

## Section 5.3.4 Application of the Energy Equation to Nonuniform Flows

**5.141** The distribution of axial direction velocity, $u$, in a pipe flow is linear from zero at the wall to maximum of $u_c$ at the centerline. Determine the average velocity, $\bar{u}$, and the kinetic energy coefficient, $\alpha$.

### ■ Lab Problems

**5.1LP** This problem involves the force that a jet of air exerts on a flat plate as the air is deflected by the plate. To proceed with this problem, go to Appendix H, which is located in *WileyPLUS* or on the book's web site, www.wiley.com/college/munson.

**5.2LP** This problem involves the pressure distribution produced on a flat plate that deflects a jet of air. To proceed with this problem, go to Appendix H, which is located in *WileyPLUS* or on the book's web site, www.wiley.com/college/munson.

**5.3LP** This problem involves the force that a jet of water exerts on a vane when the vane turns the jet through a given angle. To proceed with this problem, go to Appendix H, which is located in *WileyPLUS* or on the book's web site, www.wiley.com/college/munson.

**5.4LP** This problem involves the force needed to hold a pipe elbow stationary. To proceed with this problem, go to Appendix H, which is located in *WileyPLUS* or on the book's web site, www.wiley.com/college/munson.

### ■ Lifelong Learning Problems

**5.1LL** What are typical efficiencies associated with swimming and how can they be improved?

**5.2LL** Explain how local ionization of flowing air can accelerate it. How can this be useful?

**5.3LL** Discuss the main causes of loss of available energy in a turbo-pump and how they can be minimized. What are typical turbo-pump efficiencies?

**5.4LL** Discuss the main causes of loss of available energy in a turbine and how they can be minimized. What are typical turbine efficiencies?

### ■ FE Exam Problems

Sample FE (Fundamentals of Engineering) exam questions for fluid mechanics are provided in *WileyPLUS* or on the book's web site, www.wiley.com/college/munson.

# 6 Differential Analysis of Fluid Flow

CHAPTER OPENING PHOTO: Flow past an airfoil at angle of attack: The flow of air around a model airfoil is made visible by smoke line flow visualization. Note the flow separation on the upper surface—see Ch. 9 for more details. (Photograph courtesy of NASA.)

## Learning Objectives

After completing this chapter, you should be able to:

- determine various kinematic elements of the flow given the velocity field.
- explain the conditions necessary for a velocity field to satisfy the continuity equation.
- apply the concepts of stream function and velocity potential.
- characterize simple potential flow fields.
- analyze certain types of flows using the Navier–Stokes equations.

In the previous chapter attention is focused on the use of finite control volumes for the solution of a variety of fluid mechanics problems. This approach is very practical and useful, since it does not generally require detailed knowledge of the pressure and velocity variations within the control volume. Typically, we found that only conditions on the surface of the control volume were needed, and thus problems could be solved without detailed knowledge of the flow field. Unfortunately, many situations arise in which the details of the flow are important and the finite control volume approach will not yield the desired information. For example, we may need to know how the velocity varies over the cross section of a pipe, or how the pressure and shear stress vary along the surface of an airplane wing. In these circumstances we need to develop relationships that apply at a point, or at least in a very small infinitesimal region within a given flow field. This approach, which involves an *infinitesimal control volume*, as distinguished from a finite control volume, is commonly referred to as *differential analysis*, since (as we will soon discover) the governing equations are differential equations.

In this chapter we will provide an introduction to the differential equations that describe (in detail) the motion of fluids. Unfortunately, we will also find that these equations are rather complicated, nonlinear partial differential equations that cannot be solved exactly except in a few cases, where simplifying assumptions are made. Thus, although differential analysis has the potential for supplying very detailed information about flow fields, this information is not easily extracted. Nevertheless, this approach provides a fundamental basis for the study of fluid mechanics. We do not want to be too discouraging at this point, since there are some exact solutions for laminar flow that can be obtained, and these have proved to be very useful. A few of these are included in this chapter. In addition, by making some simplifying assumptions, many other analytical solutions can be obtained. For example, in some circumstances it may be reasonable to assume that the effect of viscosity is small and can be neglected. This rather drastic assumption greatly simplifies the analysis and provides the opportunity to obtain detailed solutions to a variety of complex flow problems. Some examples of these so-called *inviscid flow* solutions are also described in this chapter.

It is known that for certain types of flows the flow field can be conceptually divided into two regions—a very thin region near the boundaries of the system in which viscous effects are important, and a region away from the boundaries in which the flow is essentially inviscid. By making certain assumptions about the behavior of the fluid in the thin layer near the boundaries, and using the assumption of inviscid flow outside this layer, a large class of problems can be solved using differential analysis. These boundary layer problems are discussed in Chapter 9. Finally, it is to be noted that with the availability of powerful computers it is feasible to attempt to solve the differential equations using the techniques of numerical analysis. Although it is beyond the scope of this book to delve extensively into this approach, which is generally referred to as *computational fluid dynamics* (CFD), the reader should be aware of this approach to complex flow problems. CFD has become a common engineering tool, and a brief introduction can be found in Appendix A. To introduce the power of CFD, two animations based on the numerical computations are provided as shown in the margin.

We begin our introduction to differential analysis by reviewing and extending some of the ideas associated with fluid kinematics that were introduced in Chapter 4. With this background the remainder of the chapter will be devoted to the derivation of the basic differential equations (which will be based on the principle of conservation of mass and Newton's second law of motion) and to some applications.

*V6.1 Spinning football-pressure contours*

*V6.2 Spinning football-velocity vectors*

## 6.1 Fluid Element Kinematics

*Fluid element motion consists of translation, linear deformation, rotation, and angular deformation.*

In this section we will be concerned with the mathematical description of the motion of fluid elements moving in a flow field. A small fluid element in the shape of a cube that is initially in one position will move to another position during a short time interval $\delta t$ as illustrated in Fig. 6.1. Because of the generally complex velocity variation within the field, we expect the element not only to translate from one position but also to have its volume changed (linear deformation), to rotate, and to undergo a change in shape (angular deformation). Although these movements and deformations occur simultaneously, we can consider each one separately as illustrated in Fig. 6.1. Since element motion and deformation are intimately related to the velocity and variation of velocity throughout the flow field, we will briefly review the manner in which velocity and acceleration fields can be described.

■ **Figure 6.1** Types of motion and deformation for a fluid element.

### 6.1.1 Velocity and Acceleration Fields Revisited

As discussed in detail in Section 4.1, the velocity field can be described by specifying the velocity **V** at all points, and at all times, within the flow field of interest. Thus, in terms of rectangular coordinates, the notation **V** $(x, y, z, t)$ means that the velocity of a fluid particle depends on where it is located within the flow field (as determined by its coordinates, $x$, $y$, and $z$) and when it occupies the particular point (as determined by the time, $t$). As is pointed out in Section 4.1.1, this method of describing the fluid motion is called the Eulerian method. It is also convenient to express the velocity in terms of three rectangular components so that

$$\mathbf{V} = u\hat{\mathbf{i}} + v\hat{\mathbf{j}} + w\hat{\mathbf{k}} \tag{6.1}$$

where $u$, $v$, and $w$ are the velocity components in the $x$, $y$, and $z$ directions, respectively, and $\hat{\mathbf{i}}, \hat{\mathbf{j}}$, and $\hat{\mathbf{k}}$ are the corresponding unit vectors, as shown by the figure in the margin. Of course, each of these components will, in general, be a function of $x$, $y$, $z$, and $t$. One of the goals of differential analysis is to determine how these velocity components specifically depend on $x$, $y$, $z$, and $t$ for a particular problem.

With this description of the velocity field it was also shown in Section 4.2.1 that the acceleration of a fluid particle can be expressed as

$$\mathbf{a} = \frac{\partial \mathbf{V}}{\partial t} + u\frac{\partial \mathbf{V}}{\partial x} + v\frac{\partial \mathbf{V}}{\partial y} + w\frac{\partial \mathbf{V}}{\partial z} \tag{6.2}$$

and in component form:

$$a_x = \frac{\partial u}{\partial t} + u\frac{\partial u}{\partial x} + v\frac{\partial u}{\partial y} + w\frac{\partial u}{\partial z} \tag{6.3a}$$

$$a_y = \frac{\partial v}{\partial t} + u\frac{\partial v}{\partial x} + v\frac{\partial v}{\partial y} + w\frac{\partial v}{\partial z} \tag{6.3b}$$

$$a_z = \frac{\partial w}{\partial t} + u\frac{\partial w}{\partial x} + v\frac{\partial w}{\partial y} + w\frac{\partial w}{\partial z} \tag{6.3c}$$

*The acceleration of a fluid particle is described using the concept of the material derivative.*

The acceleration is also concisely expressed as

$$\mathbf{a} = \frac{D\mathbf{V}}{Dt} \tag{6.4}$$

where the operator

$$\frac{D(\ )}{Dt} = \frac{\partial(\ )}{\partial t} + u\frac{\partial(\ )}{\partial x} + v\frac{\partial(\ )}{\partial y} + w\frac{\partial(\ )}{\partial z} \tag{6.5}$$

is termed the *material derivative*, or *substantial derivative*. In vector notation

$$\frac{D(\ )}{Dt} = \frac{\partial(\ )}{\partial t} + (\mathbf{V} \cdot \boldsymbol{\nabla})(\ ) \tag{6.6}$$

where the gradient operator, $\boldsymbol{\nabla}(\ )$, is

$$\boldsymbol{\nabla}(\ ) = \frac{\partial(\ )}{\partial x}\hat{\mathbf{i}} + \frac{\partial(\ )}{\partial y}\hat{\mathbf{j}} + \frac{\partial(\ )}{\partial z}\hat{\mathbf{k}} \tag{6.7}$$

which was introduced in Chapter 2. As we will see in the following sections, the motion and deformation of a fluid element depend on the velocity field. The relationship between the motion and the forces causing the motion depends on the acceleration field.

### 6.1.2 Linear Motion and Deformation

The simplest type of motion that a fluid element can undergo is translation, as illustrated in Fig. 6.2. In a small time interval $\delta t$ a particle located at point $O$ will move to point $O'$ as is illustrated in the figure. If all points in the element have the same velocity (which is only true if there are no velocity gradients), then the element will simply translate from one position to another. However,

■ **Figure 6.2**   Translation of a fluid element.

because of the presence of velocity gradients, the element will generally be deformed and rotated as it moves. For example, consider the effect of a single velocity gradient, $\partial u/\partial x$, on a small cube having sides $\delta x$, $\delta y$, and $\delta z$. As is shown in Fig. 6.3a, if the $x$ component of velocity of $O$ and $B$ is $u$, then at nearby points $A$ and $C$ the $x$ component of the velocity can be expressed as $u + (\partial u/\partial x)\,\delta x$. This difference in velocity causes a "stretching" of the volume element by an amount $(\partial u/\partial x)(\delta x)(\delta t)$ during the short time interval $\delta t$ in which line $OA$ stretches to $OA'$ and $BC$ to $BC'$ (Fig. 6.3b). The corresponding change in the original volume, $\delta V = \delta x\,\delta y\,\delta z$, would be

$$\text{Change in } \delta V = \left(\frac{\partial u}{\partial x}\,\delta x\right)(\delta y\,\delta z)(\delta t)$$

and the *rate* at which the volume $\delta V$ is changing *per unit volume* due to the gradient $\partial u/\partial x$ is

$$\frac{1}{\delta V}\frac{d(\delta V)}{dt} = \lim_{\delta t \to 0}\left[\frac{(\partial u/\partial x)\,\delta t}{\delta t}\right] = \frac{\partial u}{\partial x} \tag{6.8}$$

*The rate of volume change per unit volume is related to the velocity gradients.*

If velocity gradients $\partial v/\partial y$ and $\partial w/\partial z$ are also present, then using a similar analysis it follows that, in the general case,

$$\frac{1}{\delta V}\frac{d(\delta V)}{dt} = \frac{\partial u}{\partial x} + \frac{\partial v}{\partial y} + \frac{\partial w}{\partial z} = \boldsymbol{\nabla}\cdot\mathbf{V} \tag{6.9}$$

This rate of change of the volume per unit volume is called the ***volumetric dilatation rate***. Thus, we see that the volume of a fluid may change as the element moves from one location to another in the flow field. However, for an *incompressible fluid* the volumetric dilatation rate is zero, since the element volume cannot change without a change in fluid density (the element mass must be conserved). Variations in the velocity in the direction of the velocity, as represented by the derivatives $\partial u/\partial x$, $\partial v/\partial y$, and $\partial w/\partial z$, simply cause a *linear deformation* of the element in the sense that the shape of the element does not change. Cross derivatives, such as $\partial u/\partial y$ and $\partial v/\partial x$, will cause the element to rotate and generally to undergo an *angular deformation*, which changes the shape of the element.

### 6.1.3  Angular Motion and Deformation

For simplicity we will consider motion in the $x$–$y$ plane, but the results can be readily extended to the more general three-dimensional case. The velocity variation that causes rotation and angular deformation is illustrated in Fig. 6.4a. In a short time interval $\delta t$ the line segments $OA$ and $OB$ will

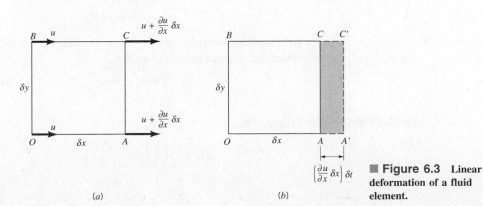

$(a)$

$(b)$

■ **Figure 6.3**   Linear deformation of a fluid element.

**Figure 6.4** Angular motion and deformation of a fluid element.

rotate through the angles $\delta\alpha$ and $\delta\beta$ to the new positions $OA'$ and $OB'$, as is shown in Fig. 6.4b. The angular velocity of line $OA$, $\omega_{OA}$, is

$$\omega_{OA} = \lim_{\delta t \to 0} \frac{\delta\alpha}{\delta t}$$

For small angles

$$\tan \delta\alpha \approx \delta\alpha = \frac{(\partial v/\partial x)\,\delta x\,\delta t}{\delta x} = \frac{\partial v}{\partial x}\,\delta t \tag{6.10}$$

so that

$$\omega_{OA} = \lim_{\delta t \to 0}\left[\frac{(\partial v/\partial x)\,\delta t}{\delta t}\right] = \frac{\partial v}{\partial x}$$

Note that if $\partial v/\partial x$ is positive, $\omega_{OA}$ will be counterclockwise. Similarly, the angular velocity of the line $OB$ is

$$\omega_{OB} = \lim_{\delta t \to 0} \frac{\delta\beta}{\delta t}$$

and

$$\tan \delta\beta \approx \delta\beta = \frac{(\partial u/\partial y)\,\delta y\,\delta t}{\delta y} = \frac{\partial u}{\partial y}\,\delta t \tag{6.11}$$

so that

$$\omega_{OB} = \lim_{\delta t \to 0}\left[\frac{(\partial u/\partial y)\,\delta t}{\delta t}\right] = \frac{\partial u}{\partial y}$$

In this instance if $\partial u/\partial y$ is positive, $\omega_{OB}$ will be clockwise. The *rotation*, $\omega_z$, of the element about the $z$ axis is defined as the average of the angular velocities $\omega_{OA}$ and $\omega_{OB}$ of the two mutually perpendicular lines $OA$ and $OB$.[1] Thus, if counterclockwise rotation is considered to be positive, it follows that

$$\omega_z = \frac{1}{2}\left(\frac{\partial v}{\partial x} - \frac{\partial u}{\partial y}\right) \tag{6.12}$$

Rotation of the field element about the other two coordinate axes can be obtained in a similar manner with the result that for rotation about the $x$ axis

$$\omega_x = \frac{1}{2}\left(\frac{\partial w}{\partial y} - \frac{\partial v}{\partial z}\right) \tag{6.13}$$

and for rotation about the $y$ axis

$$\omega_y = \frac{1}{2}\left(\frac{\partial u}{\partial z} - \frac{\partial w}{\partial x}\right) \tag{6.14}$$

*V6.3 Shear deformation*

*Rotation of fluid particles is related to certain velocity gradients in the flow field.*

---

[1]With this definition $\omega_z$ can also be interpreted to be the angular velocity of the bisector of the angle between the lines $OA$ and $OB$.

The three components, $\omega_x$, $\omega_y$, and $\omega_z$ can be combined to give the rotation vector, $\boldsymbol{\omega}$, in the form

$$\boldsymbol{\omega} = \omega_x\hat{\mathbf{i}} + \omega_y\hat{\mathbf{j}} + \omega_z\hat{\mathbf{k}} \tag{6.15}$$

An examination of this result reveals that $\boldsymbol{\omega}$ is equal to one-half the curl of the velocity vector. That is,

$$\boldsymbol{\omega} = \tfrac{1}{2}\,\text{curl }\mathbf{V} = \tfrac{1}{2}\,\boldsymbol{\nabla} \times \mathbf{V} \tag{6.16}$$

since by definition of the vector operator $\boldsymbol{\nabla} \times \mathbf{V}$

$$\frac{1}{2}\,\boldsymbol{\nabla} \times \mathbf{V} = \frac{1}{2}\begin{vmatrix} \hat{\mathbf{i}} & \hat{\mathbf{j}} & \hat{\mathbf{k}} \\ \dfrac{\partial}{\partial x} & \dfrac{\partial}{\partial y} & \dfrac{\partial}{\partial z} \\ u & v & w \end{vmatrix}$$

$$= \frac{1}{2}\left(\frac{\partial w}{\partial y} - \frac{\partial v}{\partial z}\right)\hat{\mathbf{i}} + \frac{1}{2}\left(\frac{\partial u}{\partial z} - \frac{\partial w}{\partial x}\right)\hat{\mathbf{j}} + \frac{1}{2}\left(\frac{\partial v}{\partial x} - \frac{\partial u}{\partial y}\right)\hat{\mathbf{k}}$$

*Vorticity in a flow field is related to fluid particle rotation.*

The *vorticity, $\boldsymbol{\zeta}$,* is defined as a vector that is twice the rotation vector; that is,

$$\boldsymbol{\zeta} = 2\,\boldsymbol{\omega} = \boldsymbol{\nabla} \times \mathbf{V} \tag{6.17}$$

Wing
Engine

The use of the vorticity to describe the rotational characteristics of the fluid simply eliminates the $(\tfrac{1}{2})$ factor associated with the rotation vector. The figure in the margin shows vorticity contours of the wing tip vortex flow shortly after an aircraft has passed. The lighter colors indicate stronger vorticity. (See also Fig. 4.3.)

We observe from Eq. 6.12 that the fluid element will rotate about the $z$ axis as an *undeformed* block (i.e., $\omega_{OA} = -\omega_{OB}$) only when $\partial u/\partial y = -\partial v/\partial x$. Otherwise the rotation will be associated with an angular deformation. We also note from Eq. 6.12 that when $\partial u/\partial y = \partial v/\partial x$ the rotation around the $z$ axis is zero. More generally if $\boldsymbol{\nabla} \times \mathbf{V} = 0$, then the rotation (and the vorticity) are zero, and flow fields for which this condition applies are termed *irrotational*. We will find in Section 6.4 that the condition of irrotationality often greatly simplifies the analysis of complex flow fields. However, it is probably not immediately obvious why some flow fields would be irrotational, and we will need to examine this concept more fully in Section 6.4.

---

# EXAMPLE 6.1 Vorticity

**GIVEN** For a certain two-dimensional flow field the velocity is given by the equation

$$\mathbf{V} = (x^2 - y^2)\hat{\mathbf{i}} - 2xy\hat{\mathbf{j}}$$

**FIND** Is this flow irrotational?

## SOLUTION

For an irrotational flow the rotation vector, $\boldsymbol{\omega}$, having the components given by Eqs. 6.12, 6.13, and 6.14 must be zero. For the prescribed velocity field

$$u = x^2 - y^2 \qquad v = -2xy \qquad w = 0$$

and therefore

$$\omega_x = \frac{1}{2}\left(\frac{\partial w}{\partial y} - \frac{\partial v}{\partial z}\right) = 0$$

$$\omega_y = \frac{1}{2}\left(\frac{\partial u}{\partial z} - \frac{\partial w}{\partial x}\right) = 0$$

$$\omega_z = \frac{1}{2}\left(\frac{\partial v}{\partial x} - \frac{\partial u}{\partial y}\right) = \frac{1}{2}\left[(-2y) - (-2y)\right] = 0$$

Thus, the flow is irrotational. **(Ans)**

**COMMENTS** It is to be noted that for a two-dimensional flow field (where the flow is in the $x$–$y$ plane) $\omega_x$ and $\omega_y$ will always be

zero, since by definition of two-dimensional flow $u$ and $v$ are not functions of $z$, and $w$ is zero. In this instance the condition for irrotationality simply becomes $\omega_z = 0$ or $\partial v/\partial x = \partial u/\partial y$.

The streamlines for the steady, two-dimensional flow of this example are shown in Fig. E6.1. (Information about how to calculate

■ **Figure E6.1**

streamlines for a given velocity field is given in Sections 4.1.4 and 6.2.3.) It is noted that all of the streamlines (except for the one through the origin) are curved. However, because the flow is irrotational, there is no rotation of the fluid elements. That is, lines *OA* and *OB* of Fig. 6.4 rotate with the same speed but in opposite directions.

As shown by Eq. 6.17, the condition of irrotationality is equivalent to the fact that the vorticity, $\zeta$, is zero or the curl of the velocity is zero.

In addition to the rotation associated with the derivatives $\partial u/\partial y$ and $\partial v/\partial x$, it is observed from Fig. 6.4b that these derivatives can cause the fluid element to undergo an *angular deformation*, which results in a change in shape of the element. The change in the original right angle formed by the lines *OA* and *OB* is termed the shearing strain, $\delta\gamma$, and from Fig. 6.4b

$$\delta\gamma = \delta\alpha + \delta\beta$$

where $\delta\gamma$ is considered to be positive if the original right angle is decreasing. The rate of change of $\delta\gamma$ is called the *rate of shearing strain* or the *rate of angular deformation* and is commonly denoted with the symbol $\dot{\gamma}$. The angles $\delta\alpha$ and $\delta\beta$ are related to the velocity gradients through Eqs. 6.10 and 6.11 so that

$$\dot{\gamma} = \lim_{\delta t \to 0} \frac{\delta\gamma}{\delta t} = \lim_{\delta t \to 0} \left[ \frac{(\partial v/\partial x)\,\delta t + (\partial u/\partial y)\,\delta t}{\delta t} \right]$$

and, therefore,

$$\dot{\gamma} = \frac{\partial v}{\partial x} + \frac{\partial u}{\partial y} \tag{6.18}$$

As we will learn in Section 6.8, the rate of angular deformation is related to a corresponding shearing stress that causes the fluid element to change in shape. From Eq. 6.18 we note that if $\partial u/\partial y = -\partial v/\partial x$, the rate of angular deformation is zero, and this condition corresponds to the case in which the element is simply rotating as an undeformed block (Eq. 6.12). In the remainder of this chapter we will see how the various kinematical relationships developed in this section play an important role in the development and subsequent analysis of the differential equations that govern fluid motion.

## 6.2  Conservation of Mass

*Conservation of mass requires that the mass of a system remain constant.*

As is discussed in Section 5.1, conservation of mass requires that the mass, $M$, of a system remain constant as the system moves through the flow field. In equation form this principle is expressed as

$$\frac{DM_{\text{sys}}}{Dt} = 0$$

We found it convenient to use the control volume approach for fluid flow problems, with the control volume representation of the conservation of mass written as

$$\frac{\partial}{\partial t} \int_{\text{cv}} \rho\, d\Psi + \int_{\text{cs}} \rho\, \mathbf{V} \cdot \hat{\mathbf{n}}\, dA = 0 \tag{6.19}$$

where the equation (commonly called the **continuity equation**) can be applied to a finite control volume (cv), which is bounded by a control surface (cs). The first integral on the left side of Eq. 6.19 represents the rate at which the mass within the control volume is changing, and the second integral represents the net rate at which mass is flowing out through the control surface (rate of mass outflow − rate of mass inflow). To obtain the differential form of the continuity equation, Eq. 6.19 is applied to an infinitesimal control volume.

### 6.2.1 Differential Form of Continuity Equation

We will take as our control volume the small, stationary cubical element shown in Fig. 6.5a. At the center of the element the fluid density is $\rho$, and the velocity has components $u$, $v$, and $w$. Since the element is small, the volume integral in Eq. 6.19 can be expressed as

$$\frac{\partial}{\partial t} \int_{\text{cv}} \rho\, d\Psi \approx \frac{\partial \rho}{\partial t}\, \delta x\, \delta y\, \delta z \tag{6.20}$$

■ **Figure 6.5**    A differential element for the development of conservation of mass equation.

The rate of mass flow through the surfaces of the element can be obtained by considering the flow in each of the coordinate directions separately. For example, in Fig. 6.5$b$ flow in the $x$ direction is depicted. If we let $\rho u$ represent the $x$ component of the mass rate of flow per unit area at the center of the element, then on the right face

$$\rho u|_{x+(\delta x/2)} = \rho u + \frac{\partial(\rho u)}{\partial x}\frac{\delta x}{2} \tag{6.21}$$

and on the left face

$$\rho u|_{x-(\delta x/2)} = \rho u - \frac{\partial(\rho u)}{\partial x}\frac{\delta x}{2} \tag{6.22}$$

Note that we are really using a Taylor series expansion of $\rho u$ and neglecting higher order terms such as $(\delta x)^2$, $(\delta x)^3$, and so on. This is indicated by the figure in the margin. When the right-hand sides of Eqs. 6.21 and 6.22 are multiplied by the area $\delta y\,\delta z$, the rate at which mass is crossing the right and left sides of the element is obtained as is illustrated in Fig. 6.5$b$. When these two expressions are combined, the net rate of mass flowing from the element through the two surfaces can be expressed as

$$\begin{aligned}\text{Net rate of mass} \atop \text{outflow in } x \text{ direction} &= \left[\rho u + \frac{\partial(\rho u)}{\partial x}\frac{\delta x}{2}\right]\delta y\,\delta z \\ &\quad - \left[\rho u - \frac{\partial(\rho u)}{\partial x}\frac{\delta x}{2}\right]\delta y\,\delta z = \frac{\partial(\rho u)}{\partial x}\delta x\,\delta y\,\delta z\end{aligned} \tag{6.23}$$

For simplicity, only flow in the $x$ direction has been considered in Fig. 6.5$b$, but, in general, there will also be flow in the $y$ and $z$ directions. An analysis similar to the one used for flow in the $x$ direction shows that

$$\text{Net rate of mass} \atop \text{outflow in } y \text{ direction} = \frac{\partial(\rho v)}{\partial y}\delta x\,\delta y\,\delta z \tag{6.24}$$

and

$$\text{Net rate of mass} \atop \text{outflow in } z \text{ direction} = \frac{\partial(\rho w)}{\partial z}\delta x\,\delta y\,\delta z \tag{6.25}$$

Thus,

$$\text{Net rate of} \atop \text{mass outflow} = \left[\frac{\partial(\rho u)}{\partial x} + \frac{\partial(\rho v)}{\partial y} + \frac{\partial(\rho w)}{\partial z}\right]\delta x\,\delta y\,\delta z \tag{6.26}$$

From Eqs. 6.19, 6.20, and 6.26 it now follows that the differential equation for conservation of mass is

*The continuity equation is one of the fundamental equations of fluid mechanics.*

$$\boxed{\frac{\partial\rho}{\partial t} + \frac{\partial(\rho u)}{\partial x} + \frac{\partial(\rho v)}{\partial y} + \frac{\partial(\rho w)}{\partial z} = 0} \tag{6.27}$$

As previously mentioned, this equation is also commonly referred to as the continuity equation.

The continuity equation is one of the fundamental equations of fluid mechanics and, as expressed in Eq. 6.27, is valid for steady or unsteady flow, and compressible or incompressible fluids. In vector notation, Eq. 6.27 can be written as

$$\frac{\partial \rho}{\partial t} + \nabla \cdot \rho \mathbf{V} = 0 \qquad (6.28)$$

Two special cases are of particular interest. For *steady* flow of *compressible* fluids

$$\nabla \cdot \rho \mathbf{V} = 0$$

or

$$\frac{\partial (\rho u)}{\partial x} + \frac{\partial (\rho v)}{\partial y} + \frac{\partial (\rho w)}{\partial z} = 0 \qquad (6.29)$$

*For incompressible fluids the continuity equation reduces to a simple relationship involving certain velocity gradients.*

This follows since by definition $\rho$ is not a function of time for steady flow, but could be a function of position. For *incompressible* fluids the fluid density, $\rho$, is a constant throughout the flow field so that Eq. 6.28 becomes

$$\nabla \cdot \mathbf{V} = 0 \qquad (6.30)$$

or

$$\frac{\partial u}{\partial x} + \frac{\partial v}{\partial y} + \frac{\partial w}{\partial z} = 0 \qquad (6.31)$$

Equation 6.31 applies to both steady and unsteady flow of incompressible fluids. Note that Eq. 6.31 is the same as that obtained by setting the volumetric dilatation rate (Eq. 6.9) equal to zero. This result should not be surprising since both relationships are based on conservation of mass for incompressible fluids. However, the expression for the volumetric dilatation rate was developed from a system approach, whereas Eq. 6.31 was developed from a control volume approach. In the former case the deformation of a particular differential mass of fluid was studied, and in the latter case mass flow through a fixed differential volume was studied.

---

# *E*XAMPLE 6.2    Continuity Equation

**GIVEN** The velocity components for a certain incompressible, steady-flow field are

$$u = x^2 + y^2 + z^2$$
$$v = xy + yz + z$$
$$w = ?$$

**FIND** Determine the form of the $z$ component, $w$, required to satisfy the continuity equation.

## *S*OLUTION

Any physically possible velocity distribution must for an incompressible fluid satisfy conservation of mass as expressed by the continuity equation

$$\frac{\partial u}{\partial x} + \frac{\partial v}{\partial y} + \frac{\partial w}{\partial z} = 0$$

For the given velocity distribution

$$\frac{\partial u}{\partial x} = 2x \qquad \text{and} \qquad \frac{\partial v}{\partial y} = x + z$$

so that the required expression for $\partial w/\partial z$ is

$$\frac{\partial w}{\partial z} = -2x - (x + z) = -3x - z$$

Integration with respect to $z$ yields

$$w = -3xz - \frac{z^2}{2} + f(x, y) \qquad \textbf{(Ans)}$$

**COMMENT** The third velocity component cannot be explicitly determined since the function $f(x, y)$ can have any form and conservation of mass will still be satisfied. The specific form of this function will be governed by the flow field described by these velocity components—that is, some additional information is needed to completely determine $w$.

■ **Figure 6.6**   **The representation of velocity components in cylindrical polar coordinates.**

### 6.2.2 Cylindrical Polar Coordinates

*For some problems, velocity components expressed in cylindrical polar coordinates will be convenient.*

For some problems it is more convenient to express the various differential relationships in cylindrical polar coordinates rather than Cartesian coordinates. As is shown in Fig. 6.6, with cylindrical coordinates a point is located by specifying the coordinates $r$, $\theta$, and $z$. The coordinate $r$ is the radial distance from the $z$ axis, $\theta$ is the angle measured from a line parallel to the $x$ axis (with counterclockwise taken as positive), and $z$ is the coordinate along the $z$ axis. The velocity components, as sketched in Fig. 6.6, are the radial velocity, $v_r$, the tangential velocity, $v_\theta$, and the axial velocity, $v_z$. Thus, the velocity at some arbitrary point $P$ can be expressed as

$$\mathbf{V} = v_r\hat{\mathbf{e}}_r + v_\theta\hat{\mathbf{e}}_\theta + v_z\hat{\mathbf{e}}_z \tag{6.32}$$

where $\hat{\mathbf{e}}_r$, $\hat{\mathbf{e}}_\theta$, and $\hat{\mathbf{e}}_z$ are the unit vectors in the $r$, $\theta$, and $z$ directions, respectively, as are illustrated in Fig. 6.6. The use of cylindrical coordinates is particularly convenient when the boundaries of the flow system are cylindrical. Several examples illustrating the use of cylindrical coordinates will be given in succeeding sections in this chapter.

The differential form of the continuity equation in cylindrical coordinates is

$$\frac{\partial \rho}{\partial t} + \frac{1}{r}\frac{\partial(r\rho v_r)}{\partial r} + \frac{1}{r}\frac{\partial(\rho v_\theta)}{\partial \theta} + \frac{\partial(\rho v_z)}{\partial z} = 0 \tag{6.33}$$

This equation can be derived by following the same procedure used in the preceding section. For steady, compressible flow

$$\frac{1}{r}\frac{\partial(r\rho v_r)}{\partial r} + \frac{1}{r}\frac{\partial(\rho v_\theta)}{\partial \theta} + \frac{\partial(\rho v_z)}{\partial z} = 0 \tag{6.34}$$

For incompressible fluids (for steady or unsteady flow)

$$\frac{1}{r}\frac{\partial(r v_r)}{\partial r} + \frac{1}{r}\frac{\partial v_\theta}{\partial \theta} + \frac{\partial v_z}{\partial z} = 0 \tag{6.35}$$

### 6.2.3 The Stream Function

Steady, incompressible, plane, two-dimensional flow represents one of the simplest types of flow of practical importance. By plane, two-dimensional flow we mean that there are only two velocity components, such as $u$ and $v$, when the flow is considered to be in the $x$–$y$ plane. For this flow the continuity equation, Eq. 6.31, reduces to

$$\frac{\partial u}{\partial x} + \frac{\partial v}{\partial y} = 0 \tag{6.36}$$

Velocity components in a two-dimensional flow field can be expressed in terms of a stream function.

We still have two variables, $u$ and $v$, to deal with, but they must be related in a special way as indicated by Eq. 6.36. This equation suggests that if we define a function $\psi(x, y)$, called the **stream function,** which relates the velocities shown by the figure in the margin as

$$u = \frac{\partial \psi}{\partial y} \qquad v = -\frac{\partial \psi}{\partial x} \tag{6.37}$$

then the continuity equation is identically satisfied. This conclusion can be verified by simply substituting the expressions for $u$ and $v$ into Eq. 6.36 so that

$$\frac{\partial}{\partial x}\left(\frac{\partial \psi}{\partial y}\right) + \frac{\partial}{\partial y}\left(-\frac{\partial \psi}{\partial x}\right) = \frac{\partial^2 \psi}{\partial x\,\partial y} - \frac{\partial^2 \psi}{\partial y\,\partial x} = 0$$

Thus, whenever the velocity components are defined in terms of the stream function we know that conservation of mass will be satisfied. Of course, we still do not know what $\psi(x, y)$ is for a particular problem, but at least we have simplified the analysis by having to determine only one unknown function, $\psi(x, y)$, rather than the two functions, $u(x, y)$ and $v(x, y)$.

Another particular advantage of using the stream function is related to the fact that *lines along which $\psi$ is constant are streamlines.* Recall from Section 4.1.4 that streamlines are lines in the flow field that are everywhere tangent to the velocities, as is illustrated in Fig. 6.7 and the figure in the margin. It follows from the definition of the streamline that the slope at any point along a streamline is given by

$$\frac{dy}{dx} = \frac{v}{u}$$

The change in the value of $\psi$ as we move from one point $(x, y)$ to a nearby point $(x + dx, y + dy)$ is given by the relationship:

$$d\psi = \frac{\partial \psi}{\partial x}\,dx + \frac{\partial \psi}{\partial y}\,dy = -v\,dx + u\,dy$$

Along a line of constant $\psi$ we have $d\psi = 0$ so that

$$-v\,dx + u\,dy = 0$$

and, therefore, along a line of constant $\psi$

$$\frac{dy}{dx} = \frac{v}{u}$$

which is the defining equation for a streamline. Thus, if we know the function $\psi(x, y)$, we can plot lines of constant $\psi$ to provide the family of streamlines that are helpful in visualizing the pattern

■ **Figure 6.7**  Velocity and velocity components along a streamline.

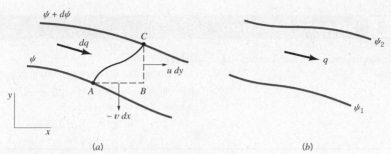

**■ Figure 6.8** The flow between two streamlines.

of flow. An infinite number of streamlines make up a particular flow field, since for each constant value assigned to $\psi$ a streamline can be drawn.

The actual numerical value associated with a particular streamline is not of particular significance, but the change in the value of $\psi$ is related to the volume rate of flow. Consider two closely spaced streamlines, shown in Fig. 6.8a. The lower streamline is designated $\psi$ and the upper one $\psi + d\psi$. Let $dq$ represent the volume rate of flow (per unit width perpendicular to the $x$–$y$ plane) passing between the two streamlines. Note that flow never crosses streamlines, since by definition the velocity is tangent to the streamline. From conservation of mass we know that the inflow, $dq$, crossing the arbitrary surface $AC$ of Fig. 6.8a must equal the net outflow through surfaces $AB$ and $BC$. Thus,

$$dq = u\,dy - v\,dx$$

or in terms of the stream function

$$dq = \frac{\partial \psi}{\partial y}\,dy + \frac{\partial \psi}{\partial x}\,dx \tag{6.38}$$

The right-hand side of Eq. 6.38 is equal to $d\psi$ so that

$$dq = d\psi \tag{6.39}$$

Thus, the volume rate of flow, $q$, between two streamlines such as $\psi_1$ and $\psi_2$ of Fig. 6.8b can be determined by integrating Eq. 6.39 to yield

$$q = \int_{\psi_1}^{\psi_2} d\psi = \psi_2 - \psi_1 \tag{6.40}$$

The relative value of $\psi_2$ with respect to $\psi_1$ determines the direction of flow, as shown by the figure in the margin.

In cylindrical coordinates the continuity equation (Eq. 6.35) for incompressible, plane, two-dimensional flow reduces to

$$\frac{1}{r}\frac{\partial (r v_r)}{\partial r} + \frac{1}{r}\frac{\partial v_\theta}{\partial \theta} = 0 \tag{6.41}$$

and the velocity components, $v_r$ and $v_\theta$, can be related to the stream function, $\psi(r, \theta)$, through the equations

$$\boxed{v_r = \frac{1}{r}\frac{\partial \psi}{\partial \theta} \qquad v_\theta = -\frac{\partial \psi}{\partial r}} \tag{6.42}$$

as shown by the figure in the margin.

Substitution of these expressions for the velocity components into Eq. 6.41 shows that the continuity equation is identically satisfied. The stream function concept can be extended to axisymmetric flows, such as flow in pipes or flow around bodies of revolution, and to two-dimensional compressible flows. However, the concept is not applicable to general three-dimensional flows using a single stream function.

*The change in the value of the stream function is related to the volume rate of flow.*

# EXAMPLE 6.3   Stream Function

**GIVEN** The velocity components in a steady, incompressible, two-dimensional flow field are

$$u = 2y$$
$$v = 4x$$

**FIND**

**(a)** Determine the corresponding stream function and

**(b)** Show on a sketch several streamlines. Indicate the direction of flow along the streamlines.

## SOLUTION

**(a)** From the definition of the stream function (Eqs. 6.37)

$$u = \frac{\partial \psi}{\partial y} = 2y$$

and

$$v = -\frac{\partial \psi}{\partial x} = 4x$$

The first of these equations can be integrated to give

$$\psi = y^2 + f_1(x)$$

where $f_1(x)$ is an arbitrary function of $x$. Similarly from the second equation

$$\psi = -2x^2 + f_2(y)$$

where $f_2(y)$ is an arbitrary function of $y$. It now follows that in order to satisfy both expressions for the stream function

$$\psi = -2x^2 + y^2 + C \qquad \textbf{(Ans)}$$

where $C$ is an arbitrary constant.

**COMMENT** Since the velocities are related to the derivatives of the stream function, an arbitrary constant can always be added to the function, and the value of the constant is actually of no consequence. Usually, for simplicity, we set $C = 0$ so that for this particular example the simplest form for the stream function is

$$\psi = -2x^2 + y^2 \qquad \textbf{(1)} \quad \textbf{(Ans)}$$

Either answer indicated would be acceptable.

**(b)** Streamlines can now be determined by setting $\psi =$ constant and plotting the resulting curve. With the preceding expression for $\psi$ (with $C = 0$) the value of $\psi$ at the origin is zero, so that the equation of the streamline passing through the origin (the $\psi = 0$ streamline) is

$$0 = -2x^2 + y^2$$

■ **Figure E6.3**

or

$$y = \pm \sqrt{2}x$$

Other streamlines can be obtained by setting $\psi$ equal to various constants. It follows from Eq. 1 that the equations of these streamlines (for $\psi \neq 0$) can be expressed in the form

$$\frac{y^2}{\psi} - \frac{x^2}{\psi/2} = 1$$

which we recognize as the equation of a hyperbola. Thus, the streamlines are a family of hyperbolas with the $\psi = 0$ streamlines as asymptotes. Several of the streamlines are plotted in Fig. E6.3. Since the velocities can be calculated at any point, the direction of flow along a given streamline can be easily deduced. For example, $v = -\partial \psi / \partial x = 4x$ so that $v > 0$ if $x > 0$ and $v < 0$ if $x < 0$. The direction of flow is indicated on the figure.

## 6.3   Conservation of Linear Momentum

To develop the differential momentum equations we can start with the linear momentum equation

$$\mathbf{F} = \frac{D\mathbf{P}}{Dt}\bigg|_{\text{sys}} \tag{6.43}$$

where $\mathbf{F}$ is the resultant force acting on a fluid mass, $\mathbf{P}$ is the linear momentum defined as

$$\mathbf{P} = \int_{\text{sys}} \mathbf{V}\, dm$$

and the operator $D(\ )/Dt$ is the material derivative (see Section 4.2.1). In the last chapter it was demonstrated how Eq. 6.43 in the form

$$\sum \mathbf{F}_{\substack{\text{contents of the} \\ \text{control volume}}} = \frac{\partial}{\partial t} \int_{cv} \mathbf{V}\rho \, d\Psi + \int_{cs} \mathbf{V}\rho \mathbf{V} \cdot \hat{\mathbf{n}} \, dA \qquad (6.44)$$

could be applied to a finite control volume to solve a variety of flow problems. To obtain the differential form of the linear momentum equation, we can either apply Eq. 6.43 to a differential system, consisting of a mass, $\delta m$, or apply Eq. 6.44 to an infinitesimal control volume, $\delta\Psi$, which initially bounds the mass $\delta m$. It is probably simpler to use the system approach since application of Eq. 6.43 to the differential mass, $\delta m$, yields

$$\delta \mathbf{F} = \frac{D(\mathbf{V} \, \delta m)}{Dt}$$

where $\delta \mathbf{F}$ is the resultant force acting on $\delta m$. Using this system approach $\delta m$ can be treated as a constant so that

$$\delta \mathbf{F} = \delta m \frac{D\mathbf{V}}{Dt}$$

But $D\mathbf{V}/Dt$ is the acceleration, $\mathbf{a}$, of the element. Thus,

$$\delta \mathbf{F} = \delta m \, \mathbf{a} \qquad (6.45)$$

which is simply Newton's second law applied to the mass $\delta_m$. This is the same result that would be obtained by applying Eq. 6.44 to an infinitesimal control volume (see Ref. 1). Before we can proceed, it is necessary to examine how the force $\delta \mathbf{F}$ can be most conveniently expressed.

### 6.3.1 Description of Forces Acting on the Differential Element

*Both surface forces and body forces generally act on fluid particles.*

In general, two types of forces need to be considered: *surface forces*, which act on the surface of the differential element, and *body forces*, which are distributed throughout the element. For our purpose, the only body force, $\delta \mathbf{F}_b$, of interest is the weight of the element, which can be expressed as

$$\delta \mathbf{F}_b = \delta m \, \mathbf{g} \qquad (6.46)$$

where $\mathbf{g}$ is the vector representation of the acceleration of gravity. In component form

$$\delta F_{bx} = \delta m \, g_x \qquad (6.47a)$$

$$\delta F_{by} = \delta m \, g_y \qquad (6.47b)$$

$$\delta F_{bz} = \delta m \, g_z \qquad (6.47c)$$

where $g_x$, $g_y$, and $g_z$ are the components of the acceleration of gravity vector in the $x$, $y$, and $z$ directions, respectively.

Surface forces act on the element as a result of its interaction with its surroundings. At any arbitrary location within a fluid mass, the force acting on a small area, $\delta A$, that lies in an arbitrary surface, can be represented by $\delta \mathbf{F}_s$, as is shown in Fig. 6.9. In general, $\delta \mathbf{F}_s$ will be inclined with respect to the surface. The force $\delta \mathbf{F}_s$ can be resolved into three components, $\delta F_n$, $\delta F_1$, and $\delta F_2$, where $\delta F_n$ is normal to the area, $\delta A$, and $\delta F_1$ and $\delta F_2$ are parallel to the area and orthogonal to each other. The *normal stress, $\sigma_n$,* is defined as

$$\sigma_n = \lim_{\delta A \to 0} \frac{\delta F_n}{\delta A}$$

Arbitrary surface

■ **Figure 6.9**  Components of force acting on an arbitrary differential area.

■ **Figure 6.10** Double subscript notation for stresses.

and the *shearing stresses* are defined as

$$\tau_1 = \lim_{\delta A \to 0} \frac{\delta F_1}{\delta A}$$

and

$$\tau_2 = \lim_{\delta A \to 0} \frac{\delta F_2}{\delta A}$$

We will use $\sigma$ for normal stresses and $\tau$ for shearing stresses. The intensity of the force per unit area at a point in a body can thus be characterized by a normal stress and two shearing stresses, if the orientation of the area is specified. For purposes of analysis it is usually convenient to reference the area to the coordinate system. For example, for the rectangular coordinate system shown in Fig. 6.10 we choose to consider the stresses acting on planes parallel to the coordinate planes. On the plane *ABCD* of Fig. 6.10*a*, which is parallel to the *y*–*z* plane, the normal stress is denoted $\sigma_{xx}$ and the shearing stresses are denoted as $\tau_{xy}$ and $\tau_{xz}$. To easily identify the particular stress component we use a double subscript notation. The first subscript indicates the direction of the *normal* to the plane on which the stress acts, and the second subscript indicates the direction of the stress. Thus, normal stresses have repeated subscripts, whereas the subscripts for the shearing stresses are always different.

It is also necessary to establish a sign convention for the stresses. We define the positive direction for the stress as the positive coordinate direction on the surfaces for which the outward normal is in the positive coordinate direction. This is the case illustrated in Fig. 6.10*a* where the outward normal to the area *ABCD* is in the positive *x* direction. The positive directions for $\sigma_{xx}$, $\tau_{xy}$, and $\tau_{xz}$ are as shown in Fig. 6.10*a*. If the outward normal points in the negative coordinate direction, as in Fig. 6.10*b* for the area *A'B'C'D'*, then the stresses are considered positive if directed in the negative coordinate directions. Thus, the stresses shown in Fig. 6.10*b* are considered to be positive when directed as shown. Note that positive normal stresses are tensile stresses; that is, they tend to "stretch" the material.

It should be emphasized that the state of stress at a point in a material is not completely defined by simply three components of a "stress vector." This follows, since any particular stress vector depends on the orientation of the plane passing through the point. However, it can be shown that the normal and shearing stresses acting on *any* plane passing through a point can be expressed in terms of the stresses acting on three orthogonal planes passing through the point (Ref. 2).

*Surface forces can be expressed in terms of the shear and normal stresses.*

We now can express the surface forces acting on a small cubical element of fluid in terms of the stresses acting on the faces of the element as shown in Fig. 6.11. It is expected that in general the stresses will vary from point to point within the flow field. Thus, through the use of Taylor series expansions we will express the stresses on the various faces in terms of the corresponding stresses at the center of the element of Fig. 6.11 and their gradients in the coordinate directions. For simplicity only the forces in the *x* direction are shown. Note that the stresses must be multiplied by the area on which they act to obtain the force. Summing all these forces in the *x* direction yields

$$\delta F_{sx} = \left( \frac{\partial \sigma_{xx}}{\partial x} + \frac{\partial \tau_{yx}}{\partial y} + \frac{\partial \tau_{zx}}{\partial z} \right) \delta x \, \delta y \, \delta z \qquad \textbf{(6.48a)}$$

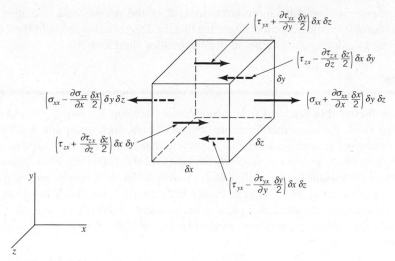

■ **Figure 6.11** Surface forces in the $x$ direction acting on a fluid element.

for the resultant surface force in the $x$ direction. In a similar manner the resultant surface forces in the $y$ and $z$ directions can be obtained and expressed as

$$\delta F_{sy} = \left(\frac{\partial \tau_{xy}}{\partial x} + \frac{\partial \sigma_{yy}}{\partial y} + \frac{\partial \tau_{zy}}{\partial z}\right) \delta x\, \delta y\, \delta z \tag{6.48b}$$

$$\delta F_{sz} = \left(\frac{\partial \tau_{xz}}{\partial x} + \frac{\partial \tau_{yz}}{\partial y} + \frac{\partial \sigma_{zz}}{\partial z}\right) \delta x\, \delta y\, \delta z \tag{6.48c}$$

The resultant surface force can now be expressed as

$$\delta \mathbf{F}_s = \delta F_{sx}\hat{\mathbf{i}} + \delta F_{sy}\hat{\mathbf{j}} + \delta F_{sz}\hat{\mathbf{k}} \tag{6.49}$$

and this force combined with the body force, $\delta \mathbf{F}_b$, yields the resultant force, $\delta \mathbf{F}$, acting on the differential mass, $\delta m$. That is, $\delta \mathbf{F} = \delta \mathbf{F}_s + \delta \mathbf{F}_b$.

### 6.3.2 Equations of Motion

The expressions for the body and surface forces can now be used in conjunction with Eq. 6.45 to develop the equations of motion. In component form Eq. 6.45 can be written as

$$\delta F_x = \delta m\, a_x$$

$$\delta F_y = \delta m\, a_y$$

$$\delta F_z = \delta m\, a_z$$

where $\delta m = \rho\, \delta x\, \delta y\, \delta z$, and the acceleration components are given by Eq. 6.3. It now follows (using Eqs. 6.47 and 6.48 for the forces on the element) that

*The motion of a fluid is governed by a set of nonlinear differential equations.*

$$\rho g_x + \frac{\partial \sigma_{xx}}{\partial x} + \frac{\partial \tau_{yx}}{\partial y} + \frac{\partial \tau_{zx}}{\partial z} = \rho\left(\frac{\partial u}{\partial t} + u\frac{\partial u}{\partial x} + v\frac{\partial u}{\partial y} + w\frac{\partial u}{\partial z}\right) \tag{6.50a}$$

$$\rho g_y + \frac{\partial \tau_{xy}}{\partial x} + \frac{\partial \sigma_{yy}}{\partial y} + \frac{\partial \tau_{zy}}{\partial z} = \rho\left(\frac{\partial v}{\partial t} + u\frac{\partial v}{\partial x} + v\frac{\partial v}{\partial y} + w\frac{\partial v}{\partial z}\right) \tag{6.50b}$$

$$\rho g_z + \frac{\partial \tau_{xz}}{\partial x} + \frac{\partial \tau_{yz}}{\partial y} + \frac{\partial \sigma_{zz}}{\partial z} = \rho\left(\frac{\partial w}{\partial t} + u\frac{\partial w}{\partial x} + v\frac{\partial w}{\partial y} + w\frac{\partial w}{\partial z}\right) \tag{6.50c}$$

where the element volume $\delta x\, \delta y\, \delta z$ cancels out.

Equations 6.50 are the general differential equations of motion for a fluid. In fact, they are applicable to any continuum (solid or fluid) in motion or at rest. However, before we can use the equations to solve specific problems, some additional information about the stresses must be obtained.

Otherwise, we will have more unknowns (all of the stresses and velocities and the density) than equations. It should not be too surprising that the differential analysis of fluid motion is complicated. We are attempting to describe, in detail, complex fluid motion.

## 6.4 Inviscid Flow

As is discussed in Section 1.6, shearing stresses develop in a moving fluid because of the viscosity of the fluid. We know that for some common fluids, such as air and water, the viscosity is small, and therefore it seems reasonable to assume that under some circumstances we may be able to simply neglect the effect of viscosity (and thus shearing stresses). Flow fields in which the shearing stresses are assumed to be negligible are said to be *inviscid*, *nonviscous*, or *frictionless*. These terms are used interchangeably. As is discussed in Section 2.1, for fluids in which there are no shearing stresses the normal stress at a point is independent of direction—that is, $\sigma_{xx} = \sigma_{yy} = \sigma_{zz}$. In this instance we define the pressure, $p$, as the negative of the normal stress so that, as indicated by the figure in the margin,

$$-p = \sigma_{xx} = \sigma_{yy} = \sigma_{zz}$$

The negative sign is used so that a *compressive* normal stress (which is what we expect in a fluid) will give a *positive* value for $p$.

In Chapter 3 the inviscid flow concept was used in the development of the Bernoulli equation, and numerous applications of this important equation were considered. In this section we will again consider the Bernoulli equation and will show how it can be derived from the general equations of motion for inviscid flow.

### 6.4.1 Euler's Equations of Motion

For an inviscid flow in which all the shearing stresses are zero, and the normal stresses are replaced by $-p$, the general equations of motion (Eqs. 6.50) reduce to

$$\rho g_x - \frac{\partial p}{\partial x} = \rho \left( \frac{\partial u}{\partial t} + u \frac{\partial u}{\partial x} + v \frac{\partial u}{\partial y} + w \frac{\partial u}{\partial z} \right) \tag{6.51a}$$

$$\rho g_y - \frac{\partial p}{\partial y} = \rho \left( \frac{\partial v}{\partial t} + u \frac{\partial v}{\partial x} + v \frac{\partial v}{\partial y} + w \frac{\partial v}{\partial z} \right) \tag{6.51b}$$

$$\rho g_z - \frac{\partial p}{\partial z} = \rho \left( \frac{\partial w}{\partial t} + u \frac{\partial w}{\partial x} + v \frac{\partial w}{\partial y} + w \frac{\partial w}{\partial z} \right) \tag{6.51c}$$

These equations are commonly referred to as ***Euler's equations of motion,*** named in honor of Leonhard Euler (1707–1783), a famous Swiss mathematician who pioneered work on the relationship between pressure and flow. In vector notation Euler's equations can be expressed as

$$\rho \mathbf{g} - \boldsymbol{\nabla} p = \rho \left[ \frac{\partial \mathbf{V}}{\partial t} + (\mathbf{V} \cdot \boldsymbol{\nabla}) \mathbf{V} \right] \tag{6.52}$$

*Euler's equations of motion apply to an inviscid flow field.*

Although Eqs. 6.51 are considerably simpler than the general equations of motion, Eqs. 6.50, they are still not amenable to a general analytical solution that would allow us to determine the pressure and velocity at all points within an inviscid flow field. The main difficulty arises from the nonlinear velocity terms ($u\, \partial u/\partial x$, $v\, \partial u/\partial y$, etc.), which appear in the convective acceleration, or the $(\mathbf{V} \cdot \boldsymbol{\nabla})\mathbf{V}$ term. Because of these terms, Euler's equations are nonlinear partial differential equations for which we do not have a general method of solving. However, under some circumstances we can use them to obtain useful information about inviscid flow fields. For example, as shown in the following section we can integrate Eq. 6.52 to obtain a relationship (the Bernoulli equation) between elevation, pressure, and velocity along a streamline.

### 6.4.2 The Bernoulli Equation

In Section 3.2 the Bernoulli equation was derived by a direct application of Newton's second law to a fluid particle moving along a streamline. In this section we will again derive this important

■ **Figure 6.12**  The nota-tion for differential length along a streamline.

equation, starting from Euler's equations. Of course, we should obtain the same result since Euler's equations simply represent a statement of Newton's second law expressed in a general form that is useful for flow problems and maintains the restriction of zero viscosity. We will restrict our attention to steady flow so Euler's equation in vector form becomes

$$\rho \mathbf{g} - \boldsymbol{\nabla} p = \rho (\mathbf{V} \cdot \boldsymbol{\nabla}) \mathbf{V} \tag{6.53}$$

We wish to integrate this differential equation along some arbitrary streamline (Fig. 6.12) and select the coordinate system with the $z$ axis vertical (with "up" being positive) so that, as shown by the figure in the margin, the acceleration of gravity vector can be expressed as

$$\mathbf{g} = -g\boldsymbol{\nabla} z$$

where $g$ is the magnitude of the acceleration of gravity vector. Also, it will be convenient to use the vector identity

$$(\mathbf{V} \cdot \boldsymbol{\nabla})\mathbf{V} = \tfrac{1}{2}\boldsymbol{\nabla}(\mathbf{V} \cdot \mathbf{V}) - \mathbf{V} \times (\boldsymbol{\nabla} \times \mathbf{V})$$

Equation 6.53 can now be written in the form

$$-\rho g\boldsymbol{\nabla} z - \boldsymbol{\nabla} p = \frac{\rho}{2}\boldsymbol{\nabla}(\mathbf{V} \cdot \mathbf{V}) - \rho\mathbf{V} \times (\boldsymbol{\nabla} \times \mathbf{V})$$

and this equation can be rearranged to yield

$$\frac{\boldsymbol{\nabla} p}{\rho} + \frac{1}{2}\boldsymbol{\nabla}(V^2) + g\boldsymbol{\nabla} z = \mathbf{V} \times (\boldsymbol{\nabla} \times \mathbf{V})$$

We next take the dot product of each term with a differential length $d\mathbf{s}$ along a streamline (Fig. 6.12). Thus,

$$\frac{\boldsymbol{\nabla} p}{\rho} \cdot d\mathbf{s} + \frac{1}{2}\boldsymbol{\nabla}(V^2) \cdot d\mathbf{s} + g\boldsymbol{\nabla} z \cdot d\mathbf{s} = [\mathbf{V} \times (\boldsymbol{\nabla} \times \mathbf{V})] \cdot d\mathbf{s} \tag{6.54}$$

Since $d\mathbf{s}$ has a direction along the streamline, the vectors $d\mathbf{s}$ and $\mathbf{V}$ are parallel. However, as shown by the figure in the margin, the vector $\mathbf{V} \times (\boldsymbol{\nabla} \times \mathbf{V})$ is perpendicular to $\mathbf{V}$ (why?), so it follows that

$$[\mathbf{V} \times (\boldsymbol{\nabla} \times \mathbf{V})] \cdot d\mathbf{s} = 0$$

Recall also that the dot product of the gradient of a scalar and a differential length gives the differential change in the scalar in the direction of the differential length. That is, with $d\mathbf{s} = dx\,\hat{\mathbf{i}} + dy\,\hat{\mathbf{j}} + dz\,\hat{\mathbf{k}}$ we can write $\boldsymbol{\nabla} p \cdot d\mathbf{s} = (\partial p/\partial x)\,dx + (\partial p/\partial y)dy + (\partial p/\partial z)dz = dp$. Thus, Eq. 6.54 becomes

$$\frac{dp}{\rho} + \frac{1}{2}\,d(V^2) + g\,dz = 0 \tag{6.55}$$

where the change in $p$, $V$, and $z$ is along the streamline. Equation 6.55 can now be integrated to give

$$\int \frac{dp}{\rho} + \frac{V^2}{2} + gz = \text{constant} \tag{6.56}$$

which indicates that the sum of the three terms on the left side of the equation must remain a constant along a given streamline. Equation 6.56 is valid for both compressible and incompressible

*V* ×(∇ × **V**)

*Euler's equations can be arranged to give the relation-ship among pres-sure, velocity, and elevation for invis-cid fluids.*

inviscid flows, but for compressible fluids the variation in $\rho$ with $p$ must be specified before the first term in Eq. 6.56 can be evaluated.

For inviscid, incompressible fluids (commonly called ***ideal fluids***) Eq. 6.56 can be written as

$$\frac{p}{\rho} + \frac{V^2}{2} + gz = \text{constant along a streamline} \qquad (6.57)$$

and this equation is the ***Bernoulli equation*** used extensively in Chapter 3. It is often convenient to write Eq. 6.57 between two points (1) and (2) along a streamline and to express the equation in the "head" form by dividing each term by $g$ so that

$$\frac{p_1}{\gamma} + \frac{V_1^2}{2g} + z_1 = \frac{p_2}{\gamma} + \frac{V_2^2}{2g} + z_2 \qquad (6.58)$$

It should be again emphasized that the Bernoulli equation, as expressed by Eqs. 6.57 and 6.58, is restricted to the following:

- inviscid flow
- steady flow
- incompressible flow
- flow along a streamline

You may want to go back and review some of the examples in Chapter 3 that illustrate the use of the Bernoulli equation.

### 6.4.3 Irrotational Flow

*The vorticity is zero in an irrotational flow field.*

If we make one additional assumption—that the flow is *irrotational*—the analysis of inviscid flow problems is further simplified. Recall from Section 6.1.3 that the rotation of a fluid element is equal to $\frac{1}{2}(\nabla \times \mathbf{V})$, and an irrotational flow field is one for which $\nabla \times \mathbf{V} = 0$ (i.e., the curl of velocity is zero). Since the vorticity, $\boldsymbol{\zeta}$, is defined as $\nabla \times \mathbf{V}$, it also follows that in an irrotational flow field the vorticity is zero. The concept of irrotationality may seem to be a rather strange condition for a flow field. Why would a flow field be irrotational? To answer this question we note that if $\frac{1}{2}(\nabla \times \mathbf{V}) = 0$, then each of the components of this vector, as are given by Eqs. 6.12, 6.13, and 6.14, must be equal to zero. Since these components include the various velocity gradients in the flow field, the condition of irrotationality imposes specific relationships among these velocity gradients. For example, for rotation about the $z$ axis to be zero, it follows from Eq. 6.12 that

$$\omega_z = \frac{1}{2}\left(\frac{\partial v}{\partial x} - \frac{\partial u}{\partial y}\right) = 0$$

and, therefore,

$$\frac{\partial v}{\partial x} = \frac{\partial u}{\partial y} \qquad (6.59)$$

Similarly from Eqs. 6.13 and 6.14

$$\frac{\partial w}{\partial y} = \frac{\partial v}{\partial z} \qquad (6.60)$$

$$\frac{\partial u}{\partial z} = \frac{\partial w}{\partial x} \qquad (6.61)$$

A general flow field would not satisfy these three equations. However, a uniform flow as is illustrated in Fig. 6.13 does. Since $u = U$ (a constant), $v = 0$, and $w = 0$, it follows that Eqs. 6.59, 6.60, and 6.61 are all satisfied. Therefore, a uniform flow field (in which there are no velocity gradients) is certainly an example of an irrotational flow.

Uniform flows by themselves are not very interesting. However, many interesting and important flow problems include uniform flow in some part of the flow field. Two examples are shown in Fig. 6.14. In Fig. 6.14*a* a solid body is placed in a uniform stream of fluid. Far away from the body the flow remains uniform, and in this far region the flow is irrotational. In Fig. 6.14*b*, flow from a large reservoir enters a pipe through a streamlined entrance where the velocity distribution is essentially uniform. Thus, at the entrance the flow is irrotational.

For an inviscid fluid there are no shearing stresses—the only forces acting on a fluid element are its weight and pressure forces. Since the weight acts through the element center of gravity, and the pressure acts in a direction normal to the element surface, neither of these forces can cause the element to rotate. Therefore, for an inviscid fluid, if some part of the flow field is irrotational, the fluid elements emanating from this region will not take on any rotation as they progress through the flow field. This phenomenon is illustrated in Fig. 6.14*a* in which fluid elements flowing far away from the body have irrotational motion, and as they flow around the body the motion remains irrotational except very near the boundary. Near the boundary the velocity changes rapidly from zero at the boundary (no-slip condition) to some relatively large value in a short distance from the boundary. This rapid change in velocity gives rise to a large velocity gradient normal to the boundary and produces significant shearing stresses, even though the viscosity is small. Of course if we had a truly inviscid fluid, the fluid would simply "slide" past the boundary and the flow would be irrotational everywhere. But this is not the case for real fluids, so we will typically have a layer (usually relatively thin) near any fixed surface in a moving stream in which shearing stresses are not negligible. This layer is called the *boundary layer*. Outside the boundary layer the flow can be treated as an irrotational flow. Another possible consequence of the boundary layer is that the main stream may "separate" from the surface and form a *wake* downstream from the body. (See the

*Flow fields involving real fluids often include both regions of negligible shearing stresses and regions of significant shearing stresses.*

■ **Figure 6.14**   **Various regions of flow: (*a*) around bodies; (*b*) through channels.**

photographs at the beginning of Chapters 7, 9, and 11.) The wake would include a region of slow, perhaps randomly moving fluid. To completely analyze this type of problem it is necessary to consider both the inviscid, irrotational flow outside the boundary layer, and the viscous, rotational flow within the boundary layer and to somehow "match" these two regions. This type of analysis is considered in Chapter 9.

As is illustrated in Fig. 6.14b, the flow in the entrance to a pipe may be uniform (if the entrance is streamlined) and thus will be irrotational. In the central core of the pipe the flow remains irrotational for some distance. However, a boundary layer will develop along the wall and grow in thickness until it fills the pipe. Thus, for this type of internal flow there will be an *entrance region* in which there is a central irrotational core, followed by a so-called *fully developed region* in which viscous forces are dominant. The concept of irrotationality is completely invalid in the fully developed region. This type of internal flow problem is considered in detail in Chapter 8.

The two preceding examples are intended to illustrate the possible applicability of irrotational flow to some "real fluid" flow problems and to indicate some limitations of the irrotationality concept. We proceed to develop some useful equations based on the assumptions of inviscid, incompressible, irrotational flow, with the admonition to use caution when applying the equations.

### 6.4.4 The Bernoulli Equation for Irrotational Flow

In the development of the Bernoulli equation in Section 6.4.2, Eq. 6.54 was integrated along a streamline. This restriction was imposed so the right side of the equation could be set equal to zero; that is,

$$[\mathbf{V} \times (\nabla \times \mathbf{V})] \cdot d\mathbf{s} = 0$$

(since $d\mathbf{s}$ is parallel to $\mathbf{V}$). However, for irrotational flow, $\nabla \times \mathbf{V} = 0$, so the right side of Eq. 6.54 is zero regardless of the direction of $d\mathbf{s}$. We can now follow the same procedure used to obtain Eq. 6.55, where the differential changes $dp$, $d(V^2)$, and $dz$ can be taken in any direction. Integration of Eq. 6.55 again yields

$$\int \frac{dp}{\rho} + \frac{V^2}{2} + gz = \text{constant} \tag{6.62}$$

where for irrotational flow the constant is the same throughout the flow field. Thus, for incompressible, irrotational flow the Bernoulli equation can be written as

$$\boxed{\frac{p_1}{\gamma} + \frac{V_1^2}{2g} + z_1 = \frac{p_2}{\gamma} + \frac{V_2^2}{2g} + z_2} \tag{6.63}$$

*The Bernoulli equation can be applied between any two points in an irrotational flow field.*

between *any two points in the flow field*. Equation 6.63 is exactly the same form as Eq. 6.58 but is not limited to application along a streamline. However, Eq. 6.63 is restricted to

■ inviscid flow            ■ incompressible flow
■ steady flow             ■ irrotational flow

It may be worthwhile to review the use and misuse of the Bernoulli equation for rotational flow as is illustrated in Example 3.18.

### 6.4.5 The Velocity Potential

For an irrotational flow the velocity gradients are related through Eqs. 6.59, 6.60, and 6.61. It follows that in this case the velocity components can be expressed in terms of a scalar function $\phi(x, y, z, t)$ as

$$u = \frac{\partial \phi}{\partial x} \qquad v = \frac{\partial \phi}{\partial y} \qquad w = \frac{\partial \phi}{\partial z} \tag{6.64}$$

where $\phi$ is called the ***velocity potential.*** Direct substitution of these expressions for the velocity components into Eqs. 6.59, 6.60, and 6.61 will verify that a velocity field defined by Eqs. 6.64 is indeed irrotational. In vector form, Eqs. 6.64 can be written as

$$\boxed{\mathbf{V} = \mathbf{\nabla}\phi} \tag{6.65}$$

so that for an irrotational flow the velocity is expressible as the gradient of a scalar function $\phi$.

The velocity potential is a consequence of the irrotationality of the flow field, whereas the stream function is a consequence of conservation of mass (see Section 6.2.3). It is to be noted, however, that the velocity potential can be defined for a general three-dimensional flow, whereas a single stream function is restricted to two-dimensional flows.

For an incompressible fluid we know from conservation of mass that

$$\mathbf{\nabla} \cdot \mathbf{V} = 0$$

and therefore for incompressible, irrotational flow (with $\mathbf{V} = \mathbf{\nabla}\phi$) it follows that

$$\nabla^2\phi = 0 \tag{6.66}$$

where $\nabla^2(\ ) = \mathbf{\nabla} \cdot \mathbf{\nabla}(\ )$ is the *Laplacian operator*. In Cartesian coordinates

$$\frac{\partial^2\phi}{\partial x^2} + \frac{\partial^2\phi}{\partial y^2} + \frac{\partial^2\phi}{\partial z^2} = 0$$

> *Inviscid, incompressible, irrotational flow fields are governed by Laplace's equation and are called potential flows.*

This differential equation arises in many different areas of engineering and physics and is called *Laplace's equation.* Thus, inviscid, incompressible, irrotational flow fields are governed by Laplace's equation. This type of flow is commonly called a ***potential flow.*** To complete the mathematical formulation of a given problem, boundary conditions have to be specified. These are usually velocities specified on the boundaries of the flow field of interest. It follows that if the potential function can be determined, then the velocity at all points in the flow field can be determined from Eq. 6.64, and the pressure at all points can be determined from the Bernoulli equation (Eq. 6.63). Although the concept of the velocity potential is applicable to both steady and unsteady flow, we will confine our attention to steady flow.

Potential flows, governed by Eqs. 6.64 and 6.66, are irrotational flows. That is, the vorticity is zero throughout. If vorticity is present (e.g., boundary layer, wake), then the flow cannot be described by Laplace's equation. The figure in the margin illustrates a flow in which the vorticity is not zero in two regions—the separated region behind the bump and the boundary layer next to the solid surface. This is discussed in detail in Chapter 9.

For some problems it will be convenient to use cylindrical coordinates, $r$, $\theta$, and $z$. In this coordinate system the gradient operator is

$\nabla^2\phi = 0$

Streamlines

$\nabla^2\phi \neq 0$

Vorticity contours

$$\mathbf{\nabla}(\ ) = \frac{\partial(\ )}{\partial r}\hat{\mathbf{e}}_r + \frac{1}{r}\frac{\partial(\ )}{\partial\theta}\hat{\mathbf{e}}_\theta + \frac{\partial(\ )}{\partial z}\hat{\mathbf{e}}_z \tag{6.67}$$

so that

$$\mathbf{\nabla}\phi = \frac{\partial\phi}{\partial r}\hat{\mathbf{e}}_r + \frac{1}{r}\frac{\partial\phi}{\partial\theta}\hat{\mathbf{e}}_\theta + \frac{\partial\phi}{\partial z}\hat{\mathbf{e}}_z \tag{6.68}$$

where $\phi = \phi(r, \theta, z)$. Since

$$\mathbf{V} = v_r\hat{\mathbf{e}}_r + v_\theta\hat{\mathbf{e}}_\theta + v_z\hat{\mathbf{e}}_z \tag{6.69}$$

it follows for an irrotational flow (with $\mathbf{V} = \mathbf{\nabla}\phi$)

$$v_r = \frac{\partial\phi}{\partial r} \qquad v_\theta = \frac{1}{r}\frac{\partial\phi}{\partial\theta} \qquad v_z = \frac{\partial\phi}{\partial z} \tag{6.70}$$

Also, Laplace's equation in cylindrical coordinates is

$$\frac{1}{r}\frac{\partial}{\partial r}\left(r\frac{\partial\phi}{\partial r}\right) + \frac{1}{r^2}\frac{\partial^2\phi}{\partial\theta^2} + \frac{\partial^2\phi}{\partial z^2} = 0 \tag{6.71}$$

## EXAMPLE 6.4 Velocity Potential and Inviscid Flow Pressure

**GIVEN** The two-dimensional flow of a nonviscous, incompressible fluid in the vicinity of the 90° corner of Fig. E6.4$a$ is described by the stream function

$$\psi = 2r^2 \sin 2\theta$$

where $\psi$ has units of m$^2$/s when $r$ is in meters. Assume the fluid density is $10^3$ kg/m$^3$ and the $x$–$y$ plane is horizontal—

that is, there is no difference in elevation between points (1) and (2).

**FIND**

**(a)** Determine, if possible, the corresponding velocity potential.

**(b)** If the pressure at point (1) on the wall is 30 kPa, what is the pressure at point (2)?

(a)

(b)

(c)

■ **Figure E6.4**

## SOLUTION

**(a)** The radial and tangential velocity components can be obtained from the stream function as (see Eq. 6.42)

$$v_r = \frac{1}{r}\frac{\partial\psi}{\partial\theta} = 4r\cos 2\theta$$

and

$$v_\theta = -\frac{\partial\psi}{\partial r} = -4r\sin 2\theta$$

Since

$$v_r = \frac{\partial\phi}{\partial r}$$

it follows that

$$\frac{\partial\phi}{\partial r} = 4r\cos 2\theta$$

and therefore by integration

$$\phi = 2r^2\cos 2\theta + f_1(\theta) \tag{1}$$

where $f_1(\theta)$ is an arbitrary function of $\theta$. Similarly

$$v_\theta = \frac{1}{r}\frac{\partial\phi}{\partial\theta} = -4r\sin 2\theta$$

and integration yields

$$\phi = 2r^2\cos 2\theta + f_2(r) \tag{2}$$

where $f_2(r)$ is an arbitrary function of $r$. To satisfy both Eqs. 1 and 2, the velocity potential must have the form

$$\phi = 2r^2\cos 2\theta + C \tag{Ans}$$

where $C$ is an arbitrary constant. As is the case for stream functions, the specific value of $C$ is not important, and it is customary to let $C = 0$ so that the velocity potential for this corner flow is

$$\phi = 2r^2\cos 2\theta \tag{Ans}$$

**COMMENT** In the statement of this problem, it was implied by the wording "if possible" that we might not be able to find a corresponding velocity potential. The reason for this concern is that we can always define a stream function for two-dimensional flow, but the flow must be *irrotational* if there is a corresponding velocity potential. Thus, the fact that we were able to determine a velocity potential means that the flow is irrotational. Several streamlines and lines of constant $\phi$ are plotted in Fig. E6.4*b*. These two sets of lines are *orthogonal*. The reason why streamlines and lines of constant $\phi$ are always orthogonal is explained in Section 6.5.

**(b)** Since we have an irrotational flow of a nonviscous, incompressible fluid, the Bernoulli equation can be applied between any two points. Thus, between points (1) and (2) with no elevation change

$$\frac{p_1}{\gamma} + \frac{V_1^2}{2g} = \frac{p_2}{\gamma} + \frac{V_2^2}{2g}$$

or

$$p_2 = p_1 + \frac{\rho}{2}(V_1^2 - V_2^2) \qquad (3)$$

Since

$$V^2 = v_r^2 + v_\theta^2$$

it follows that for any point within the flow field

$$V^2 = (4r \cos 2\theta)^2 + (-4r \sin 2\theta)^2$$
$$= 16r^2(\cos^2 2\theta + \sin^2 2\theta)$$
$$= 16r^2$$

This result indicates that the square of the velocity at any point depends only on the radial distance, $r$, to the point. Note that the constant, 16, has units of $s^{-2}$. Thus,

$$V_1^2 = (16\ s^{-2})(1\ m)^2 = 16\ m^2/s^2$$

and

$$V_2^2 = (16\ s^{-2})(0.5\ m)^2 = 4\ m^2/s^2$$

Substitution of these velocities into Eq. 3 gives

$$p_2 = 30 \times 10^3\ N/m^2 + \frac{10^3\ kg/m^3}{2}(16\ m^2/s^2 - 4\ m^2/s^2)$$
$$= 36\ kPa \qquad \textbf{(Ans)}$$

**COMMENT** The stream function used in this example could also be expressed in Cartesian coordinates as

$$\psi = 2r^2 \sin 2\theta = 4r^2 \sin\theta \cos\theta$$

or

$$\psi = 4xy$$

since $x = r \cos\theta$ and $y = r \sin\theta$. However, in the cylindrical polar form the results can be generalized to describe flow in the vicinity of a corner of angle $\alpha$ (see Fig. E6.4*c*) with the equations

$$\psi = Ar^{\pi/\alpha} \sin\frac{\pi\theta}{\alpha}$$

and

$$\phi = Ar^{\pi/\alpha} \cos\frac{\pi\theta}{\alpha}$$

where $A$ is a constant.

## 6.5 Some Basic, Plane Potential Flows

*For potential flow, basic solutions can be simply added to obtain more complicated solutions.*

A major advantage of Laplace's equation is that it is a linear partial differential equation. Since it is linear, various solutions can be added to obtain other solutions—that is, if $\phi_1(x, y, z)$ and $\phi_2(x, y, z)$ are two solutions to Laplace's equation, then $\phi_3 = \phi_1 + \phi_2$ is also a solution. The practical implication of this result is that if we have certain basic solutions we can combine them to obtain more complicated and interesting solutions. In this section several basic velocity potentials, which describe some relatively simple flows, will be determined. In the next section these basic potentials will be combined to represent complicated flows.

For simplicity, only plane (two-dimensional) flows will be considered. In this case, by using Cartesian coordinates

$$u = \frac{\partial\phi}{\partial x} \qquad v = \frac{\partial\phi}{\partial y} \qquad (6.72)$$

or by using cylindrical coordinates

$$v_r = \frac{\partial\phi}{\partial r} \qquad v_\theta = \frac{1}{r}\frac{\partial\phi}{\partial\theta} \qquad (6.73)$$

as shown by the figure in the margin. Since we can define a stream function for plane flow, we can also let

$$u = \frac{\partial\psi}{\partial y} \qquad v = -\frac{\partial\psi}{\partial x} \qquad (6.74)$$

or

$$v_r = \frac{1}{r}\frac{\partial \psi}{\partial \theta} \qquad v_\theta = -\frac{\partial \psi}{\partial r} \qquad (6.75)$$

where the stream function was previously defined in Eqs. 6.37 and 6.42. We know that by defining the velocities in terms of the stream function, conservation of mass is identically satisfied. If we now impose the condition of irrotationality, it follows from Eq. 6.59 that

$$\frac{\partial u}{\partial y} = \frac{\partial v}{\partial x}$$

and in terms of the stream function

$$\frac{\partial}{\partial y}\left(\frac{\partial \psi}{\partial y}\right) = \frac{\partial}{\partial x}\left(-\frac{\partial \psi}{\partial x}\right)$$

or

$$\frac{\partial^2 \psi}{\partial x^2} + \frac{\partial^2 \psi}{\partial y^2} = 0$$

Thus, for a plane irrotational flow we can use either the velocity potential or the stream function—both must satisfy Laplace's equation in two dimensions. It is apparent from these results that the velocity potential and the stream function are somehow related. We have previously shown that lines of constant $\psi$ are streamlines; that is,

$$\left.\frac{dy}{dx}\right|_{\text{along } \psi = \text{constant}} = \frac{v}{u} \qquad (6.76)$$

The change in $\phi$ as we move from one point $(x, y)$ to a nearby point $(x + dx, y + dy)$ is given by the relationship

$$d\phi = \frac{\partial \phi}{\partial x}\,dx + \frac{\partial \phi}{\partial y}\,dy = u\,dx + v\,dy$$

Along a line of constant $\phi$ we have $d\phi = 0$ so that

$$\left.\frac{dy}{dx}\right|_{\text{along } \phi = \text{constant}} = -\frac{u}{v} \qquad (6.77)$$

A comparison of Eqs. 6.76 and 6.77 shows that lines of constant $\phi$ (called *equipotential lines*) are orthogonal to lines of constant $\psi$ (streamlines) at all points where they intersect. (Recall that two lines are orthogonal if the product of their slopes is $-1$, as illustrated by the figure in the margin.) For any potential flow field a *"flow net"* can be drawn that consists of a family of streamlines and equipotential lines. The flow net is useful in visualizing flow patterns and can be used to obtain graphical solutions by sketching in streamlines and equipotential lines and adjusting the lines until the lines are approximately orthogonal at all points where they intersect. An example of a flow net is shown in Fig. 6.15. Velocities can be estimated from the flow net, since the velocity is inversely proportional to the streamline spacing, as shown by the figure in the margin. Thus, for example, from Fig. 6.15 we can see that the velocity near the inside corner will be higher than the velocity along the outer part of the bend. (See the photographs at the beginning of Chapters 3 and 6.)

### 6.5.1 Uniform Flow

The simplest plane flow is one for which the streamlines are all straight and parallel, and the magnitude of the velocity is constant. This type of flow is called a *uniform flow.* For example, consider a uniform flow in the positive $x$ direction as is illustrated in Fig. 6.16a. In this instance, $u = U$ and $v = 0$, and in terms of the velocity potential

$$\frac{\partial \phi}{\partial x} = U \qquad \frac{\partial \phi}{\partial y} = 0$$

Equipotential line
($\phi$ = constant)

$d_1 > d$

$V_1 < V$

$d$ → $V$

$d$ → $V$

$V_1$

$V_2$

$d_2 < d$
$V_2 > V$

Streamline
($\psi$ = constant)

■ **Figure 6.15**   Flow net for a 90° bend. (Adapted from Ref. 3.)

These two equations can be integrated to yield

$$\phi = Ux + C$$

where $C$ is an arbitrary constant, which can be set equal to zero. Thus, for a uniform flow in the positive $x$ direction

$$\phi = Ux \tag{6.78}$$

The corresponding stream function can be obtained in a similar manner, since

$$\frac{\partial \psi}{\partial y} = U \qquad \frac{\partial \psi}{\partial x} = 0$$

and, therefore,

$$\psi = Uy \tag{6.79}$$

These results can be generalized to provide the velocity potential and stream function for a uniform flow at an angle $\alpha$ with the $x$ axis, as in Fig. 6.16b. For this case

$$\phi = U(x \cos \alpha + y \sin \alpha) \tag{6.80}$$

and

$$\psi = U(y \cos \alpha - x \sin \alpha) \tag{6.81}$$

### 6.5.2 Source and Sink

Consider a fluid flowing radially outward from a line through the origin perpendicular to the $x$–$y$ plane as is shown in Fig. 6.17. Let $m$ be the volume rate of flow emanating from the line (per unit length), and therefore to satisfy conservation of mass

$$(2\pi r)v_r = m$$

or

$$v_r = \frac{m}{2\pi r}$$

$U$

$\psi = \psi_1$
$\psi = \psi_2$
$\psi = \psi_3$
$\psi = \psi_4$

$\phi = \phi_1$    $\phi = \phi_2$

(a)

$U$

$\psi = \psi_1$
$\psi = \psi_2$
$\psi = \psi_3$
$\psi = \psi_4$

$\alpha$

$\phi = \phi_2$

$\phi = \phi_1$

■ **Figure 6.16**   Uniform flow: (a) in the $x$ direction; (b) in an arbitrary direction, $\alpha$.

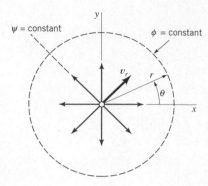

■ **Figure 6.17** The streamline pattern for a source.

*A source or sink represents a purely radial flow.*

Also, since the flow is a purely radial flow, $v_\theta = 0$, the corresponding velocity potential can be obtained by integrating the equations

$$\frac{\partial \phi}{\partial r} = \frac{m}{2\pi r} \qquad \frac{1}{r}\frac{\partial \phi}{\partial \theta} = 0$$

It follows that

$$\phi = \frac{m}{2\pi}\ln r \tag{6.82}$$

If $m$ is positive, the flow is radially outward, and the flow is considered to be a **source** flow. If $m$ is negative, the flow is toward the origin, and the flow is considered to be a **sink** flow. The flowrate, $m$, is the *strength* of the source or sink.

As shown by the figure in the margin, at the origin where $r = 0$ the velocity becomes infinite, which is of course physically impossible. Thus, sources and sinks do not really exist in real flow fields, and the line representing the source or sink is a mathematical *singularity* in the flow field. However, some real flows can be approximated at points away from the origin by using sources or sinks. Also, the velocity potential representing this hypothetical flow can be combined with other basic velocity potentials to approximately describe some real flow fields. This idea is further discussed in Section 6.6.

The stream function for the source can be obtained by integrating the relationships

$$v_r = \frac{1}{r}\frac{\partial \psi}{\partial \theta} = \frac{m}{2\pi r} \qquad v_\theta = -\frac{\partial \psi}{\partial r} = 0$$

to yield

$$\psi = \frac{m}{2\pi}\theta \tag{6.83}$$

It is apparent from Eq. 6.83 that the streamlines (lines of $\psi = $ constant) are radial lines, and from Eq. 6.82 that the equipotential lines (lines of $\phi = $ constant) are concentric circles centered at the origin.

## *E*XAMPLE 6.5 Potential Flow—Sink

**GIVEN** A nonviscous, incompressible fluid flows between wedge-shaped walls into a small opening as shown in Fig. E6.5. The velocity potential (in ft²/s), that approximately describes this flow, is

$$\phi = -2\ln r$$

**FIND** Determine the volume rate of flow (per unit length) into the opening.

■ **Figure E6.5**

## SOLUTION

The components of velocity are

$$v_r = \frac{\partial \phi}{\partial r} = -\frac{2}{r} \qquad v_\theta = \frac{1}{r}\frac{\partial \phi}{\partial \theta} = 0$$

which indicates we have a purely radial flow. The flowrate per unit width, $q$, crossing the arc of length $R\pi/6$ can thus be obtained by integrating the expression

$$q = \int_0^{\pi/6} v_r R\, d\theta = -\int_0^{\pi/6}\left(\frac{2}{R}\right)R\, d\theta$$

$$= -\frac{\pi}{3} = -1.05 \text{ ft}^2/\text{s} \qquad \textbf{(Ans)}$$

**COMMENT**  Note that the radius $R$ is arbitrary since the flowrate crossing any curve between the two walls must be the same. The negative sign indicates that the flow is toward the opening, that is, in the negative radial direction.

### 6.5.3  Vortex

We next consider a flow field in which the streamlines are concentric circles—that is, we interchange the velocity potential and stream function for the source. Thus, let

$$\phi = K\theta \qquad \textbf{(6.84)}$$

and

$$\psi = -K \ln r \qquad \textbf{(6.85)}$$

where $K$ is a constant. In this case the streamlines are concentric circles as are illustrated in Fig. 6.18, with $v_r = 0$ and

$$v_\theta = \frac{1}{r}\frac{\partial \phi}{\partial \theta} = -\frac{\partial \psi}{\partial r} = \frac{K}{r} \qquad \textbf{(6.86)}$$

*A vortex represents a flow in which the streamlines are concentric circles.*

This result indicates that the tangential velocity varies inversely with the distance from the origin, as shown by the figure in the margin, with a singularity occurring at $r = 0$ (where the velocity becomes infinite).

It may seem strange that this **vortex** motion is irrotational (and it is since the flow field is described by a velocity potential). However, it must be recalled that rotation refers to the orientation of a fluid element and not the path followed by the element. Thus, for an irrotational vortex, if a pair of small sticks were placed in the flow field at location $A$, as indicated in Fig. 6.19a, the sticks would rotate as they move to location $B$. One of the sticks, the one that is aligned along the streamline, would follow a circular path and rotate in a counterclockwise

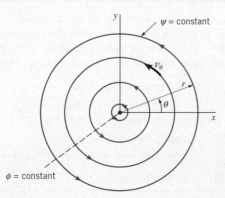

■ **Figure 6.18**  The streamline pattern for a vortex.

**■ Figure 6.19** Motion of fluid element from $A$ to $B$: ($a$) for irrotational (free) vortex; ($b$) for rotational (forced) vortex.

direction. The other stick would rotate in a clockwise direction due to the nature of the flow field—that is, the part of the stick nearest the origin moves faster than the opposite end. Although both sticks are rotating, the average angular velocity of the two sticks is zero since the flow is irrotational.

If the fluid were rotating as a rigid body, such that $v_\theta = K_1 r$ where $K_1$ is a constant, then sticks similarly placed in the flow field would rotate as is illustrated in Fig. 6.19$b$. This type of vortex motion is *rotational* and cannot be described with a velocity potential. The rotational vortex is commonly called a *forced vortex*, whereas the irrotational vortex is usually called a *free vortex*. The swirling motion of the water as it drains from a bathtub is similar to that of a free vortex, whereas the motion of a liquid contained in a tank that is rotated about its axis with angular velocity $\omega$ corresponds to a forced vortex.

> *Vortex motion can be either rotational or irrotational.*

A *combined vortex* is one with a forced vortex as a central core and a velocity distribution corresponding to that of a free vortex outside the core. Thus, for a combined vortex

$$v_\theta = \omega r \qquad r \le r_0 \tag{6.87}$$

and

$$v_\theta = \frac{K}{r} \qquad r > r_0 \tag{6.88}$$

where $K$ and $\omega$ are constants and $r_0$ corresponds to the radius of the central core. The pressure distribution in both the free and forced vortex was previously considered in Example 3.3. (See Fig. E6.6$a$ for an approximation of this type of flow.)

---

# F l u i d s   i n   t h e   N e w s

**Some hurricane facts** One of the most interesting, yet potentially devastating, naturally occurring fluid flow phenomenon is a hurricane. Broadly speaking, a hurricane is a rotating mass of air circulating around a low-pressure central core. In some respects the motion is similar to that of a *free vortex*. The Caribbean and Gulf of Mexico experience the most hurricanes, with the official hurricane season being from June 1 to November 30. Hurricanes are usually 300 to 400 miles wide and are structured around a central eye in which the air is relatively calm. The eye is surrounded by an eye wall, which is the region of strongest winds and precipitation. As one goes from the eye wall to the eye, the wind speeds decrease sharply and within the eye the air is relatively calm and clear of clouds. However, in the eye the pressure is at a minimum and may be 10% less than standard atmospheric pressure. This low pressure creates strong downdrafts of dry air from above. Hurricanes are classified into five categories based on their wind speeds:

Category one—74–95 mph

Category two—96–110 mph

Category three—111–130 mph

Category four—131–155 mph

Category five—greater than 155 mph

(See Problem 6.60.)

■ **Figure 6.20**   The notation for determining circulation around closed curve $C$.

A mathematical concept commonly associated with vortex motion is that of ***circulation***. The circulation, $\Gamma$, is defined as the line integral of the tangential component of the velocity taken around a closed curve in the flow field. In equation form, $\Gamma$ can be expressed as

$$\Gamma = \oint_C \mathbf{V} \cdot d\mathbf{s} \tag{6.89}$$

where the integral sign means that the integration is taken around a closed curve, $C$, in the counterclockwise direction, and $d\mathbf{s}$ is a differential length along the curve as is illustrated in Fig. 6.20. For an irrotational flow, $\mathbf{V} = \nabla\phi$ so that $\mathbf{V} \cdot d\mathbf{s} = \nabla\phi \cdot d\mathbf{s} = d\phi$ and, therefore,

$$\Gamma = \oint_C d\phi = 0$$

This result indicates that for an irrotational flow the circulation will generally be zero. (Chapter 9 has further discussion of circulation in real flows.) However, if there are singularities enclosed within the curve the circulation may not be zero. For example, for the free vortex with $v_\theta = K/r$ the circulation around the circular path of radius $r$ shown in Fig. 6.21 is

$$\Gamma = \int_0^{2\pi} \frac{K}{r} (r \, d\theta) = 2\pi K$$

*The numerical value of the circulation may depend on the particular closed path considered.*

which shows that the circulation is nonzero and the constant $K = \Gamma/2\pi$. However, for irrotational flows the circulation around any path that does not include a singular point will be zero. This can be easily confirmed for the closed path $ABCD$ of Fig. 6.21 by evaluating the circulation around that path.

The velocity potential and stream function for the free vortex are commonly expressed in terms of the circulation as

$$\phi = \frac{\Gamma}{2\pi}\theta \tag{6.90}$$

and

$$\psi = -\frac{\Gamma}{2\pi}\ln r \tag{6.91}$$

*V6.4 Vortex in a beaker*

The concept of circulation is often useful when evaluating the forces developed on bodies immersed in moving fluids. This application will be considered in Section 6.6.3.

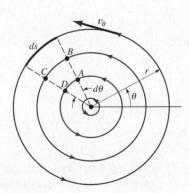

■ **Figure 6.21**   Circulation around various paths in a free vortex.

## EXAMPLE 6.6 Potential Flow—Free Vortex

**GIVEN** A liquid drains from a large tank through a small opening as illustrated in Fig. E6.6a. A vortex forms whose velocity distribution away from the tank opening can be approximated as that of a free vortex having a velocity potential

$$\phi = \frac{\Gamma}{2\pi}\theta$$

**FIND** Determine an expression relating the surface shape to the strength of the vortex as specified by the circulation $\Gamma$.

### SOLUTION

Since the free vortex represents an irrotational flow field, the Bernoulli equation

$$\frac{p_1}{\gamma} + \frac{V_1^2}{2g} + z_1 = \frac{p_2}{\gamma} + \frac{V_2^2}{2g} + z_2$$

can be written between any two points. If the points are selected at the free surface, $p_1 = p_2 = 0$, so that

$$\frac{V_1^2}{2g} = z_s + \frac{V_2^2}{2g} \qquad (1)$$

where the free surface elevation, $z_s$, is measured relative to a datum passing through point (1) as shown in Fig. E6.6b.

The velocity is given by the equation

$$v_\theta = \frac{1}{r}\frac{\partial\phi}{\partial\theta} = \frac{\Gamma}{2\pi r}$$

We note that far from the origin at point (1), $V_1 = v_\theta \approx 0$ so that Eq. 1 becomes

$$z_s = -\frac{\Gamma^2}{8\pi^2 r^2 g} \qquad \textbf{(Ans)}$$

which is the desired equation for the surface profile.

■ **Figure E6.6a**

■ **Figure E6.6b**

**COMMENT** The negative sign indicates that the surface falls as the origin is approached as shown in Fig. E6.6. This solution is not valid very near the origin since the predicted velocity becomes excessively large as the origin is approached.

### 6.5.4 Doublet

*A doublet is formed by an appropriate source–sink pair.*

The final, basic potential flow to be considered is one that is formed by combining a source and sink in a special way. Consider the equal strength, source–sink pair of Fig. 6.22. The combined stream function for the pair is

$$\psi = -\frac{m}{2\pi}(\theta_1 - \theta_2)$$

■ **Figure 6.22** The combination of a source and sink of equal strength located along the x axis.

which can be rewritten as

$$\tan\left(-\frac{2\pi\psi}{m}\right) = \tan(\theta_1 - \theta_2) = \frac{\tan\theta_1 - \tan\theta_2}{1 + \tan\theta_1 \tan\theta_2} \qquad (6.92)$$

From Fig. 6.22 it follows that

$$\tan\theta_1 = \frac{r\sin\theta}{r\cos\theta - a}$$

and

$$\tan\theta_2 = \frac{r\sin\theta}{r\cos\theta + a}$$

These results substituted into Eq. 6.92 give

$$\tan\left(-\frac{2\pi\psi}{m}\right) = \frac{2ar\sin\theta}{r^2 - a^2}$$

so that

$$\psi = -\frac{m}{2\pi}\tan^{-1}\left(\frac{2ar\sin\theta}{r^2 - a^2}\right) \qquad (6.93)$$

The figure in the margin shows typical streamlines for this flow. For small values of the distance $a$

$$\psi = -\frac{m}{2\pi}\frac{2ar\sin\theta}{r^2 - a^2} = -\frac{mar\sin\theta}{\pi(r^2 - a^2)} \qquad (6.94)$$

since the tangent of an angle approaches the value of the angle for small angles.

A doublet is formed by letting a source and sink approach one another.

The so-called **_doublet_** is formed by letting the source and sink approach one another $(a \to 0)$ while increasing the strength $m$ $(m \to \infty)$ so that the product $ma/\pi$ remains constant. In this case, since $r/(r^2 - a^2) \to 1/r$, Eq. 6.94 reduces to

$$\psi = -\frac{K\sin\theta}{r} \qquad (6.95)$$

where $K$, a constant equal to $ma/\pi$, is called the *strength* of the doublet. The corresponding velocity potential for the doublet is

$$\phi = \frac{K\cos\theta}{r} \qquad (6.96)$$

Plots of lines of constant $\psi$ reveal that the streamlines for a doublet are circles through the origin tangent to the $x$ axis as shown in Fig. 6.23. Just as sources and sinks are not physically realistic entities, neither are doublets. However, the doublet when combined with other basic potential flows

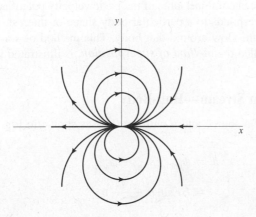

■ **Figure 6.23**  Streamlines for a doublet.

■ **Table 6.1**
**Summary of Basic, Plane Potential Flows**

| Description of Flow Field | Velocity Potential | Stream Function | Velocity Components[a] |
|---|---|---|---|
| Uniform flow at angle $\alpha$ with the $x$ axis (see Fig. 6.16$b$) | $\phi = U(x \cos \alpha + y \sin \alpha)$ | $\psi = U(y \cos \alpha - x \sin \alpha)$ | $u = U \cos \alpha$ <br> $v = U \sin \alpha$ |
| Source or sink (see Fig. 6.17) $m > 0$ source $m < 0$ sink | $\phi = \dfrac{m}{2\pi} \ln r$ | $\psi = \dfrac{m}{2\pi} \theta$ | $v_r = \dfrac{m}{2\pi r}$ <br> $v_\theta = 0$ |
| Free vortex (see Fig. 6.18) $\Gamma > 0$ counterclockwise motion $\Gamma < 0$ clockwise motion | $\phi = \dfrac{\Gamma}{2\pi} \theta$ | $\psi = -\dfrac{\Gamma}{2\pi} \ln r$ | $v_r = 0$ <br> $v_\theta = \dfrac{\Gamma}{2\pi r}$ |
| Doublet (see Fig. 6.23) | $\phi = \dfrac{K \cos \theta}{r}$ | $\psi = -\dfrac{K \sin \theta}{r}$ | $v_r = -\dfrac{K \cos \theta}{r^2}$ <br> $v_\theta = -\dfrac{K \sin \theta}{r^2}$ |

[a]Velocity components are related to the velocity potential and stream function through the relationships:
$$u = \frac{\partial \phi}{\partial x} = \frac{\partial \psi}{\partial y} \qquad v = \frac{\partial \phi}{\partial y} = -\frac{\partial \psi}{\partial x} \qquad v_r = \frac{\partial \phi}{\partial r} = \frac{1}{r}\frac{\partial \psi}{\partial \theta} \qquad v_\theta = \frac{1}{r}\frac{\partial \phi}{\partial \theta} = -\frac{\partial \psi}{\partial r}.$$

provides a useful representation of some flow fields of practical interest. For example, we will determine in Section 6.6.3 that the combination of a uniform flow and a doublet can be used to represent the flow around a circular cylinder. Table 6.1 provides a summary of the pertinent equations for the basic, plane potential flows considered in the preceding sections.

## 6.6 Superposition of Basic, Plane Potential Flows

As was discussed in the previous section, potential flows are governed by Laplace's equation, which is a linear partial differential equation. It therefore follows that the various basic velocity potentials and stream functions can be combined to form new potentials and stream functions. (Why is this true?) Whether such combinations yield useful results remains to be seen. It is to be noted that *any streamline in an inviscid flow field can be considered as a solid boundary*, since the conditions along a solid boundary and a streamline are the same—that is, there is no flow through the boundary or the streamline. Thus, if we can combine some of the basic velocity potentials or stream functions to yield a streamline that corresponds to a particular body shape of interest, that combination can be used to describe in detail the flow around that body. This method of solving some interesting flow problems, commonly called the ***method of superposition,*** is illustrated in the following three sections.

### 6.6.1 Source in a Uniform Stream—Half-Body

*Flow around a half-body is obtained by the addition of a source to a uniform flow.*

Consider the superposition of a source and a uniform flow as shown in Fig. 6.24$a$. The resulting stream function is

$$\psi = \psi_{\text{uniform flow}} + \psi_{\text{source}}$$

$$= Ur \sin \theta + \frac{m}{2\pi} \theta \qquad (6.97)$$

**Figure 6.24** The flow around a half-body: (*a*) superposition of a source and a uniform flow; (*b*) replacement of streamline $\psi = \pi bU$ with solid boundary to form a half-body.

and the corresponding velocity potential is

$$\phi = Ur \cos \theta + \frac{m}{2\pi} \ln r \tag{6.98}$$

*V6.5 Half-body*

It is clear that at some point along the negative *x* axis the velocity due to the source will just cancel that due to the uniform flow and a stagnation point will be created. For the source alone

$$v_r = \frac{m}{2\pi r}$$

so that the stagnation point will occur at $x = -b$ where

$$U = \frac{m}{2\pi b}$$

or

$$b = \frac{m}{2\pi U} \tag{6.99}$$

The value of the stream function at the stagnation point can be obtained by evaluating $\psi$ at $r = b$ and $\theta = \pi$, which yields from Eq. 6.97

$$\psi_{\text{stagnation}} = \frac{m}{2}$$

*For inviscid flow, a streamline can be replaced by a solid boundary.*

Since $m/2 = \pi bU$ (from Eq. 6.99) it follows that the equation of the streamline passing through the stagnation point is

$$\pi bU = Ur \sin \theta + bU\theta$$

or

$$r = \frac{b(\pi - \theta)}{\sin \theta} \tag{6.100}$$

where $\theta$ can vary between 0 and $2\pi$. A plot of this streamline is shown in Fig. 6.24*b*. If we replace this streamline with a solid boundary, as indicated in the figure, then it is clear that this combination of a uniform flow and a source can be used to describe the flow around a streamlined body placed in a uniform stream. The body is open at the downstream end and thus is called a **half-body**. Other streamlines in the flow field can be obtained by setting $\psi$ = constant in Eq. 6.97 and plotting the resulting equation. A number of these streamlines are shown in Fig. 6.24*b*. Although the streamlines inside the body are shown, they are actually of no interest in this case, since we are concerned with the flow field outside the body. It should be noted that the singularity in the flow field (the source) occurs inside the body, and there are no singularities in the flow field of interest (outside the body).

The width of the half-body asymptotically approaches $2\pi b = m/U$. Thus, as shown by the figure in the margin, for a given free stream velocity, *U*, the width of the half-body increases as the source strength. This follows from Eq. 6.100, which can be written as

$$y = b(\pi - \theta)$$

so that as $\theta \to 0$ or $\theta \to 2\pi$ the half-width approaches $\pm b\pi$. With the stream function (or velocity potential) known, the velocity components at any point can be obtained. For the half-body, using the stream function given by Eq. 6.97,

$$v_r = \frac{1}{r}\frac{\partial \psi}{\partial \theta} = U \cos \theta + \frac{m}{2\pi r}$$

and

$$v_\theta = -\frac{\partial \psi}{\partial r} = -U \sin \theta$$

Thus, the square of the magnitude of the velocity, $V$, at any point is

$$V^2 = v_r^2 + v_\theta^2 = U^2 + \frac{Um \cos \theta}{\pi r} + \left(\frac{m}{2\pi r}\right)^2$$

and since $b = m/2\pi U$

$$V^2 = U^2\left(1 + 2\frac{b}{r}\cos \theta + \frac{b^2}{r^2}\right) \tag{6.101}$$

With the velocity known, the pressure at any point can be determined from the Bernoulli equation, which can be written between any two points in the flow field since the flow is irrotational. Thus, applying the Bernoulli equation between a point far from the body, where the pressure is $p_0$ and the velocity is $U$, and some arbitrary point with pressure $p$ and velocity $V$, it follows that

$$p_0 + \tfrac{1}{2}\rho U^2 = p + \tfrac{1}{2}\rho V^2 \tag{6.102}$$

where elevation changes have been neglected. Equation 6.101 can now be substituted into Eq. 6.102 to obtain the pressure at any point in terms of the reference pressure, $p_0$, and the upstream velocity, $U$.

This relatively simple potential flow provides some useful information about the flow around the front part of a streamlined body, such as a bridge pier or strut placed in a uniform stream. An important point to be noted is that the velocity tangent to the surface of the body is not zero; that is, the fluid "slips" by the boundary. This result is a consequence of neglecting viscosity, the fluid property that causes real fluids to stick to the boundary, thus creating a "no-slip" condition. All potential flows differ from the flow of real fluids in this respect and do not accurately represent the velocity very near the boundary. However, outside this very thin boundary layer the velocity distribution will generally correspond to that predicted by potential flow theory if flow separation does not occur. (See Section 9.2.6.) Also, the pressure distribution along the surface will closely approximate that predicted from the potential flow theory, since the boundary layer is thin and there is little opportunity for the pressure to vary through the thin layer. In fact, as discussed in more detail in Chapter 9, the pressure distribution obtained from potential flow theory is used in conjunction with viscous flow theory to determine the nature of flow within the boundary layer.

*For a potential flow the fluid is allowed to slip past a fixed solid boundary.*

---

# $E$XAMPLE 6.7  Potential Flow—Half-body

**GIVEN**  A 40-mi/hr wind blows toward a hill arising from a plain that can be approximated with the top section of a half-body as illustrated in Fig. E6.7a. The height of the hill approaches 200 ft as shown. Assume an air density of 0.00238 slugs/ft$^3$.

**FIND**

**(a)**  What is the magnitude of the air velocity at a point on the hill directly above the origin [point (2)]?

**(b)**  What is the elevation of point (2) above the plain and what is the difference in pressure between point (1) on the plain far from the hill and point (2)?

■ **Figure E6.7a**

# SOLUTION

**(a)** The velocity is given by Eq. 6.101 as

$$V^2 = U^2\left(1 + 2\frac{b}{r}\cos\theta + \frac{b^2}{r^2}\right)$$

At point (2), $\theta = \pi/2$, and since this point is on the surface (Eq. 6.100),

$$r = \frac{b(\pi - \theta)}{\sin\theta} = \frac{\pi b}{2} \tag{1}$$

Thus,

$$V_2^2 = U^2\left[1 + \frac{b^2}{(\pi b/2)^2}\right]$$
$$= U^2\left(1 + \frac{4}{\pi^2}\right)$$

and the magnitude of the velocity at (2) for a 40 mi/hr approaching wind is

$$V_2 = \left(1 + \frac{4}{\pi^2}\right)^{1/2}(40 \text{ mi/hr}) = 47.4 \text{ mi/hr} \quad \textbf{(Ans)}$$

**(b)** The elevation at (2) above the plain is given by Eq. 1 as

$$y_2 = \frac{\pi b}{2}$$

Since the height of the hill approaches 200 ft and this height is equal to $\pi b$, it follows that

$$y_2 = \frac{200 \text{ ft}}{2} = 100 \text{ ft} \quad \textbf{(Ans)}$$

From the Bernoulli equation (with the $y$ axis the vertical axis)

$$\frac{p_1}{\gamma} + \frac{V_1^2}{2g} + y_1 = \frac{p_2}{\gamma} + \frac{V_2^2}{2g} + y_2$$

so that

$$p_1 - p_2 = \frac{\rho}{2}(V_2^2 - V_1^2) + \gamma(y_2 - y_1)$$

and with

$$V_1 = (40 \text{ mi/hr})\left(\frac{5280 \text{ ft/mi}}{3600 \text{ s/hr}}\right) = 58.7 \text{ ft/s}$$

and

$$V_2 = (47.4 \text{ mi/hr})\left(\frac{5280 \text{ ft/mi}}{3600 \text{ s/hr}}\right) = 69.5 \text{ ft/s}$$

it follows that

$$p_1 - p_2 = \frac{(0.00238 \text{ slugs/ft}^3)}{2}[(69.5 \text{ ft/s})^2 - (58.7 \text{ ft/s})^2]$$
$$+ (0.00238 \text{ slugs/ft}^3)(32.2 \text{ ft/s}^2)(100 \text{ ft} - 0 \text{ ft})$$
$$= 9.31 \text{ lb/ft}^2 = 0.0647 \text{ psi} \quad \textbf{(Ans)}$$

**COMMENTS** This result indicates that the pressure on the hill at point (2) is slightly lower than the pressure on the plain at some distance from the base of the hill with a 0.0533-psi difference due to the elevation increase and a 0.0114-psi difference due to the velocity increase.

By repeating the calculations for various values of the upstream wind speed, $U$, the results shown in Fig. E6.7b are obtained. Note that as the wind speed increases, the pressure difference increases from the calm conditions of $p_1 - p_2 = 0.0533$ psi.

The maximum velocity along the hill surface does not occur at point (2) but farther up the hill at $\theta = 63°$. At this point $V_{\text{surface}} = 1.26U$. The minimum velocity ($V = 0$) and maximum pressure occur at point (3), the stagnation point.

**■ Figure E6.7b**

## 6.6.2 Rankine Ovals

The half-body described in the previous section is a body that is "open" at one end. To study the flow around a closed body, a source and a sink of equal strength can be combined with a uniform flow as shown in Fig. 6.25a. The stream function for this combination is

$$\psi = Ur\sin\theta - \frac{m}{2\pi}(\theta_1 - \theta_2) \tag{6.103}$$

and the velocity potential is

$$\phi = Ur\cos\theta - \frac{m}{2\pi}(\ln r_1 - \ln r_2) \tag{6.104}$$

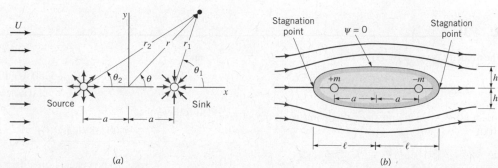

■ **Figure 6.25**   The flow around a Rankine oval: (*a*) superposition of source–sink pair and a uniform flow; (*b*) replacement of streamline $\psi = 0$ with solid boundary to form Rankine oval.

As discussed in Section 6.5.4, the stream function for the source–sink pair can be expressed as in Eq. 6.93, and, therefore, Eq. 6.103 can also be written as

$$\psi = Ur \sin \theta - \frac{m}{2\pi} \tan^{-1} \left( \frac{2ar \sin \theta}{r^2 - a^2} \right)$$

or

$$\psi = Uy - \frac{m}{2\pi} \tan^{-1} \left( \frac{2ay}{x^2 + y^2 - a^2} \right) \tag{6.105}$$

The corresponding streamlines for this flow field are obtained by setting $\psi = $ constant. If several of these streamlines are plotted, it will be discovered that the streamline $\psi = 0$ forms a closed body as is illustrated in Fig. 6.25*b*. We can think of this streamline as forming the surface of a body of length $2\ell$ and width $2h$ placed in a uniform stream. The streamlines inside the body are of no practical interest and are not shown. Note that since the body is closed, all of the flow emanating from the source flows into the sink. These bodies have an oval shape and are termed **Rankine ovals.**

*Rankine ovals are formed by combining a source and sink with a uniform flow.*

Stagnation points occur at the upstream and downstream ends of the body, as are indicated in Fig. 6.25*b*. These points can be located by determining where along the $x$ axis the velocity is zero. The stagnation points correspond to the points where the uniform velocity, the source velocity, and the sink velocity all combine to give a zero velocity. The locations of the stagnation points depend on the value of $a$, $m$, and $U$. The body half-length, $\ell$ (the value of $|x|$ that gives $\mathbf{V} = 0$ when $y = 0$), can be expressed as

$$\ell = \left( \frac{ma}{\pi U} + a^2 \right)^{1/2} \tag{6.106}$$

or

$$\frac{\ell}{a} = \left( \frac{m}{\pi U a} + 1 \right)^{1/2} \tag{6.107}$$

*Large Ua/m*

The body half-width, $h$, can be obtained by determining the value of $y$ where the $y$ axis intersects the $\psi = 0$ streamline. Thus, from Eq. 6.105 with $\psi = 0$, $x = 0$, and $y = h$, it follows that

$$h = \frac{h^2 - a^2}{2a} \tan \frac{2\pi Uh}{m} \tag{6.108}$$

or

$$\frac{h}{a} = \frac{1}{2} \left[ \left( \frac{h}{a} \right)^2 - 1 \right] \tan \left[ 2 \left( \frac{\pi Ua}{m} \right) \frac{h}{a} \right] \tag{6.109}$$

*Small Ua/m*

Equations 6.107 and 6.109 show that both $\ell/a$ and $h/a$ are functions of the dimensionless parameter, $\pi Ua/m$. Although for a given value of $Ua/m$ the corresponding value of $\ell/a$ can be determined directly from Eq. 6.107, $h/a$ must be determined by a trial-and-error solution of Eq. 6.109.

A large variety of body shapes with different length to width ratios can be obtained by using different values of $Ua/m$, as shown by the figure in the margin. As this parameter becomes large, flow around a long slender body is described, whereas for small values of the parameter, flow around a more blunt shape is obtained. Downstream from the point of maximum body width, the surface pressure increases with distance along the surface. This condition (called an adverse pressure gradient) typically leads to separation of the flow from the surface, resulting in a large low-pressure wake on the downstream side of the body. Separation is not predicted by potential theory (which simply indicates a symmetrical flow). This is illustrated by the figure in the margin for an extreme blunt shape, showing the differences between potential and viscous flow. Therefore, the potential solution for the Rankine ovals will give a reasonable approximation of the velocity outside the thin, viscous boundary layer and the pressure distribution on the front part of the body only.

Potential Flow

© S. T. Thoroddsen & Stanford University

Viscous Flow

© Stanford University, with permission

*A doublet combined with a uniform flow can be used to represent flow around a circular cylinder.*

V6.6 Circular cylinder

### 6.6.3 Flow around a Circular Cylinder

As was noted in the previous section, when the distance between the source–sink pair approaches zero, the shape of the Rankine oval becomes more blunt and in fact approaches a circular shape. Since the doublet described in Section 6.5.4 was developed by letting a source–sink pair approach one another, it might be expected that a uniform flow in the positive $x$ direction combined with a doublet could be used to represent flow around a circular cylinder. This combination gives for the stream function

$$\psi = Ur\sin\theta - \frac{K\sin\theta}{r} \tag{6.110}$$

and for the velocity potential

$$\phi = Ur\cos\theta + \frac{K\cos\theta}{r} \tag{6.111}$$

In order for the stream function to represent flow around a circular cylinder, it is necessary that $\psi$ = constant for $r = a$, where $a$ is the radius of the cylinder. Since Eq. 6.110 can be written as

$$\psi = \left(U - \frac{K}{r^2}\right)r\sin\theta$$

it follows that $\psi = 0$ for $r = a$ if

$$U - \frac{K}{a^2} = 0$$

which indicates that the doublet strength, $K$, must be equal to $Ua^2$. Thus, the stream function for flow around a circular cylinder can be expressed as

$$\psi = Ur\left(1 - \frac{a^2}{r^2}\right)\sin\theta \tag{6.112}$$

and the corresponding velocity potential is

$$\phi = Ur\left(1 + \frac{a^2}{r^2}\right)\cos\theta \tag{6.113}$$

A sketch of the streamlines for this flow field is shown in Fig. 6.26.

The velocity components can be obtained from either Eq. 6.112 or 6.113 as

$$v_r = \frac{\partial\phi}{\partial r} = \frac{1}{r}\frac{\partial\psi}{\partial\theta} = U\left(1 - \frac{a^2}{r^2}\right)\cos\theta \tag{6.114}$$

and

$$v_\theta = \frac{1}{r}\frac{\partial\phi}{\partial\theta} = -\frac{\partial\psi}{\partial r} = -U\left(1 + \frac{a^2}{r^2}\right)\sin\theta \tag{6.115}$$

On the surface of the cylinder ($r = a$) it follows from Eq. 6.114 and 6.115 that $v_r = 0$ and

$$v_{\theta s} = -2U\sin\theta$$

As shown by the figure in the margin, the maximum velocity occurs at the top and bottom of the cylinder ($\theta = \pm\pi/2$) and has a magnitude of twice the upstream velocity, $U$. As we move away from the cylinder along the ray $\theta = \pi/2$ the velocity varies, as is illustrated in Fig. 6.26.

V6.7 Ellipse

■ **Figure 6.26**   **The flow around a circular cylinder.**

*The pressure distribution on the cylinder surface is obtained from the Bernoulli equation.*

The pressure distribution on the cylinder surface is obtained from the Bernoulli equation written from a point far from the cylinder where the pressure is $p_0$ and the velocity is $U$ so that

$$p_0 + \tfrac{1}{2}\rho U^2 = p_s + \tfrac{1}{2}\rho v_{\theta s}^2$$

where $p_s$ is the surface pressure. Elevation changes are neglected. Since $v_{\theta s} = -2U \sin \theta$, the surface pressure can be expressed as

$$p_s = p_0 + \tfrac{1}{2}\rho U^2(1 - 4 \sin^2 \theta) \tag{6.116}$$

*V6.8 Circular cylinder with separation*

A comparison of this theoretical, symmetrical pressure distribution expressed in dimensionless form with a typical measured distribution is shown in Fig. 6.27. This figure clearly reveals that only on the upstream part of the cylinder is there approximate agreement between the potential flow and the experimental results. Because of the viscous boundary layer that develops on the cylinder, the main flow separates from the surface of the cylinder, leading to the large difference between the theoretical, frictionless fluid solution and the experimental results on the downstream side of the cylinder (see Chapter 9).

The resultant force (per unit length) developed on the cylinder can be determined by integrating the pressure over the surface. From Fig. 6.28 it can be seen that

$$F_x = -\int_0^{2\pi} p_s \cos \theta \, a \, d\theta \tag{6.117}$$

*V6.9 Forces on suspended ball*

■ **Figure 6.27**   **A comparison of theoretical (inviscid) pressure distribution on the surface of a circular cylinder with typical experimental distribution.**

*Potential theory*
*incorrectly predicts*
*that the drag on a*
*cylinder is zero.*

■ **Figure 6.28**   The notation for determining lift and drag on a circular cylinder.

and

$$F_y = -\int_0^{2\pi} p_s \sin \theta \, a \, d\theta \qquad (6.118)$$

where $F_x$ is the *drag* (force parallel to direction of the uniform flow) and $F_y$ is the *lift* (force perpendicular to the direction of the uniform flow). Substitution for $p_s$ from Eq. 6.116 into these two equations, and subsequent integration, reveals that $F_x = 0$ and $F_y = 0$. These results indicate that both the drag and lift as predicted by potential theory for a fixed cylinder in a uniform stream are zero. Since the pressure distribution is symmetrical around the cylinder, this is not really a surprising result. However, we know from experience that there is a significant drag developed on a cylinder when it is placed in a moving fluid. This discrepancy is known as d'Alembert's paradox. The paradox is named after Jean le Rond d'Alembert (1717–1783), a French mathematician and philosopher, who first showed that the drag on bodies immersed in inviscid fluids is zero. It was not until the latter part of the nineteenth century and the early part of the twentieth century that the role viscosity plays in the steady fluid motion was understood and d'Alembert's paradox explained (see Section 9.1).

---

# EXAMPLE 6.8   Potential Flow—Cylinder

**GIVEN**   When a circular cylinder is placed in a uniform stream, a stagnation point is created on the cylinder as is shown in Fig. E6.8a. If a small hole is located at this point, the stagnation pressure, $p_{\text{stag}}$, can be measured and used to determine the approach velocity, $U$.

**FIND**

**(a)**   Show how $p_{\text{stag}}$ and $U$ are related.

**(b)**   If the cylinder is misaligned by an angle $\alpha$ (Fig. E6.8b), but the measured pressure is still interpreted as the stagnation pressure, determine an expression for the ratio of the true velocity, $U$, to the predicted velocity, $U'$. Plot this ratio as a function of $\alpha$ for the range $-20° \le \alpha \le 20°$.

■ **Figure E6.8**

## SOLUTION

**(a)** The velocity at the stagnation point is zero, so the Bernoulli equation written between a point on the stagnation streamline upstream from the cylinder and the stagnation point gives

$$\frac{p_0}{\gamma} + \frac{U^2}{2g} = \frac{p_{\text{stag}}}{\gamma}$$

Thus,

$$U = \left[ \frac{2}{\rho} \left( p_{\text{stag}} - p_0 \right) \right]^{1/2} \qquad \textbf{(Ans)}$$

**COMMENT** A measurement of the difference between the pressure at the stagnation point and the upstream pressure can be used to measure the approach velocity. This is, of course, the same result that was obtained in Section 3.5 for Pitot-static tubes.

**(b)** If the direction of the fluid approaching the cylinder is not known precisely, it is possible that the cylinder is misaligned by some angle, $\alpha$. In this instance the pressure actually measured, $p_\alpha$, will be different from the stagnation pressure, but if the misalignment is not recognized the predicted approach velocity, $U'$, would still be calculated as

$$U' = \left[ \frac{2}{\rho} \left( p_\alpha - p_0 \right) \right]^{1/2}$$

Thus,

$$\frac{U(\text{true})}{U'(\text{predicted})} = \left( \frac{p_{\text{stag}} - p_0}{p_\alpha - p_0} \right)^{1/2} \qquad \textbf{(1)}$$

The velocity on the surface of the cylinder, $v_\theta$, where $r = a$, is obtained from Eq. 6.115 as

$$v_\theta = -2U \sin \theta$$

If we now write the Bernoulli equation between a point upstream of the cylinder and the point on the cylinder where $r = a$, $\theta = \alpha$, it follows that

$$p_0 + \tfrac{1}{2}\rho U^2 = p_\alpha + \tfrac{1}{2}\rho(-2U \sin \alpha)^2$$

and, therefore,

$$p_\alpha - p_0 = \tfrac{1}{2}\rho U^2 (1 - 4 \sin^2\alpha) \qquad \textbf{(2)}$$

Since $p_{\text{stag}} - p_0 = \tfrac{1}{2}\rho U^2$ it follows from Eqs. 1 and 2 that

$$\frac{U(\text{true})}{U'(\text{predicted})} = (1 - 4 \sin^2\alpha)^{-1/2} \qquad \textbf{(Ans)}$$

This velocity ratio is plotted as a function of the misalignment angle $\alpha$ in Fig. E6.8c.

**COMMENT** It is clear from these results that significant errors can arise if the stagnation pressure tap is not aligned with the stagnation streamline. As is discussed in Section 3.5, if two additional, symmetrically located holes are drilled on the cylinder, as are illustrated in Fig. E6.8d, the correct orientation of the cylinder can be determined. The cylinder is rotated until the pressures in the two symmetrically placed holes are equal, thus indicating that the center hole coincides with the stagnation streamline. For $\beta = 30°$ the pressure at the two holes theoretically corresponds to the upstream pressure, $p_0$. With this orientation a measurement of the difference in pressure between the center hole and the side holes can be used to determine $U$.

---

An additional, interesting potential flow can be developed by adding a free vortex to the stream function or velocity potential for the flow around a cylinder. In this case

$$\psi = Ur\left( 1 - \frac{a^2}{r^2} \right) \sin \theta - \frac{\Gamma}{2\pi} \ln r \qquad \textbf{(6.119)}$$

and

$$\phi = Ur\left( 1 + \frac{a^2}{r^2} \right) \cos \theta + \frac{\Gamma}{2\pi} \theta \qquad \textbf{(6.120)}$$

where $\Gamma$ is the circulation. We note that the circle $r = a$ will still be a streamline (and thus can be replaced with a solid cylinder), since the streamlines for the added free vortex are all circular. However, the tangential velocity, $v_\theta$, on the surface of the cylinder ($r = a$) now becomes

$$v_{\theta s} = -\frac{\partial \psi}{\partial r}\bigg|_{r=a} = -2U \sin \theta + \frac{\Gamma}{2\pi a} \qquad \textbf{(6.121)}$$

*Flow around a rotating cylinder is approximated by the addition of a free vortex.*

This type of flow field could be approximately created by placing a rotating cylinder in a uniform stream. Because of the presence of viscosity in any real fluid, the fluid in contact with the rotating cylinder would rotate with the same velocity as the cylinder, and the resulting flow field would resemble that developed by the combination of a uniform flow past a cylinder and a free vortex.

$\Gamma = 0$

(a)

$\dfrac{\Gamma}{4\pi Ua} < 1$

(b)

$\dfrac{\Gamma}{4\pi Ua} = 1$

(c)

$\dfrac{\Gamma}{4\pi Ua} > 1$

(d)

Stagnation point

■ **Figure 6.29**    The location of stagnation points on a circular cylinder: (*a*) without circulation; (*b, c, d*) with circulation.

A variety of streamline patterns can be developed, depending on the vortex strength, $\Gamma$. For example, from Eq. 6.121 we can determine the location of stagnation points on the surface of the cylinder. These points will occur at $\theta = \theta_{\text{stag}}$ where $v_\theta = 0$ and therefore from Eq. 6.121

$$\sin \theta_{\text{stag}} = \frac{\Gamma}{4\pi Ua} \tag{6.122}$$

If $\Gamma = 0$, then $\theta_{\text{stag}} = 0$ or $\pi$—that is, the stagnation points occur at the front and rear of the cylinder as are shown in Fig. 6.29*a*. However, for $-1 \le \Gamma/4\pi Ua \le 1$, the stagnation points will occur at some other location on the surface as illustrated in Figs. 6.29*b,c*. If the absolute value of the parameter $\Gamma/4\pi Ua$ exceeds 1, Eq. 6.122 cannot be satisfied, and the stagnation point is located away from the cylinder as shown in Fig. 6.29*d*.

The force per unit length developed on the cylinder can again be obtained by integrating the differential pressure forces around the circumference as in Eqs. 6.117 and 6.118. For the cylinder with circulation, the surface pressure, $p_s$, is obtained from the Bernoulli equation (with the surface velocity given by Eq. 6.121)

$$p_0 + \frac{1}{2}\rho U^2 = p_s + \frac{1}{2}\rho\left(-2U\sin\theta + \frac{\Gamma}{2\pi a}\right)^2$$

or

$$p_s = p_0 + \frac{1}{2}\rho U^2\left(1 - 4\sin^2\theta + \frac{2\Gamma\sin\theta}{\pi aU} - \frac{\Gamma^2}{4\pi^2 a^2 U^2}\right) \tag{6.123}$$

Equation 6.123 substituted into Eq. 6.117 for the drag, and integrated, again yields

$$F_x = 0$$

*Potential flow past a cylinder with circulation gives zero drag but nonzero lift.*

That is, even for the rotating cylinder no force in the direction of the uniform flow is developed. However, use of Eq. 6.123 with the equation for the lift, $F_y$ (Eq. 6.118), yields

$$F_y = -\rho U\Gamma \tag{6.124}$$

Thus, for the cylinder with circulation, lift is developed equal to the product of the fluid density, the upstream velocity, and the circulation. The negative sign means that if $U$ is positive (in the positive $x$ direction) and $\Gamma$ is positive (a free vortex with counterclockwise rotation), the direction of the $F_y$ is downward.

Of course, if the cylinder is rotated in the clockwise direction ($\Gamma < 0$), the direction of $F_y$ would be upward. This can be seen by studying the surface pressure distribution (Eq. 6.123), which is plotted in Fig. 6.30 for two situations. One has $\Gamma/4\pi Ua = 0$, which corresponds to no rotation of the cylinder. The other has $\Gamma/4\pi Ua = -0.25$, which corresponds to clockwise rotation of the cylinder. With no

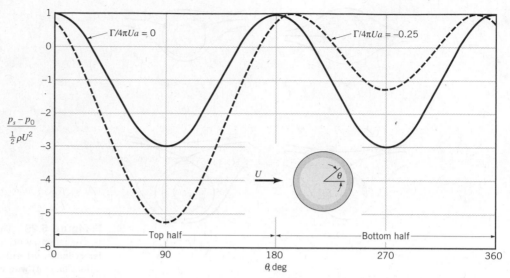

■ **Figure 6.30** Pressure distribution on a circular cylinder with and without rotation.

rotation, the flow is symmetrical both top to bottom and front to back on the cylinder. With rotation, the flow is symmetrical front to back but not top to bottom. In this case the two stagnation points [i.e., $(p_s - p_0)/(\rho U^2/2) = 1$] are located on the bottom of the cylinder, and the average pressure on the top half of the cylinder is less than that on the bottom half. The result is an upward lift force. It is this force acting in a direction perpendicular to the direction of the approach velocity that causes baseballs and golf balls to curve when they spin as they are propelled through the air. The development of this lift on rotating bodies is called the *Magnus effect.* (See Section 9.4 for further comments.)

Although Eq. 6.124 was developed for a cylinder with circulation, it gives the lift per unit length for any two-dimensional object of any cross-sectional shape placed in a uniform, inviscid stream. The circulation is determined around any closed curve containing the body. The generalized equation relating lift to fluid density, velocity, and circulation is called the *Kutta–Joukowski law* and is commonly used to determine the lift on airfoils (see Section 9.4.2 and Refs. 2–6).

## F l u i d s   i n   t h e   N e w s

**A sailing ship without sails** A sphere or cylinder spinning about its axis when placed in an airstream develops a force at right angles to the direction of the airstream. This phenomenon is commonly referred to as the *Magnus effect* and is responsible for the curved paths of baseballs and golf balls. Another lesser-known application of the Magnus effect was proposed by a German physicist and engineer, Anton Flettner, in the 1920s. Flettner's idea was to use the Magnus effect to make a ship move. To demonstrate the practicality of the "rotor-ship" he purchased a sailing schooner and replaced the ship's masts and rigging with two vertical cylinders that were 50 ft high and 9 ft in diameter. The cylinders looked like smokestacks on the ship. Their spinning motion was developed by 45-hp motors. The combination of a wind and the rotating cylinders created a force (Magnus effect) to push the ship forward. The ship, named the *Baden Baden*, made a successful voyage across the Atlantic, arriving in New York Harbor on May 9, 1926. Although the feasibility of the rotor-ship was clearly demonstrated, it proved to be less efficient and practical than more conventional vessels and the idea was not pursued. (See Problem 6.79.)

## 6.7 Other Aspects of Potential Flow Analysis

In the preceding section the method of superposition of basic potentials has been used to obtain detailed descriptions of irrotational flow around certain body shapes immersed in a uniform stream. For the cases considered, two or more of the basic potentials were combined, and the question is asked: What kind of flow does this combination represent? This approach is relatively simple and does not require the use of advanced mathematical techniques. It is, however, restrictive in its general applicability. It does not allow us to specify a priori the body shape and then determine the velocity potential or stream function that describes the flow around the particular body.

Determining the velocity potential or stream function for a given body shape is a much more complicated problem.

It is possible to extend the idea of superposition by considering a *distribution* of sources and sinks, or doublets, which when combined with a uniform flow can describe the flow around bodies of arbitrary shape. Techniques are available to determine the required distribution to give a prescribed body shape. Also, for plane potential flow problems it can be shown that complex variable theory (the use of real and imaginary numbers) can be effectively used to obtain solutions to a great variety of important flow problems. There are, of course, numerical techniques that can be used to solve not only plane two-dimensional problems but also the more general three-dimensional problems. Since potential flow is governed by Laplace's equation, any procedure that is available for solving this equation can be applied to the analysis of irrotational flow of frictionless fluids. Potential flow theory is an old and well-established discipline within the general field of fluid mechanics. The interested reader can find many detailed references on this subject, including Refs. 2–6 given at the end of this chapter.

An important point to remember is that regardless of the particular technique used to obtain a solution to a potential flow problem, the solution remains approximate because of the fundamental assumption of a frictionless fluid. Thus, "exact" solutions based on potential flow theory represent, at best, only approximate solutions to real fluid problems. The applicability of potential flow theory to real fluid problems has been alluded to in a number of examples considered in the previous section. As a rule of thumb, potential flow theory will usually provide a reasonable approximation in those circumstances when we are dealing with a low viscosity fluid moving at a relatively high velocity, in regions of the flow field in which the flow is accelerating. Under these circumstances we generally find that the effect of viscosity is confined to the thin boundary layer that develops at a solid boundary. Outside the boundary layer, the velocity distribution and the pressure distribution are closely approximated by the potential flow solution. However, in those regions of the flow field in which the flow is decelerating (for example, in the rearward portion of a bluff body or in the expanding region of a conduit), the pressure near a solid boundary will increase in the direction of flow. This so-called adverse pressure gradient can lead to flow separation, a phenomenon that causes dramatic changes in the flow field which are generally not accounted for by potential theory. However, as discussed in Chapter 9, in which boundary layer theory is developed, it is found that potential flow theory is used to obtain the appropriate pressure distribution that can then be combined with the viscous flow equations to obtain solutions near the boundary (and also to predict separation). The general differential equations that describe viscous fluid behavior and some simple solutions to these equations are considered in the remaining sections of this chapter.

*Potential flow solutions are always approximate because the fluid is assumed to be frictionless.*

*V6.11 Potential flow*

## 6.8　Viscous Flow

To incorporate viscous effects into the differential analysis of fluid motion we must return to the previously derived general equations of motion, Eqs. 6.50. Since these equations include both stresses and velocities, there are more unknowns than equations and, therefore, before proceeding it is necessary to establish a relationship between the stresses and velocities.

### 6.8.1　Stress–Deformation Relationships

For incompressible Newtonian fluids it is known that the stresses are linearly related to the rates of deformation and can be expressed in Cartesian coordinates as (for normal stresses)

$$\sigma_{xx} = -p + 2\mu \frac{\partial u}{\partial x} \tag{6.125a}$$

$$\sigma_{yy} = -p + 2\mu \frac{\partial v}{\partial y} \tag{6.125b}$$

$$\sigma_{zz} = -p + 2\mu \frac{\partial w}{\partial z} \tag{6.125c}$$

(for shearing stresses)

$$\tau_{xy} = \tau_{yx} = \mu\left(\frac{\partial u}{\partial y} + \frac{\partial v}{\partial x}\right) \tag{6.125d}$$

$$\tau_{yz} = \tau_{zy} = \mu\left(\frac{\partial v}{\partial z} + \frac{\partial w}{\partial y}\right) \tag{6.125e}$$

$$\tau_{zx} = \tau_{xz} = \mu\left(\frac{\partial w}{\partial x} + \frac{\partial u}{\partial z}\right) \tag{6.125f}$$

$p = -\frac{1}{3}(\sigma_{xx} + \sigma_{yy} + \sigma_{zz})$

where $p$ is the pressure, the negative of the average of the three normal stresses; that is, as indicated by the figure in the margin, $-p = \left(\frac{1}{3}\right)(\sigma_{xx} + \sigma_{yy} + \sigma_{zz})$. For viscous fluids in motion the normal stresses are not necessarily the same in different directions, thus, the need to define the pressure as the average of the three normal stresses. For fluids at rest, or frictionless fluids, the normal stresses are equal in all directions. (We have made use of this fact in the chapter on fluid statics and in developing the equations for inviscid flow.) Detailed discussions of the development of these stress–velocity gradient relationships can be found in Refs. 3, 7, and 8. An important point to note is that whereas for elastic solids the stresses are linearly related to the deformation (or strain), for Newtonian fluids the stresses are linearly related to the rate of deformation (or rate of strain).

> *For Newtonian fluids, stresses are linearly related to the rate of strain.*

In cylindrical polar coordinates the stresses for incompressible Newtonian fluids are expressed as (for normal stresses)

$$\sigma_{rr} = -p + 2\mu\frac{\partial v_r}{\partial r} \tag{6.126a}$$

$$\sigma_{\theta\theta} = -p + 2\mu\left(\frac{1}{r}\frac{\partial v_\theta}{\partial \theta} + \frac{v_r}{r}\right) \tag{6.126b}$$

$$\sigma_{zz} = -p + 2\mu\frac{\partial v_z}{\partial z} \tag{6.126c}$$

(for shearing stresses)

$$\tau_{r\theta} = \tau_{\theta r} = \mu\left[r\frac{\partial}{\partial r}\left(\frac{v_\theta}{r}\right) + \frac{1}{r}\frac{\partial v_r}{\partial \theta}\right] \tag{6.126d}$$

$$\tau_{\theta z} = \tau_{z\theta} = \mu\left(\frac{\partial v_\theta}{\partial z} + \frac{1}{r}\frac{\partial v_z}{\partial \theta}\right) \tag{6.126e}$$

$$\tau_{zr} = \tau_{rz} = \mu\left(\frac{\partial v_r}{\partial z} + \frac{\partial v_z}{\partial r}\right) \tag{6.126f}$$

The double subscript has a meaning similar to that of stresses expressed in Cartesian coordinates— that is, the first subscript indicates the plane on which the stress acts and the second subscript the direction. Thus, for example, $\sigma_{rr}$ refers to a stress acting on a plane perpendicular to the radial direction and in the radial direction (thus a normal stress). Similarly, $\tau_{r\theta}$ refers to a stress acting on a plane perpendicular to the radial direction but in the tangential ($\theta$ direction) and is therefore a shearing stress.

### 6.8.2 The Navier–Stokes Equations

The stresses as defined in the preceding section can be substituted into the differential equations of motion (Eqs. 6.50) and simplified by using the continuity equation for incompressible flow (Eq. 6.31). For rectangular coordinates (see the figure in the margin) the results are

($x$ direction)

$$\rho\left(\frac{\partial u}{\partial t} + u\frac{\partial u}{\partial x} + v\frac{\partial u}{\partial y} + w\frac{\partial u}{\partial z}\right) = -\frac{\partial p}{\partial x} + \rho g_x + \mu\left(\frac{\partial^2 u}{\partial x^2} + \frac{\partial^2 u}{\partial y^2} + \frac{\partial^2 u}{\partial z^2}\right) \tag{6.127a}$$

($y$ direction)

$$\rho\left(\frac{\partial v}{\partial t} + u\frac{\partial v}{\partial x} + v\frac{\partial v}{\partial y} + w\frac{\partial v}{\partial z}\right) = -\frac{\partial p}{\partial y} + \rho g_y + \mu\left(\frac{\partial^2 v}{\partial x^2} + \frac{\partial^2 v}{\partial y^2} + \frac{\partial^2 v}{\partial z^2}\right) \tag{6.127b}$$

(*z* direction)

$$\rho\left(\frac{\partial w}{\partial t} + u\frac{\partial w}{\partial x} + v\frac{\partial w}{\partial y} + w\frac{\partial w}{\partial z}\right) = -\frac{\partial p}{\partial z} + \rho g_z + \mu\left(\frac{\partial^2 w}{\partial x^2} + \frac{\partial^2 w}{\partial y^2} + \frac{\partial^2 w}{\partial z^2}\right) \qquad \textbf{(6.127c)}$$

*The Navier–Stokes equations are the basic differential equations describing the flow of Newtonian fluids.*

where $u$, $v$, and $w$ are the $x$, $y$, and $z$ components of velocity as shown in the figure in the margin of the previous page. We have rearranged the equations so that the acceleration terms are on the left side and the force terms are on the right. These equations are commonly called the **Navier–Stokes equations**, named in honor of the French mathematician L. M. H. Navier (1785–1836) and the English mechanician Sir G. G. Stokes (1819–1903), who were responsible for their formulation. These three equations of motion, when combined with the conservation of mass equation (Eq. 6.31), provide a complete mathematical description of the flow of incompressible Newtonian fluids. We have four equations and four unknowns ($u$, $v$, $w$, and $p$), and therefore the problem is "well-posed" in mathematical terms. Unfortunately, because of the general complexity of the Navier–Stokes equations (they are nonlinear, second-order, partial differential equations), they are not amenable to exact mathematical solutions except in a few instances. However, in those few instances in which solutions have been obtained and compared with experimental results, the results have been in close agreement. Thus, the Navier–Stokes equations are considered to be the governing differential equations of motion for incompressible Newtonian fluids.

In terms of cylindrical polar coordinates (see the figure in the margin), the Navier–Stokes equations can be written as

(*r* direction)

$$\rho\left(\frac{\partial v_r}{\partial t} + v_r\frac{\partial v_r}{\partial r} + \frac{v_\theta}{r}\frac{\partial v_r}{\partial \theta} - \frac{v_\theta^2}{r} + v_z\frac{\partial v_r}{\partial z}\right)$$
$$= -\frac{\partial p}{\partial r} + \rho g_r + \mu\left[\frac{1}{r}\frac{\partial}{\partial r}\left(r\frac{\partial v_r}{\partial r}\right) - \frac{v_r}{r^2} + \frac{1}{r^2}\frac{\partial^2 v_r}{\partial \theta^2} - \frac{2}{r^2}\frac{\partial v_\theta}{\partial \theta} + \frac{\partial^2 v_r}{\partial z^2}\right] \qquad \textbf{(6.128a)}$$

(*θ* direction)

$$\rho\left(\frac{\partial v_\theta}{\partial t} + v_r\frac{\partial v_\theta}{\partial r} + \frac{v_\theta}{r}\frac{\partial v_\theta}{\partial \theta} + \frac{v_r v_\theta}{r} + v_z\frac{\partial v_\theta}{\partial z}\right)$$
$$= -\frac{1}{r}\frac{\partial p}{\partial \theta} + \rho g_\theta + \mu\left[\frac{1}{r}\frac{\partial}{\partial r}\left(r\frac{\partial v_\theta}{\partial r}\right) - \frac{v_\theta}{r^2} + \frac{1}{r^2}\frac{\partial^2 v_\theta}{\partial \theta^2} + \frac{2}{r^2}\frac{\partial v_r}{\partial \theta} + \frac{\partial^2 v_\theta}{\partial z^2}\right] \qquad \textbf{(6.128b)}$$

(*z* direction)

$$\rho\left(\frac{\partial v_z}{\partial t} + v_r\frac{\partial v_z}{\partial r} + \frac{v_\theta}{r}\frac{\partial v_z}{\partial \theta} + v_z\frac{\partial v_z}{\partial z}\right)$$
$$= -\frac{\partial p}{\partial z} + \rho g_z + \mu\left[\frac{1}{r}\frac{\partial}{\partial r}\left(r\frac{\partial v_z}{\partial r}\right) + \frac{1}{r^2}\frac{\partial^2 v_z}{\partial \theta^2} + \frac{\partial^2 v_z}{\partial z^2}\right] \qquad \textbf{(6.128c)}$$

To provide a brief introduction to the use of the Navier–Stokes equations, a few of the simplest exact solutions are developed in the next section. Although these solutions will prove to be relatively simple, this is not the case in general. In fact, only a few other exact solutions have been obtained.

## 6.9    Some Simple Solutions for Laminar, Viscous, Incompressible Fluids

A principal difficulty in solving the Navier–Stokes equations is a result of their nonlinearity arising from the convective acceleration terms (i.e., $u\,\partial u/\partial x$, $w\,\partial v/\partial z$, etc.). There are no general analytical schemes for solving nonlinear partial differential equations (e.g., superposition of solutions cannot be used), and each problem must be considered individually. For most practical flow problems, fluid particles do have accelerated motion as they move from one location to another in the flow field. Thus, the convective acceleration terms are usually important. However, there are a few special cases for which the convective acceleration vanishes because of the nature of the geometry of the flow system. In these cases exact solutions are often possible. The Navier–Stokes equations apply to both laminar and turbulent flow, but for turbulent flow each velocity component fluctuates randomly with respect to time and this added complication makes an analytical solution intractable. Thus, the exact solutions referred to are for laminar flows in which the velocity is either independent of time (steady flow) or dependent on time (unsteady flow) in a well-defined manner.

### 6.9.1 Steady, Laminar Flow between Fixed Parallel Plates

*An exact solution can be obtained for steady laminar flow between fixed parallel plates.*

We first consider flow between the two horizontal, infinite parallel plates of Fig. 6.31a. For this geometry the fluid particles move in the $x$ direction parallel to the plates, and there is no velocity in the $y$ or $z$ direction—that is, $v = 0$ and $w = 0$. In this case it follows from the continuity equation (Eq. 6.31) that $\partial u/\partial x = 0$. Furthermore, there would be no variation of $u$ in the $z$ direction for infinite plates, and for steady flow $\partial u/\partial t = 0$ so that $u = u(y)$. If these conditions are used in the Navier–Stokes equations (Eqs. 6.127), they reduce to

$$0 = -\frac{\partial p}{\partial x} + \mu\left(\frac{\partial^2 u}{\partial y^2}\right) \tag{6.129}$$

$$0 = -\frac{\partial p}{\partial y} - \rho g \tag{6.130}$$

$$0 = -\frac{\partial p}{\partial z} \tag{6.131}$$

where we have set $g_x = 0$, $g_y = -g$, and $g_z = 0$. That is, the $y$ axis points up. We see that for this particular problem the Navier–Stokes equations reduce to some rather simple equations.

Equations 6.130 and 6.131 can be integrated to yield

$$p = -\rho g y + f_1(x) \tag{6.132}$$

which shows that the pressure varies hydrostatically in the $y$ direction. Equation 6.129, rewritten as

$$\frac{d^2 u}{dy^2} = \frac{1}{\mu}\frac{\partial p}{\partial x}$$

can be integrated to give

$$\frac{du}{dy} = \frac{1}{\mu}\left(\frac{\partial p}{\partial x}\right)y + c_1$$

*V6.12 No-slip boundary condition*

and integrated again to yield

$$u = \frac{1}{2\mu}\left(\frac{\partial p}{\partial x}\right)y^2 + c_1 y + c_2 \tag{6.133}$$

Note that for this simple flow the pressure gradient, $\partial p/\partial x$, is treated as constant as far as the integration is concerned, since (as shown in Eq. 6.132) it is not a function of $y$. The two constants $c_1$ and $c_2$ must be determined from the boundary conditions. For example, if the two plates are

■ **Figure 6.31**  The viscous flow between parallel plates: (*a*) coordinate system and notation used in analysis; (*b*) parabolic velocity distribution for flow between parallel fixed plates.

fixed, then $u = 0$ for $y = \pm h$ (because of the no-slip condition for viscous fluids). To satisfy this condition $c_1 = 0$ and

$$c_2 = -\frac{1}{2\mu}\left(\frac{\partial p}{\partial x}\right)h^2$$

Thus, the velocity distribution becomes

$$u = \frac{1}{2\mu}\left(\frac{\partial p}{\partial x}\right)(y^2 - h^2) \tag{6.134}$$

Equation 6.134 shows that the velocity profile between the two fixed plates is parabolic as illustrated in Fig. 6.31*b*.

The volume rate of flow, $q$, passing between the plates (for a unit width in the $z$ direction) is obtained from the relationship

$$q = \int_{-h}^{h} u \, dy = \int_{-h}^{h} \frac{1}{2\mu}\left(\frac{\partial p}{\partial x}\right)(y^2 - h^2)\, dy$$

or

$$q = -\frac{2h^3}{3\mu}\left(\frac{\partial p}{\partial x}\right) \tag{6.135}$$

*V6.13 Liquid–liquid no-slip*

The pressure gradient $\partial p/\partial x$ is negative, since the pressure decreases in the direction of flow. If we let $\Delta p$ represent the pressure *drop* between two points a distance $\ell$ apart, then

$$\frac{\Delta p}{\ell} = -\frac{\partial p}{\partial x}$$

and Eq. 6.135 can be expressed as

$$q = \frac{2h^3 \Delta p}{3\mu\ell} \tag{6.136}$$

The flow is proportional to the pressure gradient, inversely proportional to the viscosity, and strongly dependent $(\sim h^3)$ on the gap width. In terms of the mean velocity, $V$, where $V = q/2h$, Eq. 6.136 becomes

$$V = \frac{h^2 \Delta p}{3\mu\ell} \tag{6.137}$$

Equations 6.136 and 6.137 provide convenient relationships for relating the pressure drop along a parallel-plate channel and the rate of flow or mean velocity. The maximum velocity, $u_{max}$, occurs midway ($y = 0$) between the two plates, as shown in Fig. 6.31*b*, so that from Eq. 6.134

$$u_{max} = -\frac{h^2}{2\mu}\left(\frac{\partial p}{\partial x}\right)$$

or

$$u_{max} = \tfrac{3}{2}V \tag{6.138}$$

*The Navier–Stokes equations provide detailed flow characteristics for laminar flow between fixed parallel plates.*

The details of the steady laminar flow between infinite parallel plates are completely predicted by this solution to the Navier–Stokes equations. For example, if the pressure gradient, viscosity, and plate spacing are specified, then from Eq. 6.134 the velocity profile can be determined, and from Eqs. 6.136 and 6.137 the corresponding flowrate and mean velocity determined. In addition, since the pressure gradient in the $x$ direction is constant, from Eq. 6.132 it follows that

$$f_1(x) = \left(\frac{\partial p}{\partial x}\right)x + p_0$$

where $p_0$ is a reference pressure at $x = y = 0$, and the pressure variation throughout the fluid can be obtained from

$$p = -\rho gy + \left(\frac{\partial p}{\partial x}\right)x + p_0 \tag{6.139}$$

For a given fluid and reference pressure, $p_0$, the pressure at any point can be predicted. This relatively simple example of an exact solution illustrates the detailed information about the flow field which can be obtained. The flow will be laminar if the Reynolds number, $\text{Re} = \rho V(2h)/\mu$, remains below about 1400. For flow with larger Reynolds numbers the flow becomes turbulent and the preceding analysis is not valid since the flow field is complex, three-dimensional, and unsteady.

### 6.9.2 Couette Flow

*For a given flow geometry, the character and details of the flow are strongly dependent on the boundary conditions.*

Another simple parallel-plate flow can be developed by fixing one plate and letting the other plate move with a constant velocity, $U$, as is illustrated in Fig. 6.32a. The Navier–Stokes equations reduce to the same form as those in the preceding section, and the solution for the pressure and velocity distribution are still given by Eqs. 6.132 and 6.133, respectively. However, for the moving plate problem the boundary conditions for the velocity are different. For this case we locate the origin of the coordinate system at the bottom plate and designate the distance between the two plates as $b$ (see Fig. 6.32a). The two constants $c_1$ and $c_2$ in Eq. 6.133 can be determined from the no-slip boundary conditions, $u = 0$ at $y = 0$ and $u = U$ at $y = b$. It follows that

$$u = U\frac{y}{b} + \frac{1}{2\mu}\left(\frac{\partial p}{\partial x}\right)(y^2 - by) \tag{6.140}$$

or, in dimensionless form,

$$\frac{u}{U} = \frac{y}{b} - \frac{b^2}{2\mu U}\left(\frac{\partial p}{\partial x}\right)\left(\frac{y}{b}\right)\left(1 - \frac{y}{b}\right) \tag{6.141}$$

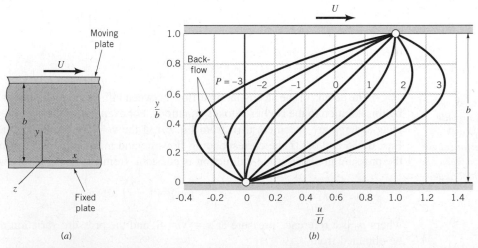

■ **Figure 6.32** The viscous flow between parallel plates with bottom plate fixed and upper plate moving (Couette flow): (a) coordinate system and notation used in analysis; (b) velocity distribution as a function of parameter, $P$, where $P = -(b^2/2\mu U)\,\partial p/\partial x$. (Data from Ref. 8.)

■ **Figure 6.33**   Flow in the narrow gap of a journal bearing.

The actual velocity profile will depend on the dimensionless parameter

$$P = -\frac{b^2}{2\mu U}\left(\frac{\partial p}{\partial x}\right)$$

*Flow between parallel plates with one plate fixed and the other moving is called Couette flow.*

Several profiles are shown in Fig. 6.32b. This type of flow is called **Couette flow**.

The simplest type of Couette flow is one for which the pressure gradient is zero; that is, the fluid motion is caused by the fluid being dragged along by the moving boundary. In this case, with $\partial p/\partial x = 0$, Eq. 6.140 simply reduces to

$$u = U\frac{y}{b} \tag{6.142}$$

which indicates that the velocity varies linearly between the two plates as shown in Fig. 6.31b for $P = 0$. This situation would be approximated by the flow between closely spaced concentric cylinders in which one cylinder is fixed and the other cylinder rotates with a constant angular velocity, $\omega$. As illustrated in Fig. 6.33, the flow in an unloaded journal bearing might be approximated by this simple Couette flow if the gap width is very small (i.e., $r_o - r_i \ll r_i$). In this case $U = r_i\omega$, $b = r_o - r_i$, and the shearing stress resisting the rotation of the shaft can be simply calculated as $\tau = \mu r_i\omega/(r_o - r_i)$. When the bearing is loaded (i.e., a force applied normal to the axis of rotation), the shaft will no longer remain concentric with the housing and the flow cannot be treated as flow between parallel boundaries. Such problems are dealt with in lubrication theory (see, for example, Ref. 9).

---

## *E*XAMPLE 6.9    Plane Couette Flow

**GIVEN**   A wide moving belt passes through a container of a viscous liquid. The belt moves vertically upward with a constant velocity, $V_0$, as illustrated in Fig. E6.9a. Because of viscous forces the belt picks up a film of fluid of thickness $h$. Gravity tends to make the fluid drain down the belt. Assume that the flow is laminar, steady, and fully developed.

**FIND**   Use the Navier–Stokes equations to determine an expression for the average velocity of the fluid film as it is dragged up the belt.

## SOLUTION

Since the flow is assumed to be fully developed, the only velocity component is in the $y$ direction (the $v$ component) so that $u = w = 0$. It follows from the continuity equation that $\partial v/\partial y = 0$, and for steady flow $\partial v/\partial t = 0$, so that $v = v(x)$. Under these conditions the Navier–Stokes equations for the $x$ direction (Eq. 6.127a) and the $z$ direction (perpendicular to the paper) (Eq. 6.127c) simply reduce to

$$\frac{\partial p}{\partial x} = 0 \qquad \frac{\partial p}{\partial z} = 0$$

■ **Figure E6.9a**

This result indicates that the pressure does not vary over a horizontal plane, and since the pressure on the surface of the film ($x = h$) is atmospheric, the pressure throughout the film must be atmospheric (or zero gage pressure). The equation of motion in the $y$ direction (Eq. 6.127b) thus reduces to

$$0 = -\rho g + \mu \frac{d^2 v}{dx^2}$$

or

$$\frac{d^2 v}{dx^2} = \frac{\gamma}{\mu} \qquad (1)$$

Integration of Eq. 1 yields

$$\frac{dv}{dx} = \frac{\gamma}{\mu} x + c_1 \qquad (2)$$

On the film surface ($x = h$) we assume the shearing stress is zero—that is, the drag of the air on the film is negligible. The shearing stress at the free surface (or any interior parallel surface) is designated as $\tau_{xy}$, where from Eq. 6.125d

$$\tau_{xy} = \mu \left( \frac{dv}{dx} \right)$$

Thus, if $\tau_{xy} = 0$ at $x = h$, it follows from Eq. 2 that

$$c_1 = -\frac{\gamma h}{\mu}$$

A second integration of Eq. 2 gives the velocity distribution in the film as

$$v = \frac{\gamma}{2\mu} x^2 - \frac{\gamma h}{\mu} x + c_2$$

At the belt ($x = 0$) the fluid velocity must match the belt velocity, $V_0$, so that

$$c_2 = V_0$$

and the velocity distribution is therefore

$$v = \frac{\gamma}{2\mu} x^2 - \frac{\gamma h}{\mu} x + V_0 \qquad (3)$$

With the velocity distribution known we can determine the flowrate per unit width, $q$, from the relationship

$$q = \int_0^h v \, dx = \int_0^h \left( \frac{\gamma}{2\mu} x^2 - \frac{\gamma h}{\mu} x + V_0 \right) dx$$

and thus

$$q = V_0 h - \frac{\gamma h^3}{3\mu}$$

The average film velocity, $V$ (where $q = Vh$), is therefore

$$V = V_0 - \frac{\gamma h^2}{3\mu} \qquad \textbf{(Ans)}$$

**COMMENT**   Equation (3) can be written in dimensionless form as

$$\frac{v}{V_0} = c \left( \frac{x}{h} \right)^2 - 2c \left( \frac{x}{h} \right) + 1$$

where $c = \gamma h^2 / 2\mu V_0$. This velocity profile is shown in Fig. E6.9b. Note that even though the belt is moving upward, for $c > 1$ (e.g., for fluids with small enough viscosity or with a small enough belt speed) there are portions of the fluid that flow downward (as indicated by $v/V_0 < 0$).

It is interesting to note from this result that there will be a net upward flow of liquid (positive $V$) only if $V_0 > \gamma h^2 / 3\mu$. It takes a relatively large belt speed to lift a small viscosity fluid.

■ **Figure E6.9b**

## 6.9.3 Steady, Laminar Flow in Circular Tubes

*An exact solution can be obtained for steady, incompressible, laminar flow in circular tubes.*

Probably the best known exact solution to the Navier–Stokes equations is for steady, incompressible, laminar flow through a straight circular tube of constant cross section. This type of flow is commonly called *Hagen–Poiseuille flow*, or simply *Poiseuille flow*. It is named in honor of J. L. Poiseuille (1799–1869), a French physician, and G. H. L. Hagen (1797–1884), a German hydraulic engineer. Poiseuille was interested in blood flow through capillaries and deduced experimentally the resistance laws for laminar flow through circular tubes. Hagen's investigation of flow in tubes was also experimental. It was actually after the work of Hagen and Poiseuille that the theoretical results presented in this section were determined, but their names are commonly associated with the solution of this problem.

Consider the flow through a horizontal circular tube of radius $R$ as is shown in Fig. 6.34a. Because of the cylindrical geometry it is convenient to use cylindrical coordinates. We assume that the flow is parallel to the walls so that $v_r = 0$ and $v_\theta = 0$, and from the continuity equation (6.34) $\partial v_z / \partial z = 0$. Also, for steady, axisymmetric flow, $v_z$ is not a function of $t$ or $\theta$, so the velocity, $v_z$,

■ **Figure 6.34**   The viscous flow in a horizontal, circular tube: (*a*) coordinate system and notation used in analysis; (*b*) flow through differential annular ring.

(*a*)                (*b*)

is only a function of the radial position within the tube—that is, $v_z = v_z(r)$. Under these conditions the Navier–Stokes equations (Eqs. 6.128) reduce to

$$0 = -\rho g \sin \theta - \frac{\partial p}{\partial r} \tag{6.143}$$

$$0 = -\rho g \cos \theta - \frac{1}{r} \frac{\partial p}{\partial \theta} \tag{6.144}$$

$$0 = -\frac{\partial p}{\partial z} + \mu \left[ \frac{1}{r} \frac{\partial}{\partial r} \left( r \frac{\partial v_z}{\partial r} \right) \right] \tag{6.145}$$

where we have used the relationships $g_r = -g \sin \theta$ and $g_\theta = -g \cos \theta$ (with $\theta$ measured from the horizontal plane).

Equations 6.143 and 6.144 can be integrated to give

$$p = -\rho g(r \sin \theta) + f_1(z)$$

or

$$p = -\rho g y + f_1(z) \tag{6.146}$$

*V6.14 Laminar flow*

Equation 6.146 indicates that the pressure is hydrostatically distributed at any particular cross section, and the $z$ component of the pressure gradient, $\partial p/\partial z$, is not a function of $r$ or $\theta$.

The equation of motion in the $z$ direction (Eq. 6.145) can be written in the form

$$\frac{1}{r} \frac{\partial}{\partial r} \left( r \frac{\partial v_z}{\partial r} \right) = \frac{1}{\mu} \frac{\partial p}{\partial z}$$

and integrated (using the fact that $\partial p/\partial z = $ constant) to give

$$r \frac{\partial v_z}{\partial r} = \frac{1}{2\mu} \left( \frac{\partial p}{\partial z} \right) r^2 + c_1$$

Integrating again we obtain

$$v_z = \frac{1}{4\mu} \left( \frac{\partial p}{\partial z} \right) r^2 + c_1 \ln r + c_2 \tag{6.147}$$

As indicated by the figure in the margin, the shape of the velocity profile depends on the values of the two constants $c_1$ and $c_2$.

Since we wish $v_z$ to be finite at the center of the tube ($r = 0$), it follows that $c_1 = 0$ [since $\ln (0) = -\infty$]. At the wall ($r = R$) the velocity must be zero so that

$$c_2 = -\frac{1}{4\mu} \left( \frac{\partial p}{\partial z} \right) R^2$$

*The velocity distribution is parabolic for steady, laminar flow in circular tubes.*

and the velocity distribution becomes

$$v_z = \frac{1}{4\mu} \left( \frac{\partial p}{\partial z} \right) (r^2 - R^2) \tag{6.148}$$

Thus, at any cross section the velocity distribution is parabolic.

To obtain a relationship between the volume rate of flow, $Q$, passing through the tube and the pressure gradient, we consider the flow through the differential, washer-shaped ring of

Fig. 6.34*b*. Since $v_z$ is constant on this ring, the volume rate of flow through the differential area $dA = (2\pi r)\, dr$ is

$$dQ = v_z(2\pi r)\, dr$$

and therefore

$$Q = 2\pi \int_0^R v_z r\, dr \tag{6.149}$$

Equation 6.148 for $v_z$ can be substituted into Eq. 6.149, and the resulting equation integrated to yield

$$Q = -\frac{\pi R^4}{8\mu}\left(\frac{\partial p}{\partial z}\right) \tag{6.150}$$

*V6.15 Complex pipe flow*

This relationship can be expressed in terms of the pressure *drop*, $\Delta p$, which occurs over a length, $\ell$, along the tube, since

$$\frac{\Delta p}{\ell} = -\frac{\partial p}{\partial z}$$

and therefore

$$Q = \frac{\pi R^4 \Delta p}{8\mu\ell} \tag{6.151}$$

*Poiseuille's law relates pressure drop and flowrate for steady, laminar flow in circular tubes.*

For a given pressure drop per unit length, the volume rate of flow is inversely proportional to the viscosity and proportional to the tube radius to the fourth power. A doubling of the tube radius produces a 16-fold increase in flow! Equation 6.151 is commonly called *Poiseuille's law.*

In terms of the mean velocity, $V$, where $V = Q/\pi R^2$, Eq. 6.151 becomes

$$V = \frac{R^2 \Delta p}{8\mu\ell} \tag{6.152}$$

The maximum velocity $v_{\max}$ occurs at the center of the tube, where from Eq. 6.148

$$v_{\max} = -\frac{R^2}{4\mu}\left(\frac{\partial p}{\partial z}\right) = \frac{R^2 \Delta p}{4\mu\ell} \tag{6.153}$$

so that

$$v_{\max} = 2V$$

$$\frac{v_z}{v_{\max}} = 1 - \left(\frac{r}{R}\right)^2$$

The velocity distribution, as shown by the figure in the margin, can be written in terms of $v_{\max}$ as

$$\frac{v_z}{v_{\max}} = 1 - \left(\frac{r}{R}\right)^2 \tag{6.154}$$

As was true for the similar case of flow between parallel plates (sometimes referred to as *plane Poiseuille flow*), a very detailed description of the pressure and velocity distribution in tube flow results from this solution to the Navier–Stokes equations. Numerous experiments performed to substantiate the theoretical results show that the theory and experiment are in agreement for the laminar flow of Newtonian fluids in circular tubes or pipes. In general, the flow remains laminar for Reynolds numbers, $\text{Re} = \rho V(2R)/\mu$, below 2100. Turbulent flow in tubes is considered in Chapter 8.

## F l u i d s   i n   t h e   N e w s

**Poiseuille's law revisited** Poiseuille's law governing *laminar flow* of fluids in tubes has an unusual history. It was developed in 1842 by a French physician, J. L. M. Poiseuille, who was interested in the flow of blood in capillaries. Poiseuille, through a series of carefully conducted experiments using water flowing through very small tubes, arrived at the formula, $Q = K\Delta p\, D^4/\ell$. In this formula $Q$ is the flowrate, $K$ an empirical constant, $\Delta p$ the pressure drop over the length $\ell$, and $D$ the tube diameter. Another formula was given for the value of $K$ as a function of the water temperature. It was not until the concept of viscosity was introduced at a later date that Poiseuille's law was derived mathematically and the constant $K$ found to be equal to $\pi/8\mu$, where $\mu$ is the fluid viscosity. The experiments by Poiseuille have long been admired for their accuracy and completeness considering the laboratory instrumentation available in the mid-nineteenth century.

■ **Figure 6.35**   The viscous flow through an annulus.

### 6.9.4 Steady, Axial, Laminar Flow in an Annulus

*An exact solution can be obtained for axial flow in the annular space between two fixed, concentric cylinders.*

The differential equations (Eqs. 6.143, 6.144, 6.145) used in the preceding section for flow in a tube also apply to the axial flow in the annular space between two fixed, concentric cylinders (Fig. 6.35). Equation 6.147 for the velocity distribution still applies, but for the stationary annulus the boundary conditions become $v_z = 0$ at $r = r_o$ and $v_z = 0$ for $r = r_i$. With these two conditions the constants $c_1$ and $c_2$ in Eq. 6.147 can be determined and the velocity distribution becomes

$$v_z = \frac{1}{4\mu}\left(\frac{\partial p}{\partial z}\right)\left[r^2 - r_o^2 + \frac{r_i^2 - r_o^2}{\ln(r_o/r_i)}\ln\frac{r}{r_o}\right] \tag{6.155}$$

The corresponding volume rate of flow is

$$Q = \int_{r_i}^{r_o} v_z(2\pi r)\,dr = -\frac{\pi}{8\mu}\left(\frac{\partial p}{\partial z}\right)\left[r_o^4 - r_i^4 - \frac{(r_o^2 - r_i^2)^2}{\ln(r_o/r_i)}\right]$$

or in terms of the pressure drop, $\Delta p$, in length $\ell$ of the annulus

$$Q = \frac{\pi\Delta p}{8\mu\ell}\left[r_o^4 - r_i^4 - \frac{(r_o^2 - r_i^2)^2}{\ln(r_o/r_i)}\right] \tag{6.156}$$

The velocity at any radial location within the annular space can be obtained from Eq. 6.155. The maximum velocity occurs at the radius $r = r_m$ where $\partial v_z/\partial r = 0$. Thus,

$$r_m = \left[\frac{r_o^2 - r_i^2}{2\ln(r_o/r_i)}\right]^{1/2} \tag{6.157}$$

An inspection of this result shows that the maximum velocity does not occur at the midpoint of the annular space, but rather it occurs nearer the inner cylinder. The specific location depends on $r_o$ and $r_i$.

These results for flow through an annulus are valid only if the flow is laminar. A criterion based on the conventional Reynolds number (which is defined in terms of the tube diameter) cannot be directly applied to the annulus, since there are really "two" diameters involved. For tube cross sections other than simple circular tubes it is common practice to use an "effective" diameter, termed the *hydraulic diameter*, $D_h$, which is defined as

$$D_h = \frac{4 \times \text{cross-sectional area}}{\text{wetted perimeter}}$$

The wetted perimeter is the perimeter in contact with the fluid. For an annulus

$$D_h = \frac{4\pi(r_o^2 - r_i^2)}{2\pi(r_o + r_i)} = 2(r_o - r_i)$$

In terms of the hydraulic diameter, the Reynolds number is $\text{Re} = \rho D_h V/\mu$ (where $V = Q/$ cross-sectional area), and it is commonly assumed that if this Reynolds number remains below 2100 the flow will be laminar. A further discussion of the concept of the hydraulic diameter as it applies to other noncircular cross sections is given in Section 8.4.3.

# EXAMPLE 6.10 | Laminar Flow in an Annulus

**GIVEN** A viscous liquid ($\rho = 1.18 \times 10^3$ kg/m$^3$; $\mu = 0.0045$ N·s/m$^2$) flows at a rate of 12 ml/s through a horizontal, 4-mm-diameter tube.

**FIND** (a) Determine the pressure drop along a 1-m length of the tube which is far from the tube entrance so that the only component of velocity is parallel to the tube axis. (b) If a 2-mm-diameter rod is placed in the 4-mm-diameter tube to form a symmetric annulus, what is the pressure drop along a 1-m length if the flowrate remains the same as in part (a)?

## SOLUTION

**(a)** We first calculate the Reynolds number, Re, to determine whether or not the flow is laminar. With the diameter $D = 4$ mm $= 0.004$ m, the mean velocity is

$$V = \frac{Q}{(\pi/4)D^2} = \frac{(12 \text{ ml/s})(10^{-6} \text{ m}^3/\text{ml})}{(\pi/4)(0.004 \text{ m})^2}$$

$$= 0.955 \text{ m/s}$$

and, therefore,

$$\text{Re} = \frac{\rho V D}{\mu} = \frac{(1.18 \times 10^3 \text{ kg/m}^3)(0.955 \text{ m/s})(0.004 \text{ m})}{0.0045 \text{ N} \cdot \text{s/m}^2}$$

$$= 1000$$

Since the Reynolds number is well below the critical value of 2100, we can safely assume that the flow is laminar. Thus, we can apply Eq. 6.151, which gives for the pressure drop

$$\Delta p = \frac{8\mu \ell Q}{\pi R^4}$$

$$= \frac{8(0.0045 \text{ N} \cdot \text{s/m}^2)(1 \text{ m})(12 \times 10^{-6} \text{ m}^3/\text{s})}{\pi(0.002 \text{ m})^4}$$

$$= 8.59 \text{ kPa} \qquad \text{(Ans)}$$

**(b)** For flow in the annulus with an outer radius $r_o = 0.002$ m and an inner radius $r_i = 0.001$ m, the mean velocity is

$$V = \frac{Q}{\pi(r_o^2 - r_i^2)} = \frac{12 \times 10^{-6} \text{ m}^3/\text{s}}{(\pi)[(0.002 \text{ m})^2 - (0.001 \text{ m})^2]}$$

$$= 1.27 \text{ m/s}$$

and the Reynolds number [based on the hydraulic diameter, $D_h = 2(r_o - r_i) = 2(0.002 \text{ m} - 0.001 \text{ m}) = 0.002$ m] is

$$\text{Re} = \frac{\rho D_h V}{\mu}$$

$$= \frac{(1.18 \times 10^3 \text{ kg/m}^3)(0.002 \text{ m})(1.27 \text{ m/s})}{0.0045 \text{ N} \cdot \text{s/m}^2}$$

$$= 666$$

This value is also well below 2100, so the flow in the annulus should also be laminar. From Eq. 6.156,

$$\Delta p = \frac{8\mu \ell Q}{\pi} \left[ r_o^4 - r_i^4 - \frac{(r_o^2 - r_i^2)^2}{\ln(r_o/r_i)} \right]^{-1}$$

so that

$$\Delta p = \frac{8(0.0045 \text{ N} \cdot \text{s/m}^2)(1 \text{ m})(12 \times 10^{-6} \text{ m}^3/\text{s})}{\pi}$$

$$\times \left\{ (0.002 \text{ m})^4 - (0.001 \text{ m})^4 \right.$$

$$\left. - \frac{[(0.002 \text{ m})^2 - (0.001 \text{ m})^2]^2}{\ln(0.002 \text{ m}/0.001 \text{ m})} \right\}^{-1}$$

$$= 68.2 \text{ kPa} \qquad \text{(Ans)}$$

**COMMENTS** The pressure drop in the annulus is much larger than that of the tube. This is not a surprising result, since to maintain the same flow in the annulus as that in the open tube, the average velocity must be larger (the cross-sectional area is smaller) and the pressure difference along the annulus must overcome the shearing stresses that develop along both an inner and an outer wall.

By repeating the calculations for various radius ratios, $r_i/r_o$, the results shown in Fig. E6.10 are obtained. It is seen that the pressure drop ratio, $\Delta p_{annulus}/\Delta p_{tube}$ (i.e., the pressure drop in the annulus compared to that in a tube with a radius equal to the outer radius of the annulus, $r_o$), is a strong function of the radius ratio. Even an annulus with a very small inner radius will have a pressure drop significantly larger than that of a tube. For example, if the inner radius is only 1/100 of the outer radius, $\Delta p_{annulus}/\Delta p_{tube} = 1.28$. As shown in the figure, for larger inner radii, the pressure drop ratio is much larger [i.e., $\Delta p_{annulus}/\Delta p_{tube} = 7.94$ for $r_i/r_o = 0.50$ as in part (b) of this example].

■ **Figure E6.10**

# 6.10    Other Aspects of Differential Analysis

In this chapter the basic differential equations that govern the flow of fluids have been developed. The Navier–Stokes equations, which can be compactly expressed in vector notation as

$$\rho\left(\frac{\partial \mathbf{V}}{\partial t} + \mathbf{V} \cdot \nabla \mathbf{V}\right) = -\nabla p + \rho \mathbf{g} + \mu \nabla^2 \mathbf{V} \tag{6.158}$$

along with the continuity equation

$$\nabla \cdot \mathbf{V} = 0 \tag{6.159}$$

*Very few practical fluid flow problems can be solved using an exact analytical approach.*

are the general equations of motion for incompressible Newtonian fluids. Although we have restricted our attention to incompressible fluids, these equations can be readily extended to include compressible fluids. It is well beyond the scope of this introductory text to consider in depth the variety of analytical and numerical techniques that can be used to obtain both exact and approximate solutions to the Navier–Stokes equations. Students, however, should be aware of the existence of these very general equations, which are frequently used as the basis for many advanced analyses of fluid motion. A few relatively simple solutions have been obtained and discussed in this chapter to indicate the type of detailed flow information that can be obtained by using differential analysis. However, it is hoped that the relative ease with which these solutions were obtained does not give the false impression that solutions to the Navier–Stokes equations are readily available. This is certainly not true, and as previously mentioned there are actually very few practical fluid flow problems that can be solved by using an exact analytical approach. In fact, there are no known analytical solutions to Eq. 6.158 for flow past any object such as a sphere, cube, or airplane.

Because of the difficulty in solving the Navier–Stokes equations, much attention has been given to various types of approximate solutions. For example, if the viscosity is set equal to zero, the Navier–Stokes equations reduce to Euler's equations. Thus, the frictionless fluid solutions discussed previously are actually approximate solutions to the Navier–Stokes equations. At the other extreme, for problems involving slowly moving fluids, viscous effects may be dominant and the nonlinear (convective) acceleration terms can be neglected. This assumption greatly simplifies the analysis, since the equations now become linear. There are numerous analytical solutions to these "*slow flow*" or "*creeping flow*" problems. Another broad class of approximate solutions is concerned with flow in the very thin boundary layer. L. Prandtl showed in 1904 how the Navier–Stokes equations could be simplified to study flow in boundary layers. Such "boundary layer solutions" play a very important role in the study of fluid mechanics. A further discussion of boundary layers is given in Chapter 9.

## 6.10.1    Numerical Methods

*V6.16 CFD example*

Numerical methods using digital computers are, of course, commonly utilized to solve a wide variety of flow problems. As discussed previously, although the differential equations that govern the flow of Newtonian fluids [the Navier–Stokes equations (6.158)] were derived many years ago, there are few known analytical solutions to them. With the advent of high-speed digital computers it has become possible to obtain numerical solutions to these (and other fluid mechanics) equations for many different types of problems. A brief introduction to computational fluid dynamics (CFD) is given in Appendix A.

## F l u i d s   i n   t h e   N e w s

**Fluids in the Academy Awards** A computer science professor at Stanford University and his colleagues were awarded a Scientific and Technical Academy Award for applying the Navier–Stokes equations for use in Hollywood movies. These researchers make use of computational algorithms to numerically solve the Navier–Stokes equations (also termed computational fluid dynam-ics, or CFD) and simulate complex liquid flows. The realism of the simulations has found application in the entertainment industry. Movie producers have used the power of these numerical tools to simulate flows from ocean waves in *Pirates of the Caribbean* to lava flows in the final duel in *Star Wars: Revenge of the Sith*. Therefore, even Hollywood has recognized the usefulness of CFD.

## 6.11 Chapter Summary and Study Guide

*volumetric dilatation rate*
*vorticity*
*irrotational flow*
*continuity equation*
*stream function*
*Euler's equations of motion*
*ideal fluid*
*Bernoulli equation*
*velocity potential*
*potential flow*
*equipotential lines*
*flow net*
*uniform flow*
*source and sink*
*vortex*
*circulation*
*doublet*
*method of superposition*
*half-body*
*Rankine oval*
*Navier–Stokes equations*
*Couette flow*
*Poiseuille's law*

Differential analysis of fluid flow is concerned with the development of concepts and techniques that can be used to provide a detailed, point-by-point, description of a flow field. Concepts related to the motion and deformation of a fluid element are introduced, including the Eulerian method for describing the velocity and acceleration of fluid particles. Linear deformation and angular deformation of a fluid element are described through the use of flow characteristics such as the volumetric dilatation rate, rate of angular deformation, and vorticity. The differential form of the conservation of mass equation (continuity equation) is derived in both rectangular and cylindrical polar coordinates.

Use of the stream function for the study of steady, incompressible, plane, two-dimensional flow is introduced. The general equations of motion are developed, and for inviscid flow these equations are reduced to the simpler Euler equations of motion. The Euler equations are integrated to give the Bernoulli equation, and the concept of irrotational flow is introduced. Use of the velocity potential for describing irrotational flow is considered in detail, and several basic velocity potentials are described, including those for a uniform flow, source or sink, vortex, and doublet. The technique of using various combinations of these basic velocity potentials, by superposition, to form new potentials is described. Flows around a half-body, a Rankine oval, and around a circular cylinder are obtained using this superposition technique.

Basic differential equations describing incompressible, viscous flow (the Navier–Stokes equations) are introduced. Several relatively simple solutions for steady, viscous, laminar flow between parallel plates and through circular tubes are included.

The following checklist provides a study guide for this chapter. When your study of the entire chapter and end-of-chapter exercises has been completed, you should be able to

- write out meanings of the terms listed here in the margin and understand each of the related concepts. These terms are particularly important and are set in *italic, **bold**, and color* type in the text.
- determine the acceleration of a fluid particle, given the equation for the velocity field.
- determine the volumetric dilatation rate, vorticity, and rate of angular deformation for a fluid element, given the equation for the velocity field.
- show that a given velocity field satisfies the continuity equation.
- use the concept of the stream function to describe a flow field.
- use the concept of the velocity potential to describe a flow field.
- use superposition of basic velocity potentials to describe simple potential flow fields.
- use the Navier–Stokes equations to determine the detailed flow characteristics of incompressible, steady, laminar, viscous flow between parallel plates and through circular tubes.

Some of the important equations in this chapter are:

| | | |
|---|---|---|
| Acceleration of fluid particle | $\mathbf{a} = \dfrac{\partial \mathbf{V}}{\partial t} + u\dfrac{\partial \mathbf{V}}{\partial x} + v\dfrac{\partial \mathbf{V}}{\partial y} + w\dfrac{\partial \mathbf{V}}{\partial z}$ | (6.2) |
| Vorticity | $\zeta = 2\,\boldsymbol{\omega} = \nabla \times \mathbf{V}$ | (6.17) |
| Conservation of mass | $\dfrac{\partial \rho}{\partial t} + \dfrac{\partial(\rho u)}{\partial x} + \dfrac{\partial(\rho v)}{\partial y} + \dfrac{\partial(\rho w)}{\partial z} = 0$ | (6.27) |
| Stream function | $u = \dfrac{\partial \psi}{\partial y} \qquad v = -\dfrac{\partial \psi}{\partial x}$ | (6.37) |
| Euler's equations of motion | $\rho g_x - \dfrac{\partial p}{\partial x} = \rho\left(\dfrac{\partial u}{\partial t} + u\dfrac{\partial u}{\partial x} + v\dfrac{\partial u}{\partial y} + w\dfrac{\partial u}{\partial z}\right)$ | (6.51a) |
| | $\rho g_y - \dfrac{\partial p}{\partial y} = \rho\left(\dfrac{\partial v}{\partial t} + u\dfrac{\partial v}{\partial x} + v\dfrac{\partial v}{\partial y} + w\dfrac{\partial v}{\partial z}\right)$ | (6.51b) |
| | $\rho g_z - \dfrac{\partial p}{\partial z} = \rho\left(\dfrac{\partial w}{\partial t} + u\dfrac{\partial w}{\partial x} + v\dfrac{\partial w}{\partial y} + w\dfrac{\partial w}{\partial z}\right)$ | (6.51c) |

| | | | |
|---|---|---|---|
| Velocity potential | | $\mathbf{V} = \nabla\phi$ | **(6.65)** |
| Laplace's equation | | $\nabla^2\phi = 0$ | **(6.66)** |

Uniform potential flow $\quad \phi = U(x\cos\alpha + y\sin\alpha) \quad \psi = U(y\cos\alpha - x\sin\alpha) \quad \begin{aligned} u &= U\cos\alpha \\ v &= U\sin\alpha \end{aligned}$

Source and sink $\qquad \phi = \dfrac{m}{2\pi}\ln r \qquad \psi = \dfrac{m}{2\pi}\theta \qquad \begin{aligned} v_r &= \dfrac{m}{2\pi r} \\ v_\theta &= 0 \end{aligned}$

Vortex $\qquad \phi = \dfrac{\Gamma}{2\pi}\theta \qquad \psi = -\dfrac{\Gamma}{2\pi}\ln r \qquad \begin{aligned} v_r &= 0 \\ v_\theta &= \dfrac{\Gamma}{2\pi r} \end{aligned}$

Doublet $\qquad \phi = \dfrac{K\cos\theta}{r} \qquad \psi = -\dfrac{K\sin\theta}{r} \qquad \begin{aligned} v_r &= -\dfrac{K\cos\theta}{r^2} \\ v_\theta &= -\dfrac{K\cos\theta}{r^2} \end{aligned}$

The Navier–Stokes equations

($x$ direction)

$$\rho\left(\frac{\partial u}{\partial t} + u\frac{\partial u}{\partial x} + v\frac{\partial u}{\partial y} + w\frac{\partial u}{\partial z}\right) = -\frac{\partial p}{\partial x} + \rho g_x + \mu\left(\frac{\partial^2 u}{\partial x^2} + \frac{\partial^2 u}{\partial y^2} + \frac{\partial^2 u}{\partial z^2}\right) \quad \textbf{(6.127a)}$$

($y$ direction)

$$\rho\left(\frac{\partial v}{\partial t} + u\frac{\partial v}{\partial x} + v\frac{\partial v}{\partial y} + w\frac{\partial v}{\partial z}\right) = -\frac{\partial p}{\partial y} + \rho g_y + \mu\left(\frac{\partial^2 v}{\partial x^2} + \frac{\partial^2 v}{\partial y^2} + \frac{\partial^2 v}{\partial z^2}\right) \quad \textbf{(6.127b)}$$

($z$ direction)

$$\rho\left(\frac{\partial w}{\partial t} + u\frac{\partial w}{\partial x} + v\frac{\partial w}{\partial y} + w\frac{\partial w}{\partial z}\right) = -\frac{\partial p}{\partial z} + \rho g_z + \mu\left(\frac{\partial^2 w}{\partial x^2} + \frac{\partial^2 w}{\partial y^2} + \frac{\partial^2 w}{\partial z^2}\right) \quad \textbf{(6.127c)}$$

# References

1. White, F. M., *Fluid Mechanics*, 5th Ed., McGraw-Hill, New York, 2003.
2. Streeter, V. L., *Fluid Dynamics*, McGraw-Hill, New York, 1948.
3. Rouse, H., *Advanced Mechanics of Fluids*, Wiley, New York, 1959.
4. Milne-Thomson, L. M., *Theoretical Hydrodynamics*, 4th Ed., Macmillan, New York, 1960.
5. Robertson, J. M., *Hydrodynamics in Theory and Application*, Prentice-Hall, Englewood Cliffs, N.J., 1965.
6. Panton, R. L., *Incompressible Flow*, 3rd Ed., Wiley, New York, 2005.
7. Li, W. H., and Lam, S. H., *Principles of Fluid Mechanics*, Addison-Wesley, Reading, Mass., 1964.
8. Schlichting, H., *Boundary-Layer Theory*, 8th Ed., McGraw-Hill, New York, 2000.
9. Fuller, D. D., *Theory and Practice of Lubrication for Engineers*, Wiley, New York, 1984.

 Problem available in *WileyPLUS* at instructor's discretion.

 Tutoring problem available in *WileyPLUS* at instructor's discretion.

Problem is related to a chapter video available in *WileyPLUS*.

\* Problem to be solved with aid of programmable calculator or computer.

† Open-ended problem that requires critical thinking. These problems require various assumptions to provide the necessary input data. There are not unique answers to these problems.

# Review Problems

Go to Appendix G (*WileyPLUS* or the book's web site, www.wiley. com/college/munson) for a set of review problems with answers. Detailed solutions can be found in the *Student Solution Manual and Study Guide for Fundamentals of Fluid Mechanics,* by Munson et al. (© 2013 John Wiley and Sons, Inc.).

# Conceptual Questions

**6.1C** Looking at Eq. 6.37, what are the units of stream function?

a) $LT^{-1}$         b) $L^2T^{-1}$

c) $MLT^{-1}$         d) $T^{-1}$

**6.2C** Two tanks filled with water are connected by two straight circular pipes that have diameters $D_1$ and $D_2$, as shown in the figure.

The water level in the left tank is twice that of the right tank. If the flow through the connection pipes is laminar and can be approximated by the fully developed Poiseuille solution, then the flow through pipe 2 will have the same velocity as the flow through pipe 1 when:

a) the diameter of pipe 2 is less than the diameter of pipe 1.

b) the diameter of pipe 2 is greater than the diameter of pipe 1.

c) the diameter of pipe 2 is equal to the diameter of pipe 1.

**6.3C** A uniform flow moving to the left (in the negative $x$-axis direction) is imposed on the doublet shown in Fig. 6.23. For the resulting flow field,

a) there are two stagnation points, one above and one below the doublet.

b) there is one stagnation point to the left of the doublet.

c) there is one stagnation point to the right of the doublet.

d) there are no stagnation points anywhere in the flow.

**6.4C** For steady, laminar flow between fixed parallel plates, the velocity profile is parabolic in shape, as shown in Fig. 6.31. If the shear stress is given by $\tau = \mu \partial u/\partial y$, what is the shear stress at the midpoint between the two plates?

a) maximum shear         b) negative shear

c) zero shear         d) equal to the wall shear

**Additional conceptual questions are available in *WileyPLUS* at the instructor's discretion.**

# Problems

**Note: Unless specific values of required fluid properties are given in the problem statement, use the values found in the tables on the inside of the front cover. Answers to the even-numbered problems are listed at the end of the book. The Lab Problems as well as the videos that accompany problems can be accessed in *Wiley-PLUS* or the book's web site, www.wiley.com/college/munson.**

### Section 6.1 Fluid Element Kinematics

**6.1** The velocity in a certain flow field is given by the equation

$$\mathbf{V} = yz\hat{\mathbf{i}} + x^2z\hat{\mathbf{j}} + x\hat{\mathbf{k}}$$

Determine the expressions for the three rectangular components of acceleration.

**6.2** *(WILEY)* The velocity in a certain two-dimensional flow field is given by the equation

$$\mathbf{V} = 2xt\hat{\mathbf{i}} - 2yt\hat{\mathbf{j}}$$

where the velocity is in ft/s when $x$, $y$, and $t$ are in feet and seconds, respectively. Determine expressions for the local and convective components of acceleration in the $x$ and $y$ directions. What is the magnitude and direction of the velocity and the acceleration at the point $x = y = 2$ ft at the time $t = 0$?

**6.3** *(WILEY)* The velocity in a certain flow field is given by the equation

$$\mathbf{V} = x\hat{\mathbf{i}} + x^2z\hat{\mathbf{j}} + yz\hat{\mathbf{k}}$$

Determine the expressions for the three rectangular components of acceleration.

**6.4** *(WILEY)* The three components of velocity in a flow field are given by

$$u = x^2 + y^2 + z^2$$
$$v = xy + yz + z^2$$
$$w = -3xz - z^2/2 + 4$$

**(a)** Determine the volumetric dilatation rate and interpret the results. **(b)** Determine an expression for the rotation vector. Is this an irrotational flow field?

**6.5** *(WILEY)* Determine the vorticity field for the following velocity vector:

$$\mathbf{V} = (x^2 - y^2)\hat{\mathbf{i}} - 2xy\hat{\mathbf{j}}$$

**6.6** *(WILEY)* Determine an expression for the vorticity of the flow field described by

$$\mathbf{V} = -xy^3\hat{\mathbf{i}} + y^4\hat{\mathbf{j}}$$

Is the flow irrotational?

**6.7** *(WILEY GO)* A two-dimensional flow field described by

$$\mathbf{V} = (2x^2y + x)\hat{\mathbf{i}} + (2xy^2 + y + 1)\hat{\mathbf{j}}$$

where the velocity is in m/s when $x$ and $y$ are in meters. Determine the angular rotation of a fluid element located at $x = 0.5$ m, $y = 1.0$ m.

**6.8** *(WILEY)* For a certain incompressible, two-dimensional flow field the velocity component in the $y$ direction is given by the equation

$$v = 3xy + x^2y$$

Determine the velocity component in the $x$ direction so that the volumetric dilatation rate is zero.

**6.9** WILEY An incompressible viscous fluid is placed between two large parallel plates as shown in Fig. P6.9. The bottom plate is fixed and the upper plate moves with a constant velocity, $U$. For these conditions the velocity distribution between the plates is linear and can be expressed as

$$u = U \frac{y}{b}$$

Determine: **(a)** the volumetric dilatation rate, **(b)** the rotation vector, **(c)** the vorticity, and **(d)** the rate of angular deformation.

■ **Figure P6.9**

**6.10** ▶ WILEY A viscous fluid is contained in the space between concentric cylinders. The inner wall is fixed, and the outer wall rotates with an angular velocity $\omega$. (See Fig. P6.10a and **Video V6.3**.) Assume that the velocity distribution in the gap is linear as illustrated in Fig. P6.10b. For the small rectangular element shown in Fig. P6.10b, determine the rate of change of the right angle $\gamma$ due to the fluid motion. Express your answer in terms of $r_o$, $r_i$, and $\omega$.

$(a)$                    $(b)$

■ **Figure P6.10**

## Section 6.2 Conservation of Mass

**6.11** For incompressible fluids the volumetric dilatation rate must be zero; that is, $\nabla \cdot \mathbf{V} = 0$. For what combination of constants, $a$, $b$, $c$, and $e$ can the velocity components

$$u = ax + by$$
$$v = cx + ey$$
$$w = 0$$

be used to describe an incompressible flow field?

**6.12** WILEY For a certain incompressible flow field it is suggested that the velocity components are given by the equations

$$u = 2xy \quad v = -x^2y \quad w = 0$$

Is this a physically possible flow field? Explain.

**6.13** WILEY The velocity components of an incompressible, two-dimensional velocity field are given by the equations

$$u = y^2 - x(1 + x)$$
$$v = y(2x + 1)$$

Show that the flow is irrotational and satisfies conservation of mass.

**6.14** WILEY For each of the following stream functions, with units of $m^2/s$, determine the magnitude and the angle the velocity vector makes with the $x$ axis at $x = 1$ m, $y = 2$ m. Locate any stagnation points in the flow field.
**(a)** $\psi = xy$
**(b)** $\psi = -2x^2 + y$

**6.15** WILEY The stream function for an incompressible, two-dimensional flow field is

$$\psi = ay - by^3$$

where $a$ and $b$ are constants. Is this an irrotational flow? Explain.

**6.16** The stream function for an incompressible, two-dimensional flow field is

$$\psi = ay^2 - bx$$

where $a$ and $b$ are constants. Is this an irrotational flow? Explain.

**6.17** The velocity components in an incompressible, two-dimensional flow field are given by the equations

$$u = x^2$$
$$v = -2xy + x$$

Determine, if possible, the corresponding stream function.

**6.18** The velocity components of an incompressible, two-dimensional velocity field are given by the equations

$$u = 2xy$$
$$v = x^2 - y^2$$

Show that the flow is irrotational and satisfies conservation of mass.

**6.19** WILEY For a certain two-dimensional flow field

$$u = 0$$
$$v = V$$

**(a)** What are the corresponding radial and tangential velocity components? **(b)** Determine the corresponding stream function expressed in Cartesian coordinates and in cylindrical polar coordinates.

**6.20** Some velocity measurements in a three-dimensional incompressible flow field indicate that $u = 6xy^2$ and $v = -4y^2z$. There is some conflicting data for the velocity component in the $z$ direction. One set of data indicates that $w = 4yz^2$ and the other set indicates that $w = 4yz^2 - 6y^2z$. Which set do you think is correct? Explain.

**6.21** WILEY A two-dimensional, incompressible flow is given by $u = -y$ and $v = x$. Show that the streamline passing through the point $x = 10$ and $y = 0$ is a circle centered at the origin.

**6.22** In a certain steady, two-dimensional flow field the fluid density varies linearly with respect to the coordinate $x$: that is, $\rho = Ax$ where $A$ is a constant. If the $x$ component of velocity $u$ is given by the equation $u = y$, determine an expression for $v$.

**6.23** WILEY In a two-dimensional, incompressible flow field, the $x$ component of velocity is given by the equation $u = 2x$. **(a)** Determine the corresponding equation for the $y$ component of velocity if $v = 0$ along the $x$ axis. **(b)** For this flow field, what is the magnitude

■ **Figure P6.23**

of the average velocity of the fluid crossing the surface *OA* of Fig. P6.23? Assume that the velocities are in feet per second when *x* and *y* are in feet.

**6.24** (WILEY) The radial velocity component in an incompressible, two-dimensional flow field ($v_z = 0$) is

$$v_r = 2r + 3r^2 \sin \theta$$

Determine the corresponding tangential velocity component, $v_\theta$, required to satisfy conservation of mass.

**6.25** (WILEY) The stream function for an incompressible flow field is given by the equation

$$\psi = 3x^2y - y^3$$

where the stream function has the units of m²/s with *x* and *y* in meters. **(a)** Sketch the streamline(s) passing through the origin. **(b)** Determine the rate of flow across the straight path *AB* shown in Fig. P6.25.

■ **Figure P6.25**

**6.26** The streamlines in a certain incompressible, two-dimensional flow field are all concentric circles so that $v_r = 0$. Determine the stream function for **(a)** $v_\theta = Ar$ and for **(b)** $v_\theta = Ar^{-1}$, where *A* is a constant.

**6.27** It is proposed that a two-dimensional, incompressible flow field be described by the velocity components

$$u = Ay$$
$$v = Bx$$

where *A* and *B* are both positive constants. **(a)** Will the continuity equation be satisfied? **(b)** Is the flow irrotational? **(c)** Determine the equation for the streamlines and show a sketch of the streamline that passes through the origin. Indicate the direction of flow along this streamline.

**\*6.28** (WILEY) The stream function for an incompressible, two-dimensional flow field is

$$\psi = 3x^2y + y$$

For this flow field, plot several streamlines.

**6.29** (WILEY GO) Consider the incompressible, two-dimensional flow of a nonviscous fluid between the boundaries shown in Fig. P6.29. The velocity potential for this flow field is

$$\phi = x^2 - y^2$$

■ **Figure P6.29**

**(a)** Determine the corresponding stream function. **(b)** What is the relationship between the discharge, *q* (per unit width normal to plane of paper) passing between the walls and the coordinates $x_i$, $y_i$ of any point on the curved wall? Neglect body forces.

## Section 6.3 Conservation of Linear Momentum

**6.30** A fluid with a density of 2000 kg/m³ flows steadily between two flat plates as shown in Fig. P6.30. The bottom plate is fixed and the top one moves at a constant speed in the *x* direction. The velocity is $V = 0.20 \, y \, \hat{\mathbf{i}}$ m/s where *y* is in meters. The acceleration of gravity is $g = -9.8 \, \hat{\mathbf{j}}$ m/s². The only nonzero shear stresses, $\tau_{yx} = \tau_{xy}$, are constant throughout the flow with a value of 5 N/m². The normal stress at the origin ($x = y = 0$) is $\sigma_{xx} = -100$ kPa. Use the *x* and *y* components of the equations of motion (Eqs. 6.50*a* and *b*) to determine the normal stress throughout the fluid. Assume that $\sigma_{xx} = \sigma_{yy}$.

■ **Figure P6.30**

**6.31** A fluid with a density of 2 slug/ft³ flows steadily between two stationary flat plates as shown in Fig. P6.31. The velocity is $V = 0.5 \left[ 1 - (y/h)^2 \right] \hat{\mathbf{i}}$ ft/s where *y* and *h* are in feet. The only nonzero shear stresses, $\tau_{yx} = \tau_{xy}$, are given by $\tau_{yx} = -4.0 \, y$ lb/ft² and the acceleration of gravity is negligible. The normal stress at the origin ($x = y = 0$) is $\sigma_{xx} = -10$ lb/ft². Use the *x* and *y* components of the equations of motion (Eqs. 6.50a and b) to determine the normal stress throughout the fluid. Assume that $\sigma_{xx} = \sigma_{yy}$.

■ **Figure P6.31**

## Section 6.4 Inviscid Flow

**6.32** Given the stream function for a flow as $\psi = 4x^2 - 4y^2$, show that the Bernoulli equation can be applied between any two points in the flow field.

**6.33** A two-dimensional flow field for a nonviscous, incompressible fluid is described by the velocity components

$$u = U_0 + 2y$$
$$v = 0$$

**Figure P6.33**

where $U_0$ is a constant. If the pressure at the origin (Fig. P6.33) is $p_0$, determine an expression for the pressure at **(a)** point $A$, and **(b)** point $B$. Explain clearly how you obtained your answer. Assume that the units are consistent and body forces may be neglected.

**6.34** In a certain two-dimensional flow field, the velocity is constant with components $u = -4$ ft/s and $v = -2$ ft/s. Determine the corresponding stream function and velocity potential for this flow field. Sketch the equipotential line $\phi = 0$ which passes through the origin of the coordinate system.

**6.35** (WILEY) The stream function for a given two-dimensional flow field is

$$\psi = 5x^2 y - (5/3)y^3$$

Determine the corresponding velocity potential.

**6.36** (WILEY) A certain flow field is described by the stream function

$$\psi = A\,\theta + B\,r\sin\theta$$

where $A$ and $B$ are positive constants. Determine the corresponding velocity potential and locate any stagnation points in this flow field.

**6.37** (WILEY) It is known that the velocity distribution for two-dimensional flow of a viscous fluid between wide parallel plates (Fig. P6.37) is parabolic; that is,

$$u = U_c\left[1 - \left(\frac{y}{h}\right)^2\right]$$

with $v = 0$. Determine, if possible, the corresponding stream function and velocity potential.

**Figure P6.37**

**6.38** (WILEY) The velocity potential for a certain inviscid flow field is

$$\phi = -(3x^2 y - y^3)$$

where $\phi$ has the units of ft$^2$/s when $x$ and $y$ are in feet. Determine the pressure difference (in psi) between the points (1, 2) and (4, 4), where the coordinates are in feet, if the fluid is water and elevation changes are negligible.

**6.39** The velocity potential for a flow is given by

$$\phi = \frac{a}{2}(x^2 - y^2)$$

where $a$ is a constant. Determine the corresponding stream function and sketch the flow pattern.

**6.40** (WILEY) The stream function for a two-dimensional, nonviscous, incompressible flow field is given by the expression

$$\psi = -2(x - y)$$

where the stream function has the units of ft$^2$/s with $x$ and $y$ in feet. **(a)** Is the continuity equation satisfied? **(b)** Is the flow field irrotational? If so, determine the corresponding velocity potential. **(c)** Determine the pressure gradient in the horizontal $x$ direction at the point $x = 2$ ft, $y = 2$ ft.

**6.41** (WILEY) The velocity potential for a certain inviscid, incompressible flow field is given by the equation

$$\phi = 2x^2 y - (\tfrac{2}{3})y^3$$

where $\phi$ has the units of m$^2$/s when $x$ and $y$ are in meters. Determine the pressure at the point $x = 2$ m, $y = 2$ m if the pressure at $x = 1$ m, $y = 1$ m is 200 kPa. Elevation changes can be neglected, and the fluid is water.

**6.42** A steady, uniform, incompressible, inviscid, two-dimensional flow makes an angle of 30° with the horizontal $x$ axis. **(a)** Determine the velocity potential and the stream function for this flow. **(b)** Determine an expression for the pressure gradient in the vertical $y$ direction. What is the physical interpretation of this result?

**6.43** (WILEY) The streamlines for an incompressible, inviscid, two-dimensional flow field are all concentric circles, and the velocity varies directly with the distance from the common center of the streamlines; that is

$$v_\theta = Kr$$

where $K$ is a constant. **(a)** For this *rotational* flow, determine, if possible, the stream function. **(b)** Can the pressure difference between the origin and any other point be determined from the Bernoulli equation? Explain.

**6.44** (WILEY) The velocity potential

$$\phi = -k(x^2 - y^2) \quad (k = \text{constant})$$

may be used to represent the flow against an infinite plane boundary, as illustrated in Fig. P6.44. For flow in the vicinity of a stagnation point, it is frequently assumed that the pressure gradient along the surface is of the form

$$\frac{\partial p}{\partial x} = Ax$$

where $A$ is a constant. Use the given velocity potential to show that this is true.

**Figure P6.44**

**6.45** In a certain steady, two-dimensional flow field the fluid may be assumed to be ideal and the weight of the fluid (specific weight $= 50$ lb/ft$^3$) is the only body force. The $x$ component of velocity is known to be $u = 6x$, which gives the velocity in ft/s when $x$ is measured in feet, and the $y$ component of velocity is known to be a function of only $y$. The $y$ axis is vertical, and at the origin the velocity is zero. **(a)** Determine the $y$ component of velocity so that the continuity equation is satisfied. **(b)** Can the difference in pressures between the points $x = 1$ ft, $y = 1$ ft and $x = 1$ ft, $y = 4$ ft be determined from the Bernoulli equation? If so, determine the value in lb/ft². If not, explain why not.

**6.46** Water is flowing between wedge-shaped walls into a small opening as shown in Fig. P6.46. The velocity potential with units m²/s for this flow is $\phi = -2 \ln r$ with $r$ in meters. Determine the pressure differential between points $A$ and $B$.

■ **Figure P6.46**

**6.47** A certain flow field is described by the velocity potential

$$\phi = A \ln r + Br \cos \theta$$

where $A$ and $B$ are positive constants. Determine the corresponding stream function and locate any stagnation points in this flow field.

**6.48** The velocity potential for a given two-dimensional flow field is

$$\phi = (\tfrac{5}{3})x^3 - 5xy^2$$

Show that the continuity equation is satisfied and determine the corresponding stream function.

**6.49** It is suggested that the velocity potential for the incompressible, nonviscous, two-dimensional flow along the wall shown in Fig. P6.49 is

$$\phi = r^{4/3} \cos \tfrac{4}{3}\theta$$

Is this a suitable velocity potential for flow along the wall? Explain.

■ **Figure P6.49**

## Section 6.5 Some Basic, Plane Potential Flows

**6.50** As illustrated in Fig. P6.50, a tornado can be approximated by a free vortex of strength $\Gamma$ for $r > R_c$, where $R_c$ is the

■ **Figure P6.50**

radius of the core. Velocity measurements at points $A$ and $B$ indicate that $V_A = 125$ ft/s and $V_B = 60$ ft/s. Determine the distance from point $A$ to the center of the tornado. Why can the free vortex model not be used to approximate the tornado throughout the flow field ($r \geq 0$)?

**6.51** The streamlines in a particular two-dimensional flow field are all concentric circles, as shown in Fig. P6.51. The velocity is given by the equation $v_\theta = \omega r$ where $\omega$ is the angular velocity of the rotating mass of fluid. Determine the circulation around the path $ABCD$.

■ **Figure P6.51**

**6.52** The motion of a liquid in an open tank is that of a combined vortex consisting of a forced vortex for $0 \leq r \leq 2$ ft and a free vortex for $r > 2$ ft. The velocity profile and the corresponding shape of the free surface are shown in Fig. P6.52. The free surface at the center of the tank is a depth $h$ below the free surface at $r = \infty$. Determine the value of $h$. Note that $h = h_{\text{forced}} + h_{\text{free}}$, where $h_{\text{forced}}$ and $h_{\text{free}}$ are the corresponding depths for the forced vortex and the free vortex, respectively. (See Section 2.12.2 for further discussion of the forced vortex.)

■ **Figure P6.52**

**6.53** A source of strength $m$ is located a distance $\ell$ from a vertical solid wall as shown in Fig. P6.53. The velocity potential for this incompressible, irrotational flow is given by

$$\phi = \frac{m}{4\pi}\{\ln[(x - \ell)^2 + y^2] + \ln[(x + \ell)^2 + y^2]\}$$

**(a)** Show that there is no flow through the wall. **(b)** Determine the velocity distribution along the wall. **(c)** Determine the pressure distribution along the wall, assuming $p = p_0$ far from the source. Neglect the effect of the fluid weight on the pressure.

**■ Figure P6.53**

**6.54**  Water flows through a two-dimensional diffuser having a 20° expansion angle as shown in Fig. P6.54. Assume that the flow in the diffuser can be treated as a radial flow emanating from a source at the origin $O$. **(a)** If the velocity at the entrance is 20 m/s, determine an expression for the pressure gradient along the diffuser walls. **(b)** What is the pressure rise between the entrance and exit?

**■ Figure P6.54**

**6.55** ▶ WILEY○ When water discharges from a tank through an opening in its bottom, a vortex may form with a curved surface profile, as shown in Fig. P6.55 and **Video V6.4.** Assume that the velocity distribution in the vortex is the same as that for a free vortex. At the same time the water is being discharged from the tank at point $A$, it is desired to discharge a small quantity of water through the pipe $B$. As the discharge through $A$ is increased, the strength of the vortex, as indicated by its circulation, is increased. Determine the maximum strength that the vortex can have in order that no air is sucked in at $B$. Express your answer in terms of the circulation. Assume that the fluid level in the tank at a large distance from the opening at $A$ remains constant and viscous effects are negligible.

**■ Figure P6.55**

**6.56** WILEY○ Water flows over a flat surface at 4 ft/s, as shown in Fig. P6.56. A pump draws off water through a narrow slit at a volume rate of 0.1 ft³/s per foot length of the slit. Assume that the fluid is incompressible and inviscid and can be represented by the combination of a uniform flow and a sink. Locate the stagnation point on the wall (point $A$) and determine the equation for the stagnation streamline. How far above the surface, $H$, must the fluid be so that it does not get sucked into the slit?

0.1 ft³/s
(per foot of length of slit)

**■ Figure P6.56**

**6.57** WILEY○ Two sources, one of strength $m$ and the other with strength $3m$, are located on the $x$ axis as shown in Fig. P6.57. Determine the location of the stagnation point in the flow produced by these sources.

**■ Figure P6.57**

**6.58** The velocity potential for a spiral vortex flow is given by $\phi = (\Gamma/2\pi)\,\theta - (m/2\pi)\ln r$, where $\Gamma$ and $m$ are constants. Show that the angle, $\alpha$, between the velocity vector and the radial direction is constant throughout the flow field (see Fig. P6.58).

**■ Figure P6.58**

**6.59** ▶ WILEY○ For a free vortex (see **Video V6.4**) determine an expression for the pressure gradient **(a)** along a streamline, and **(b)** normal to a streamline. Assume that the streamline is in a horizontal plane, and express your answer in terms of the circulation.

**6.60** WILEY○ (See Fluids in the News article titled "Some Hurricane Facts," Section 6.5.3.) Consider a category five hurricane that has a maximum wind speed of 160 mph at the eye wall, 10 mi from the center of the hurricane. If the flow in the hurricane outside of the hurricane's eye is approximated as a free vortex, determine the wind speeds at locations 20 mi, 30 mi, and 40 mi from the center of the storm.

## Section 6.6 Superposition of Basic, Plane Potential Flows

**6.61** A plane flow field is developed by the addition of a free vortex and a uniform stream in the positive $x$ direction. If the vortex is located at the origin, determine the pressure variation in this flow field. Neglect the effect of the fluid weight. Express your answer in terms of the uniform velocity, $U$, the strength of the vortex, $\Gamma$, and the pressure, $p_0$, far from the origin.

**6.62** (WILEY) Potential flow against a flat plate (Fig. P6.62a) can be described with the stream function

$$\psi = Axy$$

where $A$ is a constant. This type of flow is commonly called a stagnation point flow since it can be used to describe the flow in the vicinity of the stagnation point at $O$. By adding a source of strength $m$ at $O$, stagnation point flow against a flat plate with a "bump" is obtained as illustrated in Fig. P6.62b. Determine the relationship between the bump height, $h$, the constant, $A$, and the source strength, $m$.

(a)

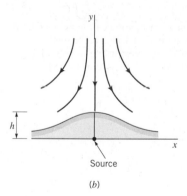

(b)

■ **Figure P6.62**

**6.63** Show on a plot several "bumps" that can be generated by combining stagnation point flow against a flat plate and a source as described in Problem 6.62.

**6.64** Consider a uniform flow in the positive $x$ direction combined with a free vortex located at the origin of the coordinate system. The streamline $\psi = 0$ passes through the point $x = 4$, $y = 0$. Determine the equation of this streamline.

**6.65** ▶ (WILEY) The combination of a uniform flow and a source can be used to describe flow around a streamlined body called a half-body. (See **Video V6.5.**) Assume that a certain body has the shape of a half-body with a thickness of 0.5 m. If this body is placed in an airstream moving at 15 m/s, what source strength is required to simulate flow around the body?

**6.66** (WILEY) A vehicle windshield is to be shaped as a portion of a half-body with the dimensions shown in Fig. P6.66. **(a)** Make a scale drawing of the windshield shape. **(b)** For a free-stream velocity of 55 mph, determine the velocity of the air at points $A$ and $B$.

■ **Figure P6.66**

**6.67** A body having the general shape of a half-body is placed in a stream of fluid. At a great distance upstream the velocity is $U$ as shown in Fig. P6.67. Show how a measurement of the differential pressure between the stagnation point and point $A$ can be used to predict the free-stream velocity, $U$. Express the pressure differential in terms of $U$ and fluid density. Neglect body forces and assume that the fluid is nonviscous and incompressible.

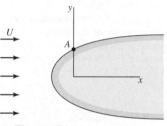

■ **Figure P6.67**

**6.68** ▶ (WILEY) One end of a pond has a shoreline that resembles a half-body as shown in Fig. P6.68. A vertical porous pipe is located near the end of the pond so that water can be pumped out. When water is pumped at the rate of 0.08 m³/s through a 3-m-long pipe, what will be the velocity at point $A$? *Hint*: Consider the flow *inside* a half-body. (See **Video V6.5.**)

■ **Figure P6.68**

**6.69** Consider two sources having equal strengths located along the $x$ axis at $x = 0$ and $x = 2$ m, and a sink located on the $y$ axis at $y = 2$ m. Determine the magnitude and direction of the fluid velocity at $x = 5$ m and $y = 0$ due to this combination if the flowrate from each of the sources is 0.5 m³/s per m and the flowrate into the sink is 1.0 m³/s per m.

**\*6.70** Consider a uniform flow with velocity $U$ in the positive $x$ direction combined with two free vortices of equal strength located along the $y$-axis. Let one vortex located at $y = a$ be a clockwise vortex ($\psi = K \ln r$) and the other at $y = -a$ be a counterclockwise vortex, where $K$ is a positive constant. It can be shown by plotting streamlines that for $Ua/K < 2$ the streamline $\psi = 0$ forms a closed contour, as shown in Fig. P6.70. Thus, this combination can be used to represent flow around a family of bodies (called *Kelvin ovals*). Show, with the aid of a graph, how the dimensionless height, $H/a$, varies with the parameter $Ua/K$ in the range $0.3 < Ua/K < 1.75$.

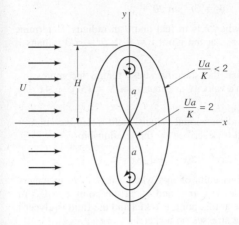

■ **Figure P6.70**

**6.71** (WILEY) A Rankine oval is formed by combining a source–sink pair, each having a strength of 36 ft²/s and separated by a distance of 12 ft along the *x* axis, with a uniform velocity of 10 ft/s (in the positive *x* direction). Determine the length and thickness of the oval.

**\*6.72** Make use of Eqs. 6.107 and 6.109 to construct a table showing how $\ell/a$, $h/a$, and $\ell/h$ for Rankine ovals depend on the parameter $\pi Ua/m$. Plot $\ell/h$ versus $\pi Ua/m$ and describe how this plot could be used to obtain the required values of *m* and *a* for a Rankine oval having a specific value of $\ell$ and $h$ when placed in a uniform fluid stream of velocity, *U*.

**6.73** (WILEY) An ideal fluid flows past an infinitely long, semicircular "hump" located along a plane boundary, as shown in Fig. P6.73. Far from the hump the velocity field is uniform, and the pressure is $p_0$. **(a)** Determine expressions for the maximum and minimum values of the pressure along the hump, and indicate where these points are located. Express your answer in terms of $\rho$, *U*, and $p_0$. **(b)** If the solid surface is the $\psi = 0$ streamline, determine the equation of the streamline passing through the point $\theta = \pi/2$, $r = 2a$.

■ **Figure P6.73**

**6.74** Assume that the flow around the long circular cylinder of Fig. P6.74 is nonviscous and incompressible. Two pressures, $p_1$ and $p_2$, are measured on the surface of the cylinder, as illustrated. It is proposed that the free-stream velocity, *U*, can be related to the pressure difference $\Delta p = p_1 - p_2$ by the equation

$$U = C\sqrt{\frac{\Delta p}{\rho}}$$

where $\rho$ is the fluid density. Determine the value of the constant *C*. Neglect body forces.

■ **Figure P6.74**

**6.75** ▶ (WILEY GO) Water flows around a 6-ft-diameter bridge pier with a velocity of 12 ft/s. Estimate the force (per unit length) that the water exerts on the pier. Assume that the flow can be approximated as an ideal fluid flow around the front half of the cylinder, but due to flow separation (see **Video V6.8**), the average pressure on the rear half is constant and approximately equal to ½ the pressure at point *A* (see Fig. P6.75).

■ **Figure P6.75**

**\*6.76** (WILEY) Consider the steady potential flow around the circular cylinder shown in Fig. 6.26. On a plot show the variation of the magnitude of the dimensionless fluid velocity, $V/U$, along the positive *y* axis. At what distance, $y/a$ (along the *y* axis), is the velocity within 1% of the free-stream velocity?

**6.77** (WILEY) The velocity potential for a cylinder (Fig. P6.77) rotating in a uniform stream of fluid is

$$\phi = Ur\left(1 + \frac{a^2}{r^2}\right)\cos\theta + \frac{\Gamma}{2\pi}\theta$$

where $\Gamma$ is the circulation. For what value of the circulation will the stagnation point be located at: **(a)** point *A*; **(b)** point *B*?

■ **Figure P6.77**

**6.78** Show that for a rotating cylinder in a uniform flow, the following pressure ratio equation is true.

$$\frac{p_{top} - p_{bottom}}{p_{stagnation}} = \frac{8q}{U}$$

Here *U* is the velocity of the uniform flow and *q* is the surface speed of the rotating cylinder.

**6.79** (WILEY) (See Fluids in the News article titled "A Sailing Ship without Sails," Section 6.6.3.) Determine the magnitude of the total force developed by the two rotating cylinders on the Flettner "rotorship" due to the Magnus effect. Assume a wind-speed relative to the ship of **(a)** 10 mph and **(b)** 30 mph. Each cylinder has a diameter of 9 ft, a length of 50 ft, and rotates at 750 rev/min. Use Eq. 6.124 and calculate the circulation by assuming the air sticks to the rotating cylinders. *Note:* This calculated force is at right angles to the direction of the wind and it is the component of this force in the direction of motion of the ship that gives the propulsive thrust. Also, due to viscous effects, the actual propulsive thrust will be smaller than that calculated from Eq. 6.124 which is based on inviscid flow theory.

**6.80** A long porous pipe runs parallel to a horizontal plane surface as shown in Fig. P6.80. The longitudinal axis of the pipe is perpendicular to the plane of the paper. Water flows radially from the pipe at a rate of $0.5\,\pi$ ft³/s per foot of pipe. Determine the difference in pressure (in lb/ft²) between point $B$ and point $A$. The flow from the pipe may be approximated by a two-dimensional source. *Hint*: To develop the stream function or velocity potential for this type of flow, place (symmetrically) another equal source on the other side of the wall. With this combination there is no flow across the $x$ axis, and this axis can be replaced with a solid boundary. This technique is called the *method of images.*

■ **Figure P6.80**

**6.81** ▶ Typical inviscid flow solutions for flow around bodies indicate that the fluid flows smoothly around the body, even for blunt bodies as shown in **Video V6.11.** However, experience reveals that due to the presence of viscosity, the main flow may actually separate from the body, creating a wake behind the body. As discussed in a later section (Section 9.2.6), whether or not separation takes place depends on the pressure gradient along the surface of the body, as calculated by inviscid flow theory. If the pressure decreases in the direction of flow (a *favorable* pressure gradient), no separation will occur. However, if the pressure increases in the direction of flow (an *adverse* pressure gradient), separation may occur. For the circular cylinder of Fig. P6.81 placed in a uniform stream with velocity, $U$, determine an expression for the pressure gradient in the direction of flow on the surface of the cylinder. For what range of values for the angle $\theta$ will an adverse pressure gradient occur?

■ **Figure P6.81**

## Section 6.8 Viscous Flow

**6.82** The velocity field for a two-dimensional source or sink flow is a solution of the Euler equations of motion for inviscid flow.

Show that this flow field is also a solution of the Navier–Stokes equations of motion for viscous flow.

**6.83** The stream function for a certain incompressible, two-dimensional flow field is

$$\psi = 3r^3 \sin 2\theta + 2\theta$$

where $\psi$ is in ft²/s when $r$ is in feet and $\theta$ in radians. Determine the shearing stress, $\tau_{r\theta}$, at the point $r = 2$ ft, $\theta = \pi/3$ radians if the fluid is water.

**6.84** (WILEY) Determine the shearing stress for an incompressible Newtonian fluid with a velocity distribution of $\mathbf{V} = (3xy^2 - 4x^3)\hat{\mathbf{i}} + (12x^2y - y^3)\hat{\mathbf{j}}$.

**6.85** (WILEY) The two-dimensional velocity field for an incompressible Newtonian fluid is described by the relationship

$$\mathbf{V} = (12xy^2 - 6x^3)\hat{\mathbf{i}} + (18x^2y - 4y^3)\hat{\mathbf{j}}$$

where the velocity has units of m/s when $x$ and $y$ are in meters. Determine the stresses $\sigma_{xx}$, $\sigma_{yy}$, and $\tau_{xy}$ at the point $x = 0.5$ m, $y = 1.0$ m if pressure at this point is 6 kPa and the fluid is glycerin at 20 °C. Show these stresses on a sketch.

**6.86** (WILEY) The velocity of a fluid particle moving along a horizontal streamline that coincides with the $x$ axis in a plane, two-dimensional, incompressible flow field was experimentally found to be described by the equation $u = x^2$. Along this streamline determine an expression for **(a)** the rate of change of the $v$ component of velocity with respect to $y$, **(b)** the acceleration of the particle, and **(c)** the pressure gradient in the $x$ direction. The fluid is Newtonian.

## Section 6.9.1 Steady, Laminar Flow between Fixed Parallel Plates

**6.87** (WILEY GO) Oil (SAE 30) at 15.6 °C flows steadily between fixed, horizontal, parallel plates. The pressure drop per unit length along the channel is 30 kPa/m, and the distance between the plates is 4 mm. The flow is laminar. Determine: **(a)** the volume rate of flow (per meter of width), **(b)** the magnitude and direction of the shearing stress acting on the bottom plate, and **(c)** the velocity along the centerline of the channel.

**6.88** (WILEY) Two fixed, horizontal, parallel plates are spaced 0.4 in. apart. A viscous liquid ($\mu = 8 \times 10^{-3}$ lb · s/ft², $SG = 0.9$) flows between the plates with a mean velocity of 0.5 ft/s. The flow is laminar. Determine the pressure drop per unit length in the direction of flow. What is the maximum velocity in the channel?

**6.89** A viscous, incompressible fluid flows between the two infinite, vertical, parallel plates of Fig. P6.89. Determine, by use of the Navier–Stokes equations, an expression for the pressure gradient in the direction of flow. Express your answer in terms of the mean velocity. Assume that the flow is laminar, steady, and uniform.

Direction of flow

■ **Figure P6.89**

**6.90** A fluid of density $\rho$ flows steadily *downward* between the two vertical, infinite, parallel plates shown in the figure for Problem 6.89. The flow is fully developed and laminar. Make use of the Navier–Stokes equation to determine the relationship between the discharge and the other parameters involved, for the case in which the change in pressure along the channel is zero.

**6.91** (WILEY) (See Fluids in the News article titled "10 Tons on 8 psi," Section 6.9.1.) A massive, precisely machined, 6-ft-diameter granite sphere rests on a 4-ft-diameter cylindrical pedestal as shown in Fig. P6.91. When the pump is turned on and the water pressure within the pedestal reaches 8 psi, the sphere rises off the pedestal, creating a 0.005-in. gap through which the water flows. The sphere can then be rotated about any axis with minimal friction. (a) Estimate the pump flowrate, $Q_0$, required to accomplish this. Assume the flow in the gap between the sphere and the pedestal is essentially viscous flow between fixed, parallel plates. (b) Describe what would happen if the pump flowrate were increased to $2Q_0$.

■ **Figure P6.91**

## Section 6.9.2 Couette Flow

**6.92** ▶ (WILEY) Two horizontal, infinite, parallel plates are spaced a distance $b$ apart. A viscous liquid is contained between the plates. The bottom plate is fixed, and the upper plate moves parallel to the bottom plate with a velocity $U$. Because of the no-slip boundary condition (see **Video V6.12**), the liquid motion is caused by the liquid being dragged along by the moving boundary. There is no pressure gradient in the direction of flow. Note that this is a so-called simple *Couette flow* discussed in Section 6.9.2. (a) Start with the Navier–Stokes equations and determine the velocity distribution between the plates. (b) Determine an expression for the flowrate passing between the plates (for a unit width). Express your answer in terms of $b$ and $U$.

**6.93** A layer of viscous liquid of constant thickness (no velocity perpendicular to plate) flows steadily down an infinite, inclined plane. Determine, by means of the Navier–Stokes equations, the relationship between the thickness of the layer and the discharge per unit width. The flow is laminar, and assume air resistance is negligible so that the shearing stress at the free surface is zero.

**6.94** (WILEY) An incompressible, viscous fluid is placed between horizontal, infinite, parallel plates as is shown in Fig. P6.94. The two plates move in opposite directions with constant velocities, $U_1$ and $U_2$, as shown. The pressure gradient in the $x$ direction is zero, and the only body force is due to the fluid weight. Use the Navier–Stokes equations to derive an expression for the velocity distribution between the plates. Assume laminar flow.

■ **Figure P6.94**

**6.95** (WILEY) Two immiscible, incompressible, viscous fluids having the same densities but different viscosities are contained between two infinite, horizontal, parallel plates (Fig. P6.95). The bottom plate is fixed, and the upper plate moves with a constant velocity $U$. Determine the velocity at the interface. Express your answer in terms of $U$, $\mu_1$, and $\mu_2$. The motion of the fluid is caused entirely by the movement of the upper plate; that is, there is no pressure gradient in the $x$ direction. The fluid velocity and shearing stress are continuous across the interface between the two fluids. Assume laminar flow.

■ **Figure P6.95**

**6.96** (WILEY) The viscous, incompressible flow between the parallel plates shown in Fig. P6.96 is caused by both the motion of the bottom plate and a pressure gradient, $\partial p/\partial x$. As noted in Section 6.9.2, an important dimensionless parameter for this type of problem is $P = -(b^2/2\ \mu U)\ (\partial p/\partial x)$ where $\mu$ is the fluid viscosity. Make a plot of the dimensionless velocity distribution (similar to that shown in Fig. 6.32b) for $P = 3$. For this case where does the maximum velocity occur?

■ **Figure P6.96**

**6.97** (WILEY) A viscous fluid (specific weight = 80 lb/ft$^3$; viscosity = 0.03 lb · s/ft$^2$) is contained between two infinite, horizontal parallel plates as shown in Fig. P6.97. The fluid moves between the plates under the action of a pressure gradient, and the upper plate moves with a velocity $U$ while the bottom plate is fixed. A U-tube manometer connected between two points along the bottom indicates a differential reading of 0.1 in. If the upper plate moves with a velocity of 0.02 ft/s, at what distance from the bottom plate does the maximum velocity in the gap between the two plates occur? Assume laminar flow.

■ **Figure P6.97**

**6.98** An infinitely long, solid, vertical cylinder of radius $R$ is located in an infinite mass of an incompressible fluid. Start with the Navier–Stokes equation in the $\theta$ direction and derive an expression for the velocity distribution for the steady flow case in which the cylinder is rotating about a fixed axis with a constant angular velocity $\omega$. You need not consider body forces. Assume that the flow is axisymmetric and the fluid is at rest at infinity.

**6.99** WILEY A vertical shaft passes through a bearing and is lubricated with an oil having a viscosity of 0.2 N · s/m² as shown in Fig. P6.99. Assume that the flow characteristics in the gap between the shaft and bearing are the same as those for laminar flow between infinite parallel plates with zero pressure gradient in the direction of flow. Estimate the torque required to overcome viscous resistance when the shaft is turning at 80 rev/min.

■ **Figure P6.99**

**6.100** ▶ WILEY A viscous fluid is contained between two long concentric cylinders. The geometry of the system is such that the flow between the cylinders is approximately the same as the laminar flow between two infinite parallel plates. **(a)** Determine an expression for the torque required to rotate the outer cylinder with an angular velocity $\omega$. The inner cylinder is fixed. Express your answer in terms of the geometry of the system, the viscosity of the fluid, and the angular velocity. **(b)** For a small, rectangular element located at the fixed wall, determine an expression for the rate of angular deformation of this element. (See **Video V6.3** and Fig. P6.10.)

*\*6.101** Oil (SAE 30) flows between parallel plates spaced 5 mm apart. The bottom plate is fixed, but the upper plate moves with a velocity of 0.2 m/s in the positive $x$ direction. The pressure gradient is 60 kPa/m, and it is negative. Compute the velocity at various points across the channel and show the results on a plot. Assume laminar flow.

### Section 6.9.3 Steady, Laminar Flow in Circular Tubes

**6.102** Ethyl alcohol flows through a horizontal tube having a diameter of 10 mm. If the mean velocity is 0.15 m/s, what is the pressure drop per unit length along the tube? What is the velocity at a distance of 2 mm from the tube axis?

**6.103** WILEY A simple flow system to be used for steady-flow tests consists of a constant head tank connected to a length of 4-mm-diameter tubing as shown in Fig. P6.103. The liquid has a viscosity of 0.015 N · s/m², a density of 1200 kg/m³, and discharges into the atmosphere with a mean velocity of 2 m/s. **(a)** Verify that the flow will be laminar. **(b)** The flow is fully developed in the last 3 m of the tube. What is the pressure at the pressure gage? **(c)** What is the magnitude of the wall shearing stress, $\tau_{rz}$, in the fully developed region?

■ **Figure P6.103**

**6.104** WILEY **(a)** Show that for Poiseuille flow in a tube of radius $R$ the magnitude of the wall shearing stress, $\tau_{rz}$, can be obtained from the relationship

$$|(\tau_{rz})_{\text{wall}}| = \frac{4\mu Q}{\pi R^3}$$

for a Newtonian fluid of viscosity $\mu$. The volume rate of flow is $Q$. **(b)** Determine the magnitude of the wall shearing stress for a fluid having a viscosity of 0.004 N · s/m² flowing with an average velocity of 130 mm/s in a 2-mm-diameter tube.

**6.105** An infinitely long, solid, vertical cylinder of radius $R$ is located in an infinite mass of an incompressible fluid. Start with the Navier–Stokes equation in the $\theta$ direction and derive an expression for the velocity distribution for the steady-flow case in which the cylinder is rotating about a fixed axis with a constant angular velocity $\omega$. You need not consider body forces. Assume that the flow is axisymmetric and the fluid is at rest at infinity.

*\*6.106** As is shown by Eq. 6.150, the pressure gradient for laminar flow through a tube of constant radius is given by the expression

$$\frac{\partial p}{\partial z} = -\frac{8\mu Q}{\pi R^4}$$

For a tube whose radius is changing very gradually, such as the one illustrated in Fig. P6.106, it is expected that this equation can be used to approximate the pressure change along the tube if the actual radius, $R(z)$, is used at each cross section. The following measurements were obtained along a particular tube.

| $z/\ell$ | 0 | 0.1 | 0.2 | 0.3 | 0.4 | 0.5 | 0.6 | 0.7 | 0.8 | 0.9 | 1.0 |
|---|---|---|---|---|---|---|---|---|---|---|---|
| $R(z)/R_o$ | 1.00 | 0.73 | 0.67 | 0.65 | 0.67 | 0.80 | 0.80 | 0.71 | 0.73 | 0.77 | 1.00 |

Compare the pressure drop over the length $\ell$ for this nonuniform tube with one having the constant radius $R_o$. *Hint:* To solve this problem you will need to numerically integrate the equation for the pressure gradient given previously.

■ **Figure P6.106**

**6.107** (WILEY GO) A liquid (viscosity = 0.002 N · s/m²; density = 1000 kg/m³) is forced through the circular tube shown in Fig. P6.107. A differential manometer is connected to the tube as shown to measure the pressure drop along the tube. When the differential reading, $\Delta h$, is 9 mm, what is the mean velocity in the tube?

**■ Figure P6.107**

## Section 6.9.4 Steady, Axial, Laminar Flow in an Annulus

**6.108** (WILEY) An incompressible Newtonian fluid flows steadily between two infinitely long, concentric cylinders as shown in Fig. P6.108. The outer cylinder is fixed, but the inner cylinder moves with a longitudinal velocity $V_0$ as shown. The pressure gradient in the axial direction is $-\Delta p/\ell$. For what value of $V_0$ will the drag on the inner cylinder be zero? Assume that the flow is laminar, axisymmetric, and fully developed.

**■ Figure P6.108**

**6.109** (WILEY) A viscous fluid is contained between two infinitely long, vertical, concentric cylinders. The outer cylinder has a radius $r_o$ and rotates with an angular velocity $\omega$. The inner cylinder is fixed and has a radius $r_i$. Make use of the Navier–Stokes equations to obtain an exact solution for the velocity distribution in the gap. Assume that the flow in the gap is axisymmetric (neither velocity nor pressure are functions of angular position $\theta$ within the gap) and

that there are no velocity components other than the tangential component. The only body force is the weight.

**6.110** For flow between concentric cylinders, with the outer cylinder rotating at an angular velocity $\omega$ and the inner cylinder fixed, it is commonly assumed that the tangential velocity ($v_\theta$) distribution in the gap between the cylinders is linear. Based on the exact solution to this problem (see Problem 6.109) the velocity distribution in the gap is not linear. For an outer cylinder with radius $r_o = 2.00$ in. and an inner cylinder with radius $r_i = 1.80$ in., show, with the aid of a plot, how the dimensionless velocity distribution, $v_\theta/r_o\omega$, varies with the dimensionless radial position, $r/r_o$, for the exact and approximate solutions.

**6.111** (WILEY) A viscous liquid ($\mu = 0.012$ lb · s/ft², $\rho = 1.79$ slugs/ft³) flows through the annular space between two horizontal, fixed, concentric cylinders. If the radius of the inner cylinder is 1.5 in. and the radius of the outer cylinder is 2.5 in., what is the pressure drop along the axis of the annulus per foot when the volume flowrate is 0.14 ft³/s?

**6.112** (WILEY) Show how Eq. 6.155 is obtained.

**6.113** (WILEY) A wire of diameter $d$ is stretched along the centerline of a pipe of diameter $D$. For a given pressure drop per unit length of pipe, by how much does the presence of the wire reduce the flowrate if **(a)** $d/D = 0.1$; **(b)** $d/D = 0.01$?

## Section 6.10 Other Aspects of Differential Analysis

**6.114** Obtain a photograph/image of a situation in which CFD has been used to solve a fluid flow problem. Print this photo and write a brief paragraph that describes the situation involved.

### ■ Lifelong Learning Problems

**6.1LL** What sometimes appear at first glance to be simple fluid flows can contain subtle, complex fluid mechanics. One such example is the stirring of tea leaves in a teacup. Obtain information about "Einstein's tea leaves" and investigate some of the complex fluid motions interacting with the leaves. Summarize your findings in a brief report.

**6.2LL** Computational fluid dynamics (CFD) has moved from a research tool to a design tool for engineering. Initially, much of the work in CFD was focused in the aerospace industry, but now has expanded into other areas. Obtain information on what other industries (e.g., automotive) make use of CFD in their engineering design. Summarize your findings in a brief report.

### ■ FE Exam Problems

Sample FE (Fundamentals of Engineering) exam questions for fluid mechanics are provided in *WileyPLUS* or on the book's web site, www.wiley.com/college/munson.

# 7 Dimensional Analysis, Similitude, and Modeling

CHAPTER OPENING PHOTO: Flow past a circular cylinder with Re $= \rho VD/\mu = 2000$: The pathlines of flow past any circular cylinder (regardless of size, velocity, or fluid) are as shown provided that the dimensionless parameter called the Reynolds number, Re, is equal to 2000. For other values of Re, the flow pattern will be different (air bubbles in water). (Photograph courtesy of ONERA, The French Aerospace Lab.)

## Learning Objectives

After completing this chapter, you should be able to:

■ apply the Buckingham pi theorem.

■ develop a set of dimensionless variables for a given flow situation.

■ discuss the use of dimensionless variables in data analysis.

■ apply the concepts of modeling and similitude to develop prediction equations.

*Experimentation and modeling are widely used techniques in fluid mechanics.*

*V7.1 Real and model flies*

Although many practical engineering problems involving fluid mechanics can be solved by using the equations and analytical procedures described in the preceding chapters, there remain a large number of problems that rely on experimentally obtained data for their solution. In fact, it is probably fair to say that very few problems involving real fluids can be solved by analysis alone. The solution to many problems is achieved through the use of a combination of theoretical and numerical analysis and experimental data. Thus, engineers working on fluid mechanics problems should be familiar with the experimental approach to these problems so that they can interpret and make use of data obtained by others, such as might appear in handbooks, or be able to plan and execute the necessary experiments in their own laboratories. In this chapter we consider some techniques and ideas that are important in the planning and execution of experiments, as well as in understanding and correlating data that may have been obtained by other experimenters.

An obvious goal of any experiment is to make the results as widely applicable as possible. To achieve this end, the concept of *similitude* is often used so that measurements made on one system (for example, in the laboratory) can be used to describe the behavior of other similar systems (outside the laboratory). The laboratory systems are usually thought of as *models* and are used to study the phenomenon of interest under carefully controlled conditions. From these model studies, empirical formulations can be developed, or specific predictions of one or more characteristics of some other similar system can be made. To do this, it is necessary to establish the relationship between the laboratory model and the "other" system. In the following sections, we find out how this can be accomplished in a systematic manner.

## 7.1 Dimensional Analysis

To illustrate a typical fluid mechanics problem in which experimentation is required, consider the steady flow of an incompressible Newtonian fluid through a long, smooth-walled, horizontal, circular pipe. An important characteristic of this system, which would be of interest to an engineer designing a pipeline, is the pressure drop per unit length that develops along the pipe as a result of friction. Although this would appear to be a relatively simple flow problem, it cannot generally be solved analytically (even with the aid of large computers) without the use of experimental data.

The first step in the planning of an experiment to study this problem would be to decide on the factors, or variables, that will have an effect on the pressure drop per unit length, $\Delta p_\ell$ [(lb/ft$^2$)/ft = lb/ft$^3$ or N/m$^3$]. We expect the list to include the pipe diameter, $D$, the fluid density, $\rho$, fluid viscosity, $\mu$, and the mean velocity, $V$, at which the fluid is flowing through the pipe. Thus, we can express this relationship as

$$\Delta p_\ell = f(D, \rho, \mu, V) \tag{7.1}$$

which simply indicates mathematically that we expect the pressure drop per unit length to be some function of the factors contained within the parentheses. At this point the nature of the function is unknown, and the objective of the experiments to be performed is to determine the nature of this function.

*It is important to develop a meaningful and systematic way to perform an experiment.*

To perform the experiments in a meaningful and systematic manner, it would be necessary to change one of the variables, such as the velocity, while holding all others constant, and measure the corresponding pressure drop. This series of tests would yield data that could be represented graphically as is illustrated in Fig. 7.1a. It is to be noted that this plot would only be valid for the specific pipe and for the specific fluid used in the tests; this certainly does not give us the general formulation we are looking for. We could repeat the process by varying each of the other variables in turn, as is illustrated in Figs. 7.1b, 7.1c, and 7.1d. This approach to determining the functional relationship between the pressure drop and the various factors that influence it, although logical in concept, is fraught with difficulties. Some of the experiments would be hard to carry out—for example, to obtain the data illustrated in Fig. 7.1c it would be necessary to vary fluid density while holding viscosity constant. How would you do this? Finally, once we obtained the various curves shown in Figs. 7.1a, 7.1b, 7.1c, and 7.1d, how could we combine these data to obtain the desired general functional relationship between $\Delta p_\ell$, $D$, $\rho$, $\mu$, and $V$ which would be valid for any similar pipe system?

Fortunately, there is a much simpler approach to this problem that will eliminate such difficulties. In the following sections we will show that rather than working with the original list of variables, as described in Eq. 7.1, we can collect these into two nondimensional combinations of variables (called **dimensionless products** or *dimensionless groups*) so that

$$\frac{D\,\Delta p_\ell}{\rho V^2} = \phi\left(\frac{\rho V D}{\mu}\right) \tag{7.2}$$

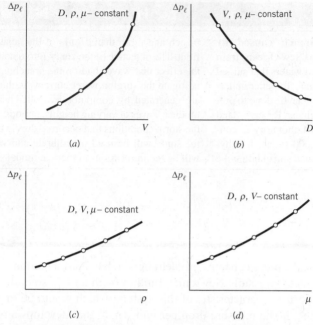

■ **Figure 7.1** Illustrative plots showing how the pressure drop in a pipe may be affected by several different factors.

*Dimensionless products are important and useful in the planning, execution, and interpretation of experiments.*

Thus, instead of having to work with five variables, we now have only two. The necessary experiment would simply consist of varying the dimensionless product $\rho VD/\mu$ and determining the corresponding value of $D\,\Delta p_\ell/\rho V^2$. The results of the experiment could then be represented by a single, universal curve as is illustrated in Fig. 7.2. This curve would be valid for any combination of smooth-walled pipe and incompressible Newtonian fluid. To obtain this curve we could choose a pipe of convenient size and a fluid that is easy to work with. Note that we wouldn't have to use different pipe sizes or even different fluids. It is clear that the experiment would be much simpler, easier to do, and less expensive (which would certainly make an impression on your boss).

The basis for this simplification lies in a consideration of the dimensions of the variables involved. As was discussed in Chapter 1, a qualitative description of physical quantities can be given in terms of **basic dimensions** such as mass, $M$, length, $L$, and time, $T$.[1] Alternatively, we could use force, $F$, $L$, and $T$ as basic dimensions, since from Newton's second law

$$F \doteq MLT^{-2}$$

■ **Figure 7.2** An illustrative plot of pressure drop data using dimensionless parameters.

---

[1]As noted in Chapter 1, we will use $T$ to represent the basic dimension of time, although $T$ is also used for temperature in thermodynamic relationships (such as the ideal gas law).

(Recall from Chapter 1 that the notation $\doteq$ is used to indicate dimensional equality.) The dimensions of the variables in the pipe flow example are $\Delta p_\ell \doteq FL^{-3}$, $D \doteq L$, $\rho \doteq FL^{-4}T^2$, $\mu \doteq FL^{-2}T$, and $V \doteq LT^{-1}$. [Note that the pressure drop per unit length has the dimensions of $(F/L^2)/L = FL^{-3}$.] A quick check of the dimensions of the two groups that appear in Eq. 7.2 shows that they are in fact *dimensionless* products; that is,

$$\frac{D\,\Delta p_\ell}{\rho V^2} \doteq \frac{L(F/L^3)}{(FL^{-4}T^2)(LT^{-1})^2} \doteq F^0L^0T^0$$

and

$$\frac{\rho VD}{\mu} \doteq \frac{(FL^{-4}T^2)(LT^{-1})(L)}{(FL^{-2}T)} \doteq F^0L^0T^0$$

Not only have we reduced the number of variables from five to two, but also the new groups are dimensionless combinations of variables, which means that the results presented in the form of Fig. 7.2 will be independent of the system of units we choose to use. This type of analysis is called *dimensional analysis*, and the basis for its application to a wide variety of problems is found in the *Buckingham pi theorem* described in the following section.

## 7.2  Buckingham Pi Theorem

A fundamental question we must answer is how many dimensionless products are required to replace the original list of variables? The answer to this question is supplied by the basic theorem of dimensional analysis that states the following:

> **If an equation involving $k$ variables is dimensionally homogeneous, it can be reduced to a relationship among $k - r$ independent dimensionless products, where $r$ is the minimum number of reference dimensions required to describe the variables.**

*Dimensional analysis is based on the Buckingham pi theorem.*

The dimensionless products are frequently referred to as *pi terms,* and the theorem is called the *Buckingham pi theorem.*[2] Edgar Buckingham used the symbol $\Pi$ to represent a dimensionless product, and this notation is commonly used. Although the pi theorem is a simple one, its proof is not so simple and we will not include it here. Many entire books have been devoted to the subject of similitude and dimensional analysis, and a number of these are listed at the end of this chapter (Refs. 1–15). Students interested in pursuing the subject in more depth (including the proof of the pi theorem) can refer to one of these books.

The pi theorem is based on the idea of dimensional homogeneity which was introduced in Chapter 1. Essentially we assume that for any physically meaningful equation involving $k$ variables, such as

$$u_1 = f(u_2, u_3, \ldots, u_k)$$

the dimensions of the variable on the left side of the equal sign must be equal to the dimensions of any term that stands by itself on the right side of the equal sign. It then follows that we can rearrange the equation into a set of dimensionless products (pi terms) so that

$$\Pi_1 = \phi(\Pi_2, \Pi_3, \ldots, \Pi_{k-r})$$

where $\phi(\Pi_2, \Pi_3, \ldots, \Pi_{k-r})$ is a function of $\Pi_2$ through $\Pi_{k-r}$.

The required number of pi terms is fewer than the number of original variables by $r$, where $r$ is determined by the minimum number of reference dimensions required to describe the original list of variables. Usually the reference dimensions required to describe the variables will be the basic dimensions $M$, $L$, and $T$ or $F$, $L$, and $T$. However, in some instances perhaps only two dimensions, such as $L$ and $T$, are required, or maybe just one, such as $L$. Also, in a few rare cases

---

[2]Although several early investigators, including Lord Rayleigh (1842–1919) in the nineteenth century, contributed to the development of dimensional analysis, Edgar Buckingham's (1867–1940) name is usually associated with the basic theorem. He stimulated interest in the subject in the United States through his publications during the early part of the twentieth century. See, for example, E. Buckingham, On Physically Similar Systems: Illustrations of the Use of Dimensional Equations, *Phys. Rev.*, 4 (1914), 345–376.

the variables may be described by some combination of basic dimensions, such as $M/T^2$ and $L$, and in this case $r$ would be equal to two rather than three. Although the use of the pi theorem may appear to be a little mysterious and complicated, we will actually develop a simple, systematic procedure for developing the pi terms for a given problem.

## 7.3    Determination of Pi Terms

Several methods can be used to form the dimensionless products, or pi terms, that arise in a dimensional analysis. Essentially we are looking for a method that will allow us to systematically form the pi terms so that we are sure that they are dimensionless and independent and that we have the right number. The method we will describe in detail in this section is called the ***method of repeating variables.***

It will be helpful to break the repeating variable method down into a series of distinct steps that can be followed for any given problem. With a little practice you will be able to readily complete a dimensional analysis for your problem.

*A dimensional analysis can be performed using a series of distinct steps.*

**Step 1**    **List all the variables that are involved in the problem.** This step is the most difficult one, and it is, of course, vitally important that all pertinent variables be included. Otherwise the dimensional analysis will not be correct! We are using the term *variable* to include any quantity, including dimensional and nondimensional constants, which play a role in the phenomenon under investigation. All such quantities should be included in the list of variables to be considered for the dimensional analysis. The determination of the variables must be accomplished by the experimenter's knowledge of the problem and the physical laws that govern the phenomenon. Typically the variables will include those that are necessary to describe the *geometry* of the system (such as a pipe diameter), to define any *fluid properties* (such as a fluid viscosity), and to indicate *external effects* that influence the system (such as a driving pressure drop per unit length). These general classes of variables are intended as broad categories that should be helpful in identifying variables. It is likely, however, that there will be variables that do not fit easily into one of these categories, and each problem needs to be carefully analyzed.

Since we wish to keep the number of variables to a minimum, so that we can minimize the amount of laboratory work, it is important that all variables be independent. For example, if in a certain problem the cross-sectional area of a pipe is an important variable, either the area or the pipe diameter could be used, but not both, since they are obviously not independent. Similarly, if both fluid density, $\rho$, and specific weight, $\gamma$, are important variables, we could list $\rho$ and $\gamma$, or $\rho$ and $g$ (acceleration of gravity), or $\gamma$ and $g$. However, it would be incorrect to use all three since $\gamma = \rho g$; that is, $\rho$, $\gamma$, and $g$ are not independent. Note that although $g$ would normally be constant in a given experiment, that fact is irrelevant as far as a dimensional analysis is concerned.

**Step 2**    **Express each of the variables in terms of basic dimensions.** For the typical fluid mechanics problem the basic dimensions will be either $M$, $L$, and $T$ or $F$, $L$, and $T$. Dimensionally these two sets are related through Newton's second law ($\mathbf{F} = m\mathbf{a}$) so that $F \doteq MLT^{-2}$. For example, $\rho \doteq ML^{-3}$ or $\rho \doteq FL^{-4}T^2$. Thus, either set can be used. The basic dimensions for typical variables found in fluid mechanics problems are listed in Table 1.1 in Chapter 1.

**Step 3**    **Determine the required number of pi terms.** This can be accomplished by means of the Buckingham pi theorem, which indicates that the number of pi terms is equal to $k - r$, where $k$ is the number of variables in the problem (which is determined from Step 1) and $r$ is the number of reference dimensions required to describe these variables (which is determined from Step 2). The reference dimensions usually correspond to the basic dimensions and can be determined by an inspection of the dimensions of the variables obtained in Step 2. As previously noted, there may be occasions (usually rare) in which the basic dimensions appear in combinations so that the number of reference dimensions is less than the number of basic dimensions. This possibility is illustrated in Example 7.2.

**Step 4**    **Select a number of repeating variables, where the number required is equal to the number of reference dimensions.** Essentially what we are doing here is selecting from the original list of variables several of which can be combined with each of the remaining variables to form a pi term. All of the required reference dimensions must be included within the group of repeating variables, and each repeating variable must be dimensionally independent of the others (i.e., the dimensions of one repeating variable cannot be reproduced by some combination of products of powers of the remaining repeating variables). This means that the repeating variables cannot themselves be combined to form a dimensionless product.

For any given problem we usually are interested in determining how one particular variable is influenced by the other variables. We would consider this variable to be the dependent variable, and we would want this to appear in only one pi term. Thus, do *not* choose the dependent variable as one of the repeating variables, since the repeating variables will generally appear in more than one pi term.

**Step 5** **Form a pi term by multiplying one of the nonrepeating variables by the product of the repeating variables, each raised to an exponent that will make the combination dimensionless.** Essentially each pi term will be of the form $u_i u_1^{a_i} u_2^{b_i} u_3^{c_i}$ where $u_i$ is one of the nonrepeating variables; $u_1$, $u_2$, and $u_3$ are the repeating variables; and the exponents $a_i$, $b_i$, and $c_i$ are determined so that the combination is dimensionless.

**Step 6** **Repeat Step 5 for each of the remaining nonrepeating variables.** The resulting set of pi terms will correspond to the required number obtained from Step 3. If not, check your work—you have made a mistake!

**Step 7** **Check all the resulting pi terms to make sure they are dimensionless.** It is easy to make a mistake in forming the pi terms. However, this can be checked by simply substituting the dimensions of the variables into the pi terms to confirm that they are all dimensionless. One good way to do this is to express the variables in terms of $M$, $L$, and $T$ if the basic dimensions $F$, $L$, and $T$ were used initially, or vice versa, and then check to make sure the pi terms are dimensionless.

**Step 8** **Express the final form as a relationship among the pi terms, and think about what it means.** Typically the final form can be written as

$$\Pi_1 = \phi(\Pi_2, \Pi_3, \ldots, \Pi_{k-r})$$

where $\Pi_1$ would contain the dependent variable in the numerator. It should be emphasized that if you started out with the correct list of variables (and the other steps were completed correctly), then the relationship in terms of the pi terms can be used to describe the problem. You need only work with the pi terms—not with the individual variables. However, it should be clearly noted that this is as far as we can go with the dimensional analysis; that is, the actual functional relationship among the pi terms must be determined by experiment.

*By using dimensional analysis, the original problem is simplified and defined with pi terms.*

To illustrate these various steps we will again consider the problem discussed earlier in this chapter which was concerned with the steady flow of an incompressible Newtonian fluid through a long, smooth-walled, horizontal circular pipe. We are interested in the pressure drop per unit length, $\Delta p_\ell$, along the pipe as illustrated by the figure in the margin. First (Step 1) we must list all of the pertinent variables that are involved based on the experimenter's knowledge of the problem. In this problem we assume that

$$\Delta p_\ell = f(D, \rho, \mu, V)$$

where $D$ is the pipe diameter, $\rho$ and $\mu$ are the fluid density and viscosity, respectively, and $V$ is the mean velocity.

Next (Step 2) we express all the variables in terms of basic dimensions. Using $F$, $L$, and $T$ as basic dimensions it follows that

$$\Delta p_\ell \doteq FL^{-3}$$

$$D \doteq L$$

$$\rho \doteq FL^{-4}T^2$$

$$\mu \doteq FL^{-2}T$$

$$V \doteq LT^{-1}$$

(1)　　　(2)

$\rho, \mu$

$V$

$D$

$\ell$

$\Delta p_\ell = (p_1 - p_2)/\ell$

We could also use $M$, $L$, and $T$ as basic dimensions if desired—the final result will be the same. Note that for density, which is a mass per unit volume ($ML^{-3}$), we have used the relationship $F \doteq MLT^{-2}$ to express the density in terms of $F$, $L$, and $T$. Do not mix the basic dimensions; that is, use either $F$, $L$, and $T$ or $M$, $L$, and $T$.

We can now apply the pi theorem to determine the required number of pi terms (Step 3). An inspection of the dimensions of the variables from Step 2 reveals that all three basic dimensions are required to describe the variables. Since there are five ($k = 5$) variables (do not forget to count the dependent variable, $\Delta p_\ell$) and three required reference dimensions ($r = 3$), then according to the pi theorem there will be ($5 - 3$), or two pi terms required.

The repeating variables to be used to form the pi terms (Step 4) need to be selected from the list $D$, $\rho$, $\mu$, and $V$. Remember, we do not want to use the dependent variable as one of the repeating variables. Since three reference dimensions are required, we will need to select three repeating variables. Generally, we would try to select as repeating variables those that are the simplest, dimensionally. For example, if one of the variables has the dimension of a length, choose it as one of the repeating variables. In this example we will use $D$, $V$, and $\rho$ as repeating variables. Note that these are dimensionally independent, since $D$ is a length, $V$ involves both length and time, and $\rho$ involves force, length, and time. This means that we cannot form a dimensionless product from this set.

*Special attention should be given to the selection of repeating variables as detailed in Step 4.*

We are now ready to form the two pi terms (Step 5). Typically, we would start with the dependent variable and combine it with the repeating variables to form the first pi term; that is,

$$\Pi_1 = \Delta p_\ell D^a V^b \rho^c$$

Since this combination is to be dimensionless, it follows that

$$(FL^{-3})(L)^a(LT^{-1})^b(FL^{-4}T^2)^c \doteq F^0L^0T^0$$

The exponents, $a$, $b$, and $c$ must be determined such that the resulting exponent for each of the basic dimensions—$F$, $L$, and $T$—must be zero (so that the resulting combination is dimensionless). Thus, we can write

$$1 + c = 0 \qquad \text{(for } F)$$
$$-3 + a + b - 4c = 0 \qquad \text{(for } L)$$
$$-b + 2c = 0 \qquad \text{(for } T)$$

The solution of this system of algebraic equations gives the desired values for $a$, $b$, and $c$. It follows that $a = 1$, $b = -2$, $c = -1$ and, therefore,

$$\Pi_1 = \frac{\Delta p_\ell D}{\rho V^2}$$

The process is now repeated for the remaining nonrepeating variables (Step 6). In this example there is only one additional variable ($\mu$) so that

$$\Pi_2 = \mu D^a V^b \rho^c$$

or

$$(FL^{-2}T)(L)^a(LT^{-1})^b(FL^{-4}T^2)^c \doteq F^0L^0T^0$$

and, therefore,

$$1 + c = 0 \qquad \text{(for } F)$$
$$-2 + a + b - 4c = 0 \qquad \text{(for } L)$$
$$1 - b + 2c = 0 \qquad \text{(for } T)$$

Solving these equations simultaneously it follows that $a = -1$, $b = -1$, $c = -1$ so that

$$\Pi_2 = \frac{\mu}{DV\rho}$$

Note that we end up with the correct number of pi terms as determined from Step 3.

At this point stop and check to make sure the pi terms are actually dimensionless (Step 7). We will check using both *FLT* and *MLT* dimensions. Thus,

$$\Pi_1 = \frac{\Delta p_\ell D}{\rho V^2} \doteq \frac{(FL^{-3})(L)}{(FL^{-4}T^2)(LT^{-1})^2} \doteq F^0 L^0 T^0$$

$$\Pi_2 = \frac{\mu}{DV\rho} \doteq \frac{(FL^{-2}T)}{(L)(LT^{-1})(FL^{-4}T^2)} \doteq F^0 L^0 T^0$$

or alternatively,

$$\Pi_1 = \frac{\Delta p_\ell D}{\rho V^2} \doteq \frac{(ML^{-2}T^{-2})(L)}{(ML^{-3})(LT^{-1})^2} \doteq M^0 L^0 T^0$$

$$\Pi_2 = \frac{\mu}{DV\rho} \doteq \frac{(ML^{-1}T^{-1})}{(L)(LT^{-1})(ML^{-3})} \doteq M^0 L^0 T^0$$

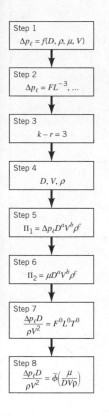

Finally (Step 8), we can express the result of the dimensional analysis as

$$\frac{\Delta p_\ell D}{\rho V^2} = \tilde{\phi}\left(\frac{\mu}{DV\rho}\right)$$

This result indicates that this problem can be studied in terms of these two pi terms, rather than the original five variables we started with. The eight steps carried out to obtain this result are summarized by the figure in the margin.

Dimensional analysis will *not* provide the form of the function $\tilde{\phi}$. This can only be obtained from a suitable set of experiments. If desired, the pi terms can be rearranged; that is, the reciprocal of $\mu/DV\rho$ could be used and, of course, the order in which we write the variables can be changed. Thus, for example, $\Pi_2$ could be expressed as

$$\Pi_2 = \frac{\rho VD}{\mu}$$

and the relationship between $\Pi_1$ and $\Pi_2$ as

$$\frac{D\,\Delta p_\ell}{\rho V^2} = \phi\left(\frac{\rho VD}{\mu}\right)$$

as shown by the figure in the margin.

This is the form we previously used in our initial discussion of this problem (Eq. 7.2). The dimensionless product $\rho VD/\mu$ is a very famous one in fluid mechanics—the Reynolds number. This number has been briefly alluded to in Chapters 1 and 6 and will be further discussed in Section 7.6.

To summarize, the steps to be followed in performing a dimensional analysis using the method of repeating variables are as follows:

**Step 1** List all the variables that are involved in the problem.

**Step 2** Express each of the variables in terms of basic dimensions.

**Step 3** Determine the required number of pi terms.

**Step 4** Select a number of repeating variables, where the number required is equal to the number of reference dimensions (usually the same as the number of basic dimensions).

**Step 5** Form a pi term by multiplying one of the nonrepeating variables by the product of repeating variables each raised to an exponent that will make the combination dimensionless.

**Step 6** Repeat Step 5 for each of the remaining nonrepeating variables.

**Step 7** Check all the resulting pi terms to make sure they are dimensionless and independent.

**Step 8** Express the final form as a relationship among the pi terms and think about what it means.

*The method of repeating variables can be most easily carried out by following a step-by-step procedure.*

# EXAMPLE 7.1 Method of Repeating Variables

**GIVEN** A thin rectangular plate having a width $w$ and a height $h$ is located so that it is normal to a moving stream of fluid as shown in Fig. E7.1. Assume the drag, $\mathcal{D}$, that the fluid exerts on the plate is a function of $w$ and $h$, the fluid viscosity and density, $\mu$ and $\rho$, respectively, and the velocity $V$ of the fluid approaching the plate.

**FIND** Determine a suitable set of pi terms to study this problem experimentally.

*V7.2 Flow past a flat plate*

## SOLUTION

From the statement of the problem we can write

$$\mathcal{D} = f(w, h, \mu, \rho, V)$$

where this equation expresses the general functional relationship between the drag and the several variables that will affect it. The dimensions of the variables (using the *MLT* system) are

$$\mathcal{D} \doteq MLT^{-2}$$
$$w \doteq L$$
$$h \doteq L$$
$$\mu \doteq ML^{-1}T^{-1}$$
$$\rho \doteq ML^{-3}$$
$$V \doteq LT^{-1}$$

■ **Figure E7.1**

We see that all three basic dimensions are required to define the six variables so that the Buckingham pi theorem tells us that three pi terms will be needed (six variables minus three reference dimensions, $k - r = 6 - 3$).

We will next select three repeating variables such as $w$, $V$, and $\rho$. A quick inspection of these three reveals that they are dimensionally independent, since each one contains a basic dimension not included in the others. Note that it would be incorrect to use both $w$ and $h$ as repeating variables since they have the same dimensions.

Starting with the dependent variable, $\mathcal{D}$, the first pi term can be formed by combining $\mathcal{D}$ with the repeating variables such that

$$\Pi_1 = \mathcal{D} w^a V^b \rho^c$$

and in terms of dimensions

$$(MLT^{-2})(L)^a(LT^{-1})^b(ML^{-3})^c \doteq M^0L^0T^0$$

Thus, for $\Pi_1$ to be dimensionless it follows that

$$1 + c = 0 \quad \text{(for } M\text{)}$$
$$1 + a + b - 3c = 0 \quad \text{(for } L\text{)}$$
$$-2 - b = 0 \quad \text{(for } T\text{)}$$

and, therefore, $a = -2$, $b = -2$, and $c = -1$. The pi term then becomes

$$\Pi_1 = \frac{\mathcal{D}}{w^2 V^2 \rho}$$

Next the procedure is repeated with the second nonrepeating variable, $h$, so that

$$\Pi_2 = h w^a V^b \rho^c$$

It follows that

$$(L)(L)^a(LT^{-1})^b(ML^{-3})^c \doteq M^0L^0T^0$$

and

$$c = 0 \quad \text{(for } M\text{)}$$
$$1 + a + b - 3c = 0 \quad \text{(for } L\text{)}$$
$$b = 0 \quad \text{(for } T\text{)}$$

so that $a = -1$, $b = 0$, $c = 0$, and therefore

$$\Pi_2 = \frac{h}{w}$$

The remaining nonrepeating variable is $\mu$ so that

$$\Pi_3 = \mu w^a V^b \rho^c$$

with

$$(ML^{-1}T^{-1})(L)^a(LT^{-1})^b(ML^{-3})^c \doteq M^0L^0T^0$$

and, therefore,

$$1 + c = 0 \quad \text{(for } M\text{)}$$
$$-1 + a + b - 3c = 0 \quad \text{(for } L\text{)}$$
$$-1 - b = 0 \quad \text{(for } T\text{)}$$

Solving for the exponents, we obtain $a = -1$, $b = -1$, $c = -1$ so that

$$\Pi_3 = \frac{\mu}{wV\rho}$$

Now that we have the three required pi terms we should check to make sure they are dimensionless. To make this check we use $F$, $L$, and $T$, which will also verify the correctness of the original dimensions used for the variables. Thus,

$$\Pi_1 = \frac{\mathcal{D}}{w^2 V^2 \rho} \doteq \frac{(F)}{(L)^2(LT^{-1})^2(FL^{-4}T^2)} \doteq F^0L^0T^0$$

$$\Pi_2 = \frac{h}{w} \doteq \frac{(L)}{(L)} \doteq F^0L^0T^0$$

$$\Pi_3 = \frac{\mu}{wV\rho} \doteq \frac{(FL^{-2}T)}{(L)(LT^{-1})(FL^{-4}T^2)} \doteq F^0L^0T^0$$

If these do not check, go back to the original list of variables and make sure you have the correct dimensions for each of the variables and then check the algebra you used to obtain the exponents $a$, $b$, and $c$.

Finally, we can express the results of the dimensional analysis in the form

$$\frac{\mathscr{D}}{w^2 V^2 \rho} = \tilde{\phi}\left(\frac{h}{w}, \frac{\mu}{wV\rho}\right) \qquad \textbf{(Ans)}$$

Since at this stage in the analysis the nature of the function $\tilde{\phi}$ is unknown, we could rearrange the pi terms if we so desire. For example, we could express the final result in the form

$$\frac{\mathscr{D}}{w^2 \rho V^2} = \phi\left(\frac{w}{h}, \frac{\rho V w}{\mu}\right) \qquad \textbf{(Ans)}$$

which would be more conventional, since the ratio of the plate width to height, $w/h$, is called the *aspect ratio*, and $\rho V w/\mu$ is the Reynolds number.

**COMMENT**   To proceed, it would be necessary to perform a set of experiments to determine the nature of the function $\phi$, as discussed in Section 7.7.

# 7.4    Some Additional Comments about Dimensional Analysis

The preceding section provides a systematic approach for performing a dimensional analysis. Other methods could be used, although we think the method of repeating variables is the easiest for the beginning student to use. Pi terms can also be formed by inspection, as is discussed in Section 7.5. Regardless of the specific method used for the dimensional analysis, there are certain aspects of this important engineering tool that must seem a little baffling and mysterious to the student (and sometimes to the experienced investigator as well). In this section we will attempt to elaborate on some of the more subtle points that, based on our experience, can prove to be puzzling to students.

## 7.4.1  Selection of Variables

One of the most important and difficult steps in applying dimensional analysis to any given problem is the selection of the variables that are involved. As noted previously, for convenience we will use the term variable to indicate any quantity involved, including dimensional and nondimensional constants. There is no simple procedure whereby the variables can be easily identified. Generally, one must rely on a good understanding of the phenomenon involved and the governing physical laws. If extraneous variables are included, then too many pi terms appear in the final solution, and it may be difficult, time consuming, and expensive to eliminate these experimentally. If important variables are omitted, then an incorrect result will be obtained; and again, this may prove to be costly and difficult to ascertain. It is, therefore, imperative that sufficient time and attention be given to this first step in which the variables are determined.

Most engineering problems involve certain simplifying assumptions that have an influence on the variables to be considered. Usually we wish to keep the problem as simple as possible, perhaps even if some accuracy is sacrificed. A suitable balance between simplicity and accuracy is a desirable goal. How "accurate" the solution must be depends on the objective of the study; that is, we may be only concerned with general trends and, therefore, some variables that are thought to have only a minor influence in the problem may be neglected for simplicity.

*It is often helpful to classify variables into three groups—geometry, material properties, and external effects.*

For most engineering problems (including areas outside of fluid mechanics), pertinent variables can be classified into three general groups—geometry, material properties, and external effects.

***Geometry.***   The geometric characteristics can usually be described by a series of lengths and angles. In most problems the geometry of the system plays an important role, and a sufficient number of geometric variables must be included to describe the system. These variables can usually be readily identified.

***Material Properties.***   Since the response of a system to applied external effects such as forces, pressures, and changes in temperature is dependent on the nature of the materials involved in the system, the material properties that relate the external effects and the responses must be included as variables. For example, for Newtonian fluids the viscosity of the fluid is the property that relates the applied forces to the rates of deformation of the fluid. As the material behavior becomes more complex, such as would be true for non-Newtonian fluids, the determination of material properties becomes difficult, and this class of variables can be troublesome to identify.

*External Effects.* This terminology is used to denote any variable that produces, or tends to produce, a change in the system. For example, in structural mechanics, forces (either concentrated or distributed) applied to a system tend to change its geometry, and such forces would need to be considered as pertinent variables. For fluid mechanics, variables in this class would be related to pressures, velocities, or gravity.

The above general classes of variables are intended as broad categories that should be helpful in identifying variables. It is likely, however, that there will be important variables that do not fit easily into one of the above categories and each problem needs to be carefully analyzed.

Since we wish to keep the number of variables to a minimum, it is important that all variables are independent. For example, if in a given problem we know that the moment of inertia of the area of a circular plate is an important variable, we could list either the moment of inertia or the plate diameter as the pertinent variable. However, it would be unnecessary to include both moment of inertia and diameter, assuming that the diameter enters the problem only through the moment of inertia. In more general terms, if we have a problem in which the variables are

$$f(p, q, r, \ldots, u, v, w, \ldots) = 0 \tag{7.3}$$

and it is known that there is an additional relationship among some of the variables, for example,

$$q = f_1(u, v, w, \ldots) \tag{7.4}$$

then $q$ is not required and can be omitted. Conversely, if it is known that the only way the variables $u$, $v$, $w$, . . . enter the problem is through the relationship expressed by Eq. 7.4, then the variables $u$, $v$, $w$, . . . can be replaced by the single variable $q$, therefore reducing the number of variables.

In summary, the following points should be considered in the selection of variables:

1. Clearly define the problem. What is the main variable of interest (the dependent variable)?
2. Consider the basic laws that govern the phenomenon. Even a crude theory that describes the essential aspects of the system may be helpful.
3. Start the variable selection process by grouping the variables into three broad classes: geometry, material properties, and external effects.
4. Consider other variables that may not fall into one of the above categories. For example, time will be an important variable if any of the variables are time dependent.
5. Be sure to include all quantities that enter the problem even though some of them may be held constant (e.g., the acceleration of gravity, $g$). For a dimensional analysis it is the dimensions of the quantities that are important—not specific values!
6. Make sure that all variables are independent. Look for relationships among subsets of the variables.

### 7.4.2 Determination of Reference Dimensions

For any given problem it is obviously desirable to reduce the number of pi terms to a minimum; therefore, we wish to reduce the number of variables to a minimum—that is, we certainly do not want to include extraneous variables. It is also important to know how many reference dimensions are required to describe the variables. As we have seen in the preceding examples, $F$, $L$, and $T$ appear to be a convenient set of basic dimensions for characterizing fluid-mechanical quantities. There is, however, really nothing "fundamental" about this set and, as previously noted, $M$, $L$, and $T$ would also be suitable. Actually any set of measurable quantities could be used as basic dimensions provided that the selected combination can be used to describe all secondary quantities. However, the use of *FLT* or *MLT* as basic dimensions is the simplest, and these dimensions can be used to describe fluid-mechanical phenomena. Of course, in some problems only one or two of these are required. In addition, we occasionally find that the number of reference dimensions needed to describe all variables is smaller than the number of basic dimensions. This point is illustrated in Example 7.2. Interesting discussions, both practical and philosophical, relative to the concept of basic dimensions can be found in the books by Huntley (Ref. 4) and by Isaacson and Isaacson (Ref. 12).

*Typically, in fluid mechanics, the required number of reference dimensions is three, but in some problems only one or two are required.*

# EXAMPLE 7.2   Determination of Pi Terms

**GIVEN**   An open, cylindrical paint can having a diameter $D$ is filled to a depth $h$ with paint having a specific weight $\gamma$. The vertical deflection, $\delta$, of the center of the bottom is a function of $D$, $h$, $d$, $\gamma$, and $E$, where $d$ is the thickness of the bottom and $E$ is the modulus of elasticity of the bottom material.

**FIND**   Determine the functional relationship between the vertical deflection, $\delta$, and the independent variables using dimensional analysis.

## SOLUTION

From the statement of the problem

$$\delta = f(D, h, d, \gamma, E)$$

and the dimensions of the variables are

$$\delta \doteq L$$
$$D \doteq L$$
$$h \doteq L$$
$$d \doteq L$$
$$\gamma \doteq FL^{-3} \doteq ML^{-2}T^{-2}$$
$$E \doteq FL^{-2} \doteq ML^{-1}T^{-2}$$

where the dimensions have been expressed in terms of both the *FLT* and *MLT* systems.

We now apply the pi theorem to determine the required number of pi terms. First, let us use $F$, $L$, and $T$ as our system of basic dimensions. There are six variables and two reference dimensions ($F$ and $L$) required so that four pi terms are needed. For repeating variables, we can select $D$ and $\gamma$ so that

$$\Pi_1 = \delta\, D^a \gamma^b$$
$$(L)(L)^a(FL^{-3})^b \doteq F^0 L^0$$

and

$$1 + a - 3b = 0 \quad \text{(for } L)$$
$$b = 0 \quad \text{(for } F)$$

Therefore, $a = -1$, $b = 0$, and

$$\Pi_1 = \frac{\delta}{D}$$

Similarly,

$$\Pi_2 = h\, D^a \gamma^b$$

and following the same procedure as above, $a = -1$, $b = 0$ so that

$$\Pi_2 = \frac{h}{D}$$

The remaining two pi terms can be found using the same procedure, with the result

$$\Pi_3 = \frac{d}{D} \qquad \Pi_4 = \frac{E}{D\gamma}$$

■ **Figure E7.2**

Thus, this problem can be studied by using the relationship

$$\frac{\delta}{D} = \phi\left(\frac{h}{D}, \frac{d}{D}, \frac{E}{D\gamma}\right) \tag{Ans}$$

**COMMENTS**   Let us now solve the same problem using the *MLT* system. Although the number of variables is obviously the same, it would seem that there are three reference dimensions required, rather than two. If this were indeed true, it would certainly be fortuitous, since we would reduce the number of required pi terms from four to three. Does this seem right? How can we reduce the number of required pi terms by simply using the *MLT* system of basic dimensions? The answer is that we cannot, and a closer look at the dimensions of the variables listed above reveals that actually only two reference dimensions, $MT^{-2}$ and $L$, are required.

This is an example of the situation in which the number of reference dimensions differs from the number of basic dimensions. It does not happen very often and can be detected by looking at the dimensions of the variables (regardless of the systems used) and making sure how many reference dimensions are actually required to describe the variables. Once the number of reference dimensions has been determined, we can proceed as before. Since the number of repeating variables must equal the number of reference dimensions, it follows that two reference dimensions are still required and we could again use $D$ and $\gamma$ as repeating variables. The pi terms would be determined in the same manner. For example, the pi term containing $E$ would be developed as

$$\Pi_4 = ED^a \gamma^b$$
$$(ML^{-1}T^{-2})(L)^a(ML^{-2}T^{-2})^b \doteq (MT^{-2})^0 L^0$$
$$1 + b = 0 \quad \text{(for } MT^{-2})$$
$$-1 + a - 2b = 0 \quad \text{(for } L)$$

and, therefore, $a = -1$, $b = -1$ so that

$$\Pi_4 = \frac{E}{D\gamma}$$

which is the same as $\Pi_4$ obtained using the *FLT* system. The other pi terms would be the same, and the final result is the same; that is,

$$\frac{\delta}{D} = \phi\left(\frac{h}{D}, \frac{d}{D}, \frac{E}{D\gamma}\right) \qquad \textbf{(Ans)}$$

This will always be true—you cannot affect the required number of pi terms by using $M$, $L$, and $T$ instead of $F$, $L$, and $T$, or vice versa.

### 7.4.3 Uniqueness of Pi Terms

A little reflection on the process used to determine pi terms by the method of repeating variables reveals that the specific pi terms obtained depend on the somewhat arbitrary selection of repeating variables. For example, in the problem of studying the pressure drop in a pipe, we selected $D$, $V$, and $\rho$ as repeating variables. This led to the formulation of the problem in terms of pi terms as

$$\frac{\Delta p_\ell D}{\rho V^2} = \phi\left(\frac{\rho V D}{\mu}\right) \qquad (7.5)$$

What if we had selected $D$, $V$, and $\mu$ as repeating variables? A quick check will reveal that the pi term involving $\Delta p_\ell$ becomes

$$\frac{\Delta p_\ell D^2}{V\mu}$$

and the second pi term remains the same. Thus, we can express the final result as

$$\frac{\Delta p_\ell D^2}{V\mu} = \phi_1\left(\frac{\rho V D}{\mu}\right) \qquad (7.6)$$

Both results are correct, and both would lead to the same final equation for $\Delta p_\ell$. Note, however, that the functions $\phi$ and $\phi_1$ in Eqs. 7.5 and 7.6 will be different because the dependent pi terms are different for the two relationships. As shown by the figure in the margin, the resulting graph of dimensionless data will be different for the two formulations. However, when extracting the physical variable, $\Delta p_\ell$, from the two results, the values will be the same.

We can conclude from this illustration that there is *not* a unique set of pi terms which arises from a dimensional analysis. However, the required *number* of pi terms is fixed, and once a correct set is determined, all other possible sets can be developed from this set by combinations of products of powers of the original set. Thus, if we have a problem involving, say, three pi terms,

$$\Pi_1 = \phi(\Pi_2, \Pi_3)$$

*Once a correct set of pi terms is obtained, any other set can be obtained by manipulation of the original set.*

we could always form a new set from this one by combining the pi terms. For example, we could form a new pi term, $\Pi_2'$, by letting

$$\Pi_2' = \Pi_2^a \, \Pi_3^b$$

where $a$ and $b$ are arbitrary exponents. Then the relationship could be expressed as

$$\Pi_1 = \phi_1(\Pi_2', \Pi_3)$$

or

$$\Pi_1 = \phi_2(\Pi_2, \Pi_2')$$

All of these would be correct. It should be emphasized, however, that the required number of pi terms cannot be reduced by this manipulation; only the form of the pi terms is altered. By using this technique, we see that the pi terms in Eq. 7.6 could be obtained from those in Eq. 7.5; that is, we multiply $\Pi_1$ in Eq. 7.5 by $\Pi_2$ so that

$$\left(\frac{\Delta p_\ell D}{\rho V^2}\right)\left(\frac{\rho V D}{\mu}\right) = \frac{\Delta p_\ell D^2}{V\mu}$$

which is the $\Pi_1$ of Eq. 7.6.

There is no simple answer to the question: Which form for the pi terms is best? Usually our only guideline is to keep the pi terms as simple as possible. Also, it may be that certain pi terms will be easier to work with in actually performing experiments. The final choice remains an arbitrary one and generally will depend on the background and experience of the investigator. It should again be emphasized, however, that although there is no unique set of pi terms for a given problem, the *number* required is fixed in accordance with the pi theorem.

## 7.5    Determination of Pi Terms by Inspection

The method of repeating variables for forming pi terms has been presented in Section 7.3. This method provides a step-by-step procedure that if executed properly will provide a correct and complete set of pi terms. Although this method is simple and straightforward, it is rather tedious, particularly for problems in which large numbers of variables are involved. Since the only restrictions placed on the pi terms are that they be (1) correct in number, (2) dimensionless, and (3) independent, it is possible to simply form the pi terms by inspection, without resorting to the more formal procedure.

To illustrate this approach, we again consider the pressure drop per unit length along a smooth pipe. Regardless of the technique to be used, the starting point remains the same—determine the variables, which in this case are

$$\Delta p_\ell = f(D, \rho, \mu, V)$$

Next, the dimensions of the variables are listed:

$$\Delta p_\ell \doteq FL^{-3}$$

$$D \doteq L$$

$$\rho \doteq FL^{-4}T^2$$

$$\mu \doteq FL^{-2}T$$

$$V \doteq LT^{-1}$$

and subsequently the number of reference dimensions determined. The application of the pi theorem then tells us how many pi terms are required. In this problem, since there are five variables and three reference dimensions, two pi terms are needed. Thus, the required number of pi terms can be easily obtained. The determination of this number should always be done at the beginning of the analysis.

*Pi terms can be formed by inspection by simply making use of the fact that each pi term must be dimensionless.*

Once the number of pi terms is known, we can form each pi term by inspection, simply making use of the fact that each pi term must be dimensionless. We will always let $\Pi_1$ contain the dependent variable, which in this example is $\Delta p_\ell$. Since this variable has the dimensions $FL^{-3}$, we need to combine it with other variables so that a nondimensional product will result. One possibility is to first divide $\Delta p_\ell$ by $\rho$ so that

$$\frac{\Delta p_\ell}{\rho} \doteq \frac{(FL^{-3})}{(FL^{-4}T^2)} \doteq \frac{L}{T^2} \qquad \text{(cancels } F\text{)}$$

The dependence on $F$ has been eliminated, but $\Delta p_\ell/\rho$ is obviously not dimensionless. To eliminate the dependence on $T$, we can divide by $V^2$ so that

$$\left(\frac{\Delta p_\ell}{\rho}\right)\frac{1}{V^2} \doteq \left(\frac{L}{T^2}\right)\frac{1}{(LT^{-1})^2} \doteq \frac{1}{L} \qquad \text{(cancels } T\text{)}$$

Finally, to make the combination dimensionless we multiply by $D$ so that

$$\left(\frac{\Delta p_\ell}{\rho V^2}\right) D \doteq \left(\frac{1}{L}\right)(L) \doteq L^0 \qquad \text{(cancels } L)$$

Thus,

$$\Pi_1 = \frac{\Delta p_\ell D}{\rho V^2}$$

Next, we will form the second pi term by selecting the variable that was not used in $\Pi_1$, which in this case is $\mu$. We simply combine $\mu$ with the other variables to make the combination dimensionless (but do not use $\Delta p_\ell$ in $\Pi_2$, since we want the dependent variable to appear only in $\Pi_1$). For example, divide $\mu$ by $\rho$ (to eliminate $F$), then by $V$ (to eliminate $T$), and finally by $D$ (to eliminate $L$). Thus,

$$\Pi_2 = \frac{\mu}{\rho V D} \doteq \frac{(FL^{-2}T)}{(FL^{-4}T^2)(LT^{-1})(L)} \doteq F^0 L^0 T^0$$

and, therefore,

$$\frac{\Delta p_\ell D}{\rho V^2} = \phi\left(\frac{\mu}{\rho V D}\right)$$

which is, of course, the same result we obtained by using the method of repeating variables.

An additional concern, when one is forming pi terms by inspection, is to make certain that they are all independent. In the pipe flow example, $\Pi_2$ contains $\mu$, which does not appear in $\Pi_1$, and therefore these two pi terms are obviously independent. In a more general case a pi term would not be independent of the others in a given problem if it can be formed by some combination of the others. For example, if $\Pi_2$ can be formed by a combination of say $\Pi_3$, $\Pi_4$, and $\Pi_5$ such as

$$\Pi_2 = \frac{\Pi_3^2 \, \Pi_4}{\Pi_5}$$

then $\Pi_2$ is not an independent pi term. We can ensure that each pi term is independent of those preceding it by incorporating a new variable in each pi term.

Although forming pi terms by inspection is essentially equivalent to the repeating variable method, it is less structured. With a little practice the pi terms can be readily formed by inspection, and this method offers an alternative to more formal procedures.

## 7.6 Common Dimensionless Groups in Fluid Mechanics

At the top of Table 7.1 is a list of variables that commonly arise in fluid mechanics problems. The list is obviously not exhaustive but does indicate a broad range of variables likely to be found in a typical problem. Fortunately, not all of these variables would be encountered in all problems. However, when combinations of these variables are present, it is standard practice to combine them into some of the common dimensionless groups (pi terms) given in Table 7.1. These combinations appear so frequently that special names are associated with them, as indicated in the table.

*A useful physical interpretation can often be given to dimensionless groups.*

It is also often possible to provide a physical interpretation to the dimensionless groups which can be helpful in assessing their influence in a particular application. For example, the Froude number is an index of the ratio of the force due to the acceleration of a fluid particle to the force due to gravity (weight). This can be demonstrated by considering a fluid particle moving along a streamline (Fig. 7.3). The magnitude of the component of inertia force $F_I$ along the streamline can be expressed as $F_I = a_s m$, where $a_s$ is the magnitude of the acceleration along the streamline for a particle having a mass $m$. From our study of particle motion along a curved path (see Section 3.1) we know that

$$a_s = \frac{dV_s}{dt} = V_s \frac{dV_s}{ds}$$

■ **Table 7.1**

**Some Common Variables and Dimensionless Groups in Fluid Mechanics**

Variables: Acceleration of gravity, $g$; Bulk modulus, $E_v$; Characteristic length, $\ell$; Density, $\rho$; Frequency of oscillating flow, $\omega$; Pressure, $p$ (or $\Delta p$); Speed of sound, $c$; Surface tension, $\sigma$; Velocity, $V$; Viscosity, $\mu$

| Dimensionless Groups | Name | Interpretation (Index of Force Ratio Indicated) | Types of Applications |
|---|---|---|---|
| $\dfrac{\rho V \ell}{\mu}$ | Reynolds number, Re | $\dfrac{\text{inertia force}}{\text{viscous force}}$ | Generally of importance in all types of fluid dynamics problems |
| $\dfrac{V}{\sqrt{g\ell}}$ | Froude number, Fr | $\dfrac{\text{inertia force}}{\text{gravitational force}}$ | Flow with a free surface |
| $\dfrac{p}{\rho V^2}$ | Euler number, Eu | $\dfrac{\text{pressure force}}{\text{inertia force}}$ | Problems in which pressure, or pressure differences, are of interest |
| $\dfrac{\rho V^2}{E_v}$ | Cauchy number,[a] Ca | $\dfrac{\text{inertia force}}{\text{compressibility force}}$ | Flows in which the compressibility of the fluid is important |
| $\dfrac{V}{c}$ | Mach number,[a] Ma | $\dfrac{\text{inertia force}}{\text{compressibility force}}$ | Flows in which the compressibility of the fluid is important |
| $\dfrac{\omega \ell}{V}$ | Strouhal number, St | $\dfrac{\text{inertia (local) force}}{\text{inertia (convective) force}}$ | Unsteady flow with a characteristic frequency of oscillation |
| $\dfrac{\rho V^2 \ell}{\sigma}$ | Weber number, We | $\dfrac{\text{inertia force}}{\text{surface tension force}}$ | Problems in which surface tension is important |

[a]The Cauchy number and the Mach number are related, and either can be used as an index of the relative effects of inertia and compressibility. See accompanying discussion.

*Special names along with physical interpretations are given to the most common dimensionless groups.*

where $s$ is measured along the streamline. If we write the velocity, $V_s$, and length, $s$, in dimensionless form, that is,

$$V_s^* = \frac{V_s}{V} \qquad s^* = \frac{s}{\ell}$$

where $V$ and $\ell$ represent some characteristic velocity and length, respectively, then

$$a_s = \frac{V^2}{\ell} V_s^* \frac{dV_s^*}{ds^*}$$

and

$$F_I = \frac{V^2}{\ell} V_s^* \frac{dV_s^*}{ds^*} m$$

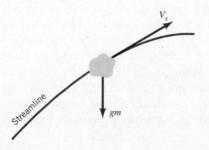

■ **Figure 7.3**   The force of gravity acting on a fluid particle moving along a streamline.

The magnitude of the weight of the particle, $F_G$, is $F_G = gm$, so the ratio of the inertia to the gravitational force is

$$\frac{F_I}{F_G} = \frac{V^2}{g\ell} V_s^* \frac{dV_s^*}{ds^*}$$

Thus, the force ratio $F_I/F_G$ is proportional to $V^2/g\ell$, and the square root of this ratio, $V/\sqrt{g\ell}$, is called the *Froude number*. We see that a physical interpretation of the Froude number is that it is a measure of, or an index of, the relative importance of inertial forces acting on fluid particles to the weight of the particle. Note that the Froude number is not really *equal* to this force ratio, but is simply some type of average measure of the influence of these two forces. In a problem in which gravity (or weight) is not important, the Froude number would not appear as an important pi term. A similar interpretation in terms of indices of force ratios can be given to the other dimensionless groups, as indicated in Table 7.1, and a further discussion of the basis for this type of interpretation is given in the last section in this chapter. Some additional details about these important dimensionless groups are given below, and the types of application or problem in which they arise are briefly noted in the last column of Table 7.1.

*V7.3 Reynolds number*

No separation
Re ≈ 0.2

Laminar boundary layer, wide turbulent wake
Re ≈ 20,000

***Reynolds Number.*** The Reynolds number is undoubtedly the most famous dimensionless parameter in fluid mechanics. It is named in honor of Osborne Reynolds (1842–1912), a British engineer who first demonstrated that this combination of variables could be used as a criterion to distinguish between laminar and turbulent flow. In most fluid flow problems there will be a characteristic length, $\ell$, and a velocity, $V$, as well as the fluid properties of density, $\rho$, and viscosity, $\mu$, which are relevant variables in the problem. Thus, with these variables the Reynolds number

$$\mathrm{Re} = \frac{\rho V \ell}{\mu}$$

arises naturally from the dimensional analysis. The Reynolds number is a measure of the ratio of the inertia force on an element of fluid to the viscous force on an element. When these two types of forces are important in a given problem, the Reynolds number will play an important role. However, if the Reynolds number is very small ($\mathrm{Re} \ll 1$), this is an indication that the viscous forces are dominant in the problem, and it may be possible to neglect the inertial effects; that is, the density of the fluid will not be an important variable. Flows at very small Reynolds numbers are commonly referred to as "creeping flows" as discussed in Section 6.10. Conversely, for large Reynolds number flows, viscous effects are small relative to inertial effects and for these cases it may be possible to neglect the effect of viscosity and consider the problem as one involving a "nonviscous" fluid. This type of problem is considered in detail in Sections 6.4 through 6.7. An example of the importance of the Reynolds number in determining the flow physics is shown in the figure in the margin for flow past a circular cylinder at two different Re values. This flow is discussed further in Chapter 9.

***Froude Number.*** The Froude number

$$\mathrm{Fr} = \frac{V}{\sqrt{g\ell}}$$

*V7.4 Froude number*

is distinguished from the other dimensionless groups in Table 7.1 in that it contains the acceleration of gravity, $g$. The acceleration of gravity becomes an important variable in a fluid dynamics problem in which the fluid weight is an important force. As discussed, the Froude number is a measure of the ratio of the inertia force on an element of fluid to the weight of the element. It will generally be important in problems involving flows with free surfaces since gravity principally affects this type of flow. Typical problems would include the study of the flow of water around ships (with the resulting wave action) or flow through rivers or open conduits. The Froude number is named in honor of William Froude (1810–1879), a British civil engineer, mathematician, and naval architect who pioneered the use of towing tanks for the study of ship design. It is to be noted that the Froude number is also commonly defined as the square of the Froude number listed in Table 7.1.

***Euler Number.*** The Euler number

$$Eu = \frac{p}{\rho V^2}$$

can be interpreted as a measure of the ratio of pressure forces to inertial forces, where $p$ is some characteristic pressure in the flow field. Very often the Euler number is written in terms of a pressure difference, $\Delta p$, so that $Eu = \Delta p / \rho V^2$. Also, this combination expressed as $\Delta p / \frac{1}{2}\rho V^2$ is called the *pressure coefficient*. Some form of the Euler number would normally be used in problems in which pressure or the pressure difference between two points is an important variable. The Euler number is named in honor of Leonhard Euler (1707–1783), a famous Swiss mathematician who pioneered work on the relationship between pressure and flow. For problems in which cavitation is of concern, the dimensionless group $(p_r - p_v)/\frac{1}{2}\rho V^2$ is commonly used, where $p_v$ is the vapor pressure and $p_r$ is some reference pressure. Although this dimensionless group has the same form as the Euler number, it is generally referred to as the *cavitation number*.

***Cauchy Number and Mach Number.*** The Cauchy number

$$Ca = \frac{\rho V^2}{E_v}$$

The Mach number is a commonly used dimensionless parameter in compressible flow problems.

and the Mach number

$$Ma = \frac{V}{c}$$

are important dimensionless groups in problems in which fluid compressibility is a significant factor. Since the speed of sound, $c$, in a fluid is equal to $c = \sqrt{E_v/\rho}$ (see Section 1.7.3), it follows that

$$Ma = V\sqrt{\frac{\rho}{E_v}}$$

and the square of the Mach number

$$Ma^2 = \frac{\rho V^2}{E_v} = Ca$$

is equal to the Cauchy number. Thus, either number (but not both) may be used in problems in which fluid compressibility is important. Both numbers can be interpreted as representing an index of the ratio of inertial forces to compressibility forces. When the Mach number is relatively small (say, less than 0.3), the inertial forces induced by the fluid motion are not sufficiently large to cause a significant change in the fluid density, and in this case the compressibility of the fluid can be neglected. The Mach number is the more commonly used parameter in compressible flow problems, particularly in the fields of gas dynamics and aerodynamics. The Cauchy number is named in honor of Augustin Louis de Cauchy (1789–1857), a French engineer, mathematician, and hydrodynamicist. The Mach number is named in honor of Ernst Mach (1838–1916), an Austrian physicist and philosopher.

***Strouhal Number.*** The Strouhal number

$$St = \frac{\omega \ell}{V}$$

*V7.5 Strouhal number*

is a dimensionless parameter that is likely to be important in unsteady, oscillating flow problems in which the frequency of the oscillation is $\omega$. It represents a measure of the ratio of inertial forces due to the unsteadiness of the flow (local acceleration) to the inertial forces due to changes in velocity from point to point in the flow field (convective acceleration). This type of unsteady flow may develop when a fluid flows past a solid body (such as a wire or cable) placed in the moving stream. For example, in a certain Reynolds number range, a periodic flow will develop downstream from a cylinder placed in a moving fluid due to a regular pattern of vortices that are shed from the body. (See the photograph at the beginning of this chapter and Fig. 9.21.) This system of vortices, called a *Kármán vortex trail* [named after Theodor von Kármán (1881–1963), a famous fluid mechanician], creates an oscillating flow at a discrete frequency, $\omega$, such that the Strouhal number can be closely correlated with the Reynolds number. When the frequency is in the audible range, a sound can be heard and the bodies appear to "sing." In fact, the Strouhal number is named in honor of Vincenz Strouhal (1850–1922), who used this parameter in his study of "singing wires." The most dramatic evidence of this phenomenon occurred in 1940 with the collapse of the Tacoma Narrows Bridge. The shedding frequency of the vortices coincided with the natural frequency of the bridge, thereby setting up a resonant condition that eventually led to the collapse of the bridge.

There are, of course, other types of oscillating flows. For example, blood flow in arteries is periodic and can be analyzed by breaking up the periodic motion into a series of harmonic components (Fourier series analysis), with each component having a frequency that is a multiple of the fundamental frequency, $\omega$ (the pulse rate). Rather than use the Strouhal number in this type of problem, a dimensionless group formed by the product of St and Re is used; that is

$$\text{St} \times \text{Re} = \frac{\rho \omega \ell^2}{\mu}$$

The square root of this dimensionless group is often referred to as the *frequency parameter*.

***Weber Number.*** The Weber number

$$\text{We} = \frac{\rho V^2 \ell}{\sigma}$$

*V7.6 Weber number*

may be important in problems in which there is an interface between two fluids. In this situation the surface tension may play an important role in the phenomenon of interest. The Weber number can be thought of as an index of the inertial force to the surface tension force acting on a fluid element. Common examples of problems in which this parameter may be important include the flow of thin films of liquid, or the formation of droplets or bubbles. Clearly, not all problems involving flows with an interface will require the inclusion of surface tension. The flow of water in a river is not affected significantly by surface tension, since inertial and gravitational effects are dominant (We $\ll$ 1). However, as discussed in a later section, for river models (which may have small depths) caution is required so that surface tension does not become important in the model, whereas it is not important in the actual river. The Weber number is named after Moritz Weber (1871–1951), a German professor of naval mechanics who was instrumental in formalizing the general use of common dimensionless groups as a basis for similitude studies.

---

### F l u i d s   i n   t h e   N e w s

**Slip at the micro scale** A goal in chemical and biological analyses is to miniaturize the experiment, which has many advantages including reduction in sample size. In recent years, there has been significant work on integrating these tests on a single microchip to form the "lab-on-a-chip" system. These devices are on the millimeter scale with complex passages for fluid flow on the micron scale (or smaller). While there are advantages to miniaturization, care must be taken in moving to smaller and smaller flow regimes, as you will eventually bump into the continuum assumption. To characterize this situation, a dimensionless number termed the Knudsen number, $\text{Kn} = \lambda/\ell$, is commonly employed. Here $\lambda$ is the mean free path and $\ell$ is the characteristic length of the system. If Kn is smaller than 0.01, then the flow can be described by the Navier–Stokes equations with no-slip at the walls. For $0.01 < \text{Kn} < 0.3$, the same equations can be used, but there can be slip between the fluid and the wall so the boundary conditions need to be adjusted. For $\text{Kn} > 10$, the continuum assumption breaks down and the Navier–Stokes equations are no longer valid.

---

## 7.7 Correlation of Experimental Data

One of the most important uses of dimensional analysis is as an aid in the efficient handling, interpretation, and correlation of experimental data. Since the field of fluid mechanics relies heavily on empirical data, it is not surprising that dimensional analysis is such an important tool in this field. As noted previously, a dimensional analysis cannot provide a complete answer to any given problem, since the analysis only provides the dimensionless groups describing the phenomenon, and not the specific relationship among the groups. To determine this relationship, suitable experimental data must be obtained. The degree of difficulty involved in this process depends on the number of pi terms, and the nature of the experiments (How hard is it to obtain the measurements?). The simplest problems are obviously those involving the fewest pi terms, and the following sections indicate how the complexity of the analysis increases with the increasing number of pi terms.

### 7.7.1 Problems with One Pi Term

Application of the pi theorem indicates that if the number of variables minus the number of reference dimensions is equal to unity, then only *one* pi term is required to describe the phenomenon. The functional relationship that must exist for one pi term is

$$\Pi_1 = C$$

*If only one pi term is involved in a problem, it must be equal to a constant.*

where $C$ is a constant. This is one situation in which a dimensional analysis reveals the specific form of the relationship and, as is illustrated by the following example, shows how the individual variables are related. The value of the constant, however, must still be determined by experiment.

---

## EXAMPLE 7.3    Flow with Only One Pi Term

**GIVEN**   As shown in Fig. E7.3, assume that the drag, $\mathscr{D}$, acting on a spherical particle that falls very slowly through a viscous fluid, is a function of the particle diameter, $D$, the particle velocity, $V$, and the fluid viscosity, $\mu$.

**FIND**   Determine, with the aid of dimensional analysis, how the drag depends on the particle velocity.

*V7.7 Stokes flow*

■ **Figure E7.3**

### SOLUTION

From the information given, it follows that

$$\mathscr{D} = f(D, V, \mu)$$

and the dimensions of the variables are

$$\mathscr{D} \doteq F$$
$$D \doteq L$$
$$V \doteq LT^{-1}$$
$$\mu \doteq FL^{-2}T$$

We see that there are four variables and three reference dimensions ($F$, $L$, and $T$) required to describe the variables. Thus, according to the pi theorem, one pi term is required. This pi term can be easily formed by inspection and can be expressed as

$$\Pi_1 = \frac{\mathscr{D}}{\mu VD}$$

Because there is only one pi term, it follows that

$$\frac{\mathscr{D}}{\mu VD} = C$$

where $C$ is a constant. Thus,

$$\mathscr{D} = C\mu VD$$

Thus, for a given particle and fluid, the drag varies directly with the velocity so that

$$\mathscr{D} \propto V \qquad \text{(Ans)}$$

**COMMENTS**   Actually, the dimensional analysis reveals that the drag not only varies directly with the velocity, but it also varies directly with the particle diameter and the fluid viscosity. We could not, however, predict the value of the drag, since the constant, $C$, is unknown. An experiment would have to be performed in which the drag and the corresponding velocity are measured for a given particle and fluid. Although in principle we would only have to run a single test, we would certainly want to repeat it several times to obtain a reliable value for $C$. It should be emphasized that once the value of $C$ is determined it is not necessary to run similar tests by using different spherical particles and fluids; that is, $C$ is a universal constant as long as the drag is a function only of particle diameter, velocity, and fluid viscosity.

An approximate solution to this problem can also be obtained theoretically, from which it is found that $C = 3\pi$ so that

$$\mathscr{D} = 3\pi\mu VD$$

This equation is commonly called *Stokes law* and is used in the study of the settling of particles. Our experiments would reveal that this result is only valid for small Reynolds numbers ($\rho VD/\mu \ll 1$). This follows, since in the original list of variables, we have neglected inertial effects (fluid density is not included as a variable). The inclusion of an additional variable would lead to another pi term so that there would be two pi terms rather than one.

Consider a free-body diagram of a sphere in Stokes flow; there would be a buoyant force in the same direction as the drag in Fig. E7.3, as well as a weight force in the opposite direction. As shown above, the drag force is proportional to the product of the diameter and fall velocity, $\mathscr{D} \propto VD$. The weight and buoyant force are

proportional to the diameter cubed, $W$ and $F_B \propto D^3$. Given equilibrium conditions, the force balance can be written as

$$\mathscr{D} = W - F_B$$

Based on the scaling laws for these terms, it follows that $VD \propto D^3$. Hence, the fall velocity will be proportional to the square of the diameter, $V \propto D^2$. Therefore, for two spheres, one having twice the diameter of the other, and falling through the same fluid, the sphere with the larger diameter will fall four times faster (see Video V7.7).

■ **Figure 7.4** The graphical presentation of data for problems involving two pi terms, with an illustration of the potential danger of extrapolation of data.

### 7.7.2 Problems with Two or More Pi Terms

If a given phenomenon can be described with two pi terms such that

$$\Pi_1 = \phi(\Pi_2)$$

*For problems involving only two pi terms, results of an experiment can be conveniently presented in a simple graph.*

the functional relationship among the variables can then be determined by varying $\Pi_2$ and measuring the corresponding values of $\Pi_1$. For this case the results can be conveniently presented in graphical form by plotting $\Pi_1$ versus $\Pi_2$ as is illustrated in Fig. 7.4. It should be emphasized that the curve shown in Fig. 7.4 would be a "universal" one for the particular phenomenon studied. This means that if the variables and the resulting dimensional analysis are correct, then there is only a single relationship between $\Pi_1$ and $\Pi_2$, as illustrated in Fig. 7.4. However, since this is an empirical relationship, we can only say that it is valid over the range of $\Pi_2$ covered by the experiments. It would be unwise to extrapolate beyond this range, since as illustrated with the dashed lines in the figure, the nature of the phenomenon could dramatically change as the range of $\Pi_2$ is extended. In addition to presenting the data graphically, it may be possible (and desirable) to obtain an empirical equation relating $\Pi_1$ and $\Pi_2$ by using a standard curve-fitting technique.

## $E$XAMPLE 7.4  Dimensionless Correlation of Experimental Data

**GIVEN** The relationship between the pressure drop per unit length along a smooth-walled, horizontal pipe and the variables that affect the pressure drop is to be determined experimentally. In the laboratory the pressure drop was measured over a 5-ft length of smooth-walled pipe having an inside diameter of 0.496 in. The fluid used was water at 60 °F ($\mu = 2.34 \times 10^{-5}$ lb · s/ft$^2$, $\rho = 1.94$ slugs/ft$^3$). Tests were run in which the velocity was varied and the corresponding pressure drop measured. The results of these tests are shown below:

| Velocity (ft/s) | Pressure drop for 5-ft length (lb/ft$^2$) |
|---|---|
| 1.17 | 6.26 |
| 1.95 | 15.6 |
| 2.91 | 30.9 |
| 5.84 | 106 |
| 11.13 | 329 |
| 16.92 | 681 |
| 23.34 | 1200 |
| 28.73 | 1730 |

**SOLUTION**

The first step is to perform a dimensional analysis during the planning stage *before* the experiments are actually run. As was discussed in Section 7.3, we will assume that the pressure drop per unit length, $\Delta p_\ell$, is a function of the pipe diameter, $D$, fluid

**FIND** Make use of these data to obtain a general relationship between the pressure drop per unit length and the other variables.

density, $\rho$, fluid viscosity, $\mu$, and the velocity, $V$. Thus,

$$\Delta p_\ell = f(D, \rho, \mu, V)$$

and application of the pi theorem yields two pi terms

$$\Pi_1 = \frac{D\,\Delta p_\ell}{\rho V^2} \quad \text{and} \quad \Pi_2 = \frac{\rho V D}{\mu}$$

Hence,

$$\frac{D\,\Delta p_\ell}{\rho V^2} = \phi\!\left(\frac{\rho V D}{\mu}\right)$$

To determine the form of the relationship, we need to vary the Reynolds number, $\text{Re} = \rho V D/\mu$, and to measure the corresponding values of $D\,\Delta p_\ell/\rho V^2$. The Reynolds number could be varied by changing any one of the variables, $\rho$, $V$, $D$, or $\mu$, or any combination of them. However, the simplest way to do this is to vary the velocity, since this will allow us to use the same fluid and pipe. Based on the data given, values for the two pi terms can be computed, with the result:

| $D\,\Delta p_\ell/\rho V^2$ | $\rho V D/\mu$ |
|---|---|
| 0.0195 | $4.01 \times 10^3$ |
| 0.0175 | $6.68 \times 10^3$ |
| 0.0155 | $9.97 \times 10^3$ |
| 0.0132 | $2.00 \times 10^4$ |
| 0.0113 | $3.81 \times 10^4$ |
| 0.0101 | $5.80 \times 10^4$ |
| 0.00939 | $8.00 \times 10^4$ |
| 0.00893 | $9.85 \times 10^4$ |

These are dimensionless groups so that their values are independent of the system of units used so long as a consistent system is used. For example, if the velocity is in ft/s, then the diameter should be in feet, not inches or meters. Note that since the Reynolds numbers are all greater than 2100, the flow in the pipe is turbulent (see Section 8.1.1).

A plot of these two pi terms can now be made with the results shown in Fig. E7.4a. The correlation appears to be quite good, and if it was not, this would suggest that either we had large experimental measurement errors or that we had perhaps omitted an important variable. The curve shown in Fig. E7.4a represents the general relationship between the pressure drop and the other factors in the range of Reynolds numbers between $4.01 \times 10^3$ and $9.85 \times 10^4$. Thus, for this range of Reynolds numbers it is *not* necessary to repeat the tests for other pipe sizes or other fluids provided the assumed independent variables ($D$, $\rho$, $\mu$, $V$) are the only important ones.

Since the relationship between $\Pi_1$ and $\Pi_2$ is nonlinear, it is not immediately obvious what form of empirical equation might be used to describe the relationship. If, however, the same data are

■ **Figure E7.4**

plotted on a logarithmic scale, as is shown in Fig. E7.4b, the data form a straight line, suggesting that a suitable equation is of the form $\Pi_1 = A\Pi_2^n$ where $A$ and $n$ are empirical constants to be determined from the data by using a suitable curve-fitting technique, such as a nonlinear regression program. For the data given in this example, a good fit of the data is obtained with the equation

$$\Pi_1 = 0.150\,\Pi_2^{-0.25} \tag{Ans}$$

**COMMENT**   In 1911, H. Blasius (1883–1970), a German fluid mechanician, established a similar empirical equation that is used widely for predicting the pressure drop in smooth pipes in the range $4 \times 10^3 < \text{Re} < 10^5$ (Ref. 16). This equation can be expressed in the form

$$\frac{D\,\Delta p_\ell}{\rho V^2} = 0.1582\!\left(\frac{\rho V D}{\mu}\right)^{-1/4}$$

The so-called Blasius formula is based on numerous experimental results of the type used in this example. Flow in pipes is discussed in more detail in the next chapter, where it is shown how pipe roughness (which introduces another variable) may affect the results given in this example (which is for smooth-walled pipes).

As the number of required pi terms increases, it becomes more difficult to display the results in a convenient graphical form and to determine a specific empirical equation that describes the phenomenon. For problems involving three pi terms

$$\Pi_1 = \phi(\Pi_2, \Pi_3)$$

it is still possible to show data correlations on simple graphs by plotting families of curves as illustrated in Fig. 7.5. This is an informative and useful way of representing the data in a general

■ **Figure 7.5** The graphical presentation of data for problems involving three pi terms.

*For problems involving more than two or three pi terms, it is often necessary to use a model to predict specific characteristics.*

way. It may also be possible to determine a suitable empirical equation relating the three pi terms. However, as the number of pi terms continues to increase, corresponding to an increase in the general complexity of the problem of interest, both the graphical presentation and the determination of a suitable empirical equation become intractable. For these more complicated problems, it is often more feasible to use models to predict specific characteristics of the system rather than to try to develop general correlations.

## 7.8 Modeling and Similitude

Prototype

Model

*V7.8 Model airplane*

Models are widely used in fluid mechanics. Major engineering projects involving structures, aircraft, ships, rivers, harbors, dams, air and water pollution, and so on, frequently involve the use of models. Although the term *model* is used in many different contexts, the "engineering model" generally conforms to the following definition. A ***model*** *is a representation of a physical system that may be used to predict the behavior of the system in some desired respect.* The physical system for which the predictions are to be made is called the ***prototype.*** Although *mathematical* or *computer* models may also conform to this definition, our interest will be in physical models, that is, models that resemble the prototype but are generally of a different size, may involve different fluids, and often operate under different conditions (pressures, velocities, etc.). As shown by the figure in the margin, usually a model is smaller than the prototype. Therefore, it is more easily handled in the laboratory and less expensive to construct and operate than a large prototype (it should be noted that variables or pi terms without a subscript will refer to the prototype, whereas the subscript *m* will be used to designate the model variables or pi terms). Occasionally, if the prototype is very small, it may be advantageous to have a model that is larger than the prototype so that it can be more easily studied. For example, large models have been used to study the motion of red blood cells, which are approximately 8 $\mu$m in diameter. With the successful development of a valid model, it is possible to predict the behavior of the prototype under a certain set of conditions. We may also wish to examine a priori the effect of possible design changes that are proposed for a hydraulic structure or fluid flow system. There is, of course, an inherent danger in the use of models in that predictions can be made that are in error and the error not detected until the prototype is found not to perform as predicted. It is, therefore, imperative that the model be properly designed and tested and that the results be interpreted correctly. In the following sections we will develop the procedures for designing models so that the model and prototype will behave in a similar fashion.

### 7.8.1 Theory of Models

The theory of models can be readily developed by using the principles of dimensional analysis. It has been shown that any given problem can be described in terms of a set of pi terms as

$$\Pi_1 = \phi(\Pi_2, \Pi_3, \ldots, \Pi_n) \tag{7.7}$$

In formulating this relationship, only a knowledge of the general nature of the physical phenomenon, and the variables involved, is required. Specific values for variables (size of components, fluid properties, and so on) are not needed to perform the dimensional analysis. Thus, Eq. 7.7 applies to any system that is governed by the same variables. If Eq. 7.7 describes the behavior of a particular prototype, a similar relationship can be written for a model of this prototype; that is,

$$\Pi_{1m} = \phi(\Pi_{2m}, \Pi_{3m}, \dots, \Pi_{nm}) \tag{7.8}$$

where the form of the function will be the same as long as the same phenomenon is involved in both the prototype and the model. Variables, or pi terms, without a subscript will refer to the prototype, whereas the subscript $m$ will be used to designate the model variables or pi terms.

The pi terms can be developed so that $\Pi_1$ contains the variable that is to be predicted from observations made on the model. Therefore, if the model is designed and operated under the following conditions,

$$\Pi_{2m} = \Pi_2$$
$$\Pi_{3m} = \Pi_3 \tag{7.9}$$
$$\vdots$$
$$\Pi_{nm} = \Pi_n$$

then with the presumption that the form of $\phi$ is the same for model and prototype, it follows that

$$\Pi_1 = \Pi_{1m} \tag{7.10}$$

Equation 7.10 is the desired **prediction equation** and indicates that the measured value of $\Pi_{1m}$ obtained with the model will be equal to the corresponding $\Pi_1$ for the prototype as long as the other pi terms are equal. The conditions specified by Eqs. 7.9 provide the **model design conditions,** also called **similarity requirements** or **modeling laws.**

As an example of the procedure, consider the problem of determining the drag, $\mathscr{D}$, on a thin rectangular plate ($w \times h$ in size) placed normal to a fluid with velocity, $V$, as shown by the figure in the margin. The dimensional analysis of this problem was performed in Example 7.1, where it was assumed that

$$\mathscr{D} = f(w, h, \mu, \rho, V)$$

Application of the pi theorem yielded

$$\frac{\mathscr{D}}{w^2 \rho V^2} = \phi\left(\frac{w}{h}, \frac{\rho V w}{\mu}\right) \tag{7.11}$$

We are now concerned with designing a model that could be used to predict the drag on a certain prototype (which presumably has a different size than the model). Since the relationship expressed by Eq. 7.11 applies to both prototype and model, Eq. 7.11 is assumed to govern the prototype, with a similar relationship

$$\frac{\mathscr{D}_m}{w_m^2 \rho_m V_m^2} = \phi\left(\frac{w_m}{h_m}, \frac{\rho_m V_m w_m}{\mu_m}\right) \tag{7.12}$$

for the model. The model design conditions, or similarity requirements, are therefore

$$\frac{w_m}{h_m} = \frac{w}{h} \quad \text{and} \quad \frac{\rho_m V_m w_m}{\mu_m} = \frac{\rho V w}{\mu}$$

The size of the model is obtained from the first requirement which indicates that

$$w_m = \frac{h_m}{h} w \tag{7.13}$$

We are free to establish the height ratio between the model and prototype, $h_m/h$, but then the model plate width, $w_m$, is fixed in accordance with Eq. 7.13.

The second similarity requirement indicates that the model and prototype must be operated at the same Reynolds number. Thus, the required velocity for the model is obtained from the relationship

$$V_m = \frac{\mu_m}{\mu} \frac{\rho}{\rho_m} \frac{w}{w_m} V \tag{7.14}$$

Note that this model design requires not only geometric scaling, as specified by Eq. 7.13, but also the correct scaling of the velocity in accordance with Eq. 7.14. This result is typical of most model designs—there is more to the design than simply scaling the geometry!

With the foregoing similarity requirements satisfied, the prediction equation for the drag is

$$\frac{\mathscr{D}}{w^2 \rho V^2} = \frac{\mathscr{D}_m}{w_m^2 \rho_m V_m^2}$$

or

$$\mathscr{D} = \left(\frac{w}{w_m}\right)^2 \left(\frac{\rho}{\rho_m}\right) \left(\frac{V}{V_m}\right)^2 \mathscr{D}_m$$

*Similarity between a model and a prototype is achieved by equating pi terms.*

Thus, a measured drag on the model, $\mathscr{D}_m$, must be multiplied by the ratio of the square of the plate widths, the ratio of the fluid densities, and the ratio of the square of the velocities to obtain the predicted value of the prototype drag, $\mathscr{D}$.

Generally, as is illustrated in this example, to achieve similarity between model and prototype behavior, *all the corresponding pi terms must be equated between model and prototype*. Usually, one or more of these pi terms will involve ratios of important lengths (such as $w/h$ in the foregoing example); that is, they are purely geometrical. Thus, when we equate the pi terms involving length ratios, we are requiring that there be complete *geometric similarity* between the model and prototype. This means that the model must be a scaled version of the prototype. Geometric scaling may extend to the finest features of the system, such as surface roughness, or small protuberances on a structure, since these kinds of geometric features may significantly influence the flow. Any deviation from complete geometric similarity for a model must be carefully considered. Sometimes complete geometric scaling may be difficult to achieve, particularly when dealing with surface roughness, since roughness is difficult to characterize and control.

*V7.9 Environmental models*

© Stanford University (with permission)

Another group of typical pi terms (such as the Reynolds number in the foregoing example) involves force ratios as noted in Table 7.1. The equality of these pi terms requires the ratio of like forces in model and prototype to be the same. Thus, for flows in which the Reynolds numbers are equal, the ratio of viscous forces in model and prototype is equal to the ratio of inertia forces. This is further illustrated by the photographs in the margin. The top photograph is of flow past a cylinder with diameter, $D$, and velocity, $V_o$, while the velocity is doubled and the diameter halved in the bottom photograph (image shown in 2x enlargement). Since the Reynolds numbers match, the flows look and behave the same. If other pi terms are involved, such as the Froude number or Weber number, a similar conclusion can be drawn; that is, the equality of these pi terms requires the ratio of like forces in model and prototype to be the same. Thus, when these types of pi terms are equal in model and prototype, we have *dynamic similarity* between model and prototype. It follows that with both geometric and dynamic similarity the streamline patterns will be the same and corresponding velocity ratios ($V_m/V$) and acceleration ratios ($a_m/a$) are constant throughout the flow field. Thus, *kinematic similarity* exists between model and prototype. To have complete similarity between model and prototype, we must maintain geometric, kinematic, and dynamic similarity between the two systems. This will automatically follow if all the important variables are included in the dimensional analysis, and if all the similarity requirements based on the resulting pi terms are satisfied.

---

## F l u i d s   i n   t h e   N e w s

**Modeling parachutes in a water tunnel** The first use of a parachute with a free-fall jump from an aircraft occurred in 1914, although parachute jumps from hot-air balloons had occurred since the late 1700s. In more modern times parachutes are commonly used by the military and for safety and sport. It is not surprising that there remains interest in the design and characteristics of parachutes, and researchers at the Worcester Polytechnic Institute have been studying various aspects of the aerodynamics associated with parachutes. An unusual part of their study is that they are using small-scale para-

chutes tested in a *water tunnel*. The *model parachutes* are reduced in size by a factor of 30 to 60 times. Various types of tests can be performed, ranging from the study of the velocity fields in the wake of the canopy with a steady free-stream velocity to the study of conditions during rapid deployment of the canopy. According to the researchers, the advantage of using water as the working fluid, rather than air, is that the velocities and deployment dynamics are slower than in the atmosphere, thus providing more time to collect detailed experimental data. (See Problem 7.57.)

# EXAMPLE 7.5   Prediction of Prototype Performance from Model Data

**GIVEN**   A long structural component of a bridge has an elliptical cross section shown in Fig. E7.5. It is known that when a steady wind blows past this type of bluff body, vortices may develop on the downwind side that are shed in a regular fashion at some definite frequency. Since these vortices can create harmful periodic forces acting on the structure, it is important to determine the shedding frequency. For the specific structure of interest, $D = 0.1$ m, $H = 0.3$ m, and a representative wind velocity is 50 km/hr. Standard air can be assumed. The shedding frequency is to be determined through the use of a small-scale model that is to be tested in a water tunnel. For the model $D_m = 20$ mm and the water temperature is 20 °C.

**FIND**   Determine the model dimension, $H_m$, and the velocity at which the test should be performed. If the shedding frequency for the model is found to be 49.9 Hz, what is the corresponding frequency for the prototype?

■ **Figure E7.5**

*V7.10 Flow past an ellipse*

## SOLUTION

We expect the shedding frequency, $\omega$, to depend on the lengths $D$ and $H$, the approach velocity, $V$, and the fluid density, $\rho$, and viscosity, $\mu$. Thus,

$$\omega = f(D, H, V, \rho, \mu)$$

where

$$\omega \doteq T^{-1}$$
$$D \doteq L$$
$$H \doteq L$$
$$V \doteq LT^{-1}$$
$$\rho \doteq ML^{-3}$$
$$\mu \doteq ML^{-1}T^{-1}$$

Since there are six variables and three reference dimensions $(MLT)$, three pi terms are required. Application of the pi theorem yields

$$\frac{\omega D}{V} = \phi\left(\frac{D}{H}, \frac{\rho VD}{\mu}\right)$$

We recognize the pi term on the left as the Strouhal number, and the dimensional analysis indicates that the Strouhal number is a function of the geometric parameter, $D/H$, and the Reynolds number. Thus, to maintain similarity between model and prototype

$$\frac{D_m}{H_m} = \frac{D}{H}$$

and

$$\frac{\rho_m V_m D_m}{\mu_m} = \frac{\rho VD}{\mu}$$

From the first similarity requirement

$$H_m = \frac{D_m}{D} H$$
$$= \frac{(20 \times 10^{-3} \text{ m})}{(0.1 \text{ m})}(0.3 \text{ m})$$
$$H_m = 60 \times 10^{-3} \text{ m} = 60 \text{ mm} \qquad \textbf{(Ans)}$$

The second similarity requirement indicates that the Reynolds number must be the same for model and prototype so that the model velocity must satisfy the condition

$$V_m = \frac{\mu_m}{\mu} \frac{\rho}{\rho_m} \frac{D}{D_m} V \qquad \textbf{(1)}$$

For air at standard conditions, $\mu = 1.79 \times 10^{-5}$ kg/m · s, $\rho = 1.23$ kg/m$^3$, and for water at 20 °C, $\mu = 1.00 \times 10^{-3}$ kg/m · s, $\rho = 998$ kg/m$^3$. The fluid velocity for the prototype is

$$V = \frac{(50 \times 10^3 \text{ m/hr})}{(3600 \text{ s/hr})} = 13.9 \text{ m/s}$$

The required velocity can now be calculated from Eq. 1 as

$$V_m = \frac{[1.00 \times 10^{-3} \text{ kg/(m · s)}] \, (1.23 \text{ kg/m}^3)}{[1.79 \times 10^{-5} \text{ kg/(m · s)}] \, (998 \text{ kg/m}^3)}$$
$$\times \frac{(0.1 \text{ m})}{(20 \times 10^{-3} \text{ m})}(13.9 \text{ m/s})$$
$$V_m = 4.79 \text{ m/s} \qquad \textbf{(Ans)}$$

This is a reasonable velocity that could be readily achieved in a water tunnel.

With the two similarity requirements satisfied, it follows that the Strouhal numbers for prototype and model will be the same so that

$$\frac{\omega D}{V} = \frac{\omega_m D_m}{V_m}$$

and the predicted prototype vortex shedding frequency is

$$\omega = \frac{V}{V_m} \frac{D_m}{D} \omega_m$$
$$= \frac{(13.9 \text{ m/s})}{(4.79 \text{ m/s})} \frac{(20 \times 10^{-3} \text{ m})}{(0.1 \text{ m})}(49.9 \text{ Hz})$$
$$\omega = 29.0 \text{ Hz} \qquad \textbf{(Ans)}$$

**COMMENT** This same model could also be used to predict the drag per unit length, $\mathcal{D}_\ell$ (lb/ft or N/m), on the prototype, since the drag would depend on the same variables as those used for the frequency. Thus, the similarity requirements would be the same and with these requirements satisfied it follows that the drag per unit length expressed in dimensionless form, such as $\mathcal{D}_\ell/D\rho V^2$, would be equal in model and prototype. The measured drag per unit length on the model could then be related to the corresponding drag per unit length on the prototype through the relationship

$$\mathcal{D}_\ell = \left(\frac{D}{D_m}\right)\left(\frac{\rho}{\rho_m}\right)\left(\frac{V}{V_m}\right)^2 \mathcal{D}_{\ell m}$$

### 7.8.2 Model Scales

It is clear from the preceding section that the ratio of like quantities for the model and prototype naturally arises from the similarity requirements. For example, if in a given problem there are two length variables $\ell_1$ and $\ell_2$, the resulting similarity requirement based on a pi term obtained from these two variables is

$$\frac{\ell_1}{\ell_2} = \frac{\ell_{1m}}{\ell_{2m}}$$

so that

$$\frac{\ell_{1m}}{\ell_1} = \frac{\ell_{2m}}{\ell_2}$$

*The ratio of a model variable to the corresponding prototype variable is called the scale for that variable.*

We define the ratio $\ell_{1m}/\ell_1$ or $\ell_{2m}/\ell_2$ as the **length scale.** For true models there will be only one length scale, and all lengths are fixed in accordance with this scale. There are, however, other scales such as the velocity scale, $V_m/V$, density scale, $\rho_m/\rho$, viscosity scale, $\mu_m/\mu$, and so on. In fact, we can define a scale for each of the variables in the problem. Thus, it is actually meaningless to talk about a "scale" of a model without specifying which scale.

We will designate the length scale as $\lambda_\ell$, and other scales as $\lambda_V$, $\lambda_\rho$, $\lambda_\mu$, and so on, where the subscript indicates the particular scale. Also, we will take the ratio of the model value to the prototype value as the scale (rather than the inverse). Length scales are often specified, for example, as $1:10$ or as a $\frac{1}{10}$ scale model. The meaning of this specification is that the model is one-tenth the size of the prototype, and the tacit assumption is that all relevant lengths are scaled accordingly so the model is geometrically similar to the prototype.

## F l u i d s   i n   t h e   N e w s

**"Galloping Gertie"** One of the most dramatic bridge collapses occurred in 1940 when the Tacoma Narrows Bridge, located near Tacoma, Washington, failed due to aerodynamic instability. The bridge had been nicknamed "Galloping Gertie" due to its tendency to sway and move in high winds. On the fateful day of the collapse the wind speed was 65 km/hr. This particular combination of a high wind and the aeroelastic properties of the bridge created large oscillations leading to its failure. The bridge was replaced in 1950, and a second bridge parallel to the existing structure was opened in 2007. To determine possible wind interference effects due to two bridges in close proximity, wind tunnel tests were run in a 9 m × 9 m wind tunnel operated by the National Research Council of Canada. *Models* of the two side-by-side bridges, each having a length scale of 1:211, were tested under various wind conditions. Since the failure of the original Tacoma Narrows Bridge, it is now common practice to use wind tunnel model studies during the design process to evaluate any bridge that is to be subjected to wind-induced vibrations. (See Problem 7.84.)

### 7.8.3 Practical Aspects of Using Models

*Validation of Model Design.* Most model studies involve simplifying assumptions with regard to the variables to be considered. Although the number of assumptions is frequently less stringent than that required for mathematical models, they nevertheless introduce some uncertainty in the model design. It is, therefore, desirable to check the design experimentally whenever possible. In some situations the purpose of the model is to predict the effects of certain proposed changes in a given prototype, and in this instance some actual prototype data may be available. The model can be designed, constructed, and tested, and the model prediction can be compared

*V7.11 Model of
fish hatchery pond*

with these data. If the agreement is satisfactory, then the model can be changed in the desired manner, and the corresponding effect on the prototype can be predicted with increased confidence.

Another useful and informative procedure is to run tests with a series of models of different sizes, where one of the models can be thought of as the prototype and the others as "models" of this prototype. With the models designed and operated on the basis of the proposed design, a necessary condition for the validity of the model design is that an accurate prediction be made between any pair of models, since one can always be considered as a model of the other. Although suitable agreement in validation tests of this type does not unequivocally indicate a correct model design (e.g., the length scales between laboratory models may be significantly different than required for actual prototype prediction), it is certainly true that if agreement between models cannot be achieved in these tests, there is no reason to expect that the same model design can be used to predict prototype behavior correctly.

***Distorted Models.***   Although the general idea behind establishing similarity requirements for models is straightforward (we simply equate pi terms), it is not always possible to satisfy all the known requirements. If one or more of the similarity requirements are not met, for example, if $\Pi_{2m} \neq \Pi_2$, then it follows that the prediction equation $\Pi_1 = \Pi_{1m}$ is not true; that is, $\Pi_1 \neq \Pi_{1m}$. Models for which one or more of the similarity requirements are not satisfied are called ***distorted models.***

*Models for which
one or more simi-
larity requirements
are not satisfied
are called distorted
models.*

Distorted models are rather commonplace, and they can arise for a variety of reasons. For example, perhaps a suitable fluid cannot be found for the model. The classic example of a distorted model occurs in the study of open channel or free-surface flows. Typically, in these problems both the Reynolds number, $\rho V \ell / \mu$, and the Froude number, $V/\sqrt{g\ell}$, are involved.

Froude number similarity requires

$$\frac{V_m}{\sqrt{g_m \ell_m}} = \frac{V}{\sqrt{g\ell}}$$

If the model and prototype are operated in the same gravitational field, then the required velocity scale is

$$\frac{V_m}{V} = \sqrt{\frac{\ell_m}{\ell}} = \sqrt{\lambda_\ell}$$

Reynolds number similarity requires

$$\frac{\rho_m V_m \ell_m}{\mu_m} = \frac{\rho V \ell}{\mu}$$

*V7.12 Distorted
river model*

and the velocity scale is

$$\frac{V_m}{V} = \frac{\mu_m}{\mu} \frac{\rho}{\rho_m} \frac{\ell}{\ell_m}$$

Since the velocity scale must be equal to the square root of the length scale, it follows that

$$\frac{\mu_m / \rho_m}{\mu / \rho} = \frac{\nu_m}{\nu} = (\lambda_\ell)^{3/2} \tag{7.15}$$

where the ratio $\mu/\rho$ is the kinematic viscosity, $\nu$. Although in principle it may be possible to satisfy this design condition, it may be quite difficult, if not impossible, to find a suitable model fluid, particularly for small length scales. For problems involving rivers, spillways, and harbors, for which the prototype fluid is water, the models are also relatively large so that the only practical model fluid is water. However, in this case (with the kinematic viscosity scale equal to unity) Eq. 7.15 will not be satisfied, and a distorted model will result. Generally, hydraulic models of this type are distorted and are designed on the basis of the Froude number, with the Reynolds number different in model and prototype.

Distorted models can be successfully used, but the interpretation of results obtained with this type of model is obviously more difficult than the interpretation of results obtained with ***true models*** for which all similarity requirements are met. There are no general rules for handling distorted models, and essentially each problem must be considered on its own merits. The success of using distorted models depends to a large extent on the skill and experience of the investigator responsible for the design of the model and in the interpretation of experimental data obtained from the model. Distorted models are widely used, and additional information can be found in the references at the end of the chapter. References 14 and 15 contain detailed discussions of several practical examples of distorted fluid flow and hydraulic models.

---

**F l u i d s    i n    t h e    N e w s**

**Old Man River in (large) miniature** One of the world's largest scale models, a Mississippi River model, resides near Jackson, Mississippi. It is a detailed, complex model that covers many acres and replicates the 1,250,000-acre Mississippi River basin. Built by the Army Corps of Engineers and used from 1943 to 1973, today it has mostly gone to ruin. As with many hydraulic models, this is a *distorted model*, with a horizontal scale of 1:2000 and a vertical scale of 1:100. One step along the model river corresponds to one mile along the river. All essential river basin elements such as geological features, levees, and railroad embankments were sculpted by hand to match the actual contours. The main purpose of the model was to predict floods. This was done by supplying specific amounts of water at prescribed locations along the model and then measuring the water depths up and down the model river. Because of the length scale, there is a difference in the time taken by the corresponding model and prototype events. Although it takes days for the actual floodwaters to travel from Sioux City, Iowa, to Omaha, Nebraska, it would take only minutes for the simulated flow in the model.

---

## 7.9    Some Typical Model Studies

Models are used to investigate many different types of fluid mechanics problems, and it is difficult to characterize in a general way all necessary similarity requirements, since each problem is unique. We can, however, broadly classify many of the problems on the basis of the general nature of the flow and subsequently develop some general characteristics of model designs in each of these classifications. In the following sections we will consider models for the study of (1) flow through closed conduits, (2) flow around immersed bodies, and (3) flow with a free surface. Turbomachine models are considered in Chapter 12.

### 7.9.1    Flow through Closed Conduits

*Geometric and Reynolds number similarity is usually required for models involving flow through closed conduits.*

Common examples of this type of flow include pipe flow and flow through valves, fittings, and metering devices. Although the conduit cross sections are often circular, they could have other shapes as well and may contain expansions or contractions. Since there are no fluid interfaces or free surfaces, the dominant forces are inertial and viscous so that the Reynolds number is an important similarity parameter. For low Mach numbers (Ma < 0.3), compressibility effects are usually negligible for both the flow of liquids and gases. For this class of problems, geometric similarity between model and prototype must be maintained. Generally the geometric characteristics can be described by a series of length terms, $\ell_1, \ell_2, \ell_3, \ldots, \ell_i$, and $\ell$, where $\ell$ is some particular length dimension for the system. Such a series of length terms leads to a set of pi terms of the form

$$\Pi_i = \frac{\ell_i}{\ell}$$

where $i = 1, 2, \ldots$, and so on. In addition to the basic geometry of the system, the roughness of the internal surface in contact with the fluid may be important. If the average height of surface roughness elements is defined as $\varepsilon$, then the pi term representing roughness will be $\varepsilon/\ell$. This parameter indicates that for complete geometric similarity, surface roughness would also have to be scaled. Note that this implies that for length scales less than 1, the model surfaces should be smoother than those in the prototype since $\varepsilon_m = \lambda_\ell \varepsilon$. To further complicate matters, the pattern of roughness elements in model and prototype would have to be similar. These are conditions that are virtually impossible to satisfy exactly. Fortunately, in some problems the surface roughness plays a minor role and can be neglected. However, in other problems (such as turbulent flow through pipes) roughness can be very important.

It follows from this discussion that for flow in closed conduits at low Mach numbers, any dependent pi term (the one that contains the particular variable of interest, such as pressure drop) can be expressed as

$$\text{Dependent pi term} = \phi\left(\frac{\ell_i}{\ell}, \frac{\varepsilon}{\ell}, \frac{\rho V \ell}{\mu}\right) \tag{7.16}$$

This is a general formulation for this type of problem. The first two pi terms of the right side of Eq. 7.16 lead to the requirement of geometric similarity so that

$$\frac{\ell_{im}}{\ell_m} = \frac{\ell_i}{\ell} \qquad \frac{\varepsilon_m}{\ell_m} = \frac{\varepsilon}{\ell}$$

or

$$\frac{\ell_{im}}{\ell_i} = \frac{\varepsilon_m}{\varepsilon} = \frac{\ell_m}{\ell} = \lambda_\ell$$

This result indicates that the investigator is free to choose a length scale, $\lambda_\ell$, but once this scale is selected, all other pertinent lengths must be scaled in the same ratio.

The additional similarity requirement arises from the equality of Reynolds numbers

$$\frac{\rho_m V_m \ell_m}{\mu_m} = \frac{\rho V \ell}{\mu}$$

From this condition the velocity scale is established so that

$$\frac{V_m}{V} = \frac{\mu_m}{\mu} \frac{\rho}{\rho_m} \frac{\ell}{\ell_m} \tag{7.17}$$

and the actual value of the velocity scale depends on the viscosity and density scales, as well as the length scale. Different fluids can be used in model and prototype. However, if the same fluid is used (with $\mu_m = \mu$ and $\rho_m = \rho$), then

$$\frac{V_m}{V} = \frac{\ell}{\ell_m}$$

Thus, $V_m = V/\lambda_\ell$, which indicates that the fluid velocity in the model will be larger than that in the prototype for any length scale less than 1. Since length scales are typically much less than unity, Reynolds number similarity may be difficult to achieve because of the large model velocities required.

With these similarity requirements satisfied, it follows that the dependent pi term will be equal in model and prototype. For example, if the dependent variable of interest is the pressure differential,[3] $\Delta p$, between two points along a closed conduit, then the dependent pi term could be expressed as

$$\Pi_1 = \frac{\Delta p}{\rho V^2}$$

The prototype pressure drop would then be obtained from the relationship

$$\Delta p = \frac{\rho}{\rho_m}\left(\frac{V}{V_m}\right)^2 \Delta p_m$$

so that from a measured pressure differential in the model, $\Delta p_m$, the corresponding pressure differential for the prototype could be predicted. Note that in general $\Delta p \neq \Delta p_m$.

---

[3]In some previous examples the pressure differential *per unit length*, $\Delta p_\ell$, was used. This is appropriate for flow in long pipes or conduits in which the pressure would vary linearly with distance. However, in the more general situation the pressure may not vary linearly with position so that it is necessary to consider the pressure differential, $\Delta p$, as the dependent variable. In this case the distance between pressure taps is an additional variable (as well as the distance of one of the taps measured from some reference point within the flow system).

# EXAMPLE 7.6 Reynolds Number Similarity

**GIVEN** Model tests are to be performed to study the flow through a large check valve having a 2-ft-diameter inlet and carrying water at a flowrate of 30 cfs as shown in Fig. E7.6a. The working fluid in the model is water at the same temperature as that in the prototype. Complete geometric similarity exists between model and prototype, and the model inlet diameter is 3 in.

**FIND** Determine the required flowrate in the model.

$Q = 30$ cfs

$(Q_m = ?)$

$D = 2$ ft
$(D_m = 3$ in.$)$

■ **Figure E7.6a**

## SOLUTION

To ensure dynamic similarity, the model tests should be run so that

$$\mathrm{Re}_m = \mathrm{Re}$$

or

$$\frac{V_m D_m}{\nu_m} = \frac{VD}{\nu}$$

where $V$ and $D$ correspond to the inlet velocity and diameter, respectively. Since the same fluid is to be used in model and prototype, $\nu = \nu_m$, and therefore

$$\frac{V_m}{V} = \frac{D}{D_m}$$

The discharge, $Q$, is equal to $VA$, where $A$ is the inlet area, so

$$\frac{Q_m}{Q} = \frac{V_m A_m}{VA} = \left(\frac{D}{D_m}\right)\frac{[(\pi/4)D_m^2]}{[(\pi/4)D^2]}$$

$$= \frac{D_m}{D}$$

and for the data given

$$Q_m = \frac{(3/12\ \mathrm{ft})}{(2\ \mathrm{ft})}\,(30\ \mathrm{ft}^3/\mathrm{s})$$

$$Q_m = 3.75\ \mathrm{cfs} \qquad \textbf{(Ans)}$$

**COMMENT** As indicated by the foregoing analysis, to maintain Reynolds number similarity using the same fluid in model and prototype, the required velocity scale is inversely proportional to the length scale, that is, $V_m/V = (D_m/D)^{-1}$. This strong influence of the length scale on the velocity scale is shown in

Fig. E7.6b. For this particular example, $D_m/D = 0.125$, and the corresponding velocity scale is 8 (see Fig. E7.6b). Thus, with the prototype velocity equal to $V = (30\ \mathrm{ft}^3/\mathrm{s})/(\pi/4)(2\ \mathrm{ft})^2 = 9.50\ \mathrm{ft/s}$, the required model velocity is $V_m = 76.4\ \mathrm{ft/s}$. Although this is a relatively large velocity, it could be attained in a laboratory facility. It is to be noted that if we tried to use a smaller model, say one with $D = 1$ in., the required model velocity is 229 ft/s, a very high velocity that would be difficult to achieve. These results are indicative of one of the difficulties encountered in maintaining Reynolds number similarity—the required model velocities may be impractical to obtain.

$V_m/V$

$D_m/D$

(0.125, 8)

■ **Figure E7.6b**

*In some problems Reynolds number similarity may be relaxed.*

Two additional points should be made with regard to modeling flows in closed conduits. First, for large Reynolds numbers, inertial forces are much larger than viscous forces, and in this case it may be possible to neglect viscous effects. The important practical consequence of this is that it would not be necessary to maintain Reynolds number similarity between model and prototype. However, *both* model and prototype would have to operate at large Reynolds numbers. Since we do not know, a priori, what a "large Reynolds number" is, the effect of Reynolds numbers would have to be determined from the model. This could be accomplished by varying the model Reynolds number to determine the range (if any) over which the dependent pi term ceases to be affected by changes in Reynolds number.

The second point relates to the possibility of cavitation in flow through closed conduits. For example, flow through the complex passages that may exist in valves may lead to local regions of high velocity (and thus low pressure), which can cause the fluid to cavitate. If the model is to be used to study cavitation phenomena, then the vapor pressure, $p_v$, becomes an important variable and an additional similarity requirement such as equality of the cavitation number $(p_r - p_v)/\frac{1}{2}\rho V^2$ is required, where $p_r$ is some reference pressure. The use of models to study cavitation is complicated, since it is not fully understood how vapor bubbles form and grow. The initiation of bubbles seems to be influenced by the microscopic particles that exist in most liquids, and how this aspect of the problem influences model studies is not clear. Additional details can be found in Ref. 17.

### 7.9.2  Flow around Immersed Bodies

*Geometric and Reynolds number similarity is usually required for models involving flow around bodies.*

Models have been widely used to study the flow characteristics associated with bodies that are completely immersed in a moving fluid. Examples include flow around aircraft, automobiles, golf balls, and buildings. (These types of models are usually tested in wind tunnels as is illustrated in Fig. 7.6.) Modeling laws for these problems are similar to those described in the preceding section; that is, geometric and Reynolds number similarity is required. Since there are no fluid interfaces, surface tension (and therefore the Weber number) is not important. Also, gravity will not affect the flow patterns, so the Froude number need not be considered. The Mach number will be important for high-speed flows in which compressibility becomes an important factor, but for incompressible fluids (such as liquids or for gases at relatively low speeds) the Mach number can be omitted as a similarity requirement. In this case, a general formulation for these problems is

$$\text{Dependent pi term} = \phi\left(\frac{\ell_i}{\ell}, \frac{\varepsilon}{\ell}, \frac{\rho V \ell}{\mu}\right) \tag{7.18}$$

*V7.13 Wind engineering models*

where $\ell$ is some characteristic length of the system and $\ell_i$ represents other pertinent lengths, $\varepsilon/\ell$ is the relative roughness of the surface (or surfaces), and $\rho V \ell/\mu$ is the Reynolds number.

Frequently, the dependent variable of interest for this type of problem is the drag, $\mathscr{D}$, developed on the body, and in this situation the dependent pi term would usually be expressed in the form of a *drag coefficient*, $C_D$, where

$$C_D = \frac{\mathscr{D}}{\frac{1}{2}\rho V^2 \ell^2}$$

The numerical factor, $\frac{1}{2}$, is arbitrary but commonly included, and $\ell^2$ is usually taken as some representative area of the object. Thus, drag studies can be undertaken with the formulation

$$\frac{\mathscr{D}}{\frac{1}{2}\rho V^2 \ell^2} = C_D = \phi\left(\frac{\ell_i}{\ell}, \frac{\varepsilon}{\ell}, \frac{\rho V \ell}{\mu}\right) \tag{7.19}$$

■ **Figure 7.6**  Model of the National Bank of Commerce, San Antonio, Texas, for measurement of peak, rms, and mean pressure distributions. The model is located in a long-test-section, meteorological wind tunnel. (Photograph courtesy of CPP Wind Engineering; www.cppwind.com.)

It is clear from Eq. 7.19 that geometric similarity

$$\frac{\ell_{im}}{\ell_m} = \frac{\ell_i}{\ell} \qquad \frac{\varepsilon_m}{\ell_m} = \frac{\varepsilon}{\ell}$$

as well as Reynolds number similarity

$$\frac{\rho_m V_m \ell_m}{\mu_m} = \frac{\rho V \ell}{\mu}$$

must be maintained. If these conditions are met, then

$$\frac{\mathcal{D}}{\frac{1}{2}\rho V^2 \ell^2} = \frac{\mathcal{D}_m}{\frac{1}{2}\rho_m V_m^2 \ell_m^2}$$

or

$$\mathcal{D} = \frac{\rho}{\rho_m}\left(\frac{V}{V_m}\right)^2 \left(\frac{\ell}{\ell_m}\right)^2 \mathcal{D}_m$$

*For flow around bodies, drag is often the dependent variable of interest.*

Measurements of model drag, $\mathcal{D}_m$, can then be used to predict the corresponding drag, $\mathcal{D}$, on the prototype from this relationship.

As was discussed in the previous section, one of the common difficulties with models is related to the Reynolds number similarity requirement which establishes the model velocity as

$$V_m = \frac{\mu_m}{\mu}\frac{\rho}{\rho_m}\frac{\ell}{\ell_m} V \tag{7.20}$$

or

$$V_m = \frac{\nu_m}{\nu}\frac{\ell}{\ell_m} V \tag{7.21}$$

where $\nu_m/\nu$ is the ratio of kinematic viscosities. If the same fluid is used for model and prototype so that $\nu_m = \nu$, then

$$V_m = \frac{\ell}{\ell_m} V$$

*V7.14 Model airplane test in water*

and, therefore, the required model velocity will be higher than the prototype velocity for $\ell/\ell_m$ greater than 1. Since this ratio is often relatively large, the required value of $V_m$ may be large. For example, for a $\frac{1}{10}$ length scale, and a prototype velocity of 50 mph, the required model velocity is 500 mph. This is a value that is unreasonably high to achieve with liquids, and for gas flows this would be in the range where compressibility would be important in the model (but not in the prototype).

As an alternative, we see from Eq. 7.21 that $V_m$ could be reduced by using a different fluid in the model such that $\nu_m/\nu < 1$. For example, the ratio of the kinematic viscosity of water to that of air is approximately $\frac{1}{10}$, so that if the prototype fluid were air, tests might be run on the model using water. This would reduce the required model velocity, but it still may be difficult to achieve the necessary velocity in a suitable test facility, such as a water tunnel.

Another possibility for wind tunnel tests would be to increase the air pressure in the tunnel so that $\rho_m > \rho$, thus reducing the required model velocity as specified by Eq. 7.20. Fluid viscosity is not strongly influenced by pressure. Although pressurized tunnels have been used, they are obviously more complicated and expensive.

*V7.15 Large scale wind tunnel*

The required model velocity can also be reduced if the length scale is modest; that is, the model is relatively large. For wind tunnel testing, this requires a large test section which greatly increases the cost of the facility. However, large wind tunnels suitable for testing very large models (or prototypes) are in use. One such tunnel, located at the NASA Ames Research Center, Moffett Field, California, has a test section that is 40 ft by 80 ft and can accommodate test speeds to 345 mph. Such a large and expensive test facility is obviously not feasible for university or industrial laboratories, so most model testing has to be accomplished with relatively small models.

# EXAMPLE 7.7 | Model Design Conditions and Predicted Prototype Performance

**GIVEN** The drag on the airplane shown in Fig. E7.7 cruising at 240 mph in standard air is to be determined from tests on a 1:10 scale model placed in a pressurized wind tunnel. To minimize compressibility effects, the airspeed in the wind tunnel is also to be 240 mph.

**FIND** Determine

**(a)** the required air pressure in the tunnel (assuming the same air temperature for model and prototype) and

**(b)** the drag on the prototype corresponding to a measured force of 1 lb on the model.

$V$ = 240 mph

■ **Figure E7.7** **(Photograph courtesy of NASA)**

## SOLUTION

**(a)** From Eq. 7.19 it follows that drag can be predicted from a geometrically similar model if the Reynolds numbers in model and prototype are the same. Thus,

$$\frac{\rho_m V_m \ell_m}{\mu_m} = \frac{\rho V \ell}{\mu}$$

For this example, $V_m = V$ and $\ell_m/\ell = \frac{1}{10}$ so that

$$\frac{\rho_m}{\rho} = \frac{\mu_m}{\mu} \frac{V}{V_m} \frac{\ell}{\ell_m}$$

$$= \frac{\mu_m}{\mu}(1)(10)$$

and therefore

$$\frac{\rho_m}{\rho} = 10\frac{\mu_m}{\mu}$$

This result shows that the same fluid with $\rho_m = \rho$ and $\mu_m = \mu$ cannot be used if Reynolds number similarity is to be maintained. One possibility is to pressurize the wind tunnel to increase the density of the air. We assume that an increase in pressure does not significantly change the viscosity so that the required increase in density is given by the relationship

$$\frac{\rho_m}{\rho} = 10$$

For an ideal gas, $p = \rho RT$ so that

$$\frac{p_m}{p} = \frac{\rho_m}{\rho}$$

for constant temperature ($T = T_m$). Therefore, the wind tunnel would need to be pressurized so that

$$\frac{p_m}{p} = 10$$

Since the prototype operates at standard atmospheric pressure, the required pressure in the wind tunnel is 10 atmospheres or

$$p_m = 10\,(14.7\ \text{psia})$$
$$= 147\ \text{psia} \qquad \textbf{(Ans)}$$

**COMMENT** Thus, we see that a high pressure would be required and this could not be achieved easily or inexpensively. However, under these conditions, Reynolds similarity would be attained.

**(b)** The drag could be obtained from Eq. 7.19 so that

$$\frac{\mathscr{D}}{\frac{1}{2}\rho V^2 \ell^2} = \frac{\mathscr{D}_m}{\frac{1}{2}\rho_m V_m^2 \ell_m^2}$$

or

$$\mathscr{D} = \frac{\rho}{\rho_m}\left(\frac{V}{V_m}\right)^2\left(\frac{\ell}{\ell_m}\right)^2 \mathscr{D}_m$$

$$= \left(\frac{1}{10}\right)(1)^2(10)^2 \mathscr{D}_m$$

$$= 10\mathscr{D}_m$$

Thus, for a drag of 1 lb on the model the corresponding drag on the prototype is

$$\mathscr{D} = 10\ \text{lb} \qquad \textbf{(Ans)}$$

*V7.16 Full Scale Race Car Model*

Fortunately, in many situations the flow characteristics are not strongly influenced by the Reynolds number over the operating range of interest. In these cases we can avoid the rather stringent similarity requirement of matching Reynolds numbers. To illustrate this point, consider the variation in the drag coefficient with the Reynolds number for a smooth sphere of diameter $d$ placed in a uniform stream with approach velocity, $V$. Some typical data are shown in Fig. 7.7. We observe that for Reynolds numbers between approximately $10^3$ and $2 \times 10^5$ the drag coefficient is relatively constant and does not strongly depend on the specific value of the Reynolds number. Thus, exact Reynolds number similarity is not required in this range. For other geometric shapes we would typically find that for high Reynolds numbers, inertial forces are dominant (rather than viscous forces), and the drag is essentially independent of the Reynolds number.

■ **Figure 7.7** The effect of Reynolds number on the drag coefficient, $C_D$, for a smooth sphere with $C_D = \mathcal{D}/\frac{1}{2}A\rho V^2$, where $A$ is the projected area of sphere, $\pi d^2/4$. (Adapted from Kurt Gramoll, www.eCourses.ou.edu. Data from Ref. 16.)

*At high Reynolds numbers the drag is often essentially independent of the Reynolds number.*

Another interesting point to note from Fig. 7.7 is the rather abrupt drop in the drag coefficient near a Reynolds number of $3 \times 10^5$. As is discussed in Section 9.3.3, this is due to a change in the flow conditions near the surface of the sphere. These changes are influenced by the surface roughness and, in fact, the drag coefficient for a sphere with a "rougher" surface will generally be less than that of the smooth sphere for high Reynolds number. For example, the dimples on a golf ball are used to reduce the drag over that which would occur for a smooth golf ball. Although this is undoubtedly of great interest to the avid golfer, it is also important to engineers responsible for fluid-flow models, since it does emphasize the potential importance of the surface roughness. However, for bodies that are sufficiently angular with sharp corners, the actual surface roughness is likely to play a secondary role compared with the main geometric features of the body.

One final note with regard to Fig. 7.7 concerns the interpretation of experimental data when plotting pi terms. For example, if $\rho$, $\mu$, and $d$ remain constant, then an increase in Re comes from an increase in $V$. Intuitively, it would seem in general that if $V$ increases, the drag would increase. However, as shown in the figure, the drag coefficient generally decreases with increasing Re. When interpreting data, one needs to be aware if the variables are nondimensional. In this case, the physical drag force is proportional to the drag coefficient times the velocity squared. Thus, as shown by the figure in the margin, the drag force does, as expected, increase with increasing velocity. The exception occurs in the Reynolds number range $2 \times 10^5 < \text{Re} < 4 \times 10^5$ where the drag coefficient decreases dramatically with increasing Reynolds number (see Fig. 7.7). This phenomena is discussed in Section 9.3.

For problems involving high velocities in which the Mach number is greater than about 0.3, the influence of compressibility, and therefore the Mach number (or Cauchy number), becomes significant. In this case complete similarity requires not only geometric and Reynolds number similarity but also Mach number similarity so that

$$\frac{V_m}{c_m} = \frac{V}{c} \tag{7.22}$$

This similarity requirement, when combined with that for Reynolds number similarity (Eq. 7.21), yields

$$\frac{c}{c_m} = \frac{\nu}{\nu_m}\frac{\ell_m}{\ell} \tag{7.23}$$

Clearly the same fluid with $c = c_m$ and $\nu = \nu_m$ cannot be used in model and prototype unless the length scale is unity (which means that we are running tests on the prototype). In high-speed aerodynamics the prototype fluid is usually air, and it is difficult to satisfy Eq. 7.23 for reasonable length scales. Thus, models involving high-speed flows are often distorted with respect to Reynolds number similarity, but Mach number similarity is maintained.

### 7.9.3  Flow with a Free Surface

*Froude number similarity is usually required for models involving free-surface flows.*

Flows in canals, rivers, spillways, and stilling basins, as well as flow around ships, are all examples of flow phenomena involving a free surface. For this class of problems, both gravitational and inertial forces are important and, therefore, the Froude number becomes an important similarity parameter. Also, since there is a free surface with a liquid–air interface, forces due to surface tension may be significant, and the Weber number becomes another similarity parameter that needs to be considered along with the Reynolds number. Geometric variables will obviously still be important. Thus, a general formulation for problems involving flow with a free surface can be expressed as

$$\text{Dependent pi term} = \phi\left(\frac{\ell_i}{\ell}, \frac{\varepsilon}{\ell}, \frac{\rho V \ell}{\mu}, \frac{V}{\sqrt{g\ell}}, \frac{\rho V^2 \ell}{\sigma}\right) \tag{7.24}$$

As discussed previously, $\ell$ is some characteristic length of the system, $\ell_i$ represents other pertinent lengths, and $\varepsilon/\ell$ is the relative roughness of the various surfaces. Since gravity is the driving force in these problems, Froude number similarity is definitely required so that

$$\frac{V_m}{\sqrt{g_m \ell_m}} = \frac{V}{\sqrt{g\ell}}$$

The model and prototype are expected to operate in the same gravitational field ($g_m = g$), and therefore it follows that

$$\frac{V_m}{V} = \sqrt{\frac{\ell_m}{\ell}} = \sqrt{\lambda_\ell} \tag{7.25}$$

*V7.17 River flow model*

Thus, when models are designed on the basis of Froude number similarity, the velocity scale is determined by the square root of the length scale. As is discussed in Section 7.8.3, to simultaneously have Reynolds and Froude number similarity it is necessary that the kinematic viscosity scale be related to the length scale as

$$\frac{\nu_m}{\nu} = (\lambda_\ell)^{3/2} \tag{7.26}$$

The working fluid for the prototype is normally either freshwater or seawater and the length scale is small. Under these circumstances it is virtually impossible to satisfy Eq. 7.26, so models involving free-surface flows are usually distorted. The problem is further complicated if an attempt is made to model surface tension effects, since this requires the equality of Weber numbers, which leads to the condition

$$\frac{\sigma_m/\rho_m}{\sigma/\rho} = (\lambda_\ell)^2 \tag{7.27}$$

for the kinematic surface tension ($\sigma/\rho$). It is again evident that the same fluid cannot be used in model and prototype if we are to have similitude with respect to surface tension effects for $\lambda_\ell \neq 1$.

*V7.18 Boat model*

Fortunately, in many problems involving free-surface flows, both surface tension and viscous effects are small and consequently strict adherence to Weber and Reynolds number similarity is not required. Certainly, surface tension is not important in large hydraulic structures and rivers. Our only concern would be if in a model the depths were reduced to the point where surface tension becomes an important factor, whereas it is not in the prototype. This is of particular importance in the design of river models, since the length scales are typically small (so that the width of the model is reasonable), but with a small length scale the required model depth may be very small. To overcome this problem, different horizontal and vertical length scales are often used for river

■ **Figure 7.8** A scale hydraulic model (1:197) of the Guri Dam in Venezuela which is used to simulate the characteristics of the flow over and below the spillway and the erosion below the spillway. (Photograph courtesy of St. Anthony Falls Hydraulic Laboratory.)

models. Although this approach eliminates surface tension effects in the model, it introduces geometric distortion that must be accounted for empirically, usually by increasing the model surface roughness. It is important in these circumstances that verification tests with the model be performed (if possible) in which model data are compared with available prototype river flow data. Model roughness can be adjusted to give satisfactory agreement between model and prototype, and then the model subsequently used to predict the effect of proposed changes on river characteristics (such as velocity patterns or surface elevations).

For large hydraulic structures, such as dam spillways, the Reynolds numbers are large so that viscous forces are small in comparison to the forces due to gravity and inertia. In this case, Reynolds number similarity is not maintained and models are designed on the basis of Froude number similarity. Care must be taken to ensure that the model Reynolds numbers are also large, but they are not required to be equal to those of the prototype. This type of hydraulic model is usually made as large as possible so that the Reynolds number will be large. A spillway model is shown in Fig. 7.8. Also, for relatively large models the geometric features of the prototype can be accurately scaled, as well as surface roughness. Note that $\varepsilon_m = \lambda_\ell \varepsilon$, which indicates that the model surfaces must be smoother than the corresponding prototype surfaces for $\lambda_\ell < 1$.

*V7.19 Dam model*

## F l u i d s   i n   t h e   N e w s

**Ice engineering** Various types of models have been studied in wind tunnels, water tunnels, and towing tanks for many years. But another type of facility is needed to study ice and ice-related problems. The U.S. Army Cold Regions Research and Engineering Laboratory has developed a unique complex that houses research facilities for studies related to the mechanical behavior of ice and ice–structure interactions. The laboratory contains three separate cold-rooms—a test basin, a flume, and a general research area. In the test basin, large-scale *model studies* of ice forces on structures

such as dams, piers, ships, and offshore platforms can be performed. Ambient temperatures can be controlled as low as −20 °F, and at this temperature a 2-mm-per-hour ice growth rate can be achieved. It is also possible to control the mechanical properties of the ice to properly match the physical scale of the model. Tests run in the recirculating flume can simulate river processes during ice formation. And in the large research area, scale models of lakes and rivers can be built and operated to model ice interactions with various types of engineering projects. (See Problem 7.85.)

## *E*XAMPLE 7.8   **Froude Number Similarity**

**GIVEN** The spillway for the dam shown in Fig. E7.8a is 20 m wide and is designed to carry 125 m³/s at flood stage. A 1:15 model is constructed to study the flow characteristics through the spillway. The effects of surface tension and viscosity are to be neglected.

**FIND**

**(a)** Determine the required model width and flowrate.

**(b)** What operating time for the model corresponds to a 24-hr period in the prototype?

■ **Figure E7.8a** (Photo by Marty Melchior.)

## SOLUTION

The width, $w_m$, of the model spillway is obtained from the length scale, $\lambda_\ell$, so that

$$\frac{w_m}{w} = \lambda_\ell = \frac{1}{15}$$

Therefore,

$$w_m = \frac{20 \text{ m}}{15} = 1.33 \text{ m} \qquad \textbf{(Ans)}$$

Of course, all other geometric features (including surface roughness) of the spillway must be scaled in accordance with the same length scale.

With the neglect of surface tension and viscosity, Eq. 7.24 indicates that dynamic similarity will be achieved if the Froude numbers are equal between model and prototype. Thus,

$$\frac{V_m}{\sqrt{g_m \ell_m}} = \frac{V}{\sqrt{g\ell}}$$

and for $g_m = g$

$$\frac{V_m}{V} = \sqrt{\frac{\ell_m}{\ell}}$$

Since the flowrate is given by $Q = VA$, where $A$ is an appropriate cross-sectional area, it follows that

$$\frac{Q_m}{Q} = \frac{V_m A_m}{VA} = \sqrt{\frac{\ell_m}{\ell}}\left(\frac{\ell_m}{\ell}\right)^2$$

$$= (\lambda_\ell)^{5/2}$$

where we have made use of the relationship $A_m/A = (\ell_m/\ell)^2$. For $\lambda_\ell = \frac{1}{15}$ and $Q = 125 \text{ m}^3/\text{s}$

$$Q_m = (\tfrac{1}{15})^{5/2}(125 \text{ m}^3/\text{s}) = 0.143 \text{ m}^3/\text{s} \qquad \textbf{(Ans)}$$

The time scale can be obtained from the velocity scale, since the velocity is distance divided by time ($V = \ell/t$), and therefore

$$\frac{V}{V_m} = \frac{\ell}{t}\frac{t_m}{\ell_m}$$

or

$$\frac{t_m}{t} = \frac{V}{V_m}\frac{\ell_m}{\ell} = \sqrt{\frac{\ell_m}{\ell}} = \sqrt{\lambda_\ell}$$

This result indicates that time intervals in the model will be smaller than the corresponding intervals in the prototype if $\lambda_\ell < 1$. For $\lambda_\ell = \frac{1}{15}$ and a prototype time interval of 24 hr

$$t_m = \sqrt{\tfrac{1}{15}}\,(24 \text{ hr}) = 6.20 \text{ hr} \qquad \textbf{(Ans)}$$

**COMMENT**    As indicated by the foregoing analysis, the time scale varies directly as the square root of the length scale. Thus, as shown in Fig. E7.8b, the model time interval, $t_m$, corresponding to a 24-hr prototype time interval can be varied by changing the length scale, $\lambda_\ell$. The ability to scale times may be very useful, since it is possible to "speed up" events in the model that may occur over a relatively long time in the prototype. There is, of course, a practical limit to how small the length scale (and the corresponding time scale) can become. For example, if the length scale is too small then surface tension effects may become important in the model whereas they are not in the prototype. In such a case the present model design, based simply on Froude number similarity, would not be adequate.

■ **Figure E7.8b**

There are, unfortunately, problems involving flow with a free surface in which viscous, inertial, and gravitational forces are all important. The drag on a ship as it moves through water is due to the viscous shearing stresses that develop along its hull, as well as a pressure-induced component of drag caused by both the shape of the hull and wave action. The shear drag is a function of the Reynolds number, whereas the pressure drag is a function of the Froude number. Since both Reynolds number and Froude number similarity cannot be simultaneously achieved by using water as the model fluid (which is the only practical fluid for ship models), some technique other than a straightforward model test must be employed. One common approach is to measure the total drag on a small, geometrically similar model as it is towed through a model basin at Froude numbers matching those of the prototype. The shear drag on the model is calculated using analytical techniques of the type described in Chapter 9. This calculated value is then subtracted from the total drag to obtain pressure drag, and using Froude number scaling the pressure drag on the prototype can then be predicted. The experimentally determined value can then be combined with a calculated value of the shear drag (again using analytical techniques) to provide the desired total drag

*V7.20 Testing of large yacht model*

■ **Figure 7.9** Instrumented, small-waterplane-area, twin hull (SWATH) model suspended from a towing carriage. (Photograph courtesy of the U.S. Navy's David W. Taylor Research Center.)

on the ship. Ship models are widely used to study new designs, but the tests require extensive facilities (see Fig. 7.9).

It is clear from this brief discussion of various types of models involving free-surface flows that the design and use of such models requires considerable ingenuity, as well as a good understanding of the physical phenomena involved. This is generally true for most model studies. Modeling is both an art and a science. Motion picture producers make extensive use of model ships, fires, explosions, and the like. It is interesting to attempt to observe the flow differences between these distorted model flows and the real thing.

## 7.10 Similitude Based on Governing Differential Equations

*Similarity laws can be directly developed from the equations governing the phenomenon of interest.*

In the preceding sections of this chapter, dimensional analysis has been used to obtain similarity laws. This is a simple, straightforward approach to modeling, which is widely used. The use of dimensional analysis requires only a knowledge of the variables that influence the phenomenon of interest. Although the simplicity of this approach is attractive, it must be recognized that omission of one or more important variables may lead to serious errors in the model design. An alternative approach is available if the equations (usually differential equations) governing the phenomenon are known. In this situation similarity laws can be developed from the governing equations, even though it may not be possible to obtain analytic solutions to the equations.

To illustrate the procedure, consider the flow of an incompressible Newtonian fluid. For simplicity we will restrict our attention to two-dimensional flow, although the results are applicable to the general three-dimensional case. From Chapter 6 we know that the governing equations are the continuity equation

$$\frac{\partial u}{\partial x} + \frac{\partial v}{\partial y} = 0 \tag{7.28}$$

and the Navier–Stokes equations

$$\rho\left(\frac{\partial u}{\partial t} + u\frac{\partial u}{\partial x} + v\frac{\partial u}{\partial y}\right) = -\frac{\partial p}{\partial x} + \mu\left(\frac{\partial^2 u}{\partial x^2} + \frac{\partial^2 u}{\partial^2 y}\right) \tag{7.29}$$

$$\rho\left(\frac{\partial v}{\partial t} + u\frac{\partial v}{\partial x} + v\frac{\partial v}{\partial y}\right) = -\frac{\partial p}{\partial y} - \rho g + \mu\left(\frac{\partial^2 v}{\partial x^2} + \frac{\partial^2 v}{\partial y^2}\right) \tag{7.30}$$

where the $y$ axis is vertical, so that the gravitational body force, $\rho g$, only appears in the "$y$ equation." To continue the mathematical description of the problem, boundary conditions are required. For example, velocities on all boundaries may be specified; that is, $u = u_B$ and $v = v_B$ at all boundary points $x = x_B$ and $y = y_B$. In some types of problems it may be necessary to specify the pressure over some part of the boundary. For time-dependent problems, initial conditions would also have to be provided, which means that the values of all dependent variables would be given at some time (usually taken at $t = 0$).

Once the governing equations, including boundary and initial conditions, are known, we are ready to proceed to develop similarity requirements. The next step is to define a new set of variables that are dimensionless. To do this we select a reference quantity for each type of variable. In this problem the variables are $u$, $v$, $p$, $x$, $y$, and $t$ so we will need a reference velocity, $V$, a reference pressure, $p_0$, a reference length, $\ell$, and a reference time, $\tau$. These reference quantities should be parameters that appear in the problem. For example, $\ell$ may be a characteristic length of a body immersed in a fluid or the width of a channel through which a fluid is flowing. The velocity, $V$, may be the free-stream velocity or the inlet velocity. The new dimensionless (starred) variables can be expressed as

*Each variable is made dimensionless by dividing by an appropriate reference quantity.*

$$u^* = \frac{u}{V} \qquad v^* = \frac{v}{V} \qquad p^* = \frac{p}{p_0}$$

$$x^* = \frac{x}{\ell} \qquad y^* = \frac{y}{\ell} \qquad t^* = \frac{t}{\tau}$$

as shown in the figure in the margin.

The governing equations can now be rewritten in terms of these new variables. For example,

$$\frac{\partial u}{\partial x} = \frac{\partial V u^*}{\partial x^*} \frac{\partial x^*}{\partial x} = \frac{V}{\ell} \frac{\partial u^*}{\partial x^*}$$

and

$$\frac{\partial^2 u}{\partial x^2} = \frac{V}{\ell} \frac{\partial}{\partial x^*} \left( \frac{\partial u^*}{\partial x^*} \right) \frac{\partial x^*}{\partial x} = \frac{V}{\ell^2} \frac{\partial^2 u^*}{\partial x^{*2}}$$

Actual

Dimensionless

The other terms that appear in the equations can be expressed in a similar fashion. Thus, in terms of the new variables the governing equations become

$$\frac{\partial u^*}{\partial x^*} + \frac{\partial v^*}{\partial y^*} = 0 \qquad (7.31)$$

and

$$\left[ \frac{\rho V}{\tau} \right] \frac{\partial u^*}{\partial t^*} + \left[ \frac{\rho V^2}{\ell} \right] \left( u^* \frac{\partial u^*}{\partial x^*} + v^* \frac{\partial u^*}{\partial y^*} \right) = -\left[ \frac{p_0}{\ell} \right] \frac{\partial p^*}{\partial x^*} + \left[ \frac{\mu V}{\ell^2} \right] \left( \frac{\partial^2 u^*}{\partial x^{*2}} + \frac{\partial^2 u^*}{\partial y^{*2}} \right) \qquad (7.32)$$

$$\underbrace{\left[ \frac{\rho V}{\tau} \right]}_{F_{I\ell}} \frac{\partial v^*}{\partial t^*} + \underbrace{\left[ \frac{\rho V^2}{\ell} \right]}_{F_{Ic}} \left( u^* \frac{\partial v^*}{\partial x^*} + v^* \frac{\partial v^*}{\partial y^*} \right)$$

$$= -\underbrace{\left[ \frac{p_0}{\ell} \right]}_{F_P} \frac{\partial p^*}{\partial y^*} - \underbrace{[\rho g]}_{F_G} + \underbrace{\left[ \frac{\mu V}{\ell^2} \right]}_{F_V} \left( \frac{\partial^2 v^*}{\partial x^{*2}} + \frac{\partial^2 v^*}{\partial y^{*2}} \right) \qquad (7.33)$$

The terms appearing in brackets contain the reference quantities and can be interpreted as indices of the various forces (per unit volume) that are involved. Thus, as is indicated in Eq. 7.33, $F_{I\ell} =$ inertia (local) force, $F_{Ic} =$ inertia (convective) force, $F_p =$ pressure force, $F_G =$ gravitational force, and $F_V =$ viscous force. As the final step in the nondimensionalization process, we will divide each term in Eqs. 7.32 and 7.33 by one of the bracketed quantities. Although any one of these quantities could be used, it is conventional to divide by the bracketed quantity $\rho V^2/\ell$ which is the index of the convective inertia force. The final nondimensional form then becomes

$$\left[ \frac{\ell}{\tau V} \right] \frac{\partial u^*}{\partial t^*} + u^* \frac{\partial u^*}{\partial x^*} + v^* \frac{\partial u^*}{\partial y^*} = -\left[ \frac{p_0}{\rho V^2} \right] \frac{\partial p^*}{\partial x^*} + \left[ \frac{\mu}{\rho V \ell} \right] \left( \frac{\partial^2 u^*}{\partial x^{*2}} + \frac{\partial^2 u^*}{\partial y^{*2}} \right) \qquad (7.34)$$

$$\left[ \frac{\ell}{\tau V} \right] \frac{\partial v^*}{\partial t^*} + u^* \frac{\partial v^*}{\partial x^*} + v^* \frac{\partial v^*}{\partial y^*} = -\left[ \frac{p_0}{\rho V^2} \right] \frac{\partial p^*}{\partial y^*} - \left[ \frac{g \ell}{V^2} \right] + \left[ \frac{\mu}{\rho V \ell} \right] \left( \frac{\partial^2 v^*}{\partial x^{*2}} + \frac{\partial^2 v^*}{\partial y^{*2}} \right) \qquad (7.35)$$

We see that bracketed terms are the standard dimensionless groups (or their reciprocals) which were developed from dimensional analysis; that is, $\ell/\tau V$ is a form of the Strouhal number, $p_0/\rho V^2$

the Euler number, $g\ell/V^2$ the reciprocal of the square of the Froude number, and $\mu/\rho V\ell$ the reciprocal of the Reynolds number. From this analysis it is now clear how each of the dimensionless groups can be interpreted as the ratio of two forces, and how these groups arise naturally from the governing equations.

*Governing equations expressed in terms of dimensionless variables lead to the appropriate dimensionless groups.*

Although we really have not helped ourselves with regard to obtaining an analytical solution to these equations (they are still complicated and not amenable to an analytical solution), the dimensionless forms of the equations, Eqs. 7.31, 7.34, and 7.35, can be used to establish similarity requirements. From these equations it follows that if two systems are governed by these equations, then the solutions (in terms of $u^*$, $v^*$, $p^*$, $x^*$, $y^*$, and $t^*$) will be the same if the four parameters $\ell/\tau V$, $p_0/\rho V^2$, $V^2/g\ell$, and $\rho V\ell/\mu$ are equal for the two systems. The two systems will be dynamically similar. Of course, boundary and initial conditions expressed in dimensionless form must also be equal for the two systems, and this will require complete geometric similarity. These are the same similarity requirements that would be determined by a dimensional analysis if the same variables were considered. However, the advantage of working with the governing equations is that the variables appear naturally in the equations, and we do not have to worry about omitting an important one, provided the governing equations are correctly specified. We can thus use this method to deduce the conditions under which two solutions will be similar even though one of the solutions will most likely be obtained experimentally.

In the foregoing analysis we have considered a general case in which the flow may be unsteady, and both the actual pressure level, $p_0$, and the effect of gravity are important. A reduction in the number of similarity requirements can be achieved if one or more of these conditions is removed. For example, if the flow is steady, the dimensionless group, $\ell/\tau V$, can be eliminated.

The actual pressure level will only be of importance if we are concerned with cavitation. If not, the flow patterns and the pressure differences will not depend on the pressure level. In this case, $p_0$ can be taken as $\rho V^2$ (or $\frac{1}{2}\rho V^2$), and the Euler number can be eliminated as a similarity requirement. However, if we are concerned about cavitation (which will occur in the flow field if the pressure at certain points reaches the vapor pressure, $p_v$), then the actual pressure level is important. Usually, in this case, the characteristic pressure, $p_0$, is defined relative to the vapor pressure such that $p_0 = p_r - p_v$ where $p_r$ is some reference pressure within the flow field. With $p_0$ defined in this manner, the similarity parameter $p_0/\rho V^2$ becomes $(p_r - p_v)/\rho V^2$. This parameter is frequently written as $(p_r - p_v)/\frac{1}{2}\rho V^2$, and in this form, as was noted previously in Section 7.6, is called the cavitation number. Thus we can conclude that if cavitation is not of concern we do not need a similarity parameter involving $p_0$, but if cavitation is to be modeled, then the cavitation number becomes an important similarity parameter.

The Froude number, which arises because of the inclusion of gravity, is important for problems in which there is a free surface. Examples of these types of problems include the study of rivers, flow through hydraulic structures such as spillways, and the drag on ships. In these situations the shape of the free surface is influenced by gravity, and therefore the Froude number becomes an important similarity parameter. However, if there are no free surfaces, the only effect of gravity is to superimpose a hydrostatic pressure distribution on the pressure distribution created by the fluid motion. The hydrostatic distribution can be eliminated from the governing equation (Eq. 7.30) by defining a new pressure, $p' = p - \rho gy$, and with this change the Froude number does not appear in the nondimensional governing equations.

We conclude from this discussion that for the steady flow of an incompressible fluid without free surfaces, dynamic and kinematic similarity will be achieved if (for geometrically similar systems) Reynolds number similarity exists. If free surfaces are involved, Froude number similarity must also be maintained. For free-surface flows we have tacitly assumed that surface tension is not important. We would find, however, that if surface tension is included, its effect would appear in the free-surface boundary condition, and the Weber number, $\rho V^2\ell/\sigma$, would become an additional similarity parameter. In addition, if the governing equations for compressible fluids are considered, the Mach number, $V/c$, would appear as an additional similarity parameter.

It is clear that all the common dimensionless groups that we previously developed by using dimensional analysis appear in the governing equations that describe fluid motion when these equations are expressed in terms of dimensionless variables. Thus, use of the governing equations to obtain similarity laws provides an alternative to dimensional analysis. This approach has the advantage that the variables are known and the assumptions involved are clearly identified. In addition, a physical interpretation of the various dimensionless groups can often be obtained.

# 7.11    Chapter Summary and Study Guide

Many practical engineering problems involving fluid mechanics require experimental data for their solution. Thus, laboratory studies and experimentation play a significant role in this field. It is important to develop good procedures for the design of experiments so they can be efficiently completed with as broad applicability as possible. To achieve this end the concept of similitude is often used in which measurements made in the laboratory can be utilized for predicting the behavior of other similar systems. In this chapter, dimensional analysis is used for designing such experiments, as an aid for correlating experimental data, and as the basis for the design of physical models. As the name implies, dimensional analysis is based on a consideration of the dimensions required to describe the variables in a given problem. A discussion of the use of dimensions and the concept of dimensional homogeneity (which forms the basis for dimensional analysis) was included in Chapter 1.

Essentially, dimensional analysis simplifies a given problem described by a certain set of variables by reducing the number of variables that need to be considered. In addition to being fewer in number, the new variables are dimensionless products of the original variables. Typically these new dimensionless variables are much simpler to work with in performing the desired experiments. The Buckingham pi theorem, which forms the theoretical basis for dimensional analysis, is introduced. This theorem establishes the framework for reducing a given problem described in terms of a set of variables to a new set of fewer dimensionless variables. A simple method, called the repeating variable method, is described for actually forming the dimensionless variables (often called pi terms). Forming dimensionless variables by inspection is also considered. It is shown how the use of dimensionless variables can be of assistance in planning experiments and as an aid in correlating experimental data.

For problems in which there are a large number of variables, the use of physical models is described. Models are used to make specific predictions from laboratory tests rather than formulating a general relationship for the phenomenon of interest. The correct design of a model is obviously imperative for the accurate predictions of other similar, but usually larger, systems. It is shown how dimensional analysis can be used to establish a valid model design. An alternative approach for establishing similarity requirements using governing equations (usually differential equations) is presented.

The following checklist provides a study guide for this chapter. When your study of the entire chapter and end-of-chapter exercises has been completed, you should be able to

*similitude*

*dimensionless product*

*basic dimensions*

*pi term*

*Buckingham pi theorem*

*method of repeating variables*

*model*

*modeling laws*

*prototype*

*prediction equation*

*model design conditions*

*similarity requirements*

*modeling laws*

*length scale*

*distorted model*

*true model*

- write out meanings of the terms listed here in the margin and understand each of the related concepts. These terms are particularly important and are set in *italic, **bold,** and color* type in the text.
- use the Buckingham pi theorem to determine the number of independent dimensionless variables needed for a given flow problem.
- form a set of dimensionless variables using the method of repeating variables.
- form a set of dimensionless variables by inspection.
- use dimensionless variables as an aid in interpreting and correlating experimental data.
- use dimensional analysis to establish a set of similarity requirements (and prediction equation) for a model to be used to predict the behavior of another similar system (the prototype).
- rewrite a given governing equation in a suitable nondimensional form and deduce similarity requirements from the nondimensional form of the equation.

Some of the important equations in this chapter are:

| | |
|---|---|
| Reynolds number | $\text{Re} = \dfrac{\rho V \ell}{\mu}$ |
| Froude number | $\text{Fr} = \dfrac{V}{\sqrt{g\ell}}$ |

| | |
|---|---|
| Euler number | $Eu = \dfrac{p}{\rho V^2}$ |
| Cauchy number | $Ca = \dfrac{\rho V^2}{E_v}$ |
| Mach number | $Ma = \dfrac{V}{c}$ |
| Strouhal number | $St = \dfrac{\omega \ell}{V}$ |
| Weber number | $We = \dfrac{\rho V^2 \ell}{\sigma}$ |

# References

1. Bridgman, P. W., *Dimensional Analysis*, Yale University Press, New Haven, Conn., 1922.
2. Murphy, G., *Similitude in Engineering*, Ronald Press, New York, 1950.
3. Langhaar, H. L., *Dimensional Analysis and Theory of Models*, Wiley, New York, 1951.
4. Huntley, H. E., *Dimensional Analysis*, Macdonald, London, 1952.
5. Duncan, W. J., *Physical Similarity and Dimensional Analysis: An Elementary Treatise*, Edward Arnold, London, 1953.
6. Sedov, K. I., *Similarity and Dimensional Methods in Mechanics*, Academic Press, New York, 1959.
7. Ipsen, D. C., *Units, Dimensions, and Dimensionless Numbers*, McGraw-Hill, New York, 1960.
8. Kline, S. J., *Similitude and Approximation Theory*, McGraw-Hill, New York, 1965.
9. Skoglund, V. J., *Similitude—Theory and Applications*, International Textbook, Scranton, Pa., 1967.
10. Baker, W. E., Westline, P. S., and Dodge, F. T., *Similarity Methods in Engineering Dynamics—Theory and Practice of Scale Modeling*, Hayden (Spartan Books), Rochelle Park, N.J., 1973.
11. Taylor, E. S., *Dimensional Analysis for Engineers*, Clarendon Press, Oxford, UK, 1974.
12. Isaacson, E. de St. Q., and Isaacson, M. de St. Q., *Dimensional Methods in Engineering and Physics*, Wiley, New York, 1975.
13. Schuring, D. J., *Scale Models in Engineering*, Pergamon Press, New York, 1977.
14. Yalin, M. S., *Theory of Hydraulic Models*, Macmillan, London, 1971.
15. Sharp, J. J., *Hydraulic Modeling*, Butterworth, London, 1981.
16. Schlichting, H., *Boundary-Layer Theory*, 7th Ed., McGraw-Hill, New York, 1979.
17. Knapp, R. T., Daily, J. W., and Hammitt, F. G., *Cavitation*, McGraw-Hill, New York, 1970.

| | |
|---|---|
|  | Problem available in *WileyPLUS* at instructor's discretion. |
|  | Tutoring problem available in *WileyPLUS* at instructor's discretion. |
| ▶ | Problem is related to a chapter video available in *WileyPLUS*. |
| ✱ | Problem to be solved with aid of programmable calculator or computer. |
| † | Open-ended problem that requires critical thinking. These problems require various assumptions to provide the necessary input data. There are not unique answers to these problems. |

# Review Problems

Go to Appendix G (*WileyPLUS* or the book's web site, www. wiley.com/college/munson) for a set of review problems with answers. Detailed solutions can be found in the *Student Solution Manual and Study Guide for Fundamentals of Fluid Mechanics,* by Munson et al. (© 2013 John Wiley and Sons, Inc.)

# Conceptual Questions

**7.1C** For flow in a pipe, the Reynolds number is defined as

$$Re = \frac{V\rho D}{\mu}$$

where $V$ is velocity, $\rho$ is density, $\mu$ is viscosity, and $D$ is diameter. Expressed this way, the Reynolds number can be interpreted as

a) the ratio of viscous to inertial forces.

b) the ratio of inertial to viscous forces.

c) the ratio of momentum flux to viscous forces.

d) the ratio of kinetic energy to viscous forces.

**7.2C** A 1/20th scale model of an airplane is used to determine forces on the actual airplane. The 1/20th scale refers to the

a) lengths.          b) velocity.          c) forces.

d) Reynolds number.      e) all of the above.

**7.3C** A 1/4th scale model of a new automobile design is tested in a wind tunnel. The Reynolds number of the model is the same as that of the full-scale prototype. Assuming the model and prototype are exposed to the same air conditions, the velocity in the wind tunnel is then

a) 1/4th that of the full-scale vehicle.

b) The same as that of the full-scale vehicle.

c) 4 times that of the full-scale vehicle.

d) 40 times that of the full-scale vehicle.

**7.4C** A 1/10th scale model of a new automobile design is tested in a wind tunnel at the same Reynolds number as that of the full-scale prototype. The force coefficient $(F/A)/(\frac{1}{2}\rho V^2)$ of the model is the same as that of the prototype. Assuming the model and prototype are both tested in air, the force on the scale model, $F_m$, is then

a) 1/10,000th that of the full-scale vehicle.

b) 1/1000th that of the full-scale vehicle.

c) 1/100th that of the full-scale vehicle.

d) 1/10th that of the full-scale vehicle.

e) The same as that of the full-scale vehicle.

**Additional conceptual questions are available in *WileyPLUS* at the instructor's discretion.**

# Problems

**Note: Unless specific values of required fluid properties are given in the problem statement, use the values found in the tables on the inside of the front cover. Answers to the even-numbered problems are listed at the end of the book. The Lab Problems as well as the videos that accompany problems can be accessed in *WileyPLUS* or the book's web site, www.wiley.com/college/munson.**

## Section 7.1 Dimensional Analysis

**7.1** What are the dimensions of density, pressure, specific weight, surface tension, and dynamic viscosity in (a) the *FLT* system, and (b) the *MLT* system? Compare your results with those given in Table 1.1 in Chapter 1.

**7.2** Verify the left-hand side of Eq. 7.2 is dimensionless using the MLT system.

**7.3** ⓦⒾⓁⒺⓎ The Reynolds number, $\rho VD/\mu$, is a very important parameter in fluid mechanics. Verify that the Reynolds number is dimensionless, using both the *FLT* system and the *MLT* system for basic dimensions, and determine its value for ethyl alcohol flowing at a velocity of 3 m/s through a 2-in.-diameter pipe.

**7.4** ⓦⒾⓁⒺⓎ What are the dimensions of acceleration of gravity, density, dynamic viscosity, kinematic viscosity, specific weight, and speed of sound in **(a)** the *FLT* system, and **(b)** the *MLT* system? Compare your results with those given in Table 1.1 in Chapter 1.

**7.5** ⓦⒾⓁⒺⓎ For the flow of a thin film of a liquid with a depth $h$ and a free surface, two important dimensionless parameters are the Froude number, $V/\sqrt{gh}$, and the Weber number, $\rho V^2 h/\sigma$. Determine the value of these two parameters for glycerin (at 20 °C) flowing with a velocity of 0.7 m/s at a depth of 3 mm.

**7.6** ⓦⒾⓁⒺⓎ The Mach number for a body moving through a fluid with velocity $V$ is defined as $V/c$, where $c$ is the speed of sound in the fluid. This dimensionless parameter is usually considered to be important in fluid dynamics problems when its value exceeds 0.3. What would be the velocity of a body at a Mach number of 0.3 if the fluid is **(a)** air at standard atmospheric pressure and 20 °C, and **(b)** water at the same temperature and pressure?

## Section 7.3 Determination of Pi Terms

**7.7** It is desired to determine the wave height when wind blows across a lake. The wave height, $H$, is assumed to be a function of the wind speed, $V$, the water density, $\rho$, the air density, $\rho_a$, the water depth, $d$, the distance from the shore, $\ell$, and the acceleration of gravity, $g$, as shown in Fig. P7.7. Use $d$, $V$, and $\rho$ as repeating variables to determine a suitable set of pi terms that could be used to describe this problem.

■ **Figure P7.7**

**7.8** ⓦⒾⓁⒺⓎ ⒼⓄ Water flows over a dam as illustrated in Fig. P7.8. Assume the flowrate, $q$, per unit length along the dam depends on the head, $H$, width, $b$, acceleration of gravity, $g$, fluid density, $\rho$, and

■ **Figure P7.8**

fluid viscosity, $\mu$. Develop a suitable set of dimensionless parameters for this problem using $b$, $g$, and $\rho$ as repeating variables.

**7.9** (WILEY) The excess pressure inside a bubble (discussed in Chapter 1) is known to be dependent on bubble radius and surface tension. After finding the pi terms, determine the variation in excess pressure if we **(a)** double the radius and **(b)** double the surface tension.

**7.10** The flowrate, $Q$, of water in an open channel is assumed to be a function of the cross-sectional area of the channel, $A$, the height of the roughness of the channel surface, $\varepsilon$, the acceleration of gravity, $g$, and the slope, $S_o$, of the hill on which the channel sits. Put this relationship into dimensionless form.

**7.11** (WILEY) At a sudden contraction in a pipe the diameter changes from $D_1$ to $D_2$. The pressure drop, $\Delta p$, which develops across the contraction is a function of $D_1$ and $D_2$, as well as the velocity, $V$, in the larger pipe, and the fluid density, $\rho$, and viscosity, $\mu$. Use $D_1$, $V$, and $\mu$ as repeating variables to determine a suitable set of dimensionless parameters. Why would it be incorrect to include the velocity in the smaller pipe as an additional variable?

**7.12** (WILEY) Water sloshes back and forth in a tank as shown in Fig. P7.12. The frequency of sloshing, $\omega$, is assumed to be a function of the acceleration of gravity, $g$, the average depth of the water, $h$, and the length of the tank, $\ell$. Develop a suitable set of dimensionless parameters for this problem using $g$ and $\ell$ as repeating variables.

■ **Figure P7.12**

**7.13** The drag, $\mathcal{D}$, on a washer-shaped plate placed normal to a stream of fluid can be expressed as

$$\mathcal{D} = f(d_1, d_2, V, \mu, \rho)$$

where $d_1$ is the outer diameter, $d_2$ the inner diameter, $V$ the fluid velocity, $\mu$ the fluid viscosity, and $\rho$ the fluid density. Some experiments are to be performed in a wind tunnel to determine the drag. What dimensionless parameters would you use to organize these data?

**7.14** (WILEY) Assume that the flowrate, $Q$, of a gas from a smokestack is a function of the density of the ambient air, $\rho_a$, the density of the gas, $\rho_g$, within the stack, the acceleration of gravity, $g$, and the height and diameter of the stack, $h$ and $d$, respectively. Use $\rho_a$, $d$, and $g$ as repeating variables to develop a set of pi terms that could be used to describe this problem.

**7.15** (WILEY) The pressure rise, $\Delta p$, across a pump can be expressed as

$$\Delta p = f(D, \rho, \omega, Q)$$

where $D$ is the impeller diameter, $\rho$ the fluid density, $\omega$ the rotational speed, and $Q$ the flowrate. Determine a suitable set of dimensionless parameters.

**7.16** (WILEY) A thin elastic wire is placed between rigid supports. A fluid flows past the wire, and it is desired to study the static deflection, $\delta$, at the center of the wire due to the fluid drag. Assume that

$$\delta = f(\ell, d, \rho, \mu, V, E)$$

where $\ell$ is the wire length, $d$ the wire diameter, $\rho$ the fluid density, $\mu$ the fluid viscosity, $V$ the fluid velocity, and $E$ the modulus

of elasticity of the wire material. Develop a suitable set of pi terms for this problem.

**7.17** The water flowrate, $Q$, in an open rectangular channel can be measured by placing a plate across the channel as shown in Fig. P7.17. This type of a device is called a *weir*. The height of the water, $H$, above the weir crest is referred to as the head and can be used to determine the flowrate through the channel. Assume that $Q$ is a function of the head, $H$, the channel width, $b$, and the acceleration of gravity, $g$. Determine a suitable set of dimensionless variables for this problem.

■ **Figure P7.17**

**7.18** ▶ (WILEY) Because of surface tension, it is possible, with care, to support an object heavier than water on the water surface as shown in Fig. P7.18. (See **Video V1.9**.) The maximum thickness, $h$, of a square of material that can be supported is assumed to be a function of the length of the side of the square, $\ell$, the density of the material, $\rho$, the acceleration of gravity, $g$, and the surface tension of the liquid, $\sigma$. Develop a suitable set of dimensionless parameters for this problem.

■ **Figure P7.18**

**7.19** ▶ (WILEY) Under certain conditions, wind blowing past a rectangular speed limit sign can cause the sign to oscillate with a frequency $\omega$. (See Fig. P7.19 and **Video V9.9**.) Assume that $\omega$ is a function of the sign width, $b$, sign height, $h$, wind velocity, $V$, air density, $\rho$, and an elastic constant, $k$, for the supporting pole. The constant, $k$, has dimensions of $FL$. Develop a suitable set of pi terms for this problem.

■ **Figure P7.19**

**7.20** ▶ (WILEY) The height, $h$, that a liquid will rise in a capillary tube is a function of the tube diameter, $D$, the specific weight of the liquid, $\gamma$, and the surface tension, $\sigma$. (See **Video V1.10.**) Perform a dimensional analysis using both the *FLT* and *MLT* systems for basic dimensions. *Note*: The results should obviously be the same regardless of the system of dimensions used. If your analysis indicates otherwise, go back and check your work, giving particular attention to the required number of reference dimensions.

**7.21** (WILEY) A cone and plate viscometer consists of a cone with a very small angle $\alpha$ that rotates above a flat surface as shown in Fig. P7.21. The torque, $\mathcal{T}$, required to rotate the cone at an angular velocity $\omega$ is a function of the radius, $R$, the cone angle, $\alpha$, and the fluid viscosity, $\mu$, in addition to $\omega$. With the aid of dimensional analysis, determine how the torque will change if both the viscosity and angular velocity are doubled.

■ **Figure P7.21**

**7.22** (WILEY) The pressure drop, $\Delta p$, along a straight pipe of diameter $D$ has been experimentally studied, and it is observed that for laminar flow of a given fluid and pipe, the pressure drop varies directly with the distance, $\ell$, between pressure taps. Assume that $\Delta p$ is a function of $D$ and $\ell$, the velocity, $V$, and the fluid viscosity, $\mu$. Use dimensional analysis to deduce how the pressure drop varies with pipe diameter.

**7.23** (WILEY) A cylinder with a diameter $D$ floats upright in a liquid as shown in Fig. P7.23. When the cylinder is displaced slightly along its vertical axis it will oscillate about its equilibrium position with a frequency, $\omega$. Assume that this frequency is a function of the diameter, $D$, the mass of the cylinder, $m$, and the specific weight, $\gamma$, of the liquid. Determine, with the aid of dimensional analysis, how the frequency is related to these variables. If the mass of the cylinder were increased, would the frequency increase or decrease?

■ **Figure P7.23**

**†7.24** Consider a typical situation involving the flow of a fluid that you encounter almost every day. List what you think are the important physical variables involved in this flow and determine an appropriate set of pi terms for this situation.

**7.25** The speed of sound in a gas, $c$, is a function of the gas pressure, $p$, and density, $\rho$. Determine, with the aid of dimensional analysis, how the velocity is related to the pressure and density. Be careful when you decide on how many reference dimensions are required.

### Section 7.5 Determination of Pi Terms by Inspection

**7.26** A viscous fluid is poured onto a horizontal plate as shown in Fig. P7.26. Assume that the time, $t$, required for the fluid to flow a certain distance, $d$, along the plate is a function of the volume of fluid poured, $\Psi$, acceleration of gravity, $g$, fluid density, $\rho$, and fluid viscosity, $\mu$. Determine an appropriate set of pi terms to describe this process. Form the pi terms by inspection.

■ **Figure P7.26**

**7.27** (WILEY) The velocity, $c$, at which pressure pulses travel through arteries (pulse-wave velocity) is a function of the artery diameter, $D$, and wall thickness, $h$, the density of blood, $\rho$, and the modulus of elasticity, $E$, of the arterial wall. Determine a set of nondimensional parameters that can be used to study experimentally the relationship between the pulse-wave velocity and the variables listed. Form the nondimensional parameters by inspection.

**7.28** ▶ (WILEY) As shown in Fig. P7.28 and **Video V5.6,** a jet of liquid directed against a block can tip over the block. Assume that the velocity, $V$, needed to tip over the block is a function of the fluid density, $\rho$, the diameter of the jet, $D$, the weight of the block, $\mathcal{W}$, the width of the block, $b$, and the distance, $d$, between the jet and the bottom of the block. **(a)** Determine a set of dimensionless parameters for this problem. Form the dimensionless parameters by inspection. **(b)** Use the momentum equation to determine an equation for $V$ in terms of the other variables. **(c)** Compare the results of parts **(a)** and **(b)**.

■ **Figure P7.28**

**7.29** (WILEY) Assume that the drag, $\mathcal{D}$, on an aircraft flying at supersonic speeds is a function of its velocity, $V$, fluid density, $\rho$, speed of sound, $c$, and a series of lengths, $\ell_1, \ldots, \ell_i$, which describe the geometry of the aircraft. Develop a set of pi terms that could be used to investigate experimentally how the drag is affected by the various factors listed. Form the pi terms by inspection.

### Section 7.6 Common Dimensionless Groups in Fluid Mechanics

**7.30** (WILEY) Shown in the following table are several flow situations and the associated characteristic velocity, size, and fluid kinematic viscosity. Determine the Reynolds number for each of the flows and indicate for which ones the inertial effects are small relative to viscous effects.

| Flow | Velocity ft/s | Size ft | Kinematic viscosity ft²/s |
|---|---|---|---|
| Airplane | 400 | 60 | $4.3 \times 10^{-4}$ |
| Mosquito | 0.02 | 0.015 | $1.6 \times 10^{-4}$ |
| Syrup on pancake | 0.03 | 0.12 | $2.5 \times 10^{-3}$ |
| Blood in capillary | 0.001 | $5 \times 10^{-5}$ | $1.2 \times 10^{-5}$ |

**7.31** (WILEY) A liquid spray nozzle is designed to produce a specific size droplet with diameter, $d$. The droplet size depends on the nozzle diameter, $D$, nozzle velocity, $V$, and the liquid properties $\rho$, $\mu$, $\sigma$. Using the common dimensionless terms found in Table 7.1, determine the functional relationship for the dependent diameter ratio of $d/D$.

### Section 7.7 Correlation of Experimental Data (also see Lab Problems 7.1LP, 7.2LP, 7.3LP, 7.4LP)

**7.32** The pressure rise, $\Delta p = p_2 - p_1$, across the abrupt expansion of Fig. P7.32 through which a liquid is flowing can be expressed as

$$\Delta p = f(A_1, A_2, \rho, V_1)$$

where $A_1$ and $A_2$ are the upstream and downstream cross-sectional areas, respectively, $\rho$ is the fluid density, and $V_1$ is the upstream velocity. Some experimental data obtained with $A_2 = 1.25$ ft², $V_1 = 5.00$ ft/s, and using water with $\rho = 1.94$ slugs/ft³ are given in the following table:

| $A_1$ (ft²) | 0.10 | 0.25 | 0.37 | 0.52 | 0.61 |
|---|---|---|---|---|---|
| $\Delta p$ (lb/ft²) | 3.25 | 7.85 | 10.3 | 11.6 | 12.3 |

Plot the results of these tests using suitable dimensionless parameters. With the aid of a standard curve fitting program determine a general equation for $\Delta p$ and use this equation to predict $\Delta p$ for water flowing through an abrupt expansion with an area ratio $A_1/A_2 = 0.35$ at a velocity $V_1 = 3.75$ ft/s.

■ **Figure P7.32**

**\*7.33** (WILEY) The pressure drop, $\Delta p$, over a certain length of horizontal pipe is assumed to be a function of the velocity, $V$, of the fluid in the pipe, the pipe diameter, $D$, and the fluid density and viscosity, $\rho$ and $\mu$. **(a)** Show that this flow can be described in dimensionless form as a "pressure coefficient," $C_p = \Delta p/(0.5 \ \rho V^2)$ that depends on the Reynolds number, Re $= \rho VD/\mu$. **(b)** The following data were obtained in an experiment involving a fluid with $\rho = 2$ slugs/ft³, $\mu = 2 \times 10^{-3}$ lb · s/ft², and $D = 0.1$ ft. Plot a dimensionless graph and use a power law equation to determine the

functional relationship between the pressure coefficient and the Reynolds number. **(c)** What are the limitations on the applicability of your equation obtained in part **(b)**?

| $V$, ft/s | $\Delta p$, lb/ft² |
|---|---|
| 3 | 192 |
| 11 | 704 |
| 17 | 1088 |
| 20 | 1280 |

**\*7.34** (WILEY GO) The pressure drop across a short hollowed plug placed in a circular tube through which a liquid is flowing (see Fig. P7.34) can be expressed as

$$\Delta p = f(\rho, V, D, d)$$

where $\rho$ is the fluid density, and $V$ is the mean velocity in the tube. Some experimental data obtained with $D = 0.2$ ft, $\rho = 2.0$ slugs/ft³, and $V = 2$ ft/s are given in the following table:

| $d$ (ft) | 0.06 | 0.08 | 0.10 | 0.15 |
|---|---|---|---|---|
| $\Delta p$ (lb/ft²) | 493.8 | 156.2 | 64.0 | 12.6 |

Plot the results of these tests, using suitable dimensionless parameters, on a log–log scale. Use a standard curve-fitting technique to determine a general equation for $\Delta p$. What are the limits of applicability of the equation?

■ **Figure P7.34**

**7.35** The pressure drop per unit length, $\Delta p_\ell$, for the flow of blood through a horizontal small-diameter tube is a function of the volume rate of flow, $Q$, the diameter, $D$, and the blood viscosity, $\mu$. For a series of tests in which $D = 2$ mm and $\mu = 0.004$ N · s/m², the following data were obtained, where the $\Delta p$ listed was measured over the length, $\ell = 300$ mm.

| $Q$ (m³/s) | $\Delta p$ (N/m²) |
|---|---|
| $3.6 \times 10^{-6}$ | $1.1 \times 10^4$ |
| $4.9 \times 10^{-6}$ | $1.5 \times 10^4$ |
| $6.3 \times 10^{-6}$ | $1.9 \times 10^4$ |
| $7.9 \times 10^{-6}$ | $2.4 \times 10^4$ |
| $9.8 \times 10^{-6}$ | $3.0 \times 10^4$ |

Perform a dimensional analysis for this problem, and make use of the data given to determine a general relationship between $\Delta p_\ell$ and $Q$ (one that is valid for other values of $D$, $\ell$, and $\mu$).

**\*7.36** ▶ (WILEY) As shown in Fig. 2.26. Fig. P7.36, and **Video V2.12,** a rectangular barge floats in a stable configuration provided the distance between the center of gravity, $CG$, of the object (boat and load) and the center of buoyancy, $C$, is less than a certain amount, $H$. If this distance is greater than $H$, the boat will tip over. Assume $H$ is a function of the boat's width, $b$, length, $\ell$, and draft, $h$. **(a)** Put this relationship into dimensionless form. **(b)** The results of a set of experiments with a model barge with a width of 1.0 m are shown in the table. Plot these data in dimensionless form and determine a power-law equation relating the dimensionless parameters.

| $\ell$, m | h, m | H, m |
|-----|------|------|
| 2.0 | 0.10 | 0.833 |
| 4.0 | 0.10 | 0.833 |
| 2.0 | 0.20 | 0.417 |
| 4.0 | 0.20 | 0.417 |
| 2.0 | 0.35 | 0.238 |
| 4.0 | 0.35 | 0.238 |

■ **Figure P7.36**

**7.37** When a sphere of diameter $d$ falls slowly in a highly viscous fluid, the settling velocity, $V$, is known to be a function of $d$, the fluid viscosity, $\mu$, and the difference, $\Delta\gamma$, between the specific weight of the sphere and the specific weight of the fluid. Due to a tight budget situation, only one experiment can be performed, and the following data were obtained: $V = 0.42$ ft/s for $d = 0.1$ in., $\mu = 0.03$ lb · s/ft², and $\Delta\gamma = 10$ lb/ft³. If possible, based on this limited amount of data, determine the general equation for the settling velocity. If you do not think it is possible, indicate what additional data would be required.

**7.38** ▶ (WILEY) The time, $t$, it takes to pour a certain volume of liquid from a cylindrical container depends on several factors, including the viscosity of the liquid. (See **Video V1.3.**) Assume that for very viscous liquids the time it takes to pour out two-thirds of the initial volume depends on the initial liquid depth, $\ell$, the cylinder diameter, $D$, the liquid viscosity, $\mu$, and the liquid specific weight, $\gamma$. The data shown in the following table were obtained in the laboratory. For these tests $\ell = 45$ mm, $D = 67$ mm, and $\gamma = 9.60$ kN/m³. **(a)** Perform a dimensional analysis, and based on the data given, determine if variables used for this problem appear to be correct. Explain how you arrived at your answer. **(b)** If possible, determine an equation relating the pouring time and viscosity for the cylinder and liquids used in these tests. If it is not possible, indicate what additional information is needed.

| $\mu$ (N · s/m²) | 11 | 17 | 39 | 61 | 107 |
|------|----|----|----|----|-----|
| $t$(s) | 15 | 23 | 53 | 83 | 145 |

**7.39** A liquid flows with a velocity $V$ through a hole in the side of a large tank. Assume that

$$V = f(h, g, \rho, \sigma)$$

where $h$ is the depth of fluid above the hole, $g$ is the acceleration of gravity, $\rho$ the fluid density, and $\sigma$ the surface tension. The following data were obtained by changing $h$ and measuring $V$, with a fluid having a density $= 10^3$ kg/m³ and surface tension $= 0.074$ N/m.

| $V$ (m/s) | 3.13 | 4.43 | 5.42 | 6.25 | 7.00 |
|-----------|------|------|------|------|------|
| $h$ (m) | 0.50 | 1.00 | 1.50 | 2.00 | 2.50 |

Plot these data by using appropriate dimensionless variables. Could any of the original variables have been omitted?

**7.40** In order to maintain uniform flight, smaller birds must beat their wings faster than larger birds. It is suggested that the relationship between the wingbeat frequency, $\omega$, beats per second, and the bird's wingspan, $\ell$, is given by a power law relationship, $\omega \sim \ell^n$. **(a)** Use dimensional analysis with the assumption that the wingbeat frequency is a function of the wingspan, the specific weight of the bird, $\gamma$, the acceleration of gravity, $g$, and the density of the air, $\rho_a$, to determine the value of the exponent $n$. **(b)** Some typical data for various birds are given in the following table. Do these data support your result obtained in part (a)? Provide appropriate analysis to show how you arrived at your conclusion.

| Bird | Wingspan, m | Wingbeat frequency, beats/s |
|------|-------------|------------------------------|
| Purple martin | 0.28 | 5.3 |
| Robin | 0.36 | 4.3 |
| Mourning dove | 0.46 | 3.2 |
| Crow | 1.00 | 2.2 |
| Canada goose | 1.50 | 2.6 |
| Great blue heron | 1.80 | 2.0 |

**7.41** A fluid flows through the horizontal curved pipe of Fig. P7.41 with a velocity $V$. The pressure drop, $\Delta p$, between the entrance and the exit to the bend is thought to be a function of the velocity, bend radius, $R$, pipe diameter, $D$, and fluid density, $\rho$. The data shown in the following table were obtained in the laboratory. For these tests $\rho = 2.0$ slugs/ft³, $R = 0.5$ ft, and $D = 0.1$ ft. Perform a dimensional analysis and based on the data given, determine if the variables used for this problem appear to be correct. Explain how you arrived at your answer.

| $V$ (ft/s) | 2.1 | 3.0 | 3.9 | 5.1 |
|------------|-----|-----|-----|-----|
| $\Delta p$ (lb/ft²) | 1.2 | 1.8 | 6.0 | 6.5 |

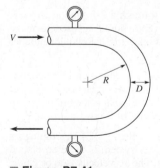

■ **Figure P7.41**

**\*7.42** (WILEY GO) The concentric cylinder device of the type shown in Fig. P7.42 is commonly used to measure the viscosity, $\mu$, of liquids by relating the angle of twist, $\theta$, of the inner cylinder to the angular velocity, $\omega$, of the outer cylinder. Assume that

$$\theta = f(\omega, \mu, K, D_1, D_2, \ell)$$

where $K$ depends on the suspending wire properties and has the dimensions $FL$. The following data were obtained in a series of tests for which $\mu = 0.01$ lb · s/ft², $K = 10$ lb · ft, $\ell = 1$ ft, and $D_1$ and $D_2$ were constant.

| $\theta$ (rad) | $\omega$ (rad/s) |
|---|---|
| 0.89 | 0.30 |
| 1.50 | 0.50 |
| 2.51 | 0.82 |
| 3.05 | 1.05 |
| 4.28 | 1.43 |
| 5.52 | 1.86 |
| 6.40 | 2.14 |

Determine from these data, with the aid of dimensional analysis, the relationship between $\theta$, $\omega$, and $\mu$ for this particular apparatus. *Hint*: Plot the data using appropriate dimensionless parameters, and determine the equation of the resulting curve using a standard curve-fitting technique. The equation should satisfy the condition that $\theta = 0$ for $\omega = 0$.

■ **Figure P7.42**

## Section 7.8 Modeling and Similitude

**7.43** (WILEY) Air at 80 °F is to flow through a 2-ft pipe at an average velocity of 6 ft/s. What size pipe should be used to move water at 60 °F and average velocity of 3 ft/s if Reynolds number similarity is enforced?

**7.44** The design of a river model is to be based on Froude number similarity, and a river depth of 3 m is to correspond to a model depth of 100 mm. Under these conditions what is the prototype velocity corresponding to a model velocity of 2 m/s?

**7.45** Glycerin at 20 °C flows with a velocity of 4 m/s through a 30-mm-diameter tube. A model of this system is to be developed using standard air as the model fluid. The air velocity is to be 2 m/s. What tube diameter is required for the model if dynamic similarity is to be maintained between model and prototype?

**7.46** To test the aerodynamics of a new prototype automobile, a scale model will be tested in a wind tunnel. For dynamic similarity, it will be required to match Reynolds number between model and prototype. Assuming that you will be testing a one-tenth-scale model and both model and prototype will be exposed to standard air pressure, will it be better for the wind tunnel air to be colder or hotter than standard sea-level air temperature of 15 °C? Why?

**7.47** ▶ (WILEY) You are to conduct wind tunnel testing of a new football design that has a smaller lace height than previous designs (see **Videos V6.1 and V6.2**). It is known that you will need to maintain Re and St similarity for the testing. Based on standard college quarterbacks, the prototype parameters are set at $V = 40$ mph and $\omega = 300$ rpm, where $V$ and $\omega$ are the speed and angular velocity of the football. The prototype football has a 7-in. diameter. Due to instrumentation required to measure pressure and shear stress on the surface of the football, the model will require a length scale of 2:1 (the model will be larger than the prototype). Determine the required model freestream velocity and model angular velocity.

**7.48** When a fluid flows slowly past a vertical plate of height $h$ and width $b$ (see Fig. P7.48), pressure develops on the face of the plate. Assume that the pressure, $p$, at the midpoint of the plate is a function of plate height and width, the approach velocity, $V$, and the fluid viscosity, $\mu$. Make use of dimensional analysis to determine how the pressure, $p$, will change when the fluid velocity, $V$, is doubled.

Plate width = $b$

■ **Figure P7.48**

**7.49** (WILEY) A model of a submarine, 1:15 scale, is to be tested at 180 ft/s in a wind tunnel with standard sea-level air, while the prototype will be operated in seawater. Determine the speed of the prototype to ensure Reynolds number similarity.

**7.50** (WILEY) SAE 30 oil at 60 °F is pumped through a 3-ft-diameter pipeline at a rate of 6400 gal/min. A model of this pipeline is to be designed using a 3-in.-diameter pipe and water at 60 °F as the working fluid. To maintain Reynolds number similarity between these two systems, what fluid velocity will be required in the model?

**7.51** (WILEY) The water velocity at a certain point along a 1:10 scale model of a dam spillway is 3 m/s. What is the corresponding prototype velocity if the model and prototype operate in accordance with Froude number similarity?

**7.52** (WILEY) The drag characteristics of a torpedo are to be studied in a water tunnel using a 1:5 scale model. The tunnel operates with fresh water at 20 °C, whereas the prototype torpedo is to be used in seawater at 15.6 °C. To correctly simulate the behavior of the prototype moving with a velocity of 30 m/s, what velocity is required in the water tunnel?

**7.53** (WILEY) For a certain fluid flow problem it is known that both the Froude number and the Weber number are important dimensionless parameters. If the problem is to be studied by using a 1:15 scale model, determine the required surface tension scale if the density scale is equal to 1. The model and prototype operate in the same gravitational field.

**7.54** (WILEY) The fluid dynamic characteristics of an airplane flying 240 mph at 10,000 ft are to be investigated with the aid of a 1:20 scale model. If the model tests are to be performed in a wind tunnel using standard air, what is the required air velocity in the wind tunnel? Is this a realistic velocity?

**7.55** The pressure drop between the entrance and exit of a 150-mm-diameter 90° elbow, through which ethyl alcohol at 20 °C is flowing, is to be determined with a geometrically similar model. The velocity of the alcohol is 5 m/s. The model fluid is to be water at 20 °C, and the model velocity is limited to 10 m/s. **(a)** What is the required diameter of the model elbow to maintain dynamic similarity? **(b)** A measured pressure drop of 20 kPa in the model will correspond to what prototype value?

**7.56** (WILEY) If an airplane travels at a speed of 1120 km/hr at an altitude of 15 km, what is the required speed at an altitude of 8 km to satisfy Mach number similarity? Assume the air properties correspond to those for the U.S. standard atmosphere.

**7.57** (WILEY) (See Fluid in the News article "Modeling Parachutes in a Water Tunnel," Section 7.8.1.) Flow characteristics for a 30-ft-diameter prototype parachute are to be determined by tests of a 1-ft-diameter model parachute in a water tunnel. Some data collected with the model parachute indicate a drag of 17 lb when the water velocity is 4 ft/s. Use the model data to predict the drag on the prototype parachute falling through air at 10 ft/s. Assume the drag to be a function of the velocity, $V$, the fluid density, $\rho$, and the parachute diameter, $D$.

**7.58** (WILEY) The lift and drag developed on a hydrofoil are to be determined through wind tunnel tests using standard air. If full-scale tests are to be run, what is the required wind tunnel velocity corresponding to a hydrofoil velocity in seawater at 15 mph? Assume Reynolds number similarity is required.

**7.59** ▶ (WILEY) A 1/50 scale model is to be used in a towing tank to study the water motion near the bottom of a shallow channel as a large barge passes over. (See **Video V7.16.**) Assume that the model is operated in accordance with the Froude number criteria for dynamic similitude. The prototype barge moves at a typical speed of 15 knots. **(a)** At what speed (in ft/s) should the model be towed? **(b)** Near the bottom of the model channel a small particle is found to move 0.15 ft in one second so that the fluid velocity at that point is approximately 0.15 ft/s. Determine the velocity at the corresponding point in the prototype channel.

**7.60** (WILEY GO) When small particles of diameter $d$ are transported by a moving fluid having a velocity $V$, they settle to the ground at some distance $\ell$ after starting from a height $h$ as shown in Fig. P7.60. The variation in $\ell$ with various factors is to be studied with a model having a length scale of $\frac{1}{10}$. Assume that

$$\ell = f(h, d, V, \gamma, \mu)$$

where $\gamma$ is the particle specific weight and $\mu$ is the fluid viscosity. The same fluid is to be used in both the model and the prototype, but $\gamma$ (model) $= 9 \times \gamma$ (prototype). **(a)** If $V = 50$ mph, at what velocity should the model tests be run? **(b)** During a certain model test it was found that $\ell$ (model) $= 0.8$ ft. What would be the predicted $\ell$ for this test?

■ **Figure P7.60**

**7.61** (WILEY) A solid sphere having a diameter $d$ and specific weight $\gamma_s$ is immersed in a liquid having a specific weight $\gamma_f (\gamma_f > \gamma_s)$ and then released. It is desired to use a model system to determine the maximum height, $h$, above the liquid surface that the sphere will rise upon release from a depth $H$. It can be assumed that the important liquid properties are the density, $\gamma_f/g$, specific weight, $\gamma_f$, and viscosity, $\mu_f$. Establish the model design conditions and the prediction equation, and determine whether the same liquid can be used in both the model and prototype systems.

**7.62** (WILEY) A thin layer of an incompressible fluid flows steadily over a horizontal smooth plate as shown in Fig. P7.62. The fluid surface is open to the atmosphere, and an obstruction having a square cross section is placed on the plate as shown. A model with a length scale of $\frac{1}{4}$ and a fluid density scale of 1.0 is to be designed to predict the depth of fluid, $y$, along the plate. Assume that inertial, gravitational, surface tension, and viscous effects are all important. What are the required viscosity and surface tension scales?

■ **Figure P7.62**

**7.63** (WILEY) The drag on a 2-m-diameter satellite dish due to an 80-km/hr wind is to be determined through a wind tunnel test using a geometrically similar 0.4-m-diameter model dish. Assume standard air for both model and prototype. **(a)** At what air speed should the model test be run? **(b)** With all similarity conditions satisfied, the measured drag on the model was determined to be 170 N. What is the predicted drag on the prototype dish?

**7.64** ▶ During a storm, a snow drift is formed behind a snow fence as shown in Fig. P7.64. (See **Video V9.9.**) Assume that the height of the drift, $h$, is a function of the number of inches of snow deposited by the storm, $d$, height of the fence, $H$, width of slats in the fence, $b$, wind speed, $V$, acceleration of gravity, $g$, air density, $\rho$, and specific weight of snow, $\gamma_s$. **(a)** If this problem is to be studied with a model, determine the similarity requirements for the model and the relationship between the drift depth for model and prototype (prediction equation). **(b)** A storm with winds of 30 mph deposits 16 in. of snow having a specific weight of 5.0 lb/ft³. A $\frac{1}{2}$-sized scale model is to be used to investigate the effectiveness of a proposed snow fence. If the air density is the same for the model and the storm, determine the required specific weight for the model snow and required wind speed for the model.

■ **Figure P7.64**

**7.65** (WILEY) A large, rigid, rectangular billboard is supported by an elastic column as shown in Fig. P7.65. There is concern about the deflection, $\delta$, of the top of the structure during a high wind of velocity $V$. A wind tunnel test is to be conducted with a 1:15 scale model. Assume the pertinent column variables are its length and cross-sectional dimensions, and the modulus of elasticity of the material used for the column. The only important "wind" variables are the air density and velocity. **(a)** Determine the model design conditions and the prediction equation for the deflection. **(b)** If the same structural materials are used for the model and prototype, and the wind tunnel operates under standard atmospheric conditions, what is the required wind tunnel velocity to match an 80 km/hr wind?

Front View                 Side View

■ **Figure P7.65**

**7.66** (WILEY) A thin flat plate having a diameter of 0.3 ft is towed through a tank of oil ($\gamma = 53$ lb/ft$^3$) at a velocity of 5 ft/s. The plane of the plate is perpendicular to the direction of motion, and the plate is submerged so that wave action is negligible. Under these conditions the drag on the plate is 1.4 lb. If viscous effects are neglected, predict the drag on a geometrically similar, 2-ft-diameter plate that is towed with a velocity of 3 ft/s through water at 60 °F under conditions similar to those for the smaller plate.

**7.67** Air bubbles discharge from the end of a submerged tube as shown in Fig. P7.67. The bubble diameter, $D$, is assumed to be a function of the air flowrate, $Q$, the tube diameter, $d$, the acceleration of gravity, $g$, the density of the liquid, $\rho$, and the surface tension of the liquid, $\sigma$. **(a)** Determine a suitable set of dimensionless variables for this problem. **(b)** Model tests are to be run on the Earth for a prototype that is to be operated on a planet where the acceleration of gravity is 10 times greater than that on Earth. The model and prototype are to use the same fluid, and the prototype tube diameter is 0.25 in. Determine the tube diameter for the model and the required model flowrate if the prototype flowrate is to be 0.001 ft$^3$/s.

■ **Figure P7.67**

**7.68** (WILEY) For a certain model study involving a 1:5 scale model it is known that Froude number similarity must be maintained. The possibility of cavitation is also to be investigated, and it is assumed that the cavitation number must be the same for model and prototype. The prototype fluid is water at 30 °C, and the model fluid is water at 70 °C. If the prototype operates at an ambient pressure of 101 kPa (abs), what is the required ambient pressure for the model system?

**7.69** (WILEY) A thin layer of particles rests on the bottom of a horizontal tube as shown in Fig. P7.69. When an incompressible fluid flows through the tube, it is observed that at some critical velocity the particles will rise and be transported along the tube. A model is to be used to determine this critical velocity. Assume the critical velocity, $V_c$, to be a function of the pipe diameter, $D$, particle diameter, $d$, the fluid density, $\rho$, and viscosity, $\mu$, the density of the particles, $\rho_p$, and the acceleration of gravity, $g$. **(a)** Determine the similarity requirements for the model, and the relationship between the critical velocity for model and prototype (the prediction equation). **(b)** For a length scale of $\frac{1}{2}$ and a fluid density scale of 1.0,

what will be the critical velocity scale (assuming all similarity requirements are satisfied)?

■ **Figure P7.69**

**7.70** ▶ (WILEY) The pressure rise, $\Delta p$, across a blast wave, as shown in Fig. P7.70, and **Video V11.8**, is assumed to be a function of the amount of energy released in the explosion, $E$, the air density, $\rho$, the speed of sound, $c$, and the distance from the blast, $d$. **(a)** Put this relationship in dimensionless form. **(b)** Consider two blasts: the prototype blast with energy release $E$ and a model blast with 1/1000th the energy release ($E_m = 0.001\ E$). At what distance from the model blast will the pressure rise be the same as that at a distance of 1 mile from the prototype blast?

■ **Figure P7.70**

**7.71** (WILEY) The drag, $\mathcal{D}$, on a sphere located in a pipe through which a fluid is flowing is to be determined experimentally (see Fig. P7.71). Assume that the drag is a function of the sphere diameter, $d$, the pipe diameter, $D$, the fluid velocity, $V$, and the fluid density, $\rho$. **(a)** What dimensionless parameters would you use for this problem? **(b)** Some experiments using water indicate that for $d = 0.2$ in., $D = 0.5$ in., and $V = 2$ ft/s, the drag is $1.5 \times 10^{-3}$ lb. If possible, estimate the drag on a sphere located in a 2-ft-diameter pipe through which water is flowing with a velocity of 6 ft/s. The sphere diameter is such that geometric similarity is maintained. If it is not possible, explain why not.

■ **Figure P7.71**

**7.72** (WILEY) An incompressible fluid oscillates harmonically ($V = V_0 \sin \omega t$, where $V$ is the velocity) with a frequency of 10 rad/s in a 4-in.-diameter pipe. A $\frac{1}{4}$ scale model is to be used to determine the pressure difference per unit length, $\Delta p_\ell$ (at any instant) along the pipe. Assume that

$$\Delta p_\ell = f(D, V_0, \omega, t, \mu, \rho)$$

where $D$ is the pipe diameter, $\omega$ the frequency, $t$ the time, $\mu$ the fluid viscosity, and $\rho$ the fluid density. **(a)** Determine the similarity requirements for the model and the prediction equation for $\Delta p_\ell$. **(b)** If the same fluid is used in the model and the prototype, at what frequency should the model operate?

**7.73** (WILEY) As shown in Fig. P7.73, a "noisemaker" B is towed behind a minesweeper A to set off enemy acoustic mines such as at C. The drag force of the noisemaker is to be studied in a water tunnel at a $\frac{1}{4}$ scale model (model ¼ the size of the prototype). The drag force is assumed to be a function of the speed of the ship, the density and viscosity of the fluid, and the diameter of the noisemaker. **(a)** If the prototype towing speed is 3 m/s, determine the water velocity in the tunnel for the model tests. **(b)** If the model tests of part (a) produced a model drag of 900 N, determine the drag expected on the prototype.

■ **Figure P7.73**

**7.74** (WILEY) The drag characteristics for a newly designed automobile having a maximum characteristic length of 20 ft are to be determined through a model study. The characteristics at both low speed (approximately 20 mph) and high speed (90 mph) are of interest. For a series of projected model tests, an unpressurized wind tunnel that will accommodate a model with a maximum characteristic length of 4 ft is to be used. Determine the range of air velocities that would be required for the wind tunnel if Reynolds number similarity is desired. Are the velocities suitable? Explain.

**7.75** (WILEY) The drag characteristics of an airplane are to be determined by model tests in a wind tunnel operated at an absolute pressure of 1300 kPa. If the prototype is to cruise in standard air at 385 km/hr, and the corresponding speed of the model is not to differ by more than 20% from this (so that compressibility effects may be ignored), what range of length scales may be used if Reynolds number similarity is to be maintained? Assume the viscosity of air is unaffected by pressure, and the temperature of air in the tunnel is equal to the temperature of the air in which the airplane will fly.

**7.76** (WILEY) Wind blowing past a flag causes it to "flutter in the breeze." The frequency of this fluttering, $\omega$, is assumed to be a function of the wind speed, $V$, the air density, $\rho$, the acceleration of gravity, $g$, the length of the flag, $\ell$, and the "area density," $\rho_A$ (with dimensions of $ML^{-2}$) of the flag material. It is desired to predict the flutter frequency of a large $\ell = 40$ ft flag in a $V = 30$ ft/s wind. To do this a model flag with $\ell = 4$ ft is to be tested in a wind tunnel. **(a)** Determine the required area density of the model flag material if the large flag has $\rho_A = 0.006$ slugs/ft$^2$. **(b)** What wind tunnel velocity is required for testing the model? **(c)** If the model flag flutters at 6 Hz, predict the frequency for the large flag.

**7.77** (WILEY) The drag on a sphere moving in a fluid is known to be a function of the sphere diameter, the velocity, and the fluid viscosity and density. Laboratory tests on a 4-in.-diameter sphere were performed in a water tunnel and some model data are plotted in

■ **Figure P7.77**

Fig. P7.77. For these tests the viscosity of the water was $2.3 \times 10^{-5}$ lb · s/ft$^2$ and the water density was 1.94 slugs/ft$^3$. Estimate the drag on an 8-ft-diameter balloon moving in air at a velocity of 3 ft/s. Assume the air to have a viscosity of $3.7 \times 10^{-7}$ lb · s/ft$^2$ and a density of $2.38 \times 10^{-3}$ slugs/ft$^3$.

### Section 7.9 Some Typical Model Studies

**7.78** (WILEY) Drag measurements were taken for a sphere with a diameter of 5 cm, moving at 4 m/s in water at 20 °C. The resulting drag on the sphere was 10 N. For a balloon with 1-m diameter rising in air with standard temperature and pressure, determine **(a)** the velocity if Reynolds number similarity is enforced and **(b)** the drag force if the drag coefficient (Eq. 7.19) is the dependent pi term.

**7.79** The pressure rise, $\Delta p$, across a centrifugal pump of a given shape can be expressed as

$$\Delta p = f(D, \omega, \rho, Q)$$

where $D$ is the impeller diameter, $\omega$ the angular velocity of the impeller, $\rho$ the fluid density, and $Q$ the volume rate of flow through the pump. A model pump having a diameter of 8 in. is tested in the laboratory using water. When operated at an angular velocity of $40\pi$ rad/s the model pressure rise as a function of $Q$ is shown in Fig. P7.79. Use this curve to predict the pressure rise across a geometrically similar pump (prototype) for a prototype flowrate of 6 ft$^3$/s. The prototype has a diameter of 12 in. and operates at an angular velocity of $60\pi$ rad/s. The prototype fluid is also water.

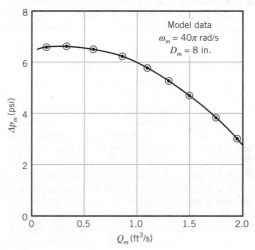

■ **Figure P7.79**

**7.80** (WILEY) A prototype automobile is designed to travel at 65 km/hr. A model of this design is tested in a wind tunnel with identical standard sea-level air properties at a 1:5 scale. The measured model drag is 400 N, enforcing dynamic similarity. Determine **(a)** the drag force on the prototype and **(b)** the power required to overcome this drag. See Eq. 7.19.

**7.81** (WILEY) A new blimp will move at 6 m/s in 20 °C air, and we want to predict the drag force. Using a 1:13-scale model in water at 20 °C and measuring a 2500-N drag force on the model, determine **(a)** the required water velocity, **(b)** the drag on the prototype blimp and, **(c)** the power that will be required to propel it through the air.

**7.82** ▶ (WILEY GO) At a large fish hatchery the fish are reared in open, water-filled tanks. Each tank is approximately square in shape with curved corners, and the walls are smooth. To create motion in the tanks, water is supplied through a pipe at the edge of the tank. The water is drained from the tank through an opening at the center. (See **Video V7.9**.) A model with a length scale of 1:13 is to be used to determine the velocity, $V$, at various locations within the tank. Assume that $V = f(\ell, \ell_i, \rho, \mu, g, Q)$ where $\ell$ is some characteristic length such as the tank width, $\ell_i$ represents a series of other pertinent lengths, such as inlet pipe diameter, fluid depth, etc., $\rho$ is the fluid density, $\mu$ is the fluid viscosity, $g$ is the acceleration of gravity, and $Q$ is the discharge through the tank. **(a)** Determine a suitable set of dimensionless parameters for this problem and the prediction equation for the velocity. If water is to be used for the model, can all of the similarity requirements be satisfied? Explain and support your answer with the necessary calculations. **(b)** If the flowrate into the full-sized tank is 250 gpm, determine the required value for the model discharge assuming Froude number similarity. What model depth will correspond to a depth of 32 in. in the full-sized tank?

**7.83** ▶ Flow patterns that develop as winds blow past a vehicle, such as a train, are often studied in low-speed environmental (meteorological) wind tunnels. (See **Video V7.16.**) Typically, the air velocities in these tunnels are in the range of 0.1 m/s to 30 m/s. Consider a cross wind blowing past a train locomotive. Assume that the local wind velocity, $V$, is a function of the approaching wind velocity (at some distance from the locomotive), $U$, the locomotive length, $\ell$, height, $h$, and width, $b$, the air density, $\rho$, and the air viscosity, $\mu$. **(a)** Establish the similarity requirements and prediction equation for a model to be used in the wind tunnel to study the air velocity, $V$, around the locomotive. **(b)** If the model is to be used for cross winds gusting to $U = 25$ m/s, explain why it is not practical to maintain Reynolds number similarity for a typical length scale 1:50.

**7.84** ▶ ⟨WILEY⟩ (See Fluids in the News article titled "Galloping Gertie," Section 7.8.2.) The Tacoma Narrows Bridge failure is a dramatic example of the possible serious effects of wind-induced vibrations. As a fluid flows around a body, vortices may be created that are shed periodically, creating an oscillating force on the body. If the frequency of the shedding vortices coincides with the natural frequency of the body, large displacements of the body can be induced as was the case with the Tacoma Narrows Bridge. To illustrate this type of phenomenon, consider fluid flow past a circular cylinder. (See **Video V7.5.**) Assume the frequency, $n$, of the shedding vortices behind the cylinder is a function of the cylinder diameter, $D$, the fluid velocity, $V$, and the fluid kinematic viscosity, $v$. **(a)** Determine a suitable set of dimensionless variables for this problem. One of the dimensionless variables should be the Strouhal number, $nD/V$. **(b)** Some results of experiments in which the shedding frequency of the vortices (in Hz) was measured, using a particular cylinder and Newtonian, incompressible fluid, are shown in Fig. P7.84. Is this a "universal curve" that can be used to predict the shedding frequency for any cylinder placed in any fluid? Explain. **(c)** A certain structural component in the form of a 1-in.-diameter, 12-ft-long rod acts as a cantilever beam

with a natural frequency of 19 Hz. Based on the data in Fig. P7.84, estimate the wind speed that may cause the rod to oscillate at its natural frequency. *Hint*: Use a trial-and-error solution.

**7.85** ⟨WILEY⟩ (See Fluids in the News article titled "Ice Engineering," Section 7.9.3.) A model study is to be developed to determine the force exerted on bridge piers due to floating chunks of ice in a river. The piers of interest have square cross sections. Assume that the force, $R$, is a function of the pier width, $b$, the thickness of the ice, $d$, the velocity of the ice, $V$, the acceleration of gravity, $g$, the density of the ice, $\rho_i$, and a measure of the strength of the ice, $E_i$, where $E_i$ has the dimensions $FL^{-2}$. **(a)** Based on these variables determine a suitable set of dimensionless variables for this problem. **(b)** The prototype conditions of interest include an ice thickness of 12 in. and an ice velocity of 6 ft/s. What model ice thickness and velocity would be required if the length scale is to be 1/10? **(c)** If the model and prototype ice have the same density, can the model ice have the same strength properties as that of the prototype ice? Explain.

**7.86** ▶ ⟨WILEY⟩ As illustrated in **Video V7.9**, models are commonly used to study the dispersion of a gaseous pollutant from an exhaust stack located near a building complex. Similarity requirements for the pollutant source involve the following independent variables: the stack gas speed, $V$, the wind speed, $U$, the density of the atmospheric air, $\rho$, the difference in densities between the air and the stack gas, $\rho - \rho_s$, the acceleration of gravity, $g$, the kinematic viscosity of the stack gas, $v_s$, and the stack diameter, $D$. **(a)** Based on these variables, determine a suitable set of similarity requirements for modeling the pollutant source. **(b)** For this type of model a typical length scale might be 1:200. If the same fluids were used in model and prototype, would the similarity requirements be satisfied? Explain and support your answer with the necessary calculations.

**7.87** ▶ A 1:40 scale model of a ship is to be tested in a towing tank. (See **Video V7.20.**) Determine the required kinematic viscosity of the model fluid so that both the Reynolds number and the Froude number are the same for model and prototype. Assume the prototype fluid to be seawater at 60 °F. Could any of the liquids with viscosities given in Fig. B.2 in Appendix B be used as the model fluid?

**7.88** ▶ ⟨WILEY⟩ River models are used to study many different types of flow situations. (See, for example, **Video V7.12.**) A certain small river has an average width and depth of 60 ft and 4 ft, respectively, and carries water at a flowrate of 700 ft³/s. A model is to be designed based on Froude number similarity so that the discharge scale is 1/250. At what depth and flowrate would the model operate?

**7.89** ▶ ⟨WILEY⟩ As winds blow past buildings, complex flow patterns can develop due to various factors such as flow separation and interactions between adjacent buildings. (See **Video V7.13.**) Assume that the local gage pressure, $p$, at a particular location on a building is a function of the air density, $\rho$, the wind speed, $V$, some characteristic length, $\ell$, and all other pertinent lengths, $\ell_i$, needed to characterize the geometry of the building or building complex. **(a)** Determine a suitable set of dimensionless parameters that can be used to study the pressure distribution. **(b)** An eight-story building that is 100 ft tall is to be modeled in a wind tunnel. If a length scale of 1:300 is to be used, how tall should the model building be? **(c)** How will a measured pressure in the model be related to the corresponding prototype pressure? Assume the same air density in model and prototype. Based on the assumed variables, does the model wind speed have to be equal to the prototype wind speed? Explain.

### Section 7.10 Similitude Based on Governing Differential Equations

**7.90** Start with the two-dimensional continuity equation and the Navier–Stokes equations (Eqs. 7.28, 7.29, and 7.30) and verify the nondimensional forms of these equations (Eqs. 7.31, 7.34, and 7.35).

■ **Figure P7.84**

**7.91** (WILEY) A viscous fluid is contained between wide, parallel plates spaced a distance $h$ apart as shown in Fig. P7.91. The upper plate is fixed, and the bottom plate oscillates harmonically with a velocity amplitude $U$ and frequency $\omega$. The differential equation for the velocity distribution between the plates is

$$\rho \frac{\partial u}{\partial t} = \mu \frac{\partial^2 u}{\partial y^2}$$

where $u$ is the velocity, $t$ is time, and $\rho$ and $\mu$ are fluid density and viscosity, respectively. Rewrite this equation in a suitable nondimensional form using $h$, $U$, and $\omega$ as reference parameters.

■ **Figure P7.91**

**7.92** (WILEY) The deflection of the cantilever beam of Fig. P7.92 is governed by the differential equation.

$$EI \frac{d^2 y}{dx^2} = P(x - \ell)$$

where $E$ is the modulus of elasticity and $I$ is the moment of inertia of the beam cross section. The boundary conditions are $y = 0$ at $x = 0$ and $dy/dx = 0$ at $x = 0$. **(a)** Rewrite the equation and boundary conditions in dimensionless form using the beam length, $\ell$, as the reference length. **(b)** Based on the results of part (a), what are the similarity requirements and the prediction equation for a model to predict deflections?

■ **Figure P7.92**

**7.93** (WILEY) A liquid is contained in a pipe that is closed at one end as shown in Fig. P7.93. Initially the liquid is at rest, but if the end is suddenly opened the liquid starts to move. Assume the pressure $p_1$ remains constant. The differential equation that describes the resulting motion of the liquid is

$$\rho \frac{\partial v_z}{\partial t} = \frac{p_1}{\ell} + \mu \left( \frac{\partial^2 v_z}{\partial r^2} + \frac{1}{r} \frac{\partial v_z}{\partial r} \right)$$

where $v_z$ is the velocity at any radial location, $r$, and $t$ is time. Rewrite this equation in dimensionless form using the liquid density, $\rho$, the viscosity, $\mu$, and the pipe radius, $R$, as reference parameters.

■ **Figure P7.93**

**7.94** (WILEY) An incompressible fluid is contained between two infinite parallel plates as illustrated in Fig. P7.94. Under the influence of a harmonically varying pressure gradient in the $x$ direction, the

■ **Figure P7.94**

fluid oscillates harmonically with a frequency $\omega$. The differential equation describing the fluid motion is

$$\rho \frac{\partial u}{\partial t} = X \cos \omega t + \mu \frac{\partial^2 u}{\partial y^2}$$

where $X$ is the amplitude of the pressure gradient. Express this equation in nondimensional form using $h$ and $\omega$ as reference parameters.

■ **Lab Problems**

**7.1LP** This problem involves the time that it takes water to drain from two geometrically similar tanks. To proceed with this problem, go to *WileyPLUS* or the book's web site, www.wiley.com/college/munson.

**7.2LP** This problem involves determining the frequency of vortex shedding from a circular cylinder as water flows past it. To proceed with this problem, go to *WileyPLUS* or the book's web site, www.wiley.com/college/munson.

**7.3LP** This problem involves the determination of the head loss for flow through a valve. To proceed with this problem, go to *WileyPLUS* or the book's web site, www.wiley.com/college/munson.

**7.4LP** This problem involves the calibration of a rotameter. To proceed with this problem, go to *WileyPLUS* or the book's web site, www.wiley.com/college/munson.

■ **Lifelong Learning Problems**

**7.1LL** Microfluidics is the study of fluid flow in fabricated devices at the micro scale. Advances in microfluidics have enhanced the ability of scientists and engineers to perform laboratory experiments using miniaturized devices known as a "lab-on-a-chip." Obtain information about a lab-on-a-chip device that is available commercially and investigate its capabilities. Summarize your findings in a brief report.

**7.2LL** For some types of aerodynamic wind tunnel testing, it is difficult to simultaneously match both the Reynolds number and Mach number between model and prototype. Engineers have developed several potential solutions to the problem including pressurized wind tunnels and lowering the temperature of the flow. Obtain information about cryogenic wind tunnels and explain the advantages and disadvantages. Summarize your findings in a brief report.

■ **FE Exam Problems**

Sample FE (Fundamental of Engineering) exam questions for fluid mechanics are provided in *WileyPLUS* or on the book's web site, www.wiley.com/college/munson.

■ **Computational Fluid Dynamics (CFD)**

The CFD problems associated with this chapter have been developed for use with the ANSYS Academic CFD software package that is associated with this text. See *WileyPLUS* or the book's web site (www.wiley.com/college/munson) for additional details.

**7.1CFD** This CFD problem involves investigation of the Reynolds number significance in fluid dynamics through the simulation of flow past a cylinder. To proceed with this problem, go to *WileyPLUS* or the book's web site, www.wiley.com/college/munson.

**There are additional CFD problems located in *WileyPLUS*.**

# 8 Viscous Flow in Pipes

CHAPTER OPENING PHOTO: Turbulent jet: The jet of water from the pipe is turbulent. The complex, irregular, unsteady structure typical of turbulent flows is apparent. (Laser-induced fluorescence of dye in water.) (© P. Dimotakis. Dimotakis, P.E., R.C. Miake-Lye, and D.A. Papantoniou 1983 Structure and Dynamics of Round Turbulent Jets. Phys. Fluids 26(11):3185–3192.)

## Learning Objectives

*V8.1 Turbulent jet*

After completing this chapter, you should be able to:

- identify and understand various characteristics of the flow in pipes.
- discuss the main properties of laminar and turbulent pipe flow and appreciate their differences.
- calculate losses in straight portions of pipes as well as those in various pipe system components.
- apply appropriate equations and principles to analyze a variety of pipe flow situations.
- predict the flowrate in a pipe by use of common flowmeters.

In the previous chapters we have considered a variety of topics concerning the motion of fluids. The basic governing principles concerning mass, momentum, and energy were developed and applied, in conjunction with rather severe assumptions, to numerous flow situations. In this chapter we will apply the basic principles to a specific, important topic—the incompressible flow of viscous fluids in pipes and ducts.

*Pipe flow is very important in our daily operations.*

    The transport of a fluid (liquid or gas) in a closed conduit (commonly called a *pipe* if it is of round cross section or a *duct* if it is not round) is extremely important in our daily operations. A brief consideration of the world around us will indicate that there is a wide variety of applications of pipe flow. Such applications range from the large, man-made Alaskan pipeline that carries crude oil almost 800 miles across Alaska, to the more complex (and certainly not less useful) natural systems of "pipes" that carry blood throughout our body and air into and out of our lungs. Other examples include the water pipes in our homes and the distribution system that delivers the water from the city well to the house. Numerous hoses and pipes carry hydraulic fluid or other fluids to various components of vehicles and machines. The air quality within our buildings is maintained at comfortable levels by the distribution of conditioned (heated, cooled, humidified/dehumidified) air through a maze of pipes and ducts. Although all of these systems are different, the fluid mechanics

■ **Figure 8.1**    **Typical pipe system components.**

principles governing the fluid motions are common. The purpose of this chapter is to understand the basic processes involved in such flows.

Some of the basic components of a typical *pipe system* are shown in Fig. 8.1. They include the pipes themselves (perhaps of more than one diameter), the various fittings used to connect the individual pipes to form the desired system, the flowrate control devices (valves), and the pumps or turbines that add energy to or remove energy from the fluid. Even the most simple pipe systems are actually quite complex when they are viewed in terms of rigorous analytical considerations. We will use an "exact" analysis of the simplest pipe flow topics (such as laminar flow in long, straight, constant diameter pipes) and dimensional analysis considerations combined with experimental results for the other pipe flow topics. Such an approach is not unusual in fluid mechanics investigations. When "real-world" effects are important (such as viscous effects in pipe flows), it is often difficult or "impossible" to use only theoretical methods to obtain the desired results. A judicious combination of experimental data with theoretical considerations and dimensional analysis often provides the desired results. The flow in pipes discussed in this chapter is an example of such an analysis.

# 8.1    General Characteristics of Pipe Flow

Before we apply the various governing equations to pipe flow examples, we will discuss some of the basic concepts of pipe flow. With these ground rules established, we can then proceed to formulate and solve various important flow problems.

Although not all conduits used to transport fluid from one location to another are round in cross section, most of the common ones are. These include typical water pipes, hydraulic hoses, and other conduits that are designed to withstand a considerable pressure difference across their walls without undue distortion of their shape. Typical conduits of noncircular cross section include heating and air conditioning ducts that are often of rectangular cross section. Normally the pressure difference between the inside and outside of these ducts is relatively small. Most of the basic principles involved are independent of the cross-sectional shape, although the details of the flow may be dependent on it. Unless otherwise specified, we will assume that the conduit is round, although we will show how to account for other shapes.

*The pipe is assumed to be completely full of the flowing fluid.*

For all flows involved in this chapter, we assume that the pipe is completely filled with the fluid being transported as is shown in Fig. 8.2*a*. Thus, we will not consider a concrete pipe through

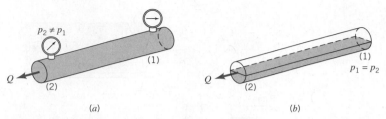

$p_2 \neq p_1$

(1)

$Q$

(2)

(a)

(1)

$p_1 = p_2$

$Q$

(2)

(b)

■ **Figure 8.2** (a) Pipe flow. (b) Open-channel flow.

*V8.2 Laminar/
turbulent pipe flow*

*A flow may be laminar, transitional, or turbulent.*

which rainwater flows without completely filling the pipe, as is shown in Fig. 8.2b. Such flows, called open-channel flow, are treated in Chapter 10. The difference between open-channel flow and the pipe flow of this chapter is in the fundamental mechanism that drives the flow. For open-channel flow, gravity alone is the driving force—the water flows down a hill. For pipe flow, gravity may be important (the pipe need not be horizontal), but the main driving force is likely to be a pressure gradient along the pipe. If the pipe is not full, it is not possible to maintain this pressure difference, $p_1 - p_2$.

### 8.1.1 Laminar or Turbulent Flow

The flow of a fluid in a pipe may be laminar flow or it may be turbulent flow. Osborne Reynolds (1842–1912), a British scientist and mathematician, was the first to distinguish the difference between these two classifications of flow by using a simple apparatus as shown by the figure in the margin, which is a sketch of Reynolds' dye experiment. Reynolds injected dye into a pipe in which water flowed due to gravity. The entrance region of the pipe is depicted in Fig. 8.3a. If water runs through a pipe of diameter $D$ with an average velocity $V$, the following characteristics are observed by injecting neutrally buoyant dye as shown. For "small enough flowrates" the dye streak (a streakline) will remain as a well-defined line as it flows along, with only slight blurring due to molecular diffusion of the dye into the surrounding water. For a somewhat larger "intermediate flowrate" the dye streak fluctuates in time and space, and intermittent bursts of irregular behavior appear along the streak. On the other hand, for "large enough flowrates" the dye streak almost immediately becomes blurred and spreads across the entire pipe in a random fashion. These three characteristics, denoted as *laminar*, *transitional*, and *turbulent* flow, respectively, are illustrated in Fig. 8.3b.

The curves shown in Fig. 8.4 represent the $x$ component of the velocity as a function of time at a point $A$ in the flow. The random fluctuations of the turbulent flow (with the associated particle mixing) are what disperse the dye throughout the pipe and cause the blurred appearance illustrated in Fig. 8.3b. For laminar flow in a pipe there is only one component of velocity,

■ **Figure 8.3** (a) Experiment to illustrate type of flow. (b) Typical dye streaks.

■ **Figure 8.4**  Time dependence of fluid velocity at a point.

$\mathbf{V} = u\hat{\mathbf{i}}$. For turbulent flow the predominant component of velocity is also along the pipe, but it is unsteady (random) and accompanied by random components normal to the pipe axis, $\mathbf{V} = u\hat{\mathbf{i}} + v\hat{\mathbf{j}} + w\hat{\mathbf{k}}$. Such motion in a typical flow occurs too fast for our eyes to follow. Slow-motion pictures of the flow can more clearly reveal the irregular, random, turbulent nature of the flow.

As was discussed in Chapter 7, we should not label dimensional quantities as being "large" or "small," such as "small enough flowrates" in the preceding paragraphs. Rather, the appropriate dimensionless quantity should be identified and the "small" or "large" character attached to it. A quantity is "large" or "small" only relative to a reference quantity. The ratio of those quantities results in a dimensionless quantity. For pipe flow the most important dimensionless parameter is the Reynolds number, Re—the ratio of the inertia to viscous effects in the flow. Hence, in the previous paragraph the term flowrate should be replaced by Reynolds number, Re $= \rho VD/\mu$, where $V$ is the average velocity in the pipe. That is, the flow in a pipe is laminar, transitional, or turbulent provided the Reynolds number is "small enough," "intermediate," or "large enough." It is not only the fluid velocity that determines the character of the flow—its density, viscosity, and the pipe size are of equal importance. These parameters combine to produce the Reynolds number. The distinction between laminar and turbulent pipe flow and its dependence on an appropriate dimensionless quantity was first pointed out by Osborne Reynolds in 1883.

The Reynolds number ranges for which laminar, transitional, or turbulent pipe flows are obtained cannot be precisely given. The actual transition from laminar to turbulent flow may take place at various Reynolds numbers, depending on how much the flow is disturbed by vibrations of the pipe, roughness of the entrance region, and the like. For general engineering purposes (i.e., without undue precautions to eliminate such disturbances), the following values are appropriate: The flow in a round pipe is laminar if the Reynolds number is less than approximately 2100. The flow in a round pipe is turbulent if the Reynolds number is greater than approximately 4000. For Reynolds numbers between these two limits, the flow may switch between laminar and turbulent conditions in an apparently random fashion (transitional flow).

*Pipe flow characteristics are dependent on the value of the Reynolds number.*

*V8.3 Intermittent turbulent burst in pipe flow*

## EXAMPLE 8.1 Laminar or Turbulent Flow

**GIVEN** Water at a temperature of 50 °F flows through a pipe of diameter $D = 0.73$ in. and into a glass as shown in Fig. E8.1a.

**FIND** Determine

**(a)** the minimum time taken to fill a 12-oz glass (volume $= 0.0125$ ft$^3$) with water if the flow in the pipe is to be laminar. Repeat the calculations if the water temperature is 140 °F.

**(b)** the maximum time taken to fill the glass if the flow is to be turbulent. Repeat the calculations if the water temperature is 140 °F.

## SOLUTION

**(a)** If the flow in the pipe is to remain laminar, the minimum time to fill the glass will occur if the Reynolds number is the maximum allowed for laminar flow, typically Re $= \rho V D/\mu = 2100$. Thus, $V = 2100 \, \mu/\rho D$, where from Table B.1, $\rho = 1.94$ slugs/ft$^3$ and $\mu = 2.73 \times 10^{-5}$ lb · s/ft$^2$ at 50 °F, while $\rho = 1.91$ slugs/ft$^3$ and $\mu = 0.974 \times 10^{-5}$ lb · s/ft$^2$ at 140 °F. Thus, the maximum average velocity for laminar flow in the pipe is

$$V = \frac{2100\mu}{\rho D} = \frac{2100(2.73 \times 10^{-5} \text{ lb} \cdot \text{s/ft}^2)}{(1.94 \text{ slugs/ft}^3)(0.73/12 \text{ ft})}$$

$$= 0.486 \text{ lb} \cdot \text{s/slug} = 0.486 \text{ ft/s}$$

Similarly, $V = 0.176$ ft/s at 140 °F. With $\Psi = $ volume of glass and $\Psi = Qt$ we obtain

$$t = \frac{\Psi}{Q} = \frac{\Psi}{(\pi/4)D^2 V} = \frac{4(0.0125 \text{ ft}^3)}{(\pi[0.73/12]^2 \text{ft}^2)(0.486 \text{ ft/s})}$$

$$= 8.85 \text{ s at } T = 50 \text{ °F} \qquad \textbf{(Ans)}$$

Similarly, $t = 24.4$ s at 140 °F. To maintain laminar flow, the less viscous hot water requires a lower flowrate than the cold water.

**(b)** If the flow in the pipe is to be turbulent, the maximum time to fill the glass will occur if the Reynolds number is the minimum allowed for turbulent flow, Re $= 4000$. Thus, $V = 4000\mu/\rho D = 0.925$ ft/s and

$$t = 4.65 \text{ s at } 50 \text{ °F} \qquad \textbf{(Ans)}$$

Similarly, $V = 0.335$ ft/s and $t = 12.8$ s at 140 °F.

**COMMENTS** Note that because water is "not very viscous," the velocity must be "fairly small" to maintain laminar flow. In general, turbulent flows are encountered more often than laminar flows because of the relatively small viscosity of most common fluids (water, gasoline, air). By repeating the calculations at various water temperatures, $T$ (i.e., with different densities and viscosities), the results shown in Fig. E8.1b are obtained. As the water temperature increases, the kinematic viscosity, $\nu = \mu/\rho$, decreases and the corresponding times to fill the glass increase as indicated. (Temperature effects on the viscosity of gases are the opposite; increase in temperature causes an increase in viscosity.)

■ **Figure E8.1a**

If the flowing fluid had been honey with a kinematic viscosity ($\nu = \mu/\rho$) 3000 times greater than that of water, the velocities given earlier would be increased by a factor of 3000 and the times reduced by the same factor. As shown in the following sections, the pressure needed to force a very viscous fluid through a pipe at such a high velocity may be unreasonably large.

■ **Figure E8.1b**

### 8.1.2 Entrance Region and Fully Developed Flow

Any fluid flowing in a pipe had to enter the pipe at some location. The region of flow near where the fluid enters the pipe is termed the *entrance region* and is illustrated in Fig. 8.5. It may be the first few feet of a pipe connected to a tank or the initial portion of a long run of a hot air duct coming from a furnace.

As is shown in Fig. 8.5, the fluid typically enters the pipe with a nearly uniform velocity profile at section (1). As the fluid moves through the pipe, viscous effects cause it to stick to the pipe wall (the no-slip boundary condition). This is true whether the fluid is relatively inviscid air or a very viscous oil. Thus, a *boundary layer* in which viscous effects are important is produced along the pipe wall such that the initial velocity profile changes with distance along the pipe, $x$, until the fluid reaches the end of the entrance length, section (2), beyond which the velocity profile does not vary with $x$. The boundary layer has grown in thickness to completely fill the pipe. Viscous effects are of considerable importance within the boundary layer. For fluid outside the boundary layer [within the *inviscid core* surrounding the centerline from (1) to (2)], viscous effects are negligible.

The shape of the velocity profile in the pipe depends on whether the flow is laminar or turbulent, as does the length of the entrance region, $\ell_e$. As with many other properties of pipe flow, the dimensionless **entrance length,** $\ell_e/D$, correlates quite well with the Reynolds number. Typical entrance lengths are given by

$$\frac{\ell_e}{D} = 0.06 \, \text{Re for laminar flow} \tag{8.1}$$

and

*The entrance length is a function of the Reynolds number.*

$$\frac{\ell_e}{D} = 4.4 \, (\text{Re})^{1/6} \text{ for turbulent flow} \tag{8.2}$$

For very low Reynolds number flows the entrance length can be quite short ($\ell_e = 0.6D$ if Re = 10), whereas for large Reynolds number flows it may take a length equal to many pipe diameters before the end of the entrance region is reached ($\ell_e = 120D$ for Re = 2000). For many practical engineering problems, $10^4 < \text{Re} < 10^5$ so that as shown by the figure in the margin, $20D < \ell_e < 30D$.

Calculation of the velocity profile and pressure distribution within the entrance region is quite complex. However, once the fluid reaches the end of the entrance region, section (2) of Fig. 8.5, the flow is simpler to describe because the velocity is a function of only the distance from the pipe centerline, $r$, and independent of $x$. This is true until the character of the pipe changes in some way, such as a change in diameter, or the fluid flows through a bend, valve, or some other component at section (3). The flow between (2) and (3) is termed *fully developed flow*. Beyond the interruption of the fully developed flow [at section (4)], the flow gradually begins its

■ **Figure 8.5** Entrance region, developing flow, and fully developed flow in a pipe system.

return to its fully developed character [section (5)] and continues with this profile until the next pipe system component is reached [section (6)]. In many cases the pipe is long enough so that there is a considerable length of fully developed flow compared with the developing flow length $[(x_3 - x_2) \gg \ell_e$ and $(x_6 - x_5) \gg (x_5 - x_4)]$. In other cases the distances between one component (bend, tee, valve, etc.) of the pipe system and the next component is so short that fully developed flow is never achieved.

### 8.1.3  Pressure and Shear Stress

Fully developed steady flow in a constant diameter pipe may be driven by gravity and/or pressure forces. For horizontal pipe flow, gravity has no effect except for a hydrostatic pressure variation across the pipe, $\gamma D$, that is usually negligible. It is the pressure difference, $\Delta p = p_1 - p_2$, between one section of the horizontal pipe and another which forces the fluid through the pipe. Viscous effects provide the restraining force that exactly balances the pressure force, thereby allowing the fluid to flow through the pipe with no acceleration. If viscous effects were absent in such flows, the pressure would be constant throughout the pipe, except for the hydrostatic variation.

In non-fully developed flow regions, such as the entrance region of a pipe, the fluid accelerates or decelerates as it flows (the velocity profile changes from a uniform profile at the entrance of the pipe to its fully developed profile at the end of the entrance region). Thus, in the entrance region there is a balance between pressure, viscous, and inertia (acceleration) forces. The result is a pressure distribution along the horizontal pipe as shown in Fig. 8.6. The magnitude of the pressure gradient, $\partial p/\partial x$, is larger in the entrance region than in the fully developed region, where it is a constant, $\partial p/\partial x = -\Delta p/\ell < 0$.

The fact that there is a nonzero pressure gradient along the horizontal pipe is a result of viscous effects. As is discussed in Chapter 3, if the viscosity were zero, the pressure would not vary with $x$. The need for the pressure drop can be viewed from two different standpoints. In terms of a force balance, the pressure force is needed to overcome the viscous forces generated. In terms of an energy balance, the work done by the pressure force is needed to overcome the viscous dissipation of energy throughout the fluid. If the pipe is not horizontal, the pressure gradient along it is due in part to the component of weight in that direction. As is discussed in Section 8.2.1, this contribution due to the weight either enhances or retards the flow, depending on whether the flow is downhill or uphill.

The nature of the pipe flow is strongly dependent on whether the flow is laminar or turbulent. This is a direct consequence of the differences in the nature of the shear stress in laminar and turbulent flows. As is discussed in some detail in Section 8.3.3, the shear stress in laminar flow is a direct result of momentum transfer among the randomly moving molecules (a microscopic phenomenon). The shear stress in turbulent flow is largely a result of momentum transfer among the randomly moving, finite-sized fluid particles (a macroscopic phenomenon). The net result is that the physical properties of the shear stress are quite different for laminar flow than for turbulent flow.

*Laminar flow characteristics are different than those for turbulent flow.*

■ **Figure 8.6**  **Pressure distribution along a horizontal pipe.**

## 8.2    Fully Developed Laminar Flow

As is indicated in the previous section, the flow in long, straight, constant diameter sections of a pipe becomes fully developed. That is, the velocity profile is the same at any cross section of the pipe. Although this is true whether the flow is laminar or turbulent, the details of the velocity profile (and other flow properties) are quite different for these two types of flow. As will be seen in the remainder of this chapter, knowledge of the velocity profile can lead directly to other useful information such as pressure drop, head loss, flowrate, and the like. Thus, we begin by developing the equation for the velocity profile in fully developed laminar flow. If the flow is not fully developed, a theoretical analysis becomes much more complex and is outside the scope of this text. If the flow is turbulent, a rigorous theoretical analysis is as yet not possible.

Although most flows are turbulent rather than laminar, and many pipes are not long enough to allow the attainment of fully developed flow, a theoretical treatment and full understanding of fully developed laminar flow is of considerable importance. First, it represents one of the few theoretical viscous analyses that can be carried out "exactly" (within the framework of quite general assumptions) without using other ad hoc assumptions or approximations. An understanding of the method of analysis and the results obtained provides a foundation from which to carry out more complicated analyses. Second, there are many practical situations involving the use of fully developed laminar pipe flow.

There are numerous ways to derive important results pertaining to fully developed laminar flow. Three alternatives include: (1) from $\mathbf{F} = m\mathbf{a}$ applied directly to a fluid element, (2) from the Navier–Stokes equations of motion, and (3) from dimensional analysis methods.

### 8.2.1    From $\mathbf{F} = m\mathbf{a}$ Applied Directly to a Fluid Element

*Steady, fully developed pipe flow experiences no acceleration.*

Velocity profiles

Streamlines

We consider the fluid element at time $t$ as is shown in Fig. 8.7. It is a circular cylinder of fluid of length $\ell$ and radius $r$ centered on the axis of a horizontal pipe of diameter $D$. Because the velocity is not uniform across the pipe, the initially flat ends of the cylinder of fluid at time $t$ become distorted at time $t + \delta t$ when the fluid element has moved to its new location along the pipe as shown in the figure. If the flow is fully developed and steady, the distortion on each end of the fluid element is the same, and no part of the fluid experiences any acceleration as it flows, as shown by the figure in the margin. The local acceleration is zero ($\partial \mathbf{V}/\partial t = 0$) because the flow is steady, and the convective acceleration is zero ($\mathbf{V} \cdot \nabla \mathbf{V} = u \, \partial u/\partial x \, \hat{\mathbf{i}} = 0$) because the flow is fully developed. Thus, every part of the fluid merely flows along its streamline parallel to the pipe walls with constant velocity, although neighboring particles have slightly different velocities. The velocity varies from one pathline to the next. This velocity variation, combined with the fluid viscosity, produces the shear stress.

If gravitational effects are neglected, the pressure is constant across any vertical cross section of the pipe, although it varies along the pipe from one section to the next. Thus, if the pressure is $p = p_1$ at section (1), it is $p_2 = p_1 - \Delta p$ at section (2) where $\Delta p$ is the pressure drop between sections (1) and (2). We anticipate the fact that the pressure decreases in the direction of flow so that $\Delta p > 0$. A shear stress, $\tau$, acts on the surface of the cylinder of fluid. This viscous stress is a function of the radius of the cylinder, $\tau = \tau(r)$.

As was done in fluid statics analysis (Chapter 2), we isolate the cylinder of fluid as is shown in Fig. 8.8 and apply Newton's second law, $F_x = ma_x$. In this case, even though the fluid is moving, it is not accelerating, so that $a_x = 0$. Thus, fully developed horizontal pipe flow is merely a

■ **Figure 8.7**  **Motion of a cylindrical fluid element within a pipe.**

■ **Figure 8.8** Free-body diagram of a cylinder of fluid.

balance between pressure and viscous forces—the pressure difference acting on the end of the cylinder of area $\pi r^2$, and the shear stress acting on the lateral surface of the cylinder of area $2\pi r\ell$. This force balance can be written as

$$(p_1)\pi r^2 - (p_1 - \Delta p)\pi r^2 - (\tau)2\pi r\ell = 0$$

which can be simplified to give

$$\frac{\Delta p}{\ell} = \frac{2\tau}{r} \tag{8.3}$$

Equation 8.3 represents the basic balance in forces needed to drive each fluid particle along the pipe with constant velocity. Since neither $\Delta p$ nor $\ell$ are functions of the radial coordinate, $r$, it follows that $2\tau/r$ must also be independent of $r$. That is, $\tau = Cr$, where $C$ is a constant. At $r = 0$ (the centerline of the pipe) there is no shear stress ($\tau = 0$). At $r = D/2$ (the pipe wall) the shear stress is a maximum, denoted $\tau_w$, the **wall shear stress**. Hence, $C = 2\tau_w/D$ and the shear stress distribution throughout the pipe is a linear function of the radial coordinate

$$\tau = \frac{2\tau_w r}{D} \tag{8.4}$$

as is indicated in Fig. 8.9. The linear dependence of $\tau$ on $r$ is a result of the pressure force being proportional to $r^2$ (the pressure acts on the end of the fluid cylinder; area $= \pi r^2$) and the shear force being proportional to $r$ (the shear stress acts on the lateral sides of the cylinder; area $= 2\pi r\ell$). If the viscosity were zero there would be no shear stress, and the pressure would be constant throughout the horizontal pipe ($\Delta p = 0$). As is seen from Eqs. 8.3 and 8.4, the pressure drop and wall shear stress are related by

$$\Delta p = \frac{4\ell\tau_w}{D} \tag{8.5}$$

*Basic horizontal pipe flow is governed by a balance between viscous and pressure forces.*

A small shear stress can produce a large pressure difference if the pipe is relatively long ($\ell/D \gg 1$).

Although we are discussing laminar flow, a closer consideration of the assumptions involved in the derivation of Eqs. 8.3, 8.4, and 8.5 reveals that these equations are valid for both laminar and turbulent flow. To carry the analysis further we must prescribe how the shear stress is related to the velocity. This is the critical step that separates the analysis of laminar from that of turbulent flow—from being able to solve for the laminar flow properties and not being able to solve for the turbulent flow properties without additional ad hoc assumptions. As is discussed in Section 8.3, the shear stress dependence for turbulent flow is very complex. However, for laminar flow of a

■ **Figure 8.9** Shear stress distribution within the fluid in a pipe (laminar or turbulent flow) and typical velocity profiles.

Newtonian fluid, the shear stress is simply proportional to the velocity gradient, "$\tau = \mu\, du/dy$" (see Section 1.6). In the notation associated with our pipe flow, this becomes

$$\tau = -\mu \frac{du}{dr} \tag{8.6}$$

The negative sign is included to give $\tau > 0$ with $du/dr < 0$ (the velocity decreases from the pipe centerline to the pipe wall).

Equations 8.3 and 8.6 represent the two governing laws for fully developed laminar flow of a Newtonian fluid within a horizontal pipe. The one is Newton's second law of motion, and the other is the definition of a Newtonian fluid. By combining these two equations, we obtain

$$\frac{du}{dr} = -\left(\frac{\Delta p}{2\mu\ell}\right) r$$

which can be integrated to give the velocity profile as follows:

$$\int du = -\frac{\Delta p}{2\mu\ell} \int r\, dr$$

or

$$u = -\left(\frac{\Delta p}{4\mu\ell}\right) r^2 + C_1$$

where $C_1$ is a constant. Because the fluid is viscous it sticks to the pipe wall so that $u = 0$ at $r = D/2$. Thus, $C_1 = (\Delta p/16\mu\ell)D^2$. Hence, the velocity profile can be written as

$$u(r) = \left(\frac{\Delta p D^2}{16\mu\ell}\right)\left[1 - \left(\frac{2r}{D}\right)^2\right] = V_c\left[1 - \left(\frac{2r}{D}\right)^2\right] \tag{8.7}$$

where $V_c = \Delta p D^2/(16\mu\ell)$ is the centerline velocity. An alternative expression can be written by using the relationship between the wall shear stress and the pressure gradient (Eqs. 8.5 and 8.7) to give

$$u(r) = \frac{\tau_w D}{4\mu}\left[1 - \left(\frac{r}{R}\right)^2\right]$$

*Under certain restrictions the velocity profile in a pipe is parabolic.*

where $R = D/2$ is the pipe radius.

This velocity profile, plotted in Fig. 8.9, is parabolic in the radial coordinate, $r$, has a maximum velocity, $V_c$, at the pipe centerline, and a minimum velocity (zero) at the pipe wall. The volume flowrate through the pipe can be obtained by integrating the velocity profile across the pipe. Since the flow is axisymmetric about the centerline, the velocity is constant on small area elements consisting of rings of radius $r$ and thickness $dr$, as shown in the figure in the margin. Thus,

$$Q = \int u\, dA = \int_{r=0}^{r=R} u(r) 2\pi r\, dr = 2\pi V_c \int_0^R \left[1 - \left(\frac{r}{R}\right)^2\right] r\, dr$$

or

$$Q = \frac{\pi R^2 V_c}{2}$$

$dA = 2\pi r\, dr$

By definition, the average velocity is the flowrate divided by the cross-sectional area, $V = Q/A = Q/\pi R^2$, so that for this flow

$$V = \frac{\pi R^2 V_c}{2\pi R^2} = \frac{V_c}{2} = \frac{\Delta p D^2}{32\mu\ell} \tag{8.8}$$

and

$$Q = \frac{\pi D^4\, \Delta p}{128\mu\ell} \tag{8.9}$$

As is indicated in Eq. 8.8, the average velocity is one-half of the maximum velocity. In general, for velocity profiles of other shapes (such as for turbulent pipe flow), the average velocity is not merely the average of the maximum ($V_c$) and minimum (0) velocities as it is for the laminar parabolic profile. The two velocity profiles indicated in Fig. 8.9 provide the same flowrate—one is the fictitious ideal ($\mu = 0$) profile; the other is the actual laminar flow profile.

*Poiseuille's law is valid for laminar flow only.*

The above results confirm the following properties of laminar pipe flow. For a horizontal pipe the flowrate is (a) directly proportional to the pressure drop, (b) inversely proportional to the viscosity, (c) inversely proportional to the pipe length, and (d) proportional to the pipe diameter to the fourth power. With all other parameters fixed, an increase in diameter by a factor of 2 will increase the flowrate by a factor of $2^4 = 16$—the flowrate is very strongly dependent on pipe size. This dependence is shown by the figure in the margin. Likewise, a small error in pipe diameter can cause a relatively large error in flowrate. For example, a 2% error in diameter gives an 8% error in flowrate ($Q \sim D^4$ or $\delta Q \sim 4D^3 \delta D$, so that $\delta Q / Q = 4 \, \delta D / D$). This flow, the properties of which were first established experimentally by two independent workers, G. Hagen (1797–1884) in 1839 and J. Poiseuille (1799–1869) in 1840, is termed *Hagen–Poiseuille flow*. Equation 8.9 is commonly referred to as **Poiseuille's law**. Recall that all of these results are restricted to laminar flow (those with Reynolds numbers less than approximately 2100) in a horizontal pipe.

The adjustment necessary to account for nonhorizontal pipes, as shown in Fig. 8.10, can be easily included by replacing the pressure drop, $\Delta p$, by the combined effect of pressure and gravity, $\Delta p - \gamma \ell \sin \theta$, where $\theta$ is the angle between the pipe and the horizontal. (Note that $\theta > 0$ if the flow is uphill, while $\theta < 0$ if the flow is downhill.) This can be seen from the force balance in the $x$ direction (along the pipe axis) on the cylinder of fluid shown in Fig. 8.10b. The method is exactly analogous to that used to obtain the Bernoulli equation (Eq. 3.6) when the streamline is not horizontal. The net force in the $x$ direction is a combination of the pressure force in that direction, $\Delta p \pi r^2$, and the component of weight in that direction, $-\gamma \pi r^2 \ell \sin \theta$. The result is a slightly modified form of Eq. 8.3 given by

$$\frac{\Delta p - \gamma \ell \sin \theta}{\ell} = \frac{2\tau}{r} \tag{8.10}$$

Thus, all of the results for the horizontal pipe are valid provided the pressure gradient is adjusted for the elevation term, that is, $\Delta p$ is replaced by $\Delta p - \gamma \ell \sin \theta$ so that

$$V = \frac{(\Delta p - \gamma \ell \sin \theta)D^2}{32\mu\ell} \tag{8.11}$$

and

$$Q = \frac{\pi(\Delta p - \gamma \ell \sin \theta)D^4}{128\mu\ell} \tag{8.12}$$

It is seen that the driving force for pipe flow can be either a pressure drop in the flow direction, $\Delta p$, or the component of weight in the flow direction, $-\gamma \ell \sin \theta$. If the flow is downhill, gravity helps the flow (a smaller pressure drop is required; $\sin \theta < 0$). If the flow is uphill, gravity works against the flow (a larger pressure drop is required; $\sin \theta > 0$). Note that $\gamma \ell \sin \theta = \gamma \Delta z$ (where

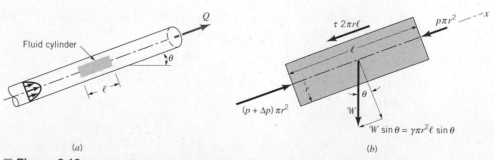

(a)                                                                                         (b)

■ **Figure 8.10**   **Free-body diagram of a fluid cylinder for flow in a nonhorizontal pipe.**

$\Delta z$ is the change in elevation) is a hydrostatic type pressure term. If there is no flow, $V = 0$ and $\Delta p = \gamma \ell \sin \theta = \gamma \Delta z$, as expected for fluid statics.

# EXAMPLE 8.2 | Laminar Pipe Flow

**GIVEN** An oil with a viscosity of $\mu = 0.40$ N · s/m$^2$ and density $\rho = 900$ kg/m$^3$ flows in a pipe of diameter $D = 0.020$ m.

**FIND** (a) What pressure drop, $p_1 - p_2$, is needed to produce a flowrate of $Q = 2.0 \times 10^{-5}$ m$^3$/s if the pipe is horizontal with $x_1 = 0$ and $x_2 = 10$ m?

**(b)** How steep a hill, $\theta$, must the pipe be on if the oil is to flow through the pipe at the same rate as in part (a), but with $p_1 = p_2$?

**(c)** For the conditions of part (b), if $p_1 = 200$ kPa, what is the pressure at section $x_3 = 5$ m, where $x$ is measured along the pipe?

## SOLUTION

**(a)** If the Reynolds number is less than 2100 the flow is laminar and the equations derived in this section are valid. Since the average velocity is $V = Q/A = (2.0 \times 10^{-5}$ m$^3$/s)/ $[\pi(0.020)^2$m$^2$/4$] = 0.0637$ m/s, the Reynolds number is Re $= \rho V D/\mu = 2.87 < 2100$. Hence, the flow is laminar and from Eq. 8.9 with $\ell = x_2 - x_1 = 10$ m, the pressure drop is

$$\Delta p = p_1 - p_2 = \frac{128 \mu \ell Q}{\pi D^4}$$

$$= \frac{128(0.40 \text{ N} \cdot \text{s/m}^2)(10.0 \text{ m})(2.0 \times 10^{-5} \text{ m}^3/\text{s})}{\pi(0.020 \text{ m})^4}$$

or

$$\Delta p = 20,400 \text{ N/m}^2 = 20.4 \text{ kPa} \quad \textbf{(Ans)}$$

**(b)** If the pipe is on a hill of angle $\theta$ such that $\Delta p = p_1 - p_2 = 0$, Eq. 8.12 gives

$$\sin \theta = -\frac{128 \mu Q}{\pi \rho g D^4} \quad \textbf{(1)}$$

or

$$\sin \theta = \frac{-128(0.40 \text{ N} \cdot \text{s/m}^2)(2.0 \times 10^{-5} \text{ m}^3/\text{s})}{\pi(900 \text{ kg/m}^3)(9.81 \text{ m/s}^2)(0.020 \text{ m})^4} \quad \textbf{(Ans)}$$

Thus, $\theta = -13.34°$.

**COMMENT** This checks with the previous horizontal result as is seen from the fact that a change in elevation of $\Delta z = \ell \sin \theta = (10 \text{ m}) \sin(-13.34°) = -2.31$ m is equivalent to a pressure change of $\Delta p = \rho g \Delta z = (900 \text{ kg/m}^3)(9.81 \text{ m/s}^2)$

$(2.31 \text{ m}) = 20,400$ N/m$^2$, which is equivalent to that needed for the horizontal pipe. For the horizontal pipe it is the work done by the pressure forces that overcomes the viscous dissipation. For the zero-pressure-drop pipe on the hill, it is the change in potential energy of the fluid "falling" down the hill that is converted to the energy lost by viscous dissipation. Note that if it is desired to increase the flowrate to $Q = 1.0 \times 10^{-4}$ m$^3$/s with $p_1 = p_2$, the value of $\theta$ given by Eq. 1 is $\sin \theta = -1.15$. Since the sine of an angle cannot be greater than 1, this flow would not be possible. The weight of the fluid would not be large enough to offset the viscous force generated for the flowrate desired. A larger-diameter pipe would be needed.

**(c)** With $p_1 = p_2$ the length of the pipe, $\ell$, does not appear in the flowrate equation (Eq. 1). This is a statement of the fact that for such cases the pressure is constant all along the pipe (provided the pipe lies on a hill of constant slope). This can be seen by substituting the values of $Q$ and $\theta$ from case (b) into Eq. 8.12 and noting that $\Delta p = 0$ for any $\ell$. For example, $\Delta p = p_1 - p_3 = 0$ if $\ell = x_3 - x_1 = 5$ m. Thus, $p_1 = p_2 = p_3$ so that

$$p_3 = 200 \text{ kPa} \quad \textbf{(Ans)}$$

**COMMENT** Note that if the fluid were gasoline ($\mu = 3.1 \times 10^{-4}$ N · s/m$^2$ and $\rho = 680$ kg/m$^3$), the Reynolds number would be Re $= 2790$, the flow would probably not be laminar, and use of Eqs. 8.9 and 8.12 would give incorrect results. Also note from Eq. 1 that the kinematic viscosity, $\nu = \mu/\rho$, is the important viscous parameter. This is a statement of the fact that with constant pressure along the pipe, it is the ratio of the viscous force ($\sim \mu$) to the weight force ($\sim \gamma = \rho g$) that determines the value of $\theta$.

### 8.2.2 From the Navier–Stokes Equations

In the previous section we obtained results for fully developed laminar pipe flow by applying Newton's second law and the assumption of a Newtonian fluid to a specific portion of the fluid—a cylinder of fluid centered on the axis of a long, round pipe. When this governing law and assumptions are applied to a general fluid flow (not restricted to pipe flow), the result is the Navier–Stokes equations as discussed in Chapter 6. In Section 6.9.3 these equations were solved for the specific geometry of fully developed laminar flow in a round pipe. The results are the same as those given in Eq. 8.7.

*Poiseuille's law can be obtained from the Navier–Stokes equations.*

We will not repeat the detailed steps used to obtain the laminar pipe flow from the Navier–Stokes equations (see Section 6.9.3) but will indicate how the various assumptions used and steps applied in the derivation correlate with the analysis used in the previous section.

General motion of an incompressible Newtonian fluid is governed by the continuity equation (conservation of mass, Eq. 6.31) and the momentum equation (Eq. 6.127), which are rewritten here for convenience:

$$\nabla \cdot \mathbf{V} = 0 \tag{8.13}$$

$$\frac{\partial \mathbf{V}}{\partial t} + (\mathbf{V} \cdot \nabla)\mathbf{V} = -\frac{\nabla p}{\rho} + \mathbf{g} + \nu \nabla^2 \mathbf{V} \tag{8.14}$$

For steady, fully developed flow in a pipe, the velocity contains only an axial component, which is a function of only the radial coordinate $[\mathbf{V} = u(r)\hat{\mathbf{i}}]$. For such conditions, the left-hand side of the Eq. 8.14 is zero. This is equivalent to saying that the fluid experiences no acceleration as it flows along. The same constraint was used in the previous section when considering $\mathbf{F} = m\mathbf{a}$ for the fluid cylinder. Thus, with $\mathbf{g} = -g\hat{\mathbf{k}}$ the Navier–Stokes equations become

$$\nabla \cdot \mathbf{V} = 0$$

$$\nabla p + \rho g\hat{\mathbf{k}} = \mu \nabla^2 \mathbf{V} \tag{8.15}$$

The flow is governed by a balance of pressure, weight, and viscous forces in the flow direction, similar to that shown in Fig. 8.10 and Eq. 8.10. If the flow were not fully developed (as in an entrance region, for example), it would not be possible to simplify the Navier–Stokes equations to that form given in Eq. 8.15 (the nonlinear term $(\mathbf{V} \cdot \nabla)\mathbf{V}$ would not be zero), and the solution would be very difficult to obtain.

Because of the assumption that $\mathbf{V} = u(r)\hat{\mathbf{i}}$, the continuity equation, Eq. 8.13, is automatically satisfied. This conservation of mass condition was also automatically satisfied by the incompressible flow assumption in the derivation in the previous section. The fluid flows across one section of the pipe at the same rate that it flows across any other section (see Fig. 8.8).

When it is written in terms of polar coordinates (as was done in Section 6.9.3), the component of Eq. 8.15 along the pipe becomes

$$\frac{\partial p}{\partial x} + \rho g \sin \theta = \mu \frac{1}{r} \frac{\partial}{\partial r}\left( r \frac{\partial u}{\partial r} \right) \tag{8.16}$$

Since the flow is fully developed, $u = u(r)$ and the right-hand side is a function of, at most, only $r$. The left-hand side is a function of, at most, only $x$. It was shown that this leads to the condition that the pressure gradient in the $x$ direction is a constant—$\partial p/\partial x = -\Delta p/\ell$. The same condition was used in the derivation of the previous section (Eq. 8.3).

It is seen from Eq. 8.16 that the effect of a nonhorizontal pipe enters into the Navier–Stokes equations in the same manner as was discussed in the previous section. The pressure gradient in the flow direction is coupled with the effect of the weight in that direction to produce an effective pressure gradient of $-\Delta p/\ell + \rho g \sin \theta$.

The velocity profile is obtained by integration of Eq. 8.16. Since it is a second-order equation, two boundary conditions are needed—(1) the fluid sticks to the pipe wall (as was also done in Eq. 8.7) and (2) either of the equivalent forms that the velocity remains finite throughout the flow (in particular $u < \infty$ at $r = 0$) or, because of symmetry, that $\partial u/\partial r = 0$ at $r = 0$. In the derivation of the previous section, only one boundary condition (the no-slip condition at the wall) was needed because the equation integrated was a first-order equation. The other condition ($\partial u/\partial r = 0$ at $r = 0$) was automatically built into the analysis because of the fact that $\tau = -\mu \, du/dr$ and $\tau = 2\tau_w r/D = 0$ at $r = 0$.

The results obtained by either applying $\mathbf{F} = m\mathbf{a}$ to a fluid cylinder (Section 8.2.1) or solving the Navier–Stokes equations (Section 6.9.3) are exactly the same. Similarly, the basic assumptions regarding the flow structure are the same. This should not be surprising because the two methods are based on the same principle—Newton's second law. One is restricted to fully developed laminar pipe flow from the beginning (the drawing of the free-body diagram), and the other starts with the general governing equations (the Navier–Stokes equations) with the appropriate restrictions concerning fully developed laminar flow applied as the solution process progresses.

### 8.2.3 From Dimensional Analysis

(1)              (2)

$\Delta p = p_1 - p_2 = F(V, \ell, D, \mu)$

Although fully developed laminar pipe flow is simple enough to allow the rather straightforward solutions discussed in the previous two sections, it may be worthwhile to consider this flow from a dimensional analysis standpoint. Thus, we assume that the pressure drop in the horizontal pipe, $\Delta p$, is a function of the average velocity of the fluid in the pipe, $V$, the length of the pipe, $\ell$, the pipe diameter, $D$, and the viscosity of the fluid, $\mu$, as shown by the figure in the margin. We have not included the density or the specific weight of the fluid as parameters because for such flows they are not important parameters. There is neither mass (density) times acceleration nor a component of weight (specific weight times volume) in the flow direction involved. Thus,

$$\Delta p = F(V, \ell, D, \mu)$$

Five variables can be described in terms of three reference dimensions $(M, L, T)$. According to the results of dimensional analysis (Chapter 7), this flow can be described in terms of $k - r = 5 - 3 = 2$ dimensionless groups. One such representation is

$$\frac{D\,\Delta p}{\mu V} = \phi\left(\frac{\ell}{D}\right) \tag{8.17}$$

where $\phi(\ell/D)$ is an unknown function of the length to diameter ratio of the pipe.

Although this is as far as dimensional analysis can take us, it seems reasonable to impose a further assumption that the pressure drop is directly proportional to the pipe length. That is, it takes twice the pressure drop to force fluid through a pipe if its length is doubled. The only way that this can be true is if $\phi(\ell/D) = C\ell/D$, where $C$ is a constant. Thus, Eq. 8.17 becomes

$$\frac{D\,\Delta p}{\mu V} = \frac{C\ell}{D}$$

which can be rewritten as

$$\frac{\Delta p}{\ell} = \frac{C\mu V}{D^2}$$

or

$$Q = AV = \frac{(\pi/4C)\,\Delta p D^4}{\mu \ell} \tag{8.18}$$

*Dimensional analysis can be used to put pipe flow parameters into dimensionless form.*

The basic functional dependence for laminar pipe flow given by Eq. 8.18 is the same as that obtained by the analysis of the two previous sections. The value of $C$ must be determined by theory (as done in the previous two sections) or experiment. For a round pipe, $C = 32$. For ducts of other cross-sectional shapes, the value of $C$ is different (see Section 8.4.3).

It is usually advantageous to describe a process in terms of dimensionless quantities. To this end we rewrite the pressure drop equation for laminar horizontal pipe flow, Eq. 8.8, as $\Delta p = 32\mu\ell V/D^2$ and divide both sides by the dynamic pressure, $\rho V^2/2$, to obtain the dimensionless form as

$$\frac{\Delta p}{\frac{1}{2}\rho V^2} = \frac{(32\mu\ell V/D^2)}{\frac{1}{2}\rho V^2} = 64\left(\frac{\mu}{\rho V D}\right)\left(\frac{\ell}{D}\right) = \frac{64}{\text{Re}}\left(\frac{\ell}{D}\right)$$

This is often written as

$$\Delta p = f\frac{\ell}{D}\frac{\rho V^2}{2}$$

where the dimensionless quantity

$$f = \Delta p(D/\ell)/(\rho V^2/2)$$

is termed the **friction factor,** or sometimes the *Darcy friction factor* [H. P. G. Darcy (1803–1858)]. (This parameter should not be confused with the less-used Fanning friction factor, which is defined to be $f/4$. In this text we will use only the Darcy friction factor.) Thus, the friction factor for laminar fully developed pipe flow is simply

$$f = \frac{64}{\text{Re}} \tag{8.19}$$

as shown by the figure in the margin.

By substituting the pressure drop in terms of the wall shear stress (Eq. 8.5), we obtain an alternate expression for the friction factor as a dimensionless wall shear stress

$$f = \frac{8\tau_w}{\rho V^2} \tag{8.20}$$

Knowledge of the friction factor will allow us to obtain a variety of information regarding pipe flow. For turbulent flow the dependence of the friction factor on the Reynolds number is much more complex than that given by Eq. 8.19 for laminar flow. This is discussed in detail in Section 8.4.

### 8.2.4 Energy Considerations

In the previous three sections we derived the basic laminar flow results from application of $\mathbf{F} = m\mathbf{a}$ or dimensional analysis considerations. It is equally important to understand the implications of energy considerations of such flows. To this end we consider the energy equation for incompressible, steady flow between two locations as is given in Eq. 5.89

$$\frac{p_1}{\gamma} + \alpha_1 \frac{V_1^2}{2g} + z_1 = \frac{p_2}{\gamma} + \alpha_2 \frac{V_2^2}{2g} + z_2 + h_L \tag{8.21}$$

Recall that the kinetic energy coefficients, $\alpha_1$ and $\alpha_2$, compensate for the fact that the velocity profile across the pipe is not uniform. For uniform velocity profiles, $\alpha = 1$, whereas for any nonuniform profile, $\alpha > 1$. The head loss term, $h_L$, accounts for any energy loss associated with the flow. This loss is a direct consequence of the viscous dissipation that occurs throughout the fluid in the pipe. For the ideal (inviscid) cases discussed in previous chapters, $\alpha_1 = \alpha_2 = 1$, $h_L = 0$, and the energy equation reduces to the familiar Bernoulli equation discussed in Chapter 3 (Eq. 3.7).

Even though the velocity profile in viscous pipe flow is not uniform, for fully developed flow it does not change from section (1) to section (2) so that $\alpha_1 = \alpha_2$. Thus, the kinetic energy is the same at any section ($\alpha_1 V_1^2/2 = \alpha_2 V_2^2/2$) and the energy equation becomes

$$\left( \frac{p_1}{\gamma} + z_1 \right) - \left( \frac{p_2}{\gamma} + z_2 \right) = h_L \tag{8.22}$$

The energy dissipated by the viscous forces within the fluid is supplied by the excess work done by the pressure and gravity forces as shown by the figure in the margin.

A comparison of Eqs. 8.22 and 8.10 shows that the head loss is given by

$$h_L = \frac{2\tau \ell}{\gamma r}$$

(recall $p_1 = p_2 + \Delta p$ and $z_2 - z_1 = \ell \sin \theta$), which, by use of Eq. 8.4, can be rewritten in the form

$$h_L = \frac{4\ell \tau_w}{\gamma D} \tag{8.23}$$

*The head loss in a pipe is a result of the viscous shear stress on the wall.*

It is the shear stress at the wall (which is directly related to the viscosity and the shear stress throughout the fluid) that is responsible for the head loss. A closer consideration of the assumptions involved in the derivation of Eq. 8.23 will show that it is valid for both laminar and turbulent flow.

# EXAMPLE 8.3    Laminar Pipe Flow Properties

**GIVEN**  The flowrate, $Q$, of corn syrup through the horizontal pipe shown in Fig. E8.3a is to be monitored by measuring the pressure difference between sections (1) and (2). It is proposed that $Q = K \Delta p$, where the calibration constant, $K$, is a function of temperature, $T$, because of the variation of the syrup's viscosity and density with temperature. These variations are given in Table E8.3.

**FIND**  (a) Plot $K(T)$ versus $T$ for $60\ °F \le T \le 160\ °F$. (b) Determine the wall shear stress and the pressure drop, $\Delta p = p_1 - p_2$, for $Q = 0.5\ ft^3/s$ and $T = 100\ °F$. (c) For the conditions of part (b), determine the net pressure force, $(\pi D^2/4)\,\Delta p$, and the net shear force, $\pi D \ell \tau_w$, on the fluid within the pipe between the sections (1) and (2).

(a)

(b)

■ **Figure E8.3**

■ **Table E8.3**

| $T$ (°F) | $\rho$ (slugs/ft$^3$) | $\mu$ (lb · s/ft$^2$) |
|----------|----------------------|------------------------|
| 60  | 2.07 | $4.0 \times 10^{-2}$ |
| 80  | 2.06 | $1.9 \times 10^{-2}$ |
| 100 | 2.05 | $3.8 \times 10^{-3}$ |
| 120 | 2.04 | $4.4 \times 10^{-4}$ |
| 140 | 2.03 | $9.2 \times 10^{-5}$ |
| 160 | 2.02 | $2.3 \times 10^{-5}$ |

# SOLUTION

**(a)**  If the flow is laminar, it follows from Eq. 8.9 that

$$Q = \frac{\pi D^4 \, \Delta p}{128 \mu \ell} = \frac{\pi (\frac{3}{12}\ ft)^4 \, \Delta p}{128 \mu (6\ ft)}$$

or

$$Q = K \, \Delta p = \frac{1.60 \times 10^{-5}}{\mu} \, \Delta p \qquad (1)$$

where the units on $Q$, $\Delta p$, and $\mu$ are ft$^3$/s, lb/ft$^2$, and lb · s/ft$^2$, respectively. Thus

$$K = \frac{1.60 \times 10^{-5}}{\mu} \qquad \textbf{(Ans)}$$

where the units of $K$ are ft$^5$/lb · s. By using values of the viscosity from Table E8.3, the calibration curve shown in Fig. E8.3b is obtained. This result is valid only if the flow is laminar.

**COMMENT**  As shown in Section 8.5, for turbulent flow the flowrate is not linearly related to the pressure drop so it would not be possible to have $Q = K \, \Delta p$. Note also that the value of $K$ is independent of the syrup density ($\rho$ was not used in the calculations) since laminar pipe flow is governed by pressure and viscous effects; inertia is not important.

**(b)**  For $T = 100\ °F$, the viscosity is $\mu = 3.8 \times 10^{-3}$ lb · s/ft$^2$ so that with a flowrate of $Q = 0.5\ ft^3/s$ the pressure drop (according to Eq. 8.9) is

$$\Delta p = \frac{128 \mu \ell Q}{\pi D^4}$$

$$= \frac{128(3.8 \times 10^{-3}\ lb \cdot s/ft^2)(6\ ft)(0.5\ ft^3/s)}{\pi (\frac{3}{12}\ ft)^4}$$

$$= 119\ lb/ft^2 \qquad \textbf{(Ans)}$$

provided the flow is laminar. For this case

$$V = \frac{Q}{A} = \frac{0.5\ ft^3/s}{\frac{\pi}{4}(\frac{3}{12}\ ft)^2} = 10.2\ ft/s$$

so that

$$Re = \frac{\rho V D}{\mu} = \frac{(2.05\ slugs/ft^3)(10.2\ ft/s)(\frac{3}{12}\ ft)}{(3.8 \times 10^{-3}\ lb \cdot s/ft^2)}$$

$$= 1380 < 2100$$

Hence, the flow is laminar. From Eq. 8.5 the wall shear stress is

$$\tau_w = \frac{\Delta p D}{4\ell} = \frac{(119\ lb/ft^2)(\frac{3}{12}\ ft)}{4(6\ ft)} = 1.24\ lb/ft^2 \qquad \textbf{(Ans)}$$

**(c)**  For the conditions of part (b), the net pressure force, $F_p$, on the fluid within the pipe between sections (1) and (2) is

$$F_p = \frac{\pi}{4} D^2 \, \Delta p = \frac{\pi}{4} \left( \frac{3}{12}\ ft \right)^2 (119\ lb/ft^2) = 5.84\ lb \qquad \textbf{(Ans)}$$

Similarly, the net viscous force, $F_v$, on that portion of the fluid is

$$F_v = 2\pi \left( \frac{D}{2} \right) \ell \tau_w$$

$$= 2\pi \left[ \frac{3}{2(12)}\ ft \right] (6\ ft)(1.24\ lb/ft^2) = 5.84\ lb \qquad \textbf{(Ans)}$$

**COMMENT**  Note that the values of these two forces are the same. The net force is zero; there is no acceleration.

## 8.3 Fully Developed Turbulent Flow

In the previous section various properties of fully developed laminar pipe flow were discussed. Since turbulent pipe flow is actually more likely to occur than laminar flow in practical situations, it is necessary to obtain similar information for turbulent pipe flow. However, turbulent flow is a very complex process. Numerous persons have devoted considerable effort in attempting to understand the variety of baffling aspects of turbulence. Although a considerable amount of knowledge about the topic has been developed, the field of turbulent flow still remains the least understood area of fluid mechanics. In this book we can provide only some of the very basic ideas concerning turbulence. The interested reader should consult some of the many books available for further reading (Refs. 1–3).

### 8.3.1 Transition from Laminar to Turbulent Flow

Flows are classified as laminar or turbulent. For any flow geometry, there is one (or more) dimensionless parameter such that with this parameter value below a particular value the flow is laminar, whereas with the parameter value larger than a certain value the flow is turbulent. The important parameters involved (i.e., Reynolds number, Mach number) and their critical values depend on the specific flow situation involved. For example, flow in a pipe and flow along a flat plate (boundary layer flow, as is discussed in Section 9.2.4) can be laminar or turbulent, depending on the value of the Reynolds number involved. As a general rule for pipe flow, the value of the Reynolds number must be less than approximately 2100 for laminar flow and greater than approximately 4000 for turbulent flow. For flow along a flat plate the transition between laminar and turbulent flow occurs at a Reynolds number of approximately 500,000 (see Section 9.2.4), where the length term in the Reynolds number is the distance measured from the leading edge of the plate.

Consider a long section of pipe that is initially filled with a fluid at rest. As the valve is opened to start the flow, the flow velocity and, hence, the Reynolds number increase from zero (no flow) to their maximum steady-state flow values, as is shown in Fig. 8.11. Assume this transient process is slow enough so that unsteady effects are negligible (quasi-steady flow). For an initial time period the Reynolds number is small enough for laminar flow to occur. At some time the Reynolds number reaches 2100, and the flow begins its transition to turbulent conditions. Intermittent spots or bursts of turbulence appear. As the Reynolds number is increased, the entire flow field becomes turbulent. The flow remains turbulent as long as the Reynolds number exceeds approximately 4000.

*Turbulent flows involve randomly fluctuating parameters.*

A typical trace of the axial component of velocity measured at a given location in the flow, $u = u(t)$, is shown in Fig. 8.12. Its irregular, random nature is the distinguishing feature of turbulent flow. The character of many of the important properties of the flow (pressure drop, heat transfer, etc.) depends strongly on the existence and nature of the turbulent fluctuations or randomness

■ **Figure 8.11**  Transition from laminar to turbulent flow in a pipe.

**■ Figure 8.12** The time-averaged, $\bar{u}$, and fluctuating, $u'$, description of a parameter for turbulent flow.

indicated. In previous considerations involving inviscid flow, the Reynolds number is (strictly speaking) infinite (because the viscosity is zero), and the flow most surely would be turbulent. However, reasonable results were obtained by using the inviscid Bernoulli equation as the governing equation. The reason that such simplified inviscid analyses gave reasonable results is that viscous effects were not very important and the velocity used in the calculations was actually the time-averaged velocity, $\bar{u}$, indicated in Fig. 8.12. Calculation of the heat transfer, pressure drop, and many other parameters would not be possible without inclusion of the seemingly small, but very important, effects associated with the randomness of the flow.

Consider flow in a pan of water placed on a stove. With the stove turned off, the fluid is stationary. The initial sloshing has died out because of viscous dissipation within the water. With the stove turned on, a temperature gradient in the vertical direction, $\partial T/\partial z$, is produced. The water temperature is greatest near the pan bottom and decreases toward the top of the fluid layer. If the temperature difference is very small, the water will remain stationary, even though the water density is smallest near the bottom of the pan because of the decrease in density with an increase in temperature. A further increase in the temperature gradient will cause a buoyancy-driven instability that results in fluid motion—the light, warm water rises to the top, and the heavy, cold water sinks to the bottom. This slow, regular "turning over" increases the heat transfer from the pan to the water and promotes mixing within the pan. As the temperature gradient increases still further, the fluid motion becomes more vigorous and eventually turns into a chaotic, random, turbulent flow with considerable mixing, vaporization (boiling) and greatly increased heat transfer rate. The flow has progressed from a stationary fluid, to laminar flow, and finally to turbulent, multiphase (liquid and vapor) flow.

Mixing processes and heat and mass transfer processes are considerably enhanced in turbulent flow compared to laminar flow. This is due to the macroscopic scale of the randomness in turbulent flow. We are all familiar with the "rolling," vigorous eddy-type motion of the water in a pan being heated on the stove (even if it is not heated to boiling). Such finite-sized random mixing is very effective in transporting energy and mass throughout the flow field, thereby increasing the various rate processes involved. Laminar flow, on the other hand, can be thought of as very small but finite-sized fluid particles flowing smoothly in layers, one over another. The only randomness and mixing take place on the molecular scale and result in relatively small heat, mass, and momentum transfer rates.

Without turbulence it would be virtually impossible to carry out life as we now know it. Mixing is one positive application of turbulence, as discussed above, but there are other situations where turbulent flow is desirable. To transfer the required heat between a solid and an adjacent fluid (such as in the cooling coils of an air conditioner or a boiler of a power plant) would require an enormously large heat exchanger if the flow were laminar. Similarly, the required mass transfer of a liquid state to a vapor state (such as is needed in the evaporated cooling system associated with sweating) would require very large surfaces if the fluid flowing past the surface were

*V8.4 Stirring color into paint*

*V8.5 Laminar and turbulent mixing*

laminar rather than turbulent. As shown in Chapter 9, turbulence can also aid in delaying flow separation.

Turbulence is also of importance in the mixing of fluids. Smoke from a stack would continue for miles as a ribbon of pollutant without rapid dispersion within the surrounding air if the flow were laminar rather than turbulent. Under certain atmospheric conditions this is observed to occur. Although there is mixing on a molecular scale (laminar flow), it is several orders of magnitude slower and less effective than the mixing on a macroscopic scale (turbulent flow). It is considerably easier to mix cream into a cup of coffee (turbulent flow) than to thoroughly mix two colors of a viscous paint (laminar flow).

In other situations laminar (rather than turbulent) flow is desirable. The pressure drop in pipes (hence, the power requirements for pumping) can be considerably lower if the flow is laminar rather than turbulent. Fortunately, the blood flow through a person's arteries is normally laminar, except in the largest arteries with high blood flowrates. The aerodynamic drag on an airplane wing can be considerably smaller with laminar flow past it than with turbulent flow.

*V8.6 Stirring cream into coffee*

### 8.3.2 Turbulent Shear Stress

The fundamental difference between laminar and turbulent flow lies in the chaotic, random behavior of the various fluid parameters. Such variations occur in the three components of velocity, the pressure, the shear stress, the temperature, and any other variable that has a field description. Turbulent flow is characterized by random, three-dimensional vorticity (i.e., fluid particle rotation or spin; see Section 6.1.3). As is indicated in Fig. 8.12, such flows can be described in terms of their mean values (denoted with an overbar) on which are superimposed the fluctuations (denoted with a prime). Thus, if $u = u(x, y, z, t)$ is the $x$ component of instantaneous velocity, then its time mean (or *time-average*) value, $\bar{u}$, is

*Turbulent flow parameters can be described in terms of mean and fluctuating portions.*

$$\bar{u} = \frac{1}{T} \int_{t_0}^{t_0+T} u(x, y, z, t)\, dt \tag{8.24}$$

where the time interval, $T$, is considerably longer than the period of the longest fluctuations, but considerably shorter than any unsteadiness of the average velocity. This is illustrated in Fig. 8.12.

The *fluctuating part* of the velocity, $u'$, is that time-varying portion that differs from the average value

$$u = \bar{u} + u' \quad \text{or} \quad u' = u - \bar{u} \tag{8.25}$$

Clearly, the time average of the fluctuations is zero, since

$$\bar{u'} = \frac{1}{T} \int_{t_0}^{t_0+T} (u - \bar{u})\, dt = \frac{1}{T} \left( \int_{t_0}^{t_0+T} u\, dt - \bar{u} \int_{t_0}^{t_0+T} dt \right)$$

$$= \frac{1}{T} (T\bar{u} - T\bar{u}) = 0$$

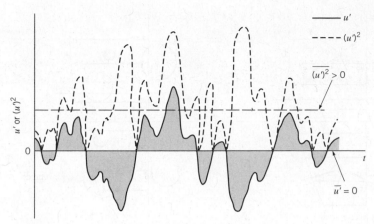

■ **Figure 8.13**   **Average of the fluctuations and average of the square of the fluctuations.**

The fluctuations are equally distributed on either side of the average. It is also clear, as is indicated in Fig. 8.13, that since the square of a fluctuation quantity cannot be negative $[(u')^2 \geq 0]$, its average value is positive. Thus,

$$\overline{(u')^2} = \frac{1}{T}\int_{t_0}^{t_0+T} (u')^2\, dt > 0$$

On the other hand, it may be that the average of products of the fluctuations, such as $\overline{u'v'}$, are zero or nonzero (either positive or negative).

The structure and characteristics of turbulence may vary from one flow situation to another. For example, the *turbulence intensity* (or the level of the turbulence) may be larger in a very gusty wind than it is in a relatively steady (although turbulent) wind. The turbulence intensity, $\mathscr{I}$, is often defined as the square root of the mean square of the fluctuating velocity divided by the time-averaged velocity, or

$$\mathscr{I} = \frac{\sqrt{\overline{(u')^2}}}{\bar{u}} = \frac{\left[\dfrac{1}{T}\displaystyle\int_{t_0}^{t_0+T} (u')^2\, dt\right]^{1/2}}{\bar{u}}$$

The larger the turbulence intensity, the larger the fluctuations of the velocity (and other flow parameters). Well-designed wind tunnels have typical values of $\mathscr{I} \approx 0.01$, although with extreme care, values as low as $\mathscr{I} = 0.0002$ have been obtained. On the other hand, values of $\mathscr{I} \gtrsim 0.1$ are found for the flow in the atmosphere and rivers. A typical atmospheric wind-speed graph is shown in the figure in the margin.

Another turbulence parameter that is different from one flow situation to another is the period of the fluctuations—the *time scale* of the fluctuations shown in Fig. 8.12. In many flows, such as the flow of water from a faucet, typical frequencies are on the order of 10, 100, or 1000 cycles per second (cps). For other flows, such as the Gulf Stream current in the Atlantic Ocean or flow of the atmosphere of Jupiter, characteristic random oscillations may have a period on the order of hours, days, or more.

It is tempting to extend the concept of viscous shear stress for laminar flow ($\tau = \mu\, du/dy$) to that of turbulent flow by replacing $u$, the instantaneous velocity, by $\bar{u}$, the time-averaged velocity. However, numerous experimental and theoretical studies have shown that such an approach leads to completely incorrect results. That is, $\tau \neq \mu\, d\bar{u}/dy$. A physical explanation for this behavior can be found in the concept of what produces a shear stress.

Laminar flow is modeled as fluid particles that flow smoothly along in layers, gliding past the slightly slower or faster ones on either side. As is discussed in Chapter 1, the fluid actually consists of numerous molecules darting about in an almost random fashion as is indicated in Fig. 8.14*a*. The motion is not entirely random—a slight bias in one direction produces the flowrate we associate

*The relationship between fluid motion and shear stress is very complex for turbulent flow.*

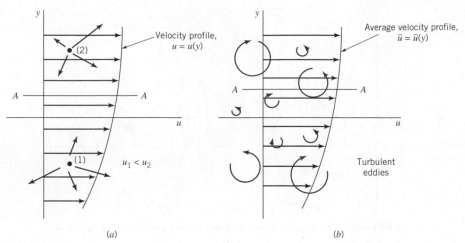

**■ Figure 8.14** (*a*) Laminar flow shear stress caused by random motion of molecules. (*b*) Turbulent flow as a series of random, three-dimensional eddies.

with the motion of fluid particles, $\bar{u}$. As the molecules dart across a given plane (plane $A{-}A$, for example), the ones moving upward have come from an area of smaller average $x$ component of velocity than the ones moving downward, which have come from an area of larger velocity.

The momentum flux in the $x$ direction across plane $A{-}A$ gives rise to a drag (to the left) of the lower fluid on the upper fluid and an equal but opposite effect of the upper fluid on the lower fluid. The sluggish molecules moving upward across plane $A{-}A$ must be accelerated by the fluid above this plane. The rate of change of momentum in this process produces (on the macroscopic scale) a shear force. Similarly, the more energetic molecules moving down across plane $A{-}A$ must be slowed down by the fluid below that plane. This shear force is present only if there is a gradient in $u = u(y)$, otherwise the average $x$ component of velocity (and momentum) of the upward and downward molecules is exactly the same. In addition, there are attractive forces between molecules. By combining these effects we obtain the well-known Newton viscosity law: $\tau = \mu \, du/dy$, where on a molecular basis $\mu$ is related to the mass and speed (temperature) of the random motion of the molecules.

*Turbulent flow shear stress is larger than laminar flow shear stress because of the irregular, random motion.*

Although the above random motion of the molecules is also present in turbulent flow, there is another factor that is generally more important. A simplistic way of thinking about turbulent flow is to consider it as consisting of a series of random, three-dimensional eddy type motions as is depicted (in one dimension only) in Fig. 8.14*b*. (See the photograph at the beginning of this chapter.) These eddies range in size from very small diameter (on the order of the size of a fluid particle) to fairly large diameter (on the order of the size of the object or flow geometry considered). They move about randomly, conveying mass with an average velocity $\bar{u} = \bar{u}(y)$. This eddy structure greatly promotes mixing within the fluid. It also greatly increases the transport of $x$ momentum across plane $A{-}A$. That is, finite particles of fluid (not merely individual molecules as in laminar flow) are randomly transported across this plane, resulting in a relatively large (when compared with laminar flow) shear force. These particles vary in size but are much larger than molecules.

## F l u i d s   i n   t h e   N e w s

**Listen to the flowrate** Sonar systems are designed to listen to transmitted and reflected sound waves in order to locate submerged objects. They have been used successfully for many years to detect and track underwater objects such as submarines and aquatic animals. Recently, sonar techniques have been refined so that they can be used to determine the flowrate in pipes. These new flowmeters work for turbulent, not laminar, pipe flows because their operation depends strictly on the existence of *turbu-*

*lent eddies* within the flow. The flowmeters contain a sonar-based array that listens to and interprets pressure fields generated by the turbulent motion in pipes. By listening to the pressure fields associated with the movement of the turbulent eddies, the device can determine the speed at which the eddies travel past an array of sensors. The flowrate is determined by using a calibration procedure that links the speed of the turbulent structures to the volumetric flowrate.

The random velocity components that account for this momentum transfer (hence, the shear force) are $u'$ (for the $x$ component of velocity) and $v'$ (for the rate of mass transfer crossing the plane). A more detailed consideration of the processes involved will show that the apparent shear stress on plane $A-A$ is given by the following (Ref. 2):

$$\tau = \mu \frac{d\bar{u}}{dy} - \rho\overline{u'v'} = \tau_{\text{lam}} + \tau_{\text{turb}} \tag{8.26}$$

*V8.7 Turbulence in a bowl*

Note that if the flow is laminar, $u' = v' = 0$, so that $\overline{u'v'} = 0$ and Eq. 8.26 reduces to the customary random molecule-motion-induced *laminar shear stress*, $\tau_{\text{lam}} = \mu\, d\bar{u}/dy$. For turbulent flow it is found that the **turbulent shear stress,** $\tau_{\text{turb}} = -\rho\overline{u'v'}$, is positive. Hence, the shear stress is greater in turbulent flow than in laminar flow. Note the units on $\tau_{\text{turb}}$ are (density)(velocity)$^2 = $ (slugs/ft$^3$)(ft/s)$^2 = $ (slugs · ft/s$^2$)/ft$^2 = $ lb/ft$^2$, or N/m$^2$, as expected. Terms of the form $-\rho\overline{u'v'}$ (or $-\rho\overline{v'w'}$, etc.) are called *Reynolds stresses* in honor of Osborne Reynolds who first discussed them in 1895.

It is seen from Eq. 8.26 that the shear stress in turbulent flow is not merely proportional to the gradient of the time-averaged velocity, $\bar{u}(y)$. It also contains a contribution due to the random fluctuations of the $x$ and $y$ components of velocity. The density is involved because of the momentum transfer of the fluid within the random eddies. Although the relative magnitude of $\tau_{\text{lam}}$ compared to $\tau_{\text{turb}}$ is a complex function dependent on the specific flow involved, typical measurements indicate the structure shown in Fig. 8.15a. (Recall from Eq. 8.4 that the shear stress is proportional to the distance from the centerline of the pipe.) In a very narrow region near the wall (the *viscous sublayer*), the laminar shear stress is dominant. Away from the wall (in the *outer layer*) the turbulent portion of the shear stress is dominant. The transition between these two regions occurs in the *overlap layer*. The corresponding typical velocity profile is shown in Fig. 8.15b.

*The shear stress is the sum of a laminar portion and a turbulent portion.*

The scale of the sketches shown in Fig. 8.15 is not necessarily correct. Typically the value of $\tau_{\text{turb}}$ is 100 to 1000 times greater than $\tau_{\text{lam}}$ in the outer region, while the converse is true in the viscous sublayer. A correct modeling of turbulent flow is strongly dependent on an accurate knowledge of $\tau_{\text{turb}}$. This, in turn, requires an accurate knowledge of the fluctuations $u'$ and $v'$, or $\rho\overline{u'v'}$. As yet it is not possible to solve the governing equations (the Navier–Stokes equations) for these details of the flow, although numerical techniques (see Appendix A) using the largest and fastest computers available have produced important information about some of the characteristics of turbulence. Considerable effort has gone into the study of turbulence. Much remains to be learned. Perhaps studies in the new areas of chaos and fractal geometry will provide the tools for a better understanding of turbulence (see Section 8.3.5).

The vertical scale of Fig. 8.15 is also distorted. The viscous sublayer is usually a very thin layer adjacent to the wall. For example, for water flow in a 3-in.-diameter pipe with an average velocity of 10 ft/s, the viscous sublayer is approximately 0.002 in. thick. Since the fluid motion within this thin layer is critical in terms of the overall flow (the no-slip condition and the wall shear stress occur in this layer), it is not surprising to find that turbulent pipe flow properties can be quite

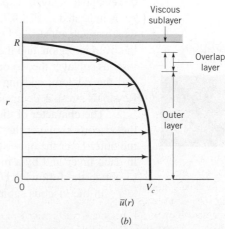

■ **Figure 8.15**   Structure of turbulent flow in a pipe. (*a*) Shear stress. (*b*) Average velocity.

dependent on the roughness of the pipe wall, unlike laminar pipe flow which is independent of roughness. Small roughness elements (scratches, rust, sand or dirt particles, etc.) can easily disturb this viscous sublayer (see Section 8.4), thereby affecting the entire flow.

An alternate form for the shear stress for turbulent flow is given in terms of the *eddy viscosity*, $\eta$, where

$$\tau_{\text{turb}} = \eta \, \frac{d\bar{u}}{dy} \tag{8.27}$$

This extension of laminar flow terminology was introduced by J. Boussinesq, a French scientist, in 1877. Although the concept of an eddy viscosity is intriguing, in practice it is not an easy parameter to use. Unlike the absolute viscosity, $\mu$, which is a known value for a given fluid, the eddy viscosity is a function of both the fluid and the flow conditions. That is, the eddy viscosity of water cannot be looked up in handbooks—its value changes from one turbulent flow condition to another and from one point in a turbulent flow to another.

The inability to accurately determine the Reynolds stress, $\overline{\rho u' v'}$, is equivalent to not knowing the eddy viscosity. Several semiempirical theories have been proposed (Ref. 3) to determine approximate values of $\eta$. L. Prandtl (1875–1953), a German physicist and aerodynamicist, proposed that the turbulent process could be viewed as the random transport of bundles of fluid particles over a certain distance, $\ell_m$, the *mixing length*, from a region of one velocity to another region of a different velocity. By the use of some ad hoc assumptions and physical reasoning, it was concluded that the eddy viscosity was given by

$$\eta = \rho \ell_m^2 \left| \frac{d\bar{u}}{dy} \right|$$

Thus, the turbulent shear stress is

$$\tau_{\text{turb}} = \rho \ell_m^2 \left( \frac{d\bar{u}}{dy} \right)^2 \tag{8.28}$$

The problem is thus shifted to that of determining the mixing length, $\ell_m$. Further considerations indicate that $\ell_m$ is not a constant throughout the flow field. Near a solid surface the turbulence is dependent on the distance from the surface. Thus, additional assumptions are made regarding how the mixing length varies throughout the flow.

The net result is that as yet there is no general, all-encompassing, useful model that can accurately predict the shear stress throughout a general incompressible, viscous turbulent flow. Without such information it is impossible to integrate the force balance equation to obtain the turbulent velocity profile and other useful information, as was done for laminar flow.

### 8.3.3 Turbulent Velocity Profile

Considerable information concerning turbulent velocity profiles has been obtained through the use of dimensional analysis, experimentation, numerical simulations, and semiempirical theoretical efforts. As is indicated in Fig. 8.15, fully developed turbulent flow in a pipe can be broken into three regions which are characterized by their distances from the wall: the viscous sublayer very near the pipe wall, the overlap region, and the outer turbulent layer throughout the center portion of the flow. Within the viscous sublayer the viscous shear stress is dominant compared with the turbulent (or Reynolds) stress, and the random, eddying nature of the flow is essentially absent. In the outer turbulent layer the Reynolds stress is dominant, and there is considerable mixing and randomness to the flow.

The character of the flow within these two regions is entirely different. For example, within the viscous sublayer the fluid viscosity is an important parameter; the density is unimportant. In the outer layer the opposite is true. By a careful use of dimensional analysis arguments for the flow in each layer and by a matching of the results in the common overlap layer, it has been possible to obtain the following conclusions about the turbulent velocity profile in a smooth pipe (Ref. 5).

In the viscous sublayer the velocity profile can be written in dimensionless form as

$$\frac{\bar{u}}{u^*} = \frac{y u^*}{\nu} \tag{8.29}$$

■ **Figure 8.16**   Typical structure of the turbulent velocity profile in a pipe.

where $y = R - r$ is the distance measured from the wall, $\bar{u}$ is the time-averaged $x$ component of velocity, and $u^* = (\tau_w/\rho)^{1/2}$ is termed the *friction velocity*. Note that $u^*$ is not an actual velocity of the fluid—it is merely a quantity that has dimensions of velocity. As is indicated in Fig. 8.16, Eq. 8.29 (commonly called the *law of the wall*) is valid very near the smooth wall, for $0 \le yu^*/\nu \lesssim 5$.

Dimensional analysis arguments indicate that in the overlap region the velocity should vary as the logarithm of $y$. Thus, the following expression has been proposed:

*A turbulent flow velocity profile can be divided into various regions.*

$$\frac{\bar{u}}{u^*} = 2.5 \ln\left(\frac{yu^*}{\nu}\right) + 5.0 \tag{8.30}$$

where the constants 2.5 and 5.0 have been determined experimentally. As is indicated in Fig. 8.16, for regions not too close to the smooth wall, but not all the way out to the pipe center, Eq. 8.30 gives a reasonable correlation with the experimental data. Note that the horizontal scale is a logarithmic scale. This tends to exaggerate the size of the viscous sublayer relative to the remainder of the flow. As is shown in Example 8.4, the viscous sublayer is usually quite thin. Similar results can be obtained for turbulent flow past rough walls (Ref. 17).

A number of other correlations exist for the velocity profile in turbulent pipe flow. In the central region (the outer turbulent layer) the expression $(V_c - \bar{u})/u^* = 2.5 \ln(R/y)$, where $V_c$ is the centerline velocity, is often suggested as a good correlation with experimental data. Another often-used (and relatively easy to use) correlation is the empirical *power-law velocity profile*

*V8.9 Laminar to turbulent flow from a pipe*

$$\frac{\bar{u}}{V_c} = \left(1 - \frac{r}{R}\right)^{1/n} \tag{8.31}$$

In this representation, the value of $n$ is a function of the Reynolds number, as is indicated in Fig. 8.17. The one-seventh power-law velocity profile ($n = 7$) is often used as a reasonable approximation for many practical flows. Typical turbulent velocity profiles based on this power-law representation are shown in Fig. 8.18.

A closer examination of Eq. 8.31 shows that the power-law profile cannot be valid near the wall, since according to this equation the velocity gradient is infinite there. In addition, Eq. 8.31 cannot be precisely valid near the centerline because it does not give $d\bar{u}/dr = 0$ at $r = 0$. However, it does provide a reasonable approximation to the measured velocity profiles across most of the pipe.

Note from Fig. 8.18 that the turbulent profiles are much "flatter" than the laminar profile and that this flatness increases with Reynolds number (i.e., with $n$). Recall from Chapter 3 that

■ **Figure 8.17** Exponent, $n$, for power-law velocity profiles. (Adapted from Ref. 1.)

*V8.10 Laminar/ turbulent velocity profiles*

■ **Figure 8.18** Typical laminar flow and turbulent flow velocity profiles.

reasonable approximate results are often obtained by using the inviscid Bernoulli equation and by assuming a fictitious uniform velocity profile. Since most flows are turbulent and turbulent flows tend to have nearly uniform velocity profiles, the usefulness of the Bernoulli equation and the uniform profile assumption is not unexpected. Of course, many properties of the flow cannot be accounted for without including viscous effects.

# $E$XAMPLE 8.4    Turbulent Pipe Flow Properties

**GIVEN** Water at $20\,°C$ ($\rho = 998\ kg/m^3$ and $\nu = 1.004 \times 10^{-6}\ m^2/s$) flows through a horizontal pipe of 0.1-m diameter with a flowrate of $Q = 4 \times 10^{-2}\ m^3/s$ and a pressure gradient of 2.59 kPa/m.

**FIND**   **(a)**   Determine the approximate thickness of the viscous sublayer.

**(b)**   Determine the approximate centerline velocity, $V_c$.

**(c)**   Determine the ratio of the turbulent to laminar shear stress, $\tau_{turb}/\tau_{lam}$, at a point midway between the centerline and the pipe wall (i.e., at $r = 0.025$ m).

# SOLUTION

**(a)**    According to Fig. 8.16, the thickness of the viscous sublayer, $\delta_s$, is approximately

$$\frac{\delta_s u^*}{\nu} = 5$$

Therefore,

$$\delta_s = 5\frac{\nu}{u^*}$$

where

$$u^* = \left(\frac{\tau_w}{\rho}\right)^{1/2} \qquad (1)$$

The wall shear stress can be obtained from the pressure drop data and Eq. 8.5, which is valid for either laminar or turbulent flow. Thus,

$$\tau_w = \frac{D\,\Delta p}{4\ell} = \frac{(0.1\ \text{m})(2.59 \times 10^3\ \text{N/m}^2)}{4(1\ \text{m})} = 64.8\ \text{N/m}^2$$

Hence, from Eq. 1 we obtain

$$u^* = \left(\frac{64.8\ \text{N/m}^2}{998\ \text{kg/m}^3}\right)^{1/2} = 0.255\ \text{m/s}$$

so that

$$\delta_s = \frac{5(1.004 \times 10^{-6}\ \text{m}^2/\text{s})}{0.255\ \text{m/s}}$$
$$= 1.97 \times 10^{-5}\ \text{m} \simeq 0.02\ \text{mm} \qquad \textbf{(Ans)}$$

**COMMENT**    As stated previously, the viscous sublayer is very thin. Minute imperfections on the pipe wall will protrude into this sublayer and affect some of the characteristics of the flow (i.e., wall shear stress and pressure drop).

**(b)**    The centerline velocity can be obtained from the average velocity and the assumption of a power-law velocity profile as follows. For this flow with

$$V = \frac{Q}{A} = \frac{0.04\ \text{m}^3/\text{s}}{\pi(0.1\ \text{m})^2/4} = 5.09\ \text{m/s}$$

the Reynolds number is

$$\text{Re} = \frac{VD}{\nu} = \frac{(5.09\ \text{m/s})(0.1\ \text{m})}{(1.004 \times 10^{-6}\ \text{m}^2/\text{s})} = 5.07 \times 10^5$$

Thus, from Fig. 8.17, $n = 8.4$ so that

$$\frac{\bar{u}}{V_c} \approx \left(1 - \frac{r}{R}\right)^{1/8.4}$$

To determine the centerline velocity, $V_c$, we must know the relationship between $V$ (the average velocity) and $V_c$. This can be

obtained by integration of the power-law velocity profile as follows. Since the flow is axisymmetric,

$$Q = AV = \int \bar{u}\, dA = V_c \int_{r=0}^{r=R} \left(1 - \frac{r}{R}\right)^{1/n} (2\pi r)\, dr$$

which can be integrated to give

$$Q = 2\pi R^2 V_c \frac{n^2}{(n+1)(2n+1)}$$

Thus, since $Q = \pi R^2 V$, we obtain

$$\frac{V}{V_c} = \frac{2n^2}{(n+1)(2n+1)}$$

With $n = 8.4$ in the present case, this gives

$$V_c = \frac{(n+1)(2n+1)}{2n^2} V = 1.186V = 1.186\,(5.09\ \text{m/s})$$
$$= 6.04\ \text{m/s} \qquad \textbf{(Ans)}$$

Recall that $V_c = 2V$ for laminar pipe flow.

**(c)**    From Eq. 8.4, which is valid for laminar or turbulent flow, the shear stress at $r = 0.025$ m is

$$\tau = \frac{2\tau_w r}{D} = \frac{2(64.8\ \text{N/m}^2)(0.025\ \text{m})}{(0.1\ \text{m})}$$

or

$$\tau = \tau_{\text{lam}} + \tau_{\text{turb}} = 32.4\ \text{N/m}^2$$

where $\tau_{\text{lam}} = -\mu\, d\bar{u}/dr$. From the power-law velocity profile (Eq. 8.31) we obtain the gradient of the average velocity as

$$\frac{d\bar{u}}{dr} = -\frac{V_c}{nR}\left(1 - \frac{r}{R}\right)^{(1-n)/n}$$

which gives

$$\frac{d\bar{u}}{dr} = -\frac{(6.04\ \text{m/s})}{8.4(0.05\ \text{m})}\left(1 - \frac{0.025\ \text{m}}{0.05\ \text{m}}\right)^{(1-8.4)/8.4}$$
$$= -26.5/\text{s}$$

Thus,

$$\tau_{\text{lam}} = -\mu\frac{d\bar{u}}{dr} = -(\nu\rho)\frac{d\bar{u}}{dr}$$
$$= -(1.004 \times 10^{-6}\ \text{m}^2/\text{s})(998\ \text{kg/m}^3)(-26.5/\text{s})$$
$$= 0.0266\ \text{N/m}^2$$

Thus, the ratio of turbulent to laminar shear stress is given by

$$\frac{\tau_{\text{turb}}}{\tau_{\text{lam}}} = \frac{\tau - \tau_{\text{lam}}}{\tau_{\text{lam}}} = \frac{32.4 - 0.0266}{0.0266} = 1220 \qquad \textbf{(Ans)}$$

**COMMENT**    As expected, most of the shear stress at this location in the turbulent flow is due to the turbulent shear stress.

The turbulent flow characteristics discussed in this section are not unique to turbulent flow in round pipes. Many of the characteristics introduced (i.e., the Reynolds stress, the viscous sublayer, the overlap layer, the outer layer, the general characteristics of the velocity profile, etc.) are found in other turbulent flows. In particular, turbulent pipe flow and turbulent flow past a solid wall (boundary layer flow) share many of these common traits. Such ideas are discussed more fully in Chapter 9.

### 8.3.4 Turbulence Modeling

Although it is not yet possible to theoretically predict the random, irregular details of turbulent flows, it would be useful to be able to predict the time-averaged flow fields (pressure, velocity, etc.) directly from the basic governing equations. To this end one can time average the governing Navier–Stokes equations (Eqs. 6.31 and 6.127) to obtain equations for the average velocity and pressure. However, because the Navier–Stokes equations are nonlinear, the resulting time-averaged differential equations contain not only the desired average pressure and velocity as variables, but also averages of products of the fluctuations—terms of the type that one tried to eliminate by averaging the equations! For example, the Reynolds stress $-\rho\overline{u'v'}$ (see Eq. 8.26) occurs in the time-averaged momentum equation.

Thus, it is not possible to merely average the basic differential equations and obtain governing equations involving only the desired averaged quantities. This is the reason for the variety of ad hoc assumptions that have been proposed to provide "closure" to the equations governing the average flow. That is, the set of governing equations must be a complete or closed set of equations—the same number of equation as unknowns.

Various attempts have been made to solve this closure problem (Refs. 1, 32). Such schemes involving the introduction of an eddy viscosity or the mixing length (as introduced in Section 8.3.2) are termed algebraic or zero-equation models. Other methods, which are beyond the scope of this book, include the one-equation model and the two-equation model. These turbulence models are based on the equation for the turbulence kinetic energy and require significant computer usage.

Turbulence modeling is an important and extremely difficult topic. Although considerable progress has been made, much remains to be done in this area.

### 8.3.5 Chaos and Turbulence

*Chaos theory may eventually provide a deeper understanding of turbulence.*

*V8.11 2D grid turbulence in soap film.*

Chaos theory is a relatively new branch of mathematical physics that may provide insight into the complex nature of turbulence. This method combines mathematics and numerical (computer) techniques to provide a new way to analyze certain problems. Chaos theory, which is quite complex and is currently under development, involves the behavior of nonlinear dynamical systems and their response to initial and boundary conditions. The flow of a viscous fluid, which is governed by the nonlinear Navier–Stokes equations (Eq. 6.127), may be such a system.

To solve the Navier–Stokes equations for the velocity and pressure fields in a viscous flow, one must specify the particular flow geometry being considered (the boundary conditions) and the condition of the flow at some particular time (the initial conditions). If, as some researchers predict, the Navier–Stokes equations allow chaotic behavior, then the state of the flow at times after the initial time may be very, very sensitive to the initial conditions. A slight variation to the initial flow conditions may cause the flow at later times to be quite different than it would have been with the original, only slightly different initial conditions. When carried to the extreme, the flow may be "chaotic," "random," or perhaps (in current terminology), "turbulent."

The occurrence of such behavior would depend on the value of the Reynolds number. For example, it may be found that for sufficiently small Reynolds numbers the flow is not chaotic (i.e., it is laminar), while for large Reynolds numbers it is chaotic with turbulent characteristics.

Thus, with the advancement of chaos theory it may be found that the numerous ad hoc turbulence ideas mentioned in previous sections (i.e., eddy viscosity, mixing length, law of the wall, etc.) may not be needed. It may be that chaos theory can provide the turbulence properties and structure directly from the governing equations. As of now we must wait until this exciting topic is developed further. The interested reader is encouraged to consult Ref. 4 for a general introduction to chaos or Ref. 33 for additional material.

## 8.4 Dimensional Analysis of Pipe Flow

As noted previously, turbulent flow can be a very complex, difficult topic—one that as yet has defied a rigorous theoretical treatment. Thus, most turbulent pipe flow analyses are based on experimental data and semi-empirical formulas. These data are expressed conveniently in dimensionless form.

It is often necessary to determine the head loss, $h_L$, that occurs in a pipe flow so that the energy equation, Eq. 5.84, can be used in the analysis of pipe flow problems. As shown in Fig. 8.1, a typical pipe system usually consists of various lengths of straight pipe interspersed with various types of components (valves, elbows, etc.). The overall head loss for the pipe system consists of the head loss due to viscous effects in the straight pipes, termed the ***major loss*** and denoted $h_{L\,major}$, and the head loss in the various pipe components, termed the ***minor loss*** and denoted $h_{L\,minor}$. That is,

$$h_L = h_{L\,major} + h_{L\,minor}$$

The head loss designations of "major" and "minor" do not necessarily reflect the relative importance of each type of loss. For a pipe system that contains many components and a relatively short length of pipe, the minor loss may actually be larger than the major loss.

### 8.4.1 Major Losses

A dimensional analysis treatment of pipe flow provides the most convenient base from which to consider turbulent, fully developed pipe flow. An introduction to this topic was given in Section 8.3. As is discussed in Sections 8.2.1 and 8.2.4, the pressure drop and head loss in a pipe are dependent on the wall shear stress, $\tau_w$, between the fluid and pipe surface. A fundamental difference between laminar and turbulent flow is that the shear stress for turbulent flow is a function of the density of the fluid, $\rho$. For laminar flow, the shear stress is independent of the density, leaving the viscosity, $\mu$, as the only important fluid property.

Thus, as indicated by the figure in the margin, the pressure drop, $\Delta p$, for steady, incompressible turbulent flow in a horizontal round pipe of diameter $D$ can be written in functional form as

$$\Delta p = F(V, D, \ell, \varepsilon, \mu, \rho) \tag{8.32}$$

where $V$ is the average velocity, $\ell$ is the pipe length, and $\varepsilon$ is a measure of the roughness of the pipe wall. It is clear that $\Delta p$ should be a function of $V$, $D$, and $\ell$. The dependence of $\Delta p$ on the fluid properties $\mu$ and $\rho$ is expected because of the dependence of $\tau$ on these parameters.

Although the pressure drop for laminar pipe flow is found to be independent of the roughness of the pipe, it is necessary to include this parameter when considering turbulent flow. As is discussed in Section 8.3.3 and illustrated in Fig. 8.19, for turbulent flow there is a relatively thin viscous sublayer formed in the fluid near the pipe wall. In many instances this layer is very thin; $\delta_s/D \ll 1$, where $\delta_s$ is the sublayer thickness. If a typical wall roughness element protrudes sufficiently far into (or even through) this layer, the structure and properties of the viscous sublayer (along with $\Delta p$ and $\tau_w$) will be different than if the wall were smooth. Thus, for turbulent flow the pressure drop is expected to be a function of the wall roughness. For laminar flow there is no thin viscous layer—viscous effects are important across the entire pipe. Thus, relatively small roughness elements have completely negligible effects on laminar pipe flow. Of course, for pipes with very large wall "roughness" ($\varepsilon/D \gtrsim 0.1$), such as that in corrugated pipes, the flowrate may be a function of the "roughness." We will consider only typical constant diameter pipes with relative roughnesses in the range $0 \le \varepsilon/D \lesssim 0.05$. Analysis of flow in corrugated pipes does not fit into the standard constant diameter pipe category, although experimental results for such pipes are available (Ref. 30).

The list of parameters given in Eq. 8.32 is apparently a complete one. That is, experiments have shown that other parameters (such as surface tension, vapor pressure, etc.) do not affect the pressure drop for the conditions stated (steady, incompressible flow; round, horizontal pipe). Since there are seven variables ($k = 7$) that can be written in terms of the three reference dimensions $MLT$ ($r = 3$), Eq. 8.32 can be written in dimensionless form in terms of $k - r = 4$ dimensionless groups. As was discussed in Section 7.9.1, one such representation is

$$\frac{\Delta p}{\frac{1}{2}\rho V^2} = \tilde{\phi}\left(\frac{\rho V D}{\mu}, \frac{\ell}{D}, \frac{\varepsilon}{D}\right)$$

This result differs from that used for laminar flow (see Eq. 8.17) in two ways. First, we have chosen to make the pressure dimensionless by dividing by the dynamic pressure, $\rho V^2/2$, rather than a characteristic viscous shear stress, $\mu V/D$. This convention was chosen in recognition of the fact that the shear stress for turbulent flow is normally dominated by $\tau_{turb}$, which is a stronger function

$\Delta p = p_1 - p_2$

(1)          (2)

$\rho, \mu$

$D$    $V$

$\varepsilon$

$\ell$

*Turbulent pipe flow properties depend on the fluid density and the pipe roughness.*

■ **Figure 8.19** Flow in the viscous sublayer near rough and smooth walls.

of the density than it is of viscosity. Second, we have introduced two additional dimensionless parameters, the Reynolds number, Re = $\rho VD/\mu$, and the **relative roughness,** $\varepsilon/D$, which are not present in the laminar formulation because the two parameters $\rho$ and $\varepsilon$ are not important in fully developed laminar pipe flow.

As was done for laminar flow, the functional representation can be simplified by imposing the reasonable assumption that the pressure drop should be proportional to the pipe length. (Such a step is not within the realm of dimensional analysis. It is merely a logical assumption supported by experiments.) The only way that this can be true is if the $\ell/D$ dependence is factored out as

$$\frac{\Delta p}{\frac{1}{2}\rho V^2} = \frac{\ell}{D}\,\phi\left(\text{Re},\frac{\varepsilon}{D}\right)$$

As was discussed in Section 8.2.3, the quantity $\Delta pD/(\ell\rho V^2/2)$ is termed the friction factor, $f$. Thus, for a horizontal pipe

$$\Delta p = f\frac{\ell}{D}\frac{\rho V^2}{2} \tag{8.33}$$

where

$$f = \phi\left(\text{Re},\frac{\varepsilon}{D}\right)$$

For laminar fully developed flow, the value of $f$ is simply $f = 64/\text{Re}$, independent of $\varepsilon/D$. For turbulent flow, the functional dependence of the friction factor on the Reynolds number and the relative roughness, $f = \phi(\text{Re}, \varepsilon/D)$, is a rather complex one that cannot, as yet, be obtained from a theoretical analysis. The results are obtained from an exhaustive set of experiments and usually presented in terms of a curve-fitting formula or the equivalent graphical form.

From Eq. 5.89 the energy equation for steady incompressible flow is

$$\frac{p_1}{\gamma} + \alpha_1\frac{V_1^2}{2g} + z_1 = \frac{p_2}{\gamma} + \alpha_2\frac{V_2^2}{2g} + z_2 + h_L$$

where $h_L$ is the head loss between sections (1) and (2). With the assumption of a constant diameter ($D_1 = D_2$ so that $V_1 = V_2$), horizontal ($z_1 = z_2$) pipe with fully developed flow ($\alpha_1 = \alpha_2$), this becomes $\Delta p = p_1 - p_2 = \gamma h_L$, which can be combined with Eq. 8.33 to give

*The major head loss in pipe flow is given in terms of the friction factor.*

$$h_{L\,\text{major}} = f\frac{\ell}{D}\frac{V^2}{2g} \tag{8.34}$$

Equation 8.34, called the *Darcy–Weisbach equation*, is valid for any fully developed, steady, incompressible pipe flow—whether the pipe is horizontal or on a hill. On the other hand, Eq. 8.33 is valid only for horizontal pipes. In general, with $V_1 = V_2$ the energy equation gives

$$p_1 - p_2 = \gamma(z_2 - z_1) + \gamma h_L = \gamma(z_2 - z_1) + f\frac{\ell}{D}\frac{\rho V^2}{2}$$

Part of the pressure change is due to the elevation change and part is due to the head loss associated with frictional effects, which are given in terms of the friction factor, $f$.

It is not easy to determine the functional dependence of the friction factor on the Reynolds number and relative roughness. Much of this information is a result of experiments conducted by J. Nikuradse in 1933 (Ref. 6) and amplified by many others since then. One difficulty lies in the determination of the roughness of the pipe. Nikuradse used artificially roughened pipes produced by gluing sand grains of known size onto pipe walls to produce pipes with sandpaper-type surfaces. The pressure drop needed to produce a desired flowrate was measured, and the data were converted into the friction factor for the corresponding Reynolds number and relative roughness. The tests were repeated numerous times for a wide range of Re and $\varepsilon/D$ to determine the $f = \phi(\text{Re}, \varepsilon/D)$ dependence.

In commercially available pipes the roughness is not as uniform and well defined as in the artificially roughened pipes used by Nikuradse. However, it is possible to obtain a measure of the effective relative roughness of typical pipes and thus to obtain the friction factor. Typical roughness values for various pipe surfaces are given in Table 8.1. Figure 8.20 shows the functional dependence of $f$ on Re and $\varepsilon/D$ and is called the **Moody chart** in honor of L. F. Moody, who, along with C. F. Colebrook, correlated the original data of Nikuradse in terms of the relative roughness of commercially available pipe materials. It should be noted that the values of $\varepsilon/D$ do not necessarily correspond to the actual values obtained by a microscopic determination of the average height of the roughness of the surface. They do, however, provide the correct correlation for $f = \phi(\text{Re}, \varepsilon/D)$.

It is important to observe that the values of relative roughness given pertain to new, clean pipes. After considerable use, most pipes (because of a buildup of corrosion or scale) may have a relative roughness that is considerably larger (perhaps by an order of magnitude) than that given. As shown by the upper figure in the margin, very old pipes may have enough scale buildup to not only alter the value of $\varepsilon$ but also to change their effective diameter by a considerable amount.

The following characteristics are observed from the data of Fig. 8.20. For laminar flow, $f = 64/\text{Re}$, which is independent of relative roughness. For turbulent flows with very large Reynolds numbers, $f = \phi(\varepsilon/D)$, which, as shown by the figure in the margin, is independent of the Reynolds number. For such flows, commonly termed *completely turbulent flow* (or *wholly turbulent flow*), the laminar sublayer is so thin (its thickness decreases with increasing Re) that the surface roughness completely dominates the character of the flow near the wall. Hence, the pressure drop required is a

**(Photograph courtesy of CorrView)**

■ **Table 8.1**

**Equivalent Roughness for New Pipes [Adapted from Moody (Ref. 7) and Colebrook (Ref. 8)]**

| Pipe | Equivalent Roughness, $\varepsilon$ | |
|---|---|---|
| | Feet | Millimeters |
| Riveted steel | 0.003–0.03 | 0.9–9.0 |
| Concrete | 0.001–0.01 | 0.3–3.0 |
| Wood stave | 0.0006–0.003 | 0.18–0.9 |
| Cast iron | 0.00085 | 0.26 |
| Galvanized iron | 0.0005 | 0.15 |
| Commercial steel or wrought iron | 0.00015 | 0.045 |
| Drawn tubing | 0.000005 | 0.0015 |
| Plastic, glass | 0.0 (smooth) | 0.0 (smooth) |

■ **Figure 8.20** Friction factor as a function of Reynolds number and relative roughness for round pipes—the Moody chart. (Data from Ref. 7 with permission.)

result of an inertia-dominated turbulent shear stress rather than the viscosity-dominated laminar shear stress normally found in the viscous sublayer. For flows with moderate values of Re, the friction factor is indeed dependent on both the Reynolds number and relative roughness—$f = \phi(\text{Re}, \varepsilon/D)$. The gap in the figure for which no values of $f$ are given (the $2100 < \text{Re} < 4000$ range) is a result of the fact that the flow in this transition range may be laminar or turbulent (or an unsteady mix of both) depending on the specific circumstances involved.

*For any pipe, even smooth ones, the head loss is not zero.*

Note that even for smooth pipes ($\varepsilon = 0$) the friction factor is not zero. That is, there is a head loss in any pipe, no matter how smooth the surface is made. This is a result of the no-slip boundary condition that requires any fluid to stick to any solid surface it flows over. There is always some microscopic surface roughness that produces the no-slip behavior (and thus $f \neq 0$) on the molecular level, even when the roughness is considerably less than the viscous sublayer thickness. Such pipes are called *hydraulically smooth*.

Various investigators have attempted to obtain an analytical expression for $f = \phi(\text{Re}, \varepsilon/D)$. Note that the Moody chart covers an extremely wide range in flow parameters. The nonlaminar region covers more than four orders of magnitude in Reynolds number—from $\text{Re} = 4 \times 10^3$ to $\text{Re} = 10^8$. Obviously, for a given pipe and fluid, typical values of the average velocity do not cover this range. However, because of the large variety in pipes ($D$), fluids ($\rho$ and $\mu$), and velocities ($V$), such a wide range in Re is needed to accommodate nearly all applications of pipe flow. In many cases the particular pipe flow of interest is confined to a relatively small region of the Moody chart, and simple semiempirical expressions can be developed for those conditions. For example, a company that manufactures cast iron water pipes with diameters between 2 and 12 in. may use a simple equation valid for their conditions only. The Moody chart, on the other hand, is universally valid for all steady, fully developed, incompressible pipe flows.

The following equation from Colebrook is valid for the entire nonlaminar range of the Moody chart

$$\frac{1}{\sqrt{f}} = -2.0 \log\left(\frac{\varepsilon/D}{3.7} + \frac{2.51}{\text{Re}\sqrt{f}}\right) \tag{8.35a}$$

In fact, the Moody chart is a graphical representation of this equation, which is an empirical fit of the pipe flow pressure drop data. Equation 8.35 is called the **Colebrook formula.** A difficulty with its use is that it is implicit in the dependence of $f$. That is, for given conditions (Re and $\varepsilon/D$), it is not possible to solve for $f$ without some sort of iterative scheme. With the use of modern computers and calculators, such calculations are not difficult. A word of caution is in order concerning the use of the Moody chart or the equivalent Colebrook formula. Because of various inherent inaccuracies involved (uncertainty in the relative roughness, uncertainty in the experimental data used to produce the Moody chart, etc.), the use of several place accuracy in pipe flow problems is usually not justified. As a rule of thumb, a 10% accuracy is the best expected. It is possible to obtain an equation that adequately approximates the Colebrook/Moody chart relationship but does not require an iterative scheme. For example, the Haaland equation (Ref. 34), which is easier to use, is given by

$$\frac{1}{\sqrt{f}} = -1.8 \log\left[\left(\frac{\varepsilon/D}{3.7}\right)^{1.11} + \frac{6.9}{\text{Re}}\right] \tag{8.35b}$$

where one can solve for $f$ explicitly.

---

## EXAMPLE 8.5    Comparison of Laminar or Turbulent Pressure Drop

**GIVEN** Air under standard conditions flows through a 4.0-mm-diameter drawn tubing with an average velocity of $V = 50$ m/s. For such conditions the flow would normally be turbulent. However, if precautions are taken to eliminate disturbances to the flow (the entrance to the tube is very smooth, the air is dust free, the tube does not vibrate, etc.), it may be possible to maintain laminar flow.

**FIND** (a) Determine the pressure drop in a 0.1-m section of the tube if the flow is laminar.

(b) Repeat the calculations if the flow is turbulent.

### SOLUTION

Under standard temperature and pressure conditions the density and viscosity are $\rho = 1.23$ kg/m$^3$ and $\mu = 1.79 \times 10^{-5}$ N·s/m$^2$. Thus, the Reynolds number is

$$\text{Re} = \frac{\rho V D}{\mu} = \frac{(1.23 \text{ kg/m}^3)(50 \text{ m/s})(0.004 \text{ m})}{1.79 \times 10^{-5} \text{ N} \cdot \text{s/m}^2} = 13{,}700$$

which would normally indicate turbulent flow.

**(a)** If the flow were laminar, then $f = 64/\text{Re} = 64/13{,}700 = 0.00467$, and the pressure drop in a 0.1-m-long horizontal section of the pipe would be

$$\Delta p = f \frac{\ell}{D} \frac{1}{2} \rho V^2$$

$$= (0.00467) \frac{(0.1 \text{ m})}{(0.004 \text{ m})} \frac{1}{2} (1.23 \text{ kg/m}^3)(50 \text{ m/s})^2$$

or

$$\Delta p = 0.179 \text{ kPa} \qquad \text{(Ans)}$$

**COMMENT** Note that the same result is obtained from Eq. 8.8:

$$\Delta p = \frac{32 \mu \ell}{D^2} V$$

$$= \frac{32(1.79 \times 10^{-5} \text{ N} \cdot \text{s/m}^2)(0.1 \text{ m})(50 \text{ m/s})}{(0.004 \text{ m})^2}$$

$$= 179 \text{ N/m}^2$$

**(b)** If the flow were turbulent, then $f = \phi(\text{Re}, \varepsilon/D)$, where from Table 8.1, $\varepsilon = 0.0015$ mm so that $\varepsilon/D = 0.0015$ mm/4.0 mm $= 0.000375$. From the Moody chart with Re $= 1.37 \times 10^4$ and $\varepsilon/D = 0.000375$ we obtain $f = 0.028$. Thus, the pressure drop in this case would be approximately

$$\Delta p = f \frac{\ell}{D} \frac{1}{2} \rho V^2 = (0.028) \frac{(0.1 \text{ m})}{(0.004 \text{ m})} \frac{1}{2} (1.23 \text{ kg/m}^3)(50 \text{ m/s})^2$$

or

$$\Delta p = 1.076 \text{ kPa} \qquad \text{(Ans)}$$

**COMMENT** A considerable savings in effort to force the fluid through the pipe could be realized (0.179 kPa rather than 1.076 kPa) if the flow could be maintained as laminar flow at this Reynolds number. In general this is very difficult to do, although laminar flow in pipes has been maintained up to Re $\approx$ 100,000 in rare instances.

An alternate method to determine the friction factor for the turbulent flow would be to use the Colebrook formula, Eq. 8.35*a*. Thus,

$$\frac{1}{\sqrt{f}} = -2.0 \log\left(\frac{\varepsilon/D}{3.7} + \frac{2.51}{\text{Re}\sqrt{f}}\right) = -2.0 \log\left(\frac{0.000375}{3.7} + \frac{2.51}{1.37 \times 10^4 \sqrt{f}}\right)$$

or

$$\frac{1}{\sqrt{f}} = -2.0 \log\left(1.01 \times 10^{-4} + \frac{1.83 \times 10^{-4}}{\sqrt{f}}\right) \quad \textbf{(1)}$$

By using a root-finding technique on a computer or calculator, the solution to Eq. 1 is determined to be $f = 0.0291$, in agreement (within the accuracy of reading the graph) with the Moody chart method of $f = 0.028$.

Equation 8.35*b* provides an alternate form to the Colebrook formula that can be used to solve for the friction factor directly.

$$\frac{1}{\sqrt{f}} = -1.8 \log\left[\left(\frac{\varepsilon/D}{3.7}\right)^{1.11} + \frac{6.9}{\text{Re}}\right] = -1.8 \log\left[\left(\frac{0.000375}{3.7}\right)^{1.11} + \frac{6.9}{1.37 \times 10^4}\right]$$

$$= 0.0289$$

This agrees with the Colebrook formula and Moody chart values obtained above.

Numerous other empirical formulas can be found in the literature (Ref. 5) for portions of the Moody chart. For example, an often-used equation, commonly referred to as the Blasius formula, for turbulent flow in smooth pipes $(\varepsilon/D = 0)$ with Re $< 10^5$ is

$$f = \frac{0.316}{\text{Re}^{1/4}}$$

For our case this gives

$$f = 0.316(13,700)^{-0.25} = 0.0292$$

which is in agreement with the previous results. Note that the value of $f$ is relatively insensitive to $\varepsilon/D$ for this particular situation. Whether the tube was smooth glass $(\varepsilon/D = 0)$ or the drawn tubing $(\varepsilon/D = 0.000375)$ would not make much difference in the pressure drop. For this flow, an increase in relative roughness by a factor of 30 to $\varepsilon/D = 0.0113$ (equivalent to a commercial steel surface; see Table 8.1) would give $f = 0.043$. This would represent an increase in pressure drop and head loss by a factor of $0.043/0.0291 = 1.48$ compared with that for the original drawn tubing.

The pressure drop of 1.076 kPa in a length of 0.1 m of pipe corresponds to a change in absolute pressure [assuming $p = 101$ kPa (abs) at $x = 0$] of approximately $1.076/101 = 0.0107$, or about 1%. Thus, the incompressible flow assumption on which the above calculations (and all of the formulas in this chapter) are based is reasonable. However, if the pipe were 2-m long the pressure drop would be 21.5 kPa, approximately 20% of the original pressure. In this case the density would not be approximately constant along the pipe, and a compressible flow analysis would be needed. Such considerations are discussed in Chapter 11.

## 8.4.2 Minor Losses

As discussed in the previous section, the head loss in long, straight sections of pipe, the major losses, can be calculated by use of the friction factor obtained from either the Moody chart or the Colebrook equation. Most pipe systems, however, consist of considerably more than straight pipes. These additional components (valves, bends, tees, and the like) add to the overall head loss of the system. Such losses are generally termed *minor losses*, with the corresponding head loss denoted $h_{L\,\text{minor}}$. In this section we indicate how to determine the various minor losses that commonly occur in pipe systems.

The head loss associated with flow through a valve is a common minor loss. The purpose of a valve is to provide a means to regulate the flowrate. This is accomplished by changing the geometry of the system (i.e., closing or opening the valve alters the flow pattern through the valve), which in turn alters the losses associated with the flow through the valve. The flow resistance or head loss through the valve may be a significant portion of the resistance in the system. In fact, with the valve closed, the resistance to the flow is infinite—the fluid cannot flow. Such minor losses may be very important indeed. With the valve wide open the extra resistance due to the presence of the valve may or may not be negligible.

The flow pattern through a typical component such as a valve is shown in Fig. 8.21. It is not difficult to realize that a theoretical analysis to predict the details of such flows to obtain the head loss for these components is not, as yet, possible. Thus, the head loss information for essentially all components is given in dimensionless form and based on experimental data. The most common method used to determine these head losses or pressure drops is to specify the *loss coefficient, $K_L$*, which is defined as

*Losses due to pipe system components are given in terms of loss coefficients.*

$$K_L = \frac{h_{L\,\text{minor}}}{(V^2/2g)} = \frac{\Delta p}{\frac{1}{2}\rho V^2}$$

■ **Figure 8.21**    Flow through a valve.

so that

$$\Delta p = K_L \tfrac{1}{2}\rho V^2$$

or

$$h_{L\text{ minor}} = K_L \frac{V^2}{2g} \qquad\qquad (8.36)$$

The pressure drop across a component that has a loss coefficient of $K_L = 1$ is equal to the dynamic pressure, $\rho V^2/2$. As shown by Eq. 8.36 and the figure in the margin, for a given value of $K_L$ the head loss is proportional to the square of the velocity.

The actual value of $K_L$ is strongly dependent on the geometry of the component considered. It may also be dependent on the fluid properties. That is,

$$K_L = \phi(\text{geometry, Re})$$

*For most flows the loss coefficient is independent of the Reynolds number.*

where Re $= \rho V D / \mu$ is the pipe Reynolds number. For many practical applications the Reynolds number is large enough so that the flow through the component is dominated by inertia effects, with viscous effects being of secondary importance. This is true because of the relatively large accelerations and decelerations experienced by the fluid as it flows along a rather curved, variable area (perhaps even torturous) path through the component (see Fig. 8.21). In a flow that is dominated by inertia effects rather than viscous effects, it is usually found that pressure drops and head losses correlate directly with the dynamic pressure. This is why the friction factor for very large Reynolds number, fully developed pipe flow is independent of the Reynolds number. The same condition is found to be true for flow through pipe components. Thus, in most cases of practical interest the loss coefficients for components are a function of geometry only, $K_L = \phi(\text{geometry})$.

Minor losses are sometimes given in terms of an *equivalent length*, $\ell_{\text{eq}}$. In this terminology, the head loss through a component is given in terms of the equivalent length of pipe that would produce the same head loss as the component. That is,

$$h_{L\text{ minor}} = K_L \frac{V^2}{2g} = f \frac{\ell_{\text{eq}}}{D} \frac{V^2}{2g}$$

or

$$\ell_{\text{eq}} = \frac{K_L D}{f}$$

■ **Figure 8.22** Entrance flow conditions and loss coefficient (Data from Refs. 28, 29). (*a*) Reentrant, $K_L = 0.8$, (*b*) sharp-edged, $K_L = 0.5$, (*c*) slightly rounded, $K_L = 0.2$ (see Fig. 8.24), (*d*) well-rounded, $K_L = 0.04$ (see Fig. 8.24).

where $D$ and $f$ are based on the pipe containing the component. The head loss of the pipe system is the same as that produced in a straight pipe whose length is equal to the pipes of the original system plus the sum of the additional equivalent lengths of all of the components of the system. Most pipe flow analyses, including those in this book, use the loss coefficient method rather than the equivalent length method to determine the minor losses.

Many pipe systems contain various transition sections in which the pipe diameter changes from one size to another. Such changes may occur abruptly or rather smoothly through some type of area change section. Any change in flow area contributes losses that are not accounted for in the fully developed head loss calculation (the friction factor). The extreme cases involve flow into a pipe from a reservoir (an entrance) or out of a pipe into a reservoir (an exit).

A fluid may flow from a reservoir into a pipe through any number of differently shaped entrance regions as are sketched in Fig. 8.22. Each geometry has an associated loss coefficient. A typical flow pattern for flow entering a pipe through a square-edged entrance is sketched in Fig. 8.23. As was discussed in Chapter 3, a vena contracta region may result because the fluid cannot turn a sharp right-angle corner. The flow is said to separate from the sharp corner. The maximum velocity at section (2) is greater than that in the pipe at section (3), and the pressure there is lower. If this high-speed fluid could slow down efficiently, the kinetic energy could be converted into pressure (the Bernoulli effect), and the ideal pressure distribution indicated in Fig. 8.23 would result. The head loss for the entrance would be essentially zero.

*Minor head losses are often a result of the dissipation of kinetic energy.*

Such is not the case. Although a fluid may be accelerated very efficiently, it is very difficult to slow down (decelerate) a fluid efficiently. Thus, the extra kinetic energy of the fluid at section (2) is partially lost because of viscous dissipation, so that the pressure does not return to the ideal value. An entrance head loss (pressure drop) is produced as is indicated in Fig. 8.23. The majority of this loss is due to inertia effects that are eventually dissipated by the shear stresses within the fluid. Only a small portion of the loss is due to the wall shear stress within the entrance region. The net effect is that the loss coefficient for a square-edged entrance is approximately $K_L = 0.50$. One-half of a velocity head is lost as the fluid enters the pipe. If the pipe protrudes into the tank (a reentrant entrance) as is shown in Fig. 8.22a, the losses are even greater.

An obvious way to reduce the entrance loss is to round the entrance region as is shown in Fig. 8.22c, thereby reducing or eliminating the vena contracta effect. Typical values for the loss coefficient for entrances with various amounts of rounding of the lip are shown in Fig. 8.24. A significant reduction in $K_L$ can be obtained with only slight rounding.

■ **Figure 8.23**   Flow pattern and pressure distribution for a sharp-edged entrance.

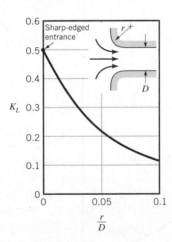

■ **Figure 8.24**   Entrance loss coefficient as a function of rounding of the inlet edge (Data from Ref. 9).

*V8.12 Entrance/exit flows*

A head loss (the exit loss) is also produced when a fluid flows from a pipe into a tank as is shown in Fig. 8.25. In these cases the entire kinetic energy of the exiting fluid (velocity $V_1$) is dissipated through viscous effects as the stream of fluid mixes with the fluid in the tank and eventually comes to rest ($V_2 = 0$). The exit loss from points (1) and (2) is therefore equivalent to one velocity head, or $K_L = 1$.

Losses also occur because of a change in pipe diameter as is shown in Figs. 8.26 and 8.27. The sharp-edged entrance and exit flows discussed in the previous paragraphs are limiting cases of this type of flow with either $A_1/A_2 = \infty$, or $A_1/A_2 = 0$, respectively. The loss coefficient for a sudden contraction, $K_L = h_L/(V_2^2/2g)$, is a function of the area ratio, $A_2/A_1$, as is shown in Fig. 8.26. The value of $K_L$ changes gradually from one extreme of a sharp-edged entrance ($A_2/A_1 = 0$ with $K_L = 0.50$) to the other extreme of no area change ($A_2/A_1 = 1$ with $K_L = 0$).

In many ways, the flow in a sudden expansion is similar to exit flow. As is indicated in Fig. 8.28, the fluid leaves the smaller pipe and initially forms a jet-type structure as it enters the larger pipe. Within a few diameters downstream of the expansion, the jet becomes dispersed across the pipe, and fully developed flow becomes established again. In this process [between sections (2) and (3)] a portion of the kinetic energy of the fluid is dissipated as a result of viscous effects. A square-edged exit is the limiting case with $A_1/A_2 = 0$.

A sudden expansion is one of the few components (perhaps the only one) for which the loss coefficient can be obtained by means of a simple analysis. To do this we consider the continuity

**■ Figure 8.25** **Exit flow conditions and loss coefficient.**
(a) Reentrant, $K_L = 1.0$, (b) sharp-edged, $K_L = 1.0$, (c) slightly rounded, $K_L = 1.0$,
(d) well-rounded, $K_L = 1.0$.

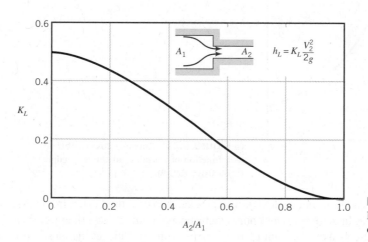

**■ Figure 8.26**
Loss coefficient for a sudden
contraction (Ref. 10).

**■ Figure 8.27**
Loss coefficient for a sudden
expansion (Ref. 10).

- - - - - - - Control volume

■ **Figure 8.28** Control volume used to calculate the loss coefficient for a sudden expansion.

and momentum equations for the control volume shown in Fig. 8.28 and the energy equation applied between (2) and (3). We assume that the flow is uniform at sections (1), (2), and (3) and the pressure is constant across the left-hand side of the control volume ($p_a = p_b = p_c = p_1$). The resulting three governing equations (mass, momentum, and energy) are

$$A_1 V_1 = A_3 V_3$$

$$p_1 A_3 - p_3 A_3 = \rho A_3 V_3 (V_3 - V_1)$$

and

*The loss coefficient for a sudden expansion can be theoretically calculated.*

$$\frac{p_1}{\gamma} + \frac{V_1^2}{2g} = \frac{p_3}{\gamma} + \frac{V_3^2}{2g} + h_L$$

These can be rearranged to give the loss coefficient, $K_L = h_L/(V_1^2/2g)$, as

$$K_L = \left(1 - \frac{A_1}{A_2}\right)^2$$

where we have used the fact that $A_2 = A_3$. This result, plotted in Fig. 8.27, is in good agreement with experimental data. As with so many minor loss situations, it is not the viscous effects directly (i.e., the wall shear stress) that cause the loss. Rather, it is the dissipation of kinetic energy (another type of viscous effect) as the fluid decelerates inefficiently.

*V8.13 Separated flow in a diffuser*

The losses may be quite different if the contraction or expansion is gradual. Typical results for a conical *diffuser* with a given area ratio, $A_2/A_1$, are shown in Fig. 8.29. (A diffuser is a device shaped to decelerate a fluid.) Clearly the included angle of the diffuser, $\theta$, is a very important parameter. For very small angles, the diffuser is excessively long and most of the head loss is due to the wall shear stress as in fully developed flow. For moderate or large angles, the flow separates from the walls and the losses are due mainly to a dissipation of the kinetic energy of the jet leaving the smaller diameter pipe. In fact, for moderate or large values of $\theta$ (i.e., $\theta > 35°$ for the case

■ **Figure 8.29** Loss coefficient for a typical conical diffuser (Ref. 5).

shown in Fig. 8.29), the conical diffuser is, perhaps unexpectedly, less efficient than a sharp-edged expansion which has $K_L = (1 - A_1/A_2)^2$. There is an optimum angle ($\theta \approx 8°$ for the case illustrated) for which the loss coefficient is a minimum. The relatively small value of $\theta$ for the minimum $K_L$ results in a long diffuser and is an indication of the fact that it is difficult to efficiently decelerate a fluid.

It must be noted that the conditions indicated in Fig. 8.29 represent typical results only. Flow through a diffuser is very complicated and may be strongly dependent on the area ratio $A_2/A_1$, specific details of the geometry, and the Reynolds number. The data are often presented in terms of a *pressure recovery coefficient*, $C_p = (p_2 - p_1)/(\rho V_1^2/2)$, which is the ratio of the static pressure rise across the diffuser to the inlet dynamic pressure. Considerable effort has gone into understanding this important topic (Refs. 11, 12).

Flow in a conical contraction (a nozzle; reverse the flow direction shown in Fig. 8.29) is less complex than that in a conical expansion. Typical loss coefficients based on the downstream (high-speed) velocity can be quite small, ranging from $K_L = 0.02$ for $\theta = 30°$, to $K_L = 0.07$ for $\theta = 60°$, for example. It is relatively easy to accelerate a fluid efficiently.

Bends in pipes produce a greater head loss than if the pipe were straight. The losses are due to the separated region of flow near the inside of the bend (especially if the bend is sharp) and the swirling secondary flow that occurs because of the imbalance of centripetal forces as a result of the curvature of the pipe centerline. These effects and the associated values of $K_L$ for large Reynolds number flows through a 90° bend are shown in Fig. 8.30. The friction loss due to the axial length of the pipe bend must be calculated and added to that given by the loss coefficient of Fig. 8.30.

For situations in which space is limited, a flow direction change is often accomplished by use of miter bends, as is shown in Fig. 8.31, rather than smooth bends. The considerable losses in such bends can be reduced by the use of carefully designed guide vanes that help direct the flow with less unwanted swirl and disturbances.

Another important category of pipe system components is that of commercially available pipe fittings such as elbows, tees, reducers, valves, and filters. The values of $K_L$ for such components depend strongly on the shape of the component and only very weakly on the Reynolds number for typical large Re flows. Thus, the loss coefficient for a 90° elbow depends on whether the pipe joints are threaded or flanged but is, within the accuracy of the data, fairly independent of the pipe diameter, flowrate, or fluid properties (the Reynolds number effect). Typical values of $K_L$ for such components are given in Table 8.2. These typical components are designed more for ease of manufacturing and costs than for reduction of the head losses that they produce. The flowrate from a faucet in a typical house is sufficient whether the value of $K_L$ for an elbow is the typical $K_L = 1.5$, or it is reduced to $K_L = 0.2$ by use of a more expensive long-radius, gradual bend (Fig. 8.30).

*V8.14 Car exhaust system*

*Extensive tables are available for loss coefficients of standard pipe components.*

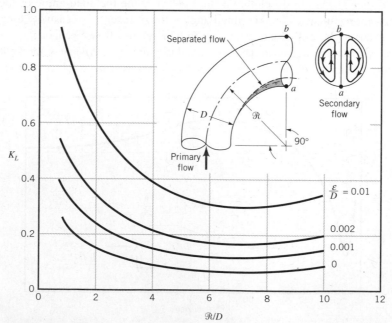

■ **Figure 8.30** Character of the flow in a 90° bend and the associated loss coefficient (Ref. 5).

■ **Figure 8.31**   Character of the flow in a 90° mitered bend and the associated loss coefficient: (*a*) without guide vanes, (*b*) with guide vanes.

■ **Table 8.2**

**Loss Coefficients for Pipe Components** $\left(h_L = K_L \dfrac{V^2}{2g}\right)$ **(Data from Refs. 5, 10, 27)**

| Component | $K_L$ |
|---|---|
| **a. Elbows** | |
| Regular 90°, flanged | 0.3 |
| Regular 90°, threaded | 1.5 |
| Long radius 90°, flanged | 0.2 |
| Long radius 90°, threaded | 0.7 |
| Long radius 45°, flanged | 0.2 |
| Regular 45°, threaded | 0.4 |
| **b. 180° return bends** | |
| 180° return bend, flanged | 0.2 |
| 180° return bend, threaded | 1.5 |
| **c. Tees** | |
| Line flow, flanged | 0.2 |
| Line flow, threaded | 0.9 |
| Branch flow, flanged | 1.0 |
| Branch flow, threaded | 2.0 |
| **d. Union, threaded** | 0.08 |
| *e. Valves** | |
| Globe, fully open | 10 |
| Angle, fully open | 2 |
| Gate, fully open | 0.15 |
| Gate, $\frac{1}{4}$ closed | 0.26 |
| Gate, $\frac{1}{2}$ closed | 2.1 |
| Gate, $\frac{3}{4}$ closed | 17 |
| Swing check, forward flow | 2 |
| Swing check, backward flow | ∞ |
| Ball valve, fully open | 0.05 |
| Ball valve, $\frac{1}{3}$ closed | 5.5 |
| Ball valve, $\frac{2}{3}$ closed | 210 |

*See Fig. 8.32 for typical valve geometry.

■ **Figure 8.32** Internal structure of various valves: (*a*) globe valve, (*b*) gate valve, (*c*) swing check valve, (*d*) stop check valve. (Courtesy of Crane Co., Fluid Handling Division.)

---

**F l u i d s   i n   t h e   N e w s**

**Safety in fluids** The following incident is taken from a NIOSH (National Institute for Occupational Safety and Health) fatality investigation report that emphasizes the dangers of oxygen-deficient environments in confined spaces. A 35-year-old male water system operator (victim) was asphyxiated after entering a valve vault at a municipal water system plant. The victim was assigned to turn on a waterline valve serving a nearby tree farm. The valve was located at the water treatment plant inside an underground valve vault that "always had normal air." The victim entered the valve vault through a ground-level manhole without

testing or ventilating the vault to atmosphere. A coworker, who had last seen the victim one hour earlier, checked the manhole and saw the victim lying on his back at the bottom. The victim did not respond to any calls. Other workers summoned from the plant building and local fire department personnel ventilated the valve vault and removed the victim. The vault atmosphere was subsequently found to be oxygen deficient. There were no witnesses to the incident, but evidence suggests that the victim lost consciousness and fell from the ladder railings to the bottom of the vault.

---

Valves control the flowrate by providing a means to adjust the overall system loss coefficient to the desired value. When the valve is closed, the value of $K_L$ is infinite and no fluid flows. Opening of the valve reduces $K_L$, producing the desired flowrate. Typical cross sections of various types of valves are shown in Fig. 8.32. Some valves (such as the conventional globe valve) are designed for general use, providing convenient control between the extremes of fully closed and fully open. Others (such as a needle valve) are designed to provide very fine control of the flowrate. The check valve provides a diode type operation that allows fluid to flow in one direction only.

Loss coefficients for typical valves are given in Table 8.2. As with many system components, the head loss in valves is mainly a result of the dissipation of kinetic energy of a high-speed portion of the flow. This high speed, $V_3$, is illustrated in Fig. 8.33.

$V_1 \rightarrow$ (1)  $V_3 \gg V_1$  $\rightarrow V_2 = V_1$ (2)

# EXAMPLE 8.6  Minor Losses

**GIVEN** The closed-circuit wind tunnel shown in Fig. E8.6a is a smaller version of that depicted in Fig. E8.6b in which air at standard conditions is to flow through the test section [between sections (5) and (6)] with a velocity of 200 ft/s. The flow is driven by a fan that essentially increases the static pressure by the amount $p_1 - p_9$ that is needed to overcome the head losses experienced by the fluid as it flows around the circuit.

**FIND** Estimate the value of $p_1 - p_9$ and the horsepower supplied to the fluid by the fan.

$V_5 = 200$ ft/s

(3)    (4)    (5)    (6)    (7)

Test section
Flow-straightening
screens

(8)

$Q$ (1)

(2)    (9)

Fan

■ **Figure E8.6b**

## SOLUTION

The maximum velocity within the wind tunnel occurs in the test section (smallest area; see Table E8.6 on the next page). Thus, the maximum Mach number of the flow is $Ma_5 = V_5/c_5$, where $V_5 = 200$ ft/s and from Eq. 1.20 the speed of sound is $c_5 = (kRT_5)^{1/2} = \{1.4 \ (1716 \ \text{ft} \cdot \text{lb/slug} \cdot °\text{R}) \ [(460 + 59)°\text{R}]\}^{1/2} = 1117$ ft/s. Thus, $Ma_5 = 200/1117 = 0.179$. As was indicated in Chapter 3 and discussed fully in Chapter 11, most flows can be considered as incompressible if the Mach number is less than about 0.3. Hence, we can use the incompressible formulas for this problem.

The purpose of the fan in the wind tunnel is to provide the necessary energy to overcome the net head loss experienced by the air as it flows around the circuit. This can be found from the energy equation between points (1) and (9) as

$$\frac{p_1}{\gamma} + \frac{V_1^2}{2g} + z_1 = \frac{p_9}{\gamma} + \frac{V_9^2}{2g} + z_9 + h_{L_{1-9}}$$

where $h_{L_{1-9}}$ is the total head loss from (1) to (9). With $z_1 = z_9$ and $V_1 = V_9$ this gives

$$\frac{p_1}{\gamma} - \frac{p_9}{\gamma} = h_{L_{1-9}} \tag{1}$$

Similarly, by writing the energy equation (Eq. 5.84) across the fan, from (9) to (1), we obtain

$$\frac{p_9}{\gamma} + \frac{V_9^2}{2g} + z_9 + h_p = \frac{p_1}{\gamma} + \frac{V_1^2}{2g} + z_1$$

where $h_p$ is the actual head rise supplied by the pump (fan) to the air. Again since $z_9 = z_1$ and $V_9 = V_1$ this, when combined with Eq. 1, becomes

$$h_p = \frac{(p_1 - p_9)}{\gamma} = h_{L_{1-9}}$$

The actual power supplied to the air (horsepower, $\mathcal{P}_a$) is obtained from the fan head by

$$\mathcal{P}_a = \gamma Q h_p = \gamma A_5 V_5 h_p = \gamma A_5 V_5 h_{L_{1-9}} \tag{2}$$

Thus, the power that the fan must supply to the air depends on the head loss associated with the flow through the wind tunnel. To obtain a reasonable, approximate answer we make the following assumptions. We treat each of the four turning corners as a mitered bend with guide vanes so that from Fig. 8.31 $K_{L_{\text{corner}}} = 0.2$. Thus, for each corner

$$h_{L_{\text{corner}}} = K_L \frac{V^2}{2g} = 0.2 \frac{V^2}{2g}$$

where, because the flow is assumed incompressible, $V = V_5 A_5/A$. The values of $A$ and the corresponding velocities throughout the tunnel are given in Table E8.6.

We also treat the enlarging sections from the end of the test section (6) to the beginning of the nozzle (4) as a conical diffuser

■ Table E8.6

| Location | Area (ft²) | Velocity (ft/s) |
|----------|-----------|-----------------|
| 1 | 22.0 | 36.4 |
| 2 | 28.0 | 28.6 |
| 3 | 35.0 | 22.9 |
| 4 | 35.0 | 22.9 |
| 5 | 4.0 | 200.0 |
| 6 | 4.0 | 200.0 |
| 7 | 10.0 | 80.0 |
| 8 | 18.0 | 44.4 |
| 9 | 22.0 | 36.4 |

(200 ft/s, 62.3 hp)

■ **Figure E8.6c**

with a loss coefficient of $K_{L_{\text{dif}}} = 0.6$. This value is larger than that of a well-designed diffuser (see Fig. 8.29, for example). Since the wind tunnel diffuser is interrupted by the four turning corners and the fan, it may not be possible to obtain a smaller value of $K_{L_{\text{dif}}}$ for this situation. Thus,

$$h_{L_{\text{dif}}} = K_{L_{\text{dif}}} \frac{V_6^2}{2g} = 0.6 \frac{V_6^2}{2g}$$

The loss coefficients for the conical nozzle between sections (4) and (5) and the flow-straightening screens are assumed to be $K_{L_{\text{noz}}} = 0.2$ and $K_{L_{\text{scr}}} = 4.0$ (Ref. 13), respectively. We neglect the head loss in the relatively short test section.

Thus, the total head loss is

$$h_{L_{1-9}} = h_{L_{\text{corner7}}} + h_{L_{\text{corner8}}} + h_{L_{\text{corner2}}} + h_{L_{\text{corner3}}}$$
$$+ h_{L_{\text{dif}}} + h_{L_{\text{noz}}} + h_{L_{\text{scr}}}$$

or

$$h_{L_{1-9}} = [0.2(V_7^2 + V_8^2 + V_2^2 + V_3^2)$$
$$+ 0.6V_6^2 + 0.2V_5^2 + 4.0V_4^2]/2g$$
$$= [0.2(80.0^2 + 44.4^2 + 28.6^2 + 22.9^2) + 0.6(200)^2$$
$$+ 0.2(200)^2 + 4.0(22.9)^2] \text{ ft}^2/\text{s}^2/[2(32.2 \text{ ft/s}^2)]$$

or

$$h_{L_{1-9}} = 560 \text{ ft}$$

Hence, from Eq. 1 we obtain the pressure rise across the fan as

$$p_1 - p_9 = \gamma h_{L_{1-9}} = (0.0765 \text{ lb/ft}^3)(560 \text{ ft})$$
$$= 42.8 \text{ lb/ft}^2 = 0.298 \text{ psi} \qquad \textbf{(Ans)}$$

From Eq. 2 we obtain the power added to the fluid as

$$\mathcal{P}_a = (0.0765 \text{ lb/ft}^3)(4.0 \text{ ft}^2)(200 \text{ ft/s})(560 \text{ ft})$$
$$= 34,300 \text{ ft} \cdot \text{lb/s}$$

or

$$\mathcal{P}_a = \frac{34,300 \text{ ft} \cdot \text{lb/s}}{550 \text{ (ft} \cdot \text{lb/s)/hp}} = 62.3 \text{ hp} \qquad \textbf{(Ans)}$$

**COMMENTS** By repeating the calculations with various test section velocities, $V_5$, the results shown in Fig. E8.6c are obtained. Since the head loss varies as $V_5^2$ and the power varies as head loss times $V_5$, it follows that the power varies as the cube of the velocity. Thus, doubling the wind tunnel speed requires an eightfold increase in power.

With a closed-return wind tunnel of this type, all of the power required to maintain the flow is dissipated through viscous effects, with the energy remaining within the closed tunnel. If heat transfer across the tunnel walls is negligible, the air temperature within the tunnel will increase in time. For steady-state operations of such tunnels, it is often necessary to provide some means of cooling to maintain the temperature at acceptable levels.

It should be noted that the actual size of the motor that powers the fan must be greater than the calculated 62.3 hp because the fan is not 100% efficient. The power calculated above is that needed by the fluid to overcome losses in the tunnel, excluding those in the fan. If the fan were 60% efficient, it would require a shaft power of $\mathcal{P} = 62.3 \text{ hp}/(0.60) = 104 \text{ hp}$ to run the fan. Determination of fan (or pump) efficiencies can be a complex problem that depends on the specific geometry of the fan. Introductory material about fan performance is presented in Chapter 12; additional material can be found in various references (Refs. 14, 15, 16, for example).

It should also be noted that the above results are only approximate. Clever, careful design of the various components (corners, diffuser, etc.) may lead to improved (i.e., lower) values of the various loss coefficients, and hence lower power requirements. Since $h_L$ is proportional to $V^2$, the components with the larger $V$ tend to have the larger head loss. Thus, even though $K_L = 0.2$ for each of the four corners, the head loss for corner (7) is $(V_7/V_3)^2 = (80/22.9)^2 = 12.2$ times greater than it is for corner (3).

### 8.4.3 Noncircular Conduits

Many of the conduits that are used for conveying fluids are not circular in cross section. Although the details of the flows in such conduits depend on the exact cross-sectional shape, many round pipe results can be carried over, with slight modification, to flow in conduits of other shapes.

Theoretical results can be obtained for fully developed laminar flow in noncircular ducts, although the detailed mathematics often becomes rather cumbersome. For an arbitrary

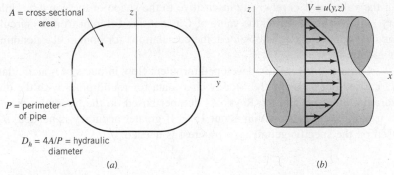

■ **Figure 8.34**   Noncircular duct.

cross section, as is shown in Fig. 8.34, the velocity profile is a function of both $y$ and $z$ $[\mathbf{V} = u(y, z)\hat{\mathbf{i}}]$. This means that the governing equation from which the velocity profile is obtained (either the Navier–Stokes equations of motion or a force balance equation similar to that used for circular pipes, Eq. 8.6) is a partial differential equation rather than an ordinary differential equation. Although the equation is linear (for fully developed flow the convective acceleration is zero), its solution is not as straightforward as for round pipes. Typically the velocity profile is given in terms of an infinite series representation (Ref. 17).

Practical, easy-to-use results can be obtained as follows. Regardless of the cross-sectional shape, there are no inertia effects in fully developed laminar pipe flow. Thus, the friction factor can be written as $f = C/\mathrm{Re}_h$, where the constant $C$ depends on the particular shape of the duct, and $\mathrm{Re}_h$ is the Reynolds number, $\mathrm{Re}_h = \rho V D_h / \mu$, based on the hydraulic diameter. The *hydraulic diameter* defined as $D_h = 4A/P$ is four times the ratio of the cross-sectional flow area divided by the wetted perimeter, $P$, of the pipe as is illustrated in Fig. 8.34. It represents a characteristic length that defines the size of a cross section of a specified shape. The factor of 4 is included in the definition of $D_h$ so that for round pipes the diameter and hydraulic diameter are equal $[D_h = 4A/P = 4(\pi D^2/4)/(\pi D) = D]$. The hydraulic diameter is also used in the definition of the friction factor, $h_L = f(\ell/D_h)V^2/2g$, and the relative roughness, $\varepsilon/D_h$.

The values of $C = f\,\mathrm{Re}_h$ for laminar flow have been obtained from theory and/or experiment for various shapes. Typical values are given in Table 8.3 along with the hydraulic diameter. Note

*The hydraulic diameter is used for noncircular duct calculations.*

■ **Table 8.3**
**Friction Factors for Laminar Flow in Noncircular Ducts (Data from Ref. 18)**

| Shape | Parameter | $C = f\,\mathrm{Re}_h$ |
|---|---|---|
| I. Concentric Annulus $\quad D_h = D_2 - D_1$ | $D_1/D_2$ | |
| | 0.0001 | 71.8 |
| | 0.01 | 80.1 |
| | 0.1 | 89.4 |
| | 0.6 | 95.6 |
| | 1.00 | 96.0 |
| II. Rectangle $\quad D_h = \dfrac{2ab}{a+b}$ | $a/b$ | |
| | 0 | 96.0 |
| | 0.05 | 89.9 |
| | 0.10 | 84.7 |
| | 0.25 | 72.9 |
| | 0.50 | 62.2 |
| | 0.75 | 57.9 |
| | 1.00 | 56.9 |

that the value of $C$ is relatively insensitive to the shape of the conduit. Unless the cross section is very "thin" in some sense, the value of $C$ is not too different from its circular pipe value, $C = 64$. Once the friction factor is obtained, the calculations for noncircular conduits are identical to those for round pipes.

Calculations for fully developed turbulent flow in ducts of noncircular cross section are usually carried out by using the Moody chart data for round pipes with the diameter replaced by the hydraulic diameter and the Reynolds number based on the hydraulic diameter. Such calculations are usually accurate to within about 15%. If greater accuracy is needed, a more detailed analysis based on the specific geometry of interest is needed.

> *The Moody chart, developed for round pipes, can also be used for noncircular ducts.*

---

# $E$XAMPLE 8.7 Noncircular Conduit

**GIVEN** Air at a temperature of 120 °F and standard pressure flows from a furnace through an 8-in.-diameter pipe with an average velocity of 10 ft/s. It then passes through a transition section similar to the one shown in Fig. E8.7 and into a square duct whose side is of length $a$. The pipe and duct surfaces are smooth ($\varepsilon = 0$). The head loss per foot is to be the same for the pipe and the duct.

**FIND** Determine the duct size, $a$.

■ **Figure E8.7**

## $S$OLUTION

We first determine the head loss per foot for the pipe, $h_L/\ell = (f/D) \, V^2/2g$, and then size the square duct to give the same value. For the given pressure and temperature we obtain (from Table B.3) $\nu = 1.89 \times 10^{-4}$ ft²/s so that

$$\text{Re} = \frac{VD}{\nu} = \frac{(10 \text{ ft/s})(\frac{8}{12} \text{ ft})}{1.89 \times 10^{-4} \text{ ft}^2/\text{s}} = 35{,}300$$

With this Reynolds number and with $\varepsilon/D = 0$ we obtain the friction factor from Fig. 8.20 as $f = 0.022$ so that

$$\frac{h_L}{\ell} = \frac{0.022}{(\frac{8}{12} \text{ ft})} \frac{(10 \text{ ft/s})^2}{2(32.2 \text{ ft/s}^2)} = 0.0512$$

Thus, for the square duct we must have

$$\frac{h_L}{\ell} = \frac{f}{D_h} \frac{V_s^2}{2g} = 0.0512 \tag{1}$$

where

$$D_h = 4A/P = 4a^2/4a = a \quad \text{and}$$

$$V_s = \frac{Q}{A} = \frac{\frac{\pi}{4}\left(\frac{8}{12} \text{ ft}\right)^2 (10 \text{ ft/s})}{a^2} = \frac{3.49}{a^2} \tag{2}$$

is the velocity in the duct.

By combining Eqs. 1 and 2 we obtain

$$0.0512 = \frac{f}{a} \frac{(3.49/a^2)^2}{2(32.2)}$$

or

$$a = 1.30 f^{1/5} \tag{3}$$

where $a$ is in feet. Similarly, the Reynolds number based on the hydraulic diameter is

$$\text{Re}_h = \frac{V_s D_h}{\nu} = \frac{(3.49/a^2)a}{1.89 \times 10^{-4}} = \frac{1.85 \times 10^4}{a} \tag{4}$$

We have three unknowns ($a$, $f$, and $\text{Re}_h$) and three equations—Eqs. 3, 4, and either in graphical form the Moody chart (Fig. 8.20) or the Colebrook equation (Eq. 8.35a).

If we use the Moody chart, we can use a trial-and-error solution as follows. As an initial attempt, assume the friction factor for the duct is the same as for the pipe. That is, assume $f = 0.022$. From Eq. 3 we obtain $a = 0.606$ ft, while from Eq. 4 we have $\text{Re}_h = 3.05 \times 10^4$. From Fig. 8.20, with this Reynolds number and the given smooth duct we obtain $f = 0.023$, which does not quite agree with the assumed value of $f$. Hence, we do not have the solution. We try again, using the latest calculated value of $f = 0.023$ as our guess. The calculations are repeated until the guessed value of $f$ agrees with the value obtained from Fig. 8.20. The final result (after only two iterations) is $f = 0.023$, $\text{Re}_h = 3.03 \times 10^4$, and

$$a = 0.611 \text{ ft} = 7.34 \text{ in.} \tag{Ans}$$

**COMMENTS** Alternatively, we can use the Colebrook equation or the Haaland equation (rather than the Moody chart) to obtain the solution as follows. For a smooth pipe ($\varepsilon/D_h = 0$) the Colebrook equation, Eq. 8.35a, becomes

$$\frac{1}{\sqrt{f}} = -2.0 \log\left(\frac{\varepsilon/D_h}{3.7} + \frac{2.51}{\mathrm{Re}_h \sqrt{f}}\right)$$

$$= -2.0 \log\left(\frac{2.51}{\mathrm{Re}_h \sqrt{f}}\right) \qquad (5)$$

where from Eq. 3,

$$f = 0.269\, a^5 \qquad (6)$$

If we combine Eqs. 4, 5, and 6 and simplify, Eq. 7 is obtained for $a$.

$$1.928\, a^{-5/2} = -2 \log\left(2.62 \times 10^{-4}\, a^{-3/2}\right) \qquad (7)$$

By using a root-finding technique on a computer or calculator, the solution to Eq. 7 is determined to be $a = 0.614$ ft, in agreement (given the accuracy of reading the Moody chart) with that obtained by the trial-and-error method given earlier.

Note that the length of the side of the equivalent square duct is $a/D = 7.34/8 = 0.918$, or approximately 92% of the diameter of the equivalent duct. It can be shown that this value, 92%, is a very good approximation for any pipe flow—laminar or turbulent. The cross-sectional area of the duct ($A = a^2 = 53.9$ in.$^2$) is greater than that of the round pipe ($A = \pi D^2/4 = 50.3$ in.$^2$). Also, it takes less material to form the round pipe (perimeter $= \pi D = 25.1$ in.) than the square duct (perimeter $= 4a = 29.4$ in.). Circles are very efficient shapes.

## 8.5    Pipe Flow Examples

*Pipe systems may contain a single pipe with components or multiple interconnected pipes.*

In the previous sections of this chapter, we discussed concepts concerning flow in pipes and ducts. The purpose of this section is to apply these ideas to the solutions of various practical problems. The application of the pertinent equations is straightforward, with rather simple calculations that give answers to problems of engineering importance. The main idea involved is to apply the energy equation between appropriate locations within the flow system, with the head loss written in terms of the friction factor and the minor loss coefficients. We will consider two classes of pipe systems: those containing a single pipe (whose length may be interrupted by various components), and those containing multiple pipes in parallel, series, or network configurations.

# F l u i d s   i n   t h e   N e w s

**New hi-tech fountains** Ancient Egyptians used fountains in their palaces for decorative and cooling purposes. Current use of fountains continues but with a hi-tech flair. Although the basic fountain still consists of a typical *pipe system* (i.e., pump, pipe, regulating valve, nozzle, filter, and basin), recent use of computer-controlled devices has led to the design of innovative fountains with special effects. For example, by using several rows of multiple nozzles, it is possible to program and activate control valves to produce water jets that resemble symbols, letters, or the time of day. Other fountains use specially designed nozzles to produce

coherent, laminar streams of water that look like glass rods flying through the air. By using fast-acting control valves in a synchronized manner it is possible to produce mesmerizing three-dimensional patterns of water droplets. The possibilities are nearly limitless. With the initial artistic design of the fountain established, the initial engineering design (i.e., the capacity and pressure requirements of the nozzles and the size of the pipes and pumps) can be carried out. It is often necessary to modify the artistic and/or engineering aspects of the design in order to obtain a functional, pleasing fountain. (See Problem 8.60.)

### 8.5.1  Single Pipes

The nature of the solution process for pipe flow problems can depend strongly on which of the various parameters are independent parameters (the "given") and which is the dependent parameter (the "determine"). The three most common types of problems are shown in Table 8.4 in terms of the parameters involved. We assume the pipe system is defined in terms of the length of pipe sections used and the number of elbows, bends, and valves needed to convey the fluid between the desired locations. In all instances we assume the fluid properties are given.

In a Type I problem we specify the desired flowrate or average velocity and determine the necessary pressure difference or head loss. For example, if a flowrate of 2.0 gal/min is required for a dishwasher that is connected to the water heater by a given pipe system as shown by the figure in the margin, what pressure is needed in the water heater?

In a Type II problem we specify the applied driving pressure (or, alternatively, the head loss) and determine the flowrate. For example, how many gal/min of hot water are supplied to the dishwasher if the pressure within the water heater is 60 psi and the pipe system details (length, diameter, roughness of the pipe; number of elbows; etc.) are specified?

■ **Table 8.4**
**Pipe Flow Types**

| Variable | Type I | Type II | Type III |
|---|---|---|---|
| **a. Fluid** | | | |
| Density | Given | Given | Given |
| Viscosity | Given | Given | Given |
| **b. Pipe** | | | |
| Diameter | Given | Given | Determine |
| Length | Given | Given | Given |
| Roughness | Given | Given | Given |
| **c. Flow** | | | |
| Flowrate or Average Velocity | Given | Determine | Given |
| **d. Pressure** | | | |
| Pressure Drop or Head Loss | Determine | Given | Given |

In a Type III problem we specify the pressure drop and the flowrate and determine the diameter of the pipe needed. For example, what diameter of pipe is needed between the water heater and dishwasher if the pressure in the water heater is 60 psi (determined by the city water system) and the flowrate is to be not less than 2.0 gal/min (determined by the manufacturer)?

Several examples of these types of problems follow.

---

## EXAMPLE 8.8  Type I, Determine Pressure Drop

**GIVEN**  Water at 60 °F flows from the basement to the second floor through the 0.75-in. (0.0625-ft)-diameter copper pipe (a drawn tubing) at a rate of $Q = 12.0$ gal/min $= 0.0267$ ft³/s and exits through a faucet of diameter 0.50 in. as shown in Fig. E8.8a.

**FIND**  Determine the pressure at point (1) if

**(a)**  all losses are neglected,

**(b)**  the only losses included are major losses, or

**(c)**  all losses are included.

■ **Figure E8.8a**

## SOLUTION

Since the fluid velocity in the pipe is given by $V_1 = Q/A_1 = Q/(\pi D^2/4) = (0.0267$ ft³/s$)/[\pi(0.0625$ ft$)^2/4] = 8.70$ ft/s, and the fluid properties are $\rho = 1.94$ slugs/ft³ and $\mu = 2.34 \times 10^{-5}$ lb · s/ft² (see Table B.1), it follows that Re $= \rho V D/\mu = (1.94$ slugs/ft³$)(8.70$ ft/s$)(0.0625$ ft$)/(2.34 \times 10^{-5}$ lb · s/ft²$) = 45,000$. Thus, the flow is turbulent. The governing equation for either case (a), (b), or (c) is the energy equation given by Eq. 8.21,

$$\frac{p_1}{\gamma} + \alpha_1 \frac{V_1^2}{2g} + z_1 = \frac{p_2}{\gamma} + \alpha_2 \frac{V_2^2}{2g} + z_2 + h_L$$

where $z_1 = 0$, $z_2 = 20$ ft, $p_2 = 0$ (free jet), $\gamma = \rho g = 62.4$ lb/ft³, and the outlet velocity is $V_2 = Q/A_2 = (0.0267$ ft³/s$)/[\pi(0.50/12)^2$ft²/4$] = 19.6$ ft/s. We assume that the kinetic energy coefficients $\alpha_1$ and $\alpha_2$ are unity. This is reasonable because turbulent velocity profiles are nearly uniform across the pipe. Thus,

$$p_1 = \gamma z_2 + \tfrac{1}{2}\rho(V_2^2 - V_1^2) + \gamma h_L \tag{1}$$

where the head loss is different for each of the three cases.

**(a)**  If all losses are neglected ($h_L = 0$), Eq. 1 gives

$$p_1 = (62.4 \text{ lb/ft}^3)(20 \text{ ft})$$
$$+ \frac{1.94 \text{ slugs/ft}^3}{2}\left[\left(19.6 \frac{\text{ft}}{\text{s}}\right)^2 - \left(8.70 \frac{\text{ft}}{\text{s}}\right)^2\right]$$
$$= (1248 + 299) \text{ lb/ft}^2 = 1547 \text{ lb/ft}^2$$

or

$$p_1 = 10.7 \text{ psi} \tag{Ans}$$

**COMMENT** Note that for this pressure drop, the amount due to elevation change (the hydrostatic effect) is $\gamma(z_2 - z_1) = 8.67$ psi and the amount due to the increase in kinetic energy is $\rho(V_2^2 - V_1^2)/2 = 2.07$ psi.

**(b)** If the only losses included are the major losses, the head loss is

$$h_L = f\frac{\ell}{D}\frac{V_1^2}{2g}$$

From Table 8.1 the roughness for a 0.75-in.-diameter copper pipe (drawn tubing) is $\varepsilon = 0.000005$ ft so that $\varepsilon/D = 8 \times 10^{-5}$. With this $\varepsilon/D$ and the calculated Reynolds number (Re = 45,000), the value of $f$ is obtained from the Moody chart as $f = 0.0215$. Note that the Colebrook equation (Eq. 8.35a) would give the same value of $f$. Hence, with the total length of the pipe as $\ell = (15 + 5 + 10 + 10 + 20)$ ft $= 60$ ft and the elevation and kinetic energy portions the same as for part (a), Eq. 1 gives

$$p_1 = \gamma z_2 + \frac{1}{2}\rho(V_2^2 - V_1^2) + \rho f\frac{\ell}{D}\frac{V_1^2}{2}$$

$$= (1248 + 299)\ \text{lb/ft}^2$$

$$+ (1.94\ \text{slugs/ft}^3)(0.0215)\left(\frac{60\ \text{ft}}{0.0625\ \text{ft}}\right)\frac{(8.70\ \text{ft/s})^2}{2}$$

$$= (1248 + 299 + 1515)\ \text{lb/ft}^2 = 3062\ \text{lb/ft}^2$$

or

$$p_1 = 21.3\ \text{psi} \qquad\qquad \textbf{(Ans)}$$

**COMMENT** Of this pressure drop, the amount due to pipe friction is approximately $(21.3 - 10.7)$ psi $= 10.6$ psi.

**(c)** If major and minor losses are included, Eq. 1 becomes

$$p_1 = \gamma z_2 + \frac{1}{2}\rho(V_2^2 - V_1^2) + f\gamma\frac{\ell}{D}\frac{V_1^2}{2g} + \sum \rho K_L\frac{V^2}{2}$$

or

$$p_1 = 21.3\ \text{psi} + \sum \rho K_L\frac{V^2}{2} \qquad\qquad \textbf{(2)}$$

where the 21.3 psi contribution is due to elevation change, kinetic energy change, and major losses [part (b)], and the last term represents the sum of all of the minor losses. The loss coefficients of the components ($K_L = 1.5$ for each elbow and $K_L = 10$ for the wide-open globe valve) are given in Table 8.2 (except for the loss coefficient of the faucet, which is given in Fig. E8.8a as $K_L = 2$). Thus,

$$\sum \rho K_L\frac{V^2}{2} = (1.94\ \text{slugs/ft}^3)\frac{(8.70\ \text{ft})^2}{2}\left[10 + 4(1.5) + 2\right]$$

$$= 1321\ \text{lb/ft}^2$$

or

$$\sum \rho K_L\frac{V^2}{2} = 9.17\ \text{psi} \qquad\qquad \textbf{(3)}$$

Note that we did not include an entrance or exit loss because points (1) and (2) are located within the fluid streams, not within an at-

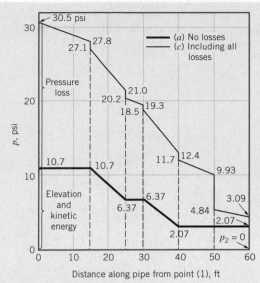

**Figure E8.8b**

taching reservoir where the kinetic energy is zero. Thus, by combining Eqs. 2 and 3 we obtain the entire pressure drop as

$$p_1 = (21.3 + 9.17)\ \text{psi} = 30.5\ \text{psi} \qquad\qquad \textbf{(Ans)}$$

This pressure drop calculated by including all losses should be the most realistic answer of the three cases considered.

**COMMENTS** More detailed calculations will show that the pressure distribution along the pipe is as illustrated in Fig. E8.8b for cases (a) and (c)—neglecting all losses or including all losses. Note that not all of the pressure drop, $p_1 - p_2$, is a "pressure loss." The pressure change due to the elevation and velocity changes is completely reversible. The portion due to the major and minor losses is irreversible.

This flow can be illustrated in terms of the energy line and hydraulic grade line concepts introduced in Section 3.7. As is shown in Fig. E8.8c, for case (a) there are no losses and the energy line (EL) is horizontal, one velocity head ($V^2/2g$) above the hydraulic grade line (HGL), which is one pressure head ($\gamma z$) above the pipe itself. For cases (b) or (c) the energy line is not horizontal. Each bit of friction in the pipe or loss in a component reduces the available

**Figure E8.8c**

energy, thereby lowering the energy line. Thus, for case (a) the total head remains constant throughout the flow with a value of

$$H = \frac{p_1}{\gamma} + \frac{V_1^2}{2g} + z_1 = \frac{(1547 \text{ lb/ft}^2)}{(62.4 \text{ lb/ft}^3)} + \frac{(8.70 \text{ ft/s})^2}{2(32.2 \text{ ft/s}^2)} + 0$$

$$= 26.0 \text{ ft.}$$

$$= \frac{p_2}{\gamma} + \frac{V_2^2}{2g} + z_2 = \frac{p_3}{\gamma} + \frac{V_3^3}{2g} + z_3 = \cdots$$

For case (c) the energy line starts at

$$H_1 = \frac{p_1}{\gamma} + \frac{V_1^2}{2g} + z_1$$

$$= \frac{(30.5 \times 144)\text{lb/ft}^2}{(62.4 \text{ lb/ft}^3)} + \frac{(8.70 \text{ ft/s})^2}{2(32.2 \text{ ft/s}^2)} + 0 = 71.6 \text{ ft}$$

and falls to a final value of

$$H_2 = \frac{p_2}{\gamma} + \frac{V_2^2}{2g} + z_2 = 0 + \frac{(19.6 \text{ ft/s})^2}{2(32.2 \text{ ft/s}^2)} + 20 \text{ ft}$$

$$= 26.0 \text{ ft}$$

The elevation of the energy line can be calculated at any point along the pipe. For example, at point (7), 50 ft from point (1),

$$H_7 = \frac{p_7}{\gamma} + \frac{V_7^2}{2g} + z_7$$

$$= \frac{(9.93 \times 144) \text{ lb/ft}^2}{(62.4 \text{ lb/ft}^3)} + \frac{(8.70 \text{ ft/s})^2}{2(32.2 \text{ ft/s}^2)} + 20 \text{ ft}$$

$$= 44.1 \text{ ft}$$

The head loss per foot of pipe is the same all along the pipe. That is,

$$\frac{h_L}{\ell} = f \frac{V^2}{2gD} = \frac{0.0215(8.70 \text{ ft/s})^2}{2(32.2 \text{ ft/s}^2)(0.0625 \text{ ft})} = 0.404 \text{ ft/ft}$$

Thus, the energy line is a set of straight-line segments of the same slope separated by steps whose height equals the head loss of the minor component at that location. As is seen from Fig. E8.8c, the globe valve produces the largest of all the minor losses.

Although the governing pipe flow equations are quite simple, they can provide very reasonable results for a variety of applications, as is shown in the next example.

---

## *E*XAMPLE 8.9    Type I, Determine Head Loss

**GIVEN** A utility worker is fixing power lines in a manhole space at a busy intersection as shown in Fig. E8.9a. A major portion of car exhaust is carbon dioxide ($CO_2$), which has a density of $3.55 \times 10^{-3}$ slugs/ft$^3$. Since the density of air at standard conditions is $2.38 \times 10^{-3}$ slugs/ft$^3$, the carbon dioxide will tend to settle into the bottom of the manhole space and displace the oxygen. This is a very relevant danger to utility workers and is further discussed in the related "Fluids in the News" article on page 440. To avoid suffocation, a fan is used to supply fresh air from the surface into the manhole space as shown in Fig. E8.9b. The air is routed through an 8-in.-diameter plastic hose that is 30 ft in length.

**FIND** Determine the horsepower needed by the fan to overcome the head loss with a required flowrate of 600 cfm, when **(a)** neglecting minor losses and **(b)** including minor losses for one sharp-edged entrance, one exit, and one 90° miter bend (no guide vanes).

■ **Figure E8.9a**
(© TIM MCCAIG/iStockphoto.)

■ **Figure E8.9b**

# SOLUTION

The purpose of the fan is to deliver air at a certain flowrate, which also requires the fan to overcome the net head loss within the pipe. This can be found from the energy equation between points (1) and (2) as

$$\frac{p_1}{\gamma} + \frac{V_1^2}{2g} + z_1 + h_p = \frac{p_2}{\gamma} + \frac{V_2^2}{2g} + z_2 + h_L$$

where $h_L$ is the total head loss from (1) to (2) and $h_p$ is the head rise supplied by the fan to the air. Points (1) and (2) are assumed to be sufficiently away from the entrance and exit of the plumbing so that $p_1 = p_2 = V_1 = V_2 = 0$. Also, since the fluid is air, the change in elevation is negligible.

Thus, the energy equation reduces to $h_p = h_L$.

**(a)**   Neglecting minor losses it follows that

$$h_P = f\frac{\ell}{D}\frac{V^2}{2g} \qquad (1)$$

where $V$ is the velocity within the hose. Therefore, we must use the Moody chart to find the friction factor, $f$. Since

$$V = \frac{Q}{A} = \frac{600 \text{ ft}^3/\text{min}}{\frac{\pi}{4}(8/12 \text{ ft})^2} = 1719 \text{ ft/min} = 28.6 \text{ ft/s}$$

it follows that

$$\text{Re} = \frac{\rho V D}{\mu} = \frac{(2.38 \times 10^{-3} \text{ slugs/ft}^3)(28.6 \text{ ft/s})(\frac{8}{12} \text{ ft})}{3.74 \times 10^{-7} \text{ lb} \cdot \text{s/ft}^2}$$

$$= 1.21 \times 10^5$$

From Table 8.1 the equivalent roughness, $\varepsilon$, for plastic is 0.0. Therefore, $\varepsilon/D = 0$. Using this information with the Moody chart, we find the friction factor, $f$, to be 0.017. We now have enough information to solve Eq. 1.

$$h_P = (0.017)\left[\frac{30 \text{ ft}}{(8/12) \text{ ft}}\right]\left[\frac{(28.6 \text{ ft/s})^2}{2(32.2 \text{ ft/s}^2)}\right] = 9.72 \text{ ft}$$

The actual power the fan must supply to the air (horsepower, $\mathcal{P}_a$) can be calculated by

$$\mathcal{P}_a = \gamma Q h_P \qquad (2)$$

$$\mathcal{P}_a = (7.65 \times 10^{-2} \text{ lb/ft}^3)(600 \text{ ft}^3/\text{min})\left(\frac{1 \text{ min}}{60 \text{ s}}\right)(9.72 \text{ ft})$$

$$= 7.44 \frac{\text{ft} \cdot \text{lb}}{\text{s}}$$

$$\mathcal{P}_a = 7.44\frac{\text{ft} \cdot \text{lb}}{\text{s}}\left(\frac{1 \text{ hp}}{550 \text{ ft} \cdot \text{lb/s}}\right) = 1.35 \times 10^{-2} \text{ hp} \qquad \textbf{(Ans)}$$

**COMMENTS**   Typically, flexible hosing is used to route air into confined spaces. Minor losses were neglected in part **(a)**, but

we need to be cautious in making this assumption with a relatively short overall pipe length.

**(b)**   For the additional minor losses, we need to include the entrance, exit, and bend. The loss coefficient for an entrance with sharp edges in 0.5, the loss coefficient for a 90° bend with no guide vanes is 1.1, and the loss coefficient for the exit is 1.0. Therefore, the total minor loss is given as

$$h_{L \text{ minor}} = \Sigma K_L \frac{V^2}{2g} = (0.5 + 1.1 + 1.0)\frac{(28.6 \text{ ft/s})^2}{2(32.2 \text{ ft/s}^2)}$$

$$h_{L \text{ minor}} = 33.0 \text{ ft}$$

The power required to account for both the major and the minor losses is

$$\mathcal{P}_a = 1.35 \times 10^{-2} \text{ hp} + (7.65 \times 10^{-2} \text{ lb/ft}^3)(600 \text{ ft}^3/\text{min})$$

$$\times \left(\frac{1 \text{ min}}{60 \text{ s}}\right)(33 \text{ ft})\left(\frac{1 \text{ hp}}{550 \text{ ft} \cdot \text{lb/s}}\right)$$

$$= 5.94 \times 10^{-2} \text{ hp} \qquad \textbf{(Ans)}$$

**COMMENTS**   Note that with the inclusion of minor losses, the horsepower required is more than double what is needed for a straight pipe. It is easy to see that in this situation neglecting minor losses would provide an underestimation of the power required to pump 600 cfm of air. The table that follows contains different power requirements for a few different pipe situations using the preceding flow properties.

| Pipe Setup | $h_{L \text{ minor}}$ (ft) | $h_{L \text{ major}}$ (ft) | $h_P$ (ft) | $\mathcal{P}_a$ (hp) |
|---|---|---|---|---|
| Straight pipe | 0 | 9.72 | 9.72 | 0.014 |
| Entrance and exit | 19.1 | 9.72 | 28.8 | 0.040 |
| Entrance, exit, 1 miter bend | 33.0 | 9.72 | 42.7 | 0.059 |
| Entrance, exit, 2 miter bends | 47.0 | 9.72 | 56.7 | 0.079 |
| Entrance, exit, 4 miter bends | 74.9 | 9.72 | 84.7 | 0.118 |

Caution also needs to be taken when placing the fan outside the confined space so that clean air is supplied and hazardous gases emitted from the manhole are not circulated back into the confined space. There is also the question of whether to have air forced into the confined space or have the contaminated air exhausted from the space and naturally draw in fresh air.

---

*Some pipe flow problems require a trial-and-error solution technique.*

Pipe flow problems in which it is desired to determine the flowrate for a given set of conditions (Type II problems) often require trial-and-error or numerical root-finding techniques. This is because it is necessary to know the value of the friction factor to carry out the calculations, but the friction factor is a function of the unknown velocity (flowrate) in terms of the Reynolds number. The solution procedure is indicated in Example 8.10.

# $E$XAMPLE 8.10  Type II, Determine Flowrate

**GIVEN** The fan shown in Fig. E8.10a is to provide airflow through the spray booth and fume hood so that workers are protected from harmful vapors and aerosols while mixing chemicals within the hood. For proper operation of the hood, the flowrate is to be between 6 ft³/s and 12 ft³/s. With the initial setup the flowrate is 9 ft³/s, the loss coefficient for the system is 5, and the duct is short enough so that major losses are negligible. It is proposed that when the factory is remodeled the 8-in.-diameter galvanized iron duct will be 100 ft long and the total loss coefficient will be 10.

**FIND** Determine if the flowrate will be within the required 6 ft³/s to 12 ft³/s after the proposed remodeling. Assume that the head added to the air by the fan remains constant.

■ **Figure E8.10a**

## Solution

We can determine the head that the fan adds to the air by considering the initial situation (i.e., before remodeling). To do this we write the energy equation between section (1) in the room and section (2) at the exit of the duct as shown in Fig. E8.10a.

$$\frac{p_1}{\gamma} + \frac{V_1^2}{2g} + z_1 + h_p = \frac{p_2}{\gamma} + \frac{V_2^2}{2g} + z_2 + h_L \qquad (1)$$

Since we are dealing with air, we can assume any change in elevation is negligible. We can also assume the pressure inside the room and at the exit of the duct is equal to atmospheric pressure and the air in the room has zero velocity. Therefore, Eq. 1 reduces to

$$h_p = \frac{V_2^2}{2g} + h_L \qquad (2)$$

The diameter of the duct is given as 8 in., so the velocity at the exit can be calculated from the flowrate, where $V = Q/A = (9 \text{ ft}^3/\text{s})/[\pi(8/12)^2/4] = 25.8$ ft/s. For the original configuration the duct is short enough to neglect major losses and only minor losses contribute to the total head loss. This head loss can be found from $h_{L,\text{minor}} = \Sigma K_L V^2/(2g) = 5(25.8 \text{ ft/s})^2/[2(32.2 \text{ ft/s}^2)] = 51.6$ ft. With this information the simplified energy equation, Eq. 2, can now be solved to find the head added to the air by the fan.

$$h_p = \frac{(25.8 \text{ ft/s})^2}{2(32.2 \text{ ft/s}^2)} + 51.6 \text{ ft} = 61.9 \text{ ft}$$

The energy equation now must be solved with the new configuration after remodeling. Using the same assumptions as before gives the same reduced energy equation as shown in Eq. 2. With the increase in duct length to 100 ft the duct is no longer short enough to neglect major losses. Thus,

$$h_p = \frac{V_2^2}{2g} + f\frac{\ell}{D}\frac{V^2}{2g} + \Sigma K_L \frac{V^2}{2g}$$

where $V_2 = V$ and $\Sigma K_L = 10$. We can now rearrange and solve for the velocity in ft/s.

$$V = \sqrt{\frac{2gh_p}{1 + f\frac{\ell}{D} + \Sigma K_L}} = \sqrt{\frac{2(32.2 \text{ ft/s}^2)(61.9 \text{ ft})}{1 + f\left(\frac{100 \text{ ft}}{8/12 \text{ ft}}\right) + 10}}$$

$$= \sqrt{\frac{3990}{11 + 150f}} \qquad (3)$$

The value of $f$ is dependent on Re, which is dependent on $V$, an unknown.

$$\text{Re} = \frac{\rho V D}{\mu} = \frac{(2.38 \times 10^{-3} \text{ slugs/ft}^3)(V)(\frac{8}{12} \text{ ft})}{3.74 \times 10^{-7} \text{ lb}\cdot\text{s/ft}^2}$$

or

$$\text{Re} = 4240V \qquad (4)$$

where again $V$ is in feet per second.

Also, since $\varepsilon/D = (0.0005 \text{ ft})/(8/12 \text{ ft}) = 0.00075$ (see Table 8.1 for the value of $\varepsilon$), we know which particular curve of the Moody chart is pertinent to this flow. Thus, we have three relationships (Eqs. 3, 4, and $\varepsilon/D = 0.00075$ curve of Fig. 8.20) from which we can solve for the three unknowns, $f$, Re, and $V$. This is done easily by an iterative scheme as follows.

It is usually simplest to assume a value of $f$, calculate $V$ from Eq. 3, calculate Re from Eq. 4, and look up the new value of $f$ in the Moody chart for this value of Re. If the assumed $f$ and the new $f$ do not agree, the assumed answer is not correct—we do not have the solution to the three equations. Although values of $f$, $V$, or Re could be assumed as starting values, it is usually simplest to assume a value of $f$ because the correct value often lies on the relatively flat portion of the Moody chart for which $f$ is quite insensitive to Re.

Thus, we assume $f = 0.019$, approximately the large Re limit for the given relative roughness. From Eq. 3 we obtain

$$V = \sqrt{\frac{3990}{11 + 150(0.019)}} = 17.0 \text{ ft/s}$$

and from Eq. 4

$$Re = 4240(17.0) = 72,100$$

With this Re and $\varepsilon/D$, Fig. 8.20 gives $f = 0.022$, which is not equal to the assumed solution of $f = 0.019$ (although it is close!). We try again, this time with the newly obtained value of $f = 0.022$, which gives $V = 16.7$ ft/s and $Re = 70,800$. With these values, Fig. 8.20 gives $f = 0.022$, which agrees with the assumed value. Thus, the solution is $V = 16.7$ ft/s, or

$$Q = VA = (16.7 \text{ ft/s})\left(\frac{\pi}{4}\right)\left(\frac{8}{12} \text{ ft}\right)^2 = 5.83 \text{ ft}^3/\text{s} \qquad \textbf{(Ans)}$$

**COMMENT**    It is seen that operation of the system after the proposed remodeling will not provide enough airflow to protect workers from inhalation hazards while mixing chemicals within the hood. By repeating the calculations for different duct lengths and different total minor loss coefficients, the results shown in Fig. E8.10b are obtained, which give flowrate as a function of

■ **Figure E8.10b**

duct length for various values of the minor loss coefficient. It will be necessary to redesign the remodeled system (e.g., larger fan, shorter ducting, larger-diameter duct) so that the flowrate will be within the acceptable range. In many companies, teams of occupational safety and health experts and engineers work together during the design phase of remodeling (or when a new operation is being planned) to consider and prevent potential negative impacts on workers' safety and health in an effort called "Prevention through Design." They also may be required to ensure that exhaust from such a system exits the building away from human activity and into an area where it will not be drawn back inside the facility.

---

# EXAMPLE 8.11    Type II, Determine Flowrate

**GIVEN**    The turbine shown in Fig. E8.11 extracts 50 hp from the water flowing through it. The 1-ft-diameter, 300-ft-long pipe is assumed to have a friction factor of 0.02. Minor losses are negligible.

**FIND**    Determine the flowrate through the pipe and turbine.

■ **Figure E8.11**

## SOLUTION

The energy equation (Eq. 8.21) can be applied between the surface of the lake [point (1)] and the outlet of the pipe as

$$\frac{p_1}{\gamma} + \frac{V_1^2}{2g} + z_1 = \frac{p_2}{\gamma} + \frac{V_2^2}{2g} + z_2 + h_L + h_T \qquad \textbf{(1)}$$

where $h_T$ is the turbine head, $p_1 = V_1 = p_2 = z_2 = 0$, $z_1 = 90$ ft, and $V_2 = V$, the fluid velocity in the pipe. The head loss is given by

$$h_L = f\frac{\ell}{D}\frac{V^2}{2g} = 0.02\frac{(300 \text{ ft})}{(1 \text{ ft})}\frac{V^2}{2(32.2 \text{ ft/s}^2)} = 0.0932V^2 \text{ ft}$$

where $V$ is in ft/s. Also, the turbine head is

$$h_T = \frac{\mathscr{P}_a}{\gamma Q} = \frac{\mathscr{P}_a}{\gamma(\pi/4)D^2V}$$

$$= \frac{(50 \text{ hp})[(550 \text{ ft} \cdot \text{lb/s})/\text{hp}]}{(62.4 \text{ lb/ft}^3)[(\pi/4)(1 \text{ ft})^2V]} = \frac{561}{V} \text{ ft}$$

Thus, Eq. 1 can be written as

$$90 = \frac{V^2}{2(32.2)} + 0.0932V^2 + \frac{561}{V}$$

or

$$0.109V^3 - 90V + 561 = 0 \quad (2)$$

where $V$ is in ft/s. The velocity of the water in the pipe is found as the solution of Eq. 2. Surprisingly, there are two real, positive roots: $V = 6.58$ ft/s or $V = 24.9$ ft/s. The third root is negative ($V = -31.4$ ft/s) and has no physical meaning for this flow. Thus, the two acceptable flowrates are

$$Q = \frac{\pi}{4} D^2 V = \frac{\pi}{4} (1 \text{ ft})^2 (6.58 \text{ ft/s}) = 5.17 \text{ ft}^3/\text{s} \quad \textbf{(Ans)}$$

or

$$Q = \frac{\pi}{4} (1 \text{ ft})^2 (24.9 \text{ ft/s}) = 19.6 \text{ ft}^3/\text{s} \quad \textbf{(Ans)}$$

**COMMENTS**  Either of these two flowrates gives the same power, $\mathscr{P}_a = \gamma Q h_T$. The reason for two possible solutions can be seen from the following. With the low flowrate ($Q = 5.17$ ft³/s), we obtain the head loss and turbine head as $h_L = 4.04$ ft and $h_T = 85.3$ ft. Because of the relatively low velocity there is a relatively small head loss and, therefore, a large head available for the turbine. With the large flowrate ($Q = 19.6$ ft³/s), we find $h_L = 57.8$ ft and $h_T = 22.5$ ft. The high-speed flow in the pipe produces a relatively large loss due to friction, leaving a relatively small head for the turbine. However, in either case the product of the turbine head times the flowrate is the same. That is, the power extracted ($\mathscr{P}_a = \gamma Q h_T$) is identical for each case. Although either flowrate will allow the extraction of 50 hp from the water, the details of the design of the turbine itself will depend strongly on which flowrate is to be used. Such information can be found in Chapter 12 and various references about turbomachines (Refs. 14, 19, 20).

If the friction factor were not given, the solution to the problem would be lengthier. A trial-and-error solution similar to that in Example 8.10 would be required, along with the solution of a cubic equation.

In pipe flow problems for which the diameter is the unknown (Type III), an iterative or numerical root-finding technique is required. This is, again, because the friction factor is a function of the diameter—through both the Reynolds number and the relative roughness. Thus, neither $\text{Re} = \rho V D/\mu = 4\rho Q/\pi\mu D$ nor $\varepsilon/D$ are known unless $D$ is known. Examples 8.12 and 8.13 illustrate this.

# $E$XAMPLE 8.12   Type III without Minor Losses, Determine Diameter

**GIVEN**  Air at standard temperature and pressure flows through a horizontal, galvanized iron pipe ($\varepsilon = 0.0005$ ft) at a rate of 2.0 ft³/s. The pressure drop is to be no more than 0.50 psi per 100 ft of pipe.

**FIND**  Determine the minimum pipe diameter.

## SOLUTION

We assume the flow to be incompressible with $\rho = 0.00238$ slugs/ft³ and $\mu = 3.74 \times 10^{-7}$ lb · s/ft². Note that if the pipe were too long, the pressure drop from one end to the other, $p_1 - p_2$, would not be small relative to the pressure at the beginning, and compressible flow considerations would be required. For example, a pipe length of 200 ft gives $(p_1 - p_2)/p_1 = [(0.50 \text{ psi})/(100 \text{ ft})](200 \text{ ft})/14.7 \text{ psia} = 0.068 = 6.8\%$, which is probably small enough to justify the incompressible assumption.

With $z_1 = z_2$ and $V_1 = V_2$ the energy equation (Eq. 8.21) becomes

$$p_1 = p_2 + f \frac{\ell}{D} \frac{\rho V^2}{2} \quad (1)$$

where $V = Q/A = 4Q/(\pi D^2) = 4(2.0 \text{ ft}^3/\text{s})/\pi D^2$, or

$$V = \frac{2.55}{D^2}$$

where $D$ is in feet. Thus, with $p_1 - p_2 = (0.5 \text{ lb/in.}^2)(144 \text{ in.}^2/\text{ft}^2)$ and $\ell = 100$ ft, Eq. 1 becomes

$$p_1 - p_2 = (0.5)(144) \text{ lb/ft}^2$$
$$= f \frac{(100 \text{ ft})}{D} (0.00238 \text{ slugs/ft}^3) \frac{1}{2} \left( \frac{2.55}{D^2} \frac{\text{ft}}{\text{s}} \right)^2$$

or

$$D = 0.404 f^{1/5} \quad (2)$$

where $D$ is in feet. Also $\text{Re} = \rho V D/\mu = (0.00238 \text{ slugs/ft}^3)[(2.55/D^2) \text{ ft/s}]D/(3.74 \times 10^{-7} \text{ lb · s/ft}^2)$, or

$$\text{Re} = \frac{1.62 \times 10^4}{D} \quad (3)$$

and

$$\frac{\varepsilon}{D} = \frac{0.0005}{D} \qquad (4)$$

Thus, we have four equations [Eqs. 2, 3, 4, and either the Moody chart, the Colebrook equation (8.35a) or the Haaland equation (8.35b)] and four unknowns ($f$, $D$, $\varepsilon/D$, and Re) from which the solution can be obtained by trial-and-error methods.

If we use the Moody chart, it is probably easiest to assume a value of $f$, use Eqs. 2, 3, and 4 to calculate $D$, Re, and $\varepsilon/D$, and then compare the assumed $f$ with that from the Moody chart. If they do not agree, try again. Thus, we assume $f = 0.02$, a typical value, and obtain $D = 0.404(0.02)^{1/5} = 0.185$ ft, which gives $\varepsilon/D = 0.0005/0.185 = 0.0027$ and Re $= 1.62 \times 10^4/0.185 = 8.76 \times 10^4$. From the Moody chart we obtain $f = 0.027$ for these values of $\varepsilon/D$ and Re. Since this is not the same as our assumed value of $f$, we try again. With $f = 0.027$, we obtain $D = 0.196$ ft, $\varepsilon/D = 0.0026$, and Re $= 8.27 \times 10^4$, which in turn give $f = 0.027$, in agreement with the assumed value. Thus, the diameter of the pipe should be

$$D = 0.196 \text{ ft} \qquad \textbf{(Ans)}$$

**COMMENT** If we use the Colebrook equation (Eq. 8.35a) with $\varepsilon/D = 0.0005/0.404\,f^{1/5} = 0.00124/f^{1/5}$ and Re $= 1.62 \times 10^4/0.404\,f^{1/5} = 4.01 \times 10^4/f^{1/5}$, we obtain

$$\frac{1}{\sqrt{f}} = -2.0 \log\left(\frac{\varepsilon/D}{3.7} + \frac{2.51}{\text{Re}\sqrt{f}}\right)$$

or

$$\frac{1}{\sqrt{f}} = -2.0 \log\left(\frac{3.35 \times 10^{-4}}{f^{1/5}} + \frac{6.26 \times 10^{-5}}{f^{3/10}}\right)$$

■ **Figure E8.12**

By using a root-finding technique on a computer or calculator, the solution to this equation is determined to be $f = 0.027$, and hence $D = 0.196$ ft, in agreement with the Moody chart method.

By repeating the calculations for various values of the flowrate, $Q$, the results shown in Fig. E8.12 are obtained. Although an increase in flowrate requires a larger diameter pipe (for the given pressure drop), the increase in diameter is minimal. For example, if the flowrate is doubled from 1 ft³/s to 2 ft³/s, the diameter increases from 0.151 ft to 0.196 ft.

In the previous example we only had to consider major losses. In some instances the inclusion of major and minor losses can cause a slightly lengthier solution procedure, even though the governing equations are essentially the same. This is illustrated in Example 8.13.

## EXAMPLE 8.13    Type III with Minor Losses, Determine Diameter

**GIVEN** Water at 60 °F ($\nu = 1.21 \times 10^{-5}$ ft²/s, see Table 1.5) is to flow from reservoir $A$ to reservoir $B$ through a pipe of length 1700 ft and roughness 0.0005 ft at a rate of $Q = 26$ ft³/s as shown in Fig. E8.13a. The system contains a sharp-edged entrance and four flanged 45° elbows.

**FIND** Determine the pipe diameter needed.

■ **Figure E8.13a**

# SOLUTION

The energy equation (Eq. 8.21) can be applied between two points on the surfaces of the reservoirs ($p_1 = p_2 = V_1 = V_2 = z_2 = 0$) as follows:

$$\frac{p_1}{\gamma} + \frac{V_1^2}{2g} + z_1 = \frac{p_2}{\gamma} + \frac{V_2^2}{2g} + z_2 + h_L$$

or

$$z_1 = \frac{V^2}{2g}\left(f\frac{\ell}{D} + \sum K_L\right) \qquad (1)$$

where $V = Q/A = 4Q/\pi D^2 = 4(26\ \text{ft}^3/\text{s})/\pi D^2$, or

$$V = \frac{33.1}{D^2} \qquad (2)$$

is the velocity within the pipe. (Note that the units on $V$ and $D$ are ft/s and ft, respectively.) The loss coefficients are obtained from Table 8.2 and Figs. 8.22 and 8.25 as $K_{L_\text{ent}} = 0.5$, $K_{L_\text{elbow}} = 0.2$, and $K_{L_\text{exit}} = 1$. Thus, Eq. 1 can be written as

$$44\ \text{ft} = \frac{V^2}{2(32.2\ \text{ft/s}^2)}\left\{\frac{1700}{D}f + [4(0.2) + 0.5 + 1]\right\}$$

or, when combined with Eq. 2 to eliminate $V$,

$$f = 0.00152\,D^5 - 0.00135\,D \qquad (3)$$

To determine $D$ we must know $f$, which is a function of Re and $\varepsilon/D$, where

$$\text{Re} = \frac{VD}{\nu} = \frac{[(33.1)/D^2]D}{1.21 \times 10^{-5}} = \frac{2.74 \times 10^6}{D} \qquad (4)$$

and

$$\frac{\varepsilon}{D} = \frac{0.0005}{D} \qquad (5)$$

where $D$ is in feet. Again, we have four equations (Eqs. 3, 4, 5, and the Moody chart or the Colebrook equation) for the four unknowns $D$, $f$, Re, and $\varepsilon/D$.

Consider the solution by using the Moody chart. Although it is often easiest to assume a value of $f$ and make calculations to determine if the assumed value is the correct one, with the inclusion of minor losses this may not be the simplest method. For example, if we assume $f = 0.02$ and calculate $D$ from Eq. 3, we would have to solve a fifth-order equation. With only major losses (see Example 8.12), the term proportional to $D$ in Eq. 3 is absent, and it is easy to solve for $D$ if $f$ is given. With both major and minor losses included, this solution for $D$ (given $f$) would require a trial-and-error or iterative technique.

Thus, for this type of problem it is perhaps easier to assume a value of $D$, calculate the corresponding $f$ from Eq. 3, and with the values of Re and $\varepsilon/D$ determined from Eqs. 4 and 5, look up the value of $f$ in the Moody chart (or the Colebrook equation). The solution is obtained when the two values of $f$ are in agreement. A few rounds of calculation will reveal that the solution is given by

$$D \approx 1.63\ \text{ft} \qquad \textbf{(Ans)}$$

**COMMENTS** Alternatively, we can use the Colebrook equation rather than the Moody chart to solve for $D$. This is easily done by using the Colebrook equation (Eq. 8.35a) with $f$ as a function of $D$ obtained from Eq. 3 and Re and $\varepsilon/D$ as functions of $D$ from Eqs. 4 and 5. The resulting single equation for $D$ can be solved by using a root-finding technique on a computer or calculator to obtain $D = 1.63\ \text{ft}$. This agrees with the solution obtained using the Moody chart.

By repeating the calculations for various pipe lengths, $\ell$, the results shown in Fig. E8.13b are obtained. As the pipe length increases it is necessary, because of the increased friction, to increase the pipe diameter to maintain the same flowrate.

It is interesting to attempt to solve this example if all losses are neglected so that Eq. 1 becomes $z_1 = 0$. Clearly from Fig. E8.13a, $z_1 = 44\ \text{ft}$. Obviously something is wrong. A fluid cannot flow from one elevation, beginning with zero pressure and velocity, and end up at a lower elevation with zero pressure and velocity unless energy is removed (i.e., a head loss or a turbine) somewhere between the two locations. If the pipe is short (negligible friction) and the minor losses are negligible, there is still the kinetic energy of the fluid as it leaves the pipe and enters the reservoir. After the fluid meanders around in the reservoir for some time, this kinetic energy is lost and the fluid is stationary. No matter how small the viscosity is, the exit loss cannot be neglected. The same result can be seen if the energy equation is written from the free surface of the upstream tank to the exit plane of the pipe, at which point the kinetic energy is still available to the fluid. In either case the energy equation becomes $z_1 = V^2/2g$ in agreement with the inviscid results of Chapter 3 (the Bernoulli equation).

■ **Figure E8.13b**

Trachea

Lung

Bronchiole

## 8.5.2 Multiple Pipe Systems

In many pipe systems there is more than one pipe involved. The complex system of tubes in our lungs (beginning as shown by the figure in the margin, with the relatively large-diameter trachea and ending in tens of thousands of minute bronchioles after numerous branchings) and the maze of pipes in a city's water distribution system are typical of such systems. The governing mechanisms for the flow in *multiple pipe systems* are the same as for the single pipe systems discussed in this chapter. However, because of the numerous unknowns involved, additional complexities may arise in solving for the flow in multiple pipe systems. Some of these complexities are discussed in this section.

---

### F l u i d s   i n   t h e   N e w s

**Deepwater pipeline** Pipelines used to transport oil and gas are commonplace. But south of New Orleans, in deep waters of the Gulf of Mexico, a not-so-common *multiple pipe system* is being built. The new so-called Mardi Gras system of pipes is being laid in water depths of 4300 to 7300 ft. It will transport oil and gas from five deepwater fields with the interesting names of Holstein, Mad Dog, Thunder Horse, Atlantis, and Na Kika. The deepwater pipelines will connect with lines at intermediate water depths to transport the oil and gas to shallow-water fixed plat-forms and shore. The steel pipe used is 28 in. in diameter with a wall thickness of 1 1/8 in. The thick-walled pipe is needed to withstand the large external pressure which is about 3250 psi at a depth of 7300 ft. The pipe is installed in 240-ft sections from a vessel the size of a large football stadium. Upon completion, the deepwater pipeline system will have a total length of more than 450 miles and the capability of transporting more than 1 million barrels of oil per day and 1.5 billion cu ft of gas per day. (See Problem 8.115.)

---

The simplest multiple pipe systems can be classified into series or parallel flows, as are shown in Fig. 8.35. The nomenclature is similar to that used in electrical circuits. Indeed, an analogy between fluid and electrical circuits is often made as follows. In a simple electrical circuit, there is a balance between the voltage ($e$), current ($i$), and resistance ($R$) as given by Ohm's law: $e = iR$. In a fluid circuit there is a balance between the pressure drop ($\Delta p$), the flowrate or velocity ($Q$ or $V$), and the flow resistance as given in terms of the friction factor and minor loss coefficients ($f$ and $K_L$). For a simple flow $[\Delta p = f(\ell/D)(\rho V^2/2)]$, it follows that $\Delta p = Q^2\tilde{R}$, where $\tilde{R}$, a measure of the resistance to the flow, is proportional to $f$.

The main differences between the solution methods used to solve electrical circuit problems and those for fluid circuit problems lie in the fact that Ohm's law is a linear equation (doubling the voltage doubles the current), while the fluid equations are generally nonlinear (doubling the pressure drop does not double the flowrate unless the flow is laminar). Thus, although some of the

(a)

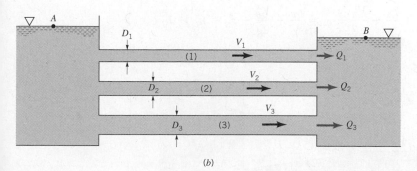

(b)

■ **Figure 8.35**  (a) Series and (b) parallel pipe systems.

standard electrical engineering methods can be carried over to help solve fluid mechanics problems, others cannot.

One of the simplest multiple pipe systems is that containing pipes in *series*, as is shown in Fig. 8.35a. Every fluid particle that passes through the system passes through each of the pipes. Thus, the flowrate (but not the velocity) is the same in each pipe, and the head loss from point $A$ to point $B$ is the sum of the head losses in each of the pipes. The governing equations can be written as follows:

$$Q_1 = Q_2 = Q_3$$

and

$$h_{L_{A-B}} = h_{L_1} + h_{L_2} + h_{L_3}$$

*Series and parallel pipe systems are often encountered.*

where the subscripts refer to each of the pipes. In general, the friction factors will be different for each pipe because the Reynolds numbers ($\mathrm{Re}_i = \rho V_i D_i/\mu$) and the relative roughnesses ($\varepsilon_i/D_i$) will be different. If the flowrate is given, it is a straightforward calculation to determine the head loss or pressure drop (Type I problem). If the pressure drop is given and the flowrate is to be calculated (Type II problem), an iteration scheme is needed. In this situation none of the friction factors, $f_i$, are known, so the calculations may involve more trial-and-error attempts than for corresponding single pipe systems. The same is true for problems in which the pipe diameter (or diameters) is to be determined (Type III problems).

Another common multiple pipe system contains pipes in *parallel*, as is shown in Fig. 8.35b. In this system a fluid particle traveling from $A$ to $B$ may take any of the paths available, with the total flowrate equal to the sum of the flowrates in each pipe. However, by writing the energy equation between points $A$ and $B$ it is found that the head loss experienced by any fluid particle traveling between these locations is the same, independent of the path taken. Thus, the governing equations for parallel pipes are

$$Q = Q_1 + Q_2 + Q_3$$

and

$$h_{L_1} = h_{L_2} = h_{L_3}$$

Again, the method of solution of these equations depends on what information is given and what is to be calculated.

Another type of multiple pipe system called a *loop* is shown in Fig. 8.36. In this case the flowrate through pipe (1) equals the sum of the flowrates through pipes (2) and (3), or $Q_1 = Q_2 + Q_3$. As can be seen by writing the energy equation between the surfaces of each reservoir, the head loss for pipe (2) must equal that for pipe (3), even though the pipe sizes and flowrates may be different for each. That is,

$$\frac{p_A}{\gamma} + \frac{V_A^2}{2g} + z_A = \frac{p_B}{\gamma} + \frac{V_B^2}{2g} + z_B + h_{L_1} + h_{L_2}$$

for a fluid particle traveling through pipes (1) and (2), while

$$\frac{p_A}{\gamma} + \frac{V_A^2}{2g} + z_A = \frac{p_B}{\gamma} + \frac{V_B^2}{2g} + z_B + h_{L_1} + h_{L_3}$$

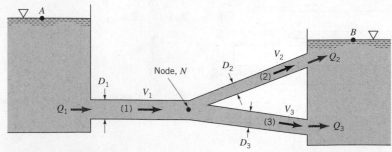

■ **Figure 8.36** Multiple-pipe loop system.

■ **Figure 8.37**  A three-reservoir system.

for fluid that travels through pipes (1) and (3). These can be combined to give $h_{L_2} = h_{L_3}$. This is a statement of the fact that fluid particles that travel through pipe (2) and particles that travel through pipe (3) all originate from common conditions at the junction (or node, $N$) of the pipes and all end up at the same final conditions.

The flow in a relatively simple looking multiple pipe system may be more complex than it appears initially. The branching system termed the *three-reservoir problem* shown in Fig. 8.37 is such a system. Three reservoirs at known elevations are connected together with three pipes of known properties (length, diameter, and roughness). The problem is to determine the flowrates into or out of the reservoirs. If valve (1) were closed, the fluid would flow from reservoir $B$ to $C$, and the flowrate could be easily calculated. Similar calculations could be carried out if valves (2) or (3) were closed with the others open.

*For some pipe systems, the direction of flow is not known a priori.*

With all valves open, however, it is not necessarily obvious which direction the fluid flows. For the conditions indicated in Fig. 8.37, it is clear that fluid flows from reservoir $A$ because the other two reservoir levels are lower. Whether the fluid flows into or out of reservoir $B$ depends on the elevation of reservoirs $B$ and $C$ and the properties (length, diameter, roughness) of the three pipes. In general, the flow direction is not obvious, and the solution process must include the determination of this direction. This is illustrated in Example 8.14.

---

# $E$XAMPLE 8.14    Three-Reservoir, Multiple Pipe System

**GIVEN**  Three reservoirs are connected by three pipes as are shown in Fig. E8.14. For simplicity we assume that the diameter of each pipe is 1 ft, the friction factor for each is 0.02, and because of the large length-to-diameter ratio, minor losses are negligible.

**FIND**  Determine the flowrate into or out of each reservoir.

■ **Figure E8.14**

## SOLUTION

It is not obvious which direction the fluid flows in pipe (2). However, we assume that it flows out of reservoir $B$, write the governing equations for this case, and check our assumption. The continuity equation requires that $Q_1 + Q_2 = Q_3$, which, since the diameters are the same for each pipe, becomes simply

$$V_1 + V_2 = V_3 \tag{1}$$

The energy equation for the fluid that flows from $A$ to $C$ in pipes (1) and (3) can be written as

$$\frac{p_A}{\gamma} + \frac{V_A^2}{2g} + z_A = \frac{p_C}{\gamma} + \frac{V_C^2}{2g} + z_C + f_1 \frac{\ell_1}{D_1} \frac{V_1^2}{2g} + f_3 \frac{\ell_3}{D_3} \frac{V_3^2}{2g}$$

By using the fact that $p_A = p_C = V_A = V_C = z_C = 0$, this becomes

$$z_A = f_1 \frac{\ell_1}{D_1} \frac{V_1^2}{2g} + f_3 \frac{\ell_3}{D_3} \frac{V_3^2}{2g}$$

For the given conditions of this problem we obtain

$$100 \text{ ft} = \frac{0.02}{2(32.2 \text{ ft/s}^2)} \frac{1}{(1 \text{ ft})} [(1000 \text{ ft})V_1^2 + (400 \text{ ft})V_3^2]$$

or

$$322 = V_1^2 + 0.4V_3^2 \qquad (2)$$

where $V_1$ and $V_3$ are in ft/s. Similarly the energy equation for fluid flowing from $B$ and $C$ is

$$\frac{p_B}{\gamma} + \frac{V_B^2}{2g} + z_B = \frac{p_C}{\gamma} + \frac{V_C^2}{2g} + z_C + f_2 \frac{\ell_2}{D_2} \frac{V_2^2}{2g} + f_3 \frac{\ell_3}{D_3} \frac{V_3^2}{2g}$$

or

$$z_B = f_2 \frac{\ell_2}{D_2} \frac{V_2^2}{2g} + f_3 \frac{\ell_3}{D_3} \frac{V_3^2}{2g}$$

For the given conditions this can be written as

$$64.4 = 0.5V_2^2 + 0.4V_3^2 \qquad (3)$$

Equations 1, 2, and 3 (in terms of the three unknowns $V_1$, $V_2$, and $V_3$) are the governing equations for this flow, provided the fluid flows from reservoir $B$. It turns out, however, that there is no solution for these equations with positive, real values of the velocities. Although these equations do not appear to be complicated, there is no simple way to solve them directly. Thus, a trial-and-error solution is suggested. This can be accomplished as follows. Assume a value of $V_1 > 0$, calculate $V_3$ from Eq. 2, and then $V_2$ from Eq. 3. It is found that the resulting $V_1$, $V_2$, $V_3$ trio does not satisfy Eq. 1 for any value of $V_1$ assumed. There is no solution to Eqs. 1, 2, and 3 with real, positive values of $V_1$, $V_2$, and $V_3$. Thus, our original assumption of flow out of reservoir $B$ must be incorrect.

To obtain the solution, assume the fluid flows into reservoirs $B$ and $C$ and out of $A$. For this case the continuity equation becomes

$$Q_1 = Q_2 + Q_3$$

or

$$V_1 = V_2 + V_3 \qquad (4)$$

Application of the energy equation between points $A$ and $B$ and $A$ and $C$ gives

$$z_A = z_B + f_1 \frac{\ell_1}{D_1} \frac{V_1^2}{2g} + f_2 \frac{\ell_2}{D_2} \frac{V_2^2}{2g}$$

and

$$z_A = z_C + f_1 \frac{\ell_1}{D_1} \frac{V_1^2}{2g} + f_3 \frac{\ell_3}{D_3} \frac{V_3^2}{2g}$$

which, with the given data, become

$$258 = V_1^2 + 0.5\, V_2^2 \qquad (5)$$

and

$$322 = V_1^2 + 0.4\, V_3^2 \qquad (6)$$

Equations 4, 5, and 6 can be solved as follows. By subtracting Eq. 5 from 6 we obtain

$$V_3 = \sqrt{160 + 1.25V_2^2}$$

Thus, Eq. 5 can be written as

$$258 = (V_2 + V_3)^2 + 0.5V_2^2$$
$$= (V_2 + \sqrt{160 + 1.25V_2^2})^2 + 0.5V_2^2$$

or

$$2V_2\sqrt{160 + 1.25V_2^2} = 98 - 2.75V_2^2 \qquad (7)$$

which, upon squaring both sides, can be written as

$$V_2^4 - 460\, V_2^2 + 3748 = 0$$

By using the quadratic formula, we can solve for $V_2^2$ to obtain either $V_2^2 = 452$ or $V_2^2 = 8.30$. Thus, either $V_2 = 21.3$ ft/s or $V_2 = 2.88$ ft/s. The value $V_2 = 21.3$ ft/s is not a root of the original equations. It is an extra root introduced by squaring Eq. 7, which with $V_2 = 21.3$ becomes "1140 = −1140." Thus, $V_2 = 2.88$ ft/s and from Eq. 5, $V_1 = 15.9$ ft/s. The corresponding flowrates are

$$Q_1 = A_1 V_1 = \frac{\pi}{4} D_1^2 V_1 = \frac{\pi}{4} (1\text{ ft})^2\, (15.9\text{ ft/s})$$
$$= 12.5\text{ ft}^3/\text{s from } A \qquad \textbf{(Ans)}$$

$$Q_2 = A_2 V_2 = \frac{\pi}{4} D_2^2 V_2 = \frac{\pi}{4} (1\text{ ft})^2\, (2.88\text{ ft/s})$$
$$= 2.26\text{ ft}^3/\text{s into } B \qquad \textbf{(Ans)}$$

and

$$Q_3 = Q_1 - Q_2 = (12.5 - 2.26)\text{ ft}^3/\text{s}$$
$$= 10.2\text{ ft}^3/\text{s into } C \qquad \textbf{(Ans)}$$

Note the slight differences in the governing equations depending on the direction of the flow in pipe (2)—compare Eqs. 1, 2, and 3 with Eqs. 4, 5, and 6.

**COMMENT** If the friction factors were not given, a trial-and-error procedure similar to that needed for Type II problems (see Section 8.5.1) would be required.

*Pipe network problems can be solved using node and loop concepts.*

The ultimate in multiple pipe systems is a *network* of pipes such as that shown in Fig. 8.38. Networks like these often occur in city water distribution systems and other systems that may have multiple "inlets" and "outlets." The direction of flow in the various pipes is by no means obvious—in fact, it may vary in time, depending on how the system is used from time to time.

The solution for pipe network problems is often carried out by use of node and loop equations similar in many ways to that done in electrical circuits. For example, the continuity equation requires that for each *node* (the junction of two or more pipes) the net flowrate is zero. What flows into a node must flow out at the same rate. In addition, the net pressure difference completely around a *loop* (starting at one location in a pipe and returning to that location) must be zero. By combining these ideas with the usual head loss and pipe flow equations, the flow throughout the entire network can

■ **Figure 8.38**    A general pipe network.

be obtained. Of course, trial-and-error solutions are usually required because the direction of flow and the friction factors may not be known. Such a solution procedure using matrix techniques is ideally suited for computer use (Refs. 21, 22).

## 8.6    Pipe Flowrate Measurement

It is often necessary to determine experimentally the flowrate in a pipe. In Chapter 3 we introduced various types of flow-measuring devices (Venturi meter, nozzle meter, orifice meter, etc.) and discussed their operation under the assumption that viscous effects were not important. In this section we will indicate how to account for the ever-present viscous effects in these flowmeters. We will also indicate other types of commonly used flowmeters.

### 8.6.1  Pipe Flowrate Meters

*Orifice, nozzle, and Venturi meters involve the concept "high velocity gives low pressure."*

Three of the most common devices used to measure the instantaneous flowrate in pipes are the orifice meter, the nozzle meter, and the Venturi meter. As was discussed in Section 3.6.3, each of these meters operates on the principle that a decrease in flow area in a pipe causes an increase in velocity that is accompanied by a decrease in pressure. Correlation of the pressure difference with the velocity provides a means of measuring the flowrate. In the absence of viscous effects and under the assumption of a horizontal pipe, application of the Bernoulli equation (Eq. 3.7) between points (1) and (2) shown in Fig. 8.39 gave

$$Q_{\text{ideal}} = A_2 V_2 = A_2 \sqrt{\frac{2(p_1 - p_2)}{\rho(1 - \beta^4)}} \tag{8.37}$$

where $\beta = D_2/D_1$. Based on the results of the previous sections of this chapter, we anticipate that there is a head loss between (1) and (2) so that the governing equations become

$$Q = A_1 V_1 = A_2 V_2$$

and

$$\frac{p_1}{\gamma} + \frac{V_1^2}{2g} = \frac{p_2}{\gamma} + \frac{V_2^2}{2g} + h_L$$

The ideal situation has $h_L = 0$ and results in Eq. 8.37. The difficulty in including the head loss is that there is no accurate expression for it. The net result is that empirical coefficients are used in the flowrate equations to account for the complex real-world effects brought on by the nonzero viscosity. The coefficients are discussed in this section.

■ **Figure 8.39**    Typical pipe flowmeter geometry.

■ **Figure 8.40** Typical orifice meter construction.

*An orifice discharge coefficient is used to account for non-ideal effects.*

A typical *orifice meter* is constructed by inserting between two flanges of a pipe a flat plate with a hole, as shown in Fig. 8.40. The pressure at point (2) within the vena contracta is less than that at point (1). Nonideal effects occur for two reasons. First, the vena contracta area, $A_2$, is less than the area of the hole, $A_o$, by an unknown amount. Thus, $A_2 = C_c A_o$, where $C_c$ is the contraction coefficient ($C_c < 1$). Second, the swirling flow and turbulent motion near the orifice plate introduce a head loss that cannot be calculated theoretically. Thus, an *orifice discharge coefficient*, $C_o$, is used to take these effects into account. That is,

$$Q = C_o Q_{\text{ideal}} = C_o A_o \sqrt{\frac{2(p_1 - p_2)}{\rho(1 - \beta^4)}} \tag{8.38}$$

where $A_o = \pi d^2/4$ is the area of the hole in the orifice plate. The value of $C_o$ is a function of $\beta = d/D$ and the Reynolds number, $\text{Re} = \rho V D/\mu$, where $V = Q/A_1$. Typical values of $C_o$ are given in Fig. 8.41. As shown by Eq. 8.38 and the figure in the margin, for a given value of $C_o$, the flowrate is proportional to the square root of the pressure difference. Note that the value of $C_o$ depends on the specific construction of the orifice meter (i.e., the placement of the pressure taps, whether the orifice plate edge is square or beveled, etc.). Very precise conditions governing the construction of standard orifice meters have been established to provide the greatest accuracy possible (Refs. 23, 24).

Another type of pipe flowmeter that is based on the same principles used in the orifice meter is the *nozzle meter,* three variations of which are shown in Fig. 8.42. This device uses a contoured nozzle (typically placed between flanges of pipe sections) rather than a simple (and less expensive) plate with a hole as in an orifice meter. The resulting flow pattern for the nozzle meter is closer to ideal than the orifice meter flow. There is only a slight vena contracta and the secondary

■ **Figure 8.41** Orifice meter discharge coefficient (Data from Ref. 24).

■ **Figure 8.42**   Typical nozzle meter construction.

*The nozzle meter is more efficient than the orifice meter.*

flow separation is less severe, but there still are viscous effects. These are accounted for by use of the *nozzle discharge coefficient*, $C_n$, where

$$Q = C_n Q_{ideal} = C_n A_n \sqrt{\frac{2(p_1 - p_2)}{\rho(1 - \beta^4)}} \tag{8.39}$$

with $A_n = \pi d^2/4$. As with the orifice meter, the value of $C_n$ is a function of the diameter ratio, $\beta = d/D$, and the Reynolds number, $Re = \rho V D/\mu$. Typical values obtained from experiments are shown in Fig. 8.43. Again, precise values of $C_n$ depend on the specific details of the nozzle design. Accepted standards have been adopted (Ref. 24). Note that $C_n > C_o$; the nozzle meter is more efficient (less energy dissipated) than the orifice meter.

The most precise and most expensive of the three obstruction-type flowmeters is the **Venturi meter** shown in Fig. 8.44 [G. B. Venturi (1746–1822)]. Although the operating principle for this device is the same as for the orifice or nozzle meters, the geometry of the Venturi meter is designed to reduce head losses to a minimum. This is accomplished by providing a relatively streamlined contraction (which eliminates separation ahead of the throat) and a very gradual expansion downstream of the throat (which eliminates separation in this decelerating portion of the device). Most of the head loss that occurs in a well-designed Venturi meter is due to friction losses along the walls rather than losses associated with separated flows and the inefficient mixing motion that accompanies such flow.

■ **Figure 8.43**   Nozzle meter discharge coefficient (Data from Ref. 24).

■ **Figure 8.44**   Typical Venturi meter construction.

$Re = \rho VD/\mu$

■ **Figure 8.45** Venturi meter discharge coefficient (Ref. 23).

Thus, the flowrate through a Venturi meter is given by

$$Q = C_v Q_{\text{ideal}} = C A_T \sqrt{\frac{2(p_1 - p_2)}{\rho(1 - \beta^4)}} \tag{8.40}$$

*The Venturi discharge coefficient is a function of the specific geometry of the meter.*

where $A_T = \pi d^2/4$ is the throat area. The range of values of $C_v$, the *Venturi discharge coefficient*, is given in Fig. 8.45. The throat-to-pipe diameter ratio ($\beta = d/D$), the Reynolds number, and the shape of the converging and diverging sections of the meter are among the parameters that affect the value of $C_v$.

Again, the precise values of $C_n$, $C_o$, and $C_v$ depend on the specific geometry of the devices used. Considerable information concerning the design, use, and installation of standard flowmeters can be found in various books (Refs. 23, 24, 25, 26, 31).

---

## $E$XAMPLE 8.15  Nozzle Flowmeter

**GIVEN**  Ethyl alcohol flows through a pipe of diameter $D = 60$ mm in a refinery. The pressure drop across the nozzle meter used to measure the flowrate is to be $\Delta p = 4.0$ kPa when the flowrate is $Q = 0.003$ m³/s.

**FIND**  Determine the diameter, $d$, of the nozzle.

### SOLUTION

From Table 1.6 the properties of ethyl alcohol are $\rho = 789$ kg/m³ and $\mu = 1.19 \times 10^{-3}$ N · s/m². Thus,

$$Re = \frac{\rho VD}{\mu} = \frac{4\rho Q}{\pi D \mu}$$

$$= \frac{4(789 \text{ kg/m}^3)(0.003 \text{ m}^3/\text{s})}{\pi(0.06 \text{ m})(1.19 \times 10^{-3} \text{ N} \cdot \text{s/m}^2)} = 42,200$$

From Eq. 8.39 the flowrate through the nozzle is

$$Q = 0.003 \text{ m}^3/\text{s} = C_n \frac{\pi}{4} d^2 \sqrt{\frac{2(4 \times 10^3 \text{ N/m}^2)}{789 \text{ kg/m}^3(1 - \beta^4)}}$$

or

$$1.20 \times 10^{-3} = \frac{C_n d^2}{\sqrt{1 - \beta^4}} \tag{1}$$

where $d$ is in meters. Note that $\beta = d/D = d/0.06$. Equation 1 and Fig. 8.43 represent two equations for the two unknowns $d$ and $C_n$ that must be solved by trial and error.

As a first approximation we assume that the flow is ideal, or $C_n = 1.0$, so that Eq. 1 becomes

$$d = (1.20 \times 10^{-3} \sqrt{1 - \beta^4})^{1/2} \tag{2}$$

In addition, for many cases $1 - \beta^4 \approx 1$, so that an approximate value of $d$ can be obtained from Eq. 2 as

$$d = (1.20 \times 10^{-3})^{1/2} = 0.0346 \text{ m}$$

Hence, with an initial guess of $d = 0.0346$ m or $\beta = d/D = 0.0346/0.06 = 0.577$, we obtain from Fig. 8.43 (using Re = 42,200) a value of $C_n = 0.972$. Clearly this does not agree with our initial assumption of $C_n = 1.0$. Thus, we do not have the solution to Eq. 1 and Fig. 8.43. Next we assume $\beta = 0.577$ and $C_n = 0.972$ and solve for $d$ from Eq. 1 to obtain

$$d = \left(\frac{1.20 \times 10^{-3}}{0.972} \sqrt{1 - 0.577^4}\right)^{1/2}$$

or $d = 0.0341$ m. With the new value of $\beta = 0.0341/0.060 = 0.568$ and Re = 42,200, we obtain (from Fig. 8.43) $C_n \approx 0.972$ in

agreement with the assumed value. Thus,

$$d = 34.1 \text{ mm} \qquad \text{(Ans)}$$

**COMMENTS**   If numerous cases are to be investigated, it may be much easier to replace the discharge coefficient data of Fig. 8.43 by the equivalent equation, $C_n = \phi(\beta, \text{Re})$, and use a computer to iterate for the answer. Such equations are available in the literature (Ref. 24). This would be similar to using the Colebrook equation rather than the Moody chart for pipe friction problems.

By repeating the calculations, the nozzle diameters, $d$, needed for the same flowrate and pressure drop but with different fluids are shown in Fig. E8.15. The diameter is a function of the fluid viscosity because the nozzle coefficient, $C_n$, is a function of the Reynolds number (see Fig. 8.43). In addition, the diameter is a function of the density because of this Reynolds number effect and, perhaps more importantly, because the density is involved directly in the flowrate equation, Eq. 8.39. These factors all combine to produce the results shown in the figure.

■ **Figure E8.15**

*There are many types of flowmeters.*

Numerous other devices are used to measure the flowrate in pipes. Many of these devices use principles other than the high-speed/low-pressure concept of the orifice, nozzle, and Venturi meters.

A quite common, accurate, and relatively inexpensive flowmeter is the *rotameter*, or variable area meter as is shown in Fig. 8.46. In this device a float is contained within a tapered, transparent metering tube that is attached vertically to the pipeline. As fluid flows through the meter (entering at the bottom), the float will rise within the tapered tube and reach an equilibrium height that is a function of the flowrate. This height corresponds to an equilibrium condition for which the net force on the float (buoyancy, float weight, fluid drag) is zero. A calibration scale in the tube provides the relationship between the float position and the flowrate.

*V8.15 Rotameter*

■ **Figure 8.46**   Rotameter-type flowmeter. (Courtesy of ABB.)

■ **Figure 8.47**
**Turbine-type flowmeter.**
**(Courtesy of FTI Flow**
**Technology, Inc.)**

Another useful pipe flowrate meter is a *turbine meter* as is shown in Fig. 8.47. A small, freely rotating propeller or turbine within the turbine meter rotates with an angular velocity that is a function of (nearly proportional to) the average fluid velocity in the pipe. This angular velocity is picked up magnetically and calibrated to provide a very accurate measure of the flowrate through the meter.

### 8.6.2 Volume Flowmeters

*Volume flowmeters measure volume rather than volume flowrate.*

In many instances it is necessary to know the amount (volume or mass) of fluid that has passed through a pipe during a given time period, rather than the instantaneous flowrate. For example, we are interested in how many gallons of gasoline are pumped into the tank in our car rather than the rate at which it flows into the tank. Numerous quantity-measuring devices provide such information.

The *nutating disk meter* shown in Fig. 8.48 is widely used to measure the net amount of water used in domestic and commercial water systems as well as the amount of gasoline delivered to your gas tank. This meter contains only one essential moving part and is relatively inexpensive and accurate. Its operating principle is very simple, but it may be difficult to understand its operation without actually inspecting the device firsthand. The device consists of a metering chamber with spherical sides and conical top and bottom. A disk passes through a central sphere and divides the chamber into two portions. The disk is constrained to be at an angle not normal to the axis of symmetry of the chamber. A radial plate (diaphragm) divides the chamber so that the entering fluid causes the disk to wobble (nutate), with fluid flowing alternately above or below the disk. The fluid exits the chamber after the disk has completed one wobble, which corresponds to a specific volume of fluid passing through the chamber. During each wobble of the disk, the pin attached to the tip

*V8.16 Water meter*

■ **Figure 8.48**  Nutating disk
flowmeter. (Courtesy of Badger
Meter, Inc. © 2011.)

**Figure 8.49**    Bellows-type flowmeter. (Based on information from BTR—Rockwell Gas Products.) (*a*) Back case emptying, back diaphragm filling. (*b*) Front diaphragm filling, front case emptying. (*c*) Back case filling, back diaphragm emptying. (*d*) Front diaphragm emptying, front case filling.

*The nutating disk meter has only one moving part; the bellows meter has a complex set of moving parts.*

of the center sphere, normal to the disk, completes one circle. The volume of fluid that has passed through the meter can be obtained by counting the number of revolutions completed.

Another quantity-measuring device that is used for gas flow measurements is the *bellows meter* as shown in Fig. 8.49. It contains a set of bellows that alternately fill and empty as a result of the pressure of the gas and the motion of a set of inlet and outlet valves. The common household natural gas meter (see photograph in the margin) is of this type. For each cycle [(*a*) through (*d*)] a known volume of gas passes through the meter.

The nutating disk meter (water meter) is an example of extreme simplicity—one cleverly designed moving part. The bellows meter (gas meter), on the other hand, is relatively complex—it contains many moving, interconnected parts. This difference is dictated by the application involved. One measures a common, safe-to-handle, relatively high-pressure liquid, whereas the other measures a relatively dangerous, low-pressure gas. Each device does its intended job very well.

There are numerous devices used to measure fluid flow, only a few of which have been discussed here. The reader is encouraged to review the literature to gain familiarity with other useful, clever devices (Refs. 25, 26).

# 8.7    Chapter Summary and Study Guide

This chapter discussed the flow of a viscous fluid in a pipe. General characteristics of laminar, turbulent, fully developed, and entrance flows are considered. Poiseuille's equation is obtained to describe the relationship among the various parameters for fully developed laminar flow.

*laminar flow*
*transitional flow*
*turbulent flow*
*entrance length*
*fully developed flow*
*wall shear stress*
*Poiseuille's law*
*friction factor*
*turbulent shear stress*
*major loss*
*minor loss*
*relative roughness*
*Moody chart*
*Colebrook formula*
*loss coefficient*
*hydraulic diameter*
*multiple pipe systems*
*orifice meter*
*nozzle meter*
*Venturi meter*

Various characteristics of turbulent pipe flow are introduced and contrasted to laminar flow. It is shown that the head loss for laminar or turbulent pipe flow can be written in terms of the friction factor (for major losses) and the loss coefficients (for minor losses). In general, the friction factor is obtained from the Moody chart or the Colebrook formula and is a function of the Reynolds number and the relative roughness. The minor loss coefficients are a function of the flow geometry for each system component.

Analysis of noncircular conduits is carried out by use of the hydraulic diameter concept. Various examples involving flow in single pipe systems and flow in multiple pipe systems are presented. The inclusion of viscous effects and losses in the analysis of orifice, nozzle, and Venturi flowmeters is discussed.

The following checklist provides a study guide for this chapter. When your study of the entire chapter and end-of-chapter exercises has been completed you should be able to

■ write out meanings of the terms listed here in the margin and understand each of the related concepts. These terms are particularly important and are set in *italic, **bold**, and color* type in the text.

■ determine which of the following types of flow will occur: entrance flow or fully developed flow; laminar flow or turbulent flow.

■ use the Poiseuille equation in appropriate situations and understand its limitations.

■ explain the main properties of turbulent pipe flow and how they are different from or similar to laminar pipe flow.

■ use the Moody chart and the Colebrook equation to determine major losses in pipe systems.

■ use minor loss coefficients to determine minor losses in pipe systems.

■ determine the head loss in noncircular conduits.

■ incorporate major and minor losses into the energy equation to solve a variety of pipe flow problems, including Type I problems (determine the pressure drop or head loss), Type II problems (determine the flow rate), and Type III problems (determine the pipe diameter).

■ solve problems involving multiple pipe systems.

■ determine the flowrate through orifice, nozzle, and Venturi flowmeters as a function of the pressure drop across the meter.

Some of the important equations in this chapter are:

| | | |
|---|---|---|
| Entrance length | $\dfrac{\ell_e}{D} = 0.06 \, \text{Re}$ for laminar flow | **(8.1)** |
| | $\dfrac{\ell_e}{D} = 4.4 \, (\text{Re})^{1/6}$ for turbulent flow | **(8.2)** |
| Pressure drop for fully developed laminar pipe flow | $\Delta p = \dfrac{4\ell\tau_w}{D}$ | **(8.5)** |
| Velocity profile for fully developed laminar pipe flow | $u(r) = \left(\dfrac{\Delta p D^2}{16\mu\ell}\right)\left[1 - \left(\dfrac{2r}{D}\right)^2\right] = V_c\left[1 - \left(\dfrac{2r}{D}\right)^2\right]$ | **(8.7)** |
| Volume flowrate for fully developed laminar horizontal pipe flow | $Q = \dfrac{\pi D^4 \, \Delta p}{128\mu\ell}$ | **(8.9)** |
| Friction factor for fully developed laminar pipe flow | $f = \dfrac{64}{\text{Re}}$ | **(8.19)** |
| Pressure drop for a horizontal pipe | $\Delta p = f\dfrac{\ell}{D}\dfrac{\rho V^2}{2}$ | **(8.33)** |
| Head loss due to major losses | $h_{L\,\text{major}} = f\dfrac{\ell}{D}\dfrac{V^2}{2g}$ | **(8.34)** |

| Colebrook formula | $\dfrac{1}{\sqrt{f}} = -2.0 \log\left(\dfrac{\varepsilon/D}{3.7} + \dfrac{2.51}{Re\sqrt{f}}\right)$ | **(8.35a)** |

| Haaland formula | $\dfrac{1}{\sqrt{f}} = -1.8 \log\left[\left(\dfrac{\varepsilon/D}{3.7}\right)^{1.11} + \dfrac{6.9}{Re}\right]$ | **(8.35b)** |

| Head loss due to minor losses | $h_{L\,\text{minor}} = K_L \dfrac{V^2}{2g}$ | **(8.36)** |

| Volume flowrate for orifice, nozzle, or Venturi meter | $Q = C_i A_i \sqrt{\dfrac{2(p_1 - p_2)}{\rho(1 - \beta^4)}}$ | **(8.38, 8.39, 8.40)** |

# References

1. Hinze, J. O., *Turbulence*, 2nd Ed., McGraw-Hill, New York, 1975.
2. Panton, R. L., *Incompressible Flow*, 3rd Ed., Wiley, New York, 2005.
3. Schlichting, H., *Boundary Layer Theory*, 8th Ed., McGraw-Hill, New York, 2000.
4. Gleick, J., *Chaos: Making a New Science*, Viking Penguin, New York, 1987.
5. White, F. M., *Fluid Mechanics*, 6th Ed., McGraw-Hill, New York, 2008.
6. Nikuradse, J., "Stomungsgesetz in Rauhen Rohren," *VDI-Forschungsch*, No. 361, 1933; or see NACA Tech Memo 1922.
7. Moody, L. F., "Friction Factors for Pipe Flow," *Transactions of the ASME*, Vol. 66, 1944.
8. Colebrook, C. F., "Turbulent Flow in Pipes with Particular Reference to the Transition Between the Smooth and Rough Pipe Laws," *Journal of the Institute of Civil Engineers London*, Vol. 11, 1939.
9. *ASHRAE Handbook of Fundamentals*, ASHRAE, Atlanta, 1981.
10. Streeter, V. L., ed., *Handbook of Fluid Dynamics*, McGraw-Hill, New York, 1961.
11. Sovran, G., and Klomp, E. D., "Experimentally Determined Optimum Geometries for Rectilinear Diffusers with Rectangular, Conical, or Annular Cross Sections," in *Fluid Mechanics of Internal Flow*, Sovran, G., ed., Elsevier, Amsterdam, 1967.
12. Runstadler, P. W., "Diffuser Data Book," Technical Note 186, Creare, Inc., Hanover, N.H., 1975.
13. Laws, E. M., and Livesey, J. L., "Flow Through Screens," *Annual Review of Fluid Mechanics*, Vol. 10, Annual Reviews, Inc., Palo Alto, Calif., 1978.
14. Balje, O. E., *Turbomachines: A Guide to Design, Selection and Theory*, Wiley, New York, 1981.
15. Wallis, R. A., *Axial Flow Fans and Ducts*, Wiley, New York, 1983.
16. Karassick, I. J. et al., *Pump Handbook*, 2nd Ed., McGraw-Hill, New York, 1985.
17. White, F. M., *Viscous Fluid Flow*, 3rd Ed., McGraw-Hill, New York, 2006.
18. Olson, R. M., *Essentials of Engineering Fluid Mechanics*, 4th Ed., Harper & Row, New York, 1980.
19. Dixon, S. L., *Fluid Mechanics of Turbomachinery*, 3rd Ed., Pergamon, Oxford, UK, 1978.
20. Finnemore, E. J., and Franzini, J. R., *Fluid Mechanics*, 10th Ed., McGraw-Hill, New York, 2002.
21. Streeter, V. L., and Wylie, E. B., *Fluid Mechanics*, 8th Ed., McGraw-Hill, New York, 1985.
22. Jeppson, R. W., *Analysis of Flow in Pipe Networks*, Ann Arbor Science Publishers, Ann Arbor, Mich., 1976.
23. Bean, H. S., ed., *Fluid Meters: Their Theory and Application*, 6th Ed., American Society of Mechanical Engineers, New York, 1971.
24. "Measurement of Fluid Flow by Means of Orifice Plates, Nozzles, and Venturi Tubes Inserted in Circular Cross Section Conduits Running Full," Int. Organ. Stand. Rep. DIS-5167, Geneva, 1976.
25. Goldstein, R. J., ed., *Flow Mechanics Measurements*, 2nd Ed., Taylor and Francis, Philadelphia, 1996.
26. Benedict, R. P., *Measurement of Temperature, Pressure, and Flow*, 2nd Ed., Wiley, New York, 1977.
27. Hydraulic Institute, *Engineering Data Book*, 1st Ed., Cleveland Hydraulic Institute, 1979.
28. Harris, C. W., *University of Washington Engineering Experimental Station Bulletin*, 48, 1928.
29. Hamilton, J. B., *University of Washington Engineering Experimental Station Bulletin*, 51, 1929.
30. Miller, D. S., *Internal Flow Systems*, 2nd Ed., BHRA, Cranfield, UK, 1990.
31. Spitzer, D. W., ed., *Flow Measurement: Practical Guides for Measurement and Control*, Instrument Society of America, Research Triangle Park, N.C., 1991.
32. Wilcox, D. C., *Turbulence Modeling for CFD*, DCW Industries, Inc., La Canada, California, 1994.
33. Mullin, T., ed., *The Nature of Chaos*, Oxford University Press, Oxford, UK, 1993.
34. Haaland, S.E., "Simple and Explicit Formulas for the Friction-Factor in Turbulent Pipe Flow," *Transactions of the ASME, Journal of Fluids Engineering*, Vol. 105, 1983.

---

 Problem available in *WileyPLUS* at instructor's discretion.

 Tutoring problem available in *WileyPLUS* at instructor's discretion.

▶ Problem is related to a chapter video available in *WileyPLUS*.

* Problem to be solved with aid of programmable calculator or computer.

† Open-ended problem that requires critical thinking. These problems require various assumptions to provide the necessary input data. There are not unique answers to these problems.

---

## Review Problems

Go to Appendix G (*WileyPLUS* or the book's web site, www.wiley.com/college/munson) for a set of review problems with answers. Detailed solutions can be found in the *Student Solution Manual and Study Guide for Fundamentals of Fluid Mechanics,* by Munson et al. (© 2013 John Wiley and Sons, Inc.).

## Conceptual Questions

**8.1C** For fully developed viscous flow in a horizontal pipe, which of the following is true?

a) Pressure forces are balanced by shear forces.

b) Pressure forces result in fluid acceleration.

c) Shear forces result in fluid deceleration.

d) Pressure forces are balanced by body forces.

**8.2C** There is a steady laminar flow of water in a pipe of length $\ell$.

As the volume flowrate increases, the pressure drop over the length $\ell$.

a) becomes larger.

b) becomes smaller.

c) stays the same.

**8.3C** Water flows steadily through a horizontal circular pipe. Which curve most correctly describes the pressure change through the pipe as the length is increased?

**8.4C** Water flows steadily through a smooth pipe. For turbulent flow, if the velocity is increased, in general

a) the friction factor increases.

b) the friction factor decreases.

c) the friction factor stays the same.

**Additional conceptual questions are available in *WileyPLUS* at the instructor's discretion.**

## Problems

**Note: Unless specific values of required fluid properties are given in the problem statement, use the values found in the tables on the inside of the front cover. Answers to the even-numbered problems are listed at the end of the book. The Lab Problems as well as the videos that accompany problems can be accessed in *WileyPLUS* or the book's web site, www.wiley.com/college/munson.**

### Section 8.1 General Characteristics of Pipe Flow (also see Lab Problem 8.5LP)

†**8.1** Under normal circumstances is the airflow though your trachea (your windpipe) laminar or turbulent? List all assumptions and show all calculations.

**8.2** ▶ Rainwater runoff from a parking lot flows through a 3-ft-diameter pipe, completely filling it. Whether flow in a pipe is laminar or turbulent depends on the value of the Reynolds number. (See **Video V8.2.**) Would you expect the flow to be laminar or turbulent? Support your answer with appropriate calculations.

**8.3** (WILEY) Blue and yellow streams of paint at 60 °F (each with a density of 1.6 slugs/ft³ and a viscosity 1000 times greater than water) enter a pipe with an average velocity of 4 ft/s as shown in Fig. P8.3. Would you expect the paint to exit the pipe as green paint or separate streams of blue and yellow paint? Explain. Repeat the problem if the paint were "thinned" so that it is only 10 times more viscous than water. Assume the density remains the same.

■ **Figure P8.3**

**8.4** (WILEY) Air at 200 °F flows at standard atmospheric pressure in a pipe at a rate of 0.08 lb/s. Determine the minimum diameter allowed if the flow is to be laminar.

**8.5** To cool a given room it is necessary to supply 4 ft³/s of air through an 8-in.-diameter pipe. Approximately how long is the entrance length in this pipe?

**\*8.6** The flow of water in a 3-mm-diameter pipe is to remain laminar. Plot a graph of the maximum flowrate allowed as a function of temperature for $0 < T < 100 \,^\circ C$.

**8.7** (WILEY) Carbon dioxide at 20 °C and a pressure of 550 kPa (abs) flows in a pipe at a rate of 0.04 N/s. Determine the maximum diameter allowed if the flow is to be turbulent.

**8.8** (WILEY) The pressure distribution measured along a straight, horizontal portion of a 50-mm-diameter pipe attached to a tank is shown in the table below. Approximately how long is the entrance length? In the fully developed portion of the flow, what is the value of the wall shear stress?

| x (m) (±0.01 m) | p (mm H₂O) (±5 mm) |
|---|---|
| 0 (tank exit) | 520 |
| 0.5 | 427 |
| 1.0 | 351 |
| 1.5 | 288 |
| 2.0 | 236 |
| 2.5 | 188 |
| 3.0 | 145 |
| 3.5 | 109 |
| 4.0 | 73 |
| 4.5 | 36 |
| 5.0 (pipe exit) | 0 |

**8.9** (WILEY) (See Fluids in the News article titled "Nanoscale Flows," Section 8.1.1.) **(a)** Water flows in a tube that has a diameter of $D = 0.1$ m. Determine the Reynolds number if the average velocity is 10 diameters per second. **(b)** Repeat the calculations if the tube is a nanoscale tube with a diameter of $D = 100$ nm.

### Section 8.2 Fully Developed Laminar Flow

**8.10** (WILEY) For fully developed laminar pipe flow in a circular pipe, the velocity profile is given by $u(r) = 2 (1 - r^2/R^2)$ in m/s, where $R$ is the inner radius of the pipe. Assuming that the pipe

diameter is 4 cm, find the maximum and average velocities in the pipe as well as the volume flow rate.

**8.11** (WILEY) The wall shear stress in a fully developed flow portion of a 12-in.-diameter pipe carrying water is 1.85 lb/ft². Determine the pressure gradient, $\partial p/\partial x$, where $x$ is in the flow direction, if the pipe is **(a)** horizontal, **(b)** vertical with flow up, or **(c)** vertical with flow down.

**8.12** (WILEY) The pressure drop needed to force water through a horizontal 1-in.-diameter pipe is 0.60 psi for every 12-ft length of pipe. Determine the shear stress on the pipe wall. Determine the shear stress at distances 0.3 and 0.5 in. away from the pipe wall.

**8.13** Repeat Problem 8.12 if the pipe is on a 20° hill. Is the flow up or down the hill? Explain.

**8.14** (WILEY) Water flows in a constant-diameter pipe with the following conditions measured: At section (a) $p_a = 32.4$ psi and $z_a = 56.8$ ft; at section (b) $p_b = 29.7$ psi and $z_b = 68.2$ ft. Is the flow from (a) to (b) or from (b) to (a)? Explain.

**\*8.15** Some fluids behave as a non-Newtonian power-law fluid characterized by $\tau = -C(du/dr)^n$, where $n = 1, 3, 5$, and so on, and $C$ is a constant. (If $n = 1$, the fluid is the customary Newtonian fluid.) **(a)** For flow in a round pipe of a diameter $D$, integrate the force balance equation (Eq. 8.3) to obtain the velocity profile

$$u(r) = \frac{-n}{(n+1)}\left(\frac{\Delta p}{2\ell C}\right)^{1/n}\left[r^{(n+1)/n} - \left(\frac{D}{2}\right)^{(n+1)/n}\right]$$

**(b)** Plot the dimensionless velocity profile $u/V_c$, where $V_c$ is the centerline velocity (at $r = 0$), as a function of the dimensionless radial coordinate $r/(D/2)$, where $D$ is the pipe diameter. Consider values of $n = 1, 3, 5,$ and 7.

**8.16** For laminar flow in a round pipe of diameter $D$, at what distance from the centerline is the actual velocity equal to the average velocity?

**8.17** Water at 20 °C flows through a horizontal 1-mm-diameter tube to which are attached two pressure taps a distance 1 m apart. **(a)** What is the maximum pressure drop allowed if the flow is to be laminar? **(b)** Assume the manufacturing tolerance on the tube diameter is $D = 1.0 \pm 0.1$ mm. Given this uncertainty in the tube diameter, what is the maximum pressure drop allowed if it must be assured that the flow is laminar?

**8.18** (WILEY GO) Glycerin at 20 °C flows upward in a vertical 75-mm-diameter pipe with a centerline velocity of 1.0 m/s. Determine the head loss and pressure drop in a 10-m length of the pipe.

**8.19** (WILEY) A large artery in a person's body can be approximated by a tube of diameter 9 mm and length 0.35 m. Also assume that blood has a viscosity of approximately $4 \times 10^{-3}$ N · s/m², a specific gravity of 1.0, and that the pressure at the beginning of the artery is equivalent to 120 mm Hg. If the flow were steady (it is not) with $V = 0.2$ m/s, determine the pressure at the end of the artery if it is oriented **(a)** vertically up (flow up) or **(b)** horizontal.

**8.20** At time $t = 0$ the level of water in tank $A$ shown in Fig. P8.20 is 2 ft above that in tank $B$. Plot the elevation of the water in tank $A$

■ **Figure P8.20**

as a function of time until the free surfaces in both tanks are at the same elevation. Assume quasisteady conditions—that is, the steady pipe flow equations are assumed valid at any time, even though the flowrate does change (slowly) in time. Neglect minor losses. *Note*: Verify and use the fact that the flow is laminar.

**8.21** ꟼWILEY A fluid flows through a horizontal 0.1-in.-diameter pipe. When the Reynolds number is 1500, the head loss over a 20-ft length of the pipe is 6.4 ft. Determine the fluid velocity.

**8.22** ꟼWILEY A viscous fluid flows in a 0.10-m-diameter pipe such that its velocity measured 0.012 m away from the pipe wall is 0.8 m/s. If the flow is laminar, determine the centerline velocity and the flowrate.

**8.23** Oil flows through the horizontal pipe shown in Fig. P8.23 under laminar conditions. All sections are the same diameter except one. Which section of the pipe (A, B, C, D, or E) is slightly smaller in diameter than the others? Explain.

■ **Figure P8.23**

**8.24** Asphalt at 120 °F, considered to be a Newtonian fluid with a viscosity 80,000 times that of water and a specific gravity of 1.09, flows through a pipe of diameter 2.0 in. If the pressure gradient is 1.6 psi/ft determine the flowrate assuming the pipe is **(a)** horizontal; **(b)** vertical with flow up.

**8.25** ꟼWILEY Oil of $SG = 0.87$ and a kinematic viscosity $\nu = 2.2 \times 10^{-4}$ m²/s flows through the vertical pipe shown in Fig. P8.25 at a rate of $4 \times 10^{-4}$ m³/s. Determine the manometer reading, $h$.

■ **Figure P8.25**

**8.26** ꟼWILEY A liquid with $SG = 0.96$, $\mu = 9.2 \times 10^{-4}$ N · s/m², and vapor pressure $p_\nu = 1.2 \times 10^4$ N/m²(abs) is drawn into the syringe as is indicated in Fig P8.26. What is the maximum flowrate if cavitation is not to occur in the syringe?

$p_{atm} = 101$ kPa (abs)

■ **Figure P8.26**

### Section 8.3 Fully Developed Turbulent Flow

**8.27** ꟼWILEY For oil ($SG = 0.86$, $\mu = 0.025$ Ns/m²) flow of 0.3 m³/s through a round pipe with diameter of 500 mm, determine the Reynolds number. Is the flow laminar or turbulent?

**8.28** For air at a pressure of 200 kPa (abs) and temperature of 15 °C, determine the maximum laminar volume flowrate for flow through a 2.0-cm-diameter tube.

**8.29** If the velocity profile for turbulent flow in a pipe is approximated by the power-law profile (Eq. 8.31), at what radial location should a Pitot tube be placed if it is to measure the average velocity in the pipe? Assume $n = 7, 8$, or 9.

**8.30** ▶ ꟼWILEY As shown in **Video V8.10** and Fig. P8.30, the velocity profile for laminar flow in a pipe is quite different from that for turbulent flow. With laminar flow the velocity profile is parabolic; with turbulent flow at Re = 10,000 the velocity profile can be approximated by the power-law profile shown in the figure. **(a)** For laminar flow, determine at what radial location you would place a Pitot tube if it is to measure the average velocity in the pipe. **(b)** Repeat part (a) for turbulent flow with Re = 10,000.

■ **Figure P8.30**

**8.31** ꟼWILEY GO Water at 10 °C flows through a smooth 60-mm-diameter pipe with an average velocity of 8 m/s. Would a layer of rust of height 0.005 mm on the pipe wall protrude through the viscous sublayer? Explain.

**8.32** ▶ (WILEY) When soup is stirred in a bowl, there is considerable turbulence in the resulting motion (see **Video V8.7**). From a very simplistic standpoint, this turbulence consists of numerous intertwined swirls, each involving a characteristic diameter and velocity. As time goes by, the smaller swirls (the fine scale structure) die out relatively quickly, leaving the large swirls that continue for quite some time. Explain why this is to be expected.

**8.33** (WILEY) Determine the thickness of the viscous sublayer in a smooth 8-in.-diameter pipe if the Reynolds number is 25,000.

**8.34** (WILEY) Water at 60 °F flows through a 6-in.-diameter pipe with an average velocity of 15 ft/s. Approximately what is the height of the largest roughness element allowed if this pipe is to be classified as smooth?

### Section 8.4.1 Major Losses (also see Lab Problem 8.1LP)

**8.35** Water flows in a constant-diameter pipe with the following conditions measured: At section (a) $p_a = 32.4$ psi and $z_a = 56.8$ ft; at section (b) $p_b = 29.7$ psi and $z_b = 68.2$ ft. Is the flow from (a) to (b) or from (b) to (a)? Explain.

**8.36** Water is pumped between two tanks as shown in Fig. P8.36. The energy line is as indicated. Is the fluid being pumped from $A$ to $B$ or $B$ to $A$? Explain. Which pipe has the larger diameter: A to the pump or B to the pump? Explain.

■ **Figure P8.36**

**8.37** (WILEY) A person with no experience in fluid mechanics wants to estimate the friction factor for 1-in.-diameter galvanized iron pipe at a Reynolds number of 8,000. The person stumbles across the simple equation of $f = 64/Re$ and uses this to calculate the friction factor. Explain the problem with this approach and estimate the error.

**8.38** (WILEY) During a heavy rainstorm, water from a parking lot completely fills an 18-in.-diameter, smooth, concrete storm sewer. If the flowrate is 10 ft³/s, determine the pressure drop in a 100-ft horizontal section of the pipe. Repeat the problem if there is a 2-ft change in elevation of the pipe per 100 ft of its length.

**8.39** Water flows through a horizontal plastic pipe with a diameter of 0.2 m at a velocity of 10 cm/s. Determine the pressure drop per meter of pipe using the Moody chart.

**8.40** (WILEY GO) For Problem 8.39, calculate the power lost to the friction per meter of pipe.

**8.41** (WILEY) Oil ($SG = 0.9$), with a kinematic viscosity of 0.007 ft²/s, flows in a 3-in.-diameter pipe at 0.01 ft³/s. Determine the head loss per unit length of this flow.

**8.42** (WILEY) Water flows through a 6-in.-diameter horizontal pipe at a rate of 2.0 cfs and a pressure drop of 4.2 psi per 100 ft of pipe. Determine the friction factor.

**8.43** Water flows downward through a vertical 10-mm-diameter galvanized iron pipe with an average velocity of 5.0 m/s and exits as a free jet. There is a small hole in the pipe 4 m above the outlet. Will water leak out of the pipe through this hole, or will air enter into the pipe through the hole? Repeat the problem if the average velocity is 0.5 m/s.

**8.44** (WILEY) Air at standard conditions flows through an 8-in.-diameter, 14.6-ft-long, straight duct with the velocity versus pressure

drop data indicated in the following table. Determine the average friction factor over this range of data.

| $V$ (ft/min) | $\Delta p$ (in. water) |
|---|---|
| 3950 | 0.35 |
| 3730 | 0.32 |
| 3610 | 0.30 |
| 3430 | 0.27 |
| 3280 | 0.24 |
| 3000 | 0.20 |
| 2700 | 0.16 |

**8.45** (WILEY) Water flows through a horizontal 60-mm-diameter galvanized iron pipe at a rate of 0.02 m³/s. If the pressure drop is 135 kPa per 10 m of pipe, do you think this pipe is **(a)** a new pipe, **(b)** an old pipe with a somewhat increased roughness due to aging, or **(c)** a very old pipe that is partially clogged by deposits? Justify your answer.

**8.46** (WILEY) Water flows at a rate of 10 gallons per minute in a new horizontal 0.75-in.-diameter galvanized iron pipe. Determine the pressure gradient, $\Delta p/\ell$, along the pipe.

**8.47** (WILEY) Two equal length, horizontal pipes, one with a diameter of 1 in., the other with a diameter of 2 in., are made of the same material and carry the same fluid at the same flow rate. Which pipe produces the larger head loss? Justify your answer.

**8.48** Carbon dioxide at a temperature of 0 °C and a pressure of 600 kPa (abs) flows through a horizontal 40-mm-diameter pipe with an average velocity of 2 m/s. Determine the friction factor if the pressure drop is 235 N/m² per 10-m length of pipe.

**8.49** Blood (assume $\mu = 4.5 \times 10^{-5}$ lb · s/ft², $SG = 1.0$) flows through an artery in the neck of a giraffe from its heart to its head at a rate of $2.5 \times 10^{-4}$ ft³/s. Assume the length is 10 ft and the diameter is 0.20 in. If the pressure at the beginning of the artery (outlet of the heart) is equivalent to 0.70 ft Hg, determine the pressure at the end of the artery when the head is **(a)** 8 ft above the heart, or **(b)** 6 ft below the heart. Assume steady flow. How much of this pressure difference is due to elevation effects, and how much is due to frictional effects?

**8.50** (WILEY) A 40-m-long, 12-mm-diameter pipe with a friction factor of 0.020 is used to siphon 30 °C water from a tank as shown in

■ **Figure P8.50**

Fig. P8.50. Determine the maximum value of $h$ allowed if there is to be no cavitation within the hose. Neglect minor losses.

**8.51** (WILEY) Gasoline flows in a smooth pipe of 40-mm diameter at a rate of 0.001 m³/s. If it were possible to prevent turbulence from occurring, what would be the ratio of the head loss for the actual turbulent flow compared to that if it were laminar flow?

**8.52** (WILEY) A 3-ft-diameter duct is used to carry ventilating air into a vehicular tunnel at a rate of 9000 ft³/min. Tests show that the pressure drop is 1.5 in. of water per 1500 ft of duct. What is the value of the friction factor for this duct and the approximate size of the equivalent roughness of the surface of the duct?

**8.53** A fluid flows in a smooth pipe with a Reynolds number of 6000. By what percent would the head loss be reduced if the flow could be maintained as laminar flow rather than the expected turbulent flow?

## Section 8.4.2 Minor Losses (also see Lab Problem 8.6LP)

**8.54** Air at standard temperature and pressure flows through a 1-in.-diameter galvanized iron pipe with an average velocity of 8 ft/s. What length of pipe produces a head loss equivalent to **(a)** a flanged 90° elbow, **(b)** a wide-open angle valve, or **(c)** a sharp-edged entrance?

**8.55** Given 90° threaded elbows used in conjunction with copper pipe (drawn tubing) of 0.75-in. diameter, convert the loss for a single elbow to equivalent length of copper pipe for wholly turbulent flow.

**8.56** Air at 80 °F and standard atmospheric pressure flows through a furnace filter with an average velocity of 2.4 ft/s. If the pressure drop across the filter is 0.06 in. of water, what is the loss coefficient for the filter?

**8.57** (WILEY) To conserve water and energy, a "flow reducer" is installed in the shower head as shown in Fig. P8.57. If the pressure at point (1) remains constant and all losses except for that in the flow reducer are neglected, determine the value of the loss coefficient (based on the velocity in the pipe) of the flow reducer if its presence is to reduce the flowrate by a factor of 2. Neglect gravity.

■ **Figure P8.57**

**8.58** (WILEY) Water flows at a rate of 0.040 m³/s in a 0.12-m-diameter pipe that contains a sudden contraction to a 0.06-m-diameter pipe. Determine the pressure drop across the contraction section. How much of this pressure difference is due to losses and how much is due to kinetic energy changes?

**8.59** (WILEY) Water flows from the container shown in Fig. P8.59. Determine the loss coefficient needed in the valve if the water is to "bubble up" 3 in. above the outlet pipe. The entrance is slightly rounded.

■ **Figure P8.59**

**8.60** (See Fluids in the News article titled "New Hi-tech Fountains," Section 8.5.) The fountain shown in Fig. P8.60 is designed to provide a stream of water that rises $h = 10$ ft to $h = 20$ ft above the nozzle exit in a periodic fashion. To do this the water from the pool enters a pump, passes through a pressure regulator that maintains a constant pressure ahead of the flow control valve. The valve is electronically adjusted to provide the desired water height. With $h = 10$ ft the loss coefficient for the valve is $K_L = 50$. Determine the valve loss coefficient needed for $h = 20$ ft. All losses except for the flow control valve are negligible. The area of the pipe is 5 times the area of the exit nozzle.

■ **Figure P8.60**

**8.61** (WILEY) Water flows through the screen in the pipe shown in Fig. P8.61 as indicated. Determine the loss coefficient for the screen.

■ **Figure P8.61**

**8.62** WILEY Air flows though the mitered bend shown in Fig. P8.62 at a rate of 5.0 cfs. To help straighten the flow after the bend, a set of 0.25-in.-diameter drinking straws is placed in the pipe as shown. Estimate the extra pressure drop between points (1) and (2) caused by these straws.

■ **Figure P8.62**

**8.63** Repeat Problem 8.62 if the straws are replaced by a piece of porous foam rubber that has a loss coefficient equal to 5.4.

**8.64** ▶ WILEY As shown in Fig. P8.64, water flows from one tank to another through a short pipe whose length is $n$ times the pipe diameter. Head losses occur in the pipe and at the entrance and exit. (See **Video V8.10.**) Determine the maximum value of $n$ if the major loss is to be no more than 10% of the minor loss and the friction factor is 0.02.

■ **Figure P8.64**

**8.65** WILEY Air flows through the fine mesh gauze shown in Fig. P8.65 with an average velocity of 1.50 m/s in the pipe. Determine the loss coefficient for the gauze.

■ **Figure P8.65**

**8.66** ▶ WILEY Water flows steadily through the 0.75-in.-diameter galvanized iron pipe system shown in **Video V8.16** and Fig. P8.66 at a rate of 0.020 cfs. Your boss suggests that friction losses in the straight pipe sections are negligible compared to losses in the threaded elbows and fittings of the system. Do you agree or disagree with your boss? Support your answer with appropriate calculations.

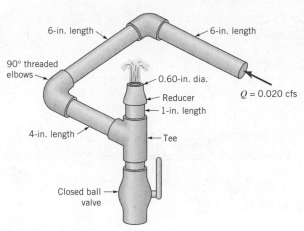

■ **Figure P8.66**

### Section 8.4.3 Noncircular Conduits

**8.67** Given two rectangular ducts with equal cross-sectional area, but different aspect ratios (width/height) of 2 and 4, which will have the greater frictional losses? Explain your answer.

**8.68** WILEY A viscous oil with a specific gravity $SG = 0.85$ and a viscosity of 0.10 Pa·s flows from tank $A$ to tank $B$ through the six rectangular slots indicated in Fig. P8.68. If the total flowrate is 30 mm³/s and minor losses are negligible, determine the pressure in tank $A$.

■ **Figure P8.68**

**8.69** WILEY Air at standard temperature and pressure flows at a rate of 7.0 cfs through a horizontal, galvanized iron duct that has a rectangular cross-sectional shape of 12 in. by 6 in. Estimate the pressure drop per 200 ft of duct.

**8.70** Air flows through a rectangular galvanized iron duct of size 0.30 m by 0.15 m at a rate of 0.068 m³/s. Determine the head loss in 12 m of this duct.

**8.71** WILEY Air at standard conditions flows through a horizontal 1 ft by 1.5 ft rectangular wooden duct at a rate of 5000 ft³/min. Determine the head loss, pressure drop, and power supplied by the fan to overcome the flow resistance in 500 ft of the duct.

### Section 8.5.1 Single Pipes—Determine Pressure Drop

**8.72** ▶ Assume a car's exhaust system can be approximated as 14 ft of 0.125-ft-diameter cast-iron pipe with the equivalent of six 90° flanged elbows and a muffler. (See **Video V8.14.**) The muffler acts as a resistor with a loss coefficient of $K_L = 8.5$. Determine the pressure at the beginning of the exhaust system if the flowrate is 0.10 cfs, the temperature is 250 °F, and the exhaust has the same properties as air.

**8.73** WILEY The pressure at section (2) shown in Fig. P8.73 is not to fall below 60 psi when the flowrate from the tank varies from 0 to 1.0 cfs and the branch line is shut off. Determine the minimum

■ **Figure P8.73**

height, $h$, of the water tank under the assumption that **(a)** minor losses are negligible, **(b)** minor losses are not negligible.

**8.74** Repeat Problem 8.73 with the assumption that the branch line is open so that half of the flow from the tank goes into the branch, and half continues in the main line.

**8.75** The $\frac{1}{2}$-in.-diameter hose shown in Fig. P8.75 can withstand a maximum pressure of 200 psi without rupturing. Determine the maximum length, $\ell$, allowed if the friction factor is 0.022 and the flowrate is 0.010 cfs. Neglect minor losses.

■ **Figure P8.75**

**8.76** The hose shown in Fig. P8.75 will collapse if the pressure within it is lower than 10 psi below atmospheric pressure. Determine the maximum length, $\ell$, allowed if the friction factor is 0.015 and the flowrate is 0.010 cfs. Neglect minor losses.

**8.77** According to fire regulations in a town, the pressure drop in a commercial steel horizontal pipe must not exceed 1.0 psi per 150 ft of pipe for flowrates up to 500 gal/min. If the water temperature is above 50 °F, can a 6-in.-diameter pipe be used?

**8.78** As shown in **Video V8.15** and Fig. P8.78, water "bubbles up" 3 in. above the exit of the vertical pipe attached to three horizontal pipe segments. The total length of the 0.75-in.-diameter galvanized iron pipe between point (1) and the exit is 21 in. Determine the pressure needed at point (1) to produce this flow.

■ **Figure P8.78**

**8.79** Water at 10 °C is pumped from a lake as shown in Fig. P8.79. If the flowrate is 0.011 m³/s, what is the maximum length inlet pipe, $\ell$, that can be used without cavitation occurring?

■ **Figure P8.79**

**8.80** At a ski resort, water at 40 °F is pumped through a 3-in.-diameter, 2000-ft-long steel pipe from a pond at an elevation of 4286 ft to a snow-making machine at an elevation of 4623 ft at a rate of 0.26 ft³/s. If it is necessary to maintain a pressure of 180 psi at the snow-making machine, determine the horsepower added to the water by the pump. Neglect minor losses.

**8.81** Water at 40 °F flows through the coils of the heat exchanger as shown in Fig. P8.81 at a rate of 0.9 gal/min. Determine the pressure drop between the inlet and outlet of the horizontal device.

■ **Figure P8.81**

**8.82** A 70-ft-long, 0.5-in.-diameter hose with a roughness of $\varepsilon = 0.0009$ ft is fastened to a water faucet where the pressure is $p_1$. Determine $p_1$ if there is no nozzle attached and the average velocity in the hose is 6 ft/s. Neglect minor losses and elevation changes.

**8.83** Repeat Problem 8.82 if there is a nozzle of diameter 0.25 in. attached to the end of the hose.

**8.84** Water flows through a 2-in.-diameter pipe with a velocity of 15 ft/s as shown in Fig. P8.84. The relative roughness of the pipe is 0.004, and the loss coefficient for the exit is 1.0. Determine the height, $h$, to which the water rises in the piezometer tube.

■ **Figure P8.84**

**8.85** Water is pumped through a 60-m-long, 0.3-m-diameter pipe from a lower reservoir to a higher reservoir whose surface is 10 m above the lower one. The sum of the minor loss coefficient for the system is $K_L = 14.5$. When the pump adds 40 kW to the water the flowrate is 0.20 m³/s. Determine the pipe roughness.

**8.86** Natural gas ($\rho = 0.0044$ slugs/ft³ and $\nu = 5.2 \times 10^{-5}$ ft²/s) is pumped through a horizontal 6-in.-diameter cast-iron pipe at a rate of 800 lb/hr. If the pressure at section (1) is 50 psi (abs), determine the pressure at section (2) 8 mi downstream if the flow is assumed incompressible. Is the incompressible assumption reasonable? Explain.

**8.87** ▶ As shown in Fig. P8.87, a standard household water meter is incorporated into a lawn irrigation system to measure the volume of water applied to the lawn. Note that these meters measure volume, not volume flowrate. (See **Video V8.16**.) With an upstream pressure of $p_1 = 50$ psi the meter registered that 120 ft³ of water was delivered to the lawn during an "on" cycle. Estimate the upstream pressure, $p_1$, needed if it is desired to have 150 ft³ delivered during an "on" cycle. List any assumptions needed to arrive at your answer.

■ **Figure P8.87**

**8.88** 〔WILEY〕 A fan is to produce a constant air speed of 40 m/s throughout the pipe loop shown in Fig. P8.88. The 3-m-diameter pipes are smooth, and each of the four 90° elbows has a loss coefficient of 0.30. Determine the power that the fan adds to the air.

■ **Figure P8.88**

### Section 8.5.1 Single Pipes—Determine Flowrate (also see Lab Problems 8.3LP and 8.4LP)

**8.89** The turbine shown in Fig. P8.89 develops 400 kW. Determine the flowrate if **(a)** head losses are negligible or **(b)** head loss due to friction in the pipe is considered. Assume $f = 0.02$. *Note*: There may be more than one solution or there may be no solution to this problem.

■ **Figure P8.89**

**8.90** 〔WILEY〕 Water flows from the nozzle attached to the spray tank shown in Fig. P8.90. Determine the flowrate if the loss coefficient for the nozzle (based on upstream conditions) is 0.75 and the friction factor for the rough hose is 0.11.

■ **Figure P8.90**

**8.91** Water flows through the pipe shown in Fig. P8.91. Determine the net tension in the bolts if minor losses are neglected and the wheels on which the pipe rests are frictionless.

■ **Figure P8.91**

**8.92** When the pump shown in Fig. P8.92 adds 0.2 horsepower to the flowing water, the pressures indicated by the two gages are equal. Determine the flowrate.

Length of pipe between gages = 60 ft
Pipe diameter = 0.1 ft
Pipe friction factor = 0.03
Filter loss coefficient = 12

■ **Figure P8.92**

**8.93** Water at 40 °F is pumped from a lake as shown in Fig. P8.93. What is the maximum flowrate possible without cavitation occurring?

■ **Figure P8.93**

**8.94** The pump shown in Fig. P8.94 adds 25 kW to the water and causes a flowrate of 0.04 m³/s. Determine the flowrate expected if the pump is removed from the system. Assume $f = 0.016$ for either case and neglect minor losses.

■ **Figure P8.94**

†**8.95** Gasoline is unloaded from the tanker truck shown in Fig. P8.95 through a 4-in.-diameter rough-surfaced hose. This is a "gravity dump" with no pump to enhance the flowrate. It is claimed that the 8800-gallon capacity truck can be unloaded in 28 minutes. Do you agree with this claim? Support your answer with appropriate calculations.

■ **Figure P8.95**

**8.96** (WILEY GO) The pump shown in Fig. P8.96 delivers a head of 250 ft to the water. Determine the power that the pump adds to the water. The difference in elevation of the two ponds is 200 ft.

■ **Figure P8.96**

**8.97** (WILEY) Water flows through two sections of the vertical pipe shown in Fig. P8.97. The bellows connection cannot support any force in the vertical direction. The 0.4-ft-diameter pipe weights 0.2 lb/ft, and the friction factor is assumed to be 0.02. At what velocity will the force, $F$, required to hold the pipe be zero?

■ **Figure P8.97**

**8.98** Water is circulated from a large tank, through a filter, and back to the tank as shown in Fig. P8.98. The power added to the water by the pump is 200 ft · lb/s. Determine the flowrate through the filter.

■ **Figure P8.98**

### Section 8.5.1 Single Pipes—Determine Diameter

**8.99** (WILEY) A certain process requires 2.3 cfs of water to be delivered at a pressure of 30 psi. This water comes from a large-diameter supply main in which the pressure remains at 60 psi. If the galvanized iron pipe connecting the two locations is 200 ft long and contains six threaded 90° elbows, determine the pipe diameter. Elevation differences are negligible.

**8.100** A flowrate of 3.5 ft³/s is to be maintained in a horizontal aluminum pipe ($\varepsilon = 5 \times 10^{-6}$ ft). The inlet and outlet pressures are 65 psi and 30 psi, respectively, and the pipe length is 500 ft. Determine the diameter of the pipe.

**8.101** Water is pumped between two large open reservoirs through 1.5 km of smooth pipe. The water surfaces in the two reservoirs are at the same elevation. When the pump adds 20 kW to the water, the flowrate is 1 m³/s. If minor losses are negligible, determine the pipe diameter.

**8.102** (WILEY) Determine the diameter of a steel pipe that is to carry 2000 gal/min of gasoline with a pressure drop of 5 psi per 100 ft of horizontal pipe.

**8.103** (WILEY) Water is to be moved from a large, closed tank in which the air pressure is 20 psi into a large, open tank through 2000 ft of smooth pipe at the rate of 3 ft³/s. The fluid level in the open tank is 150 ft below that in the closed tank. Determine the required diameter of the pipe. Neglect minor losses.

**8.104** Water flows downward through a vertical smooth pipe. When the flowrate is 0.5 ft³/s, there is no change in pressure along the pipe. Determine the diameter of the pipe.

**8.105** Rainwater flows through the galvanized iron downspout shown in Fig. P8.105 at a rate of 0.006 m³/s. Determine the size of the downspout cross section if it is a rectangle with an aspect ratio

70 mm

4 m

g

3 m

■ **Figure P8.105**

of 1.7 to 1 and it is completely filled with water. Neglect the velocity of the water in the gutter at the free surface and the head loss associated with the elbow.

**\*8.106** (WILEY) Repeat Problem 8.105 if the downspout is circular.

**8.107** For a given head loss per unit length, what effect on the flowrate does doubling the pipe diameter have if the flow is (a) laminar, or (b) completely turbulent?

## Section 8.5.2 Multiple Pipe Systems

**8.108** (WILEY) Air, assumed incompressible, flows through the two pipes shown in Fig. P8.108. Determine the flowrate if minor losses are neglected and the friction factor in each pipe is 0.015. Determine the flowrate if the 0.5-in.-diameter pipe were replaced by a 1-in.-diameter pipe. Comment on the assumption of incompressibility.

$p = 0.5$ psi
$T = 150°F$

1 in.

0.50 in.

20 ft

20 ft

■ **Figure P8.108**

**\*8.109** Repeat Problem 8.108 if the pipes are galvanized iron and the friction factors are not known a priori.

**†8.110** Estimate the power that the human heart must impart to the blood to pump it through the two carotid arteries from the heart to the brain. List all assumptions and show all calculations.

**8.111** (WILEY) The flowrate between tank $A$ and tank $B$ shown in Fig. P8.111 is to be increased by 30% (i.e., from $Q$ to $1.30Q$) by the addition of a second pipe (indicated by the dotted lines) running from node $C$ to tank $B$. If the elevation of the free surface in tank $A$ is 25 ft above that in tank $B$, determine the diameter, $D$, of this new pipe. Neglect minor losses and assume that the friction factor for each pipe is 0.02.

6-in. diameter;
600 ft long

6-in. diameter;
500 ft long

A

C

B

Diameter $D$, 500 ft long

■ **Figure P8.111**

**8.112** (WILEY) With the valve closed, water flows from tank $A$ to tank $B$ as shown in Fig. P8.112. What is the flowrate into tank $B$ when the valve is opened to allow water to flow into tank $C$ also? Neglect all minor losses and assume that the friction factor is 0.02 for all pipes.

Elevation = 15 m

Diameter of each pipe = 0.10 m

Elevations = 0

A

80 m

40 m

B

C

75 m

■ **Figure P8.112**

**\*8.113** Repeat Problem 8.112 if the friction factors are not known, but the pipes are steel pipes.

**8.114** (WILEY GO) The three water-filled tanks shown in Fig. P8.114 are connected by pipes as indicated. If minor losses are neglected, determine the flowrate in each pipe.

Elevation = 60 m

Elevation = 20 m

Elevation = 0

$D = 0.10$ m
$\ell = 200$ m
$f = 0.015$

$D = 0.08$ m
$\ell = 200$ m
$f = 0.020$

$D = 0.08$ m
$\ell = 400$ m
$f = 0.020$

■ **Figure P8.114**

**8.115** (WILEY) (See Fluids in the News article titled "Deepwater Pipeline," Section 8.5.2.) Five oil fields, each producing an output of $Q$ barrels per day, are connected to the 28-in.-diameter "mainline pipe" ($A$–$B$–$C$) by 16-in.-diameter "lateral pipes" as shown in Fig. P8.115. The friction factor is the same for each of the pipes and elevation effects are negligible. (a) For section $A$–$B$ determine the ratio of the pressure drop per mile in the mainline pipe to that in the lateral pipes. (b) Repeat the calculations for section $B$–$C$.

$Q$

Lateral

$Q$

$Q$

A

B

C

Main line

$Q$

$Q$

■ **Figure P8.115**

**†8.116** As shown in Fig. P8.116, cold water ($T = 50$ °F) flows from the water meter to either the shower or the hot water heater. In the hot water heater it is heated to a temperature of 150 °F. Thus, with equal amounts of hot and cold water, the shower is at a comfortable 100 °F. However, when the dishwasher is turned on, the

■ **Figure P8.116**

shower water becomes too cold. Indicate how you would predict this new shower temperature (assume the shower faucet is not adjusted). State any assumptions needed in your analysis.

### Section 8.6 Pipe Flowrate Measurement (also see Lab Problem 8.2LP)

**8.117** (WILEY) A 2-in.-diameter orifice plate is inserted in a 3-in.-diameter pipe. If the water flowrate through the pipe is 0.90 cfs, determine the pressure difference indicated by a manometer attached to the flowmeter.

**8.118** Air to ventilate an underground mine flows through a large 2-m-diameter pipe. A crude flowrate meter is constructed by placing a sheet metal "washer" between two sections of the pipe. Estimate the flowrate if the hole in the sheet metal has a diameter of 1.6 m and the pressure difference across the sheet metal is 8.0 mm of water.

**8.119** Water flows through a 40-mm-diameter nozzle meter in a 75-mm-diameter pipe at a rate of 0.015 m³/s. Determine the pressure difference across the nozzle if the temperature is **(a)** 10 °C, or **(b)** 80 °C.

**8.120** Gasoline flows through a 35-mm-diameter pipe at a rate of 0.0032 m³/s. Determine the pressure drop across a flow nozzle placed in the line if the nozzle diameter is 20 mm.

**8.121** Air at 200 °F and 60 psia flows in a 4-in.-diameter pipe at a rate of 0.52 lb/s. Determine the pressure at the 2-in.-diameter throat of a Venturi meter placed in the pipe.

**8.122** A 2.5-in.-diameter flow nozzle meter is installed in a 3.8-in.-diameter pipe that carries water at 160 °F. If the air–water manometer used to measure the pressure difference across the meter indicates a reading of 3.1 ft, determine the flowrate.

**8.123** A 0.064-m-diameter nozzle meter is installed in a 0.097-m-diameter pipe that carries water at 60 °C. If the inverted air–water U-tube manometer used to measure the pressure difference across the meter indicates a reading of 1 m, determine the flowrate.

**8.124** A 50-mm-diameter nozzle meter is installed at the end of a 80-mm-diameter pipe through which air flows. A manometer attached to the static pressure tap just upstream from the nozzle indicates a pressure of 7.3 mm of water. Determine the flowrate.

**8.125** Water flows through the Venturi meter shown in Fig. P8.125. The specific gravity of the manometer fluid is 1.52. Determine the flowrate.

■ **Figure P8.125**

**8.126** If the fluid flowing in Problem 8.125 were air, what would the flowrate be? Would compressibility effects be important? Explain.

**8.127** (WILEY) Water flows through the orifice meter shown in Fig. P8.127 at a rate of 0.10 cfs. If $d = 0.1$ ft, determine the value of $h$.

■ **Figure P8.127**

**8.128** Water flows through the orifice meter shown in Fig P8.127 such that $h = 1.6$ ft with $d = 1.5$ in. Determine the flowrate.

**8.129** (WILEY) Water flows through the orifice meter shown in Fig. P8.127 at a rate of 0.10 cfs. If $h = 3.8$ ft, determine the value of $d$.

**8.130** ▶ (WILEY) The scale reading on the rotameter shown in Fig. P8.130 and **Video V8.15** (also see Fig. 8.46) is directly proportional to the volumetric flowrate. With a scale reading of 2.6 the water bubbles up approximately 3 in. How far will it bubble up if the scale reading is 5.0?

■ **Figure P8.130**

### ■ Lab Problems

**8.1LP** This problem involves the determination of the friction factor in a pipe for laminar and transitional flow conditions. To proceed with this problem, go to Appendix H, which is located in *WileyPLUS* or on the book's web site, www.wiley.com/college/munson.

**8.2LP** This problem involves the calibration of an orifice meter and a Venturi meter. To proceed with this problem, go to Appendix H, which is located in *WileyPLUS* or on the book's web site, www.wiley.com/college/munson.

**8.3LP**  This problem involves the flow of water from a tank and through a pipe system. To proceed with this problem, go to Appendix H, which is located in *WileyPLUS* or on the book's web site, www.wiley.com/college/munson.

**8.4LP**  This problem involves the flow of water pumped from a tank and through a pipe system. To proceed with this problem, go to Appendix H, which is located in *WileyPLUS* or on the book's web site, www.wiley.com/college/munson.

**8.5LP**  This problem involves the pressure distribution in the entrance region of a pipe. To proceed with this problem, go to Appendix H, which is located in *WileyPLUS* or on the book's web site, www.wiley.com/college/munson.

**8.6LP**  This problem involves the power loss due to friction in a coiled pipe. To proceed with this problem, go to Appendix H, which is located in *WileyPLUS* or on the book's web site, www.wiley.com/college/munson.

■  **Lifelong Learning Problems**

**8.1LL**  The field of bioengineering has undergone significant growth in recent years. Some universities have undergraduate and graduate programs in this field. Bioengineering applies engineering principles to help solve problems in the medical field for human health. Obtain information about bioengineering applications in blood flow. Summarize your findings in a brief report.

**8.2LL**  Data used in the Moody diagram were first published in 1944. Since then, there have been many innovations in pipe material, pipe design, and measurement techniques. Investigate whether there have been any improvements or enhancements to the Moody chart. Summarize your findings in a brief report.

**8.3LL**  As discussed in Section 8.4.2, flow separation in pipes can lead to losses (we will also see in Chapter 9 that external flow separation is a significant problem). For external flows, many mechanisms have been devised to help mitigate and control flow separation from the surface, for example, from the wing of an airplane. Investigate either passive or active flow control mechanisms that can reduce or eliminate internal flow separation (e.g., flow separation in a diffuser). Summarize your findings in a brief report.

■  **FE Exam Problems**

Sample FE (Fundamentals of Engineering) exam questions for fluid mechanics are provided in *WileyPLUS* or on the book's web site, www.wiley.com/college/munson.

■  **Computational Fluid Dynamics (CFD)**

The CFD problems associated with this chapter have been developed for use with the ANSYS Academic CFD software package that is associated with this text. See *WileyPLUS* or the book's web site (www.wiley.com/college/munson) for additional details.

**8.1CFD**  This CFD problem involves the simulation of fully developed pipe flow and how the Reynolds number affects the wall friction. To proceed with this problem, go to *WileyPLUS* or the book's web site, www.wiley.com/college/munson.

**There are additional CFD problems located in *WileyPLUS*.**

# 9 Flow over Immersed Bodies

CHAPTER OPENING PHOTO: Impulsive start of flow past an array of cylinders: The complex structure of laminar flow past a relatively simple geometric structure illustrates why it is often difficult to obtain exact analytical results for external flows. (Dye in water.) (Photograph courtesy of ONERA, The French Aerospace Lab.)

## Learning Objectives

After completing this chapter, you should be able to:

- identify and discuss the features of external flow.
- explain the fundamental characteristics of a boundary layer, including laminar, transitional, and turbulent regimes.
- calculate boundary layer parameters for flow past a flat plate.
- provide a description of boundary layer separation.
- calculate the lift and drag forces for various objects.

In this chapter we consider various aspects of the flow over bodies that are immersed in a fluid. Examples include the flow of air around airplanes, automobiles, and falling snowflakes or the flow of water around submarines and fish. In these situations the object is completely surrounded by the fluid and the flows are termed *external flows*.

External flows involving air are often termed aerodynamics in response to the important external flows produced when an object such as an airplane flies through the atmosphere. Although this field of external flows is extremely important, there are many other examples that are of equal importance. The fluid force (lift and drag) on surface vehicles (cars, trucks, bicycles) has become a very important topic. By correctly designing cars and trucks, it has become possible to greatly decrease the fuel consumption and improve the handling characteristics of the vehicle. Similar efforts have resulted in improved ships, whether they are surface vessels (surrounded by two fluids, air and water) or submersible vessels (surrounded completely by water).

*Many practical situations involve flow past objects.*

Other applications of external flows involve objects that are not completely surrounded by fluid, although they are placed in some external-type flow. For example, the proper design of a building (whether it is your house or a tall skyscraper) must include consideration of the various wind effects involved.

As with other areas of fluid mechanics, various approaches (theoretical, numerical, and experimental) are used to obtain information on the fluid forces developed by external flows.

**480**

(a)

(b)

■ **Figure 9.1** (a) **Flow past a full-sized streamlined vehicle in the GM aerodynamics laboratory wind tunnel, an 18-ft by 34-ft test section facility driven by a 4000-hp, 43-ft-diameter fan. (Photograph courtesy of General Motors Corporation.)** (b) **Predicted streamlines for flow past a Formula 1 race car as obtained by using computational fluid dynamics techniques. (Courtesy of ANSYS, Inc.)**

Theoretical (i.e., analytical) techniques can provide some of the needed information about such flows. However, because of the complexities of the governing equations and the complexities of the geometry of the objects involved, the amount of information obtained from purely theoretical methods is limited.

Much of the information about external flows comes from experiments carried out, for the most part, on scale models of the actual objects. Such testing includes the obvious wind tunnel testing of model airplanes, buildings, and even entire cities. In some instances the actual device, not a model, is tested in wind tunnels. Figure 9.1a shows a test of a vehicle in a wind tunnel. Better performance of cars, bikes, skiers, and numerous other objects has resulted from testing in wind tunnels. The use of water tunnels and towing tanks also provides useful information about the flow around ships and other objects. With advancement in computational fluid dynamics, or CFD, numerical methods are also capable of predicting external flows past objects. Figure 9.1b shows streamlines around a Formula 1 car as predicted by CFD. Appendix A provides an introduction to CFD.

In this chapter we consider characteristics of external flow past a variety of objects. We investigate the qualitative aspects of such flows and learn how to determine the various forces on objects surrounded by a moving liquid.

# 9.1 General External Flow Characteristics

A body immersed in a moving fluid experiences a resultant force due to the interaction between the body and the fluid surrounding it. In some instances (such as an airplane flying through still air) the fluid far from the body is stationary and the body moves through the fluid with velocity $U$. In other instances (such as the wind blowing past a building) the body is stationary and the fluid flows past the body with velocity $U$. In any case, we can fix the coordinate system in the body and treat the situation as fluid flowing past a stationary body with velocity $U$, the *upstream velocity.* For the purposes of this book, we will assume that the upstream velocity is constant in both time and location. That is, there is a uniform, constant velocity fluid flowing past the object. In actual situations this is often not true. For example, the wind blowing past a smokestack is nearly always turbulent and gusty (unsteady) and probably not of uniform velocity from the top to the bottom of the stack. Usually the unsteadiness and nonuniformity are of minor importance.

*For external flows it is usually easiest to use a coordinate system fixed to the object.*

■ **Figure 9.2** Flow classification: (*a*) two-dimensional, (*b*) axisymmetric, (*c*) three-dimensional.

*V9.1 Space Shuttle landing*

Even with a steady, uniform upstream flow, the flow in the vicinity of an object may be unsteady. Examples of this type of behavior include the flutter that is sometimes found in the flow past airfoils (wings), the regular oscillation of telephone wires that "sing" in a wind, and the irregular turbulent fluctuations in the wake regions behind bodies.

The structure of an external flow and the ease with which the flow can be described and analyzed often depend on the nature of the body in the flow. Three general categories of bodies are shown in Fig. 9.2. They include (a) two-dimensional objects (infinitely long and of constant cross-sectional size and shape), (b) axisymmetric bodies (formed by rotating their cross-sectional shape about the axis of symmetry), and (c) three-dimensional bodies that may or may not possess a line or plane of symmetry. In practice there can be no truly two-dimensional bodies—nothing extends to infinity. However, many objects are sufficiently long so that the end effects are negligibly small.

Another classification of body shape can be made depending on whether the body is streamlined or blunt. The flow characteristics depend strongly on the amount of streamlining present. In general, *streamlined bodies* (i.e., airfoils, racing cars, etc.) have little effect on the surrounding fluid, compared with the effect that *blunt bodies* (i.e., parachutes, buildings, etc.) have on the fluid. Usually, but not always, it is easier to force a streamlined body through a fluid than it is to force a similar-sized blunt body at the same velocity. There are important exceptions to this basic rule.

### 9.1.1 Lift and Drag Concepts

*A body interacts with the surrounding fluid through pressure and shear stresses.*

When any body moves through a fluid, an interaction between the body and the fluid occurs; this effect can be given in terms of the forces at the fluid–body interface. These forces can be described in terms of the stresses—wall shear stresses on the body, $\tau_w$, due to viscous effects and normal stresses due to the pressure, $p$. Typical shear stress and pressure distributions are shown in Figs. 9.3*a* and 9.3*b*. Both $\tau_w$ and $p$ vary in magnitude and direction along the surface.

It is often useful to know the detailed distribution of shear stress and pressure over the surface of the body, although such information is difficult to obtain. Many times, however, only the integrated or resultant effects of these distributions are needed. The resultant force in the direction of the upstream velocity is termed the **drag**, $\mathscr{D}$, and the resultant force normal to the upstream

■ **Figure 9.3** Forces from the surrounding fluid on a two-dimensional object: (*a*) pressure force, (*b*) viscous force, and (*c*) resultant force (lift and drag).

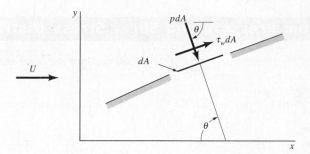

■ **Figure 9.4** Pressure and shear forces on a small element of the surface of a body.

velocity is termed the ***lift***, $\mathscr{L}$, as is indicated in Fig. 9.3c. For some three-dimensional bodies there may also be a side force that is perpendicular to the plane containing $\mathscr{D}$ and $\mathscr{L}$.

The resultant of the shear stress and pressure distributions can be obtained by integrating the effect of these two quantities on the body surface as is indicated in Fig. 9.4. The $x$ and $y$ components of the fluid force on the small area element $dA$ are

$$dF_x = (p\, dA)\cos\theta + (\tau_w\, dA)\sin\theta$$

and

$$dF_y = -(p\, dA)\sin\theta + (\tau_w\, dA)\cos\theta$$

Thus, the net $x$ and $y$ components of the force on the object are

$$\mathscr{D} = \int dF_x = \int p\cos\theta\, dA + \int \tau_w\sin\theta\, dA \tag{9.1}$$

and

$$\mathscr{L} = \int dF_y = -\int p\sin\theta\, dA + \int \tau_w\cos\theta\, dA \tag{9.2}$$

*Lift and drag on a section of a body depend on the orientation of the surface.*

Of course, to carry out the integrations and determine the lift and drag, we must know the body shape (i.e., $\theta$ as a function of location along the body) and the distribution of $\tau_w$ and $p$ along the surface. These distributions are often extremely difficult to obtain, either experimentally or theoretically. The pressure distribution can be obtained experimentally by use of a series of static pressure taps along the body surface. On the other hand, it is usually quite difficult to measure the wall shear stress distribution.

## F l u i d s   i n   t h e   N e w s

**Pressure-sensitive paint** For many years, the conventional method for measuring *surface pressure* has been to use static pressure taps consisting of small holes on the surface connected by hoses from the holes to a pressure measuring device. Pressure-sensitive paint (PSP) is now gaining acceptance as an alternative to the static surface pressure ports. The PSP material is typically a luminescent compound that is sensitive to the pressure on it and can be excited by an appropriate light that is captured by special video imaging equipment. Thus, it provides a quantitative mea-

sure of the surface pressure. One of the biggest advantages of PSP is that it is a global measurement technique, measuring pressure over the entire surface, as opposed to discrete points. PSP also has the advantage of being nonintrusive to the flow field. Although static pressure port holes are small, they do alter the surface and can slightly alter the flow, thus affecting downstream ports. In addition, the use of PSP eliminates the need for a large number of pressure taps and connecting tubes. This allows pressure measurements to be made in less time and at a lower cost.

It is seen that both the shear stress and pressure force contribute to the lift and drag, since for an arbitrary body $\theta$ is neither zero nor 90° along the entire body. The exception is a flat plate aligned either parallel to the upstream flow ($\theta = 90°$) or normal to the upstream flow ($\theta = 0$) as is discussed in Example 9.1.

## EXAMPLE 9.1 Drag from Pressure and Shear Stress Distributions

**GIVEN** Air at standard conditions flows past a flat plate as is indicated in Fig. E9.1. In case (a) the plate is parallel to the upstream flow, and in case (b) it is perpendicular to the upstream flow. The pressure and shear stress distributions on the surface are as indicated (obtained either by experiment or theory).

**FIND** Determine the lift and drag on the plate.

■ **Figure E9.1**

## SOLUTION

For either orientation of the plate, the lift and drag are obtained from Eqs. 9.1 and 9.2. With the plate parallel to the upstream flow we have $\theta = 90°$ on the top surface and $\theta = 270°$ on the bottom surface so that the lift and drag are given by

$$\mathcal{L} = -\int_{top} p \, dA + \int_{bottom} p \, dA = 0$$

and

$$\mathcal{D} = \int_{top} \tau_w \, dA + \int_{bottom} \tau_w \, dA = 2\int_{top} \tau_w \, dA \qquad (1)$$

where we have used the fact that because of symmetry the shear stress distribution is the same on the top and the bottom surfaces, as is the pressure also [whether we use gage ($p = 0$) or absolute ($p = p_{atm}$) pressure]. There is no lift generated—the plate does not know up from down. With the given shear stress distribution, Eq. 1 gives

$$\mathcal{D} = 2\int_{x=0}^{4\,ft} \left(\frac{1.24 \times 10^{-3}}{x^{1/2}} \, lb/ft^2\right)(10\,ft)\,dx$$

or

$$\mathcal{D} = 0.0992 \, lb \qquad \text{(Ans)}$$

With the plate perpendicular to the upstream flow, we have $\theta = 0°$ on the front and $\theta = 180°$ on the back. Thus, from Eqs. 9.1 and 9.2

$$\mathcal{L} = \int_{front} \tau_w \, dA - \int_{back} \tau_w \, dA = 0$$

and

$$\mathcal{D} = \int_{front} p \, dA - \int_{back} p \, dA$$

Again there is no lift because the pressure forces act parallel to the upstream flow (in the direction of $\mathcal{D}$ not $\mathcal{L}$) and the shear stress is symmetrical about the center of the plate. With the given relatively large pressure on the front of the plate (the center of the plate is a stagnation point) and the negative pressure (less than the upstream pressure) on the back of the plate, we obtain the following drag

$$\mathcal{D} = \int_{y=-2}^{2\,ft} \left[ 0.744\left(1 - \frac{y^2}{4}\right) lb/ft^2 - (-0.893)\,lb/ft^2 \right] (10\,ft)\,dy$$

or

$$\mathcal{D} = 55.6 \, lb \qquad \text{(Ans)}$$

**COMMENTS** Clearly there are two mechanisms responsible for the drag. On the ultimately streamlined body (a zero-thickness flat plate parallel to the flow) the drag is entirely due to the shear stress at the surface and, in this example, is relatively small. For the ultimately blunted body (a flat plate normal to the upstream flow) the drag is entirely due to the pressure difference between the front and back portions of the object and, in this example, is relatively large.

If the flat plate were oriented at an arbitrary angle relative to the upstream flow as indicated in Fig. E9.1c, there would be both a lift and a drag, each of which would be dependent on both the shear stress and the pressure. Both the pressure and shear stress distributions would be different for the top and bottom surfaces.

(c)

■ **Figure E9.1** *(Continued)*

Although Eqs. 9.1 and 9.2 are valid for any body, the difficulty in their use lies in obtaining the appropriate shear stress and pressure distributions on the body surface. Considerable effort has gone into determining these quantities, but because of the various complexities involved, such information is available only for certain simple situations.

*Lift coefficients and drag coefficients are dimensionless forms of lift and drag.*

Without detailed information concerning the shear stress and pressure distributions on a body, Eqs. 9.1 and 9.2 cannot be used. The widely used alternative is to define dimensionless lift and drag coefficients and determine their approximate values by means of either a simplified analysis, some numerical technique, or an appropriate experiment. The *lift coefficient, $C_L$,* and *drag coefficient, $C_D$,* are defined as

$$C_L = \frac{\mathscr{L}}{\frac{1}{2}\rho U^2 A}$$

and

$$C_D = \frac{\mathscr{D}}{\frac{1}{2}\rho U^2 A}$$

where $A$ is a characteristic area of the object (see Chapter 7). Typically, $A$ is taken to be *frontal area*—the projected area seen by a person looking toward the object from a direction parallel to the upstream velocity, $U$, as indicated by the figure in the margin. It would be the area of the shadow of the object projected onto a screen normal to the upstream velocity as formed by a light shining along the upstream flow. In other situations $A$ is taken to be the *planform area*—the projected area seen by an observer looking toward the object from a direction normal to the upstream velocity (i.e., from "above" it). Obviously, which characteristic area is used in the definition of the lift and drag coefficients must be clearly stated.

### 9.1.2  Characteristics of Flow Past an Object

External flows past objects encompass an extremely wide variety of fluid mechanics phenomena. Clearly the character of the flow field is a function of the shape of the body. Flows past relatively simple geometric shapes (i.e., a sphere or circular cylinder) are expected to have less complex flow fields than flows past a complex shape such as an airplane or a tree. However, even the simplest-shaped objects produce rather complex flows.

For a given-shaped object, the characteristics of the flow depend very strongly on various parameters such as size, orientation, speed, and fluid properties. As is discussed in Chapter 7, according to dimensional analysis arguments, the character of the flow should depend on the various dimensionless parameters involved. For typical external flows the most important of these parameters are the Reynolds number, $Re = \rho U \ell / \mu = U \ell / \nu$, the Mach number, $Ma = U/c$, and for flows with a free surface (i.e., flows with an interface between two fluids, such as the flow past a surface ship), the Froude number, $Fr = U/\sqrt{g\ell}$. (Recall that $\ell$ is some characteristic length of the object and $c$ is the speed of sound.)

For the present, we consider how the external flow and its associated lift and drag vary as a function of the Reynolds number. Recall that the Reynolds number represents the ratio of inertial effects to viscous effects. In the absence of all viscous effects ($\mu = 0$), the Reynolds number is infinite. On the other hand, in the absence of all inertial effects (negligible mass or $\rho = 0$), the Reynolds number is zero. Clearly, any actual flow will have a Reynolds number between (but not including) these two extremes. The nature of the flow past a body depends strongly on whether $Re \gg 1$ or $Re \ll 1$.

Most external flows with which we are familiar are associated with moderately sized objects with a characteristic length on the order of $0.01 \text{ m} < \ell < 10 \text{ m}$. In addition, typical upstream velocities are on the order of $0.01 \text{ m/s} < U < 100 \text{ m/s}$, and the fluids involved are typically water or air. The resulting Reynolds number range for such flows is approximately $10 < Re < 10^9$. This is shown by the figure in the margin for air. As a rule of thumb, flows with $Re > 100$ are dominated by inertial effects, whereas flows with $Re < 1$ are dominated by viscous effects. Hence, most familiar external flows are dominated by inertia.

On the other hand, there are many external flows in which the Reynolds number is considerably less than 1, indicating in some sense that viscous forces are more important than inertial forces. The gradual settling of small particles of dirt in a lake or stream is governed by low Reynolds number flow principles because of the small diameter of the particles and their small settling speed. Similarly, the Reynolds number for objects moving through large-viscosity oils is small because $\mu$ is large. The general differences between small and large Reynolds number flow past streamlined and blunt objects can be illustrated by considering flows past two objects—one a flat plate parallel to the upstream velocity and the other a circular cylinder.

Flows past three flat plates of length $\ell$ with Re $= \rho U \ell / \mu = 0.1$, 10, and $10^7$ are shown in Fig. 9.5. If the Reynolds number is small, the viscous effects are relatively strong and the plate affects the uniform upstream flow far ahead, above, below, and behind the plate. To reach that portion of the flow field where the velocity has been altered by less than 1% of its undisturbed value (i.e., $U - u < 0.01\,U$), we must travel relatively far from the plate. In low Reynolds number flows, the viscous effects are felt far from the object in all directions.

As the Reynolds number is increased (by increasing $U$, for example), the region in which viscous effects are important becomes smaller in all directions except downstream, as is shown in

*For low Reynolds number flows, viscous effects are felt far from the object.*

■ **Figure 9.5** Character of the steady, viscous flow past a flat plate parallel to the upstream velocity: (*a*) low Reynolds number flow, (*b*) moderate Reynolds number flow, (*c*) large Reynolds number flow.

Fig. 9.5*b*. One does not need to travel very far ahead, above, or below the plate to reach areas in which the viscous effects of the plate are not felt. The streamlines are displaced from their original uniform upstream conditions, but the displacement is not as great as for the Re = 0.1 situation shown in Fig. 9.5*a*.

If the Reynolds number is large (but not infinite), the flow is dominated by inertial effects and the viscous effects are negligible everywhere except in a region very close to the plate and in the relatively thin *wake region* behind the plate, as shown in Fig. 9.5*c*. Since the fluid viscosity is not zero (Re < ∞), it follows that the fluid must stick to the solid surface (the no-slip boundary condition). There is a thin *boundary layer* region of thickness $\delta = \delta(x) \ll \ell$ (i.e., thin relative to the length of the plate) next to the plate in which the fluid velocity changes from the upstream value of $u = U$ to zero velocity on the plate. The thickness of this layer increases in the direction of flow, starting from zero at the forward or leading edge of the plate. The flow within the boundary layer may be laminar or turbulent, depending on various parameters involved.

The streamlines of the flow outside of the boundary layer are nearly parallel to the plate. As we will see in the next section, the slight displacement of the external streamlines that are outside of the boundary layer is due to the thickening of the boundary layer in the direction of flow. The existence of the plate has very little effect on the streamlines outside of the boundary layer—either ahead, above, or below the plate. On the other hand, the wake region is due entirely to the viscous interaction between the fluid and the plate.

One of the great advancements in fluid mechanics occurred in 1904 as a result of the insight of Ludwig Prandtl (1875–1953), a German physicist and aerodynamicist. He conceived of the idea of the boundary layer—a thin region on the surface of a body in which viscous effects are very important and outside of which the fluid behaves essentially as if it were inviscid. Clearly the actual fluid viscosity is the same throughout; only the relative importance of the viscous effects (due to the velocity gradients) is different within or outside of the boundary layer. As is discussed in the next section, by using such a hypothesis it is possible to simplify the analysis of large Reynolds number flows, thereby allowing solution to external flow problems that are otherwise still unsolvable.

As with the flow past the flat plate described above, the flow past a blunt object (such as a circular cylinder) also varies with Reynolds number. In general, the larger the Reynolds number, the smaller the region of the flow field in which viscous effects are important. For objects that are not sufficiently streamlined, however, an additional characteristic of the flow is observed. This is termed *flow separation* and is illustrated by the figure in the margin and in Fig. 9.6.

Low Reynolds number flow (Re = $UD/\nu$ < 1) past a circular cylinder is characterized by the fact that the presence of the cylinder and the accompanying viscous effects are felt throughout a relatively large portion of the flow field. As is indicated in Fig. 9.6*a*, for Re = $UD/\nu$ = 0.1, the viscous effects are important several diameters in any direction from the cylinder. A somewhat surprising characteristic of this flow is that the streamlines are essentially symmetric about the center of the cylinder—the streamline pattern is the same in front of the cylinder as it is behind the cylinder.

As the Reynolds number is increased, the region ahead of the cylinder in which viscous effects are important becomes smaller, with the viscous region extending only a short distance ahead of the cylinder. The viscous effects are convected downstream, and the flow loses its upstream to downstream symmetry. Another characteristic of external flows becomes important—the flow separates from the body at the *separation location* as indicated in Fig. 9.6*b*. With the increase in Reynolds number, the fluid inertia becomes more important and at some location on the body, denoted the separation location, the fluid's inertia is such that it cannot follow the curved path around to the rear of the body. The result is a separation bubble behind the cylinder in which some of the fluid is actually flowing upstream, against the direction of the upstream flow. (See the photograph at the beginning of this chapter.)

At still larger Reynolds numbers, the area affected by the viscous forces is forced farther downstream until it involves only a thin ($\delta \ll D$) boundary layer on the front portion of the cylinder and an irregular, unsteady (perhaps turbulent) wake region that extends far downstream of the cylinder. The fluid in the region outside of the boundary layer and wake region flows as if it were inviscid. Of course, the fluid viscosity is the same throughout the entire flow field. Whether or not viscous effects are important depends on which region of the flow field we consider. The velocity gradients within the boundary layer and wake regions are much larger than those in the remainder of the flow

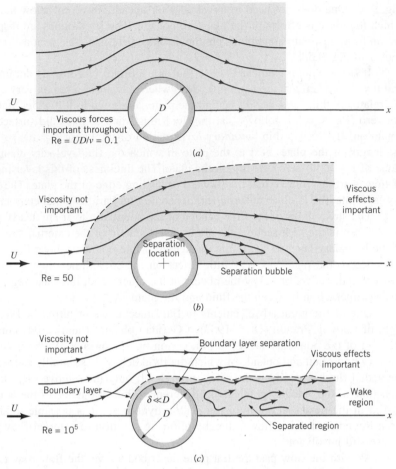

**■ Figure 9.6**   Character of the steady, viscous flow past a circular cylinder: (*a*) low Reynolds number flow, (*b*) moderate Reynolds number flow, (*c*) large Reynolds number flow.

*V9.3 Human aerodynamic wake*

*Most familiar flows involve large Reynolds numbers.*

field. Since the shear stress (i.e., viscous effect) is the product of the fluid viscosity and the velocity gradient, it follows that viscous effects are confined to the boundary layer and wake regions.

The characteristics described in Figs. 9.5 and 9.6 for flow past a flat plate and a circular cylinder are typical of flows past streamlined and blunt bodies, respectively. The nature of the flow depends strongly on the Reynolds number. (See Ref. 31 for many examples illustrating this behavior.) Most familiar flows are similar to the large Reynolds number flows depicted in Figs. 9.5*c* and 9.6*c*, rather than the low Reynolds number flow situations. (See the photograph at the beginning of Chapters 7 and 11.) In the remainder of this chapter we will investigate more thoroughly these ideas and determine how to calculate the forces on immersed bodies.

---

# EXAMPLE 9.2    Characteristics of Flow Past Objects

**GIVEN**   It is desired to experimentally determine the various characteristics of flow past a car as shown in Fig E9.2. The following tests could be carried out: (a) $U = 20$ mm/s flow of glycerin past a scale model that is 34-mm tall, 100-mm long, and 40-mm wide, (b) $U = 20$ mm/s air flow past the same scale model, or (c) $U = 25$ m/s airflow past the actual car, which is 1.7-m tall, 5-m long, and 2-m wide.

**FIND**   Would the flow characteristics for these three situations be similar? Explain.

**■ Figure E9.2**
© **Stanford University, with permission.**

# SOLUTION

The characteristics of flow past an object depend on the Reynolds number. For this instance we could pick the characteristic length to be the height, $h$, width, $b$, or length, $\ell$, of the car to obtain three possible Reynolds numbers, $Re_h = Uh/\nu$, $Re_b = Ub/\nu$, and $Re_\ell = U\ell/\nu$. These numbers will be different because of the different values of $h$, $b$, and $\ell$. Once we arbitrarily decide on the length we wish to use as the characteristic length, we must stick with it for all calculations when using comparisons between model and prototype.

With the values of kinematic viscosity for air and glycerin obtained from Tables 1.8 and 1.6 as $\nu_{air} = 1.46 \times 10^{-5}$ m$^2$/s and $\nu_{glycerin} = 1.19 \times 10^{-3}$ m$^2$/s, we obtain the following Reynolds numbers for the flows described.

| Reynolds Number | (a) Model in Glycerin | (b) Model in Air | (c) Car in Air |
|---|---|---|---|
| $Re_h$ | 0.571 | 46.6 | $2.91 \times 10^6$ |
| $Re_b$ | 0.672 | 54.8 | $3.42 \times 10^6$ |
| $Re_\ell$ | 1.68 | 137.0 | $8.56 \times 10^6$ |

Clearly, the Reynolds numbers for the three flows are quite different (regardless of which characteristic length we choose). Based on the previous discussion concerning flow past a flat plate or flow past a circular cylinder, we would expect that the flow past the actual car would behave in some way similar to the flows shown in Figs. 9.5c or 9.6c. That is, we would expect some type of boundary layer characteristic in which viscous effects would be confined to relatively thin layers near the surface of the car and the wake region behind it. Whether the car would act more like a flat plate or a cylinder would depend on the amount of streamlining incorporated into the car's design.

Because of the small Reynolds number involved, the flow past the model car in glycerin would be dominated by viscous effects, in some way reminiscent of the flows depicted in Figs. 9.5a or 9.6a. Similarly, with the moderate Reynolds number involved for the airflow past the model, a flow with characteristics similar to those indicated in Figs. 9.5b and 9.6b would be expected. Viscous effects would be important—not as important as with the glycerin flow, but more important than with the full-sized car.

It would not be a wise decision to expect the flow past the full-sized car to be similar to the flow past either of the models. The same conclusions result regardless of whether we use $Re_h$, $Re_b$, or $Re_\ell$. As is indicated in Chapter 7, the flows past the model car and the full-sized prototype will not be similar unless the Reynolds numbers for the model and prototype are the same. It is not always an easy task to ensure this condition. One (expensive) solution is to test full-sized prototypes in very large wind tunnels (see Fig. 9.1).

## 9.2    Boundary Layer Characteristics

As was discussed in the previous section, it is often possible to treat flow past an object as a combination of viscous flow in the boundary layer and inviscid flow elsewhere. If the Reynolds number is large enough, viscous effects are important only in the boundary layer regions near the object (and in the wake region behind the object). The boundary layer is needed to allow for the no-slip boundary condition that requires the fluid to cling to any solid surface that it flows past. Outside of the boundary layer the velocity gradients normal to the flow are relatively small, and the fluid acts as if it were inviscid, even though the viscosity is not zero. A necessary condition for this structure of the flow is that the Reynolds number be large.

### 9.2.1  Boundary Layer Structure and Thickness on a Flat Plate

There can be a wide variety in the size of a boundary layer and the structure of the flow within it. Part of this variation is due to the shape of the object on which the boundary layer forms. In this section we consider the simplest situation, one in which the boundary layer is formed on an infinitely long flat plate along which flows a viscous, incompressible fluid as is shown in Fig. 9.7. If the surface were curved (i.e., a circular cylinder or an airfoil), the boundary layer structure would be more complex. Such flows are discussed in Section 9.2.6.

If the Reynolds number is sufficiently large, only the fluid in a relatively thin boundary layer on the plate will feel the effect of the plate. That is, except in the region next to the plate the flow velocity will be essentially $\mathbf{V} = U\hat{\mathbf{i}}$, the upstream velocity. For the infinitely long flat plate extending from $x = 0$ to $x = \infty$, it is not obvious how to define the Reynolds number because there is no characteristic length. The plate has no thickness and is not of finite length!

For a finite length plate, it is clear that the plate length, $\ell$, can be used as the characteristic length. For an infinitely long plate we use $x$, the coordinate distance along the plate from the leading edge, as the characteristic length and define the Reynolds number as $Re_x = Ux/\nu$. Thus, for

V9.4 Laminar boundary layer

■ **Figure 9.7**   Distortion of a fluid particle as it flows within the boundary layer.

any fluid or upstream velocity the Reynolds number will be sufficiently large for boundary layer type flow (i.e., Fig. 9.5c) if the plate is long enough. Physically, this means that the flow situations illustrated in Fig. 9.5 could be thought of as occurring on the same plate but should be viewed by looking at longer portions of the plate as we step away from the plate to see the flows in Fig. 9.5a, 9.5b, and 9.5c, respectively.

If the plate is sufficiently long, the Reynolds number Re = $U\ell/\nu$ is sufficiently large so that the flow takes on its boundary layer character (except very near the leading edge). The details of the flow field near the leading edge are lost to our eyes because we are standing so far from the plate that we cannot make out these details. On this scale (Fig. 9.5c) the plate has negligible effect on the fluid ahead of the plate. The presence of the plate is felt only in the relatively thin boundary layer and wake regions. As previously noted, Prandtl in 1904 was the first to hypothesize such a concept. It has become one of the major turning points in fluid mechanics analysis.

A better appreciation of the structure of the boundary layer flow can be obtained by considering what happens to a fluid particle that flows into the boundary layer. As is indicated in Fig. 9.7, a small rectangular particle retains its original shape as it flows in the uniform flow outside of the boundary layer. Once it enters the boundary layer, the particle begins to distort because of the velocity gradient within the boundary layer—the top of the particle has a larger speed than its bottom. The fluid particles do not rotate as they flow along outside the boundary layer, but they begin to rotate once they pass through the fictitious boundary layer surface and enter the world of viscous flow. The flow is said to be irrotational outside the boundary layer and rotational within the boundary layer. (In terms of the kinematics of fluid particles as is discussed in Section 6.1, the flow outside the boundary layer has zero vorticity, and the flow within the boundary layer has nonzero vorticity.)

*V9.5 Laminar/ turbulent transition*

At some distance downstream from the leading edge, the boundary layer flow becomes turbulent and the fluid particles become greatly distorted because of the random, irregular nature of the turbulence. One of the distinguishing features of turbulent flow is the occurrence of irregular mixing of fluid particles that range in size from the smallest fluid particles up to those comparable in size with the object of interest. For laminar flow, mixing occurs only on the molecular scale. This molecular scale is orders of magnitude smaller in size than typical size scales for turbulent flow mixing. The transition from a *laminar boundary layer* to a *turbulent boundary layer* occurs at a critical value of the Reynolds number, $Re_{xcr}$, on the order of $2 \times 10^5$ to $3 \times 10^6$, depending on the roughness of the surface and the amount of turbulence in the upstream flow, as is discussed in Section 9.2.4. As shown by the figure in the margin, the location along the plate where the flow becomes turbulent, $x_{cr}$, moves toward the leading edge as the free-stream velocity increases.

The purpose of the boundary layer is to allow the fluid to change its velocity from the upstream value of U to zero on the surface. Thus, **V** = 0 at y = 0 and **V** ≈ $U\hat{\mathbf{i}}$ at the edge of the boundary layer, with the velocity profile, u = u(x, y) bridging the boundary layer thickness. This boundary layer characteristic occurs in a variety of flow situations, not just on flat plates. For example, boundary layers form on the surfaces of cars, in the water running down the gutter of the street, and in the atmosphere as the wind blows across the surface of the Earth (land or water).

■ **Figure 9.8**  Boundary layer thickness: (*a*) standard boundary layer thickness, (*b*) boundary layer displacement thickness.

*V9.6 Boundary layer thickness*

In actuality (both mathematically and physically), there is no sharp "edge" to the boundary layer; that is, $u \rightarrow U$ as we get farther from the plate. We define the ***boundary layer thickness,*** $\delta$, as that distance from the plate at which the fluid velocity is within some arbitrary value of the upstream velocity. Typically, as indicated in Fig. 9.8*a*,

$$\delta = y \qquad \text{where} \qquad u = 0.99U$$

To remove this arbitrariness (i.e., what is so special about 99%; why not 98%?), the following definitions are introduced. Shown in Fig. 9.8*b* are two velocity profiles for flow past a flat plate—one if there were no viscosity (a uniform profile) and the other if there were viscosity and zero slip at the wall (the boundary layer profile). Because of the velocity deficit, $U - u$, within the boundary layer, the flowrate across section *b–b* is less than that across section *a–a*. However, if we displace the plate at section *a–a* by an appropriate amount $\delta^*$, the *boundary layer displacement thickness*, the flowrates across each section will be identical. This is true if

*The boundary layer displacement thickness is defined in terms of volumetric flowrate.*

$$\delta^* b U = \int_0^\infty (U - u) b \, dy$$

where *b* is the plate width. Thus,

$$\delta^* = \int_0^\infty \left(1 - \frac{u}{U}\right) dy \tag{9.3}$$

The displacement thickness represents the amount that the thickness of the body must be increased so that the fictitious uniform inviscid flow has the same mass flowrate properties as the actual viscous flow. It represents the outward displacement of the streamlines caused by the viscous effects on the plate. This idea allows us to simulate the presence that the boundary layer has on the flow outside of the boundary layer by adding the displacement thickness to the actual wall and treating the flow over the thickened body as an inviscid flow. The displacement thickness concept is illustrated in Example 9.3.

---

## EXAMPLE 9.3    Boundary Layer Displacement Thickness

**GIVEN** Air flowing into a 2-ft-square duct with a uniform velocity of 10 ft/s forms a boundary layer on the walls as shown in Fig. E9.3a. The fluid within the core region (outside the boundary layers) flows as if it were inviscid. From advanced calculations it is determined that for this flow the boundary layer displacement thickness is given by

$$\delta^* = 0.0070(x)^{1/2} \qquad (1)$$

where $\delta^*$ and $x$ are in feet.

**FIND** Determine the velocity $U = U(x)$ of the air within the duct but outside of the boundary layer.

■ **Figure E9.3**

## SOLUTION

If we assume incompressible flow (a reasonable assumption because of the low velocities involved), it follows that the volume flowrate across any section of the duct is equal to that at the entrance (i.e., $Q_1 = Q_2$). That is,

$$U_1 A_1 = 10 \text{ ft/s } (2 \text{ ft})^2 = 40 \text{ ft}^3/\text{s} = \int_{(2)} u \, dA$$

According to the definition of the displacement thickness, $\delta^*$, the flowrate across section (2) is the same as that for a uniform flow with velocity $U$ through a duct whose walls have been moved inward by $\delta^*$. That is,

$$40 \text{ ft}^3/\text{s} = \int_{(2)} u \, dA = U(2 \text{ ft} - 2\delta^*)^2 \qquad (2)$$

By combining Eqs. 1 and 2 we obtain

$$40 \text{ ft}^3/\text{s} = 4U(1 - 0.0070x^{1/2})^2$$

or

$$U = \frac{10}{(1 - 0.0070x^{1/2})^2} \text{ ft/s} \qquad \textbf{(Ans)}$$

**COMMENTS** Note that $U$ increases in the downstream direction. For example, as shown in Fig. E9.3b, $U = 11.6$ ft/s at $x = 100$ ft. The viscous effects that cause the fluid to stick to the walls of the duct reduce the effective size of the duct, thereby (from conservation of mass principles) causing the fluid to accelerate. The pressure drop necessary to do this can be obtained by using the Bernoulli equation (Eq. 3.7) along the inviscid streamlines from section (1) to (2). (Recall that this equation is not valid for viscous flows within the boundary layer. It is,

however, valid for the inviscid flow outside the boundary layer.) Thus,

$$p_1 + \tfrac{1}{2}\rho U_1^2 = p + \tfrac{1}{2}\rho U^2$$

Hence, with $\rho = 2.38 \times 10^{-3}$ slugs/ft$^3$ and $p_1 = 0$ we obtain

$$p = \frac{1}{2}\rho \left(U_1^2 - U^2\right)$$

$$= \frac{1}{2}(2.38 \times 10^{-3} \text{ slugs/ft}^3)$$

$$\times \left[(10 \text{ ft/s})^2 - \frac{10^2}{(1 - 0.0079x^{1/2})^4} \text{ ft}^2/\text{s}^2\right]$$

or

$$p = 0.119 \left[1 - \frac{1}{(1 - 0.0070x^{1/2})^4}\right] \text{ lb/ft}^2$$

For example, $p = -0.0401$ lb/ft$^2$ at $x = 100$ ft.

If it were desired to maintain a constant velocity along the centerline of this entrance region of the duct, the walls could be displaced outward by an amount equal to the boundary layer displacement thickness, $\delta^*$.

Another boundary layer thickness definition, the *boundary layer momentum thickness*, $\Theta$, is often used when determining the drag on an object. Again because of the velocity deficit, $U - u$, in the boundary layer, the momentum flux across section $b-b$ in Fig. 9.8 is less than that across section $a-a$. This deficit in momentum flux for the actual boundary layer flow on a plate of width $b$ is given by

$$\int \rho u(U - u)\, dA = \rho b \int_0^\infty u(U - u)\, dy$$

which by definition is the momentum flux in a layer of uniform speed $U$ and thickness $\Theta$. That is,

$$\rho b U^2 \Theta = \rho b \int_0^\infty u(U - u)\, dy$$

or

$$\Theta = \int_0^\infty \frac{u}{U}\left(1 - \frac{u}{U}\right) dy \tag{9.4}$$

All three boundary layer thickness definitions, $\delta$, $\delta^*$, and $\Theta$, are of use in boundary layer analyses.

The boundary layer concept is based on the fact that the boundary layer is thin. For the flat plate flow this means that at any location $x$ along the plate, $\delta \ll x$. Similarly, $\delta^* \ll x$ and $\Theta \ll x$. Again, this is true if we do not get too close to the leading edge of the plate (i.e., not closer than $Re_x = Ux/\nu = 1000$ or so).

The structure and properties of the boundary layer flow depend on whether the flow is laminar or turbulent. As is illustrated in Fig. 9.9 and discussed in Sections 9.2.2 through 9.2.5, both the boundary layer thickness and the wall shear stress are different in these two regimes.

## 9.2.2 Prandtl/Blasius Boundary Layer Solution

In theory, the details of viscous, incompressible flow past any object can be obtained by solving the governing Navier–Stokes equations discussed in Section 6.8.2. For steady, two-dimensional laminar flows with negligible gravitational effects, these equations (Eqs. 6.127a, b, and c) reduce to the following:

$$u\frac{\partial u}{\partial x} + v\frac{\partial u}{\partial y} = -\frac{1}{\rho}\frac{\partial p}{\partial x} + \nu\left(\frac{\partial^2 u}{\partial x^2} + \frac{\partial^2 u}{\partial y^2}\right) \tag{9.5}$$

$$u\frac{\partial v}{\partial x} + v\frac{\partial v}{\partial y} = -\frac{1}{\rho}\frac{\partial p}{\partial y} + \nu\left(\frac{\partial^2 v}{\partial x^2} + \frac{\partial^2 v}{\partial y^2}\right) \tag{9.6}$$

which express Newton's second law. In addition, the conservation of mass equation, Eq. 6.31, for incompressible flow is

$$\frac{\partial u}{\partial x} + \frac{\partial v}{\partial y} = 0 \tag{9.7}$$

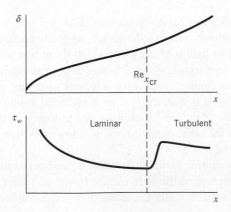

■ **Figure 9.9**  Typical characteristics of boundary layer thickness and wall shear stress for laminar and turbulent boundary layers.

The appropriate boundary conditions are that the fluid velocity far from the body is the upstream velocity and that the fluid sticks to the solid body surfaces. Although the mathematical problem is well-posed, no one has obtained an analytical solution to these equations for flow past any shaped body! Currently much work is being done to obtain numerical solutions to these governing equations for many flow geometries.

By using boundary layer concepts introduced in the previous sections, Prandtl was able to impose certain approximations (valid for large Reynolds number flows) and thereby to simplify the governing equations. In 1908, H. Blasius (1883–1970), one of Prandtl's students, was able to solve these simplified equations for the boundary layer flow past a flat plate parallel to the flow. A brief outline of this technique and the results are presented below. Additional details may be found in the literature (Refs. 1–3).

Since the boundary layer is thin, it is expected that the component of velocity normal to the plate is much smaller than that parallel to the plate and that the rate of change of any parameter across the boundary layer should be much greater than that along the flow direction. That is,

$$v \ll u \quad \text{and} \quad \frac{\partial}{\partial x} \ll \frac{\partial}{\partial y}$$

Physically, the flow is primarily parallel to the plate and any fluid property is convected downstream much more quickly than it is diffused across the streamlines.

With these assumptions it can be shown that the governing equations (Eqs. 9.5, 9.6, and 9.7) reduce to the following boundary layer equations:

$$\frac{\partial u}{\partial x} + \frac{\partial v}{\partial y} = 0 \tag{9.8}$$

$$u \frac{\partial u}{\partial x} + v \frac{\partial u}{\partial y} = \nu \frac{\partial^2 u}{\partial y^2} \tag{9.9}$$

*The Navier–Stokes equations can be simplified for boundary layer flow analysis.*

Although both these boundary layer equations and the original Navier–Stokes equations are nonlinear partial differential equations, there are considerable differences between them. For one, the $y$ momentum equation has been eliminated, leaving only the original, unaltered continuity equation and a modified $x$ momentum equation. One of the variables, the pressure, has been eliminated, leaving only the $x$ and $y$ components of velocity as unknowns. For boundary layer flow over a flat plate the pressure is constant throughout the fluid. The flow represents a balance between viscous and inertial effects, with pressure playing no role.

As shown by the figure in the margin, the boundary conditions for the governing boundary layer equations are that the fluid sticks to the plate

$$u = v = 0 \quad \text{on} \quad y = 0 \tag{9.10}$$

and that outside of the boundary layer the flow is the uniform upstream flow $u = U$. That is,

$$u \to U \quad \text{as} \quad y \to \infty \tag{9.11}$$

Mathematically, the upstream velocity is approached asymptotically as one moves away from the plate. Physically, the flow velocity is within 1% of the upstream velocity at a distance of $\delta$ from the plate.

In mathematical terms, the Navier–Stokes equations (Eqs. 9.5 and 9.6) and the continuity equation (Eq. 9.7) are elliptic equations, whereas the equations for boundary layer flow (Eqs. 9.8 and 9.9) are parabolic equations. The nature of the solutions to these two sets of equations, therefore, is different. Physically, this fact translates to the idea that what happens downstream of a given location in a boundary layer cannot affect what happens upstream of that point. That is, whether the plate shown in Fig. 9.5c ends with length $\ell$ or is extended to length $2\ell$, the flow within the first segment of length $\ell$ will be the same. In addition, the presence of the plate has no effect on the flow ahead of the plate. On the other hand, ellipticity allows flow information to propagate in all directions, including upstream.

In general, the solutions of nonlinear partial differential equations (such as the boundary layer equations, Eqs. 9.8 and 9.9) are extremely difficult to obtain. However, by applying a clever coordinate transformation and change of variables, Blasius reduced the partial differential equations to an ordinary differential equation that he was able to solve. A brief description of this process is given below. Additional details can be found in standard books dealing with boundary layer flow (Refs. 1, 2).

It can be argued that in dimensionless form the boundary layer velocity profiles on a flat plate should be similar regardless of the location along the plate. That is,

$$\frac{u}{U} = g\left(\frac{y}{\delta}\right)$$

where $g(y/\delta)$ is an unknown function to be determined. In addition, by applying an order of magnitude analysis of the forces acting on fluid within the boundary layer, it can be shown that the boundary layer thickness grows as the square root of $x$ and inversely proportional to the square root of $U$. That is,

$$\delta \sim \left(\frac{\nu x}{U}\right)^{1/2}$$

Such a conclusion results from a balance between viscous and inertial forces within the boundary layer and from the fact that the velocity varies much more rapidly in the direction across the boundary layer than along it.

*The boundary layer equations can be written in terms of a similarity variable.*

Thus, we introduce the dimensionless *similarity variable* $\eta = (U/\nu x)^{1/2}y$ and the stream function $\psi = (\nu x U)^{1/2} f(\eta)$, where $f = f(\eta)$ is an unknown function. Recall from Section 6.2.3 that the velocity components for two-dimensional flow are given in terms of the stream function as $u = \partial\psi/\partial y$ and $v = -\partial\psi/\partial x$, which for this flow become

$$u = Uf'(\eta) \tag{9.12}$$

and

$$v = \left(\frac{\nu U}{4x}\right)^{1/2} (\eta f' - f) \tag{9.13}$$

with the notation $( \; )' = d/d\eta$. We substitute Eqs. 9.12 and 9.13 into the governing equations, Eqs. 9.8 and 9.9, to obtain (after considerable manipulation) the following nonlinear, third-order ordinary differential equation:

$$2f''' + ff'' = 0 \tag{9.14a}$$

As shown by the figure in the margin, the boundary conditions given in Eqs. 9.10 and 9.11 can be written as

$$f = f' = 0 \text{ at } \eta = 0 \quad \text{and} \quad f' \to 1 \text{ as } \eta \to \infty \tag{9.14b}$$

The original partial differential equation and boundary conditions have been reduced to an ordinary differential equation by use of the similarity variable $\eta$. The two independent variables, $x$ and $y$, were combined into the similarity variable in a fashion that reduced the partial differential equation (and boundary conditions) to an ordinary differential equation. This type of reduction is not generally possible. For example, this method does not work on the full Navier–Stokes equations, although it does on the boundary layer equations (Eqs. 9.8 and 9.9).

Although there is no known analytical solution to Eq. 9.14, it is relatively easy to integrate this equation on a computer. The dimensionless boundary layer profile, $u/U = f'(\eta)$, obtained by numerical solution of Eq. 9.14 (termed the Blasius solution), is sketched in Fig. 9.10a and is tabulated in Table 9.1. The velocity profiles at different $x$ locations are similar in that there is only one curve necessary to describe the velocity at any point in the boundary layer. Because the similarity variable $\eta$ contains both $x$ and $y$, it is seen from Fig. 9.10b that the actual velocity profiles are a function of both $x$ and $y$. The profile at location $x_1$ is the same as that at $x_2$ except that the $y$ coordinate is stretched by a factor of $(x_2/x_1)^{1/2}$. You can achieve similar changes in the boundary layer velocity profile by altering the kinematic viscosity, $\nu$, as shown by the figure in the margin.

From the solution it is found that $u/U \approx 0.99$ when $\eta = 5.0$. Thus,

$$\delta = 5\sqrt{\frac{\nu x}{U}} \tag{9.15}$$

■ **Figure 9.10** Blasius boundary layer profile: (a) boundary layer profile in dimension-less form using the similarity variable $\eta$, (b) similar boundary layer profiles at different locations along the flat plate.

or

$$\frac{\delta}{x} = \frac{5}{\sqrt{\mathrm{Re}_x}}$$

where $\mathrm{Re}_x = Ux/\nu$. It can also be shown that the displacement and momentum thicknesses are given by

$$\frac{\delta^*}{x} = \frac{1.721}{\sqrt{\mathrm{Re}_x}} \tag{9.16}$$

and

$$\frac{\Theta}{x} = \frac{0.664}{\sqrt{\mathrm{Re}_x}} \tag{9.17}$$

*For large Reynolds numbers the boundary layer is relatively thin.*

As postulated, the boundary layer is thin provided that $\mathrm{Re}_x$ is large (i.e., $\delta/x \rightarrow 0$ as $\mathrm{Re}_x \rightarrow \infty$).

■ **Table 9.1**
**Laminar Flow along a Flat Plate (the Blasius Solution)**

| $\eta = y(U/\nu x)^{1/2}$ | $f'(\eta) = u/U$ | $\eta$ | $f'(\eta)$ |
|---|---|---|---|
| 0 | 0 | 3.6 | 0.9233 |
| 0.4 | 0.1328 | 4.0 | 0.9555 |
| 0.8 | 0.2647 | 4.4 | 0.9759 |
| 1.2 | 0.3938 | 4.8 | 0.9878 |
| 1.6 | 0.5168 | 5.0 | 0.9916 |
| 2.0 | 0.6298 | 5.2 | 0.9943 |
| 2.4 | 0.7290 | 5.6 | 0.9975 |
| 2.8 | 0.8115 | 6.0 | 0.9990 |
| 3.2 | 0.8761 | $\infty$ | 1.0000 |

With the velocity profile known, it is an easy matter to determine the wall shear stress, $\tau_w = \mu \left( \partial u / \partial y \right)_{y=0}$, where the velocity gradient is evaluated at the plate. The value of $\partial u / \partial y$ at $y = 0$ can be obtained from the Blasius solution to give

$$\tau_w = 0.332 U^{3/2} \sqrt{\frac{\rho \mu}{x}} \qquad (9.18)$$

As indicated by Eq. 9.18 and illustrated in the figure in the margin, the shear stress decreases with increasing $x$ because of the increasing thickness of the boundary layer—the velocity gradient at the wall decreases with increasing $x$. Also, $\tau_w$ varies as $U^{3/2}$, not as $U$ as it does for fully developed laminar pipe flow. These variations are discussed in Section 9.2.3.

### 9.2.3 Momentum Integral Boundary Layer Equation for a Flat Plate

One of the important aspects of boundary layer theory is the determination of the drag caused by shear forces on a body. As was discussed in the previous section, such results can be obtained from the governing differential equations for laminar boundary layer flow. Since these solutions are extremely difficult to obtain, it is of interest to have an alternative approximate method. The momentum integral method described in this section provides such an alternative.

We consider the uniform flow past a flat plate and the fixed control volume as shown in Fig. 9.11. In agreement with advanced theory and experiment, we assume that the pressure is constant throughout the flow field. The flow entering the control volume at the leading edge of the plate [section (1)] is uniform, while the velocity of the flow exiting the control volume [section (2)] varies from the upstream velocity at the edge of the boundary layer to zero velocity on the plate.

The fluid adjacent to the plate makes up the lower portion of the control surface. The upper surface coincides with the streamline just outside the edge of the boundary layer at section (2). It need not (in fact, does not) coincide with the edge of the boundary layer except at section (2). If we apply the $x$ component of the momentum equation (Eq. 5.22) to the steady flow of fluid within this control volume we obtain

$$\sum F_x = \rho \int_{(1)} u \mathbf{V} \cdot \hat{\mathbf{n}} \, dA + \rho \int_{(2)} u \mathbf{V} \cdot \hat{\mathbf{n}} \, dA$$

where for a plate of width $b$

$$\sum F_x = -\mathcal{D} = -\int_{\text{plate}} \tau_w \, dA = -b \int_{\text{plate}} \tau_w \, dx \qquad (9.19)$$

and $\mathcal{D}$ is the drag that the plate exerts on the fluid. Note that the net force caused by the uniform pressure distribution does not contribute to this flow. Since the plate is solid and the upper surface of the control volume is a streamline, there is no flow through these areas. Thus,

$$-\mathcal{D} = \rho \int_{(1)} U(-U) \, dA + \rho \int_{(2)} u^2 \, dA$$

*The drag on a flat plate depends on the velocity profile within the boundary layer.*

or

$$\mathcal{D} = \rho U^2 b h - \rho b \int_0^\delta u^2 \, dy \qquad (9.20)$$

■ **Figure 9.11**   Control volume used in the derivation of the momentum integral equation for boundary layer flow.

Although the height $h$ is not known, it is known that for conservation of mass the flowrate through section (1) must equal that through section (2), or

$$Uh = \int_0^\delta u \, dy$$

which can be written as

*Drag on a flat plate is related to momentum deficit within the boundary layer.*

$$\rho U^2 bh = \rho b \int_0^\delta Uu \, dy \qquad (9.21)$$

Thus, by combining Eqs. 9.20 and 9.21 we obtain the drag in terms of the deficit of momentum flux across the outlet of the control volume as

$$\mathcal{D} = \rho b \int_0^\delta u(U - u) \, dy \qquad (9.22)$$

The idea of a momentum deficit is illustrated in the figure in the margin. If the flow were inviscid, the drag would be zero, since we would have $u \equiv U$ and the right-hand side of Eq. 9.22 would be zero. (This is consistent with the fact that $\tau_w = 0$ if $\mu = 0$.) Equation 9.22 points out the important fact that boundary layer flow on a flat plate is governed by a balance between shear drag (the left-hand side of Eq. 9.22) and a decrease in the momentum of the fluid (the right-hand side of Eq. 9.22). As $x$ increases, $\delta$ increases and the drag increases. The thickening of the boundary layer is necessary to overcome the drag of the viscous shear stress on the plate. This is contrary to horizontal fully developed pipe flow in which the momentum of the fluid remains constant and the shear force is overcome by the pressure gradient along the pipe.

The development of Eq. 9.22 and its use was first put forth in 1921 by T. von Kármán (1881–1963), a Hungarian/German aerodynamicist. By comparing Eqs. 9.22 and 9.4 we see that the drag can be written in terms of the momentum thickness, $\Theta$, as

$$\mathcal{D} = \rho b U^2 \Theta \qquad (9.23)$$

Note that this equation is valid for laminar or turbulent flows.

The shear stress distribution can be obtained from Eq. 9.23 by differentiating both sides with respect to $x$ to obtain

$$\frac{d\mathcal{D}}{dx} = \rho b U^2 \frac{d\Theta}{dx} \qquad (9.24)$$

The increase in drag per length of the plate, $d\mathcal{D}/dx$, occurs at the expense of an increase of the momentum boundary layer thickness, which represents a decrease in the momentum of the fluid.

Since $d\mathcal{D} = \tau_w b dx$ (see Eq. 9.19) it follows that

$$\frac{d\mathcal{D}}{dx} = b\tau_w \qquad (9.25)$$

Hence, by combining Eqs. 9.24 and 9.25 we obtain the *momentum integral equation* for the boundary layer flow on a flat plate

$$\tau_w = \rho U^2 \frac{d\Theta}{dx} \qquad (9.26)$$

The usefulness of this relationship lies in the ability to obtain approximate boundary layer results easily by using rather crude assumptions. For example, if we knew the detailed velocity profile in the boundary layer (i.e., the Blasius solution discussed in the previous section), we could evaluate either the right-hand side of Eq. 9.23 to obtain the drag or the right-hand side of Eq. 9.26 to obtain the shear stress. Fortunately, even a rather crude guess at the velocity profile will allow us to obtain reasonable drag and shear stress results from Eq. 9.26. This method is illustrated in Example 9.4.

# EXAMPLE 9.4 Momentum Integral Boundary Layer Equation

**GIVEN** Consider the laminar flow of an incompressible fluid past a flat plate at $y = 0$. The boundary layer velocity profile is approximated as $u = Uy/\delta$ for $0 \leq y \leq \delta$ and $u = U$ for $y > \delta$, as is shown in Fig. E9.4.

**FIND** Determine the shear stress by using the momentum integral equation. Compare these results with the Blasius results given by Eq. 9.18.

■ **Figure E9.4**

## SOLUTION

From Eq. 9.26 the shear stress is given by

$$\tau_w = \rho U^2 \frac{d\Theta}{dx} \tag{1}$$

while for laminar flow we know that $\tau_w = \mu(\partial u/\partial y)_{y=0}$. For the assumed profile we have

$$\tau_w = \mu \frac{U}{\delta} \tag{2}$$

and from Eq. 9.4

$$\Theta = \int_0^\infty \frac{u}{U}\left(1 - \frac{u}{U}\right) dy = \int_0^\delta \frac{u}{U}\left(1 - \frac{u}{U}\right) dy$$

$$= \int_0^\delta \left(\frac{y}{\delta}\right)\left(1 - \frac{y}{\delta}\right) dy$$

or

$$\Theta = \frac{\delta}{6} \tag{3}$$

Note that as yet we do not know the value of $\delta$ (but suspect that it should be a function of $x$).

By combining Eqs. 1, 2, and 3 we obtain the following differential equation for $\delta$:

$$\frac{\mu U}{\delta} = \frac{\rho U^2}{6} \frac{d\delta}{dx}$$

or

$$\delta \, d\delta = \frac{6\mu}{\rho U} \, dx$$

This can be integrated from the leading edge of the plate, $x = 0$ (where $\delta = 0$) to an arbitrary location $x$ where the boundary layer thickness is $\delta$. The result is

$$\frac{\delta^2}{2} = \frac{6\mu}{\rho U} x$$

or

$$\delta = 3.46 \sqrt{\frac{\nu x}{U}} \tag{4}$$

Note that this approximate result (i.e., the velocity profile is not actually the simple straight line we assumed) compares favorably with the (much more laborious to obtain) Blasius result given by Eq. 9.15.

The wall shear stress can also be obtained by combining Eqs. 1, 3, and 4 to give

$$\tau_w = 0.289 U^{3/2} \sqrt{\frac{\rho\mu}{x}} \tag{Ans}$$

Again this approximate result is close (within 13%) to the Blasius value of $\tau_w$ given by Eq. 9.18.

---

As is illustrated in Example 9.4, the momentum integral equation, Eq. 9.26, can be used along with an assumed velocity profile to obtain reasonable, approximate boundary layer results. The accuracy of these results depends on how closely the shape of the assumed velocity profile approximates the actual profile.

Thus, we consider a general velocity profile

$$\frac{u}{U} = g(Y) \quad \text{for} \quad 0 \leq Y \leq 1$$

and

*Approximate velocity profiles are used in the momentum integral equation.*

$$\frac{u}{U} = 1 \quad \text{for} \quad Y > 1$$

where the dimensionless coordinate $Y = y/\delta$ varies from 0 to 1 across the boundary layer. The dimensionless function $g(Y)$ can be any shape we choose, although it should be a reasonable approximation to the boundary layer profile, as shown by the figure in the margin. In particular, it should certainly satisfy the boundary conditions $u = 0$ at $y = 0$ and $u = U$ at $y = \delta$. That is,

$$g(0) = 0 \quad \text{and} \quad g(1) = 1$$

The linear function $g(Y) = Y$ used in Example 9.4 is one such possible profile. Other conditions, such as $dg/dY = 0$ at $Y = 1$ (i.e., $\partial u/\partial y = 0$ at $y = \delta$), could also be incorporated into the function $g(Y)$ to more closely approximate the actual profile.

For a given $g(Y)$, the drag can be determined from Eq. 9.22 as

$$\mathcal{D} = \rho b \int_0^\delta u(U - u)\, dy = \rho b U^2 \delta \int_0^1 g(Y)[1 - g(Y)]\, dY$$

or

$$\mathcal{D} = \rho b U^2 \delta C_1 \tag{9.27}$$

where the dimensionless constant $C_1$ has the value

$$C_1 = \int_0^1 g(Y)[1 - g(Y)]\, dY$$

Also, the wall shear stress can be written as

$$\tau_w = \mu \left.\frac{\partial u}{\partial y}\right|_{y=0} = \frac{\mu U}{\delta} \left.\frac{dg}{dY}\right|_{Y=0} = \frac{\mu U}{\delta} C_2 \tag{9.28}$$

where the dimensionless constant $C_2$ has the value

$$C_2 = \left.\frac{dg}{dY}\right|_{Y=0}$$

By combining Eqs. 9.25, 9.27, and 9.28 we obtain

$$\delta\, d\delta = \frac{\mu C_2}{\rho U C_1}\, dx$$

which can be integrated from $\delta = 0$ at $x = 0$ to give

$$\delta = \sqrt{\frac{2\nu C_2 x}{U C_1}}$$

or

$$\frac{\delta}{x} = \frac{\sqrt{2C_2/C_1}}{\sqrt{\text{Re}_x}} \tag{9.29}$$

By substituting this expression back into Eqs. 9.28 we obtain

$$\tau_w = \sqrt{\frac{C_1 C_2}{2}}\; U^{3/2} \sqrt{\frac{\rho\mu}{x}} \tag{9.30}$$

*Approximate boundary layer results are obtained from the momentum integral equation.*

To use Eqs. 9.29 and 9.30 we must determine the values of $C_1$ and $C_2$. Several assumed velocity profiles and the resulting values of $\delta$ are given in Fig. 9.12 and Table 9.2. The more closely the assumed shape approximates the actual (i.e., Blasius) profile, the more accurate the final results. For any assumed profile shape, the functional dependence of $\delta$ and $\tau_w$ on the physical parameters $\rho$, $\mu$, $U$, and $x$ is the same. Only the constants are different. That is, $\delta \sim (\mu x/\rho U)^{1/2}$ or $\delta \text{Re}_x^{1/2}/x =$ constant, and $\tau_w \sim (\rho\mu U^3/x)^{1/2}$, where $\text{Re}_x = \rho U x/\mu$.

It is often convenient to use the dimensionless *local friction coefficient*, $c_f$, defined as

$$c_f = \frac{\tau_w}{\frac{1}{2}\rho U^2} \tag{9.31}$$

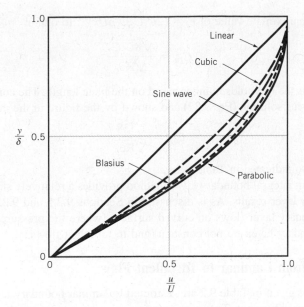

**Figure 9.12** Typical approximate boundary layer profiles used in the momentum integral equation.

to express the wall shear stress. From Eq. 9.30 we obtain the approximate value

$$c_f = \sqrt{2C_1 C_2} \sqrt{\frac{\mu}{\rho U x}} = \frac{\sqrt{2C_1 C_2}}{\sqrt{Re_x}}$$

while the Blasius solution result is given by

$$c_f = \frac{0.664}{\sqrt{Re_x}} \tag{9.32}$$

These results are also indicated in Table 9.2.

For a flat plate of length $\ell$ and width $b$, the net friction drag, $\mathcal{D}_f$, can be expressed in terms of the *friction drag coefficient*, $C_{Df}$, as

$$C_{Df} = \frac{\mathcal{D}_f}{\frac{1}{2}\rho U^2 b\ell} = \frac{b\int_0^\ell \tau_w \, dx}{\frac{1}{2}\rho U^2 b\ell}$$

*The friction drag coefficient is an integral of the local friction coefficient.*

or

$$C_{Df} = \frac{1}{\ell}\int_0^\ell c_f \, dx \tag{9.33}$$

**Table 9.2**

**Flat Plate Momentum Integral Results for Various Assumed Laminar Flow Velocity Profiles**

| Profile Character | $\delta Re_x^{1/2}/x$ | $c_f Re_x^{1/2}$ | $C_{Df}Re_\ell^{1/2}$ |
|---|---|---|---|
| a. Blasius solution | 5.00 | 0.664 | 1.328 |
| b. Linear<br>$u/U = y/\delta$ | 3.46 | 0.578 | 1.156 |
| c. Parabolic<br>$u/U = 2y/\delta - (y/\delta)^2$ | 5.48 | 0.730 | 1.460 |
| d. Cubic<br>$u/U = 3(y/\delta)/2 - (y/\delta)^3/2$ | 4.64 | 0.646 | 1.292 |
| e. Sine wave<br>$u/U = \sin[\pi(y/\delta)/2]$ | 4.79 | 0.655 | 1.310 |

We use the above approximate value of $c_f = (2C_1C_2\mu/\rho Ux)^{1/2}$ to obtain

$$C_{Df} = \frac{\sqrt{8C_1C_2}}{\sqrt{Re_\ell}}$$

where $Re_\ell = U\ell/\nu$ is the Reynolds number based on the plate length. The corresponding value obtained from the Blasius solution (Eq. 9.32) and shown by the figure in the margin gives

$$C_{Df} = \frac{1.328}{\sqrt{Re_\ell}}$$

These results are also indicated in Table 9.2.

The momentum integral boundary layer method provides a relatively simple technique to obtain useful boundary layer results. As is discussed in Sections 9.2.5 and 9.2.6, this technique can be extended to boundary layer flows on curved surfaces (where the pressure and fluid velocity at the edge of the boundary layer are not constant) and to turbulent flows.

### 9.2.4 Transition from Laminar to Turbulent Flow

*The boundary layer on a flat plate will become turbulent if the plate is long enough.*

The analytical results given in Table 9.2 are restricted to laminar boundary layer flows along a flat plate with zero pressure gradient. They agree quite well with experimental results up to the point where the boundary layer flow becomes turbulent, which will occur for any free-stream velocity and any fluid provided the plate is long enough. This is true because the parameter that governs the **transition** to turbulent flow is the Reynolds number—in this case the Reynolds number based on the distance from the leading edge of the plate, $Re_x = Ux/\nu$.

The value of the Reynolds number at the transition location is a rather complex function of various parameters involved, including the roughness of the surface, the curvature of the surface (for example, a flat plate or a sphere), and some measure of the disturbances in the flow outside the boundary layer. On a flat plate with a sharp leading edge in a typical airstream, the transition takes place at a distance $x$ from the leading edge given by $Re_{xcr} = 2 \times 10^5$ to $3 \times 10^6$. Unless otherwise stated, we will use $Re_{xcr} = 5 \times 10^5$ in our calculations.

The actual transition from laminar to turbulent boundary layer flow may occur over a region of the plate, not at a specific single location. This occurs, in part, because of the spottiness of the transition. Typically, the transition begins at random locations on the plate in the vicinity of $Re_x = Re_{xcr}$. These spots grow rapidly as they are convected downstream until the entire width of the plate is covered with turbulent flow. The photo shown in Fig. 9.13 illustrates this transition process.

The complex process of transition from laminar to turbulent flow involves the instability of the flow field. Small disturbances imposed on the boundary layer flow (i.e., from a vibration of the plate, a roughness of the surface, or a "wiggle" in the flow past the plate) will either grow (instability) or decay (stability), depending on where the disturbance is introduced into the flow. If these disturbances occur at a location with $Re_x < Re_{xcr}$ they will die out, and the boundary layer will return to laminar flow at that location. Disturbances imposed at a location with $Re_x > Re_{xcr}$ will grow and transform the boundary layer flow downstream of this location into turbulence. The study of the initiation, growth, and structure of these turbulent bursts or spots is an active area of fluid mechanics research.

*V9.7 Transition on flat plate*

■ **Figure 9.13** Turbulent spots and the transition from laminar to turbulent boundary layer flow on a flat plate. Flow from left to right. (Photograph courtesy of Brian Cantwell, Stanford University.)

■ **Figure 9.14** Typical boundary layer profiles on a flat plate for laminar, transitional, and turbulent flow (Ref. 1).

Transition from laminar to turbulent flow also involves a noticeable change in the shape of the boundary layer velocity profile. Typical profiles obtained in the neighborhood of the transition location are indicated in Fig. 9.14. The turbulent profiles are flatter, have a larger velocity gradient at the wall, and produce a larger boundary layer thickness than do the laminar profiles.

# EXAMPLE 9.5    Boundary Layer Transition

**GIVEN**    A fluid flows steadily past a flat plate with a velocity of $U = 10$ ft/s.

**FIND**    At approximately what location will the boundary layer become turbulent, and how thick is the boundary layer at that point if the fluid is (a) water at 60 °F, (b) standard air, or (c) glycerin at 68 °F?

## SOLUTION

For any fluid, the laminar boundary layer thickness is found from Eq. 9.15 as

$$\delta = 5\sqrt{\frac{\nu x}{U}} \qquad (1)$$

The boundary layer remains laminar up to

$$x_{cr} = \frac{\nu Re_{xcr}}{U}$$

Thus, if we assume $Re_{xcr} = 5 \times 10^5$ we obtain

$$x_{cr} = \frac{5 \times 10^5}{10 \text{ ft/s}}\nu = 5 \times 10^4\,\nu$$

and from Eq. (1)

$$\delta_{cr} \equiv \delta|_{x=x_{cr}} = 5\left[\frac{\nu}{10}(5 \times 10^4\,\nu)\right]^{1/2} = 354\,\nu$$

■ **Table E9.5**

| Fluid | $\nu$ (ft$^2$/s) | $x_{cr}$ (ft) | $\delta_{cr}$ (ft) |
|---|---|---|---|
| a.  Water | $1.21 \times 10^{-5}$ | 0.605 | 0.00428 |
| b.  Air | $1.57 \times 10^{-4}$ | 7.85 | 0.0556 |
| c.  Glycerin | $1.28 \times 10^{-2}$ | 640.0 | 4.53 |

(Ans)

where $\nu$ is in ft$^2$/s and $x_{cr}$ and $\delta_{cr}$ are in feet. The values of the kinematic viscosity obtained from Tables 1.5 and 1.7 are listed in Table E9.5 along with the corresponding $x_{cr}$ and $\delta_{cr}$.

**COMMENT**    Laminar flow can be maintained on a longer portion of the plate if the viscosity is increased. However, the boundary layer flow eventually becomes turbulent, provided the plate is long enough. Similarly, the boundary layer thickness is greater if the viscosity is increased.

## 9.2.5 Turbulent Boundary Layer Flow

The structure of turbulent boundary layer flow is very complex, random, and irregular. It shares many of the characteristics described for turbulent pipe flow in Section 8.3. In particular, the velocity at any given location in the flow is unsteady in a random fashion. The flow can be thought of as a jumbled mix of intertwined eddies (or swirls) of different sizes (diameters and angular velocities). The figure in the margin shows a laser-induced fluorescence visualization of a turbulent boundary layer on a flat plate (side view). The various fluid quantities involved (i.e., mass, momentum, energy) are convected downstream in the free-stream direction as in a laminar boundary layer. For turbulent flow they are also convected across the boundary layer (in the direction perpendicular to the plate) by the random transport of finite-sized fluid particles associated with the turbulent eddies. There is considerable mixing involved with these finite-sized eddies—considerably more than is associated with the mixing found in laminar flow where it is confined to the molecular scale. Although there is considerable random motion of fluid particles perpendicular to the plate, there is very little net transfer of mass across the boundary layer—the largest flowrate by far is parallel to the plate.

There is, however, a considerable net transfer of $x$ component of momentum perpendicular to the plate because of the random motion of the particles. Fluid particles moving toward the plate (in the negative $y$ direction) have some of their excess momentum (they come from areas of higher velocity) removed by the plate. Conversely, particles moving away from the plate (in the positive $y$ direction) gain momentum from the fluid (they come from areas of lower velocity). The net result is that the plate acts as a momentum sink, continually extracting momentum from the fluid. For laminar flows, such cross-stream transfer of these properties takes place solely on the molecular scale. For turbulent flow the randomness is associated with fluid particle mixing. Consequently, the shear force for turbulent boundary layer flow is considerably greater than it is for laminar boundary layer flow (see Section 8.3.2).

There are no "exact" solutions for turbulent boundary layer flow. As is discussed in Section 9.2.2, it is possible to solve the Prandtl boundary layer equations for laminar flow past a flat plate to obtain the Blasius solution (which is "exact" within the framework of the assumptions involved in the boundary layer equations). Since there is no precise expression for the shear stress in turbulent flow (see Section 8.3), solutions are not available for turbulent flow. However, considerable headway has been made in obtaining numerical (computer) solutions for turbulent flow by using approximate shear stress relationships. Also, progress is being made in the area of direct, full numerical integration of the basic governing equations, the Navier–Stokes equations.

Approximate turbulent boundary layer results can also be obtained by use of the momentum integral equation, Eq. 9.26, which is valid for either laminar or turbulent flow. What is needed for the use of this equation are reasonable approximations to the velocity profile $u = U\, g(Y)$, where $Y = y/\delta$ and $u$ is the time-averaged velocity (the overbar notation, $\bar{u}$, of Section 8.3.2 has been dropped for convenience), and a functional relationship describing the wall shear stress. For laminar flow the wall shear stress was used as $\tau_w = \mu(\partial u/\partial y)_{y=0}$. In theory, such a technique should work for turbulent boundary layers also. However, as is discussed in Section 8.3, the details of the velocity gradient at the wall are not well understood for turbulent flow. Thus, it is necessary to use some empirical relationship for the wall shear stress. This is illustrated in Example 9.6.

Plate

© J. P. Crimaldi, M. B. Wiley, & J. R. Koseff

*Random transport of finite-sized fluid particles occurs within turbulent boundary layers.*

*V9.8 Random flow but not turbulent*

# *E*XAMPLE 9.6  Turbulent Boundary Layer Properties

**GIVEN** Consider turbulent flow of an incompressible fluid past a flat plate. The boundary layer velocity profile is assumed to be $u/U = (y/\delta)^{1/7} = Y^{1/7}$ for $Y = y/\delta \leq 1$ and $u = U$ for $Y > 1$ as shown in Fig. E9.6. This is a reasonable approximation of experimentally observed profiles, except very near the plate where this formula gives $\partial u/\partial y = \infty$ at $y = 0$. Note the differences between the assumed turbulent profile and the laminar profile. Also assume that the shear stress agrees with the experimentally determined formula:

$$\tau_w = 0.0225\rho U^2 \left(\frac{\nu}{U\delta}\right)^{1/4} \qquad (1)$$

**FIND** Determine the boundary layer thicknesses $\delta$, $\delta^*$, and $\Theta$ and the wall shear stress, $\tau_w$, as a function of $x$. Determine the friction drag coefficient, $C_{Df}$.

## SOLUTION

Whether the flow is laminar or turbulent, it is true that the drag force is accounted for by a reduction in the momentum of the fluid flowing past the plate. The shear is obtained from Eq. 9.26 in terms of the rate at which the momentum boundary layer thickness, $\Theta$, increases with distance along the plate as

$$\tau_w = \rho U^2 \frac{d\Theta}{dx} \qquad (1)$$

For the assumed velocity profile, the boundary layer momentum thickness is obtained from Eq. 9.4 as

$$\Theta = \int_0^\infty \frac{u}{U}\left(1 - \frac{u}{U}\right) dy = \delta \int_0^1 \frac{u}{U}\left(1 - \frac{u}{U}\right) dY$$

or by integration

$$\Theta = \delta \int_0^1 Y^{1/7}\left(1 - Y^{1/7}\right) dY = \frac{7}{72}\delta \qquad (2)$$

where $\delta$ is an unknown function of $x$. By combining the assumed shear force dependence (Eq. 1) with Eq. 2, we obtain the following differential equation for $\delta$:

$$0.0225\rho U^2 \left(\frac{\nu}{U\delta}\right)^{1/4} = \frac{7}{72}\rho U^2 \frac{d\delta}{dx}$$

or

$$\delta^{1/4}\, d\delta = 0.231\left(\frac{\nu}{U}\right)^{1/4} dx$$

This can be integrated from $\delta = 0$ at $x = 0$ to obtain

$$\delta = 0.370\left(\frac{\nu}{U}\right)^{1/5} x^{4/5} \qquad (3) \quad \textbf{(Ans)}$$

or in dimensionless form

$$\frac{\delta}{x} = \frac{0.370}{\mathrm{Re}_x^{1/5}}$$

Strictly speaking, the boundary layer near the leading edge of the plate is laminar, not turbulent, and the precise boundary condition should be the matching of the initial turbulent boundary layer thickness (at the transition location) with the thickness of the laminar boundary layer at that point. In practice, however, the laminar boundary layer often exists over a relatively short portion of the plate, and the error associated with starting the turbulent boundary layer with $\delta = 0$ at $x = 0$ can be negligible.

The displacement thickness, $\delta^*$, and the momentum thickness, $\Theta$, can be obtained from Eqs. 9.3 and 9.4 by integrating as follows:

$$\delta^* = \int_0^\infty \left(1 - \frac{u}{U}\right) dy = \delta \int_0^1 \left(1 - \frac{u}{U}\right) dY$$

$$= \delta \int_0^1 \left(1 - Y^{1/7}\right) dY = \frac{\delta}{8}$$

Thus, by combining this with Eq. 3 we obtain

$$\delta^* = 0.0463\left(\frac{\nu}{U}\right)^{1/5} x^{4/5} \qquad \textbf{(Ans)}$$

■ **Figure E9.6**

Similarly, from Eq. 2,

$$\Theta = \tfrac{7}{72}\delta = 0.0360\left(\frac{\nu}{U}\right)^{1/5} x^{4/5} \qquad (4) \quad \textbf{(Ans)}$$

The functional dependence for $\delta$, $\delta^*$, and $\Theta$ is the same; only the constants of proportionality are different. Typically, $\Theta < \delta^* < \delta$.

By combining Eqs. 1 and 3, we obtain the following result for the wall shear stress

$$\tau_w = 0.0225\rho U^2 \left[\frac{\nu}{U(0.370)(\nu/U)^{1/5}x^{4/5}}\right]^{1/4}$$

$$= \frac{0.0288\rho U^2}{\mathrm{Re}_x^{1/5}} \qquad \textbf{(Ans)}$$

This can be integrated over the length of the plate to obtain the friction drag on one side of the plate, $\mathcal{D}_f$, as

$$\mathcal{D}_f = \int_0^\ell b\tau_w\, dx = b(0.0288\rho U^2)\int_0^\ell \left(\frac{\nu}{Ux}\right)^{1/5} dx$$

or

$$\mathcal{D}_f = 0.0360\rho U^2 \frac{A}{\mathrm{Re}_\ell^{1/5}}$$

where $A = b\ell$ is the area of the plate. (This result can also be obtained by combining Eq. 9.23 and the expression for the momentum thickness given in Eq. 4.) The corresponding friction drag coefficient, $C_{Df}$, is

$$C_{Df} = \frac{\mathcal{D}_f}{\frac{1}{2}\rho U^2 A} = \frac{0.0720}{\mathrm{Re}_\ell^{1/5}} \qquad \textbf{(Ans)}$$

**COMMENT** Note that for the turbulent boundary layer flow the boundary layer thickness increases with $x$ as $\delta \sim x^{4/5}$ and the shear stress decreases as $\tau_w \sim x^{-1/5}$. For laminar flow these dependencies are $x^{1/2}$ and $x^{-1/2}$, respectively. The random character of the turbulent flow causes a different structure of the flow.

Obviously the results presented in this example are valid only in the range of validity of the original data—the assumed velocity profile and shear stress. This range covers smooth flat plates with $5 \times 10^5 < \mathrm{Re}_\ell < 10^7$.

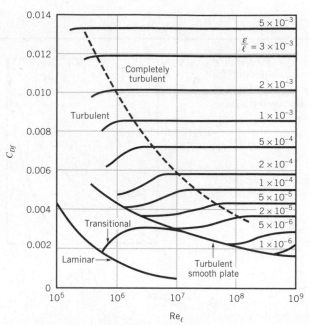

■ **Figure 9.15** Friction drag coefficient for a flat plate parallel to the upstream flow (Ref. 18, with permission).

In general, the drag coefficient for a flat plate of length $\ell$ is a function of the Reynolds number, $Re_\ell$, and the relative roughness, $\varepsilon/\ell$. The results of numerous experiments covering a wide range of the parameters of interest are shown in Fig. 9.15. For laminar boundary layer flow the drag coefficient is a function of only the Reynolds number—surface roughness is not important. This is similar to laminar flow in a pipe. However, for turbulent flow, the surface roughness does affect the shear stress and, hence, the drag coefficient. This is similar to turbulent pipe flow in which the surface roughness may protrude into or through the viscous sublayer next to the wall and alter the flow in this thin, but very important, layer (see Section 8.4.1). Values of the roughness, $\varepsilon$, for different materials can be obtained from Table 8.1.

*The flat plate drag coefficient is a function of relative roughness and Reynolds number.*

The drag coefficient diagram of Fig. 9.15 (boundary layer flow) shares many characteristics in common with the familiar Moody diagram (pipe flow) of Fig. 8.23, even though the mechanisms governing the flow are quite different. Fully developed horizontal pipe flow is governed by a balance between pressure forces and viscous forces. The fluid inertia remains constant throughout the flow. Boundary layer flow on a horizontal flat plate is governed by a balance between inertia effects and viscous forces. The pressure remains constant throughout the flow. (As is discussed in Section 9.2.6, for boundary layer flow on curved surfaces, the pressure is not constant.)

It is often convenient to have an equation for the drag coefficient as a function of the Reynolds number and relative roughness rather than the graphical representation given in Fig. 9.15. Although there is not one equation valid for the entire $Re_\ell - \varepsilon/\ell$ range, the equations presented in Table 9.3 do work well for the conditions indicated.

■ **Table 9.3**

**Empirical Equations for the Flat Plate Drag Coefficient (Ref. 1)**

| Equation | Flow Conditions |
| --- | --- |
| $C_{Df} = 1.328/(Re_\ell)^{0.5}$ | Laminar flow |
| $C_{Df} = 0.455/(\log Re_\ell)^{2.58} - 1700/Re_\ell$ | Transitional with $Re_{xcr} = 5 \times 10^5$ |
| $C_{Df} = 0.455/(\log Re_\ell)^{2.58}$ | Turbulent, smooth plate |
| $C_{Df} = [1.89 - 1.62 \log(\varepsilon/\ell)]^{-2.5}$ | Completely turbulent |

# EXAMPLE 9.7 Drag on a Flat Plate

**GIVEN** The water ski shown in Fig. E9.7a moves through 70 °F water with a velocity $U$.

**FIND** Estimate the drag caused by the shear stress on the bottom of the ski for $0 < U < 30$ ft/s.

## SOLUTION

Clearly the ski is not a flat plate, and it is not aligned exactly parallel to the upstream flow. However, we can obtain a reasonable approximation to the shear force by using the flat plate results. That is, the friction drag, $\mathscr{D}_f$, caused by the shear stress on the bottom of the ski (the wall shear stress) can be determined as

$$\mathscr{D}_f = \tfrac{1}{2}\rho U^2 \ell b C_{Df}$$

With $A = \ell b = 4 \text{ ft} \times 0.5 \text{ ft} = 2 \text{ ft}^2$, $\rho = 1.94$ slugs/ft$^3$, and $\mu = 2.04 \times 10^{-5}$ lb·s/ft$^2$ (see Table B.1) we obtain

$$\mathscr{D}_f = \tfrac{1}{2}(1.94 \text{ slugs/ft}^3)(2.0 \text{ ft}^2)U^2 C_{Df}$$
$$= 1.94 \, U^2 C_{Df} \qquad (1)$$

where $\mathscr{D}_f$ and $U$ are in pounds and ft/s, respectively.

The friction coefficient, $C_{Df}$, can be obtained from Fig. 9.15 or from the appropriate equations given in Table 9.3. As we will see, for this problem, much of the flow lies within the transition regime where both the laminar and turbulent portions of the boundary layer flow occupy comparable lengths of the plate. We choose to use the values of $C_{Df}$ from the table.

For the given conditions we obtain

$$\text{Re}_\ell = \frac{\rho U \ell}{\mu} = \frac{(1.94 \text{ slugs/ft}^3)(4 \text{ ft})U}{2.04 \times 10^{-5} \text{ lb·s/ft}^2} = 3.80 \times 10^5 \, U$$

where $U$ is in ft/s. With $U = 10$ ft/s, or $\text{Re}_\ell = 3.80 \times 10^6$, we obtain from Table 9.3 $C_{Df} = 0.455/(\log \text{Re}_\ell)^{2.58} - 1700/\text{Re}_\ell =$

0.00308. From Eq. 1 the corresponding drag is

$$\mathscr{D}_f = 1.94(10)^2(0.00308) = 0.598 \text{ lb}$$

By covering the range of upstream velocities of interest we obtain the results shown in Fig. E9.7b. **(Ans)**

**COMMENTS** If Re $\lesssim 1000$, the results of boundary layer theory are not valid—inertia effects are not dominant enough and the boundary layer is not thin compared with the length of the plate. For our problem this corresponds to $U = 2.63 \times 10^{-3}$ ft/s. For all practical purposes $U$ is greater than this value, and the flow past the ski is of the boundary layer type.

The approximate location of the transition from laminar to turbulent boundary layer flow as defined by $\text{Re}_{cr} = \rho U x_{cr}/\mu = 5 \times 10^5$ is indicated in Fig. E9.7b. Up to $U = 1.31$ ft/s the entire boundary layer is laminar. The fraction of the boundary layer that is laminar decreases as $U$ increases until only the front 0.18 ft is laminar when $U = 30$ ft/s.

For anyone who has water skied, it is clear that it can require considerably more force to be pulled along at 30 ft/s than the $2 \times 4.88 \text{ lb} = 9.76$ lb (two skis) indicated in Fig. E9.7b. As is discussed in Section 9.3, the total drag on an object such as a water ski consists of more than just the friction drag. Other components, including pressure drag and wave-making drag, add considerably to the total resistance.

(a)

(b)

■ **Figure E9.7**

## 9.2.6 Effects of Pressure Gradient

The boundary layer discussions in the previous parts of Section 9.2 have dealt with flow along a flat plate in which the pressure is constant throughout the fluid. In general, when a fluid flows past an object other than a flat plate, the pressure field is not uniform. As shown in Fig. 9.6, if the Reynolds number is large, relatively thin boundary layers will develop along the surfaces. Within

*The free-stream velocity on a curved surface is not constant.*

these layers the component of the pressure gradient in the streamwise direction (i.e., along the body surface) is not zero, although the pressure gradient normal to the surface is negligibly small. That is, if we were to measure the pressure while moving across the boundary layer from the body to the boundary layer edge, we would find that the pressure is essentially constant. However, the pressure does vary in the direction along the body surface if the body is curved, as shown by the figure in the margin. The variation in the **free-stream velocity,** $U_{fs}$, the fluid velocity at the edge of the boundary layer, is the cause of the pressure gradient in this direction. The characteristics of the entire flow (both within and outside of the boundary layer) are often highly dependent on the pressure gradient effects on the fluid within the boundary layer.

For a flat plate parallel to the upstream flow, the upstream velocity (that far ahead of the plate) and the free-stream velocity (that at the edge of the boundary layer) are equal—$U = U_{fs}$. This is a consequence of the negligible thickness of the plate. For bodies of nonzero thickness, these two velocities are different. This can be seen in the flow past a circular cylinder of diameter $D$. The upstream velocity and pressure are $U$ and $p_0$, respectively. If the fluid were completely inviscid ($\mu = 0$), the Reynolds number would be infinite (Re $= \rho UD/\mu = \infty$) and the streamlines would be symmetrical, as are shown in Fig. 9.16a. The fluid velocity along the surface would vary from $U_{fs} = 0$ at the very front and rear of the cylinder (points $A$ and $F$ are stagnation points) to a maximum of $U_{fs} = 2U$ at the top and bottom of the cylinder (point $C$). This is also indicated in the figure in the margin. The pressure on the surface of the cylinder would be symmetrical about the vertical midplane of the cylinder, reaching a maximum value of $p_0 + \rho U^2/2$ (the stagnation pressure) at both the front and back of the cylinder, and a minimum of $p_0 - 3\rho U^2/2$ at the top and bottom of the cylinder. The pressure and free-stream velocity distributions are shown in Figs. 9.16b and 9.16c. These characteristics can be obtained from potential flow analysis of Section 6.6.3.

Because of the absence of viscosity (therefore, $\tau_w = 0$) and the symmetry of the pressure distribution for inviscid flow past a circular cylinder, it is clear that the drag on the cylinder is zero. Although it is not obvious, it can be shown that the drag is zero for any object that does not produce a lift (symmetrical or not) in an inviscid fluid (Ref. 4). Based on experimental evidence, however, we know that there must be a net drag. Clearly, since there is no purely inviscid fluid, the reason for the observed drag must lie on the shoulders of the viscous effects.

To test this hypothesis, we could conduct an experiment by measuring the drag on an object (such as a circular cylinder) in a series of fluids with decreasing values of viscosity. To our initial surprise we would find that no matter how small we make the viscosity (provided it is not precisely zero) we would measure a finite drag, essentially independent of the value of $\mu$. As was noted in Section 6.6.3, this leads to what has been termed *d'Alembert's paradox*—the drag on an

■ **Figure 9.16** Inviscid flow past a circular cylinder: (a) streamlines for the flow if there were no viscous effects, (b) pressure distribution on the cylinder's surface, (c) free-stream velocity on the cylinder's surface.

object in an inviscid fluid is zero, but the drag on an object in a fluid with vanishingly small (but nonzero) viscosity is not zero.

The reason for the above paradox can be described in terms of the effect of the pressure gradient on boundary layer flow. Consider large Reynolds number flow of a real (viscous) fluid past a circular cylinder. As was discussed in Section 9.1.2, we expect the viscous effects to be confined to thin boundary layers near the surface. This allows the fluid to stick ($\mathbf{V} = 0$) to the surface—a necessary condition for any fluid, provided $\mu \neq 0$. The basic idea of boundary layer theory is that the boundary layer is thin enough so that it does not greatly disturb the flow outside the boundary layer. Based on this reasoning, for large Reynolds numbers the flow throughout most of the flow field would be expected to be as is indicated in Fig. 9.16a, the inviscid flow field.

The pressure distribution indicated in Fig. 9.16b is imposed on the boundary layer flow along the surface of the cylinder. In fact, there is negligible pressure variation across the thin boundary layer so that the pressure within the boundary layer is that given by the inviscid flow field. This pressure distribution along the cylinder is such that the stationary fluid at the nose of the cylinder ($U_{fs} = 0$ at $\theta = 0$) is accelerated to its maximum velocity ($U_{fs} = 2U$ at $\theta = 90°$) and then is decelerated back to zero velocity at the rear of the cylinder ($U_{fs} = 0$ at $\theta = 180°$). This is accomplished by a balance between pressure and inertia effects; viscous effects are absent for the inviscid flow outside the boundary layer.

Physically, in the absence of viscous effects, a fluid particle traveling from the front to the back of the cylinder coasts down the "pressure hill" from $\theta = 0$ to $\theta = 90°$ (from point $A$ to $C$ in Fig. 9.16b) and then back up the hill to $\theta = 180°$ (from point $C$ to $F$) without any loss of energy. There is an exchange between kinetic and pressure energy, but there are no energy losses. The same pressure distribution is imposed on the viscous fluid within the boundary layer. The decrease in pressure in the direction of flow along the front half of the cylinder is termed a **favorable pressure gradient**. The increase in pressure in the direction of flow along the rear half of the cylinder is termed an **adverse pressure gradient.**

Consider a fluid particle within the boundary layer indicated in Fig. 9.17a. In its attempt to flow from $A$ to $F$, it experiences the same pressure distribution as the particles in the free stream immediately outside the boundary layer—the inviscid flow field pressure. However, because of the viscous effects involved, the particle in the boundary layer experiences a loss of energy as it flows along. This loss means that the particle does not have enough energy to coast all of the way up the pressure hill (from $C$ to $F$) and to reach point $F$ at the rear of the cylinder. This kinetic energy deficit is seen in the velocity profile detail at point $C$, shown in Fig. 9.17a. Because of friction, the boundary layer fluid cannot travel from the front to the rear of the cylinder. (This conclusion can also be obtained from the concept that due to viscous effects the particle at $C$ does not have enough momentum to allow it to coast up the pressure hill to $F$.)

The situation is similar to a bicyclist coasting down a hill and up the other side of the valley. If there were no friction, the rider starting with zero speed could reach the same height from which he or she started. Clearly friction (rolling resistance, aerodynamic drag, etc.) causes a loss of energy (and momentum), making it impossible for the rider to reach the height from which he or she started without supplying additional energy (i.e., pedaling). The fluid within the boundary layer does not have such an energy supply. Thus, the fluid flows against the increasing pressure as far as it can, at which point the boundary layer separates from (lifts off) the surface. This **boundary layer separation** is indicated in Fig. 9.17a as well as the figures in the margin. (See the photographs at the beginning of Chapters 7 and 9.) Typical velocity profiles at representative locations along the surface are shown in Fig. 9.17b. At the separation location (profile $D$), the velocity gradient at the wall and the wall shear stress are zero. Beyond that location (from $D$ to $E$) there is reverse flow in the boundary layer.

As is indicated in Fig. 9.17c, because of the boundary layer separation, the average pressure on the rear half of the cylinder is considerably less than that on the front half. Thus, a large pressure drag is developed, even though (because of small viscosity) the viscous shear drag may be quite small. D'Alembert's paradox is explained. No matter how small the viscosity, provided it is not zero, there will be a boundary layer that separates from the surface, giving a drag that is, for the most part, independent of the value of $\mu$.

The pressure gradient in the external flow is imposed throughout the boundary layer fluid.

Separation location

V9.9 Snow drifts

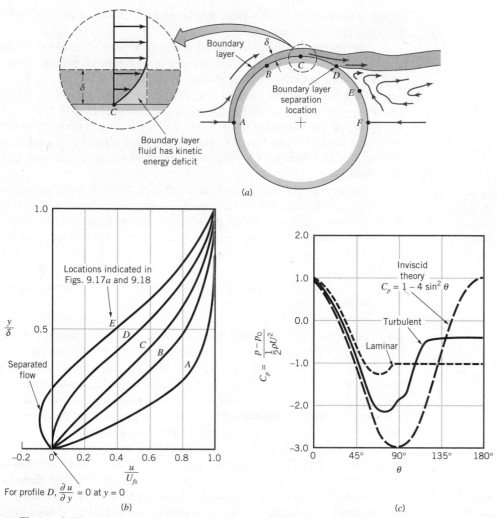

■ **Figure 9.17** Boundary layer characteristics on a circular cylinder: (*a*) boundary layer separation location, (*b*) typical boundary layer velocity profiles at various locations on the cylinder, (*c*) surface pressure distributions for inviscid flow and boundary layer flow.

**Increasing truck mpg** A large portion of the aerodynamic drag on semis (tractor-trailer rigs) is a result of the low pressure on the flat back end of the trailer. Researchers have recently developed a drag-reducing attachment that could reduce fuel costs on these big rigs by 10%. The device consists of a set of flat plates (attached to the rear of the trailer) that fold out into a box shape, thereby making the originally flat rear of the trailer a somewhat more "aerodynamic" shape. Based on thorough wind tunnel testing and actual tests conducted with a prototype design used in a series of cross-country runs, it is estimated that trucks using the device could save approximately $6000 a year in fuel costs.

*Viscous effects within the boundary layer cause boundary layer separation.*

The location of separation, the width of the wake region behind the object, and the pressure distribution on the surface depend on the nature of the boundary layer flow. Compared with a laminar boundary layer, a turbulent boundary layer flow has more kinetic energy and momentum associated with it because: (1) as is indicated in Fig. E9.6, the velocity profile is fuller, more nearly like the ideal uniform profile, and (2) there can be considerable energy associated with the swirling, random components of the velocity that do not appear in the time-averaged *x* component of velocity. Thus, as is indicated in Fig. 9.17*c*, the turbulent boundary layer can flow farther around the cylinder (farther up the pressure hill) before it separates than can the laminar boundary layer.

The structure of the flow field past a circular cylinder is completely different for a zero viscosity fluid than it is for a viscous fluid, no matter how small the viscosity is, provided it is not

(a)

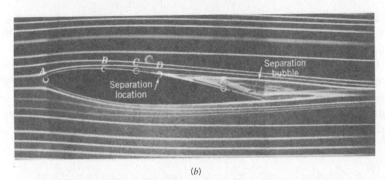

(b)

■ **Figure 9.18** Flow visualization photographs of flow past an airfoil (the boundary layer velocity profiles for the points indicated are similar to those indicated in Fig. 9.17b): (a) zero angle of attack, no separation, (b) 5° angle of attack, flow separation. Dye in water. (Photographs courtesy of ONERA, The French Aerospace Lab.)

*V9.10 Effect of body shape on pressure gradient*

*Streamlined bodies generally have no separated flow.*

zero. This is due to boundary layer separation. Similar concepts hold for other shaped bodies as well. The flow past an airfoil at zero *angle of attack* (the angle between the upstream flow and the axis of the object) is shown in Fig. 9.18a; flow past the same airfoil at a 5° angle of attack is shown in Fig. 9.18b. Over the front portion of the airfoil the pressure decreases in the direction of flow—a favorable pressure gradient. Over the rear portion the pressure increases in the direction of flow—an adverse pressure gradient. The boundary layer velocity profiles at representative locations are similar to those indicated in Fig. 9.17b for flow past a circular cylinder. If the adverse pressure gradient is not too great (because the body is not too "thick" in some sense), the boundary layer fluid can flow into the slightly increasing pressure region (i.e., from C to the trailing edge in Fig. 9.18a) without separating from the surface. However, if the pressure gradient is too adverse (because the angle of attack is too large), the boundary layer will separate from the surface as indicated in Fig. 9.18b. Such situations can lead to the catastrophic loss of lift called *stall*, which is discussed in Section 9.4.

Streamlined bodies are generally those designed to eliminate (or at least to reduce) the effects of separation, whereas nonstreamlined bodies generally have relatively large drag due to the low pressure in the separated regions (the wake). Although the boundary layer may be quite thin, it can appreciably alter the entire flow field because of boundary layer separation. These ideas are discussed in Section 9.3.

### 9.2.7 Momentum Integral Boundary Layer Equation with Nonzero Pressure Gradient

The boundary layer results discussed in Sections 9.2.2 and 9.2.3 are valid only for boundary layers with zero pressure gradients. They correspond to the velocity profile labeled C in Fig. 9.17b. Boundary layer characteristics for flows with nonzero pressure gradients can be obtained from nonlinear, partial differential boundary layer equations similar to Eqs. 9.8 and 9.9, provided the pressure gradient is appropriately accounted for. Such an approach is beyond the scope of this book (Refs. 1, 2).

An alternative approach is to extend the momentum integral boundary layer equation technique (Section 9.2.3) so that it is applicable for flows with nonzero pressure gradients. The momentum integral equation for boundary layer flows with zero pressure gradient, Eq. 9.26, is a statement of the balance between the shear force on the plate (represented by $\tau_w$) and rate of change of momentum of the fluid within the boundary layer [represented by $\rho U^2 (d\Theta/dx)$]. For such flows the free-stream velocity is constant ($U_{fs} = U$). If the free-stream velocity is not constant [$U_{fs} = U_{fs}(x)$, where $x$ is the distance measured along the curved body], the pressure will not be constant. This follows from the Bernoulli equation with negligible gravitational effects, since $p + \rho U_{fs}^2/2$ is constant along the streamlines outside the boundary layer. Thus,

$$\frac{dp}{dx} = -\rho U_{fs} \frac{dU_{fs}}{dx} \qquad (9.34)$$

For a given body the free-stream velocity and the corresponding pressure gradient on the surface can be obtained from inviscid flow techniques (potential flow) discussed in Section 6.7. (This is how the circular cylinder results of Fig. 9.16 were obtained.)

Flow in a boundary layer with nonzero pressure gradient is very similar to that shown in Fig. 9.11, except that the upstream velocity, $U$, is replaced by the free-stream velocity, $U_{fs}(x)$, and the pressures at sections (1) and (2) are not necessarily equal. By using the $x$ component of the momentum equation (Eq. 5.22) with the appropriate shear forces and pressure forces acting on the control surface indicated in Fig. 9.11, the following integral momentum equation for boundary layer flows is obtained:

$$\tau_w = \rho \frac{d}{dx}(U_{fs}^2 \Theta) + \rho\delta^* U_{fs} \frac{dU_{fs}}{dx} \qquad (9.35)$$

The derivation of this equation is similar to that of the corresponding equation for constant-pressure boundary layer flow, Eq. 9.26, although the inclusion of the pressure gradient effect brings in additional terms (Refs. 1, 2, 3). For example, both the boundary layer momentum thickness, $\Theta$, and the displacement thickness, $\delta^*$, are involved.

Equation 9.35, the general momentum integral equation for two-dimensional boundary layer flow, represents a balance between viscous forces (represented by $\tau_w$), pressure forces (represented by $\rho U_{fs} dU_{fs}/dx = -dp/dx$), and the fluid momentum (represented by $\Theta$, the boundary layer momentum thickness). In the special case of a flat plate, $U_{fs} = U = $ constant, and Eq. 9.35 reduces to Eq. 9.26.

Equation 9.35 can be used to obtain boundary layer information in a manner similar to that done for the flat plate boundary layer (Section 9.2.3). That is, for a given body shape the free-stream velocity, $U_{fs}$, is determined, and a family of approximate boundary layer profiles is assumed. Equation 9.35 is then used to provide information about the boundary layer thickness, wall shear stress, and other properties of interest. The details of this technique are not within the scope of this book (Refs. 1, 3).

## 9.3    Drag

As was discussed in Section 9.1, any object moving through a fluid will experience a drag, $\mathcal{D}$—a net force in the direction of flow due to the pressure and shear forces on the surface of the object. This net force, a combination of flow direction components of the normal and tangential forces on the body, can be determined by use of Eqs. 9.1 and 9.2, provided the distributions of pressure, $p$, and wall shear stress, $\tau_w$, are known. Only in very rare instances can these distributions be determined analytically. The boundary layer flow past a flat plate parallel to the upstream flow as is discussed in Section 9.2 is one such case. Current advances in computational fluid dynamics, CFD (i.e., the use of computers to solve the governing equations of the flow field), have provided encouraging results for more complex shapes. However, much work in this area remains.

Most of the information pertaining to drag on objects is a result of numerous experiments with wind tunnels, water tunnels, towing tanks, and other ingenious devices that are used to measure the drag on scale models. As was discussed in Chapter 7, these data can be put into dimensionless form, and the results can be appropriately ratioed for prototype calculations. Typically, the result for a given-shaped object is a drag coefficient, $C_D$, where

$$C_D = \frac{\mathcal{D}}{\frac{1}{2}\rho U^2 A} \qquad (9.36)$$

and $C_D$ is a function of other dimensionless parameters such as Reynolds number, Re, Mach number, Ma, Froude number, Fr, and relative roughness of the surface, $\varepsilon/\ell$. That is,

$$C_D = \phi(\text{shape, Re, Ma, Fr, } \varepsilon/\ell)$$

The character of $C_D$ as a function of these parameters is discussed in this section.

### 9.3.1 Friction Drag

*Friction drag,* $\mathcal{D}_f$, is that part of the drag that is due directly to the shear stress, $\tau_w$, on the object. It is a function of not only the magnitude of the wall shear stress but also of the orientation of the surface on which it acts. This is indicated by the factor $\tau_w \sin \theta$ in Eq. 9.1. If the surface is parallel to the upstream velocity, the entire shear force contributes directly to the drag. This is true for the flat plate parallel to the flow as was discussed in Section 9.2. If the surface is perpendicular to the upstream velocity, the shear stress contributes nothing to the drag. Such is the case for a flat plate normal to the upstream velocity as was discussed in Section 9.1.

In general, the surface of a body will contain portions parallel to and normal to the upstream flow, as well as any direction in between. A circular cylinder is such a body. Because the viscosity of most common fluids is small, the contribution of the shear force to the overall drag on a body is often quite small. Such a statement should be worded in dimensionless terms. That is, because the Reynolds number of most familiar flows is quite large, the percent of the drag caused directly by the shear stress is often quite small. For highly streamlined bodies or for low Reynolds number flow, however, most of the drag may be due to friction drag.

The friction drag on a flat plate of width $b$ and length $\ell$ oriented parallel to the upstream flow can be calculated from

$$\mathcal{D}_f = \tfrac{1}{2}\rho U^2 b\ell C_{Df}$$

*Friction (viscous) drag is the drag produced by viscous shear stresses.*

where $C_{Df}$ is the friction drag coefficient. The value of $C_{Df}$, given as a function of Reynolds number, $\text{Re}_\ell = \rho U\ell/\mu$, and relative surface roughness, $\varepsilon/\ell$, in Fig. 9.15 and Table 9.3, is a result of boundary layer analysis and experiments (see Section 9.2). Typical values of roughness, $\varepsilon$, for various surfaces are given in Table 8.1. As with the pipe flow discussed in Chapter 8, the flow is divided into two distinct categories—laminar or turbulent, with a transitional regime connecting them. The drag coefficient (and, hence, the drag) is not a function of the plate roughness if the flow is laminar. However, for turbulent flow the roughness does considerably affect the value of $C_{Df}$. As with pipe flow, this dependence is a result of the surface roughness elements protruding into or through the laminar sublayer (see Section 8.3).

Most objects are not flat plates parallel to the flow; instead, they are curved surfaces along which the pressure varies. As was discussed in Section 9.2.6, this means that the boundary layer character, including the velocity gradient at the wall, is different for most objects from that for a flat plate. This can be seen in the change of shape of the boundary layer profile along the cylinder in Fig. 9.17b.

The precise determination of the shear stress along the surface of a curved body is quite difficult to obtain. Although approximate results can be obtained by a variety of techniques (Refs. 1, 2), these are outside the scope of this text. As is shown by the following example, if the shear stress is known, its contribution to the drag can be determined.

---

# EXAMPLE 9.8    Drag Coefficient Based on Friction Drag

**GIVEN** A viscous, incompressible fluid flows past the circular cylinder shown in Fig. E9.8a. According to a more advanced theory of boundary layer flow, the boundary layer remains attached to the cylinder up to the separation location at $\theta \approx 108.8°$, with the dimensionless wall shear stress as is indicated in Fig. E9.8b (Ref. 1). The shear stress on the cylinder in the wake region, $108.8 < \theta < 180°$, is negligible.

**FIND** Determine $C_{Df}$, the drag coefficient for the cylinder based on the friction drag only.

## Solution

The friction drag, $\mathcal{D}_f$, can be determined from Eq. 9.1 as

$$\mathcal{D}_f = \int \tau_w \sin \theta \, dA = 2 \left( \frac{D}{2} \right) b \int_0^\pi \tau_w \sin \theta \, d\theta$$

(a)

(b)

(c)

■ **Figure E9.8**

where $b$ is the length of the cylinder. Note that $\theta$ is in radians (not degrees) to ensure the proper dimensions of $dA = 2 \, (D/2) \, b \, d\theta$. Thus,

$$C_{Df} = \frac{\mathcal{D}_f}{\frac{1}{2} \rho U^2 bD} = \frac{2}{\rho U^2} \int_0^\pi \tau_w \sin \theta \, d\theta$$

This can be put into dimensionless form by using the dimensionless shear stress parameter, $F(\theta) = \tau_w \sqrt{Re}/(\rho U^2/2)$, given in Fig. E9.8b as follows:

$$C_{Df} = \int_0^\pi \frac{\tau_w}{\frac{1}{2} \rho U^2} \sin \theta \, d\theta = \frac{1}{\sqrt{Re}} \int_0^\pi \frac{\tau_w \sqrt{Re}}{\frac{1}{2} \rho U^2} \sin \theta \, d\theta$$

where $Re = \rho UD/\mu$. Thus,

$$C_{Df} = \frac{1}{\sqrt{Re}} \int_0^\pi F(\theta) \sin \theta \, d\theta \qquad (1)$$

The function $F(\theta) \sin \theta$, obtained from Fig. E9.8b, is plotted in Fig. E9.8c. The necessary integration to obtain $C_{Df}$ from Eq. 1 can be done by an appropriate numerical technique or by an approximate graphical method to determine the area under the given curve.

The result is $\int_0^\pi F(\theta) \sin \theta \, d\theta = 5.93$, or

$$C_{Df} = \frac{5.93}{\sqrt{Re}} \qquad \textbf{(Ans)}$$

**COMMENTS** Note that the total drag must include both the shear stress (friction) drag and the pressure drag. As we will see in Example 9.9, for the circular cylinder most of the drag is due to the pressure force.

The above friction drag result is valid only if the boundary layer flow on the cylinder is laminar. As is discussed in Section 9.3.3, for a smooth cylinder this means that $Re = \rho UD/\mu < 3 \times 10^5$. It is also valid only for flows that have a Reynolds number sufficiently large to ensure the boundary layer structure to the flow. For the cylinder, this means $Re > 100$.

## 9.3.2 Pressure Drag

*Pressure (form) drag is the drag produced by normal stresses.*

**Pressure drag,** $\mathcal{D}_p$, is that part of the drag that is due directly to the pressure, $p$, on an object. It is often referred to as *form drag* because of its strong dependency on the shape or form of the object. Pressure drag is a function of the magnitude of the pressure and the orientation of the surface element on which the pressure force acts. For example, the pressure force on either side of a flat plate parallel to the flow may be very large, but it does not contribute to the drag because it acts in the direction normal to the upstream velocity. On the other hand, the pressure force on a flat plate normal to the flow provides the entire drag.

As previously noted, for most bodies, there are portions of the surface that are parallel to the upstream velocity, others normal to the upstream velocity, and the majority of which are at some angle in between, as shown by the figure in the margin. The pressure drag can be obtained from Eq. 9.1 provided a detailed description of the pressure distribution and the body shape is given. That is,

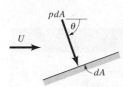

$$\mathcal{D}_p = \int p \cos \theta \, dA$$

which can be rewritten in terms of the *pressure drag coefficient*, $C_{Dp}$, as

$$C_{Dp} = \frac{\mathscr{D}_p}{\frac{1}{2}\rho U^2 A} = \frac{\int p \cos\theta \, dA}{\frac{1}{2}\rho U^2 A} = \frac{\int C_p \cos\theta \, dA}{A} \quad (9.37)$$

*The pressure coefficient is a dimensionless form of the pressure.*

Here $C_p = (p - p_0)/(\rho U^2/2)$ is the *pressure coefficient*, where $p_0$ is a reference pressure. The level of the reference pressure does not influence the drag directly because the net pressure force on a body is zero if the pressure is constant (i.e., $p_0$) on the entire surface.

For flows in which inertial effects are large relative to viscous effects (i.e., large Reynolds number flows), the pressure difference, $p - p_0$, scales directly with the dynamic pressure, $\rho U^2/2$, and the pressure coefficient is independent of Reynolds number. In such situations we expect the drag coefficient to be relatively independent of Reynolds number.

For flows in which viscous effects are large relative to inertial effects (i.e., very small Reynolds number flows), it is found that both the pressure difference and wall shear stress scale with the characteristic viscous stress, $\mu U/\ell$, where $\ell$ is a characteristic length. In such situations we expect the drag coefficient to be proportional to $1/\text{Re}$. That is, $C_D \sim \mathscr{D}/(\rho U^2/2) \sim (\mu U/\ell)/(\rho U^2/2) \sim \mu/\rho U\ell = 1/\text{Re}$. These characteristics are similar to the friction factor dependence of $f \sim 1/\text{Re}$ for laminar pipe flow and $f \sim$ constant for large Reynolds number flow (see Section 8.4).

If the viscosity were zero, the pressure drag on any shaped object (symmetrical or not) in a steady flow would be zero. There perhaps would be large pressure forces on the front portion of the object, but there would be equally large (and oppositely directed) pressure forces on the rear portion. If the viscosity is not zero, the net pressure drag may be nonzero because of boundary layer separation as is discussed in Section 9.2.6. Example 9.9 illustrates this concept.

## $E$XAMPLE 9.9   Drag Coefficient Based on Pressure Drag

**GIVEN** A viscous, incompressible fluid flows past the circular cylinder shown in Fig. E9.8a. The pressure coefficient on the surface of the cylinder (as determined from experimental measurements) is as indicated in Fig. E9.9a.

**FIND** Determine the pressure drag coefficient for this flow. Combine the results of Examples 9.8 and 9.9 to determine the drag coefficient for a circular cylinder. Compare your results with those given in Fig. 9.21.

## SOLUTION

The pressure (form) drag coefficient, $C_{Dp}$, can be determined from Eq. 9.37 as

$$C_{Dp} = \frac{1}{A}\int C_p \cos\theta \, dA = \frac{1}{bD}\int_0^{2\pi} C_p \cos\theta \, b\left(\frac{D}{2}\right) d\theta$$

or because of symmetry

$$C_{Dp} = \int_0^{\pi} C_p \cos\theta \, d\theta$$

where $b$ and $D$ are the length and diameter of the cylinder. To obtain $C_{Dp}$, we must integrate the $C_p \cos\theta$ function from $\theta = 0$ to $\theta = \pi$ radians. Again, this can be done by some numerical integration scheme or by determining the area under the curve shown in Fig. E9.9b. The result is

$$C_{Dp} = 1.17 \qquad (1) \quad \textbf{(Ans)}$$

Note that the positive pressure on the front portion of the cylinder ($0 \le \theta \le 30°$) and the negative pressure (less than the upstream value) on the rear portion ($90 \le \theta \le 180°$) produce positive contributions to the drag. The negative pressure on the front portion of the

cylinder ($30 < \theta < 90°$) reduces the drag by pulling on the cylinder in the upstream direction. The positive area under the $C_p \cos\theta$ curve is greater than the negative area—there is a net pressure drag. In the absence of viscosity, these two contributions would be equal—there would be no pressure (or friction) drag.

The net drag on the cylinder is the sum of friction and pressure drag. Thus, from Eq. 1 of Example 9.8 and Eq. 1 of this example, we obtain the drag coefficient

$$C_D = C_{Df} + C_{Dp} = \frac{5.93}{\sqrt{\text{Re}}} + 1.17 \quad (2) \quad \textbf{(Ans)}$$

This result is compared with the standard experimental value (obtained from Fig. 9.21) in Fig. E9.9c. The agreement is very good over a wide range of Reynolds numbers. For $\text{Re} < 10$ the curves diverge because the flow is not a boundary layer type flow—the shear stress and pressure distributions used to obtain Eq. 2 are not valid in this range. The drastic divergence in the curves for $\text{Re} > 3 \times 10^5$ is due to the change from a laminar to turbulent boundary layer, with the corresponding change in the pressure distribution. This is discussed in Section 9.3.3.

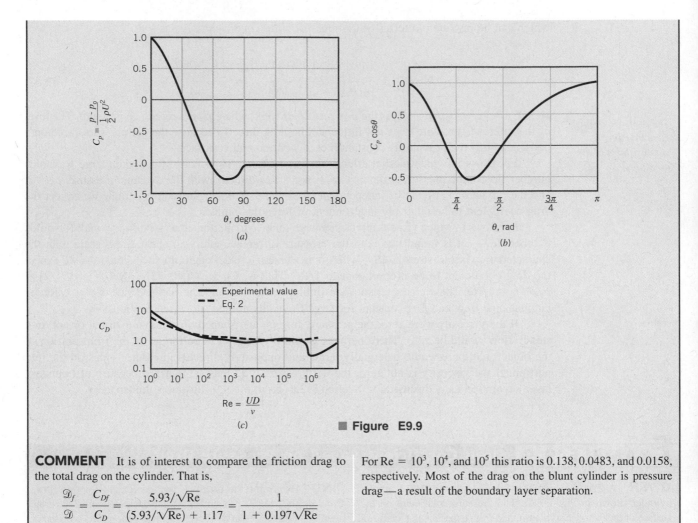

■ **Figure E9.9**

**COMMENT** It is of interest to compare the friction drag to the total drag on the cylinder. That is,

$$\frac{\mathcal{D}_f}{\mathcal{D}} = \frac{C_{Df}}{C_D} = \frac{5.93/\sqrt{Re}}{(5.93/\sqrt{Re}) + 1.17} = \frac{1}{1 + 0.197\sqrt{Re}}$$

For $Re = 10^3$, $10^4$, and $10^5$ this ratio is 0.138, 0.0483, and 0.0158, respectively. Most of the drag on the blunt cylinder is pressure drag—a result of the boundary layer separation.

### 9.3.3 Drag Coefficient Data and Examples

As was discussed in previous sections, the net drag is produced by both pressure and shear stress effects. In most instances these two effects are considered together, and an overall drag coefficient, $C_D$, as defined in Eq. 9.36 is used. There is an abundance of such drag coefficient data available in the literature. This information covers incompressible and compressible viscous flows past objects of almost any shape of interest—both man-made and natural objects. In this section we consider a small portion of this information for representative situations. Additional data can be obtained from various sources (Refs. 5, 6).

*V9.11 Skydiving practice*

*The drag coefficient may be based on the frontal area or the planform area.*

***Shape Dependence.*** Clearly the drag coefficient for an object depends on the shape of the object, with shapes ranging from those that are streamlined to those that are blunt. The drag on an ellipse with aspect ratio $\ell/D$, where $D$ and $\ell$ are the thickness and length parallel to the flow, illustrates this dependence. The drag coefficient $C_D = \mathcal{D}/(\rho U^2 \, bD/2)$, based on the frontal area, $A = bD$, where $b$ is the length normal to the flow, is as shown in Fig. 9.19. The blunter the body, the larger the drag coefficient. With $\ell/D = 0$ (i.e., a flat plate normal to the flow) we obtain the flat plate value of $C_D = 1.9$. With $\ell/D = 1$ the corresponding value for a circular cylinder is obtained. As $\ell/D$ becomes larger, the value of $C_D$ decreases.

For very large aspect ratios ($\ell/D \rightarrow \infty$) the ellipse behaves as a flat plate parallel to the flow. For such cases, the friction drag is greater than the pressure drag, and the value of $C_D$ based on the frontal area, $A = bD$, would increase with increasing $\ell/D$. (This occurs for larger $\ell/D$ values than those shown in the figure.) For such extremely thin bodies (i.e., an ellipse with $\ell/D \rightarrow \infty$, a flat plate, or very thin airfoils) it is customary to use the planform area, $A = b\ell$, in defining the drag coefficient.

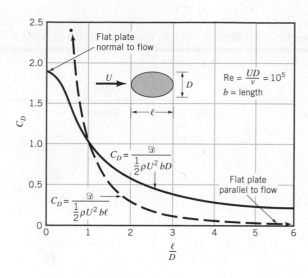

■ **Figure 9.19** Drag coefficient for an ellipse with the characteristic area either the frontal area, $A = bD$, or the planform area, $A = b\ell$ (Ref. 5).

After all, it is the planform area on which the shear stress acts, rather than the much smaller (for thin bodies) frontal area. The ellipse drag coefficient based on the planform area, $C_D = \mathscr{D}/(\rho U^2 b\ell/2)$, is also shown in Fig. 9.19. Clearly the drag obtained by using either of these drag coefficients would be the same. They merely represent two different ways to package the same information.

The amount of streamlining can have a considerable effect on the drag. Incredibly, the drag on the two two-dimensional objects drawn to scale in Fig. 9.20 is the same. The width of the wake for the streamlined strut is very thin, on the order of that for the much smaller diameter circular cylinder.

***Reynolds Number Dependence.*** Another parameter on which the drag coefficient can be very dependent is the Reynolds number. The main categories of Reynolds number dependence are (1) very low Reynolds number flow, (2) moderate Reynolds number flow (laminar boundary layer), and (3) very large Reynolds number flow (turbulent boundary layer). Examples of these three situations are discussed below.

Low Reynolds number flows (Re < 1) are governed by a balance between viscous and pressure forces. Inertia effects are negligibly small. In such instances the drag on a three-dimensional body is expected to be a function of the upstream velocity, $U$, the body size, $\ell$, and the viscosity, $\mu$. Thus, for a small grain of sand settling in a lake (see figure in the margin)

$$\mathscr{D} = f(U, \ell, \mu)$$

Re < 1

$\ell$

$\mu$

$U$

$\mathscr{D} = f(U, \ell, \mu)$

From dimensional considerations (see Section 7.7.1)

$$\mathscr{D} = C\mu\ell U \tag{9.38}$$

where the value of the constant $C$ depends on the shape of the body. If we put Eq. 9.38 into dimensionless form using the standard definition of the drag coefficient, we obtain

$$C_D = \frac{\mathscr{D}}{\frac{1}{2}\rho U^2 \ell^2} = \frac{2C\mu\ell U}{\rho U^2 \ell^2} = \frac{2C}{\mathrm{Re}}$$

where $\mathrm{Re} = \rho U\ell/\mu$. The use of the dynamic pressure, $\rho U^2/2$, in the definition of the drag coefficient is somewhat misleading in the case of creeping flows (Re < 1) because it introduces the

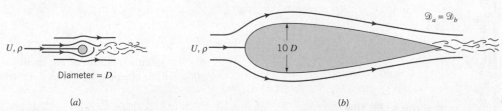

(a)                    (b)

■ **Figure 9.20** Two objects of considerably different size that have the same drag force: (a) circular cylinder $C_D = 1.2$; (b) streamlined strut $C_D = 0.12$.

■ **Table 9.4**
**Low Reynolds Number Drag Coefficients (Ref. 7)** $(Re = \rho UD/\mu, A = \pi D^2/4)$

| Object | $C_D = \mathcal{D}/(\rho U^2 A/2)$ (for Re ≲ 1) | Object | $C_D$ |
|---|---|---|---|
| a. Circular disk normal to flow | 20.4/Re | c. Sphere | 24.0/Re |
| b. Circular disk parallel to flow | 13.6/Re | d. Hemisphere | 22.2/Re |

*For very small Reynolds number flows, the drag coefficient varies inversely with the Reynolds number.*

fluid density, which is not an important parameter for such flows (inertia is not important). Use of this standard drag coefficient definition gives the 1/Re dependence for small Re drag coefficients.

Typical values of $C_D$ for low Reynolds number flows past a variety of objects are given in Table 9.4. It is of interest that the drag on a disk normal to the flow is only 1.5 times greater than that on a disk parallel to the flow. For large Reynolds number flows this ratio is considerably larger (see Example 9.1). Streamlining (i.e., making the body slender) can produce a considerable drag reduction for large Reynolds number flows; for very small Reynolds number flows it can actually increase the drag because of an increase in the area on which shear forces act. For most objects, the low Reynolds number flow results are valid up to a Reynolds number of about 1.

---

## $E$XAMPLE 9.10   Low Reynolds Number Flow Drag

**GIVEN**  While workers spray paint the ceiling of a room, numerous small paint aerosols are dispersed into the air. Eventually these particles will settle out and fall to the floor or other surfaces. Consider a small spherical paint particle of diameter $D = 1 \times 10^{-5}$ m (or 10 $\mu$m) and specific gravity $SG = 1.2$.

**FIND**  Determine the time it would take this particle to fall 8 ft from near the ceiling to the floor. Assume that the air within the room is motionless.

### SOLUTION

A free-body diagram of the particle (relative to the moving particle) is shown in Fig. E9.10a. The particle moves downward with a constant velocity $U$ that is governed by a balance between the weight of the particle, $\mathcal{W}$, the buoyancy force of the surrounding air, $F_B$, and the drag, $\mathcal{D}$, of the air on the particle.

■ **Figure  E9.10a**

From the free-body diagram, we obtain

$$\mathcal{W} = \mathcal{D} + F_B$$

where, if $V$ is the particle volume

$$\mathcal{W} = \gamma_{\text{point}} V = SG\, \gamma_{\text{H}_2\text{O}}\, \frac{\pi}{6} D^3 \qquad (1)$$

and

$$F_B = \gamma_{\text{air}} V = \gamma_{\text{air}} \frac{\pi}{6} D^3 \qquad (2)$$

We assume (because of the smallness of the object) that the flow will be creeping flow (Re < 1) with $C_D = 24/Re$ so that

$$\mathcal{D} = \frac{1}{2} \rho_{\text{air}} U^2 \frac{\pi}{4} D^2 C_D = \frac{1}{2} \rho_{\text{air}} U^2 \frac{\pi}{4} D^2 \left( \frac{24}{\rho_{\text{air}} UD/\mu_{\text{air}}} \right)$$

or

$$\mathscr{D} = 3\pi\mu_{air}UD \qquad (3)$$

We must eventually check to determine if the assumption that Re < 1 is valid. Equation 3 is called Stokes's law in honor of G. G. Stokes, a British mathematician and a physicist. By combining Eqs. 1, 2, and 3, we obtain

$$SG\,\gamma_{H_2O}\,\frac{\pi}{6}\,D^3 = 3\pi\mu_{air}UD + \gamma_{air}\,\frac{\pi}{6}\,D^3$$

Then, solving for $U$,

$$U = \frac{D^2(SG\gamma_{H_2O} - \gamma_{air})}{18\mu_{air}} \qquad (4)$$

From Tables 1.5 and 1.7 we obtain $\gamma_{H_2O} = 9800$ N/m$^3$, $\gamma_{air} = 12.0$ N/m$^3$ and $\mu_{air} = 1.79 \times 10^{-5}$ N·s/m$^2$. Thus, from Eq. 4 we obtain

$$U = \frac{(10^{-5}\,m)^2[(1.2)(9800\,N/m^3) - (12.0\,N/m^3)]}{18(1.79 \times 10^{-5}\,N \cdot s/m^2)}$$

or

$$U = 0.00365\,m/s$$

Let $t_{fall}$ denote the time it takes for the particle to fall 8 ft. Thus, since 8 ft is approximately 2.44 m,

$$t_{fall} = \frac{2.44\,m}{0.00365\,m/s} = 668\,s$$

Therefore, it will take over 11 minutes for the paint particle to fall to the floor.

Since

$$Re = \frac{\rho DU}{\mu} = \frac{(1.23\,kg/m^3)(1 \times 10^{-5}\,m)(0.00365\,m/s)}{1.79 \times 10^{-5}\,N \cdot s/m^2}$$

$$= 0.00251$$

we see that Re < 1, and the form of the drag coefficient used is valid.

**COMMENTS**   By repeating the calculations for various particle diameters, the results shown in Fig. E9.10b are obtained. Note that very small particles fall extremely slowly. In fact, particles in the general size range that can be inhaled deeply into the lungs (on the order of 5 $\mu$m or less) will remain airborne most of the working day, thus continually exposing workers to potential hazards. It is often the smaller particles that cause the greatest health concerns, particularly to individuals exposed day after day for many years. In work environments that generate particle-laden air, proper engineering controls are needed to reduce inhalation exposures (e.g., appropriate ventilation) or, in extreme cases, personal protective equipment (PPE, e.g., respirator) may be required. PPE is considered a last resort because in practice it is the least effective means to protect workers. It is more effective and usually less expensive in the long run to design out or control a problem with engineering ingenuity—prevention through design.

■ **Figure   E9.10b**

*Flow past a cylinder can take on a variety of different structures.*

Moderate Reynolds number flows tend to take on a boundary layer flow structure. For such flows past streamlined bodies, the drag coefficient tends to decrease slightly with Reynolds number. The $C_D \sim \text{Re}^{-1/2}$ dependence for a laminar boundary layer on a flat plate (see Table 9.3) is such an example. Moderate Reynolds number flows past blunt bodies generally produce drag coefficients that are relatively constant. The $C_D$ values for the spheres and circular cylinders shown in Fig. 9.21a indicate this character in the range $10^3 < \text{Re} < 10^5$.

The structure of the flow field at selected Reynolds numbers indicated in Fig. 9.21a is shown in Fig. 9.21b. For a given object there is a wide variety of flow situations, depending on the Reynolds number involved. The curious reader is strongly encouraged to study the many beautiful photographs and videos of these (and other) flow situations found in Refs. 8 and 31. (See also the photograph at the beginning of Chapter 7.)

For many shapes there is a sudden change in the character of the drag coefficient when the boundary layer becomes turbulent. This is illustrated in Fig. 9.15 for the flat plate and in Fig. 9.21 for the sphere and the circular cylinder. The Reynolds number at which this transition takes place is a function of the shape of the body.

For streamlined bodies, the drag coefficient increases when the boundary layer becomes turbulent because most of the drag is due to the shear force, which is greater for turbulent flow than for laminar flow. On the other hand, the drag coefficient for a relatively blunt object, such as a cylinder or sphere, actually decreases when the boundary layer becomes turbulent. As is discussed in Section 9.2.6, a turbulent boundary layer can travel further along the surface into the adverse

*V9.12 Karman vortex street*

$$\text{Re} = \frac{\rho U D}{\mu}$$

(a)

(A) No separation

(B) Steady separation bubble

(C) Oscillating Karman vortex street wake

(D) Laminar boundary layer, wide turbulent wake

(E) Turbulent boundary layer, narrow turbulent wake

(b)

■ **Figure 9.21** (a) **Drag coefficient as a function of Reynolds number for a smooth circular cylinder and a smooth sphere.** (b) **Typical flow patterns for flow past a circular cylinder at various Reynolds numbers as indicated in** (a).

pressure gradient on the rear portion of the cylinder before separation occurs. The result is a thinner wake and smaller pressure drag for turbulent boundary layer flow. This is indicated in Fig. 9.21 by the sudden decrease in $C_D$ for $10^5 < \text{Re} < 10^6$. In a portion of this range the actual drag (not just the drag coefficient) decreases with increasing speed. It would be very difficult to control the steady flight of such an object in this range—an increase in velocity requires a decrease in thrust (drag). In all other Reynolds number ranges the drag increases with an increase in the upstream velocity (even though $C_D$ may decrease with Re).

For extremely blunt bodies, like a flat plate perpendicular to the flow, the flow separates at the edge of the plate regardless of the nature of the boundary layer flow. Thus, the drag coefficient shows very little dependence on the Reynolds number.

The drag coefficients for a series of two-dimensional bodies of varying bluntness are given as a function of Reynolds number in Fig. 9.22. The characteristics described above are evident.

V9.16 Flow past an ellipse

$$C_D = \frac{\mathscr{D}}{\frac{1}{2}\rho U^2\, bD}$$

$b$ = length

$$\text{Re} = \frac{UD}{\nu}$$

■ **Figure 9.22** Character of the drag coefficient as a function of Reynolds number for objects with various degrees of streamlining, from a flat plate normal to the upstream flow to a flat plate parallel to the flow (two-dimensional flow) (Ref. 5).

# EXAMPLE 9.11 | Terminal Velocity of a Falling Object

**GIVEN** Hail is produced by the repeated rising and falling of ice particles in the updraft of a thunderstorm, as is indicated in Fig. E9.11a. When the hail becomes large enough, the aerodynamic drag from the updraft can no longer support the weight of the hail, and it falls from the storm cloud.

**FIND** Estimate the velocity, $U$, of the updraft needed to make $D$ = 1.5-in.-diameter (i.e., "golf ball-sized") hail.

## SOLUTION

As is discussed in Example 9.10, for steady-state conditions a force balance on an object falling through a fluid at its terminal velocity, $U$, gives

$$\mathscr{W} = \mathscr{D} + F_B$$

where $F_B = \gamma_{\text{air}} \, \mathcal{V}$ is the buoyant force of the air on the particle, $\mathscr{W} = \gamma_{\text{ice}} \, \mathcal{V}$ is the particle weight, and $\mathscr{D}$ is the aerodynamic drag. This equation can be rewritten as

$$\tfrac{1}{2}\rho_{\text{air}}U^2\,\frac{\pi}{4}\,D^2 C_D = \mathscr{W} - F_B \qquad (1)$$

With $\mathcal{V} = \pi D^3/6$ and since $\gamma_{\text{ice}} \gg \gamma_{\text{air}}$ (i.e., $\mathscr{W} \gg F_B$), Eq. 1 can be simplified to

$$U = \left(\frac{4}{3}\frac{\rho_{\text{ice}}}{\rho_{\text{air}}}\frac{gD}{C_D}\right)^{1/2} \qquad (2)$$

By using $\rho_{\text{ice}} = 1.84$ slugs/ft$^3$, $\rho_{\text{air}} = 2.38 \times 10^{-3}$ slugs/ft$^3$, and $D = 1.5$ in. $= 0.125$ ft, Eq. 2 becomes

$$U = \left[\frac{4(1.84\ \text{slugs/ft}^3)(32.2\ \text{ft/s}^2)(0.125\ \text{ft})}{3(2.38 \times 10^{-3}\ \text{slugs/ft}^3)C_D}\right]^{1/2}$$

■ **Figure E9.11a**

or

$$U = \frac{64.5}{\sqrt{C_D}} \qquad (3)$$

where $U$ is in ft/s. To determine $U$, we must know $C_D$. Unfortunately, $C_D$ is a function of the Reynolds number (see Fig. 9.21), which is not known unless $U$ is known. Thus, we must use an iterative technique similar to that done with the Moody chart for certain types of pipe flow problems (see Section 8.5).

From Fig. 9.21 we expect that $C_D$ is on the order of 0.5. Thus, we assume $C_D = 0.5$ and from Eq. 3 obtain

$$U = \frac{64.5}{\sqrt{0.5}} = 91.2\ \text{ft/s}$$

The corresponding Reynolds number (assuming $\nu = 1.57 \times 10^{-4}$ ft$^2$/s) is

$$\text{Re} = \frac{UD}{\nu} = \frac{91.2 \text{ ft/s } (0.125 \text{ ft})}{1.57 \times 10^{-4} \text{ ft}^2/\text{s}} = 7.26 \times 10^4$$

For this value of Re we obtain from Fig. 9.21, $C_D = 0.5$. Thus, our assumed value of $C_D = 0.5$ was correct. The corresponding value of $U$ is

$$U = 91.2 \text{ ft/s} = 62.2 \text{ mph} \qquad \textbf{(Ans)}$$

**COMMENTS** By repeating the calculations for various altitudes, $z$, above sea level (using the properties of the U.S. Standard Atmosphere given in Appendix C), the results shown in Fig. E9.11b are obtained. Because of the decrease in density with alti-

tude, the hail falls even faster through the upper portions of the storm than when it hits the ground.

Clearly, an airplane flying through such an updraft would feel its effects (even if it were able to dodge the hail). As seen from Eq. 2, the larger the hail, the stronger the necessary updraft. Hailstones greater than 6 in. in diameter have been reported. In reality, a hailstone is seldom spherical and often not smooth. The hailstone shown in Fig. E9.11c is larger than golf-ball sized and is clearly ellipsoidal in shape with additional smaller-scale surface roughness features. However, the calculated updraft velocities are in agreement with measured values.

■ **Figure E9.11b**

■ **Figure E9.11c** (Photograph courtesy of NOAA.)

***Compressibility Effects.*** The above discussion is restricted to incompressible flows. If the velocity of the object is sufficiently large, compressibility effects become important and the drag coefficient becomes a function of the Mach number, Ma $= U/c$, where $c$ is the speed of sound in the fluid. The introduction of Mach number effects complicates matters because the drag coefficient for a given object is then a function of both Reynolds number and Mach number— $C_D = \phi(\text{Re, Ma})$. The Mach number and Reynolds number effects are often closely connected because both are directly proportional to the upstream velocity. For example, both Re and Ma increase with increasing flight speed of an airplane. The changes in $C_D$ due to a change in $U$ are due to changes in both Re and Ma.

The precise dependence of the drag coefficient on Re and Ma is generally quite complex (Ref. 13). However, the following simplifications are often justified. For low Mach numbers, the drag coefficient is essentially independent of Ma as is indicated in Fig. 9.23. For this situation, if Ma $< 0.5$

*The drag coefficient is usually independent of Mach number for Mach numbers up to approximately 0.5.*

■ **Figure 9.23** Drag coefficient as a function of Mach number for two-dimensional objects in subsonic flow (Ref. 5).

or so, compressibility effects are unimportant. On the other hand, for larger Mach number flows, the drag coefficient can be strongly dependent on Ma, with only secondary Reynolds number effects.

For most objects, values of $C_D$ increase dramatically in the vicinity of Ma = 1 (i.e., sonic flow). This change in character, indicated by Fig. 9.24, is due to the existence of shock waves as indicated by the figure in the margin. Shock waves are extremely narrow regions in the flow field across which the flow parameters change in a nearly discontinuous manner, which are discussed in Chapter 11. Shock waves, which cannot exist in subsonic flows, provide a mechanism for the generation of drag that is not present in the relatively low-speed subsonic flows. (See the photograph at the beginning of Chapter 11.)

The character of the drag coefficient as a function of Mach number is different for blunt bodies than for sharp bodies. As is shown in Fig. 9.24, sharp-pointed bodies develop their maximum drag coefficient in the vicinity of Ma = 1 (sonic flow), whereas the drag coefficient for blunt bodies increases with Ma far above Ma = 1. This behavior is due to the nature of the shock wave structure and the accompanying flow separation. The leading edges of wings for subsonic aircraft are usually quite rounded and blunt, while those of supersonic aircraft tend to be quite pointed and sharp. More information on these important topics can be found in standard texts about compressible flow and aerodynamics (Refs. 9, 10, and 29).

***Surface Roughness.***   As is indicated in Fig. 9.15, the drag on a flat plate parallel to the flow is quite dependent on the surface roughness, provided the boundary layer flow is turbulent. In such cases the surface roughness protrudes through the laminar sublayer adjacent to the surface (see Section 8.4) and alters the wall shear stress. In addition to the increased turbulent shear stress, surface roughness can alter the Reynolds number at which the boundary layer flow becomes turbulent. Thus, a rough flat plate may have a larger portion of its length covered by a turbulent boundary layer than does the corresponding smooth plate. This also acts to increase the net drag on the plate.

In general, for streamlined bodies, the drag increases with increasing surface roughness. Great care is taken to design the surfaces of airplane wings to be as smooth as possible, since protruding rivets or screw heads can cause a considerable increase in the drag. On the other hand, for an extremely blunt body, such as a flat plate normal to the flow, the drag is independent of the surface roughness, since the shear stress is not in the upstream flow direction and contributes nothing to the drag.

*Depending on the body shape, an increase in surface roughness may increase or decrease drag.*

For blunt bodies like a circular cylinder or sphere, an increase in surface roughness can actually cause a decrease in the drag. This is illustrated for a sphere in Fig. 9.25. As is discussed in Section 9.2.6, when the Reynolds number reaches the critical value (Re = $3 \times 10^5$ for a smooth

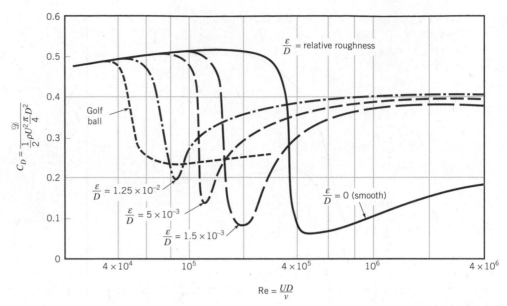

■ **Figure 9.25** **The effect of surface roughness on the drag coefficient of a sphere in the Reynolds number range for which the laminar boundary layer becomes turbulent (Ref. 5).**

sphere), the boundary layer becomes turbulent and the wake region behind the sphere becomes considerably narrower than if it were laminar (see Fig. 9.17). The result is a considerable drop in pressure drag with a slight increase in friction drag, combining to give a smaller overall drag (and $C_D$).

The boundary layer can be tripped into turbulence at a smaller Reynolds number by using a rough-surfaced sphere. For example, the critical Reynolds number for a golf ball is approximately $\text{Re} = 4 \times 10^4$. In the range $4 \times 10^4 < \text{Re} < 4 \times 10^5$, the drag on the standard rough (i.e., dimpled) golf ball is considerably less ($C_{D\text{rough}}/C_{D\text{smooth}} \approx 0.25/0.5 = 0.5$) than for the smooth ball. As is shown in Example 9.12, this is precisely the Reynolds number range for well-hit golf balls—hence, a reason for dimples on golf balls. The Reynolds number range for well-hit table tennis balls is less than $\text{Re} = 4 \times 10^4$. Thus, table tennis balls are smooth.

*Surface roughness can cause the boundary layer to become turbulent.*

# EXAMPLE 9.12 Effect of Surface Roughness

**GIVEN** A well-hit golf ball (diameter $D = 1.69$ in., weight $\mathcal{W} = 0.0992$ lb) can travel at $U = 200$ ft/s as it leaves the tee. A well-hit table tennis ball (diameter $D = 1.50$ in., weight $\mathcal{W} = 0.00551$ lb) can travel at $U = 60$ ft/s as it leaves the paddle.

**FIND** Determine the drag on a standard golf ball, a smooth golf ball, and a table tennis ball for the conditions given. Also determine the deceleration of each ball for these conditions.

*V9.17 Wake of golf ball and smooth ball*

## SOLUTION

For either ball, the drag can be obtained from

$$\mathcal{D} = \frac{1}{2} \rho U^2 \frac{\pi}{4} D^2 C_D \qquad (1)$$

where the drag coefficient, $C_D$, is given in Fig. 9.25 as a function of the Reynolds number and surface roughness. For the golf ball in standard air

$$\text{Re} = \frac{UD}{\nu} = \frac{(200 \text{ ft/s})(1.69/12 \text{ ft})}{1.57 \times 10^{-4} \text{ ft}^2/\text{s}} = 1.79 \times 10^5$$

while for the table tennis ball

$$\text{Re} = \frac{UD}{\nu} = \frac{(60 \text{ ft/s})(1.50/12 \text{ ft})}{1.57 \times 10^{-4} \text{ ft}^2/\text{s}} = 4.78 \times 10^4$$

The corresponding drag coefficients are $C_D = 0.25$ for the standard golf ball, $C_D = 0.51$ for the smooth golf ball, and $C_D = 0.50$ for the table tennis ball. Hence, from Eq. 1 for the standard golf ball

$$\mathcal{D} = \frac{1}{2}(0.00238 \text{ slugs/ft}^3)(200 \text{ ft/s})^2 \frac{\pi}{4}\left(\frac{1.69}{12} \text{ ft}\right)^2 (0.25)$$

$$= 0.185 \text{ lb} \qquad \textbf{(Ans)}$$

for the smooth golf ball

$$\mathscr{D} = \frac{1}{2}(0.00238 \text{ slugs/ft}^3)(200 \text{ ft/s})^2 \frac{\pi}{4}\left(\frac{1.69}{12}\text{ ft}\right)^2(0.51)$$

$$= 0.378 \text{ lb} \qquad \textbf{(Ans)}$$

and for the table tennis ball

$$\mathscr{D} = \frac{1}{2}(0.00238 \text{ slugs/ft}^3)(60 \text{ ft/s})^2 \frac{\pi}{4}\left(\frac{1.50}{12}\text{ ft}\right)^2(0.50)$$

$$= 0.0263 \text{ lb} \qquad \textbf{(Ans)}$$

The corresponding decelerations are $a = \mathscr{D}/m = g\mathscr{D}/\mathscr{W}$, where $m$ is the mass of the ball. Thus, the deceleration relative to the acceleration of gravity, $a/g$ (i.e., the number of $g$'s deceleration) is $a/g = \mathscr{D}/\mathscr{W}$ or

$$\frac{a}{g} = \frac{0.185 \text{ lb}}{0.0992 \text{ lb}} = 1.86 \text{ for the standard golf ball} \qquad \textbf{(Ans)}$$

$$\frac{a}{g} = \frac{0.378 \text{ lb}}{0.0992 \text{ lb}} = 3.81 \text{ for the smooth golf ball} \qquad \textbf{(Ans)}$$

$$\frac{a}{g} = \frac{0.0263 \text{ lb}}{0.00551 \text{ lb}} = 4.77 \text{ for the table tennis ball} \qquad \textbf{(Ans)}$$

**COMMENTS** Note that there is a considerably smaller deceleration for the rough golf ball than for the smooth one. Because of its much larger drag-to-mass ratio, the table tennis ball slows down relatively quickly and does not travel as far as the golf ball. Note that with $U = 60$ ft/s the standard golf ball has a drag of $\mathscr{D} = 0.0200$ lb and a deceleration of $a/g = 0.202$, considerably less than the $a/g = 4.77$ of the table tennis ball. Conversely, a table tennis ball hit from a tee at 200 ft/s would decelerate at a rate of $a = 1740$ ft/s$^2$, or $a/g = 54.1$. It would not travel nearly as far as the golf ball.

By repeating the above calculations, the drag as a function of speed for both a standard golf ball and a smooth golf ball is shown in Fig. E9.12.

The Reynolds number range for which a rough golf ball has smaller drag than a smooth one (i.e., $4 \times 10^4$ to $3.6 \times 10^5$) corresponds to a flight velocity range of $45 < U < 400$ ft/s. This is comfortably within the range of most golfers. (The fastest tee shot by top professional golfers is approximately 280 ft/s.) As discussed in Section 9.4.2, the dimples (roughness) on a golf ball also help produce a lift (due to the spin of the ball) that allows the ball to travel farther than a smooth ball.

■ **Figure E9.12**

---

# F l u i d s   i n   t h e   N e w s

**Dimpled baseball bats** For many years it has been known that dimples on golf balls can create a *turbulent boundary layer* and reduce the aerodynamic drag, allowing longer drives than with smooth balls. Thus, why not put dimples on baseball bats so that tomorrow's baseball sluggers can swing the bat faster and, therefore, hit the ball farther? MIT instructor Jeffery De Tullio pondered that question, performed experiments with dimpled bats to determine the answer, and received a patent for his dimpled bat invention. The result is that a batter can swing a dimpled bat approximately 3 to 5% faster than a smooth bat. Theoretically, this extra speed will translate to an extra 10 to 15 ft distance on a long hit. (See Problem 9.99.)

---

*V9.18 Jet ski*

*The drag coefficient for surface ships is a function of the Froude number.*

***Froude Number Effects.*** Another parameter on which the drag coefficient may be strongly dependent is the Froude number, $\text{Fr} = U/\sqrt{g\ell}$. As is discussed in Chapter 10, the Froude number is a ratio of the free-stream speed to a typical wave speed on the interface of two fluids, such as the surface of the ocean. An object moving on the surface, such as a ship, often produces waves that require a source of energy to generate. This energy comes from the ship and is manifest as a drag. [Recall that the rate of energy production (power) equals speed times force.] The nature of the waves produced often depends on the Froude number of the flow and the shape of the object—the waves generated by a water skier "plowing" through the water at a low speed (low Fr) are different than those generated by the skier "planing" along the surface at high speed (large Fr).

Thus, the drag coefficient for surface ships is a function of Reynolds number (viscous effects) and Froude number (wave-making effects); $C_D = \phi(\text{Re}, \text{Fr})$. As was discussed in Chapter 7, it is often quite difficult to run model tests under conditions similar to those of the prototype (i.e., same Re and Fr for surface ships). Fortunately, the viscous and wave effects can often be separated, with the total drag being the sum of the drag of these individual effects. A detailed account of this important topic can be found in standard texts (Ref. 11).

As is indicated in Fig. 9.26, the wave-making drag, $\mathscr{D}_w$, can be a complex function of the Froude number and the body shape. The rather "wiggly" dependence of the wave drag coefficient,

■ **Figure 9.26** Typical drag coefficient data as a function of Froude number and hull characteristics for that portion of the drag due to the generation of waves (adapted from Ref. 25).

*The drag on a complex body can be approximated as the sum of the drag on its parts.*

$C_{Dw} = \mathcal{D}_w/(\rho U^2 \ell^2/2)$, on the Froude number shown is typical. It results from the fact that the structure of the waves produced by the hull is a strong function of the ship speed or, in dimensionless form, the Froude number. This wave structure is also a function of the body shape. For example, the bow wave, which is often the major contributor to the wave drag, can be reduced by use of an appropriately designed bulb on the bow, as is indicated in Fig. 9.26. In this instance the streamlined body (hull without a bulb) has more drag than the less streamlined one.

***Composite Body Drag.*** Approximate drag calculations for a complex body can often be obtained by treating the body as a composite collection of its various parts. For example, the total force on a flag pole because of the wind (see the figure in the margin) can be approximated by adding the aerodynamic drag produced by the various components involved—the drag on the flag and the drag on the pole. In some cases considerable care must be taken in such an approach because of the interactions between the various parts. It may not be correct to merely add the drag of the components to obtain the drag of the entire object, although such approximations are often reasonable.

# EXAMPLE 9.13 | Drag on a Composite Body

**GIVEN** A 60-mph (i.e., 88-fps) wind blows past the water tower shown in Fig. E9.13a.

**FIND** Estimate the moment (torque), $M$, needed at the base to keep the tower from tipping over.

## SOLUTION

We treat the water tower as a sphere resting on a circular cylinder and assume that the total drag is the sum of the drag from these parts. The free-body diagram of the tower is shown in Fig.

$U = 60$ mph $= 88$ fps

$D_s = 40$ ft
$D_c = 15$ ft
$b = 50$ ft

*(a)*    *(b)*

■ **Figure E9.13**

E9.13b. By summing moments about the base of the tower, we obtain

$$M = \mathscr{D}_s \left( b + \frac{D_s}{2} \right) + \mathscr{D}_c \left( \frac{b}{2} \right) \qquad (1)$$

where

$$\mathscr{D}_s = \frac{1}{2} \rho U^2 \frac{\pi}{4} D_s^2 C_{Ds} \qquad (2)$$

and

$$\mathscr{D}_c = \frac{1}{2} \rho U^2 b D_c C_{Dc} \qquad (3)$$

are the drag on the sphere and cylinder, respectively. For standard atmospheric conditions, the Reynolds numbers are

$$\mathrm{Re}_s = \frac{U D_s}{\nu} = \frac{(88 \text{ ft/s})(40 \text{ ft})}{1.57 \times 10^{-4} \text{ ft}^2/\text{s}} = 2.24 \times 10^7$$

and

$$\mathrm{Re}_c = \frac{U D_c}{\nu} = \frac{(88 \text{ ft/s})(15 \text{ ft})}{1.57 \times 10^{-4} \text{ ft}^2/\text{s}} = 8.41 \times 10^6$$

The corresponding drag coefficients, $C_{Ds}$ and $C_{Dc}$, can be approximated from Fig. 9.21 as

$$C_{Ds} \approx 0.3 \quad \text{and} \quad C_{Dc} \approx 0.7$$

Note that the value of $C_{Ds}$ was obtained by an extrapolation of the given data to Reynolds numbers beyond those given (a potentially dangerous practice!). From Eqs. 2 and 3 we obtain

$$\mathscr{D}_s = 0.5(0.00238 \text{ slugs/ft}^3)(88 \text{ ft/s})^2 \frac{\pi}{4} (40 \text{ ft})^2 (0.3)$$

$$= 3470 \text{ lb}$$

and

$$\mathscr{D}_c = 0.5(0.00238 \text{ slugs/ft}^3)(88 \text{ ft/s})^2 (50 \text{ ft} \times 15 \text{ ft})(0.7)$$

$$= 4840 \text{ lb}$$

From Eq. 1 the corresponding moment needed to prevent the tower from tipping is

$$M = 3470 \text{ lb} \left( 50 \text{ ft} + \frac{40}{2} \text{ ft} \right) + 4840 \text{ lb} \left( \frac{50}{2} \text{ ft} \right)$$

$$= 3.64 \times 10^5 \text{ ft} \cdot \text{lb} \qquad \textbf{(Ans)}$$

**COMMENT** The above result is only an estimate because (a) the wind is probably not uniform from the top of the tower to the ground, (b) the tower is not exactly a combination of a smooth sphere and a circular cylinder, (c) the cylinder is not of infinite length, (d) there will be some interaction between the flow past the cylinder and that past the sphere so that the net drag is not exactly the sum of the two, and (e) a drag coefficient value was obtained by extrapolation of the given data. However, such approximate results are often quite accurate.

*V9.19 Drag on a truck*

The aerodynamic drag on automobiles provides an example of the use of adding component drag forces. The power required to move a car along a level street is used to overcome the rolling resistance and the aerodynamic drag. For speeds above approximately 30 mph, the aerodynamic drag becomes a significant contribution to the net propulsive force needed. The contribution of the drag due to various portions of car (i.e., front end, windshield, roof, rear end, windshield peak, rear roof/trunk, and cowl) have been determined by numerous model and full-sized tests as well as by

*V9.20 Automobile streamlining*

■ **Figure 9.27** The historical trend of streamlining automobiles to reduce their aerodynamic drag and increase their miles per gallon (adapted from Ref. 5).

numerical calculations. As a result, it is possible to predict the aerodynamic drag on cars of a wide variety of body styles.

As is indicated in Fig. 9.27, the drag coefficient for cars has decreased rather continuously over the years. This reduction is a result of careful design of the shape and the details (such as window molding, rear view mirrors, etc.). An additional reduction in drag has been accomplished by a reduction of the projected area. The net result is a considerable increase in the gas mileage, especially at highway speeds. Considerable additional information about the aerodynamics of road vehicles can be found in the literature (Ref. 30).

The effect of several important parameters (shape, Re, Ma, Fr, and roughness) on the drag coefficient for various objects has been discussed in this section. As stated previously, drag coefficient information for a very wide range of objects is available in the literature. Some of this information is given in Figs. 9.28, 9.29, and 9.30 below for a variety of two- and three-dimensional, natural and man-made objects. Recall that a drag coefficient of unity is equivalent to the drag produced by the dynamic pressure acting on an area of size $A$. That is, $\mathscr{D} = \frac{1}{2}\rho U^2 A C_D = \frac{1}{2}\rho U^2 A$ if $C_D = 1$. Typical nonstreamlined objects have drag coefficients on this order.

# 9.4 Lift

As is indicated in Section 9.1, any object moving through a fluid will experience a net force of the fluid on the object. For objects symmetrical perpendicular to the upstream flow, this force will be in the direction of the free stream—a drag, $\mathscr{D}$. If the object is not symmetrical (or if it does not produce a symmetrical flow field, such as the flow around a rotating sphere), there may also be a force normal to the free stream—a lift, $\mathscr{L}$. Considerable effort has been put forth to understand the various properties of the generation of lift. Some objects, such as an airfoil, are designed to generate lift. Other objects are designed to reduce the lift generated. For example, the lift on a car tends to reduce the contact force between the wheels and the ground, causing reduction in traction and cornering ability. It is desirable to reduce this lift.

### 9.4.1 Surface Pressure Distribution

The lift can be determined from Eq. 9.2 if the distributions of pressure and wall shear stress around the entire body are known. As is indicated in Section 9.1, such data are usually not known. Typically, the lift is given in terms of the lift coefficient,

*The lift coefficient is a dimensionless form of the lift.*

$$C_L = \frac{\mathscr{L}}{\frac{1}{2}\rho U^2 A} \qquad\qquad (9.39)$$

| Shape | Reference area $A$ ($b$ = length) | Drag coefficient $C_D = \dfrac{\mathscr{D}}{\frac{1}{2}\rho U^2 A}$ | Reynolds number $Re = \rho U D/\mu$ |
|---|---|---|---|
| Square rod with rounded corners | $A = bD$ | $R/D$   $C_D$<br>0   2.2<br>0.02   2.0<br>0.17   1.2<br>0.33   1.0 | $Re = 10^5$ |
| Rounded equilateral triangle | $A = bD$ | $R/D$   $C_D$ (→)   $C_D$ (←)<br>0   1.4   2.1<br>0.02   1.2   2.0<br>0.08   1.3   1.9<br>0.25   1.1   1.3 | $Re = 10^5$ |
| Semicircular shell | $A = bD$ | → 2.3<br>← 1.1 | $Re = 2 \times 10^4$ |
| Semicircular cylinder | $A = bD$ | → 2.15<br>← 1.15 | $Re > 10^4$ |
| T-beam | $A = bD$ | → 1.80<br>← 1.65 | $Re > 10^4$ |
| I-beam | $A = bD$ | 2.05 | $Re > 10^4$ |
| Angle | $A = bD$ | → 1.98<br>← 1.82 | $Re > 10^4$ |
| Hexagon | $A = bD$ | 1.0 | $Re > 10^4$ |
| Rectangle | $A = bD$ | $\ell/D$   $C_D$<br>≤0.1   1.9<br>0.5   2.5<br>0.65   2.9<br>1.0   2.2<br>2.0   1.6<br>3.0   1.3 | $Re = 10^5$ |

■ **Figure 9.28**  **Typical drag coefficients for regular two-dimensional objects (Refs. 5, 6).**

which is obtained from experiments, advanced analysis, or numerical considerations. The lift coefficient is a function of the appropriate dimensionless parameters and, as the drag coefficient, can be written as

$$C_L = \phi(\text{shape, Re, Ma, Fr, } \varepsilon/\ell)$$

*The lift coefficient is a function of other dimensionless parameters.*

The Froude number, Fr, is important only if there is a free surface present, as with an underwater "wing" used to support a high-speed hydrofoil surface ship. Often the surface roughness, $\varepsilon$, is relatively unimportant in terms of lift—it has more of an effect on the drag. The Mach number, Ma, is of importance for relatively high-speed subsonic and supersonic flows (i.e., Ma $> 0.8$), and the Reynolds number effect is often not great. The most important parameter that affects the lift coefficient is the shape of the object. Considerable effort has gone into designing optimally shaped lift-producing devices. We will emphasize the effect of the shape on lift—the effects of the other dimensionless parameters can be found in the literature (Refs. 13, 14, and 29).

| Shape | | Reference area $A$ | Drag coefficient $C_D$ | Reynolds number Re $= \rho UD/\mu$ |
|---|---|---|---|---|
| $D$ | Solid hemisphere | $A = \frac{\pi}{4}D^2$ | → 1.17 ← 0.42 | Re $> 10^4$ |
| $D$ | Hollow hemisphere | $A = \frac{\pi}{4}D^2$ | → 1.42 ← 0.38 | Re $> 10^4$ |
| $D$ | Thin disk | $A = \frac{\pi}{4}D^2$ | 1.1 | Re $> 10^3$ |
| $\ell$, $D$ | Circular rod parallel to flow | $A = \frac{\pi}{4}D^2$ | $\ell/D$   $C_D$ <br> 0.5   1.1 <br> 1.0   0.93 <br> 2.0   0.83 <br> 4.0   0.85 | Re $> 10^5$ |
| $\theta$, $D$ | Cone | $A = \frac{\pi}{4}D^2$ | $\theta$, degrees   $C_D$ <br> 10   0.30 <br> 30   0.55 <br> 60   0.80 <br> 90   1.15 | Re $> 10^4$ |
| $D$ | Cube | $A = D^2$ | 1.05 | Re $> 10^4$ |
| $D$ | Cube | $A = D^2$ | 0.80 | Re $> 10^4$ |
| $D$ | Streamlined body | $A = \frac{\pi}{4}D^2$ | 0.04 | Re $> 10^5$ |

■ **Figure 9.29**  Typical drag coefficients for regular three-dimensional objects (Ref. 5).

*Usually most lift comes from pressure forces, not viscous forces.*

Most common lift-generating devices (i.e., airfoils, fans, spoilers on cars, etc.) operate in the large Reynolds number range in which the flow has a boundary layer character, with viscous effects confined to the boundary layers and wake regions. For such cases the wall shear stress, $\tau_w$, contributes little to the lift. Most of the lift comes from the surface pressure distribution. A typical pressure distribution on a moving car is shown in Fig. 9.31. The distribution, for the most part, is consistent with simple Bernoulli equation analysis. Locations with high-speed flow (i.e., over the roof and hood) have low pressure, while locations with low-speed flow (i.e., on the grill and windshield) have high pressure. It is easy to believe that the integrated effect of this pressure distribution would provide a net upward force.

For objects operating in very low Reynolds number regimes (i.e., Re $< 1$), viscous effects are important, and the contribution of the shear stress to the lift may be as important as that of the pressure. Such situations include the flight of minute insects and the swimming of microscopic organisms. The relative importance of $\tau_w$ and $p$ in the generation of lift in a typical large Reynolds number flow is shown in Example 9.14.

| Shape | Reference area | Drag coefficient $C_D$ |
|---|---|---|
| Parachute | Frontal area $A = \frac{\pi}{4}D^2$ | 1.4 |
| Porous parabolic dish | Frontal area $A = \frac{\pi}{4}D^2$ | Porosity: 0, 0.2, 0.5<br>→ 1.42, 1.20, 0.82<br>← 0.95, 0.90, 0.80<br>Porosity = open area/total area |
| Average person | Standing<br>Sitting<br>Crouching | $C_DA = 9\ \text{ft}^2$<br>$C_DA = 6\ \text{ft}^2$<br>$C_DA = 2.5\ \text{ft}^2$ |
| Fluttering flag | $A = \ell D$ | $\ell/D$ : $C_D$<br>1 : 0.07<br>2 : 0.12<br>3 : 0.15 |
| Empire State Building | Frontal area | 1.4 |
| Six-car passenger train | Frontal area | 1.8 |
| Bikes — Upright commuter | $A = 5.5\ \text{ft}^2$ | 1.1 |
| Racing | $A = 3.9\ \text{ft}^2$ | 0.88 |
| Drafting | $A = 3.9\ \text{ft}^2$ | 0.50 |
| Streamlined | $A = 5.0\ \text{ft}^2$ | 0.12 |
| Tractor-trailer trucks — Standard | Frontal area | 0.96 |
| With fairing | Frontal area | 0.76 |
| With fairing and gap seal | Frontal area | 0.70 |
| Tree $U = 10$ m/s, $20$ m/s, $30$ m/s | Frontal area | 0.43<br>0.26<br>0.20 |
| Dolphin | Wetted area | 0.0036 at Re = $6 \times 10^6$ (flat plate has $C_{Df} = 0.0031$) |
| Large birds | Frontal area | 0.40 |

■ **Figure 9.30**   Typical drag coefficients for objects of interest (Refs. 5, 6, 15, 20).

Denotes $p > p_0$
Denotes $p < p_0$

■ **Figure 9.31** Pressure distribution on the surface of an automobile.

## EXAMPLE 9.14 | Lift from Pressure and Shear Stress Distributions

**GIVEN** When a uniform wind of velocity $U$ blows past the semicircular building shown in Fig. E9.14a,b, the wall shear stress and pressure distributions on the outside of the building are as given previously in Figs. E9.8b and E9.9a, respectively.

**FIND** If the pressure in the building is atmospheric (i.e., the value, $p_0$, far from the building), determine the lift coefficient and the lift on the roof.

## SOLUTION

From Eq. 9.2 we obtain the lift as

$$\mathcal{L} = -\int p \sin \theta \, dA + \int \tau_w \cos \theta \, dA \qquad (1)$$

As is indicated in Fig. E9.14b, we assume that on the inside of the building the pressure is uniform, $p = p_0$, and that there is no shear stress. Thus, Eq. 1 can be written as

$$\mathcal{L} = -\int_0^\pi (p - p_0) \sin \theta \, b\left(\frac{D}{2}\right) d\theta$$
$$+ \int_0^\pi \tau_w \cos \theta \, b\left(\frac{D}{2}\right) d\theta$$

or

$$\mathcal{L} = \frac{bD}{2}\left[-\int_0^\pi (p - p_0) \sin \theta \, d\theta + \int_0^\pi \tau_w \cos \theta \, d\theta\right] \qquad (2)$$

where $b$ and $D$ are the length and diameter of the building, respectively, and $dA = b(D/2)d\theta$. Equation 2 can be put into dimensionless form by using the dynamic pressure, $\rho U^2/2$, planform area, $A = bD$, and dimensionless shear stress

$$F(\theta) = \tau_w(\text{Re})^{1/2}/(\rho U^2/2)$$

to give

$$\mathcal{L} = \frac{1}{2}\rho U^2 A \left[-\frac{1}{2}\int_0^\pi \frac{(p - p_0)}{\frac{1}{2}\rho U^2} \sin \theta \, d\theta \right.$$
$$\left. + \frac{1}{2\sqrt{\text{Re}}}\int_0^\pi F(\theta) \cos \theta \, d\theta\right] \qquad (3)$$

From the data in Figs. E9.8b and E9.9a, the values of the two integrals in Eq. 3 can be obtained by determining the area under the

curves of $[(p - p_0)/(\rho U^2/2)] \sin \theta$ versus $\theta$ and $F(\theta)\cos\theta$ versus $\theta$ plotted in Figs. E9.14c and E9.14d. The results are

$$\int_0^\pi \frac{(p - p_0)}{\frac{1}{2}\rho U^2} \sin \theta \, d\theta = -1.76$$

and

$$\int_0^\pi F(\theta) \cos \theta \, d\theta = 3.92$$

Thus, the lift is

$$\mathcal{L} = \frac{1}{2}\rho U^2 A \left[\left(-\frac{1}{2}\right)(-1.76) + \frac{1}{2\sqrt{\text{Re}}}(3.92)\right]$$

or

$$\mathcal{L} = \left(0.88 + \frac{1.96}{\sqrt{\text{Re}}}\right)\left(\frac{1}{2}\rho U^2 A\right) \qquad \textbf{(Ans)}$$

and

$$C_L = \frac{\mathcal{L}}{\frac{1}{2}\rho U^2 A} = 0.88 + \frac{1.96}{\sqrt{\text{Re}}} \qquad (4) \quad \textbf{(Ans)}$$

**COMMENTS** Consider a typical situation with $D = 20$ ft, $U = 30$ ft/s, $b = 50$ ft, and standard atmospheric conditions ($\rho = 2.38 \times 10^{-3}$ slugs/ft$^3$ and $\nu = 1.57 \times 10^{-4}$ ft$^2$/s), which gives a Reynolds number of

$$\text{Re} = \frac{UD}{\nu} = \frac{(30 \text{ ft/s})(20 \text{ ft})}{1.57 \times 10^{-4} \text{ ft}^2/\text{s}} = 3.82 \times 10^6$$

Hence, the lift coefficient is

$$C_L = 0.88 + \frac{1.96}{(3.82 \times 10^6)^{1/2}} = 0.88 + 0.001 = 0.881$$

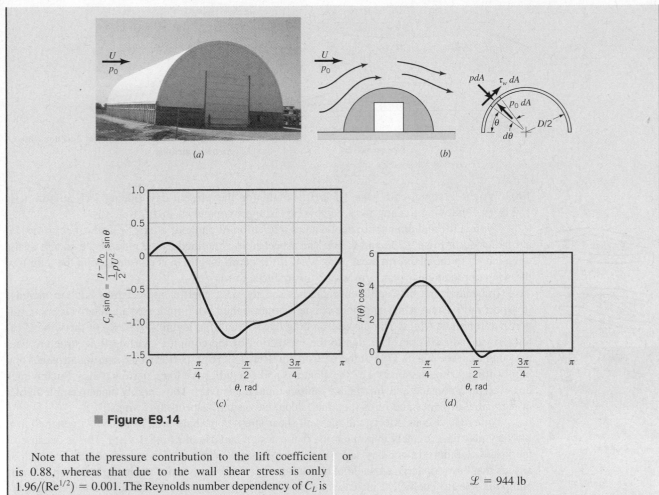

**Figure E9.14**

Note that the pressure contribution to the lift coefficient is 0.88, whereas that due to the wall shear stress is only $1.96/(\mathrm{Re}^{1/2}) = 0.001$. The Reynolds number dependency of $C_L$ is quite minor. The lift is pressure dominated. Recall from Example 9.9 that this is also true for the drag on a similar shape.

From Eq. 4 with $A = 20\ \mathrm{ft} \times 50\ \mathrm{ft} = 1000\ \mathrm{ft}^2$, we obtain the lift for the assumed conditions as

$$\mathcal{L} = \tfrac{1}{2}\rho U^2 A C_L = \tfrac{1}{2}(0.00238\ \mathrm{slugs/ft}^3)(30\ \mathrm{ft/s})^2(1000\ \mathrm{ft}^2)(0.881)$$

or

$$\mathcal{L} = 944\ \mathrm{lb}$$

There is a considerable tendency for the building to lift off the ground. Clearly this is due to the object being nonsymmetrical. The lift force on a complete circular cylinder is zero, although the fluid forces do tend to pull the upper and lower halves apart.

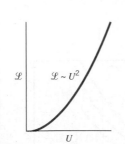

A typical device designed to produce lift does so by generating a pressure distribution that is different on the top and bottom surfaces. For large Reynolds number flows these pressure distributions are usually directly proportional to the dynamic pressure, $\rho U^2/2$, with viscous effects being of secondary importance. Hence, as indicated by the figure in the margin, for a given airfoil the lift is proportional to the square of the airspeed. Two airfoils used to produce lift are indicated in Fig. 9.32. Clearly the symmetrical one cannot produce lift unless the angle of attack, $\alpha$, is nonzero. Because of the asymmetry of the nonsymmetric airfoil, the pressure distributions on the upper and lower surfaces are different, and a lift is produced even with $\alpha = 0$. Of course, there will be a certain value of $\alpha$ (less than zero for this case) for which the lift is zero. For this situation, the pressure distributions on the upper and lower surfaces are different, but their resultant (integrated) pressure forces will be equal and opposite.

Since most airfoils are thin, it is customary to use the planform area, $A = bc$, in the definition of the lift coefficient. Here $b$ is the length of the airfoil and $c$ is the *chord length*—the length from the leading edge to the trailing edge as indicated in Fig. 9.32. Typical lift coefficients so defined are on the order of unity. That is, the lift force is on the order of the dynamic pressure times the planform area of the wing, $\mathcal{L} \approx (\rho U^2/2)A$. The *wing loading*, defined as the average lift per unit area of the wing, $\mathcal{L}/A$, therefore, increases with speed. For example, the wing loading of the

■ **Figure 9.32** Symmetrical and nonsymmetrical airfoils.

Not stalled

Stalled

**Courtesy of NASA**

*At large angles of attack the boundary layer separates and the wing stalls.*

1903 Wright Flyer aircraft was 1.5 lb/ft$^2$, while for the present-day Boeing 747 aircraft it is 150 lb/ft$^2$. The wing loading for a bumble bee is approximately 1 lb/ft$^2$ (Ref. 15).

Typical lift and drag coefficient data as a function of angle of attack, $\alpha$, and *aspect ratio*, $\mathcal{A}$, are indicated in Figs. 9.33a and 9.33b. The aspect ratio is defined as the ratio of the square of the wing length to the planform area, $\mathcal{A} = b^2/A$. If the chord length, $c$, is constant along the length of the wing (a rectangular planform wing), this reduces to $\mathcal{A} = b/c$.

In general, the lift coefficient increases and the drag coefficient decreases with an increase in aspect ratio. Long wings are more efficient because their wing tip losses are relatively more minor than for short wings. The increase in drag due to the finite length ($\mathcal{A} < \infty$) of the wing is often termed induced drag. It is due to the interaction of the complex swirling flow structure near the wing tips (see Fig. 9.37) and the free stream (Ref. 13). High-performance soaring airplanes and highly efficient soaring birds (i.e., the albatross and sea gull) have long, narrow wings. Such wings, however, have considerable inertia that inhibits rapid maneuvers. Thus, highly maneuverable fighter or acrobatic airplanes and birds (i.e., the falcon) have small-aspect-ratio wings.

Although viscous effects and the wall shear stress contribute little to the direct generation of lift, they play an extremely important role in the design and use of lifting devices. This is because of the viscosity-induced boundary layer separation that can occur on nonstreamlined bodies such as airfoils that have too large an angle of attack (see Fig. 9.18). As is indicated in Fig. 9.33, up to a certain point, the lift coefficient increases rather steadily with the angle of attack. If $\alpha$ is too large, the boundary layer on the upper surface separates, the flow over the wing develops a wide, turbulent wake region, the lift decreases, and the drag increases. This condition, as indicated by the figures in the margin, is termed *stall*. Such conditions are extremely dangerous if they occur while the airplane is flying at a low altitude where there is not sufficient time and altitude to recover from the stall.

(a)

(b)

■ **Figure 9.33** Typical lift and drag coefficient data as a function of angle of attack and the aspect ratio of the airfoil: (a) lift coefficient, (b) drag coefficient.

■ **Figure 9.34**   Two representations of the same lift and drag data for a typical airfoil: (*a*) lift-to-drag ratio as a function of angle of attack, with the onset of boundary layer separation on the upper surface indicated by the occurrence of stall, (*b*) the lift and drag polar diagram with the angle of attack indicated (Ref. 27).

In many lift-generating devices the important quantity is the ratio of the lift to drag developed, $\mathscr{L}/\mathscr{D} = C_L/C_D$. Such information is often presented in terms of $C_L/C_D$ versus $\alpha$, as is shown in Fig. 9.34a, or in a *lift-drag polar* of $C_L$ versus $C_D$ with $\alpha$ as a parameter, as is shown in Fig. 9.34b. The most efficient angle of attack (i.e., largest $C_L/C_D$) can be found by drawing a line tangent to the $C_L - C_D$ curve from the origin, as is shown in Fig. 9.34b. High-performance airfoils generate lift that is perhaps 100 or more times greater than their drag. This translates into the fact that in still air they can glide a horizontal distance of 100 m for each 1 m drop in altitude.

**F l u i d s   i n   t h e   N e w s**

**Bats feel turbulence**   Researchers have discovered that at certain locations on the wings of bats, there are special touch-sensing cells with a tiny hair poking out of the center of the cell. These cells, which are very sensitive to air flowing across the wing surface, can apparently detect turbulence in the flow over the wing. If these hairs are removed, the bats fly well in a straight line, but when maneuvering to avoid obstacles, their elevation control is erratic. When the hairs grow back, the bats regain their complete flying skills. It is proposed that these touch-sensing cells are used to detect turbulence on the wing surface and thereby tell bats when to adjust the angle of attack and curvature of their wings in order to avoid stalling out in midair.

As is indicated above, the lift and drag on an airfoil can be altered by changing the angle of attack. This actually represents a change in the shape of the object. Other shape changes can be used to alter the lift and drag when desirable. In modern airplanes it is common to utilize leading edge and trailing edge flaps as is shown in Fig. 9.35. To generate the necessary lift during the relatively low-speed landing and takeoff procedures, the airfoil shape is altered by extending special flaps on the front and/or rear portions of the wing. Use of the flaps considerably enhances the lift, although it is at the expense of an increase in the drag (the airfoil is in a "dirty" configuration). This increase in drag is not of much concern during landing and takeoff operations—the decrease in landing or takeoff speed is more important than is a temporary increase in drag. During normal flight with the flaps retracted (the "clean" configuration), the drag is relatively small, and the needed lift force is achieved with the smaller lift coefficient and the larger dynamic pressure (higher speed).

$C_L$

- ○── No flaps
- △── Trailing edge slotted flap
- □── Double slotted trailing edge flaps
- (Data not shown) Leading edge flap

V9.24 Leading edge flap

■ **Figure 9.35** Typical lift and drag alterations possible with the use of various types of flap designs (Data from Ref. 21).

## F l u i d s   i n   t h e   N e w s

**Learning from nature** For hundreds of years humans looked toward nature, particularly birds, for insight about flying. However, all early airplanes that closely mimicked birds proved to be unsuccessful. Only after much experimenting with rigid (or at least nonflapping) wings did human flight become possible. Recently, however, engineers have been turning to living systems—birds, insects, and other biological models—in an attempt to produce breakthroughs in aircraft design. Perhaps it is possible that nature's basic design concepts can be applied to airplane systems. For example, by morphing and rotating their wings in three dimensions, birds have remarkable maneuverability that to date has no technological parallel. Birds can control the airflow over their wings by moving the feathers on their wingtips and the leading edges of their wings, providing designs that are more efficient than the flaps and rigid, pivoting tail surfaces of current aircraft (Ref. 15). On a smaller scale, understanding the mechanism by which insects dynamically manage unstable flow to generate lift may provide insight into the development of microscale air vehicles. With new hi-tech materials, computers, and automatic controls, aircraft of the future may mimic nature more than was once thought possible. (See Problem 9.122.)

A wide variety of lift and drag information for airfoils can be found in standard aerodynamics books (Ref. 13, 14, 29).

## $E$XAMPLE 9.15   Lift and Power for Human-Powered Flight

**GIVEN** In 1977 the *Gossamer Condor*, shown in Fig. E9.15a, won the Kremer prize by being the first human-powered aircraft to complete a prescribed figure-of-eight course around two turning points 0.5 mi apart (Ref. 22). The following data pertain to this aircraft:

> flight speed = $U = 15$ ft/s
>
> wing size = $b = 96$ ft, $c = 7.5$ ft (average)

weight (including pilot) = $\mathcal{W} = 210$ lb

> drag coefficient = $C_D = 0.046$ (based on planform area)

power train efficiency = $\eta$

> = power to overcome drag/pilot power = 0.8

**FIND** Determine

(a) the lift coefficient, $C_L$, and

(b) the power, $\mathcal{P}$, required by the pilot.

$U$

■ **Figure E9.15a**
**(Photograph © Don Monroe.)**

## SOLUTION

**(a)**  For steady flight conditions the lift must be exactly balanced by the weight, or

$$\mathcal{W} = \mathcal{L} = \tfrac{1}{2}\rho U^2 A C_L$$

Thus,

$$C_L = \frac{2\mathcal{W}}{\rho U^2 A}$$

where $A = bc = 96 \text{ ft} \times 7.5 \text{ ft} = 720 \text{ ft}^2$, $\mathcal{W} = 210$ lb, and $\rho = 2.38 \times 10^{-3}$ slugs/ft$^3$ for standard air. This gives

$$C_L = \frac{2(210 \text{ lb})}{(2.38 \times 10^{-3} \text{ slugs/ft}^3)(15 \text{ ft/s})^2(720 \text{ ft}^2)}$$
$$= 1.09 \qquad \textbf{(Ans)}$$

a reasonable number. The overall lift-to-drag ratio for the aircraft is $C_L/C_D = 1.09/0.046 = 23.7$.

**(b)**  The product of the power that the pilot supplies and the power train efficiency equals the useful power needed to overcome the drag, $\mathcal{D}$. That is,

$$\eta\mathcal{P} = \mathcal{D}U$$

where

$$\mathcal{D} = \tfrac{1}{2}\rho U^2 A C_D$$

Thus,

$$\mathcal{P} = \frac{\mathcal{D}U}{\eta} = \frac{\tfrac{1}{2}\rho U^2 A C_D U}{\eta} = \frac{\rho A C_D U^3}{2\eta} \qquad (1)$$

or

$$\mathcal{P} = \frac{(2.38 \times 10^{-3} \text{ slugs/ft}^3)(720 \text{ ft}^2)(0.046)(15 \text{ ft/s})^3}{2(0.8)}$$

$$\mathcal{P} = 166 \text{ ft} \cdot \text{lb/s} \left( \frac{1 \text{ hp}}{550 \text{ ft} \cdot \text{lb/s}} \right) = 0.302 \text{ hp} \qquad \textbf{(Ans)}$$

**COMMENT**  This power level is obtainable by a well-conditioned athlete (as is indicated by the fact that the flight was successfully completed). Note that only 80% of the pilot's power (i.e., $0.8 \times 0.302 = 0.242$ hp, which corresponds to a drag of $\mathcal{D} = 8.86$ lb) is needed to force the aircraft through the air. The other 20% is lost because of the power train inefficiency.

By repeating the calculations for various flight speeds, the results shown in Fig. E9.15b are obtained. Note from Eq. 1 that for a constant drag coefficient, the power required increases as $U^3$—a doubling of the speed to 30 ft/s would require an eight-fold increase in power (i.e., 2.42 hp, well beyond the range of any human).

■ **Figure E9.15b**

### 9.4.2  Circulation

*Inviscid flow analysis can be used to obtain ideal flow past airfoils.*

Since viscous effects are of minor importance in the generation of lift, it should be possible to calculate the lift force on an airfoil by integrating the pressure distribution obtained from the equations governing inviscid flow past the airfoil. That is, the potential flow theory discussed in Chapter 6 should provide a method to determine the lift. Although the details are beyond the scope of this book, the following is found from such calculations (Ref. 4).

The calculation of the inviscid flow past a two-dimensional airfoil gives a flow field as indicated in Fig. 9.36. The predicted flow field past an airfoil with no lift (i.e., a symmetrical airfoil at zero angle of attack, Fig. 9.36a) appears to be quite accurate (except for the absence of thin boundary layer regions). However, as is indicated in Fig. 9.36b, the calculated flow past the same airfoil at a nonzero angle of attack (but one small enough so that boundary layer separation would not occur) is not proper near the trailing edge. In addition, the calculated lift for a nonzero angle of attack is zero—in conflict with the known fact that such airfoils produce lift.

In reality, the flow should pass smoothly over the top surface as is indicated in Fig. 9.36c, without the strange behavior indicated near the trailing edge in Fig. 9.36b. As is shown in Fig. 9.36d, the unrealistic flow situation can be corrected by adding an appropriate clockwise swirling flow around the airfoil. The results are twofold: (1) The unrealistic behavior near the trailing edge is eliminated (i.e.,

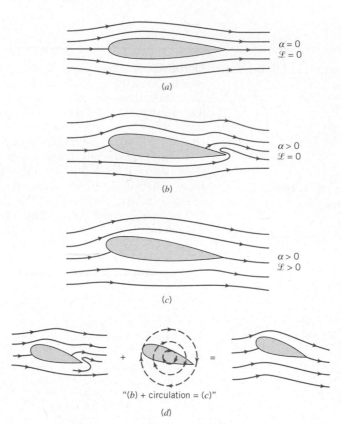

$\alpha = 0$
$\mathcal{L} = 0$

(a)

$\alpha > 0$
$\mathcal{L} = 0$

(b)

$\alpha > 0$
$\mathcal{L} > 0$

(c)

+ = 

"(b) + circulation = (c)"

(d)

■ **Figure 9.36** Inviscid flow past an airfoil: (*a*) symmetrical flow past the symmetrical airfoil at a zero angle of attack; (*b*) same airfoil at a nonzero angle of attack—no lift, flow near trailing edge not realistic; (*c*) same conditions as for (*b*) except circulation has been added to the flow—nonzero lift, realistic flow; (*d*) superposition of flows to produce the final flow past the airfoil.

the flow pattern of Fig. 9.36*b* is changed to that of Fig. 9.36*c*), and (2) the average velocity on the upper surface of the airfoil is increased while that on the lower surface is decreased. From the Bernoulli equation concepts (i.e., $p/\gamma + V^2/2g + z =$ constant), the average pressure on the upper surface is decreased and that on the lower surface is increased. The net effect is to change the original zero lift condition to that of a lift-producing airfoil.

The addition of the clockwise swirl is termed the addition of **circulation.** The amount of swirl (circulation) needed to have the flow leave the trailing edge smoothly is a function of the airfoil size and shape and can be calculated from potential flow (inviscid) theory (see Section 6.6.3 and Ref. 29). Although the addition of circulation to make the flow field physically realistic may seem artificial, it has well-founded mathematical and physical grounds. For example, consider the flow past a finite-length airfoil, as is indicated in Fig. 9.37. For lift-generating conditions the average pressure on the lower surface is greater than that on the upper surface. Near the tips of the wing this pressure difference will cause some of the fluid to attempt to migrate from the lower to the upper surface, as is indicated in Fig. 9.37*b*. At the same time, this fluid is swept downstream, forming a *trailing vortex* (swirl) from each wing tip (see Fig. 4.3). It is speculated that the reason some birds migrate in vee-formation is to take advantage of the updraft produced by the trailing vortex of the preceding bird. [It is calculated that for a given expenditure of energy, a flock of 25 birds flying in vee-formation could travel 70% farther than if each bird were to fly separately (Ref. 15).]

The trailing vortices from the right and left wing tips are connected by the *bound vortex* along the length of the wing. It is this vortex that generates the circulation that produces the lift. The combined vortex system (the bound vortex and the trailing vortices) is termed a horseshoe vortex. The strength of the trailing vortices (which is equal to the strength of the bound vortex) is proportional to the lift generated. Large aircraft (for example, a Boeing 747) can generate very strong trailing vortices that persist for a long time before viscous effects and instability mechanisms finally cause them to die out. Such vortices are strong enough to flip smaller aircraft out of control if they follow too closely behind the large aircraft. The figure in the margin clearly shows a trailing vortex produced during a wake vortex study in which an airplane flew through a column of smoke.

*V9.25 Wing tip vortices*

**(Photograph courtesy of NASA.)**

$(a)$

$(b)$

■ **Figure 9.37**　Flow past a finite-length wing: $(a)$ the horseshoe vortex system produced by the bound vortex and the trailing vortices; $(b)$ the leakage of air around the wing tips produces the trailing vortices.

## F l u i d s　i n　t h e　N e w s

**Why winglets?** Winglets, those upward turning ends of airplane wings, boost the performance by reducing drag. This is accomplished by reducing the strength of the *wing tip vortices* formed by the difference between the high pressure on the lower surface of the wing and the low pressure on the upper surface of the wing. These vortices represent an energy loss and an increase in drag. In essence, the winglet provides an effective increase in the aspect ratio of the wing without extending the wingspan. Winglets come in a variety of styles—the Airbus A320 has a very small upper and lower winglet; the Boeing 747-400 has a conventional, vertical upper winglet; and the Boeing Business Jet (a derivative of the Boeing 737) has an 8-ft winglet with a curving transition from wing to winglet. Since the airflow around the winglets is quite complicated, the winglets must be carefully designed and tested for each aircraft. In the past, winglets were more likely to be retrofitted to existing wings, but new airplanes are being designed with winglets from the start. Unlike tailfins on cars, winglets really do work. (See Problem 9.123.)

As is indicated above, the generation of lift is directly related to the production of a swirl or vortex flow around the object. A nonsymmetric airfoil, by design, generates its own prescribed amount of swirl and lift. A symmetric object like a circular cylinder or sphere, which normally provides no lift, can generate swirl and lift if it rotates.

As is discussed in Section 6.6.3, the inviscid flow past a circular cylinder has the symmetrical flow pattern indicated in Fig. 9.38$a$. By symmetry the lift and drag are zero. However, if the cylinder is rotated about its axis in a stationary real $(\mu \neq 0)$ fluid, the rotation will drag some of the fluid around, producing circulation about the cylinder as in Fig. 9.38$b$. When this circulation is combined with an ideal, uniform upstream flow, the flow pattern indicated in Fig. 9.38$c$ is obtained. The flow is no longer symmetrical about the horizontal plane through the center of the cylinder; the average pressure is greater on the lower half of the cylinder than on the upper half, and a lift is generated. This effect is called the ***Magnus effect,*** after Heinrich Magnus (1802–1870), a German chemist and physicist who first investigated this phenomenon. A similar lift is generated on a rotating sphere. It accounts for the various types of pitches in baseball (i.e., curve ball, floater, sinker, etc.), the ability of a soccer player to hook the ball, and the hook or slice of a golf ball.

Typical lift and drag coefficients for a smooth, spinning sphere are shown in Fig. 9.39. Although the drag coefficient is fairly independent of the rate of rotation, the lift coefficient is strongly

*A spinning sphere or cylinder can generate lift.*

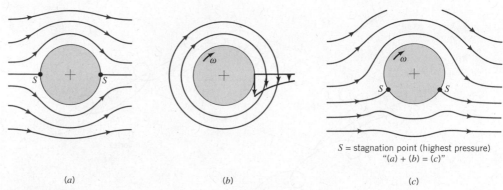

■ **Figure 9.38** Inviscid flow past a circular cylinder: (*a*) uniform upstream flow without circulation, (*b*) free vortex at the center of the cylinder, (*c*) combination of free vortex and uniform flow past a circular cylinder giving nonsymmetric flow and a lift.

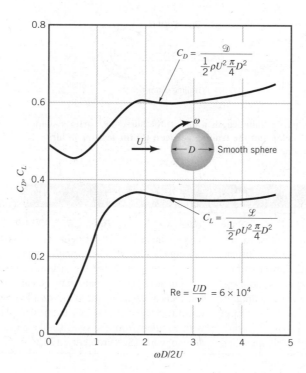

■ **Figure 9.39** Lift and drag coefficients for a spinning smooth sphere (Ref. 23).

dependent on it. In addition (although not indicated in the figure), both $C_L$ and $C_D$ are dependent on the roughness of the surface. As was discussed in Section 9.3, in a certain Reynolds number range an increase in surface roughness actually decreases the drag coefficient. Similarly, an increase in surface roughness can increase the lift coefficient because the roughness helps drag more fluid around the sphere, increasing the circulation for a given angular velocity. Thus, a rotating, rough golf ball travels farther than a smooth one because the drag is less and the lift is greater. However, do not expect a severely roughed up (cut) ball to work better—extensive testing has gone into obtaining the optimum surface roughness for golf balls.

*A dimpled golf ball has less drag and more lift than a smooth one.*

---

# $E$XAMPLE 9.16 | Lift on a Rotating Sphere

**GIVEN** A table tennis ball weighing $2.45 \times 10^{-2}$ N with diameter $D = 3.8 \times 10^{-2}$ m is hit at a velocity of $U = 12$ m/s with a back spin of angular velocity $\omega$ as is shown in Fig. E9.16.

**FIND** What is the value of $\omega$ if the ball is to travel on a horizontal path, not dropping due to the acceleration of gravity?

## SOLUTION

For horizontal flight, the lift generated by the spinning of the ball must exactly balance the weight, $\mathcal{W}$, of the ball so that

$$\mathcal{W} = \mathcal{L} = \tfrac{1}{2}\rho U^2 A C_L$$

or

$$C_L = \frac{2\mathcal{W}}{\rho U^2 (\pi/4) D^2}$$

where the lift coefficient, $C_L$, can be obtained from Fig. 9.39. For standard atmospheric conditions with $\rho = 1.23$ kg/m³ we obtain

$$C_L = \frac{2(2.45 \times 10^{-2}\,\text{N})}{(1.23\,\text{kg/m}^3)(12\,\text{m/s})^2(\pi/4)(3.8 \times 10^{-2}\,\text{m})^2}$$

$$= 0.244$$

which, according to Fig. 9.39, can be achieved if

$$\frac{\omega D}{2U} = 0.9$$

or

$$\omega = \frac{2U(0.9)}{D} = \frac{2(12\,\text{m/s})(0.9)}{3.8 \times 10^{-2}\,\text{m}} = 568\,\text{rad/s}$$

Thus,

$$\omega = (568\,\text{rad/s})(60\,\text{s/min})(1\,\text{rev}/2\pi\,\text{rad})$$

$$= 5420\,\text{rpm} \qquad \textbf{(Ans)}$$

■ **Figure E9.16**

**COMMENT**  Is it possible to impart this angular velocity to the ball? With larger angular velocities the ball will rise and follow an upward curved path. Similar trajectories can be produced by a well-hit golf ball—rather than falling like a rock, the golf ball trajectory is actually curved up and the spinning ball travels a greater distance than one without spin. However, if topspin is imparted to the ball (as in an improper tee shot) the ball will curve downward more quickly than under the action of gravity alone—the ball is "topped" and a negative lift is generated. Similarly, rotation about a vertical axis will cause the ball to hook or slice to one side or the other.

## 9.5  Chapter Summary and Study Guide

*drag*
*lift*
*lift coefficient*
*drag coefficient*
*wake region*
*boundary layer*
*laminar boundary layer*
*turbulent boundary layer*
*boundary layer thickness*
*transition*
*free-stream velocity*
*favorable pressure*
  *gradient*
*adverse pressure*
  *gradient*
*boundary layer*
  *separation*
*friction drag*
*pressure drag*
*stall*
*circulation*
*Magnus effect*

In this chapter the flow past objects is discussed. It is shown how the pressure and shear stress distributions on the surface of an object produce the net lift and drag forces on the object.

The character of flow past an object is a function of the Reynolds number. For large Reynolds number flows, a thin boundary layer forms on the surface. Properties of this boundary layer flow are discussed. These include the boundary layer thickness, whether the flow is laminar or turbulent, and the wall shear stress exerted on the object. In addition, boundary layer separation and its relationship to the pressure gradient are considered.

The drag, which contains portions due to friction (viscous) effects and pressure effects, is written in terms of the dimensionless drag coefficient. It is shown how the drag coefficient is a function of shape, with objects ranging from very blunt to very streamlined. Other parameters affecting the drag coefficient include the Reynolds number, Froude number, Mach number, and surface roughness.

The lift is written in terms of the dimensionless lift coefficient, which is strongly dependent on the shape of the object. Variation of the lift coefficient with shape is illustrated by the variation of an airfoil's lift coefficient with angle of attack.

The following checklist provides a study guide for this chapter. When your study of the entire chapter and end-of-chapter exercises has been completed, you should be able to

■ write out meanings of the terms listed here in the margin and understand each of the related concepts. These terms are particularly important and are set in *italic, **bold**, and color* type in the text.

- determine the lift and drag on an object from the given pressure and shear stress distributions on the object.
- for flow past a flat plate, calculate the boundary layer thickness, the wall shear stress, and the friction drag, and determine whether the flow is laminar or turbulent.
- explain the concept of the pressure gradient and its relationship to boundary layer separation.
- for a given object, obtain the drag coefficient from appropriate tables, figures, or equations and calculate the drag on the object.
- explain why golf balls have dimples.
- for a given object, obtain the lift coefficient from appropriate figures and calculate the lift on the object.

---

Some of the important equations in this chapter are:

| | | |
|---|---|---|
| Lift coefficient and drag coefficient | $C_L = \dfrac{\mathscr{L}}{\frac{1}{2}\rho U^2 A}, \quad C_D = \dfrac{\mathscr{D}}{\frac{1}{2}\rho U^2 A}$ | **(9.39), (9.36)** |
| Boundary layer displacement thickness | $\delta^* = \displaystyle\int_0^\infty \left(1 - \dfrac{u}{U}\right) dy$ | **(9.3)** |
| Boundary layer momentum thickness | $\Theta = \displaystyle\int_0^\infty \dfrac{u}{U}\left(1 - \dfrac{u}{U}\right) dy$ | **(9.4)** |
| Blasius boundary layer thickness, displacement thickness, and momentum thickness for flat plate | $\dfrac{\delta}{x} = \dfrac{5}{\sqrt{\mathrm{Re}_x}}, \quad \dfrac{\delta^*}{x} = \dfrac{1.721}{\sqrt{\mathrm{Re}_x}}, \quad \dfrac{\Theta}{x} = \dfrac{0.664}{\sqrt{\mathrm{Re}_x}}$ | **(9.15), (9.16), (9.17)** |
| Blasius wall shear stress for flat plate | $\tau_w = 0.332 U^{3/2} \sqrt{\dfrac{\rho\mu}{x}}$ | **(9.18)** |
| Drag on flat plate | $\mathscr{D} = \rho b U^2 \Theta$ | **(9.23)** |
| Blasius wall friction coefficient and friction drag coefficient for flat plate | $c_f = \dfrac{0.664}{\sqrt{\mathrm{Re}_x}}, \quad C_{Df} = \dfrac{1.328}{\sqrt{\mathrm{Re}_\ell}}$ | **(9.32)** |

# References

1. Schlichting, H., *Boundary Layer Theory*, 8th Ed., McGraw-Hill, New York, 2000.
2. Rosenhead, L., *Laminar Boundary Layers*, Oxford University Press, London, 1963.
3. White, F. M., *Viscous Fluid Flow*, 3rd Ed., McGraw-Hill, New York, 2005.
4. Currie, I. G., *Fundamental Mechanics of Fluids*, McGraw-Hill, New York, 1974.
5. Blevins, R. D., *Applied Fluid Dynamics Handbook*, Van Nostrand Reinhold, New York, 1984.
6. Hoerner, S. F., *Fluid-Dynamic Drag*, published by the author, Library of Congress No. 64,19666, 1965.
7. Happel, J., *Low Reynolds Number Hydrodynamics*, Prentice-Hall, Englewood Cliffs, N.J., 1965.
8. Van Dyke, M., *An Album of Fluid Motion*, Parabolic Press, Stanford, Calif., 1982.
9. Thompson, P. A., *Compressible-Fluid Dynamics*, McGraw-Hill, New York, 1972.
10. Zucrow, M. J., and Hoffman, J. D., *Gas Dynamics, Vol. I*, Wiley, New York, 1976.
11. Clayton, B. R., and Bishop, R. E. D., *Mechanics of Marine Vehicles*, Gulf Publishing Co., Houston, Tex., 1982.
12. *CRC Handbook of Tables for Applied Engineering Science*, 2nd Ed., CRC Press, Boca Raton, Fla., 1973.
13. Shevell, R. S., *Fundamentals of Flight*, 2nd Ed., Prentice-Hall, Englewood Cliffs, N.J., 1989.
14. Kuethe, A. M., and Chow, C. Y., *Foundations of Aerodynamics, Bases of Aerodynamics Design*, 4th Ed., Wiley, New York, 1986.

15. Vogel, J., *Life in Moving Fluids*, 2nd Ed., Willard Grant Press, Boston, 1994.

16. Kreider, J. F., *Principles of Fluid Mechanics*, Allyn and Bacon, Newton, Mass., 1985.

17. Dobrodzicki, G. A., Flow Visualization in the National Aeronautical Establishment's Water Tunnel, National Research Council of Canada, Aeronautical Report LR-557, 1972.

18. White, F. M., *Fluid Mechanics*, 6th Ed., McGraw-Hill, New York, 2008.

19. Vennard, J. K., and Street, R. L., *Elementary Fluid Mechanics*, 7th Ed., Wiley, New York, 1995.

20. Gross, A. C., Kyle, C. R., and Malewicki, D. J., The Aerodynamics of Human Powered Land Vehicles, *Scientific American*, Vol. 249, No. 6, 1983.

21. Abbott, I. H., and Von Doenhoff, A. E., *Theory of Wing Sections*, Dover Publications, New York, 1959.

22. MacReady, P. B., "Flight on 0.33 Horsepower: The Gossamer Condor," *Proc. AIAA 14th Annual Meeting* (Paper No. 78-308), Washington, D.C., 1978.

23. Goldstein, S., *Modern Developments in Fluid Dynamics*, Oxford Press, London, 1938.

24. Achenbach, E., Distribution of Local Pressure and Skin Friction around a Circular Cylinder in Cross-Flow Up to Re = $5 \times 10^6$, *Journal of Fluid Mechanics*, Vol. 34, Pt. 4, 1968.

25. Inui, T., Wave-Making Resistance of Ships, *Transactions of the Society of Naval Architects and Marine Engineers*, Vol. 70, 1962.

26. Sovran, G., et al., eds., *Aerodynamic Drag Mechanisms of Bluff Bodies and Road Vehicles*, Plenum Press, New York, 1978.

27. Abbott, I. H., von Doenhoff, A. E., and Stivers, L. S., Summary of Airfoil Data, NACA Report No. 824, Langley Field, Va., 1945.

28. Society of Automotive Engineers Report HSJ1566, "Aerodynamic Flow Visualization Techniques and Procedures," 1986.

29. Anderson, J. D., *Fundamentals of Aerodynamics*, 4th Ed., McGraw-Hill, New York, 2007.

30. Hucho, W. H., *Aerodynamics of Road Vehicles*, Butterworth–Heinemann, 1987.

31. Homsy, G. M., et al., *Multimedia Fluid Mechanics*, 2nd Ed., CD-ROM, Cambridge University Press, New York, 2008.

 Problem available in *WileyPLUS* at instructor's discretion.

 Tutoring problem available in *WileyPLUS* at instructor's discretion.

 Problem is related to a chapter video available in *WileyPLUS*.

\*    Problem to be solved with aid of programmable calculator or computer.

†    Open-ended problem that requires critical thinking. These problems require various assumptions to provide the necessary input data. There are not unique answers to these problems.

# Review Problems

Go to Appendix G (*WileyPLUS* or the book's web site, www. wiley.com/college/munson) for a set of review problems with answers. Detailed solutions can be found in the *Student Solution* *Manual and Study Guide for Fundamentals of Fluid Mechanics,* by Munson et al. (© 2013, John Wiley and Sons, Inc.).

# Conceptual Questions

**9.1C** The flow of two different fluids over thin flat plates is shown for two cases, A and B. The velocity profiles and boundary thicknesses are to scale.

For these two situations:

a) There is not enough information to determine the relation between the Reynolds numbers.

b) The Reynolds number for case A is lower than that for case B.

c) The Reynolds numbers for cases A and B are the same.

d) The Reynolds number for case A is higher than that for case B.

**9.2C**  Water flows over two flat plates. The free-stream velocity is the same for both plates, the flow is laminar, and both plates have the same width. The length of the first plate is $L_1$, and that of the second plate is twice as long ($L_2 = 2 L_1$).

The relation between the shear force $F_2$ for the plate of length $L_2 = 2 L_1$ relative to the shear force $F_1$ for the plate of length $L_1$ is

a) $F_2 > 2 F_1$.      b) $F_2 = 2 F_1$.      c) $F_2 < F_1$.
d) $F_2 = \sqrt{2} F_1$.      e) $F_2 = F_1$.

Free-stream velocity
$U$

Drag force $F$

Length $L$

**9.3C**  In two different experiments, air and water flow over an object that has the same shape in both experiments. It is found that the drag coefficient on each of the objects is the same for both the air and water flow. For the same velocity:

a) The drag force in the water flow is about equal to that for the airflow.

b) The drag force in the water flow is about 10 times that for the airflow.

c) The drag force in the water flow is about 100 times that for the airflow.

d) The drag force in the water flow is about 1000 times that for the airflow.

**Additional conceptual questions are available in *WileyPLUS* at the instructor's discretion.**

# Problems

**Note: Unless specific values of required fluid properties are given in the problem statement, use the values found in the tables on the inside of the front cover. Answers to the even-numbered problems are listed at the end of the book. The Lab Problems as well as the videos that accompany problems can be accessed in *WileyPLUS* or the book's web site, www.wiley.com/college/munson.**

## Section 9.1  General External Flow Characteristics

**9.1**  Assume that water flowing past the equilateral triangular bar shown in Fig. P9.1 produces the pressure distributions indicated. Determine the lift and drag on the bar and the corresponding lift and drag coefficients (based on frontal area). Neglect shear forces.

■ **Figure P9.2**

drag due to viscous effects. **(a)** Determine $\tau_{avg}$ in terms of the dynamic pressure, $\rho U^2/2$. **(b)** Determine the drag coefficient for this object.

**9.3**  Repeat Problem 9.1 if the object is a cone (made by rotating the equilateral triangle about the horizontal axis through its tip) rather than a triangular bar.

**9.4**  A small 15-mm-long fish swims with a speed of 20 mm/s. Would a boundary layer type flow be developed along the sides of the fish? Explain.

**9.5**  The average pressure and shear stress acting on the surface of the 1-m-square flat plate are as indicated in Fig. P9.5. Determine the lift and drag generated. Determine the lift and drag if the shear stress is neglected. Compare these two sets of results.

$p = -0.25 \, \rho U^2$

Linear distribution

$U = 5$ ft/s

$p = 0.5 \, \rho U^2$

0.1 ft

$b$ = length = 4 ft

■ **Figure P9.1**

**9.2**  Fluid flows past the two-dimensional bar shown in Fig. P9.2. The pressures on the ends of the bar are as shown, and the average shear stress on the top and bottom of the bar is $\tau_{avg}$. Assume that the drag due to pressure is equal to the

$p_{ave} = -1.2$ kN/m$^2$
$\tau_{ave} = 5.8 \times 10^{-2}$ kN/m$^2$

$\alpha = 7°$

$U$

$p_{ave} = 2.3$ kN/m$^2$
$\tau_{ave} = 7.6 \times 10^{-2}$ kN/m$^2$

■ **Figure P9.5**

**\*9.6** WILEY The pressure distribution on the l-m-diameter circular disk in Fig. P9.6 is given in the table. Determine the drag on the disk.

| r (m) | p (kN/m²) |
|-------|-----------|
| 0     | 4.34      |
| 0.05  | 4.28      |
| 0.10  | 4.06      |
| 0.15  | 3.72      |
| 0.20  | 3.10      |
| 0.25  | 2.78      |
| 0.30  | 2.37      |
| 0.35  | 1.89      |
| 0.40  | 1.41      |
| 0.45  | 0.74      |
| 0.50  | 0.0       |

■ **Figure P9.6**

**9.7** ▶ When you walk through still air at a rate of 1 m/s, would you expect the character of the airflow around you to be most like that depicted in Fig. 9.6a, b, or c? Explain. (See **Video 9.3**.)

**9.8** WILEY A 0.10-m-diameter circular cylinder moves through air with a speed $U$. The pressure distribution on the cylinder's surface is approximated by the three straight-line segments shown in Fig. P9.8. Determine the drag coefficient on the cylinder. Neglect shear forces.

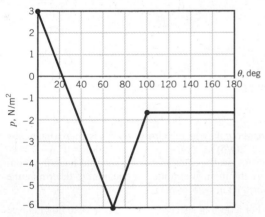

■ **Figure P9.8**

**9.9** WILEY Typical values of the Reynolds number for various animals moving through air or water are listed below. For which cases is inertia of the fluid important? For which cases do viscous effects dominate? For which cases would the flow be laminar; turbulent? Explain.

| Animal | Speed | Re |
|--------|-------|-----|
| (a) large whale | 10 m/s | 300,000,000 |
| (b) flying duck | 20 m/s | 300,000 |
| (c) large dragonfly | 7 m/s | 30,000 |
| (d) invertebrate larva | 1 mm/s | 0.3 |
| (e) bacterium | 0.01 mm/s | 0.00003 |

**†9.10** Estimate the Reynolds numbers associated with the following objects moving through water. **(a)** a kayak, **(b)** a minnow, **(c)** a submarine, **(d)** a grain of sand settling to the bottom, **(e)** you swimming.

## Section 9.2 Boundary Layer Characteristics (also see Lab Problems 9.1LP and 9.2LP.)

**9.11** A 12-ft-long kayak moves with a speed of 5 ft/s. Would a boundary layer type flow be developed along the sides of the boat? Explain.

**9.12** WILEY Water flows past a flat plate that is oriented parallel to the flow with an upstream velocity of 0.5 m/s. Determine the approximate location downstream from the leading edge where the boundary layer becomes turbulent. What is the boundary layer thickness at this location?

**9.13** WILEY A viscous fluid flows past a flat plate such that the boundary layer thickness at a distance 1.3 m from the leading edge is 12 mm. Determine the boundary layer thickness at distances of 0.20, 2.0, and 20 m from the leading edge. Assume laminar flow.

**9.14** WILEY If the upstream velocity of the flow in Problem 9.13 is $U = 1.5$ m/s, determine the kinematic viscosity of the fluid.

**9.15** WILEY Water flows past a flat plate with an upstream velocity of $U = 0.02$ m/s. Determine the water velocity a distance of 10 mm from the plate at distances of $x = 1.5$ m and $x = 15$ m from the leading edge.

**9.16** WILEY Approximately how fast can the wind blow past a 0.25-in.-diameter twig if viscous effects are to be of importance throughout the entire flow field (i.e., Re < 1)? Explain. Repeat for a 0.004-in.-diameter hair and a 6-ft-diameter smokestack.

**9.17** The typical shape of small cumulus clouds is as indicated in Fig. P9.17. Based on boundary layer ideas, explain why it is clear that the wind is blowing from right to left as indicated.

■ **Figure P9.17**

**9.18** Consider flow past the flat plate described in Problem 9.15. Based on the nature of the boundary layer equations that govern this flow, Eqs. 9.8 and 9.9, explain why the answer to Problem 9.15 is independent of the plate length.

**9.19** WILEY Because of the velocity deficit, $U - u$, in the boundary layer, the streamlines for flow past a flat plate are not exactly parallel to the plate. This deviation can be determined by use of the displacement thickness, $\delta^*$. For air blowing past the flat plate shown in Fig. P9.19, plot the streamline $A-B$ that passes through the edge of the boundary layer ($y = \delta_B$ at $x = \ell$) at point $B$. That is, plot $y = y(x)$ for streamline $A-B$. Assume laminar boundary layer flow.

■ **Figure P9.19**

**9.20** *(WILEY GO)* Air enters a square duct through a 1-ft opening as is shown in Fig. P9.20. Because the boundary layer displacement thickness increases in the direction of flow, it is necessary to increase the cross-sectional size of the duct if a constant $U = 2$ ft/s velocity is to be maintained outside the boundary layer. Plot a graph of the duct size, $d$, as a function of $x$ for $0 \leq x \leq 10$ ft if $U$ is to remain constant. Assume laminar flow.

■ **Figure P9.20**

**9.21** *(WILEY)* A smooth, flat plate of length $\ell = 6$ m and width $b = 4$ m is placed in water with an upstream velocity of $U = 0.5$ m/s. Determine the boundary layer thickness and the wall shear stress at the center and the trailing edge of the plate. Assume a laminar boundary layer.

**9.22** *(WILEY)* An atmospheric boundary layer is formed when the wind blows over the Earth's surface. Typically, such velocity profiles can be written as a power law: $u = ay^n$, where the constants $a$ and $n$ depend on the roughness of the terrain. As is indicated in Fig. P9.22, typical values are $n = 0.40$ for urban areas, $n = 0.28$ for woodland or suburban areas, and $n = 0.16$ for flat open country (Ref. 23). **(a)** If the velocity is 20 ft/s at the bottom of the sail on your boat ($y = 4$ ft), what is the velocity at the top of the mast ($y = 30$ ft)? **(b)** If the average velocity is 10 mph on the tenth floor of an urban building, what is the average velocity on the sixtieth floor?

■ **Figure P9.22**

**9.23** *(WILEY)* A 30-story office building (each story is 12 ft tall) is built in a suburban industrial park. Plot the dynamic pressure, $\rho u^2/2$, as a function of elevation if the wind blows at hurricane strength (75 mph) at the top of the building. Use the atmospheric boundary layer information of Problem 9.22.

**9.24** Show that by writing the velocity in terms of the similarity variable $\eta$ and the function $f(\eta)$, the momentum equation for boundary layer flow on a flat plate (Eq. 9.9) can be written as the ordinary differential equation given by Eq. 9.14.

**\*9.25** Integrate the Blasius equation (Eq. 9.14) numerically to determine the boundary layer profile for laminar flow past a flat plate. Compare your results with those of Table 9.1.

**9.26** *(WILEY GO)* An airplane flies at a speed of 400 mph at an altitude of 10,000 ft. If the boundary layers on the wing surfaces behave as those on a flat plate, estimate the extent of laminar boundary layer flow along the wing. Assume a transitional Reynolds number of $Re_{xcr} = 5 \times 10^5$. If the airplane maintains its 400-mph speed but descends to sea-level elevation, will the portion of the wing covered by a laminar boundary layer increase or decrease compared with its value at 10,000 ft? Explain.

**†9.27** If the boundary layer on the hood of your car behaves as one on a flat plate, estimate how far from the front edge of the hood the boundary layer becomes turbulent. How thick is the boundary layer at this location?

**9.28** *(WILEY)* A laminar boundary layer velocity profile is approximated by $u/U = [\, 2 - (y/\delta)]\,(y/\delta)$ for $y \leq \delta$, and $u = U$ for $y > \delta$. **(a)** Show that this profile satisfies the appropriate boundary conditions. **(b)** Use the momentum integral equation to determine the boundary layer thickness, $\delta = \delta(x)$.

**9.29** *(WILEY)* A laminar boundary layer velocity profile is approximated by the two straight-line segments indicated in Fig. P9.29. Use the momentum integral equation to determine the boundary layer thickness, $\delta = \delta(x)$, and wall shear stress, $\tau_w = \tau_w(x)$. Compare these results with those in Table 9.2.

■ **Figure P9.29**

**9.30** A laminar boundary layer velocity profile is approximated by $u/U = 2(y/\delta) - 2(y/\delta)^3 + (y/\delta)^4$ for $y \leq \delta$, and $u = U$ for $y > \delta$. **(a)** Show that this profile satisfies the appropriate boundary conditions. **(b)** Use the momentum integral equation to determine the boundary layer thickness, $\delta = \delta(x)$.

**\*9.31** *(WILEY GO)* For a fluid of specific gravity $SG = 0.86$ flowing past a flat plate with an upstream velocity of $U = 5$ m/s, the wall shear stress on the flat plate was determined to be as indicated in the table below. Use the momentum integral equation to determine the boundary layer momentum thickness, $\Theta = \Theta(x)$. Assume $\Theta = 0$ at the leading edge, $x = 0$.

| $x$ (m) | $\tau_w$ (N/m²) |
|---|---|
| 0 | — |
| 0.2 | 13.4 |
| 0.4 | 9.25 |
| 0.6 | 7.68 |
| 0.8 | 6.51 |
| 1.0 | 5.89 |
| 1.2 | 6.57 |
| 1.4 | 6.75 |
| 1.6 | 6.23 |
| 1.8 | 5.92 |
| 2.0 | 5.26 |

## Section 9.3 Drag

**9.32** Should a canoe paddle be made rough to get a "better grip on the water" for paddling purposes? Explain.

**9.33** (WILEY) Two different fluids flow over two identical flat plates with the same laminar free-stream velocity. Both fluids have the same viscosity, but one is twice as dense as the other. What is the relationship between the drag forces for these two plates?

**9.34** (WILEY) Fluid flows past a flat plate with a drag force $\mathcal{D}_1$. If the free-stream velocity is doubled, will the new drag force, $\mathcal{D}_2$, be larger or smaller than $\mathcal{D}_1$ and by what amount?

**9.35** A model is placed in an airflow with a given velocity and then placed in water flow with the same velocity. If the drag coefficients are the same between these two cases, how do the drag forces compare between the two fluids?

**9.36** (WILEY) The drag coefficient for a newly designed hybrid car is predicted to be 0.21. The cross-sectional area of the car is 30 ft$^2$. Determine the aerodynamic drag on the car when it is driven through still air at 55 mph.

**9.37** (WILEY) A 5-m-diameter parachute of a new design is to be used to transport a load from flight altitude to the ground with an average vertical speed of 3 m/s. The total weight of the load and parachute is 200 N. Determine the approximate drag coefficient for the parachute.

**9.38** (WILEY) A 50-mph wind blows against an outdoor movie screen that is 70 ft wide and 20 ft tall. Estimate the wind force on the screen.

**9.39** (WILEY) The aerodynamic drag on a car depends on the "shape" of the car. For example, the car shown in Fig. P9.39 has a drag coefficient of 0.35 with the windows and roof closed. With the windows and roof open, the drag coefficient increases to 0.45. With the windows and roof open, at what speed is the amount of power needed to overcome aerodynamic drag the same as it is at 65 mph with the windows and roof closed? Assume the frontal area remains the same. Recall that power is force times velocity.

Windows and roof closed: $C_D = 0.35$    Windows open; roof open: $C_D = 0.45$

■ **Figure P9.39**

**9.40** (WILEY) A rider on a bike with the combined mass of 100 kg attains a terminal speed of 15 m/s on a 12% slope. Assuming that the only forces affecting the speed are the weight and the drag, calculate the drag coefficient. The frontal area is 0.9 m$^2$. Speculate whether the rider is in the upright or racing position.

**9.41** (WILEY) A baseball is thrown by a pitcher at 95 mph through standard air. The diameter of the baseball is 2.82 in. Estimate the drag force on the baseball.

**9.42** (WILEY) A logging boat is towing a log that is 2 m in diameter and 8 m long at 4 m/s through water. Estimate the power required if the axis of the log is parallel to the tow direction.

**9.43** A sphere of diameter $D$ and density $\rho_s$ falls at a steady rate through a liquid of density $\rho$ and viscosity $\mu$. If the Reynolds number, Re $= \rho D U/\mu$, is less than 1, show that the viscosity can be determined from $\mu = gD^2(\rho_s - \rho)/18\ U$.

**9.44** (WILEY GO) Determine the drag on a small circular disk of 0.01-ft diameter moving 0.01 ft/s through oil with a specific gravity of 0.87 and a viscosity 10,000 times that of water. The disk is oriented normal to the upstream velocity. By what percent is the drag reduced if the disk is oriented parallel to the flow?

**9.45** (WILEY) The square, flat plate shown in Fig. P9.45a is cut into four equal-sized pieces and arranged as shown in Fig. P9.45b. Determine the ratio of the drag on the original plate [case (**a**)] to the drag on the plates in the configuration shown in (**b**). Assume laminar boundary flow. Explain your answer physically.

(a)

(b)

■ **Figure P9.45**

**9.46** (WILEY) If the drag on one side of a flat plate parallel to the upstream flow is $\mathcal{D}$ when the upstream velocity is $U$, what will the drag be when the upstream velocity is $2U$; or $U/2$? Assume laminar flow.

**9.47** (WILEY) Water flows past a triangular flat plate oriented parallel to the free stream as shown in Fig. P9.47. Integrate the wall shear stress over the plate to determine the friction drag on one side of the plate. Assume laminar boundary layer flow.

■ **Figure P9.47**

**9.48** For small Reynolds number flows, the drag coefficient of an object is given by a constant divided by the Reynolds number (see Table 9.4). Thus, as the Reynolds number tends to zero, the drag coefficient becomes infinitely large. Does this mean that for small velocities (hence, small Reynolds numbers) the drag is very large? Explain.

**9.49** (WILEY) A rectangular cartop carrier of 1.6-ft height, 5.0-ft length (front to back), and 4.2-ft width is attached to the top of a car. Estimate the additional power required to drive the car with the carrier at 60 mph through still air compared with the power required to drive only the car at 60 mph.

**9.50** ▶ (WILEY) As shown in **Video V9.2** and Fig. P9.50*a*, a kayak is a relatively streamlined object. As a first approximation in calculating the drag on a kayak, assume that the kayak acts as if it were a smooth, flat plate 17 ft long and 2 ft wide. Determine the drag as a function of speed and compare your results with the measured values given in Fig. P9.50*b*. Comment on reasons why the two sets of values may differ.

(a)

(b)

■ **Figure P9.50**

**9.51** A three-bladed helicopter blade rotates at 200 rpm. If each blade is 12 ft long and 1.5 ft wide, estimate the torque needed to overcome the friction on the blades if they act as flat plates.

**9.52** A ceiling fan consists of five blades of 0.80-m length and 0.10-m width which rotate at 100 rpm. Estimate the torque needed to overcome the friction on the blades if they act as flat plates.

**9.53** A thin smooth sign is attached to the side of a truck as is indicated in Fig. P9.53. Estimate the friction drag on the sign when the truck is driven at 55 mph.

■ **Figure P9.53**

**9.54** (WILEY) A 38.1-mm-diameter, 0.0245-N table tennis ball is released from the bottom of a swimming pool. With what velocity does it rise to the surface? Assume it has reached its terminal velocity.

**9.55** (WILEY) A hot-air balloon roughly spherical in shape has a volume of 70,000 ft³ and a weight of 500 lb (including passengers, basket, balloon fabric, etc.). If the outside air temperature is 80 °F

and the temperature within the balloon is 165 °F, estimate the rate at which it will rise under steady-state conditions if the atmospheric pressure is 14.7 psi.

**9.56** It is often assumed that "sharp objects can cut through the air better than blunt ones." Based on this assumption, the drag on the object shown in Fig. P9.56 should be less when the wind blows from right to left than when it blows from left to right. Experiments show that the opposite is true. Explain.

■ **Figure P9.56**

**9.57** An object falls at a rate of 100 ft/s immediately prior to the time that the parachute attached to it opens. The final descent rate with the chute open is 10 ft/s. Calculate and plot the speed of falling as a function of time from when the chute opens. Assume that the chute opens instantly, that the drag coefficient and air density remain constant, and that the flow is quasisteady.

**9.58** Estimate the velocity with which you would contact the ground if you jumped from an airplane at an altitude of 5,000 ft and **(a)** air resistance is negligible, **(b)** air resistance is important, but you forgot your parachute, or **(c)** you use a 25-ft-diameter parachute.

**9.59** As is discussed in Section 9.3, the drag on a rough golf ball is less than that on an equal-sized smooth ball. Does it follow that a 10-m-diameter spherical water tank resting on a 20-m-tall support should have a rough surface so as to reduce the moment needed at the base of the support when a wind blows? Explain.

**9.60** (WILEY) A 12-mm-diameter cable is strung between a series of poles that are 50 m apart. Determine the horizontal force this cable puts on each pole if the wind velocity is 30 m/s.

**9.61** (WILEY) How fast do small water droplets of 0.06-μm (6 × $10^{-8}$ m) diameter fall through the air under standard sea-level conditions? Assume the drops do not evaporate. Repeat the problem for standard conditions at 5000-m altitude.

**9.62** (WILEY) A strong wind can blow a golf ball off the tee by pivoting it about point 1 as shown in Fig. P9.62. Determine the wind speed necessary to do this.

■ **Figure P9.62**

**9.63** ▶ (WILEY) A 22 in. by 34 in. speed limit sign is supported on a 3-in.-wide, 5-ft-long pole. Estimate the bending moment in the pole at ground level when a 30-mph wind blows against the sign. (See **Video V9.14**.) List any assumptions used in your calculations.

**9.64** (WILEY) Determine the moment needed at the base of a 20-m-tall, 0.12-m-diameter flag pole to keep it in place in a 20-m/s wind.

**9.65** (WILEY) Repeat Problem 9.64 if a 2-m by 2.5-m flag is attached to the top of the pole. See Fig. 9.30 for drag coefficient data for flags.

†**9.66** (WILEY) During a flash flood, water rushes over a road as shown in Fig. P9.66 with a speed of 12 mph. Estimate the maximum water depth, $h$, that would allow a car to pass without being swept away. List all assumptions and show all calculations.

$U = 12$ mph

$h$

■ **Figure P9.66**

**9.67** (WILEY) How much more power is required to pedal a bicycle at 15 mph into a 20-mph head-wind than at 15 mph through still air? Assume a frontal area of 3.9 ft² and a drag coefficient of $C_D = 0.88$.

†**9.68** (WILEY) Estimate the wind velocity necessary to knock over a 20-lb garbage can that is 3 ft tall and 2 ft in diameter. List your assumptions.

**9.69** (WILEY) On a day without any wind, your car consumes $x$ gallons of gasoline when you drive at a constant speed, $U$, from point $A$ to point $B$ and back to point $A$. Assume that you repeat the journey, driving at the same speed, on another day when there is a steady wind blowing from $B$ to $A$. Would you expect your fuel consumption to be less than, equal to, or greater than $x$ gallons for this windy round-trip? Support your answer with appropriate analysis.

**9.70** (WILEY) The structure shown in Fig. P9.70 consists of three cylindrical support posts to which an elliptical flat plate sign is attached. Estimate the drag on the structure when a 50-mph wind blows against it.

16 ft

LAUREN'S PEANUT SHOP

5 ft

0.6 ft

15 ft

0.8 ft

15 ft

1 ft

15 ft

■ **Figure P9.70**

**9.71** A 25-ton (50,000-lb) truck coasts down a steep 7% mountain grade without brakes, as shown in Fig. P9.71. The truck's ultimate steady-state speed, $V$, is determined by a balance between weight, rolling resistance, and aerodynamic drag. Determine $V$ if the rolling resistance for a truck on concrete is 1.2% of the weight and the drag coefficient based on frontal area is 0.76.

12 ft

$V$

100

7

Truck width = 10 ft

■ **Figure P9.71**

**9.72** ▶ (WILEY) As shown in **Video V9.19** and Fig. P9.72, the aerodynamic drag on a truck can be reduced by the use of appropriate air deflectors. A reduction in drag coefficient from $C_D = 0.96$ to $C_D = 0.70$ corresponds to a reduction of how many horsepower needed at a highway speed of 65 mph?

$b$ = width = 10 ft

12 ft

Lily 2011

(a) $C_D = 0.70$

Lily 2011

(b) $C_D = 0.96$

■ **Figure P9.72**

**9.73** ▶ (WILEY) As shown in **Video V9.11** and Fig. P9.73, a vertical wind tunnel can be used for skydiving practice. Estimate the vertical wind speed needed if a 150-lb person is to be able to "float" motionless when the person (**a**) curls up as in a crouching position or (**b**) lies flat. See Fig. 9.30 for appropriate drag coefficient data.

$U$

■ **Figure P9.73**

**9.74** Compare the rise velocity of an $\frac{1}{8}$-in.-diameter air bubble in water to the fall velocity of an $\frac{1}{8}$-in.-diameter water drop in air. Assume each to behave as a solid sphere.

**9.75** A 50-lb box shaped like a 1-ft cube falls from the cargo hold of an airplane at an altitude of 30,000 ft. If the drag coefficient of the falling box is 1.2, determine the time it takes for the box to hit the ocean. Assume that it falls at the terminal velocity corresponding to its current altitude and use a standard atmosphere (see Table C.1).

**9.76** (WILEY GO) A 500-N cube of specific gravity $SG = 1.8$ falls through water at a constant speed $U$. Determine $U$ if the cube falls (a) as oriented in Fig. P9.76a, (b) as oriented in Fig. P9.76b.

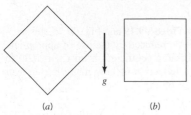

(a)        (b)

■ **Figure P9.76**

**\*9.77** (WILEY) The helium-filled balloon shown in Fig P9.77 is to be used as a wind-speed indicator. The specific weight of the helium is $y = 0.011$ lb/ft$^3$, the weight of the balloon material is 0.20 lb, and the weight of the anchoring cable is negligible. Plot a graph of $\theta$ as a function of $U$ for $1 \leq U \leq 50$ mph. Would this be an effective device over the range of $U$ indicated? Explain.

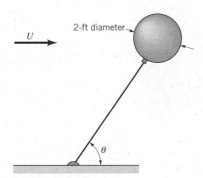

2-ft diameter

$U$

$\theta$

■ **Figure P9.77**

**9.78** (WILEY) A 0.30-m-diameter cork ball ($SG = 0.21$) is tied to an object on the bottom of a river as is shown in Fig. P9.78. Estimate the speed of the river current. Neglect the weight of the cable and the drag on it.

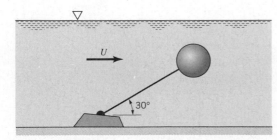

$U$

30°

■ **Figure P9.78**

**9.79** (WILEY) A shortwave radio antenna is constructed from circular tubing, as is illustrated in Fig. P9.79. Estimate the wind force on the antenna in a 100-km/hr wind.

10-mm diameter
1 m long

0.6 m

20-mm diameter
1.5 m long

0.5 m

0.25 m

40-mm diameter
5 m long

■ **Figure P9.79**

**†9.80** (WILEY) Estimate the wind force on your hand when you hold it out of your car window while driving 55 mph. Repeat your calculations if you were to hold your hand out of the window of an airplane flying 550 mph.

**†9.81** Estimate the energy that a runner expends to overcome aerodynamic drag while running a complete marathon race. This expenditure of energy is equivalent to climbing a hill of what height? List all assumptions and show all calculations.

**9.82** (WILEY) A 2-mm-diameter meteor of specific gravity 2.9 has a speed of 6 km/s at an altitude of 50,000 m where the air density is $1.03 \times 10^{-3}$ kg/m$^3$. If the drag coefficient at this large Mach number condition is 1.5, determine the deceleration of the meteor.

**9.83** (WILEY) Air flows past two equal sized spheres (one rough, one smooth) that are attached to the arm of a balance as is indicated in Fig. P9.83. With $U = 0$ the beam is balanced. What is the minimum air velocity for which the balance arm will rotate clockwise?

$D = 0.1$ m

Rough sphere
$\epsilon/D = 1.25 \times 10^{-2}$

Smooth sphere

0.3 m        0.5 m

$U$

■ **Figure P9.83**

**9.84** ▶ (WILEY) A 2-in.-diameter sphere weighing 0.14 lb is suspended by the jet of air shown in Fig. P9.84 and **Video V3.2.** The drag coefficient for the sphere is 0.5. Determine the reading on the pressure gage if friction and gravity effects can be neglected for the flow between the pressure gage and the nozzle exit.

Area = 0.3 ft$^2$

Air

Area = 0.6 ft$^2$

Pressure
gage

■ **Figure P9.84**

**9.85** A 60 mph wind blows against a football stadium scoreboard that is 36 ft tall, 80 ft wide, and 8 ft thick (parallel to the wind). Estimate the wind force on the scoreboard. See Fig. 9.28 for drag coefficient data.

**9.86** (WILEY) The United Nations Building in New York is approximately 87.5 m wide and 154 m tall. (a) Determine the drag on this building if the drag coefficient is 1.3 and the wind speed is a uniform 20 m/s. (b) Repeat your calculations if the velocity profile against the building is a typical profile for an urban area (see Problem 9.22) and the wind speed halfway up the building is 20 m/s.

**9.87** (WILEY) A regulation football is 6.78 in. in diameter and weighs 0.91 lb. If its drag coefficient is $C_D = 0.2$, determine its deceleration if it has a speed of 20 ft/s at the top of its trajectory.

**9.88** (WILEY) An airplane tows a banner that is $b = 0.8$ m tall and $\ell = 25$ m long at a speed of 150 km/hr. If the drag coefficient based on the area $b\ell$ is $C_D = 0.06$, estimate the power required to tow the banner. Compare the drag force on the banner with that on a rigid flat plate of the same size. Which has the larger drag force and why?

**9.89** (WILEY) The paint stirrer shown in Fig. P9.89 consists of two circular disks attached to the end of a thin rod that rotates at 80 rpm. The specific gravity of the paint is $SG = 1.1$ and its viscosity is $\mu = 2 \times 10^{-2}$ lb · s/ft$^2$. Estimate the power required to drive the mixer if the induced motion of the liquid is neglected.

■ **Figure P9.89**

**†9.90** If the wind becomes strong enough, it is "impossible" to paddle a canoe into the wind. Estimate the wind speed at which this will happen. List all assumptions and show all calculations.

**9.91** (WILEY) A fishnet consists of 0.10-in.-diameter strings tied into squares 4 in. per side. Estimate the force needed to tow a 15-ft by 30-ft section of this net through seawater at 5 ft/s.

**9.92** It is suggested that the power, $\mathcal{P}$, needed to overcome the aerodynamic drag on a vehicle traveling at a speed $U$ varies as $\mathcal{P} \sim U^n$. What is an appropriate value for the constant $n$? Explain.

**9.93** Estimate the power needed to overcome the aerodynamic drag of a person who runs at a rate of 100 yds in 10 s in still air. Repeat the calculations if the race is run into a 20-mph headwind; a 20-mph tailwind. Explain.

**9.94** (WILEY) By appropriate streamlining, the drag coefficient for an airplane is reduced by 12% while the frontal area remains the same. For the same power output, by what percentage is the flight speed increased?

**9.95** Two bicycle racers ride 30 km/hr through still air. By what percentage is the power required to overcome aerodynamic drag for the second cyclist reduced if she drafts closely behind the first cyclist rather than riding alongside her? Neglect any forces other than aerodynamic drag. (See Fig. 9.30.)

**9.96** (WILEY) As indicated in Fig. P9.96, the orientation of leaves on a tree is a function of the wind speed, with the tree becoming "more streamlined" as the wind increases. The resulting drag coefficient for the tree (based on the frontal area of the tree, $HW$) as a function of Reynolds number (based on the leaf length, $L$) is approxi-

■ **Figure P9.96**

mated as shown. Consider a tree with leaves of length $L = 0.3$ ft. What wind speed will produce a drag on the tree that is 6 times greater than the drag on the tree in a 15 ft/s wind?

**9.97** (WILEY) The blimp shown in Fig. P9.97 is used at various athletic events. It is 128 ft long and has a maximum diameter of 33 ft. If its drag coefficient (based on the frontal area) is 0.060, estimate the power required to propel it (a) at its 35-mph cruising speed, or (b) at its maximum 55-mph speed.

■ **Figure P9.97**

**9.98** If for a given vehicle it takes 20 hp to overcome aerodynamic drag while being driven at 65 mph, estimate the horsepower required at 75 mph.

**9.99** (WILEY) (See Fluids in the News article "Dimpled Baseball Bats," Section 9.3.3.) How fast must a 3.5-in.-diameter, dimpled baseball bat move through the air in order to take advantage of drag reduction produced by the dimples on the bat? Although there are differences, assume the bat (a cylinder) acts the same as a golf ball in terms of how the dimples affect the transition from a laminar to a turbulent boundary layer.

**9.100** (WILEY) (See Fluids in the News article "At 12,600 mpg It Doesn't Cost Much to 'Fill 'er Up,'" Section 9.3.3.) (a) Determine the power it takes to overcome aerodynamic drag on a small (6 ft$^2$ cross section), streamlined ($C_D = 0.12$) vehicle traveling 15 mph. (b) Compare the power calculated in part (a) with that for a large (36 ft$^2$ cross-sectional area), nonstreamlined ($C_D = 0.48$) SUV traveling 65 mph on the interstate.

## Section 9.4 Lift

**9.101** (WILEY GO) A rectangular wing with an aspect ratio of 6 is to generate 1000 lb of lift when it flies at a speed of 200 ft/s. Determine the length of the wing if its lift coefficient is 1.0.

**9.102** A 1.2-lb kite with an area of 6 ft$^2$ flies in a 20-ft/s wind such that the weightless string makes an angle of 55° relative to the horizontal. If the pull on the string is 1.5 lb, determine the lift and drag coefficients based on the kite area.

**9.103** (WILEY) A Piper Cub airplane has a gross weight of 1750 lb, a cruising speed of 115 mph, and a wing area of 179 ft$^2$. Determine the lift coefficient of this airplane for these conditions.

**9.104** (WILEY) A light aircraft with a wing area of 200 ft$^2$ and a weight of 2000 lb has a lift coefficient of 0.40 and a drag coefficient of 0.05. Determine the power required to maintain level flight.

**9.105** ▶ (WILEY) As shown in **Video V9.25** and Fig. P9.105, a spoiler is used on race cars to produce a negative lift, thereby giving a better tractive force. The lift coefficient for the airfoil shown is $C_L = 1.1$, and the coefficient of friction between the wheels and the pavement is 0.6. At a speed of 200 mph, by how much would use of the spoiler increase the maximum tractive force that could be generated between the wheels and ground? Assume the airspeed past the spoiler equals the car speed and that the airfoil acts directly over the drive wheels.

■ **Figure P9.105**

**9.106** (WILEY) The wings of old airplanes are often strengthened by the use of wires that provided cross-bracing as shown in Fig. P9.106. If the drag coefficient for the wings was 0.020 (based on the planform area), determine the ratio of the drag from the wire bracing to that from the wings.

Speed: 70 mph
Wing area: 148 ft²
Wire: length = 160 ft
        diameter = 0.05 in.

■ **Figure P9.106**

**9.107** For a given airplane, compare the power to maintain level flight at a 5000-ft altitude with that at 30,000 ft at the same velocity. Assume $C_D$ remains constant.

**9.108** (WILEY) A wing generates a lift $\mathcal{L}$ when moving through sea-level air with a velocity $U$. How fast must the wing move through the air at an altitude of 10,000 m with the same lift coefficient if it is to generate the same lift?

**9.109** (WILEY) Air blows over the flat-bottomed, two-dimensional object shown in Fig. P9.109. The shape of the object, $y = y(x)$, and the fluid speed along the surface, $u = u(x)$, are given in the table. Determine the lift coefficient for this object.

■ **Figure P9.109**

| $x(\% c)$ | $y(\% c)$ | $u/U$ |
|-----------|-----------|-------|
| 0         | 0         | 0     |
| 2.5       | 3.72      | 0.971 |
| 5.0       | 5.30      | 1.232 |
| 7.5       | 6.48      | 1.273 |
| 10        | 7.43      | 1.271 |
| 20        | 9.92      | 1.276 |
| 30        | 11.14     | 1.295 |
| 40        | 11.49     | 1.307 |
| 50        | 10.45     | 1.308 |
| 60        | 9.11      | 1.195 |
| 70        | 6.46      | 1.065 |
| 80        | 3.62      | 0.945 |
| 90        | 1.26      | 0.856 |
| 100       | 0         | 0.807 |

***9.110** (WILEY) When air flows past the airfoil shown in Fig. P9.110, the velocity just outside the boundary layer, $u$, is as indicated. Estimate the lift coefficient for these conditions.

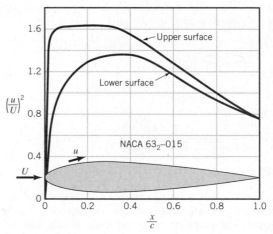

■ **Figure P9.110**

**9.111** (WILEY) A Boeing 747 aircraft weighing 580,000 lb when loaded with fuel and 100 passengers takes off with an airspeed of 140 mph. With the same configuration (i.e., angle of attack, flap settings, etc.), what is its takeoff speed if it is loaded with 372 passengers? Assume each passenger with luggage weighs 200 lb.

**9.112** Show that for unpowered flight (for which the lift, drag, and weight forces are in equilibrium) the glide slope angle, $\theta$, is given by $\tan \theta = C_D/C_L$.

**9.113** A sail plane with a lift-to-drag ratio of 25 flies with a speed of 50 mph. It maintains or increases its altitude by flying in thermals, columns of vertically rising air produced by buoyancy effects of nonuniformly heated air. What vertical airspeed is needed if the sail plane is to maintain a constant altitude?

**9.114** (WILEY) If the lift coefficient for a Boeing 777 aircraft is 15 times greater than its drag coefficient, can it glide from an altitude of 30,000 ft to an airport 80 mi away if it loses power from its engines? Explain. (See Problem 9.112.)

**9.115** (WILEY) On its final approach to the airport, an airplane flies on a flight path that is 3.0° relative to the horizontal. What lift-to-drag ratio is needed if the airplane is to land with its engines idled back to zero power? (See Problem 9.112.)

**9.116** (WILEY) Over the years there has been a dramatic increase in the flight speed ($U$), altitude ($h$), weight ($W$), and wing loading ($W/A$ = weight divided by wing area) of aircraft. Use the data given in the table below to determine the lift coefficient for each of the aircraft listed.

| Aircraft | Year | $W$, lb | $U$, mph | $W/A$, lb/ft$^2$ | $h$, ft |
|---|---|---|---|---|---|
| Wright Flyer | 1903 | 750 | 35 | 1.5 | 0 |
| Douglas DC-3 | 1935 | 25,000 | 180 | 25.0 | 10,000 |
| Douglas DC-6 | 1947 | 105,000 | 315 | 72.0 | 15,000 |
| Boeing 747 | 1970 | 800,000 | 570 | 150.0 | 30,000 |

**9.117** If the takeoff speed of a particular airplane is 120 mi/hr at sea level, what will it be at Denver (elevation 5000 ft)? Use properties of the U.S. Standard Atmosphere.

**9.118** ▶ (WILEY) The landing speed of an airplane such as the Space Shuttle is dependent on the air density. (See **Video V9.1.**) By what percent must the landing speed be increased on a day when the temperature is 110 °F compared to a day when it is 50 °F? Assume that the atmospheric pressure remains constant.

**9.119** (WILEY) Commercial airliners normally cruise at relatively high altitudes (30,000 to 35,000 ft). Discuss how flying at this high altitude (rather than 10,000 ft, for example) can save fuel costs.

**9.120** A pitcher can pitch a "curve ball" by putting sufficient spin on the ball when it is thrown. A ball that has absolutely no spin will follow a "straight" path. A ball that is pitched with a very small amount of spin (on the order of one revolution during its flight between the pitcher's mound and home plate) is termed a knuckle ball. A ball pitched this way tends to "jump around" and "zig-zag" back and forth. Explain this phenomenon. *Note*: A baseball has seams.

**9.121** (WILEY) For many years, hitters have claimed that some baseball pitchers have the ability to actually throw a rising fastball. Assuming that a top major leaguer pitcher can throw a 95-mph pitch and impart an 1800-rpm spin to the ball, is it possible for the ball to actually rise? Assume the baseball diameter is 2.9 in. and its weight is 5.25 oz.

**9.122** (WILEY) (See Fluids in the News article "Learning from Nature," Section 9.4.1.) As indicated in Fig. P9.122, birds can significantly alter their body shape and increase their planform area, $A$, by spreading their wing and tail feathers, thereby reducing their flight speed. If during landing the planform area is increased by 50% and the lift coefficient increased by 30% while all other parameters are held constant, by what percent is the flight speed reduced?

■ **Figure P9.122**

**9.123** (WILEY) (See Fluids in the News article "Why Winglets?," Section 9.4.2.) It is estimated that by installing appropriately designed winglets on a certain airplane the drag coefficient will be reduced by 5%. For the same engine thrust, by what percent will the aircraft speed be increased by use of the winglets?

■ **Lab Problems**

**9.1LP** This problem involves measuring the boundary layer profile on a flat plate. To proceed with this problem, go to Appendix H, which is located in *WileyPLUS* or on the book's web site, www.wiley.com/college/munson.

**9.2LP** This problem involves measuring the pressure distribution on a circular cylinder. To proceed with this problem, go to Appendix H, which is located in *WileyPLUS* or on the book's web site, www.wiley.com/college/munson.

■ **Lifelong Learning Problems**

**9.1LL** One of the "Fluids in the News" articles in this chapter discusses pressure-sensitive paint—a new technique of measuring surface pressure. There have been other advances in fluid measurement techniques, particularly in velocity measurements. One such technique is particle image velocimetry, or PIV. Obtain information about PIV and its advantages. Summarize your findings in a brief report.

**9.2LL** For typical aircraft flying at cruise conditions, it is advantageous to have as much laminar flow over the wing as possible because there is an increase in friction drag once the flow becomes turbulent. Various techniques have been developed to help promote laminar flow over the wing, both in airfoil geometry configurations as well as active flow control mechanisms. Obtain information on one of these techniques. Summarize your findings in a brief report.

**9.3LL** We have seen in this chapter that streamlining an automobile can help to reduce the drag coefficient. One of the methods of reducing the drag has been to reduce the projected area. However, it is difficult for some road vehicles, such as a tractor-trailer, to reduce this projected area due to the storage volume needed to haul the required load. Over the years, work has been done to help minimize some of the drag on this type of vehicle. Obtain information on a method that has been developed to reduce drag on a tractor-trailer. Summarize your findings in a brief report.

■ **FE Exam Problems**

Sample FE (Fundamentals of Engineering) exam questions for fluid mechanics are provided in *WileyPLUS* or on the book's web site, www.wiley.com/college/munson.

■ **Computational Fluid Dynamics (CFD)**

The CFD problems associated with this chapter have been developed for use with the ANSYS Academic CFD software package that is associated with this text. See *WileyPLUS* or the book's web site (www.wiley.com/college/munson) for additional details.

**9.1CFD** This CFD problem involves simulating the pressure distribution for flow past a cylinder and investigating the differences between inviscid and viscous flows. To proceed with this problem, go to *WileyPLUS* or the book's web site, www.wiley.com/college/munson.

**There are additional CFD problems located in *WileyPLUS*.**

# $10$ Open-Channel Flow

CHAPTER OPENING PHOTO: **Hydraulic jump:** Under certain conditions, when water flows in an open channel, even if it has constant geometry, the depth of the water may increase considerably over a short distance along the channel. This phenomenon is termed a hydraulic jump (water flow from left to right).

## Learning Objectives

After completing this chapter, you should be able to:

■ discuss the general characteristics of open-channel flow.

■ use a specific energy diagram.

■ apply appropriate equations to analyze open-channel flow with uniform depth.

■ calculate key properties of a hydraulic jump.

■ determine flowrates based on open-channel flow-measuring devices.

*V10.1 Offshore oil drilling platform.*

*Open-channel flow* involves the flow of a liquid in a channel or conduit that is not completely filled. A free surface exists between the flowing fluid (usually water) and fluid above it (usually the atmosphere). The main driving force for such flows is the fluid weight—gravity forces the fluid to flow downhill. Most open-channel flow results are based on correlations obtained from model and full-scale experiments. Additional information can be gained from various analytical and numerical efforts.

Open-channel flows are essential to the world as we know it. The natural drainage of water through the numerous creek and river systems is a complex example of open-channel flow. Although the flow geometry for these systems is extremely complex, the resulting flow properties are of considerable economic, ecological, and recreational importance. Other examples of open-channel flows include the flow of rainwater in the gutters of our houses; the flow in canals, drainage ditches, sewers, and gutters along roads; the flow of small rivulets and sheets of water on a road during a rain (as shown by the photograph in the margin); and the flow in the chutes of water rides in amusement parks.

Open-channel flow involves the existence of a free surface which can distort into various shapes. Thus, a brief introduction into the properties and characteristics of surface waves is included.

The purpose of this chapter is to investigate the concepts of open-channel flow. Because of the amount and variety of material available, only a brief introduction to the topic can be presented. Further information can be obtained from the references indicated.

# 10.1    General Characteristics of Open-Channel Flow

*Open-channel flow can have a variety of characteristics.*

In our study of pipe flow (Chapter 8), we found that there are many ways to classify a flow—developing, fully developed, laminar, turbulent, and so on. For open-channel flow, the existence of a free surface allows additional types of flow. The extra freedom that allows the fluid to select its free-surface location and configuration (because it does not completely fill a pipe or conduit) allows important phenomena in open-channel flow that cannot occur in pipe flow. Some of the classifications of the flows are described below.

The manner in which the fluid depth, $y$, varies with time, $t$, and distance along the channel, $x$, is used to partially classify a flow. For example, the flow is *unsteady* or *steady* depending on whether the depth at a given location does or does not change with time. Some unsteady flows can be viewed as steady flows if the reference frame of the observer is changed. For example, a tidal bore (difference in water level) moving up a river is unsteady to an observer standing on the bank, but steady to an observer moving along the bank with the speed of the wave front of the bore. Other flows are unsteady regardless of the reference frame used. The complex, time-dependent, wind-generated waves on a lake are in this category. In this book we will consider only steady open-channel flows.

An open-channel flow is classified as *uniform flow* (UF) if the depth of flow does not vary along the channel ($dy/dx = 0$). Conversely, it is *nonuniform flow* or *varied flow* if the depth varies with distance ($dy/dx \neq 0$). Nonuniform flows are further classified as *rapidly varying flow* (RVF) if the flow depth changes considerably over a relatively short distance; $dy/dx \sim 1$. *Gradually varying flows* (GVF) are those in which the flow depth changes slowly with distance along the channel; $dy/dx \ll 1$. Examples of these types of flow are illustrated in Fig. 10.1 and the photographs in the margin. The relative importance of the various types of forces involved (pressure, weight, shear, inertia) is different for the different types of flows.

As for any flow geometry, open-channel flow may be *laminar*, *transitional*, or *turbulent*, depending on various conditions involved. Which type of flow occurs depends on the Reynolds number, $\text{Re} = \rho V R_h / \mu$, where $V$ is the average velocity of the fluid and $R_h$ is the hydraulic radius of the channel (see Section 10.4). A general rule is that open-channel flow is laminar if $\text{Re} < 500$, turbulent if $\text{Re} > 12{,}500$, and transitional otherwise. The values of these dividing Reynolds numbers are only approximate—a precise knowledge of the channel geometry is necessary to obtain specific values. Since most open-channel flows involve water (which has a fairly small viscosity) and have relatively large characteristic lengths, it is rare to have laminar open-channel flows. For example, flow of 50 °F water ($\nu = 1.41 \times 10^{-5}$ ft$^2$/s) with an average velocity of $V = 1$ ft/s in a river with a hydraulic radius of $R_h = 10$ ft has $\text{Re} = V R_h / \nu = 7.1 \times 10^5$. The flow is turbulent. However, flow of a thin sheet of water down a driveway with an average velocity of $V = 0.25$ ft/s such that $R_h = 0.02$ ft (in such cases the hydraulic radius is approximately equal to the fluid depth; see Section 10.4) has $\text{Re} = 355$. The flow is laminar.

In some cases *stratified flows* are important. In such situations layers of two or more fluids of different densities flow in a channel. A layer of oil on water is one example of this type of flow. All of the open-channel flows considered in this book are *homogeneous flows*. That is, the fluid has uniform properties throughout.

Open-channel flows involve a free surface that can deform from its undisturbed, relatively flat configuration to form waves. Such waves move across the surface at speeds that depend on

Uniform flow

Rapidly varying flow
**(Photo by Marty Melchior)**

UF uniform flow
GVF gradually varying flow
RVF rapidly varying flow

| RVF | UF | RVF | UF | RVF | GVF | RVF | UF |

■ **Figure 10.1**    Classification of open-channel flow.

$$\mathrm{Fr} = \frac{V}{\sqrt{gy}}$$

Supercritical

1 — Critical

Subcritical

0

their size (height, length) and properties of the channel (depth, fluid velocity, etc.). The character of an open-channel flow may depend strongly on how fast the fluid is flowing relative to how fast a typical wave moves relative to the fluid. The dimensionless parameter that describes this behavior is termed the **Froude number,** $\mathrm{Fr} = V/(g\ell)^{1/2}$, where $\ell$ is an appropriate characteristic length of the flow. This dimensionless parameter was introduced in Chapter 7 and is discussed more fully in Section 10.2. As shown by the figure in the margin, the special case of a flow with a Froude number of unity, $\mathrm{Fr} = 1$, is termed a *critical flow.* If the Froude number is less than 1, the flow is **subcritical** (or *tranquil*). A flow with the Froude number greater than 1 is termed **supercritical** (or *rapid*).

## 10.2 Surface Waves

The distinguishing feature of flows involving a free surface (as in open-channel flows) is the opportunity for the free surface to distort into various shapes. The surface of a lake or the ocean is seldom "smooth as a mirror." It is usually distorted into ever-changing patterns associated with surface waves as shown in the photos in the margin. Some of these waves are very high, some barely ripple the surface; some waves are very long (the distance between wave crests), some are short; some are breaking waves that form whitecaps, others are quite smooth. Although a general study of this wave motion is beyond the scope of this book, an understanding of certain fundamental properties of simple waves is necessary for open-channel flow considerations. The interested reader is encouraged to use some of the excellent references available for further study about wave motion (Refs. 1, 2, 3).

### 10.2.1 Wave Speed

*V10.2 Filling your car's gas tank.*

Consider the situation illustrated in Fig. 10.2*a* in which a single elementary wave of small height, $\delta y$, is produced on the surface of a channel by suddenly moving the initially stationary end wall with speed $\delta V$. The water in the channel was stationary at the initial time, $t = 0$. A stationary observer will observe a single wave move down the channel with a **wave speed** $c$, with no fluid motion ahead of the wave and a fluid velocity of $\delta V$ behind the wave. The motion is unsteady for such an observer. For an observer moving along the channel with speed $c$, the flow will appear steady as shown in Fig. 10.2*b*. To this observer, the fluid velocity will be $\mathbf{V} = -c\hat{\mathbf{i}}$ on the observer's right and $\mathbf{V} = (-c + \delta V)\hat{\mathbf{i}}$ to the left of the observer.

The relationship between the various parameters involved for this flow can be obtained by application of the continuity and momentum equations to the control volume shown in Fig. 10.2*b* as follows. With the assumption of uniform one-dimensional flow, the continuity equation (Eq. 5.12) becomes

$$-cyb = (-c + \delta V)(y + \delta y)b$$

where $b$ is the channel width. This simplifies to

$$c = \frac{(y + \delta y)\delta V}{\delta y}$$

■ **Figure 10.2**  (*a*) **Production of a single elementary wave in a channel as seen by a stationary observer. (*b*) Wave as seen by an observer moving with a speed equal to the wave speed.**

or in the limit of small-amplitude waves with $\delta y \ll y$

$$c = y \frac{\delta V}{\delta y} \tag{10.1}$$

Similarly, the momentum equation (Eq. 5.22) is

$$\tfrac{1}{2} \gamma y^2 b - \tfrac{1}{2} \gamma (y + \delta y)^2 b = \rho b c y [(c - \delta V) - c]$$

where we have written the mass flowrate as $\dot{m} = \rho b c y$ and have assumed that the pressure variation is hydrostatic within the fluid. That is, the pressure forces on the channel cross sections (1) and (2) are $F_1 = \gamma y_{c1} A_1 = \gamma (y + \delta y)^2 b/2$ and $F_2 = \gamma y_{c2} A_2 = \gamma y^2 b/2$, respectively. If we again impose the assumption of small-amplitude waves [i.e., $(\delta y)^2 \ll y \, \delta y$], the momentum equation reduces to

*The wave speed can be obtained from the continuity and momentum equations.*

$$\frac{\delta V}{\delta y} = \frac{g}{c} \tag{10.2}$$

Combination of Eqs. 10.1 and 10.2 gives the wave speed

$$c = \sqrt{gy} \tag{10.3}$$

as indicated by the figure in the margin.

The speed of a small-amplitude solitary wave as is indicated in Fig. 10.2 is proportional to the square root of the fluid depth, $y$, and independent of the wave amplitude, $\delta y$. The fluid density is not an important parameter, although the acceleration of gravity is. This is a result of the fact that such wave motion is a balance between inertial effects (proportional to $\rho$) and weight or hydrostatic pressure effects (proportional to $\gamma = \rho g$). A ratio of these forces eliminates the common factor $\rho$ but retains $g$. For very small waves (like those produced by insects on water as shown in **Video V10.3**), Eq. 10.3 is not valid because the effects of surface tension are significant.

The wave speed can also be calculated by using the energy and continuity equations rather than the momentum and continuity equations as is done above. A simple wave on the surface is shown in Fig. 10.3. As seen by an observer moving with the wave speed, $c$, the flow is steady. Since the pressure is constant at any point on the free surface, the Bernoulli equation for this frictionless flow is simply

$$\frac{V^2}{2g} + y = \text{constant}$$

or by differentiating

■ **Figure 10.3** Stationary simple wave in a flowing fluid.

*V10.3 Water strider*

$$\frac{V\,\delta V}{g} + \delta y = 0$$

Also, by differentiating the continuity equation, $Vy =$ constant, we obtain

$$y\,\delta V + V\,\delta y = 0$$

We combine these two equations to eliminate $\delta V$ and $\delta y$ and use the fact that $V = c$ for this situation (the observer moves with speed $c$) to obtain the wave speed given by Eq. 10.3.

The above results are restricted to waves of small amplitude because we have assumed one-dimensional flow. That is, $\delta y/y \ll 1$. More advanced analysis and experiments show that the wave speed for finite-sized solitary waves exceeds that given by Eq. 10.3. To a first approximation, one obtains (Ref. 4)

$$c \approx \sqrt{gy}\left(1 + \frac{\delta y}{y}\right)^{1/2}$$

As indicated by the figure in the margin, the larger the amplitude, the faster the wave travels.

A more general description of wave motion can be obtained by considering continuous (not solitary) waves of sinusoidal shape as is shown in Fig. 10.4. By combining waves of various wavelengths, $\lambda$, and amplitudes, $\delta y$, it is possible to describe very complex surface patterns found in nature, such as the wind-driven waves on a lake. Mathematically, such a process consists of using a Fourier series (each term of the series represented by a wave of different wavelength and amplitude) to represent an arbitrary function (the free-surface shape).

A more advanced analysis of such sinusoidal surface waves of small amplitude shows that the wave speed varies with both the wavelength and fluid depth as (Ref. 1)

$$c = \left[\frac{g\lambda}{2\pi}\tanh\left(\frac{2\pi y}{\lambda}\right)\right]^{1/2} \tag{10.4}$$

*V10.4 Sinusoidal waves*

where $\tanh(2\pi y/\lambda)$ is the hyperbolic tangent of the argument $2\pi y/\lambda$. The result is plotted in Fig. 10.5. For conditions for which the water depth is much greater than the wavelength ($y \gg \lambda$, as in the ocean), the wave speed is independent of $y$ and given by

$$c = \sqrt{\frac{g\lambda}{2\pi}}$$

■ **Figure 10.4** Sinusoidal surface wave.

■ **Figure 10.5**   Wave speed as a function of wavelength.

This result, shown in the figure in the margin, follows from Eq. 10.4, since $\tanh(2\pi y/\lambda) \to 1$ as $y/\lambda \to \infty$. Note that waves with very long wavelengths [e.g., waves created by a tsunami ("tidal wave") with wavelengths on the order of several kilometers] travel very rapidly. On the other hand, if the fluid layer is shallow ($y \gg \lambda$, as often happens in open channels), the wave speed is given by $c = (gy)^{1/2}$, as derived for the solitary wave in Fig. 10.2. This result also follows from Eq. 10.4, since $\tanh(2\pi y/\lambda) \to 2\pi y/\lambda$ as $y/\lambda \to 0$. These two limiting cases are shown in Fig. 10.5. For moderate depth layers ($y \sim \lambda$), the results are given by the complete Eq. 10.4. Note that for a given fluid depth, the long wave travels the fastest. Hence, for our purposes we will consider the wave speed to be this limiting situation, $c = (gy)^{1/2}$.

# F l u i d s    i n    t h e    N e w s

**Tsunami, the nonstorm wave** A tsunami, often miscalled a "tidal wave," is a wave produced by a disturbance (for example, an earthquake, volcanic eruption, or meteorite impact) that vertically displaces the water column. Tsunamis are characterized as shallow-water waves, with long periods, very long wavelengths, and extremely large wave speeds. For example, the waves of the great December 2005, Indian Ocean tsunami traveled with speeds of 500–1000 m/s. Typically, these waves were of small amplitude in deep water far from land. Satellite radar measured the wave height as less than 1 m in these areas. However, as the waves approached shore and moved into shallower water, they slowed down considerably and reached heights up to 30 m. Because the rate at which a wave loses its energy is inversely related to its wavelength, tsunamis, with their wavelengths on the order of 100 km, not only travel at high speeds, but they also travel great distances with minimal energy loss. The furthest reported death from the Indian Ocean tsunami occurred approximately 8000 km from the epicenter of the earthquake that produced it. A more recent example is the tsunami that struck Japan in 2011. (See Problem 10.14.)

## 10.2.2   Froude Number Effects

Consider an elementary wave traveling on the surface of a fluid, as is shown in the figure in the margin and Fig. 10.2a. If the fluid layer is stationary, the wave moves to the right with speed $c$ relative to the fluid and the stationary observer. If the fluid is flowing to the left with velocity $V < c$, the wave (which travels with speed $c$ relative to the fluid) will travel to the right with a speed of $c - V$ relative to a fixed observer. If the fluid flows to the left with $V = c$, the wave will remain stationary, but if $V > c$ the wave will be washed to the left with speed $V - c$.

The above ideas can be expressed in dimensionless form by use of the Froude number, $\text{Fr} = V/(gy)^{1/2}$, where we take the characteristic length to be the fluid depth, $y$. Thus, the Froude number, $\text{Fr} = V/(gy)^{1/2} = V/c$, is the ratio of the fluid velocity to the wave speed.

The following characteristics are observed when a wave is produced on the surface of a moving stream, as happens when a rock is thrown into a river. If the stream is not flowing, the wave spreads equally in all directions. If the stream is nearly stationary or moving in a tranquil manner (i.e., $V < c$), the wave can move upstream. Upstream locations are said to be in hydraulic communication with the downstream locations. That is, an observer upstream of a disturbance can tell that there has been a disturbance on the surface because that disturbance can propagate upstream to the observer. Viscous effects, which have been neglected in this discussion, will eventually damp out such waves far upstream. Such flow conditions, $V < c$, or Fr $< 1$, are termed *subcritical*.

On the other hand, if the stream is moving rapidly so that the flow velocity is greater than the wave speed (i.e., $V > c$), no upstream communication with downstream locations is possible. Any disturbance on the surface downstream from the observer will be washed farther downstream. Such conditions, $V > c$ or Fr $> 1$, are termed *supercritical*. For the special case of $V = c$ or Fr $= 1$, the upstream propagating wave remains stationary and the flow is termed *critical*.

---

# $E$XAMPLE 10.1 Surface Waves

**GIVEN** At a certain location along the Rock River shown in Fig. E10.1a, the velocity, $V$, of the flow is a function of the depth, $y$, of the river as indicated in Fig. E10.1b. A reasonable approximation to these experimental results is

$$V = 5\, y^{2/3} \qquad (1)$$

where $V$ is in ft/s and $y$ is in ft.

**FIND** For what range of water depth will a surface wave on the river be able to travel upstream?

■ **Figure E10.1a**

## $S$OLUTION

While the river travels to the left with speed $V$, the surface wave travels upstream (to the right) with speed $c = (g\,y)^{1/2}$ relative to the water (not relative to the ground). Hence relative to the stationary ground, the wave travels to the right with speed

$$c - V = (g\,y)^{1/2} - 5\, y^{2/3}$$
$$= (32.2 \text{ ft/s}^2\, y)^{1/2} - 5\, y^{2/3} \qquad (2)$$

For the wave to travel upstream, $c - V > 0$ so that from Eq. 2,

$$(32.2\, y)^{1/2} > 5\, y^{2/3}$$

or

$$y < 2.14 \text{ ft} \qquad \textbf{(Ans)}$$

**COMMENT** As shown above, if the river depth is less than 2.14 ft, its velocity is less than the wave speed and the wave can travel upstream against the current. This is consistent with the fact that if a wave is to travel upstream, the flow must be subcritical (i.e., Fr $= V/c < 1$). For this flow

$$\text{Fr} = V/c = (5\, y^{2/3})/(g\,y)^{1/2}$$
$$= 5\, y^{1/6}/(32.2 \text{ ft/s}^2)^{1/2}$$
$$= 0.881\, y^{1/6}$$

This result is plotted in Fig. E10.1c. Note that in agreement with the above answer, for $y < 2.14$ the flow is subcritical; the wave can travel upstream.

■ **Figure E10.1b**

■ **Figure E10.1c**

---

The character of an open-channel flow may depend strongly on whether the flow is subcritical or supercritical. The characteristics of the flow may be completely opposite for subcritical flow than for supercritical flow. For example, as is discussed in Section 10.3, a "bump" on the bottom

*V10.5 Bicycle through a puddle*

of a river (such as a submerged log) may cause the surface of the river to dip below the level it would have had if the log were not there, or it may cause the surface level to rise above its undisturbed level. Which situation will happen depends on the value of Fr. Similarly, for supercritical flows it is possible to produce steplike discontinuities in the fluid depth (called a hydraulic jump; see Section 10.6.1). For subcritical flows, however, changes in depth must be smooth and continuous. Certain open-channel flows, such as the broad-crested weir (Section 10.6.3), depend on the existence of critical flow conditions for their operation.

As strange as it may seem, there exist many similarities between the open-channel flow of a liquid and the compressible flow of a gas. The governing dimensionless parameter in each case is the fluid velocity, $V$, divided by a wave speed, the surface wave speed for open-channel flow or sound wave speed for compressible flow. Many of the differences between subcritical (Fr $<$ 1) and supercritical (Fr $>$ 1) open-channel flows have analogs in subsonic (Ma $<$ 1) and supersonic (Ma $>$ 1) compressible gas flow, where Ma is the Mach number. Some of these similarities are discussed in this chapter and in Chapter 11.

## 10.3    Energy Considerations

*The slope of the bottom of most open channels is very small; the bottom is nearly horizontal.*

A typical segment of an open-channel flow is shown in Fig. 10.6. The slope of the channel bottom (or *bottom slope*), $S_0 = (z_1 - z_2)/\ell$, is assumed constant over the segment shown. The fluid depths and velocities are $y_1$, $y_2$, $V_1$, and $V_2$ as indicated. Note that the fluid depth is measured in the vertical direction and the distance $x$ is horizontal. For most open-channel flows the value of $S_0$ is very small (the bottom is nearly horizontal). For example, the Mississippi River drops a distance of 1470 ft in its 2350-mi length to give an average value of $S_0 = 0.000118$. In such circumstances the values of $x$ and $y$ are often taken as the distance along the channel bottom and the depth normal to the bottom, with negligibly small differences introduced by the two coordinate schemes.

With the assumption of a uniform velocity profile across any section of the channel, the one-dimensional energy equation for this flow (Eq. 5.84) becomes

$$\frac{p_1}{\gamma} + \frac{V_1^2}{2g} + z_1 = \frac{p_2}{\gamma} + \frac{V_2^2}{2g} + z_2 + h_L \tag{10.5}$$

where $h_L$ is the head loss due to viscous effects between sections (1) and (2) and $z_1 - z_2 = S_0\ell$. Since the pressure is essentially hydrostatic at any cross section, we find that $p_1/\gamma = y_1$ and $p_2/\gamma = y_2$ so that Eq. 10.5 becomes

$$y_1 + \frac{V_1^2}{2g} + S_0\ell = y_2 + \frac{V_2^2}{2g} + h_L \tag{10.6}$$

One of the difficulties of analyzing open-channel flow, similar to that discussed in Chapter 8 for pipe flow, is associated with the determination of the head loss in terms of other physical parameters. Without getting into such details at present, we write the head loss in terms of the slope of the energy line, $S_f = h_L/\ell$ (often termed the *friction slope*), as indicated in Fig. 10.6. Recall from

■ **Figure 10.6    Typical open-channel geometry.**

Chapter 3 that the energy line is located a distance $z$ (the elevation from some datum to the channel bottom) plus the pressure head $(p/\gamma)$ plus the velocity head $(V^2/2g)$ above the datum. Therefore, Eq. 10.6 can be written as

$$y_1 - y_2 = \frac{(V_2^2 - V_1^2)}{2g} + (S_f - S_0)\ell \tag{10.7}$$

If there is no head loss, the energy line is horizontal $(S_f = 0)$, and the total energy of the flow is free to shift between kinetic energy and potential energy in a conservative fashion. In the specific instance of a horizontal channel bottom $(S_0 = 0)$ and negligible head loss $(S_f = 0)$, Eq. 10.7 simply becomes

$$y_1 - y_2 = \frac{(V_2^2 - V_1^2)}{2g}$$

### 10.3.1 Specific Energy

The concept of the *specific energy* or specific head, $E$, defined as

*The specific energy is the sum of potential energy and kinetic energy (per unit weight).*

$$E = y + \frac{V^2}{2g} \tag{10.8}$$

is often useful in open-channel flow considerations. The energy equation, Eq. 10.7, can be written in terms of $E$ as

$$E_1 = E_2 + (S_f - S_0)\ell \tag{10.9}$$

If head losses are negligible, then $S_f = 0$ so that $(S_f - S_0)\ell = -S_0\ell = z_2 - z_1$ and the sum of the specific energy and the elevation of the channel bottom remains constant (i.e., $E_1 + z_1 = E_2 + z_2$, a statement of the Bernoulli equation).

If we consider a simple channel whose cross-sectional shape is a rectangle of width $b$, the specific energy can be written in terms of the flowrate per unit width, $q = Q/b = Vyb/b = Vy$, as

$$E = y + \frac{q^2}{2gy^2} \tag{10.10}$$

which is illustrated by the figure in the margin.

For a given channel of constant width, the value of $q$ remains constant along the channel, although the depth, $y$, may vary. To gain insight into the flow processes involved, we consider the *specific energy diagram*, a graph of $E = E(y)$, with $q$ fixed, as shown in Fig. 10.7. The relationship between the flow depth, $y$, and the velocity head, $V^2/2g$, as given by Eq. 10.8 is indicated in the figure.

■ **Figure 10.7** Specific energy diagram.

*For a given value of specific energy, a flow may have alternate depths.*

For given $q$ and $E$, Eq. 10.10 is a cubic equation $[y^3 - Ey^2 + (q^2/2g) = 0]$ with three solutions, $y_{sup}$, $y_{sub}$, and $y_{neg}$. If the specific energy is large enough (i.e., $E > E_{min}$, where $E_{min}$ is a function of $q$), two of the solutions are positive and the other, $y_{neg}$, is negative. The negative root, represented by the curved dashed line in Fig. 10.7, has no physical meaning and can be ignored. Thus, for a given flowrate and specific energy there are two possible depths, unless the vertical line from the $E$ axis does not intersect the specific energy curve corresponding to the value of $q$ given (i.e., $E < E_{min}$). These two depths are termed *alternate depths*.

For large values of $E$ the upper and lower branches of the specific energy diagram ($y_{sub}$ and $y_{sup}$) approach $y = E$ and $y = 0$, respectively. These limits correspond to a very deep channel flowing very slowly ($E = y + V^2/2g \rightarrow y$ as $y \rightarrow \infty$ with $q = Vy$ fixed), or a very high-speed flow in a shallow channel ($E = y + V^2/2g \rightarrow V^2/2g$ as $y \rightarrow 0$).

As is indicated in Fig. 10.7, $y_{sup} < y_{sub}$. Thus, since $q = Vy$ is constant along the curve, it follows that $V_{sup} > V_{sub}$, where the subscripts "sub" and "sup" on the velocities correspond to the depths so labeled. The specific energy diagram consists of two portions divided by the $E_{min}$ "nose" of the curve. We will show that the flow conditions at this location correspond to critical conditions (Fr = 1), those on the upper portion of the curve correspond to subcritical conditions (hence, the "sub" subscript), and those on the lower portion of the curve correspond to supercritical conditions (hence, the "sup" subscript).

To determine the value of $E_{min}$, we use Eq. 10.10 and set $dE/dy = 0$ to obtain

$$\frac{dE}{dy} = 1 - \frac{q^2}{gy^3} = 0$$

or

$$y_c = \left(\frac{q^2}{g}\right)^{1/3} \tag{10.11}$$

where the subscript "$c$" denotes conditions at $E_{min}$. By substituting this back into Eq. 10.10, we obtain

$$E_{min} = \frac{3y_c}{2}$$

By combining Eq. 10.11 and $V_c = q/y_c$, we obtain

$$V_c = \frac{q}{y_c} = \frac{(y_c^{3/2} g^{1/2})}{y_c} = \sqrt{gy_c}$$

or $\text{Fr}_c \equiv V_c/(gy_c)^{1/2} = 1$. Thus, critical conditions (Fr = 1) occur at the location of $E_{min}$. Since the layer is deeper and the velocity smaller for the upper part of the specific energy diagram (compared with the conditions at $E_{min}$), such flows are subcritical (Fr < 1). Conversely, flows for the lower part of the diagram are supercritical. This is shown by the figure in the margin. Thus, for a given flowrate, $q$, if $E > E_{min}$ there are two possible depths of flow, one subcritical and the other supercritical.

It is often possible to determine various characteristics of a flow by considering the specific energy diagram. Example 10.2 illustrates this for a situation in which the channel bottom elevation is not constant.

## $E$XAMPLE 10.2    Specific Energy Diagram—Quantitative

**GIVEN**   Water flows up a 0.5-ft-tall ramp in a constant width, rectangular channel at a rate $q = 5.75$ ft²/s as is shown in Fig. E10.2a. (For now disregard the "bump.") The upstream depth is 2.3 ft, and viscous effects are negligible.

**FIND**   Determine the elevation of the water surface downstream of the ramp, $y_2 + z_2$.

## SOLUTION

With $S_0\ell = z_1 - z_2$ and $h_L = 0$, conservation of energy (Eq. 10.6 which, under these conditions, is actually the Bernoulli equation) requires that

$$y_1 + \frac{V_1^2}{2g} + z_1 = y_2 + \frac{V_2^2}{2g} + z_2$$

For the conditions given ($z_1 = 0$, $z_2 = 0.5$ ft, $y_1 = 2.3$ ft, and $V_1 = q/y_1 = 2.5$ ft/s), this becomes

$$1.90 = y_2 + \frac{V_2^2}{64.4} \tag{1}$$

where $V_2$ and $y_2$ are in ft/s and ft, respectively. The continuity equation provides the second equation

$$y_2 V_2 = y_1 V_1$$

or

$$y_2 V_2 = 5.75 \text{ ft}^2/\text{s} \tag{2}$$

Equations 1 and 2 can be combined to give

$$y_2^3 - 1.90 y_2^2 + 0.513 = 0$$

which has solutions

$$y_2 = 1.72 \text{ ft}, \qquad y_2 = 0.638 \text{ ft}, \quad \text{or} \quad y_2 = -0.466 \text{ ft}$$

Note that two of these solutions are physically realistic, but the negative solution is meaningless. This is consistent with the previous discussions concerning the specific energy (recall the three roots indicated in Fig. 10.7). The corresponding elevations of the free surface are either

$$y_2 + z_2 = 1.72 \text{ ft} + 0.50 \text{ ft} = 2.22 \text{ ft}$$

or

$$y_2 + z_2 = 0.638 \text{ ft} + 0.50 \text{ ft} = 1.14 \text{ ft}$$

Which of these two flows is to be expected? This question can be answered by use of the specific energy diagram obtained from Eq. 10.10, which for this problem is

$$E = y + \frac{0.513}{y^2}$$

where $E$ and $y$ are in feet. The diagram is shown in Fig. E10.2b. The upstream condition corresponds to subcritical flow; the downstream condition is either subcritical or supercritical, corresponding to points 2 or 2'. Note that since $E_1 = E_2 + (z_2 - z_1) = E_2 + 0.5$ ft, it follows that the downstream conditions are located 0.5 ft to the left of the upstream conditions on the diagram.

With a constant-width channel, the value of $q$ remains the same for any location along the channel. That is, all points for the flow from (1) to (2) or (2') must lie along the $q = 5.75$ ft²/s curve shown. Any deviation from this curve would imply either a change in $q$ or a relaxation of the one-dimensional flow assumption. To stay on the curve and go from (1) around the critical point (point c) to point (2') would require a reduction in

(a)

(b)

■ **Figure E10.2**

specific energy to $E_{min}$. As is seen from Fig. E10.2a, this would require a specified elevation (bump) in the channel bottom so that critical conditions would occur above this bump. The height of this bump can be obtained from the energy equation (Eq. 10.9) written between points (1) and (c) with $S_f = 0$ (no viscous effects) and $S_0\ell = z_1 - z_c$. That is, $E_1 = E_{min} - z_1 + z_c$. In particular, since $E_1 = y_1 + 0.513/y_1^2 = 2.40$ ft and $E_{min} = 3y_c/2 = 3(q^2/g)^{1/3}/2 = 1.51$ ft, the top of this bump would need to be $z_c - z_1 = E_1 - E_{min} = 2.40$ ft $- 1.51$ ft $= 0.89$ ft above the channel bottom at section (1). The flow could then accelerate to supercritical conditions (Fr$_{2'}$ > 1) as is shown by the free surface represented by the dashed line in Fig. E10.2a.

Since the actual elevation change (a ramp) shown in Fig. E10.2a does not contain a bump, the downstream conditions will correspond to the subcritical flow denoted by (2), not the supercritical condition (2'). Without a bump on the channel bottom, the state (2') is inaccessible from the upstream condition state (1). Such considerations are often termed the *accessibility of flow regimes*. Thus, the surface elevation is

$$y_2 + z_2 = 2.22 \text{ ft} \tag{Ans}$$

Note that since $y_1 + z_1 = 2.30$ ft and $y_2 + z_2 = 2.22$ ft, the elevation of the free surface decreases as it goes across the ramp.

**COMMENT**   If the flow conditions upstream of the ramp were supercritical, the free-surface elevation and fluid depth would increase as the fluid flows up the ramp. This is indicated in Fig. E10.2c along with the corresponding specific energy diagram, as is shown in Fig. E10.2d. For this case the flow starts at (1) on the lower (supercritical) branch of the specific energy curve and ends at (2) on the same branch with $y_2 > y_1$. Since both $y$ and $z$ increase from (1) to (2), the surface elevation, $y + z$, also increases. Thus, flow up a ramp is different for subcritical than it is for supercritical conditions.

(c)                                      (d)

■ **Figure E10.2**   (*Continued*)

### 10.3.2  Channel Depth Variations

By using the concepts of the specific energy and critical flow conditions ($\text{Fr} = 1$), it is possible to determine how the depth of a flow in an open channel changes with distance along the channel. In some situations the depth change is very rapid so that the value of $dy/dx$ is of the order of 1. Complex effects involving two- or three-dimensional flow phenomena are often involved in such flows.

In this section we consider only gradually varying flows. For such flows, $dy/dx \ll 1$ and it is reasonable to impose the one-dimensional velocity assumption. At any section the total head is $H = V^2/2g + y + z$ and the energy equation (Eq. 10.5) becomes

$$H_1 = H_2 + h_L$$

where $h_L$ is the head loss between sections (1) and (2).

As is discussed in the previous section, the slope of the energy line is $dH/dx = dh_L/dx = S_f$ and the slope of the channel bottom is $dz/dx = S_0$. Thus, since

$$\frac{dH}{dx} = \frac{d}{dx}\left(\frac{V^2}{2g} + y + z\right) = \frac{V}{g}\frac{dV}{dx} + \frac{dy}{dx} + \frac{dz}{dx}$$

we obtain

$$\frac{dh_L}{dx} = \frac{V}{g}\frac{dV}{dx} + \frac{dy}{dx} + S_0$$

or

$$\frac{V}{g}\frac{dV}{dx} + \frac{dy}{dx} = S_f - S_0 \tag{10.12}$$

For a given flowrate per unit width, $q$, in a rectangular channel of constant width $b$, we have $V = q/y$ or by differentiation

$$\frac{dV}{dx} = -\frac{q}{y^2}\frac{dy}{dx} = -\frac{V}{y}\frac{dy}{dx}$$

so that the kinetic energy term in Eq. 10.12 becomes

$$\frac{V}{g}\frac{dV}{dx} = -\frac{V^2}{gy}\frac{dy}{dx} = -\text{Fr}^2\frac{dy}{dx} \tag{10.13}$$

where $\text{Fr} = V/(gy)^{1/2}$ is the local Froude number of the flow. Substituting Eq. 10.13 into Eq. 10.12 and simplifying gives

$$\frac{dy}{dx} = \frac{(S_f - S_0)}{(1 - \text{Fr}^2)} \tag{10.14}$$

It is seen that the rate of change of fluid depth, $dy/dx$, depends on the local slope of the channel bottom, $S_0$, the slope of the energy line, $S_f$, and the Froude number, Fr. As shown by the figure in the margin, the value of $dy/dx$ can be either negative, zero, or positive, depending on the values of these three parameters. That is, the channel flow depth may be constant or it may increase or decrease in the flow direction, depending on the values of $S_0$, $S_f$, and Fr. The behavior of subcritical flow may be the opposite of that for supercritical flow, as seen by the denominator, $1 - \text{Fr}^2$, of Eq. 10.14.

Although in the derivation of Eq. 10.14 we assumed $q$ is constant (i.e., a rectangular channel), Eq. 10.14 is valid for channels of any constant cross-sectional shape, provided the Froude number is interpreted properly (Ref. 3). In this book we will consider only rectangular cross-sectional channels when using this equation.

## 10.4    Uniform Depth Channel Flow

*V10.6 Merging channels*

Many channels are designed to carry fluid at a uniform depth all along their length. Irrigation canals are frequently of uniform depth and cross section for considerable lengths. Natural channels such as rivers and creeks are seldom of uniform shape, although a reasonable approximation to the flowrate in such channels can often be obtained by assuming uniform flow. In this section we will discuss various aspects of such flows.

*Uniform depth flow* $(dy/dx = 0)$ can be accomplished by adjusting the bottom slope, $S_0$, so that it precisely equals the slope of the energy line, $S_f$. That is, $S_0 = S_f$. This can be seen from Eq. 10.14. From an energy point of view, uniform depth flow is achieved by a balance between the potential energy lost by the fluid as it coasts downhill and the energy that is dissipated by viscous effects (head loss) associated with shear stresses throughout the fluid. Similar conclusions can be reached from a force balance analysis as discussed in the following section.

### 10.4.1  Uniform Flow Approximations

*The wall shear stress acts on the wetted perimeter of the channel.*

We consider fluid flowing in an open channel of constant cross-sectional size and shape such that the depth of flow remains constant as is indicated in Fig. 10.8. The area of the section is $A$, and the *wetted perimeter* (i.e., the length of the perimeter of the cross section in contact with the fluid) is $P$. The interaction between the fluid and the atmosphere at the free surface is assumed negligible so that this portion of the perimeter is not included in the definition of the wetted perimeter.

Since the fluid must adhere to the solid surfaces, the actual velocity distribution in an open channel is not uniform. Some typical velocity profiles measured in channels of various shapes are indicated in Fig. 10.9a. The maximum velocity is often found somewhat below the free surface,

■ **Figure 10.8**
Uniform flow in an open channel.

Lines of constant velocity

Centerline velocity profiles

——— Actual
– – – Uniform

$\tau_w$ = shear stress distribution

**Figure 10.9**  Typical velocity and shear stress distributions in an open channel: (*a*) velocity distribution throughout the cross section, (*b*) shear stress distribution on the wetted perimeter.

and the fluid velocity is zero on the wetted perimeter, where a wall shear stress, $\tau_w$, is developed. This shear stress is seldom uniform along the wetted perimeter, with typical variations as are indicated in Fig. 10.9*b*.

Fortunately, reasonable analytical results can be obtained by assuming a uniform velocity profile, $V$, and a constant wall shear stress, $\tau_w$. Similar assumptions were made for pipe flow situations (Chapter 8), with the friction factor being used to obtain the head loss.

### 10.4.2  The Chezy and Manning Equations

The basic equations used to determine the uniform flowrate in open channels were derived many years ago. Continual refinements have taken place to obtain better values of the empirical coefficients involved. The result is a semiempirical equation that provides reasonable engineering results. A more refined analysis is perhaps not warranted because of the complexity and uncertainty of the flow geometry (i.e., channel shape and the irregular makeup of the wetted perimeter, particularly for natural channels).

Under the assumptions of steady uniform flow, the $x$ component of the momentum equation (Eq. 5.22) applied to the control volume indicated in Fig. 10.10 simply reduces to

$$\Sigma F_x = \rho Q(V_2 - V_1) = 0$$

*For steady, uniform depth flow in an open channel there is no fluid acceleration.*

since $V_1 = V_2$. There is no acceleration of the fluid, and the momentum flux across section (1) is the same as that across section (2). The flow is governed by a simple balance between the forces in the direction of the flow. Thus, $\Sigma F_x = 0$, or

$$F_1 - F_2 - \tau_w P\ell + \mathcal{W} \sin \theta = 0 \tag{10.15}$$

■ **Figure 10.10**   **Control volume for uniform flow in an open channel.**

where $F_1$ and $F_2$ are the hydrostatic pressure forces across either end of the control volume, as shown by the figure in the margin. Because the flow is at a uniform depth $(y_1 = y_2)$, it follows that $F_1 = F_2$ so that these two forces do not contribute to the force balance. The term $\mathcal{W}\sin\theta$ is the component of the fluid weight that acts down the slope, and $\tau_w P\ell$ is the shear force on the fluid, acting up the slope as a result of the interaction of the water and the channel's wetted perimeter. Thus, Eq. 10.15 becomes

$$\tau_w = \frac{\mathcal{W}\sin\theta}{P\ell} = \frac{\mathcal{W}\,S_0}{P\ell}$$

where we have used the approximation that $\sin\theta \approx \tan\theta = S_0$, since the bottom slope is typically very small (i.e., $S_0 \ll 1$). Since $\mathcal{W} = \gamma A\ell$ and the **hydraulic radius** is defined as $R_h = A/P$, the force balance equation becomes

$$\tau_w = \frac{\gamma A\ell S_0}{P\ell} = \gamma R_h S_0 \tag{10.16}$$

*For uniform depth, channel flow is governed by a balance between friction and weight.*

Most open-channel flows are turbulent rather than laminar. In fact, typical Reynolds numbers are quite large, well above the transitional value and into the wholly turbulent regime. As was discussed in Chapter 8, and shown by the figure in the margin, for very large Reynolds number pipe flows (wholly turbulent flows), the friction factor, $f$, is found to be independent of Reynolds number, dependent only on the relative roughness, $\varepsilon/D$, of the pipe surface. For such cases, the wall shear stress is proportional to the dynamic pressure, $\rho V^2/2$, and independent of the viscosity. That is,

$$\tau_w = K\rho\,\frac{V^2}{2}$$

where $K$ is a constant dependent upon the roughness of the pipe.

It is not unreasonable that similar shear stress dependencies occur for the large Reynolds number open-channel flows. In such situations, Eq. 10.16 becomes

$$K\rho\,\frac{V^2}{2} = \gamma R_h S_0$$

or

$$V = C\sqrt{R_h S_0} \tag{10.17}$$

where the constant $C$ is termed the Chezy coefficient and Eq. 10.17 is termed the **Chezy equation.** This equation, one of the oldest in the area of fluid mechanics, was developed in 1768 by A. Chezy (1718–1798), a French engineer who designed a canal for the Paris water supply. The value of the Chezy coefficient, which must be determined by experiments, is not dimensionless but has the dimensions of $(\text{length})^{1/2}$ per time (i.e., the square root of the units of acceleration).

From a series of experiments it was found that the slope dependence of Eq. 10.17 ($V \sim S_0^{1/2}$) is reasonable but that the dependence on the hydraulic radius is more nearly $V \sim R_h^{2/3}$ rather than $V \sim R_h^{1/2}$. In 1889, R. Manning (1816–1897), an Irish engineer, developed the following somewhat modified equation for open-channel flow to more accurately describe the $R_h$ dependence:

$$V = \frac{R_h^{2/3} S_0^{1/2}}{n} \tag{10.18}$$

■ **Table 10.1**
**Values of the Manning Coefficient, $n$ (Ref. 6)**

| Wetted Perimeter | $n$ | Wetted Perimeter | $n$ |
|---|---|---|---|
| **A. *Natural channels*** | | **D. *Artificially lined channels*** | |
| Clean and straight | 0.030 | Glass | 0.010 |
| Sluggish with deep pools | 0.040 | Brass | 0.011 |
| Major rivers | 0.035 | Steel, smooth | 0.012 |
| | | Steel, painted | 0.014 |
| **B. *Floodplains*** | | Steel, riveted | 0.015 |
| Pasture, farmland | 0.035 | Cast iron | 0.013 |
| Light brush | 0.050 | Concrete, finished | 0.012 |
| Heavy brush | 0.075 | Concrete, unfinished | 0.014 |
| Trees | 0.15 | Planed wood | 0.012 |
| | | Clay tile | 0.014 |
| **C. *Excavated earth channels*** | | Brickwork | 0.015 |
| Clean | 0.022 | Asphalt | 0.016 |
| Gravelly | 0.025 | Corrugated metal | 0.022 |
| Weedy | 0.030 | Rubble masonry | 0.025 |
| Stony, cobbles | 0.035 | | |

Equation 10.18 is termed the ***Manning equation***, and the parameter $n$ is the ***Manning resistance coefficient***. Its value is dependent on the surface material of the channel's wetted perimeter and is obtained from experiments. It is not dimensionless, having the units of $s/m^{1/3}$ or $s/ft^{1/3}$.

As is discussed in Chapter 7, any correlation should be expressed in dimensionless form, with the coefficients that appear being dimensionless coefficients, such as the friction factor for pipe flow or the drag coefficient for flow past objects. Thus, Eq. 10.18 should be expressed in dimensionless form. Unfortunately, the Manning equation is so widely used and has been used for so long that it will continue to be used in its dimensional form with a coefficient, $n$, that is not dimensionless. The values of $n$ found in the literature (such as Table 10.1) were developed for SI units. Standard practice is to use the same value of $n$ when using the BG system of units and to insert a conversion factor into the equation.

*The Manning equation is used to obtain the velocity or flowrate in an open channel.*

Thus, uniform flow in an open channel is obtained from the Manning equation written as

$$V = \frac{\kappa}{n} R_h^{2/3} S_0^{1/2} \tag{10.19}$$

and

$$Q = \frac{\kappa}{n} A R_h^{2/3} S_0^{1/2} \tag{10.20}$$

where $\kappa = 1$ if SI units are used, and $\kappa = 1.49$ if BG units are used. The value 1.49 is the cube root of the number of feet per meter: $(3.281 \text{ ft/m})^{1/3} = 1.49$. Thus, by using $R_h$ in meters, $A$ in $m^2$, and $\kappa = 1$, the average velocity is m/s and the flowrate $m^3$/s. By using $R_h$ in feet, $A$ in $ft^2$, and $\kappa = 1.49$, the average velocity is ft/s and the flowrate $ft^3$/s.

Typical values of the Manning coefficient are indicated in Table 10.1. As expected, the rougher the wetted perimeter, the larger the value of $n$. For example, the roughness of floodplain surfaces increases from pasture to brush to tree conditions. So does the corresponding value of the Manning coefficient. Thus, for a given depth of flooding, the flowrate varies with floodplain roughness as indicated by the figure in the margin.

Precise values of $n$ are often difficult to obtain. Except for artificially lined channel surfaces like those found in new canals or flumes, the channel surface structure may be quite complex and variable. There are various methods used to obtain a reasonable estimate of the value of $n$ for a given situation (Ref. 5). For the purpose of this book, the values from Table 10.1 are sufficient. Note that the error in $Q$ is directly proportional to the error in $n$. A 10%

error in the value of $n$ produces a 10% error in the flowrate. Considerable effort has been put forth to obtain the best estimate of $n$, with extensive tables of values covering a wide variety of surfaces (Ref. 7). It should be noted that the values of $n$ given in Table 10.1 are valid only for water as the flowing fluid.

*V10.7 Uniform channel flow*

Both the friction factor for pipe flow and the Manning coefficient for channel flow are parameters that relate the wall shear stress to the makeup of the bounding surface. Thus, various results are available that describe $n$ in terms of the equivalent pipe friction factor, $f$, and the surface roughness, $\varepsilon$ (Ref. 8). For our purposes we will use the values of $n$ from Table 10.1.

### 10.4.3 Uniform Depth Examples

A variety of interesting and useful results can be obtained from the Manning equation. The following examples illustrate some of the typical considerations.

The main parameters involved in uniform depth open-channel flow are the size and shape of the channel cross section $(A, R_h)$, the slope of the channel bottom $(S_0)$, the character of the material lining the channel bottom and walls $(n)$, and the average velocity or flowrate $(V$ or $Q)$. Although the Manning equation is a rather simple equation, the ease of using it depends in part on which variables are given and which are to be determined.

Determination of the flowrate of a given channel with flow at a given depth (often termed the *normal flowrate* for *normal depth*, sometimes denoted $y_n$) is obtained from a straightforward calculation as is shown in Example 10.3.

---

## EXAMPLE 10.3 | Uniform Flow, Determine Flowrate

**GIVEN** Water flows in the canal of trapezoidal cross section shown in Fig. E10.3*a*. The bottom drops 1.4 ft per 1000 ft of length. The canal is lined with new finished concrete.

**FIND** Determine

**(a)** the flowrate and

**(b)** the Froude number for this flow.

■ **Figure E10.3*a***

### SOLUTION

**(a)** From Eq. 10.20,

$$Q = \frac{1.49}{n} A R_h^{2/3} S_0^{1/2} \qquad (1)$$

where we have used $\kappa = 1.49$, since the dimensions are given in BG units. For a depth of $y = 5$ ft, the flow area is

$$A = 12 \text{ ft} (5 \text{ ft}) + 5 \text{ ft} \left( \frac{5}{\tan 40°} \text{ ft} \right) = 89.8 \text{ ft}^2$$

so that with a wetted perimeter of $P = 12 \text{ ft} + 2(5/\sin 40° \text{ ft}) = 27.6 \text{ ft}$, the hydraulic radius is determined to be $R_h = A/P = 3.25 \text{ ft}$. Note that even though the channel is quite wide (the free-surface width is 23.9 ft), the hydraulic radius is only 3.25 ft, which is less than the depth.

Thus, with $S_0 = 1.4 \text{ ft}/1000 \text{ ft} = 0.0014$, Eq. 1 becomes

$$Q = \frac{1.49}{n} (89.8 \text{ ft}^2)(3.25 \text{ ft})^{2/3}(0.0014)^{1/2} = \frac{10.98}{n}$$

where $Q$ is in ft³/s.

From Table 10.1, we obtain $n = 0.012$ for the finished concrete. Thus,

$$Q = \frac{10.98}{0.012} = 915 \text{ cfs} \qquad \textbf{(Ans)}$$

**COMMENT** The corresponding average velocity, $V = Q/A$, is 10.2 ft/s. It does not take a very steep slope ($S_0 = 0.0014$ or $\theta = \tan^{-1}(0.0014) = 0.080°$) for this velocity.

By repeating the calculations for various surface types (i.e., various Manning coefficient values), the results shown in Fig. E10.3*b* are obtained. Note that the increased roughness causes a decrease in the flowrate. This is an indication that for the turbulent flows involved, the wall shear stress increases with surface roughness. [For water at 50 °F, the Reynolds number based on the 3.25-ft hydraulic radius of the channel and a smooth concrete surface is $Re = R_h V/\nu = 3.25 \text{ ft} (10.2 \text{ ft/s})/ (1.41 \times 10^{-5} \text{ ft}^2/\text{s}) = 2.35 \times 10^6$, well into the turbulent regime.]

**(b)** The Froude number based on the maximum depth for the flow can be determined from $Fr = V/(gy)^{1/2}$. For the finished

**Figure E10.3b**

concrete case,

$$\text{Fr} = \frac{10.2 \text{ ft/s}}{(32.2 \text{ ft/s}^2 \times 5 \text{ ft})^{1/2}} = 0.804 \qquad \textbf{(Ans)}$$

The flow is subcritical.

**COMMENT**   The same results would be obtained for the channel if its size were given in meters. We would use the same value of $n$ but set $\kappa = 1$ for this SI units situation.

In some instances a trial-and-error or iteration method must be used to solve for the dependent variable. This is often encountered when the flowrate, channel slope, and channel material are given, and the flow depth is to be determined as illustrated in the following examples.

---

## EXAMPLE 10.4   Uniform Flow, Determine Flow Depth

**GIVEN**   Water flows in the channel shown in Fig. E10.3a at a rate of $Q = 10.0$ m$^3$/s. The canal lining is weedy.

**FIND**   Determine the depth of the flow.

### SOLUTION

In this instance neither the flow area nor the hydraulic radius is known, although they can be written in terms of the depth, $y$. Since the flowrate is given in m$^3$/s, we will solve this problem using SI units. Hence, the bottom width is (12 ft) (1 m/3.281 ft) = 3.66 m and the area is

$$A = y\left(\frac{y}{\tan 40°}\right) + 3.66y = 1.19y^2 + 3.66y$$

where $A$ and $y$ are in square meters and meters, respectively. Also, the wetted perimeter is

$$P = 3.66 + 2\left(\frac{y}{\sin 40°}\right) = 3.11y + 3.66$$

so that

$$R_h = \frac{A}{P} = \frac{1.19y^2 + 3.66y}{3.11y + 3.66}$$

where $R_h$ and $y$ are in meters. Thus, with $n = 0.030$ (from Table 10.1), Eq. 10.20 can be written as

$$Q = 10 = \frac{\kappa}{n} A R_h^{2/3} S_0^{1/2}$$

$$= \frac{1.0}{0.030}(1.19y^2 + 3.66y)\left(\frac{1.19y^2 + 3.66y}{3.11y + 3.66}\right)^{2/3}$$

$$\times (0.0014)^{1/2}$$

which can be rearranged into the form

$$(1.19y^2 + 3.66y)^5 - 515(3.11y + 3.66)^2 = 0 \qquad \textbf{(1)}$$

where $y$ is in meters. The solution of Eq. 1 can be easily obtained by use of a simple rootfinding numerical technique or by trial-

and-error methods. The only physically meaningful root of Eq. 1 (i.e., a positive, real number) gives the solution for the normal flow depth at this flowrate as

$$y = 1.50 \text{ m} \qquad \textbf{(Ans)}$$

**COMMENT**   By repeating the calculations for various flowrates, the results shown in Fig. E10.4 are obtained. Note that the water depth is not linearly related to the flowrate. That is, if the flowrate is doubled, the depth is not doubled.

**Figure E10.4**

In Example 10.4 we found the flow depth for a given flowrate. Since the equation for this depth is a nonlinear equation, it may be that there is more than one solution to the problem. For a given channel there may be two or more depths that carry the same flowrate. Although this is not normally so, it can and does happen, as is illustrated by Example 10.5.

## $E$XAMPLE 10.5    Uniform Flow, Maximum Flowrate

**GIVEN** Water flows in a round pipe of diameter $D$ at a depth of $0 \leq y \leq D$, as is shown in Fig. E10.5a. The pipe is laid on a constant slope of $S_0$, and the Manning coefficient is $n$.

**FIND**    **(a)**    At what depth does the maximum flowrate occur?

**(b)**    Show that for certain flowrates there are two depths possible with the same flowrate. Explain this behavior.

■ **Figure E10.5**

## SOLUTION

**(a)** According to the Manning equation (Eq. 10.20), the flowrate is

$$Q = \frac{\kappa}{n} A R_h^{2/3} S_0^{1/2} \qquad (1)$$

where $S_0$, $n$, and $\kappa$ are constants for this problem. From geometry it can be shown that

$$A = \frac{D^2}{8}(\theta - \sin \theta)$$

where $\theta$, the angle indicated in Fig. E10.5a, is in radians. Similarly, the wetted perimeter is

$$P = \frac{D\theta}{2}$$

so that the hydraulic radius is

$$R_h = \frac{A}{P} = \frac{D(\theta - \sin \theta)}{4\theta}$$

Therefore, Eq. 1 becomes

$$Q = \frac{\kappa}{n} S_0^{1/2} \frac{D^{8/3}}{8(4)^{2/3}} \left[ \frac{(\theta - \sin \theta)^{5/3}}{\theta^{2/3}} \right]$$

This can be written in terms of the flow depth by using $y = (D/2)[1 - \cos(\theta/2)]$.

A graph of flowrate versus flow depth, $Q = Q(y)$, has the characteristic indicated in Fig. E10.5b. In particular, the maximum flowrate, $Q_{max}$, does not occur when the pipe is full;

$Q_{full} = 0.929 Q_{max}$. It occurs when $y = 0.938D$, or $\theta = 5.28$ rad $= 303°$. Thus,

$$Q = Q_{max} \text{ when } y = 0.938D \qquad \textbf{(Ans)}$$

**(b)** For any $0.929 < Q/Q_{max} < 1$ there are two possible depths that give the same $Q$. The reason for this behavior can be seen by considering the gain in flow area, $A$, compared to the increase in wetted perimeter, $P$, for $y \approx D$. The flow area increase for an increase in $y$ is very slight in this region, whereas the increase in wetted perimeter, and hence the increase in shear force holding back the fluid, is relatively large. The net result is a decrease in flowrate as the depth increases.

**COMMENT** For most practical problems, the slight difference between the maximum flowrate and full pipe flowrates is negligible, particularly in light of the usual inaccuracy of the value of $n$.

## F l u i d s   i n   t h e   N e w s

**Done without GPS or lasers** Two thousand years before the invention of such tools as the Global Positioning System (GPS) or laser surveying equipment, Roman engineers were able to design and construct structures that made a lasting contribution to Western civilization. For example, one of the best surviving examples of Roman aqueduct construction is the Pont du Gard, an aqueduct that spans the Gardon River near Nîmes, France.

This aqueduct is part of a circuitous, 50-km-long open channel that transported water to Rome from a spring located 20 km from Rome. The spring is only 14.6 m above the point of delivery, giving an average *bottom slope* of only $3 \times 10^{-4}$. It is obvious that to carry out such a project, the Roman understanding of hydraulics, surveying, and construction was well advanced. (See Problem 10.65.)

In many man-made channels and in most natural channels, the surface roughness (and, hence, the Manning coefficient) varies along the wetted perimeter of the channel. A drainage ditch, for example, may have a rocky bottom surface with concrete side walls to prevent erosion. Thus, the effective $n$ will be different for shallow depths than for deep depths of flow. Similarly, a river channel may have one value of $n$ appropriate for its normal channel and another very different value of $n$ during its flood stage when a portion of the flow occurs across fields or through floodplain woods. An ice-covered channel usually has a different value of $n$ for the ice than for the remainder of the wetted perimeter (Ref. 7). (Strictly speaking, such ice-covered channels are not "open" channels, although analysis of their flow is often based on open-channel flow equations. This is acceptable, since the ice cover is often thin enough so that it represents a fixed boundary in terms of the shear stress resistance, but it cannot support a significant pressure differential as in pipe flow situations.)

*For many open channels, the surface roughness varies across the channel.*

A variety of methods have been used to determine an appropriate value of the effective roughness of channels that contain subsections with different values of $n$. Which method gives the most accurate, easy-to-use results is not firmly established, since the results are nearly the same for each method (Ref. 5). A reasonable approximation is to divide the channel cross section into $N$ subsections, each with its own wetted perimeter, $P_i$, area, $A_i$, and Manning coefficient, $n_i$. The $P_i$ values do not include the imaginary boundaries between the different subsections. The total flowrate is assumed to be the sum of the flowrates through each section. This technique is illustrated by Example 10.7.

---

## $E$XAMPLE 10.6 | Uniform Flow, Variable Roughness

**GIVEN** Water flows along the drainage canal having the properties shown in Fig. E10.6a. The bottom slope is $S_0 = 1 \text{ ft}/500 \text{ ft} = 0.002$.

**FIND** Estimate the flowrate when the depth is $y = 0.8 \text{ ft} + 0.6 \text{ ft} = 1.4 \text{ ft}$.

■ Figure E10.6a

## $S$OLUTION

We divide the cross section into three subsections as is indicated in Fig. E10.6a and write the flowrate as $Q = Q_1 + Q_2 + Q_3$, where for each section

$$Q_i = \frac{1.49}{n_i} A_i R_{h_i}^{2/3} S_0^{1/2}$$

The appropriate values of $A_i, P_i, R_{hi}$, and $n_i$ are listed in Table E10.6. Note that the imaginary portions of the perimeters between sections (denoted by the vertical dashed lines in Fig. E10.6a) are not included in the $P_i$. That is, for section (2)

$$A_2 = 2 \text{ ft} (0.8 + 0.6) \text{ ft} = 2.8 \text{ ft}^2$$

and

$$P_2 = 2 \text{ ft} + 2(0.8 \text{ ft}) = 3.6 \text{ ft}$$

so that

$$R_{h_2} = \frac{A_2}{P_2} = \frac{2.8 \text{ ft}^2}{3.6 \text{ ft}} = 0.778 \text{ ft}$$

Thus, the total flowrate is

$$Q = Q_1 + Q_2 + Q_3 = 1.49(0.002)^{1/2}$$
$$\times \left[ \frac{(1.8 \text{ ft}^2)(0.500 \text{ ft})^{2/3}}{0.020} + \frac{(2.8 \text{ ft}^2)(0.778 \text{ ft})^{2/3}}{0.015} \right.$$
$$\left. + \frac{(1.8 \text{ ft}^2)(0.500 \text{ ft})^{2/3}}{0.030} \right]$$

or

$$Q = 16.8 \text{ ft}^3/\text{s} \qquad \textbf{(Ans)}$$

**COMMENTS** If the entire channel cross section were considered as one flow area, then $A = A_1 + A_2 + A_3 = 6.4 \text{ ft}^2$ and $P = P_1 + P_2 + P_3 = 10.8 \text{ ft}$, or $R_h = A/P = 6.4 \text{ ft}^2/10.8 \text{ ft} = 0.593 \text{ ft}$. The flowrate is given by Eq. 10.20, which can be written as

$$Q = \frac{1.49}{n_{\text{eff}}} A R_h^{2/3} S_0^{1/2}$$

■ Table E10.6

| $i$ | $A_i$ (ft$^2$) | $P_i$ (ft) | $R_{hi}$ (ft) | $n_i$ |
|---|---|---|---|---|
| 1 | 1.8 | 3.6 | 0.500 | 0.020 |
| 2 | 2.8 | 3.6 | 0.778 | 0.015 |
| 3 | 1.8 | 3.6 | 0.500 | 0.030 |

where $n_{eff}$ is the effective value of $n$ for this channel. With $Q = 16.8$ ft³/s as determined above, the value of $n_{eff}$ is found to be

$$n_{eff} = \frac{1.49 A R_h^{2/3} S_0^{1/2}}{Q}$$

$$= \frac{1.49(6.4)(0.593)^{2/3}(0.002)^{1/2}}{16.8} = 0.0179$$

As expected, the effective roughness (Manning $n$) is between the minimum ($n_2 = 0.015$) and maximum ($n_3 = 0.030$) values for the individual subsections.

By repeating the calculations for various depths, $y$, the results shown in Fig. E10.6$b$ are obtained. Note that there are two distinct portions of the graph—one when the water is contained entirely within the main, center channel ($y < 0.8$ ft) and the other when the water overflows into the side portions of the channel ($y > 0.8$ ft).

**■ Figure E10.6$b$**

One type of problem often encountered in open-channel flows is that of determining the *best hydraulic cross section,* defined as the section of the minimum area for a given flowrate, $Q$, slope, $S_0$, and roughness coefficient, $n$. By using $R_h = A/P$ we can write Eq. 10.20 as

$$Q = \frac{\kappa}{n} A \left(\frac{A}{P}\right)^{2/3} S_0^{1/2} = \frac{\kappa}{n} \frac{A^{5/3} S_0^{1/2}}{P^{2/3}}$$

which can be rearranged as

$$A = \left(\frac{nQ}{\kappa S_0^{1/2}}\right)^{3/5} P^{2/5}$$

*For a given flowrate, the channel of minimum area is denoted as the best hydraulic cross section.*

where the quantity in the parentheses is a constant. Thus, a channel with minimum $A$ is one with a minimum $P$, so that both the amount of excavation needed and the amount of material to line the surface are minimized by the best hydraulic cross section.

The best hydraulic cross section possible is that of a semicircular channel. No other shape has as small a wetted perimeter for a given area. It is often desired to determine the best shape for a class of cross sections. The results for rectangular, trapezoidal (with 60° sides), and triangular shapes are shown in Fig. 10.11. For example, the best hydraulic cross section for a rectangle is one whose depth is half its width; for a triangle it is a 90° triangle.

## EXAMPLE 10.7    Uniform Flow—Best Hydraulic Cross Section

**GIVEN** Water flows uniformly in a rectangular channel of width $b$ and depth $y$.

**FIND** Determine the aspect ratio, $b/y$, for the best hydraulic cross section.

### SOLUTION

The uniform flow is given by Eq. 10.20 as

$$Q = \frac{\kappa}{n} A R_h^{2/3} S_0^{1/2} \qquad (1)$$

where $A = by$ and $P = b + 2y$, so that $R_h = A/P = by/(b + 2y)$. We rewrite the hydraulic radius in terms of $A$ as

$$R_h = \frac{A}{(2y + b)} = \frac{A}{(2y + A/y)} = \frac{Ay}{(2y^2 + A)}$$

so that Eq. 1 becomes

$$Q = \frac{\kappa}{n} A \left(\frac{Ay}{2y^2 + A}\right)^{2/3} S_0^{1/2}$$

This can be rearranged to give

$$A^{5/2} y = K(2y^2 + A) \qquad (2)$$

where $K = (nQ/\kappa S_0^{1/2})^{3/2}$ is a constant. The best hydraulic section is the one that gives the minimum $A$ for all $y$. That is, $dA/dy = 0$. By differentiating Eq. 2 with respect to $y$, we obtain

$$\frac{5}{2} A^{3/2} \frac{dA}{dy} y + A^{5/2} = K\left(4y + \frac{dA}{dy}\right)$$

which, with $dA/dy = 0$, reduces to

$$A^{5/2} = 4Ky \tag{3}$$

With $K = A^{5/2} y/(2y^2 + A)$ from Eq. 2, Eq. 3 can be written in the form

$$A^{5/2} = \frac{4A^{5/2}y^2}{(2y^2 + A)}$$

which simplifies to $y = (A/2)^{1/2}$. Thus, because $A = by$, the best hydraulic cross section for a rectangular shape has a width $b$ and a depth

$$y = \left(\frac{A}{2}\right)^{1/2} = \left(\frac{by}{2}\right)^{1/2}$$

or

$$2y^2 = by$$

That is, the rectangle with the best hydraulic cross section is twice as wide as it is deep, or

$$b/y = 2 \tag{Ans}$$

**COMMENTS**    A rectangular channel with $b/y = 2$ will give the smallest area (and smallest wetted perimeter) for a given flowrate. Conversely, for a given area, the largest flowrate in a rectangular channel will occur when $b/y = 2$. For $A = by = $ constant, if $y \to 0$ then $b \to \infty$, and the flowrate is small because of

the large wetted perimeter $P = b + 2y \to \infty$. The maximum $Q$ occurs when $y = b/2$. However, as seen in Fig. E10.7, the maximum represented by this optimal configuration is a rather weak one. For example, for aspect ratios between 1 and 4, the flowrate is within 96% of the maximum flowrate obtained with the same area and $b/y = 2$.

An alternate but equivalent method to obtain the aforementioned answer is to use the fact that $dR_h/dy = 0$, which follows from Eq. 1 using $dQ/dy = 0$ (constant flowrate) and $dA/dy = 0$ (best hydraulic cross section has minimum area). Differentiation of $R_h = Ay/(2y^2 + A)$ with constant $A$ gives $b/y = 2$ when $dR_h/dy = 0$.

The best hydraulic cross section can be calculated for other shapes in a similar fashion. The results (given here without proof) for rectangular, trapezoidal (with 60° sides), and triangular shapes are shown in Fig. 10.11.

■ **Figure E10.7**

## 10.5    Gradually Varied Flow

In many situations the flow in an open channel is not of uniform depth ($y = $ constant) along the channel. This can occur because of several reasons: The bottom slope is not constant, the cross-sectional shape and area vary in the flow direction, or there is some obstruction across a portion of the channel. Such flows are classified as gradually varying flows if $dy/dx \ll 1$.

If the bottom slope and the energy line slope are not equal, the flow depth will vary along the channel, either increasing or decreasing in the flow direction. In such cases $dy/dx \neq 0$, $dV/dx \neq 0$, and the right-hand side of Eq. 10.10 is not zero. Physically, the difference between the

■ **Figure 10.11**    Best hydraulic cross sections for a rectangle, a 60° trapezoid, and a triangle.

component of weight and the shear forces in the direction of flow produces a change in the fluid momentum that requires a change in velocity and, from continuity considerations, a change in depth. Whether the depth increases or decreases depends on various parameters of the flow, with a variety of surface profile configurations [flow depth as a function of distance, $y = y(x)$] possible (Refs. 5 and 9).

## 10.6    Rapidly Varied Flow

In many open channels, flow depth changes occur over a relatively short distance so that $dy/dx \sim 1$. Such ***rapidly varied flow*** conditions are often quite complex and difficult to analyze in a precise fashion. Fortunately, many useful approximate results can be obtained by using a simple one-dimensional model along with appropriate experimentally determined coefficients when necessary. In this section we discuss several of these flows.

*In many cases the flow depth may change significantly in a short distance.*

Some rapidly varied flows occur in constant area channels for reasons that are not immediately obvious. The hydraulic jump is one such case. As is indicated in Fig. 10.12, the flow may change from a relatively shallow, high-speed condition into a relatively deep, low-speed condition within a horizontal distance of just a few channel depths. Other rapidly varied flows may be due to a sudden change in the channel geometry such as the flow in an expansion or contraction section of a channel as is indicated in Fig. 10.13.

In such situations the flow field is often two- or three-dimensional in character. There may be regions of flow separation, flow reversal, or unsteady oscillations of the free surface. For the purpose of some analyses, these complexities can be neglected and a simplified analysis can be undertaken. In other cases, however, it is the complex details of the flow that are the most important property of the flow; any analysis must include their effects. The scouring of a river bottom in the neighborhood of a bridge pier, as is indicated in Fig. 10.14, is such an example. A one- or two-dimensional model of this flow would not be sufficient to describe the complex structure of the flow that is responsible for the erosion near the foot of the bridge pier.

Many open-channel flow-measuring devices are based on principles associated with rapidly varied flows. Among these devices are broad-crested weirs, sharp-crested weirs, critical flow flumes, and sluice gates. The operation of such devices is discussed in the following sections.

▶ *V10.8 Erosion in a channel*

■ **Figure 10.12**  **Hydraulic jump.**

▶ *V10.9 Bridge pier scouring*

■ **Figure 10.13**  **Rapidly varied flow may occur in a channel transition section.**

■ **Figure 10.14** **The complex three-dimensional flow structure around a bridge pier.**

### 10.6.1 The Hydraulic Jump

Observations of flows in open channels show that under certain conditions it is possible that the fluid depth will change very rapidly over a short length of the channel without any change in the channel configuration. Such changes in depth can be approximated as a discontinuity in the free-surface elevation ($dy/dx = \infty$). For reasons discussed below, this step change in depth is always from a shallow to a deeper depth—always a step up, never a step down.

*A hydraulic jump is a steplike increase in fluid depth in an open channel.*

Physically, this near discontinuity, called a ***hydraulic jump***, may result when there is a conflict between the upstream and downstream influences that control a particular section (or reach) of a channel. For example, a sluice gate may require that the conditions at the upstream portion of the channel (downstream of the gate) be supercritical flow, while obstructions in the channel on the downstream end of the reach may require that the flow be subcritical. The hydraulic jump provides the mechanism (a nearly discontinuous one at that) to make the transition between the two types of flow.

The simplest type of hydraulic jump occurs in a horizontal, rectangular channel as is indicated in Fig. 10.15. Although the flow within the jump itself is extremely complex and agitated, it is reasonable to assume that the flow at sections (1) and (2) is nearly uniform, steady, and one-dimensional. In addition, we neglect any wall shear stresses, $\tau_w$, within the relatively short segment between these two sections. Under these conditions the $x$ component of the momentum equation (Eq. 5.22) for the control volume indicated can be written as

$$F_1 - F_2 = \rho Q(V_2 - V_1) = \rho V_1 y_1 b(V_2 - V_1)$$

where, as indicated by the figure in the margin, the pressure force at either section is hydrostatic. That is, $F_1 = p_{c1}A_1 = \gamma y_1^2 b/2$ and $F_2 = p_{c2}A_2 = \gamma y_2^2 b/2$, where $p_{c1} = \gamma y_1/2$ and $p_{c2} = \gamma y_2/2$ are the pressures at the centroids of the channel cross sections and $b$ is the channel width (see Sect. 2.8). Thus, the momentum equation becomes

$$\frac{y_1^2}{2} - \frac{y_2^2}{2} = \frac{V_1 y_1}{g}(V_2 - V_1) \tag{10.21}$$

In addition to the momentum equation, we have the conservation of mass equation (Eq. 5.12)

$$y_1 b V_1 = y_2 b V_2 = Q \tag{10.22}$$

■ **Figure 10.15** **Hydraulic jump geometry.**

and the energy equation (Eq. 5.84)

$$y_1 + \frac{V_1^2}{2g} = y_2 + \frac{V_2^2}{2g} + h_L \tag{10.23}$$

The head loss, $h_L$, in Eq. 10.23 is due to the violent turbulent mixing and dissipation that occur within the jump itself. We have neglected any head loss due to wall shear stresses.

Clearly Eqs. 10.21, 10.22, and 10.23 have a solution $y_1 = y_2$, $V_1 = V_2$, and $h_L = 0$. This represents the trivial case of no jump. Since these are nonlinear equations, it may be possible that more than one solution exists. The other solutions can be obtained as follows. By combining Eqs. 10.21 and 10.22 to eliminate $V_2$ we obtain

$$\frac{y_1^2}{2} - \frac{y_2^2}{2} = \frac{V_1 y_1}{g}\left(\frac{V_1 y_1}{y_2} - V_1\right) = \frac{V_1^2 y_1}{g y_2}(y_1 - y_2)$$

which can be simplified by factoring out a common nonzero factor $y_1 - y_2$ from each side to give

$$\left(\frac{y_2}{y_1}\right)^2 + \left(\frac{y_2}{y_1}\right) - 2\,\text{Fr}_1^2 = 0$$

where $\text{Fr}_1 = V_1/\sqrt{gy_1}$ is the upstream Froude number. By using the quadratic formula we obtain

$$\frac{y_2}{y_1} = \frac{1}{2}\left(-1 \pm \sqrt{1 + 8\text{Fr}_1^2}\right)$$

Clearly the solution with the minus sign is not possible (it would give a negative $y_2/y_1$). Thus,

$$\frac{y_2}{y_1} = \frac{1}{2}\left(-1 + \sqrt{1 + 8\text{Fr}_1^2}\right) \tag{10.24}$$

This depth ratio, $y_2/y_1$, across the hydraulic jump is shown as a function of the upstream Froude number in Fig. 10.16. The portion of the curve for $\text{Fr}_1 < 1$ is dashed in recognition of the fact that to have a hydraulic jump the flow must be supercritical. That is, the solution as given by Eq. 10.24 must be restricted to $\text{Fr}_1 \geq 1$, for which $y_2/y_1 \geq 1$. This can be shown by consideration of the energy equation, Eq. 10.23, as follows. The dimensionless head loss, $h_L/y_1$, can be obtained from Eq. 10.23 as

$$\frac{h_L}{y_1} = 1 - \frac{y_2}{y_1} + \frac{\text{Fr}_1^2}{2}\left[1 - \left(\frac{y_1}{y_2}\right)^2\right] \tag{10.25}$$

where, for given values of $\text{Fr}_1$, the values of $y_2/y_1$ are obtained from Eq. 10.24. As is indicated in Fig. 10.16, the head loss is negative if $\text{Fr}_1 < 1$. Since negative head losses violate the second law

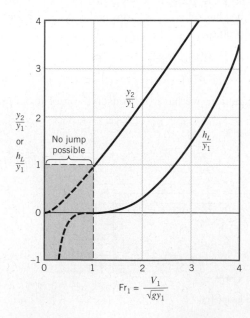

■ **Figure 10.16** Depth ratio and dimensionless head loss across a hydraulic jump as a function of upstream Froude number.

of thermodynamics (viscous effects dissipate energy, they cannot create energy; see Section 5.3), it is not possible to produce a hydraulic jump with $Fr_1 < 1$. The head loss across the jump is indicated by the lowering of the energy line shown in Fig. 10.15.

A flow must be supercritical (Froude number > 1) to produce the discontinuity called a hydraulic jump. This is analogous to the compressible flow ideas discussed in Chapter 11 in which it is shown that the flow of a gas must be supersonic (Mach number > 1) to produce the discontinuity called a normal shock wave. However, the fact that a flow is supercritical (or supersonic) does not guarantee the production of a hydraulic jump (or shock wave). The trivial solution $y_1 = y_2$ and $V_1 = V_2$ is also possible.

The fact that there is an energy loss across a hydraulic jump is useful in many situations. For example, the relatively large amount of energy contained in the fluid flowing down the spillway of a dam like that shown in the figure in the margin could cause damage to the channel below the dam. By placing suitable flow control objects in the channel downstream of the spillway, it is possible (if the flow is supercritical) to produce a hydraulic jump on the apron of the spillway and thereby dissipate a considerable portion of the energy of the flow. That is, the dam spillway produces supercritical flow, and the channel downstream of the dam requires subcritical flow. The resulting hydraulic jump provides the means to change the character of the flow.

(Photograph courtesy of U.S. Army Corps of Engineers.)

*Hydraulic jumps dissipate energy.*

---

# F l u i d s   i n   t h e   N e w s

**Grand Canyon rapids building** Virtually all of the rapids in the Grand Canyon were formed by rock debris carried into the Colorado River from side canyons. Severe storms wash large amounts of sediment into the river, building debris fans that narrow the river. This debris forms crude dams that back up the river to form quiet pools above the rapids. Water exiting the pool through the narrowed channel can reach supercritical conditions and produce *hydraulic jumps* downstream. Since the configuration of the jumps is a function of the flowrate, the difficulty in running the rapids can change from day to day. Also, rapids change over the years as debris is added to or removed from the rapids. For example, Crystal Rapid, one of the notorious rafting stretches of the river, changed very little between the first photos of 1890 and those of 1966. However, a debris flow from a severe winter storm in 1966 greatly constricted the river. Within a few minutes the configuration of Crystal Rapid was completely changed. The new, immature rapid was again drastically changed by a flood in 1983. While Crystal Rapid is now considered full grown, it will undoubtedly change again, perhaps in 100 or 1000 years. (See Problem 10.107.)

---

# $E$XAMPLE 10.8   Hydraulic Jump

**GIVEN**   Water on the horizontal apron of the 100-ft-wide spillway shown in Fig. E10.8a has a depth of 0.60 ft and a velocity of 18 ft/s.

**FIND**   Determine the depth, $y_2$, after the jump, the Froude numbers before and after the jump, $Fr_1$ and $Fr_2$, and the power dissipated, $\mathcal{P}_d$, within the jump.

## SOLUTION

Conditions across the jump are determined by the upstream Froude number

$$Fr_1 = \frac{V_1}{\sqrt{gy_1}} = \frac{18 \text{ ft/s}}{[(32.2 \text{ ft/s}^2)(0.60 \text{ ft})]^{1/2}} = 4.10 \quad \textbf{(Ans)}$$

Thus, the upstream flow is supercritical, and it is possible to generate a hydraulic jump as sketched.

From Eq. 10.24 we obtain the depth ratio across the jump as

$$\frac{y_2}{y_1} = \frac{1}{2}(-1 + \sqrt{1 + 8 \, Fr_1^2})$$

$$= \frac{1}{2}[-1 + \sqrt{1 + 8(4.10)^2}] = 5.32$$

or

$$y_2 = 5.32 \, (0.60 \text{ ft}) = 3.19 \text{ ft} \quad \textbf{(Ans)}$$

Since $Q_1 = Q_2$, or $V_2 = (y_1 V_1)/y_2 = 0.60 \text{ ft} (18 \text{ ft/s})/3.19 \text{ ft} = 3.39 \text{ ft/s}$, it follows that

$$Fr_2 = \frac{V_2}{\sqrt{gy_2}} = \frac{3.39 \text{ ft/s}}{[(32.2 \text{ ft/s}^2)(3.19 \text{ ft})]^{1/2}} = 0.334 \quad \textbf{(Ans)}$$

As is true for any hydraulic jump, the flow changes from supercritical to subcritical flow across the jump.

The power (energy per unit time) dissipated, $\mathcal{P}_d$, by viscous effects within the jump can be determined from the head loss as (see Eq. 5.85)

$$\mathcal{P}_d = \gamma Q h_L = \gamma b y_1 V_1 h_L \qquad (1)$$

where $h_L$ is obtained from Eq. 10.23 or 10.25 as

$$h_L = \left(y_1 + \frac{V_1^2}{2g}\right) - \left(y_2 + \frac{V_2^2}{2g}\right) = \left[0.60 \text{ ft} + \frac{(18.0 \text{ ft/s})^2}{2(32.2 \text{ ft/s}^2)}\right]$$
$$- \left[3.19 \text{ ft} + \frac{(3.39 \text{ ft/s})^2}{2(32.2 \text{ ft/s}^2)}\right]$$

or

$$h_L = 2.26 \text{ ft}$$

Thus, from Eq. 1,

$$\mathcal{P}_d = (62.4 \text{ lb/ft}^3)(100 \text{ ft})(0.60 \text{ ft})(18.0 \text{ ft/s})(2.26 \text{ ft})$$
$$= 1.52 \times 10^5 \text{ ft} \cdot \text{lb/s}$$

or

$$\mathcal{P}_d = \frac{1.52 \times 10^5 \text{ ft} \cdot \text{lb/s}}{550[(\text{ft} \cdot \text{lb/s})/\text{hp}]} = 277 \text{ hp} \qquad \textbf{(Ans)}$$

**COMMENTS** This power, which is dissipated within the highly turbulent motion of the jump, is converted into an increase in water temperature, $T$. That is, $T_2 > T_1$. Although the power dissipated is considerable, the difference in temperature is not great because the flowrate is quite large.

By repeating the calculations for the given flowrate $Q_1 = A_1 V_1 = b_1 y_1 V_1 = 100 \text{ ft } (0.6 \text{ ft})(18 \text{ ft/s}) = 1080 \text{ ft}^3/\text{s}$ but with various upstream depths, $y_1$, the results shown in Fig. E10.8b are obtained. Note that a slight change in water depth can produce a considerable change in energy dissipated. Also, if $y_1 > 1.54$ ft the flow is subcritical (Fr$_1 < 1$) and no hydraulic jump can occur.

The hydraulic jump flow process can be illustrated by use of the specific energy concept introduced in Section 10.3 as follows. Equation 10.23 can be written in terms of the specific energy, $E = y + V^2/2g$, as $E_1 = E_2 + h_L$, where $E_1 = y_1 + V_1^2/2g = 5.63$ ft and $E_2 = y_2 + V_2^2/2g = 3.37$ ft. As is discussed in Section 10.3, the specific energy diagram for this flow can be obtained by using $V = q/y$, where

$$q = q_1 = q_2 = \frac{Q}{b} = y_1 V_1 = 0.60 \text{ ft } (18.0 \text{ ft/s})$$
$$= 10.8 \text{ ft}^2/\text{s}$$

Thus,

$$E = y + \frac{q^2}{2gy^2} = y + \frac{(10.8 \text{ ft}^2/\text{s})^2}{2(32.2 \text{ ft/s}^2)y^2} = y + \frac{1.81}{y^2}$$

where $y$ and $E$ are in feet. The resulting specific energy diagram is shown in Fig. E10.8c. Because of the head loss across the

(b)

(c)

■ **Figure E10.8**

jump, the upstream and downstream values of $E$ are different. In going from state (1) to state (2), the fluid does not proceed along the specific energy curve and pass through the critical condition at state 2′. Rather, it jumps from (1) to (2), as is represented by the dashed line in the figure. From a one-dimensional consideration, the jump is a discontinuity. In actuality, the jump is a complex three-dimensional flow incapable of being represented on the one-dimensional specific energy diagram.

The actual structure of a hydraulic jump is a complex function of Fr$_1$, even though the depth ratio and head loss are given quite accurately by a simple one-dimensional flow analysis (Eqs. 10.24 and 10.25). A detailed investigation of the flow indicates that there are essentially five types of surface and jump conditions. The classification of these jumps is indicated in Table 10.2, along with sketches of the structure of the jump. For flows that are barely supercritical, the jump is more like a standing wave, without a nearly step change in depth. In some Froude number ranges the jump is

**■ Table 10.2**
**Classification of Hydraulic Jumps (Ref. 12)**

| Fr$_1$ | y$_2$/y$_1$ | Classification | Sketch |
|---|---|---|---|
| <1 | 1 | Jump impossible | |
| 1 to 1.7 | 1 to 2.0 | Standing wave or undulant jump | |
| 1.7 to 2.5 | 2.0 to 3.1 | Weak jump | |
| 2.5 to 4.5 | 3.1 to 5.9 | Oscillating jump | |
| 4.5 to 9.0 | 5.9 to 12 | Stable, well-balanced steady jump; insensitive to downstream conditions | |
| >9.0 | >12 | Rough, somewhat intermittent strong jump | |

*The actual structure of a hydraulic jump depends on the Froude number.*

*V10.12 Hydraulic jump in a sink*

unsteady, with regular periodic oscillations traveling downstream. (Recall that the wave cannot travel upstream against the supercritical flow.)

The length of a hydraulic jump (the distance between the nearly uniform upstream and downstream flows) may be of importance in the design of channels. Although its value cannot be determined theoretically, experimental results indicate that over a wide range of Froude numbers, the jump is approximately seven downstream depths long (Ref. 5).

Hydraulic jumps can occur in a variety of channel flow configurations, not just in horizontal, rectangular channels as discussed above. Jumps in nonrectangular channels (i.e., circular pipes, trapezoidal canals) behave in a manner quite like those in rectangular channels, although the details of the depth ratio and head loss are somewhat different from jumps in rectangular channels.

Other common types of hydraulic jumps include those that occur in sloping channels, as is indicated in Fig. 10.17$a$ and the submerged hydraulic jumps that can occur just downstream of a

**■ Figure 10.17   Hydraulic jump variations:** ($a$) **jump caused by a change in channel slope,** ($b$) **submerged jump.**

**Figure 10.18** **Sharp-crested weir geometry.**

sluice gate, as is indicated in Fig. 10.17b. Details of these and other jumps can be found in standard open-channel flow references (Refs. 3 and 5).

### 10.6.2 Sharp-Crested Weirs

A weir is an obstruction on a channel bottom over which the fluid must flow. It provides a convenient method of determining the flowrate in an open channel in terms of a single depth measurement. A *sharp-crested weir* is essentially a vertical sharp-edged flat plate placed across the channel in a way such that the fluid must flow across the sharp edge and drop into the pool downstream of the weir plate, as is shown in Fig. 10.18. The specific shape of the flow area in the plane of the weir plate is used to designate the type of weir. Typical shapes include the rectangular weir, the triangular weir, and the trapezoidal weir, as indicated in Fig. 10.19.

> *A sharp-crested weir can be used to determine the flowrate.*

The complex nature of the flow over a weir makes it impossible to obtain precise analytical expressions for the flow as a function of other parameters, such as the weir height, $P_w$, **weir head,** $H$, the fluid depth upstream, and the geometry of the weir plate (angle $\theta$ for triangular weirs or aspect ratio, $b/H$, for rectangular weirs). The flow structure is far from one-dimensional, with a variety of interesting flow phenomena obtained.

The main mechanisms governing flow over a weir are gravity and inertia. From a highly simplified point of view, gravity accelerates the fluid from its free-surface elevation upstream of the weir to larger velocity as it flows down the hill formed by the nappe. Although viscous and surface tension effects are usually of secondary importance, such effects cannot be entirely neglected. Generally, appropriate experimentally determined coefficients are used to account for these effects.

As a first approximation, we assume that the velocity profile upstream of the weir plate is uniform and that the pressure within the nappe is atmospheric. In addition, we assume that the fluid flows horizontally over the weir plate with a nonuniform velocity profile, as indicated in Fig. 10.20. With $p_B = 0$ the Bernoulli equation for flow along the arbitrary streamline $A-B$ indicated can be written as

$$\frac{p_A}{\gamma} + \frac{V_1^2}{2g} + z_A = (H + P_w - h) + \frac{u_2^2}{2g} \tag{10.26}$$

■ **Figure 10.19** **Sharp-crested weir plate geometry:** (*a*) **rectangular,** (*b*) **triangular,** (*c*) **trapezoidal.**

■ **Figure 10.20**   Assumed flow structure over a weir.

where $h$ is the distance that point $B$ is below the free surface. We do not know the location of point $A$ from which came the fluid that passes over the weir at point $B$. However, since the total head for any particle along the vertical section (1) is the same, $z_A + p_A/\gamma + V_1^2/2g = H + P_w + V_1^2/2g$, the specific location of $A$ (i.e., $A$ or $A'$ shown in the figure in the margin) is not needed, and the velocity of the fluid over the weir plate is obtained from Eq. 10.26 as

$$u_2 = \sqrt{2g\left(h + \frac{V_1^2}{2g}\right)}$$

The flowrate can be calculated from

$$Q = \int_{(2)} u_2 \, dA = \int_{h=0}^{h=H} u_2 \ell \, dh \tag{10.27}$$

where $\ell = \ell(h)$ is the cross-channel width of a strip of the weir area, as is indicated in Fig. 10.20$b$. For a rectangular weir, $\ell$ is constant. For other weirs, such as triangular or circular weirs, the value of $\ell$ is known as a function of $h$.

For a rectangular weir, $\ell = b$, and the flowrate becomes

$$Q = \sqrt{2g}\, b \int_0^H \left(h + \frac{V_1^2}{2g}\right)^{1/2} dh$$

or

$$Q = \frac{2}{3}\sqrt{2g}\, b \left[\left(H + \frac{V_1^2}{2g}\right)^{3/2} - \left(\frac{V_1^2}{2g}\right)^{3/2}\right] \tag{10.28}$$

Equation 10.28 is a rather cumbersome expression that can be simplified by using the fact that with $P_w \gg H$ (as often happens in practical situations) the upstream velocity is negligibly small. That is, $V_1^2/2g \ll H$ and Eq. 10.28 simplifies to the basic rectangular weir equation

$$Q = \tfrac{2}{3}\sqrt{2g}\, b H^{3/2} \tag{10.29}$$

Note that the weir head, $H$, is the height of the upstream free surface above the crest of the weir. As is indicated in Fig. 10.18, because of the drawdown effect, $H$ is not the distance of the free surface above the weir crest as measured directly above the weir plate.

*A weir coefficient is used to account for nonideal conditions excluded in the simplified analysis.*

Because of the numerous approximations made to obtain Eq. 10.29, it is not unexpected that an experimentally determined correction factor must be used to obtain the actual flowrate as a function of weir head. Thus, the final form is

$$Q = C_{\text{wr}}\tfrac{2}{3}\sqrt{2g}\, b H^{3/2} \tag{10.30}$$

where $C_{wr}$ is the rectangular weir coefficient. From dimensional analysis arguments, it is expected that $C_{wr}$ is a function of Reynolds number (viscous effects), Weber number (surface tension effects), and $H/P_w$ (geometry). In most practical situations, the Reynolds and Weber number effects are negligible, and the following correlation, shown in the figure in the margin, can be used (Refs. 4, 7):

$$C_{wr} = 0.611 + 0.075\left(\frac{H}{P_w}\right) \tag{10.31}$$

More precise values of $C_{wr}$ can be found in the literature, if needed (Refs. 3 and 14).

The triangular sharp-crested weir is often used for flow measurements, particularly for measuring flowrates over a wide range of values. For small flowrates, the head, $H$, for a rectangular weir would be very small and the flowrate could not be measured accurately. However, with the triangular weir, the flow width decreases as $H$ decreases so that even for small flowrates, reasonable heads are developed. Accurate results can be obtained over a wide range of $Q$.

*V10.13 Triangular weir*

The triangular weir equation can be obtained from Eq. 10.27 by using

$$\ell = 2(H - h)\tan\left(\frac{\theta}{2}\right)$$

where $\theta$ is the angle of the V-notch (see Figs. 10.19 and 10.20). After carrying out the integration and again neglecting the upstream velocity ($V_1^2/2g \ll H$), we obtain

$$Q = \frac{8}{15}\tan\left(\frac{\theta}{2}\right)\sqrt{2g}\,H^{5/2}$$

An experimentally determined triangular weir coefficient, $C_{wt}$, is used to account for the real-world effects neglected in the analysis so that

$$Q = C_{wt}\frac{8}{15}\tan\left(\frac{\theta}{2}\right)\sqrt{2g}\,H^{5/2} \tag{10.32}$$

*V10.14 Low-head dam*

Typical values of $C_{wt}$ for triangular weirs are in the range of 0.58 to 0.62, as is shown in Fig. 10.21. Note that although $C_{wt}$ and $\theta$ are dimensionless, the value of $C_{wt}$ is given as a function of the weir head, $H$, which is a dimensional quantity. Although using dimensional parameters is not recommended (see the dimensional analysis discussion in Chapter 7), such parameters are often used for open-channel flow.

■ **Figure 10.21** Weir coefficient for triangular sharp-crested weirs (Ref. 10, with permission from ASCE).

■ **Figure 10.22**   Flow conditions over a weir without a free nappe: (*a*) plunging nappe, (*b*) submerged nappe.

The above results for sharp-crested weirs are valid provided the area under the nappe is ventilated to atmospheric pressure. Although this is not a problem for triangular weirs, for rectangular weirs it is sometimes necessary to provide ventilation tubes to ensure atmospheric pressure in this region. In addition, depending on downstream conditions, it is possible to obtain the submerged weir operation, as is indicated in Fig. 10.22. Clearly the flowrate will be different for these situations from that given by Eqs. 10.30 and 10.32.

### 10.6.3   Broad-Crested Weirs

$H/L_w = 0.08$

$H/L_w = 0.50$

A *broad-crested weir* is a structure in an open channel that has a horizontal crest above which the fluid pressure may be considered hydrostatic. A typical configuration is shown in Fig. 10.23. Generally, to ensure proper operation, these weirs are restricted to the range $0.08 < H/L_w < 0.50$. These conditions are drawn to scale in the figure in the margin. For long weir blocks ($H/L_w$ less than 0.08), head losses across the weir cannot be neglected. On the other hand, for short weir blocks ($H/L_w$ greater than 0.50), the streamlines of the flow over the weir block are not horizontal. Although broad-crested weirs can be used in channels of any cross-sectional shape, we restrict our attention to rectangular channels.

The operation of a broad-crested weir is based on the fact that nearly uniform critical flow is achieved in the short reach above the weir block. (If $H/L_w < 0.08$, viscous effects are important, and the flow is subcritical over the weir.) If the kinetic energy of the upstream flow is negligible, then $V_1^2/2g \ll y_1$ and the upstream specific energy is $E_1 = V_1^2/2g + y_1 \approx y_1$. Observations show that as the flow passes over the weir block, it accelerates and reaches critical conditions, $y_2 = y_c$ and $\mathrm{Fr}_2 = 1$ (i.e., $V_2 = c_2$), corresponding to the nose of the specific energy curve (see Fig. 10.7). The flow does not accelerate to supercritical conditions ($\mathrm{Fr}_2 > 1$). To do so would require the ability of the downstream fluid to communicate with the upstream fluid to let it know that there is an end of the weir block. Since waves cannot propagate upstream against a critical flow, this information cannot be transmitted. The flow remains critical, not supercritical, across the weir block.

The Bernoulli equation can be applied between point (1) upstream of the weir and point (2) over the weir where the flow is critical to obtain

$$H + P_w + \frac{V_1^2}{2g} = y_c + P_w + \frac{V_c^2}{2g}$$

or, if the upstream velocity head is negligible

$$H - y_c = \frac{(V_c^2 - V_1^2)}{2g} = \frac{V_c^2}{2g}$$

■ **Figure 10.23**   Broad-crested weir geometry.

However, since $V_2 = V_c = (gy_c)^{1/2}$, we find that $V_c^2 = gy_c$ so that we obtain

$$H - y_c = \frac{y_c}{2}$$

or

$$y_c = \frac{2H}{3}$$

Thus, the flowrate is

$$Q = by_2 V_2 = by_c V_c = by_c(gy_c)^{1/2} = b \sqrt{g}\, y_c^{3/2}$$

or

$$Q = b \sqrt{g} \left(\frac{2}{3}\right)^{3/2} H^{3/2}$$

Again an empirical weir coefficient is used to account for the various real-world effects not included in the above simplified analysis. That is

$$Q = C_{wb}\, b \sqrt{g} \left(\frac{2}{3}\right)^{3/2} H^{3/2} \qquad (10.33)$$

where approximate values of $C_{wb}$, the broad-crested weir coefficient shown in the figure in the margin, can be obtained from the equation (Ref. 6)

$$C_{wb} = 1.125 \left(\frac{1 + H/P_w}{2 + H/P_w}\right)^{1/2} \qquad (10.34)$$

## EXAMPLE 10.9  Sharp-Crested and Broad-Crested Weirs

**GIVEN** Water flows in a rectangular channel of width $b = 2$ m with flowrates between $Q_{min} = 0.02$ m³/s and $Q_{max} = 0.60$ m³/s. This flowrate is to be measured by using either (a) a rectangular sharp-crested weir, (b) a triangular sharp-crested weir with $\theta = 90°$, or (c) a broad-crested weir. In all cases the bottom of the flow area over the weir is a distance $P_w = 1$ m above the channel bottom.

**FIND** Plot a graph of $Q = Q(H)$ for each weir and comment on which weir would be best for this application.

### SOLUTION

**(a)** For the rectangular weir with $P_w = 1$ m, Eqs. 10.30 and 10.31 give

$$Q = C_{wr} \frac{2}{3} \sqrt{2g}\, bH^{3/2}$$

$$= \left(0.611 + 0.075 \frac{H}{P_w}\right) \frac{2}{3} \sqrt{2g}\, bH^{3/2}$$

Thus,

$$Q = (0.611 + 0.075H) \frac{2}{3} \sqrt{2(9.81 \text{ m/s}^2)}\, (2 \text{ m})\, H^{3/2}$$

or

$$Q = 5.91(0.611 + 0.075H)H^{3/2} \qquad (1)$$

where $H$ and $Q$ are in meters and m³/s, respectively. The results from Eq. 1 are plotted in Fig. E10.9.

■ **Figure E10.9**

**(b)**    Similarly, for the triangular weir, Eq. 10.32 gives

$$Q = C_{wt} \frac{8}{15} \tan\left(\frac{\theta}{2}\right) \sqrt{2g} \ H^{5/2}$$

$$= C_{wt} \frac{8}{15} \tan(45°) \sqrt{2(9.81 \ \text{m/s}^2)} \ H^{5/2}$$

or

$$Q = 2.36 C_{wt} \ H^{5/2} \qquad (2)$$

where $H$ and $Q$ are in meters and m³/s and $C_{wt}$ is obtained from Fig. 10.21. For example, with $H = 0.20$ m $= 0.656$ ft, we find $C_{wt} = 0.584$, or $Q = 2.36 \ (0.584)(0.20)^{5/2} = 0.0247$ m³/s. The triangular weir results are also plotted in Fig. E10.9.

**(c)**    For the broad-crested weir, Eqs. 10.28 and 10.29 give

$$Q = C_{wb} \, b \sqrt{g} \left(\frac{2}{3}\right)^{3/2} H^{3/2}$$

$$= 1.125 \left(\frac{1 + H/P_w}{2 + H/P_w}\right)^{1/2} b\sqrt{g} \left(\frac{2}{3}\right)^{3/2} H^{3/2}$$

Thus, with $P_w = 1$ m

$$Q = 1.125 \left(\frac{1 + H}{2 + H}\right)^{1/2} (2 \ \text{m}) \ \sqrt{9.81 \ \text{m/s}^2} \left(\frac{2}{3}\right)^{3/2} H^{3/2}$$

or

$$Q = 3.84 \left(\frac{1 + H}{2 + H}\right)^{1/2} H^{3/2} \qquad (3)$$

where, again, $H$ and $Q$ are in meters and m³/s. This result is also plotted in Fig. E10.9.

**COMMENTS**    Although it appears as though any of the three weirs would work well for the upper portion of the flowrate range, neither the rectangular nor the broad-crested weir would be very accurate for small flowrates near $Q = Q_{min}$ because of the small head, $H$, at these conditions. The triangular weir, however, would allow reasonably large values of $H$ at the lowest flowrates. The corresponding heads with $Q = Q_{min} = 0.02$ m³/s for rectangular, triangular, and broad-crested weirs are 0.0312, 0.182, and 0.0375 m, respectively.

In addition, as discussed in this section, for proper operation the broad-crested weir geometry is restricted to $0.08 < H/L_w < 0.50$, where $L_w$ is the weir block length. From Eq. 3 with $Q_{max} = 0.60$ m³/s, we obtain $H_{max} = 0.349$. Thus, we must have $L_w > H_{max}/0.5 = 0.698$ m to maintain proper critical flow conditions at the largest flowrate in the channel. However, with $Q = Q_{min} = 0.02$ m³/s, we obtain $H_{min} = 0.0375$ m. Thus, we must have $L_w < H_{min}/0.08 = 0.469$ m to ensure that frictional effects are not important. Clearly, these two constraints on the geometry of the weir block, $L_w$, are incompatible.

A broad-crested weir will not function properly under the wide range of flowrates considered in this example. The sharp-crested triangular weir would be the best of the three types considered, provided the channel can handle the $H_{max} = 0.719$-m head.

### 10.6.4 Underflow Gates

(Photograph courtesy of Pend Oreille Public Utility District.)

A variety of **underflow gate** structures is available for flowrate control at the crest of an overflow spillway (as shown by the figure in the margin), or at the entrance of an irrigation canal or river from a lake. Three types are illustrated in Fig. 10.24. Each has certain advantages and disadvantages in terms of costs of construction, ease of use, and the like, although the basic fluid mechanics involved are the same in all instances.

The flow under a gate is said to be free outflow when the fluid issues as a jet of supercritical flow with a free surface open to the atmosphere as shown in Fig. 10.24. In such cases it is customary to write this flowrate as the product of the distance, $a$, between the channel bottom and the bottom of the gate times the convenient reference velocity $(2gy_1)^{1/2}$. That is,

$$q = C_d a \sqrt{2gy_1} \qquad (10.35)$$

where $q$ is the flowrate per unit width. The discharge coefficient, $C_d$, is a function of the contraction coefficient, $C_c = y_2/a$, and the depth ratio $y_1/a$. Typical values of the discharge coefficient for free

V10.15 Spillway gate

(a)    (b)    (c)

■ **Figure 10.24**    Three variations of underflow gates: (*a*) vertical gate, (*b*) radial gate, (*c*) drum gate.

■ **Figure 10.25** Typical discharge coefficients for underflow gates (Ref. 3).

■ **Figure 10.26** Drowned outflow from a sluice gate.

*The flowrate from an underflow gate depends on whether the outlet is free or drowned.*

V10.16 Unsteady under and over

outflow (or free discharge) from a vertical sluice gate are on the order of 0.55 to 0.60 as indicated by the top line in Fig. 10.25 (Ref. 3).

As indicated in Fig. 10.26, in certain situations the depth downstream of the gate is controlled by some downstream obstacle and the jet of water issuing from under the gate is overlaid by a mass of water that is quite turbulent.

The flowrate for a submerged (or drowned) gate can be obtained from the same equation that is used for free outflow (Eq. 10.35), provided the discharge coefficient is modified appropriately. Typical values of $C_d$ for drowned outflow cases are indicated as the series of lower curves in Fig. 10.25. Consider flow for a given gate and upstream conditions (i.e., given $y_1/a$) corresponding to a vertical line in the figure. With $y_3/a = y_1/a$ (i.e., $y_3 = y_1$) there is no head to drive the flow so that $C_d = 0$ and the fluid is stationary. For a given upstream depth ($y_1/a$ fixed), the value of $C_d$ increases with decreasing $y_3/a$ until the maximum value of $C_d$ is reached. This maximum corresponds to the free discharge conditions and is represented by the free outflow line so labeled in Fig. 10.25. For values of $y_3/a$ that give $C_d$ values between zero and its maximum, the jet from the gate is overlaid (drowned) by the downstream water and the flowrate is therefore reduced when compared with a free discharge situation. Similar results are obtained for the radial gate and drum gate.

# $E$XAMPLE 10.10   Sluice Gate

**GIVEN** Water flows under the sluice gate shown in Fig. E10.10. The channel width is $b = 20$ ft, the upstream depth is $y_1 = 6$ ft, and the gate is $a = 1.0$ ft off the channel bottom.

**FIND** Plot a graph of flowrate, $Q$, as a function of $y_3$.

## SOLUTION

From Eq. 10.35 we have

$$Q = bq = baC_d \sqrt{2gy_1}$$
$$= 20 \text{ ft } (1.0 \text{ ft}) \, C_d \sqrt{2(32.2 \text{ ft/s}^2)(6.0 \text{ ft})}$$

or

$$Q = 393C_d \text{ cfs} \qquad (1)$$

The value of $C_d$ is obtained from Fig. 10.25 along the vertical line $y_1/a = 6 \text{ ft}/1 \text{ ft} = 6$. For $y_3 = 6 \text{ ft}$ (i.e., $y_3/a = 6 = y_1/a$) we obtain $C_d = 0$, indicating that there is no flow when there is no head difference across the gate. The value of $C_d$ increases as $y_3/a$ decreases, reaching a maximum of $C_d = 0.56$ when $y_3/a = 3.2$. Thus, with $y_3 = 3.2a = 3.2 \text{ ft}$

$$Q = 393 (0.56) \text{ cfs} = 220 \text{ cfs}$$

The flowrate for $3.2 \text{ ft} \le y_3 \le 6 \text{ ft}$ is obtained from Eq. 1 and the $C_d$ values of Fig. 10.24 with the results as indicated in Fig. E10.10.

**Figure E10.10**

**COMMENT**  For $y_3 < 3.2 \text{ ft}$ the flowrate is independent of $y_3$, and the outflow is a free (not submerged) outflow. For such cases the inertia of the water flowing under the gate is sufficient to produce free outflow even with $y_3 > a$.

## 10.7    Chapter Summary and Study Guide

*open-channel flow*
*Froude number*
*critical flow*
*subcritical flow*
*supercritical flow*
*wave speed*
*specific energy*
*specific energy diagram*
*uniform depth flow*
*wetted perimeter*
*hydraulic radius*
*Chezy equation*
*Manning equation*
*Manning coefficient*
*rapidly varied flow*
*hydraulic jump*
*sharp-crested weir*
*weir head*
*broad-crested weir*
*underflow gate*

This chapter discussed various aspects of flows in an open channel. A typical open-channel flow is driven by the component of gravity in the direction of flow. The character of such flows can be a strong function of the Froude number, which is the ratio of the fluid speed to the free-surface wave speed. The specific energy diagram is used to provide insight into the flow processes involved in open-channel flow.

Uniform depth channel flow is achieved by a balance between the potential energy lost by the fluid as it coasts downhill and the energy dissipated by viscous effects. Alternately, it represents a balance between weight and friction forces. The relationship among the flowrate, the slope of the channel, the geometry of the channel, and the roughness of the channel surfaces is given by the Manning equation. Values of the Manning coefficient used in the Manning equation are dependent on the surface material roughness.

The hydraulic jump is an example of nonuniform depth open-channel flow. If the Froude number of a flow is greater than 1, the flow is supercritical, and a hydraulic jump may occur. The momentum and mass equations are used to obtain the relationship between the upstream Froude number and the depth ratio across the jump. The energy dissipated in the jump and the head loss can then be determined by use of the energy equation.

The use of weirs to measure the flowrate in an open channel is discussed. The relationships between the flowrate and the weir head are given for both sharp-crested and broad-crested weirs.

The following checklist provides a study guide for this chapter. When your study of the entire chapter and end-of-chapter exercises has been completed, you should be able to

- write out meanings of the terms listed here in the margin and understand each of the related concepts. These terms are particularly important and are set in ***italic, bold, and color*** type in the text.
- determine the Froude number for a given flow and explain the concepts of subcritical, critical, and supercritical flows.

- plot and interpret the specific energy diagram for a given flow.
- use the Manning equation to analyze uniform depth flow in an open channel.
- calculate properties such as the depth ratio and the head loss for a hydraulic jump.
- determine the flowrates over sharp-crested weirs, broad-crested weirs, and under underflow gates.

Some of the important equations in this chapter are:

| | | |
|---|---|---|
| Froude number | $\text{Fr} = V/(gy)^{1/2}$ | |
| Wave speed | $c = \sqrt{gy}$ | (10.3) |
| Specific energy | $E = y + \dfrac{V^2}{2g}$ | (10.8) |
| Manning equation | $V = \dfrac{\kappa}{n} R_h^{2/3} S_0^{1/2}$ | (10.19) |
| Hydraulic jump depth ratio | $\dfrac{y_2}{y_1} = \dfrac{1}{2}\left(-1 + \sqrt{1 + 8\text{Fr}_1^2}\right)$ | (10.24) |
| Hydraulic jump head loss | $\dfrac{h_L}{y_1} = 1 - \dfrac{y_2}{y_1} + \dfrac{\text{Fr}_1^2}{2}\left[1 - \left(\dfrac{y_1}{y_2}\right)^2\right]$ | (10.25) |
| Rectangular sharp-crested weir | $Q = C_{wr}\dfrac{2}{3}\sqrt{2g}\, b\, H^{3/2}$ | (10.30) |
| Triangular sharp-crested weir | $Q = C_{wt}\dfrac{8}{15}\tan\left(\dfrac{\theta}{2}\right)\sqrt{2g}\, H^{5/2}$ | (10.32) |
| Broad-crested weir | $Q = C_{wb}\, b\, \sqrt{g}\left(\dfrac{2}{3}\right)^{3/2} H^{3/2}$ | (10.33) |
| Underflow gate | $q = C_d a\sqrt{2gy_1}$ | (10.35) |

# References

1. Currie, C. G., and Currie, I. G., *Fundamental Mechanics of Fluids*, 3rd Ed., Marcel Dekker, New York, 2003.
2. Stoker, J. J., *Water Waves*, Interscience, New York, 1957.
3. Henderson, F. M., *Open Channel Flow*, Macmillan, New York, 1966.
4. Rouse, H., *Elementary Fluid Mechanics*, Wiley, New York, 1946.
5. French, R. H., *Open Channel Hydraulics*, McGraw-Hill, New York, 1992.
6. Chow, V. T., *Open Channel Hydraulics*, McGraw-Hill, New York, 1959.
7. Blevins, R. D., *Applied Fluid Dynamics Handbook*, Van Nostrand Reinhold, New York, 1984.
8. Daugherty, R. L., and Franzini, J. B., *Fluid Mechanics with Engineering Applications*, McGraw-Hill, New York, 1977.
9. Vennard, J. K., and Street, R. L., *Elementary Fluid Mechanics*, 7th Ed., Wiley, New York, 1995.
10. Lenz, A. T., "Viscosity and Surface Tension Effects on V-Notch Weir Coefficients," *Transactions of the American Society of Civil Engineers*, Vol. 108, 759–802, 1943.
11. White, F. M., *Fluid Mechanics*, 5th Ed., McGraw-Hill, New York, 2003.
12. U.S. Bureau of Reclamation, Research Studies on Stilling Basins, Energy Dissipators, and Associated Appurtenances, Hydraulic Lab Report Hyd.-399, June 1, 1955.
13. Wallet, A., and Ruellan, F., *Houille Blanche*, Vol. 5, 1950.
14. Spitzer, D. W., ed., *Flow Measurement: Practical Guides for Measurement and Control*, Instrument Society of America, Research Triangle Park, N.C., 1991.

Problem available in *WileyPLUS* at instructor's discretion.

 Tutoring problem available in *WileyPLUS* at instructor's discretion.

▶ Problem is related to a chapter video available in *WileyPLUS*.

* Problem to be solved with aid of programmable calculator or computer.

† Open-ended problem that requires critical thinking. These problems require various assumptions to provide the necessary input data. There are not unique answers to these problems.

# Review Problems

Go to Appendix G (*WileyPLUS* or the book's web site, www.wiley.com/college/munson) for a set of review problems with answers. Detailed solutions can be found in the *Student Solution Manual and Study Guide for Fundamentals of Fluid Mechanics*, by Munson et al. (© 2013 John Wiley and Sons, Inc.).

# Conceptual Questions

**10.1C** When water flows steadily in a channel that is sloped downward,

a) the flow depth will decrease only if the slope is reduced.

b) the flow depth could increase, decrease, or stay the same even if the slope stays constant.

c) the flow depth remains uniform unless the slope of the bottom changes.

d) the flow depth will increase only if the slope is reduced.

**10.2C** A surface wave produced by a disturbance can travel upstream,

a) when the flow is subcritical.

b) when the flow is critical.

c) when the flow is either subcritical or supercritical, depending on the Froude number.

d) when the flow is supercritical.

**10.3C** For a hydraulic jump to occur when water flows down a channel.

a) the upstream flow must be critical.

b) the upstream flow must be subcritical.

c) the upstream flow must be supercritical.

d) the Froude number must be unity.

**10.4C** When a hydraulic jump occurs in a channel,

a) the energy of the flow upstream and that downstream of the jump are equal.

b) the depth of the flow upstream and that downstream of the jump are equal.

c) the momentum of the flow upstream and that downstream of the jump are equal.

d) the Froude number of the flow upstream and that downstream of the jump are equal.

**Additional conceptual questions are available in *WileyPLUS* at the instructor's discretion.**

# Problems

**Note: Unless specific values of required fluid properties are given in the problem statement, use the values found in the tables on the inside of the front cover. Answers to the even-numbered problems are listed at the end of the book. The Lab Problems as well as the videos that accompany problems can be accessed in *WileyPLUS* or the book's web site, www.wiley.com/college/munson.**

### Section 10.2 Surface Waves

**10.1** On a distant planet small-amplitude waves travel across a 1-m-deep pond with a speed of 5 m/s. Determine the acceleration of gravity on the surface of that planet.

**10.2** The flowrate in a 50-ft-wide, 2-ft-deep river is $Q = 190$ cfs. Is the flow subcritical or supercritical?

**10.3** The flowrate per unit width in a wide channel is $q = 2.3$ m²/s. Is the flow subcritical or supercritical if the depth is **(a)** 0.2 m, **(b)** 0.8 m, or **(c)** 2.5 m?

**10.4** A rectangular channel 3 m wide carries 10 m³/s at a depth of 2 m. Is the flow subcritical or supercritical? For the same flowrate, what depth will give critical flow?

**10.5** Determine the critical depth for a flow of 200 m³/s through a rectangular channel of 10-m width. If the water flows 3.8-m deep, is the flow supercritical? Explain.

**10.6** ▶ Consider waves made by dropping objects (one after another from a fixed location) into a stream of depth $y$ that is moving with speed $V$ as shown in Fig. P10.6 (see **Video V10.5**). The circular wave crests that are produced travel with speed $c = (gy)^{1/2}$ relative to the moving water. Thus, as the circular waves are washed downstream, their diameters increase and the center of each circle is fixed relative to the moving water. **(a)** Show that if the flow is supercritical, lines tangent to the waves generate a

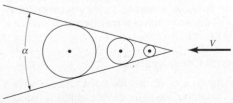

■ **Figure P10.6**

wedge of half-angle $\alpha/2 = \arcsin(1/Fr)$, where $Fr = V/(gy)^{1/2}$ is the Froude number. **(b)** Discuss what happens to the wave pattern when the flow is subcritical, $Fr < 1$.

**10.7** (WILEY) Waves on the surface of a tank are observed to travel at a speed of 2 m/s. How fast would these waves travel if **(a)** the tank were in an elevator accelerating downward at a rate of 4 m/s², **(b)** the tank accelerates horizontally at a rate of 9.81 m/s², **(c)** the tank were aboard the orbiting Space Shuttle? Explain.

**10.8** (WILEY) In flowing from section (1) to section (2) along an open channel, the water depth decreases by a factor of 2 and the Froude number changes from a subcritical value of 0.5 to a supercritical value of 3.0. Determine the channel width at (2) if it is 12 ft wide at (1).

**10.9** Water flows in a canal at a depth of 2.8 ft and a velocity of 5.3 ft/s. Will waves produced by throwing a stick into the canal travel both upstream and downstream, or will they all be washed downstream? Explain.

**10.10** A trout jumps, producing waves on the surface of a 0.8-m-deep mountain stream. If it is observed that the waves do not travel upstream, what is the minimum velocity of the current?

**10.11** (WILEY) Observations at a shallow sandy beach show that even though the waves several hundred yards out from the shore are not parallel to the beach, the waves often "break" on the beach nearly parallel to the shore as indicated in Fig. P10.11. Explain this behavior based on the wave speed $c = (gy)^{1/2}$.

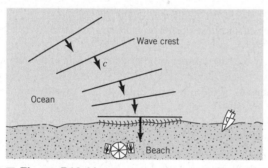

**■ Figure P10.11**

**10.12** (WILEY) (See Fluids in the News article titled "Tsunami, the Nonstorm Wave," Section 10.2.1.) Often when an earthquake shifts a segment of the ocean floor, a relatively small-amplitude wave of very long wavelength is produced. Such waves go unnoticed as they move across the open ocean; only when they approach the shore do they become dangerous (e.g., a tsunami). Determine the wave speed if the wavelength, $\lambda$, is 6000 ft and the ocean depth is 15,000 ft.

**10.13** (WILEY) Determine the minimum depth in a 3-m-wide rectangular channel if the flow is to be subcritical with a flowrate of $Q = 60$ m³/s.

**10.14** (WILEY) (See Fluids in the News article titled "Tsunami, the Nonstorm Wave," Section 10.2.1.) An earthquake causes a shift in the ocean floor that produces a tsunami with a wavelength of 100 km. How fast will this wave travel across the ocean surface if the ocean depth is 3000 m?

## Section 10.3 Energy Considerations

**10.15** (WILEY) Water flows in a 10-m-wide open channel with a flowrate of 5 m³/s. Determine the two possible depths if the specific energy of the flow is $E = 0.6$ m.

**10.16** (WILEY) Water flows in a rectangular channel with a flowrate per unit width of $q = 2.5$ m²/s. Plot the specific energy diagram for this flow. Determine the two possible depths of flow if $E = 2.5$ m.

**\*10.17** Water flows in a rectangular channel with a specific energy of $E = 5$ ft. If the flowrate per unit width is $q = 30$ ft²/s, determine the two possible flow depths and the corresponding Froude numbers. Plot the specific energy diagram for this flow. Repeat the problem for $E = 1, 2, 3$, and 4 ft.

**10.18** (WILEY) Water flows in a 10-ft-wide rectangular channel with a flowrate of 200 ft³/s. Plot the specific energy diagram for this flow. Determine the two possible flowrates when the specific energy is 6 ft.

**10.19** (WILEY GO) Water flows in a rectangular channel at a rate of $q = 20$ cfs/ft. When a Pitot tube is placed in the stream, water in the tube rises to a level of 4.5 ft above the channel bottom. Determine the two possible flow depths in the channel. Illustrate this flow on a specific energy diagram.

**10.20** (WILEY) Water flows in a 5-ft-wide rectangular channel with a flowrate of $Q = 30$ ft³/s and an upstream depth of $y_1 = 2.5$ ft as is shown in Fig. P10.20. Determine the flow depth and the surface elevation at section (2).

**■ Figure P10.20**

**10.21** (WILEY) Repeat Problem 10.20 if the upstream depth is $y_1 = 0.5$ ft.

**\*10.22** (WILEY) Water flows over the bump in the bottom of the rectangular channel shown in Fig. P10.22 with a flowrate per unit width of $q = 4$ m²/s. The channel bottom contour is given by $z_B = 0.2e^{-x^2}$, where $z_B$ and $x$ are in meters. The water depth far upstream of the bump is $y_1 = 2$ m. Plot a graph of the water depth, $y = y(x)$, and the surface elevation, $z = z(x)$, for $-4$ m $\leq x \leq 4$ m. Assume one-dimensional flow.

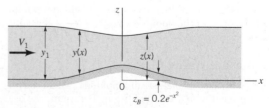

**■ Figure P10.22**

**\*10.23** (WILEY) Repeat Problem 10.22 if the upstream depth is 0.4 m.

**10.24** (WILEY) Water in a rectangular channel flows into a gradual contraction section as is indicated in Fig. P10.24. If the flowrate is $Q = 25$ ft³/s and the upstream depth is $y_1 = 2$ ft, determine the downstream depth, $y_2$.

**■ Figure P10.24**

**10.25** Sketch the specific energy diagram for the flow of Problem 10.24 and indicate its important characteristics. Note that $q_1 \neq q_2$.

**10.26** (WILEY) Repeat Problem 10.24 if the upstream depth is $y_1 = 0.5$ ft. Assume that there are no losses between sections (1) and (2).

**10.27** (WILEY) Water flows in a rectangular channel with a flowrate per unit width of $q = 1.5$ m²/s and a depth of 0.5 m at section (1). The head loss between sections (1) and (2) is 0.03 m. Plot the specific energy diagram for this flow and locate states (1) and (2) on this diagram. Is it possible to have a head loss of 0.06 m? Explain.

**10.28** (WILEY) Water flows in a horizontal rectangular channel with a flowrate per unit width of $q = 10$ ft²/s and a depth of 1.0 ft at the downstream section (2). The head loss between section (1) upstream and section (2) is 0.2 ft. Plot the specific energy diagram for this flow and locate states (1) and (2) on this diagram.

**10.29** (WILEY) Water flows in a horizontal, rectangular channel with an initial depth of 1 m and an initial velocity of 4 m/s. Determine the depth downstream if losses are negligible. Note that there may be more than one solution.

**10.30** Water flows in a horizontal, rectangular channel with an initial depth of 2 ft and initial velocity of 12 ft/s. Determine the depth downstream if losses are negligible. Note that there may be more than one solution. Repeat the problem if the initial depth remains the same, but the initial velocity is 6 ft/s.

**10.31** (WILEY) A smooth transition section connects two rectangular channels as shown in Fig. P10.31. The channel width increases from 6.0 to 7.0 ft, and the water surface elevation is the same in each channel. If the upstream depth of flow is 3.0 ft, determine $h$, the amount the channel bed needs to be raised across the transition section to maintain the same surface elevation.

**■ Figure P10.31**

**10.32** Water flows over a bump of height $h = h(x)$ on the bottom of a wide rectangular channel as is indicated in Fig. P10.32. If energy losses are negligible, show that the slope of the water surface is given by $dy/dx = -(dh/dx)/[1 - (V^2/gy)]$, where $V = V(x)$ and $y = y(x)$ are the local velocity and depth of flow. Comment on the sign (i.e., $<0$, $= 0$, or $>0$) of $dy/dx$ relative to the sign of $dh/dx$.

**■ Figure P10.32**

**10.33** Water flows in a rectangular channel with a bottom slope of 4.2 ft/mi and a head loss of 2.3 ft/mi. At a section where the depth is 5.8 ft and the average velocity 5.9 ft/s, does the flow depth increase or decrease in the direction of flow? Explain.

**10.34** (WILEY) Water flows in the river shown in Fig. P10.34 with a uniform bottom slope. The total head at each section is measured by using Pitot tubes as indicated. Determine the value of $dy/dx$ at the location where the Froude number is 0.357.

**■ Figure P10.34**

**10.35** (WILEY) Repeat Problem 10.34 if the Froude number is 2.75.

**10.36** (WILEY) Assume that the conditions given in Fig. P10.34 are as indicated except that the value of $z_4$ is not known. Determine the value of $z_4$ if the flow is uniform depth.

**10.37** (WILEY) Water flows in a horizontal rectangular channel at a depth of 0.5 ft and a velocity of 8 ft/s. Determine the two possible depths at a location slightly downstream. Viscous effects between the water and the channel surface are negligible.

## Section 10.4.2 The Manning Equation

**10.38** A viscous oil flows down a wide plate with a uniform depth of 8 mm and an average velocity of 50 mm/s. The plate is on a 3° hill and the specific gravity of the oil is 0.85. Determine the average shear stress between the oil and the plate.

**10.39** (WILEY) Water flows in a 5-m-wide channel with a speed of 2 m/s and a depth of 1 m. The channel bottom slopes at a rate of 1 m per 1000 m. Determine the Manning coefficient for this channel.

**10.40** Fluid properties such as viscosity or density do not appear in the Manning equation (Eq. 10.20). Does this mean that this equation is valid for any open-channel flow such as that involving mercury, water, oil, or molasses? Explain.

**10.41** (WILEY) The following data are taken from measurements on Indian Fork Creek: $A = 26$ m², $P = 16$ m, and $S_0 = 0.02$ m/62 m. Determine the average shear stress on the wetted perimeter of this channel.

**10.42** (WILEY) The following data are obtained for a particular reach of the Provo River in Utah: $A = 183$ ft², free-surface width = 55 ft, average depth = 3.3 ft, $R_h = 3.32$ ft, $V = 6.56$ ft/s, length of reach = 116 ft, and elevation drop of reach = 1.04 ft. Determine **(a)** the average shear stress on the wetted perimeter, **(b)** the Manning coefficient, $n$, and **(c)** the Froude number of the flow.

**10.43** (WILEY) At a particular location, the cross section of the Columbia River is as indicated in Fig. P10.43. If on a day without wind it takes 5 min to float 0.5 mi along the river, which drops 0.46 ft in that distance, determine the value of the Manning coefficient, $n$.

■ **Figure P10.43**

**10.44** (WILEY) If the free surface of the Columbia River shown in Fig. P10.43 were 20 ft above the bottom rather than 44 ft, as is indicated in the figure, how long would it take to float the 0.5-mi stretch considered in Problem 10.43? Assume the elevation change remains 0.46 ft.

### Section 10.4.3 Uniform Depth Examples—Determine Flowrate

**10.45** (WILEY) A 2-m-diameter pipe made of finished concrete lies on a slope of 1-m elevation change per 1000-m horizontal distance. Determine the flowrate when the pipe is half full.

**10.46** (WILEY) By what percent is the flowrate reduced in the rectangular channel shown in Fig. P10.46 because of the addition of the thin center board? All surfaces are of the same material.

■ **Figure P10.46**

**10.47** (WILEY) The great Kings River flume in Fresno County, California, was used from 1890 to 1923 to carry logs from an elevation of 4500 ft where trees were cut to an elevation of 300 ft at the railhead. The flume was 54 miles long, constructed of wood, and had a V-cross section as indicated in Fig. P10.47. It is claimed

■ **Figure P10.47**

that logs would travel the length of the flume in 15 hours. Do you agree with this claim? Provide appropriate calculations to support your answer.

**10.48** (WILEY) Water flows in a channel as shown in Fig. P10.48. The velocity is 4.0 ft/s when the channel is half full with depth $d$. Determine the velocity when the channel is completely full, depth $2d$.

■ **Figure P10.48**

**10.49** A natural riverbed has an approximate parabolic cross section such that the free-surface width is 200 ft when the center depth is 5 ft. Determine the flowrate if $n = 0.035$ and the slope is 1.5 ft/mi.

**10.50** (WILEY) A trapezoidal channel with a bottom width of 3.0 m and sides with a slope of 2 : 1 (horizontal:vertical) is lined with fine gravel ($n = 0.020$) and is to carry 10 m³/s. Can this channel be built with a slope of $S_0 = 0.00010$ if it is necessary to keep the velocity below 0.75 m/s to prevent scouring of the bottom? Explain.

**10.51** (WILEY GO) Water flows in a 2-m-diameter finished concrete pipe so that it is completely full and the pressure is constant all along the pipe. If the slope is $S_0 = 0.005$, determine the flowrate by using open-channel flow methods. Compare this result with that obtained by using pipe flow methods of Chapter 8.

**10.52** (WILEY) Water flows in a weedy earthen channel at a rate of 30 m³/s. What flowrate can be expected if the weeds are removed and the depth remains constant?

**10.53** Water flows in an unfinished concrete channel at a rate of 30 m³/s. What flowrate can be expected if the concrete were finished and the depth remains constant?

**10.54** The Blue Ridge Flume, constructed in Shasta County, California, in 1872 to supply water for gold mining operations, was built of planed wood and had a cross section as shown in Fig. P10.54. Determine the water velocity in a section of this flume if it was full and the elevation change was 27 ft per mile.

■ **Figure P10.54**

**10.55** (WILEY) A round concrete storm sewer pipe used to carry rainfall runoff from a parking lot is designed to be half full when the rainfall rate is a steady 1 in./hr. Will this pipe be able to handle the flow from a 2-in./hr rainfall without water backing up into the parking lot? Support your answer with appropriate calculations.

**10.56** (WILEY) A 10-ft-wide rectangular channel is built to bypass a dam so that fish can swim upstream during their migration. During normal conditions when the water depth is 4 ft, the water velocity is 5 ft/s. Determine the velocity during a flood when the water depth is 8 ft.

**†10.57** Overnight a thin layer of ice forms on the surface of a river. Estimate the percent reduction in flowrate caused by this condition. List all assumptions and show all calculations.

**\*10.58** The cross section of a long tunnel carrying water through a mountain is as indicated in Fig. P10.58. Plot a graph of flowrate as a function of water depth, $y$, for $0 \leq y \leq 18$ ft. The slope is 2 ft/mi and the surface of the tunnel is rough rock (equivalent to rubble masonry). At what depth is the flowrate maximum? Explain.

■ **Figure P10.58**

**10.59** The cross section of an ancient Roman aqueduct is drawn to scale in Fig. P10.59. When it was new the channel was essentially rectangular and for a flowrate of 100,000 m³/day, the water depth was as indicated. Archaeological evidence indicates that after many years of use, calcium carbonate deposits on the sides and bottom modified the shape to that shown in the figure. Estimate the flowrate for the modified shape if the slope and surface roughness did not change.

Modified water surface

Original water surface

Calcium carbonate deposits

0  0.2  0.4  0.6  0.8  1.0 m

■ **Figure P10.59**

**10.60** (WILEY) The smooth, concrete-lined channel shown in Fig. P10.60 is built on a slope of 2 m/km. Determine the flowrate if the depth is $y = 1.5$ m.

6 m

1.0 m

0.5 m

3 m

Concrete

■ **Figure P10.60**

**\*10.61** (WILEY) At a given location, under normal conditions a river flows with a Manning coefficient of 0.030, and a cross section as indicated in Fig. P10.61a. During flood conditions at this location, the river has a Manning coefficient of 0.040 (because of trees and brush in the floodplain) and a cross section as shown in Fig. P10.61b. Determine the ratio of the flowrate during flood conditions to that during normal conditions.

12 ft

800 ft

(a)

20 ft

8 ft

800 ft

1000 ft

(b)

■ **Figure P10.61**

**\*10.62** The cross section of a creek valley is shown in Fig. P10.62. Plot a graph of flowrate as a function of depth, $y$, for $0 \leq y \leq 10$ ft. The slope is 5 ft/mi.

$z = 530$ ft

$z = 526$ ft

$z = 530$ ft   $z = 522$ ft

90 ft    8 ft   30 ft    50 ft

Pasture   Clean   Light   Trees
          straight  brush

■ **Figure P10.62**

**10.63** (WILEY) Repeat Problem 10.60 if the surfaces are smooth concrete as is indicated, except for the diagonal surface, which is gravelly with $n = 0.025$.

**10.64** (WILEY) Determine the flowrate for the symmetrical channel shown in Fig. P10.88 if the bottom is smooth concrete and the sides are weedy. The bottom slope is $S_0 = 0.001$.

**10.65** (WILEY) (See Fluids in the News article titled "Done without GPS or Lasers," Section 10.4.3.) Determine the number of gallons of water delivered per day by a rubble masonry, 1.2-m-wide aqueduct laid on an average slope of 14.6 m per 50 km if the water depth is 1.8 m.

### Section 10.4.3 Uniform Depth Examples—Determine Depth or Size

**10.66** (WILEY) Water flows in a rectangular, finished concrete channel at a rate of 2 m³/s. The bottom slope is 0.001. Determine the channel width if the water depth is to be equal to its width.

**10.67** (WILEY) An old, rough-surfaced, 2-m-diameter concrete pipe with a Manning coefficient of 0.025 carries water at a rate of 5.0 m³/s when it is half full. It is to be replaced by a new pipe with a Manning coefficient of 0.012 that is also to flow half full at the same flowrate. Determine the diameter of the new pipe.

**10.68** (WILEY) Four sewer pipes of 0.5-m diameter join to form one pipe of diameter $D$. If the Manning coefficient, $n$, and the slope are the same for all of the pipes, and if each pipe flows half full, determine $D$.

**10.69** (WILEY) The flowrate in the clay-lined channel ($n = 0.025$) shown in Fig. P10.69 is to be 300 ft³/s. To prevent erosion of the sides, the velocity must not exceed 5 ft/s. For this maximum velocity, determine the width of the bottom, $b$, and the slope, $S_0$.

■ **Figure P10.69**

**10.70** Water flows in a rectangular, brick-lined aqueduct of width 1.2 m at a rate of 73,000 m³/day. Determine the water depth if the change in elevation over the 16-km length of this channel is 9.6 m.

**10.71** (WILEY) A rectangular, unfinished concrete channel of 28-ft width is laid on a slope of 8 ft/mi. Determine the flow depth and Froude number of the flow if the flowrate is 400 ft³/s.

**10.72** (WILEY) An engineer is to design a channel lined with planed wood to carry water at a flowrate of 2 m³/s on a slope of 10 m/800 m. The channel cross section can be either a 90° triangle or a rectangle with a cross section twice as wide as its depth. Which would require less wood and by what percent?

**10.73** A circular finished concrete culvert is to carry a discharge of 50 ft³/s on a slope of 0.0010. It is to flow not more than half full. The culvert pipes are available from the manufacturer with diameters that are multiples of 1 ft. Determine the smallest suitable culvert diameter.

**10.74** (WILEY) At what depth will 50 ft³/s of water flow in a 6-ft-wide rectangular channel lined with rubble masonry set on a slope of 1 ft in 500 ft? Is a hydraulic jump possible under these conditions? Explain.

**10.75** (WILEY) The rectangular canal shown in Fig. P10.75 changes to a round pipe of diameter $D$ as it passes through a tunnel in a mountain. Determine $D$ if the surface material and slope remain the same and the round pipe is to flow completely full.

■ **Figure P10.75**

**10.76** (WILEY) An 8-ft-diameter concrete drainage pipe that flows half full is to be replaced by a concrete-lined V-shaped open channel having an interior angle of 90°. Determine the depth of fluid that will exist in the V-shaped channel if it is laid on the same slope and carries the same discharge as the drainage pipe.

**10.77** Water flows in the symmetrical, unfinished concrete trapezoidal channel shown in Fig. P10.77 at a rate of 120 ft³/s. The slope is 4.2 ft/2000 ft. Determine the number of cubic yards of concrete needed to line each 1000 ft of the channel.

■ **Figure P10.77**

**10.78** (WILEY) When the channel of triangular cross section shown in Fig. P10.78 was new, a flowrate of $Q$ caused the water to reach $L = 2$ m up the side as indicated. After considerable use, the walls of the channel became rougher and the Manning coefficient, $n$, doubled. Determine the new value of $L$ if the flowrate stayed the same.

■ **Figure P10.78**

**10.79** Water flows in a channel with an equilateral triangle cross section as shown in Fig. P10.79. For a given Manning coefficient, $n$, and channel slope, determine the depth that gives the maximum flowrate.

■ **Figure P10.79**

**10.80** ▶ Two canals join to form a larger canal as shown in **Video V10.6** and Fig. P10.80. Each of the three rectangular canals is lined with the same material and has the same bottom slope. The water depth in each is to be 2 m. Determine the width of the merged canal, $b$. Explain physically (i.e., without using any equations) why it is expected that the width of the merged canal is less than the combined widths of the two original canals (i.e., $b < 4$ m $+ 8$ m $= 12$ m).

■ **Figure P10.80**

**10.81** (WILEY) Water flows uniformly at a depth of 1 m in a channel that is 5 m wide as shown in Fig. P10.81. Further downstream, the channel cross section changes to that of a square of width and height $b$. Determine the value of $b$ if the two portions of this channel are made of the same material and are constructed with the same bottom slope.

Width = 5 m

■ **Figure P10.81**

**10.82** (WILEY) Determine the flow depth for the channel shown in Fig. P10.60 if the flowrate is 15 m³/s.

**10.83** (WILEY) Rainwater runoff from a 200-ft by 500-ft parking lot is to drain through a circular concrete pipe that is laid on a slope of 3 ft/mi. Determine the pipe diameter if it is to be full with a steady rainfall of 1.5 in./hr.

**10.84** (WILEY) (See Fluids in the News article titled "Plumbing the Everglades," Section 10.4.1.) The canal shown in Fig. P10.84 is to be widened so that it can carry twice the amount of water. Determine the additional width, $L$, required if all other parameters (i.e., flow depth, bottom slope, surface material, side slope) are to remain the same.

■ **Figure P10.84**

## Section 10.4.3 Uniform Depth Examples—Determine Slope

**10.85** (WILEY) Water flows 1 m deep in a 2-m-wide finished concrete channel. Determine the slope if the flowrate is 3 m³/s.

**10.86** (WILEY) Water flows in the channel shown in Fig. P10.86 at a rate of 90 ft³/s. Determine the minimum slope that this channel can have so that the water does not overflow the sides. The Manning coefficient for this channel is $n = 0.014$.

■ **Figure P10.86**

**10.87** It is claimed that on steep portions of the Blue Ridge Flume (see Problem 10.54 for the flume cross section), the water would rush along at 50 mph. Is this possible? If so, determine the slope; if not, explain why not. *Note:* Use a flowrate of 52 ft³/s, which is that obtained for the conditions (slope) of Problem 10.54.

**10.88** (WILEY) To prevent weeds from growing in a clean earthen-lined canal, it is recommended that the velocity be no less than 2.5 ft/s. For the symmetrical canal shown in Fig. P10.88, determine the minimum slope needed.

■ **Figure P10.88**

**10.89** ▶ (WILEY) The smooth, concrete-lined, symmetrical channel shown in **Video V10.7** and Fig. P10.88 carries water from the silt-laden Colorado River. If the velocity must be 4.0 ft/s to prevent the silt from settling out (and eventually clogging the channel), determine the minimum slope needed.

**10.90** (WILEY) The symmetrical channel shown in Fig. P10.88 is dug in sandy loam soil with $n = 0.020$. For such surface material it is recommended that to prevent scouring of the surface the average velocity be no more than 1.75 ft/s. Determine the maximum slope allowed.

**10.91** (WILEY) The depth downstream of a sluice gate in a rectangular wooden channel of width 5 m is 0.60 m. If the flowrate is 18 m³/s, determine the channel slope needed to maintain this depth. Will the depth increase or decrease in the flow direction if the slope is **(a)** 0.02; **(b)** 0.01?

**10.92** (WILEY) A 50-ft-long aluminum gutter (Manning coefficient $n = 0.011$) on a section of a roof is to handle a flowrate of 0.15 ft³/s during a heavy rainstorm. The cross section of the gutter is shown in Fig. P10.92. Determine the vertical distance that this gutter must be pitched (i.e., the difference in elevation between the two ends of the gutter) so that the water does not overflow the gutter. Assume uniform depth channel flow.

■ **Figure P10.92**

**10.93** Water flows in a rectangular channel of width $b$ and depth $y$ with a Froude number of unity. The slope, $S_{0c}$, of the channel needed to produce this critical flow is a function of $y$. Show that as $y \to \infty$ the slope becomes proportional to $y$ (i.e., $S_{0c} = C_1 y$, where $C_1$ is a constant) and that as $y \to 0$ the slope becomes proportional to $y^{-1/3}$ (i.e., $S_{0c} = C_2/y^{1/3}$, where $C_2$ is a constant). Show that the channel with an aspect ratio of $b/y = 6$ gives the minimum value of $S_{0c}$.

### Section 10.6.1 The Hydraulic Jump (also see Lab Problems 10.3LP and 10.4LP)

**10.94** (WILEY) Water flows upstream of a hydraulic jump with a depth of 0.5 m and a velocity of 6 m/s. Determine the depth of the water downstream of the jump.

**10.95** (WILEY) A 2.0-ft standing wave is produced at the bottom of the rectangular channel in an amusement park water ride. If the water depth upstream of the wave is estimated to be 1.5 ft, determine how fast the boat is traveling when it passes through this standing wave (hydraulic jump) for its final "splash."

**10.96** (WILEY GO) The water depths upstream and downstream of a hydraulic jump are 0.3 and 1.2 m, respectively. Determine the upstream velocity and the power dissipated if the channel is 50 m wide.

**10.97** ▶ (WILEY) Under appropriate conditions, water flowing from a faucet, onto a flat plate, and over the edge of the plate can produce a circular hydraulic jump as shown in Fig. P10.97 and **Video V10.12.** Consider a situation where a jump forms 3.0 in. from the center of the plate with depths upstream and downstream of the jump of 0.05 in. and 0.20 in., respectively. Determine the flowrate from the faucet.

■ **Figure P10.97**

**10.98** Show that for a hydraulic jump in a rectangular channel, the Froude number upstream, $\text{Fr}_1$, and the Froude number downstream, $\text{Fr}_2$, are related by

$$\text{Fr}_2^2 = \frac{8\text{Fr}_1^2}{[(1 + 8\,\text{Fr}_1^2)^{1/2} - 1]^3}$$

Plot $\text{Fr}_2$ as a function of $\text{Fr}_1$ and show that the flow downstream of a jump is subcritical.

**10.99** (WILEY) Water flows in a 2-ft-wide rectangular channel at a rate of 10 ft³/s. If the water depth downstream of a hydraulic jump is 2.5 ft, determine **(a)** the water depth upstream of the jump, **(b)** the upstream and downstream Froude numbers, and **(c)** the head loss across the jump.

**10.100** ▶ (WILEY) A hydraulic jump at the base of a spillway of a dam is such that the depths upstream and downstream of the jump are 0.90 and 3.6 m, respectively (see **Video V10.11**). If the spillway is 10 m wide, what is the flowrate over the spillway?

**10.101** (WILEY) Determine the head loss and power dissipated by the hydraulic jump of Problem 10.100.

**10.102** (WILEY) A hydraulic jump occurs in a 4-m-wide rectangular channel at a point where the slope changes from 3 m per 100 m upstream of the jump to $h$ m per 100 m downstream of the jump. The depth and velocity of the uniform flow upstream of the jump are 0.5 m and 8 m/s, respectively. Determine the value of $h$ if the flow downstream of the jump is to be uniform flow.

**10.103** (WILEY) At a given location in a 12-ft-wide rectangular channel the flowrate is 900 ft³/s and the depth is 4 ft. Is this location upstream or downstream of the hydraulic jump that occurs in this channel? Explain.

**10.104** Water flows in a wide, finished concrete channel as is shown in Fig. P10.104 such that a hydraulic jump occurs at the transition of the change in slope of the channel bottom. If the upstream Froude number and depth are 4.0 and 0.2 ft, respectively, determine the slopes upstream, $S_{01}$, and downstream, $S_{02}$, of the jump to maintain uniform flows in those regions. The jump can be treated as a jump on a horizontal surface.

■ **Figure P10.104**

**10.105** Water flows in a rectangular channel at a depth of $y = 1$ ft and a velocity of $V = 20$ ft/s. When a gate is suddenly placed across the end of the channel, a wave (a moving hydraulic jump) travels upstream with velocity $V_w$ as is indicated in Fig. P10.105. Determine $V_w$. Note that this is an unsteady problem for a stationary observer. However, for an observer moving to the left with velocity $V_w$, the flow appears as a steady hydraulic jump.

■ **Figure P10.105**

**10.106** (WILEY) Water flows in a rectangular channel with velocity $V = 6$ m/s. A gate at the end of the channel is suddenly closed so that a wave (a moving hydraulic jump) travels upstream with velocity $V_w = 2$ m/s as is indicated in Fig. P10.105. Determine the depths ahead of and behind the wave. Note that this is an unsteady problem for a stationary observer. However, for an observer moving to the left with velocity $V_w$, the flow appears as a steady hydraulic jump.

**10.107** (WILEY) (See Fluids in the News article titled "Grand Canyon Rapids Building." Section 10.6.1.) During the flood of 1983, a large hydraulic jump formed at "Crystal Rapid" on the Colorado River. People rafting the river at that time report "entering the rapid at almost 30 mph, hitting a 20-ft-tall wall of water, and exiting at about 10 mph." Is this information (i.e., upstream and downstream velocities and change in depth) consistent with the principles of a hydraulic jump? Show calculations to support your answer.

### Section 10.6.2, 3 Sharp-Crested and Broad-Crested Weirs (also see Lab Problems 10.1LP and 10.2LP)

**10.108** (WILEY) Water flows over a 2-m-wide rectangular sharp-crested weir. Determine the flowrate if the weir head is 0.1 m and the upstream channel depth is 1 m.

**10.109** (WILEY) Water flows over a 5-ft-wide, rectangular sharp-crested weir that is $P_w = 4.5$ ft tall. If the depth upstream is 5 ft, determine the flowrate.

**10.110** (WILEY) A rectangular sharp-crested weir is used to measure the flowrate in a channel of width 10 ft. It is desired to have the upstream channel flow depth be 6 ft when the flowrate is 50 cfs. Determine the height, $P_w$, of the weir plate.

**10.111** ▶ Water flows from a storage tank, over two triangular weirs, and into two irrigation channels as shown in **Video V10.13** and Fig. P10.111. The head for each weir is 0.4 ft, and the flowrate in the channel fed by the 90°-V-notch weir is to be twice the flowrate in the other channel. Determine the angle $\theta$ for the second weir.

■ **Figure P10.111**

**10.112** A basin at a water treatment plant is 60 ft long, 10 ft wide, and 5 ft deep. Water flows from the basin over a 3-ft-long, rectangular weir whose crest is 4 ft above the bottom of the basin. Estimate how long it will take for the depth of the water in the basin to change from 4.5 ft to 4.4 ft if there is no flow into the basin.

**10.113** Water flows over the rectangular sharp-crested weir in a wide channel as shown in Fig. P10.113. If the channel is lined with unfinished concrete with a bottom slope of 2 m/300 m, will it be possible to produce a hydraulic jump in the channel downstream of the weir? Explain.

■ **Figure P10.113**

**10.114** Water flows over a sharp-crested triangular weir with $\theta = 90°$. The head range covered is $0.20 \leq H \leq 1.0$ ft and the accuracy in the measurement of the head, $H$, is $\delta H = \pm 0.01$ ft. Plot a graph of the percent error expected in $Q$ as a function of $Q$.

**10.115** (WILEY GO) **(a)** The rectangular sharp-crested weir shown in Fig. P10.115a is used to maintain a relatively constant depth in the channel upstream of the weir. How much deeper will the water be upstream of the weir during a flood when the flowrate is 45 ft³/s compared to normal conditions when the flowrate is 30 ft³/s? Assume the weir coefficient remains constant at $C_{wr} = 0.62$.

(a)                              (b)

■ **Figure P10.115**

**(b)** Repeat the calculations if the weir of part (a) is replaced by a rectangular sharp-crested "duck bill" weir that is oriented at an angle of 30° relative to the channel centerline as shown in Fig. P10.115b. The weir coefficient remains the same.

**10.116** Water flows over the sharp-crested weir shown in Fig. P10.116. The weir plate cross section consists of a semicircle and a rectangle. Plot a graph of the estimated flowrate, $Q$, as a function of head, $H$. List all assumptions and show all calculations.

■ **Figure P10.116**

**10.117** (WILEY) Water flows over a broad-crested weir that has a width of 4 m and a height of $P_w = 1.5$ m. The free-surface well upstream of the weir is at a height of 0.5 m above the surface of the weir. Determine the flowrate in the channel and the minimum depth of the water above the weir block.

**10.118** Determine the flowrate per unit width, $q$, over a broad-crested weir that is 2.0 m tall if the head, $H$, is 0.50 m.

## Section 10.6.4 Underflow Gates

**10.119** (WILEY) Water flows under a sluice gate in a 60-ft-wide finished concrete channel as is shown in Fig. P10.119. Determine the flowrate. If the slope of the channel is 2.5 ft/200 ft, will the water depth increase or decrease downstream of the gate? Assume $C_c = y_2/a = 0.65$. Explain.

■ **Figure P10.119**

**10.120** Water flows under a sluice gate in a channel of 10-ft width. If the upstream depth remains constant at 5 ft, plot a graph of flowrate as a function of the distance between the gate and the channel bottom as the gate is slowly opened. Assume free outflow.

**10.121** A water-level regulator (not shown) maintains a depth of 2.0 m downstream from a 10-m-wide drum gate as shown in Fig. P10.121. Plot a graph of flowrate, $Q$, as a function of water depth upstream of the gate, $y_1$, for $2.0 \leq y_1 \leq 5.0$ m.

■ **Figure P10.121**

### ■ Lab Problems

**10.1LP** This problem involves the calibration of a triangular weir. To proceed with this problem, go to Appendix H, which is located in *WileyPLUS* or on the book's web site, www.wiley.com/college/munson.

**10.2LP** This problem involves the calibration of a rectangular weir. To proceed with this problem, go to Appendix H, which is located in *WileyPLUS* or on the book's web site, www.wiley.com/college/munson.

**10.3LP** This problem involves the depth ratio across a hydraulic jump. To proceed with this problem, go to Appendix H, which is located in *WileyPLUS* or on the book's web site, www.wiley.com/college/munson.

**10.4LP** This problem involves the head loss across a hydraulic jump. To proceed with this problem, go to Appendix H, which is located in *WileyPLUS* or on the book's web site, www.wiley.com/college/munson.

### ■ Lifelong Learning Problems

**10.1LL** With the increased usage of low-lying coastal areas and the possible rise in ocean levels because of global warming, the potential for widespread damage from tsunamis is increasing. Obtain information about new and improved methods available to predict the occurrence of these damaging waves and how to better use coastal areas so that massive loss of life and property does not occur. Summarize your findings in a brief report.

**10.2LL** Recent photographs from NASA's Mars Orbiter Camera on the Mars Global Surveyor provide new evidence that water may still flow on the surface of Mars. Obtain information about the possibility of current or past open-channel flows on Mars and other planets or their satellites. Summarize your findings in a brief report.

**10.3LL** Hydraulic jumps are normally associated with water flowing in rivers, gullies, and other such relatively high-speed open channels. However, recently, hydraulic jumps have been used in various manufacturing processes involving fluids other than water (such as liquid metal solder) in relatively small-scale flows. Obtain information about new manufacturing processes that involve hydraulic jumps as an integral part of the process. Summarize your findings in a brief report.

### ■ FE Exam Problems

Sample FE (Fundamentals of Engineering) exam questions for fluid mechanics are provided in *WileyPLUS* or on the book's web site, www.wiley.com/college/munson.

# 11 Compressible Flow

CHAPTER OPENING PHOTO: Supersonic flow past a scale model of the X-15 experimental aircraft. An object moving through a fluid at supersonic speed (Mach number greater than one) creates shock waves (a discontinuity in flow conditions shown by the light and dark lines), which would be heard as a sonic boom as the object passes overhead. (Photograph courtesy of NASA)

## Learning Objectives

After completing this chapter, you should be able to:

- distinguish between incompressible and compressible flows, and know when the approximations associated with assuming fluid incompressibility are acceptable.

- understand some important features of different categories of compressible flows of ideal gases.

- explain speed of sound and Mach number and their practical significance.

- solve useful problems involving isentropic and nonisentropic flows including flows across normal shock waves.

- appreciate the compelling similarities between compressible flows of gases and open-channel flows of liquids.

- move on to understanding more advanced concepts about compressible flows.

Most first courses in fluid mechanics concentrate on constant density (incompressible) flows. In earlier chapters of this book, we mainly considered incompressible flow behavior. In a few instances, variable density (compressible) flow effects were covered briefly. The notion of an incompressible fluid is convenient because when constant density and constant (including zero) viscosity are assumed, problem solutions are greatly simplified. Also, fluid incompressibility allows us to build on the Bernoulli equation as was done, for example, in Chapter 5. Preceding examples should have convinced us that nearly incompressible flows are common in everyday experiences.

Any study of fluid mechanics would, however, be incomplete without a brief introduction to compressible flow behavior. Fluid compressibility is a very important consideration in numerous engineering applications of fluid mechanics. For example, the measurement of high-speed flow velocities requires compressible flow theory. The flows in gas turbine engine components are generally compressible. Many aircraft fly fast enough to involve compressible flow.

The variation of fluid density for compressible flows requires attention to density and other fluid property relationships. The fluid equation of state, often unimportant for incompressible flows, is vital in the analysis of compressible flows. Also, temperature variations for compressible flows are usually significant, and thus the energy equation is important. Curious phenomena can occur with compressible flows. For example, with compressible flows we can have fluid acceleration because of friction, fluid deceleration in a converging duct, fluid temperature decrease with heating, and the formation of abrupt discontinuities in flows across which fluid properties change appreciably.

For simplicity, in this introductory study of compressibility effects we mainly consider the steady, one-dimensional, constant (including zero) viscosity, compressible flow of an ideal gas. We limit our study to compressibility due to high-speed flow. In this chapter, one-dimensional flow refers to flow involving uniform distributions of fluid properties over any flow cross-sectional area. Both frictionless ($\mu = 0$) and frictional ($\mu \neq 0$) compressible flows are considered. If the change in volume associated with a change of pressure is considered a measure of compressibility, our experience suggests that gases and vapors are much more compressible than liquids. We focus our attention on the compressible flow of a gas because such flows occur often. We limit our discussion to ideal gases, since the equation of state for an ideal gas is uncomplicated, yet representative of actual gases at pressures and temperatures of engineering interest, and because the flow trends associated with an ideal gas are generally applicable to other compressible fluids.

An excellent film about compressible flow is available (see Ref. 1). This resource is a useful supplement to the material covered in this chapter.

## 11.1 Ideal Gas Relationships

*V11.1 Lighter flame*

Before we can proceed to develop **compressible flow** equations, we need to become more familiar with the fluid we will work with, the ideal gas. Specifically, we must learn how to evaluate ideal gas property changes. The equation of state for an **ideal gas** is

$$\rho = \frac{p}{RT} \tag{11.1}$$

We have already discussed fluid pressure, $p$, density, $\rho$, and temperature, $T$, in earlier chapters. The gas constant, $R$, represents a constant for each distinct ideal gas or mixture of ideal gases, where

$$R = \frac{\lambda}{M_{\text{gas}}} \tag{11.2}$$

With this notation, $\lambda$ is the universal gas constant and $M_{\text{gas}}$ is the molecular weight of the ideal gas or gas mixture. Listed in Tables 1.7 and 1.8 are values of the gas constants of some commonly used gases. Knowing the pressure and temperature of a gas, we can estimate its density. Nonideal gas state equations are beyond the scope of this text, and those interested in this topic are directed to texts on engineering thermodynamics, for example, Ref. 2. Note that the trends of ideal gas flows are generally good indicators of what nonideal gas flow behavior is like.

*We consider ideal gas flows only.*

For an ideal gas, **internal energy**, $\breve{u}$, is part of the stored energy of the gas as explained in Section 5.3 and is considered to be a function of temperature only (Ref. 2). Thus, the ideal gas specific heat at constant volume, $c_v$, can be expressed as

$$c_v = \left(\frac{\partial \breve{u}}{\partial T}\right)_v = \frac{d\breve{u}}{dT} \tag{11.3}$$

where the subscript $v$ on the partial derivative refers to differentiation at constant specific volume, $v = 1/\rho$. From Eq. 11.3 we conclude that for a particular ideal gas, $c_v$ is a function of temperature only. Equation 11.3 can be rearranged to yield

$$d\breve{u} = c_v \, dT$$

Thus,

$$\check{u}_2 - \check{u}_1 = \int_{T_1}^{T_2} c_v \, dT \tag{11.4}$$

Equation 4 is useful because it allows us to evaluate the change in internal energy, $\check{u}_2 - \check{u}_1$, associated with ideal gas flow from section (1) to section (2) in a flow. For simplicity, we can assume that $c_v$ is constant for a particular ideal gas and obtain from Eq. 11.4

$$\boxed{\check{u}_2 - \check{u}_1 = c_v(T_2 - T_1)} \tag{11.5}$$

Actually, $c_v$ for a particular gas varies with temperature (see Ref. 2). However, for moderate changes in temperature, the constant $c_v$ assumption is reasonable.

The fluid property **enthalpy**, $\check{h}$, is defined as

$$\check{h} = \check{u} + \frac{p}{\rho} \tag{11.6}$$

It combines internal energy, $\check{u}$, and pressure energy, $p/\rho$, and is useful when dealing with the energy equation (Eq. 5.69). For an ideal gas, we have already stated that

$$\check{u} = \check{u}(T)$$

From the equation of state (Eq. 11.1)

$$\frac{p}{\rho} = RT$$

Thus, it follows that

$$\check{h} = \check{h}(T)$$

Since for an ideal gas, enthalpy is a function of temperature only, the ideal gas specific heat at constant pressure, $c_p$, can be expressed as

$$c_p = \left(\frac{\partial \check{h}}{\partial T}\right)_p = \frac{d\check{h}}{dT} \tag{11.7}$$

where the subscript $p$ on the partial derivative refers to differentiation at constant pressure, and $c_p$ is a function of temperature only. The rearrangement of Eq. 11.7 leads to

$$d\check{h} = c_p \, dT$$

and

$$\check{h}_2 - \check{h}_1 = \int_{T_1}^{T_2} c_p \, dT \tag{11.8}$$

Equation 11.8 is useful because it allows us to evaluate the change in enthalpy, $\check{h}_2 - \check{h}_1$, associated with ideal gas flow from section (1) to section (2) in a flow. For simplicity, we can assume that $c_p$ is constant for a specific ideal gas and obtain from Eq. 11.8

$$\boxed{\check{h}_2 - \check{h}_1 = c_p(T_2 - T_1)} \tag{11.9}$$

As is true for $c_v$, the value of $c_p$ for a given gas varies with temperature. Nevertheless, for moderate changes in temperature, the constant $c_p$ assumption is reasonable.

*For moderate temperature changes, specific heat values can be considered constant.*

From Eqs. 11.5 and 11.9 we see that changes in internal energy and enthalpy are related to changes in temperature by values of $c_v$ and $c_p$. We turn our attention now to developing useful relationships for determining $c_v$ and $c_p$. Combining Eqs. 11.6 and 11.1 we get

$$\check{h} = \check{u} + RT \tag{11.10}$$

Differentiating Eq. 11.10 leads to

$$d\check{h} = d\check{u} + R\, dT$$

or

$$\frac{d\check{h}}{dT} = \frac{d\check{u}}{dT} + R \tag{11.11}$$

From Eqs. 11.3, 11.7, and 11.11 we conclude that

$$\boxed{c_p - c_v = R} \tag{11.12}$$

Equation 11.12 indicates that the difference between $c_p$ and $c_v$ is constant for each ideal gas regardless of temperature. Also $c_p > c_v$. If the *specific heat ratio*, $k$, is defined as

$$k = \frac{c_p}{c_v} \tag{11.13}$$

then combining Eqs. 11.12 and 11.13 leads to

$$\boxed{c_p = \frac{Rk}{k-1}} \tag{11.14}$$

and

$$\boxed{c_v = \frac{R}{k-1}} \tag{11.15}$$

*The gas constant is related to the specific heat values.*

Actually, $c_p$, $c_v$, and $k$ are all somewhat temperature dependent for any ideal gas. We will assume constant values for these variables in this book. Values of $k$ and $R$ for some commonly used gases at nominal temperatures are listed in Tables 1.7 and 1.8. These tabulated values can be used with Eqs. 11.13 and 11.14 to determine the values of $c_p$ and $c_v$. Example 11.1 demonstrates how internal energy and enthalpy changes can be calculated for a flowing ideal gas having constant $c_p$ and $c_v$.

## $E$XAMPLE 11.1 Internal Energy, Enthalpy, and Density for an Ideal Gas

**GIVEN** Air flows steadily between two sections in a long, straight portion of 4-in.-diameter pipe as is indicated in Fig. E11.1. The uniformly distributed temperature and pressure at each section are $T_1 = 540\,°R$, $p_1 = 100\,psia$, and $T_2 = 453\,°R$, $p_2 = 18.4\,psia$.

■ **Figure E11.1**

**FIND** Calculate the (**a**) change in internal energy between sections (1) and (2), (**b**) change in enthalpy between sections (1) and (2), and (**c**) change in density between sections (1) and (2).

## SOLUTION

(**a**) Assuming air behaves as an ideal gas, we can use Eq. 11.5 to evaluate the change in internal energy between sections (1) and (2). Thus

$$\check{u}_2 - \check{u}_1 = c_v(T_2 - T_1) \tag{1}$$

From Eq. 11.15 we have

$$c_v = \frac{R}{k-1} \tag{2}$$

and from Table 1.7, $R = 1716\,(ft \cdot lb)/(slug \cdot °R)$ and $k = 1.4$. Throughout this book, we use the nominal values of $k$ for common gases listed in Tables 1.7 and 1.8 and consider these values as being representative. Since English Engineering System

units are used more often than British Gravitational System units in compressible flow discussions, we use

$$1\,slug = 32.174\,lbm$$

to get

$$R = \frac{1716\,(ft \cdot lb)/(slug \cdot °R)}{32.174\,(lbm/slug)} = 53.3\,(ft \cdot lb)/(lbm \cdot °R)$$

From Eq. 2 we obtain

$$c_v = \frac{53.3}{(1.4 - 1)} \text{ (ft} \cdot \text{lb)/(lbm} \cdot \text{°R)}$$
$$= 133 \text{ (ft} \cdot \text{lb)/(lbm} \cdot \text{°R)} \tag{3}$$

Combining Eqs. 1 and 3 yields

$$\breve{u}_2 - \breve{u}_1 = c_v(T_2 - T_1) = 133 \text{ (ft} \cdot \text{lb)/(lbm} \cdot \text{°R)}$$
$$\times (453 \text{ °R} - 540 \text{ °R})$$
$$= -11{,}600 \text{ ft} \cdot \text{lb/lbm} \qquad \textbf{(Ans)}$$

Or, if Btu are wanted as energy units, we note that

$$778 \text{ ft} \cdot \text{lb} = 1 \text{ Btu}$$

and so

$$\breve{u}_2 - \breve{u}_1 = -\frac{11{,}600 \text{ (ft} \cdot \text{lb)/lbm}}{778 \text{ (ft} \cdot \text{lb)/Btu}} = -14.9 \text{ Btu/lbm}$$

**(b)**   For enthalpy change we use Eq. 11.9. Thus

$$\breve{h}_2 - \breve{h}_1 = c_p(T_2 - T_1) \tag{4}$$

where since $k = c_p/c_v$ we obtain

$$c_p = kc_v = (1.4)[133 \text{ (ft} \cdot \text{lb)/(lbm} \cdot \text{°R)}]$$
$$= 186 \text{ (ft} \cdot \text{lb)/(lbm} \cdot \text{°R)} \tag{5}$$

From Eqs. 4 and 5 we obtain

$$\breve{h}_2 - \breve{h}_1 = c_p(T_2 - T_1) = 186 \text{ (ft} \cdot \text{lb)/(lbm} \cdot \text{°R)}$$
$$\times (453 \text{ °R} - 540 \text{ °R})$$
$$= -16{,}200 \text{ ft} \cdot \text{lb/lbm} \qquad \textbf{(Ans)}$$

**(c)**   For density change we use the ideal gas equation of state (Eq. 11.1) to get

$$\rho_2 - \rho_1 = \frac{p_2}{RT_2} - \frac{p_1}{RT_1} = \frac{1}{R}\left(\frac{p_2}{T_2} - \frac{p_1}{T_1}\right) \tag{6}$$

Using the pressures and temperatures given in the problem statement we calculate from Eq. 6

$$\rho_2 - \rho_1 = \frac{1}{53.3 \text{ (ft} \cdot \text{lb)/(lbm} \cdot \text{°R)}}$$
$$\times \left[\frac{(18.4 \text{ psia})(144 \text{ in.}^2/\text{ft}^2)}{453 \text{ °R}}\right.$$
$$\left. - \frac{(100 \text{ psia})(144 \text{ in.}^2/\text{ft}^2)}{540 \text{ °R}}\right]$$

or

$$\rho_2 - \rho_1 = -0.389 \text{ lbm/ft}^3 \qquad \textbf{(Ans)}$$

**COMMENT**   This is a significant change in density when compared with the upstream density

$$\rho_1 = \frac{p_1}{RT_1} = \frac{(100 \text{ psia})(144 \text{ in.}^2/\text{ft}^2)}{[53.3 \text{ (ft} \cdot \text{lb)/(lbm} \cdot \text{°R)}](540 \text{ °R})}$$
$$= 0.499 \text{ lbm/ft}^3$$

Compressibility effects are important for this flow.

---

*Changes in entropy are important because they are related to loss of available energy.*

For compressible flows, changes in the thermodynamic property ***entropy***, $s$, are important. For any pure substance including ideal gases, the "first $T \, ds$ equation" is (see Ref. 2)

$$T \, ds = d\breve{u} + p \, d\left(\frac{1}{\rho}\right) \tag{11.16}$$

where $T$ is absolute temperature, $s$ is entropy, $\breve{u}$ is internal energy, $p$ is absolute pressure, and $\rho$ is density. Differentiating Eq. 11.6 leads to

$$d\breve{h} = d\breve{u} + p \, d\left(\frac{1}{\rho}\right) + \left(\frac{1}{\rho}\right) dp \tag{11.17}$$

By combining Eqs. 11.16 and 11.17, we obtain

$$T \, ds = d\breve{h} - \left(\frac{1}{\rho}\right) dp \tag{11.18}$$

Equation 11.18 is often referred to as the "second $T \, ds$ equation." For an ideal gas, Eqs. 11.1, 11.3, and 11.16 can be combined to yield

$$ds = c_v \frac{dT}{T} + \frac{R}{1/\rho} d\left(\frac{1}{\rho}\right) \tag{11.19}$$

and Eqs. 11.1, 11.7, and 11.18 can be combined to yield

$$ds = c_p \frac{dT}{T} - R \frac{dp}{p} \tag{11.20}$$

If $c_p$ and $c_v$ are assumed to be constant for a given gas, Eqs. 11.19 and 11.20 can be integrated to get

*Changes in entropy are related to changes in temperature, pressure, and density.*

$$s_2 - s_1 = c_v \ln \frac{T_2}{T_1} + R \ln \frac{\rho_1}{\rho_2} \tag{11.21}$$

and

$$s_2 - s_1 = c_p \ln \frac{T_2}{T_1} - R \ln \frac{p_2}{p_1} \tag{11.22}$$

Equations 11.21 and 11.22 allow us to calculate the change of entropy of an ideal gas flowing from one section to another with constant specific heat values ($c_p$ and $c_v$).

---

## EXAMPLE 11.2  Entropy for an Ideal Gas

**GIVEN** Consider the airflow of Example 11.1.

**FIND** Calculate the change in entropy, $s_2 - s_1$, between sections (1) and (2).

**SOLUTION**

Assuming that the flowing air in Fig. E11.1 behaves as an ideal gas, we can calculate the entropy change between sections by using either Eq. 11.21 or Eq. 11.22. We use both to demonstrate that the same result is obtained either way.

From Eq. 11.21,

$$s_2 - s_1 = c_v \ln \frac{T_2}{T_1} + R \ln \frac{\rho_1}{\rho_2} \tag{1}$$

To evaluate $s_2 - s_1$ from Eq. 1 we need the density ratio, $\rho_1/\rho_2$, which can be obtained from the ideal gas equation of state (Eq. 11.1) as

$$\frac{\rho_1}{\rho_2} = \left(\frac{p_1}{T_1}\right)\left(\frac{T_2}{p_2}\right) \tag{2}$$

and thus from Eqs. 1 and 2,

$$s_2 - s_1 = c_v \ln \frac{T_2}{T_1} + R \ln \left[\left(\frac{p_1}{T_1}\right)\left(\frac{T_2}{p_2}\right)\right] \tag{3}$$

By substituting values already identified in the Example 11.1 problem statement and solution into Eq. 3 with

$$\left(\frac{p_1}{T_1}\right)\left(\frac{T_2}{p_2}\right) = \left(\frac{100 \text{ psia}}{540 \text{ °R}}\right)\left(\frac{453 \text{ °R}}{18.4 \text{ psia}}\right) = 4.56$$

we get

$$s_2 - s_1 = [133 \text{ (ft·lb)/(lbm·°R)}] \ln\left(\frac{453 \text{ °R}}{540 \text{ °R}}\right)$$
$$+ [53.3 \text{ (ft·lb)/(lbm·°R)}] \ln 4.56$$

or

$$s_2 - s_1 = 57.5 \text{ (ft·lb)/(lbm·°R)} \tag{Ans}$$

From Eq. 11.22,

$$s_2 - s_1 = c_p \ln \frac{T_2}{T_1} - R \ln \frac{p_2}{p_1} \tag{4}$$

By substituting known values into Eq. 4 we obtain

$$s_2 - s_1 = [186 \text{ (ft·lb)/(lbm·°R)}] \ln\left(\frac{453 \text{ °R}}{540 \text{ °R}}\right)$$
$$- [53.3 \text{ (ft·lb)/(lbm·°R)}] \ln\left(\frac{18.4 \text{ psia}}{100 \text{ psia}}\right)$$

or

$$s_2 - s_1 = 57.5 \text{ (ft·lb)/(lbm·°R)} \tag{Ans}$$

**COMMENT** As anticipated, both Eqs. 11.21 and 11.22 yield the same result for the entropy change, $s_2 - s_1$.

Note that since the ideal gas equation of state was used in the derivation of the entropy difference equations, both the pressures and temperatures used must be absolute.

---

If internal energy, enthalpy, and entropy changes for ideal gas flow with variable specific heats are desired, Eqs. 11.4, 11.8, and 11.19 or 11.20 must be used as explained in Ref. 2. Detailed tables (see, for example, Ref. 3) are available for variable specific heat calculations.

The second law of thermodynamics requires that the *adiabatic* and frictionless flow of any fluid results in $ds = 0$ or $s_2 - s_1 = 0$. Constant entropy flow is called *isentropic* flow. For the isentropic flow of an ideal gas with constant $c_p$ and $c_v$, we get from Eqs. 11.21 and 11.22

$$c_v \ln \frac{T_2}{T_1} + R \ln \frac{\rho_1}{\rho_2} = c_p \ln \frac{T_2}{T_1} - R \ln \frac{p_2}{p_1} = 0 \tag{11.23}$$

By combining Eq. 11.23 with Eqs. 11.14 and 11.15 we obtain

$$\left(\frac{T_2}{T_1}\right)^{k/(k-1)} = \left(\frac{\rho_2}{\rho_1}\right)^k = \left(\frac{p_2}{p_1}\right) \tag{11.24}$$

which is a useful relationship between temperature, density, and pressure for the isentropic flow of an ideal gas. From Eq. 11.24 we can conclude that

$$\boxed{\frac{p}{\rho^k} = \text{constant}} \tag{11.25}$$

for an ideal gas with constant $c_p$ and $c_v$ flowing isentropically, a result already used without proof earlier in Chapters 1, 3, and 5.

---

### F l u i d s   i n   t h e   N e w s

**Hilsch tube (Ranque vortex tube)** Years ago (around 1930) a French physics student (George Ranque) discovered that appreciably warmer and colder portions of *rapidly swirling airflow* could be separated in a simple apparatus consisting of a tube open at both ends into which was introduced, somewhere in between the two openings, swirling air at *high pressure.* Warmer air near the outer portion of the swirling air flowed out one open end of the tube through a simple valve and colder air near the inner portion of the swirling air flowed out the opposite end of the tube. Rudolph Hilsch, a German physicist, improved on this discovery (ca. 1947). Hot air temperatures of 260 °F (127 °C) and cold air temperatures of −50 °F (−46 °C) have been claimed in an optimized version of this apparatus. Thus far the inefficiency of the process has prevented it from being widely adopted. (See Problem 11.4LL.)

---

## 11.2    Mach Number and Speed of Sound

The **Mach number**, Ma, was introduced in Chapters 1 and 7 as a dimensionless measure of compressibility in a fluid flow. In this and subsequent sections, we develop some useful relationships involving the Mach number. The Mach number is defined as the ratio of the value of the local flow velocity, $V$, to the local **speed of sound**, $c$. In other words,

$$\text{Ma} = \frac{V}{c}$$

*Mach number is the ratio of local flow and sound speeds.*

What we perceive as sound generally consists of weak pressure pulses that move through air with a Mach number of one. When our eardrums respond to a succession of moving pressure pulses, we hear sounds.

To better understand the notion of speed of sound, we analyze the one-dimensional fluid mechanics of an infinitesimally thin, weak pressure pulse moving at the speed of sound through a fluid at rest (see Fig. 11.1a). Ahead of the pressure pulse, the fluid velocity is zero, and the fluid pressure and density are $p$ and $\rho$. Behind the pressure pulse, the fluid velocity has changed by an amount $\delta V$, and the pressure and density of the fluid have also changed by amounts $\delta p$ and $\delta \rho$. We select an infinitesimally thin control volume that moves with the pressure pulse as is sketched

■ **Figure 11.1**   (*a*) **Weak pressure pulse moving through a fluid at rest.**
(*b*) **The flow relative to a control volume containing a weak pressure pulse.**

in Fig. 11.1$a$. The speed of the weak pressure pulse is considered constant and in one direction only; thus, our control volume is inertial.

For an observer moving with this control volume (Fig. 11.1$b$), it appears as if fluid is entering the control volume through surface area $A$ with speed $c$ at pressure $p$ and density $\rho$ and leaving the control volume through surface area $A$ with speed $c - \delta V$, pressure $p + \delta p$, and density $\rho + \delta \rho$. When the continuity equation (Eq. 5.16) is applied to the flow through this control volume, the result is

$$\rho A c = (\rho + \delta\rho)A(c - \delta V) \tag{11.26}$$

*The changes in fluid properties across a sound wave are very small compared to their local values.*

or

$$\rho c = \rho c - \rho\,\delta V + c\,\delta\rho - (\delta\rho)(\delta V) \tag{11.27}$$

Since $(\delta\rho)(\delta V)$ is much smaller than the other terms in Eq. 11.27, we drop it from further consideration and keep

$$\rho\,\delta V = c\,\delta\rho \tag{11.28}$$

The linear momentum equation (Eq. 5.29) can also be applied to the flow through the control volume of Fig. 11.1$b$. The result is

$$-c\rho c A + (c - \delta V)(\rho + \delta\rho)(c - \delta V)A = pA - (p + \delta p)A \tag{11.29}$$

Note that any frictional forces are considered as being negligibly small. We again neglect higher order terms [such as $(\delta V)^2$ compared to $c\,\delta V$, for example] and combine Eqs. 11.26 and 11.29 to get

$$-c\rho c A + (c - \delta V)\rho A c = -\delta p A$$

or

$$\rho\delta V = \frac{\delta p}{c} \tag{11.30}$$

From Eqs. 11.28 (continuity) and 11.30 (linear momentum) we obtain

$$c^2 = \frac{\delta p}{\delta\rho}$$

or

$$c = \sqrt{\frac{\delta p}{\delta\rho}} \tag{11.31}$$

This expression for the speed of sound results from application of the conservation of mass and conservation of linear momentum principles to the flow through the control volume of Fig. 11.1$b$. These principles were similarly used in Section 10.2.1 to obtain an expression for the speed of surface waves traveling on the surface of fluid in a channel.

The conservation of energy principle can also be applied to the flow through the control volume of Fig. 11.1$b$. If the energy equation (Eq. 5.103) is used for the flow through this control volume, the result is

$$\frac{\delta p}{\rho} + \delta\left(\frac{V^2}{2}\right) + g\,\delta z = \delta(\text{loss}) \tag{11.32}$$

For gas flow we can consider $g\,\delta z$ as being negligibly small in comparison to the other terms in the equation. Also, if we assume that the flow is frictionless, then $\delta(\text{loss}) = 0$ and Eq. 11.32 becomes

$$\frac{\delta p}{\rho} + \frac{(c - \delta V)^2}{2} - \frac{c^2}{2} = 0$$

or, neglecting $(\delta V)^2$ compared to $c\,\delta V$, we obtain

$$\rho\,\delta V = \frac{\delta p}{c} \tag{11.33}$$

By combining Eqs. 11.28 (continuity) and 11.33 (energy) we again find that

$$c = \sqrt{\frac{\delta p}{\delta \rho}}$$

which is identical to Eq. 11.31. Thus, the conservation of linear momentum and the conservation of energy principles lead to the same result. If we further assume that the frictionless flow through the control volume of Fig. 11.1*b* is adiabatic (no heat transfer), then the flow is isentropic. In the limit, as $\delta p$ becomes vanishingly small ($\delta p \rightarrow \partial p \rightarrow 0$)

$$c = \sqrt{\left(\frac{\partial p}{\partial \rho}\right)_s} \qquad (11.34)$$

where the subscript $s$ is used to designate that the partial differentiation occurs at constant entropy.

Equation 11.34 suggests to us that we can calculate the speed of sound by determining the partial derivative of pressure with respect to density at constant entropy. For the isentropic flow of an ideal gas (with constant $c_p$ and $c_v$), we learned earlier (Eq. 11.25) that

$$p = (\text{constant})(\rho^k)$$

and thus

$$\left(\frac{\partial p}{\partial \rho}\right)_s = (\text{constant}) \, k\rho^{k-1} = \frac{p}{\rho^k} k\rho^{k-1} = \frac{p}{\rho} k = RTk \qquad (11.35)$$

Thus, for an ideal gas

$$c = \sqrt{RTk} \qquad (11.36)$$

From Eq. 11.36 and the charts in the margin we conclude that for a given temperature, the speed of sound, $c$, in hydrogen and in helium, is higher than in air.

More generally, the bulk modulus of elasticity, $E_v$, of any fluid including liquids is defined as (see Section 1.7.1)

$$E_v = \frac{dp}{d\rho/\rho} = \rho\left(\frac{\partial p}{\partial \rho}\right)_s \qquad (11.37)$$

*Speed of sound is larger in fluids that are more difficult to compress.*

Thus, in general, from Eqs. 11.34 and 11.37,

$$c = \sqrt{\frac{E_v}{\rho}} \qquad (11.38)$$

Values of the speed of sound are tabulated in Tables B.1 and B.2 for water and in Tables B.3 and B.4 for air. From experience we know that air is more easily compressed than water. Note from the values of $c$ in Tables B.1 through B.4 and the graph in the margin that the speed of sound in air is much less than it is in water. From Eq. 11.37, we can conclude that if a fluid is truly incompressible, its bulk modulus would be infinitely large, as would be the speed of sound in that fluid. Thus, an incompressible flow must be considered an idealized approximation of reality.

---

# F l u i d s   i n   t h e   N e w s

**Sonification** The normal human ear is capable of detecting even very subtle sound patterns produced by *sound waves.* Most of us can distinguish the bark of a dog from the meow of a cat or the roar of a lion, or identify a person's voice on the telephone before they identify who is calling. The number of "things" we can identify from subtle sound patterns is enormous. Combine this ability with the power of computers to transform the information from sensor transducers into variations in pitch, rhythm, and volume and you have *sonification,* the representation of data in the form of sound. With this emerging technology, pathologists may soon learn to "hear" abnormalities in tissue samples, engineers may "hear" flaws in gas turbine engine blades being inspected, and scientists may "hear" a desired attribute in a newly invented material. Perhaps the concept of hearing the trends in data sets may become as commonplace as seeing them. Analysts may listen to the stock market and make decisions. Of course, none of this can happen in a vacuum.

**EXAMPLE 11.3** Speed of Sound

**GIVEN** Consider the data in Table B.4.

**FIND** Verify the speed of sound for air at 0 °C.

**SOLUTION**

In Table B.4, we find the speed of sound of air at 0 °C given as 331.4 m/s. Assuming that air behaves as an ideal gas, we can calculate the speed of sound from Eq. 11.36 as

$$c = \sqrt{RTk} \qquad (1)$$

The value of the gas constant is obtained from Table 1.8 as

$$R = 286.9 \text{ J/(kg} \cdot \text{K)}$$

and the specific heat ratio is listed in Table B.4 as

$$k = 1.401$$

By substituting values of $R$, $k$, and $T$ into Eq. 1 we obtain

$$c = \sqrt{[(286.9) \text{ J/(kg} \cdot \text{K)}](273.15 \text{ K})(1.401)}$$
$$= 331.4 \text{ (J/kg)}^{1/2}$$

Thus, since $1 \text{ J/kg} = 1 \text{ N} \cdot \text{m/kg} = 1 \text{ (kg} \cdot \text{m/s}^2) \cdot \text{m/kg} = 1 \text{ (m/s)}^2$, we obtain

$$c = 331.4 \text{ m/s} \qquad \textbf{(Ans)}$$

**COMMENT** The value of the speed of sound calculated with Eq. 11.36 agrees very well with the value of $c$ listed in Table B.4. The ideal gas approximation does not compromise this result significantly.

## 11.3 Categories of Compressible Flow

*Compressibility effects are more important at higher Mach numbers.*

In Section 3.8.1, we learned that the effects of compressibility become more significant as the Mach number increases. For example, the error associated with using $\rho V^2/2$ in calculating the *stagnation pressure* of an ideal gas increases at larger Mach numbers. From Fig. 3.24 we can conclude that incompressible flows can only occur at low Mach numbers.

Experience has also demonstrated that compressibility can have a large influence on other important flow variables. For example, in Fig. 11.2 the variation of drag coefficient with Reynolds

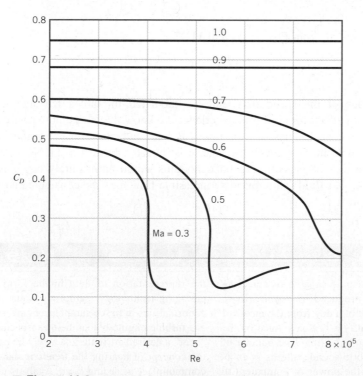

■ **Figure 11.2** **The variation of the drag coefficient of a sphere with Reynolds number and Mach number. (Data from Fig. 1.8 in Ref. 1 of Chapter 9.)**

*The wave pattern from a moving source is not symmetrical.*

number and Mach number is shown for airflow over a sphere. Compressibility effects can be of considerable importance.

To further illustrate some curious features of compressible flow, a simplified example is considered. Imagine the emission of weak pressure pulses from a point source. These pressure waves are spherical and expand radially outward from the point source at the speed of sound, $c$. If a pressure wave is emitted at different times, $t_{wave}$, we can determine where several waves will be at a common instant of time, $t$, by using the relationship

$$r = (t - t_{wave})c$$

where $r$ is the radius of the sphere-shaped wave emitted at time $= t_{wave}$. For a stationary point source, the symmetrical wave pattern shown in Fig. 11.3a is involved.

When the point source moves to the left with a constant velocity, $V$, the wave pattern is no longer symmetrical. In Figs. 11.3b, 11.3c, and 11.3d are illustrated the wave patterns at $t = 3$ s for different values of $V$. Also shown with a "+" are the positions of the moving point source at values of time, $t$, equal to 0 s, 1 s, 2 s, and 3 s. Knowing where the point source has been at different instances is important because it indicates to us where the different waves originated.

From the pressure wave patterns of Fig. 11.3, we can draw some useful conclusions. Before doing this we should recognize that if instead of moving the point source to the left, we held the point source stationary and moved the fluid to the right with velocity $V$, the resulting pressure wave patterns would be identical to those indicated in Fig. 11.3.

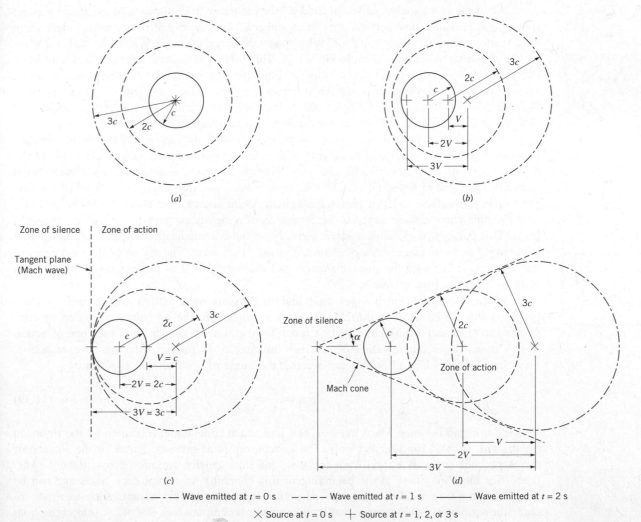

■ **Figure 11.3**   (a) Pressure waves at $t = 3$ s, $V = 0$; (b) pressure waves at $t = 3$ s, $V < c$; (c) pressure waves at $t = 3$ s, $V = c$; (d) pressure waves at $t = 3$ s, $V > c$.

When the point source and the fluid are stationary, the pressure wave pattern is symmetrical (Fig. 11.3*a*) and an observer anywhere in the pressure field would hear the same sound frequency from the point source. When the velocity of the point source (or the fluid) is very small in comparison with the speed of sound, the pressure wave pattern will still be nearly symmetrical. The speed of sound in an incompressible fluid is infinitely large. Thus, the stationary point source and stationary fluid situation are representative of incompressible flows. For truly incompressible flows, the communication of pressure information throughout the flow field is unrestricted and instantaneous ($c = \infty$).

---

*V11.3 Jet noise*

When the point source moves in fluid at rest (or when fluid moves past a stationary point source), the pressure wave patterns vary in asymmetry, with the extent of asymmetry depending on the ratio of the point source (or fluid) velocity and the speed of sound. When $V/c < 1$, the wave pattern is similar to the one shown in Fig. 11.3*b*. This flow is considered **subsonic** and compressible. A stationary observer will hear a different sound frequency coming from the point source depending on where the observer is relative to the source because the wave pattern is asymmetrical. We call this phenomenon the Doppler effect. Pressure information can still travel unrestricted throughout the flow field, but not symmetrically or instantaneously.

When $V/c = 1$, pressure waves are not present ahead of the moving point source. The flow is **sonic.** If you were positioned to the left of the moving point source, you would not hear the point source until it was coincident with your location. For flow moving past a stationary point source at the speed of sound ($V/c = 1$), the pressure waves are all tangent to a plane that is perpendicular to the flow and that passes through the point source. The concentration of pressure waves in this tangent plane suggests the formation of a significant pressure variation across the plane. This plane is often called a **Mach wave.** Note that communication of pressure information is restricted to the region of flow downstream of the Mach wave. The region of flow upstream of the Mach wave is called the *zone of silence,* and the region of flow downstream of the tangent plane is called the *zone of action.*

When $V > c$, the flow is **supersonic,** and the pressure wave pattern resembles the one depicted in Fig. 11.3*d*. A cone (**Mach cone**) that is tangent to the pressure waves can be constructed to represent the Mach wave that separates the zone of silence from the zone of action in this case. The communication of pressure information is restricted to the zone of action. From the sketch of Fig. 11.3*d*, we can see that the angle of this cone, $\alpha$, is given by

$$\sin \alpha = \frac{c}{V} = \frac{1}{\text{Ma}} \tag{11.39}$$

*V11.4 Speed boat*

This relationship between Mach number, Ma, and Mach cone angle, $\alpha$, shown by the figure in the margin, is valid for $V/c > 1$ only. The concentration of pressure waves at the surface of the Mach cone suggests a significant pressure, and thus density, variation across the cone surface. (See the photograph at the beginning of this chapter.) An abrupt density change can be visualized in a flow field by using special optics. Examples of flow visualization methods include the schlieren, shadowgraph, and interferometer techniques (see Ref. 4). A schlieren photo of a flow for which $V > c$ is shown in Fig. 11.4. The airflow through the row of compressor blade airfoils is as shown with the arrow. The flow enters supersonically ($\text{Ma}_1 = 1.14$) and

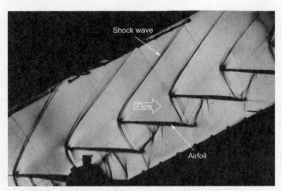

■ **Figure 11.4**  The Schlieren visualization of flow (supersonic to subsonic) through a row of compressor airfoils. (Photograph courtesy of TAGT-1.2 Cascade, R. Fuchs, H. A. Schreiber, W. Steinert, German Aerospace Center (DLR), Institute of Propulsion Technology, 1996.)

*V11.5 Compressible flow visualization*

leaves subsonically ($\mathrm{Ma}_2 = 0.86$). The center two airfoils have pressure tap hoses connected to them. Regions of significant changes in fluid density appear in the supersonic portion of the flow. Also, the region of separated flow on each airfoil is visible.

This discussion about pressure wave patterns suggests the following categories of fluid flow:

1. Incompressible flow: $\mathrm{Ma} \leq 0.3$. Unrestricted, nearly symmetrical and instantaneous pressure communication.
2. Compressible subsonic flow: $0.3 < \mathrm{Ma} < 1.0$. Unrestricted but noticeably asymmetrical pressure communication.
3. Compressible supersonic flow: $\mathrm{Ma} \geq 1.0$. Formation of Mach wave; pressure communication restricted to zone of action.

In addition to the above-mentioned categories of flows, two other regimes are commonly referred to: namely, *transonic flows* ($0.9 \leq \mathrm{Ma} \leq 1.2$) and *hypersonic flows* ($\mathrm{Ma} > 5$). Modern aircraft are mainly powered by gas turbine engines that involve transonic flows. When a space shuttle reenters the Earth's atmosphere, the flow is hypersonic. Future aircraft may be expected to operate from subsonic to hypersonic flow conditions.

### F l u i d s   i n   t h e   N e w s

**Supersonic and compressible flows in gas turbines**  Modern gas turbine engines commonly involve compressor and turbine blades that are moving so fast that the fluid flows over the blades are locally *supersonic*. Density varies considerably in these flows so they are also considered to be *compressible*. Shock waves can form when these supersonic flows are sufficiently decelerated. Shocks formed at blade leading edges or on blade surfaces can interact with other blades and shocks and seriously affect blade aerodynamic and structural performance. It is possible to have supersonic flows past blades near the outer diameter of a rotor with *subsonic flows* near the inner diameter of the same rotor. These rotors are considered to be *transonic* in their operation. Very large aero gas turbines can involve thrust levels exceeding 100,000 lb. Two of these engines are sufficient to carry over 350 passengers halfway around the world at high subsonic speed. (See Problem 11.5LL.)

### *E*XAMPLE 11.4    Mach Cone

**GIVEN**  An aircraft cruising at 1000-m elevation, $z$, above you moves past in a flyby. It is moving with a Mach number equal to 1.5 and the ambient temperature is 20 °C.

**FIND**  How many seconds after the plane passes overhead do you expect to wait before you hear the aircraft?

#### SOLUTION

Since the aircraft is moving supersonically ($\mathrm{Ma} > 1$), we can imagine a Mach cone originating from the forward tip of the craft

as is illustrated in Fig. E11.4a. A photograph of this phenomenon is shown in Fig. E11.4b. When the surface of the cone reaches the

■ **Figure E11.4a**

■ **Figure E11.4b** NASA Schlieren photograph of shock waves from a T-38 aircraft at Mach 1.1, 13,000 ft. (Photo courtesy of NASA)

observer, the "sound" of the aircraft is perceived. The angle $\alpha$ in Fig. E11.4 is related to the elevation of the plane, $z$, and the ground distance, $x$, by

$$\alpha = \tan^{-1}\frac{z}{x} = \tan^{-1}\frac{1000}{Vt} \qquad (1)$$

Also, assuming negligible change of Mach number with elevation, we can use Eq. 11.39 to relate Mach number to the angle $\alpha$. Thus,

$$\text{Ma} = \frac{1}{\sin \alpha} \qquad (2)$$

Combining Eqs. 1 and 2 we obtain

$$\text{Ma} = \frac{1}{\sin \left[\tan^{-1}\left(1000/Vt\right)\right]} \qquad (3)$$

The speed of the aircraft can be related to the Mach number with

$$V = (\text{Ma})c \qquad (4)$$

where $c$ is the speed of sound. From Table B.4, $c = 343.3$ m/s. Using Ma = 1.5, we get from Eqs. 3 and 4

$$1.5 = \frac{1}{\sin\left\{\tan^{-1}\left[\dfrac{1000\ \text{m}}{(1.5)(343.3\ \text{m/s})t}\right]\right\}}$$

or

$$t = 2.17\ \text{s} \qquad \textbf{(Ans)}$$

**COMMENT** By repeating the calculations for various values of Mach number, Ma, the results shown in Fig. E11.4c are obtained. Note that for subsonic flight (Ma < 1) there is no delay since the sound travels faster than the aircraft. You can hear a subsonic aircraft approaching.

■ **Figure E11.4c**

## 11.4 Isentropic Flow of an Ideal Gas

In this section, we consider in further detail the steady, one-dimensional, isentropic flow of an ideal gas with constant specific heat values ($c_p$ and $c_v$). Because the flow is steady throughout, shaft work cannot be involved. Also, as explained earlier, the one-dimensionality of flows we discuss in this chapter implies velocity and fluid property changes in the streamwise direction only. We consider flows through finite control volumes with uniformly distributed velocities and fluid properties at each section of flow. Much of what we develop can also apply to the flow of a fluid particle along its pathline.

*An important class of isentropic flow involves no heat transfer and zero friction.*

Isentropic flow involves constant entropy and was discussed earlier in Section 11.1, where we learned that adiabatic and frictionless (reversible) flow is one form of isentropic flow. Some ideal gas relationships for isentropic flows were developed in Section 11.1. An isentropic flow is not achievable with actual fluids because of friction. Nonetheless, the study of isentropic flow trends is useful because it helps us to gain an understanding of actual compressible flow phenomena including choked flow, shock waves, acceleration from subsonic to supersonic flow, and deceleration from supersonic to subsonic flow.

### 11.4.1  Effect of Variations in Flow Cross-Sectional Area

When fluid flows steadily through a conduit that has a flow cross-sectional area that varies with axial distance, the conservation of mass (continuity) equation

$$\dot{m} = \rho A V = \text{constant} \tag{11.40}$$

can be used to relate the flow rates at different sections. For incompressible flow, the fluid density remains constant and the flow velocity from section to section varies inversely with cross-sectional area. However, when the flow is compressible, density, cross-sectional area, and flow velocity can all vary from section to section. We proceed to determine how fluid density and flow velocity change with axial location in a variable area duct when the fluid is an ideal gas and the flow through the duct is steady and isentropic.

In Chapter 3, Newton's second law was applied to the inviscid (frictionless) and steady flow of a fluid particle. For the streamwise direction, the result (Eq. 3.5) for either compressible or incompressible flows is

$$dp + \tfrac{1}{2}\rho \, d(V^2) + \gamma \, dz = 0 \tag{11.41}$$

The frictionless flow from section to section through a finite control volume is also governed by Eq. 11.41, if the flow is one-dimensional, because every particle of fluid involved will have the same experience. For ideal gas flow, the potential energy difference term, $\gamma \, dz$, can be dropped because of its small size in comparison to the other terms, namely, $dp$ and $d(V^2)$. Thus, an appropriate equation of motion in the streamwise direction for the steady, one-dimensional, and isentropic (adiabatic and frictionless) flow of an ideal gas is obtained from Eq. 11.41 as

$$\frac{dp}{\rho V^2} = -\frac{dV}{V} \tag{11.42}$$

If we form the logarithm of both sides of the continuity equation (Eq. 11.40), the result is

$$\ln \rho + \ln A + \ln V = \text{constant} \tag{11.43}$$

*Density, cross-sectional area, and velocity may all vary for a compressible flow.*

Differentiating Eq. 11.43 we get

$$\frac{d\rho}{\rho} + \frac{dA}{A} + \frac{dV}{V} = 0$$

or

$$-\frac{dV}{V} = \frac{d\rho}{\rho} + \frac{dA}{A} \tag{11.44}$$

Now we combine Eqs. 11.42 and 11.44 to obtain

$$\frac{dp}{\rho V^2}\left(1 - \frac{V^2}{dp/d\rho}\right) = \frac{dA}{A} \tag{11.45}$$

Since the flow being considered is isentropic, the speed of sound is related to variations of pressure with density by Eq. 11.34, repeated here for convenience as

$$c = \sqrt{\left(\frac{\partial p}{\partial \rho}\right)_s}$$

Equation 11.34, combined with the definition of Mach number

$$\text{Ma} = \frac{V}{c} \tag{11.46}$$

and Eq. 11.45 yields

$$\frac{dp}{\rho V^2}(1 - \text{Ma}^2) = \frac{dA}{A} \tag{11.47}$$

Equations 11.42 and 11.47 merge to form

$$\frac{dV}{V} = -\frac{dA}{A}\frac{1}{(1 - \mathrm{Ma}^2)} \tag{11.48}$$

We can use Eq. 11.48 to conclude that when the flow is subsonic ($\mathrm{Ma} < 1$), velocity and section area changes are in opposite directions. In other words, the area increase associated with subsonic flow through a diverging duct like the one shown in Fig. 11.5a is accompanied by a velocity decrease. Subsonic flow through a converging duct (see Fig. 11.5b) involves an increase of velocity. These trends are consistent with incompressible flow behavior, which we described earlier in this book, for instance, in Chapters 3 and 8.

*A converging duct will decelerate a supersonic flow and accelerate a subsonic flow.*

Equation 11.48 also serves to show us that when the flow is supersonic ($\mathrm{Ma} > 1$), velocity and area changes are in the same direction. A diverging duct (Fig. 11.5a) will accelerate a supersonic flow. A converging duct (Fig. 11.5b) will decelerate a supersonic flow. These trends are the opposite of what happens for incompressible and subsonic compressible flows.

To better understand why subsonic and supersonic duct flows are so different, we combine Eqs. 11.44 and 11.48 to form

$$\frac{d\rho}{\rho} = \frac{dA}{A}\frac{\mathrm{Ma}^2}{(1 - \mathrm{Ma}^2)} \tag{11.49}$$

Using Eq. 11.49, we can conclude that for subsonic flows ($\mathrm{Ma} < 1$), density and area changes are in the same direction, whereas for supersonic flows ($\mathrm{Ma} > 1$), density and area changes are in opposite directions. Since $\rho AV$ must remain constant (Eq. 11.40), when the duct diverges and the flow is subsonic, density and area both increase and thus flow velocity must decrease. However, for supersonic flow through a diverging duct, when the area increases, the density decreases enough so that the flow velocity has to increase to keep $\rho AV$ constant.

By rearranging Eq. 11.48, we can obtain

$$\frac{dA}{dV} = -\frac{A}{V}(1 - \mathrm{Ma}^2) \tag{11.50}$$

Equation 11.50 gives us some insight into what happens when $\mathrm{Ma} = 1$. For $\mathrm{Ma} = 1$, Eq. 11.50 requires that $dA/dV = 0$. This result suggests that the area associated with $\mathrm{Ma} = 1$ is either a minimum or a maximum amount.

A *converging–diverging duct* (Fig. 11.6a and margin photograph) involves a minimum area. If the flow entering such a duct were subsonic, Eq. 11.48 discloses that the fluid velocity would increase in the converging portion of the duct, and achievement of a sonic condition ($\mathrm{Ma} = 1$) at the minimum area location appears possible. If the flow entering the converging–diverging duct is supersonic, Eq. 11.48 states that the fluid velocity would decrease in the converging portion of the duct and the sonic condition at the minimum area is possible.

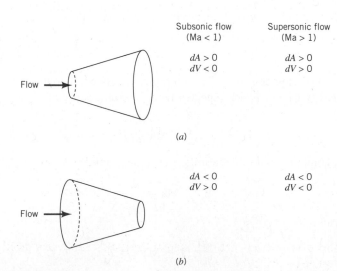

■ **Figure 11.5** (a) A diverging duct. (b) A converging duct.

**■ Figure 11.6**  (*a*) A converging–diverging duct. (*b*) A diverging–converging duct.

A diverging–converging duct (Fig. 11.6*b*), on the other hand, would involve a maximum area. If the flow entering this duct were subsonic, the fluid velocity would decrease in the diverging portion of the duct and the sonic condition could not be attained at the maximum area location. For supersonic flow in the diverging portion of the duct, the fluid velocity would increase and thus Ma = 1 at the maximum area is again impossible.

For the steady isentropic flow of an ideal gas, we conclude that the sonic condition (Ma = 1) can be attained in a converging–diverging duct at the minimum area location. This minimum area location is often called the ***throat*** of the converging–diverging duct. Furthermore, to achieve supersonic flow from a subsonic state in a duct, a converging–diverging area variation is necessary. For this reason, we often refer to such a duct as a *converging–diverging nozzle*. Note that a converging–diverging duct can also decelerate a supersonic flow to subsonic conditions. Thus, a converging–diverging duct can be a nozzle or a diffuser depending on whether the flow in the converging portion of the duct is subsonic or supersonic. A supersonic wind tunnel test section is generally preceded by a converging–diverging nozzle and followed by a converging–diverging diffuser (see Ref. 1). Further details about steady, isentropic, ideal gas flow through a converging–diverging duct are discussed in the next section.

*A converging–diverging duct is required to accelerate a flow from subsonic to supersonic flow conditions.*

### 11.4.2  Converging–Diverging Duct Flow

In the preceding section, we discussed the variation of density and velocity of the steady isentropic flow of an ideal gas through a variable area duct. We proceed now to develop equations that help us determine how other important flow properties vary in these flows.

It is convenient to use the stagnation state of the fluid as a reference state for compressible flow calculations. The stagnation state is associated with zero flow velocity and an entropy value that corresponds to the entropy of the flowing fluid. The subscript 0 is used to designate the stagnation state. Thus, stagnation temperature and pressure are $T_0$ and $p_0$. For example, if the fluid flowing through the converging–diverging duct of Fig. 11.6*a* were drawn isentropically from the atmosphere, the atmospheric pressure and temperature would represent the stagnation state of the flowing fluid. The stagnation state can also be achieved by isentropically decelerating a flow to zero velocity. This can be accomplished with a diverging duct for subsonic flows or a converging–diverging duct for supersonic flows. Also, as discussed earlier in Chapter 3, an approximately isentropic deceleration can be accomplished with a Pitot-static tube (see Fig. 3.6). It is thus possible to measure, with only a small amount of uncertainty, values of stagnation pressure, $p_0$, and stagnation temperature, $T_0$, of a flowing fluid.

In Section 11.1, we demonstrated that for the isentropic flow of an ideal gas (see Eq. 11.25)

$$\frac{p}{\rho^k} = \text{constant} = \frac{p_0}{\rho_0^k}$$

The streamwise equation of motion for steady and frictionless flow (Eq. 11.41) can be expressed for an ideal gas as

$$\frac{dp}{\rho} + d\left(\frac{V^2}{2}\right) = 0 \tag{11.51}$$

since the potential energy term, $\gamma\, dz$, can be considered as being negligibly small in comparison with the other terms involved.

By incorporating Eq. 11.25 into Eq. 11.51 we obtain

$$\frac{p_0^{1/k}}{\rho_0}\frac{dp}{(p)^{1/k}} + d\left(\frac{V^2}{2}\right) = 0 \tag{11.52}$$

Consider the steady, one-dimensional, isentropic flow of an ideal gas with constant $c_p$ and $c_v$ through the converging–diverging nozzle of Fig. 11.6a. Equation 11.52 is valid for this flow and can be integrated between the common stagnation state of the flowing fluid to the state of the gas at any location in the converging–diverging duct to give

$$\frac{k}{k-1}\left(\frac{p_0}{\rho_0} - \frac{p}{\rho}\right) - \frac{V^2}{2} = 0 \tag{11.53}$$

By using the ideal gas equation of state (Eq. 11.1) with Eq. 11.53, we obtain

$$\frac{kR}{k-1}(T_0 - T) - \frac{V^2}{2} = 0 \tag{11.54}$$

It is of interest to note that combining Eqs. 11.14 and 11.54 leads to

$$c_p(T_0 - T) - \frac{V^2}{2} = 0$$

which, when merged with Eq. 11.9, results in

$$\check{h}_0 - \left(\check{h} + \frac{V^2}{2}\right) = 0 \tag{11.55}$$

where $\check{h}_0$ is the stagnation enthalpy. If the steady-flow energy equation (Eq. 5.69) is applied to the flow situation we are presently considering, the resulting equation will be identical to Eq. 11.55. Further, we conclude that the stagnation enthalpy is constant. The conservation of momentum and energy principles lead to the same equation (Eq. 11.55) for steady isentropic flows.

The definition of Mach number (Eq. 11.46) and the speed of sound relationship for ideal gases (Eq. 11.36) can be combined with Eq. 11.54 to yield

*For isentropic flows the temperature, pressure, and density ratios are functions of the Mach number.*

$$\boxed{\frac{T}{T_0} = \frac{1}{1 + [(k-1)/2]\text{Ma}^2}} \tag{11.56}$$

which is graphed in the margin for air. With Eq. 11.56 we can calculate the temperature of an ideal gas anywhere in the converging–diverging duct of Fig. 11.6a if the flow is steady, one-dimensional, and isentropic, provided we know the value of the local Mach number and the stagnation temperature.

We can also develop an equation for pressure variation. Since $p/\rho = RT$, then

$$\left(\frac{p}{p_0}\right)\left(\frac{\rho_0}{\rho}\right) = \frac{T}{T_0} \tag{11.57}$$

From Eqs. 11.57 and 11.25 we obtain

$$\left(\frac{p}{p_0}\right) = \left(\frac{T}{T_0}\right)^{k/(k-1)} \tag{11.58}$$

Combining Eqs. 11.58 and 11.56 leads to

$$\boxed{\frac{p}{p_0} = \left\{\frac{1}{1 + [(k-1)/2]\text{Ma}^2}\right\}^{k/(k-1)}} \tag{11.59}$$

For density variation we consolidate Eqs. 11.56, 11.57, and 11.59 to get

$$\boxed{\frac{\rho}{\rho_0} = \left\{\frac{1}{1 + [(k-1)/2]\text{Ma}^2}\right\}^{1/(k-1)}} \tag{11.60}$$

These relationships are graphed in the margin for air.

A very useful means of keeping track of the states of an isentropic flow of an ideal gas involves a *temperature–entropy (T–s) diagram,* as is shown in Fig. 11.7. Experience has shown (see, for example, Refs. 2 and 3) that lines of constant pressure are generally as are sketched in Fig. 11.7. An isentropic flow is confined to a vertical line on a $T–s$ diagram. The vertical line in Fig. 11.7 is representative of flow between the stagnation state and any state within the converging–diverging nozzle. Equation 11.56 shows that fluid temperature decreases with an increase in Mach number. Thus, the lower temperature levels on a $T–s$ diagram correspond to higher Mach numbers. Equation 11.59 suggests that fluid pressure also decreases with an increase in Mach number. Thus, lower fluid temperatures and pressures are associated with higher Mach numbers in our isentropic converging–diverging duct example.

One way to produce flow through a converging–diverging duct like the one in Fig. 11.6a is to connect the downstream end of the duct to a vacuum pump. When the pressure at the downstream end of the duct (the back pressure) is decreased slightly, air will flow from the atmosphere through the duct and vacuum pump. Neglecting friction and heat transfer and considering the air to act as an ideal gas, Eqs. 11.56, 11.59, and 11.60 and a $T–s$ diagram can be used to describe steady flow through the converging–diverging duct.

If the pressure in the duct is only slightly less than atmospheric pressure, we predict with Eq. 11.59 that the Mach number levels in the duct will be low. Thus, with Eq. 11.60 we conclude that the variation of fluid density in the duct is also small. The continuity equation (Eq. 11.40) leads us to state that there is a small amount of fluid flow acceleration in the converging portion of the duct followed by flow deceleration in the diverging portion of the duct. We considered this type of flow when we discussed the Venturi meter in Section 3.6.3. The $T–s$ diagram for this flow is sketched in Fig. 11.8.

We next consider what happens when the back pressure is lowered further. Since the flow starts from rest upstream of the converging portion of the duct of Fig. 11.6a, Eqs. 11.48 and 11.50 reveal to us that flow up to the nozzle throat can be accelerated to a maximum allowable Mach number of 1 at the throat. Thus, when the duct back pressure is lowered sufficiently, the Mach number at the throat of the duct will be 1. Any further decrease of the back pressure will not affect the flow in the converging portion of the duct because, as is discussed in Section 11.3, information about pressure cannot move upstream when Ma = 1. When Ma = 1 at the throat of the converging–diverging duct, we have a condition called *choked flow.* Some useful equations for choked flow are developed below.

*Choked flow occurs when the Mach number is 1.0 at the minimum cross-sectional area.*

We have already used the stagnation state for which Ma = 0 as a reference condition. It will prove helpful to us to use the state associated with Ma = 1 and the same entropy level as the flowing fluid as another reference condition we shall call the *critical state,* denoted ( )*.

The ratio of pressure at the converging–diverging duct throat for choked flow, $p^*$, to stagnation pressure, $p_0$, is referred to as the *critical pressure ratio.* By substituting Ma = 1 into Eq. 11.59 we obtain

$$\frac{p^*}{p_0} = \left(\frac{2}{k+1}\right)^{k/(k-1)} \tag{11.61}$$

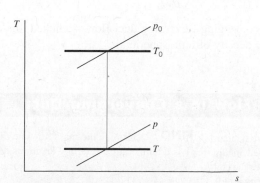

■ **Figure 11.7** The (*T–s*) diagram relating stagnation and static states.

■ **Figure 11.8** The *T–s* diagram for Venturi meter flow.

■ **Figure 11.9** The relationship between the stagnation and critical states.

For $k = 1.4$, the nominal value of $k$ for air, Eq. 11.61 yields

$$\left(\frac{p^*}{p_0}\right)_{k=1.4} = 0.528 \qquad (11.62)$$

Because the stagnation pressure for our converging–diverging duct example is the atmospheric pressure, $p_{\text{atm}}$, the throat pressure for choked air flow is, from Eq. 11.62

$$p^*_{k=1.4} = 0.528 p_{\text{atm}}$$

We can get a relationship for the critical temperature ratio, $T^*/T_0$, by substituting Ma = 1 into Eq. 11.56. Thus,

$$\frac{T^*}{T_0} = \frac{2}{k+1} \qquad (11.63)$$

or for $k = 1.4$

$$\left(\frac{T^*}{T_0}\right)_{k=1.4} = 0.833 \qquad (11.64)$$

For the duct of Fig. 11.6a, Eq. 11.64 yields

$$T^*_{k=1.4} = 0.833 T_{\text{atm}}$$

The stagnation and critical pressures and temperatures are shown on the $T–s$ diagram of Fig. 11.9.

*The stagnation and critical states are at the same entropy level.*

When we combine the ideal gas equation of state (Eq. 11.1) with Eqs. 11.61 and 11.63, for Ma = 1 we get

$$\frac{\rho^*}{\rho_0} = \left(\frac{p^*}{T^*}\right)\left(\frac{T_0}{p_0}\right) = \left(\frac{2}{k+1}\right)^{k/(k-1)}\left(\frac{k+1}{2}\right) = \left(\frac{2}{k+1}\right)^{1/(k-1)} \qquad (11.65)$$

For air ($k = 1.4$), Eq. 11.65 leads to

$$\left(\frac{\rho^*}{\rho_0}\right)_{k=1.4} = 0.634 \qquad (11.66)$$

and we see that when the converging–diverging duct flow is choked, the density of the air at the duct throat is 63.4% of the density of atmospheric air.

---

## $E$XAMPLE 11.5    Isentropic Flow in a Converging Duct

**GIVEN** A converging duct passes air steadily from standard atmospheric conditions to a receiver pipe as illustrated in Fig. E11.5a. The throat (minimum) flow cross-sectional area of the converging duct is $1 \times 10^{-4}$ m$^2$. The receiver pressure is **(a)** 80 kPa (abs), **(b)** 40 kPa (abs).

**FIND** Determine the mass flowrate through the duct and sketch temperature–entropy diagrams for situations **(a)** and **(b)**.

# SOLUTION

To determine the mass flowrate through the converging duct, we use Eq. 11.40. Thus,

$$\dot{m} = \rho A V = \text{constant}$$

or in terms of the given throat area, $A_{th}$,

$$\dot{m} = \rho_{th} A_{th} V_{th} \tag{1}$$

We assume that the flow through the converging duct is isentropic and that the air behaves as an ideal gas with constant $c_p$ and $c_v$. Then, from Eq. 11.60

$$\frac{\rho_{th}}{\rho_0} = \left\{ \frac{1}{1 + [(k-1)/2]\text{Ma}_{th}^2} \right\}^{1/(k-1)} \tag{2}$$

The stagnation density, $\rho_0$, for the standard atmosphere is 1.23 kg/m³, and the specific heat ratio is 1.4. To determine the throat Mach number, $\text{Ma}_{th}$, we can use Eq. 11.59,

$$\frac{p_{th}}{p_0} = \left\{ \frac{1}{1 + [(k-1)/2]\text{Ma}_{th}^2} \right\}^{k/(k-1)} \tag{3}$$

The critical pressure, $p^*$, is obtained from Eq. 11.62 as

$$p^* = 0.528 p_0 = 0.528 p_{atm}$$

$$= (0.528)[101 \text{ kPa(abs)}] = 53.3 \text{ kPa(abs)}$$

If the receiver pressure, $p_{re}$, is greater than or equal to $p^*$, then $p_{th} = p_{re}$. If $p_{re} < p^*$, then $p_{th} = p^*$ and the flow is choked. With $p_{th}$, $p_0$, and $k$ known, $\text{Ma}_{th}$ can be obtained from Eq. 3, and $\rho_{th}$ can be determined from Eq. 2.

The flow velocity at the throat can be obtained from Eqs. 11.36 and 11.46 as

$$V_{th} = \text{Ma}_{th} c_{th} = \text{Ma}_{th} \sqrt{R T_{th} k} \tag{4}$$

The value of temperature at the throat, $T_{th}$, can be calculated from Eq. 11.56,

$$\frac{T_{th}}{T_0} = \frac{1}{1 + [(k-1)/2]\text{Ma}_{th}^2} \tag{5}$$

Since the flow through the converging duct is assumed to be isentropic, the stagnation temperature is considered constant at the standard atmosphere value of $T_0 = 15 \text{ K} + 273 \text{ K} = 288 \text{ K}$. Note that absolute pressures and temperatures are used.

**(a)** For $p_{re} = 80 \text{ kPa(abs)} > 53.3 \text{ kPa(abs)} = p^*$, we have $p_{th} = 80 \text{ kPa(abs)}$. Then from Eq. 3

$$\frac{80 \text{ kPa(abs)}}{101 \text{ kPa(abs)}} = \left\{ \frac{1}{1 + [(1.4-1)/2]\text{Ma}_{th}^2} \right\}^{1.4/(1.4-1)}$$

or

$$\text{Ma}_{th} = 0.587$$

From Eq. 2

$$\frac{\rho_{th}}{1.23 \text{ kg/m}^3} = \left\{ \frac{1}{1 + [(1.4-1)/2](0.587)^2} \right\}^{1/(1.4-1)}$$

or

$$\rho_{th} = 1.04 \text{ kg/m}^3$$

**■ Figure E11.5**

From Eq. 5

$$\frac{T_{th}}{288 \text{ K}} = \frac{1}{1 + [(1.4-1)/2](0.587)^2}$$

or

$$T_{th} = 269 \text{ K}$$

Substituting $\text{Ma}_{th} = 0.587$ and $T_{th} = 269 \text{ K}$ into Eq. 4 we obtain

$$V_{th} = 0.587 \sqrt{[286.9 \text{ J/(kg} \cdot \text{K)}](269 \text{ K})(1.4)}$$

$$= 193 \text{ (J/kg)}^{1/2}$$

Thus, since $1 \text{ J/kg} = 1 \text{ N} \cdot \text{m/kg} = 1 \text{ (kg} \cdot \text{m/s}^2) \cdot \text{m/kg} = \text{(m/s)}^2$, we obtain

$$V_{th} = 193 \text{ m/s}$$

Finally from Eq. 1 we have

$$\dot{m} = (1.04 \text{ kg/m}^3)(1 \times 10^{-4} \text{ m}^2)(193 \text{ m/s})$$
$$= 0.0201 \text{ kg/s} \qquad \textbf{(Ans)}$$

**(b)** For $p_{re} = 40 \text{ kPa(abs)} < 53.3 \text{ kPa(abs)} = p^*$, we have $p_{th} = p^* = 53.3 \text{ kPa(abs)}$ and $\text{Ma}_{th} = 1$. The converging duct is choked. From Eq. 2 (see also Eq. 11.66)

$$\frac{\rho_{th}}{1.23 \text{ kg/m}^3} = \left\{ \frac{1}{1 + [(1.4 - 1)/2](1)^2} \right\}^{1/(1.4 - 1)}$$

or

$$\rho_{th} = 0.780 \text{ kg/m}^3$$

From Eq. 5 (see also Eq. 11.64),

$$\frac{T_{th}}{288 \text{ K}} = \frac{1}{1 + [(1.4 - 1)/2](1)^2}$$

or

$$T_{th} = 240 \text{ K}$$

From Eq. 4,

$$V_{th} = (1) \sqrt{[286.9 \text{ J/(kg} \cdot \text{K)}](240 \text{ K})(1.4)}$$
$$= 310 \text{ (J/kg)}^{1/2} = 310 \text{ m/s}$$

since $1 \text{ J/kg} = 1 \text{ N} \cdot \text{m/kg} = 1 (\text{kg} \cdot \text{m/s}^2) \cdot \text{m/kg} = (\text{m/s})^2$. Finally from Eq. 1

$$\dot{m} = (0.780 \text{ kg/m}^3)(1 \times 10^{-4} \text{ m}^2)(310 \text{ m/s})$$
$$= 0.0242 \text{ kg/s} \qquad \textbf{(Ans)}$$

From the values of throat temperature and throat pressure calculated above for flow situations (a) and (b), we can construct the temperature–entropy diagram shown in Fig. E11.5b.

**COMMENT** Note that the flow from standard atmosphere to the receiver for receiver pressure, $p_{re}$, greater than or equal to the critical pressure, $p^*$, is isentropic. When the receiver pressure is less than the critical pressure as in situation (b) above, what is the flow like downstream from the exit of the converging duct? Experience suggests that this flow, when $p_{re} < p^*$, is three-dimensional and nonisentropic and involves a drop in pressure from $p_{th}$ to $p_{re}$, a drop in temperature, and an increase of entropy as are indicated in Fig. E11.5c.

---

Isentropic flow Eqs. 11.56, 11.59, and 11.60 have been used to construct Fig. D.1 in Appendix D for air ($k = 1.4$). Examples 11.6 and 11.7 illustrate how these graphs of $T/T_0$, $p/p_0$, and $\rho/\rho_0$ as a function of Mach number, Ma, can be used to solve compressible flow problems.

---

## EXAMPLE 11.6   Use of Compressible Flow Graphs in Solving Problems

**GIVEN**   Consider the flow described in Example 11.5.

**FIND**   Solve Example 11.5 using Fig. D.1 of Appendix D.

### SOLUTION

We still need the density and velocity of the air at the converging duct throat to solve for mass flowrate from

$$\dot{m} = \rho_{th} A_{th} V_{th} \qquad (1)$$

**(a)** Since the receiver pressure, $p_{re} = 80 \text{ kPa(abs)}$, is greater than the critical pressure, $p^* = 53.3 \text{ kPa(abs)}$, the throat pressure, $p_{th}$, is equal to the receiver pressure. Thus

$$\frac{p_{th}}{p_0} = \frac{80 \text{ kPa(abs)}}{101 \text{ kPa(abs)}} = 0.792$$

From Fig. D.1, for $p/p_0 = 0.79$, we get from the graph

$$\text{Ma}_{th} = 0.59$$

$$\frac{T_{th}}{T_0} = 0.94 \qquad (2)$$

$$\frac{\rho_{th}}{\rho_0} = 0.85 \qquad (3)$$

Thus, from Eqs. 2 and 3

$$T_{th} = (0.94)(288 \text{ K}) = 271 \text{ K}$$

and

$$\rho_{th} = (0.85)(1.23 \text{ kg/m}^3) = 1.04 \text{ kg/m}^3$$

Furthermore, using Eqs. 11.36 and 11.46 we get

$$V_{th} = \text{Ma}_{th} \sqrt{RT_{th}k}$$
$$= (0.59) \sqrt{[286.9 \text{ J/(kg} \cdot \text{K)}](269 \text{ K})(1.4)}$$
$$= 194 \text{ (J/kg)}^{1/2} = 194 \text{ m/s}$$

since $1 \text{ J/kg} = 1 \text{ N} \cdot \text{m/kg} = 1 (\text{kg} \cdot \text{m/s}^2) \cdot \text{m/kg} = (\text{m/s})^2$. Finally, from Eq. 1

$$\dot{m} = (1.04 \text{ kg/m}^3)(1 \times 10^{-4} \text{ m}^2)(194 \text{ m/s})$$
$$= 0.0202 \text{ kg/s} \qquad \textbf{(Ans)}$$

**(b)**  For $p_{re} = 40$ kPa(abs) $< 53.3$ kPa(abs) $= p^*$, the throat pressure is equal to 53.3 kPa(abs) and the duct is choked with $Ma_{th} = 1$. From Fig. D.1, for $Ma = 1$ we get

$$\frac{T_{th}}{T_0} = 0.83 \tag{4}$$

and

$$\frac{\rho_{th}}{\rho_0} = 0.64 \tag{5}$$

From Eqs. 4 and 5 we obtain

$$T_{th} = (0.83)(288 \text{ K}) = 240 \text{ K}$$

and

$$\rho_{th} = (0.64)(1.23 \text{ kg/m}^3) = 0.79 \text{ kg/m}^3$$

Also, from Eqs. 11.36 and 11.46 we conclude that

$$\begin{aligned}
V_{th} &= Ma_{th} \sqrt{RT_{th}k} \\
&= (1) \sqrt{[286.9 \text{ J/(kg} \cdot \text{K)}](240 \text{ K})(1.4)} \\
&= 310 \text{ (J/kg)}^{1/2} = 310 \text{ m/s}
\end{aligned}$$

Then, from Eq. 1

$$\begin{aligned}
\dot{m} &= (0.79 \text{ kg/m}^3)(1 \times 10^{-4} \text{ m}^2)(310 \text{ m/s}) \\
&= 0.024 \text{ kg/s} \quad \text{(Ans)}
\end{aligned}$$

**COMMENT**    The values from Fig. D.1 resulted in answers for mass flowrate that are close (within the accuracy of reading data from the figures) to those using the ideal gas equations (see Example 11.5).

The temperature–entropy diagrams remain the same as those provided in the solution of Example 11.5.

---

# EXAMPLE 11.7    Static to Stagnation Pressure Ratio

**GIVEN**    The static pressure to stagnation pressure ratio at a point in a flow stream is measured with a Pitot-static tube (see Fig. 3.6) as being equal to 0.82. The stagnation temperature of the fluid is 68 °F.

**FIND**    Determine the flow velocity if the fluid is **(a)** air, **(b)** helium.

## SOLUTION

We consider both air and helium, flowing as described above, to act as ideal gases with constant specific heats. Then, we can use any of the ideal gas relationships developed in this chapter. To determine the flow velocity, we can combine Eqs. 11.36 and 11.46 to obtain

$$V = Ma \sqrt{RTk} \tag{1}$$

By knowing the value of static to stagnation pressure ratio, $p/p_0$, and the specific heat ratio we can obtain the corresponding Mach number from Eq. 11.59, or for air, from Fig. D.1. Figure D.1 cannot be used for helium, since $k$ for helium is 1.66 and Fig. D.1 is for $k = 1.4$ only. With Mach number, specific heat ratio, and stagnation temperature known, the value of static temperature can be subsequently ascertained from Eq. 11.56 (or Fig. D.1 for air).

**(a)**    For air, $p/p_0 = 0.82$; thus from Fig. D.1,

$$Ma = 0.54 \tag{2}$$

and

$$\frac{T}{T_0} = 0.94 \tag{3}$$

Then, from Eq. 3

$$T = (0.94)[(68 + 460) \text{ °R}] = 496 \text{ °R} \tag{4}$$

and using Eqs. 1, 2, and 4 we get

$$\begin{aligned}
V &= (0.54) \sqrt{[53.3 \text{ (ft} \cdot \text{lb)/(lbm} \cdot \text{°R)}](496 \text{ °R})(1.4)} \\
&= 104 \text{ (ft} \cdot \text{lb/lbm)}^{1/2}
\end{aligned}$$

Thus, since 1 lb = 32.2 lbm $\cdot$ ft/s$^2$, it follows that

$$\begin{aligned}
V &= 104 \text{ (ft} \cdot \text{lb/lbm)}^{1/2} [(32.2 \text{ lbm} \cdot \text{ft/s}^2)/\text{lb}]^{1/2} \\
&= 590 \text{ ft/s} \quad \text{(Ans)}
\end{aligned}$$

**(b)**    For helium, $p/p_0 = 0.82$ and $k = 1.66$. By substituting these values into Eq. 11.59 we get

$$0.82 = \left\{ \frac{1}{1 + [(1.66 - 1)/2] \, Ma^2} \right\}^{1.66/(1.66-1)}$$

or

$$Ma = 0.499$$

From Eq. 11.56 we obtain

$$\frac{T}{T_0} = \frac{1}{1 + [(k - 1)/2]Ma^2}$$

Thus,

$$\begin{aligned}
T &= \left\{ \frac{1}{1 + [(1.66 - 1)/2](0.499)^2} \right\} [(68 + 460) \text{ °R}] \\
&= 488 \text{ °R}
\end{aligned}$$

From Eq. 1 we obtain

$$V = (0.499) \sqrt{[386 \,(\text{ft} \cdot \text{lb})/(\text{lbm} \cdot °\text{R})](488 \,°\text{R})(1.66)}$$
$$= 279 \,(\text{ft} \cdot \text{lb/lbm})^{1/2}$$

or, using $1 \,\text{lb} = 32.2 \,\text{lbm} \cdot \text{ft/s}^2$,

$$V = 279 \,(\text{ft} \cdot \text{lb/lbm})^{1/2} \,[(32.2 \,\text{lbm} \cdot \text{ft/s}^2)/\text{lb}]^{1/2}$$
$$= 1580 \,\text{ft/s} \qquad \textbf{(Ans)}$$

**COMMENT** Note that the isentropic flow equations and Fig. D.1 for $k = 1.4$ were used presently to describe fluid particle isentropic flow along a pathline in a stagnation process. Even though these equations and graph were developed for one-dimensional duct flows, they can be used for frictionless, adiabatic pathline flows also.

Furthermore, while the Mach numbers calculated above are of similar size for the air and helium flows, the flow speed is much larger for helium than for air because the speed of sound in helium is much larger than it is in air.

Also included in Fig. D.1 is a graph of the ratio of local area, $A$, to critical area, $A^*$, for different values of local Mach number. The importance of this area ratio is clarified below.

For choked flow through the converging–diverging duct of Fig. 11.6$a$, the conservation of mass equation (Eq. 11.40) yields

$$\rho A V = \rho^* A^* V^*$$

or

$$\frac{A}{A^*} = \left(\frac{\rho^*}{\rho}\right)\left(\frac{V^*}{V}\right) \qquad (11.67)$$

From Eqs. 11.36 and 11.46, we obtain

$$V^* = \sqrt{RT^*k} \qquad (11.68)$$

and

$$V = \text{Ma} \sqrt{RTk} \qquad (11.69)$$

By combining Eqs. 11.67, 11.68, and 11.69 we get

$$\frac{A}{A^*} = \frac{1}{\text{Ma}}\left(\frac{\rho^*}{\rho_0}\right)\left(\frac{\rho_0}{\rho}\right)\sqrt{\frac{(T^*/T_0)}{(T/T_0)}} \qquad (11.70)$$

The incorporation of Eqs. 11.56, 11.60, 11.63, 11.65, and 11.70 results in

$$\frac{A}{A^*} = \frac{1}{\text{Ma}}\left\{\frac{1 + [(k-1)/2]\text{Ma}^2}{1 + [(k-1)/2]}\right\}^{(k+1)/[2(k-1)]} \qquad (11.71)$$

*The ratio of flow area to the critical area is a useful concept for isentropic duct flow.*

Equation 11.71 was used to generate the values of $A/A^*$ for air ($k = 1.4$) in Fig. D.1. These values of $A/A^*$ are graphed as a function of Mach number in Fig. 11.10. As is demonstrated in the following examples, whether or not the critical area, $A^*$, is physically present in the flow, the area ratio, $A/A^*$, is still a useful concept for the isentropic flow of an ideal gas through a converging–diverging duct.

■ **Figure 11.10** **The variation of area ratio with Mach number for isentropic flow of an ideal gas ($k = 1.4$, linear coordinate scales).**

# EXAMPLE 11.8   Isentropic Choked Flow in a Converging–Diverging Duct with Subsonic Entry

**GIVEN** Air enters subsonically from standard atmosphere and flows isentropically through a choked converging–diverging duct having a circular cross-sectional area, $A$, that varies with axial distance from the throat, $x$, according to the formula

$$A = 0.1 + x^2$$

where $A$ is in square meters and $x$ is in meters. The duct extends from $x = -0.5$ m to $x = +0.5$ m.

**FIND** For this flow situation, sketch the side view of the duct and graph the variation of Mach number, static temperature to stagnation temperature ratio, $T/T_0$, and static pressure to stagnation pressure ratio, $p/p_0$, through the duct from $x = -0.5$ m to $x = +0.5$ m. Also show the possible fluid states at $x = -0.5$ m, 0 m, and $+0.5$ m using temperature–entropy coordinates.

## SOLUTION

The side view of the converging–diverging duct is a graph of radius $r$ from the duct axis as a function of axial distance. For a circular flow cross section we have

$$A = \pi r^2 \tag{1}$$

where

$$A = 0.1 + x^2 \tag{2}$$

Thus, combining Eqs. 1 and 2, we have

$$r = \left(\frac{0.1 + x^2}{\pi}\right)^{1/2} \tag{3}$$

and a graph of radius as a function of axial distance can be easily constructed (see Fig. E11.8a).

Since the converging–diverging duct in this example is choked, the throat area is also the critical area, $A^*$. From Eq. 2 we see that

$$A^* = 0.1 \text{ m}^2 \tag{4}$$

For any axial location, from Eqs. 2 and 4 we get

$$\frac{A}{A^*} = \frac{0.1 + x^2}{0.1} \tag{5}$$

(a)

(b)

(c)

(d)

**■ Figure E11.8**

Values of $A/A^*$ from Eq. 5 can be used in Eq. 11.71 to calculate corresponding values of Mach number, Ma. For air with $k = 1.4$, we could enter Fig. D.1 with values of $A/A^*$ and read off values of the Mach number. With values of Mach number ascertained, we could use Eqs. 11.56 and 11.59 to calculate related values of $T/T_0$ and $p/p_0$. For air with $k = 1.4$, Fig. D.1 could be entered with $A/A^*$ or Ma to get values of $T/T_0$ and $p/p_0$. To solve this example, we elect to use values from Fig. D.1.

The following table was constructed by using Eqs. 3 and 5 and Fig. D.1.

With the air entering the choked converging–diverging duct subsonically, only one isentropic solution exists for the converging portion of the duct. This solution involves an accelerating flow that becomes sonic ($Ma = 1$) at the throat of the passage. Two isentropic flow solutions are possible for the diverging portion of the duct—one subsonic, the other supersonic. If the pressure ratio, $p/p_0$, is set at 0.98 at $x = +0.5$ m (the outlet), the subsonic flow will occur. Alternatively, if $p/p_0$ is set at 0.04 at $x = +0.5$ m, the supersonic flow field will exist. These conditions are illustrated in Fig. E11.8. An unchoked subsonic flow through the converging–diverging duct of this example is discussed in Example 11.10. Choked flows involving flows other than the two isentropic flows in the diverging portion of the duct of this example are discussed after Example 11.10.

**COMMENT**  Note that if the diverging portion of this duct is extended, larger values of $A/A^*$ and Ma are achieved. From Fig. D1, note that further increases of $A/A^*$ result in smaller changes of Ma after $A/A^*$ values of about 10. The ratio of $p/p_0$

| $x$ (m) | From Eq. 3, $r$ (m) | From Eq. 5, $A/A^*$ | From Fig. D.1 Ma | $T/T_0$ | $p/p_0$ | State |
|---|---|---|---|---|---|---|
| *Subsonic Solution* | | | | | | |
| −0.5 | 0.334 | 3.5 | 0.17 | 0.99 | 0.98 | $a$ |
| −0.4 | 0.288 | 2.6 | 0.23 | 0.99 | 0.97 | |
| −0.3 | 0.246 | 1.9 | 0.32 | 0.98 | 0.93 | |
| −0.2 | 0.211 | 1.4 | 0.47 | 0.96 | 0.86 | |
| −0.1 | 0.187 | 1.1 | 0.69 | 0.91 | 0.73 | |
| 0 | 0.178 | 1 | 1.00 | 0.83 | 0.53 | $b$ |
| +0.1 | 0.187 | 1.1 | 0.69 | 0.91 | 0.73 | |
| +0.2 | 0.211 | 1.4 | 0.47 | 0.96 | 0.86 | |
| +0.3 | 0.246 | 1.9 | 0.32 | 0.98 | 0.93 | |
| +0.4 | 0.288 | 2.6 | 0.23 | 0.99 | 0.97 | |
| +0.5 | 0.344 | 3.5 | 0.17 | 0.99 | 0.98 | $c$ |
| *Supersonic Solution* | | | | | | |
| +0.1 | 0.187 | 1.1 | 1.37 | 0.73 | 0.33 | |
| +0.2 | 0.211 | 1.4 | 1.76 | 0.62 | 0.18 | |
| +0.3 | 0.246 | 1.9 | 2.14 | 0.52 | 0.10 | |
| +0.4 | 0.288 | 2.6 | 2.48 | 0.45 | 0.06 | |
| +0.5 | 0.334 | 3.5 | 2.80 | 0.39 | 0.04 | $d$ |

becomes vanishingly small, suggesting a practical limit to the expansion.

## EXAMPLE 11.9  Isentropic Choked Flow in a Converging–Diverging Duct with Supersonic Entry

**GIVEN**  Air enters supersonically with $T_0$ and $p_0$ equal to standard atmosphere values and flows isentropically through the choked converging–diverging duct described in Example 11.8.

**FIND**  Graph the variation of Mach number, Ma, static temperature to stagnation temperature ratio, $T/T_0$, and static pressure to stagnation pressure ratio, $p/p_0$, through the duct from $x = -0.5$ m to $x = +0.5$ m. Also show the possible fluid states at $x = -0.5$ m, 0 m, and $+0.5$ m by using temperature–entropy coordinates.

## SOLUTION

With the air entering the converging–diverging duct of Example 11.8 supersonically instead of subsonically, a unique isentropic flow solution is obtained for the converging portion of the duct. Now, however, the flow decelerates to the sonic condition at the throat. The two solutions obtained previously in Example 11.8 for the diverging portion are still valid. Since the area variation in the duct is symmetrical with respect to the duct throat, we can use the supersonic flow values obtained from Example 11.8 for the supersonic flow in the converging portion of the duct. The supersonic flow solution for the converging passage is summarized in the following table. The solution values for the entire duct are graphed in Fig. E11.9.

| $x$ (m) | $A/A^*$ | From Fig. D.1 Ma | $T/T_0$ | $p/p_0$ | State |
|---|---|---|---|---|---|
| −0.5 | 3.5 | 2.8 | 0.39 | 0.04 | $e$ |
| −0.4 | 2.6 | 2.5 | 0.45 | 0.06 | |
| −0.3 | 1.9 | 2.1 | 0.52 | 0.10 | |
| −0.2 | 1.4 | 1.8 | 0.62 | 0.18 | |
| −0.1 | 1.1 | 1.4 | 0.73 | 0.33 | |
| 0 | 1 | 1.0 | 0.83 | 0.53 | $b$ |

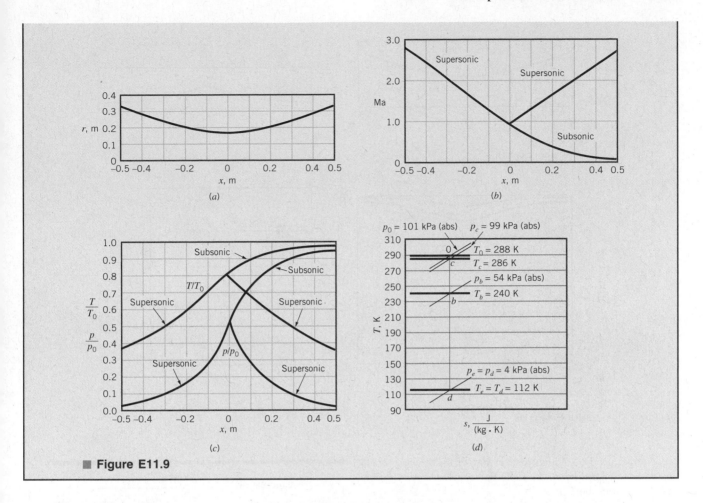

■ **Figure E11.9**

# EXAMPLE 11.10 | Isentropic Unchoked Flow in a Converging–Diverging Duct

**GIVEN** Air flows subsonically and isentropically through the converging–diverging duct of Example 11.8.

**FIND** Graph the variation of Mach number, Ma, static temperature to stagnation temperature ratio, $T/T_0$, and the static pressure to stagnation pressure ratio, $p/p_0$, through the duct from $x = -0.5$ m to $x = +0.5$ m for Ma = 0.48 at $x = 0$ m. Show the corresponding temperature–entropy diagram.

## SOLUTION

Since for this example, Ma = 0.48 at $x = 0$ m, the isentropic flow through the converging–diverging duct will be entirely subsonic and not choked. For air ($k = 1.4$) flowing isentropically through the duct, we can use Fig. D.1 for flow field quantities. Entering Fig. D.1 with Ma = 0.48 we read off $p/p_0 = 0.85$, $T/T_0 = 0.96$, and $A/A^* = 1.4$. Even though the duct flow is not choked in this example and $A^*$ does not therefore exist physically, it still represents a valid reference. For a given isentropic flow, $p_0$, $T_0$, and $A^*$ are constants. Since $A$ at $x = 0$ m is equal to 0.10 m² (from Eq. 2 of Example 11.8),

$A^*$ for this example is

$$A^* = \frac{A}{(A/A^*)} = \frac{0.10 \text{ m}^2}{1.4} = 0.07 \text{ m}^2 \tag{1}$$

With known values of duct area at different axial locations, we can calculate corresponding area ratios, $A/A^*$, knowing $A^* = 0.07$ m². Having values of the area ratio, we can use Fig. D.1 and obtain related values of Ma, $T/T_0$, and $p/p_0$. The following table summarizes flow quantities obtained in this manner. The results are graphed in Fig. E11.10.

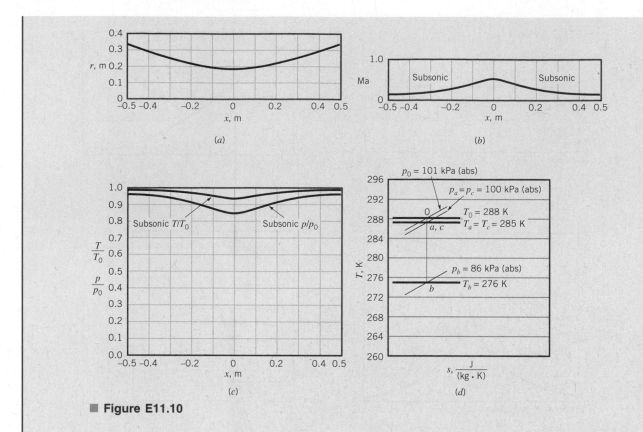

**■ Figure E11.10**

A more precise solution for the flow of this example could have been obtained with isentropic flow equations by following the steps outlined below.

1. Use Eq. 11.59 to get $p/p_0$ at $x = 0$ knowing $k$ and Ma = 0.48.
2. From Eq. 11.71, obtain value of $A/A^*$ at $x = 0$ knowing $k$ and Ma.
3. Determine $A^*$ knowing $A$ and $A/A^*$ at $x = 0$.
4. Determine $A/A^*$ at different axial locations, $x$.
5. Use Eq. 11.71 and $A/A^*$ from step 4 above to get values of Mach numbers at different axial locations.
6. Use Eqs. 11.56 and 11.59 and Ma from step 5 above to obtain $T/T_0$ and $p/p_0$ at different axial locations, $x$.

**COMMENT** There are an infinite number of subsonic, isentropic flow solutions for the converging–diverging duct considered in this example (one for any given Ma < 1 at $x = 0$).

| x (m) | Calculated, A/A* | From Fig. D.1 | | | State |
|---|---|---|---|---|---|
| | | Ma | $T/T_0$ | $p/p_0$ | |
| −0.5 | 5.0 | 0.12 | 0.99 | 0.99 | a |
| −0.4 | 3.7 | 0.16 | 0.99 | 0.98 | |
| −0.3 | 2.7 | 0.23 | 0.99 | 0.96 | |
| −0.2 | 2.0 | 0.31 | 0.98 | 0.94 | |
| −0.1 | 1.6 | 0.40 | 0.97 | 0.89 | |
| 0 | 1.4 | 0.48 | 0.96 | 0.85 | b |
| +0.1 | 1.6 | 0.40 | 0.97 | 0.89 | |
| +0.2 | 2.0 | 0.31 | 0.98 | 0.94 | |
| +0.3 | 2.7 | 0.23 | 0.99 | 0.96 | |
| +0.4 | 3.7 | 0.16 | 0.99 | 0.98 | |
| +0.5 | 5.0 | 0.12 | 0.99 | 0.99 | c |

The isentropic flow behavior for the converging–diverging duct discussed in Examples 11.8, 11.9, and 11.10 is summarized in the area ratio–Mach number graphs sketched in Fig. 11.11. The points $a$, $b$, and $c$ represent states at axial distance $x = -0.5$ m, 0 m, and $+0.5$ m. In Fig. 11.11$a$, the isentropic flow through the converging–diverging duct is subsonic without choking at the throat. This situation was discussed in Example 11.10. Figure 11.11$b$ represents subsonic to subsonic choked flow (Example 11.8), and Fig. 11.11$c$ is for subsonic to supersonic choked flow (also Example 11.8). The states in Fig. 11.11$d$ are related to the supersonic to supersonic choked flow of Example 11.9; the states in Fig. 11.11$e$ are for the supersonic to subsonic choked flow of Example 11.9. Not covered by an example but also possible are the isentropic flow states $a$, $b$, and $c$ shown in Fig. 11.11$f$ for supersonic to supersonic flow without choking. These six categories generally represent the possible kinds of isentropic, ideal gas flow through a converging–diverging duct.

For a given stagnation state (i.e., $T_0$ and $p_0$ fixed), ideal gas ($k = $ constant), and converging–diverging duct geometry, an infinite number of isentropic subsonic to subsonic (not choked) and supersonic to supersonic (not choked) flow solutions exist. In contrast, the isentropic subsonic to supersonic (choked), subsonic to subsonic (choked), supersonic to subsonic (choked), and supersonic to supersonic (choked) flow solutions are each unique. The above-mentioned isentropic

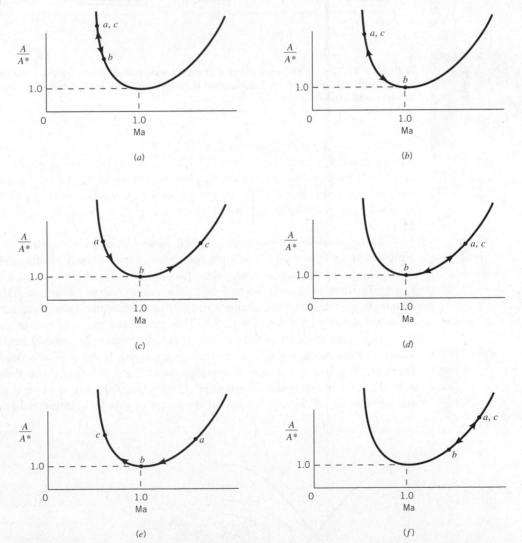

■ **Figure 11.11**  (*a*) Subsonic to subsonic isentropic flow (not choked). (*b*) Subsonic to subsonic isentropic flow (choked). (*c*) Subsonic to supersonic isentropic flow (choked). (*d*) Supersonic to supersonic isentropic flow (choked). (*e*) Supersonic to subsonic isentropic flow (choked). (*f*) Supersonic to supersonic isentropic flow (not choked).

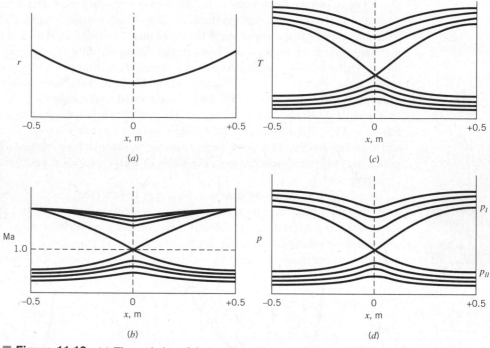

■ **Figure 11.12** (*a*) The variation of duct radius with axial distance. (*b*) The variation of Mach number with axial distance. (*c*) The variation of temperature with axial distance. (*d*) The variation of pressure with axial distance.

**Photographs courtesy of NASA.**

*V11.7 Supersonic nozzle flow*

flow solutions are represented in Fig. 11.12. When the pressure at $x = +0.5$ (exit) is greater than or equal to $p_I$ indicated in Fig. 11.12*d*, an isentropic flow is possible. When the pressure at $x = +0.5$ is equal to or less than $p_{II}$, isentropic flows in the duct are possible. However, when the exit pressure is less than $p_I$ and greater than $p_{III}$ as indicated in Fig. 11.13, isentropic flows are no longer possible in the duct. Determination of the value of $p_{III}$ is discussed in Example 11.19.

Some possible nonisentropic choked flows through our converging–diverging duct are represented in Fig. 11.13. Each abrupt pressure rise shown within and at the exit of the flow passage occurs across a very thin discontinuity in the flow called a ***normal shock wave***. Except for flow across the normal shock wave, the flow is isentropic. The nonisentropic flow equations that describe the changes in fluid properties that take place across a normal shock wave are developed in Section 11.5.3. The less abrupt pressure rise or drop that occurs after the flow leaves the duct is nonisentropic and attributable to three-dimensional ***oblique shock waves*** or **expansion waves** (see margin photograph). If the pressure rises downstream of the duct exit, the flow is considered ***overexpanded***. If the pressure drops downstream of the duct exit, the flow is called ***underexpanded***. Further details about over- and underexpanded flows and oblique shock waves are beyond the scope of this text. Interested readers are referred to

■ **Figure 11.13** Shock formation in converging–diverging duct flows.

Constant area duct

Fluid flow $\longrightarrow$

■ **Figure 11.14**   **Constant area duct flow.**

texts on compressible flows and gas dynamics (for example, Refs. 4, 5, and 6) for additional material on this subject.

### 11.4.3 Constant Area Duct Flow

For steady, one-dimensional, isentropic flow of an ideal gas through a constant area duct (see Fig. 11.14), Eq. 11.50 suggests that $dV = 0$ or that flow velocity remains constant. With the energy equation (Eq. 5.69) we can conclude that since flow velocity is constant, the fluid enthalpy and thus temperature are also constant for this flow. This information and Eqs. 11.36 and 11.46 indicate that the Mach number is constant for this flow also. This being the case, Eqs. 11.59 and 11.60 tell us that fluid pressure and density also remain unchanged. Thus, we see that a steady, one-dimensional, isentropic flow of an ideal gas does not involve varying velocity or fluid properties unless the flow cross-sectional area changes.

In Section 11.5 we discuss nonisentropic, steady, one-dimensional flows of an ideal gas through a constant area duct and also a normal shock wave. We learn that friction and/or heat transfer can also accelerate or decelerate a fluid.

## 11.5   Nonisentropic Flow of an Ideal Gas

*Fanno flow involves wall friction with no heat transfer and constant cross-sectional area.*

Actual fluid flows are generally nonisentropic. An important example of **nonisentropic flow** involves adiabatic (no heat transfer) flow with friction. Flows with heat transfer (diabatic flows) are generally nonisentropic also. In this section we consider the adiabatic flow of an ideal gas through a constant area duct with friction. This kind of flow is often referred to as **Fanno flow**. We also analyze the diabatic flow of an ideal gas through a constant area duct without friction (**Rayleigh flow**). The concepts associated with Fanno and Rayleigh flows lead to further discussion of normal shock waves.

### 11.5.1 Adiabatic Constant Area Duct Flow with Friction (Fanno Flow)

Consider the steady, one-dimensional, and adiabatic flow of an ideal gas through the constant area duct shown in Fig. 11.15. This is Fanno flow. For the control volume indicated, the energy equation (Eq. 5.69) leads to

$$\dot{m}\left[\check{h}_2 - \check{h}_1 + \frac{V_2^2 - V_1^2}{2} + g(z_2 - z_1)\right] = \dot{Q}_{\substack{\text{net} \\ \text{in.}}} + \dot{W}_{\substack{\text{shaft} \\ \text{net in}}}$$

$0$(negligibly small for gas flow)    $0$(flow is adiabatic)    $0$(flow is steady throughout)

**■ Figure 11.15** Adiabatic constant area flow.

or

$$\check{h} + \frac{V^2}{2} = \check{h}_0 = \text{constant} \tag{11.72}$$

where $h_0$ is the stagnation enthalpy. For an ideal gas we gather from Eq. 11.9 that

$$\check{h} - \check{h}_0 = c_p(T - T_0) \tag{11.73}$$

so that by combining Eqs. 11.72 and 11.73 we get

$$T + \frac{V^2}{2c_p} = T_0 = \text{constant}$$

or

$$T + \frac{(\rho V)^2}{2c_p\rho^2} = T_0 = \text{constant} \tag{11.74}$$

By substituting the ideal gas equation of state (Eq. 11.1) into Eq. 11.74, we obtain

$$T + \frac{(\rho V)^2 T^2}{2c_p(p^2/R^2)} = T_0 = \text{constant} \tag{11.75}$$

From the continuity equation (Eq. 11.40) we can conclude that the density–velocity product, $\rho V$, is constant for a given Fanno flow since the area, $A$, is constant. Also, for a particular Fanno flow, the stagnation temperature, $T_0$, is fixed. Thus, Eq. 11.75 allows us to calculate values of fluid temperature corresponding to values of fluid pressure in the Fanno flow. We postpone our discussion of how pressure is determined until later.

As with earlier discussions in this chapter, it is helpful to describe Fanno flow with a temperature–entropy diagram. From the second $T\,ds$ relationship, an expression for entropy variation was already derived (Eq. 11.22). If the temperature, $T_1$, pressure, $p_1$, and entropy, $s_1$, at the entrance of the Fanno flow duct are considered as reference values, then Eq. 11.22 yields

$$s - s_1 = c_p \ln\frac{T}{T_1} - R \ln\frac{p}{p_1} \tag{11.76}$$

*Entropy increases in Fanno flows because of wall friction.*

Equations 11.75 and 11.76 taken together result in a curve with $T$–$s$ coordinates as is illustrated in Fig. 11.16. This curve involves a given gas ($c_p$ and $R$) with fixed values of stagnation temperature, density–velocity product, and inlet temperature, pressure, and entropy. Curves like the one sketched in Fig. 11.16 are called Fanno lines.

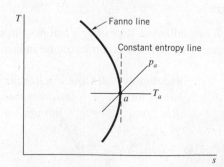

**■ Figure 11.16** The $T$–$s$ diagram for Fanno flow.

# EXAMPLE 11.11    Compressible Flow with Friction (Fanno Flow)

**GIVEN**  Air ($k = 1.4$) enters [section (1)] an insulated, constant cross-sectional area duct with the following properties:

$$T_0 = 518.67\ {}^\circ R$$

$$T_1 = 514.55\ {}^\circ R$$

$$p_1 = 14.3\ \text{psia}$$

**FIND**  For Fanno flow, determine corresponding values of fluid temperature and entropy change for various values of downstream pressures and plot the related Fanno line.

## SOLUTION

To plot the Fanno line we can use Eq. 11.75

$$T + \frac{(\rho V)^2 T^2}{2c_p p^2 / R^2} = T_0 = \text{constant} \tag{1}$$

and Eq. 11.76

$$s - s_1 = c_p \ln \frac{T}{T_1} - R \ln \frac{p}{p_1} \tag{2}$$

to construct a table of values of temperature and entropy change corresponding to different levels of pressure in the Fanno flow.

We need values of the ideal gas constant and the specific heat at constant pressure to use in Eqs. 1 and 2. From Table 1.7 we read for air

$$R = 1716\ (\text{ft} \cdot \text{lb})/(\text{slug} \cdot {}^\circ R) = 53.3\ (\text{ft} \cdot \text{lb})/(\text{lbm} \cdot {}^\circ R)$$

From Eq. 11.14 we obtain

$$c_p = \frac{Rk}{k - 1} \tag{3}$$

or

$$c_p = \frac{[53.3\ (\text{ft} \cdot \text{lb})/(\text{lbm} \cdot {}^\circ R)](1.4)}{1.4 - 1}$$

$$= 187\ (\text{ft} \cdot \text{lb})/(\text{lbm} \cdot {}^\circ R)$$

From Eqs. 11.1 and 11.69 we obtain

$$\rho V = \frac{p}{RT}\, \text{Ma} \sqrt{RTk}$$

and $\rho V$ is constant for this flow

$$\rho V = \rho_1 V_1 = \frac{p_1}{RT_1}\, \text{Ma}_1 \sqrt{RT_1 k} \tag{4}$$

But

$$\frac{T_1}{T_0} = \frac{514.55\ {}^\circ F}{518.67\ {}^\circ R} = 0.992$$

and from Eq. 11.56, $\text{Ma}_1 = \{(T_0/T_1 - 1)/[(k-1)/2]\}^{1/2}$, or

$$\text{Ma}_1 = \sqrt{\left(\frac{1}{0.992} - 1\right) / 0.2} = 0.2$$

Thus, with

$$\sqrt{RT_1 k} = \sqrt{(1.4)[53.3\ (\text{ft} \cdot \text{lb})/(\text{lbm} \cdot {}^\circ R)](514.55\ {}^\circ R)}$$

$$= 196\ [(\text{ft} \cdot \text{lb})/\text{lbm}]^{1/2}[(32.2\ \text{lbm} \cdot \text{ft/s}^2)/\text{lb}]^{1/2}$$

$$= 1112\ \text{ft/s}$$

Eq. 4 becomes

$$\rho V = \frac{(14.3\ \text{psia})(144\ \text{in.}^2/\text{ft}^2)0.2(1112\ \text{ft/s})}{[53.3\ (\text{ft} \cdot \text{lb})/(\text{lbm} \cdot {}^\circ R)](514.55\ {}^\circ R)}$$

or

$$\rho V = 16.7\ \text{lbm}/(\text{ft}^2 \cdot \text{s})$$

For $p = 7$ psia we have from Eq. 1

$$T + \frac{[16.7\ \text{lbm}/(\text{ft}^2 \cdot \text{s})]^2 T^2}{2[187\ (\text{ft} \cdot \text{lb})/(\text{lbm} \cdot {}^\circ R)]\dfrac{(7\ \text{psia})^2(144\ \text{in.}^2/\text{ft}^2)^2}{[53.3\ (\text{ft} \cdot \text{lb})/(\text{lbm} \cdot {}^\circ R)]^2}}$$

$$= 518.67\ {}^\circ R$$

or

$$2.08 \times 10^{-3}\ [(\text{lbm} \cdot \text{ft/s}^2)/(\text{lb} \cdot {}^\circ R)]\, T^2 + T - 518.67\ {}^\circ R = 0$$

Thus, since 1 lb = 32.2 lbm $\cdot$ ft/s$^2$ we obtain

$$6.5 \times 10^{-5} T^2 + T - 518.67 = 0$$

Hence,

$$T = 502.3\ {}^\circ R \tag{Ans}$$

where $T$ is in ${}^\circ R$.

From Eq. 2, we obtain

$$s - s_1 = [187\ (\text{ft} \cdot \text{lb})/(\text{lbm} \cdot {}^\circ R)] \ln\left(\frac{502.3\ {}^\circ R}{514.55\ {}^\circ R}\right)$$

$$- [53.3\ (\text{ft} \cdot \text{lb})/(\text{lbm} \cdot {}^\circ R)] \ln\left(\frac{7\ \text{psia}}{14.3\ \text{psia}}\right)$$

or

$$s - s_1 = 33.6\ (\text{ft} \cdot \text{lb})/(\text{lbm} \cdot {}^\circ R) \tag{Ans}$$

Proceeding as outlined above, we construct the table of values shown below and graphed as the Fanno line in Fig. E11.11. The

**■ Figure E11.11**

maximum entropy difference occurs at a pressure of 2.62 psia and a temperature of 432.1 °R.

**COMMENT** Note that for Fanno flow the entropy must increase in the direction of flow. Hence, as discussed in the material that follows this example, this flow can proceed either from subsonic conditions upstream to a sonic condition (Ma = 1) downstream or from supersonic conditions upstream to a sonic condition downstream. The arrows in Fig. 11.11 indicate in which direction a Fanno flow can proceed.

| $p$ (psia) | $T$ (°R) | $s - s_1$ [(ft · lb)/(lbm · °R)] |
|---|---|---|
| 7 | 502.3 | 33.6 |
| 6 | 496.8 | 39.8 |
| 5 | 488.3 | 46.3 |
| 4 | 474.0 | 52.6 |
| 3 | 447.7 | 57.3 |
| 2.62 | 432.1 | 57.9 |
| 2 | 394.7 | 55.4 |
| 1.8 | 378.1 | 53.0 |
| 1.5 | 347.6 | 47.0 |
| 1.4 | 335.6 | 44.2 |

We can learn more about Fanno lines by further analyzing the equations that describe the physics involved. For example, the second $T\,ds$ equation (Eq. 11.18) is

$$T\,ds = d\breve{h} - \frac{dp}{\rho} \tag{11.18}$$

For an ideal gas

$$d\breve{h} = c_p\,dT \tag{11.7}$$

and

$$\rho = \frac{p}{RT} \tag{11.1}$$

or

$$\frac{dp}{p} = \frac{d\rho}{\rho} + \frac{dT}{T} \tag{11.77}$$

*Fanno flow properties can be obtained from the second T ds equation combined with the continuity and energy equations.*

Thus, consolidating Eqs. 11.1, 11.7, 11.18, and 11.77 we obtain

$$T\,ds = c_p\,dT - RT\left(\frac{d\rho}{\rho} + \frac{dT}{T}\right) \tag{11.78}$$

Also, from the continuity equation (Eq. 11.40), we get for Fanno flow $\rho V = $ constant, or

$$\frac{d\rho}{\rho} = -\frac{dV}{V} \tag{11.79}$$

Substituting Eq. 11.79 into Eq. 11.78 yields

$$T\,ds = c_p\,dT - RT\left(-\frac{dV}{V} + \frac{dT}{T}\right)$$

or

$$\frac{ds}{dT} = \frac{c_p}{T} - R\left(-\frac{1}{V}\frac{dV}{dT} + \frac{1}{T}\right) \tag{11.80}$$

By differentiating the energy equation (11.74) obtained earlier, we obtain

$$\frac{dV}{dT} = -\frac{c_p}{V} \tag{11.81}$$

■ **Figure 11.17**   *(a)* Subsonic Fanno flow. *(b)* Supersonic Fanno flow. *(c)* Normal shock occurrence in Fanno flow.

which, when substituted into Eq. 11.80, results in

$$\frac{ds}{dT} = \frac{c_p}{T} - R\left(\frac{c_p}{V^2} + \frac{1}{T}\right) \tag{11.82}$$

The Fanno line in Fig. 11.16 goes through a state (labeled state *a*) for which $ds/dT = 0$. At this state, we can conclude from Eqs. 11.14 and 11.82 that

$$V_a = \sqrt{RT_a k} \tag{11.83}$$

However, by comparing Eqs. 11.83 and 11.36 we see that the Mach number at state *a* is 1. Since the stagnation temperature is the same for all points on the Fanno line [see energy equation (Eq. 11.74)], the temperature at point *a* is the critical temperature, $T^*$, for the entire Fanno line. Thus, Fanno flow corresponding to the portion of the Fanno line above the critical temperature must be subsonic, and Fanno flow on the line below $T^*$ must be supersonic.

The second law of thermodynamics states that, based on all past experience, entropy can only remain constant or increase for adiabatic flows. For Fanno flow to be consistent with the second law of thermodynamics, flow can only proceed along the Fanno line toward state *a*, the critical state. The critical state may or may not be reached by the flow. If it is, the Fanno flow is *choked*. Some examples of Fanno flow behavior are summarized in Fig. 11.17. A case involving subsonic Fanno flow that is accelerated by friction to a higher Mach number without choking is illustrated in Fig. 11.17*a*. A supersonic flow that is decelerated by friction to a lower Mach number without choking is illustrated in Fig. 11.17*b*. In Fig. 11.17*c*, an abrupt change from supersonic to subsonic flow in the Fanno duct is represented. This sudden deceleration occurs across a standing *normal shock wave* that is described in more detail in Section 11.5.3.

The qualitative aspects of Fanno flow that we have already discussed are summarized in Table 11.1 and Fig. 11.18. To quantify Fanno flow behavior we need to combine a relationship that represents the linear momentum law with the set of equations already derived in this chapter.

> *Friction accelerates a subsonic Fanno flow.*

■ **Table 11.1**
**Summary of Fanno Flow Behavior**

| Parameter | Flow | |
| --- | --- | --- |
| | Subsonic Flow | Supersonic Flow |
| Stagnation temperature | Constant | Constant |
| Ma | Increases (maximum is 1) | Decreases (minimum is 1) |
| Friction | Accelerates flow | Decelerates flow |
| Pressure | Decreases | Increases |
| Temperature | Decreases | Increases |

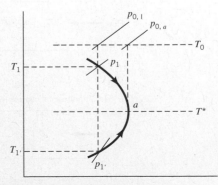

■ **Figure 11.18**   Fanno flow.

If the linear momentum equation (Eq. 5.22) is applied to the Fanno flow through the control volume sketched in Fig. 11.19a, the result is

$$p_1 A_1 - p_2 A_2 - R_x = \dot{m}(V_2 - V_1)$$

where $R_x$ is the frictional force exerted by the inner pipe wall on the fluid. Since $A_1 = A_2 = A$ and $\dot{m} = \rho A V = \text{constant}$ (i.e. $\rho V = \rho_1 V_1 = \rho_2 V_2$), we obtain

$$p_1 - p_2 - \frac{R_x}{A} = \rho V(V_2 - V_1) \tag{11.84}$$

The differential form of Eq. 11.84, which is valid for Fanno flow through the semi-infinitesimal control volume shown in Fig. 11.19b, is

$$-dp - \frac{\tau_w \pi D \, dx}{A} = \rho V \, dV \tag{11.85}$$

The wall shear stress, $\tau_w$, is related to the wall friction factor, $f$, by Eq. 8.20 as

$$f = \frac{8\tau_w}{\rho V^2} \tag{11.86}$$

By substituting Eq. 11.86 and $A = \pi D^2/4$ into Eq. 11.85, we obtain

$$-dp - f\rho \frac{V^2}{2} \frac{dx}{D} = \rho V \, dV \tag{11.87}$$

or

$$\frac{dp}{p} + \frac{f}{p} \frac{\rho V^2}{2} \frac{dx}{D} + \frac{\rho}{p} \frac{d(V^2)}{2} = 0 \tag{11.88}$$

Combining the ideal gas equation of state (Eq. 11.1), the ideal gas speed-of-sound equation (Eq. 11.36), and the Mach number definition (Eq. 11.46) with Eq. 11.88 leads to

$$\frac{dp}{p} + \frac{fk}{2} \text{Ma}^2 \frac{dx}{D} + k \frac{\text{Ma}^2}{2} \frac{d(V^2)}{V^2} = 0 \tag{11.89}$$

Since $V = \text{Ma} \, c = \text{Ma} \sqrt{RTk}$, then

$$V^2 = \text{Ma}^2 RTk$$

or

$$\frac{d(V^2)}{V^2} = \frac{d(\text{Ma}^2)}{\text{Ma}^2} + \frac{dT}{T} \tag{11.90}$$

(a)

(b)

■ **Figure 11.19**  *(a)* Finite control volume. *(b)* Semi-infinitesimal control volume.

The application of the energy equation (Eq. 5.69) to Fanno flow gave Eq. 11.74. If Eq. 11.74 is differentiated and divided by temperature, the result is

$$\frac{dT}{T} + \frac{d(V^2)}{2c_p T} = 0 \tag{11.91}$$

Substituting Eqs. 11.14, 11.36, and 11.46 into Eq. 11.91 yields

$$\frac{dT}{T} + \frac{k-1}{2}\,\mathrm{Ma}^2\,\frac{d(V^2)}{V^2} = 0 \tag{11.92}$$

which can be combined with Eq. 11.90 to form

$$\frac{d(V^2)}{V^2} = \frac{d(\mathrm{Ma}^2)/\mathrm{Ma}^2}{1 + [(k-1)/2]\mathrm{Ma}^2} \tag{11.93}$$

We can merge Eqs. 11.77, 11.79, and 11.90 to get

$$\frac{dp}{p} = \frac{1}{2}\frac{d(V^2)}{V^2} - \frac{d(\mathrm{Ma}^2)}{\mathrm{Ma}^2} \tag{11.94}$$

Consolidating Eqs. 11.94 and 11.89 leads to

$$\frac{1}{2}\,(1 + k\mathrm{Ma}^2)\,\frac{d(V^2)}{V^2} - \frac{d(\mathrm{Ma}^2)}{\mathrm{Ma}^2} + \frac{fk}{2}\,\mathrm{Ma}^2\,\frac{dx}{D} = 0 \tag{11.95}$$

Finally, incorporating Eq. 11.93 into Eq. 11.95 yields

$$\frac{(1 - \mathrm{Ma}^2)\,d(\mathrm{Ma}^2)}{\{1 + [(k-1)/2]\mathrm{Ma}^2\}k\mathrm{Ma}^4} = f\frac{dx}{D} \tag{11.96}$$

Equation 11.96 can be integrated from one section to another in a Fanno flow duct. We elect to use the critical (*) state as a reference and to integrate Eq. 11.96 from an upstream state to the critical state. Thus

$$\int_{\mathrm{Ma}}^{\mathrm{Ma}^*=1} \frac{(1 - \mathrm{Ma}^2)\,d(\mathrm{Ma}^2)}{\{1 + [(k-1)/2]\,\mathrm{Ma}^2\}k\mathrm{Ma}^4} = \int_{\ell}^{\ell^*} f\frac{dx}{D} \tag{11.97}$$

where $\ell$ is length measured from an arbitrary but fixed upstream reference location to a section in the Fanno flow. For an approximate solution, we can assume that the friction factor is constant at an average value over the integration length, $\ell^* - \ell$. We also consider a constant value of $k$. Thus, we obtain from Eq. 11.97

$$\boxed{\frac{1}{k}\frac{(1 - \mathrm{Ma}^2)}{\mathrm{Ma}^2} + \frac{k+1}{2k}\ln\left\{\frac{[(k+1)/2]\mathrm{Ma}^2}{1 + [(k-1)/2]\mathrm{Ma}^2}\right\} = \frac{f(\ell^* - \ell)}{D}} \tag{11.98}$$

*For Fanno flow, the Mach number is a function of the distance to the critical state.*

For a given gas, values of $f(\ell^* - \ell)/D$ can be tabulated as a function of Mach number for Fanno flow. For example, values of $f(\ell^* - \ell)/D$ for air ($k = 1.4$) Fanno flow are graphed as a function of Mach number in Fig. D.2 in Appendix D and in the figure in the margin. Note that the critical state does not have to exist in the actual Fanno flow being considered, since for any two sections in a given Fanno flow

$$\boxed{\frac{f(\ell^* - \ell_2)}{D} - \frac{f(\ell^* - \ell_1)}{D} = \frac{f}{D}\,(\ell_1 - \ell_2)} \tag{11.99}$$

The sketch in Fig. 11.20 illustrates the physical meaning of Eq. 11.99.

For a given Fanno flow (constant specific heat ratio, duct diameter, and friction factor) the length of duct required to change the Mach number from $\mathrm{Ma}_1$ to $\mathrm{Ma}_2$ can be determined from Eqs. 11.98 and 11.99 or a graph such as Fig. D.2. To get the values of other fluid properties in the Fanno flow field we need to develop more equations.

■ **Figure 11.20** *(a) Unchoked Fanno flow. (b) Choked Fanno flow.*

*For Fanno flow, the length of duct needed to produce a given change in Mach number can be determined.*

By consolidating Eqs. 11.90 and 11.92 we obtain

$$\frac{dT}{T} = -\frac{(k-1)}{2\{1 + [(k-1)/2]\mathrm{Ma}^2\}}\, d(\mathrm{Ma}^2) \tag{11.100}$$

Integrating Eq. 11.100 from any state upstream in a Fanno flow to the critical (*) state leads to

$$\boxed{\frac{T}{T^*} = \frac{(k+1)/2}{1 + [(k-1)/2]\mathrm{Ma}^2}} \tag{11.101}$$

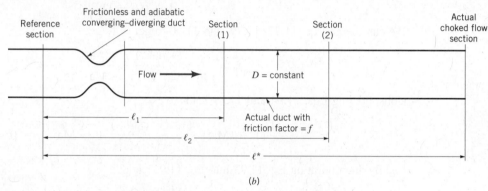

Equations 11.68 and 11.69 allow us to write

$$\frac{V}{V^*} = \frac{\mathrm{Ma}\,\sqrt{RTk}}{\sqrt{RT^*k}} = \mathrm{Ma}\,\sqrt{\frac{T}{T^*}} \tag{11.102}$$

Substituting Eq. 11.101 into Eq. 11.102 yields

$$\boxed{\frac{V}{V^*} = \left\{\frac{[(k+1)/2]\mathrm{Ma}^2}{1 + [(k-1)/2]\mathrm{Ma}^2}\right\}^{1/2}} \tag{11.103}$$

Equations 11.101 and 11.103 are graphed in the margin for air.
From the continuity equation (Eq. 11.40) we get for Fanno flow

$$\frac{\rho}{\rho^*} = \frac{V^*}{V} \tag{11.104}$$

Combining 11.104 and 11.103 results in

$$\frac{\rho}{\rho^*} = \left\{\frac{1 + [(k-1)/2]\mathrm{Ma}^2}{[(k+1)/2]\mathrm{Ma}^2}\right\}^{1/2} \tag{11.105}$$

The ideal gas equation of state (Eq. 11.1) leads to

$$\frac{p}{p^*} = \frac{\rho}{\rho^*}\frac{T}{T^*}$$

(11.106)

and merging Eqs. 11.106, 11.105, and 11.101 gives

$$\frac{p}{p^*} = \frac{1}{\text{Ma}}\left\{\frac{(k+1)/2}{1+[(k-1)/2]\text{Ma}^2}\right\}^{1/2}$$

(11.107)

This relationship is graphed in the margin for air.

Finally, the stagnation pressure ratio can be written as

$$\frac{p_0}{p_0^*} = \left(\frac{p_0}{p}\right)\left(\frac{p}{p^*}\right)\left(\frac{p^*}{p_0^*}\right)$$

(11.108)

*For Fanno flow, thermodynamic and flow properties can be calculated as a function of Mach number.*

which by use of Eqs. 11.59 and 11.107 yields

$$\frac{p_0}{p_0^*} = \frac{1}{\text{Ma}}\left[\left(\frac{2}{k+1}\right)\left(1+\frac{k-1}{2}\text{Ma}^2\right)\right]^{[(k+1)/2(k-1)]}$$

(11.109)

Values of $f(\ell^* - \ell)/D$, $T/T^*$, $V/V^*$, $p/p^*$, and $p_0/p_0^*$ for Fanno flow of air $(k = 1.4)$ are graphed as a function of Mach number (using Eqs. 11.99, 11.101, 11.103, 11.107, and 11.109) in Fig. D.2 of Appendix D. The usefulness of Fig. D.2 is illustrated in Examples 11.12, 11.13, and 11.14. See Ref. 7 for additional compressible internal flow material.

---

## EXAMPLE 11.12    Choked Fanno Flow

**GIVEN** Standard atmospheric air $[T_0 = 288 \text{ K}, p_0 = 101$ kPa(abs)$]$ is drawn steadily through a frictionless, adiabatic converging nozzle into an adiabatic, constant area duct as shown in Fig. E11.12a. The duct is 2 m long and has an inside diameter of 0.1 m. The average friction factor for the duct is estimated as being equal to 0.02.

**FIND** What is the maximum mass flowrate through the duct? For this maximum flowrate, determine the values of static temperature, static pressure, stagnation temperature, stagnation pressure, and velocity at the inlet [section (1)] and exit [section (2)] of the constant area duct. Sketch a temperature–entropy diagram for this flow.

## SOLUTION

We consider the flow through the converging nozzle to be isentropic and the flow through the constant area duct to be Fanno flow. A decrease in the pressure at the exit of the constant area duct (back pressure) causes the mass flowrate through the nozzle and the duct to increase. The flow throughout is subsonic. The maximum flowrate will occur when the back pressure is lowered to the extent that the constant area duct chokes and the Mach number at the duct exit is equal to 1. Any further decrease of back pressure will not affect the flowrate through the nozzle–duct combination.

■ **Figure E11.12**

For the maximum flowrate condition, the constant area duct must be choked, and

$$\frac{f(\ell^* - \ell_1)}{D} = \frac{f(\ell_2 - \ell_1)}{D} = \frac{(0.02)(2 \text{ m})}{(0.1 \text{ m})} = 0.4 \qquad (1)$$

With $k = 1.4$ for air and the above calculated value of $f(\ell^* - \ell_1)/D = 0.4$, we could use Eq. 11.98 to determine a value of Mach number at the entrance of the duct [section (1)]. With $k = 1.4$ and $Ma_1$ known, we could then rely on Eqs. 11.101, 11.103, 11.107, and 11.109 to obtain values of $T_1/T^*$, $V_1/V^*$, $p_1/p^*$, and $p_{0,1}/p_0^*$. Alternatively, for air ($k = 1.4$), we can use Fig. D.2 with $f(\ell^* - \ell_1)/D = 0.4$ and read off values of $Ma_1$, $T_1/T^*$, $V_1/V^*$, $p_1/p^*$, and $p_{0,1}/p_0^*$.

The pipe entrance Mach number, $Ma_1$, also represents the Mach number at the throat (and exit) of the isentropic, converging nozzle. Thus, the isentropic flow equations of Section 11.4 or Fig. D.1 can be used with $Ma_1$. We use Fig. D.1 in this example.

With $Ma_1$ known, we can enter Fig. D.1 and get values of $T_1/T_0$, $p_1/p_0$, and $\rho_1/\rho_0$. Through the isentropic nozzle, the values of $T_0$, $p_0$, and $\rho_0$ are each constant, and thus $T_1$, $p_1$, and $\rho_1$ can be readily obtained.

Since $T_0$ also remains constant through the constant area duct (see Eq. 11.75), we can use Eq. 11.63 to get $T^*$. Thus,

$$\frac{T^*}{T_0} = \frac{2}{k+1} = \frac{2}{1.4+1} = 0.8333 \qquad (2)$$

Since $T_0 = 288$ K, we get from Eq. 2,

$$T^* = (0.8333)(288 \text{ K}) = 240 \text{ K} = T_2 \qquad (3) \quad \textbf{(Ans)}$$

With $T^*$ known, we can calculate $V^*$ from Eq. 11.36 as

$$\begin{aligned}
V^* &= \sqrt{RT^*k} \\
&= \sqrt{[(286.9 \text{ J})/(\text{kg} \cdot \text{K})](240 \text{ K})(1.4)} \\
&= 310 \ (\text{J/kg})^{1/2}
\end{aligned}$$

Thus, since $1 \text{ J/kg} = 1 \text{ N} \cdot \text{m/kg} = 1 \ (\text{kg} \cdot \text{m/s}^2) \cdot \text{m/kg} = (\text{m/s})^2$, we obtain

$$V^* = 310 \text{ m/s} = V_2 \qquad (4) \quad \textbf{(Ans)}$$

Now $V_1$ can be obtained from $V^*$ and $V_1/V^*$. Having $A_1$, $\rho_1$, and $V_1$ we can get the mass flowrate from

$$\dot{m} = \rho_1 A_1 V_1 \qquad (5)$$

Values of the other variables asked for can be obtained from the ratios mentioned.

Entering Fig. D.2 with $f(\ell^* - \ell)/D = 0.4$ we read

$$Ma_1 = 0.63 \qquad (7)$$

$$\frac{T_1}{T^*} = 1.1 \qquad (8)$$

$$\frac{V_1}{V^*} = 0.66 \qquad (9)$$

$$\frac{p_1}{p^*} = 1.7 \qquad (10)$$

$$\frac{p_{0,1}}{p_0^*} = 1.16 \qquad (11)$$

Entering Fig. D.1 with $Ma_1 = 0.63$ we read

$$\frac{T_1}{T_0} = 0.93 \qquad (12)$$

$$\frac{p_1}{p_{0,1}} = 0.76 \qquad (13)$$

$$\frac{\rho_1}{\rho_{0,1}} = 0.83 \qquad (14)$$

Thus, from Eqs. 4 and 9 we obtain

$$V_1 = (0.66)(310 \text{ m/s}) = 205 \text{ m/s} \qquad \textbf{(Ans)}$$

From Eq. 14 we get

$$\rho_1 = 0.83\rho_{0,1} = (0.83)(1.23 \text{ kg/m}^3) = 1.02 \text{ kg/m}^3$$

and from Eq. 5 we conclude that

$$\begin{aligned}
\dot{m} &= (1.02 \text{ kg/m}^3)\left[\frac{\pi(0.1 \text{ m})^2}{4}\right](206 \text{ m/s}) \\
&= 1.65 \text{ kg/s} \qquad \textbf{(Ans)}
\end{aligned}$$

From Eq. 12, it follows that

$$T_1 = (0.93)(288 \text{ K}) = 268 \text{ K} \qquad \textbf{(Ans)}$$

Equation 13 yields

$$p_1 = (0.76)[101 \text{ kPa(abs)}] = 77 \text{ kPa(abs)} \qquad \textbf{(Ans)}$$

The stagnation temperature, $T_0$, remains constant through this adiabatic flow at a value of

$$T_{0,1} = T_{0,2} = 288 \text{ K} \qquad \textbf{(Ans)}$$

The stagnation pressure, $p_0$, at the entrance of the constant area duct is the same as the constant value of stagnation pressure through the isentropic nozzle. Thus

$$p_{0,1} = 101 \text{ kPa(abs)} \qquad \textbf{(Ans)}$$

To obtain the duct exit pressure ($p_2 = p^*$) we can use Eqs. 10 and 13. Thus,

$$\begin{aligned}
p_2 &= \left(\frac{p^*}{p_1}\right)\left(\frac{p_1}{p_{0,1}}\right)(p_{0,1}) = \left(\frac{1}{1.7}\right)(0.76)[101 \text{ kPa(abs)}] \\
&= 45 \text{ kPa(abs)} \qquad \textbf{(Ans)}
\end{aligned}$$

For the duct exit stagnation pressure ($p_{0,2} = p_0^*$) we can use Eq. 11 as

$$\begin{aligned}
p_{0,2} &= \left(\frac{p_0^*}{p_{0,1}}\right)(p_{0,1}) = \left(\frac{1}{1.16}\right)[101 \text{ kPa(abs)}] \\
&= 87.1 \text{ kPa(abs)} \qquad \textbf{(Ans)}
\end{aligned}$$

The stagnation pressure, $p_0$, decreases in a Fanno flow because of friction.

**COMMENT** Use of graphs such as Figs. D.1 and D.2 illustrates the solution of a problem involving Fanno flow. The $T-s$ diagram for this flow is shown in Fig. E.11.12b, where the entropy difference, $s_2 - s_1$, is obtained from Eq. 11.22.

# EXAMPLE 11.13  Effect of Duct Length on Choked Fanno Flow

**GIVEN**  The duct in Example 11.12 is shortened by 50%, but the duct discharge pressure is maintained at the choked flow value for Example 11.12, namely,

$$p_d = 45 \text{ kPa(abs)}$$

**FIND**  Will shortening the duct cause the mass flowrate through the duct to increase or decrease? Assume that the average friction factor for the duct remains constant at a value of $f = 0.02$.

## SOLUTION

We guess that the shortened duct will still choke and check our assumption by comparing $p_d$ with $p^*$. If $p_d < p^*$, the flow is choked; if not, another assumption has to be made. For choked flow we can calculate the mass flowrate just as we did for Example 11.12. For unchoked flow, we will have to devise another strategy.

For choked flow

$$\frac{f(\ell^* - \ell_1)}{D} = \frac{(0.02)(1 \text{ m})}{0.1 \text{ m}} = 0.2$$

and from Fig. D.2, we read the values $Ma_1 = 0.70$ and $p_1/p^* = 1.5$. With $Ma_1 = 0.70$, we use Fig. D.1 and get

$$\frac{p_1}{p_0} = 0.72$$

Now the duct exit pressure ($p_2 = p^*$) can be obtained from

$$p_2 = p^* = \left(\frac{p^*}{p_1}\right)\left(\frac{p_1}{p_{0,1}}\right)(p_{0,1})$$

$$= \left(\frac{1}{1.5}\right)(0.72)[101 \text{ kPa(abs)}] = 48.5 \text{ kPa(abs)}$$

and we see that $p_d < p^*$. Our assumption of choked flow is justified. The pressure at the exit plane is greater than the surrounding pressure outside the duct exit. The final drop of pressure from 48.5 kPa(abs) to 45 kPa(abs) involves complicated three-dimensional flow downstream of the exit.

To determine the mass flowrate we use

$$\dot{m} = \rho_1 A_1 V_1 \tag{1}$$

The density at section (1) is obtained from

$$\frac{\rho_1}{\rho_{0,1}} = 0.79 \tag{2}$$

which is read in Fig. D.1 for $Ma_1 = 0.7$. Thus,

$$\rho_1 = (0.79)(1.23 \text{ kg/m}^3) = 0.97 \text{ kg/m}^3 \tag{3}$$

We get $V_1$ from

$$\frac{V_1}{V^*} = 0.73 \tag{4}$$

from Fig. D.2 for $Ma_1 = 0.7$. The value of $V^*$ is the same as it was in Example 11.12, namely,

$$V^* = 310 \text{ m/s} \tag{5}$$

Thus, from Eqs. 4 and 5 we obtain

$$V_1 = (0.73)(310) = 226 \text{ m/s} \tag{6}$$

and from Eqs. 1, 3, and 6 we get

$$\dot{m} = (0.97 \text{ kg/m}^3)\left[\frac{\pi(0.1\text{m})^2}{4}\right](226 \text{ m/s})$$

$$= 1.73 \text{ kg/s} \tag{Ans}$$

The mass flowrate associated with a shortened tube is larger than the mass flowrate for the longer tube, $\dot{m} = 1.65$ kg/s. This trend is general for subsonic Fanno flow.

**COMMENT**  For the same upstream stagnation state and downstream pressure, the mass flowrate for the Fanno flow will decrease with increase in length of duct for subsonic flow. Equivalently, if the length of the duct remains the same but the wall friction is increased, the mass flowrate will decrease.

# EXAMPLE 11.14  Unchoked Fanno Flow

**GIVEN**  The same flowrate obtained in Example 11.12 ($\dot{m} = 1.65$ kg/s) is desired through the shortened duct of Example 11.13 ($\ell_2 - \ell_1 = 1$ m). Assume $f$ remains constant at a value of 0.02.

**FIND**  Determine the Mach number at the exit of the duct, $M_2$, and the back pressure, $p_2$, required.

## SOLUTION

Since the mass flowrate of Example 11.12 is desired, the Mach number and other properties at the entrance of the constant area duct remain at the values determined in Example 11.12. Thus,

from Example 11.12, $Ma_1 = 0.63$ and from Fig. D.2

$$\frac{f(\ell^* - \ell_1)}{D} = 0.4$$

For this example,

$$\frac{f(\ell_2 - \ell_1)}{D} = \frac{f(\ell^* - \ell_1)}{D} - \frac{f(\ell^* - \ell_2)}{D}$$

or

$$\frac{(0.02)(1 \text{ m})}{0.1 \text{ m}} = 0.4 - \frac{f(\ell^* - \ell_2)}{D}$$

so that

$$\frac{f(\ell^* - \ell_2)}{D} = 0.2 \qquad (1)$$

By using the value from Eq. 1 and Fig. D.2, we get

$$\text{Ma}_2 = 0.70 \qquad \textbf{(Ans)}$$

and

$$\frac{p_2}{p^*} = 1.5 \qquad (2)$$

We obtain $p_2$ from

$$p_2 = \left(\frac{p_2}{p^*}\right)\left(\frac{p^*}{p_1}\right)\left(\frac{p_1}{p_{0,1}}\right)(p_{0,1})$$

where $p_2/p^*$ is given in Eq. 2 and $p^*/p_1$, $p_1/p_{0,1}$, and $p_{0,1}$ are the same as they were in Example 11.12. Thus,

$$p_2 = (1.5)\left(\frac{1}{1.7}\right)(0.76)[101 \text{ kPa(abs)}]$$

$$= 68.0 \text{ kPa(abs)} \qquad \textbf{(Ans)}$$

**COMMENT** A larger back pressure [68.0 kPa(abs)] than the one associated with choked flow through a Fanno duct [45 kPa(abs)] will maintain the same flowrate through a shorter Fanno duct with the same friction coefficient. The flow through the shorter duct is not choked. It would not be possible to maintain the same flowrate through a Fanno duct longer than the choked one with the same friction coefficient, regardless of what back pressure is used.

### 11.5.2 Frictionless Constant Area Duct Flow with Heat Transfer (Rayleigh Flow)

*Rayleigh flow involves heat transfer with no wall friction and constant cross-sectional area.*

Consider the steady, one-dimensional, and frictionless flow of an ideal gas through the constant area duct with heat transfer illustrated in Fig. 11.21. This is **Rayleigh flow.** Application of the linear momentum equation (Eq. 5.22) to the Rayleigh flow through the finite control volume sketched in Fig. 11.21 results in

$$p_1 A_1 + \dot{m} V_1 = p_2 A_2 + \dot{m} V_2 + R_x \overset{\nearrow 0 \text{(frictionless flow)}}{}$$

or

$$p + \frac{(\rho V)^2}{\rho} = \text{constant} \qquad \textbf{(11.110)}$$

Use of the ideal gas equation of state (Eq. 11.1) in Eq. 11.110 leads to

$$p + \frac{(\rho V)^2 RT}{p} = \text{constant} \qquad \textbf{(11.111)}$$

Since the flow cross-sectional area remains constant for Rayleigh flow, from the continuity equation (Eq. 11.40) we conclude that

$$\rho V = \text{constant}$$

For a given Rayleigh flow, the constant in Eq. 11.111, the density–velocity product, $\rho V$, and the ideal gas constant are all fixed. Thus, Eq. 11.111 can be used to determine values of fluid temperature corresponding to the local pressure in a Rayleigh flow.

To construct a temperature–entropy diagram for a given Rayleigh flow, we can use Eq. 11.76, which was developed earlier from the second $T \, ds$ relationship. Equations 11.111 and 11.76 can be solved simultaneously to obtain the curve sketched in Fig. 11.22. Curves like the one in Fig. 11.22 are called *Rayleigh lines.*

■ **Figure 11.21** **Rayleigh flow.**

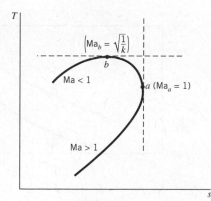

■ **Figure 11.22**   Rayleigh line.

# EXAMPLE 11.15   Frictionless, Constant Area Compressible Flow with Heat Transfer (Rayleigh Flow)

**GIVEN**   Air $(k = 1.4)$  enters [section (1)] a frictionless, constant flow cross-sectional area duct with the following properties (the same as in Example 11.11):

$$T_0 = 518.67\ °R$$
$$T_1 = 514.55\ °R$$
$$p_1 = 14.3\ \text{psia}$$

**FIND**   For Rayleigh flow, determine corresponding values of fluid temperature and entropy change for various levels of downstream pressure and plot the related Rayleigh line.

## SOLUTION

To plot the Rayleigh line asked for, use Eq. 11.111

$$p + \frac{(\rho V)^2\, RT}{p} = \text{constant} \qquad (1)$$

and Eq. 11.76

$$s - s_1 = c_p \ln \frac{T}{T_1} - R \ln \frac{p}{p_1} \qquad (2)$$

to construct a table of values of temperature and entropy change corresponding to different levels of pressure downstream in a Rayleigh flow.

Use the value of ideal gas constant for air from Table 1.7

$$R = 1716\ (\text{ft} \cdot \text{lb})/(\text{slug} \cdot °R)$$

or in EE system units

$$R = 53.3\ (\text{ft} \cdot \text{lb})/(\text{lbm} \cdot °R)$$

and the value of specific heat at constant pressure for air from Example 11.11, namely,

$$c_p = 187\ (\text{ft} \cdot \text{lb})/(\text{lbm} \cdot °R)$$

Also, from Example 11.11, $\rho V = 16.7\ \text{lbm}/(\text{ft}^2 \cdot \text{s})$. For the given inlet [section (1)] conditions, we get

$$\frac{RT_1}{p_1} = \frac{[53.3\ (\text{ft} \cdot \text{lb})/(\text{lbm} \cdot °R)](514.55\ °R)}{14.3\ \text{psia}\ (144\ \text{in.}^2/\text{ft}^2)}$$
$$= 13.3\ \text{ft}^3/\text{lbm}$$

Thus, from Eq. 1 we get

$$p + \frac{(\rho V)^2\, RT}{p} = 14.3\ \text{psia} + [16.7\ \text{lbm}/(\text{ft}^2 \cdot \text{s})]^2(13.3\ \text{ft}^3/\text{lbm})$$
$$= 14.3\ \text{psia} + 3720\ \text{lbm}/(\text{ft} \cdot \text{s}^2) = \text{constant}$$

or,   since   $1\ \text{lbm}/(\text{ft} \cdot \text{s}^2) = [(1/32.2)(\text{lb} \cdot \text{s}^2/\text{ft})]/(\text{ft} \cdot \text{s}^2) = (1/32.2)\ \text{lb/ft}^2$,

$$p + \frac{(\rho V)^2\, RT}{p} = 14.3\ \text{psia} + [(3720/32.2)\text{lb/ft}^2](1\ \text{ft}^2/144\ \text{in.}^2)$$
$$= 15.10\ \text{psia} = \text{constant} \qquad (3)$$

With the downstream pressure of $p = 13.5$ psia, we can obtain the downstream temperature by using Eq. 3 with the fact that

$$\frac{(\rho V)^2 R}{p} = \frac{[16.7\ \text{lbm}/(\text{ft}^2 \cdot \text{s})]^2[53.3\ (\text{ft} \cdot \text{lb})/(\text{lbm} \cdot °R)]}{(144\ \text{in.}^2/\text{ft}^2)\ 13.5\ \text{psia}}$$
$$= 7.65\ [\text{lbm}/(\text{ft} \cdot \text{s}^2)]/°R\ [1\ \text{lb}/(32.2\ \text{lbm} \cdot \text{ft/s}^2)]$$
$$= 0.238\ (\text{lb/ft}^2)/°R(1\ \text{ft}^2/144\ \text{in.}^2)$$
$$= 1.65 \times 10^{-3}\ (\text{lb/in.}^2)/°R$$

Hence, from Eq. 3,

$$13.5\ \text{psia} + [1.65 \times 10^{-3}\ (\text{lb/in.}^2)/°R]\,T = 15.10\ \text{psia}$$

or

$$T = 969\ °R$$

From Eq. 2 with the downstream pressure $p = 13.5$ psia and temperature $T = 969\ °R$ we get

$$s - s_1 = [187\ (\text{ft} \cdot \text{lb})/(\text{lbm} \cdot °R)] \ln\!\left(\frac{969\ °R}{514.55\ °R}\right)$$
$$\quad - [53.3\ (\text{ft} \cdot \text{lb})/(\text{lbm} \cdot °R)] \ln\!\left(\frac{13.5\ \text{psia}}{14.3\ \text{psia}}\right)$$
$$s - s_1 = 121\ (\text{ft} \cdot \text{lb})/(\text{lbm} \cdot °R)$$

By proceeding as outlined above, we can construct the table of values shown below and graph the Rayleigh line of Fig. E11.15.

| $p$ (psia) | $T$ (°R) | $s - s_1$ [(ft · lb)/(lbm · °R)] |
|---|---|---|
| 13.5 | 969 | 9.32 |
| 12.5 | 1459 | 202 |
| 11.5 | 1859 | 251 |
| 10.5 | 2168 | 285 |
| 9.0 | 2464 | 317 |
| 8.0 | 2549 | 330 |
| 7.6 | 2558 | 333 |
| 7.5 | 2558 | 334 |
| 7.0 | 2544 | 336 |
| 6.3 | 2488 | 338 |
| 6.0 | 2450 | 338 |
| 5.5 | 2369 | 336 |
| 5.0 | 2266 | 333 |
| 4.5 | 2140 | 328 |
| 4.0 | 1992 | 321 |
| 2.0 | 1175 | 259 |
| 1.0 | 633 | 181 |

■ **Figure E11.15**

**COMMENT** Depending on whether the flow is being heated or cooled, it can proceed in either direction along the curve.

At point $a$ on the Rayleigh line of Fig. 11.22, $ds/dT = 0$. To determine the physical importance of point $a$, we analyze further some of the governing equations. By differentiating the linear momentum equation for Rayleigh flow (Eq. 11.110), we obtain

$$dp = -\rho V \, dV$$

or

$$\frac{dp}{\rho} = -V \, dV \tag{11.112}$$

Combining Eq. 11.112 with the second $T \, ds$ equation (Eq. 11.18) leads to

$$T \, ds = d\breve{h} + V \, dV \tag{11.113}$$

For an ideal gas (Eq. 11.7) $d\breve{h} = c_p \, dT$. Thus, substituting Eq. 11.7 into Eq. 11.113 gives

$$T \, ds = c_p \, dT + V \, dV$$

or

$$\frac{ds}{dT} = \frac{c_p}{T} + \frac{V}{T} \frac{dV}{dT} \tag{11.114}$$

Consolidation of Eqs. 11.114, 11.112 (linear momentum), 11.1, 11.77 (differentiated equation of state), and 11.79 (continuity) leads to

$$\frac{ds}{dT} = \frac{c_p}{T} + \frac{V}{T} \frac{1}{[(T/V) - (V/R)]} \tag{11.115}$$

*The maximum entropy state on the Rayleigh line corresponds to sonic conditions.*

Hence, at state $a$ where $ds/dT = 0$, Eq. 11.115 reveals that

$$V_a = \sqrt{RT_a k} \tag{11.116}$$

Comparison of Eqs. 11.116 and 11.36 tells us that the Mach number at state $a$ is equal to 1,

$$\text{Ma}_a = 1 \tag{11.117}$$

At point $b$ on the Rayleigh line of Fig. 11.22, $dT/ds = 0$. From Eq. 11.115 we get

$$\frac{dT}{ds} = \frac{1}{ds/dT} = \frac{1}{(c_p/T) + (V/T)[(T/V) - (V/R)]^{-1}}$$

which for $dT/ds = 0$ (point $b$) gives

$$Ma_b = \sqrt{\frac{1}{k}} \tag{11.118}$$

The flow at point $b$ is subsonic ($Ma_b < 1.0$). Recall that $k > 1$ for any gas.

To learn more about Rayleigh flow, we need to consider the energy equation in addition to the equations already used. Application of the energy equation (Eq. 5.69) to the Rayleigh flow through the finite control volume of Fig. 11.21 yields

$$\dot{m}\left[\check{h}_2 - \check{h}_1 + \frac{V_2^2 - V_1^2}{2} + g(z_2 - z_1)\right] = \overset{0(\text{negligibly small for gas flow})}{\dot{Q}_{\text{net}}} + \overset{0(\text{flow is steady throughout})}{\dot{W}_{\text{shaft}}}$$

or in differential form for Rayleigh flow through the semi-infinitesimal control volume of Fig. 11.21

$$d\check{h} + V\,dV = \delta q \tag{11.119}$$

where $\delta q$ is the heat transfer per unit mass of fluid in the semi-infinitesimal control volume.

By using $d\check{h} = c_p\,dT = Rk\,dT/(k-1)$ in Eq. 11.119, we obtain

$$\frac{dV}{V} = \frac{\delta q}{c_p T}\left[\frac{V}{T}\frac{dT}{dV} + \frac{V^2(k-1)}{kRT}\right]^{-1} \tag{11.120}$$

Thus, by combining Eqs. 11.36 (ideal gas speed of sound), 11.46 (Mach number), 11.1 and 11.77 (ideal gas equation of state), 11.79 (continuity), and 11.112 (linear momentum) with Eq. 11.120 (energy) we get

$$\frac{dV}{V} = \frac{\delta q}{c_p T}\frac{1}{(1 - Ma^2)} \tag{11.121}$$

With the help of Eq. 11.121, we see clearly that when the Rayleigh flow is subsonic ($Ma < 1$), fluid heating ($\delta q > 0$) increases fluid velocity while fluid cooling ($\delta q < 0$) decreases fluid velocity. When Rayleigh flow is supersonic ($Ma > 1$), fluid heating decreases fluid velocity and fluid cooling increases fluid velocity.

The second law of thermodynamics states that, based on experience, entropy increases with heating and decreases with cooling. With this additional insight provided by the conservation of energy principle and the second law of thermodynamics, we can say more about the Rayleigh line in Fig. 11.22. A summary of the qualitative aspects of Rayleigh flow is outlined in Table 11.2 and Fig. 11.23. Along the upper portion of the line, which includes point $b$, the flow is subsonic. Heating the fluid results in flow acceleration to a maximum Mach number of 1 at point $a$. Note that between points $b$ and $a$ along the Rayleigh line, heating the fluid results in a temperature decrease and cooling the fluid leads to a temperature increase. This trend is not surprising if we consider the stagnation temperature and fluid velocity changes that occur between points $a$ and $b$ when the fluid is heated or cooled. Along the lower portion of the Rayleigh curve the flow is supersonic. Rayleigh flows may or may not be choked. The amount of heating or cooling involved determines what will happen in a specific instance. As with Fanno flows, an abrupt deceleration from supersonic flow to subsonic flow across a normal shock wave can also occur in Rayleigh flows.

*Fluid temperature reduction can accompany heating a subsonic Rayleigh flow.*

■ **Table 11.2**

**Summary of Rayleigh Flow Characteristics**

|          | Heating      | Cooling      |
|----------|--------------|--------------|
| $Ma < 1$ | Acceleration | Deceleration |
| $Ma > 1$ | Deceleration | Acceleration |

**■ Figure 11.23** (*a*) **Subsonic Rayleigh flow.** (*b*) **Supersonic Rayleigh flow.** (*c*) **Normal shock in a Rayleigh flow.**

To quantify Rayleigh flow behavior we need to develop appropriate forms of the governing equations. We elect to use the state of the Rayleigh flow fluid at point *a* of Fig. 11.22 as the reference state. As shown earlier, the Mach number at point *a* is 1. Even though the Rayleigh flow being considered may not choke and state *a* is not achieved by the flow, this reference state is useful.

If we apply the linear momentum equation (Eq. 11.110) to Rayleigh flow between any upstream section and the section, actual or imagined, where state *a* is attained, we get

$$p + \rho V^2 = p_a + \rho_a V_a^2$$

or

$$\frac{p}{p_a} + \frac{\rho V^2}{p_a} = 1 + \frac{\rho_a}{p_a} V_a^2 \qquad (11.122)$$

By substituting the ideal gas equation of state (Eq. 11.1) into Eq. 11.122 and making use of the ideal gas speed-of-sound equation (Eq. 11.36) and the definition of Mach number (Eq. 11.46), we obtain

$$\boxed{\frac{p}{p_a} = \frac{1 + k}{1 + k\mathrm{Ma}^2}} \qquad (11.123)$$

This relationship is graphed in the margin for air.

From the ideal gas equation of state (Eq. 11.1) we conclude that

$$\frac{T}{T_a} = \frac{p}{p_a} \frac{\rho_a}{\rho} \qquad (11.124)$$

Conservation of mass (Eq. 11.40) with constant *A* gives

$$\frac{\rho_a}{\rho} = \frac{V}{V_a} \qquad (11.125)$$

which when combined with Eqs. 11.36 (ideal gas speed of sound) and 11.46 (Mach number definition) gives

$$\frac{\rho_a}{\rho} = \mathrm{Ma} \sqrt{\frac{T}{T_a}} \qquad (11.126)$$

Combining Eqs. 11.124 and 11.126 leads to

$$\frac{T}{T_a} = \left( \frac{p}{p_a} \mathrm{Ma} \right)^2 \qquad (11.127)$$

which when combined with Eq. 11.123 gives (also see margin figure)

$$\boxed{\frac{T}{T_a} = \left[ \frac{(1 + k)\mathrm{Ma}}{1 + k\mathrm{Ma}^2} \right]^2} \qquad (11.128)$$

From Eqs. 11.125, 11.126, and 11.128 we see that

$$\frac{\rho_a}{\rho} = \frac{V}{V_a} = \mathrm{Ma}\left[\frac{(1 + k)\mathrm{Ma}}{1 + k\mathrm{Ma}^2}\right] \tag{11.129}$$

This relationship is graphed in the margin for air.

The energy equation (Eq. 5.69) tells us that because of the heat transfer involved in Rayleigh flows, the stagnation temperature varies. We note that

$$\frac{T_0}{T_{0,a}} = \left(\frac{T_0}{T}\right)\left(\frac{T}{T_a}\right)\left(\frac{T_a}{T_{0,a}}\right) \tag{11.130}$$

*Unlike Fanno flow, the stagnation temperature in Rayleigh flow varies.*

We can use Eq. 11.56 (developed earlier for steady, isentropic, ideal gas flow) to evaluate $T_0/T$ and $T_a/T_{0,a}$ because these two temperature ratios, by definition of the stagnation state, involve isentropic processes. Equation 11.128 can be used for $T/T_a$. Thus, consolidating Eqs. 11.130, 11.56, and 11.128 we obtain

$$\frac{T_0}{T_{0,a}} = \frac{2(k + 1)\mathrm{Ma}^2\left(1 + \dfrac{k - 1}{2}\mathrm{Ma}^2\right)}{(1 + k\mathrm{Ma}^2)^2} \tag{11.131}$$

This relationship is graphed in the margin for air.

Finally, we observe that

$$\frac{p_0}{p_{0,a}} = \left(\frac{p_0}{p}\right)\left(\frac{p}{p_a}\right)\left(\frac{p_a}{p_{0,a}}\right) \tag{11.132}$$

We can use Eq. 11.59 developed earlier for steady, isentropic, ideal gas flow to evaluate $p_0/p$ and $p_a/p_{0,a}$ because these two pressure ratios, by definition, involve isentropic processes. Equation 11.123 can be used for $p/p_a$. Together, Eqs. 11.59, 11.123, and 11.132 give

$$\frac{p_0}{p_{0,a}} = \frac{(1 + k)}{(1 + k\mathrm{Ma}^2)}\left[\left(\frac{2}{k + 1}\right)\left(1 + \frac{k - 1}{2}\mathrm{Ma}^2\right)\right]^{k/(k-1)} \tag{11.133}$$

This relationship is graphed in the margin for air.

Values of $p/p_a$, $T/T_a$, $\rho/\rho_a$ or $V/V_a$, $T_0/T_{0,a}$, and $p_0/p_{0,a}$ are graphed in Fig. D.3 of Appendix D as a function of Mach number for Rayleigh flow of air ($k = 1.4$). The values in Fig. D.3 were calculated from Eqs. 11.123, 11.128, 11.129, 11.131, and 11.133. The usefulness of Fig. D.3 is illustrated in Example 11.16.

See Ref. 7 for a more advanced treatment of internal flows with heat transfer.

---

**$E$XAMPLE  11.16**  **Effect of Mach Number and Heating/Cooling for Rayleigh Flow**

**GIVEN**  The information in Table 11.2 shows us that subsonic Rayleigh flow accelerates when heated and decelerates when cooled. Supersonic Rayleigh flow behaves just opposite to subsonic Rayleigh flow; it decelerates when heated and accelerates when cooled.

**FIND**  Using Fig. D.3 for air ($k = 1.4$), state whether velocity, Mach number, static temperature, stagnation temperature, static pressure, and stagnation pressure increase or decrease as subsonic and supersonic Rayleigh flow is (a) heated, (b) cooled.

## SOLUTION

Acceleration occurs when $V/V_a$ in Fig. D.3 increases. For deceleration, $V/V_a$ decreases. From Fig. D.3 and Table 11.2 the following chart can be constructed.

From the Rayleigh flow trends summarized in the table below, we note that heating affects Rayleigh flows much like friction affects Fanno flows. Heating and friction both accelerate subsonic flows and decelerate supersonic flows. More importantly, both heating and friction cause the stagnation pressure to decrease. Since stagnation pressure loss is considered undesirable in terms of fluid mechanical efficiency, heating a fluid flow must be accomplished with this loss in mind.

**COMMENT**   Note that for a small range of Mach numbers cooling actually results in a rise in temperature, $T$.

|       | Heating | | Cooling | |
|-------|---------|------------|---------|------------|
|       | **Subsonic** | **Supersonic** | **Subsonic** | **Supersonic** |
| $V$   | Increase | Decrease | Decrease | Increase |
| Ma    | Increase | Decrease | Decrease | Increase |
| $T$   | Increase for $0 \le$ Ma $\le \sqrt{1/k}$ <br> Decrease for $\sqrt{1/k} \le$ Ma $\le 1$ | Increase | Decrease for $0 \le$ Ma $\le \sqrt{1/k}$ <br> Increase for $\sqrt{1/k} \le$ Ma $\le 1$ | Decrease |
| $T_0$ | Increase | Increase | Decrease | Decrease |
| $p$   | Decrease | Increase | Increase | Decrease |
| $p_0$ | Decrease | Decrease | Increase | Increase |

### 11.5.3 Normal Shock Waves

*V11.8 Blast waves*

As mentioned earlier, normal shock waves can occur in supersonic flows through converging–diverging and constant area ducts. Past experience suggests that normal shock waves involve deceleration from a supersonic flow to a subsonic flow, a pressure rise, and an increase of entropy. To develop the equations that verify this observed behavior of flows across a normal shock, we apply first principles to the flow through a control volume that completely surrounds a normal shock wave (see Fig. 11.24). We consider the normal shock and thus the control volume to be infinitesimally thin and stationary.

For steady flow through the control volume of Fig. 11.24, the conservation of mass principle yields

$$\rho V = \text{constant} \tag{11.134}$$

*Normal shock waves are assumed to be infinitesimally thin discontinuities.*

because the flow cross-sectional area remains essentially constant within the infinitesimal thickness of the normal shock. Note that Eq. 11.134 is identical to the continuity equation used for Fanno and Rayleigh flows considered earlier.

The friction force acting on the contents of the infinitesimally thin control volume surrounding the normal shock is considered to be negligibly small. Also for ideal gas flow, the effect of gravity is neglected. Thus, the linear momentum equation (Eq. 5.22) describing steady gas flow through the control volume of Fig. 11.24 is

$$p + \rho V^2 = \text{constant}$$

or for an ideal gas for which $p = \rho RT$,

$$p + \frac{(\rho V)^2 RT}{p} = \text{constant} \tag{11.135}$$

Equation 11.135 is the same as the linear momentum equation for Rayleigh flow, which was derived earlier (Eq. 11.111).

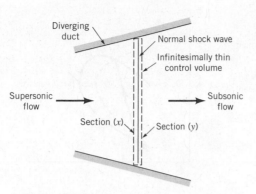

**Figure 11.24**   Normal shock control volume.

**Figure 11.25**   The relationship between a normal shock and Fanno and Rayleigh lines.

For the control volume containing the normal shock, no shaft work is involved and the heat transfer is assumed negligible. Thus, the energy equation (Eq. 5.69) can be applied to steady gas flow through the control volume of Fig. 11.24 to obtain

$$\check{h} + \frac{V^2}{2} = \check{h}_0 = \text{constant}$$

or, for an ideal gas, since $\check{h} - \check{h}_0 = c_p(T - T_0)$ and $p = \rho R T$

$$T + \frac{(\rho V)^2 T^2}{2 c_p (p^2/R^2)} = T_0 = \text{constant} \tag{11.136}$$

Equation 11.136 is identical to the energy equation for Fanno flow analyzed earlier (Eq. 11.75).

The $T\,ds$ relationship previously used for ideal gas flow (Eq. 11.22) is valid for the flow through the normal shock (Fig. 11.24) because it (Eq. 11.22) is an ideal gas property relationship.

From the analyses in the previous paragraphs, it is apparent that the steady flow of an ideal gas across a normal shock is governed by some of the same equations used for describing both Fanno and Rayleigh flows (energy equation for Fanno flows and momentum equation for Rayleigh flow). Thus, for a given density–velocity product ($\rho V$), gas ($R$, $k$), and conditions at the inlet of the normal shock ($T_x$, $p_x$, and $s_x$), the conditions downstream of the shock (state $y$) will be on both a Fanno line and a Rayleigh line that pass through the inlet state (state $x$), as is illustrated in Fig. 11.25. To conform with common practice we designate the states upstream and downstream of the normal shock with $x$ and $y$ instead of numerals 1 and 2. The Fanno and Rayleigh lines describe more of the flow field than just in the vicinity of the normal shock when Fanno and Rayleigh flows are actually involved (solid lines in Figs. 11.26$a$ and 11.26$b$). Otherwise, these lines (dashed lines in Figs. 11.26$a$, 11.26$b$, and 11.26$c$) are useful mainly as a way to better visualize how the governing equations combine to yield a solution to the normal shock flow problem.

The second law of thermodynamics requires that entropy must increase across a normal shock wave. This law and sketches of the Fanno line and Rayleigh line intersections, like those of Figs. 11.25 and 11.26, persuade us to conclude that flow across a normal shock can only proceed from supersonic to subsonic flow. Similarly, in open-channel flows (see Chapter 10) the flow across a hydraulic jump proceeds from supercritical to subcritical conditions.

Since the states upstream and downstream of a normal shock wave are represented by the supersonic and subsonic intersections of actual and/or imagined Fanno and Rayleigh lines, we should be able to use equations developed earlier for Fanno and Rayleigh flows to quantify normal shock flow. For example, for the Rayleigh line of Fig. 11.26$b$

$$\frac{p_y}{p_x} = \left(\frac{p_y}{p_a}\right)\left(\frac{p_a}{p_x}\right) \tag{11.137}$$

*The energy equation for Fanno flow and the momentum equation for Rayleigh flow are valid for flow across normal shocks.*

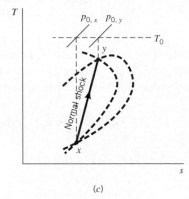

■ **Figure 11.26** (*a*) The normal shock in a Fanno flow. (*b*) The normal shock in a Rayleigh flow. (*c*) The normal shock in a frictionless and adiabatic flow.

But from Eq. 11.123 for Rayleigh flow we get

$$\frac{p_y}{p_a} = \frac{1 + k}{1 + k\text{Ma}_y^2} \tag{11.138}$$

and

$$\frac{p_x}{p_a} = \frac{1 + k}{1 + k\text{Ma}_x^2} \tag{11.139}$$

Thus, by combining Eqs. 11.137, 11.138, and 11.139 we get

*Ratios of thermodynamic properties across a normal shock are functions of the Mach numbers.*

$$\frac{p_y}{p_x} = \frac{1 + k\text{Ma}_x^2}{1 + k\text{Ma}_y^2} \tag{11.140}$$

Equation 11.140 can also be derived starting with

$$\frac{p_y}{p_x} = \left(\frac{p_y}{p^*}\right)\left(\frac{p^*}{p_x}\right)$$

and using the Fanno flow equation (Eq. 11.107)

$$\frac{p}{p^*} = \frac{1}{\text{Ma}}\left\{\frac{(k + 1)/2}{1 + [(k - 1)/2]\text{Ma}^2}\right\}^{1/2}$$

As might be expected, Eq. 11.140 can be obtained directly from the linear momentum equation

$$p_x + \rho_x V_x^2 = p_y + \rho_y V_y^2$$

since $\rho V^2/p = V^2/RT = kV^2/RTk = k\,\text{Ma}^2$.

For the Fanno flow of Fig. 11.26*a*,

$$\frac{T_y}{T_x} = \left(\frac{T_y}{T^*}\right)\left(\frac{T^*}{T_x}\right) \tag{11.141}$$

From Eq. 11.101 for Fanno flow we get

$$\frac{T_y}{T^*} = \frac{(k+1)/2}{1 + [(k-1)/2]\mathrm{Ma}_y^2}$$

(11.142)

and

$$\frac{T_x}{T^*} = \frac{(k+1)/2}{1 + [(k-1)/2]\mathrm{Ma}_x^2}$$

(11.143)

A consolidation of Eqs. 11.141, 11.142, and 11.143 gives

$$\frac{T_y}{T_x} = \frac{1 + [(k-1)/2]\mathrm{Ma}_x^2}{1 + [(k-1)/2]\mathrm{Ma}_y^2}$$

(11.144)

We seek next to develop an equation that will allow us to determine the Mach number downstream of the normal shock, $\mathrm{Ma}_y$, when the Mach number upstream of the normal shock, $\mathrm{Ma}_x$, is known. From the ideal gas equation of state (Eq. 11.1), we can form

$$\frac{p_y}{p_x} = \left(\frac{T_y}{T_x}\right)\left(\frac{\rho_y}{\rho_x}\right)$$

(11.145)

Using the continuity equation

$$\rho_x V_x = \rho_y V_y$$

with Eq. 11.145 we obtain

$$\frac{p_y}{p_x} = \left(\frac{T_y}{T_x}\right)\left(\frac{V_x}{V_y}\right)$$

(11.146)

When combined with the Mach number definition (Eq. 11.46) and the ideal gas speed-of-sound equation (Eq. 11.36), Eq. 11.146 becomes

$$\frac{p_y}{p_x} = \left(\frac{T_y}{T_x}\right)^{1/2}\left(\frac{\mathrm{Ma}_x}{\mathrm{Ma}_y}\right)$$

(11.147)

Thus, Eqs. 11.147 and 11.144 lead to

$$\frac{p_y}{p_x} = \left\{\frac{1 + [(k-1)/2]\mathrm{Ma}_x^2}{1 + [(k-1)/2]\mathrm{Ma}_y^2}\right\}^{1/2} \frac{\mathrm{Ma}_x}{\mathrm{Ma}_y}$$

(11.148)

which can be merged with Eq. 11.140 to yield

$$\mathrm{Ma}_y^2 = \frac{\mathrm{Ma}_x^2 + [2/(k-1)]}{[2k/(k-1)]\mathrm{Ma}_x^2 - 1}$$

(11.149)

*The flow changes from supersonic to subsonic across a normal shock.*

This relationship is graphed in the margin for air.

Thus, we can use Eq. 11.149 to calculate values of Mach number downstream of a normal shock from a known Mach number upstream of the shock. As suggested by Fig. 11.26, to have a normal shock we must have $\mathrm{Ma}_x > 1$. From Eq. 11.149 we find that $\mathrm{Ma}_y < 1$.

If we combine Eqs. 11.149 and 11.140, we get

$$\frac{p_y}{p_x} = \frac{2k}{k+1}\mathrm{Ma}_x^2 - \frac{k-1}{k+1}$$

(11.150)

This relationship is graphed in the margin for air.

This equation allows us to calculate the pressure ratio across a normal shock from a known upstream Mach number. Similarly, taking Eqs. 11.149 and 11.144 together we obtain

$$\frac{T_y}{T_x} = \frac{\{1 + [(k-1)/2]\mathrm{Ma}_x^2\}\{[2k/(k-1)]\mathrm{Ma}_x^2 - 1\}}{\{(k+1)^2/[2(k-1)]\}\mathrm{Ma}_x^2} \qquad (11.151)$$

This relationship is graphed in the margin for air.

From the continuity equation (Eq. 11.40), we have for flow across a normal shock

$$\frac{\rho_y}{\rho_x} = \frac{V_x}{V_y} \qquad (11.152)$$

and from the ideal gas equation of state (Eq. 11.1)

$$\frac{\rho_y}{\rho_x} = \left(\frac{p_y}{p_x}\right)\left(\frac{T_x}{T_y}\right) \qquad (11.153)$$

Thus, by combining Eqs. 11.152, 11.153, 11.150, and 11.151, we get

$$\frac{\rho_y}{\rho_x} = \frac{V_x}{V_y} = \frac{(k+1)\mathrm{Ma}_x^2}{(k-1)\mathrm{Ma}_x^2 + 2} \qquad (11.154)$$

This relationship is graphed in the margin for air.

The stagnation pressure ratio across the shock can be determined by combining

$$\frac{p_{0,y}}{p_{0,x}} = \left(\frac{p_{0,y}}{p_y}\right)\left(\frac{p_y}{p_x}\right)\left(\frac{p_x}{p_{0,x}}\right) \qquad (11.155)$$

with Eqs. 11.59, 11.149, and 11.150 to get

$$\frac{p_{0,y}}{p_{0,x}} = \frac{\left(\frac{k+1}{2}\mathrm{Ma}_x^2\right)^{k/(k-1)}\left(1 + \frac{k-1}{2}\mathrm{Ma}_x^2\right)^{k/(1-k)}}{\left(\frac{2k}{k+1}\mathrm{Ma}_x^2 - \frac{k-1}{k+1}\right)^{1/(k-1)}} \qquad (11.156)$$

This relationship is graphed in the margin for air.

Figure D.4 in Appendix D graphs values of downstream Mach numbers, $\mathrm{Ma}_y$, pressure ratio, $p_y/p_x$, temperature ratio, $T_y/T_x$, density ratio, $\rho_y/\rho_x$, or velocity ratio, $V_x/V_y$, and stagnation pressure ratio, $p_{0,y}/p_{0,x}$, as a function of upstream Mach number, $\mathrm{Ma}_x$, for the steady flow across a normal shock wave of an ideal gas having a specific heat ratio $k = 1.4$. These values were calculated from Eqs. 11.149, 11.150, 11.151, 11.154, and 11.156.

Important trends associated with the steady flow of an ideal gas across a normal shock wave can be determined by studying Fig. D.4. These trends are summarized in Table 11.3.

Examples 11.17 and 11.18 illustrate how Fig. D.4 can be used to solve fluid flow problems involving normal shock waves.

*Across a normal shock the values of some parameters increase, some remain constant, and some decrease.*

■ **Table 11.3**

**Summary of Normal Shock Wave Characteristics**

| Variable | Change Across Normal Shock Wave |
|---|---|
| Mach number | Decrease |
| Static pressure | Increase |
| Stagnation pressure | Decrease |
| Static temperature | Increase |
| Stagnation temperature | Constant |
| Density | Increase |
| Velocity | Decrease |

# EXAMPLE 11.17    Stagnation Pressure Drop across a Normal Shock

**GIVEN**  Designers involved with fluid mechanics work hard at minimizing loss of available energy in their designs. Adiabatic, frictionless flows involve no loss in available energy. Entropy remains constant for these idealized flows. Adiabatic flows with friction involve available energy loss and entropy increase. Generally, larger entropy increases imply larger losses.

**FIND**  For normal shocks, show that the stagnation pressure drop (and thus loss) is larger for higher Mach numbers.

## SOLUTION

We assume that air ($k = 1.4$) behaves as a typical gas and use Fig. D.4 to respond to the above-stated requirements. Since

$$1 - \frac{p_{0,y}}{p_{0,x}} = \frac{p_{0,x} - p_{0,y}}{p_{0,x}}$$

we can construct the following table with values of $p_{0,y}/p_{0,x}$ from Fig. D.4.

**COMMENT**  When the Mach number of the flow entering the shock is low, say $Ma_x = 1.2$, the flow across the shock is nearly isentropic and the loss in stagnation pressure is small. However, as shown in Fig. E11.17, at larger Mach numbers, the entropy change across the normal shock rises dramatically and the stagna-

tion pressure drop across the shock is appreciable. If a shock occurs at $Ma_x = 2.5$, only about 50% of the upstream stagnation pressure is recovered.

In devices where supersonic flows occur, for example, high-performance aircraft engine inlet ducts and high-speed wind tunnels, designers attempt to prevent shock formation, or if shocks must occur, they design the flow path so that shocks are positioned where they are weak (small Mach number).

Of interest also is the static pressure rise that occurs across a normal shock. These static pressure ratios, $p_y/p_x$, obtained from Fig. D.4 are shown in the table for a few Mach numbers. For a developing boundary layer, any pressure rise in the flow direction is considered as an adverse pressure gradient that can possibly cause flow separation (see Section 9.2.6). Thus, shock–boundary layer interactions are of great concern to designers of high-speed flow devices.

| $Ma_x$ | $p_{0,y}/p_{0,x}$ | $\dfrac{p_{0,x} - p_{0,y}}{p_{0,x}}$ | $Ma_x$ | $p_y/p_x$ |
|---|---|---|---|---|
| 1.0 | 1.0 | 0 | 1.0 | 1.0 |
| 1.2 | 0.99 | 0.01 | 1.2 | 1.5 |
| 1.5 | 0.93 | 0.07 | 1.5 | 2.5 |
| 2.0 | 0.72 | 0.28 | 2.0 | 4.5 |
| 2.5 | 0.50 | 0.50 | 3.0 | 10 |
| 3.0 | 0.33 | 0.67 | 4.0 | 18 |
| 3.5 | 0.21 | 0.79 | 5.0 | 29 |
| 4.0 | 0.14 | 0.86 | | |
| 5.0 | 0.06 | 0.94 | | |

■ **Figure E11.17**

# EXAMPLE 11.18    Supersonic Flow Pitot Tube

**GIVEN**  A total pressure probe is inserted into a supersonic air flow. A shock wave forms just upstream of the impact hole and head as illustrated in Fig. E11.18. The probe measures a total pressure of 60 psia. The stagnation temperature at the probe head is 1000 °R. The static pressure upstream of the shock is measured with a wall tap to be 12 psia.

**FIND**  Determine the Mach number and velocity of the flow.

■ **Figure E11.18**

## SOLUTION

We assume that the flow along the stagnation pathline is isentropic except across the shock. Also, the shock is treated as a normal shock. Thus, in terms of the data we have

$$\frac{p_{0,y}}{p_x} = \left(\frac{p_{0,y}}{p_{0,x}}\right)\left(\frac{p_{0,x}}{p_x}\right) \tag{1}$$

where $p_{0,y}$ is the stagnation pressure measured by the probe, and $p_x$ is the static pressure measured by the wall tap. The stagnation pressure upstream of the shock, $p_{0,x}$, is not measured.

Combining Eqs. 1, 11.156, and 11.59 we obtain

$$\frac{p_{0,y}}{p_x} = \frac{\{[(k+1)/2]\mathrm{Ma}_x^2\}^{k/(k-1)}}{\{[2k/(k+1)]\mathrm{Ma}_x^2 - [(k-1)/(k+1)]\}^{1/(k-1)}} \tag{2}$$

which is called the *Rayleigh Pitot-tube formula*. Values of $p_{0,y}/p_x$ from Eq. 2 are considered important enough to be included in Fig. D.4 for $k = 1.4$. Thus, for $k = 1.4$ and

$$\frac{p_{0,y}}{p_x} = \frac{60 \text{ psia}}{12 \text{ psia}} = 5$$

we use Fig. D.4 (or Eq. 2) to ascertain that

$$\mathrm{Ma}_x = 1.9 \tag{Ans}$$

To determine the flow velocity we need to know the static temperature upstream of the shock, since Eqs. 11.36 and 11.46 can be used to yield

$$V_x = \mathrm{Ma}_x c_x = \mathrm{Ma}_x \sqrt{RT_x k} \tag{3}$$

The stagnation temperature downstream of the shock was measured and found to be

$$T_{0,y} = 1000 \text{ °R}$$

Since the stagnation temperature remains constant across a normal shock (see Eq. 11.136),

$$T_{0,x} = T_{0,y} = 1000 \text{ °R}$$

For the isentropic flow upstream of the shock, Eq. 11.56 or Fig. D.1 can be used. For $\mathrm{Ma}_x = 1.9$,

$$\frac{T_x}{T_{0,x}} = 0.59$$

or

$$T_x = (0.59)(1000 \text{ °R}) = 590 \text{ °R}$$

With Eq. 3 we obtain

$$V_x = 1.87 \sqrt{[53.3 \text{ (ft} \cdot \text{lb)/(lbm} \cdot \text{°R})](590 \text{ °R})(1.4)}$$
$$= 392 \text{ (ft} \cdot \text{lb/lbm)}^{1/2}[(32.2 \text{ lbm} \cdot \text{ft/s}^2)/\text{lb}]^{1/2}$$
$$= 2220 \text{ ft/s} \tag{Ans}$$

**COMMENT** Application of the incompressible flow Pitot tube results (see Section 3.5) would give highly inaccurate results because of the large pressure and density changes involved.

---

# EXAMPLE 11.19 Normal Shock in a Converging–Diverging Duct

**GIVEN** Consider the converging–diverging duct of Example 11.8.

**FIND** Determine the ratio of back pressure to inlet stagnation pressure, $p_{III}/p_{0,x}$ (see Fig. 11.13), that will result in a standing normal shock at the exit ($x = +0.5$ m) of the duct. What value of the ratio of back pressure to inlet stagnation pressure would be required to position the shock at $x = +0.3$ m? Show related temperature–entropy diagrams for these flows.

## SOLUTION

For supersonic, isentropic flow through the nozzle to just upstream of the standing normal shock at the duct exit, we have from the table of Example 11.8 at $x = +0.5$ m

$$\mathrm{Ma}_x = 2.8$$

and

$$\frac{p_x}{p_{0,x}} = 0.04$$

From Fig. D.4 for $\mathrm{Ma}_x = 2.8$ we obtain

$$\frac{p_y}{p_x} = 9.0$$

Thus,

$$\frac{p_y}{p_{0,x}} = \left(\frac{p_y}{p_x}\right)\left(\frac{p_x}{p_{0,x}}\right) = (9.0)(0.04)$$

$$= 0.36 = \frac{p_{III}}{p_{0,x}} \tag{Ans}$$

When the ratio of duct back pressure to inlet stagnation pressure, $p_{III}/p_{0,x}$, is set equal to 0.36, the air will accelerate through the converging–diverging duct to a Mach number of 2.8 at the duct exit. The air will subsequently decelerate to a subsonic flow across a normal shock at the duct exit. The stagnation pressure ratio across the normal shock, $p_{0,y}/p_{0,x}$, is 0.38 (Fig. D.4 for

$Ma_x = 2.8$). A considerable amount of available energy is lost across the shock.

For a normal shock at $x = +0.3$ m, we note from the table of Example 11.8 that $Ma_x = 2.14$ and

$$\frac{p_x}{p_{0,x}} = 0.10 \qquad (1)$$

From Fig. D.4 for $Ma_x = 2.14$ we obtain $p_y/p_x = 5.2$, $Ma_y = 0.56$, and

$$\frac{p_{0,y}}{p_{0,x}} = 0.66 \qquad (2)$$

From Fig. D.1 for $Ma_y = 0.56$ we get

$$\frac{A_y}{A^*} = 1.24 \qquad (3)$$

For $x = +0.3$ m, the ratio of duct exit area to local area $(A_2/A_y)$ is, using the area equation from Example 11.8,

$$\frac{A_2}{A_y} = \frac{0.1 + (0.5)^2}{0.1 + (0.3)^2} = 1.842 \qquad (4)$$

Using Eqs. 3 and 4 we get

$$\frac{A_2}{A^*} = \left(\frac{A_y}{A^*}\right)\left(\frac{A_2}{A_y}\right) = (1.24)(1.842) = 2.28$$

Note that for the isentropic flow upstream of the shock, $A^* = 0.10$ m$^2$ (the actual throat area), while for the isentropic flow downstream of the shock, $A^* = A_2/2.28 = 0.35$ m$^2/2.28 = 0.15$ m$^2$. With $A_2/A^* = 2.28$ we use Fig. D.1 and find $Ma_2 = 0.26$ and

$$\frac{p_2}{p_{0,y}} = 0.95 \qquad (5)$$

Combining Eqs. 2 and 5 we obtain

$$\frac{p_2}{p_{0,x}} = \left(\frac{p_2}{p_{0,y}}\right)\left(\frac{p_{0,y}}{p_{0,x}}\right) = (0.95)(0.66) = 0.63 \qquad \textbf{(Ans)}$$

When the back pressure, $p_2$, is set equal to 0.63 times the inlet stagnation pressure, $p_{0,x}$, the normal shock will be positioned at $x = +0.3$ m. The corresponding $T-s$ diagrams are shown in Figs. E11.19a and E11.19b.

**COMMENT**   Note that $p_2/p_{0,x} = 0.63$ is less than the value of this ratio for subsonic isentropic flow through the converging–diverging duct, $p_2/p_0 = 0.98$ (from Example 11.8) and is larger than $p_{III}/p_{0,x} = 0.36$, for duct flow with a normal shock at the exit (see Fig. 11.13). Also the stagnation pressure ratio with the shock at $x = +0.3$ m, $p_{0,y}/p_{0,x} = 0.66$, is much greater than the stagnation pressure ratio, 0.38, when the shock occurs at the exit $(x = +0.5$ m) of the duct.

■ **Figure E11.19**

## 11.6   Analogy between Compressible and Open-Channel Flows

During a first course in fluid mechanics, students rarely study both open-channel flows (Chapter 10) and compressible flows. This is unfortunate because these two kinds of flows are strikingly similar in several ways. Furthermore, the analogy between open-channel and compressible flows is useful because important two-dimensional compressible flow phenomena can be simply and inexpensively demonstrated with a shallow, open-channel flow field in a *ripple tank* or *water table*.

The propagation of weak pressure pulses (sound waves) in a compressible flow can be considered to be comparable to the movement of small-amplitude waves on the surface of an open-channel flow. In each case—two-dimensional compressible flow and open-channel flow— the influence of flow velocity on wave pattern is similar. When the flow velocity is less than the wave speed, wave fronts can move upstream of the wave source and the flow is subsonic (compressible flow) or subcritical (open-channel flow). When the flow velocity is equal to the wave speed, wave fronts cannot move upstream of the wave source and the flow is sonic (compressible flow) or critical (open-channel flow). When the flow velocity is greater than the wave speed, the flow is supersonic (compressible flow) or supercritical (open-channel flow). Normal shocks can occur in supersonic compressible flows. Hydraulic jumps can occur in supercritical open-channel flows. Comparison of the characteristics of normal shocks (Section 11.5.3) and hydraulic jumps (Section 10.6.1) suggests a strong resemblance and thus analogy between the two phenomena.

For compressible flows a meaningful dimensionless variable is the Mach number, where

> *Compressible gas flows and open-channel liquid flows are strikingly similar in several ways.*

$$\text{Ma} = \frac{V}{c} \tag{11.46}$$

In open-channel flows, an important dimensionless variable is the Froude number, where

$$\text{Fr} = \frac{V_{oc}}{\sqrt{gy}} \tag{11.157}$$

The velocity of the channel flow is $V_{oc}$, the acceleration of gravity is $g$, and the depth of the flow is $y$. Since the speed of a small-amplitude wave on the surface of an open-channel flow, $c_{oc}$, is (see Section 10.2.1)

$$c_{oc} = \sqrt{gy} \tag{11.158}$$

we conclude that

$$\text{Fr} = \frac{V_{oc}}{c_{oc}} \tag{11.159}$$

From Eqs. 11.46 and 11.159 we see the similarity between Mach number (compressible flow) and Froude number (open-channel flow).

For compressible flow, the continuity equation is

$$\rho AV = \text{constant} \tag{11.160}$$

where $V$ is the flow velocity, $\rho$ is the fluid density, and $A$ is the flow cross-sectional area. For an open-channel flow, conservation of mass leads to

$$ybV_{oc} = \text{constant} \tag{11.161}$$

where $V_{oc}$ is the flow velocity, and $y$ and $b$ are the depth and width of the open-channel flow. Comparing Eqs. 11.160 and 11.161 we note that if flow velocities are considered similar and flow area, $A$, and channel width, $b$, are considered similar, then compressible flow density, $\rho$, is analogous to open-channel flow depth, $y$.

It should be pointed out that the similarity between Mach number and Froude number is generally not exact. If compressible flow and open-channel flow velocities are considered to be similar, then it follows that for Mach number and Froude number similarity the wave speeds $c$ and $c_{oc}$ must also be similar.

From the development of the equation for the speed of sound in an ideal gas (see Eqs. 11.34 and 11.35) we have for the compressible flow

$$c = \sqrt{(\text{constant})\,k\rho^{k-1}} \tag{11:162}$$

From Eqs. 11.162 and 11.158, we see that if $y$ is to be similar to $\rho$ as suggested by comparing Eqs. 11.160 and 11.161, then $k$ should be equal to 2. Typically $k = 1.4$ or 1.67, not 2. This limitation to exactness is, however, usually not serious enough to compromise the benefits of the analogy between compressible and open-channel flows.

## 11.7 Two-Dimensional Compressible Flow

A brief introduction to two-dimensional compressible flow is included here for those who are interested. We begin with a consideration of supersonic flow over a wall with a small change of direction as sketched in Fig. 11.27.

We apply the component of the linear momentum equation (Eq. 5.22) parallel to the Mach wave to the flow across the Mach wave. (See Eq. 11.39 for the definition of a Mach wave.) The result is that the component of velocity parallel to the Mach wave is constant across the Mach wave. That is, $V_{t1} = V_{t2}$. Thus, from the simple velocity triangle construction indicated in Fig. 11.27, we conclude that the flow accelerates because of the change in direction of the flow. If several changes in wall direction are involved as shown in Fig. 11.28, then the supersonic flow accelerates (expands) because of the changes in flow direction across the Mach waves (also called *expansion* waves). Each Mach wave makes an appropriately smaller angle $\alpha$ with the upstream wall because of the increase in Mach number that occurs with each direction change (see Section 11.3). A rounded expansion corner may be considered as a series of infinitesimal changes in direction. Conversely, even sharp corners are actually rounded when viewed on a small enough scale. Thus, expansion fans as illustrated in Fig. 11.29 are commonly used for supersonic flow around a "sharp" corner. If the flow across the Mach waves is considered to be isentropic, then Eq. 11.42 suggests that the increase in flow speed is accompanied by a decrease in static pressure.

When the change in supersonic flow direction involves the change in wall orientation sketched in Fig. 11.30, compression rather than expansion occurs. The flow decelerates and the static pressure increases across the Mach wave. For several changes in wall direction, as indicated in Fig. 11.31, several Mach waves occur, each at an appropriately larger angle $\alpha$ with the upstream wall. A rounded compression corner may be considered as a series of infinitesimal changes in

*Supersonic flows accelerate across expansion Mach waves.*

*V11.9 Wedge oblique shocks and expansions*

■ **Figure 11.27** Flow acceleration across a Mach wave.

■ **Figure 11.28** Flow acceleration across Mach waves.

■ **Figure 11.29** Corner expansion fan.

■ **Figure 11.30** Flow deceleration across a Mach wave.

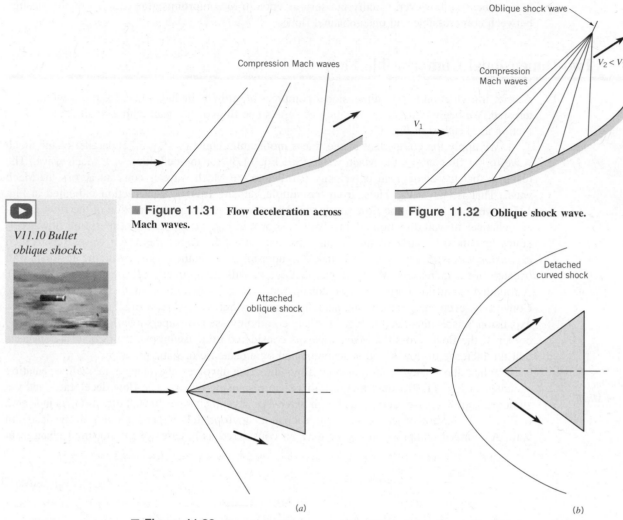

■ **Figure 11.31** **Flow deceleration across Mach waves.**

■ **Figure 11.32** **Oblique shock wave.**

■ **Figure 11.33** **Supersonic flow over a wedge:** (*a*) **Smaller wedge angle results in attached oblique shock.** (*b*) **Large wedge angle results in detached curved shock.**

direction and even sharp corners are actually rounded. Mach waves or compression waves can coalesce to form an oblique shock wave as shown in Fig. 11.32.

The above discussion of compression waves can be usefully extended to supersonic flow impinging on an object. For example, for supersonic flow incident on a wedge-shaped leading edge (see Fig. 11.33), an attached oblique shock can form as suggested in Fig. 11.33*a*. For the same incident Mach number but with a larger wedge angle, a detached curved shock as sketched in Fig. 11.33*b* can result. In Example 11.19, we considered flow along a stagnation pathline across a detached curved shock to be identical to flow across a normal shock wave.

From this brief look at two-dimensional supersonic flow, one can easily conclude that the extension of these concepts to flows over immersed objects and within ducts can be exciting, especially if three-dimensional effects are considered. Reference 6 provides much more on this subject than could be included here.

# 11.8    Chapter Summary and Study Guide

In this chapter, consideration is given to the flow of gas involving substantial changes in fluid density caused mainly by high speeds. While the flow of liquids may most often be considered of constant density or incompressible over a wide range of speeds, the flow of gases and vapors

*compressible flow*
*ideal gas*
*internal energy*
*enthalpy*
*specific heat ratio*
*entropy*
*adiabatic*
*isentropic*
*Mach number*
*speed of sound*
*stagnation pressure*
*subsonic*
*sonic*
*Mach wave*
*supersonic*
*Mach cone*
*transonic flows*
*hypersonic flows*
*converging–diverging*
    *duct*
*throat*
*temperature–entropy*
    *(T–s) diagram*
*choked flow*
*critical state*
*critical pressure ratio*
*normal shock wave*
*oblique shock wave*
*expansion wave*
*overexpanded*
*underexpanded*
*nonisentropic flow*
*Fanno flow*
*Rayleigh flow*

may involve substantial fluid density changes at higher speeds. At lower speeds, gas and vapor density changes are not appreciable, and so these flows may be treated as incompressible.

Since fluid density and other fluid property changes are significant in compressible flows, property relationships are important. An ideal gas, with well-defined fluid property relationships, is used as an approximation of an actual gas. This profound simplification still allows useful conclusions to be made about compressible flows.

The Mach number is a key variable in compressible flow theory. Most easily understood as the ratio of the local speed of flow and the speed of sound in the flowing fluid, it is a measure of the extent to which the flow is compressible or not. It is used to define categories of compressible flows, which range from subsonic (Mach number less than 1) to supersonic (Mach number greater than 1). The speed of sound in a truly incompressible fluid is infinite, so the Mach numbers associated with liquid flows are generally low.

The notion of an isentropic or constant entropy flow is introduced. The most important isentropic flow is one that is adiabatic (no heat transfer to or from the flowing fluid) and frictionless (zero viscosity). This simplification, like the one associated with approximating real gases with an ideal gas, leads to useful results including trends associated with accelerating and decelerating flows through converging, diverging, and converging–diverging flow paths. Phenomena including flow choking, acceleration in a diverging passage, deceleration in a converging passage, and the achievement of supersonic flows are discussed.

Three major nonisentropic compressible flows considered in this chapter are Fanno flows, Rayleigh flows, and flows across normal shock waves. Unusual outcomes include the conclusions that friction can accelerate a subsonic Fanno flow, heating can result in fluid temperature reduction in a subsonic Rayleigh flow, and a flow can decelerate from supersonic flow to subsonic flow across a very small distance. The value of temperature–entropy ($T$–$s$) diagramming of flows to better understand them is demonstrated.

Numerous formulas describing a variety of ideal gas compressible flows are derived. These formulas can be easily solved with computers. However, to provide the learner with a better grasp of the details of a compressible flow process, a graphical approach, albeit approximate, is used.

The striking analogy between compressible and open-channel flows leads to a brief discussion of the usefulness of a ripple tank or water table to simulate compressible flows.

Expansion and compression Mach waves associated with two-dimensional compressible flows are introduced, as is the formation of oblique shock waves from compression Mach waves.

The following checklist provides a study guide for this chapter. When your study of the entire chapter and end-of-chapter exercises is completed, you should be able to

■ write out the meanings of the terms listed here in the margin and understand each of the related concepts. These terms are particularly important and are set in *__italic, bold, and color__* type in the text.

■ estimate the change in ideal gas properties in a compressible flow.

■ calculate Mach number value for a specific compressible flow.

■ estimate when a flow may be considered incompressible and when it must be considered compressible to preserve accuracy.

■ estimate details of isentropic flows of an ideal gas though converging, diverging, and converging–diverging passages.

■ estimate details of nonisentropic Fanno and Rayleigh flows and flows across normal shock waves.

■ explain the analogy between compressible and open-channel flows.

Some of the important equations in this chapter are:

| | | |
|---|---|---|
| Ideal gas equation of state | $\rho = \dfrac{p}{RT}$ | **(11.1)** |
| Internal energy change | $\check{u}_2 - \check{u}_1 = c_v(T_2 - T_1)$ | **(11.5)** |
| Enthalpy | $\check{h} = \check{u} + \dfrac{p}{\rho}$ | **(11.6)** |

| | | |
|---|---|---|
| Enthalpy change | $\check{h}_2 - \check{h}_1 = c_p(T_2 - T_1)$ | **(11.9)** |
| Specific heat difference | $c_p - c_v = R$ | **(11.12)** |
| Specific heat ratio | $k = \dfrac{c_p}{c_v}$ | **(11.13)** |
| Specific heat at constant pressure | $c_p = \dfrac{Rk}{k-1}$ | **(11.14)** |
| Specific heat at constant volume | $c_v = \dfrac{R}{k-1}$ | **(11.15)** |
| First $T ds$ equation | $T\,ds = d\check{u} + pd\left(\dfrac{1}{\rho}\right)$ | **(11.16)** |
| Second $T ds$ equation | $T\,ds = d\check{h} - \left(\dfrac{1}{\rho}\right)dp$ | **(11.18)** |
| Entropy change | $s_2 - s_1 = c_v \ln \dfrac{T_2}{T_1} + R \ln \dfrac{\rho_1}{\rho_2}$ | **(11.21)** |
| Entropy change | $s_2 - s_1 = c_p \ln \dfrac{T_2}{T_1} - R \ln \dfrac{p_2}{p_1}$ | **(11.22)** |
| Isentropic flow | $\dfrac{p}{\rho^k} = \text{constant}$ | **(11.25)** |
| Speed of sound | $c = \sqrt{\left(\dfrac{\partial p}{\partial \rho}\right)_s}$ | **(11.34)** |
| Speed of sound in gas | $c = \sqrt{RTk}$ | **(11.36)** |
| Speed of sound in liquid | $c = \sqrt{\dfrac{E_v}{\rho}}$ | **(11.38)** |
| Mach cone angle | $\sin \alpha = \dfrac{c}{V} = \dfrac{1}{\text{Ma}}$ | **(11.39)** |
| Mach number | $\text{Ma} = \dfrac{V}{c}$ | **(11.46)** |
| Isentropic flow | $\dfrac{dV}{V} = -\dfrac{dA}{A}\dfrac{1}{(1 - \text{Ma}^2)}$ | **(11.48)** |
| Isentropic flow | $\dfrac{d\rho}{\rho} = \dfrac{dA}{A}\dfrac{\text{Ma}^2}{(1 - \text{Ma}^2)}$ | **(11.49)** |
| Isentropic flow | $\dfrac{T}{T_0} = \dfrac{1}{1 + [(k-1)/2]\text{Ma}^2}$ | **(11.56)** |
| Isentropic flow | $\dfrac{p}{p_0} = \left\{\dfrac{1}{1 + [(k-1)/2]\text{Ma}^2}\right\}^{k/(k-1)}$ | **(11.59)** |
| Isentropic flow | $\dfrac{\rho}{\rho_0} = \left\{\dfrac{1}{1 + [(k-1)/2]\text{Ma}^2}\right\}^{1/(k-1)}$ | **(11.60)** |
| Isentropic flow-critical pressure ratio | $\dfrac{p^*}{p_0} = \left(\dfrac{2}{k+1}\right)^{k/(k-1)}$ | **(11.61)** |
| Isentropic flow-critical temperature ratio | $\dfrac{T^*}{T_0} = \dfrac{2}{k+1}$ | **(11.63)** |

| | | |
|---|---|---|
| Isentropic flow | $$\frac{A}{A^*} = \frac{1}{\text{Ma}}\left\{\frac{1 + [(k-1)/2]\text{Ma}^2}{1 + [(k-1)/2]}\right\}^{(k+1)/[2(k-1)]}$$ | **(11.71)** |
| Fanno flow | $$\frac{1}{k}\frac{(1-\text{Ma}^2)}{\text{Ma}^2} + \frac{k+1}{2k}\ln\left\{\frac{[(k+1)/2]\text{Ma}^2}{1+[(k-1)/2]\text{Ma}^2}\right\} = \frac{f(\ell^* - \ell)}{D}$$ | **(11.98)** |
| Fanno flow | $$\frac{T}{T^*} = \frac{(k+1)/2}{1+[(k-1)/2]\text{Ma}^2}$$ | **(11.101)** |
| Fanno flow | $$\frac{V}{V^*} = \left\{\frac{[(k+1)/2]\text{Ma}^2}{1+[(k-1)/2]\text{Ma}^2}\right\}^{1/2}$$ | **(11.103)** |
| Fanno flow | $$\frac{p}{p^*} = \frac{1}{\text{Ma}}\left\{\frac{(k+1)/2}{1+[(k-1)/2]\text{Ma}^2}\right\}^{1/2}$$ | **(11.107)** |
| Fanno flow | $$\frac{p_0}{p_0^*} = \frac{1}{\text{Ma}}\left[\left(\frac{2}{k+1}\right)\left(1+\frac{k-1}{2}\text{Ma}^2\right)\right]^{[(k+1)/2(k-1)]}$$ | **(11.109)** |
| Rayleigh flow | $$\frac{p}{p_a} = \frac{1+k}{1+k\text{Ma}^2}$$ | **(11.123)** |
| Rayleigh flow | $$\frac{T}{T_a} = \left[\frac{(1+k)\text{Ma}}{1+k\text{Ma}^2}\right]^2$$ | **(11.128)** |
| Rayleigh flow | $$\frac{\rho_a}{\rho} = \frac{V}{V_a} = \text{Ma}\left[\frac{(1+k)\text{Ma}}{1+k\text{Ma}^2}\right]$$ | **(11.129)** |
| Rayleigh flow | $$\frac{T_0}{T_{0,a}} = \frac{2(k+1)\text{Ma}^2\left(1+\dfrac{k-1}{2}\text{Ma}^2\right)}{(1+k\text{Ma}^2)^2}$$ | **(11.131)** |
| Rayleigh flow | $$\frac{p_0}{p_{0,a}} = \frac{(1+k)}{(1+k\text{Ma}^2)}\left[\left(\frac{2}{k+1}\right)\left(1+\frac{k-1}{2}\text{Ma}^2\right)\right]^{k/(k-1)}$$ | **(11.133)** |
| Normal shock | $$\text{Ma}_y^2 = \frac{\text{Ma}_x^2 + [2/(k-1)]}{[2k/(k-1)]\text{Ma}_x^2 - 1}$$ | **(11.149)** |
| Normal shock | $$\frac{p_y}{p_x} = \frac{2k}{k+1}\text{Ma}_x^2 - \frac{k-1}{k+1}$$ | **(11.150)** |
| Normal shock | $$\frac{T_y}{T_x} = \frac{\{1+[(k-1)/2]\text{Ma}_x^2\}\{[2k/(k-1)]\text{Ma}_x^2 - 1\}}{\{(k+1)^2/[2(k-1)]\}\text{Ma}_x^2}$$ | **(11.151)** |
| Normal shock | $$\frac{\rho_y}{\rho_x} = \frac{V_x}{V_y} = \frac{(k+1)\text{Ma}_x^2}{(k-1)\text{Ma}_x^2 + 2}$$ | **(11.154)** |
| Normal shock | $$\frac{p_{0,y}}{p_{0,x}} = \frac{\left(\dfrac{k+1}{2}\text{Ma}_x^2\right)^{k/(k-1)}\left(1+\dfrac{k-1}{2}\text{Ma}_x^2\right)^{k/(1-k)}}{\left(\dfrac{2k}{k+1}\text{Ma}_x^2 - \dfrac{k-1}{k+1}\right)^{1/(k-1)}}$$ | **(11.156)** |

# References

1. Coles, D., "Channel Flow of a Compressible Fluid," Summary description of film in *Illustrated Experiments in Fluid Mechanics, The NCFMF Book of Film Notes,* MIT Press, Cambridge, Mass., 1972.
2. Moran, M. J., and Shapiro, H. N., *Fundamentals of Engineering Thermodynamics,* 6th Ed., Wiley, New York, 2008.

3. Keenan, J. H., Chao, J., and Kaye, J., *Gas Tables,* 2nd Ed., Wiley, New York, 1980.

4. Shapiro, A. H., *The Dynamics and Thermodynamics of Compressible Fluid Flow,* Vol. 1, Wiley, New York, 1953.

5. Liepmann, H. W., and Roshko, A., *Elements of Gasdynamics,* Dover Publications, 2002.

6. Anderson, J. D., Jr., *Modern Compressible Flow with Historical Perspective,* 3rd Ed., McGraw-Hill, New York, 2003.

7. Greitzer, E. M., Tan, C. S., and Graf, M. B., *Internal Flow Concepts and Applications,* Cambridge University Press, UK, 2004.

---

| | |
|---|---|
|  | Problem available in *WileyPLUS* at instructor's discretion. |
|  | Tutoring problem available in *WileyPLUS* at instructor's discretion. |
| ▶ | Problem is related to a chapter video available in *WileyPLUS*. |
| * | Problem to be solved with aid of programmable calculator or computer. |
| † | Open-ended problem that requires critical thinking. These problems require various assumptions to provide the necessary input data. There are not unique answers to these problems. |

## Review Problems

Go to Appendix G (*WileyPLUS* or the book's web site, www.wiley.com/college/munson) for a set of review problems with answers. Detailed solutions can be found in the *Student Solution Manual and Study Guide for Fundamentals of Fluid Mechanics,* by Munson et al. (© 2013 John Wiley and Sons, Inc.).

## Conceptual Questions

**11.1C** You are watching fireworks on the Fourth of July and notice that you hear the sound of a firework exploding 2 seconds after you see the flash of the explosion. Approximately how far away are the fireworks?

a) About the length of a football field.

b) About one-half of a mile.

c) About a mile.

d) About two miles.

**11.2C** An object, shown as the dark circle below, is moving along and emitting sound waves. The sound waves that were emitted by the object at times −1 second (one second ago), −2 seconds, and −3 seconds are shown below.

If the Mach number of the object is equal to unity, which is the correct picture for the sound waves?

a) Picture A.  b) Picture B.  c) Picture C.  d) Picture D.

**Additional conceptual questions are available in *WileyPLUS* at the instructor's discretion.**

## Problems

**Note: Unless specific values of required fluid properties are given in the problem statement, use the values found in the tables on the inside of the front cover. Answers to the even-numbered problems are listed at the end of the book. The Lab Problems as well as the videos that accompany problems can be accessed in *WileyPLUS* or the book's web site, www.wiley.com/college/munson.**

### Section 11.1 Ideal Gas Relationships

**11.1**  Distinguish between flow of an ideal gas and inviscid flow of a fluid.

**11.2** (WILEY) Compare the density of standard air listed in Table 1.8 with the value of standard air calculated with the ideal gas equation of state, and comment on what you discover.

**11.3** (WILEY) Five pounds mass of air are heated in a closed, rigid container from 80 °F, 15 psia to 500 °F. Estimate the final pressure of the air and the entropy rise involved.

**11.4** (WILEY) Air flows steadily between two sections in a duct. At section (1), the temperature and pressure are $T_1 = 80\ °C$, $p_1 = 301$ kPa(abs), and at section (2), the temperature and pressure are $T_2 = 180\ °C$, $p_2 = 181$ kPa(abs). Calculate the **(a)** change in internal energy between sections (1) and (2), **(b)** change in enthalpy between sections (1) and (2), **(c)** change in density between sections (1) and (2), **(d)** change in entropy between sections (1) and (2). How would you estimate the loss of available energy between the two sections of this flow?

**11.5** Determine, in SI units, nominal values of $c_p$ and $c_v$ for: **(a)** air, **(b)** carbon dioxide, **(c)** helium. Use information provided in Table 1.8.

**11.6** ▶ As demonstrated in **Video V11.1** fluid density differences in a flow may be seen with the help of a schlieren optical system. Discuss what variables affect fluid density and the different ways in which a variable density flow can be achieved.

**11.7** ▶ Describe briefly how a schlieren optical visualization system (**Videos V11.1** and **V11.5**, also Fig. 11.4) works. How else might density changes in a fluid flow be made visible to the eye?

**11.8** Three kilograms of hydrogen contained in a nondeforming sealed vessel are cooled from 400 °C, 400 kPa(abs) until the hydrogen pressure is 100 kPa(abs). Calculate the change in internal energy, enthalpy, and entropy associated with this cooling process.

**11.9** WILEY° GO Helium is compressed isothermally from 121 kPa(abs) to 301 kPa(abs). Determine the entropy change associated with this process.

**11.10** WILEY° Air at 14.7 psia and 70 °F is compressed adiabatically by a centrifugal compressor to a pressure of 100 psia. What is the minimum temperature rise possible? Explain.

**11.11** WILEY° Methane is compressed adiabatically from 100 kPa(abs) and 25 °C to 200 kPa(abs). What is the minimum compressor exit temperature possible? Explain.

**11.12** WILEY° Air expands adiabatically through a turbine from a pressure and temperature of 180 psia, 1600 °R to a pressure of 14.7 psia. If the actual temperature change is 85% of the ideal temperature change, determine the actual temperature of the expanded air and the actual enthalpy and entropy differences across the turbine.

**11.13** WILEY° An expression for the value of $c_p$ for carbon dioxide as a function of temperature is

$$c_p = 286 - \frac{1.15 \times 10^5}{T} + \frac{2.49 \times 10^6}{T^2}$$

where $c_p$ is in (ft · lb)/(lbm · °R) and $T$ is in °R. Compare the change in enthalpy of carbon dioxide using the constant value of $c_p$ (see Table 1.7) with the change in enthalpy of carbon dioxide using the expression above, for $T_2 - T_1$ equal to **(a)** 10 °R, **(b)** 1000 °R, **(c)** 3000 °R. Set $T_1 = 540$ °R.

## Section 11.2 Mach Number and Speed of Sound

**11.14** WILEY° Confirm the speed of sound for air at 70 °F listed in Table B.3.

**11.15** WILEY° From Table B.1 we can conclude that the speed of sound in water at 60 °F is 4814 ft/s. Is this value of $c$ consistent with the value of bulk modulus, $E_v$, listed in Table 1.5?

**11.16** WILEY° If the observed speed of sound in steel is 5300 m/s, determine the bulk modulus of elasticity of steel in N/m³. The density of steel is nominally 7790 kg/m³. How does your value of $E_v$ for steel compare with $E_v$ for water at 15.6 °C? Compare the speeds of sound in steel, water, and air at standard atmospheric pressure and 15 °C and comment on what you observe.

**11.17** WILEY° Using information provided in Table C.1, develop a table of speed of sound in ft/s as a function of elevation for U.S. standard atmosphere.

**11.18** If a high-performance aircraft is able to cruise at a Mach number of 3.0 at an altitude 80,000 ft, how fast is this in **(a)** mph, **(b)** ft/s, **(c)** m/s?

**11.19** Compare values of the speed of sound in m/s at 20 °C in the following gases: **(a)** air, **(b)** carbon dioxide, **(c)** helium.

**11.20** WILEY° Determine the Mach number of a car moving in standard air at a speed of **(a)** 25 mph, **(b)** 55 mph, and **(c)** 100 mph.

**11.21** The flow of an ideal gas may be considered incompressible if the Mach number is less than 0.3. Determine the velocity level in ft/s and in m/s for Ma = 0.3 in the following gases: **(a)** standard air, **(b)** hydrogen at 68 °F.

## Section 11.3 Categories of Compressible Flow

**11.22** ▶ A schlieren photo of a bullet moving through air (see **Video V11.10**) at 14.7 psia and 68 °F indicates a Mach cone angle of 28°. How fast was the bullet moving in: **(a)** m/s, **(b)** ft/s, **(c)** mph?

**11.23** WILEY° At a given instant of time, two pressure waves, each moving at the speed of sound, emitted by a point source moving with constant velocity in a fluid at rest are shown in Fig. P11.23. Determine the Mach number involved and indicate with a sketch the instantaneous location of the point source.

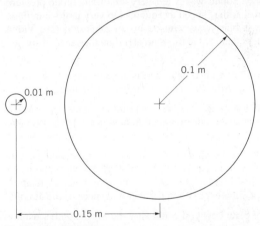

0.1 m

0.01 m

0.15 m

■ **Figure P11.23**

**11.24** At a given instant of time, a pressure wave moving at the speed of sound, which is emitted by a point source moving with constant velocity along the line $A$–$B$ in a fluid at rest, is shown in Fig. P11.24. If the point source Mach number is 0.8, sketch, for the given instant, the pressure wave emitted by the point source when it was at location $B$.

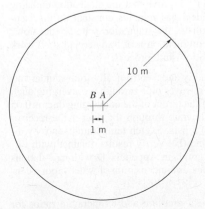

10 m

$B$ $A$

1 m

■ **Figure P11.24**

**11.25** (WILEY) At a given instant of time, two pressure waves, each moving at the speed of sound, emitted by a point source moving with constant velocity in a fluid at rest, are shown in Fig. P11.25. Determine the Mach number involved and indicate with a sketch the instantaneous location of the point source.

■ **Figure P11.25**

**11.26** ▶ (WILEY) Sound waves are very small-amplitude pressure pulses that travel at the "speed of sound." Do very large-amplitude waves such as a blast wave caused by an explosion (see **Video V11.8**) travel less than, equal to, or greater than the speed of sound? Explain.

**11.27** How would you estimate the distance between you and an approaching storm front involving lightning and thunder?

**11.28** (WILEY) If a new Boeing 787 Dreamliner cruises at a Mach number of 0.87 at an altitude of 30,000 ft, how fast is this in **(a)** mph, **(b)** ft/s, **(c)** m/s?

**11.29** (WILEY) At the seashore, you observe a high-speed aircraft moving overhead at an elevation of 10,000 ft. You hear the plane 8 s after it passes directly overhead. Using a nominal air temperature of 40 °F, estimate the Mach number and speed of the aircraft.

**11.30** (WILEY) Explain how you could vary the Mach number but not the Reynolds number in airflow past a sphere. For a constant Reynolds number of 300,000, estimate how much the drag coefficient will increase as the Mach number is increased from 0.3 to 1.0.

## Section 11.4 Isentropic Flow of an Ideal Gas

**11.31** Starting with the enthalpy form of the energy equation (Eq. 5.69), show that for isentropic flows, the stagnation temperature remains constant. Why is this important?

**11.32** Starting with Eq. 11.52, prove that the stagnation enthalpy (and temperature) of an ideal gas remains constant during isentropic flow. Does this result fix the stagnation state for the flow? Explain. Using Eq. 5.69, comment on heat transfer and shaft work during any constant stagnation state process.

**11.33** (WILEY) Air flows steadily and isentropically from standard atmospheric conditions to a receiver pipe through a converging duct. The cross-sectional area of the throat of the converging duct is 0.05 ft². Determine the mass flowrate through the duct if the receiver pressure is **(a)** 10 psia, **(b)** 5 psia. Sketch temperature–entropy diagrams for situations **(a)** and **(b)**. Verify results obtained with values from the appropriate graph in Appendix D with calculations involving ideal gas equations. Is condensation of water vapor a concern? Explain.

**11.34** Determine the critical pressure and temperature ratios for **(a)** air, **(b)** carbon dioxide, **(c)** helium.

**11.35** (WILEY) Determine the static pressure to stagnation pressure ratio associated with the following motion in standard air: **(a)** a runner moving at the rate of 10 mph, **(b)** a cyclist moving at the rate of 40 mph, **(c)** a car moving at the rate of 65 mph, **(d)** an airplane moving at the rate of 500 mph.

**11.36** The static pressure to stagnation pressure ratio at a point in a gas flow field is measured with a Pitot-static probe as being equal to 0.6. The stagnation temperature of the gas is 20 °C. Determine the flow speed in m/s and the Mach number if the gas is air. What error would be associated with assuming that the flow is incompressible?

**11.37** (WILEY) The stagnation pressure and temperature of air flowing past a probe are 120 kPa(abs) and 100 °C, respectively. The air pressure is 80 kPa(abs). Determine the airspeed and the Mach number considering the flow to be **(a)** incompressible, **(b)** compressible.

**11.38** (WILEY) The stagnation pressure indicated by a Pitot tube mounted on an airplane in flight is 45 kPa(abs). If the aircraft is cruising in standard atmosphere at an altitude of 10,000 m, determine the speed and Mach number involved.

**11.39** (WILEY GO) Helium at 68 °F and 14.7 psia in a large tank flows steadily and isentropically through a converging nozzle to a receiver pipe. The cross-sectional area of the throat of the converging passage is 0.05 ft². Determine the mass flowrate through the duct if the receiver pressure is **(a)** 10 psia, **(b)** 5 psia. Sketch temperature–entropy diagrams for situations **(a)** and **(b)**.

*****11.40** ▶ (WILEY) An ideal gas enters subsonically and flows isentropically through a choked converging–diverging duct having a circular cross-sectional area $A$ that varies with axial distance from the throat, $x$, according to the formula

$$A = 0.1 + x^2$$

where $A$ is in square feet and $x$ is in feet. For this flow situation, sketch the side view of the duct and graph the variation of Mach number, static temperature to stagnation temperature ratio, $T/T_0$, and static pressure to stagnation pressure ratio, $p/p_0$, through the duct from $x = -0.6$ ft to $x = +0.6$ ft. Also show the possible fluid states at $x = -0.6$ ft, 0 ft, and +0.6 ft using temperature-entropy coordinates. Consider the gas as being helium (use $0.051 \le$ Ma $\le 5.193$). Sketch on your pressure variation graph the nonisentropic paths that would occur with over- and underexpanded duct exit flows (see **Video V11.7**) and explain when they will occur. When will isentropic supersonic duct exit flow occur?

*****11.41** ▶ (WILEY) An ideal gas enters supersonically and flows isentropically through the choked converging–diverging duct described in Problem 11.40. Graph the variation of Ma, $T/T_0$, and $p/p_0$ from the entrance to the exit sections of the duct for helium (use $0.051 \le$ Ma $\le 5.193$). Show the possible fluid states at $x = -0.6$ ft, 0 ft, and +0.6 ft using temperature–entropy coordinates. Sketch on your pressure variation graph the nonisentropic paths that would occur with over- and underexpanded duct exit flows (see **Video V11.7**) and explain when they will occur. When will isentropic supersonic duct exit flow occur?

**11.42** (WILEY) An ideal gas is to flow isentropically from a large tank where the air is maintained at a temperature and pressure of 59 °F and 80 psia to standard atmospheric discharge conditions. Describe in general terms the kind of duct involved and determine the duct exit Mach number and velocity in ft/s if the gas is air.

**11.43** (WILEY) An ideal gas flows isentropically through a converging-diverging nozzle. At a section in the converging portion of the nozzle, $A_1 = 0.1$ m², $p_1 = 600$ kPa(abs), $T_1 = 20$ °C, and $Ma_1 = 0.6$. For section (2) in the diverging part of the nozzle, determine $A_2$, $p_2$, and $T_2$ if $Ma_2 = 3.0$ and the gas is air.

**11.44** (WILEY) Upstream of the throat of an isentropic converging–diverging nozzle at section (1), $V_1 = 150$ m/s, $p_1 = 100$ kPa(abs), and $T_1 = 20$ °C. If the discharge flow is supersonic and the throat area is 0.1 m², determine the mass flowrate in kg/s for the flow of air.

**11.45** (WILEY) The flow blockage associated with the use of an intrusive probe can be important. Determine the percentage increase in section velocity corresponding to a 0.5% reduction in flow area due to probe blockage for airflow if the section area is 1.0 m², $T_0 = 20$ °C, and the unblocked flow Mach numbers are **(a)** Ma = 0.2, **(b)** Ma = 0.8, **(c)** Ma = 1.5, **(d)** Ma = 30.

**11.46** (See Fluids in the News article titled "Rocket Nozzles," Section 11.4.2.) Comment on the practical limits of area ratio for the diverging portion of a converging–diverging nozzle designed to achieve supersonic exit flow.

### Section 11.5.1 Adiabatic Constant Area Duct Flow with Friction (Fanno Flow)

**11.47** (WILEY) An ideal gas enters [section (1)] an insulated, constant cross-sectional area duct with the following properties:

$$T_0 = 293 \text{ K}$$
$$p_0 = 101 \text{ kPa(abs)}$$
$$Ma_1 = 0.2$$

For Fanno flow, determine corresponding values of fluid temperature and entropy change for various levels of pressure and plot the Fanno line if the gas is helium.

**11.48** For Fanno flow, prove that

$$\frac{dV}{V} = \frac{fk(Ma^2/2)(dx/D)}{1 - Ma^2}$$

and in so doing show that when the flow is subsonic, friction accelerates the fluid, and when the flow is supersonic, friction decelerates the fluid.

**11.49** (WILEY GO) Standard atmospheric air ($T_0 = 59$ °F, $p_0 = 14.7$ psia) is drawn steadily through a frictionless and adiabatic converging nozzle into an adiabatic, constant cross-sectional area duct. The duct is 10 ft long and has an inside diameter of 0.5 ft. The average friction factor for the duct may be estimated as being equal to 0.03. What is the maximum mass flowrate in slugs/s through the duct? For this maximum flowrate, determine the values of static temperature, static pressure, stagnation temperature, stagnation pressure, and velocity at the inlet [section (1)] and exit [section (2)] of the constant area duct. Sketch a temperature–entropy diagram for this flow.

**11.50** (WILEY) The duct in Problem 11.49 is shortened by 50%. The duct discharge pressure is maintained at the choked flow value determined in Problem 11.49. Determine the change in mass flowrate through the duct associated with the 50% reduction in length. The average friction factor remains constant at a value of 0.03.

**11.51** (WILEY) If the same mass flowrate of air obtained in Problem 11.49 is desired through the shortened duct of Problem 11.50, determine the back pressure, $p_2$, required. Assume $f$ remains constant at a value of 0.03.

**11.52** (WILEY) If the average friction factor of the duct of Example 11.12 is changed to **(a)** 0.01 or **(b)** 0.03, determine the maximum mass flowrate of air through the duct associated with each new friction factor; compare with the maximum mass flowrate value of Example 11.12.

**11.53** If the length of the constant area duct of Example 11.12 is changed to **(a)** 1 m or **(b)** 3 m, and all other specifications in the problem statement remain the same, determine the maximum mass flowrate of air through the duct associated with each new length; compare with the maximum mass flowrate of Example 11.12.

**11.54** The duct of Example 11.12 is lengthened by 50%. If the duct discharge pressure is set at a value of $p_d = 46.2$ kPa(abs), determine the mass flowrate of air through the lengthened duct. The average friction factor for the duct remains constant at a value 0.02.

**11.55** (WILEY) Air flows adiabatically between two sections in a constant area pipe. At upstream section (1), $p_{0,1} = 100$ psia, $T_{0,1} = 600$ °R, and $Ma_1 = 0.5$. At downstream section (2), the flow is choked. Estimate the magnitude of the force per unit cross-sectional area exerted by the inside wall of the pipe on the fluid between sections (1) and (2).

### Section 11.5.2 Frictionless Constant Area Duct Flow with Heat Transfer (Rayleigh Flow)

**11.56** Cite an example of an actual subsonic flow of practical importance that may be approximated with a Rayleigh flow.

**11.57** (WILEY) Standard atmospheric air [$T_0 = 288$ K, $p_0 = 101$ kPa(abs)] is drawn steadily through an isentropic converging nozzle into a frictionless diabatic ($q = 500$ kJ/kg) constant area duct. For maximum flow, determine the values of static temperature, static pressure, stagnation temperature, stagnation pressure, and flow velocity at the inlet [section (1)] and exit [section (2)] of the constant area duct. Sketch a temperature–entropy diagram for this flow.

**11.58** (WILEY) Air enters a 0.5-ft inside diameter duct with $p_1 = 20$ psia, $T_1 = 80$ °F, and $V_1 = 200$ ft/s. What frictionless heat addition rate in Btu/s is necessary for an exit gas temperature $T_2 = 1500$ =°F? Determine $p_2$, $V_2$, and $Ma_2$ also.

**11.59** (WILEY) Air enters a length of constant area pipe with $p_1 = 200$ kPa(abs), $T_1 = 500$ K, and $V_1 = 400$ m/s. If 500 kJ/kg of energy is removed from the air by frictionless heat transfer between sections (1) and (2), determine $p_2$, $T_2$, and $V_2$. Sketch a temperature–entropy diagram for the flow between sections (1) and (2).

**11.60** An ideal gas enters [section (1)] a frictionless, constant area duct with the following properties:

$$T_0 = 293 \text{ K}$$
$$p_0 = 101 \text{ kPa(abs)}$$
$$Ma_1 = 0.2$$

For Rayleigh flow, determine corresponding values of fluid temperature and entropy change for various levels of pressure and plot the Rayleigh line if the gas is helium.

**11.61** Air flows through a constant area pipe. At an upstream section (1), $p_1 = 15$ psia, $T_1 = 530$ °R, and $V_1 = 200$ ft/s. Downstream at section (2), $p_2 = 10$ psia and $T_2 = 1760$ °R. For this flow, determine the stagnation temperature and pressure ratios, $T_{0,2}/T_{0,1}$ and $p_{0,2}/p_{0,1}$, and the heat transfer per unit mass of air flowing between sections (1) and (2). Is the flow between sections (1) and (2) frictionless? Explain.

**11.62** Describe what happens to Fanno flow when heat transfer is allowed to occur. Is this the same as a Rayleigh flow with friction considered?

### Section 11.5.3 Normal Shock Waves

**11.63** (WILEY) The Mach number and stagnation pressure of air are 2.0 and 200 kPa(abs) just upstream of a normal shock. Estimate the stagnation pressure loss across the shock.

**11.64** (WILEY) The stagnation pressure ratio across a normal shock in an airflow is 0.6. Estimate the Mach number of the flow entering the shock.

**11.65** (WILEY) Just upstream of a normal shock in an airflow, $Ma = 3.0$, $T = 600$ °R, and $p = 30$ psia. Estimate values of Ma, $T_0$, $T$, $p_0$, $p$, and $V$ downstream of the shock.

**11.66** ▶ (WILEY) A total pressure probe like the one shown in **Video V3.8** is inserted into a supersonic airflow. A shock wave forms just upstream of the impact hole. The probe measures a total pressure of 500 kPa(abs). The stagnation temperature at the probe head is 500 K. The static pressure upstream of the shock is measured with a wall tap to be 100 kPa(abs). From these data, estimate the Mach number and velocity of the flow.

**11.67** ▶ (WILEY) The Pitot tube on a supersonic aircraft (see **Video V3.8**) cruising at an altitude of 30,000 ft senses a stagnation pressure of 12 psia. If the atmosphere is considered standard, determine the airspeed and Mach number of the aircraft. A shock wave is present just upstream of the probe impact hole.

**11.68** (WILEY) An aircraft cruises at a Mach number of 2.0 at an altitude of 15 km. Inlet air is decelerated to a Mach number of 0.4 at the engine compressor inlet. A normal shock occurs in the inlet diffuser upstream of the compressor inlet at a section where the Mach number is 1.2. For isentropic diffusion, except across the shock, and for standard atmosphere, determine the stagnation temperature and pressure of the air entering the engine compressor.

**11.69** (WILEY) Determine, for the airflow through the frictionless and adiabatic converging–diverging duct of Example 11.8, the ratio of duct exit pressure to duct inlet stagnation pressure that will result in a standing normal shock at: **(a)** $x = +0.1$ m, **(b)** $x = +0.2$ m, **(c)** $x = +0.4$ m. How large is the stagnation pressure loss in each case?

**11.70** (WILEY) A normal shock is positioned in the diverging portion of a frictionless, adiabatic, converging–diverging airflow duct where the cross-sectional area is 0.1 ft² and the local Mach number is 2.0. Upstream of the shock, $p_0 = 200$ psia and $T_0 = 1200$ °R. If the duct exit area is 0.15 ft², determine the exit area temperature and pressure and the duct mass flowrate.

**11.71** (WILEY GO) Supersonic airflow enters an adiabatic, constant area (inside diameter = 1 ft) 30-ft-long pipe with $Ma_1 = 3.0$. The pipe friction factor is estimated to be 0.02. What ratio of pipe exit pressure to pipe inlet stagnation pressure would result in a normal shock wave standing at **(a)** $x = 5$ ft, or **(b)** $x = 10$ ft, where $x$ is the distance downstream from the pipe entrance? Determine also the duct exit Mach number and sketch the temperature–entropy diagram for each situation.

**11.72** (WILEY) Supersonic airflow enters an adiabatic, constant area pipe (inside diameter = 0.1 m) with $Ma_1 = 2.0$. The pipe friction factor is 0.02. If a standing normal shock is located right at the pipe exit, and the Mach number just upstream of the shock is 1.2, determine the length of the pipe.

**11.73** (WILEY) Air enters a frictionless, constant area duct with $Ma_1 = 2.0$, $T_{0,1} = 59$ °F, and $p_{0,1} = 14.7$ psia. The air is decelerated by heating until a normal shock wave occurs where the local Mach number is 1.5. Downstream of the normal shock, the subsonic flow is accelerated with heating until it chokes at the duct exit. Determine the static temperature and pressure, the stagnation temperature and pressure, and the fluid velocity at the duct entrance, just upstream and downstream of the normal shock, and at the duct exit. Sketch the temperature–entropy diagram for this flow.

**11.74** (WILEY) Air enters a frictionless, constant area duct with $Ma = 2.5$, $T_0 = 20$ °C, and $p_0 = 101$ kPa(abs). The gas is decelerated by heating until a normal shock occurs where the local Mach number is 1.3. Downstream of the shock, the subsonic flow is accelerated with heating until it exits with a Mach number of 0.9. Determine the static temperature and pressure, the stagnation temperature and pressure, and the fluid velocity at the duct entrance, just upstream and downstream of the normal shock, and at the duct exit. Sketch the temperature–entropy diagram for this flow.

■ **Lifelong Learning Problems**

**11.1LL** Is there a limit to how fast an object can move through the atmosphere? Explain.

**11.2LL** Discuss the similarities between hydraulic jumps in open-channel flow and shock waves in compressible flow. Explain how this knowledge can be useful.

**11.3LL** Estimate the surface temperature associated with the reentry of the Space Shuttle into the Earth's atmosphere. Why is knowing this important?

**11.4LL** [See Fluids in the News article titled "Hilsch Tube (Ranque Vortex Tube)," Section 11.1.] Explain why a Hilsch tube works and cite some high and low gas temperatures actually achieved. What is the most important limitation of a Hilsch tube, and how can it be overcome?

**11.5LL** [See Fluids in the News article titled "Supersonic and Compressible Flows in Gas Turbines," Section 11.3.] Using typical physical dimensions and rotation speeds of manufactured gas turbine rotors, consider the possibility that supersonic fluid velocities relative to blade surfaces are possible. How do designers use this knowledge?

**11.6LL** Develop useful equations describing the constant temperature (isothermal) flow of an ideal gas through a constant cross-sectional area pipe. What important practical flow situations would these equations be useful for? How are real gas effects estimated?

■ **FE Exam Problems**

Sample FE (Fundamentals of Engineering) exam questions for fluid mechanics are provided in *WileyPLUS* or on the book's web site, www.wiley.com/college/munson.

# $12$ *Turbomachines*

CHAPTER OPENING PHOTO: A mixed-flow, transonic compressor stage. (Photograph courtesy of Concepts NREC.)

## Learning Objectives

After completing this chapter, you should be able to:

- explain how and why a turbomachine works.
- know the basic differences between a turbine and a pump.
- recognize the importance of minimizing loss in a turbomachine.
- select an appropriate class of turbomachines for a particular application.
- understand why turbomachine blades are shaped like they are.
- appreciate the basic fundamentals of sensibly scaling turbomachines that are larger or smaller than a prototype.
- move on to more advanced engineering work involving the fluid mechanics of turbomachinery (e.g., design, development, research).

In previous chapters we often used generic "black boxes" to represent fluid machines such as pumps or turbines. The purpose of this chapter is to understand the fluid mechanics of these devices when they are turbomachines.

Pumps and turbines (often turbomachines) occur in a wide variety of configurations. In general, pumps add energy to the fluid—they do work on the fluid to move and/or increase the pressure of the fluid; turbines extract energy from the fluid—the fluid does work on them. The term "pump" will be used to generically refer to all pumping machines, including *pumps*, *fans*, *blowers*, and *compressors*.

*Turbomachines are dynamic fluid machines that add (for pumps) or extract (for turbines) flow energy.*

Turbomachines involve a collection of blades, buckets, flow channels, or passages arranged around an axis of rotation to form a rotor. A fluid that is moving can force rotation and produce shaft power. In this case we have a turbine. On the other hand, we can exert a shaft torque, typically with a motor, and by using blades, flow channels, or passages force the fluid to move. In this case we have a pump. In Fig. 12.1 are shown the turbine and compressor (pump) rotors of an automobile turbocharger. Examples of turbomachine-type pumps include simple window fans, propellers on ships or airplanes, squirrel-cage fans on home furnaces, axial-flow water

Exhaust gas flow

Exhaust gas flow

Exhaust gas flow

Turbine rotor

Exhaust gas flow

Shaft rotation

Compressed airflow

Airflow

Airflow

Compressor rotor

Compressed airflow

■ **Figure 12.1** **Automotive turbocharger turbine and compressor rotors. (Photograph courtesy of Concepts NREC.)**

pumps used in deep wells, and compressors in automobile turbochargers. Examples of turbines include the turbine portion of gas turbine engines on aircraft, steam turbines used to drive generators at electrical generation stations, and the small, high-speed air turbines that power dentist drills.

Turbomachines serve in an enormous array of applications in our daily lives and thus play an important role in modern society. These machines can have a high-power density (large-power transfer per size), relatively few moving parts, and reasonable efficiency. The following sections provide an introduction to the fluid mechanics of these important machines. References 1–3 are a few examples of the many books that offer much more knowledge about turbomachines.

# 12.1 Introduction

*Turbomachines involve the related parameters of force, torque, work, and power.*

**(Photograph courtesy of MidAmerican Energy.)**

*Turbomachines* are mechanical devices that either extract energy from a fluid (turbine) or add energy to a fluid (pump) as a result of dynamic interactions between the device and the fluid. While the actual design and construction of these devices often require considerable insight and effort, their basic operating principles are quite simple.

Using a food blender to make a fruit drink is an example of turbo-pump action. The blender blades are forced to rotate around an axis by a motor. The moving blades pulverize fruit and ice and mix them with a base liquid to form a "smoothie."

Conversely, the dynamic effect of the wind blowing past the sail on a boat creates pressure differences on the sail. The wind force on the moving sail in the direction of the boat's motion provides power to propel the boat. The sail and boat act as a machine extracting energy from the air. Turbine blades are like sails. See, for example, the enormous wind turbine blades in the figure in the margin.

The fluid involved can be either a gas (as with a window fan or a gas turbine engine) or a liquid (as with the water pump on a car or a turbine at a hydroelectric power plant). While the basic operating principles are the same whether the fluid is a liquid or a gas, important differences in the fluid dynamics involved can occur. For example, cavitation may be an important design consideration when liquids are involved if the pressure at any point within the flow is reduced to the vapor pressure. Compressibility effects may be important when gases are involved if the Mach number becomes large enough.

Many turbomachines contain some type of housing or casing that surrounds the rotating blades or rotor, thus forming an internal flow passageway through which the fluid flows (see Fig. 12.2). Others, such as a windmill or a window fan, are unducted. Some turbomachines

Inlet

Rotor

Outlet

Housing or casing

(a) Radial-flow fan

Rotor

Inlet

Stator
or strut

Outlet

Housing
or casing

(b) Axial-flow fan

■ **Figure 12.2**  (*a*) A radial-flow turbomachine, (*b*) an axial-flow turbomachine.
(Photographs courtesy of Comair Rotron, Inc.)

*A group of blades moving with or against a lift force is the essence of a turbomachine.*

include stationary blades or vanes in addition to rotor blades. These stationary vanes can be arranged to accelerate the flow and thus serve as nozzles. Or these vanes can be set to diffuse the flow and act as diffusers.

Turbomachines are classified as ***axial-flow***, ***mixed-flow***, or ***radial-flow*** machines depending on the predominant direction of the fluid motion relative to the rotor's axis as the fluid passes the blades (see Fig. 12.2). For an axial-flow machine the fluid maintains a significant axial-flow direction component from the inlet to outlet of the rotor. For a radial-flow machine the flow across the blades involves a substantial radial-flow component at the rotor inlet, exit, or both. In other machines, designated as mixed-flow machines, there may be significant radial- and axial-flow velocity components for the flow through the rotor row. Each type of machine has advantages and disadvantages for different applications and in terms of fluid-mechanical performance.

## 12.2    Basic Energy Considerations

An understanding of the work transfer in turbomachines can be obtained by considering the basic operation of a household fan (pump) and a windmill (turbine). Although the actual flows in such devices are very complex (i.e., three-dimensional and unsteady), the essential phenomena can be illustrated by use of simplified flow considerations and velocity triangles.

Consider a fan blade driven at constant angular velocity, $\omega$, by a motor as is shown in Fig. 12.3a. We denote the blade speed as $U = \omega r$, where $r$ is the radial distance from the axis of the fan. The absolute fluid velocity (that seen by a person sitting stationary at the table on which the fan rests) is denoted $\mathbf{V}$, and the relative velocity (that seen by a person riding on the fan blade) is denoted $\mathbf{W}$. As shown by the figure in the margin, the actual (absolute) fluid velocity is the vector sum of the relative velocity and the blade velocity

$$\mathbf{V} = \mathbf{W} + \mathbf{U} \tag{12.1}$$

A simplified sketch of the fluid velocity as it "enters" and "exits" the fan at radius $r$ is shown in Fig. 12.3b. The shaded surface labeled $a-b-c-d$ is a portion of the cylindrical surface (including a "slice" through the blade) shown in Fig. 12.3a. We assume for simplicity that the

W

U

V

*(a)*

*(b)*

flow moves smoothly along the blade so that relative to the moving blade the velocity is parallel to the leading and trailing edges (points 1 and 2) of the blade. For now we assume that the fluid enters and leaves the fan at the same distance from the axis of rotation; thus, $U_1 = U_2 = \omega r$. In actual turbomachines, the entering and leaving flows are not necessarily tangent to the blades, and the fluid pathlines can involve changes in radius. These considerations are important at design and off-design operating conditions. Interested readers are referred to Refs. 1, 2, and 3 for more information about these aspects of turbomachine flows.

With this information we can construct the ***velocity triangles*** shown in Fig. 12.3*b*. Note that this view is from the top of the fan, looking radially down toward the axis of rotation. The motion of the blade is up; the motion of the incoming air is assumed to be directed along the axis of rotation. The important concept to grasp from this sketch is that the fan blade (because of its shape and motion) "pushes" the fluid, causing it to change direction. The absolute velocity vector, **V**, is turned during its flow across the blade from section (1) to section (2). Initially the fluid had no component of absolute velocity in the direction of the motion of the blade, the $\theta$ (or tangential) direction. When the fluid leaves the blade, this tangential component of absolute velocity is nonzero. For this to occur, the blade must push on the fluid in the tangential direction. That is, the blade exerts a tangential force component on the fluid in the direction of the motion of the blade. This tangential force component and the blade motion are in the same direction—the blade does work on the fluid. This device is a pump.

On the other hand, consider the windmill shown in Fig. 12.4*a* (also see **Video V12.1**). Rather than the rotor being driven by a motor, the blades move in the direction of the lift force (compared to the fan in Fig. 12.3) exerted on each blade by the wind blowing through the rotor. We again note that because of the blade shape and motion, the absolute velocity vectors at sections (1) and (2), **V**₁

*When blades move
because of the fluid
force, we have a
turbine; when
blades are forced to
move fluid, we have
a pump.*

*V12.1 Windmills*

■ **Figure 12.4** **Idealized flow through a windmill:** (*a*) **windmill;** (*b*) **windmill blade geometry;** (*c*) **absolute velocity, V; relative velocity, W; and blade velocity, U; at the inlet and exit of the windmill blade section.**

and $\mathbf{V}_2$, have different directions. For this to happen, the blades must have pushed up on the fluid—opposite to the direction of blade motion. Alternatively, because of equal and opposite forces (action/reaction) the fluid must have pushed on the blades in the direction of their motion—the fluid does work on the blades. This extraction of energy from the fluid is the purpose of a turbine.

These examples involve work transfer to or from a flowing fluid in two axial-flow turbomachines. Similar concepts hold for other turbomachines including mixed-flow and radial-flow configurations.

---

**F l u i d s     i n     t h e     N e w s**

**Current from currents** The use of large, efficient wind *turbines* to generate electrical power is becoming more commonplace throughout the world. "Wind farms" containing numerous turbines located at sites that have proper wind conditions can produce a significant amount of electrical power. Recently, researchers in the United States, the United Kingdom, and Canada have been investigating the possibility of harvesting the power of ocean currents and tides by using current turbines that function much like wind turbines. Rather than being driven by wind, they derive energy from ocean currents that occur at many locations in the 70% of the Earth's surface that is water. Clearly, a 4-knot (2.5 m/s) tidal current is not as fast as a 40-mph (70 km/hr) wind driving a wind turbine. However, since turbine power output is proportional to the fluid density, and since seawater is more than 800 times as dense as air, significant power can be extracted from slow, but massive, ocean currents. One promising configuration involves *blades* twisted in a helical pattern. This technology may provide electrical power that is both ecologically and economically sound.

# EXAMPLE 12.1 Basic Difference between a Pump and a Turbine

**GIVEN**  The rotor shown in Fig. E12.1a rotates at a constant angular velocity of $\omega = 100$ rad/s. Although the fluid initially approaches the rotor in an axial direction, the flow across the blades is primarily outward (see Fig. 12.2a). Measurements indicate that the absolute velocity at the inlet and outlet are $V_1 = 12$ m/s and $V_2 = 15$ m/s, respectively.

**FIND**  Is this device a pump or a turbine?

## SOLUTION

To answer this question, we need to know if the tangential component of the force of the blade on the fluid is in the direction of the blade motion (a pump) or opposite to it (a turbine). We assume that the blades are tangent to the incoming relative velocity and that the relative flow leaving the rotor is tangent to the blades as shown in Fig. E12.1b. We can also calculate the inlet and outlet blade speeds as

$$U_1 = \omega r_1 = (100 \text{ rad/s})(0.1 \text{ m}) = 10 \text{ m/s}$$

and

$$U_2 = \omega r_2 = (100 \text{ rad/s})(0.2 \text{ m}) = 20 \text{ m/s}$$

With the known, absolute fluid velocity and blade velocity at the inlet, we can draw the velocity triangle (the graphical representation of Eq. 12.1) at that location as shown in Fig. E12.1c.

Note that we have assumed that the absolute flow at the blade row inlet is radial (i.e., the direction of $\mathbf{V}_1$ is radial). At the outlet we know the blade velocity, $\mathbf{U}_2$, the outlet speed, $V_2$, and the relative velocity direction, $\beta_2$ (because of the blade geometry). Therefore, we can graphically (or trigonometrically) construct the outlet velocity triangle as shown in the figure. By comparing the velocity triangles at the inlet and outlet, it can be seen that as the fluid flows across the blade row, the absolute velocity vector turns in the direction of the blade motion. At the inlet there is no component of absolute velocity in the direction of rotation; at the outlet this component is not zero. That is, the blade pushes and turns the fluid in the direction of the blade motion, thereby doing work on the fluid, adding energy to it.

This device is a pump.  **(Ans)**

■ **Figure E12.1**

**COMMENT**   On the other hand, by reversing the direction of flow from larger to smaller radii, this device can become a radial-flow turbine. In this case (Fig. E12.1*d*) the flow direction is reversed (compared to that in Figs. E12.1*a*, *b*, and *c*) and the velocity triangles are as indicated. Stationary vanes around the perimeter of the rotor would be needed to achieve $\mathbf{V}_1$ as shown. Note that the component of the absolute velocity, $\mathbf{V}$, in the di- rection of the blade motion is smaller at the outlet than at the inlet. The blade must push against the fluid in the direction opposite the motion of the blade to cause this. Hence (by equal and opposite forces), the fluid pushes against the blade in the direction of blade motion, thereby doing work on the blade. There is a transfer of work from the fluid to the blade—a turbine operation.

## 12.3   Basic Angular Momentum Considerations

In the previous section we indicated how work transfer to or from a fluid flowing through a pump or a turbine occurs by interaction between moving rotor blades and the fluid. Since all of these turbomachines involve the rotation of an impeller or a rotor about a central axis, it is appropriate to discuss their performance in terms of torque and angular momentum.

Recall that work can be written as force times distance or as torque times angular displacement. Hence, if the shaft torque (the torque that the shaft applies to the rotor) and the rotation of the rotor are in the same direction, energy is transferred from the shaft to the rotor and from the rotor to the fluid—the machine is a pump. Conversely, if the torque exerted by the shaft on the rotor is opposite to the direction of rotation, the energy transfer is from the fluid to the rotor—a turbine. The amount of shaft torque (and, hence, shaft work) can be obtained from the moment-of-momentum equation derived formally in Section 5.2.3 and discussed as follows.

*When shaft torque and rotation are in the same direction, we have a pump; otherwise we have a turbine.*

Consider a fluid particle traveling outward through the rotor in the radial-flow machine shown in Figs. E12.1*a*, *b*, and *c*. For now, assume that the particle enters the rotor with a radial velocity only (i.e., no "swirl"). After being acted upon by the rotor blades during its passage from the inlet [section (1)] to the outlet [section (2)], this particle exits with radial ($r$) and circumferential ($\theta$) components of velocity. Thus, the particle enters with no angular momentum about the rotor axis of rotation but leaves with nonzero angular momentum about that axis. (Recall that the axial component of angular momentum for a particle is its mass times the distance from the axis times the $\theta$ component of absolute velocity.)

*V12.2 Self- propelled lawn sprinkler*

A similar experience can occur at the neighborhood playground. Consider yourself as a particle and a merry-go-round as a rotor. Walk from the center to the edge of the spinning merry- go-round and note the forces involved. The merry-go-round does work on you—there is a "side- ward force" on you. Another person must apply a torque (and power) to the merry-go-round to maintain a constant angular velocity; otherwise the angular momentum of the system (you and the merry-go-round) is conserved, and the angular velocity decreases as you increase your dis- tance from the axis of rotation. (Similarly, if the motor driving a pump is turned off, the pump will obviously slow down and stop.) Your friend is the motor supplying energy to the rotor that is transferred to you. Is the amount of energy your friend expends to keep the angular velocity constant dependent on what path you follow along the merry-go-round (i.e., the blade shape); on how fast and in what direction you walk off the edge (i.e., the exit velocity); on how much you weigh (i.e., the density of the fluid)? What happens if you walk from the outside edge toward the center of the rotating merry-go-round? Recall that the opposite of a pump is a turbine.

In a turbomachine a series of particles (a continuum) passes through the rotor. Thus, the moment-of-momentum equation applied to a control volume as derived in Section 5.2.3 is valid. For steady flow (or for turbomachine rotors with steady-in-the-mean or steady-on-average cycli- cal flow), Eq. 5.42 gives

$$\sum (\mathbf{r} \times \mathbf{F}) = \int_{cs} (\mathbf{r} \times \mathbf{V}) \rho \mathbf{V} \cdot \hat{\mathbf{n}} \, dA$$

Recall that the left-hand side of this equation represents the sum of the external torques (moments) acting on the contents of the control volume, and the right-hand side is the net rate of flow of moment-of-momentum (***angular momentum***) through the control surface.

The axial component of this equation applied to the one-dimensional simplification of flow through a turbomachine rotor with section (1) as the inlet and section (2) as the outlet results in

$$T_{\text{shaft}} = -\dot{m}_1(r_1 V_{\theta 1}) + \dot{m}_2(r_2 V_{\theta 2}) \tag{12.2}$$

where $T_{\text{shaft}}$ is the **shaft torque** applied to the contents of the control volume. The "$-$" is associated with mass flowrate into the control volume, and the "$+$" is used with the outflow. The sign of the $V_{\theta}$ component depends on the direction of $V_{\theta}$ and the blade motion, $U$. If $V_{\theta}$ and $U$ are in the same direction, then $V_{\theta}$ is positive. The sign of the torque exerted by the shaft on the rotor, $T_{\text{shaft}}$, is positive if $T_{\text{shaft}}$ is in the same direction as rotation, and negative otherwise.

As seen from Eq. 12.2, the shaft torque is directly proportional to the mass flowrate, $\dot{m} = \rho Q$. (It takes considerably more torque and power to pump water than to pump air with the same volume flowrate.) The torque also depends on the tangential component of the absolute velocity, $V_{\theta}$. Equation 12.2 is often called the **Euler turbomachine equation**.

Also recall that the **shaft power**, $\dot{W}_{\text{shaft}}$, is related to the shaft torque and angular velocity by

$$\dot{W}_{\text{shaft}} = T_{\text{shaft}}\, \omega \tag{12.3}$$

By combining Eqs. 12.2 and 12.3 and using the fact that $U = \omega r$, we obtain

$$\dot{W}_{\text{shaft}} = -\dot{m}_1(U_1 V_{\theta 1}) + \dot{m}_2(U_2 V_{\theta 2}) \tag{12.4}$$

Again, the value of $V_{\theta}$ is positive when $V_{\theta}$ and $U$ are in the same direction and negative otherwise. Also, $\dot{W}_{\text{shaft}}$ is positive when the shaft torque and $\omega$ are in the same direction and negative otherwise. Thus, $\dot{W}_{\text{shaft}}$ is positive when power is supplied to the contents of the control volume (pumps) and negative otherwise (turbines). This outcome is consistent with the sign convention involving the work term in the energy equation considered in Chapter 5 (see Eq. 5.67).

Finally, in terms of work per unit mass, $w_{\text{shaft}} = \dot{W}_{\text{shaft}}/\dot{m}$, we obtain

$$w_{\text{shaft}} = -U_1 V_{\theta 1} + U_2 V_{\theta 2} \tag{12.5}$$

where we have used the fact that by conservation of mass, $\dot{m}_1 = \dot{m}_2$. Equations 12.3, 12.4, and 12.5 are the basic governing equations for pumps or turbines whether the machines are radial-, mixed-, or axial-flow devices and for compressible and incompressible flows. Note that neither the axial nor the radial component of velocity enter into the specific work (work per unit mass) equation. [In the above merry-go-round example the amount of work your friend does is independent of how fast you jump "up" (axially) or "out" (radially) as you exit. The only thing that counts is your $\theta$ component of velocity.]

Another useful but more laborious form of Eq. 12.5 can be obtained by writing the right-hand side in a slightly different form based on the velocity triangles at the entrance or exit as shown generically in Fig. 12.5. The velocity component $V_x$ is the generic through-flow component of velocity and it can be axial, radial, or in-between depending on the rotor configuration. From the large right triangle we note that

$$V^2 = V_{\theta}^2 + V_x^2$$

or

$$V_x^2 = V^2 - V_{\theta}^2 \tag{12.6}$$

*The Euler turbomachine equation is the axial component of the moment-of-momentum equation.*

■ **Figure 12.5** Velocity triangle: $V$ = absolute velocity, $W$ = relative velocity, $U$ = blade velocity.

From the small right triangle we note that

$$V_x^2 + (V_\theta - U)^2 = W^2 \tag{12.7}$$

By combining Eqs. 12.6 and 12.7 we obtain

$$V_\theta U = \frac{V^2 + U^2 - W^2}{2}$$

which when written for the inlet and exit and combined with Eq. 12.5 gives

$$w_{\text{shaft}} = \frac{V_2^2 - V_1^2 + U_2^2 - U_1^2 - (W_2^2 - W_1^2)}{2} \tag{12.8}$$

*Turbomachine work is related to changes in absolute, relative, and blade velocities.*

Thus, the power and the shaft work per unit mass can be obtained from the speed of the blade, $U$, the absolute fluid speed, $V$, and the fluid speed relative to the blade, $W$. This is an alternative to using fewer components of the velocity as suggested by Eq. 12.5. Equation 12.8 contains more terms than Eq. 12.5; however, it is an important concept equation because it shows how the work transfer is related to absolute, relative, and blade velocity changes. Because of the general nature of the velocity triangle in Fig. 12.5, Eq. 12.8 is applicable for axial-, radial-, and mixed-flow rotors.

## F l u i d s   i n   t h e   N e w s

**1948 Buick Dynaflow started it** Prior to 1948 almost all cars had manual transmissions, which required the use of a clutch pedal to shift gears. The 1948 Buick Dynaflow was the first automatic transmission to use the hydraulic *torque* converter and was the model for present-day automatic transmissions. Currently, in the United States over 84% of the cars have automatic transmissions. The torque converter replaces the clutch found on manual shift vehicles and allows the engine to continue running when the vehicle comes to a stop. In principle, but certainly not in detail or complexity, operation of a torque converter is similar to blowing air from a *fan* onto another fan that is unplugged. One can hold the *blade* of the unplugged fan and keep it from turning, but as soon as it is let go, it will begin to speed up until it comes close to the speed of the powered fan. The torque converter uses transmission *fluid* (not air) and consists of a *pump* (the powered fan) driven by the engine driveshaft, a *turbine* (the unplugged fan) connected to the input shaft of the transmission, and a *stator* (absent in the fan model) to efficiently direct the flow between the pump and turbine.

## 12.4    The Centrifugal Pump

*V12.3 Windshield washer pump*

One of the most common radial-flow turbomachines is the ***centrifugal pump***. This type of pump has two main components: an *impeller* attached to a rotating shaft, and a stationary *casing*, *housing*, or *volute* enclosing the impeller. The impeller consists of a number of blades (usually curved), also sometimes called *vanes*, arranged in a regular pattern around the shaft. A sketch showing the essential features of a centrifugal pump is shown in Fig. 12.6. As the impeller rotates, fluid is sucked in through the *eye* of the casing and flows radially outward. Energy is added to the fluid by the rotating blades, and both pressure and absolute velocity are increased as the fluid flows from the eye to the periphery of the blades. For the simplest type of centrifugal pump, the

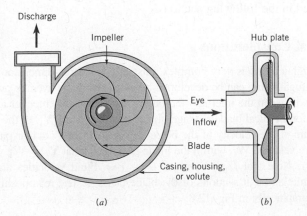

*(a)*                    *(b)*

■ **Figure 12.6**  Schematic diagram of basic elements of a centrifugal pump.

■ **Figure 12.7** (*a*) **Open impeller,** (*b*) **enclosed or shrouded impeller.** (**Courtesy of Flowserve Corporation**)

(*a*)  (*b*)

fluid discharges directly into a volute-shaped casing. The casing shape is designed to reduce the velocity as the fluid leaves the impeller, and this decrease in kinetic energy is converted into an increase in pressure. The volute-shaped casing, with its increasing area in the direction of flow, is used to produce an essentially uniform velocity distribution as the fluid moves around the casing into the discharge opening. For large centrifugal pumps, a different design is often used in which diffuser guide vanes surround the impeller. The diffuser vanes decelerate the flow as the fluid is directed into the pump casing. This type of centrifugal pump is referred to as a *diffuser* pump.

Impellers are generally of two types. For one configuration the blades are arranged on a hub or backing plate and are open on the other (casing or shroud) side. A typical *open impeller* is shown in Fig. 12.7*a*. For the second type of impeller, called an *enclosed* or *shrouded* impeller, the blades are covered on both hub and shroud ends as shown in Fig. 12.7*b*.

Pump impellers can also be *single* or *double suction*. For the single-suction impeller the fluid enters through the eye on only one side of the impeller, whereas for the double-suction impeller the fluid enters the impeller along its axis from both sides. The double-suction arrangement reduces end thrust on the shaft, and also, since the net inlet flow area is larger, inlet velocities are reduced.

Pumps can be *single* or *multistage*. For a single-stage pump, only one impeller is mounted on the shaft, whereas for multistage pumps, several impellers are mounted on the same shaft. The stages operate in series; that is, the discharge from the first stage flows into the eye of the second stage, the discharge from the second stage flows into the eye of the third stage, and so on. The flowrate is the same through all stages, but each stage develops an additional pressure rise. Thus, a very large discharge pressure, or head, can be developed by a multistage pump.

*Centrifugal pumps involve radially outward flows.*

Centrifugal pumps come in a variety of arrangements (open or shrouded impellers, volute or diffuser casings, single- or double-suction, single- or multistage), but the basic operating principle remains the same. Work is done on the fluid by the rotating blades (centrifugal action and tangential blade force acting on the fluid over a distance), creating a large increase in kinetic energy of the fluid flowing through the impeller. This kinetic energy is converted into an increase in pressure as the fluid flows from the impeller into the casing enclosing the impeller. A simplified theory describing the behavior of the centrifugal pump was introduced in the previous section and is expanded in the following section.

### 12.4.1 Theoretical Considerations

Although flow through a pump is very complex (unsteady and three-dimensional), the basic theory of operation of a centrifugal pump can be developed by considering the average one-dimensional flow of the fluid as it passes between the inlet and the outlet sections of the impeller as the blades rotate. As shown in Fig. 12.8, for a typical blade passage, the absolute velocity, $V_1$, of the fluid entering the passage is the vector sum of the velocity of the blade, $U_1$, rotating in a circular path with angular velocity $\omega$, and the relative velocity, $W_1$, within the blade passage so that $V_1 = W_1 + U_1$. Similarly, at the exit $V_2 = W_2 + U_2$. Note that $U_1 = r_1\omega$ and $U_2 = r_2\omega$. Fluid velocities are taken to be average velocities over the inlet and exit sections of the blade passage. The relationship between the various velocities is shown graphically in Fig. 12.8.

As discussed in Section 12.3, the moment-of-momentum equation indicates that the shaft torque, $T_{shaft}$, required to rotate the pump impeller is given by equation Eq. 12.2 applied to a pump with $\dot{m}_1 = \dot{m}_2 = \dot{m}$. That is,

$$T_{shaft} = \dot{m}(r_2 V_{\theta 2} - r_1 V_{\theta 1}) \tag{12.9}$$

or

$$T_{shaft} = \rho Q(r_2 V_{\theta 2} - r_1 V_{\theta 1}) \tag{12.10}$$

where $V_{\theta 1}$ and $V_{\theta 2}$ are the tangential components of the absolute velocities, $\mathbf{V}_1$ and $\mathbf{V}_2$ (see Figs. 12.8b,c).

For a rotating shaft, the power transferred, $\dot{W}_{shaft}$, is given by

$$\dot{W}_{shaft} = T_{shaft}\omega$$

and, therefore, from Eq. 12.10

$$\dot{W}_{shaft} = \rho Q \omega (r_2 V_{\theta 2} - r_1 V_{\theta 1})$$

*Centrifugal pump impellers involve an increase in blade velocity along the flow path.*

Since $U_1 = r_1 \omega$ and $U_2 = r_2 \omega$ we obtain

$$\dot{W}_{shaft} = \rho Q(U_2 V_{\theta 2} - U_1 V_{\theta 1}) \tag{12.11}$$

Equation 12.11 shows how the power supplied to the shaft of the pump is transferred to the flowing fluid. It also follows that the shaft power per unit mass of flowing fluid is

$$w_{shaft} = \frac{\dot{W}_{shaft}}{\rho Q} = U_2 V_{\theta 2} - U_1 V_{\theta 1} \tag{12.12}$$

For incompressible pump flow, we get from Eq. 5.82

$$w_{shaft} = \left(\frac{p_{out}}{\rho} + \frac{V_{out}^2}{2} + gz_{out}\right) - \left(\frac{p_{in}}{\rho} + \frac{V_{in}^2}{2} + gz_{in}\right) + loss$$

■ **Figure 12.8**  **Velocity diagrams at the inlet and exit of a centrifugal pump impeller.**

Combining Eq. 12.12 with this, we get

$$U_2 V_{\theta 2} - U_1 V_{\theta 1} = \left( \frac{p_{out}}{\rho} + \frac{V_{out}^2}{2} + g z_{out} \right) - \left( \frac{p_{in}}{\rho} + \frac{V_{in}^2}{2} + g z_{in} \right) + loss$$

Dividing both sides of this equation by the acceleration of gravity, $g$, we obtain

$$\frac{U_2 V_{\theta 2} - U_1 V_{\theta 1}}{g} = H_{out} - H_{in} + h_L$$

where $H$ is total head defined by

$$H = \frac{p}{\rho g} + \frac{V^2}{2g} + z$$

and $h_L = loss/g$ is head loss.

From this equation we see that $(U_2 V_{\theta 2} - U_1 V_{\theta 1})/g$ is the shaft work head added to the fluid by the pump. Head loss, $h_L$, reduces the actual head rise, $H_{out} - H_{in}$, achieved by the fluid. Thus, the ideal head rise possible, $h_i$, is

$$h_i = \frac{U_2 V_{\theta 2} - U_1 V_{\theta 1}}{g} \tag{12.13}$$

*The pump actual head rise is less than the pump ideal head rise by an amount equal to the head loss in the pump.*

The actual head rise, $H_{out} - H_{in} = h_a$, is always less than the ideal head rise, $h_i$, by an amount equal to the head loss, $h_L$, in the pump. Some additional insight into the meaning of Eq. 12.13 can be obtained by using the following alternate version (see Eq. 12.8).

$$h_i = \frac{1}{2g} \left[ (V_2^2 - V_1^2) + (U_2^2 - U_1^2) + (W_1^2 - W_2^2) \right] \tag{12.14}$$

A detailed examination of the physical interpretation of Eq. 12.14 would reveal the following. The first term in brackets on the right-hand side represents the increase in the kinetic energy of the fluid, and the other two terms represent the pressure head rise that develops across the impeller due to the centrifugal effect, $U_2^2 - U_1^2$, and the diffusion of relative flow in the blade passages, $W_1^2 - W_2^2$.

An appropriate relationship between the flowrate and the pump ideal head rise can be obtained as follows. Often the fluid has no tangential component of velocity $V_{\theta 1}$, or *swirl*, as it enters the impeller; that is, the angle between the absolute velocity and the tangential direction is 90° ($\alpha_1 = 90°$ in Fig. 12.8). In this case, Eq. 12.13 reduces to

$$h_i = \frac{U_2 V_{\theta 2}}{g} \tag{12.15}$$

From Fig. 12.8c

$$\cot \beta_2 = \frac{U_2 - V_{\theta 2}}{V_{r2}}$$

so that Eq. 12.15 can be expressed as

$$h_i = \frac{U_2^2}{g} - \frac{U_2 V_{r2} \cot \beta_2}{g} \tag{12.16}$$

The flowrate, $Q$, is related to the radial component of the absolute velocity through the equation

$$Q = 2\pi r_2 b_2 V_{r2} \tag{12.17}$$

where $b_2$ is the impeller blade height at the radius $r_2$. Thus, combining Eqs. 12.16 and 12.17 yields

$$h_i = \frac{U_2^2}{g} - \frac{U_2 \cot \beta_2}{2\pi r_2 b_2 g} Q \tag{12.18}$$

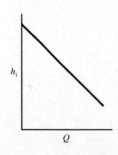

This equation is graphed in the margin and shows that the ideal or maximum head rise for a centrifugal pump varies linearly with $Q$ for a given blade geometry and angular velocity. For actual pumps, the blade angle $\beta_2$ falls in the range of $15°-35°$, with a normal range of $20° < \beta_2 < 25°$, and with $15° < \beta_1 < 50°$ (Ref. 10). Blades with $\beta_2 < 90°$ are called *backward curved*, whereas blades with $\beta_2 > 90°$ are called *forward curved*. Pumps are not usually designed with forward-curved vanes since such pumps tend to suffer unstable flow conditions.

---

## EXAMPLE 12.2    Centrifugal Pump Performance Based on Inlet/Outlet Velocities

**GIVEN**  Water is pumped at the rate of 1400 gpm through a centrifugal pump operating at a speed of 1750 rpm. The impeller has a uniform blade height, $b$, of 2 in. with $r_1 = 1.9$ in. and $r_2 = 7.0$ in., and the exit blade angle $\beta_2$ is $23°$ (see Fig. 12.8). Assume ideal flow conditions and that the tangential velocity component, $V_{\theta 1}$, of the water entering the blade is zero ($\alpha_1 = 90°$).

**FIND**  Determine (a) the tangential velocity component, $V_{\theta 2}$, at the exit, (b) the ideal head rise, $h_i$, and (c) the power, $\dot{W}_{\text{shaft}}$, transferred to the fluid. Discuss the difference between ideal and actual head rise. Is the power, $\dot{W}_{\text{shaft}}$, ideal or actual? Explain.

## SOLUTION

**(a)**  At the exit the velocity diagram is as shown in Fig. 12.8c, where $\mathbf{V}_2$ is the absolute velocity of the fluid, $\mathbf{W}_2$ is the relative velocity, and $\mathbf{U}_2$ is the tip velocity of the impeller with

$$U_2 = r_2\omega = (7/12 \text{ ft})(2\pi \text{ rad/rev})\frac{(1750 \text{ rpm})}{(60 \text{ s/min})}$$

$$= 107 \text{ ft/s}$$

Since the flowrate is given, it follows that

$$Q = 2\pi r_2 b_2 V_{r2}$$

or

$$V_{r2} = \frac{Q}{2\pi r_2 b_2}$$

$$= \frac{1400 \text{ gpm}}{(7.48 \text{ gal/ft}^3)(60 \text{ s/min})(2\pi)(7/12 \text{ ft})(2/12 \text{ ft})}$$

$$= 5.11 \text{ ft/s}$$

From Fig. 12.8c we see that

$$\cot \beta_2 = \frac{U_2 - V_{\theta 2}}{V_{r2}}$$

so that

$$V_{\theta 2} = U_2 - V_{r2} \cot \beta_2$$
$$= (107 - 5.11 \cot 23°) \text{ ft/s}$$
$$= 95.0 \text{ ft/s} \qquad \textbf{(Ans)}$$

**(b)**  From Eq. 12.15 the ideal head rise is given by

$$h_i = \frac{U_2 V_{\theta 2}}{g} = \frac{(107 \text{ ft/s})(95.0 \text{ ft/s})}{32.2 \text{ ft/s}^2}$$

$$= 316 \text{ ft} \qquad \textbf{(Ans)}$$

Alternatively, from Eq. 12.16, the ideal head rise is

$$h_i = \frac{U_2^2}{g} - \frac{U_2 V_{r2} \cot \beta_2}{g}$$

$$= \frac{(107 \text{ ft/s})^2}{32.2 \text{ ft/s}^2} - \frac{(107 \text{ ft/s})(5.11 \text{ ft/s}) \cot 23°}{32.2 \text{ ft/s}^2}$$

$$= 316 \text{ ft} \qquad \textbf{(Ans)}$$

**(c)**  From Eq. 12.11, with $V_{\theta 1} = 0$, the power transferred to the fluid is given by the equation

$$\dot{W}_{\text{shaft}} = \rho Q U_2 V_{\theta 2}$$

$$= \frac{(1.94 \text{ slugs/ft}^3)(1400 \text{ gpm})(107 \text{ ft/s})(95.0 \text{ ft/s})}{[1(\text{slug} \cdot \text{ft/s}^2)/\text{lb}](7.48 \text{ gal/ft}^3)(60 \text{ s/min})}$$

$$= (61,500 \text{ ft} \cdot \text{lb/s})(1 \text{ hp}/550 \text{ ft} \cdot \text{lb/s}) = 112 \text{ hp} \qquad \textbf{(Ans)}$$

Note that the ideal head rise and the power transferred to the fluid are related through the relationship

$$\dot{W}_{\text{shaft}} = \rho g Q h_i$$

**COMMENT**  It should be emphasized that results given in the previous equation involve the ideal head rise. The actual head-rise performance characteristics of a pump are usually determined by experimental measurements obtained in a testing laboratory. The actual head rise is always less than the ideal head rise for a specific flowrate because of the loss of available energy associated with actual flows. Also, it is important to note that even if actual values of $U_2$ and $V_{r2}$ are used in Eq. 12.16, the ideal head rise is calculated. The only idealization used in this example problem is that the exit flow angle is identical to the blade angle at the exit. If the actual exit flow angle was made available in this example, it could have been used in Eq. 12.16 to calculate the ideal head rise.

The pump power, $\dot{W}_{\text{shaft}}$, is the actual power required to achieve a blade speed of 107 ft/s, a flowrate of 1400 gpm, and the tangential velocity, $V_{\theta 2}$, associated with this example. If pump losses could somehow be reduced to zero (every pump designer's dream), the actual and ideal head rise would have been identical at 316 ft. As is, the ideal head rise is 316 ft and the actual head rise something less.

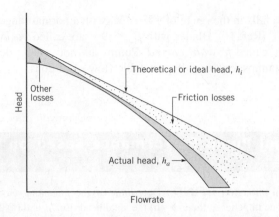

■ **Figure 12.9** Effect of losses on the pump head–flowrate curve.

*Ideal and actual head-rise levels differ by the head loss.*

Figure 12.9 shows the ideal head versus flowrate curve (Eq. 12.18) for a centrifugal pump with backward-curved vanes ($\beta_2 < 90°$). Since there are simplifying assumptions (i.e., zero losses) associated with the equation for $h_i$, we would expect that the actual rise in head of fluid, $h_a$, would be less than the ideal head rise, and this is indeed the case. As shown in Fig. 12.9, the $h_a$ versus $Q$ curve lies below the ideal head-rise curve and shows a nonlinear variation with $Q$. The differences between the two curves (as represented by the shaded areas between the curves) arise from several sources. These differences include losses due to fluid skin friction in the blade passages, which vary as $Q^2$, and other losses due to such factors as flow separation, impeller blade-casing clearance flows, and other three-dimensional flow effects. Near the design flowrate, some of these other losses are minimized.

Centrifugal pump design is a highly developed field, with much known about pump theory and design procedures (see, for example, Refs. 4–6). However, due to the general complexity of flow through a centrifugal pump, the actual performance of the pump cannot be accurately predicted on a completely theoretical basis as indicated by the data of Fig. 12.9. Actual pump performance is determined experimentally through tests on the pump. From these tests, pump characteristics are determined and presented as ***pump performance curves.*** It is this information that is most helpful to the engineer responsible for incorporating pumps into a given flow system.

### 12.4.2 Pump Performance Characteristics

The actual head rise, $h_a$, gained by fluid flowing through a pump can be determined with an experimental arrangement of the type shown in Fig. 12.10, using the energy equation (Eq. 5.84 with $h_a = h_s - h_L$ where $h_s$ is the shaft work head and is identical to $h_i$, and $h_L$ is the pump head loss)

$$h_a = \frac{p_2 - p_1}{\gamma} + z_2 - z_1 + \frac{V_2^2 - V_1^2}{2g} \tag{12.19}$$

with sections (1) and (2) at the pump inlet and exit, respectively. Typically, the differences in elevations and velocities are small so that

$$h_a \approx \frac{p_2 - p_1}{\gamma} \tag{12.20}$$

The power, $\mathcal{P}_f$, gained by the fluid is given by the equation

$$\mathcal{P}_f = \gamma Q h_a \tag{12.21}$$

■ **Figure 12.10** Typical experimental arrangement for determining the head rise gained by a fluid flowing through a pump.

and this quantity, expressed in terms of horsepower is traditionally called the *water horsepower*. Thus,

$$\mathcal{P}_f = \text{water horsepower} = \frac{\gamma Q h_a}{550} \tag{12.22}$$

with $\gamma$ expressed in lb/ft$^3$, $Q$ in ft$^3$/s, and $h_a$ in ft. Note that if the pumped fluid is not water, the $\gamma$ appearing in Eq. 12.22 must be the specific weight of the fluid moving through the pump.

In addition to the head or power added to the fluid, the ***overall efficiency,*** $\eta$, is of interest, where

$$\eta = \frac{\text{power gained by the fluid}}{\text{shaft power driving the pump}} = \frac{\mathcal{P}_f}{\dot{W}_{\text{shaft}}}$$

The denominator of this relationship represents the total power applied to the shaft of the pump and is often referred to as *brake horsepower* (bhp). Thus,

$$\eta = \frac{\gamma Q h_a/550}{\text{bhp}} \tag{12.23}$$

*Overall pump efficiency is the ratio of power actually gained by the fluid to the shaft power supplied.*

The overall pump efficiency is affected by the *hydraulic losses* in the pump, as previously discussed, and in addition, by the *mechanical losses* in the bearings and seals. There may also be some power loss due to leakage of the fluid between the back surface of the impeller hub plate and the casing, or through other pump components. This leakage contribution to the overall efficiency is called the *volumetric loss.* Thus, the overall efficiency arises from three sources, the *hydraulic efficiency,* $\eta_h$, the *mechanical efficiency,* $\eta_m$, and the *volumetric efficiency,* $\eta_v$, so that $\eta = \eta_h \eta_m \eta_v$.

Performance characteristics for a given pump geometry and operating speed are usually given in the form of plots of $h_a$, $\eta$, and bhp versus $Q$ (commonly referred to as *capacity*) as illustrated in Fig. 12.11. Actually, only two curves are needed since $h_a$, $\eta$, and bhp are related through Eq. 12.23. For convenience, all three curves are usually provided. Note that for the pump characterized by the data of Fig. 12.11, the head curve continuously rises as the flowrate decreases, and in this case the pump is said to have a *rising head* curve. As shown by the figure in the margin, pumps may also have $h_a - Q$ curves that initially rise as $Q$ is decreased from the design value and then fall with a continued decrease in $Q$. These pumps have a *falling head* curve. The head developed by the pump at zero discharge is called the shutoff head, and it represents the rise in pressure head across the pump with the discharge valve closed. Since there is no flow with the valve closed, the related efficiency is zero, and the power supplied by the pump (bhp at $Q = 0$) is simply dissipated as heat. Although centrifugal pumps can be operated for short periods of time with the discharge valve closed, damage will occur due to overheating and large mechanical stress with any extended operation with the valve closed.

As can be seen from Fig. 12.11, as the discharge is increased from zero the brake horsepower increases, with a subsequent fall as the maximum discharge is approached. As previously noted, with $h_a$ and bhp known, the efficiency can be calculated. As shown in Fig. 12.11, the efficiency is a function

■ **Figure 12.11**  Typical performance characteristics for a centrifugal pump of a given size operating at a constant impeller speed.

■ **Figure 12.12** Performance curves for a two-stage centrifugal pump operating at 3500 rpm. Data given for three different impeller diameters.

of the flowrate and reaches a maximum value at some particular value of the flowrate, commonly referred to as the *normal* or *design* flowrate or capacity for the pump. The points on the various curves corresponding to the maximum efficiency are denoted as the *best efficiency points* (BEP). It is apparent that when selecting a pump for a particular application, it is usually desirable to have the pump operate near its maximum efficiency. Thus, performance curves of the type shown in Fig. 12.11 are very important to the engineer responsible for the selection of pumps for a particular flow system. Matching the pump to a particular flow system is discussed in Section 12.4.4.

Pump performance characteristics are also presented in charts of the type shown in Fig. 12.12. Since impellers with different diameters may be used in a given casing, performance characteristics for several impeller diameters can be provided with corresponding lines of constant efficiency and brake horsepower as illustrated in Fig. 12.12. Thus, the same information can be obtained from this type of graph as from the curves shown in Fig. 12.11.

It is to be noted that an additional curve is given in Fig. 12.12, labeled $\text{NPSH}_R$, which stands for *required net positive suction head*. As discussed in the following section, the significance of this curve is related to conditions on the suction side of the pump, which must also be carefully considered when selecting and positioning a pump.

### 12.4.3 Net Positive Suction Head (NPSH)

On the suction side of a pump, low pressures are commonly encountered, with the concomitant possibility of cavitation occurring within the pump. As discussed in Section 1.8, cavitation occurs when the liquid pressure at a given location is reduced to the vapor pressure of the liquid. When this occurs, vapor bubbles form (the liquid starts to "boil"); this phenomenon can cause a loss in efficiency as well as structural damage to the pump. To characterize the potential for cavitation, the difference between the total head on the suction side, near the pump impeller inlet, $p_s/\gamma + V_s^2/2g$, and the liquid vapor pressure head, $p_v/\gamma$, is used. The position reference for the elevation head passes through the centerline of the pump impeller inlet. This difference is called the net positive suction head (NPSH) so that

*Cavitation, which may occur when pumping a liquid, is usually avoided.*

$$\text{NPSH} = \frac{p_s}{\gamma} + \frac{V_s^2}{2g} - \frac{p_v}{\gamma} \tag{12.24}$$

There are actually two values of NPSH of interest. The first is the *required* NPSH, denoted $\text{NPSH}_R$, that must be maintained, or exceeded, so that cavitation will not occur. Since pressures lower than those in the suction pipe will develop in the impeller eye, it is usually necessary to determine experimentally, for a given pump, the required $\text{NPSH}_R$. This is the curve shown in Fig. 12.12. Pumps are tested to determine the value for $\text{NPSH}_R$, as defined by Eq. 12.24, by either directly detecting cavitation or observing a change in the head–flowrate curve (Ref. 7). The second value for NPSH of concern is the *available* NPSH, denoted $\text{NPSH}_A$, which represents the head that actually occurs for the particular flow system. This value can be determined experimentally, or calculated if the system parameters are known. For example, a typical flow system is shown in

■ **Figure 12.13**   **Schematic of a pump installation in which the pump must lift fluid from one level to another.**

Fig. 12.13. The energy equation applied between the free liquid surface, where the pressure is atmospheric, $p_{atm}$, and a point on the suction side of the pump near the impeller inlet yields

$$\frac{p_{atm}}{\gamma} - z_1 = \frac{p_s}{\gamma} + \frac{V_s^2}{2g} + \sum h_L$$

where $\sum h_L$ represents head losses between the free surface and the pump impeller inlet. Thus, the head available at the pump impeller inlet is

$$\frac{p_s}{\gamma} + \frac{V_s^2}{2g} = \frac{p_{atm}}{\gamma} - z_1 - \sum h_L$$

so that

$$\text{NPSH}_A = \frac{p_{atm}}{\gamma} - z_1 - \sum h_L - \frac{p_v}{\gamma} \tag{12.25}$$

*For proper pump operation, the available net positive suction head must be greater than the required net positive suction head.*

For this calculation, absolute pressures are normally used since the vapor pressure is usually specified as an absolute pressure. For proper pump operation it is necessary that

$$\text{NPSH}_A \geq \text{NPSH}_R$$

It is noted from Eq. 12.25 that as the height of the pump impeller above the fluid surface, $z_1$, is increased, the NPSH$_A$ is decreased. Therefore, there is some critical value for $z_1$ above which the pump cannot operate without cavitation. The specific value depends on the head losses and the value of the vapor pressure. It is further noted that if the supply tank or reservoir is *above* the pump, $z_1$ will be negative in Eq. 12.25, and the NPSH$_A$ will increase as this height is increased.

---

## EXAMPLE 12.3   Net Positive Suction Head

**GIVEN**   A centrifugal pump is to be placed above a large, open water tank, as shown in Fig. 12.13, and is to pump water at a rate of 0.5 ft³/s. At this flowrate the required net positive suction head, NPSH$_R$, is 15 ft, as specified by the pump manufacturer. The water temperature is 80 °F and atmospheric pressure is 14.7 psi. Assume that the major head loss between the tank and the pump inlet is due to filter at the pipe inlet having a minor loss coefficient

$K_L = 20$. Other losses can be neglected. The pipe on the suction side of the pump has a diameter of 4 in.

**FIND**   Determine the maximum height, $z_1$, that the pump can be located above the water surface without cavitation. If you were required to place a valve in the flow path, would you place it upstream or downstream of the pump? Why?

## SOLUTION

From Eq. 12.25 the available net positive suction head, NPSH$_A$, is given by the equation

$$\text{NPSH}_A = \frac{p_{atm}}{\gamma} - z_1 - \sum h_L - \frac{p_v}{\gamma}$$

and the maximum value for $z_1$ will occur when NPSH$_A$ = NPSH$_R$. Thus,

$$(z_1)_{max} = \frac{p_{atm}}{\gamma} - \sum h_L - \frac{p_v}{\gamma} - \text{NPSH}_R \tag{1}$$

Since the only head loss to be considered is the loss

$$\sum h_L = K_L \frac{V^2}{2g}$$

with

$$V = \frac{Q}{A} = \frac{0.5 \text{ ft}^3/\text{s}}{(\pi/4)(4/12 \text{ ft})^2} = 5.73 \text{ ft/s}$$

it follows that

$$\sum h_L = \frac{(20)(5.73 \text{ ft/s})^2}{2(32.2 \text{ ft/s}^2)} = 10.2 \text{ ft}$$

From Table B.1 the water vapor pressure at 80 °F is 0.5069 psia and $\gamma = 62.22$ lb/ft$^3$. Equation (1) can now be written as

$$(z_1)_{max} = \frac{(14.7 \text{ lb/in.}^2)(144 \text{ in.}^2/\text{ft}^2)}{62.22 \text{ lb/ft}^3} - 10.2 \text{ ft}$$
$$- \frac{(0.5069 \text{ lb/in.}^2)(144 \text{ in.}^2/\text{ft}^2)}{62.22 \text{ lb/ft}^3} - 15 \text{ ft}$$
$$= 7.65 \text{ ft} \qquad \textbf{(Ans)}$$

Thus, to prevent cavitation, with its accompanying poor pump performance, the pump should not be located higher than 7.65 ft above the water surface.

**COMMENT**   If the valve is placed upstream of the pump, not only would the pump have to operate with an additional loss in the system, it would now operate with a lower inlet pressure because of this additional upstream loss and could now suffer cavitation with its usually negative consequences. If the valve is placed downstream of the pump, the pump would need to operate with more loss in the system and with higher back pressure than without the valve. Depending on the stability of the pump at higher back pressures, this could be inconsequential or important. Usually, pumps are stable even with higher back pressures. So, placing the valve on the downstream side of the pump is normally the better choice.

### 12.4.4 System Characteristics and Pump Selection

*The system equation relates the actual head gained by the fluid to the flowrate.*

A typical flow system in which a pump is used is shown in Fig. 12.14. The energy equation applied between points (1) and (2) indicates that

$$h_a = z_2 - z_1 + \sum h_L \qquad \textbf{(12.26)}$$

where $h_a$ is the actual head gained by the fluid from the pump, and $\sum h_L$ represents all friction losses in the pipe and minor losses for pipe fittings and valves. From our study of pipe flow, we know that typically $h_L$ varies approximately as the flowrate squared; that is, $h_L \propto Q^2$ (see Section 8.4). Thus, Eq. 12.26 can be written in the form

$$h_a = z_2 - z_1 + KQ^2 \qquad \textbf{(12.27)}$$

where $K$ depends on the pipe sizes and lengths, friction factors, and minor loss coefficients. Equation 12.27, which is shown in the figure in the margin, is the *system equation* and shows how the actual head gained by the fluid from the pump is related to the system parameters. In this case the parameters include the change in elevation head, $z_2 - z_1$, and the losses due to friction as expressed by $KQ^2$. Each flow system has its own specific system equation. If the flow is laminar, the frictional losses will be proportional to $Q$ rather than $Q^2$ (see Section 8.2).

■ **Figure 12.14**   Typical flow system.

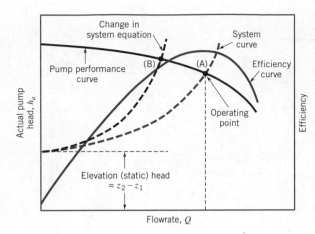

■ **Figure 12.15** Utilization of the system curve and the pump performance curve to obtain the operating point for the system.

*The intersection of the pump performance curve and the system curve is the operating point.*

There is also a unique relationship between the actual pump head gained by the fluid and the flowrate, which is governed by the pump design (as indicated by the pump performance curve). To select a pump for a particular application, it is necessary to utilize both the *system curve*, as determined by the system equation, and the pump performance curve. If both curves are plotted on the same graph, as illustrated in Fig. 12.15, their intersection (point A) represents the operating point for the system. That is, this point gives the head and flowrate that satisfy both the system equation and the pump equation. On the same graph the pump efficiency is shown. Ideally, we want the operating point to be near the best efficiency point (BEP) for the pump. For a given pump, it is clear that as the system equation changes, the operating point will shift. For example, if the pipe friction increases due to pipe wall fouling, the system curve changes, resulting in the operating point A shifting to point B in Fig. 12.15 with a reduction in flowrate and efficiency. The following example shows how the system and pump characteristics can be used to decide if a particular pump is suitable for a given application.

## F l u i d s   i n   t h e   N e w s

**Space Shuttle fuel pumps** The fuel *pump* of your car engine is vital to its operation. Similarly, the fuels (liquid hydrogen and oxygen) of each Space Shuttle main engine (there are three per shuttle) rely on multistage *turbopumps* to get from storage tanks to main combustors. High pressures are utilized throughout the pumps to avoid cavitation. The pumps, some *centrifugal* and some *axial*, are driven by axial-flow, *multistage turbines*. Pump speeds are as high as 35,360 rpm. The liquid oxygen is pumped from 100 to 7420 psia, the liquid hydrogen from 30 to 6515 psia. Liquid hydrogen and oxygen flowrates of about 17,200 gpm and 6100 gpm, respectively, are achieved. These pumps could empty your home swimming pool in seconds. The hydrogen goes from $-423$ °F in storage to $+6000$ °F in the combustion chamber!

## *E*XAMPLE 12.4    Use of Pump Performance Curves

**GIVEN** Water is to be pumped from one large, open tank to a second large, open tank as shown in Fig. E12.4*a*. The pipe diameter throughout is 6 in., and the total length of the pipe between the pipe entrance and exit is 200 ft. Minor loss coefficients for the entrance, exit, and the elbow are shown, and the friction factor for the pipe can be assumed constant and equal to 0.02. A certain centrifugal pump having the performance characteristics shown in Fig. E12.4*b* is suggested as a good pump for this flow system.

**FIND** With this pump, what would be the flowrate between the tanks? Do you think this pump would be a good choice?

■ **Figure E12.4a, b**

## SOLUTION

Application of the energy equation between the two free surfaces, points (1) and (2) as indicated, gives

$$\frac{p_1}{\gamma} + \frac{V_1^2}{2g} + z_1 + h_a = \frac{p_2}{\gamma} + \frac{V_2^2}{2g} + z_2$$

$$+ f\frac{\ell}{D}\frac{V^2}{2g} + \sum K_L \frac{V^2}{2g} \qquad (1)$$

Thus, with $p_1 = p_2 = 0$, $V_1 = V_2 = 0$, $\Delta z = z_2 - z_1 = 10$ ft, $f = 0.02$, $D = 6/12$ ft, and $\ell = 200$ ft, Eq. 1 becomes

$$h_a = 10 + \left[0.02\,\frac{(200\text{ ft})}{(6/12\text{ ft})}\right.$$

$$\left. + (0.5 + 1.5 + 1.0)\right]\frac{V^2}{2(32.2\text{ ft/s}^2)} \qquad (2)$$

where the given minor loss coefficients have been used. Since

$$V = \frac{Q}{A} = \frac{Q(\text{ft}^3/\text{s})}{(\pi/4)(6/12\text{ ft})^2}$$

Eq. 2 can be expressed as

$$h_a = 10 + 4.43\,Q^2 \qquad (3)$$

where $Q$ is in ft$^3$/s, or with $Q$ in gallons per minute

$$h_a = 10 + 2.20 \times 10^{-5}Q^2 \qquad (4)$$

Equation 3 or 4 represents the system equation for this particular flow system and reveals how much actual head the fluid will need to gain from the pump to maintain a certain flowrate. Performance data shown in Fig. E12.4b indicate the actual head the fluid will gain from this particular pump when it operates at a certain flowrate. Thus, when Eq. 4 is plotted on the same graph with performance data, the intersection of the two curves represents the operating point for the pump and the system. This combination is shown in Fig. E12.4c with the intersection (as obtained graphically) occurring at

$$Q = 1600\text{ gal/min} \qquad \text{(Ans)}$$

with the corresponding actual head gained equal to 66.5 ft.

Another concern is whether the pump is operating efficiently at the operating point. As can be seen from Fig. E12.4c, although this is not peak efficiency, which is about 86%, it is close (about 84%). Thus, this pump would be a satisfactory choice, assuming the 1600 gal/min flowrate is at or near the desired flowrate.

**■ Figure E12.4c** *(Continued)*

**■ Figure E12.4d**

The amount of pump head needed at the pump shaft is 66.5 ft/0.84 = 79.2 ft. The power needed to drive the pump is

$$\dot{W}_{shaft} = \frac{\gamma Q h_a}{\eta}$$

$$= \frac{(62.4 \text{ lb/ft}^3)[(1600 \text{ gal/min})/(7.48 \text{ gal/ft}^3)(60 \text{ s/min})](66.5 \text{ ft})}{0.84}$$

$$= 17{,}600 \text{ ft·lb/s} = 32.0 \text{ hp}$$

**COMMENT**  By repeating the calculations for $\Delta z = z_2 - z_1 = 80$ ft and 100 ft (rather than the given 10 ft), the results shown in Fig. E12.4d are obtained. Although the given pump could be used with $\Delta z = 80$ ft (provided that the 500 gal/min flowrate produced is acceptable), it would not be an ideal pump for this application since its efficiency would be only 36%.

Energy could be saved by using a different pump with a performance curve that more nearly matches the new system requirements (i.e., higher efficiency at the operating condition). On the other hand, the given pump would not work at all for $\Delta z = 100$ ft since its maximum head ($h_a = 88$ ft when $Q = 0$) is not enough to lift the water 100 ft, let alone overcome head losses. This is shown in Fig. E12.4d by the fact that for $\Delta z = 100$ ft the system curve and the pump performance curve do not intersect.

Note that head loss within the pump itself was accounted for with the pump efficiency, $\eta$. Thus, $h_s = h_a/\eta$, where $h_s$ is the pump shaft work head and $h_a$ is the actual head rise experienced by the flowing fluid.

*For two pumps in series, add heads; for two in parallel, add flowrates.*

Pumps can be arranged in series or in parallel to provide for additional head or flow capacity. When two pumps are placed in *series*, the resulting pump performance curve is obtained by adding heads at the same flowrate. As illustrated in Fig. 12.16a, for two identical pumps in series, both the actual head gained by the fluid and the flowrate are increased, but neither will be doubled if the system curve remains the same. The operating point is at (A) for one pump and moves to (B) for two pumps in series. For two identical pumps in *parallel*, the combined performance curve is obtained by adding flowrates at the same head, as shown in Fig. 12.16b. As illustrated, the flowrate for the system will not be doubled with the addition of two pumps in parallel (if the same system

*(a)*

*(b)*

**■ Figure 12.16** **Effect of operating pumps in** (*a*) **series and** (*b*) **in parallel.**

curve applies). However, for a relatively flat system curve, as shown in Fig. 12.16b, a significant increase in flowrate can be obtained as the operating point moves from point (A) to point (B).

## 12.5 Dimensionless Parameters and Similarity Laws

As discussed in Chapter 7, dimensional analysis is particularly useful in the planning and execution of experiments. Since the characteristics of pumps are usually determined experimentally, it is expected that dimensional analysis and similitude considerations will prove to be useful in the study and documentation of these characteristics.

From the previous section we know that the principal, dependent pump variables are the actual head rise, $h_a$, shaft power, $\dot{W}_{\text{shaft}}$, and efficiency, $\eta$. We expect that these variables will depend on the geometrical configuration, which can be represented by some characteristic diameter, $D$, other pertinent lengths, $\ell_i$, and surface roughness, $\varepsilon$. In addition, the other important variables are flowrate, $Q$, the pump shaft rotational speed, $\omega$, fluid viscosity, $\mu$, and fluid density, $\rho$. We will only consider incompressible fluids presently, so compressibility effects need not concern us yet. Thus, any one of the dependent variables $h_a$, $\dot{W}_{\text{shaft}}$, and $\eta$ can be expressed as

$$\text{Dependent variable} = f(D, \ell_i, \varepsilon, Q, \omega, \mu, \rho)$$

and a straightforward application of dimensional analysis leads to

$$\text{Dependent pi term} = \phi\left(\frac{\ell_i}{D}, \frac{\varepsilon}{D}, \frac{Q}{\omega D^3}, \frac{\rho \omega D^2}{\mu}\right) \tag{12.28}$$

*Dimensionless pi terms and similarity laws are important pump considerations.*

The dependent pi term involving the head is usually expressed as $C_H = gh_a/\omega^2 D^2$, where $gh_a$ is the actual head rise in terms of energy per unit mass, rather than simply $h_a$, which is energy per unit weight. This dimensionless parameter is called the **head rise coefficient.** The dependent pi term involving the shaft power is expressed as $C_{\mathscr{P}} = \dot{W}_{\text{shaft}}/\rho \omega^3 D^5$, and this standard dimensionless parameter is termed the **power coefficient.** The power appearing in this dimensionless parameter is commonly based on the shaft (brake) horsepower, bhp, so that in BG units, $\dot{W}_{\text{shaft}} = 550 \times (\text{bhp})$. The rotational speed, $\omega$, which appears in these dimensionless groups is expressed in rad/s. The final dependent pi term is the efficiency, $\eta$, which is already dimensionless. Thus, in terms of dimensionless parameters the performance characteristics are expressed as

$$C_H = \frac{gh_a}{\omega^2 D^2} = \phi_1\left(\frac{\ell_i}{D}, \frac{\varepsilon}{D}, \frac{Q}{\omega D^3}, \frac{\rho \omega D^2}{\mu}\right)$$

$$C_{\mathscr{P}} = \frac{\dot{W}_{\text{shaft}}}{\rho \omega^3 D^5} = \phi_2\left(\frac{\ell_i}{D}, \frac{\varepsilon}{D}, \frac{Q}{\omega D^3}, \frac{\rho \omega D^2}{\mu}\right)$$

$$\eta = \frac{\rho g Q h_a}{\dot{W}_{\text{shaft}}} = \phi_3\left(\frac{\ell_i}{D}, \frac{\varepsilon}{D}, \frac{Q}{\omega D^3}, \frac{\rho \omega D^2}{\mu}\right)$$

The last pi term in each of the above equations is a form of Reynolds number that represents the relative influence of viscous effects. When the pump flow involves high Reynolds numbers, as is usually the case, experience has shown that the effect of the Reynolds number can be neglected. For simplicity, the relative roughness, $\varepsilon/D$, can also be neglected in pumps since the highly irregular shape of the pump chamber is usually the dominant geometric factor rather than the surface roughness. Thus, with these simplifications and for *geometrically similar* pumps (all pertinent dimensions, $\ell_i$, scaled by a common length scale), the dependent pi terms are functions of only $Q/\omega D^3$, so that

$$\frac{gh_a}{\omega^2 D^2} = \phi_1\left(\frac{Q}{\omega D^3}\right) \tag{12.29}$$

$$\frac{\dot{W}_{\text{shaft}}}{\rho \omega^3 D^5} = \phi_2\left(\frac{Q}{\omega D^3}\right) \tag{12.30}$$

$$\eta = \phi_3\left(\frac{Q}{\omega D^3}\right) \tag{12.31}$$

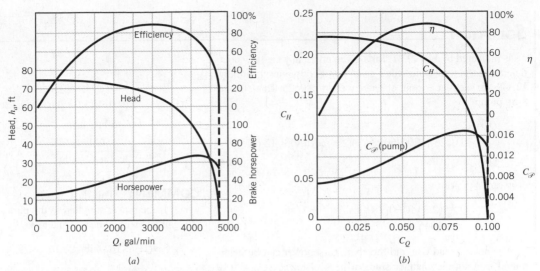

■ **Figure 12.17** Typical performance data for a centrifugal pump: (*a*) characteristic curves for a 12-in. centrifugal pump operating at 1000 rpm, (*b*) dimensionless characteristic curves. (Data from Ref. 8, IIHR-Hydroscience & Engineering, University of Iowa College of Engineering, used by permission.)

The dimensionless parameter $C_Q = Q/\omega D^3$ is called the *flow coefficient*. These three equations provide the desired similarity relationships among a family of geometrically similar pumps. If two pumps from the family are operated at the same value of flow coefficient

$$\left(\frac{Q}{\omega D^3}\right)_1 = \left(\frac{Q}{\omega D^3}\right)_2 \qquad (12.32)$$

it then follows that

$$\left(\frac{gh_a}{\omega^2 D^2}\right)_1 = \left(\frac{gh_a}{\omega^2 D^2}\right)_2 \qquad (12.33)$$

$$\left(\frac{\dot{W}_{\text{shaft}}}{\rho\omega^3 D^5}\right)_1 = \left(\frac{\dot{W}_{\text{shaft}}}{\rho\omega^3 D^5}\right)_2 \qquad (12.34)$$

$$\eta_1 = \eta_2 \qquad (12.35)$$

where the subscripts 1 and 2 refer to any two pumps from the family of geometrically similar pumps.

With these so-called *pump scaling laws* it is possible to experimentally determine the performance characteristics of one pump in the laboratory and then use these data to predict the corresponding characteristics for other pumps within the family under different operating conditions. Figure 12.17*a* shows some typical curves obtained for a centrifugal pump. Figure 12.17*b* shows the results plotted in terms of the dimensionless coefficients, $C_Q$, $C_H$, $C_{\mathscr{P}}$, and $\eta$. From these curves the performance of different-sized, geometrically similar pumps can be predicted, as can the effect of changing speeds on the performance of the pump from which the curves were obtained. It is to be noted that the efficiency, $\eta$, is related to the other coefficients through the relationship $\eta = C_Q C_H C_{\mathscr{P}}^{-1}$. This follows directly from the definition of $\eta$.

*Pump scaling laws relate geometrically similar pumps.*

---

# EXAMPLE 12.5 Use of Pump Scaling Laws

**GIVEN** An 8-in.-diameter centrifugal pump operating at 1200 rpm is geometrically similar to the 12-in.-diameter pump having the performance characteristics of Figs. 12.17*a* and 12.17*b* while operating at 1000 rpm. The working fluid is water at 60 °F.

**FIND** For peak efficiency, predict the discharge, actual head rise, and shaft horsepower for this smaller pump.

## SOLUTION

As is indicated by Eq. 12.31, for a given efficiency the flow coefficient has the same value for a given family of pumps. From Fig. 12.17b we see that at peak efficiency $C_Q = 0.0625$. Thus, for the 8-in. pump

$$Q = C_Q \omega D^3$$
$$= (0.0625)(1200/60 \text{ rev/s})(2\pi \text{ rad/rev})(8/12 \text{ ft})^3$$
$$Q = 2.33 \text{ ft}^3/\text{s} \qquad \textbf{(Ans)}$$

or in terms of gpm

$$Q = (2.33 \text{ ft}^3/\text{s})(7.48 \text{ gal/ft}^3)(60 \text{ s/min})$$
$$= 1046 \text{ gpm} \qquad \textbf{(Ans)}$$

The actual head rise and the shaft horsepower can be determined in a similar manner since at peak efficiency $C_H = 0.19$ and $C_{\mathscr{P}} = 0.014$, so that with $\omega = 1200$ rev/min (1 min/60 s) $(2\pi \text{ rad/rev}) = 126$ rad/s

$$h_a = \frac{C_H \omega^2 D^2}{g} = \frac{(0.19)(126 \text{ rad/s})^2(8/12 \text{ ft})^2}{32.2 \text{ ft/s}^2} = 41.6 \text{ ft} \quad \textbf{(Ans)}$$

and

$$\dot{W}_{\text{shaft}} = C_{\mathscr{P}} \rho \omega^3 D^5$$
$$= (0.014)(1.94 \text{ slugs/ft}^3)(126 \text{ rad/s})^3(8/12 \text{ ft})^5$$
$$= 7150 \text{ ft} \cdot \text{lb/s}$$
$$\dot{W}_{\text{shaft}} = \frac{7150 \text{ ft} \cdot \text{lb/s}}{550 \text{ ft} \cdot \text{lb/s/hp}} = 13.0 \text{ hp} \qquad \textbf{(Ans)}$$

**COMMENT**   This last result gives the shaft horsepower, which is the power supplied to the pump shaft. The power actually gained by the fluid is equal to $\gamma Q h_a$, which in this example is

$$\mathscr{P}_f = \gamma Q h_a = (62.4 \text{ lb/ft}^3)(2.33 \text{ ft}^3/\text{s})(41.6 \text{ ft}) = 6050 \text{ ft} \cdot \text{lb/s}$$

Thus, the efficiency, $\eta$, is

$$\eta = \frac{\mathscr{P}_f}{\dot{W}_{\text{shaft}}} = \frac{6050}{7150} = 85\%$$

which checks with the efficiency curve of Fig. 12.17b.

### 12.5.1 Special Pump Scaling Laws

*Effects of changes in pump operating speed and impeller diameter are often of interest.*

Two special cases related to pump similitude commonly arise. In the first case we are interested in how a change in the operating speed, $\omega$, for a *given pump*, affects pump characteristics. It follows from Eq. 12.32 that for the same flow coefficient (and therefore the same efficiency) with $D_1 = D_2$ (the same pump)

$$\frac{Q_1}{Q_2} = \frac{\omega_1}{\omega_2} \qquad \textbf{(12.36)}$$

The subscripts 1 and 2 now refer to the same pump operating at two different speeds at the same flow coefficient. Also, from Eqs. 12.33 and 12.34 it follows that

$$\frac{h_{a1}}{h_{a2}} = \frac{\omega_1^2}{\omega_2^2} \qquad \textbf{(12.37)}$$

and

$$\frac{\dot{W}_{\text{shaft1}}}{\dot{W}_{\text{shaft2}}} = \frac{\omega_1^3}{\omega_2^3} \qquad \textbf{(12.38)}$$

Thus, for a given pump operating at a given flow coefficient, the flow varies directly with speed, the head varies as the speed squared, and the power varies as the speed cubed. These effects of angular velocity variation are illustrated in the sketch in the margin. These scaling laws are useful in estimating the effect of changing pump speed when some data are available from a pump test obtained by operating the pump at a particular speed.

In the second special case we are interested in how a change in the impeller diameter, $D$, of a geometrically similar family of pumps, operating at a *given speed*, affects pump characteristics. As before, it follows from Eq. 12.32 that for the same flow coefficient with $\omega_1 = \omega_2$

$$\frac{Q_1}{Q_2} = \frac{D_1^3}{D_2^3} \qquad \textbf{(12.39)}$$

Similarly, from Eqs. 12.33 and 12.34

$$\frac{h_{a1}}{h_{a2}} = \frac{D_1^2}{D_2^2} \qquad \textbf{(12.40)}$$

and

$$\frac{\dot{W}_{shaft1}}{\dot{W}_{shaft2}} = \frac{D_1^5}{D_2^5} \qquad (12.41)$$

Thus, for a family of geometrically similar pumps operating at a given speed and the same flow coefficient, the flow varies as the diameter cubed, the head varies as the diameter squared, and the power varies as the diameter raised to the fifth power. These strong effects of diameter variation are illustrated in the sketch in the margin. These scaling relationships are based on the condition that, as the impeller diameter is changed, all other important geometric variables are properly scaled to maintain geometric similarity. This type of geometric scaling is not always possible due to practical difficulties associated with manufacturing the pumps. It is common practice for manufacturers to put impellers of different diameters in the same pump casing. In this case, complete geometric similarity is not maintained, and the scaling relationships expressed in Eqs. 12.39, 12.40, and 12.41 will not, in general, be valid. However, experience has shown that if the impeller diameter change is not too large, less than about 20%, these scaling relationships can still be used to estimate the effect of a change in the impeller diameter. The pump similarity laws expressed by Eqs. 12.36 through 12.41 are sometimes referred to as the *pump affinity laws*.

*Pump affinity laws relate the same pump at different speeds or geometrically similar pumps at the same speed.*

The effects of viscosity and surface roughness have been neglected in the foregoing similarity relationships. However, it has been found that as the pump size decreases, these effects more significantly influence efficiency because of smaller clearances and blade size. An approximate, empirical relationship to estimate the influence of diminishing size on efficiency is (Ref. 9)

$$\frac{1 - \eta_2}{1 - \eta_1} \approx \left(\frac{D_1}{D_2}\right)^{1/5} \qquad (12.42)$$

In general, it is to be expected that the similarity laws will not be very accurate if tests on a model pump with water are used to predict the performance of a prototype pump with a highly viscous fluid, such as oil, because at the much smaller Reynolds number associated with the oil flow, the fluid physics involved is different from the higher Reynolds number flow associated with water.

### 12.5.2 Specific Speed

A useful pi term can be obtained by eliminating diameter $D$ between the flow coefficient and the head rise coefficient. This is accomplished by raising the flow coefficient to an appropriate exponent $(1/2)$ and dividing this result by the head coefficient raised to another appropriate exponent $(3/4)$ so that

$$\frac{(Q/\omega D^3)^{1/2}}{(gh_a/\omega^2 D^2)^{3/4}} = \frac{\omega \sqrt{Q}}{(gh_a)^{3/4}} = N_s \qquad (12.43)$$

The dimensionless parameter $N_s$ is called the **specific speed**. Specific speed varies with flow coefficient just as the other coefficients and efficiency discussed earlier do. However, for any pump it is customary to specify a value of specific speed at the flow coefficient corresponding to peak efficiency only. For pumps with low $Q$ and high $h_a$, the specific speed is low compared to a pump with high $Q$ and low $h_a$. Centrifugal pumps typically are low-capacity, high-head pumps, and therefore have low specific speeds.

Specific speed as defined by Eq. 12.43 is dimensionless, and therefore independent of the system of units used in its evaluation as long as a consistent unit system is used. However, in the United States a modified, dimensional form of specific speed, $N_{sd}$, is commonly used, where

$$N_{sd} = \frac{\omega(\text{rpm})\sqrt{Q(\text{gpm})}}{[h_a(\text{ft})]^{3/4}} \qquad (12.44)$$

In this case $N_{sd}$ is said to be expressed in *U.S. customary units*. Typical values of $N_{sd}$ are in the range $500 < N_{sd} < 4000$ for centrifugal pumps. Both $N_s$ and $N_{sd}$ have the same physical meaning, but their magnitudes will differ by a constant conversion factor $(N_{sd} = 2733 N_s)$ when $\omega$ in Eq. 12.43 is expressed in rad/s.

Each family or class of pumps has a particular range of values of specific speed associated with it. Thus, pumps that have low-capacity, high-head characteristics will have specific speeds that are smaller than pumps that have high-capacity, low-head characteristics. The concept of specific speed is very useful to engineers and designers, since if the required head, flowrate, and speed are specified, it is possible to select an appropriate (most efficient) type of pump for a particular application. For example, as shown by the figure in the margin, as the specific speed, $N_{sd}$, increases beyond about 2000 the peak efficiency, $\eta$, of the purely radial-flow centrifugal pump starts to fall off, and other types of more efficient pump design are preferred. In addition to the centrifugal pump, the *axial-flow* pump is widely used. As discussed in Section 12.6, in an axial-flow pump the direction of flow is primarily parallel to the rotating shaft rather than radial as in the centrifugal pump. Axial-flow pumps are essentially high-capacity, low-head pumps, and therefore have large specific speeds ($N_{sd} > 9000$) compared to centrifugal pumps. *Mixed-flow* pumps combine features of both radial-flow and axial-flow pumps and have intermediate values of specific speed. Figure 12.18 illustrates how the specific speed changes as the configuration of the pump changes from centrifugal or radial to axial.

### 12.5.3 Suction Specific Speed

With an analysis similar to that used to obtain the specific speed pi term, the *suction specific speed*, $S_s$, can be expressed as

$$S_s = \frac{\omega\sqrt{Q}}{[g(\text{NPSH}_R)]^{3/4}} \tag{12.45}$$

where $h_a$ in Eq. 12.43 has been replaced by the required net positive suction head ($\text{NPSH}_R$). This dimensionless parameter is useful in determining the required operating conditions on the suction side of the pump. As was true for the specific speed, $N_s$, the value for $S_s$ commonly used is for peak efficiency. For a family of geometrically similar pumps, $S_s$ should have a fixed value. If this value is known, then the $\text{NPSH}_R$ can be estimated for other pumps within the same family operating at different values of $\omega$ and $Q$.

*Specific speed may be used to approximate what general pump geometry (axial, mixed, or radial) to use for maximum efficiency.*

As noted for $N_s$, the suction specific speed as defined by Eq. 12.45 is also dimensionless, and the value for $S_s$ is independent of the system of units used. However, as was the case for specific speed, in the United States a modified dimensional form for the suction specific speed, designated as $S_{sd}$, is commonly used, where

$$S_{sd} = \frac{\omega(\text{rpm})\sqrt{Q(\text{gpm})}}{[\text{NPSH}_R\,(\text{ft})]^{3/4}} \tag{12.46}$$

For double-suction pumps the discharge, $Q$, in Eq. 12.46 is one-half the total discharge.

Typical values for $S_{sd}$ fall in the range of 7000 to 12,000 (Ref. 11). If $S_{sd}$ is specified, Eq. 12.46 can be used to estimate the $\text{NPSH}_R$ for a given set of operating conditions. However, this calculation would generally only provide an approximate value for the $\text{NPSH}_R$, and the actual determination of the $\text{NPSH}_R$ for a particular pump should be made through a direct measurement whenever possible. Note that $S_{sd} = 2733\,S_s$, with $\omega$ expressed in rad/s in Eq. 12.45.

■ **Figure 12.18** Variation in specific speed at maximum efficiency with type of pump. (Adapted from Ref. 10, used with permission. Courtesy of Hydraulic Institute, Parsippany, NJ 07054 www.Pumps.org.)

## 12.6    Axial-Flow and Mixed-Flow Pumps

As noted previously, centrifugal pumps are radial-flow machines that operate most efficiently for applications requiring high heads at relatively low flowrates. This head–flowrate combination typically yields specific speeds ($N_{sd}$) that are less than approximately 4000. For many applications, such as those associated with drainage and irrigation, high flowrates at low heads are required and centrifugal pumps are not suitable. In this case, axial-flow pumps are commonly used. This type of pump consists essentially of a propeller confined within a cylindrical casing. Axial-flow pumps are often called *propeller pumps*. For this type of pump the flow is primarily in the axial direction (parallel to the axis of rotation of the shaft), as opposed to the radial flow found in the centrifugal pump. Whereas the head developed by a centrifugal pump includes a contribution due to centrifugal action, the head developed by an axial-flow pump is due primarily to the tangential force exerted by the rotor blades on the fluid. A schematic of an axial-flow pump arranged for vertical operation is shown in Fig. 12.19. The rotor is connected to a motor through a shaft, and as it rotates (usually at a relatively high speed) the fluid is sucked in through the inlet. Typically the fluid discharges through a row of fixed stator (guide) vanes used to straighten the flow leaving the rotor. Some axial-flow pumps also have inlet guide vanes upstream of the rotor row, and some are multistage in which pairs (*stages*) of rotating blades (*rotor blades*) and fixed vanes (*stator blades*) are arranged in series. Axial-flow pumps usually have specific speeds ($N_{sd}$) in excess of 9000.

*Axial-flow pumps often have alternating rows of stator blades and rotor blades.*

The definitions and broad concepts that were developed for centrifugal pumps are also applicable to axial-flow pumps. The actual flow characteristics, however, are quite different. In Fig. 12.20 typical head, power, and efficiency characteristics are compared for a centrifugal pump and an axial-flow pump. It is noted that at design capacity (maximum efficiency) the head and brake horsepower are the same for the two pumps selected. But as the flowrate decreases, the power input to the centrifugal pump falls to 180 hp at shutoff, whereas for the axial-flow pump the power input increases to 520 hp at shutoff. This characteristic of the axial-flow pump can cause overloading of the drive motor if the flowrate is reduced significantly from the design capacity. It is also noted that the head curve for the axial-flow pump is much steeper than that for the centrifugal pump. Thus, with axial-flow pumps there will be a large change in head with a small change in the flowrate, whereas for the centrifugal pump, with its relatively flat head curve, there will be only a small change in head with large changes in the flowrate. It is further observed from Fig. 12.20 that, except at design capacity, the efficiency of the axial-flow pump is lower than that of the centrifugal pump. To improve operating characteristics, some axial-flow pumps are constructed with adjustable blades.

For applications requiring specific speeds intermediate to those for centrifugal and axial-flow pumps, mixed-flow pumps have been developed that operate efficiently in the specific speed range $4000 < N_{sd} < 9000$. As the name implies, the flow in a mixed-flow pump has both a radial and an axial component. Figure 12.21 shows some typical data for centrifugal, mixed-flow, and axial-flow pumps, each operating with the same flowrate. These data indicate that as we proceed from the centrifugal pump to the mixed-flow pump to the axial-flow pump, the specific speed increases, the

■ **Figure 12.19** Schematic diagram of an axial-flow pump arranged for vertical operation.

■ **Figure 12.20** Comparison of performance characteristics for a centrifugal pump and an axial-flow pump, each rated 42,000 gal/min at a 17-ft head. (Data from Ref. 12, used with permission.)

head decreases, the speed increases, the impeller diameter decreases, and the eye diameter increases. These general trends are commonly found when these three types of pumps are compared.

The dimensionless parameters and scaling relationships developed in the previous sections apply to all three types of pumps—centrifugal, mixed-flow, and axial-flow—since the dimensional analysis used is not restricted to a particular type of pump. Additional information about pumps can be found in Refs. 4, 7, 9, 12, and 13.

# F l u i d s   i n   t h e   N e w s

**Mechanical heart assist devices** As with any pump, the human heart can suffer various malfunctions and problems during its useful life. Recent developments in artificial heart technology may be able to provide help to those whose pumps have broken down beyond repair. One of the more promising techniques is use of a left-ventricular assist device (LVAD), which supplements a diseased heart. Rather than replacing a diseased heart, an LVAD pump is implanted alongside the heart and works in parallel with the cardiovascular system to assist the pumping function of the heart's left ventricle. (The left ventricle supplies oxygenated blood to the entire body and performs about 80% of the heart's work.) Some LVADs are *centrifugal or axial-flow pumps* that provide a continuous flow of blood. The continuous-flow devices may take some adjustment on the part of patients who do not hear a pulse or a heartbeat. Despite advances in artificial heart technology, it is probably still several years before fully implantable, quiet, and reliable devices will be considered for widespread use.

| Type | Centrifugal | Centrifugal | Mixed flow | Axial flow |
|---|---|---|---|---|
| $N_{sd}$ | 1,250 | 2,200 | 6,200 | 13,500 |
| Gal/min | 2,400 | 2,400 | 2,400 | 2,400 |
| Head, ft | 70 | 48 | 33 | 20 |
| Rpm | 870 | 1,160 | 1,750 | 2,600 |
| $D$, in. | 19 | 12 | 10 | 7 |
| $D_{eye}/D$ | 0.5 | 0.7 | 0.9 | 1.0 |

■ **Figure 12.21** Comparison of different types of impellers. Specific speed for centrifugal pumps based on single suction and identical flowrate. (Adapted from Ref. 12, used with permission.)

# 12.7    Fans

*Fans are used to pump air and other gases and vapors.*

When the fluid to be moved is air, or some other gas or vapor, *fans* are commonly used. Types of fans vary from the small fan used for cooling desktop computers to large fans used in many industrial applications such as ventilating of large buildings. Fans typically operate at relatively low rotation speeds and are capable of moving large volumes of gas. Although the fluid of interest is a gas, the change in gas density through the fan does not usually exceed 7%, which for air represents a change in pressure of only about 1 psi (Ref. 14). Thus, in dealing with fans, the gas density is treated as a constant, and the flow analysis is based on incompressible flow concepts. Because of the low-pressure rise involved, fans are often constructed of lightweight sheet metal. Fans are also called *blowers*, *boosters*, and *exhausters*, depending on the location within the system; that is, blowers are located at the system entrance, exhausters are at the system exit, and boosters are at some intermediate position within the system. Turbomachines used to produce larger changes in gas density and pressure than is possible with fans are called *compressors* (see Section 12.9.1).

As is the case for pumps, fan designs include centrifugal (radial-flow) fans, as well as mixed-flow and axial-flow (propeller) fans, and the analysis of fan performance closely follows that previously described for pumps. The shapes of typical performance curves for centrifugal and axial-flow fans are quite similar to those shown in Fig. 12.20 for centrifugal and axial-flow pumps. However, fan head-rise data are often given in terms of pressure rise, either static or total, rather than the more conventional head rise commonly used for pumps.

Scaling relationships for fans are the same as those developed for pumps, that is, Eqs. 12.32 through 12.35 apply to fans as well as pumps. As noted above, for fans it is common to replace the head, $h_a$, in Eq. 12.33 with pressure head, $p_a/\rho g$, so that Eq. 12.33 becomes

$$\left(\frac{p_a}{\rho \omega^2 D^2}\right)_1 = \left(\frac{p_a}{\rho \omega^2 D^2}\right)_2 \tag{12.47}$$

where, as before, the subscripts 1 and 2 refer to any two fans from the family of geometrically similar fans. Equations 12.47, 12.32, and 12.34 are called the *fan laws* and can be used to scale performance characteristics between members of a family of geometrically similar fans. Additional information about fans can be found in Refs. 14–17.

---

### F l u i d s   i n   t h e   N e w s

**Hi-tech ceiling fans** Energy savings of up to 25% can be realized if thermostats in air-conditioned homes are raised by a few degrees. This can be accomplished by using ceiling *fans* and taking advantage of the increased sensible cooling brought on by air moving over skin. If the energy used to run the fans can be reduced, additional energy savings can be realized. Most ceiling fans use flat, fixed pitch, nonaerodynamic *blades* with uniform chord length. Because the tip of a paddle moves through air faster than its root does, airflow over such fan blades is lowest near the hub and highest at the tip. By making the fan blade more propeller-like, it is possible to have a more uniform, efficient distribution. However, since ceiling fans are restricted by law to operate at less than 200 rpm, ordinary airplane propeller design is not appropriate. After considerable design effort, a highly efficient ceiling fan capable of delivering the same airflow as the conventional design with only half the power has been successfully developed and marketed. The fan blades are based on the slowly turning prop used in the *Gossamer Albatross*, the human-powered aircraft that flew across the English Channel in 1979. (See Problem 12.46.)

---

# 12.8    Turbines

As discussed in Section 12.2, turbines are devices that extract energy from a flowing fluid. The geometry of turbines is such that the fluid exerts a torque on the rotor in the direction of its rotation. The shaft power generated is available to drive generators or other devices.

In the following sections we discuss mainly the operation of hydraulic turbines (those for which the working fluid is water) and to a lesser extent gas and steam turbines (those for which the density of the working fluid may be much different at the inlet than at the outlet).

Although there are numerous ingenious hydraulic turbine designs, most of these turbines can be classified into two basic types—*impulse turbines* and *reaction turbines*. (Reaction is related to

■ **Figure 12.22** (*a*) Schematic diagram of a Pelton wheel turbine, (*b*) photograph of a Pelton wheel turbine. (Courtesy of Voith Hydro, York, PA.)

*The two basic types of hydraulic turbines are impulse and reaction.*

the ratio of static pressure drop that occurs across the rotor to static pressure drop across the turbine stage, with larger rotor pressure drop corresponding to larger reaction.) For hydraulic impulse turbines, the pressure drop across the rotor is zero; all of the pressure drop across the turbine stage occurs in the nozzle row. The ***Pelton wheel*** shown in Fig. 12.22 is a classical example of an impulse turbine. In these machines the total head of the incoming fluid (the sum of the pressure head, velocity head, and elevation head) is converted into a large-velocity head at the exit of the supply nozzle (or nozzles if a multiple nozzle configuration is used). Both the pressure drop across the bucket (blade) and the change in relative speed (i.e., fluid speed relative to the moving bucket) of the fluid across the bucket are negligible. The space surrounding the rotor is not completely filled with fluid. It is the impulse of the individual jets of fluid striking the buckets that generates the torque.

For reaction turbines, on the other hand, the rotor is surrounded by a casing (or volute), which is completely filled with the working fluid. There is both a pressure drop and a fluid relative speed change across the rotor. As shown for the radial-inflow turbine in Fig 12.23, guide vanes act as nozzles to accelerate the flow and turn it in the appropriate direction as the fluid enters the rotor. Thus, part of the pressure drop occurs across the guide vanes and part occurs across the rotor. In many respects the operation of a reaction turbine is similar to that of a pump "flowing backward," although such oversimplification can be quite misleading.

Both impulse and reaction turbines can be analyzed using the moment-of-momentum principles discussed in Section 12.3. In general, impulse turbines are high-head, low-flowrate devices, while reaction turbines are low-head, high-flowrate devices.

### 12.8.1 Impulse Turbines

Although there are various types of impulse turbine designs, perhaps the easiest to understand is the Pelton wheel (see Fig. 12.24). Lester Pelton (1829–1908), an American mining engineer during the

■ **Figure 12.23**   (*a*) Schematic diagram of a reaction turbine, (*b*) photograph of a reaction turbine. (Courtesy of Voith Hydro, York, PA.)

California gold mining days, is responsible for many of the still-used features of this type of turbine. It is most efficient when operated with a large head (for example, a water source from a lake located significantly above the turbine nozzle), which is converted into a relatively large velocity at the exit of the nozzle. Among the many design considerations for such a turbine are the head loss that occurs in the pipe (the penstock) transporting the water to the turbine, the design of the nozzle, and the design of the buckets on the rotor.

■ **Figure 12.24**   Details of Pelton wheel turbine bucket.

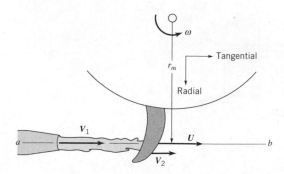

■ **Figure 12.25** Ideal fluid velocities for a Pelton wheel turbine.

As shown in Fig. 12.24, a high-speed jet of water strikes the Pelton wheel buckets and is deflected. The water enters and leaves the control volume surrounding the wheel as free jets (atmospheric pressure). In addition, a person riding on the bucket would note that the speed of the water does not change as it slides across the buckets (assuming viscous effects are negligible). That is, the magnitude of the relative velocity does not change, but its direction does. The change in direction of the velocity of the fluid jet causes a torque on the rotor, resulting in a power output from the turbine.

Design of the optimum, complex shape of the buckets to obtain maximum power output is a very difficult matter. Ideally, the fluid enters and leaves the control volume shown in Fig. 12.25 with no radial component of velocity. (In practice there often is a small but negligible radial component.) In addition, the buckets would ideally turn the relative velocity vector through a 180° turn, but physical constraints dictate that $\beta$, the angle of the exit edge of the blade, is less than 180°. Thus, the fluid leaves with an axial component of velocity as shown in Fig. 12.26.

The inlet and exit velocity triangles at the arithmetic mean radius, $r_m$, are assumed to be as shown in Fig. 12.27. To calculate the torque and power, we must know the tangential components of the absolute velocities at the inlet and exit. (Recall from the discussion in Section 12.3 that neither the radial nor the axial components of velocity enter into the torque or power equations.) From Fig. 12.27 we see that

$$V_{\theta 1} = V_1 = W_1 + U \tag{12.48}$$

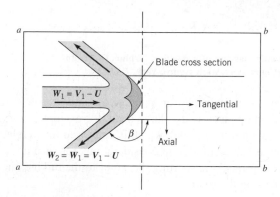

■ **Figure 12.26** Flow as viewed by an observer riding on the Pelton wheel—relative velocities.

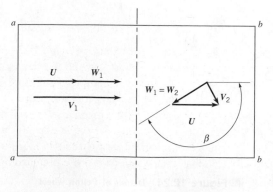

■ **Figure 12.27** Inlet and exit velocity triangles for a Pelton wheel turbine.

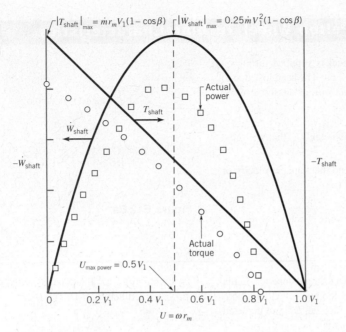

■ **Figure 12.28** Typical theoretical and experimental power and torque for a Pelton wheel turbine as a function of bucket speed.

and

$$V_{\theta 2} = W_2 \cos \beta + U \tag{12.49}$$

Thus, with the assumption that $W_1 = W_2$ (i.e., the relative speed of the fluid does not change as it is deflected by the buckets), we can combine Eqs. 12.48 and 12.49 to obtain

$$V_{\theta 2} - V_{\theta 1} = (U - V_1)(1 - \cos \beta) \tag{12.50}$$

This change in tangential component of velocity combined with the torque and power equations developed in Section 12.3 (i.e., Eqs. 12.2 and 12.4) gives

$$T_{\text{shaft}} = \dot{m} r_m (U - V_1)(1 - \cos \beta)$$

where $\dot{m} = \rho Q$ is the mass flowrate through the turbine. Since $U = \omega R_m$, it follows that

$$\dot{W}_{\text{shaft}} = T_{\text{shaft}} \omega = \dot{m} U (U - V_1)(1 - \cos \beta) \tag{12.51}$$

These results are plotted in Fig. 12.28 along with typical experimental results. Note that $V_1 > U$ (i.e., the jet impacts the bucket), and $\dot{W}_{\text{shaft}} < 0$ (i.e., the turbine extracts power from the fluid).

Several interesting points can be noted from the above results. First, the power is a function of $\beta$. However, a typical value of $\beta = 165°$ (rather than the optimum 180°) results in a relatively small (less than 2%) reduction in power since $1 - \cos 165° = 1.966$, compared to $1 - \cos 180° = 2$. Second, although the torque is maximum when the wheel is stopped ($U = 0$), there is no power under this condition—to extract power one needs force and motion. On the other hand, the power output is a maximum when

$$U|_{\text{max power}} = \frac{V_1}{2} \tag{12.52}$$

This can be shown by using Eq. 12.51 and solving for $U$ that gives $d\dot{W}_{\text{shaft}}/dU = 0$. A bucket speed of one-half the speed of the fluid coming from the nozzle gives the maximum power. Third, the maximum speed occurs when $T_{\text{shaft}} = 0$ (i.e., the load is completely removed from the turbine, as would happen if the shaft connecting the turbine to the generator were to break and frictional torques were negligible). For this case $U = \omega R = V_1$, the turbine is "freewheeling," and the water simply passes across the rotor without putting any force on the buckets.

Although the actual flow through a Pelton wheel is considerably more complex than assumed in the above simplified analysis, reasonable results and trends are obtained by this simple application of the moment-of-momentum principle.

*In Pelton wheel analyses, we assume the relative speed of the fluid is constant (no friction).*

## EXAMPLE 12.6 Pelton Wheel Turbine Characteristics

**GIVEN** Water to drive a Pelton wheel is supplied through a pipe from a lake as indicated in Fig. E12.6a. The head loss due to friction in the pipe is important, but minor losses can be neglected.

**FIND** (a) Determine the nozzle diameter, $D_1$, that will give the maximum power output.

(b) Determine the maximum power and the angular velocity of the rotor at the conditions found in part (a).

■ **Figure E12.6a**

## SOLUTION

(a) As indicated by Eq. 12.51, the power output depends on the flowrate, $Q = \dot{m}/\rho$, and the jet speed at the nozzle exit, $V_1$, both of which depend on the diameter of the nozzle, $D_1$, and the head loss associated with the supply pipe. That is

$$\dot{W}_{shaft} = \rho Q U (U - V_1)(1 - \cos \beta) \tag{1}$$

The nozzle exit speed, $V_1$, can be obtained by applying the energy equation (Eq. 5.85) between a point on the lake surface (where $V_0 = p_0 = 0$) and the nozzle outlet (where $z_1 = p_1 = 0$) to give

$$z_0 = \frac{V_1^2}{2g} + h_L \tag{2}$$

where the head loss is given in terms of the friction factor, $f$, as (see Eq. 8.34)

$$h_L = f \frac{\ell}{D} \frac{V^2}{2g}$$

The speed, $V$, of the fluid in the pipe of diameter $D$ is obtained from the continuity equation

$$V = \frac{A_1 V_1}{A} = \left(\frac{D_1}{D}\right)^2 V_1$$

We have neglected minor losses associated with the pipe entrance and the nozzle. With the given data, Eq. 2 becomes

$$z_0 = \left[1 + f \frac{\ell}{D} \left(\frac{D_1}{D}\right)^4\right] \frac{V_1^2}{2g} \tag{3}$$

or

$$V_1 = \left[\frac{2gz_0}{1 + f\dfrac{\ell}{D}\left(\dfrac{D_1}{D}\right)^4}\right]^{1/2}$$

$$= \left[\frac{2(32.2 \text{ ft/s}^2)(200 \text{ ft})}{1 + 0.02\left(\dfrac{1000 \text{ ft}}{8/12 \text{ ft}}\right)\left(\dfrac{D_1}{8/12}\right)^4}\right]^{1/2}$$

$$= \frac{113.5}{\sqrt{1 + 152 D_1^4}} \tag{4}$$

where $D_1$ is in feet.

By combining Eqs. 1 and 4 and using $Q = \pi D_1^2 V_1/4$ we obtain the power as a function of $D_1$ and $U$ as

$$\dot{W}_{shaft} = \frac{323 U D_1^2}{\sqrt{1 + 152 D_1^4}} \left[U - \frac{113.5}{\sqrt{1 + 152 D_1^4}}\right] \tag{5}$$

where $U$ is in feet per second and $\dot{W}_{shaft}$ is in ft · lb/s. These results are plotted as a function of $U$ for various values of $D_1$ in Fig. E12.6b.

As shown by Eq. 12.52, the maximum power (in terms of its variation with $U$) occurs when $U = V_1/2$, which, when used with Eqs. 4 and 5, gives

$$\dot{W}_{shaft} = -\frac{1.04 \times 10^6 D_1^2}{(1 + 152 D_1^4)^{3/2}} \tag{6}$$

The maximum power possible occurs when $d\dot{W}_{shaft}/dD_1 = 0$, which according to Eq. 6 can be found as

$$\frac{d\dot{W}_{shaft}}{dD_1} = -1.04 \times 10^6 \left[\frac{2 D_1}{(1 + 152 D_1^4)^{3/2}}\right.$$

$$\left. - \left(\frac{3}{2}\right) \frac{4(152) D_1^5}{(1 + 152 D_1^4)^{5/2}}\right] = 0$$

■ **Figure E12.6b**

or

$$304 D_1^4 = 1$$

Thus, the nozzle diameter for maximum power output is

$$D_1 = 0.239 \text{ ft} \qquad \text{(Ans)}$$

**(b)** The corresponding maximum power can be determined from Eq. 6 as

$$\dot{W}_{\text{shaft}} = -\frac{1.04 \times 10^6 (0.239)^2}{[1 + 152(0.239)^4]^{3/2}} = -3.25 \times 10^4 \text{ ft} \cdot \text{lb/s}$$

or

$$\dot{W}_{\text{shaft}} = -3.25 \times 10^4 \text{ ft} \cdot \text{lb/s} \times \frac{1 \text{ hp}}{550 \text{ ft} \cdot \text{lb/s}}$$

$$= -59.0 \text{ hp} \qquad \text{(Ans)}$$

The rotor speed at the maximum power condition can be obtained from

$$U = \omega R = \frac{V_1}{2}$$

where $V_1$ is given by Eq. 4. Thus,

$$\omega = \frac{V_1}{2R} = \frac{\dfrac{113.5}{\sqrt{1 + 152(0.239)^4}} \text{ ft/s}}{2\left(\dfrac{3}{2}\text{ ft}\right)}$$

$$= 30.9 \text{ rad/s} \times 1 \text{ rev}/2\pi \text{ rad} \times 60 \text{ s/min}$$

$$= 295 \text{ rpm} \qquad \text{(Ans)}$$

**COMMENT** The reason that an optimum diameter nozzle exists can be explained as follows. A larger-diameter nozzle will allow a larger flowrate but will produce a smaller jet velocity because of the head loss within the supply side. A smaller-diameter nozzle will reduce the flowrate but will produce a larger jet velocity. Since the power depends on a product combination of flowrate and jet velocity (see Eq. 1), there is an optimum-diameter nozzle that gives the maximum power.

These results can be generalized (i.e., without regard to the specific parameter values of this problem) by considering Eqs. 1 and 3 and the condition that $U = V_1/2$ to obtain

$$\dot{W}_{\text{shaft}|U = V_1/2} = -\frac{\pi}{16}\, \rho(1 - \cos\beta)$$

$$\times (2gz_0)^{3/2}\, D_1^2 \Big/ \left(1 + f\frac{\ell}{D^5}D_1^4\right)^{3/2}$$

By setting $d\dot{W}_{\text{shaft}}/dD_1 = 0$, it can be shown (see Problem 12.61) that the maximum power occurs when

$$D_1 = D\Big/\left(2f\frac{\ell}{D}\right)^{1/4}$$

which gives the same results obtained earlier for the specific parameters of the example problem. Note that the optimum condition depends only on the friction factor and the length-to-diameter ratio of the supply pipe. What happens if the supply pipe is frictionless or of essentially zero length?

In previous chapters we mainly treated turbines (and pumps) as "black boxes" in the flow that removed (or added) energy to the fluid. We treated these devices as objects that removed a certain shaft work head from or added a certain shaft work head to the fluid. The relationship between the shaft work head and the power output as described by the moment-of-momentum considerations is illustrated in Example 12.7.

---

## *E*XAMPLE 12.7   Maximum Power Output for a Pelton Wheel Turbine

**GIVEN** Water flows through the Pelton wheel turbine shown in Fig. 12.24. For simplicity we assume that the water is turned 180° by the blade.

**FIND** Show, based on the energy equation (Eq. 5.84), that the maximum power output occurs when the absolute velocity of the fluid exiting the turbine is zero.

### *S*OLUTION

As indicated by Eq. 12.51, the shaft power of the turbine is given by

$$\dot{W}_{\text{shaft}} = \rho Q U(U - V_1)(1 - \cos\beta)$$

$$= 2\rho Q(U^2 - V_1 U) \qquad (1)$$

For this impulse turbine with $\beta = 180°$, the velocity triangles simplify into the diagram types shown in Fig. E12.7. Three possibilities are indicated:

**(a)** the exit absolute velocity, $\mathbf{V}_2$, is directed back toward the nozzle,

**(b)** the absolute velocity at the exit is zero, or

**(c)** the exiting stream flows in the direction of the incoming stream.

According to Eq. 12.52, the maximum power occurs when $U = V_1/2$, which corresponds to the situation shown in Fig. E12.7b, that is, $U = V_1/2 = W_1$. If viscous effects are negligible, then $W_1 = W_2$ and we have $U = W_2$, which gives

$$V_2 = 0 \qquad \text{(Ans)}$$

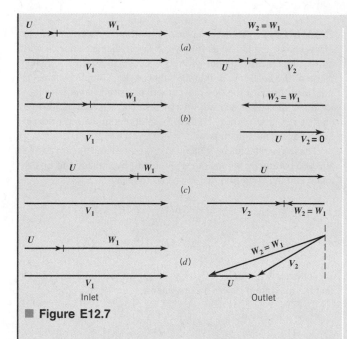

**■ Figure E12.7**

If we consider the energy equation (Eq. 5.84) for flow across the rotor, we have

$$\frac{p_1}{\gamma} + \frac{V_1^2}{2g} + z_1 + h_S = \frac{p_2}{\gamma} + \frac{V_2^2}{2g} + z_2 + h_L$$

where $h_S$ is the shaft work head. This simplifies to

$$h_S = \frac{V_2^2 - V_1^2}{2g} + h_L \qquad (2)$$

since $p_1 = p_2$ and $z_1 = z_2$. Note that the impulse turbine obtains its energy from a reduction in the velocity head. The largest shaft work head possible (and therefore the largest power) occurs when all of the kinetic energy available is extracted by the turbine, giving

$$V_2 = 0 \qquad \textbf{(Ans)}$$

This is consistent with the maximum power condition represented by Fig. E12.7b.

**COMMENT**   As indicated by Eq. 1, if the exit absolute velocity is not in the plane of the rotor (i.e., $\beta < 180°$), there is a reduction in the power available (by a factor of $1 - \cos\beta$). This is also supported by the energy equation, Eq. 2, as follows. For $\beta < 180°$ the inlet and exit velocity triangles are as shown in Fig. E12.7d. Regardless of the bucket speed, $U$, it is not possible to reduce the value of $V_2$ to zero—there is always a component in the axial direction. Thus, according to Eq. 2, the turbine cannot extract the entire velocity head; the exiting fluid has some kinetic energy left in it.

A second type of impulse turbine that is widely used (most often with air as the working fluid) is indicated in Fig. 12.29. A circumferential series of fluid jets strikes the rotating blades which, as with the Pelton wheel, alter both the direction and magnitude of the absolute velocity. As with the Pelton wheel, the inlet and exit pressures (i.e., on either side of the rotor) are equal, and the magnitude of the relative velocity is unchanged as the fluid slides across the blades (if frictional effects are negligible).

Typical inlet and exit velocity triangles (absolute, relative, and blade velocities) are shown in Fig. 12.30. As discussed in Section 12.2, in order for the absolute velocity of the fluid to be changed

*Dentist drill turbines are usually of the impulse class.*

**■ Figure 12.29**   (a) A multinozzle, non-Pelton wheel impulse turbine commonly used with air as the working fluid, (b) dental drill.

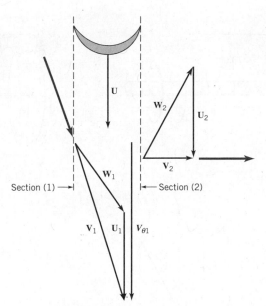

■ **Figure 12.30**  **Inlet and exit velocity triangles for the impulse turbine shown in Fig. 12.29.**

as indicated during its passage across the blade, the blade must push on the fluid in the direction opposite of the blade motion. Hence, the fluid pushes on the blade in the direction of the blade's motion—the fluid does work on the blade (a turbine).

---

## EXAMPLE 12.8 | Non-Pelton Wheel Impulse Turbine (Dental Drill)

**GIVEN**  An air turbine used to drive the high-speed drill used by your dentist is shown in Figs. 12.29 and E12.8a (also see **Video V12.5**). Air exiting from the upstream nozzle holes forces the turbine blades to move in the direction shown. The turbine rotor speed is 300,000 rpm, the tangential component of velocity out of the nozzle is twice the blade speed, and the tangential component of the absolute velocity out of the rotor is zero.

**FIND**  Estimate the shaft energy per unit mass of air flowing through the turbine.

### SOLUTION

We use the fixed, nondeforming control volume that includes the turbine rotor and the fluid in the rotor blade passages at an instant of time (see Fig. E12.8b). The only torque acting on this control volume is the shaft torque. For simplicity we analyze this problem using an arithmetic mean radius, $r_m$, where

$$r_m = \frac{1}{2}(r_0 + r_i)$$

A sketch of the velocity triangles at the rotor entrance and exit is shown in Fig. E12.8c.

Application of Eq. 12.5 (a form of the moment-of-momentum equation) gives

$$w_{shaft} = -U_1 V_{\theta 1} + U_2 V_{\theta 2} \qquad (1)$$

where $w_{shaft}$ is shaft energy per unit of mass flowing through the turbine. From the problem statement, $V_{\theta 1} = 2U$ and $V_{\theta 2} = 0$, where

$$U = \omega r_m = (300{,}000 \text{ rev/min})(1 \text{ min/60 s})(2\pi \text{ rad/rev})$$
$$\times (0.168 \text{ in.} + 0.133 \text{ in.})/2(12 \text{ in./ft}) \qquad (2)$$
$$= 394 \text{ ft/s}$$

is the mean-radius blade velocity. Thus, Eq. (1) becomes

$$
\begin{aligned}
w_{shaft} &= -U_1 V_{\theta 1} = -2U^2 = -2(394 \text{ ft/s})^2 \\
&= -310{,}000 \text{ ft}^2/\text{s}^2 \\
&= (-310{,}000 \text{ ft}^2/\text{s}^2)/(32.174(\text{ft} \cdot \text{lbm})/(\text{lb} \cdot \text{s}^2)) \\
&= -9640 \text{ ft} \cdot \text{lb/lbm} \qquad \textbf{(Ans)}
\end{aligned}
$$

**COMMENT**  For each lbm of air passing through the turbine there is 9640 ft · lb of energy available at the shaft to drive the drill. However, because of fluid friction, the actual amount of energy given up by each slug of air will be greater than the amount available at the shaft. How much greater depends on the efficiency of the fluid-mechanical energy transfer between the fluid and the turbine blades.

Recall that the shaft power, $\dot{W}_{shaft}$, is given by

$$\dot{W}_{shaft} = \dot{m} w_{shaft}$$

Hence, to determine the power, we need to know the mass flowrate, $\dot{m}$, which depends on the size and number of the nozzles. Although the energy per unit mass is large (i.e., 9640 ft · lb/lbm), the flowrate is small, so the power is not "large."

V12.5 Dental drill

(a)

(b)

(c)

■ **Figure E12.8**

### 12.8.2 Reaction Turbines

*Reaction turbines are best suited for higher flowrate and lower head situations.*

As indicated in the previous section, impulse turbines are best suited (i.e., most efficient) for lower-flowrate and higher-head operations. Reaction turbines, on the other hand, are best suited for higher-flowrate and lower-head situations such as are often encountered in hydroelectric power plants associated with a dammed river, for example.

In a reaction turbine the working fluid completely fills the passageways through which it flows (unlike an impulse turbine, which contains one or more individual unconfined jets of fluid). The angular momentum, pressure, and velocity of the fluid decrease as it flows through the turbine rotor—the turbine rotor extracts energy from the fluid.

As with pumps, turbines are manufactured in a variety of configurations—radial-flow, mixed-flow, and axial-flow types. Typical radial- and mixed-flow hydraulic turbines are called *Francis turbines*, named after James B. Francis, an American engineer. At very low heads the most efficient type of turbine is the axial-flow or propeller turbine. The *Kaplan turbine*, named after Victor Kaplan, a German professor, is an efficient axial-flow hydraulic turbine with adjustable blades. Cross sections of these different turbine types are shown in Fig. 12.31.

As shown in Fig. 12.31a, flow across the rotor blades of a radial-inflow turbine has a major component in the radial direction. Inlet guide vanes (which may be adjusted to allow optimum performance) direct the water into the rotor with a tangential component of velocity. The absolute velocity of the water leaving the rotor is essentially without tangential velocity. Hence, the rotor decreases the angular momentum of the fluid, the fluid exerts a torque on the rotor in the direction of rotation, and the rotor extracts energy from the fluid. The Euler turbomachine equation (Eq. 12.2) and the corresponding power equation (Eq. 12.4) are equally valid for this turbine as they are for the centrifugal pump discussed in Section 12.4.

As shown in Fig. 12.31b, for an axial-flow Kaplan turbine, the fluid flows through the inlet guide vanes and achieves a tangential velocity in a vortex (swirl) motion before it reaches the

■ **Figure 12.31**   (*a*) **Typical radial-flow Francis turbine,** (*b*) **typical axial-flow Kaplan turbine.**

rotor. Flow across the rotor contains a major axial component. Both the inlet guide vanes and the turbine blades can be adjusted by changing their setting angles to produce the best match (optimum output) for the specific operating conditions. For example, the operating head available may change from season to season and/or the flowrate through the rotor may vary.

Pumps and turbines are often thought of as the "inverse" of each other. Pumps add energy to the fluid; turbines remove energy. The propeller on an outboard motor (a pump) and the propeller on a Kaplan turbine are in some ways geometrically similar, but they perform opposite tasks. Similar comparisons can be made for centrifugal pumps and Francis turbines. In fact, some large turbomachines at hydroelectric power plants are designed to be run as turbines during high-power demand periods (i.e., during the day) and as pumps to resupply the upstream reservoir from the downstream reservoir during low-demand times (i.e., at night). Thus, a pump type often has its corresponding turbine type. However, is it possible to have the "inverse" of a Pelton wheel turbine— an impulse pump?

As with pumps, incompressible flow turbine performance is often specified in terms of appropriate dimensionless parameters. The flow coefficient, $C_Q = Q/\omega D^3$, the head coefficient, $C_H = gh_a/\omega^2 D^2$, and the power coefficient, $C_{\mathscr{P}} = \dot{W}_{\text{shaft}}/\rho\omega^3 D^5$, are defined in the same way for pumps and turbines. On the other hand, turbine efficiency, $\eta$, is the inverse of pump efficiency. That is, the efficiency is the ratio of the shaft power output to the power available in the flowing fluid, or

*Actual head available for a turbine, $h_a$, is always greater than shaft work head, $h_s$, because of head loss, $h_L$, in the turbine.*

$$\eta = \frac{\dot{W}_{\text{shaft}}}{\rho g Q h_a}$$

For geometrically similar turbines and for negligible Reynolds number and surface roughness difference effects, the relationships between the dimensionless parameters are given functionally by that shown in Eqs. 12.29, 12.30, and 12.31. That is,

$$C_H = \phi_1(C_Q), \quad C_{\mathscr{P}} = \phi_2(C_Q), \quad \text{and} \quad \eta = \phi_3(C_Q)$$

where the functions $\phi_1$, $\phi_2$, and $\phi_3$ are dependent on the type of turbine involved. Also, for turbines the efficiency, $\eta$, is related to the other coefficients according to $\eta = C_{\mathscr{P}}/C_H C_Q$.

As indicated above, the design engineer has a variety of turbine types available for any given application. It is necessary to determine which type of turbine would best fit the job (i.e., be most efficient) before detailed design work is attempted. As with pumps, the use of a specific speed parameter can help provide this information. For hydraulic turbines, the rotor diameter $D$ is eliminated between the flow coefficient and the power coefficient to obtain the *power specific speed*, $N'_s$, where

$$N'_s = \frac{\omega \sqrt{\dot{W}_{\text{shaft}}/\rho}}{(gh_a)^{5/4}}$$

We use the more common, but not dimensionless, definition of specific speed

$$N'_{sd} = \frac{\omega(\text{rpm}) \sqrt{\dot{W}_{\text{shaft}} \, (\text{bhp})}}{[h_a \, (\text{ft})]^{5/4}} \tag{12.53}$$

*Specific speed may be used to approximate what kind of turbine geometry (axial to radial) would operate most efficiently.*

That is, $N'_{sd}$ is calculated with angular velocity, $\omega$, in rpm; shaft power, $\dot{W}_{\text{shaft}}$, in brake horsepower; and actual head available, $h_a$, in feet. Optimum turbine efficiency (for large turbines) as a function of specific speed is indicated in Fig. 12.32. Also shown are representative rotor and casing cross sections. Note that impulse turbines are best at low specific speeds; that is, when operating with large heads and small flowrate. The other extreme is axial-flow turbines, which are the most efficient type if the head is low and if the flowrate is large. For intermediate values of specific speeds, radial- and mixed-flow turbines offer the best performance.

The data shown in Fig. 12.32 are meant only to provide a guide for turbine-type selection. The actual turbine efficiency for a given turbine depends very strongly on the detailed design of the turbine. Considerable analysis, testing, and experience are needed to produce an efficient turbine. However, the data of Fig. 12.32 are representative. Much additional information can be found in the literature.

■ **Figure 12.32** Typical turbine cross sections and maximum efficiencies as a function of specific speed.

**Cavitation damage in hydraulic turbines** The occurrence of *cavitation* in hydraulic *pumps* seems to be an obvious possibility since low-suction *pressures* are expected. Cavitation damage can also occur in hydraulic *turbines,* even though they do not seem obviously prone to this kind of problem. Local *acceleration* of liquid over *blade* surfaces can be sufficient to result in local pressures low enough to cause fluid vaporization or cavitation.

Further along the flow path, the fluid can decelerate rapidly enough with accompanying increase in local pressure to make cavitation bubbles collapse with enough intensity to cause blade surface damage in the form of material erosion. Over time, this erosion can be severe enough to require blade repair or replacement which is very expensive. (See Problem 12.4LL.)

---

# EXAMPLE 12.9    Use of Specific Speed to Select Turbine Type

**GIVEN**  A hydraulic turbine is to operate at an angular velocity of 6 rev/s, a flowrate of 10 ft³/s, and a head of 20 ft.

**FIND**  What type of turbine should be selected? Explain.

## SOLUTION

The most efficient type of turbine to use can be obtained by calculating the specific speed, $N'_{sd}$, and using the information of Fig. 12.32. To use the dimensional form of the specific speed indicated in Fig. 12.32, we must convert the given data into the appropriate units. For the rotor speed we get

$$\omega = 6 \text{ rev/s} \times 60 \text{ s/min} = 360 \text{ rpm}$$

To estimate the shaft power, we assume all of the available head is converted into power and multiply this amount by an assumed efficiency (94%).

$$\dot{W}_{\text{shaft}} = \gamma Q z \eta = (62.4 \text{ lb/ft}^3)(10 \text{ ft}^3/\text{s})\left[\frac{20 \text{ ft}(0.94)}{550 \text{ ft} \cdot \text{lb/s} \cdot \text{hp}}\right]$$

$$\dot{W}_{\text{shaft}} = 21.3 \text{ hp}$$

Thus for this turbine,

$$N'_{sd} = \frac{\omega \sqrt{\dot{W}_{\text{shaft}}}}{(h_a)^{5/4}} = \frac{(360 \text{ rpm}) \sqrt{21.3 \text{ hp}}}{(20 \text{ ft})^{5/4}} = 39.3$$

According to the information of Fig. 12.32,

A mixed-flow Francis turbine would probably give the highest efficiency and an assumed efficiency of 0.94 is appropriate.    **(Ans)**

**COMMENT**  What would happen if we wished to use a Pelton wheel for this application? Note that with only a 20-ft head, the maximum jet velocity, $V_1$, obtainable (neglecting viscous effects) would be

$$V_1 = \sqrt{2\,gz} = \sqrt{2 \times 32.2 \text{ ft/s}^2 \times 20 \text{ ft}} = 35.9 \text{ ft/s}$$

As shown by Eq. 12.52, for maximum efficiency of a Pelton wheel the jet velocity is ideally two times the blade velocity. Thus, $V_1 = 2\omega R$, or the wheel diameter, $D = 2R$, is

$$D = \frac{V_1}{\omega} = \frac{35.9 \text{ ft/s}}{(6 \text{ rev/s} \times 2\pi \text{ rad/rev})} = 0.952 \text{ ft}$$

To obtain a flowrate of $Q = 10$ ft³/s at a velocity of $V_1 = 35.9$ ft/s, the jet diameter, $d_1$, must be given by

$$Q = \frac{\pi}{4} d_1^2 V_1$$

or

$$d_1 = \left[\frac{4Q}{\pi V_1}\right]^{1/2} = \left[\frac{4(10 \text{ ft}^3/\text{s})}{\pi(35.9 \text{ ft/s})}\right]^{1/2} = 0.596 \text{ ft}$$

A Pelton wheel with a diameter of $D = 0.952$ ft supplied with water through a nozzle of diameter $d_1 = 0.596$ ft is not a practical design. Typically $d_1 \ll D$ (see Fig. 12.22). By using multiple jets it would be possible to reduce the jet diameter. However, even with 8 jets, the jet diameter would be 0.211 ft, which is still too large (relative to the wheel diameter) to be practical. Hence, the above calculations reinforce the results presented in Fig. 12.32—a Pelton wheel would not be practical for this application. If the flowrate were considerably smaller, the specific speed could be reduced to the range where a Pelton wheel would be the type to use (rather than a mixed-flow reaction turbine).

---

## 12.9    Compressible Flow Turbomachines

*Compressible flow turbomachines* are in many ways similar to the incompressible flow pumps and turbines described in previous portions of this chapter. The main difference is that the density of the fluid (a gas or vapor) changes significantly from the inlet to the outlet of the compressible flow machines. This added feature has interesting consequences, benefits, and complications.

Compressors are pumps that add energy to the fluid, causing a significant pressure rise and a corresponding significant increase in density. Compressible flow turbines, on the other hand, remove energy from the fluid, causing a lower pressure and a smaller density at the outlet than at the inlet. The information provided earlier about basic energy considerations (Section 12.2) and basic angular momentum considerations (Section 12.3) is directly applicable to these turbomachines in the ways demonstrated earlier.

As discussed in Chapter 11, compressible flow study requires an understanding of the principles of thermodynamics. Similarly, an in-depth analysis of compressible flow turbomachines requires use of various thermodynamic concepts. In this section we provide only a brief discussion of some of the general properties of compressors and compressible flow turbines. The interested reader is encouraged to read some of the excellent references available for further information (e.g., Refs. 1–3, 18–20).

### 12.9.1 Compressors

Turbocompressors operate with the continuous compression of gas flowing through the device. Since there is a significant pressure and density increase, there is also a considerable temperature increase.

Radial-flow (or centrifugal) compressors are essentially centrifugal pumps (see Section 12.4) that use a gas (rather than a liquid) as the working fluid. They are typically high pressure rise, low flowrate, and axially compact turbomachines. A photograph of the rotor of a centrifugal compressor rotor is shown in Fig. 12.33.

The amount of compression is typically given in terms of the *total pressure ratio*, $PR = p_{02}/p_{01}$, where the pressures are absolute. Thus, a radial-flow compressor with $PR = 3.0$ can compress standard atmospheric air from 14.7 psia to $3.0 \times 14.7 = 44.1$ psia.

Higher pressure ratios can be obtained by using *multiple-stage* devices in which flow from the outlet of the preceding stage proceeds to the inlet of the following stage. If each stage has the same pressure ratio, $PR$, the overall pressure ratio after $n$ stages is $PR^n$. Thus, as shown by the figure in the margin, a four-stage compressor with individual stage $PR = 2.0$ can compress standard air from $p_{0\,in} = 14.7$ psia to $p_{0\,out} = 2^4 \times 14.7 = 235$ psia. Adiabatic (i.e., no heat transfer) compression of a gas causes an increase in temperature and requires more work than isothermal (constant temperature) compression of a gas. An interstage cooler (i.e., an intercooler heat exchanger) as shown in Fig. 12.34 can be used to reduce the compressed gas temperature and thus the work required.

*Multistaging is common in high-pressure ratio compressors.*

Relative to centrifugal water pumps, radial compressors of comparable size rotate at much higher speeds. It is not uncommon for the rotor blade exit speed and the speed of the absolute flow leaving the impeller to be greater than the speed of sound. That such large speeds are necessary for compressors can be seen by noting that the large pressure rise designed for is related to the differences of several squared speeds (see Eq. 12.14).

The axial-flow compressor is the other widely used configuration. This type of turbomachine has a lower pressure rise per stage and a higher flowrate, and is more radially compact than a centrifugal compressor. As shown in Fig. 12.35, axial-flow compressors usually consist of several stages, with each stage containing a rotor/stator row pair. For an 11-stage compressor, a compression ratio of $PR = 1.2$ per stage gives an overall pressure ratio of $p_{02}/p_{01} = 1.2^{11} = 7.4$. As the

■ **Figure 12.33** Centrifugal compressor rotor. (Photograph courtesy of Concepts NREC.)

■ **Figure 12.34**  Two-stage centrifugal compressor with an intercooler.

■ **Figure 12.35**  Enthalpy, velocity, and pressure distribution in an axial-flow compressor.

*Axial-flow compressor multistaging requires less space than centrifugal compressors.*

*V12.6 Flow in a compressor stage*

gas is compressed and its density increases, a smaller annulus cross-sectional area is required and the flow channel size decreases from the inlet to the outlet of the compressor. The typical jet aircraft engine uses an axial-flow compressor as one of its main components (see Fig. 12.36 and Ref. 21).

An axial-flow compressor can include a set of *inlet guide vanes* upstream of the first rotor row. These guide vanes optimize the size of the relative velocity into the first rotor row by directing the flow away from the axial direction. *Rotor blades* push on the gas in the direction of blade motion and to the rear, adding energy (like in an axial pump) and moving the gas through the compressor. The *stator blade* rows act as diffusers, turning the fluid back toward the axial direction and increasing the static pressure. The stator blades cannot add energy to the fluid because they are stationary. Typical pressure, velocity, and enthalpy distributions along the axial direction are shown in Fig. 12.35. [If you

■ **Figure 12.36**  Rolls-Royce Trent 900 three-shaft propulsion system. (Courtesy of Rolls-Royce plc.)

are not familiar with the thermodynamic concept of enthalpy (see Section 11.1), you may replace "enthalpy" by temperature as an approximation.] The reaction of the compressor stage is equal to the ratio of the rise in static enthalpy or temperature achieved across the rotor to the enthalpy or temperature rise across the stage. Most modern compressors involve 50% or higher reaction.

The blades in an axial-flow compressor are airfoils carefully designed to produce appropriate lift and drag forces on the flowing gas. As occurs with airplane wings, compressor blades can stall (see Section 9.4). When the flowrate is decreased from the design amount, the velocity triangle at the entrance of the rotor row indicates that the relative flow meets the blade leading edge at larger angles of incidence than the design value. When the angle of incidence becomes too large, blade stall can occur and the result is *compressor surge* or *stall*—unstable flow conditions that can cause excessive vibration, noise, poor performance, and possible damage to the machine. The lower flowrate bound of compressor operation is related to the beginning of these instabilities (see Fig. 12.37).

Other important compressible flow phenomena such as variations of the Mach cone (see Section 11.3), shock waves (see Section 11.5.3), and choked flow (see Section 11.4.2) occur commonly in compressible flow turbomachines. They must be carefully designed for. These phenomena are very sensitive to even very small changes or variations of geometry. Shock strength is kept low to minimize shock loss, and choked flows limit the upper flowrate boundary of machine operation (see Fig. 12.37).

The experimental performance data for compressors are systematically summarized with parameters prompted by dimensional analysis. As mentioned earlier, total pressure ratio, $p_{02}/p_{01}$, is used instead of the head-rise coefficient associated with pumps, blowers, and fans.

Either isentropic or polytropic efficiencies are used to characterize compressor performance. A detailed explanation of these efficiencies is beyond the scope of this text. Those interested in learning more about these parameters should study any of several available books on turbomachines (for example, Refs. 2 and 3). Basically, each of these compressor efficiencies involves a ratio of ideal work to actual work required to accomplish the compression. The isentropic efficiency involves a ratio of the ideal work required with an adiabatic and frictionless (no loss) compression process to the actual work required to achieve the same total pressure rise. The polytropic efficiency involves a ratio of the ideal work required to achieve the actual end state of the compression with a polytropic and frictionless process between the actual beginning and end stagnation states across the compressor and the actual work involved between these same states.

The flow parameter commonly used for compressors is based on the following dimensionless grouping from dimensional analysis

$$\frac{R\dot{m}\sqrt{kRT_{01}}}{D^2 p_{01}}$$

■ **Figure 12.37** Performance characteristics of an axial-flow compressor (Ref. 19).

where $R$ is the gas constant, $\dot{m}$ the mass flowrate, $k$ the specific heat ratio, $T_{01}$ the stagnation temperature at the compressor inlet, $D$ a characteristic length, and $p_{01}$ the stagnation pressure at the compressor inlet.

To account for variations in test conditions, the following strategy is employed. We set

$$\left(\frac{R\dot{m}\sqrt{kRT_{01}}}{D^2 p_{01}}\right)_{\text{test}} = \left(\frac{R\dot{m}\sqrt{kRT_{01}}}{D^2 p_{01}}\right)_{\text{std}}$$

where the subscript "test" refers to a specific test condition and "std" refers to the standard atmosphere ($p_0 = 14.7$ psia, $T_0 = 518.7\ °R$) condition. When we consider a given compressor operating on a given working fluid (so that $R$, $k$, and $D$ are constant), the above equation reduces to

$$\dot{m}_{\text{std}} = \frac{\dot{m}_{\text{test}}\sqrt{T_{01\,\text{test}}/T_{0\,\text{std}}}}{p_{01\,\text{test}}/p_{0\,\text{std}}} \tag{12.54}$$

In essence, $\dot{m}_{\text{std}}$ is the compressor-test mass flowrate "corrected" to the standard atmosphere inlet condition. *The corrected compressor mass flowrate*, $\dot{m}_{\text{std}}$, is used instead of flow coefficient. Often, $\dot{m}_{\text{std}}$ is divided by $A$, the frontal area of the compressor flow path.

While for pumps, blowers, and fans, rotor speed was accounted for in the flow coefficient, it is not in the corrected mass flowrate derived above. Thus, for compressors, rotor speed needs to be accounted for with an additional group. This dimensionless group is

$$\frac{ND}{\sqrt{kRT_{01}}}$$

For the same compressor operating on the same gas, we eliminate $D$, $k$, and $R$ and, as with corrected mass flowrate, obtain a corrected speed, $N_{\text{std}}$, where

$$N_{\text{std}} = \frac{N}{\sqrt{T_{01}/T_{\text{std}}}} \tag{12.55}$$

Often, the percentage of the corrected speed design value is used.

An example of how compressor performance data are typically summarized is shown in Fig. 12.37.

### 12.9.2　Compressible Flow Turbines

Turbines that use a gas or vapor as the working fluid are in many respects similar to hydraulic turbines (see Section 12.8). Compressible flow turbines may be impulse or reaction turbines, and mixed-, radial-, or axial-flow turbines. The fact that the gas may expand (compressible flow) in coursing through the turbine can introduce some important phenomena that do not occur in hydraulic turbines. (*Note:* It is tempting to label turbines that use a gas as the working fluid as gas turbines. However, the terminology "gas turbine" is commonly used to denote a *gas turbine engine*, as employed, for example, for aircraft propulsion or stationary power generation. As shown in Fig. 12.36, these engines typically contain a compressor, combustion chamber, and turbine.)

Although for compressible flow turbines the axial-flow type is common, the radial-inflow type is also used for various purposes. As shown in Fig. 12.33, the turbine that drives the typical automobile turbocharger compressor is a radial-inflow type. The main advantages of the radial-inflow turbine are: (1) It is robust and durable, (2) it is axially compact, and (3) it can be relatively inexpensive. A radial-flow turbine usually has a lower efficiency than an axial-flow turbine, but lower initial costs may be the compelling incentive in choosing a radial-flow turbine over an axial-flow one.

Axial-flow turbines are widely used compressible flow turbines. Steam engines used in electrical generating plants and marine propulsion and the turbines used in gas turbine engines are usually of the axial-flow type. Often they are multistage turbomachines, although single-stage compressible turbines are also produced. They may be either an impulse type or a reaction type. With

*A gas turbine engine generally consists of a compressor, a combustor, and a turbine.*

compressible flow turbines, the ratio of static enthalpy or temperature drop across the rotor to this drop across the stage, rather than the ratio of static pressure differences, is used to determine reaction. Strict impulse (zero pressure drop) turbines have slightly negative reaction; the static enthalpy or temperature actually increases across the rotor. Zero-reaction turbines involve no change of static enthalpy or temperature across the rotor but do involve a slight pressure drop.

A two-stage, axial-flow impulse turbine is shown in Fig. 12.38a. The gas accelerates through the supply nozzles, has some of its energy removed by the first-stage rotor blades, accelerates again through the second-stage nozzle row, and has additional energy removed by the second-stage rotor blades. As shown in Fig. 12.38b, the static pressure remains constant across the rotor rows. Across the second-stage nozzle row, the static pressure decreases, absolute velocity increases, and the stagnation enthalpy (temperature) is constant. Flow across the second rotor is similar to flow across the first rotor. Since the working fluid is a gas, the significant decrease in static pressure across the turbine results in a significant decrease in density—the flow is compressible. Hence, more detailed analysis of this flow must incorporate various compressible flow concepts developed in Chapter 11. Interesting phenomena such as shock waves and choking due to sonic conditions at the "throat" of the flow passage between blades can occur because of compressibility effects. The interested reader is encouraged to consult the various references available (e.g., Refs. 2, 3, 20) for fascinating applications of compressible flow principles in turbines.

The rotor and nozzle blades in a three-stage, axial-flow reaction turbine are shown in Fig. 12.39a. The axial variations of pressure and velocity are shown in Fig. 12.39c. Both the stationary and rotor blade (passages) act as flow-accelerating nozzles. That is, the static pressure and enthalpy (temperature) decrease in the direction of flow for both the fixed and the rotating blade rows. This distinguishes the reaction turbine from the impulse turbine (see Fig. 12.38b). Energy is removed from the fluid by the rotors only (the stagnation enthalpy or temperature is constant across the adiabatic flow stators).

■ **Figure 12.38** Enthalpy, velocity, and pressure distribution in a two-stage impulse turbine.

■ **Figure 12.39** Enthalpy, pressure, and velocity distribution in a three-stage reaction turbine.

■ **Figure 12.40** Typical compressible flow turbine performance "map." (Ref. 20)

Because of the reduction of static pressure in the downstream direction, the gas expands, and the flow passage area must increase from the inlet to the outlet of this turbine. This is seen in Fig. 12.39b.

Performance data for compressible flow turbines are summarized with the help of parameters derived from dimensional analysis. Isentropic and polytropic efficiencies (see Refs. 2, 3, and 20) are commonly used, as are inlet-to-outlet total pressure ratios ($p_{01}/p_{02}$), corrected rotor speed (see Eq. 12.55), and corrected mass flowrate (see Eq. 12.54). Figure 12.40 shows a compressible flow turbine performance "map."

> *Turbine performance maps are used to display complex turbine characteristics.*

## 12.10    Chapter Summary and Study Guide

> *turbomachine*
> *axial-, mixed-, and radial-flow*
> *velocity triangle*
> *angular momentum*
> *shaft torque*
> *Euler turbomachine equation*
> *shaft power*
> *centrifugal pump*
> *pump performance curve*
> *overall efficiency*
> *system equation*
> *head-rise coefficient*
> *power coefficient*
> *flow coefficient*
> *pump scaling laws*
> *specific speed*
> *impulse turbine*
> *reaction turbine*
> *Pelton wheel*

Various aspects of turbomachine flow are considered in this chapter. The connection between fluid angular momentum change and shaft torque is key to understanding how turbo-pumps and turbines operate.

The shaft torque associated with change in the axial component of angular momentum of a fluid as it flows through a pump or turbine is described in terms of the inlet and outlet velocity triangle diagrams. Such diagrams indicate the relationship among absolute, relative, and blade velocities.

Performance characteristics for centrifugal pumps are discussed. Standard dimensionless pump parameters, similarity laws, and the concept of specific speed are presented for use in pump analysis. How to use pump performance curves and the system curve for proper pump selection is presented. A brief discussion of axial-flow and mixed-flow pumps is given.

An analysis of impulse turbines is provided, with emphasis on the Pelton wheel turbine. For impulse turbines there is negligible pressure difference across the blade; the torque is a result of the change in direction of the fluid jet striking the blade. Radial-flow and axial-flow reaction turbines are also briefly discussed.

The following checklist provides a study guide for this chapter. When your study of the entire chapter and end-of-chapter exercises has been completed, you should be able to

■ write out meanings of the terms listed here in the margin and understand each of the related concepts. These terms are particularly important and are set in *italic, **bold**, and color* type in the text.

■ draw appropriate velocity triangles for flows entering and leaving given pump or turbine configurations.

- estimate the actual shaft torque, actual shaft power, and ideal pump head rise for a given centrifugal pump configuration.
- use pump performance curves and the system curve to predict pump performance in a given system.
- predict the performance characteristics for one pump based on the performance of another pump of the same family using the pump scaling laws.
- use specific speed to determine whether a radial-flow, mixed-flow, or axial-flow pump would be most appropriate for a given situation.
- estimate the actual shaft torque and actual shaft power for flow through an impulse turbine configuration.
- estimate the actual shaft torque and actual shaft power for a given reaction turbine.
- use specific speed to determine whether an impulse or a reaction turbine would be most appropriate for a given situation.

Some of the important equations in this chapter are:

| | | |
|---|---|---|
| Vector addition of velocities | $\mathbf{V} = \mathbf{W} + \mathbf{U}$ | (12.1) |
| Shaft torque | $T_{\text{shaft}} = -\dot{m}_1(r_1 V_{\theta 1}) + \dot{m}_2(r_2 V_{\theta 2})$ | (12.2) |
| Shaft power | $\dot{W}_{\text{shaft}} = T_{\text{shaft}}\,\omega$ | (12.3) |
| Shaft power | $\dot{W}_{\text{shaft}} = -\dot{m}_1(U_1 V_{\theta 1}) + \dot{m}_2(U_2 V_{\theta 2})$ | (12.4) |
| Shaft work | $w_{\text{shaft}} = U_2 V_{\theta 2} - U_1 V_{\theta 1}$ | (12.12) |
| Pump ideal head rise | $h_i = \dfrac{U_2 V_{\theta 2} - U_1 V_{\theta 1}}{g}$ | (12.13) |
| Pump actual head rise | $h_a = \dfrac{p_2 - p_1}{\gamma} + z_2 - z_1 + \dfrac{V_2^2 - V_1^2}{2g}$ | (12.19) |
| Pump similarity relationship | $\dfrac{g h_a}{\omega^2 D^2} = \phi_1\left(\dfrac{Q}{\omega D^3}\right)$ | (12.29) |
| Pump similarity relationship | $\dfrac{\dot{W}_{\text{shaft}}}{\rho \omega^3 D^5} = \phi_2\left(\dfrac{Q}{\omega D^3}\right)$ | (12.30) |
| Pump similarity relationship | $\eta = \phi_3\left(\dfrac{Q}{\omega D^3}\right)$ | (12.31) |
| Pump scaling law | $\left(\dfrac{Q}{\omega D^3}\right)_1 = \left(\dfrac{Q}{\omega D^3}\right)_2$ | (12.32) |
| Pump scaling law | $\left(\dfrac{g h_a}{\omega^2 D^2}\right)_1 = \left(\dfrac{g h_a}{\omega^2 D^2}\right)_2$ | (12.33) |
| Pump scaling law | $\left(\dfrac{\dot{W}_{\text{shaft}}}{\rho \omega^3 D^5}\right)_1 = \left(\dfrac{\dot{W}_{\text{shaft}}}{\rho \omega^3 D^5}\right)_2$ | (12.34) |
| Pump scaling law | $\eta_1 = \eta_2$ | (12.35) |
| Specific speed (pumps) | $N_{sd} = \dfrac{\omega(\text{rpm})\sqrt{Q(\text{gpm})}}{[h_a(\text{ft})]^{3/4}}$ | (12.44) |
| Specific speed (turbines) | $N'_{sd} = \dfrac{\omega(\text{rpm})\sqrt{\dot{W}_{\text{shaft}}(\text{bhp})}}{[h_a(\text{ft})]^{5/4}}$ | (12.53) |

# References

1. Cumpsty, N. A., *Jet Propulsion*, 2nd Ed., Cambridge University Press, Cambridge, UK, 2003.
2. Saravanamuttoo, H. I. H., Rogers, G. F. C., and Cohen, H., *Gas Turbine Theory*, 5th Ed., Prentice-Hall, Saddle River, N.J., 2001.
3. Wilson, D. G., and Korakianitis, T., *The Design of High-Efficiency Turbomachinery and Gas Turbines*, 2nd Ed., Prentice-Hall, Saddle River, N.J., 1998.
4. Stepanoff, H. J., *Centrifugal and Axial Flow Pumps*, 2nd Ed., Wiley, New York, 1957.
5. Wislicenus, G. F., *Preliminary Design of Turbopumps and Related Machinery*, NASA Reference Publication 1170, 1986.
6. Neumann, B., *The Interaction Between Geometry and Performance of a Centrifugal Pump*, Mechanical Engineering Publications Limited, London, 1991.
7. Garay, P. N., *Pump Application Desk Book*, Fairmont Press, Lilburn, Ga., 1990.
8. Rouse, H., *Elementary Mechanics of Fluids*, Wiley, New York, 1946.
9. Moody, L. F., and Zowski, T., "Hydraulic Machinery," in *Handbook of Applied Hydraulics*, 3rd Ed., by C. V. Davis and K. E. Sorensen, McGraw-Hill, New York, 1969.
10. Hydraulic Institute, *Hydraulic Institute Standards*, 14th Ed., Hydraulic Institute, Cleveland, Ohio, 1983.
11. Heald, C. C., ed., *Cameron Hydraulic Data*, 17th Ed., Ingersoll-Rand, Woodcliff Lake, N.J., 1988.
12. Kristal, F. A., and Annett, F. A., *Pumps: Types, Selection, Installation, Operation, and Maintenance*, McGraw-Hill, New York, 1953.
13. Karassick, I. J., et al., *Pump Handbook*, 3rd Ed., McGraw-Hill, New York, 2000.
14. Stepanoff, A. J., *Turboblowers*, Wiley, New York, 1955.
15. Berry, C. H., *Flow and Fan Principles of Moving Air Through Ducts*, Industrial Press, New York, 1954.
16. Wallis, R. A., *Axial Flow Fans and Ducts*, Wiley, New York, 1983.
17. Reason, J., "Fans," *Power*, Vol. 127, No. 9, 103–128, 1983.
18. Cumpsty, N. A., *Compressor Aerodynamics*, Longman Scientific & Technical, Essex, UK, and Wiley, New York, 1989.
19. Johnson, I. A., and Bullock, R. D., eds., *Aerodynamic Design of Axial-Flow Compressors*, NASA SP-36, National Aeronautics and Space Administration, Washington, D.C., 1965.
20. Glassman, A. J., ed., *Turbine Design and Application*, Vol. 3, NASA SP-290, National Aeronautics and Space Administration, Washington, D.C., 1975.
21. Saeed Farokhi, *Aircraft Propulsion*, Wiley, New York, 2009.

---

 Problem available in *WileyPLUS* at instructor's discretion.

 Tutoring problem available in *WileyPLUS* at instructor's discretion.

 Problem is related to a chapter video available in *WileyPLUS*.

\* Problem to be solved with aid of programmable calculator or computer.

† Open-ended problem that requires critical thinking. These problems require various assumptions to provide the necessary input data. There are not unique answers to these problems.

# Review Problems

Go to Appendix G (*WileyPLUS* or the book's web site, www.wiley.com/college/munson) for a set of review problems with answers. Detailed solutions can be found in the *Student Solution Manual and Study Guide for Fundamentals of Fluid Mechanics*, by Munson et al. (© 2013 John Wiley and Sons, Inc.).

# Conceptual Questions

**12.1C** For a turbomachine pump, such as a window fan or a propeller,

a) rotation of the fan or propeller results in movement of fluid.

b) rotation of the fan or propeller results in energy being transferred to the fluid.

c) rotation of the fan or propeller requires work input to the fan or propeller shaft.

d) All of the above.

**12.2C** A purpose of a water pump is to turn the pump shaft work into an increase in

a) the pressure of the fluid.    b) the volume of the fluid.

c) the density of the fluid.    d) the enthalpy of the fluid.

**12.3C** When wind is acting on a windmill, the power produced will depend on

a) the velocity of the wind relative to the blade.

b) the velocity of the blade relative to the ground.

c) the velocity of the ground relative to the blade.

d) the velocity of the wind relative to the ground.

**Additional conceptual questions are available in *WileyPLUS* at the instructor's discretion.**

# Problems

**Note: Unless specific values of required fluid properties are given in the problem statement, use the values found in the tables on the inside of the front cover. Answers to the even-numbered problems are listed at the end of the book. The Lab Problems as well as the videos that accompany problems can be accessed in *WileyPLUS* or the book's web site, www.wiley.com/college/munson.**

## Section 12.1 Introduction and Section 12.2 Basic Energy Considerations

**12.1** (WILEY) The rotor shown in Fig. P12.1 rotates clockwise. Assume that the fluid enters in the radial direction and the relative velocity is tangent to the blades and remains constant across the entire rotor. Is the device a pump or a turbine? Explain.

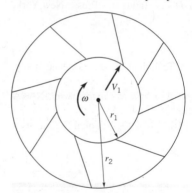

■ **Figure P12.1**

**12.2** Air (assumed incompressible) flows across the rotor shown in Fig. P12.2 such that the magnitude of the absolute velocity increases from 15 m/s to 25 m/s. Measurements indicate that the absolute velocity at the inlet is in the direction shown. Determine the direction of the absolute velocity at the outlet if the fluid puts no torque on the rotor. Is the rotation CW or CCW? Is this device a pump or a turbine?

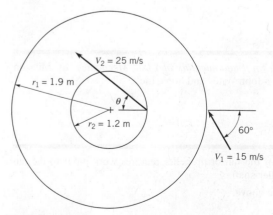

■ **Figure P12.2**

**12.3** The measured shaft torque on the turbomachine shown in Fig. P12.3 is $-60$ N·m when the absolute velocities are as indicated. Determine the mass flowrate. What is the angular velocity if the magnitude of the shaft power is 1800 N·m/s? Is this machine a pump or a turbine? Explain.

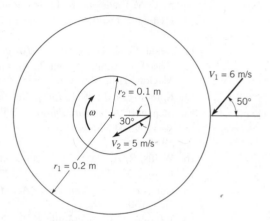

■ **Figure P12.3**

## Section 12.3 Basic Angular Momentum Considerations

**12.4** ▶ (WILEY) Water flows through a rotating sprinkler arm as shown in Fig. P12.4 and **Video V12.2.** Estimate the minimum water pressure necessary for an angular velocity of 150 rpm. Is this a turbine or a pump?

■ **Figure P12.4**

**12.5** WILEY Water is supplied to a dishwasher through the manifold shown in Fig. P12.5. Determine the rotational speed of the manifold if bearing friction and air resistance are neglected. The total flowrate of 2.3 gpm is divided evenly among the six outlets, each of which produces a 5/16-in.-diameter stream.

■ **Figure P12.5**

**12.6** WILEY Water flows axially up the shaft and out through the two sprinkler arms as sketched in Fig. P12.4 and as shown in **Video V12.2.** With the help of the moment-of-momentum equation explain why only at a threshold amount of water flow, the sprinkler arms begin to rotate. What happens when the flowrate increases above this threshold amount? If the exit nozzle could be varied, what would happen for a set flowrate above the threshold amount, when the angle is increased to 90°? Decreased to 0°?

**12.7** WILEY GO Uniform horizontal sheets of water of 3-mm thickness issue from the slits on the rotating manifold shown in Fig. P12.7. The velocity relative to the arm is a constant 3 m/s along each slit. Determine the torque needed to hold the manifold stationary. What would the angular velocity of the manifold be if the resisting torque is negligible?

■ **Figure P12.7**

**12.8** WILEY At a given radial location, a 15-mph wind against a windmill (**see Video V12.1**) results in the upstream (1) and downstream (2) velocity triangles shown in Fig. P12.8. Sketch an appropriate blade section at that radial location and determine the energy transferred per unit mass of fluid.

■ **Figure P12.8**

**12.9** WILEY Sketch how you would arrange four 3-in.-wide by 12-in.-long thin but rigid strips of sheet metal on a hub to create a windmill like the one shown in **Video V12.1**. Discuss, with the help of velocity triangles, how you would arrange each blade on the hub and how you would orient your windmill in the wind.

**12.10** WILEY Sketched in Fig. P12.10 are the upstream [section (1)] and downstream [section (2)] velocity triangles at the arithmetic mean radius for flow through an axial-flow turbomachine rotor. The axial component of velocity is 50 ft/s at sections (1) and (2). **(a)** Label each velocity vector appropriately. Use **V** for absolute velocity, **W** for relative velocity, and **U** for blade velocity. **(b)** Are you dealing with a turbine or a fan? **(c)** Calculate the work per unit mass involved. **(d)** Sketch a reasonable blade section. Do you think that the actual blade exit angle will need to be less or greater than 15°? Why?

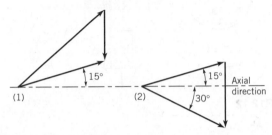

■ **Figure P12.10**

**12.11** WILEY Shown in Fig. P12.11 is a toy "helicopter" powered by air escaping from a balloon. The air from the balloon flows radially through each of the three propeller blades and out small nozzles at the tips of the blades. The nozzles (along with the rotating propeller blades) are tilted at a small angle as indicated. Sketch the velocity triangle (i.e., blade, absolute, and relative velocities) for the flow from the nozzles. Explain why this toy tends to move upward. Is this a turbine? Pump?

■ **Figure P12.11**

## Section 12.4 The Centrifugal Pump and Section 12.4.1 Theoretical Considerations

**12.12** (WILEY) The radial component of velocity of water leaving the centrifugal pump sketched in Fig. P12.12 is 45 ft/s. The magnitude of the absolute velocity at the pump exit is 90 ft/s. The fluid enters the pump rotor radially. Calculate the shaft work required per unit mass flowing through the pump.

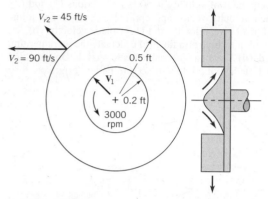

■ **Figure P12.12**

**12.13** (WILEY) A centrifugal water pump having an impeller diameter of 0.5 m operates at 900 rpm. The water enters the pump parallel to the pump shaft. If the exit blade angle, $\beta_2$ (see Fig. 12.8), is 25°, determine the shaft power required to turn the impeller when the flow through the pump is 0.16 m³/s. The uniform blade height is 50 mm.

**12.14** (WILEY GO) A centrifugal pump impeller is rotating at 1200 rpm in the direction shown in Fig. P12.14. The flow enters parallel to the axis of rotation and leaves at an angle of 30° to the radial direction. The absolute exit velocity, $V_2$, is 90 ft/s. (a) Draw the velocity triangle for the impeller exit flow. (b) Estimate the torque necessary to turn the impeller if the fluid is water. What will the impeller rotation speed become if the shaft breaks?

■ **Figure P12.14**

**12.15** (WILEY) A centrifugal radial water pump has the dimensions shown in Fig. P12.15. The volume rate of flow is 0.25 ft³/s, and the absolute inlet velocity is directed radially outward. The angular velocity of the impeller is 960 rpm. The exit velocity as seen from a coordinate system attached to the impeller can be assumed to be tangent to the vane at its trailing edge. Calculate the power required to drive the pump.

■ **Figure P12.15**

## Section 12.4.2 Pump Performance Characteristics

**12.16** (WILEY) Water is pumped with a centrifugal pump, and measurements made on the pump indicate that for a flowrate of 240 gpm the required input power is 6 hp. For a pump efficiency of 62%, what is the actual head rise of the water being pumped?

**12.17** (WILEY) The performance characteristics of a certain centrifugal pump are determined from an experimental setup similar to that shown in Fig. 12.10. When the flowrate of a liquid ($SG = 0.9$) through the pump is 120 gpm, the pressure gage at (1) indicates a vacuum of 95 mm of mercury and the pressure gage at (2) indicates a pressure of 80 kPa. The diameter of the pipe at the inlet is 110 mm and at the exit it is 55 mm. If $z_2 - z_1 = 0.5$ m, what is the actual head rise across the pump? Explain how you would estimate the pump motor power requirement.

**12.18** (WILEY) The performance characteristics of a certain centrifugal pump having a 9-in.-diameter impeller and operating at 1750 rpm are determined using an experimental setup similar to that shown in Fig. 12.10. The following data were obtained during a series of tests in which $z_2 - z_1 = 0$, $V_2 = V_1$, and the fluid was water.

| $Q$ (gpm) | 20 | 40 | 60 | 80 | 100 | 120 | 140 |
|---|---|---|---|---|---|---|---|
| $p_2 - p_1$ (psi) | 40.2 | 40.1 | 38.1 | 36.2 | 33.5 | 30.1 | 25.8 |
| Power input (hp) | 1.58 | 2.27 | 2.67 | 2.95 | 3.19 | 3.49 | 4.00 |

Based on these data, show or plot how the actual head rise, $h_a$, and the pump efficiency, $\eta$, vary with the flowrate. What is the design flowrate for this pump?

## Section 12.4.3 Net Positive Suction Head (NPSH)

**12.19** (WILEY) In Example 12.3, how will the maximum height, $z_1$, that the pump can be located above the water surface change if the water temperature is decreased to 40 °F?

**12.20** In Example 12.3, how will the maximum height, $z_1$, that the pump can be located above the water surface change if (a) the water temperature is increased to 120 °F, or (b) the fluid is changed from water to gasoline at 60 °F?

**12.21** (WILEY) A centrifugal pump with a 7-in.-diameter impeller has the performance characteristics shown in Fig. 12.12. The pump is used to pump water at 100 °F, and the pump inlet is located 12 ft above the open water surface. When the flowrate is 200 gpm, the head loss between the water surface and the pump inlet is 6 ft of water. Would you expect cavitation in the pump to be a problem? Assume standard atmospheric pressure. Explain how you arrived at your answer.

**12.22** (WILEY) Water at 40 °C is pumped from an open tank through 200 m of 50-mm-diameter smooth horizontal pipe as shown in Fig. P12.22 and discharges into the atmosphere with a velocity of 3 m/s. Minor losses are negligible. **(a)** If the efficiency of the pump is 70%, how much power is being supplied to the pump? **(b)** What is the NPSH$_A$ at the pump inlet? Neglect losses in the short section of pipe connecting the pump to the tank. Assume standard atmospheric pressure.

■ **Figure P12.22**

**12.23** A small model of a pump is tested in the laboratory and found to have a specific speed, $N_{sd}$, equal to 1000 when operating at peak efficiency. Predict the discharge of a larger, geometrically similar pump operating at peak efficiency at a speed of 1800 rpm across an actual head rise of 200 ft.

**12.24** (WILEY) The centrifugal pump shown in Fig. P12.24 is not self-priming. That is, if the water is drained from the pump and pipe as shown in Fig. P12.24(a), the pump will not draw the water into the pump and start pumping when the pump is turned on. However, if the pump is primed [i.e., filled with water as in Fig. P12.24(b)], the pump does start pumping water when turned on. Explain this behavior.

(a)        (b)

■ **Figure P12.24**

## Section 12.4.4 System Characteristics and Pump Selection

**12.25** Contrast the advantages and disadvantages of using pumps in parallel and in series.

**12.26** (WILEY) Owing to fouling of the pipe wall, the friction factor for the pipe of Example 12.4 increases from 0.02 to 0.03. Determine the new flowrate, assuming all other conditions remain the same. What is the pump efficiency at this new flowrate? Explain how a line valve could be used to vary the flowrate through the pipe of Example 12.4. Would it be better to place the valve upstream or downstream of the pump? Why?

**12.27** (WILEY) A centrifugal pump having a head-capacity relationship given by the equation $h_a = 180 - 6.10 \times 10^{-4} Q^2$, with $h_a$ in feet when $Q$ is in gpm, is to be used with a system similar to that shown in Fig. 12.14. For $z_2 - z_1 = 50$ ft, what is the expected flowrate if the total length of constant diameter pipe is 600 ft and the fluid is water? Assume the pipe diameter to be 4 in. and the friction factor to be equal to 0.02. Neglect all minor losses.

**12.28** (WILEY) A centrifugal pump having a 6-in.-diameter impeller and the characteristics shown in Fig. 12.12 is to be used to pump gasoline through 4000 ft of commercial steel 3-in.-diameter pipe. The pipe connects two reservoirs having open surfaces at the same elevation. Determine the flowrate. Do you think this pump is a good choice? Explain.

**12.29** (WILEY) Determine the new flowrate for the system described in Problem 12.28 if the pipe diameter is increased from 3 in. to 4 in. Is this pump still a good choice? Explain.

**12.30** (WILEY) A centrifugal pump having the characteristics shown in Example 12.4 is used to pump water between two large open tanks through 100 ft of 8-in.-diameter pipe. The pipeline contains four regular flanged 90° elbows, a check valve, and a fully open globe valve. Other minor losses are negligible. Assume the friction factor $f = 0.02$ for the 100-ft section of pipe. If the static head (difference in height of fluid surfaces in the two tanks) is 30 ft, what is the expected flowrate? Do you think this pump is a good choice? Explain.

**12.31** (WILEY) In a chemical processing plant a liquid is pumped from an open tank, through a 0.1-m-diameter vertical pipe, and into another open tank as shown in Fig. P12.31(a). A valve is located in the pipe, and the minor loss coefficient for the valve as a function of the valve setting is shown in Fig. P12.31(b). The pump head-capacity relationship is given by the equation $h_a = 52.0 - 1.01 \times 10^3 Q^2$ with $h_a$ in meters when $Q$ is in m³/s. Assume the friction factor $f = 0.02$ for the pipe, and all minor losses, except for the valve, are negligible. The fluid levels in the two tanks can be assumed to remain constant. **(a)** Determine the flowrate with the valve wide open. **(b)** Determine the required valve setting (percent open) to reduce the flowrate by 50%.

(a)

(b)

■ **Figure P12.31**

†**12.32** Water is pumped between the two tanks described in Example 12.4 once a day, 365 days a year, with each pumping period lasting two hours. The water levels in the two tanks remain essentially constant. Estimate the annual cost of the electrical power needed to operate the pump if it were located in your city. You will have to make a reasonable estimate for the efficiency of the motor used to drive the pump. Due to aging, it can be expected that the overall resistance of the system will increase with time. If the operating point shown in Fig. E12.4c changes to a point where the flowrate has been reduced to 1000 gpm, what will be the new annual cost of operating the pump? Assume that the cost of electrical power remains the same.

## Section 12.5 Dimensionless Parameters and Similarity Laws

**12.33** What is the rationale for operating two geometrically similar pumps differing in feature size at the same flow coefficient?

**12.34** (WILEY) A centrifugal pump having an impeller diameter of 1 m is to be constructed so that it will supply a head rise of 200 m at a flowrate of 4.1 $m^3$/s of water when operating at a speed of 1200 rpm. To study the characteristics of this pump, a 1/5 scale, geometrically similar model operated at the same speed is to be tested in the laboratory. Determine the required model discharge and head rise. Assume that both model and prototype operate with the same efficiency (and therefore the same flow coefficient).

**12.35** (WILEY) A centrifugal pump with a 12-in.-diameter impeller requires a power input of 60 hp when the flowrate is 3200 gpm against a 60-ft head. The impeller is changed to one with a 10-in. diameter. Determine the expected flowrate, head, and input power if the pump speed remains the same.

**12.36** Do the head–flowrate data shown in Fig. 12.12 appear to follow the similarity laws as expressed by Eqs. 12.39 and 12.40? Explain.

**12.37** (WILEY) A centrifugal pump has the performance characteristics of the pump with the 6-in.-diameter impeller described in Fig. 12.12. Note that the pump in this figure is operating at 3500 rpm. What is the expected head gained if the speed of this pump is reduced to 2800 rpm while operating at peak efficiency?

**12.38** (WILEY) A centrifugal pump provides a flowrate of 500 gpm when operating at 1750 rpm against a 200-ft head. Determine the pump's flowrate and developed head if the pump speed is increased to 3500 rpm.

**12.39** (WILEY) Use the data given in Problem 12.18 and plot the dimensionless coefficients $C_H$, $C_{\mathscr{P}}$, $\eta$ versus $C_Q$ for this pump. Calculate a meaningful value of specific speed, discuss its usefulness, and compare the result with data of Fig. 12.18.

**12.40** (WILEY) In a certain application, a pump is required to deliver 5000 gpm against a 300-ft head when operating at 1200 rpm. What type of pump would you recommend?

## Section 12.6 Axial-Flow and Mixed-Flow Pumps

**12.41** Explain how a marine propeller and an axial-flow pump are similar in the main effect they produce.

**12.42** (WILEY) A certain axial-flow pump has a specific speed of $N_s$ = 5.0. If the pump is expected to deliver 3000 gpm when operating against a 15-ft head, at what speed (rpm) should the pump be run?

**12.43** (WILEY) A certain pump is known to have a capacity of 3 $m^3$/s when operating at a speed of 60 rad/s against a head of 20 m. Based on the information in Fig. 12.18, would you recommend a radial-flow, mixed-flow, or axial-flow pump?

**12.44** (WILEY) Fuel oil (sp. wt = 48.0 lb/$ft^3$, viscosity = 2.0 × $10^{-5}$ lb · s/$ft^2$) is pumped through the piping system of Fig. P12.44 with a velocity of 4.6 ft/s. The pressure 200 ft upstream from the pump is 5 psi. Pipe losses downstream from the pump are negligible, but minor losses are not (minor loss coefficients are given on the figure). **(a)** For a pipe diameter of 2 in. with a relative roughness $\varepsilon/D$ = 0.001, determine the head that must be added by the pump. **(b)** For a pump operating speed of 1750 rpm, what type of pump (radial-flow, mixed-flow, or axial-flow) would you recommend for this application?

■ **Figure P12.44**

**12.45** (WILEY) The axial-flow pump shown in Fig. 12.19 is designed to move 5000 gal/min of water over a head rise of 5 ft of water. Estimate the motor power requirement and the $U_2V_{\theta 2}$ needed to achieve this flowrate on a continuous basis. Comment on any cautions associated with where the pump is placed vertically in the pipe.

## Section 12.7 Fans

**12.46** (See Fluids in the News Article titled "Hi-tech Ceiling Fans," Section 12.7.) Explain why reversing the direction of rotation of a ceiling fan results in airflow in the opposite direction.

**12.47** For the fan of both Examples 5.19 and 5.28 discuss what fluid flow properties you would need to measure to estimate fan efficiency.

## Section 12.8 Turbines (also see Sec. 12.3)

**12.48** An inward-flow radial turbine (see Fig. P12.48) involves a nozzle angle, $\alpha_1$, of 60° and an inlet rotor tip speed, $U_1$, of 3 m/s.

■ **Figure P12.48**

The ratio of rotor inlet to outlet diameters is 2.0. The absolute velocity leaving the rotor at section (2) is radial with a magnitude of 6 m/s. Determine the energy transfer per unit mass of fluid flowing through this turbine if the fluid is **(a)** air, **(b)** water.

**12.49** WILEY GO A simplified sketch of a hydraulic turbine runner is shown in Fig. P12.49. Relative to the rotating runner, water enters at section (1) (cylindrical cross section area $A_1$ at $r_1 = 1.5$ m) at an angle of 100° from the tangential direction and leaves at section (2) (cylindrical cross section area $A_2$ at $r_2 = 0.85$ m) at an angle of 50° from the tangential direction. The blade height at sections (1) and (2) is 0.45 m, and the volume flowrate through the turbine is 30 m³/s. The runner speed is 130 rpm in the direction shown. Determine the shaft power developed. Is the shaft power greater or less than the power lost by the fluid? Explain.

■ **Figure P12.49**

**12.50** WILEY A water turbine wheel rotates at the rate of 100 rpm in the direction shown in Fig. P12.50. The inner radius, $r_2$, of the blade row is 1 ft, and the outer radius, $r_1$, is 2 ft. The absolute velocity vector at the turbine rotor entrance makes an angle of 20° with the tangential direction. The inlet blade angle is 60° relative to the tangential direction. The blade outlet angle is 120°. The flowrate is 10 ft³/s. For the flow tangent to the rotor blade surface at inlet and outlet, determine an appropriate constant blade height, $b$, and the corresponding power available at the rotor shaft. Is the shaft power greater or less than the power lost by the fluid? Explain.

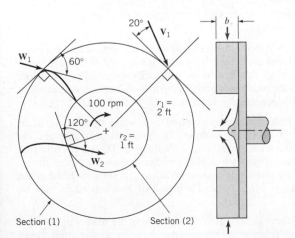

■ **Figure P12.50**

**12.51** A sketch of the arithmetic mean radius blade sections of an axial-flow water turbine stage is shown in Fig. P12.51. The rotor speed is 1500 rpm. **(a)** Sketch and label velocity triangles for the flow entering and leaving the rotor row. Use **V** for absolute velocity, **W** for relative velocity, and **U** for blade velocity. Assume flow enters and leaves each blade row at the blade angles shown. **(b)** Calculate the work per unit mass delivered at the shaft.

■ **Figure P12.51**

**12.52** WILEY An inward flow radial turbine (see Fig. P12.52) involves a nozzle angle, $\alpha_1$, of 60° and an inlet rotor tip speed, $U_1$, of 9 m/s. The ratio of rotor inlet to outlet diameters is 2.0. The radial component of velocity remains constant at 6 m/s through the rotor, and the flow leaving the rotor at section (2) is without angular momentum. If the flowing fluid is water and the stagnation pressure drop across the rotor is 110 kPa, determine the loss of available energy across the rotor and the efficiency involved.

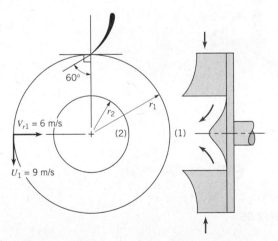

■ **Figure P12.52**

**12.53** ▶ Consider the Pelton wheel turbine illustrated in Figs. 12.24, 12.25, 12.26, and 12.27. This kind of turbine is used to drive the oscillating sprinkler shown in **Video V12.4.** Explain how this kind of sprinkler is started, and subsequently operated at constant oscillating speed. What is the physical significance of the zero torque condition with the Pelton wheel rotating?

**12.54** ▶ ⟨WILEY⟩ A small Pelton wheel is used to power an oscillating lawn sprinkler as shown in **Video V12.4** and Fig. P12.54. The arithmetic mean radius of the turbine is 1 in., and the exit angle of the blade is 135° relative to the blade motion. Water is supplied through a single 0.20-in.-diameter nozzle at a speed of 50 ft/s. Determine the flowrate, the maximum torque developed, and the maximum power developed by this turbine.

■ **Figure P12.54**

**12.55** A Pelton wheel turbine is illustrated in Fig. P12.55. The radius to the line of action of the tangential reaction force on each vane is 1 ft. Each vane deflects fluid by an angle of 135° as indicated. Assume all of the flow occurs in a horizontal plane. Each of the four jets shown strikes a vane with a velocity of 100 ft/s and a stream diameter of 1 in. The magnitude of velocity of the jet remains constant along the vane surface.

**(a)** How much torque is required to hold the wheel stationary?

**(b)** How fast will the wheel rotate if shaft torque is negligible, and what practical situation is simulated by this condition?

■ **Figure P12.55**

**12.56** ⟨WILEY⟩ The single-stage, axial-flow turbomachine shown in Fig. P12.56 involves water flow at a volumetric flowrate of 9 m³/s. The rotor revolves at 600 rpm. The inner and outer radii of the annular flow path through the stage are 0.46 and 0.61 m, and $\beta_2 = 60°$. The flow entering the rotor row and leaving the stator row is axial when viewed from the stationary casing. Is this device a turbine or a pump? Estimate the amount of power transferred to or from the fluid.

■ **Figure P12.56**

**12.57** ▶ ⟨WILEY⟩ For an air turbine of a dentist's drill like the one shown in Fig. E12.8 and **Video V12.5,** calculate the average blade speed associated with a rotational speed of 350,000 rpm. Estimate the air pressure needed to run this turbine.

**12.58** ⟨WILEY⟩ Water for a Pelton wheel turbine flows from the headwater and through the penstock as shown in Fig. P12.58. The effective friction factor for the penstock, control valves, and the like is 0.032, and the diameter of the jet is 0.20 m. Determine the maximum power output.

■ **Figure P12.58**

**12.59** A 10.9-ft-diameter Pelton wheel operates at 500 rpm with a total head just upstream of the nozzle of 5330 ft. Estimate the diameter of the nozzle of the single-nozzle wheel if it develops 25,000 horsepower.

**12.60** A Pelton wheel has a diameter of 2 m and develops 500 kW when rotating 180 rpm. What is the average force of the water against the blades? If the turbine is operating at maximum efficiency, determine the speed of the water jet from the nozzle and the mass flowrate.

**12.61** ⟨WILEY⟩ Water to run a Pelton wheel is supplied by a penstock of length $\ell$ and diameter $D$ with a friction factor $f$. If the only losses associated with the flow in the penstock are due to pipe friction, show that the maximum power output of the turbine occurs when the nozzle diameter, $D_1$, is given by $D_1 = D/(2f\,\ell/D)^{1/4}$.

**12.62** A Pelton wheel is supplied with water from a lake at an elevation $H$ above the turbine. The penstock that supplies the water to the wheel is of length $\ell$, diameter $D$, and friction factor $f$. Minor losses are negligible. Show that the power developed by the turbine is maximum when the velocity head at the nozzle exit is $2H/3$. *Note:* The result of Problem 12.61 may be of use.

**12.63** If there is negligible friction along the blades of a Pelton wheel, the relative speed remains constant as the fluid flows across the blades, and the maximum power output occurs when the blade speed is one-half the jet speed (see Eq.12.52). Consider the case where friction is not negligible and the relative speed leaving the blade is some fraction, c, of the relative speed entering the blade. That is, $W_2 = cW_1$. Show that Eq. 12.52 is valid for this case also.

**12.64** A 1-m-diameter Pelton wheel rotates at 300 rpm. Which of the following heads (in meters) would be best suited for this turbine: (**a**) 2, (**b**) 5, (**c**) 40, (**d**) 70, or (**e**) 140? Explain.

**12.65** ⟨WILEY⟩ A hydraulic turbine operating at 180 rpm with a head of 100 ft develops 20,000 horsepower. Estimate the power if the same turbine were to operate under a head of 50 ft.

**12.66** Draft tubes as shown in Fig. P12.66 are often installed at the exit of Kaplan and Francis turbines. Explain why such draft tubes are advantageous.

■ **Figure P12.66**

**12.67** ⟨WILEY⟩ Turbines are to be designed to develop 30,000 horsepower while operating under a head of 70 ft and an angular velocity of 60 rpm. What type of turbine is best suited for this purpose? Estimate the flowrate needed.

**12.68** Water at 400 psi is available to operate a turbine at 1750 rpm. What type of turbine would you suggest to use if the turbine should have an output of approximately 200 hp?

**12.69** It is desired to produce 50,000 hp with a head of 50 ft and an angular velocity of 100 rpm. How many turbines would be needed if the specific speed is to be (**a**) 50, (**b**) 100?

**12.70** ▶ ⟨WILEY⟩ Show how you would estimate the relationship between feature size and power production for a wind turbine like the one shown in **Video V12.1.**

**12.71** ⟨WILEY⟩ Test data for the small Francis turbine shown in Fig. P12.71 is given in the following table. The test was run at a constant 32.8-ft head just upstream of the turbine. The Prony brake on the turbine output shaft was adjusted to give various angular velocities, and the force on the brake arm, F, was recorded. Use the given data to plot curves of torque as a function of angular velocity and turbine efficiency as a function of angular velocity.

| $\omega$ (rpm) | $Q$ (ft³/s) | $F$ (lb) |
|---|---|---|
| 0 | 0.129 | 2.63 |
| 1000 | 0.129 | 2.40 |
| 1500 | 0.129 | 2.22 |
| 1870 | 0.124 | 1.91 |
| 2170 | 0.118 | 1.49 |
| 2350 | 0.0942 | 0.876 |
| 2580 | 0.0766 | 0.337 |
| 2710 | 0.068 | 0.089 |

■ **Figure P12.71**

†**12.72** ▶ It is possible to generate power by using the water from your garden hose to drive a small Pelton wheel turbine (see **Video V12.4**). Provide a preliminary design of such a turbine and estimate the power output expected. List all assumptions and show calculations.

**12.73** The device shown in Fig. P12.73 is used to investigate the power produced by a Pelton wheel turbine. Water supplied at a constant flowrate issues from a nozzle and strikes the turbine buckets as indicated. The angular velocity, $\omega$, of the turbine wheel is varied by adjusting the tension on the Prony brake spring, thereby varying the torque, $T_{shaft}$, applied to the output shaft. This torque can be determined from the measured force, $R$, needed to keep the brake arm stationary as $T_{shaft} = F\ell$, where $\ell$ is the moment arm of the brake force.

■ **Figure P12.73**

Experimentally determined values of $\omega$ and $R$ are shown in the following table. Use these results to plot a graph of torque as a function of the angular velocity. On another graph plot the power output, $W_{shaft} = T_{shaft}\,\omega$, as a function of the angular velocity. On each of these graphs plot the theoretical curves for this turbine, assuming 100% efficiency.

Compare the experimental and theoretical results and discuss some possible reasons for any differences between them.

| $\omega$ (rpm) | $R$ (lb) |
|----------------|----------|
| 0    | 2.47 |
| 360  | 1.91 |
| 450  | 1.84 |
| 600  | 1.69 |
| 700  | 1.55 |
| 940  | 1.17 |
| 1120 | 0.89 |
| 1480 | 0.16 |

## Section 12.9 Compressible Flow Turbomachines

**12.74** Obtain photographs/images of a variety of turbo-compressor rotors and categorize them as axial-flow or radial-flow compressors. Explain briefly how they are used. Note any unusual features.

**12.75** Obtain photographs/images of a variety of compressible flow turbines and categorize them as axial-flow or radial-flow turbines. Explain briefly how they are used. Note any unusual features.

## ■ Lifelong Learning Problems

**12.1LL** What do you think are the major unresolved fluid dynamics problems associated with gas turbine engine compressors? For gas turbine engine high-pressure and low-pressure turbines? For gas turbine engine fans?

**12.2LL** Outline the steps associated with the preliminary design of a turbomachine rotor.

**12.3LL** What are current efficiencies achieved by the following categories of turbomachines? **(a)** Wind turbines; **(b)** hydraulic turbines; **(c)** power plant steam turbines; **(d)** aircraft gas turbine engines; **(e)** natural gas pipeline compressors; **(f)** home vacuum cleaner blowers; **(g)** laptop computer cooling fan; **(h)** irrigation pumps; **(i)** dentist drill air turbines. What is being done to improve these devices?

**12.4LL** (See Fluids in the News Article titled "Cavitation Damage in Hydraulic Turbines," Section 12.8.2.) How are cavitation and, more importantly, the damage it can cause detected in hydraulic turbines? How can this damage be minimized?

# A Computational Fluid Dynamics

## A.1    Introduction

Numerical methods using digital computers are, of course, commonly utilized to solve a wide variety of flow problems. As discussed in Chapter 6, although the differential equations that govern the flow of Newtonian fluids [the Navier–Stokes equations (Eq. 6.127)] were derived many years ago, there are few known analytical solutions to them. However, with the advent of high-speed digital computers it has become possible to obtain approximate numerical solutions to these (and other fluid mechanics) equations for a wide variety of circumstances.

*Computational fluid dynamics* (CFD) involves replacing the partial differential equations with discretized algebraic equations that approximate the partial differential equations. These equations are then numerically solved to obtain flow field values at the discrete points in space and/or time. Since the Navier–Stokes equations are valid everywhere in the flow field of the fluid continuum, an analytical solution to these equations provides the solution for an infinite number of points in the flow. However, analytical solutions are available for only a limited number of simplified flow geometries. To overcome this limitation, the governing equations can be discretized and put in algebraic form for the computer to solve. The CFD simulation solves for the relevant flow variables only at the discrete points, which make up the grid or mesh of the solution (discussed in more detail below). Interpolation schemes are used to obtain values at non-grid point locations.

CFD can be thought of as a numerical experiment. In a typical fluids experiment, an experimental model is built, measurements of the flow interacting with that model are taken, and the results are analyzed. In CFD, the building of the model is replaced with the formulation of the governing equations and the development of the numerical algorithm. The process of obtaining measurements is replaced with running an algorithm on the computer to simulate the flow interaction. Of course, the analysis of results is common ground to both techniques.

CFD can be classified as a subdiscipline to the study of fluid dynamics. However, it should be pointed out that a thorough coverage of CFD topics is well beyond the scope of this textbook. This appendix highlights some of the more important topics in CFD, but is only intended as a brief introduction. The topics include discretization of the governing equations, grid generation, boundary conditions, application of CFD, and some representative examples.

VA.1 Pouring a liquid

## A.2    Discretization

The process of *discretization* involves developing a set of algebraic equations (based on discrete points in the flow domain) to be used in place of the partial differential equations. Of the various discretization techniques available for the numerical solution of the governing differential equations, the following three types are most common: (1) the finite difference method, (2) the finite element (or finite volume) method, and (3) the boundary element method. In each of these methods, the continuous flow field (i.e., velocity or pressure as a function of space and time) is described in terms of discrete (rather than continuous) values at prescribed locations. Through this technique the differential equations are replaced by a set of algebraic equations that can be solved on the computer.

For the *finite element* (or *finite volume*) method, the flow field is broken into a set of small fluid elements (usually triangular areas if the flow is two-dimensional, or small volume elements if the flow is three-dimensional). The conservation equations (i.e., conservation of mass, momentum, and energy) are written in an appropriate form for each element, and the set of resulting

**■ Figure A.1**   **Panel method for flow past an airfoil.**

algebraic equations for the flow field is solved numerically. The number, size, and shape of elements are dictated in part by the particular flow geometry and flow conditions for the problem at hand. As the number of elements increases (as is necessary for flows with complex boundaries), the number of simultaneous algebraic equations that must be solved increases rapidly. Problems involving one million to ten million (or more) grid cells are not uncommon in today's CFD community, particularly for complex three-dimensional geometries. Further information about this method can be found in Refs. 1 and 2.

For the *boundary element method*, the boundary of the flow field (not the entire flow field as in the finite element method) is broken into discrete segments (Ref. 3) and appropriate singularities such as sources, sinks, doublets, and vortices are distributed on these boundary elements. The strengths and type of the singularities are chosen so that the appropriate boundary conditions of the flow are obtained on the boundary elements. For points in the flow field not on the boundary, the flow is calculated by adding the contributions from the various singularities on the boundary. Although the details of this method are rather mathematically sophisticated, it may (depending on the particular problem) require less computational time and space than the finite element method. Typical boundary elements and their associated singularities (vortices) for two-dimensional flow past an airfoil are shown in Fig. A.1. Such use of the boundary element method in aerodynamics is often termed the *panel method* in recognition of the fact that each element plays the role of a panel on the airfoil surface (Ref. 4).

The *finite difference method* for computational fluid dynamics is perhaps the most easily understood of the three methods listed above. For this method the flow field is dissected into a set of grid points and the continuous functions (velocity, pressure, etc.) are approximated by discrete values of these functions calculated at the grid points. Derivatives of the functions are approximated by using the differences between the function values at local grid points divided by the grid spacing. The standard method for converting the partial differential equations to algebraic equations is through the use of Taylor series expansions. (See Ref. 5.) For example, assume a standard rectangular grid is applied to a flow domain as shown in Fig. A.2.

This grid stencil shows five grid points in $x$–$y$ space with the center point being labeled as $i, j$. This index notation is used as subscripts on variables to signify location. For example, $u_{i+1, j}$ is the $u$ component of velocity at the first point to the right of the center point $i, j$. The grid spacing in the $i$ and $j$ directions is given as $\Delta x$ and $\Delta y$, respectively.

To find an algebraic approximation to a first derivative term such as $\partial u/\partial x$ at the $i, j$ grid point, consider a Taylor series expansion written for $u$ at $i + 1$ as

$$u_{i+1, j} = u_{i, j} + \left(\frac{\partial u}{\partial x}\right)_{i, j} \frac{\Delta x}{1!} + \left(\frac{\partial^2 u}{\partial x^2}\right)_{i, j} \frac{(\Delta x)^2}{2!} + \left(\frac{\partial^3 u}{\partial x^3}\right)_{i, j} \frac{(\Delta x)^3}{3!} + \cdots \tag{A.1}$$

**■ Figure A.2**   **Standard rectangular grid.**

Solving for the underlined term in the above equation results in the following:

$$\left(\frac{\partial u}{\partial x}\right)_{i,j} = \frac{u_{i+1,j} - u_{i,j}}{\Delta x} + O(\Delta x) \tag{A.2}$$

where $O(\Delta x)$ contains higher order terms proportional to $\Delta x$, $(\Delta x)^2$, and so forth. Equation A.2 represents a forward difference equation to approximate the first derivative using values at $i + 1, j$ and $i, j$ along with the grid spacing in the $x$ direction. Obviously in solving for the $\partial u/\partial x$ term we have ignored higher order terms such as the second and third derivatives present in Eq. A.1. This process is termed *truncation* of the Taylor series expansion. The lowest order term that was truncated included $(\Delta x)^2$. Notice that the first derivative term contains $\Delta x$. When solving for the first derivative, all terms on the right-hand side were divided by $\Delta x$. Therefore, the term $O(\Delta x)$ signifies that this equation has error of "order $(\Delta x)$," which is due to the neglected terms in the Taylor series and is called truncation error. Hence, the forward difference is termed first-order accurate.

Thus, we can transform a partial derivative into an algebraic expression involving values of the variable at neighboring grid points. This method of using the Taylor series expansions to obtain discrete algebraic equations is called the finite difference method. Similar procedures can be used to develop approximations termed backward difference and central difference representations of the first derivative. The central difference makes use of both the left and right points (i.e., $i - 1, j$ and $i + 1, j$) and is second-order accurate. In addition, finite difference equations can be developed for the other spatial directions (i.e., $\partial u/\partial y$) as well as for second derivatives ($\partial^2 u/\partial x^2$), which are also contained in the Navier–Stokes equations (see Ref. 5 for details).

Applying this method to all terms in the governing equations transfers the differential equations into a set of algebraic equations involving the physical variables at the grid points (i.e., $u_{i,j}, p_{i,j}$ for $i = 1, 2, 3, \ldots$ and $j = 1, 2, 3, \ldots$, etc.). This set of equations is then solved by appropriate numerical techniques. The larger the number of grid points used, the larger the number of equations that must be solved.

A student of CFD should realize that the discretization of the continuum governing equations involves the use of algebraic equations that are an approximation to the original partial differential equation. Along with this approximation comes some amount of error. This type of error is termed truncation error because the Taylor series expansion used to represent a derivative is "truncated" at some reasonable point and the higher order terms are ignored. The truncation errors tend to zero as the grid is refined by making $\Delta x$ and $\Delta y$ smaller, so grid refinement is one method of reducing this type of error. Another type of unavoidable numerical error is the so-called round-off error. This type of error is due to the limit of the computer on the number of digits it can retain in memory. Engineering students can run into round-off errors from their calculators if they plug values into the equations at an early stage of the solution process. Fortunately, for most CFD cases, if the algorithm is set up properly, round-off errors are usually negligible.

# A.3    Grids

CFD computations using the finite difference method provide the flow field at discrete points in the flow domain. The arrangement of these discrete points is termed the *grid* or the *mesh*. The type of grid developed for a given problem can have a significant impact on the numerical simulation, including the accuracy of the solution. The grid must represent the geometry correctly and accurately, since an error in this representation can have a significant effect on the solution.

The grid must also have sufficient grid resolution to capture the relevant flow physics, otherwise they will be lost. This particular requirement is problem dependent. For example, if a flow field has small-scale structures, the grid resolution must be sufficient to capture these structures. It is usually necessary to increase the number of grid points (i.e., use a finer mesh) where large gradients are to be expected, such as in the boundary layer near a solid surface. The same can also be

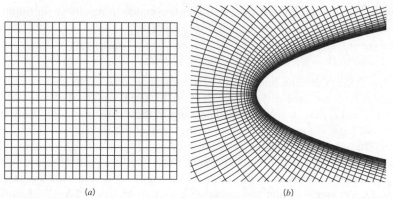

■ **Figure A.3**   Structured grids. (*a*) Rectangular grid. (*b*) Grid around a parabolic surface.

*VA.2 Dynamic grid*

said for the temporal resolution. The time step, $\Delta t$, used for unsteady flows must be smaller than the smallest time scale of the flow features being investigated.

Generally, the types of grids fall into two categories: structured and unstructured, depending on whether or not there exists a systematic pattern of connectivity of the grid points with their neighbors. As the name implies, a *structured grid* has some type of regular, coherent structure to the mesh layout that can be defined mathematically. The simplest structured grid is a uniform rectangular grid, as shown in Fig. A.3*a*. However, structured grids are not restricted to rectangular geometries. Figure A.3*b* shows a structured grid wrapped around a parabolic surface. Notice that grid points are clustered near the surface (i.e., grid spacing in normal direction increases as one moves away from the surface) to help capture the steep flow gradients found in the boundary layer region. This type of variable grid spacing is used wherever there is a need to increase grid resolution and is termed grid stretching.

For the *unstructured grid*, the grid cell arrangement is irregular and has no systematic pattern. The grid cell geometry usually consists of various-sized triangles for two-dimensional problems and tetrahedrals for three-dimensional grids. An example of an unstructured grid is shown in Fig. A.4. Unlike structured grids, for an unstructured grid each grid cell and the connection information to neighboring cells is defined separately. This produces an increase in the computer code complexity as well as a significant computer storage requirement. The advantage to an unstructured grid is that it can be applied to complex geometries, where structured grids would have severe difficulty. The finite difference method is usually restricted to structured grids whereas the finite volume (or finite element) method can use either structured or unstructured grids.

Other grids include hybrid, moving, and adaptive grids. A grid that uses a combination of grid elements (rectangles, triangles, etc.) is termed a *hybrid grid*. As the name implies, the *moving grid*

■ **Figure A.4**   Anisotropic adaptive mesh for flow induced by the rotor of a helicopter in hover, with ground effect. Left: flow; Right: grid. (From Newmerical Technologies International, Montreal, Canada. Used by permission.)

is helpful for flows involving a time-dependent geometry. If, for example, the problem involves simulating the flow within a pumping heart or the flow around a flapping wing, a mesh that moves with the geometry is desired. The nature of the adaptive grid lies in its ability to literally adapt itself during the simulation. For this type of grid, while the CFD code is trying to reach a converged solution, the grid will adapt itself to place additional grid resources in regions of high-flow gradients. Such a grid is particularly useful when a new problem arises and the user is not quite sure where to refine the grid due to high-flow gradients.

## A.4  Boundary Conditions

The same governing equations, the Navier–Stokes equations (Eq. 6.127), are valid for all incompressible Newtonian fluid flow problems. Thus, if the same equations are solved for all types of problems, how is it possible to achieve different solutions for different types of flows involving different flow geometries? The answer lies in the *boundary conditions* of the problem. The boundary conditions are what allow the governing equations to differentiate between different flow fields (for example, flow past an automobile and flow past a person running) and produce a solution unique to the given flow geometry.

It is critical to specify the correct boundary conditions so that the CFD simulation is a well-posed problem and is an accurate representation of the physical problem. Poorly defined boundary conditions can ultimately affect the accuracy of the solution. One of the most common boundary conditions used for simulation of viscous flow is the no-slip condition, as discussed in Section 1.6. Thus, for example, for two-dimensional external or internal flows, the $x$ and $y$ components of velocity ($u$ and $v$) are set to zero at the stationary wall to satisfy the no-slip condition. Other boundary conditions that must be appropriately specified involve inlets, outlets, far-field, wall gradients, etc. It is important to not only select the correct physical boundary condition for the problem, but also to correctly implement this boundary condition into the numerical simulation.

## A.5  Basic Representative Examples

A very simple one-dimensional example of the finite difference technique is presented in the following example.

---

**EXAMPLE A.1**  **Flow from a Tank**

A viscous oil flows from a large, open tank and through a long, small-diameter pipe as shown in Fig. EA.1a. At time $t = 0$ the fluid depth is $H$. Use a finite difference technique to determine the liquid depth as a function of time, $h = h(t)$. Compare this result with the exact solution of the governing equation.

**SOLUTION**

Although this is an unsteady flow (i.e., the deeper the oil, the faster it flows from the tank) we assume that the flow is "quasisteady" and apply steady flow equations as follows.

As shown by Eq. 6.152, the mean velocity, $V$, for steady laminar flow in a round pipe of diameter $D$ is given by

$$V = \frac{D^2 \Delta p}{32 \mu \ell} \quad (1)$$

where $\Delta p$ is the pressure drop over the length $\ell$. For this problem the pressure at the bottom of the tank (the inlet of the pipe) is $\gamma h$ and that at the pipe exit is zero. Hence, $\Delta p = \gamma h$ and Eq. 1 becomes

$$V = \frac{D^2 \gamma h}{32 \mu \ell} \quad (2)$$

Conservation of mass requires that the flowrate from the tank, $Q = \pi D^2 V/4$, is related to the rate of change of depth of oil in the tank, $dh/dt$, by

$$Q = -\frac{\pi}{4} D_T^2 \frac{dh}{dt}$$

where $D_T$ is the tank diameter. Thus,

$$\frac{\pi}{4} D^2 V = -\frac{\pi}{4} D_T^2 \frac{dh}{dt}$$

or

$$V = -\left(\frac{D_T}{D}\right)^2 \frac{dh}{dt} \quad (3)$$

**Figure EA.1**

By combining Eqs. 2 and 3 we obtain

$$\frac{D^2 \gamma h}{32 \mu \ell} = -\left(\frac{D_T}{D}\right)^2 \frac{dh}{dt}$$

or

$$\frac{dh}{dt} = -Ch$$

where $C = \gamma D^4 / 32 \mu \ell D_T^2$ is a constant. For simplicity we assume the conditions are such that $C = 1$. Thus, we must solve

$$\frac{dh}{dt} = -h \quad \text{with} \quad h = H \text{ at } t = 0 \tag{4}$$

The exact solution to Eq. 4 is obtained by separating the variables and integrating to obtain

$$h = He^{-t} \tag{5}$$

However, assume this solution was not known. The following finite difference technique can be used to obtain an approximate solution.

As shown in Fig. EA.1b, we select discrete points (nodes or grid points) in time and approximate the time derivative of $h$ by the expression

$$\left. \frac{dh}{dt} \right|_{t=t_i} \approx \frac{h_i - h_{i-1}}{\Delta t} \tag{6}$$

where $\Delta t$ is the time step between the different node points on the time axis and $h_i$ and $h_{i-1}$ are the approximate values of $h$ at nodes $i$ and $i - 1$. Equation 6 is called the backward-difference approximation to $dh/dt$. We are free to select whatever value of $\Delta t$ that we wish. (Although we do not need to space the nodes at equal distances, it is often convenient to do so.) Since the governing equation (Eq. 4) is an ordinary differential equation, the "grid" for the finite difference method is a one-dimensional grid as shown in Fig. EA.1b rather than a two-dimensional grid (which occurs for partial differential equations) as shown in Fig. EA.2b, or a three-dimensional grid.

Thus, for each value of $i = 2, 3, 4, \ldots$ we can approximate the governing equation, Eq. 4, as

$$\frac{h_i - h_{i-1}}{\Delta t} = -h_i$$

or

$$h_i = \frac{h_{i-1}}{(1 + \Delta t)} \tag{7}$$

We cannot use Eq. 7 for $i = 1$ since it would involve the non-existing $h_0$. Rather we use the initial condition (Eq. 4), which gives

$$h_1 = H$$

The result is the following set of $N$ algebraic equations for the $N$ approximate values of $h$ at times $t_1 = 0, t_2 = \Delta t, \ldots, t_N = (N - 1)\Delta t$.

$$h_1 = H$$
$$h_2 = h_1/(1 + \Delta t)$$
$$h_3 = h_2/(1 + \Delta t)$$
$$\vdots \qquad \vdots$$
$$h_N = h_{N-1}/(1 + \Delta t)$$

For most problems the corresponding equations would be more complicated than those just given, and a computer would be used to solve for the $h_i$. For this problem the solution is simply

$$h_2 = H/(1 + \Delta t)$$
$$h_3 = H/(1 + \Delta t)^2$$
$$\vdots \qquad \vdots$$

or in general

$$h_i = H/(1 + \Delta t)^{i-1}$$

The results for $0 < t < 1$ are shown in Fig. EA.1c. Tabulated values of the depth for $t = 1$ are listed in the table below.

| $\Delta t$ | $i$ for $t = 1$ | $h_i$ for $t = 1$ |
|---|---|---|
| 0.2 | 6 | 0.4019H |
| 0.1 | 11 | 0.3855H |
| 0.01 | 101 | 0.3697H |
| 0.001 | 1001 | 0.3681H |
| Exact (Eq. 5) | — | 0.3678H |

It is seen that the approximate results compare quite favorably with the exact solution given by Eq. 5. It is expected that the finite difference results would more closely approximate the exact results as $\Delta t$ is decreased since in the limit of $\Delta t \to 0$ the finite difference approximation for the derivatives (Eq. 6) approaches the actual definition of the derivative.

For most CFD problems the governing equations to be solved are partial differential equations [rather than an ordinary differential equation as in the above example (Eq. A.1)] and the finite difference method becomes considerably more involved. The following example illustrates some of the concepts involved.

# EXAMPLE A.2    Flow Past a Cylinder

Consider steady, incompressible flow of an inviscid fluid past a circular cylinder as shown in Fig. EA.2a. The stream function, $\psi$, for this flow is governed by the Laplace equation (see Section 6.5)

$$\frac{\partial^2 \psi}{\partial x^2} + \frac{\partial^2 \psi}{\partial y^2} = 0 \tag{1}$$

The exact analytical solution is given in Section 6.6.3.

Describe a simple finite difference technique that can be used to solve this problem.

## SOLUTION

The first step is to define a flow domain and set up an appropriate grid for the finite difference scheme. Since we expect the flow field to be symmetrical both above and below and in front of and behind the cylinder, we consider only one-quarter of the entire flow domain as indicated in Fig. EA.2b. We locate the upper boundary and right-hand boundary far enough from the cylinder so that we expect the flow to be essentially uniform at these locations. It is not always clear how far from the object these boundaries must be located. If they are not far enough, the solution obtained will be incorrect because we have imposed

artificial, uniform flow conditions at a location where the actual flow is not uniform. If these boundaries are farther than necessary from the object, the flow domain will be larger than necessary and excessive computer time and storage will be required. Experience in solving such problems is invaluable!

Once the flow domain has been selected, an appropriate grid is imposed on this domain (see Fig. EA.2b). Various grid structures can be used. If the grid is too coarse, the numerical solution may not be capable of capturing the fine scale structure of the actual flow field. If the grid is too fine, excessive computer time and

(c)          **Figure EA.2**

storage may be required. Considerable work has gone into forming appropriate grids (Ref. 6). We consider a grid that is uniformly spaced in the $x$ and $y$ directions, as shown in Fig. EA.2$b$.

As shown in Eq. 6.112, the exact solution to Eq. 1 (in terms of polar coordinates $r$, $\theta$ rather than Cartesian coordinates $x$, $y$) is $\psi = Ur(1 - a^2/r^2) \sin \theta$. The finite difference solution approximates these stream function values at a discrete (finite) number of locations (the grid points) as $\psi_{i,j}$, where the $i$ and $j$ indices refer to the corresponding $x_i$ and $y_j$ locations.

The derivatives of $\psi$ can be approximated as follows:

$$\frac{\partial \psi}{\partial x} \approx \frac{1}{\Delta x}(\psi_{i+1,j} - \psi_{i,j})$$

and

$$\frac{\partial \psi}{\partial y} \approx \frac{1}{\Delta y}(\psi_{i,j+1} - \psi_{i,j})$$

This particular approximation is called a forward-difference approximation. Other approximations are possible. By similar reasoning, it is possible to show that the second derivatives of $\psi$ can be written as follows:

$$\frac{\partial^2 \psi}{\partial x^2} \approx \frac{1}{(\Delta x)^2}(\psi_{i+1,j} - 2\psi_{i,j} + \psi_{i-1,j}) \quad \textbf{(2)}$$

and

$$\frac{\partial^2 \psi}{\partial y^2} \approx \frac{1}{(\Delta y)^2}(\psi_{i,j+1} - 2\psi_{i,j} + \psi_{i,j-1}) \quad \textbf{(3)}$$

Thus, by combining Eqs. 1, 2, and 3 we obtain

$$\frac{\partial^2 \psi}{\partial x^2} + \frac{\partial^2 \psi}{\partial y^2} \approx \frac{1}{(\Delta x)^2}(\psi_{i+1,j} + \psi_{i-1,j}) + \frac{1}{(\Delta y)^2}(\psi_{i,j+1}$$
$$+ \psi_{i,j-1}) - 2\left(\frac{1}{(\Delta x)^2} + \frac{1}{(\Delta y)^2}\right)\psi_{i,j} = 0 \quad \textbf{(4)}$$

Equation 4 can be solved for the stream function at $x_i$ and $y_j$ to give

$$\psi_{i,j} = \frac{1}{2[(\Delta x)^2 + (\Delta y)^2]}[(\Delta y)^2(\psi_{i+1,j} + \psi_{i-1,j})$$
$$+ (\Delta x)^2(\psi_{i,j+1} + \psi_{i,j-1})] \quad \textbf{(5)}$$

Note that the value of $\psi_{i,j}$ depends on the values of the stream function at neighboring grid points on either side and above and below the point of interest (see Eq. 5 and Fig. EA. 2$c$).

To solve the problem (either exactly or by the finite difference technique), it is necessary to specify boundary conditions for points located on the boundary of the flow domain (see Section 6.6.3). For example, we may specify that $\psi = 0$ on the lower boundary of the domain (see Fig. EA.2$b$) and $\psi = C$, a constant, on the upper boundary of the domain. Appropriate boundary conditions on the two vertical ends of the flow domain can also be specified. Thus, for points interior to the boundary Eq. 5 is valid; similar equations or specified values of $\psi_{i,j}$ are valid for boundary points. The result is an equal number of equations and unknowns, $\psi_{i,j}$, one for every grid point. For this problem, these equations represent a set of linear algebraic equations for $\psi_{i,j}$, the solution of which provides the finite difference approximation for the stream function at discrete grid points in the flow field. Streamlines (lines of constant $\psi$) can be obtained by interpolating values of $\psi_{i,j}$ between the grid points and "connecting the dots" of $\psi = $ constant. The velocity field can be obtained from the derivatives of the stream function according to Eq. 6.74. That is,

$$u = \frac{\partial \psi}{\partial y} \approx \frac{1}{\Delta y}\left(\psi_{i,j+1} - \psi_{i,j}\right)$$

and

$$v = -\frac{\partial \psi}{\partial x} \approx -\frac{1}{\Delta x}\left(\psi_{i+1,j} - \psi_{i,j}\right)$$

Further details of the finite difference technique can be found in standard references on the topic (Refs. 5, 7, 8). Also, see the completely solved viscous flow CFD problem in Appendix I.

The preceding two examples are rather simple because the governing equations are not too complex. A finite difference solution of the more complicated, nonlinear Navier–Stokes equation (Eq. 6.127) requires considerably more effort and insight and larger and faster computers. A typical finite difference grid for a more complex flow, the flow past a turbine blade, is shown in Fig. A.5. Note that the mesh is much finer in regions where large gradients are to be expected (i.e., near the leading and trailing edges of the blade) and coarser away from the blade.

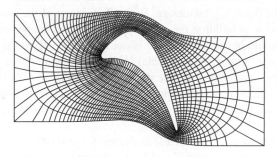

■ **Figure A.5**  Finite difference grid for flow past a turbine blade. (From Ref. 9, used by permission.)

## A.6    Methodology

In general, most applications of CFD take the same basic approach. Some of the differences include problem complexity, available computer resources, available expertise in CFD, and whether a commercially available CFD package is used, or a problem-specific CFD algorithm is developed. In today's market, there are many commercial CFD codes available to solve a wide variety of problems. However, if the intent is to conduct a thorough investigation of a specific fluid flow problem such as in a research environment, it is possible that taking the time to develop a problem-specific algorithm may be most efficient in the long run. The features common to most CFD applications can be summarized in the flow chart shown in Fig. A.6. A complete, detailed CFD solution for a viscous flow obtained by using the steps summarized in the flow chart can be accessed from the book's web site at www.wiley.com/college/munson.

■ **Figure A.6**  Flow chart of general CFD methodology.

The Algorithm Development box is grayed because this step is required only when developing your own CFD code. When using a commercial CFD code, this step is not necessary. This chart represents a generalized methodology to CFD. There are other more complex components that are hidden in the above steps, which are beyond the scope of a brief introduction to CFD.

# A.7    Application of CFD

In the early stages of CFD, research and development was primarily driven by the aerospace industry. Today, CFD is still used as a research tool, but it also has found a place in industry as a design tool. There is now a wide variety of industries that make at least some use of CFD, including automotive, industrial, HVAC, naval, civil, chemical, biological, and others. Industries are using CFD as an added engineering tool that complements the experimental and theoretical work in fluid dynamics.

## A.7.1  Advantages of CFD

There are many advantages to using CFD for simulation of fluid flow. One of the most important advantages is the realizable savings in time and cost for engineering design. In the past, coming up with a new engineering design meant somewhat of a trial-and-error method of building and testing multiple prototypes prior to finalizing the design. With CFD, many of the issues dealing with fluid flow can be flushed out prior to building the actual prototype. This translates to a significant savings in time and cost. It should be noted that CFD is not meant to replace experimental testing, but rather to work in conjunction with it. Experimental testing will always be a necessary component of engineering design. Other advantages include the ability of CFD to: (1) obtain flow information in regions that would be difficult to test experimentally, (2) simulate real flow conditions, (3) conduct large parametric tests on new designs in a shorter time, and (4) enhance visualization of complex flow phenomena.

*VA.3 Tornado simulation*

A good example of the advantages of CFD is shown in Figure A.7. Researchers use a type of CFD approach called "large-eddy simulation" or LES to simulate the fluid dynamics of a tornado as it encounters a debris field and begins to pick up sand-sized particles. A full animation of this tornado simulation can be accessed by visiting the book web site. The motivation for this work is to investigate whether there are significant differences in the fluid mechanics when debris particles are present. Historically it has been difficult to get comprehensive experimental data throughout a tornado, so CFD is helping to shine some light on the complex fluid dynamics involved in such a flow.

## A.7.2  Difficulties in CFD

One of the key points that a beginning CFD student should understand is that one cannot treat the computer as a "magic black box" when performing flow simulations. It is quite possible to obtain a fully converged solution for the CFD simulation, but this is no guarantee that the results are physically correct. This is why it is important to have a good understanding of the flow physics and how they are modeled. Any numerical technique (including those discussed above), no matter how simple in concept, contains many hidden subtleties and potential problems. For example, it may seem reasonable that a finer grid would ensure a more accurate numerical solution. While this may be true (as Example A.1), it is not always so straightforward; a variety of stability or convergence problems may occur. In such cases the numerical "solution" obtained may exhibit unreasonable oscillations or the numerical result may "diverge" to an unreasonable (and incorrect) result. Other problems that may arise include (but are not limited to): (1) difficulties in dealing with the nonlinear terms of the Navier–Stokes equations, (2) difficulties in modeling or capturing turbulent flows, (3) convergence issues, (4) difficulties in obtaining a quality grid for complex geometries, and (5) managing resources, both time and computational, for complex problems such as unsteady three-dimensional flows.

## A.7.3  Verification and Validation

Verification and validation of the simulation are critical steps in the CFD process. This is a necessary requirement for CFD, particularly since it is possible to have a converged solution that is nonphysical. Figure A.8 shows the streamlines for viscous flow past a circular cylinder at a given instant

■ **Figure A.7**  Results from a large-eddy simulation showing the visual appearance of the debris and funnel cloud from a simulated medium swirl F3-F4 tornado. The funnel cloud is translating at 15 m/s and is ingesting 1-mm-diameter "sand" from the surface as it encounters a debris field. Please visit the book web site to access a full animation of this tornado simulation. (Photographs and animation courtesy of Dr. David Lewellen (Ref. 10) and Paul Lewellen, West Virginia University.)

after it was impulsively started from rest. The lower half of the figure represents the results of a finite difference calculation; the upper half of the figure represents the photograph from an experiment of the same flow situation. It is clear that the numerical and experimental results agree quite well. For any CFD simulation, several levels of testing need to be accomplished before one can have confidence in the solution. The most important verification to be performed is grid convergence testing. In its simplest form, it consists of proving that further refinement of the grid (i.e., increasing the number of grid points) does not alter the final solution. When this has been achieved, you have a grid-independent solution. Other verification factors that need to be investigated include

■ **Figure A.8**  Streamlines for flow past a circular cylinder at a short time after the flow was impulsively started. The upper half is a photograph from a flow visualization experiment. The lower half is from a finite difference calculation. (See the photograph at the beginning of Chapter 9.) (From Ref. 9, used by permission.)

the suitability of the convergence criterion, whether the time step is adequate for the time scale of the problem, and comparison of CFD solutions to existing data, at least for baseline cases. Even when using a commercial CFD code that has been validated on many problems in the past, the CFD practitioner still needs to verify the results through such measures as grid-dependence testing.

### A.7.4 Summary

In CFD, there are many different numerical schemes, grid techniques, etc. They all have their advantages and disadvantages. A great deal of care must be used in obtaining approximate numerical solutions to the governing equations of fluid motion. The process is not as simple as the often-heard "just let the computer do it." Remember that CFD is a tool and as such needs to be used appropriately to produce meaningful results. The general field of computational fluid dynamics, in which computers and numerical analysis are combined to solve fluid flow problems, represents an extremely important subject area in advanced fluid mechanics. Considerable progress has been made in the past relatively few years, but much remains to be done. The reader is encouraged to consult some of the available literature.

## References

1. Baker, A. J., *Finite Element Computational Fluid Mechanics*, McGraw-Hill, New York, 1983.
2. Carey, G. F., and Oden, J. T., *Finite Elements: Fluid Mechanics*, Prentice-Hall, Englewood Cliffs, N.J., 1986.
3. Brebbia, C. A., and Dominguez, J., *Boundary Elements: An Introductory Course*, McGraw-Hill, New York, 1989.
4. Moran, J., *An Introduction to Theoretical and Computational Aerodynamics*, Wiley, New York, 1984.
5. Anderson, J. D., *Computational Fluid Dynamics: The Basics with Applications*, McGraw-Hill, New York, 1995.
6. Thompson, J. F., Warsi, Z. U. A., and Mastin, C. W., *Numerical Grid Generation: Foundations and Applications*, North-Holland, New York, 1985.
7. Peyret, R., and Taylor, T. D., *Computational Methods for Fluid Flow*, Springer-Verlag, New York, 1983.
8. Tannehill, J. C., Anderson, D. A., and Pletcher, R. H., *Computational Fluid Mechanics and Heat Transfer*, 2nd Ed., Taylor and Francis, Washington, D.C., 1997.
9. Hall, E. J., and Pletcher, R. H., *Simulation of Time Dependent, Compressible Viscous Flow Using Central and Upwind-Biased Finite-Difference Techniques*, Technical Report HTL-52, CFD-22, College of Engineering, Iowa State University, 1990.
10. Lewellen, D. C., Gong, B., and Lewellen, W. S., *Effects of Debris on Near-Surface Tornado Dynamics*, 22nd Conference on Severe Local Storms, Paper 15.5, American Meteorological Society, 2004.

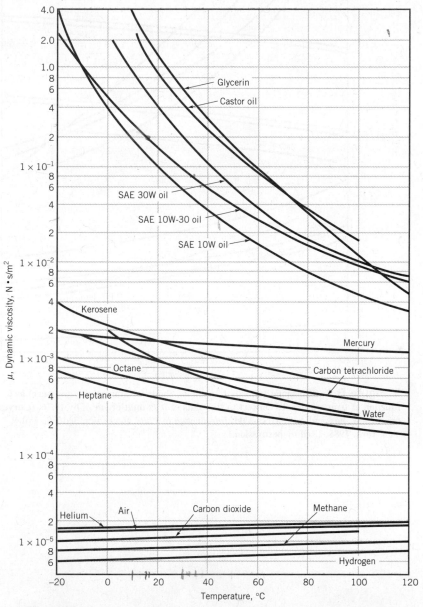

■ **Figure B.1**  Dynamic (absolute) viscosity of common fluids as a function of temperature. To convert to BG units of lb · s/ft² multiply N · s/m² by 2.089 × 10⁻². (Curves from R. W. Fox and A. T. McDonald, *Introduction to Fluid Mechanics*, 3rd Ed., Wiley, New York, 1985. Used by permission.)

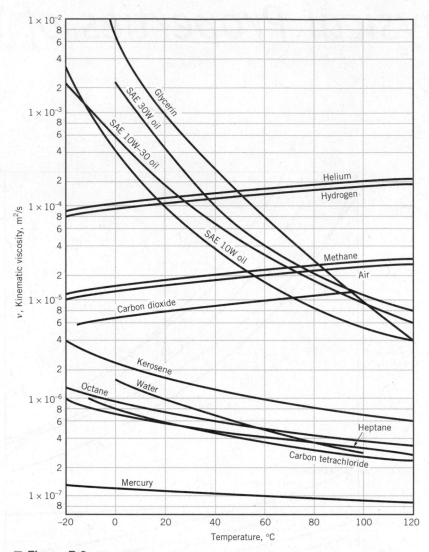

■ **Figure B.2** Kinematic viscosity of common fluids (at atmospheric pressure) as a function of temperature. To convert to BG units of ft²/s multiply m²/s by 10.76. (Curves from R. W. Fox and A. T. McDonald, *Introduction to Fluid Mechanics*, 3rd Ed., Wiley, New York, 1985. Used by permission.)

■ **Table B.1**
**Physical Properties of Water (BG Units)[a]**

| Temperature (°F) | Density, $\rho$ (slugs/ft³) | Specific Weight[b], $\gamma$ (lb/ft³) | Dynamic Viscosity, $\mu$ (lb·s/ft²) | Kinematic Viscosity, $\nu$ (ft²/s) | Surface Tension[c], $\sigma$ (lb/ft) | Vapor Pressure, $p_v$ [lb/in.²(abs)] | Speed of Sound[d], $c$ (ft/s) |
|---|---|---|---|---|---|---|---|
| 32 | 1.940 | 62.42 | 3.732 E − 5 | 1.924 E − 5 | 5.18 E − 3 | 8.854 E − 2 | 4603 |
| 40 | 1.940 | 62.43 | 3.228 E − 5 | 1.664 E − 5 | 5.13 E − 3 | 1.217 E − 1 | 4672 |
| 50 | 1.940 | 62.41 | 2.730 E − 5 | 1.407 E − 5 | 5.09 E − 3 | 1.781 E − 1 | 4748 |
| 60 | 1.938 | 62.37 | 2.344 E − 5 | 1.210 E − 5 | 5.03 E − 3 | 2.563 E − 1 | 4814 |
| 70 | 1.936 | 62.30 | 2.037 E − 5 | 1.052 E − 5 | 4.97 E − 3 | 3.631 E − 1 | 4871 |
| 80 | 1.934 | 62.22 | 1.791 E − 5 | 9.262 E − 6 | 4.91 E − 3 | 5.069 E − 1 | 4819 |
| 90 | 1.931 | 62.11 | 1.500 E − 5 | 8.233 E − 6 | 4.86 E − 3 | 6.979 E − 1 | 4960 |
| 100 | 1.927 | 62.00 | 1.423 E − 5 | 7.383 E − 6 | 4.79 E − 3 | 9.493 E − 1 | 4995 |
| 120 | 1.918 | 61.71 | 1.164 E − 5 | 6.067 E − 6 | 4.67 E − 3 | 1.692 E + 0 | 5049 |
| 140 | 1.908 | 61.38 | 9.743 E − 6 | 5.106 E − 6 | 4.53 E − 3 | 2.888 E + 0 | 5091 |
| 160 | 1.896 | 61.00 | 8.315 E − 6 | 4.385 E − 6 | 4.40 E − 3 | 4.736 E + 0 | 5101 |
| 180 | 1.883 | 60.58 | 7.207 E − 6 | 3.827 E − 6 | 4.26 E − 3 | 7.507 E + 0 | 5195 |
| 200 | 1.869 | 60.12 | 6.342 E − 6 | 3.393 E − 6 | 4.12 E − 3 | 1.152 E + 1 | 5089 |
| 212 | 1.860 | 59.83 | 5.886 E − 6 | 3.165 E − 6 | 4.04 E − 3 | 1.469 E + 1 | 5062 |

[a]Based on data from *Handbook of Chemistry and Physics*, 69th Ed., CRC Press, 1988. Where necessary, values obtained by interpolation.
[b]Density and specific weight are related through the equation $\gamma = \rho g$. For this table, $g = 32.174$ ft/s².
[c]In contact with air.
[d]Based on data from R. D. Blevins, *Applied Fluid Dynamics Handbook*, Van Nostrand Reinhold Co., Inc., New York, 1984.

■ **Table B.2**
**Physical Properties of Water (SI Units)[a]**

| Temperature (°C) | Density, $\rho$ (kg/m³) | Specific Weight[b], $\gamma$ (kN/m³) | Dynamic Viscosity, $\mu$ (N·s/m²) | Kinematic Viscosity, $\nu$ (m²/s) | Surface Tension[c], $\sigma$ (N/m) | Vapor Pressure, $p_v$ [N/m²(abs)] | Speed of Sound[d], $c$ (m/s) |
|---|---|---|---|---|---|---|---|
| 0 | 999.9 | 9.806 | 1.787 E − 3 | 1.787 E − 6 | 7.56 E − 2 | 6.105 E + 2 | 1403 |
| 5 | 1000.0 | 9.807 | 1.519 E − 3 | 1.519 E − 6 | 7.49 E − 2 | 8.722 E + 2 | 1427 |
| 10 | 999.7 | 9.804 | 1.307 E − 3 | 1.307 E − 6 | 7.42 E − 2 | 1.228 E + 3 | 1447 |
| 20 | 998.2 | 9.789 | 1.002 E − 3 | 1.004 E − 6 | 7.28 E − 2 | 2.338 E + 3 | 1481 |
| 30 | 995.7 | 9.765 | 7.975 E − 4 | 8.009 E − 7 | 7.12 E − 2 | 4.243 E + 3 | 1507 |
| 40 | 992.2 | 9.731 | 6.529 E − 4 | 6.580 E − 7 | 6.96 E − 2 | 7.376 E + 3 | 1526 |
| 50 | 988.1 | 9.690 | 5.468 E − 4 | 5.534 E − 7 | 6.79 E − 2 | 1.233 E + 4 | 1541 |
| 60 | 983.2 | 9.642 | 4.665 E − 4 | 4.745 E − 7 | 6.62 E − 2 | 1.992 E + 4 | 1552 |
| 70 | 977.8 | 9.589 | 4.042 E − 4 | 4.134 E − 7 | 6.44 E − 2 | 3.116 E + 4 | 1555 |
| 80 | 971.8 | 9.530 | 3.547 E − 4 | 3.650 E − 7 | 6.26 E − 2 | 4.734 E + 4 | 1555 |
| 90 | 965.3 | 9.467 | 3.147 E − 4 | 3.260 E − 7 | 6.08 E − 2 | 7.010 E + 4 | 1550 |
| 100 | 958.4 | 9.399 | 2.818 E − 4 | 2.940 E − 7 | 5.89 E − 2 | 1.013 E + 5 | 1543 |

[a]Based on data from *Handbook of Chemistry and Physics*, 69th Ed., CRC Press, 1988.
[b]Density and specific weight are related through the equation $\gamma = \rho g$. For this table, $g = 9.807$ m/s².
[c]In contact with air.
[d]Based on data from R. D. Blevins, *Applied Fluid Dynamics Handbook*, Van Nostrand Reinhold Co., Inc., New York, 1984.

■ **Table B.3**
**Physical Properties of Air at Standard Atmospheric Pressure (BG Units)**[a]

| Temperature (°F) | Density, $\rho$ (slugs/ft³) | | Specific Weight[b], $\gamma$ (lb/ft³) | | Dynamic Viscosity, $\mu$ (lb·s/ft²) | | Kinematic Viscosity, $\nu$ (ft²/s) | | Specific Heat Ratio, $k$ (—) | Speed of Sound, $c$ (ft/s) |
|---|---|---|---|---|---|---|---|---|---|---|
| −40 | 2.939 | E − 3 | 9.456 | E − 2 | 3.29 | E − 7 | 1.12 | E − 4 | 1.401 | 1004 |
| −20 | 2.805 | E − 3 | 9.026 | E − 2 | 3.34 | E − 7 | 1.19 | E − 4 | 1.401 | 1028 |
| 0 | 2.683 | E − 3 | 8.633 | E − 2 | 3.38 | E − 7 | 1.26 | E − 4 | 1.401 | 1051 |
| 10 | 2.626 | E − 3 | 8.449 | E − 2 | 3.44 | E − 7 | 1.31 | E − 4 | 1.401 | 1062 |
| 20 | 2.571 | E − 3 | 8.273 | E − 2 | 3.50 | E − 7 | 1.36 | E − 4 | 1.401 | 1074 |
| 30 | 2.519 | E − 3 | 8.104 | E − 2 | 3.58 | E − 7 | 1.42 | E − 4 | 1.401 | 1085 |
| 40 | 2.469 | E − 3 | 7.942 | E − 2 | 3.60 | E − 7 | 1.46 | E − 4 | 1.401 | 1096 |
| 50 | 2.420 | E − 3 | 7.786 | E − 2 | 3.68 | E − 7 | 1.52 | E − 4 | 1.401 | 1106 |
| 60 | 2.373 | E − 3 | 7.636 | E − 2 | 3.75 | E − 7 | 1.58 | E − 4 | 1.401 | 1117 |
| 70 | 2.329 | E − 3 | 7.492 | E − 2 | 3.82 | E − 7 | 1.64 | E − 4 | 1.401 | 1128 |
| 80 | 2.286 | E − 3 | 7.353 | E − 2 | 3.86 | E − 7 | 1.69 | E − 4 | 1.400 | 1138 |
| 90 | 2.244 | E − 3 | 7.219 | E − 2 | 3.90 | E − 7 | 1.74 | E − 4 | 1.400 | 1149 |
| 100 | 2.204 | E − 3 | 7.090 | E − 2 | 3.94 | E − 7 | 1.79 | E − 4 | 1.400 | 1159 |
| 120 | 2.128 | E − 3 | 6.846 | E − 2 | 4.02 | E − 7 | 1.89 | E − 4 | 1.400 | 1180 |
| 140 | 2.057 | E − 3 | 6.617 | E − 2 | 4.13 | E − 7 | 2.01 | E − 4 | 1.399 | 1200 |
| 160 | 1.990 | E − 3 | 6.404 | E − 2 | 4.22 | E − 7 | 2.12 | E − 4 | 1.399 | 1220 |
| 180 | 1.928 | E − 3 | 6.204 | E − 2 | 4.34 | E − 7 | 2.25 | E − 4 | 1.399 | 1239 |
| 200 | 1.870 | E − 3 | 6.016 | E − 2 | 4.49 | E − 7 | 2.40 | E − 4 | 1.398 | 1258 |
| 300 | 1.624 | E − 3 | 5.224 | E − 2 | 4.97 | E − 7 | 3.06 | E − 4 | 1.394 | 1348 |
| 400 | 1.435 | E − 3 | 4.616 | E − 2 | 5.24 | E − 7 | 3.65 | E − 4 | 1.389 | 1431 |
| 500 | 1.285 | E − 3 | 4.135 | E − 2 | 5.80 | E − 7 | 4.51 | E − 4 | 1.383 | 1509 |
| 750 | 1.020 | E − 3 | 3.280 | E − 2 | 6.81 | E − 7 | 6.68 | E − 4 | 1.367 | 1685 |
| 1000 | 8.445 | E − 4 | 2.717 | E − 2 | 7.85 | E − 7 | 9.30 | E − 4 | 1.351 | 1839 |
| 1500 | 6.291 | E − 4 | 2.024 | E − 2 | 9.50 | E − 7 | 1.51 | E − 3 | 1.329 | 2114 |

[a]Based on data from R. D. Blevins, *Applied Fluid Dynamics Handbook*, Van Nostrand Reinhold Co., Inc., New York, 1984.
[b]Density and specific weight are related through the equation $\gamma = \rho g$. For this table $g = 32.174$ ft/s².

■ **Table B.4**
**Physical Properties of Air at Standard Atmospheric Pressure (SI Units)[a]**

| Temperature (°C) | Density, $\rho$ (kg/m³) | Specific Weight[b], $\gamma$ (N/m³) | Dynamic Viscosity, $\mu$ (N·s/m²) | Kinematic Viscosity, $\nu$ (m²/s) | Specific Heat Ratio, $k$ (—) | Speed of Sound, $c$ (m/s) |
|---|---|---|---|---|---|---|
| −40 | 1.514 | 14.85 | 1.57 E − 5 | 1.04 E − 5 | 1.401 | 306.2 |
| −20 | 1.395 | 13.68 | 1.63 E − 5 | 1.17 E − 5 | 1.401 | 319.1 |
| 0 | 1.292 | 12.67 | 1.71 E − 5 | 1.32 E − 5 | 1.401 | 331.4 |
| 5 | 1.269 | 12.45 | 1.73 E − 5 | 1.36 E − 5 | 1.401 | 334.4 |
| 10 | 1.247 | 12.23 | 1.76 E − 5 | 1.41 E − 5 | 1.401 | 337.4 |
| 15 | 1.225 | 12.01 | 1.80 E − 5 | 1.47 E − 5 | 1.401 | 340.4 |
| 20 | 1.204 | 11.81 | 1.82 E − 5 | 1.51 E − 5 | 1.401 | 343.3 |
| 25 | 1.184 | 11.61 | 1.85 E − 5 | 1.56 E − 5 | 1.401 | 346.3 |
| 30 | 1.165 | 11.43 | 1.86 E − 5 | 1.60 E − 5 | 1.400 | 349.1 |
| 40 | 1.127 | 11.05 | 1.87 E − 5 | 1.66 E − 5 | 1.400 | 354.7 |
| 50 | 1.109 | 10.88 | 1.95 E − 5 | 1.76 E − 5 | 1.400 | 360.3 |
| 60 | 1.060 | 10.40 | 1.97 E − 5 | 1.86 E − 5 | 1.399 | 365.7 |
| 70 | 1.029 | 10.09 | 2.03 E − 5 | 1.97 E − 5 | 1.399 | 371.2 |
| 80 | 0.9996 | 9.803 | 2.07 E − 5 | 2.07 E − 5 | 1.399 | 376.6 |
| 90 | 0.9721 | 9.533 | 2.14 E − 5 | 2.20 E − 5 | 1.398 | 381.7 |
| 100 | 0.9461 | 9.278 | 2.17 E − 5 | 2.29 E − 5 | 1.397 | 386.9 |
| 200 | 0.7461 | 7.317 | 2.53 E − 5 | 3.39 E − 5 | 1.390 | 434.5 |
| 300 | 0.6159 | 6.040 | 2.98 E − 5 | 4.84 E − 5 | 1.379 | 476.3 |
| 400 | 0.5243 | 5.142 | 3.32 E − 5 | 6.34 E − 5 | 1.368 | 514.1 |
| 500 | 0.4565 | 4.477 | 3.64 E − 5 | 7.97 E − 5 | 1.357 | 548.8 |
| 1000 | 0.2772 | 2.719 | 5.04 E − 5 | 1.82 E − 4 | 1.321 | 694.8 |

[a]Based on data from R. D. Blevins, *Applied Fluid Dynamics Handbook*, Van Nostrand Reinhold Co., Inc., New York, 1984.
[b]Density and specific weight are related through the equation $\gamma = \rho g$. For this table $g = 9.807$ m/s².

# C Properties of the U.S. Standard Atmosphere

■ Table C.1

Properties of the U.S. Standard Atmosphere (BG Units)[a]

| Altitude (ft) | Temperature (°F) | Acceleration of Gravity, $g$ (ft/s$^2$) | Pressure, $p$ [lb/in.$^2$(abs)] | Density, $\rho$ (slugs/ft$^3$) | | Dynamic Viscosity, $\mu$ (lb·s/ft$^2$) | |
|---|---|---|---|---|---|---|---|
| −5,000 | 76.84 | 32.189 | 17.554 | 2.745 | E − 3 | 3.836 | E − 7 |
| 0 | 59.00 | 32.174 | 14.696 | 2.377 | E − 3 | 3.737 | E − 7 |
| 5,000 | 41.17 | 32.159 | 12.228 | 2.048 | E − 3 | 3.637 | E − 7 |
| 10,000 | 23.36 | 32.143 | 10.108 | 1.756 | E − 3 | 3.534 | E − 7 |
| 15,000 | 5.55 | 32.128 | 8.297 | 1.496 | E − 3 | 3.430 | E − 7 |
| 20,000 | −12.26 | 32.112 | 6.759 | 1.267 | E − 3 | 3.324 | E − 7 |
| 25,000 | −30.05 | 32.097 | 5.461 | 1.066 | E − 3 | 3.217 | E − 7 |
| 30,000 | −47.83 | 32.082 | 4.373 | 8.907 | E − 4 | 3.107 | E − 7 |
| 35,000 | −65.61 | 32.066 | 3.468 | 7.382 | E − 4 | 2.995 | E − 7 |
| 40,000 | −69.70 | 32.051 | 2.730 | 5.873 | E − 4 | 2.969 | E − 7 |
| 45,000 | −69.70 | 32.036 | 2.149 | 4.623 | E − 4 | 2.969 | E − 7 |
| 50,000 | −69.70 | 32.020 | 1.692 | 3.639 | E − 4 | 2.969 | E − 7 |
| 60,000 | −69.70 | 31.990 | 1.049 | 2.256 | E − 4 | 2.969 | E − 7 |
| 70,000 | −67.42 | 31.959 | 0.651 | 1.392 | E − 4 | 2.984 | E − 7 |
| 80,000 | −61.98 | 31.929 | 0.406 | 8.571 | E − 5 | 3.018 | E − 7 |
| 90,000 | −56.54 | 31.897 | 0.255 | 5.610 | E − 5 | 3.052 | E − 7 |
| 100,000 | −51.10 | 31.868 | 0.162 | 3.318 | E − 5 | 3.087 | E − 7 |
| 150,000 | 19.40 | 31.717 | 0.020 | 3.658 | E − 6 | 3.511 | E − 7 |
| 200,000 | −19.78 | 31.566 | 0.003 | 5.328 | E − 7 | 3.279 | E − 7 |
| 250,000 | −88.77 | 31.415 | 0.000 | 6.458 | E − 8 | 2.846 | E − 7 |

[a]Data abridged from *U.S. Standard Atmosphere*, 1976, U.S. Government Printing Office, Washington, D.C.

■ Table C.2
**Properties of the U.S. Standard Atmosphere (SI Units)**[a]

| Altitude (m) | Temperature (°C) | Acceleration of Gravity, $g$ (m/s²) | Pressure, $p$ [N/m²(abs)] | | Density, $\rho$ (kg/m³) | | Dynamic Viscosity, $\mu$ (N·s/m²) | |
|---|---|---|---|---|---|---|---|---|
| −1,000 | 21.50 | 9.810 | 1.139 | E + 5 | 1.347 | E + 0 | 1.821 | E − 5 |
| 0 | 15.00 | 9.807 | 1.013 | E + 5 | 1.225 | E + 0 | 1.789 | E − 5 |
| 1,000 | 8.50 | 9.804 | 8.988 | E + 4 | 1.112 | E + 0 | 1.758 | E − 5 |
| 2,000 | 2.00 | 9.801 | 7.950 | E + 4 | 1.007 | E + 0 | 1.726 | E − 5 |
| 3,000 | −4.49 | 9.797 | 7.012 | E + 4 | 9.093 | E − 1 | 1.694 | E − 5 |
| 4,000 | −10.98 | 9.794 | 6.166 | E + 4 | 8.194 | E − 1 | 1.661 | E − 5 |
| 5,000 | −17.47 | 9.791 | 5.405 | E + 4 | 7.364 | E − 1 | 1.628 | E − 5 |
| 6,000 | −23.96 | 9.788 | 4.722 | E + 4 | 6.601 | E − 1 | 1.595 | E − 5 |
| 7,000 | −30.45 | 9.785 | 4.111 | E + 4 | 5.900 | E − 1 | 1.561 | E − 5 |
| 8,000 | −36.94 | 9.782 | 3.565 | E + 4 | 5.258 | E − 1 | 1.527 | E − 5 |
| 9,000 | −43.42 | 9.779 | 3.080 | E + 4 | 4.671 | E − 1 | 1.493 | E − 5 |
| 10,000 | −49.90 | 9.776 | 2.650 | E + 4 | 4.135 | E − 1 | 1.458 | E − 5 |
| 15,000 | −56.50 | 9.761 | 1.211 | E + 4 | 1.948 | E − 1 | 1.422 | E − 5 |
| 20,000 | −56.50 | 9.745 | 5.529 | E + 3 | 8.891 | E − 2 | 1.422 | E − 5 |
| 25,000 | −51.60 | 9.730 | 2.549 | E + 3 | 4.008 | E − 2 | 1.448 | E − 5 |
| 30,000 | −46.64 | 9.715 | 1.197 | E + 3 | 1.841 | E − 2 | 1.475 | E − 5 |
| 40,000 | −22.80 | 9.684 | 2.871 | E + 2 | 3.996 | E − 3 | 1.601 | E − 5 |
| 50,000 | −2.50 | 9.654 | 7.978 | E + 1 | 1.027 | E − 3 | 1.704 | E − 5 |
| 60,000 | −26.13 | 9.624 | 2.196 | E + 1 | 3.097 | E − 4 | 1.584 | E − 5 |
| 70,000 | −53.57 | 9.594 | 5.221 | E + 0 | 8.283 | E − 5 | 1.438 | E − 5 |
| 80,000 | −74.51 | 9.564 | 1.052 | E + 0 | 1.846 | E − 5 | 1.321 | E − 5 |

[a]Data abridged from *U.S. Standard Atmosphere*, 1976, U.S. Government Printing Office, Washington, D.C.

# D Compressible Flow Graphs for an Ideal Gas (k = 1.4)

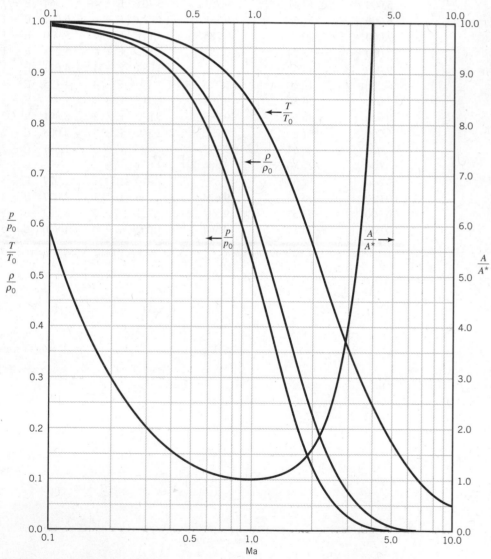

■ **Figure D.1** Isentropic flow of an ideal gas with *k* = 1.4. (Graph provided by Dr. Bruce A. Reichert.)

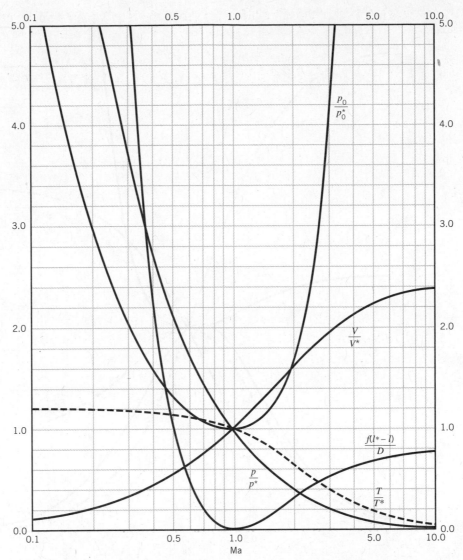

■ **Figure D.2** Fanno flow of an ideal gas with $k = 1.4$. (Graph provided by Dr. Bruce A. Reichert.)

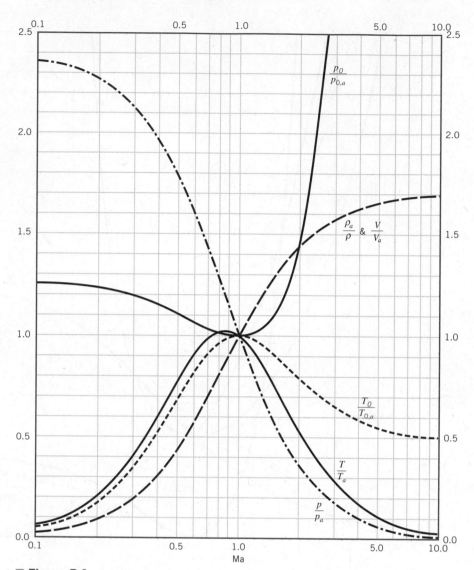

■ **Figure D.3**   **Rayleigh flow of an ideal gas with $k = 1.4$. (Graph provided by Dr. Bruce A. Rcichcrt.)**

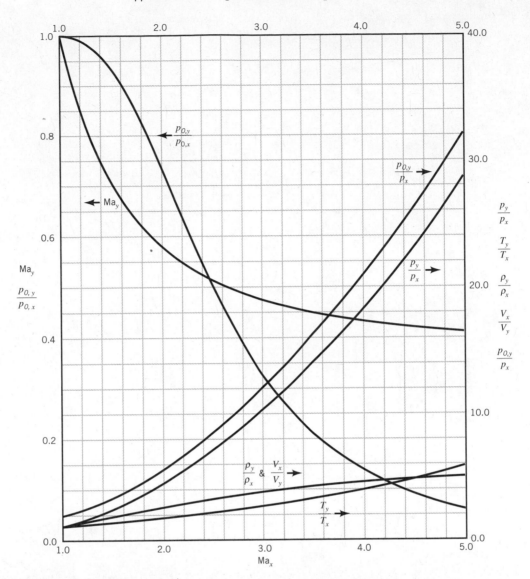

■ **Figure D.4**  Normal shock flow of an ideal gas with $k = 1.4$. (Graph provided by Dr. Bruce A. Reichert.)

# Answers to Selected Even-Numbered Homework Problems

## Chapter 1

| | |
|---|---|
| **1.2** | (a) $MLT^{-1}$; (b) $ML^4 L^{-2}$; (c) $MT^{-2}$ |
| **1.4** | (a) $FLT^{-2}$; (b) $FL^{-1}T^{-1}$; (c) $FL^{-3} T$ |
| **1.10** | (a) $FL^{-1}$; (b) $FL^{-3}$; FL |
| **1.12** | $LT^{-1}$; $F^0 L^0 T^0$; $LT^{-1}$ |
| **1.14** | Yes |
| **1.16** | $1/2$, $-1/2$ |
| **1.18** | Dimensionless |
| **1.20** | (a) 4.32 mm/s; (b) 70.2 kg; (c) 13.4 N; (d) $22.3$ m/s$^2$; (e) $1.12$ N $\cdot$ s/m$^2$ |
| **1.22** | (a) $6.47 \times 10^5$ m$^2$; (b) $56.8 \times 10^{-2}$ m$^3$; (c) $3.86 \times 10^5$ m; (d) $5.90 \times 10^4$ W; (e) 289 K |
| **1.30** | (a) 0.240 mi$^3$; (b) $4.41 \times 10^5$ lb |
| **1.32** | 30.6 kg; 37.3 N |
| **1.34** | 1150 kg/m$^3$; 11.3 kN/m$^3$ |
| **1.36** | 0.0186 ft$^3$; No |
| **1.38** | 0.9971 lb |
| **1.40** | 991.5 kg/m$^3$ |
| **1.42** | 16.0 kN/m$^3$; $1.63 \times 10^3$ kg/m$^3$; 1.63 |
| **1.44** | 4.76 kg |
| **1.46** | (a) 0.0214 kg/m$^3$; (b) $\rho_{Mars}/\rho_{earth} = 1.75\%$ |
| **1.48** | $6.44 \times 10^{-3}$ slugs/ft$^3$; 0.622 lb |
| **1.50** | 98.7 psia |
| **1.52** | 668 lb |
| **1.56** | 0.6 N $\cdot$ s/m$^2$; $1.3 \times 10^{-2}$ lb $\cdot$ s/ft$^2$ |
| **1.58** | 0.727 N $\cdot$ s/m$^2$ |
| **1.60** | 23.7, $2.55 \times 10^{-2}$ |
| **1.62** | 0.277 N $\cdot$ s/m$^2$ |
| **1.64** | $2.05 \times 10^{-5}$ N $\cdot$ s/m$^2$ |
| **1.66** | (a) No; (b) Not correct |
| **1.68** | 184 |
| **1.70** | $C = 1.43 \times 10^{-6}$ kg/(m $\cdot$ s $\cdot$ K$^{1/2}$); $S = 107$ K |
| **1.72** | $D = 1.767 \times 10^{-6}$ N $\cdot$ s/m$^2$; $B = 1.870 \times 10^3$ K; $5.76 \times 10^{-4}$ N $\cdot$ s/m$^2$ |
| **1.74** | 0.300 N $\cdot$ s/m$^2$ |
| **1.76** | non-Newtonian |
| **1.78** | 286 N |
| **1.80** | 0.0883 m/s |
| **1.82** | 0.268 ft/s |
| **1.84** | $3.43 \times 10^{-4}$ lb |
| **1.86** | $9.53 \times 10^{-4}$ ft $\cdot$ lb |
| **1.88** | (a) 12.7 ft$^2$/rev; (b) $4.73 \times 10^{-3}$ lb $\cdot$ s/ft$^2$ |
| **1.94** | $2.03 \times 10^3$ psi |
| **1.96** | $4.14 \times 10^3$ psi |

| | |
|---|---|
| **1.98** | (a) 343 m/s; (b) 1010 m/s; (c) 446 m/s |
| **1.100** | 104 psi (gage) |
| **1.102** | 2.88 kg/m$^3$ |
| **1.104** | $4.25 \times 10^{-3}$ slugs/ft$^3$; 305 °F |
| **1.108** | 1.06 |
| **1.110** | 14.1% |
| **1.112** | 4.74 psi (abs) |
| **1.114** | 13 kPa (abs) |
| **1.116** | 5.81 kPa (abs); 0.842 psi (abs) |
| **1.118** | 0.060 N/m |
| **1.120** | 538 Pa |
| **1.122** | 97.9 Pa |
| **1.124** | (a) 24.5 deg |
| **1.126** | 0.0614 in.; Yes |
| **1.128** | $1.80 \times 10^{-2}$ ft |
| **1.130** | 7.49 mm |

## Chapter 2

| | |
|---|---|
| **2.2** | 59.2 kPa |
| **2.6** | 34.7 psi |
| **2.8** | 404 kPa |
| **2.10** | (a) $p = -E_v \ln (1 - \rho_0 gh/E_v)$; (b) 61.4 MPa; (c) 60.6 MPa |
| **2.12** | $p = Kh^2/2 + \gamma_0 h$ |
| **2.18** | 464 mm |
| **2.20** | 62.9% |
| **2.24** | 4,250 ft |
| **2.26** | 543 m |
| **2.28** | 60 kPa |
| **2.32** | 14.4 psia; 99.3 kPa (abs) |
| **2.34** | (a) 18.2 ft; (b) 8.73 psi; 21.7 psia |
| **2.36** | 0.317 ft |
| **2.38** | 6.28 ft |
| **2.40** | $-3.32$ kPa |
| **2.42** | (a) 4.00 ft; (b) 2.08 ft |
| **2.44** | 1.45 ft |
| **2.46** | (a) 26.9 kPa; (b) 0.202 m |
| **2.48** | 1.55 slugs/ft$^3$ |
| **2.50** | 94.9 kPa |
| **2.54** | 21.6 ft |
| **2.56** | 575 lb/ft$^2$ |
| **2.58** | 0.100 m |
| **2.60** | 27.8° |
| **2.62** | 0.212 m |

**2.64**  0.304 ft

**2.66**  $\ell = [d + 11.31 - (d^2 + 18.61d + 128)^{1/2}]/2$

**2.68**  36,000 lb

**2.70**  $p = 0.416\,\theta$

**2.74**  2.90 lb

**2.76**  12.19 ft

**2.78**  7200 lb

**2.80**  665 lb

**2.82**  107 kPa

**2.84**  7.11 m; No

**2.86**  $F_B = \gamma\pi R^3/4$

**2.88**  (a) 9.00 ft; (b) 11,200 lb; (c) 0 lb

**2.90**  33,900 lb

**2.92**  373 kN

**2.94**  314 kN; 497 kN

**2.96**  1680 lb

**2.100**  3.63 ft

**2.102**  (a) 2.11 m; (b) 941 kN

**2.104**  599 lb

**2.106**  426 kN, 2.46 m below surface

**2.108**  2.34 ft

**2.110**  0.146

**2.116**  392 kN; 437 kN; Yes

**2.118**  581 lb; 0.289 psi

**2.120**  $7.77 \times 10^9$ lb acting 406 ft above base

**2.122**  64.4 kN

**2.124**  3360 psi

**2.126**  203 kN down

**2.128**  3370 lb; 882 lb

**2.130**  485 kN

**2.132**  46.8 lb/ft$^3$

**2.134**  1.22 ft; no change

**2.136**  786,000 lb; 315,000 lb

**2.138**  54,600 lb

**2.142**  127 lb

**2.144**  18.9 kPa; 0.208 m$^3$

**2.146**  681 lb at 6.22 ft to right of A

**2.148**  89.5%

**2.150**  8.63 ft/s$^2$

**2.152**  (a) 68.9 lb/ft$^2$; (b) 57.4 lb

**2.154**  30°

**2.156**  $h = a\ell/g$

**2.158**  6.04 rad/s

**2.160**  5.76 ft

**2.162**  28.8 kPa

## Chapter 3

**3.2**  13.7 m/s

**3.4**  −30.0 kPa/m

**3.6**  −0.838 psi/ft; 0.0292 psi/ft

**3.8**  (a) $-2\rho a^2 V_0^2\,[1 - (a/x)^2]/x^3$
    (b) $p_0 + \rho V_0^2\,[(a/x)^2 - (a/x)^4/2]$

**3.12**  12.0 kPa; −20.1 kPa

**3.14**  (a) 4.97 lb/ft$^3$; 0.681 lb/ft$^3$

**3.20**  3.21 N/m$^2$

**3.24**  203 ft/s

**3.28**  46.5 m/s

**3.32**  20.9 ft/s

**3.34**  45.7 ft

**3.36**  1248 lb/ft$^2$

**3.40**  20.8 lb/s

**3.42**  0.0119 ft$^3$/s

**3.44**  15.4 m

**3.46**  0.223 ft; 9.22 lb/ft$^2$

**3.48**  $2.45 \times 10^5$ kN/m$^3$; $5.50 \times 10^{-6}$ m$^3$/s

**3.50**  3.19 ft

**3.52**  0.0156 m$^3$/s for any D

**3.54**  6.59 ft; B above A

**3.56**  $D = D_1/[1 - (\pi^2 g D_1^4\, z/8Q^2]^{1/4}$

**3.58**  3.13 ft

**3.60**  2.94 m

**3.62**  0.0758 ft$^3$/s; 499 lb/ft$^2$; 488 lb/ft$^2$; −11.7 lb/ft$^2$

**3.64**  (a) 3.98 ft; (b) 36.0 ft/s; −510 lb/ft$^2$

**3.66**  19.63 psi

**3.68**  6.26 m

**3.70**  (a) $1.06 \times 10^{-3}$; (b) $3.02 \times 10^{-3}$; 0.118

**3.72**  29.5 Pa

**3.74**  0 (gage); 0.146 m

**3.76**  (a) 0.0696 m$^3$/s; (b) 0.574 m

**3.78**  $H/H_0 = 1/(1 + cx/L)^{1/2}$, where
    $c = 2\gamma_{H_2O}\, d_{max}/\rho V_0^2$

**3.80**  0.0132 m$^3$/s

**3.82**  2.50 in.

**3.84**  4

**3.86**  13.0 psi

**3.88**  0.607 ft$^3$/s; (b) 21.3 ft

**3.90**  0.123 ft$^3$/s

**3.92**  0.400 ft

**3.94**  $R = 0.998\, z^{1/4}$

**3.100**  404.5 kPa

**3.102**  252 kPa; 114 kPa

**3.104**  $9.10 \times 10^{-3}$ m$^3$/s; 57.9 kPa

**3.106**  3120 ft$^3$/s; 2080 ft$^3$/s; 2200 ft$^3$/s

**3.108**  155 N/m$^2$

**3.112**  0.630 ft; 4.48 ft

**3.116**  $6.10 \times 10^{-3}$ m$^3$/s

**3.118**  1.016 in.

**3.120**  1.73 m$^2$/s

**3.122**  25.4 m; 6.51 m; −9.59 m

## Chapter 4

**4.2**  $(9y^2 + 12y + 68)^{1/2}$ ft/s

**4.4**  $x = y + y^2/2$

**4.8**  $x = -2, y = 2$

**4.10**  Everywhere; $x^2 + y^2 = $ constant

**4.14**  $x = -h\,(u_0/v_0)\ln(1 - y/h)$

**4.18**  10, 9, 6, 4 ft/s

**4.20**  $2c^2x^3$; $2c^2y^3$; $x = y = 0$

**4.22**  $4x\hat{\mathbf{i}} + y\hat{\mathbf{j}} + z\hat{\mathbf{k}}$

**4.24**  (a) 4ft/s$^2$; 2 ft/s$^2$; (b) negative

**4.26**  10 ft/s; 10 ft/s$^2$

**4.28**  $-1.65 \times 10^{11}$ ft/s$^2$; $-5.12 \times 10^9$

**4.32** 720 m/s$^2$; 1440 m/s$^2$; 2160 m/s$^2$
**4.34** $-33.8$ ft/s$^2$; 1.05
**4.36** $(225x + 150)$ m/s$^2$; 0; 150 m/s$^2$; 375 m/s$^2$
**4.42** 8.02 ft/s
**4.46** 8860 m/s$^2$; 7440 m/s$^2$
**4.48** (a) 10 ft/s$^2$; (b) 20 ft/s$^2$, 10 ft/s$^2$; (c) 22.4 ft/s$^2$
**4.50** $a_x = 2c^2x \, (x^2 + y^2)$; $a_y = 2c^2y \, (x^2 + y^2)$
**4.52** 3.20 m/s$^2$; 1.48 m/s$^2$
**4.54** $-225$ ft/s$^2$; $-28.1$ ft/s$^2$; $-8.33$ ft/s$^2$
**4.56** 525 °F/s
**4.58** $-28.4$ °C/s; $-25.1$ °C/s
**4.60** (a) 200 °C/s; 100 °C/s
**4.62** 5.0 m$^3$/s for each line
**4.64** 132 ft$^3$/s
**4.68** (b) 2 $V_0 \, hb/3$
**4.72** (a) 3000 kg/s; (b) 3.00 m$^3$/s

## Chapter 5
**5.2** 20 m/s
**5.4** 3.66 slugs/s
**5.6** 1.57 in.
**5.8** 4.99 ft/s
**5.10** 314 m/s
**5.12** $-0.0189$ kg/s
**5.14** 0.0125 lb$_m$/ft$^3$
**5.16** 1.00 m$^3$/s
**5.18** 150 liters/s
**5.20** 3.63 ft/s
**5.22** 0.711; 0.791; 0.837; 0.866
**5.24** (a) 6280 kg/s
**5.26** (a) 0.00456 slug/s increasing;
(b) $2.28 \times 10^{-4}$ slug/ft$^3 \cdot$ s
**5.28** 1260 min
**5.30** 14.8 m/s
**5.32** (a) 15.6 gal/min; (b) 62.4 gal/min
**5.34** 120 N
**5.36** $2.66 \times 10^{-4}$ m$^3$/s
**5.38** 15.5 lb
**5.40** 7.01 ft$^3$/s; 674 lb
**5.42** 352 lb
**5.44** 2.69 lb
**5.46** (a) 6.26 m/s; (b) 392 N; (c) 675 N
**5.48** 0 N
**5.50** 482 N downward
**5.52** 17,900 lb
**5.54** 28.7 ft$^3$/s
**5.56** 4.02 lb
**5.58** (a) $W_1 - 9.8$ lb; (b) $W_2 + 14.7$ lb
**5.60** $-185$ kN; 45.8 kN
**5.62** 1290 lb/ft; 3120 lb/ft
**5.64** 3580 lb
**5.66** 2.96 lb
**5.68** 66.6 N
**5.70** 13.3 lb
**5.72** non-uniform = (4/3) uniform
**5.74** (a) 181 lb; (b) 146 lb

**5.76** (b) 234 lb
**5.78** 9170 lb
**5.80** 78.5 kN
**5.88** (a) 231 N $\cdot$ m, 185 rad/s; (b) 200 N $\cdot$ m, 160 rad/s;
(c) 116 N $\cdot$ m, 92.5 rad/s
**5.90** 39.6 hp
**5.92** (a) 43 deg; (b) 53.4 ft $\cdot$ lb/s
**5.92** $-36.8$ ft $\cdot$ lb/slug
**5.96** $-3130$ ft $\cdot$ lb/slug
**5.100** 0.32 ft; right to left
**5.102** 0.042 m$^3$/s
**5.104** $4.58 \times 10^{-3}$ m$^3$/s
**5.106** 5660 ft $\cdot$ lb/slug
**5.108** 926 ft $\cdot$ lb/slug; 200 ft $\cdot$ lb/slug
**5.110** 734 ft
**5.112** $2.22 \times 10^6$ W
**5.114** 5.48 hp
**5.116** Will work for 2 atm., not for 3 atm.
**5.118** 0.052 m$^3$/s
**5.120** 1610 ft $\cdot$ lb/slug; 1110 ft $\cdot$ lb/s
**5.122** (a) 4.08 hp; (b) 9.94 ft
**5.124** 0.58
**5.126** 1.78 kW
**5.128** (a) 0.56; (b) 11.7 lb
**5.132** 2.36 ft
**5.134** (a) 17.2 deg, 4.29 m/s; 558 N $\cdot$ m/s
**5.136** (a) 3.43 ft $\cdot$ lb/lb$_m$; (b) 3.36 ft $\cdot$ lb/lb$_m$
**5.138** 11.6 N $\cdot$ m/kg; 0.796
**5.140** 148 ft $\cdot$ lb/slug; 0.875

## Chapter 6
**6.2** $2x$, $4xt^2$; $-2y$, $4yt^2$; $V = 0$; $a = 4\hat{\mathbf{i}} - 4\hat{\mathbf{j}}$ ft/s$^2$;
$a = 5.66$ ft/s$^2$
**6.4** (a) 0; (b) $\omega = -(y/2 + z)\hat{\mathbf{i}} + (5z/2)\hat{\mathbf{j}} - (y/2)\hat{\mathbf{k}}$;
No
**6.6** $\zeta = 3xy^2\hat{\mathbf{k}}$; No
**6.8** $u = -\dfrac{3}{2}x^2 + \dfrac{x^3}{3} + f(y)$
**6.10** $\dot{\gamma} = \dfrac{-r_0\omega}{r_0 - r_i}$
**6.12** No
**6.14** (a) 2.24 m/s, $-63.4$ deg, $x = y = 0$; (b) 4.12 m/s,
76.0 deg, none
**6.16** No
**6.18** Second one
**6.22** $v = \dfrac{-y^2}{2x} + f(x)$
**6.24** $v_\theta = -4r\theta - 9r^2 \cos \theta + f(r)$
**6.26** (a) $\psi = \dfrac{-Ar^2}{2} + C$; (b) $\psi = -A \ln r + C$
**6.30** $19.6y - 100$ kPa, where $y$ is m
**6.34** $\psi = 2x - 4y + C$; $\phi = -4x - 2y + C$
**6.36** $\phi = A \ln r + Br \cos \theta + C$; $\theta = \pi$, $r = A/B$
**6.38** 60.5 psi
**6.40** (a) Yes; (b) Yes, $\phi = 2(x + y) + C$; (c) 0

**6.42**  (a) $\phi = U(0.866x + 0.500y)$;

$\psi = U(0.866y - 0.500x)$; (b) $\dfrac{\partial p}{\partial y} = -\gamma$

**6.46**  $-7.10$ kPa

**6.48**  $\psi = 5x^2y - 5y^3/3 + C$

**6.50**  92.3 ft

**6.52**  3.10 ft

**6.54**  (a) $1.60 \times 10^3/r^3$ kPa/m; (b) 184 kPa

**6.56**  0.00796 ft to right of slit; 0.0250 ft

**6.60**  80 mph; 53.3 mph; 40 mph

**6.62**  $h^2 = m/2\pi A$

**6.64**  $\sin\theta = (\Gamma/2\pi rU)\ln(r/4)$

**6.66**  (b) 0; 68.2 mph

**6.68**  $8.49 \times 10^{-4}$ m/s

**6.74**  $(2/3)^{1/2}$

**6.76**  10

**6.80**  0.00620 lb/ft$^2$

**6.84**  $\tau_{xy} = 30\mu xy$; $\tau_{yz} = \tau_{xz} = 0$

**6.86**  (a) $\partial v/\partial y = -2x$; (b) $\mathbf{a} = 2x^3\hat{\mathbf{i}}$;
(c) $\partial p/\partial x = 2\mu - 2\rho x^3$

**6.88**  43.2 lb/ft$^3$; 0.75 ft/s

**6.90**  $q = -2\,\rho g h^3/3\mu$

**6.92**  (a) $u = Uy/b$; (b) $Ub/2$

**6.94**  $u = [(U_1 + U_2)/b]y - U_2$

**6.96**  $y/b = 1/3$

**6.98**  $v_\theta = R^2\omega/r$

**6.100**  (a) $\mathcal{T} = 2\pi r_0^3\,\mu\omega\ell/(r_0 - r_i)$; (b) $\dot{\gamma} = -U/b$

**6.102**  57.1 N/m$^3$; 0.252 m/s

**6.106**  $\Delta p$ (nonuniform)/$\Delta p$ (uniform) $= 3.52$

**6.108**  $V_0 = (\Delta p/\ell)[2r_i^2\ln(r_i/r_0) - (r_i^2 - r_0^2)]/4\mu$

## Chapter 7

**7.6**  (a) 103 m/s; (b) 444 m/s

**7.8**  $q/(b^{3/2}g^{1/2}) = \phi\,(H/b,\ \mu/b^{3/2}g^{1/2}\rho)$

**7.10**  $Q/A^{5/4}\,g^{1/2} = \phi(\varepsilon/A^{1/2},\ S_0)$

**7.12**  $\omega\ell^{1/2}/g^{1/2} = \phi(h/\ell)$

**7.14**  $Q/d^{5/2}\,g^{1/2} = \phi(\rho_g/\rho_a,\ h/d)$

**7.16**  $\delta/d = \phi\,(\ell/d,\ \rho V^2/E,\ \mu V/dE)$

**7.18**  $h/\ell = \phi(\sigma/\ell^2\,g\rho)$

**7.20**  $h/D = \phi(\sigma/\gamma D^2)$

**7.22**  $\Delta p \propto 1/D^2$ (for a given velocity)

**7.26**  $t\sqrt{g/d} = \phi(\mathcal{V}/d^3,\ \rho\sqrt{g}\,d^{3/2}/\mu)$

**7.28**  (a) $VD\sqrt{\rho/\mathcal{W}} = \phi(b/d,\ d/D)$;
(b) $V = \sqrt{2\mathcal{W}b/\pi\rho dD^2}$

**7.30**  $5.58 \times 10^7$; 1.88; 1.44; $4.17 \times 10^{-3}$

**7.32**  $\Delta p/\rho V_1^2 = -1.10\,(A_1/A_2)^2 + 1.07\,(A_1/A_2)$
$-\ 0.0103$; 6.26 lb/ft$^2$

**7.34**  $\Delta p/(\rho V^2) = 0.505\,(D/d)^{3.99}$

**7.36**  (a) $H/b = \phi(h/b,\ \ell/b)$;
(b) $H/b = 0.0833\,(h/b)^{-1.00}$

**7.38**  (a) Correct; (b) $t = 1.36\,\mu$

**7.40**  (a) $-1/2$; (b) Yes

**7.42**  (a) $\theta = 2.98 \times 10^3\,(\omega\mu\ell^3/K)$

**7.44**  11.0 m/s

**7.46**  Colder

**7.48**  Double pressure

**7.50**  $6.52 \times 10^{-2}$ ft/s

**7.52**  129 m/s

**7.54**  3750 mph; No

**7.56**  1170 km/hr

**7.58**  187 mph

**7.60**  (a) 4.50 mph; (b) 8.00 ft

**7.62**  0.125; 0.0625

**7.64**  (a) $h/H = \phi(d/H,\ b/H,\ \rho g/\gamma_s,\ V/\sqrt{gH})$;
(b) 21.2 mph

**7.66**  26.4 lb

**7.68**  50.2 kPa (abs)

**7.70**  (a) $\Delta p/\rho c^2 = \phi(E/\rho c^2 d^3)$

**7.72**  (a) $V_{0m}\,t_m/D_m = V_0\,t/D$, $\omega_m\,t_m = \omega t$,
$\rho_m\,V_{0m}\,D_m/\mu_m = \rho V_0\,D/\mu$, $D\,\Delta p_\ell/\rho V_0^2 =$
$D_m\,\Delta p_{\ell m}/\rho_m\,V_{0m}^2$; (b) 160 rad/s

**7.74**  100 mph to 450 mph

**7.76**  (a) 0.0006 slugs/ft$^3$; (b) 9.49 ft/s; (c) 1.90 Hz

**7.78**  (a) 2.91 m/s; (b) 2.61 N

**7.80**  (a) 400 N; (b) 7220 W

**7.82**  (a) $V\ell^2/Q = \phi(\ell_i/\ell,\ Q^2/\ell^5 g,\ \rho Q/\ell\mu)$; no
(b) 0.410 gpm; 2.46 in.

**7.84**  (a) $nD/V = \phi(VD/\nu)$; (b) Yes; (c) 7.54 ft/s

**7.86**  (a) $V_m/U_m = V/U$, $V_m\,D_m/\nu_{sm} = VD/\nu_s$,
$V_m^2/g_m\,D_m = V^2/gD$, $(\rho - \rho_s)_m/\rho_m = (\rho - \rho_s)/\rho$;
(b) No

**7.86**  0.440 ft; 2.80 ft$^3$/s

**7.92**  (a) $d^2y^*/dx^{*2} = (P\ell^2/EI)(x^* - 1)$

**7.94**  $\partial u^*/\partial t^* = (X/\rho h\omega^2)\cos t^* + (\mu/\rho h^2\omega)\partial^2 u^*/\partial y^{*2}$

## Chapter 8

**8.2**  Turbulent

**8.4**  3.36 ft

**8.6**  $4.95 \times 10^{-6}$ m$^3$/s at 20 °C

**8.8**  3 m; 8.83 N/m$^2$

**8.10**  2 m/s; 1 m/s; $1.26 \times 10^{-3}$ m$^3$/s

**8.12**  0.15 lb/ft$^2$; 0.06 lb/ft$^2$; 0 lb/ft$^2$

**8.14**  (b) to (a)

**8.16**  0.354 D

**8.18**  3.43 m; 166 kPa

**8.20**  $z = 1 + e^{-(t/2.8 \times 10^5)}$ ft, where $t$ is seconds

**8.22**  1.89 m/s; $7.42 \times 10^{-3}$ m$^3$/s

**8.24**  (a) $4.69 \times 10^{-3}$ ft$^3$/s; $3.30 \times 10^{-3}$ ft$^3$/s

**8.26**  $8.88 \times 10^{-8}$ m$^3$/s

**8.28**  $2.4 \times 10^{-4}$ m$^3$/s

**8.30**  (a) 0.707 $R$; (b) 0.750 $R$

**8.34**  $1.02 \times 10^{-4}$ ft

**8.36**  B to A; A to pump

**8.38**  (a) 0.266 psi; (b) 1.13 psi; (c) $-0.601$ psi

**8.40**  $2.04 \times 10^{-3}$ W

**8.42**  0.0300

**8.44**  0.0162

**8.46**  0.211 psi/ft

**8.48**  0.0404

**8.50**  19.6 m

**8.52**  0.0132 ft

**8.54**  (a) 0.556 ft; (b) 3.70 ft; (c) 0.926 ft

| | | | | |
|---|---|---|---|---|
| **8.56** | 47.5 | | **9.46** | 2.83 $\mathcal{D}_U$; 0.354 $\mathcal{D}_U$ |
| **8.58** | 39.7 kPa; 39.7 kPa, 93.0 kPa | | **9.52** | 0.0438 N · m |
| **8.60** | 13 | | **9.54** | 1.06 m/s |
| **8.62** | 0.223 lb/ft$^2$ | | **9.58** | (a) 567 ft/s; (b) 118 ft/s; (c) 13.5 ft/s |
| **8.64** | 9 | | **9.60** | 133 N |
| **8.66** | Disagree | | **9.62** | 77.0 ft/s |
| **8.68** | 1.54 kPa | | **9.64** | 7080 N · m |
| **8.70** | 0.188 m | | **9.68** | 30.6 ft/s |
| **8.72** | 0.899 lb/ft$^2$ | | **9.70** | 378 lb |
| **8.74** | (a) 135 ft; (b) 137 ft | | **9.72** | 58.4 hp |
| **8.76** | 84.0 ft | | **9.74** | 0.946 ft/s in water; 30.2 ft/s in air |
| **8.78** | 0.750 psi | | **9.76** | (a) 2.44 m/s; (b) 2.13 m/s |
| **8.80** | 24.4 hp | | **9.78** | 5.31 m/s |
| **8.82** | 21.2 psi | | **9.80** | 1.42 lb; 142 lb |
| **8.84** | 16.5 ft | | **9.82** | 7220 m/s$^2$ |
| **8.86** | 48.0 psi | | **9.84** | 9.65 lb/ft$^2$ |
| **8.88** | 379 kW | | **9.86** | (a) 4.31 MN; (b) 4.17 MN |
| **8.90** | $5.46 \times 10^{-4}$ m$^3$/s | | **9.88** | 53.5 kW; 4.46 kW |
| **8.92** | 0.0615 ft$^3$/s | | **9.92** | 3 |
| **8.94** | 0.0289 m$^3$/s | | **9.94** | 4.35% |
| **8.96** | 155 hp | | **9.96** | 41.6 ft/s |
| **8.98** | 0.0494 ft$^3$/s | | **9.98** | 30.8 hp |
| **8.100** | 0.476 ft | | **9.100** | (a) 0.0166 hp; (b) 32.4 hp |
| **8.102** | 0.535 ft | | **9.102** | 0.851; 0.301 |
| **8.104** | 0.155 ft | | **9.104** | 65.9 hp |
| **8.106** | 0.0445 m | | **9.106** | 0.225 |
| **8.108** | 0.500 ft$^3$/s | | **9.108** | 1.72 $U$ |
| **8.112** | 0.0180 m$^3$/s | | **9.110** | 0.206 |
| **8.114** | 0.0284 m$^3$/s; 0.0143 m$^3$/s; 0.0141 m$^3$/s | | **9.114** | Yes |
| **8.118** | 18.0 m$^3$/s | | **9.116** | 0.480; 0.409; 0.451; 0.482 |
| **8.120** | 32.4 kPa | | **9.118** | 5.72% |
| **8.122** | 0.528 ft$^3$/s | | **9.122** | 28.4% |
| **8.124** | 0.0221 m$^3$/s | | | |
| **8.126** | 5.69 ft$^3$/s | | **Chapter 10** | |
| **8.128** | 0.0936 ft$^3$/s | | **10.2** | Subcritical |
| **8.130** | 11.1 in. | | **10.4** | Subcritical; 1.04 m |
| | | | **10.8** | 5.66 ft |
| **Chapter 9** | | | **10.10** | 2.80 m/s |
| **9.2** | (a) 0.06 ($\rho U^2/2$); (b) 2.40 | | **10.12** | 175 ft/s |
| **9.4** | No | | **10.14** | 616 km/hr |
| **9.6** | 5.43 kN | | **10.16** | 2.45 m or 0.388 m |
| **9.8** | 0.142 | | **10.18** | 5.82 ft or 1.129 ft |
| **9.12** | 1.12 m; $7.92 \times 10^{-3}$ m | | **10.20** | 1.774 ft or 1.974 ft |
| **9.14** | $6.65 \times 10^{-6}$ m$^2$/s | | **10.24** | 1.828 ft |
| **9.16** | $7.54 \times 10^{-3}$ ft/s; 0.471 ft/s; $2.62 \times 10^{-5}$ ft/s | | **10.26** | 0.694 ft |
| **9.20** | $d = 1 + 0.0304\sqrt{x}$ ft where $x$ is ft | | **10.30** | 2 ft or 3.51 ft; 2 ft or 1.38 ft |
| **9.22** | (a) 27.6 ft/s; (b) 20.5 mph | | **10.34** | 0.000532 |
| **9.26** | (a) 0.171 ft; (b) 0.134 ft | | **10.36** | 626.9 ft |
| **9.28** | (b) $\delta/x = 5.48/\sqrt{Re_x}$ | | **10.38** | 3.49 N/m$^2$ |
| **9.30** | (b) $\delta = 5.83\sqrt{\nu x/U}$ | | **10.42** | (a) 1.80 lb/ft$^2$; (b) 0.0469; (c) 0.636 |
| **9.34** | 0.707 | | **10.44** | 7.42 min |
| **9.36** | 48.8 lb | | **10.46** | 23.7% |
| **9.38** | 10,300 lb | | **10.48** | 5.73 ft/s |
| **9.40** | 1.4; upright | | **10.50** | Yes |
| **9.42** | 85.4 kW | | **10.52** | 40.9 m$^3$/s |
| **9.44** | $1.88 \times 10^4$ lb; 33.3% | | **10.54** | 8.42 ft/s |

**10.56**   6.22 ft/s
**10.58**   17.1 ft
**10.60**   18.2 m$^3$/s
**10.64**   119 ft$^3$/s
**10.66**   1.19 m
**10.68**   0.841 m
**10.70**   0.861 m
**10.72**   Each requires the same.
**10.74**   No
**10.76**   5.16 ft
**10.78**   2.59 m
**10.80**   10.66 m
**10.82**   1.22 m
**10.84**   5.94 ft
**10.86**   0.00226
**10.88**   0.000664
**10.90**   0.000269
**10.92**   0.452 in.
**10.94**   1.68 m
**10.96**   401 kW
**10.100**   84.5 m$^3$/s
**10.102**   0.0378 m
**10.104**   0.0577; 0.000240
**10.106**   0.652 m; 2.61 m
**10.108**   0.116 m$^3$/s
**10.110**   4.70 ft
**10.112**   20.1 s
**10.118**   0.350 m$^2$/s

## Chapter 11
**11.4**   (a) 71,700 J/kg; (b) 100,000 J/kg;
(c) $-1.58$ kg/m$^3$; (d) 396 J/kg $\cdot$ K
**11.8**   $-5.08 \times 10^6$ N $\cdot$ m/kg; $-7.16 \times 10^6$ N $\cdot$ m/kg;
$-1.40 \times 10^4$ N $\cdot$ m/kg
**11.10**   387 °R
**11.12**   905 °R; $-4.17 \times 10^6$ ft $\cdot$ lb/slug; 877 ft $\cdot$ lb/slug
**11.16**   5300 m/s; 1470 m/s; 340 m/s
**11.18**   (a) 2000 ft/s; (b) 2930; (c) 893 m/s
**11.20**   (a) 0.0328; (b) 0.0722; (c) 0.131
**11.22**   (a) 732 m/s; (b) 2400 ft/s; (c) 1636 mph
**11.24**   8.75 m
**11.28**   (a) 590 mph; (b) 866 ft/s; (c) 264 m/s
**11.30**   From 0.47 to 0.75
**11.32**   No
**11.34**   (a) 0.528, 0.833; (b) 0.546, 0.869; (c) 0.488, 0.752
**11.36**   283 m/s; 0.89
**11.38**   269 m/s; 0.90
**11.42**   1.8; 1580 ft/s
**11.44**   26.5 kg/s

**11.50**   9.8%
**11.52**   (a) 1.70 kg/s; (b) 1.52 kg/s
**11.54**   1.55 kg/s
**11.58**   17 psia; 868 ft/s; 1390 Btu/s
**11.64**   2.29
**11.66**   1.9; 648 m/s
**11.68**   92 kPa (abs); 82 kPa (abs)
**11.70**   1160 °R; 132 psia; 0.81 slug/s
**11.72**   1.35 m
**11.74**   at duct entrance $T = 130$ K; $p = 6.0$ kPa (abs);
$T_0 = 293$ K; $p_0 = 101$ kPa (abs); $V = 571$ m/s;
just upstream of shock $T = 296$ K;
$p = 17$ kPa (abs); $T_0 = 396$ K;
$p_0 = 47.7$ kPa (abs); $V = 448$ m/s;
just downstream of shock $T = 354$ K;
$p = 30.8$ kPa (abs); $T_0 = 396$ K;
$p_0 = 46.4$ kPa (abs); $V = 299$ m/s;
at duct exit $T = 351$ K; $p = 26.6$ kPa (abs);
$T_0 = 407$ K; $p_0 = 45.9$ kPa (abs); $V = 337$ m/s

## Chapter 12
**12.2**   55.4 deg
**12.4**   10.8 lb/ft$^2$
**12.8**   $-71$ ft $\cdot$ lb/slug
**12.10**   (b) fan; (c) 1780 ft$^2$/s$^2$
**12.12**   379 ft $\cdot$ lb/lb$_m$
**12.14**   (b) 891 ft $\cdot$ lb
**12.16**   61.3 ft
**12.18**   107 gpm
**12.20**   (a) 5.15 ft; (b) $-2.49$ ft
**12.22**   (a) 2.07 kW; (b) 12.6 m
**12.26**   1400 gal/min
**12.28**   158 gal/min; Yes
**12.30**   1740 gal/min; Yes
**12.34**   0.0328 m$^3$/s; 8.00 m
**12.36**   Yes
**12.38**   1000 gpm; 800 ft
**12.40**   Centrifugal
**12.42**   1900 rpm
**12.44**   (a) 18.3 ft; radial-flow
**12.48**   (a) $-15.6$ m$^2$/s$^2$; (b) $-15.6$ m$^2$/s$^2$
**12.50**   $-19.8$ hp
**12.52**   16.6 m$^2$/s$^2$; 0.849
**12.54**   0.0109 ft$^3$/s; 0.0409 hp
**12.56**   816 kW pump
**12.58**   23,200 kW
**12.60**   26,600 N; 37.6 m/s; 707 kg/s
**12.64**   70 m
**12.68**   impulse

# Index

## A

Absolute pressure, 13, 50
Absolute temperature:
    in British Gravitational (BG) System, 8
    in English Engineering (EE) System, 8
    in ideal gas law, 12, 13
    in International System (SI), 7
Absolute velocity:
    defined, 208, 669
    in rotating sprinkler example, 230–231
    in turbomachine example, 670, 671, 672–673
    *vs.* relative velocity, 186–187
Absolute viscosity:
    of air, 740, 741
    of common gases, 14, 737
    of common liquids, 12, 737
    defined, 15
    of U.S. Standard Atmosphere, 742, 743
    of water, 739
Acceleration:
    along streamline, 168–169
    centrifugal, 174, video V3.4
    convective, 170, 174
    defined, 166
    as field, 166, 278
    gravitational, 7, 8, 742, 743
    local, 169
    normal, 103, 173–174, video V3.1,
        video V3.4
    particle, 103, 166, 278
    role of material derivative, 166–168, 278
    streamwise, 103, 174, video V3.1, video V5.3
Accessibility of flow, 564
Acoustic velocity:
    of air, 740, 741
    defined, 22
    of water, 739
Adiabatic flow, 237, 606
    with friction, *see* Fanno flow
    reversible, *see* Isentropic flow
Adverse pressure gradient, 313, 509
Aerodynamics, in history of fluid mechanics, 28
Affinity laws, pump, 691
Air, table of properties, 740, 741
Airfoils, 533–534, video V9.21, video V9.23,
    video V9.24
Alternate depths, 563
Andrade's equation, 18
Aneroid barometers, 57
Angle of attack, 511, 534, 535, 537, 538
Angle of contact, 25
Angular deformation, 277, 279–282, video V6.3
Angular momentum, *see* Moment-
    of-momentum
Angular momentum flow, 228, video V5.12
Annulus, flow in, 329–330
Apparent viscosity, 16–17

Archimedes, 27, 28, 71
Archimedes' principle, 70–73, video V2.7
Area:
    centroid, 61, 62
    moment of inertia, 61, 62
    product of inertia, 62
Area ratio, critical, isentropic flow, 624
Aspect ratio:
    airfoil, 534
    plate, 355
Atmosphere, standard:
    defined, 49
    pressure variation, 49–50
    properties of, 49, 742, 743
Atmospheric pressure, 13, 49, video V2.2
Available energy, 244–245, 249, 605, 653
Average velocity, in mass flowrate, 202
Axial-flow machines:
    compressors, 708–710
    defined, 669
    pumps, 693–694
Axisymmetric flow, 482

## B

Backward curved blades, 679, 680
Backward difference, 727, 730
Barometers:
    aneroid, 57
    mercury, 51
Basic dimensions, 4, 348
Bellows meters, 465
Bends, 438
Bernoulli, D., 27, 28, 107
Bernoulli equation:
    cautions on use of, 107–108, 294
    derivation from Euler's equations, 293–294
    derivation from first law of thermodynamics,
        243–249
    derivation from F=*ma*, 107, 292–293,
        video V3.1
    examples of use, 117–131
    extended, 245
    irrotational flow, 138–139, 296
    outside boundary layer, 492
    physical interpretation, 110–113
    restrictions on use of, 134–139, 294,
        video V3.12
    unsteady flow, 136–138
    use of, video V3.2
Best efficiency points, 682
Best hydraulic cross section, 574–575
BG units, *see* British Gravitational (BG) System
    of units
Bingham plastic, 17
Blade velocity, 669, 670, 671
Blasius, H., 29, 367, 494–497

Blasius formula, 367, 432
Blasius solution, 495–497
Blood pressure, measuring, 2, 45, 52,
    video V2.3
Blowers, 667, 695
Blunt bodies, 482, video V9.2
Body forces, 42, 214, 289
Boiling, 23–24, 124–125
Boosters, 695
Bottom slope, 561–562
Bound vortex, 538, 539, video V4.6
Boundary conditions:
    in Bernoulli equation, 107
    inviscid flow, 308, 309
    no-slip, 15, 310, 323, video V1.4,
        video V6.12, video V6.13
    in solving CFD simulation problems, 729
Boundary element method, 725, 726
Boundary layer:
    concept in history of fluid mechanics, 27–28
    concept of, 295–296, 487, video V1.4,
        video V6.12, video V9.12
    displacement thickness, 491–492
    effect of wall roughness, 523
    effects of pressure gradient, 507–511,
        video V9.10
    equations, 494–497
    flat plate, 489–493
    and flow separation, 487
    laminar flow
        approximate solution, 501–502
        Blasius solution, 495–497
    momentum integral equation for, 497–502,
        511–512
    momentum thickness, 493
    pipe, 405
    separation, 509, video V6.8
    thickness, 490–492, video V9.6
    transition from laminar to turbulent flow,
        490, 502–503, video V9.7
    turbulent flow, 490, 504–505
    velocity profiles, 503, 509, 510, video V9.4
    vorticity, 490
Bourdon pressure gage, 57, video V2.4
Boussinesq, J., 422
Brake horsepower, 681
British Gravitational (BG) System of units, 8
Broad-crested weirs, 585–587, video V10.14
Buckingham, E., 29, 349
Buckingham pi theorem, 349–350
Bulk modulus, video V1.7
    characterizing compressibility, 20, 21
    of common liquids, 12
    defined, 20
    of elasticity, 609
Buoyant force, 70–73, video V2.7, video V2.9,
    video V11.5

**C**

Canals, 566
Capillary action, 25, video V1.10
Capillary tube viscometer, 33, video V1.5
Casings, 675, 676
Cauchy, A., 28, 363
Cauchy number, 361, 363
Cavitation:
    defined, 24, 124
    number, 363, 377, 386
    producing in flowing liquid, 125
Center of buoyancy, 71–72, video V2.12
Center of pressure, 62
Central difference, 727
Centrifugal pumps, 675–688, video V12.3
    performance curves, 685–687
    system curve, 685
    theoretical considerations, 676–679
Centroid, 61, 62
CFD, *see* Computational fluid dynamics (CFD)
Chaos, 426–427
Chezy, A., 28, 568
Chezy coefficient, 568
Chezy equation, 568
Choked flow, 619, 635
Chord length, 533, 534
Circulation:
    flow past airfoils, 537–540
    origin of, 538
    and vortex motion, 305
Colebrook formula, 431
Completely turbulent flow, 429–430
Composite body drag, 526–528
Compressibility criterion, 135
Compressibility of fluids, 20–23, 47–48
Compressible flow, video V11.2, video V11.5
    analogy with open-channel flow, 655–657
    with area change, 615–617, video V11.6,
        video V11.7
    categories, 610–613
    effect on use of Bernoulli equation, 134–136
    with friction, 631, 633–634
    graphs of, 744–747
    with heat transfer, 642–649
    introduction, 601–602
    as reversible process, 134–135
    turbomachines, 707–713
    two-dimensional, 657–658
Compression waves, 657–658, video V11.8
Compressors, video V12.6
    axial-flow, 708–710
    corrected speed, 711
    defined, 695, 708
    multi-stage, 708
    as pumps, 667
    radial-flow, 708
    stalling, 710
    surging, 710
Computational fluid dynamics (CFD),
        video V2.11, video V6.2, video V6.16,
        video V11.6, video VA.1
    advantages for flow simulation studies, 734
    difficulties in using, 734
    importance of verification and validation
        steps, 734–736
    methodology, 733–734
    overview, 277, 331, 725
    role of boundary conditions, 729
    use of finite difference method, 727–728
Conservation of energy, *see* First law of
    thermodynamics
Conservation of linear momentum, 288–292,
        video V5.4, video V5.5, video V5.6
Conservation of mass, 120, 199, 200–212,
        282–288
Contact angle, 25, video V1.11
Continuity equation, *see also* Conservation of
    mass
    cylindrical coordinates, 285
    defined, 283–284
    differential form, 282–284
    finite control volume, 120–121
        compressible flow, 615
        incompressible flow, 207, 208, video V5.1,
        video V5.2
        unsteady flow, 201, 284
    rectangular coordinates, 282–283
Continuum hypothesis, 4
Contraction coefficient, 119, 120, 129, 587
Control surfaces, 175
Control volume, video V4.14
    defined, 175
    deforming, 176, 210–212
    in derivation of continuity equation, 201
    and first law of thermodynamics, 237–238
    fixed, 175, 202–208
    guidelines for selection, 187–188, 212
    infinitesimal, 276
    moving, 186–187, 208–210, 226–228
    velocity, 208
Convective acceleration, 170, 174, 321
Convective derivative, 169–170
Converging-diverging diffusers, 617
Converging-diverging duct flow, 616–620,
        video V11.6, video V11.7
Converging-diverging nozzles, 617,
        video V11.6, video V11.7
Conversion factors, for units of measure, 9
Core flow, 296
Couette flow, 324–326
Creeping flow, 331, 362
Critical depth, 563
Critical flow, open-channel, 556, 560, 656
Critical pressure ratio, 619
Critical Reynolds number, 490, 502
Critical state, compressible flow, 619, 637
Curl operator, 281
Curved surfaces, forces on, 68–70
Cylinders:
    circular, 313–316
    drag coefficient for, 520, video V9.13
    inviscid flow around, 314, video V6.6
    large *vs.* low Reynolds number flow around,
        487, video V6.8, video V6.16
    pressure distribution on surface, 314, 510–511
    rotating, 316–318, 539
    separation location, 487, 509
Cylindrical coordinates, 77, 285, 297

**D**

D'Alembert, J., 27, 28, 315
D'Alembert's paradox, 315, 508–509
Darcy, H., 28, 413
Darcy friction factor, 413–414
Darcy-Weisbach equation, 429
DaVinci, L., 27, 28
Deepwater waves, 559
Deformation:
    angular, 277, 279–282, video V6.3
    linear, 277, 278–279
    rate of, 279
Deforming control volume, 176, 210–212
Del (gradient) operator:
    cylindrical coordinates, 297
    rectangular coordinates, 43
Density:
    of air, 740, 741
    of common gases, 14
    of common liquids, 12
    as fluid characteristic, 11, video V2.10
    of U.S. Standard Atmosphere, 742, 743
    of water, 11, 739
Dental drills, 702–704
Derivative:
    convective, 169–170
    local, 169
    material, 166–169, 183–184, 278
    substantial, 166–169, 278
Detached curved shock, 658
Differential analysis:
    and conservation of linear momentum,
        288–292
    and conservation of mass, 282–288
    determination of pi terms by inspection,
        359–360
    determination of pi terms by example,
        357–358
    and fluid element kinematics, 277–282
    introduction, 276–277
    and inviscid flow, 292–299
    and plane potential flows, 299–318
Differential equations, *see* Differential analysis
Diffuser pumps, 676, video V8.13
Diffusers, 437–438, 617
Dimensional analysis:
    and Buckingham pi theorem, 349–355
    concept of dimensions, 4–5
    determination of reference dimensions,
        356–358
    for fully developed laminar pipe flow, 413–414
    overview, 347–349
    selection of variables, 355–356
Dimensional homogeneity, 4–5, 349
Dimensionless groups:
    common in fluid mechanics, 360–365
    defined, 347–348
Dimensionless products:
    defined, 347–348
    pi terms as, 349, 360–361
Dimensions:
    basic, 4, 348
    of common physical quantities, 5
    reference, 349–350, 356–358
Discharge coefficient:
    nozzle, 461
    orifice, 460
    underflow gates, 587–588
    Venturi meter, 462
    weir, 583, 585
Discretization, 725–727
Displacement thickness, 491–492

Distorted models, 373–374, video V7.8,
    video V7.11, video V7.12, video V7.17,
    video V7.18
Diverging-converging duct flow, 617
Doppler effect, 612
Doublet, 306–308
Drag:
    on circular cylinder, 315
    defined, 482–483
    example, 484
    form, 514–516, video V7.2, video V9.11
    friction, 513–514
    friction coefficient, 501, 506
    overview, 512–513
    pressure, 514–516
    on rotating cylinder, 317
Drag coefficient:
    airfoil, 538
    automobile, 527–528
    based on friction drag, 513–514
    composite body, 526–528
    cylinder, 520, video V9.13
    defined, 485
    effect of compressibility on, 522–523, 610–611
    effect of Froude number on, 525–526,
        video V9.18
    effect of Mach number on, 522–523,
        610–611
    effect of Reynolds number on, 517–518,
        610–611
    effect of shape on, 516–517, 521,
        video V9.11, video V9.19
    effect of surface roughness on, 523–524
    ellipse, 516–517
    flat plate normal to flow, 516, 517, 518, 521,
        video V7.2
    flat plate parallel to flow, 506, 516, 517, 518,
        521
    form, 514
    friction, 501, 506
    in model studies of flow around bodies, 377
    pressure, 514
    Reynolds number dependence, 517–521
    selected three-dimensional objects, 528, 530
    selected two-dimensional objects, 528, 529
    sphere, 379–380, 518, 520, 523–524, 540,
        610–611, video V9.17
    streamlined body, 517
    vehicle, 527–528
Ducts:
    constant area, flow, 631
    converging-diverging, flow, 616–620
    dimensional analysis, 442–445
    vs. pipes, 400
Dynamic pressure, 113
Dynamic similarity, 370
Dynamic viscosity:
    of air, 740, 741
    of common gases, 14, 737
    of common liquids, 12, 737
    defined, 15
    of U.S. Standard Atmosphere, 742, 743
    of water, 739

**E**

Eddies, 420
Eddy viscosity, 422, video V8.7

EE units, *see* English Engineering (EE) System
    of units
Efficiency:
    centrifugal pump, 681
    fan performance example, 252
    hydraulic, 681
    isentropic, 710
    mechanical, 681
    overall, pump, 681
    polytropic, 710
    pumps *vs.* turbines, 705–706
    volumetric, 681
Elbows, losses in, 438
Elevation head, 111, 132
Energy:
    available, 244–245
    internal, 236, 602–603
    kinetic, 111, 236
    per unit mass, 236
    per unit volume, 247
    per unit weight, *see* Head
    potential, 111, 236
    specific, 562–565
    stored, 236
Energy equation:
    application to nonuniform flows, 249–252
    application to one-dimensional flows,
        239–243
    combined with moment-of-momentum
        equation, 252
    comparison with Bernoulli equation,
        243–249
    defined, 239
    derivation, 236–239
Energy line, 131–133, 447–448, 561–562
English Engineering (EE) System of
    units, 8–9
Enthalpy, 240, 603, 606
Entrance length, 405, video V8.12
Entrance loss, 434, 435
Entrance region, 296, 405, 434
Entropy, 605, 606
Equation of state, 12–13, 602
Equations:
    general homogeneous, 5
    of motion, 291–292
    restricted homogeneous, 5
Equipotential lines, 300
Equivalent length, 433, 434
Euler, L., 27, 28, 292, 363
Euler number, 361, 363, 386
Euler turbomachine equation, 674
Eulerian description of motion, 160–161,
    video V6.2
Euler's equations, 102
Euler's equations of motion, 292
Exhausters, 695
Exit pipe loss, 435, video V8.12
Expansion waves, 630, 657, video V11.6
Extensive property, 176–177
External flow:
    background, 480–481
    boundary layer characteristics, 489–512
    drag, 512–528
    general characteristics, 481–489
    lift, 528–541
Eyes, in pump casings, 675

**F**

Falling head curve, 681
Fanning friction factor, 413
Fanno flow:
    basic equations of, 631–632
    choking length for, 635
    defined, 631
    example of nonisentropic flow of ideal gas,
        631–642
    graph for computations of, 745
    normal shock in, 635
    *T—s* diagram, 632, 635
Fans, 667, 695
Favorable pressure gradient, 509
Field representation, 158, video V4.2
Finite control volume analysis, 199–253
Finite difference method, 725, 726–728, 733
Finite element method, 725–726, 728
Finite volume, 725
First law of thermodynamics, 200, 236–252,
        video V5.15, video V5.16
First moment of area, 61
Fittings, losses in, 438
Flaps, airfoil, 535–536, video V9.23, video V9.24
Flat plates:
    Blasius solution, 495–497, 496
    drag coefficient
        normal to flow, 516, 517, 518, 521,
            video V7.2
        parallel to flow, 506, 516, 517, 518, 521
    momentum integral equation, 497–502
        laminar flow, 499
        turbulent flow, 504
    roughhouse effect, 506
Floating bodies, video V2.7
    Archimedes' principle, 70–73
    stability, 73–74
Flow coefficient, 689, 705
Flow measurement:
    external flow
        Pitot tube, 114–115, 132, 653–654,
            video V3.8
    internal flow
        bellows meter, 465
        nozzle meter, 127, 459, 460–461, 462–463
        orifice meter, 128, 459, 460
        rotameter, 463, video V8.15
        rotating disk meter, 465, video V8.16
        turbine meter, 464
        Venturi meter, 127, 128, 459, 461–462,
            video V3.10
    open-channel flow
        sluice gate, 128–130, 587–589
        weir, 130–131, 582–587, video V10.13
Flow meters, *see* Flow measurement
Flow net, 300
Flow separation, 487, 511, video V9.9,
        video V9.19
Flow visualization, 161, 162–166, video V4.1,
        video V4.6, video V4.9, video V7.16,
        video V9.19, video V11.5
Flowrate:
    mass, 120, 201–202
    volume, 120, 202
Fluid dynamics *vs.* fluid statics, 11
Fluid machines, 667–668, videos
        V12.1–V12.5

Fluid mechanics:
    defined, 1
    flow parameters, 1–2
    history of, 27–29
Fluid statics:
    equilibrium in, 40
    forces in, 59–65
    pressure height relation, 44, 47–48, 49–50
    *vs.* fluid dynamics, 11
Fluids:
    defined, 3
    describing characteristics, 4–5
    flow parameters, 1–2
    ideal, 294
    inviscid, 292–299
    molecular structure, 4
    Newtonian, 15
    non-Newtonian, 16, video V1.6
    properties, 12, 14, 737–741
Force:
    anchoring, 214, 218
    body, 42, 214, 289
    buoyant, 70–73, video V2.7, video V2.9
    compressibility, 361, 363
    drag, 315, 482–483, 484, 512–528
    gravity, 361
    hydrostatic
        on curved surfaces, 68–70, video V2.6
        on plane surfaces, 59–65, video V2.5
    inertia, 361
    lift, 315, 483
    pressure, 57, 66, 361
    shear, 14
    surface, 42–43, 214, 289
    surface tension, 24–26, 361, video V1.9
    as unit of measure, 7, 8–9
    viscous, 15, 361
Forced vortex, 304
Form drag, 514–516
Forward curved blades, 679, 680
Forward difference, 727, 732
Francis, J., 704
Francis turbine, 704, 705
Free jets, 118–119, video V3.9
Free surface, 60
Free vortex, 109, 304, 306, video V3.6, video V6.4
Free-stream velocity, 508
Free-surface flow, 381–382
Frequency parameter, 364
Friction coefficient, 500–501
Friction drag, *see* Drag
Friction factor:
    data correlation for, 430
    in dimensional analysis of laminar pipe flow,
        413–414
    Moody chart, 429, 430–431
    noncircular conduits, 443
    smooth pipe, 430
    in wholly turbulent flow, 429–430
Friction slope, 561
Frictionless flow:
    compressible adiabatic, *see* Isentropic flow
    compressible with heat transfer, *see* Rayleigh
        flow
    incompressible, 107, 294
    open channel, *see* Open-channel flow
Fricton velocity, 423

Frontal area, 485, 516
Frontinus, S., 27, 28
Froude, W., 28, 362
Froude number, 361, 362, 382–384, 386, 556,
    559–560, video V7.4, video V10.5
Fully developed flow:
    applying Navier-Stokes equations, 411–412
    laminar, 407–415, video V8.10
    in pipes, 405–406
    turbulent, 405, 416–420

**G**
$G_c$, 8–9
Gage fluid, 53
Gage pressure, 13, 50
Galilei, G., 27, 28
Gas constants:
    for common gases, 14
    defined, 14
    in ideal gas equation of state, 12, 602
    universal, 602
Gas dynamics, *see* Compressible flow
Gas turbine engines, 711–713
General homogeneous equations, 5
Geometric similarity, 370
Grade line:
    energy, 131–133, 561–562
    hydraulic, 131–133
Gradient, pressure, *see* Pressure gradient
Gradient operator, 167, 278, 297
Gradually varied flow (GVF), 555, 565, 575–576
Gravity, acceleration of, 7, 8, 742, 743
Grids, 725, 727–729, 733, video VA.2
Guide vanes, 438–439, 676, 692, 693, 704–705,
    709

**H**
Hagen, G., 28, 326, 410
Hagen-Poiseuille flow, 326, 410, video V6.14
Half-body, 308–311, video V6.5
Head:
    elevation, 111, 132
    in energy line concept, 131–132
    loss, 247
    piezometric, 132
    pressure, 45, 111, 132
    pump, 678–679, 680, 684–685
    shaft, 247
    shaft work head, 247
    total, 132
    velocity, 111, 132
    weir, 582–587
Head coefficient, 705
Head curve:
    falling, 681
    rising, 681
Head loss:
    defined, 247
    in diffusers, 437–438
    in enlargements and contractions, 435–438
    in entrances, 434, 435, video V8.12
    in exits, 435, video V8.12
    in gradual contractions, 437–438
    in hydraulic jumps, 578, video V10.11,
        video V10.12
    major, 427–431
    minor, 427, 432–442, video V8.14

    in mitered bends, 438, 439
    in nozzles, 438
    in open-channel flow, 562
    in pipe bends, 438, 439
    in pipe entrances, 434
    in pipes, 414, 438
    in sudden area changes, 434
    in valves and fittings, 438, 439
Head loss coefficient, 432–433, 434
Head rise coefficient, 688
Heat transfer:
    net rate, 236
    rate, expressed in first law of
        thermodynamics, 236–237
    Rayleigh flow, 631, 642–649
    relationship between head loss, internal
        energy and, 245
History of fluid mechanics, 27–29
Homogeneous flow, 555
Horseshoe vortex, 538, 539, 577, video V4.6,
    video V9.1, video V9.25
Hybrid grids, 728–729
Hydraulic diameter, 329, 443
Hydraulic efficiency, 681
Hydraulic grade line, 131–133
Hydraulic jumps, 577–582, video V10.11,
    video V10.12
    classification of, 580–581
    comparison of compressible and open-channel
        flows, 656
    depth ratio across, 578
    as example of rapidly varied flow, 576
    head loss across, 578
Hydraulic radius, 568
Hydraulically smooth wall, 430
Hydraulics, in history of fluid mechanics, 27
Hydrodynamics, in history of fluid mechanics,
    27
Hydrometers, 72, video V2.8
Hydrostatic force:
    on curved surfaces, 68–70, video V2.6
    on plane surfaces, 59–65, video V2.5
Hydrostatic pressure distribution, 44–47
Hypersonic flow, 613

**I**
Ideal fluid, 294
Ideal gas flow, 602, video V11.1
Ideal gas law, 12–13, 602, video V11.1
Ideal head rise, 678–679, 680
Impellers, 675, 676
Impulse turbines, video V12.5
    defined, 696
    dental drills as, 702–704
    non-Pelton wheel example, 702–704
    Pelton wheel example, 696–702
    *vs.* reaction turbines, 695–696
Inclined tube manometer, 56
Incompressible flow, 107, 284
Incompressible fluids, 20, 21, 44–47, 48,
    video V1.7
Induced drag, 534
Inertial control volume, 214
Intensity of turbulence, 419
Intensive property, 176–177
Internal energy, 236, 602–603
International System (SI) of units, 7–8

Inviscid core, 295, 405
Inviscid flow:
　applying differential analysis, 292–299
　applying Newton's second law, 102–104
　defined, 102, 292
Irreversible flow, 253
Irrotational flow, 138–139, 281–282, 294–297,
　　video V3.6, video V6.4
Irrotational vortex, 303, 304
Isentropic flow:
　with area change, 615–617
　basic equations for, 615–616
　and compressible flow condition, 134–135
　in converging nozzle, 639–640
　in converging-diverging duct, 616–617,
　　video V11.7
　defined, 606
　effect of area variation on, 615–617
　graph for calculation of, 744
　ideal gas, 614–631
　reference condition for, 617, 619
Isentropic process, 21, 607
Isentropic stagnation properties, 618
Isothermal atmosphere, 48
Isothermal process, 21


**J**

Jet exit pressure, 118–119
Joule, 7
Jumps, hydraulic, *see* Hydraulic jumps


**K**

Kaplan turbines, 704–705
Kármán vortex trail, 363–364, 520, video V4.12,
　　video V6.16, video V9.12
Kilogram, 7
Kinematic similarity, 370
Kinematic viscosity:
　of air, 740, 741
　of common gases, 14, 738
　of common liquids, 12
　defined, 20
　of water, 739
Kinematics of fluid flow, 157–188, 277–282
Kinetic energy:
　coefficient, 249–252
　in first law of thermodynamics, 236
　in work-energy principle, 111
Kutta-Joukowski law, 318


**L**

Lagrangian description of flow, 160–161,
　　video V4.4, video V4.5
Laminar boundary layer:
　Blasius solution, 495–497
　description of, 487, video V9.4
　effect of pressure gradient, 507–511
　friction drag coefficient for, 501
　thickness, 492, video V9.4
Laminar flow, video V4.7
　in an annulus, 329–330
　applying dimensional analysis, 413–414
　in boundary layers, 490
　between parallel plates, 322–326
　past cylinders, 510
　past spheres, 519–520

　in pipes, 326–328, 402–403, 404, 407–415,
　　video V6.14, video V8.2
　*vs.* turbulent flow, 162
Laplace's equation, 297, 299
Laplacian operator, 297
Lapse rate, 50
Law of the wall, 423
Length scale, 372
Lift, 315, 317–318, 483, 528
Lift coefficient:
　airfoil, 533–534
　defined, 485
　rotating cylinder or sphere, 539–540
　writing, 529
Lift/drag polar diagram, 535
Lift/drag ratio, 535
Linear deformation, 277
Linear momentum, *see* Newton's second law
　　of motion
Linear momentum equation:
　application, 214–228
　derivation, 213–214
Linear momentum flow, 213, video V5.4, video
　　V5.5, video V5.6, video V5.8, video V5.9
Linear partial differential equations, *see*
　　Laplace's equation
Local acceleration, 169
Local derivative, 169
Local friction coefficient, 500–501
Logarithmic velocity profile, 423
Loss, major and minor, *see* Head loss
Loss coefficient, 432–433, 434
Loss of available energy, 244–245, 249, 605, 653
Low Reynolds number flow, 485–488, 517–518,
　　video V1.3, video V7.7, video V8.4


**M**

Mach, E., 28, 363
Mach cone, 612–614, video V1.8, video V11.3,
　　video V11.4
Mach number, 23, 361, 363, 386, 607, 610–612
Mach waves, 612, 657–658, video V1.8,
　　video V11.3, video V11.7
Magnus, H., 539
Magnus effect, 318, 539
Major loss, *see* Head loss
Manning, R., 28, 568
Manning coefficient, 569–570
Manning equation, 568–569
Manometers:
　defined, 52
　differential, 54–55
　inclined tube, 56
　piezometer tube, 52–53, 113, video V2.3
　U-tube, 53–56
Mass, *see* Conservation of mass
Mass, units of:
　in British Gravitational (BG) System, 8
　in English Engineering (EE) System, 8
　in International System (SI), 7
Mass flowrate, 120, 201–202
Mass *vs.* weight, 7, 8–9
Material derivative, 166–169, 183–184, 278
Measurement, flow, *see* Flow measurement
Mechanical efficiency, 681
Mechanical energy equation, 245
Meniscus, 53, 55

Mesh, *see* Grids
Meters, flow, *see* Flow measurement
Method of repeating variables, 350, 354–355
Method of superposition, 308–318, video V6.5
Methods of description:
　Eulerian, 160–161
　Lagrangian, 160–161
Minor loss, 427, 432–442
Minor loss coefficient, 432–433, 434
Mixed-flow machines, 669
Mixed-flow pumps, 693–694
Mixing length, 422
Modeling laws, 369
Models:
　defined, 368
　design conditions, 369
　distorted, 373–374, video V7.8, video V7.12,
　　video V7.17, video V7.18
　scales for, 372
　similitude concept, 346
　theory of, 368–370
　true, 373
　typical studies, 374–384, video V7.1,
　　video V7.9, video V7.18, video V7.19,
　　video V7.20
Moment of inertia, 61
Moment-of-momentum:
　application of equation, 229–236, video V5.13
　derivation, 228–229
　relationship to Newton's second law of
　　motion, 228–229
　turbomachine considerations, 234, 673–675
Momentum:
　angular, *see* Moment-of-momentum
　linear, *see* Newton's second law of motion
Momentum coefficient, 223
Momentum equation, *see also* Linear
　　momentum equation
　for control volume, 213, 225, 289
　for inviscid flow, 292–299
Momentum flux, 223, 224, video V5.7
Momentum integral equation:
　boundary layer flow on flat plate, 497–502
　boundary layer with nonzero pressure gradient,
　　511–512
　boundary layer with zero pressure gradient,
　　498
Momentum thickness, 493
Moody, L., 29, 429
Moody chart, 429, 430–431
Moving grids, 728–729
Multiple pipe system, *see* Pipe systems


**N**

Nappe, 582, 585
Navier, L., 28, 321
Navier-Stokes equation:
　applying to laminar flow, 322–324, 411–412
　cautions on use of, 331
　cylindrical coordinates, 321
　defined, 320–321, 331
　rectangular coordinates, 320–321
Net positive suction head, 682–684
Newton (unit), 7
Newton, I., 27, 28
Newtonian fluids, 15, 319–321
Newton's law of viscosity, 420

Newton's second law of motion, 41, 101–104, 199, 212–236, 289
Nikuradse, J., 429
Noncircular conduits, 442–445
Noninertial reference frame, 214
Nonisentropic flow of an ideal gas:
    Fanno flow, 631–642
    normal shock waves, 648–655
    Rayleigh flow, 642–649
Non-Newtonian fluids, 16, video V1.6
Nonuniform flow, 248–252, 555
Normal acceleration, 103, 173–174, video V3.1, video V3.4
Normal depth, 570
Normal shock waves, 648–655, video V11.7, video V11.8
    basic equations, 648–649
    defined, 630
    examples, 653–655
    graph for calculation, 747
    *T—s* diagram, 649
Normal stress:
    in conservation of linear momentum, 289
    in Newtonian fluids, 15, 319, 320
    and work transfer, 238
No-slip condition, 15, 310, 323, video V1.4, video V6.12, video V6.13
Nozzle discharge coefficient, 461
Nozzle meters, 127, 459, 460–461, 462–463
Nozzles:
    choked flow in, 619
    converging, 617
    converging-diverging, 617, video V11.6, video V11.7
    defined, 118
    loss coefficients, 438
    normal shock waves in, 630
    overexpanded flow, 630, video V11.6, video V11.7
    thrust, video V5.11
    underexpanded flow, 630, video V11.7
Numerical methods, 331, video V6.15

O
Oblique shock waves, 630, 658, video V1.8, video V11.6, video V11.7
One-dimensional flow, 161–162
One-equation model, 426
Open channels:
    characteristics of, 555–556, video V10.6
    geometric properties of, 566–570, video V10.7
    optimum cross sections, 574–575
    regimes of flow in, 556
    shear stress in, 567
    velocity contours in, 566–567
Open-channel flow:
    analogy with compressible flow, 655–657
    critical flow, 556, 560
    defined, 554
    depth variations, 565–566
    energy equation for, 561–562
    hydraulic jump, 576, 577–582, video V10.11, video V10.12
    laminar, 555
    Manning equation for, 568–569
    most efficient, 574–575
    normal depth, 570

surface wave propagation, 556–561
transitional, 555
turbulent, 555
uniform flow, 555, 566–575
varied flow, 555, 565, 575–577, video V10.6, video V10.8
Orifice discharge coefficient, 460
Orifice meters, 127, 459, 460
Orifices, flow through, 119
Outer layer, turbulent pipe flow, 421, 422
Overexpanded flow, 630, video V11.6, video V11.7
Overlap layer, turbulent pipe flow, 421, 422

P
Panel method, 726
Parallel plates:
    Couette flow, 324–326
    steady laminar flow between, 322–324
Pascal (unit), 13
Pascal, B., 28, 41
Pascal's law, 41
Pathlines, 164–166, video V4.11
Pelton wheels, 696–698, video V5.6, video V12.2, video V12.4
Perfect gas law, *see* Ideal gas law
Physical properties of fluids:
    of air, 740, 741
    of common gases, 14, 737–738, 740, 741
    of common liquids, 12, 737–738, 739
    of U.S. Standard Atmosphere, 742, 743
    of water, 739
Pi terms:
    defined, 349
    determination by inspection, 359–360
    determination by method of repeating variables, 350–355
    determination example, 357–358
    uniqueness, 358–359
Pi theorem, 349–350
Piezometers, 52–53, 113, video V2.3
Piezometric head, 132, video V2.3
Pipe systems, video V6.15
    loop, 456–457
    multiple pipes, 455–459
    networks, 458–459
    parallel, 455, 456
    series, 455, 456
    single pipes, 445–454
Pipes:
    aging, 429–430
    compressible adiabatic flow in, *see* Fanno flow
    developing flow in, 405–406
    fittings, 438–439
    flow characteristics, 401–406
    flow meters, 458–464
    flow rate, 458–464
    fully developed flow in, 326–328, 405–406, 407–415
    head loss, *see* Head loss
    heat transfer in, *see* Rayleigh flow
    hydraulically smooth, 430
    laminar flow in, 326–328, 402–403, 404, video V6.14
    laminar *vs.* turbulent flow in, 405
    noncircular, 442–445
    relative roughness of, 428–431

smooth, 430
standard components, 438–440
transitional flow in, 402, 416–418, video V8.2, video V8.3, video V8.9
turbulent flow in, 402–403, 404, 416–420, video V8.2, video V8.3
typical system components, 401
*vs.* ducts, 400
Pitot, H., 28
Pitot tubes, 114–117, 132, 653–654, video V3.8
Plane Poiseuille flow, 328
Planform area, 485, 516–517, 533, 534
Poise, 20
Poiseuille, J., 28, 326, 410
Poiseuille flow, 326, 328, 410, video V6.14
Poiseuille's law, 328, 410
Polar coordinates, 77
Polar diagram, lift-drag, 535
Position vector, 158
Potential, velocity, 296–297
Potential energy, 111, 236
Potential flow theory, video V6.9, video V6.10
    basic plane flows, 299–300, 308, video V6.6, video V6.7, video V6.11
        doublet, 306–308
        sink, 302–303
        source, 302
        uniform flow, 300–301
        vortex motion, 303–304
    defined, 297
    finding solutions to problems, 318–319
    singularity, 302
    superposition of basic plane flows, 308–318, video V6.5
        flow past circular cylinder, 313–316
        flow past half-body, 308–311, video V6.5
        flow past Rankine oval, 311–313
        flow past rotating cylinder, 316–318
Potential function, 297
Pound, 8
Power:
    as expression of work transfer rate, 237
    shaft, 674
    units of, 7
Power coefficient, 688, 705
Power specific speed, 706
Power-law velocity profile, 423
Prandtl, L., 27–28, 29, 422, 487
Prandtl boundary layer equations, 493–497
Prediction equation, 369
Pressure:
    absolute, 13, 50
    atmospheric, 13, 742, 743
    center of, 62
    coefficient, 363, 515
    defined, 13, 40
    dynamic, 113
    gage, 13, 50
    hydrostatic, 44–47, 113
    and ideal gas law, 12–13
    isentropic stagnation, *see* Isentropic stagnation properties
    measurement of, 50–52, 117, video V2.3, video V2.4
    at a point, 40–42, 102, 103, 653–654, video V3.7
    static, 113

suction, 50
thermodynamic, 113
total, 114
vacuum, 50
vapor, 23–24
Pressure coefficient, 363, 515
Pressure distribution:
airfoils, 482
automobile, 530, 532, video V2.1
converging-diverging nozzle, 630
cylinder, inviscid flow, 314
cylinder, viscous effects, 510–511
entrance of pipe, 405, 434
hydrostatic, 44–47
in rigid-body motion, 74–79
in rotating fluids, 77
standard atmosphere, 49–50
Pressure drag, see Drag
Pressure drag coefficient, 515
Pressure forces, 42–43
Pressure gradient:
adverse, 313, 509
effect on surface forces, 43
effects on boundary layer, 507–511,
video V9.10
favorable, 509
Pressure head, 45, 111, 132
Pressure at a jet exit, 118–119
Pressure prism, 65–68
Pressure recovery coefficient, 438
Pressure tap, 113, 116, 117, 132
Pressure transducers:
defined, 57
illustrated, 58, 59
role of strain gages, 58, 59
Primary quantities, 4
Product of inertia, 62
Profile, velocity, see Velocity profile
Propeller pumps, 693
Properties:
of air, 740, 741
of common gases, 14, 737–738, 740, 741
of common liquids, 12, 737–738, 739
of U.S. Standard Atmosphere, 49, 742, 743
of water, 739
Prototype, 368–370
Pumps:
affinity laws, 691
axial-flow, 693–694
blades, 675
centrifugal, 675–688, video V12.3
comparison with fans, 695
defined, 667
dimensionless parameters, 688–689
double-stage, 676
efficiency, 248
geometrically similar, 688–689
head, 678, 682–684
ideal vs. actual head rise, 678–679, 680
mixed-flow, 693–694
multistage, 676
parallel, 687
performance characteristics, 680–682
propeller, 693
scaling laws, 689–691
selection, 684–685
series, 687

similarity laws, 688–689
single-stage, 676
specific speed, 691–692
stages in, 693
system equation, 684
vs. turbines, 673

Q
Quasi-steady flow, 138

R
Radial-flow machines, defined, 669
Rankine ovals, 311–313
Rapidly varied flow (RVF), 555, 576–577,
video V10.9, video V10.10
Rate of angular deformation, 282, video V6.3
Rate of heat transfer, 236–237
Rate of shearing strain, 15, 282
Rayleigh, Lord, 29
Rayleigh flow:
basic equations for, 642–643
defined, 631, 642
example of nonisentropic flow of ideal gas,
642–649
graph for calculation of, 746
normal shock wave in, 645, 646
T—s diagram, 643, 646
Rayleigh Pitot tube formula, 654
Reaction turbines, 696, 704–705
Rectangular weirs, 130–131, 582, 583
Reentrant entrance, 434
Reference dimensions, 349–350, 356–358
Relative roughness, 428–431, 506
Relative velocity:
defined, 208, 669
and Reynolds transport theorem, 187
in rotating sprinkler example, 231
as turbomachine energy consideration, 669,
670, 671, 672
vs. absolute velocity, 186–187
Repeating variable method, 350, 354–355
Resistance coefficient, Manning, 569–570
Restricted homogeneous equations, 5
Reynolds, O., 29, 362, 402, 421
Reynolds number:
as common variable in fluid mechanics, 361
critical, see Transition to turbulence
defined, 18, 362
large vs. low, properties of flows, 485–488,
video V1.1, video V1.2, video V7.3,
video V7.7, video V9.15
similarity, 386
Reynolds pipe flow experiment, 403, video V8.2
Reynolds stress, 421, 422
Reynolds transport theorem:
defined, 176
derivation, 178–180
example of use, 180
general form, 183
obtaining for moving control volume, 186–187
physical interpretation, 183
relationship to material derivative, 183–184
role in selection of control volume, 187–188
steady effects, 184
unsteady effects, 184–186
Rheology, 3

Right-hand rule, 231
Rigid-body motion of fluids, 74–79
Ripple tank, comparison of compressible and
open-channel flows, 656
Rising head curve, 681
Rotameters, 463, video V8.15
Rotating disk meters, 464–465, video V8.16
Rotation, 138–139, 277, 279–282
Rotors, turbomachine, 230, 234, 235, 669, 670,
709
Roughness:
effect on models, 374–375, 381, 382,
video V7.13
on flat plates, 506
in open channels, 568, 569
in pipes, 427–431
relative, 428–431
typical values of, 429
Round-off errors, 727

S
Scales:
length, 372
types for models, 372
Scaling laws for pumps, 689–691
Second law of thermodynamics, 200, 253
Second moment of area, 61
Secondary flow, in pipes, 438
Secondary quantities, 4
Separation, video V6.8, video V9.9,
video V9.16, video V9.19
in boundary layers, 295–296, 487, 509
in pipes, 434, 438
with shock waves, 653
Separation location:
on airfoils, 511
on cylinders, 487, 509
defined, 509, 520
illustrated, 488
Shaft power, 232, 674
Shaft torque, 232, 674
Shaft work, 232, 239, 245, 247, 673, 675
Shaft work head, 247, 678, 680, 701, 705
Shallow water waves, 559
Sharp-crested weirs, 130–131, 582–585,
video V10.13
Shear thickening fluids, 17
Shear thinning fluids, 16
Shearing strain:
defined, 14
rate of, 15–16, 282, video V6.3
Shearing stress:
defined, 3
distribution in pipes, 408, 414
effect on fluids, 3, 14, 15–17, 19
as force acting on differential element, 290, 320
in laminar flow, 419–420
in Newtonian fluid, 319, 320
in open-channel flow, 567
in turbulent flow, 418–422
in work transfer, 238
Shock waves, videos V11.6–V11.11
curved, 658
detached, 658
normal, see Normal shock waves
oblique, 630, 658, video V11.6, video V11.7,
videos V11.9–V11.11

SI (International System) units, 7–8
Similarity requirements, 369–370
Similarity variable, 495, 496
Similitude:
    based on differential equations, 384–387
    as concept in experimentation, 346
Sink flow:
    defined, 302
    example, 302–303
    strength of, 302
Sinusoidal waves, 558–559
Siphon, 125–126
Slope:
    of channel bottom, 561–562
    energy line, 131–133, 561–562
    friction, 561
Slug, 8
Sluice gates:
    discharge coefficient, 587–588
    flow rate under, 128–130, 586
    force on, 224–225
    free outflow, 587–588
    submerged, 588
Smooth pipe, 430
Sonic flow, 612
Sound, speed of, video V11.3
    defined, 22–23
    ideal gas, 22, 609
    and Mach number, 607–609
    table of values for air at standard atmospheric
        pressure, 740–741
    table of values for water at standard
        atmospheric pressure, 739
Source flow, 302, video V6.5
Specific energy, 562–565
Specific energy diagram, 562–563
Specific gravity, 12
Specific head, see specific energy
Specific heat:
    constant pressure, 21, 603
    constant volume, 21, 602
Specific heat ratio:
    of air, 740, 741
    of common gases, 14
    defined, 21, 604
Specific speed:
    specifying for pumps, 691–692
    and turbine efficiency, 706, 707
Specific volume, 11
Specific weight:
    of air, 740, 741
    of common gases, 14
    of common liquids, 12
    as fluid characteristic, 12
    of water, 739
Speed of sound, video V11.3
    defined, 22–23
    ideal gas, 22, 609
    and Mach number, 607–609
    table of values for air at standard atmospheric
        pressure, 740–741
    table of values for water at standard atmos-
        pheric pressure, 739
Spheres:
    compressible flow past, 523, 610–611
    drag coefficient for, 379–380, 518, 520,
        523–524, 540, 610–611, video V9.17

effect of surface roughness, 523–524
    low Reynolds number flow past, 518
    rotating, 539–540
Spillways, 579, video V10.15, video V10.16
Stability, 73–74, video V2.11, video V2.12
Stagnation enthalpy, 618
Stagnation point, 113, video V3.7
Stagnation pressure, 114, 610, video V3.8
    isentropic, see Isentropic stagnation properties
Stagnation pressure probe, see Pitot tubes
Stagnation properties, see Isentropic stagnation
        properties
Stagnation state, 617, 618, 619
Stagnation streamline, 114, video V3.7,
        video V4.9
Stagnation temperature, 617
Stall condition, 511, 534, 710, video V9.21
Standard atmosphere:
    defined, 49
    pressure variation, 49–50
    properties of, 49, 742, 743
State, equation of, 12–13, 602
Static fluids, pressure variation in, 43–48
Static pressure, 113
Static pressure tap, 113, 116, 117, 132
Stators, turbomachine, 693, 709–710
Steady flow, video V4.7
    defined, 162
    fluid particles, 102
    and Reynolds transport theorem, 184
    streaklines in, 164
    streamlines in, 163
Stoke, 20
Stokes, G., 28, 321, 519
Stokes' law, 365
Stratified flow, 555
Stratosphere, 49
Streaklines, 164–166, video V4.10, video V7.14
Stream function, 285–288
Streamline, video V3.1, video V4.9
Streamline bodies, video V2.1, video V9.2,
        video V9.20
Streamline coordinates, video V4.13
Streamline curvature, 102–103
Streamlined bodies, 482
Streamlines:
    defined, 102–103
    equation, 286
    use of coordinate systems, 173–174
    in visualization and analysis of flow fields,
        163, 164–166
Streamtube, 239–240
Streamwise acceleration, 103
Stress:
    components, 319–320
    Newtonian fluid, 15, 319–320
    normal, 238, 289, 319, 320
    notation, 290
    shearing, 3, 14, 238, 290, 320
    sign convention, 290
    tangential, 238
    turbulent, 418–422
Strouhal, V., 29, 364
Strouhal number, 361, 363, 364, 385, video V7.5
Structured grids, 728
Subcritical flow, 556, 560–561, 563, 656,
        video V10.5

Submerged area, forces on, 59–62
Subsonic flow, 23, 612, 656
Substantial derivative, 166–169, 278
Suction, 50
Suction specific speed, 692
Sudden contraction, 435–437
Sudden expansion, 435–437
Supercritical flow, 556, 560–561, 563, 656,
        video V10.5
Superposition, 308–319, video V6.5
Supersonic flow, 23, 612–613, 656, video V1.8,
        video V11.6, video V11.7
Surface forces, 42–43, 214, 289, 290–291
Surface tension, video V1.9, video V7.6,
        video V10.3
    of common liquids, 12
    defined, 24–26
    effect on models, 381–382
    of water, 739
Surface waves, 556–561, video V7.4,
        video V9.18, videos V10.1–V10.5
Sutherland equation, 18
System curve, 685
System equation, 685
Systems, 175, 176
Systems of units:
    British Gravitational (BG) System, 8
    conversion factors, 9
    English Engineering (EE) System, 8–9
    International System (SI), 7–8
    need for consistency, 9, 11

T
T ds equations, 605
T—s diagrams, 619
Tangential stress work, 238
Temperature:
    absolute, 7, 8, 12, 13
    Celsius, 7
    centigrade, 7
    Fahrenheit, 8
    Kelvin, 7
    Rankine, 8
    stagnation, 617
Thermodynamic pressure, see Pressure
Thermodynamics:
    first law, 200, 236–252
    second law, 200, 253
Three-dimensional flow, 161
Three-dimensional objects, 482, 530
Three-reservoir problem, 457–458
Throat, converging-diverging duct, 617, 619
Thrust, 223–224, video V5.11
Tidal waves, see Tsunamis
Time-averaged equations, 426
Time-averaged velocity, 417, 418–419
Torque:
    in moment-of-momentum equation, 228, 229,
        230, 231–233, 234
    shaft, 232, 674
Torricelli, E., 28, 51, 118
Torricelli's formula, 118
Total head, 132
Total pressure, 114
Total pressure ratio, 708
Trailing vortex, 538, 539, video V4.6,
        video V9.1, video V9.25

Tranquil flow, 556
Transition to turbulence, video V9.5
  in boundary layer flows, 490, 502–503, video V9.7
  in fully developed flows, 416–418
  in open-channel flows, 555
  in pipe flows, 402, 416–418, video V8.2, video V8.3
Translation, in fluid element flow, 277, 278–279
Transonic flow, 613
Triangular weirs, 131, 583, 584–585, video V10.13
Troposphere, 49–50
True models, 373
Truncation errors, 727
Tsunamis, 559
Turbine flow meters, 464
Turbines:
  axial-flow, 669
  compressible flow, 707, 711–713
  defined, 695
  Francis, 704, 705
  gas, 711–713
  impulse, 695–696, 696–704, video V12.4, video V12.5
  Kaplan, 704
  Pelton wheel, 696–698, 699, 700–702, video V5.6, video V5.14, video V12.4
  performance maps, 713
  radial-flow, 669
  reaction, 696, 704
  specific speed, 691–692
  vs. pumps, 673
Turbomachines:
  angular momentum considerations, 234, 673–675
  axial-flow, 669, 693–694, 708–710
  basic energy considerations, 669–673
  centrifugal pumps, 675–688, video V12.3
  compressible flow, 707–713
  defined, 668–669
  flow coefficient, 689, 705
  head coefficient, 691
  mixed-flow, 669, 693–694
  radial-flow, 669, 708
  rotary sprinkler, 230–233, video V4.11, video V5.12, video V12.2
  specific speed, 691–692
Turbulence:
  chaos, 426–427
  characteristics of, 162, 418–420, video V1.1, video V8.1, video V8.7, video V8.8, video V8.11
  intensity of, 419
  mixing in, 418, 420, video V1.1, video V8.5, video V8.6, video V8.7
  modeling, 426
  Reynolds stress in, 421
  spots in, 502
Turbulent boundary layer:
  approximate solution for flat plate, 504
  effect on separation, 510–511
  properties, 504–505, video V9.8
  transition from laminar boundary layer, 490
  velocity profiles in, 503
Turbulent jet, video V8.1

Turbulent pipe flow, video V8.2, video V8.3, video V8.9
  fully developed, 416–420
  general characteristics, 402–403, 404
  mean vs. fluctuating velocity, 418–419
  overlap layer, 421, 422
  shear stress distribution, 408, 419, 421
  velocity profile, 422–424, video V8.2
    logarithmic, 423
    power law, 423
    viscous sublayer, 421–422
Turbulent stresses, 418–422
Two-dimensional flow, 161, 657–658
Two-dimensional objects, 482, 529
Two-equation model, 426

U
U-tube manometers, 53–56
Underexpanded flow, 630, video V11.7
Underflow gates, 587–589
Uniform depth flow, 566–575
Uniform flow:
  in open channels, 555, 566–575
  potential, 300–301
  in sections, 202
Uniform flow field, 294
Units:
  conversion factors, 9
  defined, 4
  need for consistency, 9, 11
  systems of, 7–9
Universal gas constant, 602
Unsteady Bernoulli equation, 136–137
Unsteady flow, 136–138, 162, 163, 164, 169, 184–185, 206–207, video V3.11, video V4.7, video V5.10
Unstructured grids, 728
Upstream velocity, 481
U.S. standard atmosphere, see Standard atmosphere
Useful energy, see Available energy

V
Vacuum pressure, 50
Valve losses, 439
Valves, 433, 440
Vanes, 669, 675, see also Guide vanes
Vapor pressure:
  in barometer tubes, 51
  boiling, 23–24
  of common liquids, 12
  defined, 23–24
  effect on cavitation, 24
  of water, 739
Varied flow, see Gradually varied flow (GVF); Rapidly varied flow (RVF)
Velocity:
  absolute, 208, 230–231, 669, 670, 671
  absolute vs. relative, 186–187
  average, 202
  friction, 423
  particle, 102, 158
  relative, 208, 231, 669, 670, 671
  time-averaged, 417, 418–419
Velocity distribution, see Velocity profile
Velocity field, 158–166, 278, video V4.2, video V4.3, video V6.1, video V6.2

Velocity gradient, 14–15
Velocity head, 111, 132
Velocity measurement, see Flow measurement
Velocity potential, 296–297
Velocity profile:
  in boundary layers
    effect of pressure gradient on, 509
    laminar flow, 501
    in transition from laminar to turbulent flow, 503
    turbulent flow, 504, 505
  between flat plates, 322–324
  in open channels, 566–567
  in pipes
    laminar flow, 327, 409–410, video V8.10
    power law, 423
    turbulent flow, 422–424, video V8.10
Velocity triangle, 669, 670, 674–675
Vena contracta, 119, 129, 434
Venturi, G., 28
Venturi discharge coefficient, 462
Venturi meters, 127, 128, 459, 461–462, video V3.10
Viscometers, capillary tube, 34, video V1.5
Viscosity, video V1.3
  absolute or dynamic, 15
  of air, 740, 741
  apparent, 16, 17
  of common gases, 14, 737, 738
  of common liquids, 12, 737, 738
  defined, 14
  eddy, 422
  effect of temperature on, 17–18
  kinematic, 20
  of U.S. Standard Atmosphere, 742, 743
  of water, 739
Viscous dissipation, 414
Viscous flow, 319–321
Viscous stresses, 320
Viscous sublayer, turbulent pipe flow, 421–422
Visualization flow, 161, 162–166, video V4.6, video V4.9, video V7.16, video V9.19
Volume flow meters, 464–465
Volume flowrate, 120
Volumetric dilatation, 279
Volumetric efficiency, 681
Volutes, 675
Von Kármán, T., 29, 363–364, 498
Vortex:
  bound, 538, 539, video V4.6, video V9.25
  combined, 304, video V4.8
  defined, 303–304
  forced, 304
  free, 109, 304, 306, video V3.6, video V6.4
  horseshoe, 538, 539, 577
  irrotational, 303, 304
  Kármán trail, 520, video V6.16, video V9.14
  ring, video V4.1
  rotational, 304
  strength of, 304
  tornado, video VA.3
  trailing, 538, video V3.5, video V4.6, video V9.1, video V9.25
Vortex shedding, 363–364, 371, video V4.12, video V7.5, video V7.10, video V9.14

Vortex street, 519, 520, video V6.16,
    video V7.5, video V9.12
Vorticity, 281–282, 294, 297, 490

**W**

Wake, video V9.3, video V9.9
    and boundary layer, 295–296
    defined, 487
    flat plate, 487
    streamlined object, 517
    vortex street, 520, video V6.16,
        video V9.14
Wall shear stress, 408, 414, 497
Water, table of properties, 739
Water horsepower, 681
Water table, comparison of compressible and
        open-channel flows, 656
Watt, 7
Wave motion, 558–559
Wave speed, 556–559, video V11.3
Waves:
    calculating speed, 556–559

gravity, video V10.4, video V11.4
    role of gravity, 557
    shock, *see* Shock waves
    surface, 556–561, video V9.18, video V10.3,
        video V10.4, video V10.5, video V11.4
Weber, M., 29, 364
Weber number, 361, 364, 386, video V7.6
Weight *vs.* mass, 7, 8–9
Weir coefficient, 583, 584, 585, 586
Weir head, 582
Weirs:
    broad-crested, 585–587, video V10.14
    defined, 130, 582
    rectangular, 130–131, 582, 583
    sharp-crested, 130–131, 582–585,
        video V10.13
    triangular, 131, 583, 584–585, video V10.13
Weisbach, J., 28
Wetted perimeter, 329, 443, 566–567
Wholly turbulent flow, 429–430
Wind tunnels, 441–442, video V3.3,
        video V7.15, video V9.11, video V9.22

Windmills, 670–671, video V12.1
Wing loading, 533–534
Wing tip vortices, 281, 538–539,
        video V3.5
Work:
    expressed in first law of thermodynamics,
        237
    normal stress, 238
    rate of, 232, 237
    shaft, 232, 239, 245, 247, 673, 675
    shear stress, 238
    sign convention for, 237
    tangential stress, 238
    and turbomachines, 674–675
    unit of, 7
Work-energy equation, 111

**Z**

Zero-equation model, 426
Zone:
    of action, 612
    of silence, 612

# ndex of Fluids Phenomena Videos

**V1.1**
Mount St. Helens eruption, 2
*Courtesy of USGS EROS Data Center*

**V1.2**
E. coli swimming, 2
*© L. Turner, W. Ryu, & H. Berg*

**V1.3**
Viscous fluids, 14
*© B. R. Munson*

**V1.4**
No-slip condition, 14
*© B. R. Munson*

**V1.5**
Capillary tube viscometer, 15
*© B. R. Munson*

**V1.6**
Non-Newtonian behavior, 17
*© B. R. Munson*

**V1.7**
Water balloon, 20
*© S. T. Thoroddsen & Stanford University*

**V1.8**
As fast as a speeding bullet, 22
*Courtesy of Vision Research, Inc.*

**V1.9**
Floating razor blade, 24
*© B. R. Munson*

**V1.10**
Capillary rise, 25
*© The Regents of the University of California*

**V1.11**
Contact angle, 25
*© 2004 Gas Dynamics Laboratory, The Pennsylvania State University with a grant from the National Science Foundation*

**V2.1**
Pressure on a car, 44
*© Stanford University*

**V2.2**
Demonstration of atmospheric pressure, 44
*© 2004 Gas Dynamics Laboratory, The Pennsylvania State University with a grant from the National Science Foundation*

**V2.3**
Blood pressure measurement, 53
*© B. R. Munson*

**V2.4**
Bourdon gage, 57
*© B. R. Munson*

**V2.5**
Hoover Dam, 59
*Courtesy of U.S. Bureau of Reclamation*

**V2.6**
Pop bottle, 68
*© B. R. Munson*

**V2.7**
Cartesian diver, 71
*© B. R. Munson*

**V2.8**
Hydrometer, 72
*© B. R. Munson*

**V2.9**
Atmospheric buoyancy, 73
*Courtesy of U.S. Department of Energy*

**V2.10**
Density differences in fluids, 73
*© 2004 Gas Dynamics Laboratory, The Pennsylvania State University with a grant from the National Science Foundation*

**V2.11**
Stability of a floating cube, 74
*© ANSYS, Inc. with permission*

**V2.12**
Stability of a model barge, 74
*© B. R. Munson*

**V3.1**
Streamlines past an airfoil, 103
*© S.T. Thoroddsen & Stanford University*

**V3.2**
Balancing ball, 107
*© B. R. Munson*

**V3.3**
Flow past a biker, 107
*Courtesy of A2 Wind Tunnel, www.A2WT.com*

**V3.4**
Hydrocyclone separator, 108
*© ANSYS, Inc. with permission*

**V3.5**
Aircraft wing tip vortex, 108
*Courtesy of NASA Dryden Flight Research Center*

**V3.6**
Free vortex, 109
*© B. R. Munson*

**V3.7**
Stagnation point flow, 114
*© B. R. Munson*

**V3.8**
Airspeed indicator, 115
*© B. R. Munson*

**V3.9**
Flow from a tank, 118
*© B. R. Munson*

**V3.10**
Venturi channel, 124
*© B. R. Munson*

**V3.11**
Oscillations in a U-tube, 137
*© B. R. Munson*

**V3.12**
Flow over a cavity, 139
*© ONERA, The French Aerospace Lab*

**V4.1**
Streaklines, 157
*© ONERA, The French Aerospace Lab*

**V4.2**
Velocity field, 158
*© Chris Griffin*

**V4.3**
Cylinder velocity vectors, 158
*© ANSYS, Inc. with permission*

**V4.4**
Follow the particles (experiment), 161
*© Dr. Hui Hu*

**V4.5** Follow the particles (computer), 161
*Courtesy of NASA*

**V4.6** Flow past a wing, 161
*© ONERA, The French Aerospace Lab*

**V4.7** Flow types, 162
*© B. R. Munson*

**V4.8** Jupiter red spot, 162
*Courtesy of NASA*

**V4.9** Streamlines, 163
*© B. R. Munson*

**V4.10** Streaklines, 164
*© Stanford University*

**V4.11** Pathlines, 166
*© B. R. Munson*

**V4.12** Unsteady flow, 169
*© B. R. Munson & Stanford University*

**V4.13** Streamline coordinates, 173
*© ONERA, The French Aerospace Lab*

**V4.14** Control volume, 187
*© 2004 Gas Dynamics Laboratory, The Pennsylvania State University with a grant from the National Science Foundation*

**V5.1** Sink flow, 202
*© B. R. Munson*

**V5.2** Shop vac filter, 207
*© B. R. Munson*

**V5.3** Flow through a contraction, 207
*© Education Development Center, Inc.*

**V5.4** Smokestack plume momentum, 213
*© B. R. Munson*

**V5.5** Marine propulsion, 213
*© Education Development Center, Inc.*

**V5.6** Force due to a water jet, 213
*© B. R. Munson*

**V5.7** Air jet momentum flux, 218
*© 2004 Gas Dynamics Laboratory, The Pennsylvania State University with a grant from the National Science Foundation*

**V5.8** Fire hose, 221
*© Critical Information Network (CiNet) Fire and Emergency Training Network (FETN)*

**V5.9** Jellyfish, 225
*© S. P. Colin & J. H. Costello*

**V5.10** Running on water, 226
*Footage provided by Thought Equity Motion*

**V5.11** Thrust from water nozzle, 227
*© 2004 Gas Dynamics Laboratory, The Pennsylvania State University with a grant from the National Science Foundation*

**V5.12** Rotating lawn sprinkler, 230
*© B. R. Munson*

**V5.13** Impulse-type lawn sprinkler, 232
*© B. R. Munson*

**V5.14** Pelton wheel turbine, 242
*© B. R. Munson*

**V5.15** Energy transfer, 245
*© B. R. Munson*

**V5.16** Water plant aerator, 247
*© B. R. Munson*

**V6.1** Spinning football-pressure contours, 277
*© ANSYS, Inc. with permission*

**V6.2** Spinning football-velocity vectors, 277
*© ANSYS, Inc. with permission*

**V6.3** Shear deformation, 280
*© B. R. Munson*

**V6.4** Vortex in a beaker, 305
*© B. R. Munson*

**V6.5** Half-body, 309
*© B. R. Munson*

**V6.6** Circular cylinder, 313
*© S. T. Thoroddsen & Stanford Univers*

**V6.7** Ellipse, 313
*© S. T. Thoroddsen & Stanford Univers*

**V6.8** Circular cylinder with separation, 314
*© Stanford University*

**V6.9** Forces on suspended ball, 314
*© 2004 Gas Dynamics Laboratory, The Pennsylvania State University with a gr from the National Science Foundation*

**V6.10** Potential and viscous flow, 315
*© S. T. Thoroddsen & Stanford Univers*

**V6.11** Potential flow, 319
*© B. R. Munson*

**V6.12** No-slip boundary condition, 32
*© B. R. Munson*

**V6.13** Liquid-liquid no-slip, 323
*© B. R. Munson & Stanford University*

**V6.14** Laminar flow, 327
*© B. R. Munson*

**V6.15** Complex pipe flow, 328
*© R. W. Metcalfe & S. Lundin, Universi of Houston*

**V6.16** CFD example, 331
*© 2004 Gas Dynamics Laboratory, The Pennsylvania State University with a gr from the National Science Foundation*

**V7.1**
Real and model flies, 346
*© Will Dickinson & Michael Dickinson, Caltech*

**V7.2**
Flow past a flat plate, 354
*© B. R. Munson & Stanford University*

**V7.3**
Reynolds number, 362
*© B. R. Munson*

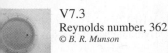

**V7.4**
Froude number, 362
*© B. R. Munson*

**V7.5**
Strouhal number, 363
*© ONERA, The French Aerospace Lab*

**V7.6**
Weber number, 364
*© The Regents of the University of California*

**V7.7**
Stokes flow, 365
*© Education Development Center, Inc.*

**V7.8**
Model airplane, 368
*© Dennis Goodrich*

**V7.9**
Environmental models, 370
*Courtesy of CPP Wind Engineering, www.cppwind.com*

**V7.10**
Flow past an ellipse, 371
*© B. R. Munson & Stanford University*

**V7.11**
Model of fish hatchery pond, 373
*© B. R. Munson*

**V7.12**
Distorted river model, 373
*Courtesy of U.S. Army Corps of Engineers Coastal and Hydraulics Lab*

**V7.13**
Wind engineering models, 377
*Courtesy of CPP Wind Engineering, www.cppwind.com*

**V7.14**
Model airplane test in water, 378
*Courtesy of NASA Dryden Flow Visualization Facility*

**V7.15**
Large scale wind tunnel, 378
*Courtesy of NASA Langley Center for Distance Learning*

**V7.16**
Full Scale Race Car Model, 379
*© 2004 Gas Dynamics Laboratory, The Pennsylvania State University with a grant from the National Science Foundation*

**V7.17**
River flow model, 381
*Courtesy of the U.S. Army Engineer Waterways Experiment Station*

**V7.18**
Boat model, 381
*Courtesy of the U.S. Army Engineer Waterways Experiment Station*

**V7.19**
Dam model, 382
*Courtesy of St. Anthony Falls Lab, University of Minnesota*

**V7.20**
Testing of large yacht model, 383
*Courtesy of Marine Hydrodynamics, Laboratory University of Michigan*

**V8.1**
Turbulent jet, 400
*© A. M. Fincham*

**V8.2**
Laminar/turbulent pipe flow, 402
*© B. R. Munson*

**V8.3**
Intermittent turbulent burst in pipe flow, 403
*© K. S. Breuer & Stanford University*

**V8.4**
Stirring color into paint, 417
*© B. R. Munson*

**V8.5**
Laminar and turbulent mixing, 417
*© Stanford University*

**V8.6**
Stirring cream into coffee, 418
*© B. R. Munson & Stanford University*

**V8.7**
Turbulence in a bowl, 421
*© B. R. Munson*

**V8.8**
Turbulent channel flow, 422
*Courtesy of NASA*

**V8.9**
Laminar to turbulent flow from a pipe, 423
*© Education Development Center, Inc.*

**V8.10**
Laminar/turbulent velocity profiles, 424
*© B. R. Munson*

**V8.11**
2D grid turbulence in soap film, 426
*© M. Gharib & M. Beizaei, California Institute of Technology*

**V8.12**
Entrance/exit flows, 435
*© B. R. Munson*

**V8.13**
Separated flow in a diffuser, 437
*© Education Development Center, Inc.*

**V8.14**
Car exhaust system, 438
*© B. R. Munson*

**V8.15**
Rotameter, 463
*© B. R. Munson*

**V8.16**
Water meter, 464
*© B. R. Munson*

**V9.1**
Space Shuttle landing, 482
*Courtesy of NASA*

**V9.2**
Streamlined and blunt bodies, 487
*© B. R. Munson*

**V9.3**
Human aerodynamic wake, 488
*© 2004 Gas Dynamics Laboratory, The Pennsylvania State University with a grant from the National Science Foundation*

**V9.4**
Laminar boundary layer, 489
*© Education Development Center, Inc.*

**V9.5**
Laminar/turbulent transition, 490
*© 2004 Gas Dynamics Laboratory, The Pennsylvania State University with a grant from the National Science Foundation*

**V9.6**
Boundary layer thickness, 491
*© K. S. Breuer & Stanford University*

**V9.7**
Transition on flat plate, 502
*© ONERA, The French Aerospace Lab*

**V9.8**
Random flow but not turbulent, 504
*© J. M. Wallace, J.-L. Balint & M. Ayrault, Ecole Centrale de Lyon*

**V9.9**
Snow drifts, 509
*© B. R. Munson*

**V9.10**
Effect of body shape on pressure gradient, 511
*© ONERA, The French Aerospace Lab*

**V9.11**
Skydiving practice, 516
*Courtesy of Vegas Indoor Skydiving*

**V9.12**
Karman vortex street, 519
*© B. R. Munson & Stanford University*

**V9.13**
Flow past cylinder, 520
*© 2004 Gas Dynamics Laboratory, The Pennsylvania State University with a grant from the National Science Foundation*

**V9.14**
Oscillating sign, 520
*© B. R. Munson*

**V9.15**
Flow past a flat plate, 520
*© B. R. Munson & Stanford University*

**V9.16**
Flow past an ellipse, 521
*© Stanford University*

**V9.17**
Wake of golf ball and smooth ball, 524
*© Stanford University*

**V9.18**
Jet ski, 525
*© B. R. Munson*

**V9.19**
Drag on a truck, 527
*Courtesy of the National Research Council of Canada Institute for Aerospace Research*

**V9.20**
Automobile streamlining, 527
*© Stanford University*

**V9.21**
Stalled airfoil, 535
*© ONERA, The French Aerospace Lab*

**V9.22**
Bat flying, 535
*© K. Breuer & S. Swartz*

**V9.23**
Trailing edge flap, 535
*© ANSYS, Inc. with permission*

**V9.24**
Leading edge flap, 536
*© ONERA, The French Aerospace Lab*

**V9.25**
Wing tip vortices, 538
*Courtesy of the National Research Council of Canada Institute for Aerospace Research*

**V10.1**
Offshore oil drilling platform, 554
*© ANSYS, Inc. with permission*

**V10.2**
Filling your car's gas tank, 556
*© ANSYS, Inc. with permission*

**V10.3**
Water strider, 558
*© C. Loudon, University of Kansas*

**V10.4**
Sinusoidal waves, 558
*Courtesy of St. Anthony Falls, Lab University of Minnesota*

**V10.5**
Bicycle through a puddle, 561
*© B. R. Munson*

**V10.6**
Merging channels, 566
*Courtesy of U.S. Army Engineer Waterways Experiment Station*

**V10.7**
Uniform channel flow, 570
*© B. R. Munson*

**V10.8**
Erosion in a channel, 576
*Courtesy of U.S. Army Engineer Waterways Experiment Station*

**V10.9**
Bridge pier scouring, 576
*© ONERA, The French Aerospace Lab*

**V10.10**
Big Sioux River Bridge collapse, 577
*Courtesy of St. Anthony Falls Lab, University of Minnesota*

**V10.11**
Hydraulic jump in a river, 578
*© B. R. Munson*

**V10.12**
Hydraulic jump in a sink, 581
*© B. R. Munson*

**V10.13**
Triangular weir, 584
*© B. R. Munson*

**V10.14**
Low-head dam, 584
*Courtesy of U.S. Army Engineer Waterways Experiment Station*

**V10.15**
Spillway gate, 587
*Courtesy of St. Anthony Falls Lab, University of Minnesota*

**V10.16**
Unsteady under and over, 588
*© ANSYS, Inc. with permission*

**V11.1**
Lighter flame, 602
*Courtesy of Professor W. J. Cook*

**V11.2**
Waves on a water surface, 611
*© 2004 Gas Dynamics Laboratory, The Pennsylvania State University with a grant from the National Science Foundation*

**V11.3**
Jet noise, 612
*© J. B. Freund*

**V11.4**
Speed boat, 612
*Footage provided by Thought Equity Motio*

**V11.5**
Compressible flow visualization,
*© 2004 Gas Dynamics Laboratory, The Pennsylvania State University with a grant from the National Science Foundation*

**V11.6**
Rocket engine start-up, 629
*Courtesy of Mike Massee/XCOR Aerospac*

**V11.7**
Supersonic nozzle flow, 630
*© 2004 Gas Dynamics Laboratory, The Pennsylvania State University with a grant from the National Science Foundation*

**V11.8**
Blast waves, 648
*Courtesy of the Department of Energy*

**V11.9**
Wedge oblique shocks and
expansions, 657
*© 2004 Gas Dynamics Laboratory, The
Pennsylvania State University with a grant
from the National Science Foundation*

**V11.10**
Bullet oblique shocks, 658
*Courtesy of Vision Research, Inc.*

**V11.11**
Space shuttle oblique shocks, 658
*© 2004 Gas Dynamics Laboratory, The
Pennsylvania State University with a grant
from the National Science Foundation*

**V12.1**
Windmills, 671
*© B. R. Munson*

**V12.2**
Self-propelled lawn
sprinkler, 673
*© B. R. Munson*

**V12.3**
Windshield washer
pump, 675
*© B. R. Munson*

**V12.4**
Pelton wheel lawn
sprinkler, 698
*© B. R. Munson*

**V12.5**
Dental drill, 704
*© B. R. Munson*

**V12.6**
Flow in a compressor stage, 709
*© J. U. Schluter, X. Wu, S. Kim, J.J. Alonso,
& H. Pitsch, CTR; Courtesy of NASA*

**VA.1**
Pouring a liquid, 725
*© ANSYS, Inc. with permission*

**VA.2**
Dynamic grid, 728
*© ANSYS, Inc. with permission*

**VA.3**
Tornado simulation, 734
*Courtesy of Dr. David Lewellen and Paul
Lewellen, West Virginia University*

# Photo Credits

For the stills in the video icons in the text margins, the copyright holders are indicated in the foregoing *Index of Fluids Phenomena Videos*.

The following images are in the public domain:

**Chapter 1**
Fig. E1.2a
Margin figure Pg. 28, Leonardo da Vinci
Margin figure Pg. 28, Isaac Newton
Margin figure Pg. 28, Daniel Bernoulli
Margin figure Pg. 28, Ernst Mach
Margin figure Pg. 29, Ludwig Prandtl
Margin figure Pg. 29, Osborne Reynolds

**Chapter 4**
Margin figure Pg. 175, Space Shuttle
Margin figure Pg. 184, Space Shuttle

**Chapter 5**
Fig. E5.5a

**Chapter 6**
Chapter 6 opening photo

**Chapter 7**
Fig. E7.7
Fig. 7.9

**Chapter 8**
Margin figure Pg. 402, sketch

**Chapter 9**
Fig. E9.11c
Margin figure Pg. 534, stalled wing
Margin figure Pg. 538, trailing vortex

**Chapter 10**
Margin figure Pg. 579, dam spillway

**Chapter 11**
Chapter opening photo
Fig. E11.4b
Margin figures (2) Pg. 630, Space Shuttle

For all other third party photo provisions in the text, the copyright holders are indicated near the photo.

The following photos were provided by the authors, and are copyright John Wiley & Sons, Inc.:

**Chapter 1:**
Fig. E1.1a
Margin figure Pg. 15, streamlines past solid body
Margin figure Pg. 17, knife and toast
Fig. E1.4
Fig. E1.6a

**Chapter 2:**
Margin figure Pg. 41, beaker
Fig. 2.4
Fig. 2.5
Margin figure Pg. 52, sphygmomanometer
Fig. 2.15a
Margin figure Pg. 68, wine bottle
Margin figure Pg. 74, kayak
Fig. 2.29a

**Chapter 3:**
Chapter opening photo
Fig. 3.1a
Fig. E3.1a
Fig. E3.2
Fig. E3.4a
Fig. 3.5
Margin figure Pg. 118, coffee urn
Fig. 3.12b
Margin figure Pg. 119, sprinkling can

Fig. 3.6b
Fig. 3.15a
Fig. E3.7a
Fig. E3.10a

**Chapter 4:**
Margin figure Pg. 167, flame
Margin figure Pg. 169, garden hose nozzle
Fig. 4.11a
Margin figure Pg. 183, sink faucet

**Chapter 5:**
Fig. E5.3a
Fig. E5.12a
Margin figure Pg. 240, hair dryer
Fig. E5.21a

**Chapter 6:**
Fig. 6.7
Fig. 6.12
Fig. E6.6a
Fig. 6.34a

**Chapter 7:**
Fig. E7.2

**Chapter 8:**
Fig. E8.1a
Fig. 8.21

Table 8.2 figures
Fig. E8.7
Margin figure Pg. 465, gas meter

**Chapter 9:**
Margin figure Pg. 517, rock
Margin figure Pg. 526, flag
Fig. E9.13a
Fig. E9.14a

**Chapter 10:**
Chapter opening photo
Margin figure Pg. 554, water on a road
Margin figure Pg. 555, uniform flow
Margin figures (2) Pg. 556, waves
Fig. E10.1a
Fig. 10.12

**Chapter 11:**
Margin figure Pg. 616, converging-diverging duct

**Chapter 12:**
Fig. 12.4a
Fig. 12.29b

**Appendices:**
None

Conversion Factors from BG and EE Units to SI Units[a]

| | To Convert from | to | Multiply by |
|---|---|---|---|
| Acceleration | ft/s$^2$ | m/s$^2$ | 3.048 E − 1 |
| Area | ft$^2$ | m$^2$ | 9.290 E − 2 |
| Density | lbm/ft$^3$ | kg/m$^3$ | 1.602 E + 1 |
| | slugs/ft$^3$ | kg/m$^3$ | 5.154 E + 2 |
| Energy | Btu | J | 1.055 E + 3 |
| | ft · lb | J | 1.356 |
| Force | lb | N | 4.448 |
| Length | ft | m | 3.048 E − 1 |
| | in. | m | 2.540 E − 2 |
| | mile | m | 1.609 E + 3 |
| Mass | lbm | kg | 4.536 E − 1 |
| | slug | kg | 1.459 E + 1 |
| Power | ft · lb/s | W | 1.356 |
| | hp | W | 7.457 E + 2 |
| Pressure | in. Hg (60 °F) | N/m$^2$ | 3.377 E + 3 |
| | lb/ft$^2$ (psf) | N/m$^2$ | 4.788 E + 1 |
| | lb/in.$^2$ (psi) | N/m$^2$ | 6.895 E + 3 |
| Specific weight | lb/ft$^3$ | N/m$^3$ | 1.571 E + 2 |
| Temperature | °F | °C | $T_C = (5/9)(T_F - 32°)$ |
| | °R | K | 5.556 E − 1 |
| Velocity | ft/s | m/s | 3.048 E − 1 |
| | mi/hr (mph) | m/s | 4.470 E − 1 |
| Viscosity (dynamic) | lb · s/ft$^2$ | N · s/m$^2$ | 4.788 E + 1 |
| Viscosity (kinematic) | ft$^2$/s | m$^2$/s | 9.290 E − 2 |
| Volume flowrate | ft$^3$/s | m$^3$/s | 2.832 E − 2 |
| | gal/min (gpm) | m$^3$/s | 6.309 E − 5 |

[a]If more than four-place accuracy is desired, refer to Appendix E.